INDUSTRIAL SOLVENTS HANDBOOK

INDUSTRIAL SOLVENTS HANDBOOK

Third Edition

Edited by

Ernest W. Flick

NOYES DATA CORPORATION

Park Ridge, New Jersey, U.S.A.

Published in the United States of America by
Noyes Data Corporation
Mill Road, Park Ridge, New Jersey 07656

10 9 8 7 6 5 4 3

Library of Congress Cataloging in Publication Data
Main entry under title:

Industrial solvents handbook.

    Bibliography: p.
    Includes index.
    1. Solvents--Handbooks, manuals, etc. I. Flick,
Ernest W.
TP247.5.I53   1985   660.2'9482'0212   84-22637
ISBN 0-8155-1010-1

Foreword

Completely revised, and vastly expanded, this well-established and successful reference volume is designed principally for the chemical and other process industries, but will be found useful by anyone needing the latest pertinent data on industrial solvents.

This Third Edition is uniquely helpful when it becomes necessary to select a new solvent on a competitive or comparative basis; when the customary solvent, employed hitherto, might no longer be available, or can no longer be used because of environmental reasons; or when prices have risen to such an extent that an existing process must be redesigned to make it economically feasible again.

The more than 1,100 tables in this book contain basic data on the physical properties of most solvents and on the solubilities of a variety of materials in these solvents. Even phase diagrams for multicomponent systems are included.

The contents of the tables were selected by the editor mainly from manufacturers' literature at no cost to, nor influence from, the manufacturers or distributors of these solvents. The source of each table is indicated by a reference number following the title. A complete set of references is found at the end of the book.

The vast amount of information contained in the book is evidenced at once in the large table of contents, which is organized by chemical groups and also serves as the index. An abbreviated summary of the contents is given below.

Advanced composition and production methods developed by Noyes Data Corporation are employed to bring this durably bound book to you in a minimum of time. Special techniques are used to close the gap between "manuscript" and "completed book." In order to keep the price of the book to a reasonable level, it has been partially reproduced by photo-offset directly from the original material and the cost saving passed on to the reader. Due to this method of publishing, certain portions of the book may be less legible than desired.

To my wife, Ruth, for the last 30 years

Contents and Index

Introduction

A solution may be defined as a mixture of two or more substances which has uniform chemical and physical properties throughout. It may also be defined as a system whose component parts are two or more molecular species, there being no boundary surfaces between these parts larger than molecules. There are two components to every solution—the solvent and the solute. As a matter of convenience we designate the part of a solution which is in excess as the solvent; the solute is the component which is in smaller proportion.

The purpose of solvents is to convert substances into a form suitable for a particular use. The importance of the role of solvents is brought out most clearly by the fact that many substances exhibit their greatest usefulness when in solution. Lacquer solvents, for example, are selected to produce homogeneous combinations and so selected as to impart the most desirable mechanical properties. The physical properties of a fabricated solution can be regulated at will by the proper choice of solvents, thus adapting them to the most varied uses and methods of applications. Some of the more important uses for solvents are in lacquers of all kinds, oils, paints, varnishes, polishes, fat extraction and cleaning of metals and fabrics.

Solvents vary in their dissolving power, so that the line of demarcation between solvents, latent solvents and nonsolvents is difficult to define. Some of the factors which influence solvency are atmospheric conditions, purity and molecular association. Molecular aggregation is the explanation for increased, attenuated, or decreased solvent power or, more concisely, eccentric solvency. Any substance that will dissolve another is called a solvent. Thus, we have a gaseous solution when a liquid or a solid is dissolved in a gas, a liquid solution when any one of these is dissolved in a liquid, and a solid solution when any one of them is dissolved in a solid.

Mixing of solvents, diluents and thinners often results in change of solvent properties. Some chlorinated compounds become good solvents for cellulose esters when mixed with an alcohol. On the other hand, some active solvents for esters of cellulose lose some of their solvent power when mixed with hydrocarbons. Alcohols are added to lacquers to improve flow and to prevent blushing, although they vary considerably in these respects. Alcohols are not true or active solvents for nitrocellulose as are the active dissolvents like ethyl lactate or n-butyl acetate. The alcohol group, however, cannot be classed as nonsolvents like toluene or naphtha. When an alcohol is added to a true solvent, the solvent power of the latter is not reduced but, on the contrary, this active solvent activates the alcohol to such an extent that it too becomes a solvent. Therefore, alcohols are referred to as latent solvents, whose hidden solvent qualities are brought out by the addition of an active solvent. The presence of a latent solvent increases the tolerance of an active solvent for a nonsolvent. This group of latent solvents is also called extenders, because they increase the volume of a mixture without decreasing the solvent power.

In general, simple esters and ketones activate alcohols so that they too become solvents and are capable of tolerating various proportions of diluents. This is due to the molecular aggregates formed. Two-type solvents containing both an alcohol and an active solvent group, such as an ester, ether or ketone, activate alcohol to a lesser degree. Unit volumes of a solvent will activate only a limited amount of alcohol, indicating that definite molecular aggregates are formed. A mixture of 50% n-butyl acetate and 50% n-butyl alcohol will not lose its solvent power until 85 to 95% of the volume is evaporated, contributing further evidence of the validity of the theory of molecular aggregates. Plasticizers, which are the high-boiling solvents, also activate alcohols.

Liquids vary in their rate of evaporation. Naturally, in a mixture of liquids, some evaporate more rapidly than others. For example, if the solvent constituent of a lacquer evaporates more rapidly than the diluent, the limit of tolerance of the residual mixture is exceeded and gelling or precipitation occurs. As evaporation goes on, gigantic molecular reactions take place. Vast numbers of molecules change places as the new aggregates are formed. Some are replaced and some are repelled, causing immiscibility, precipitation, blushing, or one or more of the many lacquer faults. It follows that dilution ratios do not indicate tolerance during the change of solvent-nonsolvent balance which occurs during drying.

In the theory of molecular aggregation, higher concentrations of cellulose derivatives contain fewer secondary-valence bonds. Consequently, smaller amounts of diluent can be tolerated. This condition occurs during film drying. Hydroxyl-containing solvents show greater tolerance for toluene than do the simpler esters. In the case of naphtha the condition is reversed. There are, however, exceptions to this statement, among which are butyl lactate and butyl "Cellosolve," which have very high naphtha tolerance. Simple esters will tolerate 50 to 100% more naphtha than will such materials as ethyl lactate, ethyl ether, ethylene glycol, diacetone alcohol, and so forth. Ethers of glycols generally have higher dilution ratios than do the butyl esters with respect to benzene, toluene, and xylene.

Solutions of nitrocellulose tolerate larger quantities of nonsolvents than solutions of cellulose acetate. The "solvent-power number" is influenced by both the nature of the diluent and the mixing of two or more solvents. Frequently, when two or more nonsolvents are mixed, they may exhibit the qualities of a good solvent. This is especially true when one of the ingredients is an alcohol. The ether-alcohol solvent mixture for collodion is an example with which we are most familiar. Another example of acquired solubility is the mixing of butyl acetate with amyl or ethyl alcohol for the less highly polymerized forms of glyceryl phthalate resins. Some of the chlorinated hydrocarbons will dissolve nitrocellulose when mixed with an alcohol. A mixture of benzene and alcohol will dissolve nitrocellulose containing up to 11% nitrogen. A toluene-ethyl alcohol solution of alkyd resin will dissolve nitrocellulose. In many cases the solvent property of esters for resins and nitrocellulose is increased by the addition of an alcohol. On the other hand, when active solvents for cellulose esters are mixed with aliphatic or aromatic hydrocarbons, the solvent power of these active solvents is decreased.

These facts bring to light reasons why many of the old-type solvents have been valued for their impurities. For example, methyl acetone, made from the distillation of wood, had particularly valuable solvent properties. Actually, it is a mixed solvent which consists of methanol, acetone, esters and higher ketones. This mixture has certain desirable properties not obtained by any of its component ingredients when used separately. For this reason the "synthetic methyl acetone" is made to simulate it. For this same reason commercial grades of butyl and amyl acetate contain 85% ester and the remaining portion is the corresponding alcohol. Amyl acetate, containing its characteristic impurities when manufactured from fusel oil, is also valued for its solvent properties. The synthetic product is different because it lacks these impurities. It is made from the pentane fraction of gasoline by chlorination; the chloropentane is hydrolyzed to form amyl alcohol, and is finally esterified to the acetate.

Because of growing concern with atmospheric pollution, chemical composition limitations of solvent formulations have been adopted by many state and local governmental agencies in the more highly industrialized areas of the country. These rules and regulations seriously affect the use of many solvents, and solvent blends must be reformulated to conform to the maximum allowable concentrations of the restricted solvents. It is necessary for the solvent user to acquaint himself with the governmental regulations of solvent use in his particular locale.

Hydrocarbon Solvents

PARAFFINS

Table 2.1: Methane *(4)*

FORMULA	CH$_4$	
PROPERTIES	RESEARCH GRADE	PURE GRADE
Composition, mol per cent		
Nitrogen	0.01	0.61
Carbon Dioxide		0.24
Methane	99.98	99.08
Ethylene		
Ethane	0.01	0.06
Propylene		
Propane		0.01
Freezing point, triple point, F	−296.46*	
Boiling point, F	−258.68*	
Specific gravity of liquid at 60/60 F		
at 20/4 C		
Density of liquid at 60 F, lb/gal		
Vapor pressure at 70 F, psia		
Specific gravity of real gas at 60 F and 14.7 psia (Air = 1)	0.55491*	
Specific volume of real gas at 60 F and 14.7 psia, cu ft/lb	23.6113*	
Density of real gas at 60 F and 14.7 psia, lbs/cu ft	0.04235	
Liquid volume, cu ft/lb at −260 F and 13.8 psia	0.03766*	
Critical temperature, F	−115.78*	
Critical pressure, psia	673.1*	
Flash point, approximate, F	−306*	
Flammability limits, volume % in air		
Lower	5.0*	
Higher	15.0*	
Heating value for real gas at 60 F and 30 in Hg, saturated basis BTU/ cu ft		994
Heating value for ideal gas at 60 F and 14.7 psia, BTU/cu ft,		
Dry basis	1010*	
Saturated basis		985

*Literature values.

Table 2.2: Ethane *(4)*

FORMULA	CH$_3$−CH$_3$	
PROPERTIES	RESEARCH GRADE	PURE GRADE
Composition, mol per cent		
Nitrogen		
Carbon Dioxide		
Methane		trace
Ethylene	trace	0.05
Ethane	99.97	99.35
Propylene	0.01	0.25
Propane	0.02	0.35
Freezing point, triple point, F	−297.89*	
Boiling point, F	−127.53*	
Specific gravity of liquid at 60/60 F	0.3771*	
at 20/4 C	0.362*	
Density of liquid at 60 F, lb/gal	3.144*	
Vapor pressure at 70 F, psia	560*	
Specific gravity of real gas at 60 F and 14.7 psia (Air = 1)	1.0469*	
Specific volume of real gas at 60 F and 14.7 psia, cu ft/lb	12.515*	
Density of real gas at 60 F and 14.7 psia, lbs/cu ft		
Liquid volume, cu ft/lb at −260 F and 13.8 psia	0.04252 (60 F)	
Critical temperature, F	90.32*	
Critical pressure, psia	707.8*	
Flash point, approximate, F	−211*	
Flammability limits, volume % in air		
Lower	2.9*	
Higher	13.0*	
Heating value for real gas at 60 F and 30 in Hg, saturated basis BTU/ cu ft		
Heating value for ideal gas at 60 F and 14.7 psia, BTU/cu ft,		
Dry basis	1769*	
Saturated basis		

*Literature values.

Table 2.3: Propane *(4)*

FORMULA	$CH_3-CH_2-CH_3$		
PROPERTIES	**RESEARCH GRADE**	**PURE GRADE**	**TECHNICAL GRADE**
Composition, weight per cent			
Ethane		0.07	0.01
Propylene		0.01	0.01
Propane	99.98	99.35	97.50
Isobutane	0.02	0.52	2.38
Normal Butane		0.05	0.10
Butene-2			
Neopentane			
Isopentane			
Normal Pentane			
Purity by freezing point, mol percent			
Freezing point, F	−305.84* (triple point)		
Boiling point, F	−43.73*		
Specific gravity of liquid at 60/60 F	0.5077*	0.508	0.510
20/4 C	0.5005*	0.501	
API gravity at 60 F		147.0	145.9
Density of liquid at 60 F, lb/gal		4.22	4.24
Vapor pressure at 70 F, psia		123	123
100 F, psia		189	189
130 F, psia		271	271
Sulfur content, weight per cent		< 0.0005	< 0.0005
Specific gravity of real gas at 60 F and 14.7 psia (Air = 1)	1.5503*		
Specific volume of real gas at 60 F and 14.7 psia, cu ft /lb	8.4515*		
Flash point, approximate, F	−156*		
Flammability limits, volume % in air			
Lower	2.1*		
Higher	9.5*		
Heating value for ideal gas at 60 F and 14.7 psia, dry basis BTU/cu ft	2517*		

*Literature values.

Table 2.4: Isobutane *(4)*

FORMULA	$\begin{array}{c}CH_3\\ \vert\\ CH_3-CH-CH_3\end{array}$		
PROPERTIES	**RESEARCH GRADE**	**PURE GRADE**	**TECHNICAL GRADE**
Composition, weight per cent			
Ethane			
Propylene			
Propane		0.1	0.4
Isobutane	99.98	99.5	96.8
Normal Butane	0.02	0.4	2.8
Butene-2			
Neopentane			
Isopentane			
Normal Pentane			
Purity by freezing point, mol percent	99.96	99.5	
Freezing point, F	−255.28*		
Boiling point, F	10.89*		
Specific gravity of liquid at 60/60 F	0.5631*	0.563	0.563
20/4 C	0.5572*	0.557	0.557
API gravity at 60 F		119.8	119.8
Density of liquid at 60 F, lb/gal		4.68	4.68
Vapor pressure at 70 F, psia		45.8	45.4
100 F, psia		72.2	72.2
130 F, psia		111.5	111.5
Sulfur content, weight per cent		< 0.0005	< 0.0005
Specific gravity of real gas at 60 F and 14.7 psia (Air = 1)	2.06805*		
Specific volume of real gas at 60 F and 14.7 psia, cu ft /lb	6.3355*		
Flash point, approximate, F		−117	−117
Flammability limits, volume % in air			
Lower	1.8*		
Higher	8.4*		
Heating value for ideal gas at 60 F and 14.7 psia, dry basis BTU/cu ft	3253*		

*Literature values.

Table 2.5: n-Butane *(4)*

FORMULA	$CH_3-CH_2-CH_2-CH_3$		
PROPERTIES	**RESEARCH GRADE**	**PURE GRADE**	**TECHNICAL GRADE**
Composition, weight per cent			
Ethane			
Propylene			
Propane			0.6
Isobutane	0.05	0.3	1.0
Normal Butane	99.95	99.4	97.6 95.0 min
Butene-2			0.1
Neopentane			0.2
Isopentane		0.2	0.3
Normal Pentane		0.1	0.2
Purity by freezing point, mol percent	99.95	99.4 99.0 min	
Freezing point, F	−217.03*		
Boiling point, F	31.10*		
Specific gravity of liquid at 60/60 F	0.5844*	0.584	0.584
20/4 C	0.5788*	0.579	0.579
API gravity at 60 F		110.8	110.8
Density of liquid at 60 F, lb/gal		4.86	4.86
Vapor pressure at 70 F, psia		31.6	32.0
100 F, psia		51.6	52.0
130 F, psia		82.2	83.0
Sulfur content, weight per cent		< 0.0005	< 0.0005
Specific gravity of real gas at 60 F and 14.7 psia (Air = 1)	2.0757*		
Specific volume of real gas at 60 F and 14.7 psia, cu ft /lb	6.3120*		
Flash point, approximate, F		−100	−100
Flammability limits, volume % in air			
Lower	1.8*		
Higher	8.4*		
Heating value for ideal gas at 60 F and 14.7 psia, dry basis BTU/cu ft	3262*		

*Literature values.

Table 2.6: 2,2-Dimethylpropane *(4)*

Neopentane

FORMULA	$CH_3-\overset{\displaystyle CH_3}{\underset{\displaystyle CH_3}{C}}-CH_3$		
PROPERTIES	**RESEARCH GRADE**	**PURE GRADE**	**TECHNICAL GRADE**
Composition, weight per cent			
Normal Butane	trace	0.1	1.7
cis-Butene-2		trace	0.1
2,2,-Dimethylpropane	99.99+	99.6	97.8
Isopentane			
Normal Pentane			0.4
Pentene-2			
Cyclopentane			
Purity by freezing point, mol per cent	99.99	99.3	
Freezing point, F	2.21*		
Boiling point, F	49.10*		
Distillation range, F			
Initial boiling point			
10% Condensed			
50% Condensed			
90% Condensed			
Dry point			
Specific gravity of liquid at 60/60 F	0.5967*	0.597	0.597
at 20/4 C	0.5910*	0.591	0.591

(continued)

Table 2.6: (continued)

PROPERTIES	RESEARCH GRADE	PURE GRADE	TECHNICAL GRADE
API gravity at 60 F		105.5	105.5
Density of liquid at 60 F, lb/gal		4.96	4.96
Vapor pressure at 70 F, psia	21.9*	21.9	22.0
100 F, psia		35.9	36.7
130 F, psia		57.4	57.7
Refractive index, 20/D			
Color, Saybolt (unless indicated)	+30	+30	+30
Acidity, distillation residue		neutral	netural
Nonvolatile matter, grams/100 ml		0.0005	0.0005
Sulfur content, weight per cent		0.005	0.005
Copper corrosion		1	1
Doctor test		negative	negative
Kinematic viscosity, cs at 32 F	0.532*		
Specific gravity of real gas at 60 F and 14.7 psia (Air = 1)	2.622*		
Specific volume of real gas at 60 F and 14.7 psia, cu ft/lb	4.997*		
Flash point, approximate, F		−85	−85
Flammability limits, volume % in air			
Lower	1.4*		
Higher	8.3*		

*Literature values.

Table 2.7: Isopentane *(4)*

FORMULA	CH_3 $CH_3{-}CH{-}CH_2{-}CH_3$		
PROPERTIES	RESEARCH GRADE	PURE GRADE	TECHNICAL GRADE
Composition, weight per cent			
Normal Butane		0.1	0.2
cis-Butene-2			
2,2,-Dimethylpropane		0.1	0.1
Isopentane	99.99	99.4	97.1
Normal Pentane	0.01	0.4	2.6
Pentene-2			
Cyclopentane			
Purity by freezing point, mol per cent	99.99	99.4	
Freezing point, F	−255.82*		
Boiling point, F	82.13*		
Distillation range, F			
Initial boiling point			82
10% Condensed			83
50% Condensed			83
90% Condensed			84
Dry point			86
Specific gravity of liquid at 60/60 F	0.6248*	0.625	0.625
at 20/4 C	0.61967*	0.620	0.620
API gravity at 60 F		94.9	94.9
Density of liquid at 60 F, lb/gal		5.20	5.20
Vapor pressure at 70 F, psia	11.57*	11.5	11.4
100 F, psia	20.44*	20.4	20.2
130 F, psia			33.5
Refractive index, 20/D	1.35373*		
Color, Saybolt (unless indicated)	+30	+30	+30
Acidity, distillation residue		neutral	neutral
Nonvolatile matter, grams/100 ml		0.0005	0.0005
Sulfur content, weight per cent		0.005	0.005
Copper corrosion		1	1
Doctor test		negative	negative
Kinematic viscosity, cs at 32 F	0.433*		
Specific gravity of real gas at 60 F and 14.7 psia (Air = 1)	2.6269*		
Specific volume of real gas at 60 F and 14.7 psia, cu ft/lb	4.9876*		
Flash point, approximate, F		−70	−70
Flammability limits, volume % in air			
Lower	1.4*		
Higher	8.3*		

*Literature values.

Table 2.8: n-Pentane *(4)*

FORMULA	$CH_3-CH_2-CH_2-CH_2-CH_3$		
PROPERTIES	RESEARCH GRADE	PURE GRADE	TECHNICAL GRADE
Composition, weight per cent			
Normal Butane			
cis-Butene-2			
2,2,-Dimethylpropane			
Isopentane	0.01	0.2	0.5
Normal Pentane	99.99	99.4	98.8
Pentene-2		0.1	0.2
Cyclopentane		0.3	0.5
Purity by freezing point, mol per cent	99.98	99.2	
Freezing point, F	−201.50*		
Boiling point, F	96.93*		
Distillation range, F			
Initial boiling point			96
10% Condensed			97
50% Condensed			97
90% Condensed			97
Dry point			99
Specific gravity of liquid at 60/60 F	0.6312*	0.631	0.633
at 20/4 C	0.62624*	0.626	
API gravity at 60 F		92.7	92.0
Density of liquid at 60 F, lb/gal		5.25	5.27
Vapor pressure at 70 F, psia	8.56*	8.6	
100 F, psia	15.57*	15.6	
130 F, psia	26.4*	26.3	
Refractive index, 20/D	1.35748*		
Color, Saybolt (unless indicated)	+30	+30	+30
Acidity, distillation residue		neutral	neutral
Nonvolatile matter, grams/100 ml		0.0005	0.0005
Sulfur content, weight per cent		0.005	0.005
Copper corrosion		1	1
Doctor test		negative	negative
Kinematic viscosity, cs at 32 F	0.431*		
Specific gravity of real gas at 60 F and 14.7 psia (Air = 1)	2.6400*		
Specific volume of real gas at 60 F and 14.7 psia, cu ft/lb	4.9629*		
Flash point, approximate, F		−57	−50
Flammability limits, volume % in air			
Lower	1.4*		
Higher	8.3*		

*Literature values.

Table 2.9: 2,2-Dimethylbutane *(4)*

Neohexane

FORMULA	$CH_3-\overset{\overset{\displaystyle CH_3}{\vert}}{\underset{\underset{\displaystyle CH_3}{\vert}}{C}}-CH_2-CH_3$		
PROPERTIES	RESEARCH GRADE	PURE GRADE	TECHNICAL GRADE
Composition, weight per cent			
Isopentane			
Cyclopentane		0.2	0.1
2,2-Dimethylbutane	99.98	99.5	96.4 95.0 min.
2,3-Dimethylbutane	0.01	0.2	2.2
2-Methylpentane	0.01	0.1	0.3
3-Methylpentane			
Purity by freezing point, mol per cent	99.97	99.4 99.0 min.	
Freezing point, F	−147.77*		
Boiling point, F	121.53*		
Distillation range, F			
Initial boiling point			120.5
Dry point			122.2

(continued)

Table 2.9: (continued)

PROPERTIES	RESEARCH GRADE	PURE GRADE	TECHNICAL GRADE
Specific gravity of liquid at 60/60 F	0.6540*	0.655	0.659
at 20/4 C	0.64916*	0.650	0.654
API gravity at 60 F		84.5	83.2
Density of liquid at 60 F, lbs/gal		5.45	5.49
Vapor pressure at 70 F, psia	5.30*	5.3	5.3
100 F, psia	9.86*	9.9	9.9
130 F, psia	17.04*	16.8	16.8
Refractive index, 20/D	1.36876*	1.369	1.369
Color, Saybolt	+30	+30	+30
Acidity, distillation residue		neutral	neutral
Nonvolatile matter, grams/100 ml		0.0005	0.0005
Sulfur content, weight per cent		0.005	0.005
Copper corrosion		1	1
Doctor test		negative	negative
Flash point, approximate, F		−25	−25
Flammability limits, volume % in air			
Lower	1.2*		
Higher	7.7*		

*Literature values.

Table 2.10: 2,3-Dimethylbutane *(4)*

Diisopropyl

FORMULA	$CH_3-CH-CH-CH_3$ with CH_3 CH_3		
PROPERTIES	RESEARCH GRADE	PURE GRADE	TECHNICAL GRADE
Composition, weight per cent			
Isopentane			0.6
Cyclopentane			
2,2-Dimethylbutane	0.07	0.2	1.1
2,3-Dimethylbutane	99.88	99.7	98.0 95.0 min.
2-Methylpentane	0.04	0.1	0.2
3-Methylpentane	0.01		0.1
Purity by freezing point, mol per cent	99.88	99.3 99.0 min.	
Freezing point, F	−199.37*		
Boiling point, F	136.37*		
Distillation range, F			
Initial boiling point			135
Dry point			136
Specific gravity of liquid at 60/60 F	0.6664*	0.666	0.666
at 20/4 C	0.66164*	0.662	0.661
API gravity at 60 F		81.0	81.0
Density of liquid at 60 F, lbs/gal		5.54	5.54
Vapor pressure at 70 F, psia	3.87*	3.8	3.8
100 F, psia	7.40*	7.3	7.3
130 F, psia	13.12*	12.9	12.9
Refractive index, 20/D	1.37495*	1.375	1.375
Color, Saybolt	+30	+30	+30
Acidity, distillation residue		neutral	neutral
Nonvolatile matter, grams/100 ml		0.0005	0.0005
Sulfur content, weight per cent		0.005	0.005
Copper corrosion		1	1
Doctor test		negative	negative
Flash point, approximate, F		−20	−20
Flammability limits, volume % in air			
Lower	1.2*		
Higher	7.7*		

*Literature values.

Table 2.11: 2-Methylpentane *(4)*

FORMULA	$CH_3-CH-CH_2-CH_2-CH_3$ (with CH_3 branch)		
PROPERTIES	RESEARCH GRADE	PURE GRADE	TECHNICAL GRADE
Composition, weight per cent			
Isopentane			
Cyclopentane			0.2
2,2-Dimethylbutane			
2,3-Dimethylbutane	0.01	0.5	3.8
2-Methylpentane	99.98	99.3	95.4 95.0 min.
3-Methylpentane	0.01	0.2	0.6
Purity by freezing point, mol per cent	99.98	99.2 99.0 min.	
Freezing point, F	−244.61*		
Boiling point, F	140.49*		
Distillation range, F			
Initial boiling point			140
Dry point			141
Specific gravity of liquid at 60/60 F	0.6579*	0.658	0.658
at 20/4 C	0.65315*	0.653	0.653
API gravity at 60 F		85.2	85.2
Density of liquid at 60 F, lbs/gal		5.44	5.44
Vapor pressure at 70 F, psia	3.48*	3.5	3.5
100 F, psia	6.77*	6.8	6.8
130 F, psia	13.32*	13.0	13.0
Refractive index, 20/D	1.37145*	1.371	1.371
Color, Saybolt	+30	+30	+30
Acidity, distillation residue		neutral	neutral
Nonvolatile matter, grams/100 ml		0.0005	0.0005
Sulfur content, weight per cent		0.005	0.005
Copper corrosion		1	1
Doctor test		negative	negative
Flash point, approximate, F		−10	−10
Flammability limits, volume % in air			
Lower	1.2*		
Higher	7.7*		

*Literature values.

Table 2.12: 3-Methylpentane *(4)*

FORMULA	$CH_3-CH_2-CH-CH_2-CH_3$ (with CH_3 branch)		
PROPERTIES	RESEARCH GRADE	PURE GRADE	TECHNICAL GRADE
Composition, weight per cent			
2,3-Dimethylbutane			0.1
2-Methylpentane	0.01	0.6	3.8
3-Methylpentane	99.99	99.4 99.0 min	0.02
Normal Hexane			
Methylcyclopentane			
2,2-Dimethylpentane			
2,4-Dimethylpentane			
Cyclohexane			
2,3-Dimethylpentane			
2-Methylhexane			
3-Methylhexane			
Purity by freezing point, mol per cent	**		
Freezing point, F	**		
Boiling point, F	145.91*		
Distillation range, F			
Initial boiling point			145
10% Condensed			
50% Condensed			
90% Condensed			
Dry point			146

(continued)

Table 2.12: (continued)

PROPERTIES	RESEARCH GRADE	PURE GRADE	TECHNICAL GRADE
Specific gravity of liquid at 60/60F	0.6690*	0.669	0.669
at 20/4 C	0.66431*	0.664	0.664
API gravity at 60 F		80.0	80.0
Density of liquid at 60 F, lbs/gal		5.57	5.57
Vapor pressure at 70 F, psia	3.11*	3.1	3.1
100 F, psia	6.10*	6.1	6.0
130 F, psia	11.03*	11.0	10.9
Refractive index, 20/D	1.37652*	1.376	1.376
Color, Saybolt	+30	+30	+30
Acidity, distillation residue		neutral	neutral
Nonvolatile matter, grams/100 ml		0.0005	0.0005
Sulfur content, weight per cent		0.005	0.005
Copper corrosion		1	1
Doctor test		negative	negative
Flash point, approximate, F		−25	25
Flammability limits, volume % in air			
Lower	1.2*		
Higher	7.7*		

*Literature values. **Forms a glass.

Table 2.13: n-Hexane *(4)*

FORMULA	$CH_3-CH_2-CH_2-CH_2-CH_2-CH_3$		
PROPERTIES	RESEARCH GRADE	PURE GRADE	TECHNICAL GRADE
Composition, weight per cent			
2,3-Dimethylbutane			
2-Methylpentane	trace	trace	trace
3-Methylpentane	0.02	0.1	0.2
Normal Hexane	99.98	99.5	97.7 95.0 min
Methylcyclopentane	trace	0.4	2.1
2,2-Dimethylpentane			
2,4-Dimethylpentane			
Cyclohexane			
2,3-Dimethylpentane			
2-Methylhexane			
3-Methylhexane			
Purity by freezing point, mol per cent	99.98	99.4	
Freezing point, F	−139.63*		
Boiling point, F	155.73*		
Distillation range, F			
Initial boiling point			155.1
10% Condensed			155.3
50% Condensed			155.3
90% Condensed			155.7
Dry point			156.4
Specific gravity of liquid at 60/60F	0.6640*	0.664	0.666
at 20/4 C	0.65937*	0.660	0.661
API gravity at 60 F		81.6	81.0
Density of liquid at 60 F, lbs/gal		5.53	5.54
Vapor pressure at 70 F, psia	2.46*	2.5	2.5
100 F, psia	4.96*	5.0	4.9
130 F, psia	9.17*	9.2	9.1
Refractive index, 20/D	1.37486*	1.375	1.375
Color, Saybolt	+30	+30	+30
Acidity, distillation residue		neutral	neutral
Nonvolatile matter, grams/100 ml		0.0005	0.0005
Sulfur content, weight per cent		0.005	0.005
Copper corrosion		1	1
Doctor test		negative	negative
Flash point, approximate, F		−10	−10
Flammability limits, volume % in air			
Lower	1.2*		
Higher	7.7*		

*Literature values. **Forms a glass.

Table 2.14: 2,4-Dimethylpentane *(4)*

FORMULA	CH_3 CH_3 $CH_3-CH-CH_2-CH-CH_3$		
PROPERTIES	**RESEARCH GRADE**	**PURE GRADE**	**TECHNICAL GRADE**
Composition, weight per cent			
2,3-Dimethylbutane			
2-Methylpentane			
3-Methylpentane			
Normal Hexane		trace	0.1
Methylcyclopentane			
2,2-Dimethylpentane	0.01	0.1	2.9
2,4-Dimethylpentane	99.99	99.7	96.0 95.0 min
Cyclohexane		0.1	0.5
2,3-Dimethylpentane		0.1	0.5
2-Methylhexane			
3-Methylhexane			
Purity by freezing point, mol per cent	99.77	99.2 99.0 min	
Freezing point, F	−182.64*		
Boiling point, F	176.90*		
Distillation range, F			
Initial boiling point			175
10% Condensed			
50% Condensed			
90% Condensed			
Dry point			176
Specific gravity of liquid at 60/60F	0.6772*	0.677	0.678
at 20/4 C	0.67270*	0.673	0.673
API gravity at 60 F		77.4	77.2
Density of liquid at 60 F, lbs/gal		5.64	5.64
Vapor pressure at 70 F, psia	1.59*	1.6	1.6
100 F, psia	3.29*	3.3	3.3
130 F, psia	6.24*	6.2	6.2
Refractive index, 20/D	1.38145*	1.381	1.381
Color, Saybolt	+30	+30	+30
Acidity, distillation residue		neutral	neutral
Nonvolatile matter, grams/100 ml		0.0005	0.0005
Sulfur content, weight per cent		0.005	0.005
Copper corrosion		1	1
Doctor test		negative	negative
Flash point, approximate, F		10	10
Flammability limits, volume % in air			
Lower	1.0*		
Higher	7.0*		

*Literature values.

Table 2.15: 2,3-Dimethylpentane *(4)*

FORMULA	CH_3 CH_3 $CH_3-CH-CH-CH_2-CH_3$	PROPERTIES	90% GRADE
PROPERTIES	**90% GRADE**	50% Condensed	
		90% Condensed	
Composition, weight per cent		Dry point	194
2,3-Dimethylbutane		Specific gravity of liquid at 60/60F	0.6990
2-Methylpentane		at 20/4 C	0.6943
3-Methylpentane		API gravity at 60 F	70.9
Normal Hexane		Density of liquid at 60 F, lbs/gal	5.82
Methylcyclopentane		Vapor pressure at 70 F, psia	1.2
2,2-Dimethylpentane		100 F, psia	3.6
2,4-Dimethylpentane		130 F, psia	
Cyclohexane		Refractive index, 20/D	1.3922
2,3-Dimethylpentane	90.4 90.0 min	Color, Saybolt	+30
2-Methylhexane	3.4	Acidity, distillation residue	neutral
3-Methylhexane	6.2	Nonvolatile matter, grams/100 ml	0.0005
Purity by freezing point, mol per cent		Sulfur content, weight per cent	
Freezing point, F		Copper corrosion	
Boiling point, F		Doctor test	negative
Distillation range, F		Flash point, approximate, F	< 10
Initial boiling point	193	Flammability limits, volume % in air	
10% Condensed		Lower	
		Higher	

Table 2.16: 3-Methylhexane (4)

FORMULA	$CH_3-CH_2-\overset{\overset{CH_3}{|}}{CH}-(CH_2)_2-CH_3$
PROPERTIES	**TECHNICAL GRADE**
Composition, weight percent	
2,3-Dimethylpentane	0.1
2-Methylhexane	1.4
3-Methylhexane	97.2 95.0 min
3-Ethylpentane	0.8
Normal Heptane	
Dimethylcyclopentane	0.5
Methylcyclohexane	
2,2-Dimethylhexane	
2,4-Dimethylhexane	
2,5-Dimethylhexane	
Other Dimethylhexanes	
2,2,4-Trimethylpentane	
2,2,3-Trimethylpentane	
2,3,4-Trimethylpentane	
2,3,3-Trimethylpentane	

PROPERTIES	TECHNICAL GRADE
Boiling point, F	
Distillation range, F	
Initial boiling point	195
Dry point	196
Specific gravity of liquid at 60/60 F	0.692
20/4 C	0.688
API gravity at 60 F	73.0
Density of liquid at 60 F, lbs/gal	5.76
Vapor pressure at 70 F, psia	
100 F, psia	2.1
130 F, psia	
Refractive index, 20/D	1.388
Color, Saybolt	+30
Acidity, distillation residue	neutral
Nonvolatile matter, grams/100 ml	0.0005
Sulfur content, weight percent	0.005
Copper corrosion	1
Doctor test	negative
Flash point, approximate, F	25 Estimated
Flammabilty limits, volume % in air	
Lower	1
Higher	7

Table 2.17: n-Heptane (4)

FORMULA	$CH_3-(CH_2)_5-CH_3$	
PROPERTIES	**RESEARCH GRADE**	**PURE GRADE**
Composition, weight percent		
2,3-Dimethylpentane		
2-Methylhexane		
3-Methylhexane		
3-Ethylpentane		trace
Normal Heptane	99.99	99.8
Dimethylcyclopentane	0.01	0.2
Methylcyclohexane		trace
2,2-Dimethylhexane		
2,4-Dimethylhexane		
2,5-Dimethylhexane		
Other Dimethylhexanes		
2,2,4-Trimethylpentane		
2,2,3-Trimethylpentane		
2,3,4-Trimethylpentane		
2,3,3-Trimethylpentane		
Purity by freezing point, mol %	99.92	99.7 99.0 min
Freezing point, F	−131.10*	
Boiling point, F	209.17*	
Distillation range, F		
Initial boiling point		
Dry point		
Specific gravity of liquid at 60/60 F	0.6882*	0.688
20/4 C	0.68376*	0.684
API gravity at 60 F		74.1
Density of liquid at 60 F, lbs/gal		5.73
Vapor pressure at 70 F, psia		
100 F, psia	1.62*	1.6
130 F, psia		
Refractive index, 20/D	1.38764*	1.388
Color, Saybolt	+30	+30
Acidity, distillation residue		neutral
Nonvolatile matter, grams/100 ml		0.0005
Sulfur content, weight percent		0.005
Copper corrosion		1
Doctor test		negative
Flash point, approximate, F		25
Flammabilty limits, volume % in air		
Lower		1.0*
Higher		7.0*

*Literature values.

Table 2.18: 2,2,4-Trimethylpentane (4)

Isooctane

FORMULA	$CH_3-\overset{\overset{CH_3}{|}}{\underset{\underset{CH_3}{|}}{C}}-CH_2-\overset{\overset{CH_3}{|}}{CH}-CH_3$	
PROPERTIES	**RESEARCH GRADE**	**PURE GRADE**
Composition, weight percent		
2,3-Dimethylpentane		
2-Methylhexane		
3-Methylhexane		
3-Ethylpentane		
Normal Heptane		trace
Dimethylcyclopentane		
Methylcyclohexane		
2,2-Dimethylhexane	0.01	0.2
2,4-Dimethylhexane		
2,5-Dimethylhexane		
Other Dimethylhexanes		
2,2,4-Trimethylpentane	99.99	99.8
2,2,3-Trimethylpentane		
2,3,4-Trimethylpentane		
2,3,3-Trimethylpentane		
Purity by freezing point, mol %	99.98	99.7 99.0 min
Freezing point, F	−161.28*	
Boiling point, F	210.63*	
Distillation range, F		
Initial boiling point		
Dry point		
Specific gravity of liquid at 60/60 F	0.6963*	0.696
20/4 C	0.69193*	0.692
API gravity at 60 F		71.7
Density of liquid at 60 F, lbs/gal		5.80
Vapor pressure at 70 F, psia	0.79*	0.8
100 F, psia	1.71*	1.7
130 F, psia	3.37*	3.3
Refractive index, 20/D	1.39145*	1.391
Color, Saybolt	+30	+30
Acidity, distillation residue		neutral
Nonvolatile matter, grams/100 ml		0.0005
Sulfur content, weight percent		0.005
Copper corrosion		1
Doctor test		negative
Flash point, approximate, F		18
Flammabilty limits, volume % in air		
Lower		
Higher		

*Literature values.

Table 2.19: 2,3,4-Trimethylpentane (4)

FORMULA	$CH_3-\overset{\underset{\displaystyle |}{CH_3}}{CH}-\overset{\underset{\displaystyle |}{CH_3}}{CH}-\overset{\underset{\displaystyle |}{CH_3}}{CH}-CH_3$		
PROPERTIES	RESEARCH GRADE	PURE GRADE	TECHNICAL GRADE
Composition, weight percent			
2,3-Dimethylpentane			
2-Methylhexane			
3-Methylhexane			
3-Ethylpentane			
Normal Heptane			
Dimethylcyclopentane			
Methylcyclohexane			
2,2-Dimethylhexane			
2,4-Dimethylhexane			
2,5-Dimethylhexane			
Other Dimethylhexanes	trace	trace	0.6
2,2,4-Trimethylpentane			
2,2,3-Trimethylpentane			
2,3,4-Trimethylpentane	99.99+	99.8	98.0 95.0 min
2,3,3-Trimethylpentane	trace	0.2	1.4
Purity by freezing point, mol %		99.1 99.0 min	
Freezing point, F	−164.58*		
Boiling point, F	236.24*		
Distillation range, F			
Initial boiling point			
Dry point			
Specific gravity of liquid at 60/60 F	0.7233*	0.723	0.723
20/4 C	0.71906*	0.719	0.719
API gravity at 60 F		64.1	64.1
Density of liquid at 60 F, lbs/gal		6.02	6.02
Vapor pressure at 70 F, psia			
100 F, psia	0.98*	1.0	1.0
130 F, psia			
Refractive index, 20/D	1.40422*	1.404	1.404
Color, Saybolt	+30	+30	+30
Acidity, distillation residue		neutral	neutral
Nonvolatile matter, grams/100 ml		0.0005	0.0005
Sulfur content, weight percent		0.005	0.005
Copper corrosion		1	1
Doctor test		negative	negative
Flash point, approximate, F	41 (D 56)	41 (D 56)	41 (D 56)
Flammabilty limits, volume % in air			
Lower			
Higher			

*Literature values.

Table 2.20: Mixed Trimethylpentanes (4)

FORMULA	C_8H_{18}
PROPERTIES	TECHNICAL GRADE
Composition, weight percent	
2,3-Dimethylpentane	
2-Methylhexane	
3-Methylhexane	
3-Ethylpentane	
Normal Heptane	
Dimethylcyclopentane	
Methylcyclohexane	
2,2-Dimethylhexane	0.3
2,4-Dimethylhexane	0.1
2,5-Dimethylhexane	0.1
Other Dimethylhexanes	3.5
2,2,4-Trimethylpentane	
2,2,3-Trimethylpentane	0.1
2,3,4-Trimethylpentane	80.9 } 95.0 min
2,3,3-Trimethylpentane	15.0
Purity by freezing point, mol %	
Freezing point, F	
Boiling point, F	
Distillation range, F	
Initial boiling point	235
Dry point	236
Specific gravity of liquid at 60/60 F	0.723
20/4 C	0.719
API gravity at 60 F	64.2
Density of liquid at 60 F, lbs/gal	6.02
Vapor pressure at 70 F, psia	
100 F, psia	1.0
130 F, psia	
Refractive index, 20/D	1.404
Color, Saybolt	+30
Acidity, distillation residue	neutral
Nonvolatile matter, grams/100 ml	0.0005
Sulfur content, weight percent	0.005
Copper corrosion	1
Doctor test	negative
Flash point, approximate, F	50
Flammabilty limits, volume % in air	
Lower	
Higher	

Table 2.21: Mixed Dimethylhexanes (4)

FORMULA	C_8H_{18}
PROPERTIES	TECHNICAL GRADE
Composition, weight percent	
2,3-Dimethylpentane	
2-Methylhexane	
3-Methylhexane	
3-Ethylpentane	
Normal Heptane	
Dimethylcyclopentane	
Methylcyclohexane	
2,2-Dimethylhexane	4.3
2,4-Dimethylhexane	36.7 } 95.0 min
2,5-Dimethylhexane	53.9
Other Dimethylhexanes	
2,2,4-Trimethylpentane	1.6
2,2,3-Trimethylpentane	3.5
2,3,4-Trimethylpentane	
2,3,3-Trimethylpentane	
Purity by freezing point, mol %	
Freezing point, F	
Boiling point, F	

PROPERTIES	TECHNICAL GRADE
Distillation range, F	
Initial boiling point	228.6
Dry point	228.8
Specific gravity of liquid at 60/60 F	0.704
20/4 C	0.700
API gravity at 60 F	69.4
Density of liquid at 60 F, lbs/gal	5.86
Vapor pressure at 70 F, psia	
100 F, psia	1.0
130 F, psia	
Refractive index, 20/D	1.394
Color, Saybolt	+30
Acidity, distillation residue	neutral
Nonvolatile matter, grams/100 ml	0.0005
Sulfur content, weight percent	0.005
Copper corrosion	1
Doctor test	negative
Flash point, approximate, F	50
Flammabilty limits, volume % in air	
Lower	
Higher	

*Literature values.

Table 2.22: n-Octane (4)

FORMULA	$CH_3-(CH_2)_6-CH_3$		
PROPERTIES	RESEARCH GRADE	PURE GRADE	TECHNICAL GRADE
Composition, weight per cent			
Isooctanes	0.02	0.3	0.7
Normal Octane	99.92	99.6	98.7
2,2,5-Trimethylhexane			
2,2,4-Trimethylhexane			
Isononanes	0.06	0.1	0.6
Isoparaffins			
Normal Nonane			
Purity by freezing point, mol %	99.88	99.2 99.0 min	96.2 99.0 min
Freezing point, F	−70.23*		
Boiling point, F	258.20*		
Distillation range, F			
Initial boiling point			
Dry point			
Specific gravity of liquid at 60/60 F	0.7068*	0.707	0.707
at 20/4 C	0.70252*	0.702	0.702
API gravity at 60 F		68.6	68.2
Density of liquid at 60 F, lbs/gal		5.89	5.89
Vapor pressure at 70 F, psia			
100 F, psia	0.54*	0.5	0.5
130 F, psia			
Refractive index, 20/D	1.39743*	1.397	1.397
Color, Saybolt	+30	+30	+30
Acidity, distillation residue		neutral	neutral
Nonvolatile matter, grams/100 ml		0.0005	0.0005
Sulfur content, weight percent		0.005	0.005
Copper corrosion		1	1
Doctor test		negative	negative
Flash point, approximate, F		72	72

*Literature values.

Table 2.23: 2,2,5-Trimethylhexane (4)

FORMULA	$CH_3-\overset{\overset{CH_3}{\mid}}{\underset{\underset{CH_3}{\mid}}{C}}-CH_2-CH_2-\overset{\overset{CH_3}{\mid}}{CH}-CH_3$		
PROPERTIES	RESEARCH GRADE	PURE GRADE	TECHNICAL GRADE
Composition, weight per cent			
Isooctanes			
Normal Octane			
2,2,5-Trimethylhexane	99.99	99.6	97.2 95.0 min
2,2,4-Trimethylhexane	0.01	0.4	2.8
Isononanes			
Isoparaffins			
Normal Nonane			
Purity by freezing point, mol %	99.80	99.3 99.0 min	
Freezing point, F	−158.40*		
Boiling point, F	255.35		
Distillation range, F			
Initial boiling point			
Dry point			
Specific gravity of liquid at 60/60 F	0.7174*	0.717	0.717
at 20/4 C	0.70721*	0.707	0.707
API gravity at 60 F		65.7	65.7
Density of liquid at 60 F, lbs/gal		5.97	5.97
Vapor pressure at 70 F, psia	0.26*	0.3	0.3
100 F, psia	0.62*	0.6	0.6
130 F, psia	1.34*	1.3	1.3
Refractive index, 20/D	1.39972*	1.400	1.400
Color, Saybolt	+30	+30	+30
Acidity, distillation residue		neutral	netural
Nonvolatile matter, grams/100 ml		0.0005	0.0005
Sulfur content, weight percent		0.005	0.005
Copper corrosion		1	1
Doctor test		negative	negative
Flash point, approximate, F		55	55

*Literature values.

Table 2.24: n-Nonane *(4)*

FORMULA	$CH_3-(CH_2)_7-CH_3$		
PROPERTIES	RESEARCH GRADE	PURE GRADE	TECHNICAL GRADE
Composition, weight per cent			
Isooctanes			
Normal Octane			
2,2,5-Trimethylhexane			
2,2,4-Trimethylhexane			
Isononanes			
Isoparaffins	0.1	0.4	0.5
Normal Nonane	99.9	99.6	99.5
Purity by freezing point, mol %	99.67	99.2 99.0 min	95.9 95.0 min
Freezing point, F	−64.33*		
Boiling point, F	303.44		
Distillation range, F			
Initial boiling point			303.4
Dry point			304.0
Specific gravity of liquid at 60/60 F	0.7217*	0.722	0.722
at 20/4 C	0.71763*	0.718	0.718
API gravity at 60 F		64.4	64.4
Density of liquid at 60 F, lbs/gal		6.01	6.01
Vapor pressure at 70 F, psia			
100 F, psia	0.18*	0.2	0.2
130 F, psia			
Refractive index, 20/D	1.40542*	1.405	1.397
Color, Saybolt	+30	+30	+30
Acidity, distillation residue		neutral	neutral
Nonvolatile matter, grams/100 ml		0.0005	0.0005
Sulfur content, weight percent			
Copper corrosion			
Doctor test			
Flash point, approximate, F		86	86

*Literature values.

Table 2.25: n-Decane *(4)*

FORMULA	$CH_3-(CH_2)_8-CH_3$		
PROPERTIES	RESEARCH GRADE	PURE GRADE	TECHNICAL GRADE
Composition, weight per cent			
Normal Nonane	0.05		
Normal Decane	99.94	99.5	99.0
Normal Undecane			
Normal Dodecane			
Normal Tridecane			
Isoparaffins	0.01	0.5	1
Purity by freezing point, mol %	99.55	99.1 99.0 min	96.5 95.0 min
Freezing point, F	−21.39*		
Boiling point, F	345.42*		
Distillation range, F			
Initial boiling point			344.9
Dry Point			345.4
Specific gravity of liquid at 60/60 F	0.7341*	0.734	0.734
at 20/4 C	0.73005*	0.730	0.730
API gravity at 60 F		61.3	61.3
Density of liquid at 60 F, lbs/gal		6.11	6.11
Vapor pressure at 100 F, psia		0.1	0.1
Refractive index, 20/D	1.41189*	1.412	1.412
Color, Saybolt	+30	+30	+30
Acidity, distillation residue		neutral	neutral
Nonvolatile matter, grams/100 ml		0.0005	0.0005
Flash point, approximate, F		111	111

*Literature values.

Table 2.26: n-Undecane *(4)*

FORMULA	$CH_3-(CH_2)_9-CH_3$		
PROPERTIES	**RESEARCH GRADE**	**PURE GRADE**	**TECHNICAL GRADE**
Composition, weight per cent			
Normal Nonane			
Normal Decane			
Normal Undecane	99.8	99.6	99.1
Normal Dodecane			
Normal Tridecane			
Isoparaffins	0.2	0.4	0.9
Purity by freezing point, mol %	99.64	99.1 99.0 min	96.7 95.0 min
Freezing point, F	−14.07*		
Boiling point, F	384.60*		
Distillation range, F			
Initial boiling point			384
Dry Point			385
Specific gravity of liquid at 60/60 F	0.7443*	0.744	0.744
at 20/4 C	0.74024*	0.740	0.739
API gravity at 60 F		58.7	58.7
Density of liquid at 60 F, lbs/gal		6.19	6.19
Vapor pressure at 100 F, psia			
Refractive index, 20/D	1.41725*	1.417	1.419
Color, Saybolt	+30	+30	+30
Acidity, distillation residue		neutral	neutral
Nonvolatile matter, grams/100 ml		0.0005	0.0005
Flash point, approximate, F		149	149

*Literature values.

Table 2.27: n-Dodecane *(4)*

FORMULA	$CH_3-(CH_2)_{10}-CH_3$		
PROPERTIES	**RESEARCH GRADE**	**PURE GRADE**	**TECHNICAL GRADE**
Composition, weight per cent			
Normal Nonane			
Normal Decane			
Normal Undecane	0.05	0.1	0.3
Normal Dodecane	99.95	99.9	99.7
Normal Tridecane			trace
Isoparaffins			
Purity by freezing point, mol %	99.70	99.3 99.0 min	95.5 95.0 min
Freezing point, F	14.74*		
Boiling point, F	421.30*		
Distillation range, F			
Initial boiling point		419	418
Dry Point		424	424
Specific gravity of liquid at 60/60 F	0.7526*	0.753	0.753
at 20/4 C	0.74869*	0.749	
API gravity at 60 F		56.4	56.4
Density of liquid at 60 F, lbs/gal		6.27	6.26
Vapor pressure at 100 F, psia			
Refractive index, 20/D	1.42160*	1.422	1.422
Color, Saybolt	+30	+30	+30
Acidity, distillation residue		neutral	neutral
Nonvolatile matter, grams/100 ml		0.0005	0.0005
Flash point, approximate, F		160	160

*Literature values.

Table 2.28: n-Tridecane *(4)*

FORMULA	$CH_3-(CH_2)_{11}-CH_3$		
PROPERTIES	**RESEARCH GRADE**	**PURE GRADE**	**TECHNICAL GRADE**
Composition, weight percent			
Normal Tridecane	99.9	99.8	99.2
Normal Tetradecane			
Normal Pentadecane			
Normal Hexadecane			
Normal Heptadecane			
Isoparaffins	0.1	0.2	0.8
Purity by freezing point, mol %	99.80	99.49 99.0 min	96.81 95.0 min
Freezing point, F	22.29*		
Boiling point, F	455.78*		
Distillation range, F			
Initial boiling point			452
10% Condensed			
50% Condensed			
90% Condensed			
Dry point			458
Specific gravity of liquid at 60/60 F	0.7601*	0.760	0.762
at 20/4 C	0.75622*	0.756	0.758
API gravity at 80 F			
API gravity at 60 F		54.7	54.2
Density of liquid at 60 F, lbs/gal		6.33	6.34
Refractive index, 20/D	1.42560*	1.426	1.427
Color, Gardner			
Acidity, distillation residue		neutral	neutral
Sulfur content, weight percent		0.005	0.005
Bromine number			
Kinematic viscosity, cs at 77 F			2.25
Flash point, approximate, F		175 (D-56)	175 (D-56)

*Literature values

Table 2.29: n-Tetradecane *(4)*

FORMULA	$CH_3-(CH_2)_{12}-CH_3$	
PROPERTIES	**PURE GRADE**	**TECHNICAL GRADE**
Composition, weight percent		
Normal Tridecane		
Normal Tetradecane	99.6	99
Normal Pentadecane		
Normal Hexadecane		
Normal Heptadecane		
Isoparaffins	0.4	1
Purity by freezing point, mol %	99.14	95.8 95.0 min
Freezing point, F	42.55*	
Boiling point, F	488.33*	
Distillation range, F		
Initial boiling point		485
10% Condensed		
50% Condensed		
90% Condensed		
Dry point		492
Specific gravity of liquid at 60/60 F	0.7667*	0.769
at 20/4 C	0.76276*	0.765
API gravity at 80 F		
API gravity at 60 F		52.5
Density of liquid at 60 F, lbs/gal		6.40
Refractive index, 20/D	1.42892*	1.430
Color, Gardner	1	1
Acidity, distillation residue		
Sulfur content, weight percent		
Bromine number		
Kinematic viscosity, cs at 77 F		
Flash point, approximate, F	250**	250**

*Literature values

Table 2.30: n-Pentadecane *(4)*

FORMULA	$CH_3-(CH_2)_{13}-CH_3$
PROPERTIES	**TECHNICAL GRADE**
Composition, weight percent	
Normal Tridecane	
Normal Tetradecane	
Normal Pentadecane	99.7
Normal Hexadecane	
Normal Heptadecane	
Isoparaffins	0.3
Purity by freezing point, mol %	96.80 95.0 min
Freezing point, F	48.74
Boiling point, F	
Distillation range, F	
Initial boiling point	502
10% Condensed	512
50% Condensed	514
90% Condensed	516
Dry point	
Specific gravity of liquid at 60/60 F	0.7721*
at 20/4 C	0.76830*
API gravity at 80 F	
API gravity at 60 F	51.77*
Density of liquid at 60 F, lbs/gal	6.43*
Refractive index, 20/D	1.4332
Color, Gardner	< 1
Acidity, distillation residue	
Sulfur content, weight percent	
Bromine number	0.10
Kinematic viscosity, cs at 77 F	
Flash point, approximate, F	270

*Literature values

Table 2.31: n-Hexadecane *(4)*

FORMULA	$CH_3-(CH_2)_{14}-CH_3$
PROPERTIES	**TECHNICAL GRADE**
Composition, weight percent	
Normal Tridecane	
Normal Tetradecane	
Normal Pentadecane	0.2
Normal Hexadecane	99.6
Normal Heptadecane	
Isoparaffins	0.2
Purity by freezing point, mol %	96.35 95.0 min
Freezing point, F	63.79
Boiling point, F	
Distillation range, F	
Initial boiling point	521
10% Condensed	531
50% Condensed	531
90% Condensed	533
Dry point	540
Specific gravity of liquid at 60/60 F at 20/4 C	
API gravity at 80 F	51.8
API gravity at 60 F	49.9†
Density of liquid at 60 F, lbs/gal	6.49
Refractive index, 20/D	1.4352
Color, Gardner	< 1
Acidity, distillation residue	
Sulfur content, weight percent	
Bromine number	0.21
Kinematic viscosity, cs at 77 F	
Flash point, approximate, F	275

*Literature values

Table 2.32: n-Heptadecane *(4)*

FORMULA	$CH_3-(CH_2)_{15}-CH_3$
PROPERTIES	**TECHNICAL GRADE**
Composition, weight percent	
Normal Tridecane	
Normal Tetradecane	
Normal Pentadecane	
Normal Hexadecane	0.3
Normal Heptadecane	99.4
Isoparaffins	0.3
Purity by freezing point, mol %	96.60 95.0 min
Freezing point, F	70.47
Boiling point, F	
Distillation range, F	5 mm Hg
Initial boiling point	289
10% Condensed	291
50% Condensed	292
90% Condensed	292
Dry point	
Specific gravity of liquid at 60/60 F at 20/4 C	
API gravity at 80 F	50.8
API gravity at 60 F	48.9†
Density of liquid at 60 F, lbs/gal	6.53
Refractive index, 20/D	
Color, Gardner	< 1
Acidity, distillation residue	
Sulfur content, weight percent	
Bromine number	0.43
Kinematic viscosity, cs at 77 F	
Flash point, approximate, F	300

*Literature values

Table 2.33: n-Octadecane *(4)*

FORMULA	$CH_3-(CH_2)_{16}-CH_3$
PROPERTIES	**TECHNICAL GRADE**
Composition, weight percent	
Normal Hexadecane	0.1
Normal Heptadecane	0.3
Normal Octadecane	99.2
Normal Nonadecane	
Normal Eicosane	
Isoparaffins	0.4
Purity by freezing point, mol %	95.95 95.0 min
Freezing point, F	81.82
Distillation range, F	5 mm Hg
Initial boiling point	302
10% Condensed	310
50% Condensed	312
90% Condensed	312
95% Condensed	313
API gravity at 100 F	51.8
API gravity at 60 F	48.0†
Density of liquid at 60 F, lbs/gal	6.56
Color, Gradner	1
Bromine number	0.48
Flash point, approximate, F	330

†API gravity at 60 F is corrected from 100F.

Table 2.34: n-Nonadecane *(4)*

FORMULA	$CH_3-(CH_2)_{17}-CH_3$
PROPERTIES	**TECHNICAL GRADE**
Composition, weight percent	
Normal Hexadecane	
Normal Heptadecane	
Normal Octadecane	0.5
Normal Nonadecane	99.3
Normal Eicosane	
Isoparaffins	0.2
Purity by freezing point, mol %	95.37 95.0 min
Freezing point, F	87.98
Distillation range, F	5 mm Hg
Initial boiling point	320
10% Condensed	333
50% Condensed	336
90% Condensed	336
95% Condensed	336
API gravity at 100 F	51.0
API gravity at 60 F	47.3†
Density of liquid at 60 F, lbs/gal	6.59
Color, Gradner	1
Bromine number	0.53
Flash point, approximate, F	335

†API gravity at 60 F is corrected from 100F.

Table 2.35: n-Eicosane *(4)*

FORMULA	$CH_3-(CH_2)_{18}-CH_3$
PROPERTIES	**90% GRADE**
Composition, weight percent	
Normal Hexadecane	
Normal Heptadecane	
Normal Octadecane	
Normal Nonadecane	1.25
Normal Eicosane	98.75
Isoparaffins	
Purity by freezing point, mol %	91.83 90.0 min
Freezing point, F	95.83

PROPERTIES	90% GRADE
Distillation range, F	5 mm Hg
Initial boiling point	340
10% Condensed	352
50% Condensed	354
90% Condensed	355
95% Condensed	356
API gravity at 100 F	49.7
API gravity at 60 F	46.1†
Density of liquid at 60 F, lbs/gal	6.63
Color, Gradner	1
Bromine number	0.74
Flash point, approximate, F	360

†API gravity at 60 F is corrected from 100F.

CYCLOPARAFFINS

Table 2.36: Cyclopentane *(4)*

FORMULA	$\begin{array}{l}CH_2-CH_2\\ |\qquad\quad > CH_2\\ CH_2-CH_2\end{array}$			
PROPERTIES	**RESEARCH GRADE**	**PURE GRADE**	**TECHNICAL GRADE**	**90% GRADE**
Composition, weight percent				
Normal Pentane		0.2	2.0	
Cyclopentane	99.97	99.5	97.4 95.0 min	93 90 min
2,2-Dimethylbutane	0.03	0.1	0.1	**
Normal Hexane			0.2	
Methylcyclopentane		0.2	0.3	
Purity by freezing point, mol %	99.97	99.5 99.0 min		
Freezing point, F	−136.96*			
Boiling point, F	120.67*			
Distillation range, F				
Initial boiling point			120.6	120.4
10% Condensed				120.9
50% Condensed				120.9
90% Condensed				121.1
Dry point			120.8	121.5
Specific gravity of liquid at 60/60 F	0.7505*	0.750	0.749	0.744
at 20/4 C	0.74538*	0.745	0.745	
API gravity at 60 F		57.2	57.2	58.8
Density of liquid at 60 F, lbs/gal		6.24	6.24	6.19
Vapor pressure at 70 F, psia	5.25*	5.3	5.3	
100 F, psia	9.91*	9.9	9.9	10.0
130 F, psia	17.37*	17.4	17.4	
Refractive index, 20/D	1.40645*	1.406	1.405	1.404
Color, Saybolt	+30	+30	+30	
Acidity, distillation residue		neutral	neutral	neutral
Nonvolatile matter, grams/100 ml		0.0005	0.0005	0.0005
Sulfur content, weight percent				0.006
Kauri Butanol value				53.4
Aniline point, F				70.5
Copper corrosion				1
Doctor test		negative	negative	negative
Flash point, approximate, F		−35	−35	−35

*Literature values.

**Major impurities are 2,2-Dimethylbutane and 2,3-Dimethylbutane.

Table 2.37: Methylcyclopentane *(4)*

FORMULA	$\begin{array}{c} CH_2-CH_2 \\ | \qquad\qquad CH-CH_3 \\ CH_2-CH_2 \end{array}$		
PROPERTIES	**RESEARCH GRADE**	**PURE GRADE**	**TECHNICAL GRADE**
Composition, weight percent			
Normal Hexane	0.06	0.4	2.0
Methylcyclopentane	99.94	99.5	96.5 95.0 min
2,4-Dimethylpentane		0.1	
Cyclohexane			1.5
Isoheptanes			
3,3-Dimethylpentane			
Benzene & Toluene, ppm			
1,1-Dimethylcyclopentane			
1,2 & 1,3-Dimethylcyclopentane			
Purity by freezing point, mol %	99.94	99.3 99.0 min	
Freezing point, F	−224.42*		
Boiling point, F	161.26*		
Distillation range, F			
Initial boiling point			161
10% Condensed			
50% Condensed			
90% Condensed			
Dry point			162
Specific gravity of liquid at 60/60 F	0.7535*	0.754	0.754
at 20/4 C	0.74864*	0.749	0.749
API gravity at 60 F		56.2	56.2
Density of liquid at 60 F, lbs/gal		6.28	6.28
Vapor pressure at 70 F, psia	2.24*	2.2	2.3
100 F, psia	4.50*	4.5	4.5
130 F, psia	8.33*	8.3	8.3
Refractive index, 20/D	1.40970*	1.410	1.410
Color, Saybolt	+30	+30	+30
Acidity, distillation residue		neutral	neutral
Nonvolatile matter, grams/100 ml		0.0005	0.0005
Sulfur content, weight percent		0.005	0.005
Aniline point, F			
Kauri Butanol value			
Copper corrosion		1	1
Doctor test		negative	negative
Kinematic viscosity, cs at 32 F		−	
Flash point, approximate, F		−17	−17

*Literature values.

Table 2.38: Cyclohexane *(4)*

FORMULA	$\begin{array}{c} CH_2 \\ CH_2 \quad\quad CH_2 \\ | \qquad\qquad | \\ CH_2 \quad CH_2 \\ CH_2 \end{array}$		
PROPERTIES	**RESEARCH GRADE**	**99.5% GRADE**	**98.0% GRADE**
Composition, weight percent			
Normal Hexane			
Methylcyclopentane			0.5
2,4-Dimethylpentane	0.01	0.1	0.1
Cyclohexane	99.98	99.8 99.5 min	98.8 98.0 min
Isoheptanes		0.1	0.4
3,3-Dimethylpentane	0.01		0.2
Benzene & Toluene, ppm		193 500 max	200 500 max
1,1-Dimethylcyclopentane			
1,2 & 1,3-Dimethylcyclopentane			
Purity by freezing point, mol %	99.98		98.8
Freezing point, F	43.80*		

(continued)

Table 2.38: (continued)

PROPERTIES	RESEARCH GRADE	99.5% GRADE	98.0% GRADE
Boiling point, F	177.33*		
Distillation range, F			
Initial boiling point		177.3 175.1 min	177.3 175.1 min
10% Condensed			
50% Condensed			
90% Condensed			
Dry point		177.8 179.6 max	177.8 179.6 max
Specific gravity of liquid at 60/60 F	0.7834*	0.783	0.781
at 20/4 C	0.77855*	0.779	0.778
API gravity at 60 F	49.1*	49.3	49.6
Density of liquid at 60 F, lbs/gal	6.53*	6.52	6.51
Vapor pressure at 70 F, psia			
100 F, psia	3.26*	3.3 3.5 max	3.3 3.5 max
130 F, psia			
Refractive index, 20/D	1.42623*	1.426	1.424
Color, Saybolt	+30	+30 +30 min	+30 +30 min
Acidity, distillation residue		neutral	neutral
Nonvolatile matter, grams/100 ml		0.0007 0.0010 max	0.0007 0.0010 max
Sulfur content, weight percent		1 ppm 5 ppm max	1 ppm 5 ppm max
Aniline point, F			
Kauri Butanol value		56	55.1
Copper corrosion		1 1 max	1 1 max
Doctor test		neg. neg.	neg. neg.
Kinematic viscosity, cs at 32 F			0.94
Flash point, approximate, F		10	−1

*Literature values.

Table 2.39: 1,1-Dimethylcyclopentane *(4)*

FORMULA	CH_2-CH_2 $\mid$ $C-(CH_3)_2$ CH_2-CH_2
PROPERTIES	90% GRADE
Composition, weight percent	
Normal Hexane	
Methylcyclopentane	
2,4-Dimethylpentane	
Cyclohexane	
Isoheptanes	
3,3-Dimethylpentane	
Benzene & Toluene, ppm	
1,1-Dimethylcyclopentane	92*
1,2 & 1,3-Dimethylcyclopentane	
Purity by freezing point, mol %	
Freezing point, F	
Boiling point, F	
Distillation range, F	
Initial boiling point	189
10% Condensed	190
50% Condensed	190
90% Condensed	190
Dry point	190

PROPERTIES	90% GRADE
Specific gravity of liquid at 60/60 F	0.754
at 20/4 C	0.749
API gravity at 60 F	
Density of liquid at 60 F, lbs/gal	
Vapor pressure at 70 F, psia	
100 F, psia	
130 F, psia	
Refractive index, 20/D	
Color, Saybolt	
Acidity, distillation residue	
Nonvolatile matter, grams/100 ml	
Sulfur content, weight percent	
Aniline point, F	117
Kauri Butanol value	42.9
Copper corrosion	
Doctor test	
Kinematic viscosity, cs at 32 F	
Flash point, approximate, F	< 70

*Major impurities are: Cyclohexane, 3,3-Dimethylpentane and 2-Methylhexane.

Table 2.40: 1,2- and 1,3-Dimethylcyclopentane *(4)*

FORMULA	CH₃ CH–CH | H CH₂–CH₂ CH–CH₃
PROPERTIES	**90% GRADE**
Composition, weight percent	
Normal Hexane	
Methylcyclopentane	
2,4-Dimethylpentane	
Cyclohexane	
Isoheptanes	
3,3-Dimethylpentane	
Benzene & Toluene, ppm	
1,1-Dimethylcyclopentane	
1,2 & 1,3-Dimethylcyclopentane	92†
Purity by freezing point, mol %	
Freezing point, F	
Boiling point, F	
Distillation range, F	
Initial boiling point	196
10% Condensed	197
50% Condensed	197
90% Condensed	197
Dry point	197

PROPERTIES	90% GRADE
Specific gravity of liquid at 60/60 F	0.748
at 20/4 C	0.744
API gravity at 60 F	
Density of liquid at 60 F, lbs/gal	
Vapor pressure at 70 F, psia	
100 F, psia	
130 F, psia	
Refractive index, 20/D	
Color, Saybolt	
Acidity, distillation residue	
Nonvolatile matter, grams/100 ml	
Sulfur content, weight percent	
Aniline point, F	120
Kauri Butanol value	40.5
Copper corrosion	
Doctor test	
Kinematic viscosity, cs at 32 F	
Flash point, approximate, F	< 70

†Major impurity is 3-Methylhexane.

Table 2.41: Methylcyclohexane *(4)*

FORMULA		CH₃–CH CH₂–CH₂ CH₂ CH₂–CH₂	
PROPERTIES	**RESEARCH GRADE**	**PURE GRADE**	**TECHNICAL GRADE**
Composition, weight percent			
1,2-Dimethylcyclopentane	0.08	0.4	1.1
Normal Heptane		trace	0.4
Methylcyclohexane	99.90	99.4	97.9 95.0 min
Ethylcyclopentane		trace	0.1
Toluene	0.02	0.2	0.5
trans-1,4-Dimethylcyclohexane			
cis-1,4-Dimethylcyclohexane			
Other Dimethylcyclohexanes			
trans-1,2-Dimethylcyclohexane			
cis-1,2-Dimethylcyclohexane			
ortho-Xylene			
Unidentified Impurities			
Purity by freezing point, mol %	99.86	99.3 99.0 min	
Freezing point, F	−195.87*	−196.20	
Boiling point, F	213.68*		
Distillation range, F			
Initial boiling point			211
50% Condensed			
Dry point			213
Specific gravity of liquid at 60/60 F	0.7740*	0.774	0.774
at 20/4 C	0.76939*	0.769	0.769
API gravity at 60 F		51.3	51.3
Density of liquid at 60 F, lbs/gal		6.44	6.44
Vapor pressure at 70 F, psia			
100 F, psia	1.61*	1.6	1.6
Refractive index, 20/D	1.42312*	1.423	1.423
Color, Saybolt	+30	+30	+30
Acidity, distillation residue		neutral	neutral
Nonvolatile matter, grams/100 ml		0.0005	0.0005
Sulfur content, weight percent		0.005	0.005
Copper corrosion		1	1
Doctor test		negative	negative
Flash point, approximate, F		22	22

*Literature values.

Table 2.42: trans-1,4-Dimethylcyclohexane *(4)*

PROPERTIES	TECHNICAL GRADE
FORMULA (CH₃–CH / CH₂ CH₂ / CH₂ CH₂ / CH–CH₃)	
Composition, weight percent	
1,2-Dimethylcyclopentane	
Normal Heptane	
Methylcyclohexane	
Ethylcyclopentane	
Toluene	
trans-1,4-Dimethylcyclohexane	99.6
cis-1,4-Dimethylcyclohexane	
Other Dimethylcyclohexanes	
trans-1,2-Dimethylcyclohexane	
cis-1,2-Dimethylcyclohexane	
ortho-Xylene	
Unidentified Impurities	0.4
Purity by freezing point, mol %	95.03 95.0 min
Freezing point, F	−37.97
Boiling point, F	
Distillation range, F	
Initial boiling point	245
50% Condensed	
Dry point	248
Specific gravity of liquid at 60/60 F	0.7704
at 20/4 C	0.7661
API gravity at 60 F	52.2
Density of liquid at 60 F, lbs/gal	6.41
Vapor pressure at 70 F, psia	0.4
100 F, psia	2.0
Refractive index, 20/D	1.4229
Color, Saybolt	+28
Acidity, distillation residue	neutral
Nonvolatile matter, grams/100 ml	0.0005
Sulfur content, weight percent	
Copper corrosion	
Doctor test	negative
Flash point, approximate, F	40 (D 56)

Table 2.43: cis-1,4-Dimethylcyclohexane *(4)*

PROPERTIES	TECHNICAL GRADE
FORMULA (CH₃–CH / CH₂ CH₂ / CH₂ CH₂ / CH₃–CH)	
Composition, weight percent	
1,2-Dimethylcyclopentane	
Normal Heptane	
Methylcyclohexane	
Ethylcyclopentane	
Toluene	
trans-1,4-Dimethylcyclohexane	0.11
cis-1,4-Dimethylcyclohexane	99.89
Other Dimethylcyclohexanes	
trans-1,2-Dimethylcyclohexane	
cis-1,2-Dimethylcyclohexane	
ortho-Xylene	
Unidentified Impurities	
Purity by freezing point, mol %	97.4 95.0 min
Freezing point, F	−125.38*
Boiling point, F	255.78*
Distillation range, F	
Initial boiling point	255
50% Condensed	
Dry point	256
Specific gravity of liquid at 60/60 F	0.7872
at 20/4 C	0.7825
API gravity at 60 F	48.2
Density of liquid at 60 F, lbs/gal	6.56
Vapor pressure at 70 F, psia	
100 F, psia	0.7
Refractive index, 20/D	1.4297
Color, Saybolt	+30
Acidity, distillation residue	netural
Nonvolatile matter, grams/100 ml	0.0005
Sulfur content, weight percent	
Copper corrosion	
Doctor test	
Flash point, approximate, F	60

*Literature values.

Table 2.44: Mixed 1,4-Dimethylcyclohexanes *(4)*

PROPERTIES	TECHNICAL GRADE
FORMULA	C₈H₁₆
Composition, weight percent	
1,2-Dimethylcyclopentane	
Normal Heptane	
Methylcyclohexane	
Ethylcyclopentane	
Toluene	
trans-1,4-Dimethylcyclohexane	44.0 } 95.0 min
cis-1,4-Dimethylcyclohexane	54.9
Other Dimethylcyclohexanes	1.1
trans-1,2-Dimethylcyclohexane	
cis-1,2-Dimethylcyclohexane	
ortho-Xylene	
Unidentified Impurities	
Freezing point, F	
Boiling point, F	
Distillation range, F	
Initial boiling point	250
50% Condensed	252
Dry point	253
Specific gravity of liquid at 60/60 F	0.7784
at 20/4 C	0.7739
API gravity at 60 F	50.3
Density of liquid at 60 F, lbs/gal	6.48
Vapor pressure at 70 F, psia	0.4
100 F, psia	2.0
Refractive index, 20/D	1.4257
Color, Saybolt	+30
Acidity, distillation residue	neutral
Nonvolatile matter, grams/100 ml	0.0005
Sulfur content, weight percent	
Copper corrosion	
Doctor test	negative
Flash point, approximate, F	45 (D 56)

Table 2.45: trans-1,2-Dimethylcyclohexane *(4)*

FORMULA	CH₂ structure		

PROPERTIES	RESEARCH GRADE	PURE GRADE	TECHNICAL GRADE
Composition, weight percent			
1,2-Dimethylcyclopentane			
Normal Heptane			
Methylcyclohexane			0.7
Ethylcyclopentane			
Toluene			
trans-1,4-Dimethylcyclohexane			
cis-1,4-Dimethylcyclohexane			
Other Dimethylcyclohexanes	0.02	0.2	0.2
trans-1,2-Dimethylcyclohexane	99.90	99.6	96.9 95.0 min
cis-1,2-Dimethylcyclohexane	0.08	0.2	2.0
ortho-Xylene			0.2
Unidentified Impurities			
Purity by freezing point, mol %	99.73	99.3 99.0 min	
Freezing point, F	−126.75*		
Boiling point, F	254.15*		
Distillation range, F			
Initial boiling point			252
50% Condensed			
Dry point			253
Specific gravity of liquid at 60/60 F	0.7803*	0.780	0.780
at 20/4 C	0.77601*	0.776	0.776
API gravity at 60 F		49.9	49.9
Density of liquid at 60 F, lbs/gal		6.49	6.49
Vapor pressure at 70 F, psia			
100 F, psia	0.71*	0.7	0.7
Refractive index, 20/D	1.42695*	1.427	1.427
Color, Saybolt	+30	+30	+30
Acidity, distillation residue		neutral	neutral
Nonvolatile matter, grams/100 ml		0.0005	0.0005
Sulfur content, weight percent		0.005	0.005
Copper corrosion		1	1
Doctor test		negative	negative
Flash point, approximate, F		51 (D 56)	51 (D 56)

*Literature values.

Table 2.46: cis-1,2-Dimethylcyclohexane *(4)*

FORMULA	CH₂ structure	

PROPERTIES	RESEARCH GRADE	PURE GRADE
Composition, weight percent		
Methylcyclohexane		
trans-1,2-Dimethylcyclohexane	0.03	0.1
cis-1,2-Dimethylcyclohexane	99.96	99.7
Ethylcyclohexane		
Ethylbenzene		
Xylenes	0.01	0.2
Isopropylbenzene		
Isopropylcyclohexane		
Unidentified		
Purity by freezing point, mol %	99.91	99.5 99.0 min
Freezing point, F	−58.04*	
Boiling point, F	265.51*	
Distillation range, F		
Initial boiling point		

(continued)

Table 2.46: (continued)

PROPERTIES	RESEARCH GRADE	PURE GRADE
Dry point		
Specific gravity of liquid at 60/60 F	0.8006*	0.801
at 20/4 C	0.79627*	0.796
API gravity at 60 F		45.2
Density of liquid at 60 F, lbs/gal		6.67
Vapor pressure at 70 F, psia	0.23*	0.2
100 F, psia	0.54*	0.5
Refractive index, 20/D	1.43596*	1.436
Color, Saybolt	+30	+30
Acidity, distillation residue		neutral
Nonvolatile matter, grams/100 ml		0.0005
Sulfur content, weight percent		0.005
Copper corrosion		1
Doctor test		negative
Flash point, approximate, F		60 (D 56)

*Literature values.

Table 2.47: Mixed 1,2-Dimethylcyclohexane *(4)*

FORMULA	C_8H_{16}
PROPERTIES	**PURE GRADE**
Composition, weight percent	
Methylcyclohexane	trace
trans-1,2-Dimethylcyclohexane	34 } 99.0 min
cis-1,2-Dimethylcyclohexane	66 }
Ethylcyclohexane	
Ethylbenzene	
Xylenes	trace
Isopropylbenzene	
Isopropylcyclohexane	
Unidentified	
Purity by freezing point, mol %	
Freezing point, F	
Boiling point, F	260

PROPERTIES	PURE GRADE
Distillation range, F	
Initial boiling point	
Dry point	
Specific gravity of liquid at 60/60 F	0.792
at 20/4 C	0.789
API gravity at 60 F	47.2
Density of liquid at 60 F, lbs/gal	6.59
Vapor pressure at 70 F, psia	
100 F, psia	0.6
Refractive index, 20/D	1.432
Color, Saybolt	+30
Acidity, distillation residue	neutral
Nonvolatile matter, grams/100 ml	0.0005
Sulfur content, weight percent	0.005
Copper corrosion	1
Doctor test	negative
Flash point, approximate, F	55 (D 56)

*Literature values.

Table 2.48: Ethylcyclohexane *(4)*

FORMULA			
PROPERTIES	**RESEARCH GRADE**	**PURE GRADE**	**TECHNICAL GRADE**
Composition, weight percent			
Methylcyclohexane			
trans-1,2-Dimethylcyclohexane			
cis-1,2-Dimethylcyclohexane			2.0
Ethylcyclohexane	99.98	99.5	96.9
Ethylbenzene	0.02	0.4	0.8
Xylenes			
Isopropylbenzene			
Isopropylcyclohexane			
Unidentified	trace	0.1	0.3

(continued)

Table 2.48: (continued)

PROPERTIES	RESEARCH GRADE	PURE GRADE	TECHNICAL GRADE
Purity by freezing point, mol %	99.66	99.19 99.0 min	96.06 95.0 min
Freezing point, F	−168.38*		
Boiling point, F	269.21*		
Distillation range, F			
Initial boiling point			266
Dry point			269
Specific gravity of liquid at 60/60 F	0.7922*	0.793	0.793
at 20/4 C	0.78792*	0.788	0.788
API gravity at 60 F		46.9	46.9
Density of liquid at 60 F, lbs/gal		6.60	6.60
Vapor pressure at 70 F, psia			
100 F, psia	0.48*	0.5	0.5
Refractive index, 20/D	1.43304*	1.433	1.433
Color, Saybolt	+30	+30	+30
Acidity, distillation residue		neutral	neutral
Nonvolatile matter, grams/100 ml		0.0005	0.0005
Sulfur content, weight percent			
Copper corrosion			
Doctor test			
Flash point, approximate, F	66	66	66

*Literature values.

Table 2.49: Isopropylcyclohexane *(4)*

FORMULA			
PROPERTIES	RESEARCH GRADE	PURE GRADE	TECHNICAL GRADE
Composition, weight percent			
Methylcyclohexane			
trans-1,2-Dimethylcyclohexane			
cis-1,2-Dimethylcyclohexane			
Ethylcyclohexane			
Ethylbenzene			
Xylenes		0.02	
Isopropylbenzene	0.03	0.02	0.79
Isopropylcyclohexane	99.97	99.90	99.05
Unidentified		0.06	0.16
Purity by freezing point, mol %	99.67	99.4 99.0 min	95.2 95.0 min
Freezing point, F	−128.9*		
Boiling point, F	310.57*		
Distillation range, F			
Initial boiling point			307
Dry point			310
Specific gravity of liquid at 60/60 F	0.8064*	0.807	0.807
at 20/4 C	0.8024*	0.803	0.803
API gravity at 60 F		43.8	43.8
Density of liquid at 60 F, lbs/gal		6.72	6.72
Vapor pressure at 70 F, psia			
100 F, psia			
Refractive index, 20/D	1.44087*	1.441	1.441
Color, Saybolt	+30	+30	+30
Acidity, distillation residue		neutral	neutral
Nonvolatile matter, grams/100 ml		0.0005	0.0005
Sulfur content, weight percent			
Copper corrosion			
Doctor test			
Flash point, approximate, F	96	96	96

*Literature values.

OLEFINS

Table 2.50: Ethylene *(4)*

FORMULA	$CH_2 = CH_2$	
PROPERTIES	RESEARCH GRADE	99.8% GRADE
Composition, weight percent		
Propane		
Propylene		
Ethylene	99.97	99.88 99.8 min
Ethane	0.01	0.04
Methane	0.02	0.08
Carbon Dioxide, ppm		1 15 max
Acetylene, ppm (liquid)		1 5 max
Carbonyl, ppm (liquid)		
Carbon Monoxide, ppm		1 5 max
Oxygen, ppm		20
Hydrogen, ppm		1 5 max
Freezing point, triple point, F	−272.47*	
Boiling point, F	−154.68*	
Specific gravity of liquid at 60/60 F		
at 20/4 C		
API gravity at 60 F		
Density of liquid at 60 F, lbs/gal		
Vapor pressure at 70 F, psia		
100 F, psia		
130 F, psia		
Sulfur content, ppm		3 10 max
Specific gravity of real gas at 60 F and 14.7 psia (Air = 1)	0.9740*	
Specific volume of real gas at 60 F and 14.7 psia, cu ft/lb	13.4524*	
Critical temperature, F	49.82*	
Critical pressure, psia	742.1*	
Density of real gas at 60 F and 14.7 psia, lbs/cu ft		0.0743
Flash point, approximate, F		−213
Flammability limits, volume % in air		
Lower	2.7*	
Higher	34*	
Heating value for ideal gas at 60 F and 14.7 psia, BTU/cu ft, dry basis	1599*	

*Literature values.

Table 2.51: Propylene *(4)*

FORMULA	$CH_2 = CH-CH_3$	
PROPERTIES	RESEARCH GRADE	POLYMERIZATION GRADE
Composition, weight percent		
Propane	0.01	0.5
Propylene	99.99	99.5 99.0 min
Ethylene		
Ethane		trace
Methane		
Carbon Dioxide, ppm		
Acetylene, ppm (liquid)		10
Carbonyl, ppm (liquid)		20
Carbon Monoxide, ppm		
Oxygen, ppm		
Hydrogen, ppm		
Freezing point, triple point, F	−301.45*	
Boiling point, F	−53.86*	
Specific gravity of liquid at 60/60 F	0.5220*	0.522
at 20/4 C	0.5139*	0.514
API gravity at 60 F		139.6
Density of liquid at 60 F, lbs/gal		4.35
Vapor pressure at 70 F, psia		151
100 F, psia		242
130 F, psia		328
Sulfur content, ppm		4
Specific gravity of real gas at 60 F and 14.7 psia (Air = 1)	1.4765*	
Specific volume of real gas at 60 F and 14.7 psia, cu ft/lb	8.8736*	
Critical temperature, F	197.4*	
Critical pressure, psia	667*	
Density of real gas at 60 F and 14.7 psia, lbs/cu ft		0.1127
Flash point, approximate, F		−162
Flammability limits, volume % in air		
Lower	2.0*	
Higher	10*	
Heating value for ideal gas at 60 F and 14.7 psia, BTU/cu ft, dry basis	2334*	

*Literature values.

Table 2.52: Isobutylene *(4)*

FORMULA	$CH_3-\overset{CH_3}{\underset{|}{C}} = CH_2$	
PROPERTIES	RESEARCH GRADE	PURE GRADE
Composition, weight percent		
Isobutane	0.06	0.1
Isobutylene	99.81	99.3 99.0 min
Butene-1	0.09	0.4
Butadiene-1,3		
Normal Butane	0.04	0.2
Butene-2	trace	trace
Acetylene (as Methylacetylene) ppm, wt.		
Water, ppm, weight		177
Carbonyl (as Acetaldehyde) ppm, weight		nil
Propadiene, ppm, weight		

PROPERTIES	RESEARCH GRADE	PURE GRADE
Freezing point, F	−220.63*	
Boiling point, F	19.58*	
Specific gravity of liquid at 60/60 F	0.6004*	0.600
at 20/4 C	0.5942*	
API gravity at 60 F		104.3
Density of liquid at 60 F, lbs/gal		4.99
Vapor pressure at 70 F, psia		
100 F, psia		63.4
130 F, psia		
Sulfur content, ppm		8
Specific gravity of real gas at 60 F and 14.7 psia (Air = 1)		1.997
Specific volume of real gas at 60 F and 14.7 psia, cu ft/lb		6.561
Flash point, approximate, F		−105
Flammability limits, volume % in air		
Lower		
Higher		

*Literature values.

Table 2.53: Butene-1 *(4)*

FORMULA	$CH_3-CH_2-CH=CH_2$	
PROPERTIES	**RESEARCH GRADE**	**POLYMERIZATION GRADE**
Composition, weight percent		
Isobutane		0.1
Isobutylene	0.2	0.3
Butene-1	99.8	99.4 99.0 min
Butadiene-1,3		trace
Normal Butane		0.2
Butene-2		trace
Acetylene (as Methylacetylene) ppm, wt.		15 25 max
Water, ppm, weight		
Carbonyl (as Acetaldehyde) ppm, weight		10 20 max
Propadiene, ppm, weight		4

PROPERTIES	RESEARCH GRADE	POLYMERIZATION GRADE
Freezing point, F	−301.63*	
Boiling point, F	20.73*	
Specific gravity of liquid at 60/60 F	0.6013*	0.601
at 20/4 C	0.5951*	
API gravity at 60 F		103.9
Density of liquid at 60 F, lbs/gal		5.00
Vapor pressure at 70 F, psia		37.5
100 F, psia		67.5 (105F)
130 F, psia		99.7
Sulfur content, ppm		1 10 max
Specific gravity of real gas at 60 F and 14.7 psia (Air = 1)		
Specific volume of real gas at 60 F and 14.7 psia, cu ft/lb		
Flash point, approximate, F		−112
Flammability limits, volume % in air		
Lower	1.6*	
Higher	9.3*	

*Literature values.

Table 2.54: trans-Butene-2 *(4)*

FORMULA	$CH_3-\overset{H}{\underset{H}{C}}=C-CH_3$		
PROPERTIES	**RESEARCH GRADE**	**PURE GRADE**	**TECHNICAL GRADE**
Composition, weight percent			
Butene-1	0.03		trace
Normal Butane	0.07	0.2	1.3
trans-Butene-2	99.80	99.6	97.7 95.0 min
cis-Butene-2	0.10	0.2	1.0
Purity by freezing point, mol %	99.76	99.2 99.0 min	
Freezing point, F	−157.99*		
Boiling point, F	33.58*		
Specific gravity of liquid at 60/60 F	0.6100*	0.610	0.609
at 20/4 C	0.6042*		
API gravity at 60 F		100.5	100.8
Density of liquid at 60 F, lbs/gal		5.07	5.07
Vapor pressure at 70 F, psia	29.94*	29.9	30
105 F, psia		52.2	52
130 F, psia		76.4	76
Flash point, approximate, F		−100	−100

*Literature values.

Table 2.55: cis-Butene-2 *(4)*

FORMULA	$CH_3-\overset{H}{C}=\overset{H}{C}-CH_3$		
PROPERTIES	**RESEARCH GRADE**	**PURE GRADE**	**TECHNICAL GRADE**
Composition, weight percent			
Butene-1			
Normal Butane			
trans-Butene-2	0.03	0.5	4.3
cis-Butene-2	99.97	99.5	95.7 95.0 min

(continued)

Table 2.55: (continued)

PROPERTIES	RESEARCH GRADE	PURE GRADE	TECHNICAL GRADE
Purity by freezing point, mol %	99.92	99.4 99.0 min	
Freezing point, F	−218.04*		
Boiling point, F	38.70*		
Specific gravity of liquid at 60/60 F	0.6271*	0.627	0.632
at 20/4 C	0.6213*		
API gravity at 60 F		94.2	92.4
Density of liquid at 60 F, lbs/gal		5.22	5.26
Vapor pressure at 70 F, psia	27.29*	27.3	27.8
105 F, psia		49.8	50.8
130 F, psia		73.2	74.8
Flash point, approximate, F		−100	−100

*Literature values.

Table 2.56: Mixed 2-Butenes *(4)*

FORMULA	$CH_3-CH = CH-CH_3$	
PROPERTIES	**PURE GRADE**	**TECHNICAL GRADE**
Composition, weight percent		
Butene-1		0.3
Normal Butane		2.7
trans-Butene-2	45.0	52.0
cis-Butene-2	55.0 } 99.0 min	45.0 } 95.0 min
Purity by freezing point, mol %		
Freezing point, F		
Boiling point, F		
Specific gravity of liquid at 60/60 F	0.619	0.618
at 20/4 C	0.615	0.614
API gravity at 60 F	97.1	97.5
Density of liquid at 60 F, lbs/gal	5.15	5.14
Vapor pressure at 70 F, psia	28.0	28.1
105 F, psia	51.0	51.2
130 F, psia	76.5	76.7
Flash point, approximate, F	−95	−100

*Literature values.

Table 2.57: 3-Methylbutene-1 *(4)*

FORMULA	$CH_2 = CH-\overset{\overset{\displaystyle CH_3}{\displaystyle |}}{CH}-CH_3$		
PROPERTIES	**RESEARCH GRADE**	**PURE GRADE**	**TECHNICAL GRADE**
Composition, weight percent			
3-Methylbutene-1	99.97	99.8	96.3 95.0 min
2-Methylbutene-1		0.1	2.0
2-Methylbutene-2			
Pentene-1			0.2
Pentenes-2	0.02	0.1	1.0
Isopentane	0.01	trace	0.5
Normal Pentane			

(continued)

Table 2.57: (continued)

PROPERTIES	RESEARCH GRADE	PURE GRADE	TECHNICAL GRADE
Purity by freezing point, mol %	99.93	99.4 99.0 min	
Freezing point, F	−271.29*		
Boiling point, F	68.11*		
Distillation range, F			
Initial boiling point			
Dry point			
Specific gravity of liquid at 60/60 F	0.6325*	0.633	0.633
at 20/4 C	0.6272*	0.628	0.628
API gravity at 60 F		91.9	91.9
Density of liquid at 60 F, lbs/gal		5.27	5.27
Vapor pressure at 70 F, psia	15.25*	15.2	15.0
100 F, psia	26.41*	26.4	26.0
130 F, psia			
Refractive index, 20/D	1.3643*	1.364	1.364
Color, Saybolt	+30	+30	+30
Acidity, distillation residue			
Nonvolatile matter, grams/100 ml			
Flash point, approximate, F		−70	−70

*Literature values.

Table 2.58: 2-Methylbutene-1 *(4)*

FORMULA	$CH_2=C-CH_2-CH_3$ with CH_3		
PROPERTIES	**RESEARCH GRADE**	**PURE GRADE**	**TECHNICAL GRADE**
Composition, weight percent			
3-Methylbutene-1		trace	0.3
2-Methylbutene-1	99.99	99.8	97.3 95.0 min
2-Methylbutene-2		trace	0.1
Pentene-1	0.01	0.1	1.9
Pentenes-2		trace	0.2
Isopentane			
Normal Pentane		0.1	0.2
Purity by freezing point, mol %	99.85	99.5 99.0 min	
Freezing point, F	−215.61*		
Boiling point, F	88.09*		
Distillation range, F			
Initial boiling point			87
Dry point			88
Specific gravity of liquid at 60/60 F	0.6557*	0.656	0.656
at 20/4 C	0.6504*	0.650	0.650
API gravity at 60 F		84.2	84.2
Density of liquid at 60 F, lbs/gal		5.46	5.46
Vapor pressure at 70 F, psia	10.21*	10.3	10.3
100 F, psia	18.40*	18.8	18.8
130 F, psia		32.0	32.0
Refractive index, 20/D	1.3778*	1.378	1.378
Color, Saybolt	+30	+30	+30
Acidity, distillation residue		neutral	neutral
Nonvolatile matter, grams/100 ml		0.0005	0.0005
Flash point, approximate, F		−55	−55

*Literature values.

Table 2.59: Methylbutene-2 *(4)*

FORMULA	$CH_3-\underset{\underset{CH_3}{\mid}}{C}=CH-CH_3$			
PROPERTIES	**RESEARCH GRADE**	**PURE GRADE**	**TECHNICAL GRADE**	**COMMERCIAL GRADE**
Composition, weight percent				
3-Methylbutene-1				
2-Methylbutene-1	trace	0.2	0.2	10.3
2-Methylbutene-2	99.99	99.5	97.4 95.0 min	87.8
Pentene-1	trace		0.1	
Pentenes-2	0.01	0.3	2.3	0.8
Isopentane				1.1
Normal Pentane				
Purity by freezing point, mol %	99.78	99.3 99.0 min		
Freezing point, F	−208.78*			
Boiling point, F	101.42*			
Distillation range, F				
Initial boiling point			101	100.7
Dry point			102	101.3
Specific gravity of liquid at 60/60 F	0.6676*	0.668	0.667	0.668
at 20/4 C	0.6623*	0.662	0.662	0.663
API gravity at 60 F		80.3	80.6	
Density of liquid at 60 F, lbs/gal		5.56	5.55	5.56
Vapor pressure at 70 F, psia	7.76*	7.8	7.8	
100 F, psia	14.30*	14.3	14.3	
130 F, psia	24.56*	24.6	24.6	
Refractive index, 20/D	1.3874*	1.387	1.387	1.387
Color, Saybolt	+30	+30	+30	
Acidity, distillation residue		neutral	neutral	
Nonvolatile matter, grams/100 ml		0.0005	0.0005	
Flash point, approximate, F		−50	−50	−60

*Literature values.

Table 2.60: Pentene-1 *(4)*

FORMULA	$CH_2=CH-CH_2-CH_2-CH_3$		
PROPERTIES	**RESEARCH GRADE**	**PURE GRADE**	**TECHNICAL GRADE**
Composition, weight percent			
Isopentane		trace	0.1
Pentene-1	99.9	99.4 99.0 min	97.0 95.0 min
2-Methylbutene-1		0.3	1.4
Normal Pentane		0.1	0.5
trans-Pentene-2	0.1	0.2	0.5
cis-Pentene-2			
2-Methylbutene-2		trace	0.5
Purity by freezing point, mol %			
Freezing point, F	−265.40*		
Boiling point, F	85.94*		
Distillation range, F			
Initial boiling point			85
Dry Point			87
Specific gravity of liquid at 60/60 F	0.6457*	0.646	0.646
at 20/4 C	0.64050*	0.641	0.641
API gravity at 60 F		87.5	87.5
Density of liquid at 60 F, lbs/gal		5.38	5.38
Vapor pressure at 70 F, psia	10.70*	10.7	10.6
100 F, psia	19.12*	19.1	19.0
130 F, psia		33.0	32.8
Refractive index, 20/D	1.37148*	1.372	1.372
Color, Saybolt	+30	+30	+30
Acidity, distillation residue		neutral	neutral
Nonvolatile matter, grams/100 ml		0.0005	0.0005
Flash point, approximate, F		−60	−60
Flammability limits, volume % in air			
Lower	1.4*		
Higher	8.7*		

* Literature values.

Table 2.61: cis-Pentene-2 *(4)*

FORMULA	$CH_3 - \overset{H}{\underset{}{C}} = \overset{H}{\underset{}{C}} - CH_2 - CH_3$	
PROPERTIES	**RESEARCH GRADE**	**TECHNICAL GRADE**
Composition, weight percent		
Isopentane		
Pentene-1		
2-Methylbutene-1		
Normal Pentane		
trans-Pentene-2	0.1	3.2
cis-Pentene-2	99.9	96.8 95.0 min
2-Methylbutene-2		
Purity by freezing point, mol %	99.8	
Freezing point, F	−240.50*	
Boiling point, F	98.50*	
Distillation range, F		
Initial boiling point		
Dry Point		
Specific gravity of liquid at 60/60 F	0.6608*	0.660
at 20/4 C	0.6556*	0.655
API gravity at 60 F		82.9
Density of liquid at 60 F, lbs/gal		5.49
Vapor pressure at 70 F, psia	8.24*	8.3
100 F, psia	15.12	15.1
130 F, psia	25.84*	26.6
Refractive index, 20/D	1.3830*	1.383
Color, Saybolt		+30
Acidity, distillation residue		neutral
Nonvolatile matter, grams/100 ml		0.0005
Flash point, approximate, F		−50
Flammability limits, volume % in air		
Lower		
Higher		

* Literature values.

Table 2.62: trans-Pentene-2 *(4)*

FORMULA	$CH_3 - \overset{H}{\underset{H}{C}} = C - CH_2 - CH_3$	
PROPERTIES	**RESEARCH GRADE**	
Composition, weight percent		
Isopentane		
Pentene-1		
2-Methylbutene-1		
Normal Pentane	0.02	
trans-Pentene-2	99.63	
cis-Pentene-2	0.35	
2-Methylbutene-2		
Purity by freezing point, mol %	99.53	
Freezing point, F	−220.44*	
Boiling point, F	97.44*	
Distillation range, F		
Initial boiling point		
Dry Point		
Specific gravity of liquid at 60/60 F	0.6533*	
at 20/4 C	0.6482*	
API gravity at 60 F		
Density of liquid at 60 F, lbs/gal	5.447*	
Vapor pressure at 70 F, psia	10.2*	
100 F, psia	15.4*	
130 F, psia	26.3*	
Refractive index, 20/D	1.3793*	
Color, Saybolt	+30	
Acidity, distillation residue		
Nonvolatile matter, grams/100 ml		
Flash point, approximate, F		
Flammability limits, volume % in air		
Lower		
Higher		

* Literature values.

Table 2.63: Mixed 2-Pentenes *(4)*

FORMULA	$CH_3 - CH = CH - CH_2 - CH_3$	
PROPERTIES	**PURE GRADE**	**TECHNICAL GRADE**
Composition, weight percent		
Isopentane	0.1	0.1
Pentene-1	0.1	0.3
2-Methylbutene-1	0.1	1.6
Normal Pentane		
trans-Pentene-2	46.6⎱ 99.0 min	48.8⎱ 95.0 min
cis-Pentene-2	53.0⎰	48.2⎰
2-Methylbutene-2	0.1	1.0
Purity by freezing point, mol %		
Freezing point, F		
Boiling point, F		
Distillation range, F		
Initial boiling point		97
Dry Point		99

PROPERTIES	PURE GRADE	TECHNICAL GRADE
Specific gravity of liquid at 60/60 F	0.656	0.658
at 20/4 C	0.652	0.654
API gravity at 60 F	84.2	83.5
Density of liquid at 60 F, lbs/gal	5.46	5.48
Vapor pressure at 70 F, psia	8.4	8.3
100 F, psia	15.4	15.2
130 F, psia	27.0	26.7
Refractive index, 20/D	1.380	1.381
Color, Saybolt	+30	+30
Acidity, distillation residue	neutral	neutral
Nonvolatile matter, grams/100 ml	0.0005	0.0005
Flash point, approximate, F	−50	−50
Flammability limits, volume % in air		
Lower		
Higher		

Table 2.64: 3,3-Dimethylbutene-1 *(4)*

Neohexene

FORMULA	$CH_2 = CH - \overset{\overset{\displaystyle CH_3}{\displaystyle |}}{\underset{\underset{\displaystyle CH_3}{\displaystyle |}}{C}} - CH_3$
PROPERTIES	**TECHNICAL GRADE**
Composition, weight percent	
3,3-Dimethylbutene-1	98.9
2,3-Dimethylbutane	
4-Methylpentene-1	
3-Methylpentene-1	
2,3-Dimethylbutene-1	
2,3-Dimethylbutene-2	
cis-4-Methylpentene-2	
trans-4-Methylpentene-2	
2-Methylpentene-2	
Other Olefins	1.1
Purity by freezing point, mol %	
Freezing point, F	−184.43
Boiling point, F	
Distillation range, F	
Initial boiling point	106
10% Condensed	106
50% Condensed	107
90% Condensed	111
Dry point	114
Specific gravity of liquid at 60/60 F	0.6582
at 20/4 C	0.6533
API gravity at 60 F	83.5
Density of liquid at 60 F, lbs/gal	5.48
Vapor pressure at 70 F, psia	7.2
100 F, psia	14.6
130 F, psia	
Refractive index, 20/D	1.3766
Color, Saybolt	+30
Acidity, distillation residue	neutral
Nonvolatile matter, grams/100 ml	0.0005
Doctor test	negative
Flash point, approximate, F	45

*Literature values.

Table 2.65: Mixed 2,3-Dimethylbutenes *(4)*

FORMULA	C_6H_{12}
PROPERTIES	**TECHNICAL GRADE**
Composition, weight percent	
3,3-Dimethylbutene-1	
2,3-Dimethylbutane	2.0
4-Methylpentene-1	
3-Methylpentene-1	
2,3-Dimethylbutene-1	32.0 } 95.0 min
2,3-Dimethylbutene-2	63.9 }
cis-4-Methylpentene-2	
trans-4-Methylpentene-2	
2-Methylpentene-2	2.1
Other Olefins	
Purity by freezing point, mol %	
Freezing point, F	
Boiling point, F	
Distillation range, F	
Initial boiling point	140
10% Condensed	
50% Condensed	
90% Condensed	
Dry point	169
Specific gravity of liquid at 60/60 F	0.703
at 20/4 C	
API gravity at 60 F	69.8
Density of liquid at 60 F, lbs/gal	5.85
Vapor pressure at 70 F, psia	
100 F, psia	5.7
130 F, psia	
Refractive index, 20/D	
Color, Saybolt	+30
Acidity, distillation residue	neutral
Nonvolatile matter, grams/100 ml	0.0005
Doctor test	
Flash point, approximate, F	−35

*Literature values.

Table 2.66: 4-Methylpentene-1 *(4)*

FORMULA	$CH_2 = CH - CH_2 - \overset{\overset{\displaystyle CH_3}{\displaystyle |}}{CH} - CH_3$		
PROPERTIES	**RESEARCH GRADE**	**PURE GRADE**	**TECHNICAL GRADE**
Composition, weight percent			
3,3-Dimethylbutene-1			
2,3-Dimethylbutane			
4-Methylpentene-1	99.94	99.6	99.1
3-Methylpentene-1	0.02	0.1	0.2
2,3-Dimethylbutene-1			
2,3-Dimethylbutene-2			
cis-4-Methylpentene-2	0.02	0.3	0.6
trans-4-Methylpentene-2	0.02		0.1
2-Methylpentene-2			
Other Olefins			
Purity by freezing point, mol %	99.81	99.3 99.0 min	97.5 95.0 min
Freezing point, F		−244.53*	
Boiling point, F		128.96*	
Distillation range, F			
Initial boiling point			129
10% Condensed			
50% Condensed			
90% Condensed			
Dry point			130
Specific gravity of liquid at 60/60 F	0.6686*	0.669	0.669
at 20/4 C	0.66370*	0.664	0.664

(continued)

Table 2.66: (continued)

PROPERTIES	RESEARCH GRADE	PURE GRADE	TECHNICAL GRADE
API gravity at 60 F		80.0	80.0
Density of liquid at 60 F, lbs/gal		5.57	5.57
Vapor pressure at 70 F, psia	4.48*	4.5	4.5
100 F, psia	8.50*	8.5	8.5
130 F, psia	14.97*	15.0	15.0
Refractive index, 20/D	1.38267*	1.383	1.383
Color, Saybolt	+30	+30	+30
Acidity, distillation residue		neutral	neutral
Nonvolatile matter, grams/100 ml		0.0005	0.0005
Doctor test			
Flash point, approximate, F		−25	−25

*Literature values.

Table 2.67: cis-4-Methylpentene-2 *(4)*

FORMULA	$\begin{array}{ccc} & H & H\ CH_3 \\ & | & |\ | \\ CH_3-C & = & C-CH-CH_3 \end{array}$		
PROPERTIES	**RESEARCH GRADE**	**PURE GRADE**	**TECHNICAL GRADE**
Composition, weight percent			
3,3-Dimethylbutene-1			
2,3-Dimethylbutane			
4-Methylpentene-1	0.06	0.1	0.2
3-Methylpentene-1			
2,3-Dimethylbutene-1			
2,3-Dimethylbutene-2			
cis-4-Methylpentene-2	99.87	99.8	97.1
trans-4-Methylpentene-2	0.07	0.1	2.4
2-Methylpentene-2			
Other Olefins			0.3
Purity by freezing point, mol %	99.71	99.52 99.0 min	96.2 95.0 min
Freezing point, F	−209.97*		
Boiling point, F	133.50*		
Distillation range, F			
Initial boiling point			130
10% Condensed			
50% Condensed			
90% Condensed			
Dry point			133
Specific gravity of liquid at 60/60 F	0.6741*	0.674	0.674
at 20/4 C	0.66918*	0.669	0.669
API gravity at 60 F		78.4	78.4
Density of liquid at 60 F, lbs/gal		5.61	5.61
Vapor pressure at 70 F, psia	4.01*	4.0	4.0
100 F, psia	7.73*	7.7	7.7
130 F, psia	13.80*	13.8	13.8
Refractive index, 20/D	1.38793*	1.388	1.388
Color, Saybolt	+30	+30	+30
Acidity, distillation residue		neutral	neutral
Nonvolatile matter, grams/100 ml		0.0005	0.0005
Doctor test			
Flash point, approximate, F		−25	−25

*Literature values.

Table 2.68: trans-4-Methylpentene-2 *(4)*

FORMULA	$CH_3-\underset{\underset{H}{\vert}}{\overset{\overset{H}{\vert}}{C}}=C-\underset{\underset{CH_3}{\vert}}{\overset{\overset{CH_3}{\vert}}{CH}}-CH_3$		
PROPERTIES	**RESEARCH GRADE**	**PURE GRADE**	**TECHNICAL GRADE**
Composition, weight percent			
4-Methylpentene-1			
cis-4-Methylpentene-2	trace	0.1	0.6
trans-4-Methylpentene-2	99.98	99.9	96.5
2-Methylpentene-1			
2-Methylpentene-2			
Isoolefins	0.02	trace	2.9
Purity by freezing point, mol %	99.94	99.2 99.0 min	95.6 95.0 min
Freezing point, F	−221.46*		
Boiling point, F	137.50*		
Distillation range, F			
Initial boiling point			137.1
Dry point			137.8
Specific gravity of liquid at 60/60F	0.6736*	0.673	0.674
at 20/4 C	0.66862*	0.669	0.670
API gravity at 60 F		78.7	78.4
Density of liquid at 60 F, lbs/gal		5.60	5.61
Vapor pressure at 70 F, psia	3.66*	3.7	3.7
100 F, psia	7.12*	7.1	7.1
130 F, psia	12.82*	12.8	12.8
Refractive index, 20/D	1.38878*	1.389	1.389
Color, Saybolt	+30	+30	+30
Acidity, distillation residue		neutral	neutral
Nonvolatile matter, grams/100 ml		0.0005	0.0005
Flash point, approximate, F		−20	−20

*Literature values.

Table 2.69: Mixed 4-Methyl-2-Pentenes *(4)*

FORMULA	$CH_3-CH=CH-\underset{\underset{CH_3}{\vert}}{\overset{\overset{CH_3}{\vert}}{CH}}-CH_3$	
PROPERTIES	**PURE GRADE**	**TECHNICAL GRADE**
Composition, weight percent		
4-Methylpentene-1	0.1	2.8
cis-4-Methylpentene-2	76.5 } 99.0 min	76.2 } 95.0 min
trans-4-Methylpentene-2	23.4	20.8
2-Methylpentene-1		
2-Methylpentene-2		
Isoolefins		0.2
Purity by freezing point, mol %		
Freezing point, F		
Boiling point, F	135.0	
Distillation range, F		
Initial boiling point		136.0
Dry point		137.2
Specific gravity of liquid at 60/60F	0.673	0.673
at 20/4 C	0.669	0.669
API gravity at 60 F	78.8	78.8
Density of liquid at 60 F, lbs/gal	5.60	5.60
Vapor pressure at 70 F, psia	3.8	3.8
100 F, psia	7.5	7.5
130 F, psia	13.0	13.0
Refractive index, 20/D	1.388	1.388
Color, Saybolt	+30	+30
Acidity, distillation residue	neutral	neutral
Nonvolatile matter, grams/100 ml	0.0005	0.0005
Flash point, approximate, F	−20	−20

Table 2.70: 2-Methylpentene-1 *(4)*

FORMULA	$CH_2 = \overset{\overset{\textstyle CH_3}{\textstyle |}}{C} - CH_2 - CH_2 - CH_3$		
PROPERTIES	RESEARCH GRADE	PURE GRADE	TECHNICAL GRADE
Composition, weight percent			
4-Methylpentene-1			0.6
cis-4-Methylpentene-2			0.5
trans-4-Methylpentene-2		0.1	0.2
2-Methylpentene-1	99.90	99.8	95.8 95.0 min
2-Methylpentene-2			
Isoolefins	0.10	0.1	2.9
Purity by freezing point, mol %	99.84	99.65 99.0 min	
Freezing point, F	−212.30*		
Boiling point, F	143.80*		
Distillation range, F			
Initial boiling point			142.8
Dry point			143.6
Specific gravity of liquid at 60/60F	0.6848*	0.685	0.685
at 20/4 C	0.67987*	0.680	0.680
API gravity at 60 F		75.4	75.4
Density of liquid at 60 F, lbs/gal		5.69	5.69
Vapor pressure at 70 F, psia	3.20*	3.2	3.2
100 F, psia	6.30*	6.3	6.3
130 F, psia	11.43*	11.4	11.4
Refractive index, 20/D	1.39200*	1.392	1.392
Color, Saybolt		+30	+30
Acidity, distillation residue		neutral	neutral
Nonvolatile matter, grams/100 ml		0.0005	0.0005
Flash point, approximate, F		−15	−15

*Literature values.

Table 2.71: 2-Methylpentene-2 *(4)*

FORMULA	$CH_3 - \overset{\overset{\textstyle CH_3}{\textstyle |}}{C} = CH - CH_2 - CH_3$	
PROPERTIES	PURE GRADE	TECHNICAL GRADE
Composition, weight percent		
4-Methylpentene-1		
cis-4-Methylpentene-2		
trans-4-Methylpentene-2		
2-Methylpentene-1	0.1	1.6
2-Methylpentene-2	99.8	96.0 95.0 min
Isoolefins	0.1	2.4
Purity by freezing point, mol %		
Freezing point, F	−211.13*	
Boiling point, F	153.15*	
Distillation range, F		
Initial boiling point		152
Dry point		158
Specific gravity of liquid at 60/60F	0.6913*	0.692
at 20/4 C	0.68650*	0.687
API gravity at 60 F		73.1
Density of liquid at 60 F, lbs/gal		5.76
Vapor pressure at 70 F, psia	2.57*	2.6
100 F, psia	5.17*	5.1
130 F, psia	9.58*	9.6
Refractive index, 20/D	1.40030*	1.400
Color, Saybolt	+30	+30
Acidity, distillation residue	neutral	neutral
Nonvolatile matter, grams/100 ml	0.0005	0.0005
Flash point, approximate, F	−10	−10

*Literature values.

Table 2.72: Hexene-1 *(4)*

FORMULA	$CH_2 = CH-CH_2-CH_2-CH_2-CH_3$		
PROPERTIES	RESEARCH GRADE	PURE GRADE	TECHNICAL GRADE
Composition, weight percent			
Hexene-1	99.98	99.7	96.8
trans-Hexene-2	} 0.02	} 0.2	} 0.1
cis-Hexene-2			
Hexenes-3			0.3
Normal Hexane		0.1	1.2
Isoolefins			1.6
Heptene-1			
trans-Heptene-3			
cis-Heptene-3			
trans-Heptene-2			
cis-Heptene-2			
Purity by freezing point, mol %	99.97	99.14 99.0 min	95.8 95.0 min
Freezing point, F	−219.67*		
Boiling point, F	146.27*		
Distillation range, F			
Initial boiling point			146.2
Dry point			146.3
Specific gravity of liquid at 60/60 F	0.6780*	0.678	0.677
at 20/4 C	0.67317*	0.673	0.674
API gravity at 60 F		77.2	77.5
Density of liquid at 60 F, lbs/gal		5.64	5.64
Vapor pressure at 70 F, psia	3.04*	3.0	3.0
100 F, psia	6.01*	6.0	6.0
130 F, psia	10.93*	10.9	10.9
Refractive index, 20/D	1.38788*	1.388	1.388
Color, Saybolt	+30	+30	+30
Acidity, distillation residue		neutral	neutral
Nonvolatile matter, grams/100 ml		0.0005	0.0005
Flash point, approximate, F		−15	−15

*Literature values.

Table 2.73: cis-Hexene-2 *(4)*

FORMULA	$CH_3-\overset{H}{C} = \overset{H}{C}-CH_2-CH_2-CH_3$
PROPERTIES	RESEARCH GRADE
Composition, weight percent	
Hexene-1	0.1
trans-Hexene-2	0.2
cis-Hexene-2	99.6
Hexenes-3	
Normal Hexane	
Isoolefins	0.1
Heptene-1	
trans-Heptene-3	
cis-Heptene-3	
trans-Heptene-2	
cis-Heptene-2	
Purity by freezing point, mol %	99.28
Freezing point, F	−222.04*
Boiling point, F	156.00*
Distillation range, F	
Initial boiling point	
Dry point	
Specific gravity of liquid at 60/60 F	0.6920*
at 20/4 C	0.68720*
API gravity at 60 F	
Density of liquid at 60 F, lbs/gal	5.760*
Vapor pressure at 70 F, psia	2.4*
100 F, psia	4.9*
130 F, psia	9.1*
Refractive index, 20/D	1.39761*
Color, Saybolt	+30
Acidity, distillation residue	
Nonvolatile matter, grams/100 ml	
Flash point, approximate, F	

*Literature values.

Table 2.74: Mixed 2-Hexenes *(4)*

FORMULA	$CH_3-CH = CH-CH_2-CH_2-CH_3$	
PROPERTIES	PURE GRADE	TECHNICAL GRADE
Composition, weight percent		
Hexene-1	trace	0.3
trans-Hexene-2	35.6 } 99.0 min	34.1 } 95.0 min
cis-Hexene-2	63.6	63.5
Hexenes-3	0.8	2.1
Normal Hexane		
Isoolefins		
Heptene-1		
trans-Heptene-3		
cis-Heptene-3		
trans-Heptene-2		
cis-Heptene-2		
Purity by freezing point, mol %		
Freezing point, F		
Boiling point, F		
Distillation range, F		
Initial boiling point	155.0	155.0
Dry point	155.1	155.1
Specific gravity of liquid at 60/60 F	0.684	0.686
at 20/4 C		
API gravity at 60 F	75.4	74.8
Density of liquid at 60 F, lbs/gal	5.69	5.71
Vapor pressure at 70 F, psia	2.4	2.4
100 F, psia	5.0	5.0
130 F, psia	9.2	9.2
Refractive index, 20/D	1.396	1.396
Color, Saybolt	+30	+30
Acidity, distillation residue	neutral	neutral
Nonvolatile matter, grams/100 ml	0.0005	0.0005
Flash point, approximate, F	−5	−5

Table 2.75: Mixed 2- and 3-Hexenes *(4)*

FORMULA	C_6H_{12}
PROPERTIES	**TECHNICAL GRADE**
Composition, weight percent	
Hexene-1	2.3
trans-Hexene-2	71.1
cis-Hexene-2	15.8 } 95.0 min
Hexenes-3	10.8
Normal Hexane	
Isoolefins	
Heptene-1	
trans-Heptene-3	
cis-Heptene-3	
trans-Heptene-2	
cis-Heptene-2	
Purity by freezing point, mol %	
Freezing point, F	
Boiling point, F	
Distillation range, F	
Initial boiling point	152.2
Dry point	155.4
Specific gravity of liquid at 60/60 F	0.685
at 20/4 C	
API gravity at 60 F	75.1
Density of liquid at 60 F, lbs/gal	5.70
Vapor pressure at 70 F, psia	2.5
100 F, psia	5.2
130 F, psia	9.6
Refractive index, 20/D	1.396
Color, Saybolt	+30
Acidity, distillation residue	neutral
Nonvolatile matter, grams/100 ml	0.0005
Flash point, approximate, F	−10

Table 2.76: Heptene-1 *(4)*

FORMULA	$CH_2 = CH-(CH_2)_4-CH_3$
PROPERTIES	**TECHNICAL GRADE**
Composition, weight percent	
Hexene-1	
trans-Hexene-2	
cis-Hexene-2	
Hexenes-3	
Normal Hexane	
Isoolefins	1.3
Heptene-1	97.6
trans-Heptene-3	0.5
cis-Heptene-3	0.5
trans-Heptene-2	0.1
cis-Heptene-2	
Purity by freezing point, mol %	95.4 95.0 min
Freezing point, F	−183.0
Boiling point, F	
Distillation range, F	
Initial boiling point	199
Dry point	202
Specific gravity of liquid at 60/60 F	0.7032
at 20/4 C	0.6982
API gravity at 60 F	70.0
Density of liquid at 60 F, lbs/gal	5.85
Vapor pressure at 70 F, psia	0.9
100 F, psia	2.0
130 F, psia	3.9
Refractive index, 20/D	1.4003
Color, Saybolt	+30
Acidity, distillation residue	neutral
Nonvolatile matter, grams/100 ml	
Flash point, approximate, F	25 (D 56)

Table 2.77: cis-Heptene-2 *(4)*

FORMULA	$CH_3-\overset{H}{C} = \overset{H}{C}-(CH_2)_3-CH_3$
PROPERTIES	**TECHNICAL GRADE**
Composition, weight percent	
Hexene-1	
trans-Hexene-2	
cis-Hexene-2	
Hexenes-3	
Normal Hexane	
Isoolefins	
Heptene-1	
trans-Heptene-3	
cis-Heptene-3	
trans-Heptene-2	4.0
cis-Heptene-2	96.0 95.0 min

PROPERTIES	TECHNICAL GRADE
Freezing point, F	
Boiling point, F	209.3
Distillation range, F	
Initial boiling point	
Dry point	
Specific gravity of liquid at 60/60 F	0.717
at 20/4 C	
API gravity at 60 F	67.3
Density of liquid at 60 F, lbs/gal	5.94
Vapor pressure at 70 F, psia	
100 F, psia	
130 F, psia	
Refractive index, 20/D	1.406
Color, Saybolt	+30
Acidity, distillation residue	neutral
Nonvolatile matter, grams/100 ml	0.0005
Flash point, approximate, F	0

Table 2.78: Mixed 2-Heptenes *(4)*

FORMULA	$CH_3-CH = CH-(CH_2)_3-CH_3$	
PROPERTIES	PURE GRADE	TECHNICAL GRADE
Composition, weight percent		
Heptene-1	trace	1.2
trans-Heptene-3	} 0.5	} 1.4
cis-Heptene-3		
trans-Heptene-2	35.0 } 99.0 min	52.1 } 95.0 min
cis-Heptene-2	64.5	45.0
2,4,4-Trimethylpentene-1		
2,4,4-Trimethylpentene-2		
2,3,3-Trimethylpentene-1		
Isoolefins	trace	0.3
Purity by freezing point, mol %		
Freezing point, F		
Boiling point, F		
Distillation range, F		
Initial boiling point	208.4	208.6
10% Condensed		
50% Condensed		
90% Condensed		
Dry point	212.0	217.1
Specific gravity of liquid at 60/60 F	0.711	0.709
at 20/4 C		
API gravity at 60 F	67.5	68.0
Density of liquid at 60 F, lbs/gal	5.92	5.91
Vapor pressure at 100 F, psia		
Refractive index, 20/D	1.406	1.405
Color, Saybolt	+30	+30
Acidity, distillation residue	neutral	neutral
Nonvolatile matter, grams/100 ml	0.0005	0.0005
Bromine number		
Kauri Butanol value		
Copper corrosion		
Doctor test		
Flash point, approximate, F	28	28

Table 2.79: Mixed 3-Heptenes *(4)*

FORMULA	$CH_3-CH_2-CH = CH-(CH_2)_2-CH_3$
PROPERTIES	TECHNICAL GRADE
Composition, weight percent	
Heptene-1	2.0
trans-Heptene-3	66.5 } 95.0 min
cis-Heptene-3	29.3
trans-Heptene-2	0.5
cis-Heptene-2	1.7
2,4,4-Trimethylpentene-1	
2,4,4-Trimethylpentene-2	
2,3,3-Trimethylpentene-1	
Isoolefins	
Purity by freezing point, mol %	
Freezing point, F	
Boiling point, F	
Distillation range, F	
Initial boiling point	204.0
10% Condensed	
50% Condensed	
90% Condensed	
Dry point	204.4
Specific gravity of liquid at 60/60 F	0.705
at 20/4 C	
API gravity at 60 F	69.2
Density of liquid at 60 F, lbs/gal	5.87
Vapor pressure at 100 F, psia	
Refractive index, 20/D	1.405
Color, Saybolt	+30
Acidity, distillation residue	neutral
Nonvolatile matter, grams/100 ml	0.0005
Bromine number	
Kauri Butanol value	
Copper corrosion	
Doctor test	
Flash point, approximate, F	21

Table 2.80: 2,4,4-Trimethylpentene-1 *(4)*

α–Diisobutylene

FORMULA	$CH_2 = C-CH_2-C-CH_3$ with CH_3 groups		
PROPERTIES	RESEARCH GRADE	PURE GRADE	TECHNICAL GRADE
Composition, weight percent			
Heptene-1			
trans-Heptene-3			
cis-Heptene-3			
trans-Heptene-2			
cis-Heptene-2			
2,4,4-Trimethylpentene-1	99.86	99.39	98.7
2,4,4-Trimethylpentene-2	0.05	0.08	0.1
2,3,3-Trimethylpentene-1			
Isoolefins	0.09	0.53	1.2
Purity by freezing point, mol %	99.58	99.0 99.0 min	97.6 95.0 min
Freezing point, F	−136.26*		
Boiling point, F	214.59*		

(continued)

Table 2.80: (continued)

PROPERTIES	RESEARCH GRADE	PURE GRADE	TECHNICAL GRADE
Distillation range, F			
Initial boiling point			214.3
10% Condensed			
50% Condensed			
90% Condensed			
Dry point			214.6
Specific gravity of liquid at 60/60 F	0.7194*	0.719	0.720
at 20/4 C	0.7150*	0.715	0.716
API gravity at 60 F		65.3	65.0
Density of liquid at 60 F, lbs/gal		5.98	5.99
Vapor pressure at 100 F, psia		1.6	1.6
Refractive index, 20/D	1.4086*	1.409	1.409
Color, Saybolt		+30	+30
Acidity, distillation residue		neutral	neutral
Nonvolatile matter, grams/100 ml		0.0005	0.0005
Bromine number			
Kauri Butanol value			
Copper corrosion			
Doctor test			
Flash point, approximate, F		< 20 (Est.)	< 20 (Est.)

*Literature values.

Table 2.81: 2,4,4-Trimethylpentene-2 (4)

β–Diisobutylene

FORMULA	$CH_3-C(CH_3)=CH-C(CH_3)_2-CH_3$
PROPERTIES	**TECHNICAL GRADE**
Composition, weight percent	
Heptene-1	
trans-Heptene-3	
cis-Heptene-3	
trans-Heptene-2	
cis-Heptene-2	
2,4,4-Trimethylpentene-1	1.9
2,4,4-Trimethylpentene-2	97.1
2,3,3-Trimethylpentene-1	
Isoolefins	1.0
Purity by freezing point, mol %	95.1 95.0 min
Freezing point, F	
Boiling point, F	
Distillation range, F	
Initial boiling point	219
10% Condensed	
50% Condensed	
90% Condensed	
Dry point	230
Specific gravity of liquid at 60/60 F	0.724
at 20/4 C	
API gravity at 60 F	64.0
Density of liquid at 60 F, lbs/gal	6.03
Vapor pressure at 100 F, psia	1.5
Refractive index, 20/D	1.416
Color, Saybolt	+30
Acidity, distillation residue	neutral
Nonvolatile matter, grams/100 ml	0.0005
Bromine number	
Kauri Butanol value	
Copper corrosion	
Doctor test	
Flash point, approximate, F	35 (D 1310)

Table 2.82: Mixed Diisobutylenes (4)

FORMULA	C_8H_{16}
PROPERTIES	**90% GRADE**
Composition, weight percent	
Heptene-1	
trans-Heptene-3	
cis-Heptene-3	
trans-Heptene-2	
cis-Heptene-2	
2,4,4-Trimethylpentene-1	73.2
2,4,4-Trimethylpentene-2	17.0
2,3,3-Trimethylpentene-1	2.9
Isoolefins	6.9
Purity by freezing point, mol %	
Freezing point, F	
Boiling point, F	
Distillation range, F	
Initial boiling point	216 200 min
10% Condensed	217
50% Condensed	218
90% Condensed	220
Dry point	224 260 max
Specific gravity of liquid at 60/60 F	0.723
at 20/4 C	
API gravity at 60 F	64.2
Density of liquid at 60 F, lbs/gal	6.02
Vapor pressure at 100 F, psia	2.0
Refractive index, 20/D	
Color, Saybolt	+30
Acidity, distillation residue	neutral
Nonvolatile matter, grams/100 ml	0.0005
Bromine number	140.3 130 min
Kauri Butanol value	38.3
Copper corrosion	1 1 max
Doctor test	neg. neg.
Flash point, approximate, F	35 (Est.)

Table 2.83: Octene-1 *(4)*

FORMULA	$CH_2 = CH-(CH_2)_5-CH_3$		
PROPERTIES	**RESEARCH GRADE**	**PURE GRADE**	**TECHNICAL GRADE**
Composition, weight percent			
Octene-1	99.95	99.8	98.0
trans-Octene-2			
cis-Octene-2			
mixed-Octenes-3	0.01	0.1	0.5
trans-Octene-4			
Nonene-1			
Decene-1			
Isoolefins	0.04	0.1	1.5
Purity by freezing point, mol %	99.73	99.3 99.0 min	95.6 95.0 min
Freezing point, F	−151.12*		
Boiling point, F	250.30*		
Distillation range, F			
Initial boiling point			250.0
Dry point			250.3
Specific gravity of liquid at 60/60 F	0.7194*	0.719	0.718
at 20/4 C	0.71492*	0.715	0.714
API gravity at 60 F		65.3	65.6
Density of liquid at 60 F, lbs/gal		5.98	5.98
Vapor pressure at 70 F, psia	0.23*	0.2	0.2
100 F, psia	0.66*	0.7	0.7
130 F, psia	1.42*	1.4	1.4
Refractive index, 20/D	1.40870*	1.409	1.409
Color, Saybolt	+30	+30	+30
Acidity, distillation residue		neutral	neutral
Nonvolatile matter, grams/100 ml		0.0005	0.0005
Flash point, approximate, F		70	70

*Literature values.

Table 2.84: cis-Octene-2 *(4)*

FORMULA	$\begin{array}{c} H\ \ H \\ |\ \ | \\ CH_3-C = C-(CH_2)_4-CH_3 \end{array}$
PROPERTIES	**TECHNICAL GRADE**
Composition, weight percent	
Octene-1	trace
trans-Octene-2	3.6
cis-Octene-2	95.6
mixed-Octenes-3	0.1
trans-Octene-4	0.1
Nonene-1	
Decene-1	
Isoolefins	0.6
Purity by freezing point, mol %	95.0 95.0 min
Freezing point, F	
Boiling point, F	
Distillation range, F	
Initial boiling point	257
Dry point	259
Specific gravity of liquid at 60/60 F	0.728
at 20/4 C	
API gravity at 60 F	62.4
Density of liquid at 60 F, lbs/gal	6.07
Vapor pressure at 70 F, psia	
100 F, psia	
130 F, psia	
Refractive index, 20/D	1.414
Color, Saybolt	
Acidity, distillation residue	neutral
Nonvolatile matter, grams/100 ml	0.0005
Flash point, approximate, F	

Table 2.85: Mixed 2-Octenes *(4)*

FORMULA	$CH_3-CH = CH-(CH_2)_4-CH_3$	
PROPERTIES	**PURE GRADE**	**TECHNICAL GRADE**
Composition, weight percent		
Octene-1		
trans-Octene-2	50.3 } 99.0 min	61.4 } 95.0 min
cis-Octene-2	49.2 }	36.5 }
mixed-Octenes-3	0.4	1.8
trans-Octene-4		
Nonene-1		
Decene-1		
Isoolefins	0.1	0.3
Purity by freezing point, mol %		
Freezing point, F		
Boiling point, F		
Distillation range, F		
Initial boiling point	257.0	256.9
Dry point	258.0	257.5
Specific gravity of liquid at 60/60 F	0.731	0.730
at 20/4 C		
API gravity at 60 F	62.1	62.3
Density of liquid at 60 F, lbs/gal	6.08	6.08
Vapor pressure at 70 F, psia		
100 F, psia		
130 F, psia		
Refractive index, 20/D	1.414	1.414
Color, Saybolt	+30	+30
Acidity, distillation residue	neutral	neutral
Nonvolatile matter, grams/100 ml	0.0005	0.0005
Flash point, approximate, F	70	70

Table 2.86: Mixed Octenes *(4)*   **Table 2.87: Nonene-1** *(4)*   **Table 2.88: Decene-1** *(4)*

FORMULA	C_8H_{16}	FORMULA	$CH_2 = CH - (CH_2)_6 - CH_3$	FORMULA	$CH_2 = CH - (CH_2)_7 - CH_3$
PROPERTIES	TECHNICAL GRADE	PROPERTIES	TECHNICAL GRADE	PROPERTIES	TECHNICAL GRADE
Composition, weight percent		Composition, weight percent		Composition, weight percent	
Octene-1	34.6	Octene-1	0.7	Octene-1	
trans-Octene-2	20.7	trans-Octene-2		trans-Octene-2	
cis-Octene-2	42.3 95.0 min	cis-Octene-2		cis-Octene-2	
mixed-Octenes-3	2.3	mixed-Octenes-3		mixed-Octenes-3	
trans-Octene-4		trans-Octene-4		trans-Octene-4	
Nonene-1		Nonene-1	98.7	Nonene-1	
Decene-1		Decene-1		Decene-1	98.9
Isoolefins		Isoolefins	0.6	Isoolefins	1.1
Purity by freezing point, mol %		Purity by freezing point, mol %	97.1 95.0 min	Purity by freezing point, mol %	96.0 95.0 min
Freezing point, F		Freezing point, F	−115.04	Freezing point, F	−89.25
Boiling point, F		Boiling point, F		Boiling point, F	
Distillation range, F		Distillation range, F		Distillation range, F	
Initial boiling point	250.0	Initial boiling point	293	Initial boiling point	336
Dry point	255.0	Dry point	297	Dry point	342
Specific gravity of liquid at 60/60 F	0.724	Specific gravity of liquid at 60/60 F	0.7352	Specific gravity of liquid at 60/60 F	0.7452
at 20/4 C	0.720	at 20/4 C	0.7306	at 20/4 C	0.7408
API gravity at 60 F	63.8	API gravity at 60 F	61.2	API gravity at 60 F	59.70
Density of liquid at 60 F, lbs/gal	6.08	Density of liquid at 60 F, lbs/gal	6.12	Density of liquid at 60 F, lbs/gal	6.16
Vapor pressure at 70 F, psia		Vapor pressure at 70 F, psia		Vapor pressure at 70 F, psia	
100 F, psia	0.5	100 F, psia		100 F, psia	
130 F, psia		130 F, psia		130 F, psia	
Refractive index, 20/D	1.412	Refractive index, 20/D	1.4161	Refractive index, 20/D	1.4216
Color, Saybolt	+30	Color, Saybolt	+30	Color, Saybolt	+30
Acidity, distillation residue	neutral	Acidity, distillation residue	neutral	Acidity, distillation residue	neutral
Nonvolatile matter, grams/100 ml	0.0005	Nonvolatile matter, grams/100 ml	0.0005	Nonvolatile matter, grams/100 ml	
Flash point, approximate, F	70	Flash point, approximate, F	115 (Est.)	Flash point, approximate, F	120 (Est.)

Table 2.89: Undecene-1 *(4)*   **Table 2.90: Dodecene-1** *(4)*

FORMULA	$CH_2 = CH - (CH_2)_8 - CH_3$	FORMULA	$CH_2 = CH - (CH_2)_9 - CH_3$
PROPERTIES	TECHNICAL GRADE	PROPERTIES	TECHNICAL GRADE
Composition, weight percent		Composition, weight percent	
Undecene-1	99.0	Undecene-1	
Dodecene-1		Dodecene-1	99.2
Tridecene-1		Tridecene-1	
Tetradecene-1		Tetradecene-1	0.1
Pentadecene-1		Pentadecene-1	
Hexadecene-1		Hexadecene-1	0.2
Isoolefins	1.0	Isoolefins	0.5
Purity by freezing point, mol %	95.7 95.0 min	Purity by freezing point, mol %	95.4 95.0 min
Freezing point, F	−58.27	Freezing point, F	−33.39
Distillation range, F		Distillation range, F	
Initial boiling point	372	Initial boiling point	410
Dry point	377	Dry point	416
Specific gravity of liquid at 60/60 F	0.7563	Specific gravity of liquid at 60/60 F	0.7624
at 20/4 C	0.7519	at 20/4 C	0.7584
API gravity at 60 F	56.0	API gravity at 60 F	54.10
Density of liquid at 60 F, lbs/gal	6.31	Density of liquid at 60 F, lbs/gal	6.347
Refractive index, 20/D	1.4266	Refractive index, 20/D	1.4300
Color, Saybolt	+30	Color, Saybolt	+30
Acidity, distillation residue	neutral	Acidity, distillation residue	neutral
Flash point, approximate, F	160	Flash point, approximate, F	174

Table 2.91: Tridecene-1 *(4)*

FORMULA	$CH_2 = CH - (CH_2)_{10} - CH_3$
PROPERTIES	TECHNICAL GRADE
Composition, weight percent	
Undecene-1	
Dodecene-1	0.1
Tridecene-1	99.7
Tetradecene-1	
Pentadecene-1	
Hexadecene-1	
Isoolefins	0.2
Purity by freezing point, mol %	96.6 95.0 min
Freezing point, F	−10.95
Distillation range, F	
Initial boiling point	442.6
Dry point	450.7
Specific gravity of liquid at 60/60 F	0.7704
at 20/4 C	0.7662
API gravity at 60 F	52.7
Density of liquid at 60 F, lbs/gal	6.41
Refractive index, 20/D	1.4336
Color, Saybolt	+30
Acidity, distillation residue	neutral
Flash point, approximate, F	175

Table 2.92: Tetradecene-1 *(4)*

FORMULA	$CH_2 = CH - (CH_2)_{11} - CH_3$
PROPERTIES	TECHNICAL GRADE
Composition, weight percent	
Undecene-1	
Dodecene-1	0.1
Tridecene-1	0.3
Tetradecene-1	99.6
Pentadecene-1	trace
Hexadecene-1	
Isoolefins	
Purity by freezing point, mol %	95.5 95.0 min
Freezing point, F	7.05
Distillation range, F	
Initial boiling point	474
Dry point	485
Specific gravity of liquid at 60/60 F	0.7779
at 20/4 C	0.7737
API gravity at 60 F	50.4
Density of liquid at 60 F, lbs/gal	6.48
Refractive index, 20/D	1.4373
Color, Saybolt	+30
Acidity, distillation residue	neutral
Flash point, approximate, F	240

Table 2.93: Butadiene-1,3 *(4)*

FORMULA	$CH_2 = CH-CH = CH_2$		
PROPERTIES	RESEARCH GRADE	SPECIAL PURITY	RUBBER GRADE
Composition, weight percent			
Isobutylene	0.02	0.05	0.1
Butene-1	0.02	0.10	0.2
Butadiene-1,3	99.95	99.70	99.5
trans-Butene-2	0.01	0.10	0.1
Butadiene Dimer		0.05	0.1
Purity by freezing point, mol %	99.89	99.6 99.5 min	99.4 99.0 min
Freezing point, F	−164.05*		
Boiling point, F	24.06*		
Specific gravity of liquid at 60/60 F	0.6272*	0.627	0.627
at 20/4 C	0.6211*	0.621	0.621
API gravity at 60 F		94.2	94.2
Density of liquid at 60 F, lbs/gal		5.22	5.22
Vapor pressure at 70 F, psia		35.6	35.6
100 F, psia		64.0	64.0
130 F, psia		92.2	92.2
Specific gravity of real gas at			
60 F and 14.7 psia (Air = 1)	1.9153*		
Specific volume of real gas at			
60 F and 14.7 psia, cu ft/lb	6.841*		
Flash point, approximate, F		−105	−105
Flammability limits, volume % in air			
Lower	2.0*		
Higher	11.5*		

*Literature values.

Table 2.94: Isoprene *(4)*

FORMULA	CH₂ = C–CH = CH₂ with CH₃	
PROPERTIES	**RESEARCH GRADE**	**POLYMERIZATION GRADE**
Composition, weight percent		
2-Methylbutene-1	trace	0.1
2-Methylbutadiene-1,3	99.99	99.8
Pentenes-2	0.01	0.1
2-Methybutene-2	trace	trace
trans-Pentadiene-1,3		
cis-Pentadiene-1,3		
Cyclopentene		
Cyclooctadiene-1,5		
4-Vinylcyclohexene-1		
1-Methylcyclohexene-1		
3-Methylcyclohexene-1		
4-Methylcyclohexane-1		
Unidentified		
Purity by freezing point, mol %	99.98	99.6 99.0 min
Freezing point, F	−230.71*	
Boiling point, F	93.32*	
Distillation range, F		
Initial boiling point		
10% Condensed		
50% Condensed		
90% Condensed		
Dry point		
Specific gravity of liquid at 60/60 F	0.6861*	0.686
at 20/4 C	0.68095*	0.681
API gravity at 60 F		74.8
Density of Liquid at 60 F, lbs/gal	5.71*	5.71
Vapor pressure at 70 F, psia	9.19*	9.2
100 F, psia	16.67*	16.7
130 F, psia	28.23*	28.2
Refractive index, 20/D	1.42194*	1.422
Color, Saybolt	+30	+30
Acidity, distillation residue		
Nonvolatile matter, grams/100 ml		
Doctor test		
Flash point, approximate, F		− 55 (Est.)

*Literature values.

Table 2.95: Piperylene *(4)*

FORMULA	CH₂ = CH–CH = CH – CH₃
PROPERTIES	**90% GRADE**
Composition, weight percent	
2-Methylbutene-1	
2-Methylbutadiene-1,3	
Pentenes-2	0.1
2-Methylbutene-2	0.7
trans-Pentadiene-1,3	57 } 90.0 min**
cis-Pentadiene-1,3	34
Cyclopentene	8.2
Cyclooctadiene-1,5	
4-Vinylcyclohexene-1	
1-Methylcyclohexene-1	
3-Methylcyclohexene-1	
4-Methylcyclohexene-1	
Unidentified	
Purity by freezing point, mol %	
Freezing point, F	
Boiling point, F	
Distillation range, F	
Initial boiling point	107
10% Condensed	108
50% Condensed	108
90% Condensed	109
Dry point	113
Specific gravity of liquid at 60/60 F	0.690
at 20/4 C	
API gravity at 60 F	73.5
Density of Liquid at 60 F, lbs/gal	5.75
Vapor pressure at 70 F, psia	
100 F, psia	12.7
130 F, psia	
Refractive index, 20/D	
Color, Saybolt	
Acidity, distillation residue	
Nonvolatile matter, grams/100 ml	
Doctor test	
Flash point, approximate, F	− 20 (Est.)

*Literature values.
**Distribution of isomer content varies.

CYCLOOLEFINS

Table 2.96: Cyclopentene *(4)*

FORMULA	CH₂ – CH₂, CH = CH, CH₂		
PROPERTIES	**RESEARCH GRADE**	**PURE GRADE**	**TECHNICAL GRADE**
Composition, weight percent			
Pentenes-2	0.02	0.2	3.6
2-Methylbutene-2	0.03	0.2	0.2
Cyclopentene	99.95	99.6	95.7 95.0 min
2-Methylbutene-1			0.1
Pentene-1			0.1
Cyclopentane			0.4
Cyclohexene			
Cyclohexane			

(continued)

Table 2.96: (continued)

PROPERTIES	RESEARCH GRADE	PURE GRADE	TECHNICAL GRADE
Unidentified			
Benzene			
Toluene			
Ethylbenzene			
Xylenes			
Purity by freezing point, mol %	99.93	99.5 99.0 min	
Freezing point, F	−211.14*		
Boiling point, F	111.64*		
Distillation range, F			
Initial boiling point			111
Dry point			112
Specific gravity of liquid at 60/60 F	0.7775*	0.778	0.778
at 20/4 C	0.77199*	0.772	0.772
API gravity at 60 F		50.4	50.4
Density of liquid at 60 F, lbs/gal		6.48	6.48
Vapor pressure at 70 F, psia			
100 F, psia			
130 F, psia			
Refractive index, 20/D	1.42246*	1.422	1.422
Color, Saybolt	+30	+30	+30
Acidity, distillation residue		neutral	neutral
Nonvolatile matter, grams/100 ml		0.0005	0.0005
Copper corrosion			
Doctor test			
Flash point, approximate, F		−35	
Flammability limits, volume % in air			
Lower			
Higher			

*Literature values.
Cyclopentene and Cyclohexene are sometimes inhibited with 2,6-ditertiarybutyl-4-methylphenol which can be removed by distillation.

Table 2.97: Cyclohexene (4)

FORMULA	CH_2-CH_2 / CH_2 \ CH_2 / CH_2 $CH=CH$	
PROPERTIES	**RESEARCH GRADE**	**PURE GRADE**
Composition, weight percent		
Pentenes-2		
2-Methylbutene-2		
Cyclopentene		
2-Methylbutene-1		
Pentene-1		
Cyclopentane		
Cyclohexene	99.99	99.5
Cyclohexane	0.01	0.2
Unidentified		0.3
Benzene		
Toluene		
Ethylbenzene		
Xylenes		
Purity by freezing point, mol %	99.92	99.4 99.0 min
Freezing point, F	−154.32*	
Boiling point, F	181.36*	

PROPERTIES	RESEARCH GRADE	PURE GRADE
Distillation range, F		
Initial boiling point		181
Dry point		182
Specific gravity of liquid at 60/60 F	0.8159*	0.816
at 20/4 C	0.81096*	0.811
API gravity at 60 F		41.9
Density of liquid at 60 F, lbs/gal		6.79
Vapor pressure at 70 F, psia		
100 F, psia		3.1
130 F, psia		
Refractive index, 20/D	1.44654*	1.446
Color, Saybolt	+30	+30
Acidity, distillation residue		neutral
Nonvolatile matter, grams/100 ml		0.0005
Copper corrosion		
Doctor test		
Flash point, approximate, F		10
Flammability limits, volume % in air		
Lower		
Higher		

*Literature values.
Cyclopentene and Cyclohexene are sometimes inhibited with 2,6-ditertiarybutyl-4-methylphenol which can be removed by distillation.

Table 2.98: 4-Vinylcyclohexene-1 *(4)*

FORMULA			
	CH₂—CH / CH₂ CH₂ CH₂ CH₂ CH—CH = CH₂		
PROPERTIES	**RESEARCH GRADE**	**PURE GRADE**	**TECHNICAL GRADE**
Composition, weight percent			
2-Methylbutene-1			
2-Methylbutadiene-1,3			
Pentenes-2			
2-Methybutene-2			
trans-Pentadiene-1,3			
cis-Pentadiene-1,3			
Cyclopentene			
Cyclooctadiene-1,5	0.01	0.1	1.5
4-Vinylcyclohexene-1	99.99	99.9	98.5
1-Methylcyclohexene-1			
3-Methylcyclohexene-1			
4-Methylcyclohexane-1			
Unidentified			
Purity by freezing point, mol %	99.88	99.3 99.0 min	97.0 95.0 min
Freezing point, F	−164.07*		
Boiling point, F	262.4*		
Distillation range, F			
Initial boiling point			262
10% Condensed			
50% Condensed			
90% Condensed			
Dry point			265
Specific gravity of liquid at 60/60 F	0.834*	0.836	0.836
at 20/4 C	0.8303*	0.830	0.833
API gravity at 60 F		37.8	37.8
Density of Liquid at 60 F, lbs/gal		6.96	6.96
Vapor pressure at 70 F, psia			
100 F, psia	0.5*	0.5	0.5
130 F, psia			
Refractive index, 20/D		1.464	1.464
Color, Saybolt		+30	+30
Acidity, distillation residue		neutral	neutral
Nonvolatile matter, grams/100 ml		0.0005	0.0005
Doctor test			
Flash point, approximate, F		70	70

*Literature values.

Table 2.99: Mixed Methylcyclohexenes *(4)*

FORMULA	C₇H₁₂
PROPERTIES	**TECHNICAL GRADE**
Composition, weight percent	
2-Methylbutene-1	
2-Methylbutadiene-1,3	
Pentenes-2	
2-Methybutene-2	
trans-Pentadiene-1,3	
cis-Pentadiene-1,3	
Cyclopentene	
Cyclooctadiene-1,5	
4-Vinylcyclohexene-1	
1-Methylcyclohexene-1	0.4 ⎫
3-Methylcyclohexene-1	45.5 ⎬ 95.0 min
4-Methylcyclohexane-1	52.5 ⎭
Unidentified	1.6
Purity by freezing point, mol %	
Freezing point, F	
Boiling point, F	
Distillation range, F	
Initial boiling point	215

PROPERTIES	TECHNICAL GRADE
10% Condensed	
50% Condensed	218
90% Condensed	
Dry point	222
Specific gravity of liquid at 60/60 F	0.8086
at 20/4 C	0.8041
API gravity at 60 F	43.5
Density of Liquid at 60 F, lbs/gal	6.73
Vapor pressure at 70 F, psia	0.6
100 F, psia	2.6
130 F, psia	
Refractive index, 20/D	1.4431
Color, Saybolt	+30
Acidity, distillation residue	neutral
Nonvolatile matter, grams/100 ml	
Doctor test	negative
Flash point, approximate, F	30

Table 2.100: Cyclooctadiene-1,5 *(4)*

FORMULA	CH = CH / CH₂ CH₂ / CH₂ CH₂ / CH = CH	
PROPERTIES	PURE GRADE	TECHNICAL GRADE
Composition, weight percent		
2-Methylbutene-1		
2-Methylbutadiene-1,3		
Pentenes-2		
2-Methybutene-2		
trans-Pentadiene-1,3		
cis-Pe..adiene-1,3		
Cyclopentene		
Cyclooctadiene-1,5	99.8	96.4 95.0 min
4-Vinylcyclohexene-1	0.2	3.6
1-Methylcyclohexene-1		
3-Methylcyclohexene-1		
4-Methylcyclohexane-1		
Unidentified		
Purity by freezing point, mol %	99.5 99.0 min	
Freezing point, F	−69.53*	
Boiling point, F		

PROPERTIES	PURE GRADE	TECHNICAL GRADE
Distillation range, F		
Initial boiling point		298
10% Condensed		
50% Condensed		
90% Condensed		
Dry point		304
Specific gravity of liquid at 60/60 F	0.8865*	0.886
at 20/4 C	0.8833*	0.883
API gravity at 60 F		28.2
Density of Liquid at 60 F, lbs/gal		7.38
Vapor pressure at 70 F, psia		
100 F, psia	0.5*	0.5
130 F, psia		
Refractive index, 20/D	1.4933*	1.493
Color, Saybolt	+30	+30
Acidity, distillation residue		
Nonvolatile matter, grams/100 ml		
Doctor test		
Flash point, approximate, F	100	96

*Literature values.

AROMATICS

Table 2.101: Benzene *(4)*

FORMULA	CH − CH / CH CH / CH = CH	
PROPERTIES	RESEARCH GRADE	PURE GRADE
Composition, weight percent		
Pentenes-2		
2-Methylbutene-2		
Cyclopentene		
2-Methylbutene-1		
Pentene-1		
Cyclopentane		
Cyclohexene		
Cyclohexane		
Unidentified		
Benzene	99.99	99.8
Toluene	0.01	0.1
Ethylbenzene		0.1
Xylenes		
Purity by freezing point, mol %	99.90	99.7 99.0 min
Freezing point, F	41.96*	
Boiling point, F	176.18*	
Distillation range, F		
Initial boiling point		175
Dry point		177
Specific gravity of liquid at 60/60 F	0.8845*	0.884
at 20/4 C	0.87901*	0.879
API gravity at 60 F		28.6
Density of liquid at 60 F, lbs/gal		7.36
Vapor pressure at 70 F, psia	1.53*	1.5
100 F, psia	3.22*	3.2
130 F, psia	6.20*	6.2
Refractive index, 20/D	1.50112*	1.501
Color, Saybolt	+30	+30

PROPERTIES	RESEARCH GRADE	PURE GRADE
Acidity, distillation residue		neutral
Nonvolatile matter, grams/100 ml		0.0005
Copper corrosion		
Doctor test		negative
Flash point, approximate, F		10
Flammability limits, volume % in air		
Lower	1.3*	
Higher	7.9*	

*Literature values.

Table 2.102: Toluene *(4)*

FORMULA: toluene structure

PROPERTIES	RESEARCH GRADE	PURE GRADE
Composition, weight percent		
Pentenes-2		
2-Methylbutene-2		
Cyclopentene		
2-Methylbutene-1		
Pentene-1		
Cyclopentane		
Cyclohexene		
Cyclohexane		
Unidentified		
Benzene	0.01	0.1
Toluene	99.99	99.8
Ethylbenzene		
Xylenes		0.1
Purity by freezing point, mol %	99.90	99.7 99.0 min
Freezing point, F	−138.98*	
Boiling point, F	231.12*	
Distillation range, F		
Initial boiling point		230
Dry point		231

PROPERTIES	RESEARCH GRADE	PURE GRADE
Specific gravity of liquid at 60/60 F	0.8719*	0.872
at 20/4 C	0.86696*	0.867
API gravity at 60 F		30.8
Density of liquid at 60 F, lbs/gal		7.26
Vapor pressure at 70 F, psia	0.45*	0.4
100 F, psia	1.03*	1.0
130 F, psia	2.15*	2.2
Refractive index, 20/D	1.49693*	1.497
Color, Saybolt	+30	+30
Acidity, distillation residue		neutral
Nonvolatile matter, grams/100 ml		0.0005
Copper corrosion		1
Doctor test		
Flash point, approximate, F		40 (D 56)
Flammability limits, volume % in air		
Lower	1.2* 212F	
Higher	7.1* 212F	

*Literature values.

Table 2.103: Ethylbenzene *(4)*

FORMULA: ethylbenzene structure

PROPERTIES	RESEARCH GRADE	PURE GRADE	TECHNICAL GRADE
Composition, weight percent			
Benzene	0.01	0.3	0.6
Toluene	0.01	0.2	0.4
Ethylbenzene	99.98	99.5	99.0
para-Xylene			
meta-Xylene			
ortho-Xylene			
Purity by freezing point, mol %	99.92	99.2 99.0 min	98.5 95.0 min
Freezing point, F	−138.96*		
Boiling point, F	277.13*		
Distillation range, F			
Initial boiling point			277
Dry point			278
Specific gravity of liquid at 60/60 F	0.8717*	0.872	0.872
at 20/4 C	0.86702*	0.867	0.867
API gravity at 60 F		30.8	30.8
Density of liquid at 60 F, lbs/gal		7.26	7.26
Vapor pressure at 100 F, psia	0.37*	0.4	0.4
130 F, psia	0.84*	0.8	0.8
Refractive index, 20/D	1.49588*	1.496	1.496
Color, Saybolt	+30	+30	+30
Acidity, distillation residue		neutral	neutral
Nonvolatile matter, grams/100 ml		0.0005	0.0005
Acid wash color			
Color			
Doctor test			
Flash point, approximate, F		59 (D 1310)	59 (D 1319)
Flammability limits, volume % in air			
Lower	1.0*		
Higher	6.7*		

*Literature values.
Ethylbenzene is sometimes stablized with 2,6-ditertiarybutyl-4-methylphenol which can be removed by distillation.

Table 2.104: p-Xylene *(4)*

FORMULA	CH_3 ring structure		
PROPERTIES	RESEARCH GRADE	PURE GRADE	TECHNICAL GRADE
Composition, weight percent			
Benzene			
Toluene			
Ethylbenzene			
para-Xylene	99.99	99.8	99.0
meta-Xylene	0.01	0.2	0.6
ortho-Xylene		trace	0.4
Purity by freezing point, mol %	99.94	99.5 99.0 min	98.0 95.0 min
Freezing point, F	55.87*		
Boiling point, F	281.03*		
Distillation range, F			
Initial boiling point			280
Dry point			281
Specific gravity of liquid at 60/60 F	0.8657*	0.866	0.866
at 20/4 C	0.86105*	0.861	0.861
API gravity at 60 F		31.9	31.9
Density of liquid at 60 F, lbs/gal		7.21	7.21
Vapor pressure at 100 F, psia	0.34*	0.3	0.3
130 F, psia	0.77*	0.8	0.8
Refractive index, 20/D	1.49582*	1.496	1.496
Color, Saybolt	+30	+30	+30
Acidity, distillation residue		neutral	neutral
Nonvolatile matter, grams/100 ml		0.0005	0.0005
Acid wash color		1	1
Color		Pass	Pass
Doctor test			
Flash point, approximate, F		81	81
Flammability limits, volume % in air			
Lower	1.1*		
Higher	6.6*		

*Literature values.

Table 2.105: m-Xylene *(4)*

FORMULA	CH_3 ring structure		
PROPERTIES	RESEARCH GRADE	PURE GRADE	TECHNICAL GRADE
Composition, weight percent			
Benzene			
Toluene			
Ethylbenzene			0.1
para-Xylene	0.01	0.1	0.4
meta-Xylene	99.99	99.9	99.2
ortho-Xylene			0.3
Purity by freezing point, mol %	99.94	99.2 99.0 min	98.2 95.0 min
Freezing point, F	−54.17*		
Boiling point, F	282.39*		
Distillation range, F			
Initial boiling point			280
Dry point			281
Specific gravity of liquid at 60/60 F	0.8687*	0.869	0.869
at 20/4 C	0.86417*	0.864	0.864
API gravity at 60 F		31.3	31.3

(continued)

Table 2.105: (continued)

PROPERTIES	RESEARCH GRADE	PURE GRADE	TECHNICAL GRADE
Density of liquid at 60 F, lbs/gal		7.24	7.24
Vapor pressure at 100 F, psia	0.33*	0.3	0.3
130 F, psia	0.74*	0.7	0.7
Refractive index, 20/D	1.49722*	1.497	1.497
Color, Saybolt	+30	+30	+30
Acidity, distillation residue		neutral	neutral
Nonvolatile matter, grams/100 ml		0.0005	0.0005
Acid wash color			
Color			
Doctor test		negative	negative
Flash point, approximate, F		84	84
Flammability limits, volume % in air			
Lower	1.1*		
Higher	6.4*		

*Literature values.

Table 2.106: o-Xylene *(4)*

FORMULA	CH₃ / CH-C-C-CH₃ / CH / CH-CH	
PROPERTIES	RESEARCH GRADE	PURE GRADE
Composition, weight percent		
meta-Xylene	0.01	0.3
ortho-Xylene	99.99	99.7
Ethylbenzene		
Isopropylbenzene		
Normal Propylbenzene		
Methylethylbenzenes		
1,2,4-Trimethylbenzene		
1,3,5-Trimethylbenzene		
1,2,3-Trimethylbenzene		
Purity by freezing point, mol %	99.96	99.4 99.0 min
Freezing point, F	−13.33*	
Boiling point, F	291.94*	
Distillation range, F		
Initial boiling point		289
Dry point		291
Specific gravity of liquid at 60/60 F	0.8848*	0.885
at 20/4 C	0.88020*	0.880
API gravity at 60 F		28.4
Density of liquid at 60 F, lbs/gal		7.37
Vapor pressure at 100 F, psia	0.26*	0.3
130 F, psia	0.61*	0.6
Refractive index, 20/D	1.50545*	1.505
Color, Saybolt	+30	+30
Acidity, distillation residue		neutral
Nonvolatile matter, grams/100 ml		0.0005
Flash point, approximate, F		88 (D 56)
Flammability limits, volume % in air		
Lower	1.1*	
Higher	6.4*	

*Literature values.

Table 2.107: Cumene *(4)*

FORMULA	CH₃-CH-CH₃ / CH-C-CH / CH-CH		
PROPERTIES	RESEARCH GRADE	PURE GRADE	TECHNICAL GRADE
Composition, weight percent			
meta-Xylene			
ortho-Xylene	0.01	0.1	0.2
Ethylbenzene	0.03	0.3	0.6
Isopropylbenzene	99.96	99.6	99.2
Normal Propylbenzene			
Methylethylbenzenes			
1,2,4-Trimethylbenzene			
1,3,5-Trimethylbenzene			
1,2,3-Trimethylbenzene			
Purity by freezing point, mol %	99.92	99.3 99.0 min	98.5 95.0 min
Freezing point, F	−140.86*		
Boiling point, F	306.31*		
Distillation range, F			
Initial boiling point		306	
Dry point		307	
Specific gravity of liquid at 60/60 F	0.8663*	0.866	0.866
at 20/4 C	0.86179*	0.862	0.862
API gravity at 60 F		31.9	31.9
Density of liquid at 60 F, lbs/gal		7.21	7.21
Vapor pressure at 100 F, psia	0.19*	0.2	0.2
130 F, psia	0.45*	0.4	0.4
Refractive index, 20/D	1.49145*	1.491	1.491
Color, Saybolt	+30	+30	+30
Acidity, distillation residue		neutral	neutral
Nonvolatile matter, grams/100 ml		0.0005	0.0005
Flash point, approximate, F		111	111
Flammability limits, volume % in air			
Lower	0.9*		
Higher	6.5*		

*Literature values.
Isopropylbenzene and 1,2,4-Trimethylbenzene are sometimes stabilized
 with 2,6-ditertiary butyl-4-methylphenol which can be removed by distillation.

Table 2.108: n-Propylbenzene *(4)*

FORMULA	CH₂ – CH₂ – CH₃ structure
PROPERTIES	**TECHNICAL GRADE**
Composition, weight percent	
meta-Xylene	
ortho-Xylene	
Ethylbenzene	
Isopropylbenzene	1.5
Normal Propylbenzene	96.6 95.0 min
Methylethylbenzenes	1.9
1,2,4-Trimethylbenzene	
1,3,5-Trimethylbenzene	
1,2,3-Trimethylbenzene	
Purity by freezing point, mol %	
Freezing point, F	
Boiling point, F	
Distillation range, F	
Initial boiling point	315
Dry point	319
Specific gravity of liquid at 60/60 F	0.8669
at 20/4 C	0.8621
API gravity at 60 F	31.7
Density of liquid at 60 F, lbs/gal	7.22
Vapor pressure at 100 F, psia	
130 F, psia	
Refractive index, 20/D	1.4915
Color, Saybolt	+30
Acidity, distillation residue	
Nonvolatile matter, grams/100 ml	0.0005
Flash point, approximate, F	114
Flammability limits, volume % in air	
Lower	
Higher	

Table 2.109: Pseudocumene *(4)*

FORMULA	CH₃ structure		
PROPERTIES	**RESEARCH GRADE**	**PURE GRADE**	**TECHNICAL GRADE**
Composition, weight percent			
meta-Xylene			
ortho-Xylene			
Ethylbenzene			
Isopropylbenzene			
Normal Propylbenzene			
Methylethylbenzenes		0.2	1.6
1,2,4-Trimethylbenzene	99.99	99.7	96.1
1,3,5-Trimethylbenzene	0.01	0.1	1.4
1,2,3-Trimethylbenzene			0.9
Purity by freezing point, mol %	99.90	99.5 99.0 min	95.5 95.0 min
Freezing point, F	−46.84*		
Boiling point, F	336.83*		
Distillation range, F			
Initial boiling point		336	
Dry point		337	
Specific gravity of liquid at 60/60 F	0.8802*	0.880	0.880
at 20/4 C	0.87582*	0.876	0.876
API gravity at 60 F		29.3	29.3
Density of liquid at 60 F, lbs/gal		7.33	7.33
Vapor pressure at 100 F, psia			
130 F, psia	0.22*	0.2	0.2
Refractive index, 20/D	1.50484*	1.505	1.505
Color, Saybolt	+30	+30	+30
Acidity, distillation residue		neutral	neutral
Nonvolatile matter, grams/100 ml		0.0005	0.0005
Flash point, approximate, F		130	130
Flammability limits, volume % in air			
Lower			
Higher			

*Literature values.

Isopropylbenzene and 1,2,4-Trimethylbenzene are sometimes stabilized with 2,6-ditertiary butyl-4-methylphenol which can be removed by distillation.

Table 2.110: n-Butylbenzene *(4)*

FORMULA	CH₂ – CH₂ – CH₂ – CH₃ structure		
PROPERTIES	**RESEARCH GRADE**	**PURE GRADE**	**TECHNICAL GRADE**
Composition, weight percent			
secondary-Butylbenzene		0.2	3.8
normal-Butylbenzene	99.8	99.4	95.6 95.0 min
1-Phenylbutene-2			0.6
other Alkylbenzenes	0.2	0.4	
other Phenylbutenes			
secondary-Amylbenzene			
3-Phenylpentane			
2-Phenyl-2-methylbutane			
Light Amylbenzenes			
Alkylbenzenes			
secondary-Butyl Chloride			
Butenes			
Purity by freezing point, mol %	99.50	99.2 99.0 min	
Freezing point, F	−126.35*		
Boiling point, F	361.89*		
Distillation range, F			
Initial boiling point			
10% Condensed			
50% Condensed			
90% Condensed			
Dry point			

(continued)

Table 2.110: (continued)

PROPERTIES	RESEARCH GRADE	PURE GRADE	TECHNICAL GRADE
Specific gravity of liquid at 60/60 F	0.8646*	0.865	0.865
at 20/4 C	0.86013*	0.860	0.860
API gravity at 60 F		32.1	32.1
Density of liquid at 60 F, lbs/gal		7.20	7.20
Refractive index, 20/D	1.48979*	1.490	1.490
Color, Saybolt	+30	+30	+30
Acidity, distillation residue		neutral	neutral
Nonvolatile matter, grams/100 ml		0.0005	0.0005
Color Alpha			
Flash point, approximate, F		160	160
Flammability limits, volume % in air			
Lower	0.8		
Higher	5.8		

*Literature values.
Normal Butylbenzene is sometimes stablized with tertiary-butylcatechol (TBC) which can be removed by distillation.

Table 2.111: Isobutylbenzene *(4)*

FORMULA	$CH_2 - CH\genfrac{}{}{0pt}{}{CH_3}{CH_3}$ $CH \cdot C \cdot CH$ $CH \cdot CH \cdot CH$		
PROPERTIES	**RESEARCH GRADE**	**PURE GRADE**	**TECHNICAL GRADE**
Composition, weight percent			
Toluene	0.01	0.2	0.5
Isopropylbenzene			
tertiary-Butylbenzene			
Isobutylbenzene	99.97	99.6	99.1
secondary-Butylbenzene			
normal-Butylbenzene	0.02	0.2	0.4
Water, ppm, weight		<100	<100
Purity by freezing point, mol %	99.80	99.3 99.0 min	98.5 95.0 min
Freezing point, F	−60.66*		
Boiling point, F	342.97*		
Distillation range, F			
Initial boiling point		340	337
Dry point		343	344
Specific gravity of liquid at 60/60 F	0.8576*	0.858	0.858
at 20/4 C	0.85321*	0.853	0.853
API gravity at 60 F		33.4	33.4
Density of liquid at 60 F, lbs/gal		7.14	7.14
Vapor pressure at 130 F, psia	0.21	0.2	0.2
Refractive index, 20/D	1.48646*	1.486	1.486
Color, Saybolt	+30	+30	+30
Acidity, distillation residue		neutral	neutral
Nonvolatile matter, grams/100 ml			
Aniline point, F			
Bromine number			
Flash point, approximate, F		140	140

*Literature values.

Table 2.112: sec-Butylbenzene *(4)*

FORMULA	CH₃ – CH – CH₂ – CH₃ (ring structure)		
PROPERTIES	**RESEARCH GRADE**	**PURE GRADE**	**TECHNICAL GRADE**
Composition, weight percent			
Toluene			
Isopropylbenzene			
tertiary-Butylbenzene	0.02	0.6	0.9
Isobutylbenzene			0.1
secondary-Butylbenzene	99.98	99.4	98.9
normal-Butylbenzene			0.1
Water, ppm, weight			
Purity by freezing point, mol %	99.93	99.2 99.0 min	96.7 95.0 min
Freezing point, F	−103.85*		
Boiling point, F	343.95*		
Distillation range, F			
Initial boiling point			338
Dry point			343
Specific gravity of liquid at 60/60 F	0.8664*	0.866	0.866
at 20/4 C	0.86207*	0.862	0.862
API gravity at 60 F		31.9	31.9
Density of liquid at 60 F, lbs/gal		7.21	7.21
Vapor pressure at 130 F, psia	0.20	0.2	0.2
Refractive index, 20/D	1.49020*	1.490	1.490
Color, Saybolt	+30	+30	+30
Acidity, distillation residue		neutral	neutral
Nonvolatile matter, grams/100 ml		0.0005	0.0005
Aniline point, F		−20	
Bromine number		0.5	
Flash point, approximate, F		126	126

*Literature values.

Secondary-Butylbenzene is sometimes stablized with tertiary-butylcatechol (TBC) which can be removed by distillation.

Table 2.113: tert-Butylbenzene *(4)*

FORMULA	CH₃ – C(CH₃) – CH₃ (ring structure)		
PROPERTIES	**RESEARCH GRADE**	**PURE GRADE**	**TECHNICAL GRADE**
Composition, weight percent			
Toluene			
Isopropylbenzene		0.1	0.5
tertiary-Butylbenzene	99.99	99.8	98.2
Isobutylbenzene			
secondary-Butylbenzene	0.01	0.1	1.1
normal-Butylbenzene			0.2
Water, ppm, weight			
Purity by freezing point, mol %	99.82	99.4 99.0 min	97.0 95.0 min
Freezing point, F	−72.13*		
Boiling point, F	336.41*		
Distillation range, F			
Initial boiling point		336	331
Dry point		337	336
Specific gravity of liquid at 60/60 F	0.8710*	0.871	0.871
at 20/4 C	0.86650*	0.866	0.866
API gravity at 60 F		31.0	31.0
Density of liquid at 60 F, lbs/gal		7.25	7.25
Vapor pressure at 130 F, psia	0.23	0.2	0.2
Refractive index, 20/D	1.49266*	1.493	1.493
Color, Saybolt	+30	+30	+30
Acidity, distillation residue		neutral	neutral
Nonvolatile matter, grams/100 ml		0.0005	0.0005
Aniline point, F			
Bromine number			
Flash point, approximate, F		140	140

*Literature values.

Table 2.114: 1-Phenylbutene-2 *(4)*

FORMULA	$CH_2-CH=CH-CH_3$ $CH \cdot C \cdot CH$ $CH \cdot CH$ $\cdot CH$
PROPERTIES	**TECHNICAL GRADE**
Composition, weight percent	
secondary-Butylbenzene	
normal-Butylbenzene	
1-Phenylbutene-2	96.4 95.0 min
other Alkylbenzenes	
other Phenylbutenes	3.6
secondary-Amylbenzene	
3-Phenylpentane	
2-Phenyl-2-methylbutane	
Light Amylbenzenes	
Alkylbenzenes	
secondary-Butyl Chloride	
Butenes	
Purity by freezing point, mol %	
Freezing point, F	
Boiling point, F	
Distillation range, F	
Initial boiling point	360
10% Condensed	
50% Condensed	
90% Condensed	
Dry point	367
Specific gravity of liquid at 60/60 F	0.888
at 20/4 C	
API gravity at 60 F	27.8
Density of liquid at 60 F, lbs/gal	7.40
Refractive index, 20/D	1.511
Color, Saybolt	+30
Acidity, distillation residue	neutral
Nonvolatile matter, grams/100 ml	0.0005
Color Alpha	
Flash point, approximate, F	160
Flammability limits, volume % in air	
Lower	
Higher	

*Literature values.
1-Phenylbutene-2 is sometimes stabilized with tertiary-butylcatechol (TBC) which can be removed by distillation.

Table 2.115: sec-Amylbenzene *(4)*

FORMULA	$CH_3-CH-CH_2-CH_2-CH_3$ $CH \cdot C \cdot CH$ $CH \cdot CH$ $\cdot CH$	
PROPERTIES	**PURE GRADE**	**TECHNICAL GRADE**
Composition, weight percent		
secondary-Butylbenzene		
normal-Butylbenzene		
1-Phenylbutene-2		
other Alkylbenzenes		
other Phenylbutenes		
secondary-Amylbenzene	99.5 99.0 min	97.3 95.0 min
3-Phenylpentane	0.1	0.8
2-Phenyl-2-methylbutane	0.4	1.9
Light Amylbenzenes		
Alkylbenzenes		
secondary-Butyl Chloride		
Butenes		
Purity by freezing point, mol %		
Freezing point, F		
Boiling point, F	379.4*	
Distillation range, F		
Initial boiling point	374	370
10% Condensed		
50% Condensed		
90% Condensed		
Dry point	380	380
Specific gravity of liquid at 60/60 F	0.8628*	0.863
at 20/4 C	0.8585	0.858
API gravity at 60 F		32.5
Density of liquid at 60 F, lbs/gal		7.18
Refractive index, 20/D	1.4876*	1.488
Color, Saybolt	+30	+30
Acidity, distillation residue	neutral	neutral
Nonvolatile matter, grams/100 ml		
Color Alpha		
Flash point, approximate, F	155	155
Flammability limits, volume % in air		
Lower		
Higher		

*Literature values.

Table 2.116: Mixed Amylbenzenes *(4)*

FORMULA	C_6H_{11} $CH \cdot C \cdot CH$ $CH \cdot CH$ $\cdot CH$
PROPERTIES	**PURE GRADE**
Composition, weight percent	
secondary-Butylbenzene	
normal-Butylbenzene	
1-Phenylbutene-2	
other Alkylbenzenes	
other Phenylbutenes	
secondary-Amylbenzene	38.8
3-Phenylpentane	39.2 } 99.0 min
2-Phenyl-2-methylbutane	21.7
Light Amylbenzenes	0.2
Alkylbenzenes	0.1
secondary-Butyl Chloride	
Butenes	

PROPERTIES	PURE GRADE
Distillation range, F	
Initial boiling point	369.6
10% Condensed	370.4
50% Condensed	371.6
90% Condensed	372.0
Dry point	372.4
Specific gravity of liquid at 60/60 F	
at 20/4 C	0.864
API gravity at 60 F	
Density of liquid at 60 F, lbs/gal	7.25
Refractive index, 20/D	1.490
Color, Saybolt	+30
Acidity, distillation residue	neutral
Nonvolatile matter, grams/100 ml	
Color Alpha	
Flash point, approximate, F	155
Flammability limits, volume % in air	
Lower	
Higher	

*Literature values.

TERPENES

Table 2.117: "Dipentene" No. 122 *(28)*

Dipentene No. 122 terpene solvent meets the Federal specification for commercial dipentene. It is used mainly in oleoresin-based protective coatings as a component of the solvent system and as an antiskinning agent. Other uses include the production of dipping finishes and various chemical specialties. Dipentene No. 122 is a near-colorless liquid with a pleasant terpene odor. Obtained by fractionation of oils extracted from pinewood, it is a blend of related monocyclic terpene hydrocarbons, predominantly dipentene.

Product Specifications

Color (Hercules terpene)	0.5 max
Specific gravity at 15.6/15.6°C	0.845-0.865
Refractive index at 20°	1.472-1.477
Distillation range, °C	
5%	168 min
95%	188 max

Typical Properties

Specific gravity at 15.6/15.6°C	0.853
Refractive index at 20°C	1.475
Distillation range, °C	
5%	175
95%	183
Freezing point, °C	<-40
Flash point, TCC, °C (°F)	49 (120)
Kauri-butanol value	62
Aniline point, °C	<0
Unpolymerized residue, %	1.5

Outstanding Characteristics

High clarity; near colorlessness; pleasant odor; high solvency; good anti-skinning properties; good wetting and dispersing properties for pigments.

Solvent for synthetic and natural resins, rubber, waxes, raw and polymerized oils, and metallic driers.

Table 2.118: "Solvenol" No. 1 *(28)*

Solvenol 1 is a pale yellow to near colorless liquid with a pleasant terpene odor and high solvency for resins, waxes, and greases. Of pinewood origin, it is a mixture of terpenes, mainly monocyclic hydrocarbons similar to and including dipentene. Solvenol 1, like Dipentene No. 122 and turpentine, can be used as a component of solvent systems for oleoresinous and oil-based paints and varnishes. In these applications, Solvenol 1 has equivalent or slightly greater solvent power; it is also a stronger solvent than turpentine for resins and waxes used in polishes and other specialty products.

Product Specifications

Specific gravity at 15.6/15.6°C	0.845-0.865
Refractive index at 20°C	1.472-1.485
Unpolymerized residue, %	3 max
Distillation range, °C, first cm³	168 min
95%	188 max

Typical Properties

Specific gravity at 15.6/15.6°C	0.858
Refractive index at 20°C	1.477
Unpolymerized residue, %	1.5
Distillation range, °C, 5%	172
95%	184
Color, Hazen	30
Freezing point, °C	<−40
Flash point, Tag closed cup, °F (°C)	120 (49)
Kauri-butanol value	70
Aniline point, °F (°C)	<23 (<−5)

Outstanding Characteristics

Clear, near colorless liquid; pleasant, terpenaceous odor; high solvent power; antiskinning properties; slow, uniform evaporation; wetting and dispersing action on pigments.

Table 2.119: "Solvenol" No. 2 *(28)*

Solvenol 2 terpene solvent is a pale yellow to near colorless liquid that has high solvency for resins, waxes, and greases. It is exceptionally effective as a softening and swelling agent for rubber. Of pinewood origin, it is a mixture of monocyclic terpenes similar to those composing Solvenol 1, but in different proportions to one another and slightly broader in distillation range. It is comparable in solvent power to Solvenol 1 and, like the latter, is a stronger solvent than turpentine for waxes and resins.

Product Specifications

Specific gravity at 15.6/15.6°C	0.845-0.870
Distillation range, °C, first cm³	165 min
95%	195 max

Typical Properties

Specific gravity at 15.6/15.6°C	0.860
Distillation range, °C, 5%	174
95%	183
Color, Hazen	45
Freezing point, °C	<-40
Flashpoint, Tag closed cup, °F (°C)	115 (46)
Kauri-butanol value	80
Aniline point, °F (°C)	<23 (<-5)

Outstanding Characteristics

Clear, near colorless liquid; high solvent power; highly effective softening and swelling agent for natural and synthetic rubbers.

Table 2.120: Hercules Steam-Distilled Wood Turpentine *(28)*

Hercules steam-distilled wood turpentine is a clear, water-white liquid that complies with all requirements of Federal and ASTM specifications for pure spirits of turpentine. It is used as a solvent and thinner for paints and varnishes; as a constituent of cleaning compounds and wax and liquid polishes; and in various other chemical specialties.

Product Specifications

Specific gravity at 15.6/15.6°C	0.860-0.866
Refractive index at 20°C	1.465-1.469
ASTM distillation range, °C, 5%	154.0
95%	170.0

Typical Properties

Specific gravity at 15.6/15.6°C	0.861
Refractive index at 20°C	1.468
Unpolymerized residue, %	1.3
Initial boiling point, °C	150
ASTM distillation below 170°C, %	98
Freezing point, °C	<-40
Aniline point, °C	21
Kauri-butanol value	56
Moisture	trace
Flash point, TCC, °C (°F)	36 (97)

Outstanding Characteristics

Clarity; water-white color; typical turpentine odor; high solvency power; excellent wetting and penetrating properties; uniform purity.

Solvent for raw and bodied drying oils, and for natural and synthetic resins and waxes.

Table 2.121: Hercules Alpha-Pinene *(28)*

Hercules alpha-pinene is a clear, water-white product obtained by fractional distillation of steam-distilled wood turpentine. It consists predominantly of the bicyclic terpene hydrocarbon, alpha-pinene. Hercules alpha-pinene can be used wherever a high-purity-grade alpha-pinene is required.

Product Specifications

Specific gravity at 15.6/15.6°C	0.8620-0.8645
Refractive index at 20°C	1.4655-1.4670
Distillation range, °C, 5%, min	155
95%, max	159

(continued)

Table 2.121: (continued)

Typical Properties

Specific gravity at 15.6/15.6°C	0.863
Refractive index at 20°C	1.466
Components, %	
Low boilers	0.3
alpha-pinene	84.7
Camphene	13.9
beta-pinene	0.5
Monocyclics	0.6
Distillation range, °C. 5%	156
95%	158
Color, Hazen	10
Freezing point, °C	<-40
Flashpoint, Tag closed cup, °F (°C)	91 (33)
Kauri-butanol value	52
Density at 60°F (15.6°C), lbs/gal (kg/L)	7.2 (0.86)

Outstanding Characteristics

Clear, water-white, high purity, chemically reactive, excellent solvent, narrow distillation range.

Table 2.122: Typical Composition of Turpentines Produced in the United States *(5)*

	TYPE OF TURPENTINE		
	Gum	**Wood**	**Sulfate**
Alpha-pinene, %	60-65	75-80	55-60
Beta-pinene, %	25-30	0-2	20-25
Camphene, %	1-2	4-8	1-2
Dipentene, %	3-4	5-7	7-8
Other Terpenes, %	2-3	10-13	9-10

COMPARATIVE DATA

Table 2.123: Amoco "Panasol" Solvents *(20)*

Panasol solvents are high boiling point aromatics containing substituted mono- and di-alkylnaphthalenes. Panasol solvents are the superior choice for reducers and solvents for coating resins, high flash point solvents for pesticides and herbicides and raw materials for the petroleum naphthalene market.

Specification Properties	Test Method	PANASOL AN-2K	PANASOL AN-2L	PANASOL AN-3N
Gravity @ 16°C	ASTM D 287			
min. °API		19	22	9
max. °API		20	25	13
Distillation	ASTM D 86			
min. IBP, °C (°F)		177 (350)	140 (284)	210 (410)
max. EP, °C (°F)		288 (550)	288 (550)	288 (550)
Flash point				
min. TCC, °C (°F)	ASTM D 56	63 (145)	38 (100)	82 (180)
Color				
max. ASTM	ASTM D 1500	2	– – –	2
Appearance	visual	clear, no free suspended matter	– – –	clear, no free suspended matter
Aromatics				
min. vol. %	ASTM D 1319	78	– – –	95
Lighter than naphthalene				
max. wt. %	chromatograph	– – –	– – –	9.5
Mixed aniline point				
max. °C	ASTM D 611	31.0	– – –	– – –
Copper corrosion	ASTM D 849	pass	– – –	pass

Table 2.124: Amsco Aliphatic and Aromatic Solvents (13)

	Chemical Abstracts Service Registry Number	Gravity, 60°F (15.56°C) API	Specific	lb. per gal	Distillation Range, °F(°C) IBP	50%	DP	Vapor Pressure (mm Hg at 20°C)	Relative Evap. Rate n-BuAc = 1	Aniline Cloud Point °F(°C)	Kauri-Butanol Value	Flash Point TCC °F	Hydrocarbon Compositions, % Aromatics	Paraffins	Cyclo-paraffins
Aliphatic Hydrocarbons.															
Amsco pentane	109-66-0	99.2	0.633	5.26	95(35.0)	98(36.7)	102(38.9)	439	—	—	26	<0	Nil	95	5
Amsco petroleum ether (ligroin)	8032-32-4	88.2	0.644	5.77	100(37.8)	106(41.7)	140(60.0)	400	—	—	26	<0	2	81	17
Amsco rubber solvent	64742-89-8	71.6	0.700	5.83	107(41.7)	189(87.2)	275(135.0)	180	6.1	142(61.1)	34	<0	4	82	14
Amsco textile spirits	64741-84-0	76.0	0.682	5.68	146(63.6)	155(68.3)	176(80.0)	155	8.8	146(63.3)	31	<0	<1	90	10
Amsco hexane	110-54-3	78.3	0.675	5.61	149(65.0)	151(66.2)	157(69.7)	140	8.1	150(65.6)	30	<0	Nil	88	12
Amsco heptane	142-82-5	71.2	0.698	5.81	198(92.4)	201(93.8)	206(96.7)	45	4.5	151(66.1)	30	<20	Nil	90	10
Amsco lactol spirits 9300	64742-89-8	62.0	0.731	6.09	201(93.9)	206(96.7)	223(106.1)	66	3.9	126(52.2)	39	<20	7	59	34
Amsco lactol spirits 9500	64742-89-8	57.3	0.749	6.24	206(96.7)	213(100.6)	225(107.2)	60	3.9	109(42.8)	43	<20	15	41	44
Amsco roto solv	64742-89-8	61.1	0.735	6.12	241(116.1)	246(118.9)	256(124.4)	17	1.7	145(62.8)	34	45	5	60	35
Amsco special naphtholite 66/3 9300	64742-48-9	59.2	0.753	6.27	248(120.0)	265(129.4)	296(146.7)	5.6	1.0	148(64.4)	33	54	<1	51	49
Amsco special naphtholite 66/3 9500	64742-48-9	56.0	0.755	6.28	260(126.7)	269(131.7)	288(142.2)	5.2	1.0	143(61.7)	35	65	<1	42	58
Amsco naphthol spirits 66/3 9300	64742-48-9	54.9	0.759	6.32	313(156.1)	328(164.4)	353(178.3)	3.0	0.21	160(71.1)	31	106	<1	64	36
Amsco naphthol spirits 66/3 9500	64742-48-9	51.9	0.772	6.42	315(157.2)	328(164.4)	355(179.4)	2.9	0.21	152(66.7)	33	104	<1	44	56
Amsco mineral spirits 66/3 9300	64742-47-8	53.3	0.766	6.38	314(156.7)	341(171.7)	388(197.8)	2.9	0.13	161(71.7)	31	108	<1	48	51
Amsco mineral spirits 66/3 9500	64742-47-8	51.7	0.772	6.43	318(158.9)	339(170.6)	386(196.7)	2.6	0.13	155(68.3)	33	108	<1	47	53
Amsco mineral spirits 75*	64741-41-9	52.0	0.771	6.42	319(159.4)	345(173.9)	397(202.8)	2.6	0.13	150(65.6)	32	104	7	48	45
Amsco regular mineral spirits	64741-41-9	48.3	0.787	6.55	307(152.8)	340(171.1)	389(198.3)	3.4	0.12	132(55.6)	37	107	15	48	37
Amsco retardsol*	8008-20-6	43.3	0.810	6.74	328(164.4)	425(218.3)	510(265.6)	1.8	—	144(62.2)	34	133	18	41	41
Amsco odorless mineral spirits	64741-65-7	54.4	0.761	6.34	354(178.9)	362(183.3)	386(196.7)	1.2	0.17	184(84.4)	27	128	<1	86	14
Amsco 140 solvent 66/3 9300	64742-47-8	50.9	0.776	6.46	361(182.8)	381(193.9)	407(208.3)	<1	0.08	168(75.6)	29	145	1	66	33
Amsco 140 solvent 66/3 9500	64742-47-8	48.3	0.787	6.55	374(190.0)	386(196.7)	397(202.8)	<1	0.08	162(72.2)	31	147	<1	45	55
Amsco odorless base*	—	47.6	0.790	6.58	370(187.8)	441(227.2)	485(251.7)	—	—	171(77.5)	28	160	7	66	27
Amsco 460 solvent*	—	45.1	0.801	6.67	372(188.9)	405(207.2)	503(261.7)	<1	—	153(67.2)	33	150	8	56	36
Amsco mineral seal oil	64741-44-2	42.0	0.813	6.77	518(270.0)	560(293.3)	610(321.1)	<1	—	189(87.2)	27	275(TOC)	3	—	—
Aromatic Hydrocarbons.															
Amsco benzene	71-43-2	28.5	0.884	7.36	176(79.8)	176(80.1)	177(80.4)	75	4.50	—	110	<0	100	Nil	Nil
Amsco toluene	108-88-3	30.9	0.871	7.26	231(110.6)	231(110.6)	232(111.1)	22	1.90	47(8.3)**	105	35	100	Nil	Nil
Amsco xylene	1330-20-7	31.0	0.871	7.25	281(138.5)	283(139.6)	287(141.4)	6.6	0.80	50(10.0)**	98	81	100	Nil	Nil
Amsco super high flash naphtha	64742-95-6	30.3	0.875	7.28	318(158.9)	325(162.8)	350(176.6)	2.6	0.37	56(13.3)**	93	113	100	Nil	Nil
Amsco-Solv D	—	33.3	0.859	7.15	324(162.2)	345(173.9)	394(201.1)	2.1	—	84(28.9)**	72	119	77	10	13
Amsco-Solv F	64742-94-5	30.0	0.876	7.30	339(170.6)	358(181.1)	399(203.9)	1.3	0.13	70(21.1)**	84	120	70	17	13
Amsco-Solv G	—	25.8	0.900	7.49	362(183.3)	377(191.7)	407(208.3)	<1	—	60(15.6)**	95	147	100	Nil	Nil
Amsco-Solv H-SB	—	20.2	0.934	7.77	357(180.6)	456(235.6)	537(280.6)	<1	—	76(24.4)**	105	152	89	4	7

*Specifications may vary according to location.   **Mixed aniline point.

Table 2.125: Ashland Aliphatic and Aromatic Solvents (69)

ALIPHATIC SOLVENTS

Listed By IBP	API Gravity @ 60°F	Dist. Range °F IBP	Dist. Range °F DP	Flash Pt °F TCC	KB Value
n-Pentane	92.7	95	102	−50	25
Rubber Solvent	71.0	115	278	−40	33
Hexane	78.5	150	157	−10	29
Heptane	69.0	190	210	20	31
LACOLENE	69.0	195	230	20	30
Super LACOLENE	56.2	195	230	22	44
VM & P Naphtha	61.0	245	280	52	32
90 Solvent	52.0	285	320	90	33
KWIK-DRI	50.9	308	355	101	32
Mineral Spirits	48.3	310	400	105	35
Rule 66 Mineral Spirits	51.5	315	390	105	35
Odorless Mineral Spirits	54.0	355	400	128	28
Odorless Solvent LR	50.5	360	580	136	26
140 Solvent	48.6	368	400	141	31
Low Odor Base Solvent	45.0	370	500	145	30
Mineral Seal Oil	35.5	490	610	265	22

AROMATIC SOLVENTS

Listed By IBP	API Gravity @ 60°F	Dist Range °F IBP	Dist Range °F DP	Flash Pt °F TCC	KB Value
Toluene	30.6	230	232	45	105
Ethylbenzene	31.0	276	278	59	96
Xylene	31.1	280	288	81	98
HI-SOL XB	28.2	310	450	105	96
HI-SOL 10	30.5	315	350	110	90
HI-SOL 70	33.6	325	395	112	70
HI-SOL 3	32.5	347	425	141	71
HI-SOL 15	28.0	360	400	140	90

Table 2.126: Charter Hydrocarbon Solvents (21)

Chartersol® Aromatic Solvents

	DISTILLATION RANGE °C			DISTILLATION RANGE °F			GRAVITY			SOLVENCY MIXED ANILINE POINT			FLASH T.C.C.		AROMATICS	
	IBP	50%	DP	IBP	50%	DP	API	Rel. Den.	Lbs. per Gal.	°C	°F	KB	°C	°F	C8+ EXCEPT EB	TOTAL
Benzene	79.7	80.1	80.4	175	176	177	28.7	.883	7.36	13	55	105	−11	12	0	99+
Ethylbenzene	135.6	136.2	137.0	276	277	279	31.1	.870	7.25	10	50	95	20	68	0	99+
Toluene	110.1	110.6	110.9	230	231	232	31.1	.870	7.25	10	50	105	4	40	0	99+
Xylene	37.7	138.7	139.5	280	282	283	31.1	.870	7.25	12	54	98	29	85	94	99+
Chartersol 1	162	166	175	323	331	347	30.5	.874	7.27	15	59	90	46	114	96	96
Chartersol 2	179	193	213	354	380	415	29.0	.882	7.34	17	63	86	61	142	98	98
Chartersol 12-B	161	171	204	323	340	400	31.5	.868	7.23	19	67	86	41	105	87	87
Chartersol 12-M	164	172	201	327	341	394	29.7	.878	7.31	16	60	90	49	121	97	97
Chartersol 20	178	188	212	352	370	413	29.1	.881	7.34	18	64	85	61	142	93	93

High Purity and Specialty Chemicals

	DISTILLATION RANGE °C			DISTILLATION RANGE °F			GRAVITY			SOLVENCY ANILINE POINT			FLASH T.C.C.		VOLUME %	
	IBP	50%	DP	IBP	50%	DP	API	Rel. Den.	Lbs. per Gal.	°C	°F	KB	°C	°F	Purity	Aromatics
H.P.N. Hexane	68	68	69	155	156	156	80.4	.668	5.56	69	156	28	−32	−26	95.1	0.1
H.P.N. Heptane	98	98	99	208	209	211	73.4	.691	5.75	68	155	28	− 4	25	95.1	0.5
H.P.N. Octane	125	125	126	257	258	259	68.3	.708	5.90	72	161	28	16	61	95.4	0.5

(continued)

Table 2.126: (continued)

High Purity and Specialty Chemicals

Chartersol 3-AG	150	161	271	302	321	520	27.3	.891	7.42	14*	58*	89	41	105	96	96
Chartersol 5-AG	162	166	175	323	331	347	30.5	.874	7.27	15*	59*	90	46	114	96	96

*MIXED ANILINE POINT

Chartersol® Aliphatic Solvents

These solvents contain less than 0.1% benzene

		DISTILLATION RANGE						GRAVITY			SOLVENCY			FLASH T.C.C.		AROMATICS	
		°C			°F						ANILINE POINT						
		IBP	50%	DP	IBP	50%	DP	°API	Rel. Den.	Lbs. per Gal.	°C	°F	KB	C	°F	C_8+ EXCEPT EB	TOTAL
LACQUER Diluent; Rubber & Adhesive Solvents	Chartersol Heptane LF	82	90	98	180	194	209	73.8	.689	5.74	68	154	29	−13	8	0	0.2
LACQUER Diluent; Rubber & Adhesive Solvents	Chartersol 100-66-WRS	86	97	110	187	206	230	69.2	.705	5.87	64	147	31	−11	13	0	3
LACQUER Diluent; Rubber & Adhesive Solvents	Chartersol 206-66	102	106	114	216	223	237	59.2	.742	6.18	48	118	40	−2	29	0.3	18.5
LACQUER Diluent; Rubber & Adhesive Solvents	Chartersol 210-66	102	106	114	215	223	238	64.0	.724	6.03	61	141	34	−2	29	0.3	6.0
LACQUER Diluent; VM & P Naphthas	Chartersol 230-66	112	115	117	234	239	243	66.0	.717	5.96	68	155	29.5	4	40	1.0	1.5
LACQUER Diluent; VM & P Naphthas	Chartersol 250	121	127	140	249	261	284	57.9	.747	6.22	55	131	36	12	53	14	16
LACQUER Diluent; VM & P Naphthas	Chartersol 250-C	121	127	141	249	261	285	52.3	.770	6.41	36	97	46	12	53	29	31
LACQUER Diluent; VM & P Naphthas	Chartersol 250-66	121	127	139	249	261	283	61.6	.733	6.10	66	151	31	12	53	5	6
LACQUER Diluent; VM & P Naphthas	Chartersol 260-H	132	134	141	268	274	282	60.8	.736	6.13	66	151	31	18	65	4	5
VM & P Naphthas	Chartersol 280-S	139	143	160	283	289	320	61.5	.733	6.10	69	156	31	27	80	3	3
VM & P Naphthas	Chartersol 286	140	162	176	284	324	348	54.0	.763	6.35	66	150	32.5	27	81	3.0	3.2
MINERAL Spirits	Chartersol 300	160	178	199	320	353	390	47.5	.791	6.58	53	127	38	43	110	18	18
MINERAL Spirits	Chartersol 300-66	160	179	199	320	355	390	51.5	.773	6.44	67	152	33	42	108	4	4
MINERAL Spirits	Chartersol 300-S	161	176	198	321	349	388	44.0	.806	6.71	38	100	46	43	109	34	34
MINERAL Spirits	Chartersol 306-66	160	178	199	320	353	390	51.0	.775	6.46	63	146	34	43	110	7.6	7.6
MINERAL Spirits	Chartersol 310-66	156	167	175	312	332	347	52.8	.768	6.39	68	154	33	39	103	2.5	2.5
MINERAL Spirits	Chartersol 350-66	184	192	199	364	378	390	48.5	.786	6.55	67	152	32.2	61	141	5	5
MINERAL Spirits	Chartersol 380-66	188	221	270	370	430	518	44.0	.806	6.71	71	160	31.5	63	145	0.1	0.1

Table 2.127: Chevron Alpha Olefins *(6)*

· Typical Alpha Olefins Inspections ·

	C_6-C_7	C_8-C_9	C_6-C_9	C_8-C_{10}	C_{10}	C_{11}-C_{12}	C_{13}-C_{14}	C_{11}-C_{14}	C_{13}-C_{18}	C_{15}-C_{18}	C_{18}-C_{20}	C_{15}-C_{20}
Straight chain mono alpha olefins, wt %	83.3	88.2	86	83	90	88	88	89	89	89	86	88
Diolefins, wt %	3.2	3	4	4	5	5	5	5	5	6	4	5
Paraffins, wt %	1.5	2.3	3	2	2	2	2	2	2	2	9	5
Peroxide content, meq/ℓ	2	2	1	3	1	2	2	1	1	1	1	1
Appearance	· · · · · · · · · · · · · · · · · Clear and bright and free of sediment · · · · · · · · · · · · · · ·											
Color, Saybolt	+30	+19	+18	+20	+17	+12	+11	+14	+14	+7	–16	–12
Density (20°/4°C) g/mℓ	0.689	0.732	0.713	0.737	0.751	0.76	0.768	0.764	0.781	0.787	0.796	0.794
Density (60°/60°F) lb/gal	5.75	6.11	5.95	6.15	6.27	6.34	6.41	6.38	6.52	6.57	6.64	6.63
Flash point, TOC, °F	<30	<60	<30	<80	103	160	210	162	215	260	330	280
Pour point, °F	<-60	<-60	<-60	<-60	<-60	-50	-10	-20	+25	+40	+70	+55
Bromine No. g/100 g	189	141	165	135	118	106	92	98	82	73	57	67
Water content, ppm	110	107	89	101	162	76	61	66	46	53	52	66
Sulfur content, ppm	7	7	5	7	8	10	10	10	10	15	10	11
Carbon number distribution, wt %												
C_5	1	—	2	—	—	—	—	—	—	—	—	—
C_6	59	—	39	—	—	—	—	—	—	—	—	—
C_7	38	1.8	24	2	—	—	—	—	—	—	—	—
C_8	2	49.9	17	30.6	—	—	—	—	—	—	—	—
C_9	—	45.9	16	32	4	—	—	—	—	—	—	—
C_{10}	—	2.4	2	34.6	95	2	—	1	—	—	—	—
C_{11}	—	—	—	0.8	1	51	—	27	—	—	—	—
C_{12}	—	—	—	—	—	45	2	24	0.6	—	—	—
C_{13}	—	—	—	—	—	2	47	24	14.1	—	—	—
C_{14}	—	—	—	—	—	—	48	23	15.8	2	—	1
C_{15}	—	—	—	—	—	—	3	1	21.9	30	—	17
C_{16}	—	—	—	—	—	—	—	—	21.0	30	—	18
C_{17}	—	—	—	—	—	—	—	—	19.6	28	1	17
C_{18}	—	—	—	—	—	—	—	—	7.0	10	23	17
C_{19}	—	—	—	—	—	—	—	—	0.2	0.1	37	15
C_{20}	—	—	—	—	—	—	—	—	—	—	30	12
C_{21}	—	—	—	—	—	—	—	—	—	—	9	3
Average molecular weight	90	119	100	123	140	161	189	174	214	226	269	244

· Distillation Data of Typical Chevron Alpha Olefins ·

	C_6-C_7	C_8-C_9	C_6-C_9	C_8-C_{10}	C_{10}	C_{11}-C_{12}	C_{13}-C_{14}	C_{11}-C_{14}	C_{13}-C_{18}	C_{15}-C_{18}	C_{18}-C_{20}	C_{15}-C_{20}
Distillation °F, ASTM method	· · · · · · · · · · · · D-86 at 760 mm Hg · · · · · · · · · · · · ·								D-1160 at 10 mm Hg			
Start	152	245	148	245	314	365	420	388	462	286	352	288
5%	159	257	161	260	325	385	449	398	475	293	360	295
95%	201	299	284	340	328	410	472	480	545	343	395	367
End point	252	317	315	350	343	427	494	494	567	350	406	406

· Chevron Alpha Olefin C_5 ·

	Typical Tests
Condition at 70°F	Clear and bright, free of sediment
Color, Saybolt	+30
2-Methyl-1, 3-butadiene, wt %	0.9
1,4-Pentadiene, wt %	2.5
cis- and trans-1,3-Pentadiene, wt %	0.2
1-Pentene, wt %	96.4
Vapor pressure at 100°F, psia	19
Flash Point, TOC, °F	-40
Density at 20°C, g/mℓ	0.64
Density at 60°F, lb/gal	5.385

(continued)

Table 2.127: (continued)

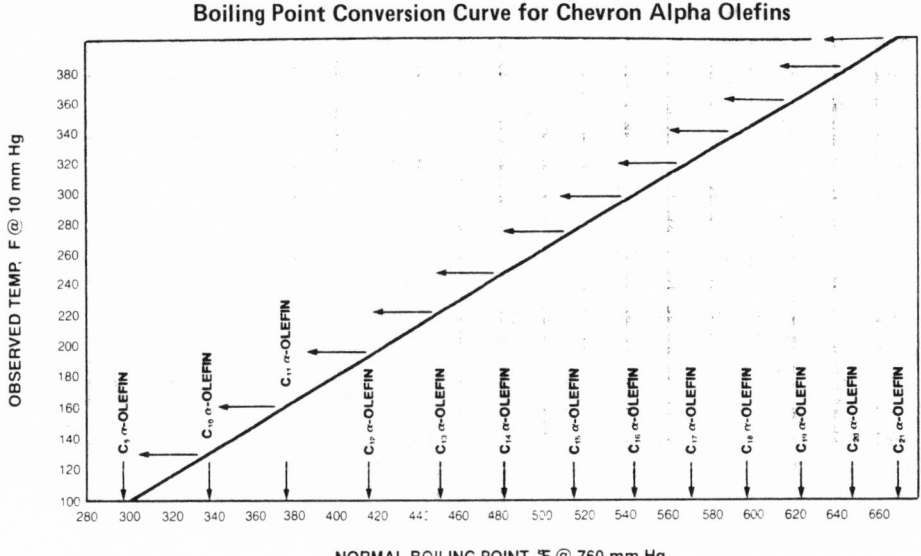

Boiling Point Conversion Curve for Chevron Alpha Olefins

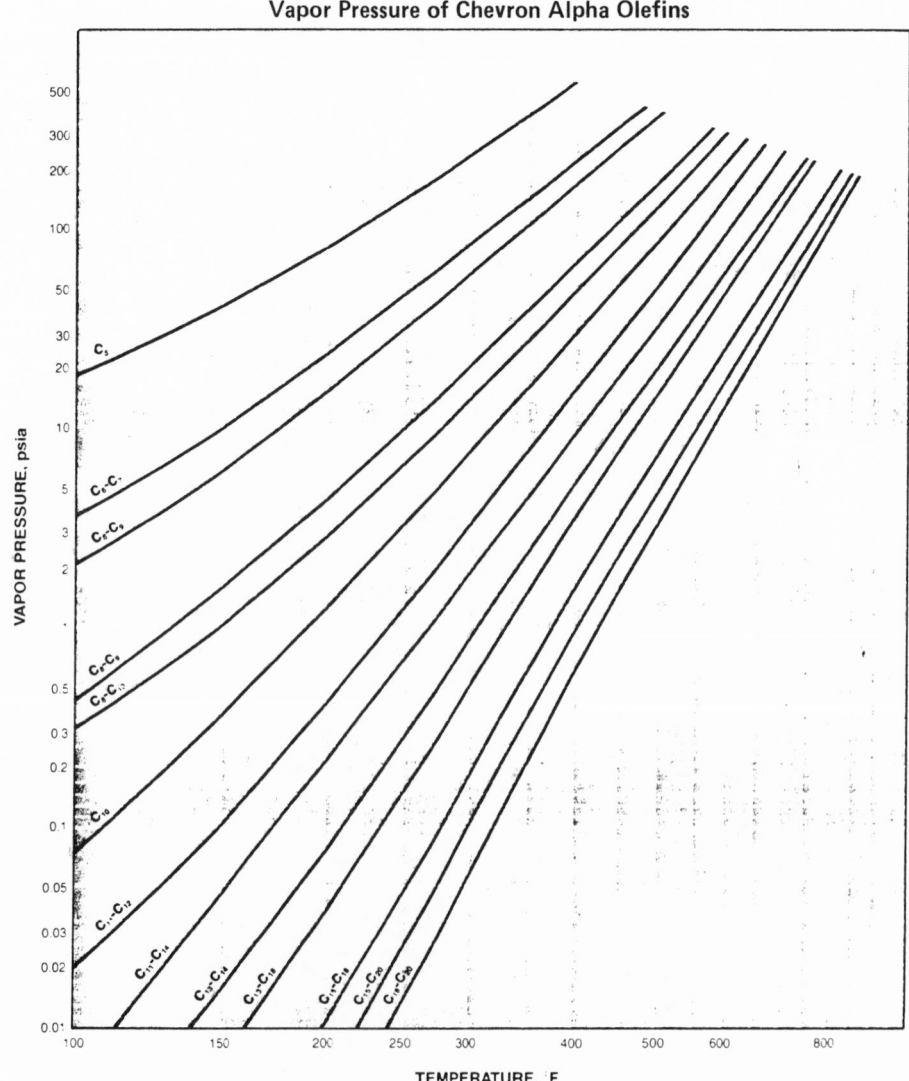

Vapor Pressure of Chevron Alpha Olefins

Table 2.128: Conoco "LPA" Solvent *(40)*

Description:

Conoco LPA Solvent is a very high purity aliphatic hydrocarbon having a molecular weight range similar to kerosene. It is a colorless liquid having a mild odor, low viscosity and extremely low aromatic and olefin content.

Properties	Specification	Typical
Test method ASTM-D-86, under distillation range, °F.		
IBP	350 Min.	370
10	—	395
20	—	405
50	—	430
90	—	475
95	—	480
EP	530 Max.	518
Aromatics, Wt. %	1.0 Max.	0.1
Flash point, Pensky-Martens, °F.	140 Min.	144
Color, Saybolt	25 + Min.	30 +
Water, ppm	100 Max.	20
UV Absorbance	Passes Test*	Passes
Specific Gravity, 60°F.	—	0.8063
Bromine Number	—	<0.2
Sulfur, ppm	—	<1
Nitrogen, ppm	—	<1
Carbonyl, as C = O, ppm	—	10
Aniline Point, °F.	—	160
Kauri Butanol Value	—	31.5

*Method to meet FDA requirement—CFR Title 21, Sections 172.884 and 178.3650

Table 2.129: Eastman Chemical Diluents *(41)*

	Evaporation Rate, nBuOAc = 1	Formula	Aromatic, Vol %	Kauri Butanol Value	Specific Gravity at 60°/60°F	Pounds Per Gallon at 60°F	Flash Point, T.C.C., °F
TOLUENE	1.9	$C_6H_5CH_3$	100	105	0.871	7.25	45
VM & P NAPHTHA	1.6	Mixture	11.6	39	0.753	6.27	44
XYLENE	0.6	$C_6H_4(CH_3)_2$	100	98	0.871	7.25	83
HIGH-FLASH AROMATIC NAPHTHA 100	0.2	Mixture	98	92	0.873	7.26	113
HIGH-FLASH AROMATIC NAPHTHA 150	0.04	Mixture	98	94	0.891	7.41	151

	Solvent Constants			Aniline Point, °F	Azeotrope		Boiling Range at 760 torr, °F	Gram Molecular Weight	Electrical Resistance,* Megohms
	Solubility Parameter	Hydrogen Bonding	Dipole Moment		BP, °C	Wt % Water°			
TOLUENE	8.9	3.0	0.4	46, mixed	84.6	18	228 - 233	92.13	>20
VM & P NAPHTHA	7.6	0.3	0	126	—	—	244 - 282	—	>20
XYLENE	8.8	3.3	0.4	52, mixed	94.5	40	275 - 290	106.16	>20
HIGH-FLASH AROMATIC NAPHTHA 100	8.8	5.0	0	57.2, mixed	—	—	311 - 344	—	>20
HIGH-FLASH AROMATIC NAPHTHA 150	8.7	5.3	0	64.4, mixed	—	—	379 - 410	—	>20

*Electrical resistance was measured by Ransburg paint resistance meter, part no. 7924, model no. 234. Values shown are typical for commercial production. Certain solvents vary in resistivity with age.

Table 2.130: Eastman Urethane Grade Solvents *(41)*

Typical Properties and Sales Specifications for Eastman Urethane-Grade Solvents

Solvent	Evaporation Rate[b]	Sales Specifications[a]									
		Purity, Wt % (Min)	Acid as Acetic, Wt % (Max)	Water, Wt % (Max)	Color[c] (Max)	Specific Gravity at 20°/20°C	Initial Boiling Point, °C (Min)	Dry Point, °C (Max)	Alcohol Wt % (Max)	Odor	Appearance
Ethyl acetate	4.2	99.5	0.005	0.05	10	0.900-0.905	75.5	78.0	0.2	d	e
Isopropyl acetate	3.0	99.0	0.01	0.05	10	0.871-0.874	86.0	90.0	0.5	d	e
Methyl isobutyl ketone (MIBK)	1.6	99.0	0.01	0.05	10	0.800-0.803	114.0	117.0	0.5	d	e
Isobutyl acetate	1.4	99.0	0.01	0.05	10	0.868-0.873	112.0	119.0	0.5	d	e
n-Butyl acetate	1.0	99.0	0.01	0.05	10	0.881-0.884	124.0	129.0	0.5	d	e
Methyl n-butyl ketone (MBK)	1.0	95.0[f]	0.05	0.05	10	0.810-0.813	124.5	129.5	0.5	d	e
Methyl isoamyl ketone (MIAK)	0.5	99.0	0.02	0.05	10	0.812-0.815	141.0	148.0	0.5	d	e
Methyl n-amyl ketone (MAK)	0.4	98.0[g]	0.02	0.05	10	0.815-0.818	147.0	152.0	0.5	d	e
EKTASOLVE® EE acetate	0.2	99.0	0.02	0.05	15	0.971-0.976	150.0	160.0	0.5	mild	e

[a]*Specifications are subject to change without notice.*

[b]*Relative to: n-butyl acetate = 1.0*

[c]*Platinum-Cobalt Scale*

[d]*Characteristic, nonresidual*

[e]*Free from insoluble matter or haze*

[f]*Purity as MBK; total ketone 99% (min)*

[g]*Purity as MAK; total ketone 99% (min)*

Table 2.131: Ethyl Alpha Olefins *(82)*

"Ethyl" Alpha Olefins are clear, colorless, water-white, mobile liquids. They are 99 + % olefinic.

Products	Applications	Typical Properties			
		Specific Gravity 25°C/25°C	Pour Point, °C	Flash Point, °F. (closed cup)	Boiling Range, °C
Butene-1 (C4)	Comonomer for polymer impact modification	0.595 (Liquid)	Liquefied gas	Flammable gas	−6.1 B.P.
Hexene-1 (C6)	Comonomer, and for epoxides, amines, oxo-alcohols, syn. fatty acids.	0.678	−140	−15	63-64 (5%-95%)
Octene-1 (C8)	Same as above, also for oligomers.	0.711	−102	50	121-123 (5%-95%)
Decene-1 (C10)	Same as above, also for lubricants.	0.734	−66	114	170-171 (5%-95%)
Dodecene-1 (C12)	Same as above, plus mercaptans, and alpha olefin sulfonates.	0.755	−37	171	213-216 (5%-95%)
Tetradecene-1 (C14)	Same as above	0.769	−18	225	245-250 (5%-95%)
Hexadecene-1 (C16)	Same as above	0.778	−2	263	276-283 (5%-95%)
Octadecene-1 (C18)	Same as above	0.784	5	290	298-316 (5%-95%)
Dodecene-1/Tetradecene-1 Blend (approx. 2/1 by weight) (C12-C14)	Same as above, plus alkylated aromatics	0.760	−35	178	216-250 (5%-95%)
Tetradecene-1/Hexadecene-1 Blend (approx. 5/3 by weight) (C14-C16)	Same as above. Well suited for AOS applications	0.773	−14	235	245-279 (5%-95%)
Hexadecene-1/Octadecene-1 Blend (Var. proportions) (C16-C18)	Same as above, plus sulfated olefins for leather conditioning.	0.782	−2	275	285-316 (5%-95%)

Coproducts

"Ethyl" Coproducts are mixtures of olefins and paraffins as described in the table below. Composition may vary somewhat from those shown. *Primary Applications* are in alkylated aromatics and sulfonated products.

Products	Approximate Distribution, wt %			Typical Properties		
	Paraffin	Olefin*	Alcohol	Density lb/gal	Pour Point °F	Flash Point closed cup, °F
C12-26 Olefin/Paraffin	22	72	6	6.5	35	130
C14-26 Olefin/Paraffin	14	83	3	6.5	45	250
C18+ Olefin	5	85	10	6.5	85	240
C18-24 Olefin	2	93	5	6.5	75	335

*Mixture of vinyl, internal and branched

Table 2.132: Exxon Petroleum Solvents (8)

Test Method	Aniline Point °C (ASTM D 611)	°F	Kauri-Butanol Number (ASTM D 1133)	Flash Point TCC °C (ASTM D 56)	°F	Distillation IBP °C	°F	50% °C	°F	Dry Point °C	°F	Inhalation TLV(3) ppm	Evap 25 Mass%	50 Mass%	75 Mass%	95 Mass%	Rule 102 (formerly 66) Exempt	Total Saturates	Total Aromatics	C₉ and Higher(2)	Sulfur ppm Microcoulometer	Bromine Index ASTM D 2710	Specific Gravity @ 15.6°C (60°F) ASTM D 1250	API Gravity ASTM D 287	Viscosity cp at 25°C ASTM D 445	CAS Number(9)
EXSOL HEXANE	66	150	31	<-18	<0	66(b)	150	-	-	70	158	100	14	31	51	72	Yes	100	<0.01	-	<1	10	0.672	79.1	0.31	64742-49-0*
HEXANE	62	143	31	<-18	<0	65(b)	149	-	-	69	157	100	16	30	47	66	Yes	100	0.05	-	1	100	0.682	76.0	0.32	110-54-3 or 64742-49-0*
HEPTANE	53	127	38	~-8	~18	94(b)	201	96	204	99	210	400	29	59	92	130	Yes	96.0	4.0	-	1	60	0.731	62.0	0.47	64742-89-8*
LAKTANE	30	86	51	~-6	~21	101(b)	214	103	218	106	223	400(4)	42	84	128	180	No	80	20	-	2	80	0.772	51.8	0.54	64742-89-8*
VM&P NAPHTHA	47	117	40	7	44	117(b)	242	122	251	139	282	300	75	150	236	337	Yes	85.0	15.0	7.8	5	130	0.759	54.9	0.56	64742-89-8*
VARSOL 1	54	130	37	42	108	156(a)	313	175	347	202	395	100	690	1737	3300	5700	No	82.0	18.0	18.0	2	200	0.791	47.5	0.95	64742-48-9*
VARSOL 3	54	130	38	41	105	154(a)	310	163	325	176	348	200(4)	572	1206	1944	2772	No	81.5	18.5	18.5	2	200	0.784	49.0	0.83	64742-48-9*
VARSOL 18	63	146	34	41	106	156(a)	313	176	348	202	395	200(4)	720	1750	3348	5640	Yes	92.3	7.7	7.5	2	200	0.780	49.9	0.90	64742-48-9*
LOW ODOR PARAFFIN SOLVENT	76	169	29	68	154	191(a)	375	216	420	244	472	200(4)	8496	28080	79920	-	Yes	96.0	4.0	4.0	2	50	0.794	46.7	1.69	64741-92-0* & 64742-48-9*
TOLUENE	8.8(5)	47.8(5)	105	~7	~45	110.2(c)	230.4	-	-	111.0	231.8	100	73	142	210	274	No	0.12	99.88	0.02	2	-	0.872	30.8	0.57	108-88-3*
XYLENE	10.4(5)	50.7(5)	98	27	80	138(c)	280.4	-	-	140.0	284.0	100	198	407	606	816	No	0.1	99.9	75.0	2	-	0.871	31.0	0.62	1330-20-7
AROMATIC 100	13.4(5)	56.1(5)	91	42	108	155(a)	311	160	320	173	344	50(4)	648	1344	2117	2976	No	2.0	98.0	98.0	3	-	0.872	30.7	0.78	64742-95-6*
AROMATIC 150(1)	15.4(5)	59.7(5)	95	66	150	183(a)	362	191	375	210	410	100(4)	3312	7320	12120	17520	No	2.0	98.0	98.0	2	-	0.902	25.4	1.20	64742-94-5*
HAN(1)	29(5)	84(5)	78	57(6)	135(6)	177(a)	350	218	425	277(7)	530(7)	100(4)	6351	27432	70200	-	No	20	80	80	0.3(8)	800	0.893	26.9	2.0	64742-06-9*
ISOPAR C	77	171	27	~-7	~19	98(b)	208	99	211	106	222	400(4)	36	68	107	151	Yes	100	0.02	-	1	6	0.701	70.5	0.48	64742-48-9*
ISOPAR E	74	165	29	<7	<45	116(a)	240	118	245	138	280	400(4)	69	148	220	310	Yes	100	0.02	-	1	8	0.721	64.7	0.62	64742-48-9*
ISOPAR G	82	179	28	40	104	157(a)	315	163	326	176	348	300(4)	615	1200	1872	2625	Yes	100	0.08	-	1	4	0.748	57.6	0.99	64742-48-9*
ISOPAR H	84	184	27	53	127	176(a)	348	181	358	191	375	300(4)	1098	2820	4710	6636	Yes	100	0.04	-	1	4	0.759	55.0	1.29	64742-48-9*
ISOPAR K	84	184	27	53	127	177(a)	350	183	361	197	386	300(4)	1974	4100	6660	9600	Yes	99.9	0.06	-	1	4	0.761	54.5	1.39	64742-48-9*
ISOPAR L	84	184	27	61	142	188(a)	370	194	382	206	403	300(4)	4176	8400	13260	18840	Yes	99.9	0.06	-	1	4	0.767	53.0	1.61	64742-48-9*
ISOPAR M	89	192	23	80(6)	176(6)	207(a)	405	223	434	254	490	300(4)	10080	27900	-	-	Yes	99.8	0.2	-	1	50	0.783	49.2	2.46	64742-47-8*
NORPAR 12	82	180	23	69(6)	156(6)	188(a)	370	201	393	217	422	300(4)	5640	13700	-	-	Yes	99.4	0.6	-	<1	-	0.751	57.0	1.26	64742-47-8*
NORPAR 13	87	189	22	93(6)	200(6)	226(a)	438	231	448	242	468	300(4)	57500	126000	261400	-	Yes	99.8	0.2	-	<1	-	0.764	53.8	1.84	64771-72-8*
NORPAR 15	92	198	18	118(6)	244(6)	252(a)	486	261	502	272	522	300(4)	>115200	-	-	-	Yes	100	0.01	-	5	500	0.771	52.0	2.50	64742-14-9*

IBP footnotes: (a) ASTM D 86 (b) ASTM D 1076 (c) ASTM D 850

(1) Note: All solvents test 30 Saybolt color except HAN (1.5 ASTM) and AROMATIC 150 (+ 27 Saybolt).

(2) Excluding ethylbenzene.

(3) TLV is a registered trademark of the American Conference of Governmental Industrial Hygienists. It is the threshold limit value or occupational exposure limit - the time weighted average concentration for a normal 8-hour workday 40-hour work-week, to which nearly all workers may be exposed repeatedly without adverse effect. Refer to the most recent Material Safety Data Sheet for the latest recommended maximum exposure limit for each solvent.

(4) A TLV has not been established for this product. The value shown has been recommended by Exxon Corporation Medical Research based on consideration of available toxicological data. Additional data are being obtained to help define a recommended occupational exposure limit more conclusively.

(5) Mixed aniline point.

(6) Pensky-Martens method. ASTM D 93.

(7) Final Boiling Point.

(8) Mass %.

(9) The asterisk (*) is part of the Chemical Abstract Service (CAS) Registry Number and identifies substances that appear in the UVCB Index and that are described in Appendix A. "Chemical Substances Definitions," of the Toxic Substances Control Act Chemical Substance Inventory.

Table 2.133: Getty Aliphatic and Aromatic Solvents (16)

GETTYSOLVE B
A highly refined commercial hexane with an extremely low benzene content. Its typical boiling range is 4°F and offers a low evaporation residue and balanced composition. Uses include vegetable oil extraction, pharmaceutical extraction, compounding rubber cements and sealants, and polyolefin production.

GETTYSOLVE C
A mixed heptane fraction with low aromatic content and low evaporation residue. Uses include adhesives, rubber, cements, sealants, and animal tallow extraction.

GETTYSOLVE F
Analytical grade petroleum ether that meets USP, AOCS, and APS specifications. Uses include lab reagent.

GETTYSOLVE H
This hexane-heptane fraction offers fast evaporation and extremely low evaporation residue. Uses include rubber tire manufacture, rubber cements, sealants, and adhesives.

GETTYSOLVE SPECIAL H
A light naphtha with consistent low non-volatile content and fast evaporation needed for special applications. Uses include rubber cements, sealants, and ink diluent.

GETTYSOLVE L
This lacquer diluent's aromatic content of approximately 20% gives high solvent power. Evaporation rate is comparable to that of toluene, but with a lower boiling range. Uses include protective coating formulation, rubber cements, sealants, and ink formulations.

GETTYSOLVE R
A rubber solvent with excellent solvency throughout evaporation. Also uniform evaporation rate. High IBP helps eliminate excessive evaporation losses in storage and use. Uses include compounding rubber cements in tire manufacture, rubber belts, rubber hose.

GETTYSOLVE S
This mineral spirit meets or exceeds all requirements of both mineral spirits and Stoddard Solvent. Uniformity and pleasant odor allow varied applications. Uses include paint and protective coatings, dry cleaning, degreasing, wood treating, and charcoal lighter fluid.

GETTYSOLVE S-66
Exempt mineral spirits meet or exceed all requirements of both mineral spirits and Stoddard Solvent. Maximum aromatic content (C_8 and higher) of 7%, to meet air quality standards, while retaining solvency. Uses include paint and protective coatings, dry cleaning, degreasing, wood treating, and charcoal lighter fluid.

GETTYSOLVE S-1
120°F flash mineral spirits manufactured for applications where high flash and specific evaporation are needed. Offers carefully controlled "front end" distillation and narrow boiling range. Uses include degreasing, charcoal lighter fluid, and absorption oil.

GETTYSOLVE S-2
This quick-dry mineral spirit has a narrow boiling range, fast evaporation, no residual odor, and a flash point in excess of 100°F. Its low dry point makes it unique. Also known as Stoddard Solvent. Uses include paint and protective coatings, dry cleaning, other quick drying mineral spirit applications.

GETTYSOLVE T
A 140°F flash naphtha, commonly referred to as "140 solvent." Has a pleasant odor, low sulfur content, and low residual odor. Uses include dry cleaning, paint and protective coatings.

GETTYSOLVE V
A VM&P naphtha with a pleasant odor, high flash point and uniform evaporation. Uses include paint and protective coatings, spot dry cleaners, lighter fluid and rubber cement.

SKELLITE
A clean burning naphtha free of residue and gum. Manufactured especially for gasoline pressure applications. Uses include stove and lamp fuel.

ALIPHATIC SOLVENTS

	Description	Specific Gravity At 60°F	API Gravity At 60°F	Pounds Per Gallon At 60°F	Distillation, °F					Evaporation Residue gm/100 ml	Flash Point, °F	Aniline Point, °F
					IBP	10%	50%	90%	Dry Point			
Test Method		ASTM D287			(a) ASTM D1078 (b) ASTM D86 (c) ASTM D850 (d) ASTM 216					ASTM D1353	ASTM D56	ASTM D611
Gettysolve—B	Commercial Hexane	0.679	76.8	5.65	152 (a)	153	154	155	156	0.0001	−25	144
Gettysolve—C	Commercial Heptane	0.724	64.0	6.01	193 (a)	197	200	204	208	0.0002	13	134
Gettysolve—F	Analytical Grade Petroleum Ether	0.643	88.6	5.35	96 (d)	—	—	—	136	0.0001	−70	158
Gettysolve—H	Hexane-Heptane Fraction	0.700	70.6	5.83	158 (b)	165	176	196	205	0.0003	−20	138
Gettysolve Special—H	Light Naphtha	0.682	75.9	5.68	150 (a)	152	153	157	175	0.0006	−20	142
Gettysolve—L	Lacquer Diluent	0.749	57.5	6.23	205 (b)	211	213	220	228	0.0005	17	105
Gettysolve—R	Rubber Solvent Naphtha	0.713	67.0	5.92	134 (b)	164	196	232	245	0.0009	−18	131
Gettysolve—S	Mineral Spirits	0.780	50.0	6.49	313 (b)	322	335	350	375	0.005	107	138
Gettysolve—S-66	Mineral Spirits	0.778	50.3	6.48	313 (b)	322	335	350	375	0.005	107	142
Gettysolve—S-1	High Flash Mineral Spirits	0.785	48.7	6.54	335 (b)	341	346	365	380	0.005	120	139

(continued)

Table 2.133: (continued)

ALIPHATIC SOLVENTS

	Description	Specific Gravity At 60°F	API Gravity At 60°F	Pounds Per Gallon At 60°F	IBP	10%	50%	90%	Dry Point	Evaporation Residue gm/100 ml	Flash Point, °F	Aniline Point, °F
Test Method		ASTM D287			(a) ASTM D1078 (b) ASTM D86 (c) ASTM D850 (d) ASTM 216					ASTM D1353	ASTM D56	ASTM D611
Gettysolve—S-2	Quick Dry Mineral Spirits	0.778	50.4	6.46	310 (b)	316	321	326	330	0.003	102	131
Gettysolve—T	140 Flash Type Naphtha	0.790	47.6	6.58	363 (b)	366	372	381	400	0.015	141	142
Gettysolve—V	VM&P	0.755	56.0	6.28	242 (b)	251	259	266	275	0.001	54	127
Skellite	Stove & Lantern Solvent	0.720 0.724	64.0 65.0	6.01	112 (b)	152	220	282	343	—	—	—

	Kauri Butanol No.	Color (Saybolt)	Corrosion Cu Strip	Sulfur % By Weight	Benzene, Vol. %	Reidvapor Pressure (PSIA at 100°F)	Paraffins	Olefins	Naphthene	Aromatic	C₈ And Above
Test Method	ASTM D1133	ASTM D156	ASTM D130	ASTM D3120	ASTM D2267	ASTM D323		Method Dependent On Particular Product Characteristics Used In Analysis			
Gettysolve—B	30.0	+30	1A	0.0002	0.02	5.0	83.6	—	16.0	0.2	—
Gettysolve—C	36.0	30	1A	0.002	0.1	2.1	61.7	0.3	35.1	2.9	—
Gettysolve—F	26.4	30	1A	0.003	NIL	13.5	—	—	—	—	—
Gettysolve—H	33.5	30	1A	0.001	0.05	4	72.9	0.3	25.3	1.5	—
Gettysolve Special—H	32.0	+30	1A	0.001	0.05	4.7	82.5	—	17.4	0.1	—
Gettysolve—L	46.0	30	1A	0.005	NIL	2.0	48.3	0.3	35.3	16.1	—
Gettysolve—R	35.0	30	1A	0.005	0.30	4.4	52.4	0.7	45.3	1.6	0.3
Gettysolve—S	37.0	30	1B	0.001	NIL	0.2	43.4	0.7	44.1	11.8	11.8
Gettysolve—S-66	36.0	30	1B	0.001	NIL	0.2	43.4	0.4	49.4	6.8	6.8
Gettysolve—S-1	36.5	23	1B	0.02	NIL	0.1	35.7	0.5	51.7	12.1	12.1
Gettysolve—S-2	37	30	1B	0.016	NIL	0.15	39.5	0.6	48.2	11.7	11.7
Gettysolve—T	34.5	+20	1B	0.03	NIL	0.05	32.1	0.9	54.5	9.3	9.3
Gettysolve—V	38.5	30	1A	0.009	NIL	0.7	47.0	0.4	43.6	9.0	3.9
Skellite	—	28	1A	—	0.3	5.6	74.4	—	18.3	7.3	—

(continued)

Table 2.133: (continued)

GETTY TOLUENE
A toluene that meets or exceeds all ASTM specifications for nitration grade. Uses include paint and protective coatings, and as an intermediate for demanding chemical manufacture.

GETTY A-150
A 150°F flash aromatic naphtha with a narrow distillation range, high solvency power, high flash point, and low sulfur content. Uses include paint and

protective coatings, herbicide and pesticide carrier, synthetic resin manufacturing, degreasing.

GETTY A-400
A heavy aromatic naphtha consisting of 100% aromatic hydrocarbons offers high solvency power, high flash point, and low pour point. Uses include degreasing, pesticide and herbicide carrier, fuel additive treatments.

AROMATIC SOLVENTS

	Description	Specific Gravity At 60°F	API Gravity At 60°F	Pounds Per Gallon At 60°F	Distillation, °F IBP	10%	50%	90%	Dry Point	Flash Point, °F	Aniline Point, °F
Test Method		ASTM D891					(a) D850 (b) D86 ASTM			(a) ASTM D56 (b) ASTM D93	ASTM D1012
Toluene	Nitration Grade	.871	30.9	7.26	110.4 (a)	110.5	110.6	110.7	110.8	35 (a)	51
A-150	Alkyl Benzenes	0.905	24.9	7.53	386 (a)	391	394	403	411	154 (a)	62
A-400	Alkyl Benzene Isomer Mix	0.969	14.6	8.07	416 (b)	427	442	493	627	198 (b)	56.5

	Kauri Butanol No.	Color	Corrosion Cu Strip	Sulfur % By Weight	Benzene, Vol. %	Composition Vol. % Paraffins	Olefins	Naphthene	Aromatic	C_8 And Above
Test Method	ASTM D1133	(a) ASTM 1209 (APHA) (b) ASTM D156 (Saybolt)	ASTM D849	ASTM (a) D853 (b) D2785	ASTM (a) D2600 (b) D3257	Method Dependent On Particular Product Characteristics Used In Analysis				
Toluene	105	5 (a)	1A	Free (a)	NIL (a)	—	—	—	99.9	—
A-150	89	+ 30 (b)	No Discoloration	—	NIL (b)	1	—	2	97	97
A-400	107	< 3 (b)	—	0.00043 (b)	NIL (b)	—	—	—	100	100

Table 2.134: Kendall Special Solvents *(76)*

KENSOL 10

TYPICAL ANALYSIS

Gravity		64.8
Flash Point °C		-23°C (<-10°F)
Aniline Point, °C		60°C
Kauri Butanol Value		34.2
Pounds per Gallon (U.S.)		6.000

Distillation (D-86)	°C	°F
Initial Boiling Point	40	105
5%	53	128
50%	103	218
95%	144	291
End Boiling Point	166	330

(continued)

Table 2.134: (continued)

KENSOL 30

TYPICAL ANALYSIS

ASTM METHOD

D-287	*Gravity, °API*	*52.7*
D-445	*Pounds Per Gallon (U.S.)*	*6.396*
D-56	*Flash Point, °F., TCC*	*102 (39°C.)*
D-611	*Aniline Point, °F.*	*156 (69°C.)*
D-1133	*Kauri Butanol Value*	*31.6*
D-156	*Color, Saybolt*	*+ 27*
	Odor	*Mild Petroleum Solvent*

D-86	*Distillation*	*°F.*	*(°C.)*
	Initial Boiling Point	*301*	*(149)*
	5%	*310*	*(154)*
	10%	*314*	*(157)*
	50%	*333*	*(167)*
	90%	*363*	*(184)*
	95%	*372*	*(189)*
	End Boiling Point	*391*	*(199)*

D-1319 *Fluorescent Indicator Adsorption Analysis*

Aromatics, Volume %	*11.3*
Olefins, Volume %	*1.8*
Saturates, Volume %	*86.9*

COMBUSTIBLE - Keep away from heat or flame - Use adequate ventilation - Close container when not in use.

KENSOL 48T

TYPICAL ANALYSIS

Gravity, °API	*47.8*	
Flash Point, °C., COC	*79.8*	*(175°F.)*
Pour Point, °C.	*-42.8*	*(-45°F.)*
Viscosity @ 40°C., CST	*1.56*	
Viscosity @ 100°F., SUS	*31.2*	
Viscosity @ 70°F., SUS	*33.0*	
Viscosity @ 130°F., SUS	*30.2*	
Aniline Point, °C.	*74*	*(165°F.)*
Kauri Butanol Value	*29.3*	
Color, Saybolt	*+22*	
Odor	*Mild Distillate*	
Pounds Per Gallon (U.S.)	*6.570*	

Boiling Range, °C.

Initial Boiling Point	*195.8*	*(385°F.)*
5%	*205.7*	*(403°F.)*
50%	*217.1*	*(422°F.)*
95%	*233.7*	*(453°F.)*
End Boiling Point	*245.8*	*(475°F.)*

FIA Analysis, Vol. %

Aromatics	*9.6*
Olefins	*3.3*
Saturates	*87.1*

COMBUSTIBLE - Keep away from heat or flame - Use adequate ventilation - Close container when not in use.

(continued)

Table 2.134: (continued)

KENSOL 50 T

TYPICAL ANALYSIS

Gravity, °API	43.7
Flash Point, °C., COC	110 (230°F.)
Pour Point, °C.	−20.5 (−5°F.)
Viscosity @ 40°C., CST	2.38
Viscosity @ 100°F., SUS	34.5
Viscosity @ 70°F., SUS	37.4
Viscosity @ 130°F., SUS	32.1
Aniline Point, °C.	76.3
Kauri Butanol Value	27.0
Color, Saybolt	+16
Odor	Mild
Pound per Gallon (U.S.)	6.724

Boiling Range, °C.

Initial Boiling Point	234.8	(455°F.)
5%	243.0	(469°F.)
50%	253.5	(488°F.)
95%	268.2	(514°F.)
End Boiling Point	273.8	(525°F.)

KENSOL 51

TYPICAL ANALYSIS

Gravity, °API	41.9
Flash Point, °C., COC	132 (270°F.)
Pour Point, °C.	−6.7 (20°F.)
Viscosity @ 40°C., CST	3.46
Viscosity @ 100°F., SUS	38.0
Viscosity @ 70°F., SUS	44.0
Viscosity @ 130°F., SUS	34.8
Aniline Point, °C.	83.5
Kauri Butanol Value	24.0
Color, Saybolt	+16
Odor	Mild Distillate
Pounds Per Gallon (U.S.)	6.794

Boiling Range, °C.

Initial Boiling Point	268.2	(514°F.)
5%	271.0	(520°F.)
50%	282.0	(540°F.)
95%	293.0	(560°F.)
End Boiling Point	302.3	(576°F.)

FIA Analysis, Volume %

Aromatics	11.7
Olefins	4.1
Saturates	84.2

KENSOL 53

TYPICAL ANALYSIS

Gravity, °API	39.0
Flash Point, °C., COC	154.0 (310°F.)
Fire Point, °C., COC	162.8 (325°F.)
Pour Point, °C.	10.0 (+50°F.)
Viscosity @ 40°C., CST	5.9
Viscosity @ 100°F., SUS	45.9
Viscosity @ 70°F., SUS	64.5
Aniline Point, °C.	89.0
Kauri Butanol Value	21.0
Pounds per Gallon (U.S.)	6.910

Boiling Range, °C.

Initial Boiling Point	299.0	(570°F.)
5%	305.1	(582°F.)
50%	321.0	(610°F.)
95%	334.8	(635°F.)
End Boiling Point	340.8	(645°F.)

KENSOL 61

TYPICAL ANALYSIS

Gravity, °API	40.5
Flash Point, °C., COC	134.8 (275°F.)
Pour Point, °C.	−1.1 (30°F.)
Viscosity @ 40°C., CST	4.36
Viscosity @ 100°F., SUS	40.9
Viscosity @ 70°F., SUS	48.7
Viscosity @ 130°F., SUS	36.4
Aniline Point, °C.	84.1
Kauri Butanol Value	24.4
Odor	Mild
Pounds Per Gallon (U.S.)	6.850

Boiling Range, °C.

Initial Boiling Point	272.1	(522°F.)
5%	281.4	(538°F.)
50%	295.2	(564°F.)
95%	321.0	(610°F.)
End Boiling Point	329.8	(625°F.)

FIA Analysis

Vol. % Saturates	78.7
Olefins	5.6
Aromatics	15.7

(continued)

Table 2.134: (continued)

KENSOL 80

TYPICAL ANALYSIS

Gravity, °API	*60.8*	
Pounds per Gallon (U.S.)	*6.126*	
Flash Point, °C. TCC	*- 2.2*	*(28°F.)*
Aniline Point, °C.	*57.1*	*(135°F.)*
Kauri Butanol Value	*35.4*	
Color, Saybolt	*+30*	
Odor	*Mild Solvent*	

Distillation	*°C.*	*°F.*
Initial Boiling Point	*90*	*194*
5%	*99*	*210*
10%	*103*	*218*
50%	*119*	*245*
90%	*140*	*283*
95%	*148*	*298*
End Boiling Point	*157*	*315*

Florescent Indicator Analysis

Aromatics, Volume %	*7.8*
Olefins, Volume %	*0.5*
Saturates, Volume %	*91.7*

Flammable - Keep away from heat or flame - Use adequate ventilation -
Close container when not in use.

Table 2.135: Mobil Oil Aromatic Solvents *(74)*

CERTREX 50C

TEST	
Gravity, API	19.5 Typical
Color ASTM	3.0 Max.
Flash Point, °F, COC	150 Min.
Viscosity, SUS @ 100°F	36 Min.
cst @ 100°F	3.0 Min
Solvency, wt. % pentachlorophenol	10 Min
Distillation, °F @ 50%	490 Min.
°F @ 90%	585 Min.
AWPA Sludge test, wet %	0.9 Typical
Water and sediment, vol %	0.5 Max.

CERTREX 409

TEST	
Gravity, API	16 - 22
Viscosity, SUS @ 100 °F	35 - 45
Pour Point, °F	0 Max.
Flash Pt., °F, PMCC	170 Min.
Color, ASTM	6.0 Typical
Total Acid Absorption, ASTM D-1019, Wt. %	65 Min.
Water & Sediment, %	0.05 Max.
Distillation, F.B.P., %F	700 Max.

Table 2.136: Penreco Hydrocarbon Solvents (75)

	2251 Oil	2263 Oil	2257 Oil	2259 Oil	2260 Oil
Specifications:					
API Gravity at 60°F	46/50	46/50	44/47	43/47	40/44
Specific Gravity at 60/60°F	.779/.797	.779/.797	.793/.806	.793/.811	.806/.825
Distillation, ASTM D86					
IBP, °F, Min.	375	375	430	445	500
End Point, °F. Max.	500	500	510	535	610
Typical Properties:					
Viscosity at 100°F, SUS	30.5	30.5	33.2	33.6	40.2
Pounds per Gallon at 60°F	6.56	6.56	6.64	6.69	6.79
Flash Point, COC°F	165	165	220	240	280
Pour Point, ASTM D97. °F	-40	-40	-10	-10	25
CAS or TSCA Inventory Numbers	64742-14-9*	64742-47-8*	64742-46-7*	64742-46-7*	64742-46-7*

Table 2.137: Phillips Hydrocarbon Solvents (4)

	ISOBUTANES	PENTANES				HEXANES			HEPTANES	
TYPICAL PROPERTIES	POLY GRADE ISOBUTANE	ISOPENTANE	NORMAL PENTANE	SPECIAL GRADE PENTANE	USP PETROLEUM ETHER (30°-60°C)	ISOHEXANES	NORMAL HEXANE(S)	HIGH PURITY NORMAL HEXANE	ISOHEPTANES	NORMAL HEPTANE
Distillation range. F at 760 mm Hg		(D 216)	(D 216)	(D 216)	(D 216)	(D 216)	(D 1078)	(D 1078)	(D 216)	
Initial boiling point		82	93	99	97	125	152	152	191	202
10% recovered		83	96	100	102	133	153	153	193	205
50% recovered		89	97	101	107	137	154	154	195	206
90% recovered		98	100	102	122	140	155	155	196	207
Dry point		102	104	105	138	141	156	156	198	209
API gravity at 60 F		95.0	92.6	91.5	98.9	83.0	77.6	79.5	66.8	65.4
Specific gravity of liquid at 60/60 F		0.624	0.633	0.63	0.642	0.660	0.677	0.672	0.714	0.719
Density of liquid at 60 F, lb/gal	4.69	5.19	5.27	5.28	5.34	5.49	5.63	5.59	5.94	5.98
Vapor pressure at 100 F, psia	72.2	20.1	15.5			8.4	5.3	5.0	2.5	2.2
Flash point, approximate, F	-117	-70	-55		-50	-25	-15	-10	+14	26
Kinematic viscosity cs at 32 F						0.565	0.576	0.576		0.83
cs at 100 F		0.486	0.443			0.406	0.406	0.417		0.56
Purity, percent	96.4	97	99.4			99+Paraffins	66	87.8		
Aromatic content (Total), percent	nil	nil	nil	nil	nil	nil	<0.01	<0.02	<0.9	1.3
Benzene content, weight percent	nil		nil	nil					<0.1	<0.1
Sulfur content, weight percent	nil	<0.0005	<0.0005	nil		0.0008	0.0001	0.0001	0.0004	0.003
Bromine number		0.1	<0.0005		<0.5			0.01	0.3	0.2
Nonvolatile matter, grams/100 ml	nil	<0.0005			<0.0005	<0.0005	<0.0005	<0.0005	<0.0005	<0.0005
Odor, residual	None							None		
Color, Saybolt	+30	+30	+30	+30	+30	+30	+30	+30	+30	+30
Acidity, distillation residue	neutral	neutral	neutral		neutral	neutral	neutral	neutral	neutral	neutral
Copper corrosion, 2 hrs at 212 F	1	1	1	1	1	1	1	1	1	1
Unsulfonated residue, volume percent										
Doctor test	negative	negative	negative	negative	negative	negative	negative	negative	negative	negative
Kauri-Butanol value			25.4		27.7	27	29.7	27.8	37	34
Aniline point, F							146	151	140	135

(continued)

Table 2.137: (continued)

TYPICAL PROPERTIES	ISOPARAFFINS							CYCLOPARAFFINS*			TEST METHOD
	SOLTROL 10	SOLTROL 50	SOLTROL 100 "ODORLESS"	SOLTROL 130 "ODORLESS"	SOLTROL 170 "ODORLESS"	SOLTROL 220	SOLTROL 145	POLY GRADE CYCLOHEXANE	98% CYCLOHEXANE	99.5% CYCLOHEXANE	
Distillation range, F at 760 mm Hg											ASTM D 86 (Unless indicated)
Initial boiling point	203	241	320	365	423	471	347	174 (D 1078)	177.3 (D 1078)	177.3 (D 1078)	
10% recovered	204	246	326	369	431	488	358	176	177.3	177.3	
50% recovered	206	252	332	374	438	502	374	177			
90% recovered	212	278	340	391	453	522	437	178			
Dry point	218	297	347	407	461	538	535	178	177.8	177.8	
API gravity at 60 F	71.0	64.0	58.0	55.2	50.5		53.0	52.3	49.6	49.2	ASTM D 287
Specific gravity of liquid at 60/60 F	0.699	0.724	0.735	0.762	0.785	.809	0.767	0.770	0.782	0.783	ASTM D 1298
Density of liquid at 60 F, lb/gal	5.82	6.02	6.22	6.31	6.48	6.67	6.38	6.41	6.51	6.52	ASTM D 1250
Vapor pressure at 100 F, psia	2.1							3.4	3.25	3.3	ASTM D 323
Flash point, approximate, F	16	50	105	139	185	222 (D-93)	123	+6	-1	+10	ASTM D-56 (Unless indicated)
Kinematic viscosity, cs at 32 F	0.818			3.21	6.65		3.85	1.54	0.94		
cs at 100 F	0.574			1.55	2.47	4.68	1.75	0.85			ASTM D 445
Purity, percent	99+Isoparaffins	99+Isoparaffins	99+Isoparaffins	99+Isoparaffins	99+Isoparaffins	99+Isoparaffins	99+Isoparaffins	86.5	98.8	99.8	Gas Liquid Chromatography
Aromatic content (Total), percent	nil (Est.)	nil (Est.)	nil (Est.)	nil (Est.)	nil (Est.)	nil (Est.)	nil (Est.)	0.010 (D 1017)	0.003 (D 1017)	0.003 (D 1017)	ASTM D 1319 (Unless indicated)
Benzene content, weight percent											ASTM D 1017
Sulfur content, weight percent	0.0005	0.0004	0.0004	0.0002	0.001	0.0007	0.0008	0.0002	0.0001	0.0001	ASTM D 1266
Bromine number			0.3	0.6	1.0	4.2	2.5				ASTM D 1159
Nonvolatile matter grams/100 ml	<0.0005	<0.0005	<0.0005	<0.0005				<0.0005	<0.0005	<0.0005	ASTM D 1353
Odor, residual			None	None	None						Panel
Color, Saybolt	+30	+30	+30	+30	+30	+30	+25	+30	+30	+30	ASTM D 156
Acidity, distillation residue	neutral	neutral	neutral	neutral	neutral	neutral	neutral	neutral	neutral	neutral	ASTM D 1093
Copper corrosion, 2 hrs at 212 F	1	1	1	1	1	1	1	1	1	1	ASTM D 130
Unsulfonated residue, volume percent			99.0	99.0	98.5		96.2				ASTM D 483
Doctor test	negative	negative	negative	negative	negative	negative	negative	negative	negative	negative	ASTM D 484
Kauri-Butanol value	28.3	29.1		26	24		28	52	55	56	ASTM D 1133
Aniline point, F	167	168	179.8	184	192	194.5	184	93.4			ASTM D 611

*Naphthenes

Table 2.138: Phillips Hydrocarbon Propellants *(4)*

Phillips normal butane, isobutane and propane, and their various blends have excellent characteristics for aerosol propellant application. They are non-corrosive, odorless hydrocarbons of high purity, high stability, low toxicity, and low cost. They are soluble in alcohol, chloroform, methylene chloride, and ether, as well as many other organic diluents. These hydrocarbon propellants are non-polar and not hydrolyzed by water (Figures 1 and 2). Their Kauri-Butanol value is approximately 18. Their dispersive properties are unique. The low surface tension of these propellants aids in dispersion as less energy is required to separate the particles in a spray mist (Figure 3).

Figures 4, 5 and 6 represent ternary diagrams of water, alcohol, and hydrocarbon propellant. These diagrams show percentages of ingredients (water, alcohol, and hydrocarbon propellant) that can be used to retain either a one-phase or a two-phase system. Figures 7 and 10 provide vapor pressures of blends.

Curves illustrating both the change in liquid density with temperature and the change in vapor density with temperature are shown. These data are useful in calculating the filling volume and head space (Figures 8 and 9). Figure 11 gives formulas for conversion between mol percent, liquid-volume percent, and weight percent in a propane-isobutane mixture.

TYPICAL PROPERTIES OF PHILLIPS HYDROCARBON PROPELLANTS

Property	Hydrocarbon Propellant A-108* (Propane)	Hydrocarbon Propellant A-31* (Isobutane)	Hydrocarbon Propellant A-17* (N-Butane)
Molecular Formula	C_3H_8	C_4H_{10}	C_4H_{10}
Molecular Weight	44.10	58.12	58.12
Distillation Range. °F			
Initial Boiling Point	– 46	9	28
Dry Point	– 42	15	33
Specific Gravity. Gas. 60/60 F	0.509	0.564	0.585
API Gravity, 60° F	146.6	119.6	110.5
Density of Liq., 60° F (lb/gal)	4.24	4.70	4.88
Vapor Pressure			
70° F psia	123	48	32
100° F psia	189	72	51
105° F psia	198	80	56
130° F psia	271	111	82
Coefficient of Liquid Expansion, 60° F	0.0015	0.0012	0.0011
Heat of Vaporization Btu/lb at 14.696 psia and normal boiling point	183.1	157.5	165.6
Specific Heat of Gas CP, 60° F Btu/lb ° F	0.389	0.387	0.391
Viscosity. Saturated Gas, 70° F, centipoise	0.0082	0.0078	0.0074
Critical Temperature, ° F	206.3	274.9	305.6
Critical Pressure. psia	617.4	529.2	550.7
Flash Point, approx. ° F	–156	–117	–101
Purity, Mol %	98	96	98
Sulfur content, ppm	2.0	2.0	2.0
Residue, g/100 ml	0.0005	0.0005	0.0005
Acidity of Residue	Neutral	Neutral	Neutral
Odor	Pass	Pass	Pass

(The "A" designates aerosol grade, the number following indicates the pressure in psig at 70° F)
*Trademark of Phillips Petroleum Company

AEROSOL GRADE PROPELLANTS SPECIFICATIONS

Property	Hydrocarbon Propellant A-108* Propane	Hydrocarbon Propellant A-31* Isobutane	Hydrocarbon Propellant A-17* Normal Butane	Test Method
Composition, mol %				Chromatography
Ethane	max 0.2	—	—	
Propane	min 98.0	max 1.0	max 1.0	
Isobutane	max 2.0	min 95.0	max 3.0	
Normal Butane	max 0.5	max 5.0	min 97.0	
Pentanes	—	max 0.1	max 2.0	
Total Saturates	min 99.9	min 99.9	min 99.9	Chromatography
Vapor Pressure				
psig at 70 F	108 ± 4	31 ± 2	17 ± 2	ASTM D 1267 (mod. 70 F)
Sulfur, ppm	max 5	max 5	max 5	ASTM D 1266
Residue, g/100 ml	max 0.0005	max 0.0005	max 0.0005	ASTM D 1351 (mod.)
Acidity of Residue	neutral	neutral	neutral	ASTM D 1093
Odor	pass	pass	pass	Panel

*Trademark of Phillips Petroleum Company

Figure 1.

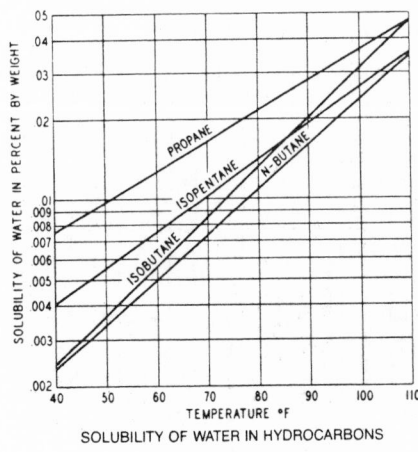

SOLUBILITY OF WATER IN HYDROCARBONS

Figure 2.

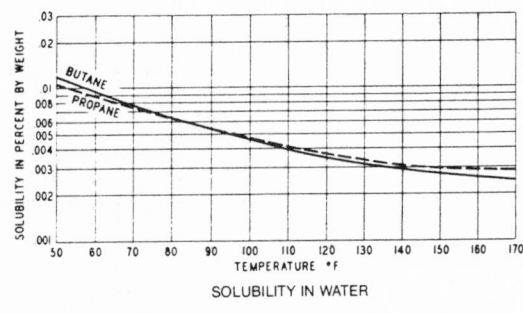

SOLUBILITY IN WATER

(continued)

Table 2.138: (continued)

Figure 3.

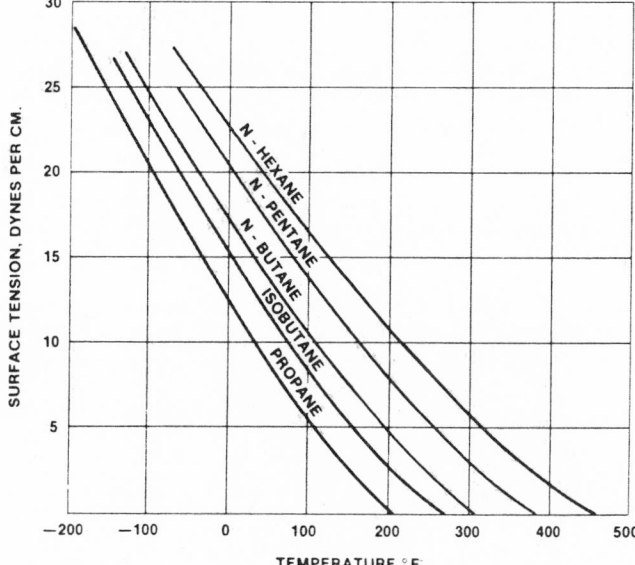

SURFACE TENSION OF PARAFFIN HYDROCARBONS

Figure 4.

WEIGHT PER CENT @ 72°F		
ETHANOL	WATER	N-BUTANE
29.7	69.3	1.0
64.5	30.6	4.8
63.1	9.7	27.2
38.2	3.4	58.3

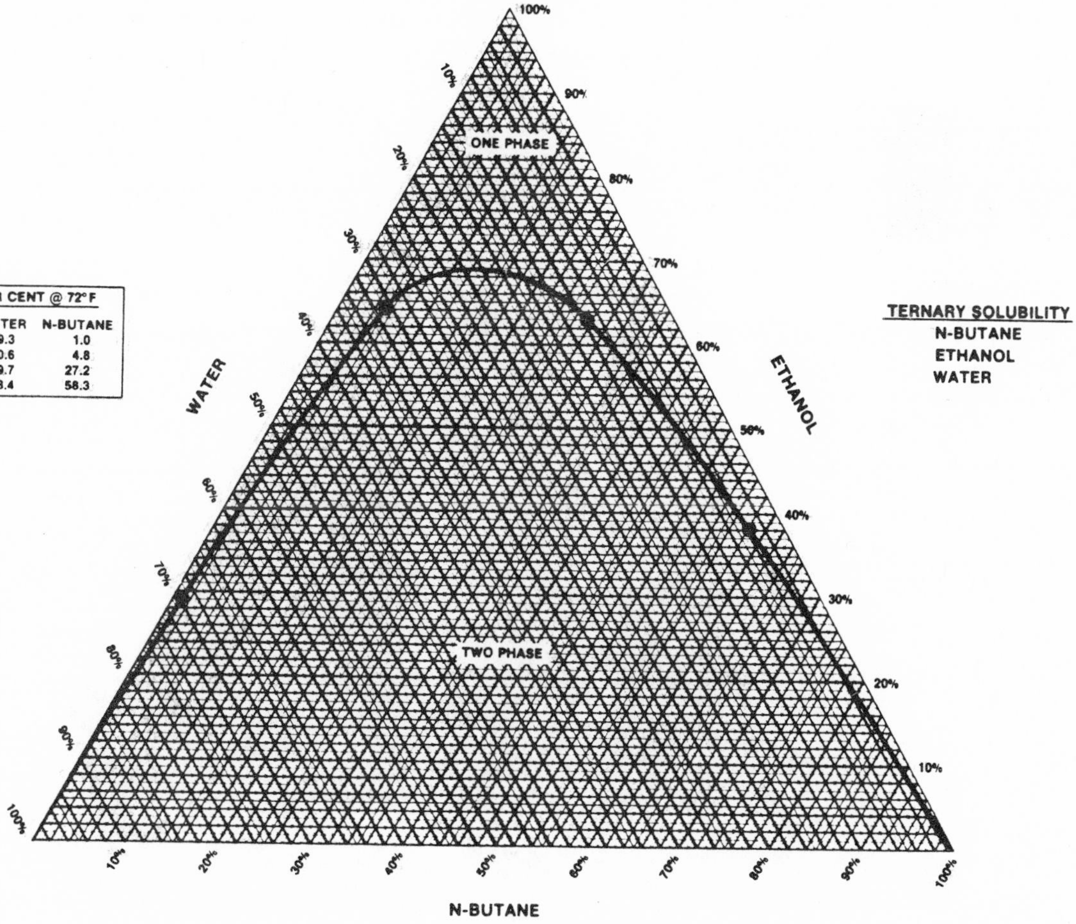

TERNARY SOLUBILITY
N-BUTANE
ETHANOL
WATER

ONE PHASE

TWO PHASE

WATER

ETHANOL

N-BUTANE

(continued)

Table 2.138: (continued)

Figure 5.

WEIGHT PER CENT @ 72° F		
ETHANOL	WATER	ISOBUTANE
19.8	79.2	1.0
49.2	49.2	1.5
65.3	27.8	6.9
68.0	19.4	12.6
66.7	13.0	20.3
62.5	8.6	28.9
41.2	3.8	55.0

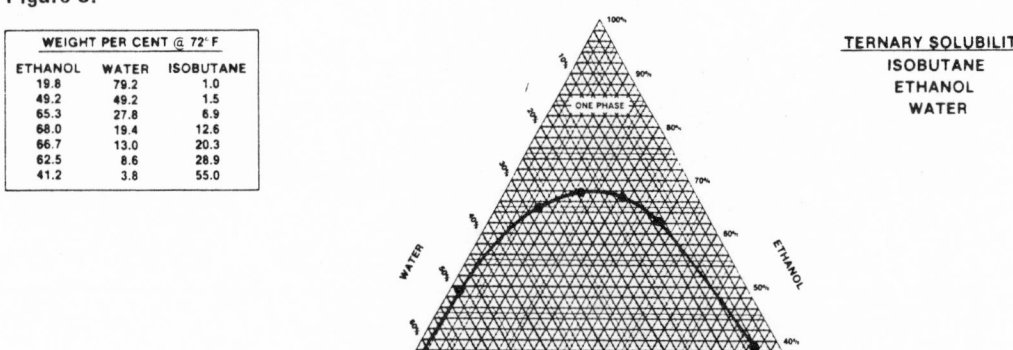

TERNARY SOLUBILITY
ISOBUTANE
ETHANOL
WATER

Figure 6.

WEIGHT PERCENT @ 72° F		
ETHANOL	WATER	PROPANE
9.9	90	~.1
30	69.6	~.4
49	49	2
66	28	6
44	5	51

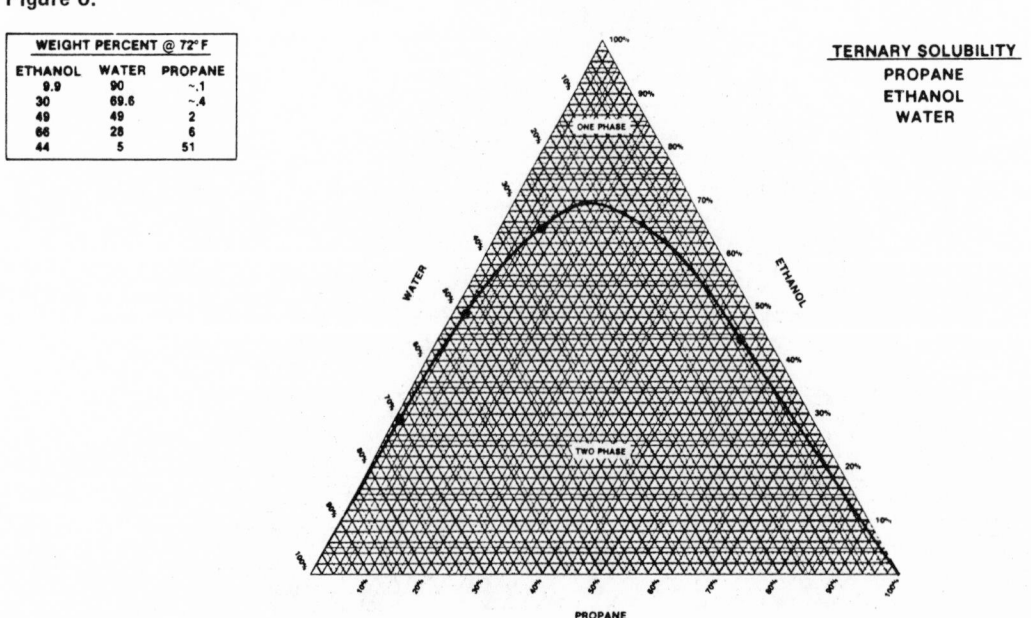

TERNARY SOLUBILITY
PROPANE
ETHANOL
WATER

(continued)

Table 2.138: (continued)

Figure 7.

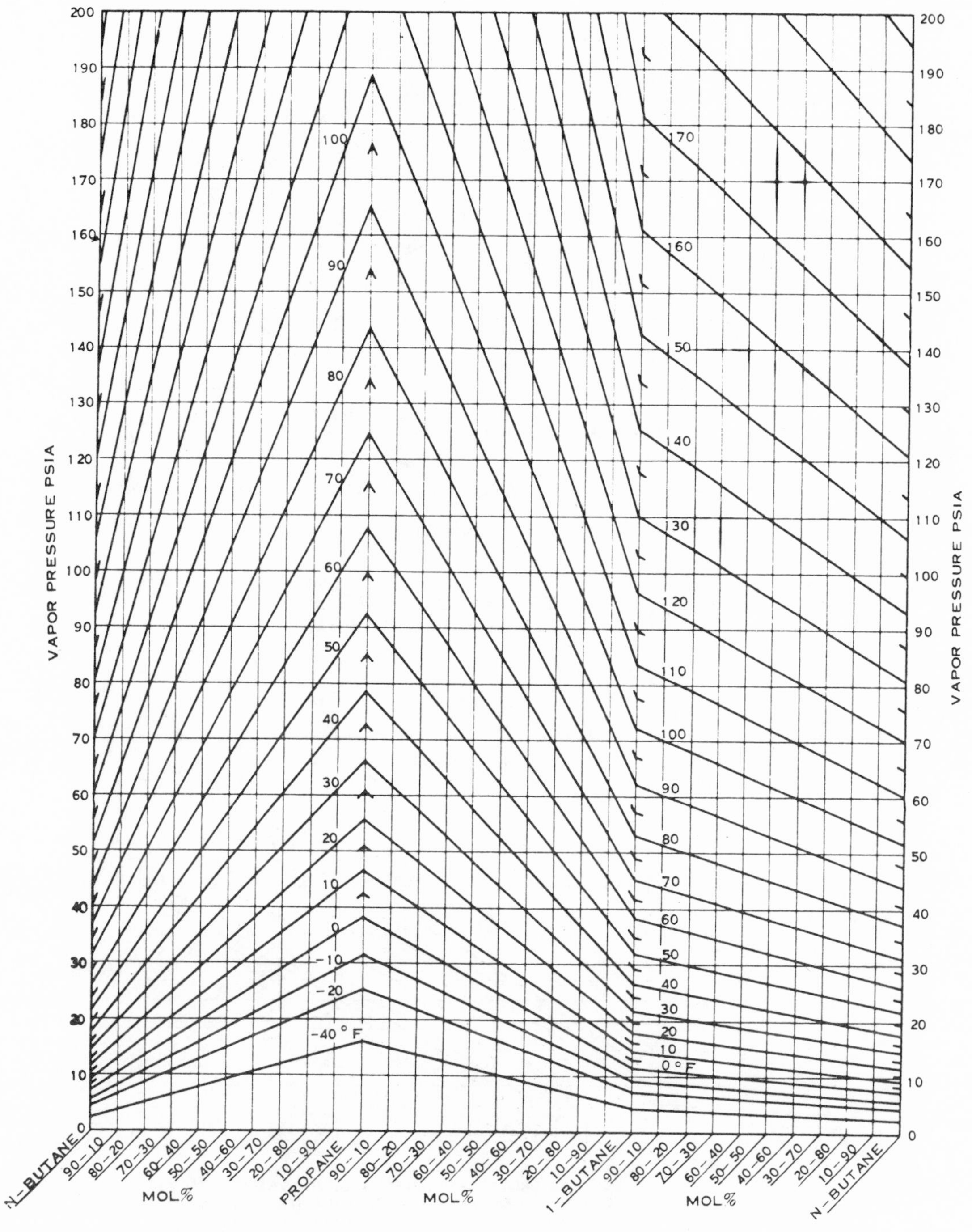

VAPOR PRESSURES OF HYDROCARBON PROPELLANT BLENDS

(continued)

Table 2.138: (continued)

Figure 8.

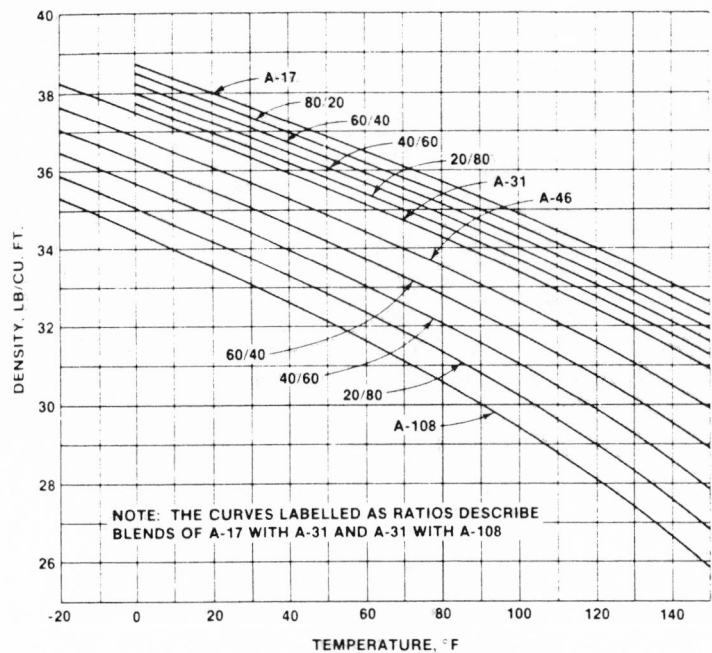

PHILLIPS
HYDROCARBON PROPELLANTS
DENSITY OF LIQUID PHASE

Figure 9.

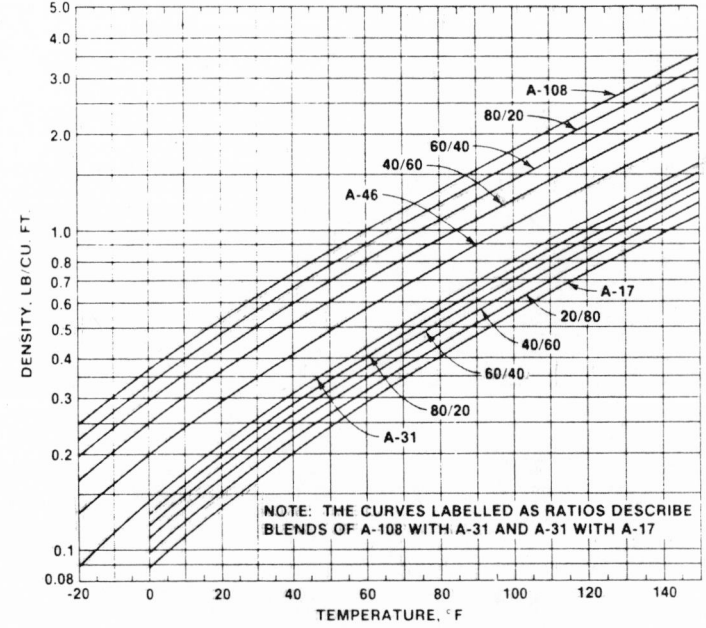

PHILLIPS
HYDROCARBON PROPELLANTS
DENSITY OF VAPOR ABOVE SATURATED LIQUID

(continued)

Table 2.138: (continued)

Figure 10.

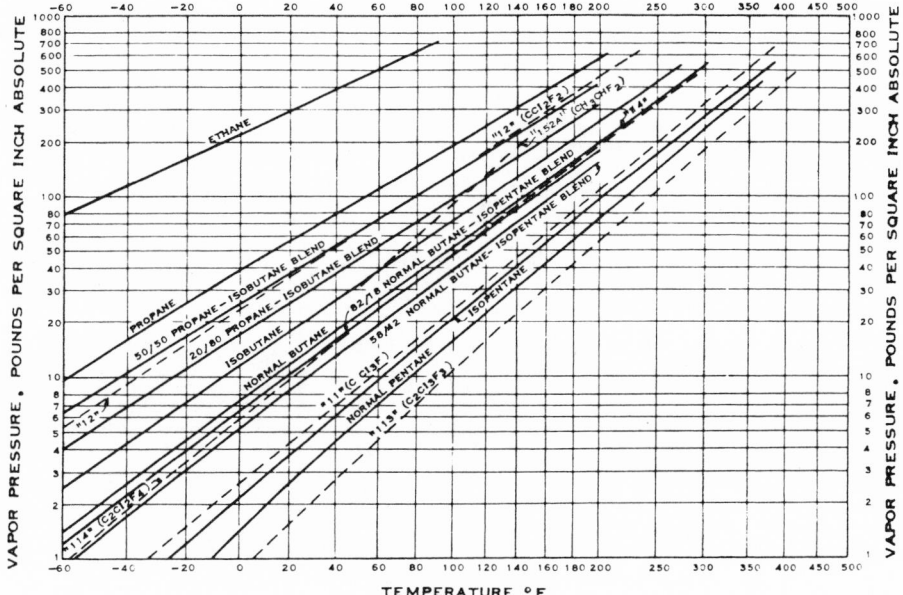

VAPOR PRESSURES OF LIGHT HYDROCARBONS

Figure 11.

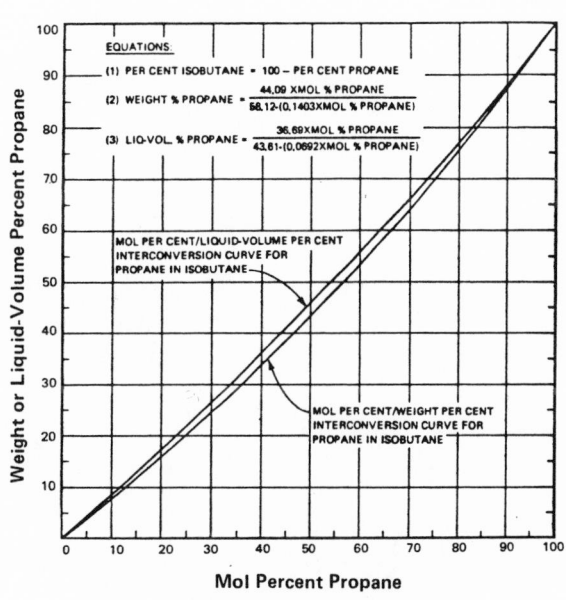

Propane-Isobutane Mixtures
Interconversion of Mol (Gas-Volume),
Liquid-Volume, or Weight Percent

Table 2.139: Shell Hydrocarbon Solvents (East of the Rockies) (14)

Property	ASTM test method	Shell Sol B	Shell Sol B-8	Shell rubber solvent	Shell Sol BT-4	Shell Sol BT-67-EC	Shell Tolu-Sol 5	Shell Sol BJ-10	Shell Sol BT-9	Shell Tolu-Sol 25
Specific gravity, 60/60°	D 1250	0.672	0.692	0.695	0.694	0.694	0.702	0.704	0.706	0.740
Gravity, °API	D 287	79.0	73.0	72.0	72.5	72.3	70.0	69.5	68.9	58.8
Pounds per gallon @ 60/F	D 1250	5.60	5.76	5.79	5.78	5.78	5.85	5.86	5.88	6.16
Color, Saybolt	D 156	+30	+30	+30	+30	+30	+30	+30	+30	+30
Viscosity, cs @ 77°F		0.5	0.5	0.5	-	0.55	0.6	-	-	0.6
Kauri-butanol number	D 1133	29	33	34	30	30	30	35	30	42
Aniline point, °F	D 611/D 1012	153	137	135	145	152	152	131	139	103
Mixed aniline point, °F	D 611/D 1012	-	-	-	-	-	-	-	-	-
Flash point, TCC, °F	D 56	<0	<0	<0	<10	<10	20	<10	<10	24
Autoignition temp, °F	D 2155	639	-	-	-	-	608	-	-	-
Reid vapor pressure, psia	D 323	6.2	6.0	5.5	4.3	3.3	2.3	4.2	3.0	1.8
Distillation, °F	ASTM	D 86	D 86	D 86	D 86	D 1078	D 1078	D 86	D 1078	D 1078
IBP		146	146	147	150	156	193	147	168	198
10% recovered		149	150	152	164	175	195	160	184	205
30%		151	152	155	171	181	197	168	191	207
50%		153	155	160	177	188	198	178	197	209
70%		160	163	170	188	196	200	193	203	211
90%		171	177	196	202	206	204	224	211	218
Dry point		190	225	238	215	217	217	276	220	225
End point		-	-	-	-	-	-	-	-	-
Evaporation time, seconds	D 3539									
10%		6	6	7	8	9	10	8	10	11
30%		16	18	20	22	26	29	24	29	34
50%		28	32	35	40	46	51	43	50	60
70%		42	48	52	61	67	74	66	74	87
90%		**60**	**68**	**75**	**68**	**93**	**102**	**103**	**103**	**118**
100%		89	111	135	131	130	139	164	143	198
Typical composition,[1] %v										
Paraffins		88.2	82.0	79.6	86.8	88.7	89.0	79.7	82.6	69.0
Naphthenes		11.2	10.4	11.5	9.2	8.6	7.3	8.7	7.7	5.7
Olefins		0.1	0.1	0.1	0.4	0.5	0.7	0.3	0.5	0.5
Total aromatics		0.5	7.5	8.8	3.6	2.2	3.0	11.3	9.2	24.8
Benzene		0.5	0.5	0.4	0.3	0.2	.04	0.3	0.1	0.03
Toluene/ethylbenzene		0.0	7.0	8.1	0.3	0.2	2.96	5.1	9.1	24.77
C8+ aromatics		-	-	0.3	3.3	2.0	-	5.9	9.1	-

(continued)

Table 2.139: (continued)

Property	ASTM test method	Shell Tolu-Sol 19-EC	Shell Tolu-Sol 40	Shell toluene	Shell super VM&P naphtha	Shell VM&P naphtha EC	Shell xylene	Shell Sol 340	Shell Cyclo Sol® 53	Shell mineral spirits 143-EC
Specific gravity, 60/60°	D 1250	0.730	0.760	0.871	0.754	0.750	0.871	0.769	0.874	0.778
Gravity, °API	D 287	62.3	54.7	31.0	56.3	57.2	31.0	52.5	30.5	50.5
Pounds per gallon @ 60/F	D 1250	6.08	6.33	7.25	6.27	6.24	7.25	6.40	7.27	6.47
Color, Saybolt	D 156	+30	+30	+30	+30	+30	+30	+30	+30	+30
Viscosity, cs @ 77°F		0.6	0.6	0.64	-	0.7	0.75	1.2	1.0	-
Kauri-butanol number	D 1133	39	52	105	39	38	95	33	92	34
Aniline point, °F	D 611/D 1012	116	76	-	125	129	-	150	-	144
Mixed aniline point, °F	D 611/D 1012	-	-	51	-	-	51	-	56	-
Flash point, TCC, °F	D 56	24	26	41	55	55	81	104	116	109
Autoignition temp, °F	D 2155	-	-	996	-	608	1,013	567	968	-
Reid vapor pressure, psia	D 323	2.0	1.7	1.2	<0.1	0.5	0.7	0.3	<0.4	<0.1
Distillation, °F	ASTM	D 86	D 1078	D 850	D 86	D 86	D 850	D 86	D 86	D 86
IBP		195	198	230	246	246	281	316	320	326
10% recovered		199	203	-	250	247	-	320	321	329
30%		201	205	-	252	248	-	322	322	332
50%		202	207	-	252	250	-	324	324	336
70%		203	210	-	255	255	-	329	327	341
90%		208	216	-	261	260	-	337	333	360
Dry point		220	226	231	272	270	285	-	-	392
End point		-	-	-	-	-	-	358	345	-

Evaporation time, seconds   D 3539

10%		13	15	25	29	29	69	154	210	180
30%		35	44	76	87	87	200	475	655	570
50%		60	75	127	148	148	333	825	1110	1000
70%		88	109	179	211	212	469	1220	1600	1540
90%		**119**	**149**	**235**	**285**	**288**	**610**	**1680**	**2140**	**2290**
100%		178	225	297	378	390	740	2200	2710	3310

Typical composition, %v

Paraffins (saturates)		74.3	57.0	(<0.01)	48.3	49.8	0.5	47.2	(0.6)	46.3
Naphthenes		6.1	4.7	-	39.8	41.0	0.0	50.2	-	46.2
Olefins		0.6	0.4	0.0	<0.1	0.1	0.0	0.1	>0.1	0.1
Total aromatics		19.0	37.9	99.9+	11.9	9.2	99.5	2.6	99.4	7.5
Benzene		.04	0.03	0.005	NIL	NIL	NIL	NIL	NIL	NIL
Toluene/ethylbenzene		18.96	37.87	99.9+	2.7	2.3	15.0	-	-	-
C₈+ aromatics		-	-	0.0	9.2	6.9	84.5	2.6	99.4	7.5

(continued)

Table 2.139: (continued)

Property	ASTM test method	Shell mineral spirits 145-EC	Shell TS-28 solvent	Shell mineral spirits 150-EC	Shell minerals spirits 135 (WR)	Shell Sol 71	Shell Sol 140	Shell Cyclo Sol 63	Shell 460 solvent EC
Specific gravity, 60/60°	D 1250	0.779	0.852	0.777	0.783	0.759	0.791	0.887	0.795
Gravity, ° API	D 287	50.1	34.6	50.6	49.2	55.0	47.5	28.0	46.4
Pounds per gallon @ 60/F	D 1250	6.49	7.09	6.47	6.52	6.32	6.58	7.39	6.62
Color, Saybolt	D 156	+30	+30	+30	+30	+30	+27	+27	+25
Viscosity, cs @ 77°F	D 1133	–	–	–	–	–	–	–	–
Kauri-butanol number	D 611/D 1012	34	73	33	37	26	31	89	33
Aniline point, °F	D 611/D 1012	145	–	151	133	184	153	–	151
Mixed aniline point, °F		–	86	–	–	–	–	57	–
Flash point, TCC, °F	D 56	112	122	111	110	125	143	142	144
Autoignition temp, °F	D 2155	–	–	–	581	586	563	996	–
Reid vapor pressure, psia	D 323	<0.1	<0.1	<0.1	0.3	<0.1	<0.1	<0.1	<0.1
Distillation, °F	ASTM	D 86	D 86	D 86	D 86	D 86	D 86	D 86	D 86
IBP		317	320	323	315	355	371	354	366
10% recovered		327	323	331	327	360	375	360	370
30%		332	324	336	–	362	377	–	374
50%		336	327	342	331	364	380	364	378
70%		343	335	351	–	370	387	–	384
90%		358	370	369	359	376	396	398	408
Dry point		–	–	–	–	–	–	–	–
End point		388	400	390	393	400	416	420	475

Evaporation time, seconds	D 3539								
10%		183	210	200	240	330	615	615	555
30%		625	705	665	750	1080	1910	1890	1860
50%		1140	1250	1210	1350	1960	3340	3330	3420
70%		1800	1880	1990	2110	3010	4980	4500	5430
90%		**2720**	**2740**	**2950**	**3270**	**4510**	**7060**	**7120**	**9180**
100%		4470	4260	4430	4940	6870	9260	9260	25000

Typical composition, %v									
Paraffins (saturates)		47.5	14.2	49.5	45.7	95.9	55.7	(2.5)	62.6
Naphthenes		45.1	9.8	46.9	37.3		39.7		27.4
Olefins		0.1	0.1	0.1	0.5	4.1	<0.1	<0.1	0.8
Total aromatics		7.4	76.0	3.6	16.5	<0.1	4.6	97.5	9.2
Benzene		NIL	NIL	NIL	NIL	NIL	NIL	NIL	NIL
Toluene/ethylbenzene		–	–	–	0.3	–	–	–	–
C_8+ aromatics		7.4	76.0	3.6	16.2	<0.1	4.6	97.5	9.2

Compositions are subject to variation due to feedstock makeup and should be considered as representative only.
NIL: <10 ppm; a product which in normal manufacture would not contain benzene and therefore is not routinely analyzed for benzene content.

Table 2.140: Shell Hydrocarbon Solvents (West of the Rockies) (14)

Property	ASTM test method	Shell Tolu-Sol® 6 EC	Shell Tolu-Sol 20 EC	Shell Sol® M-95 EC	Shell Cyclosol® 27	Shell Sol M-75 EC	Shell Toluene	Shell Cyclosol 28	Shell Super VM&P Naphtha EC	Shell Cyclosol 37
Specific gravity, 60/60°F	D 891/D 1298	0.752	0.752	0.752	0.841	0.753	0.871	0.841	0.754	0.857
Gravity, °API	D 287/D 1298	56.7	54.1	56.7	36.8	56.5	31.0	36.8	56.3	33.7
Pounds per gallon @ 60°F	D 1250	6.26	6.35	6.26	7.00	6.27	7.25	7.00	6.27	7.13
Color, Saybolt	D 156	+30	+30	+30	+30	+30	+30	+30	+30	+30
Viscosity, cs @ 77°F	D 445	0.64	—	—	—	—	0.64	—	0.80	—
Kauri-butanol number	D 1133	43	46	45	91	42	105	90	38	92
Aniline point, °F	D 611/D 1012	112	100	111	—	120	—	—	132	—
Mixed aniline point, °F	D 611/D 1012	—	—	—	72	—	51	73	—	63
Flash point, closed cup, °F	D 56/D 3278	21	29	—	—	—	43	36	55	76
Autoignition temp., °F	D 2155	635	—	—	—	—	996	—	612	—
Reid vapor pressure, psia	D 323	—	—	—	—	—	1.2	—	—	—
Distillation, °F	ASTM	D 86	D 86	D 86	D 86	D 86	D 850	D 86	D 86	D 86
IBP		206	208	208	216	220	230	210	254	274
10% recovered		208	210	211	220	224	—	227	260	277
30		209	210	214	220	232	—	230	—	277
50		210	211	215	222	239	—	235	265	278
70		213	212	220	224	250	—	243	—	280
90		218	216	236	226	270	—	260	277	284
Dry point		224	225	268	230	282	232	277	285	300
End point		—	—	—	—	—	—	—	—	—
Evaporation time, seconds D 3539										
10%		12	12	13	18	17	25	21	30	59
30%		38	37	40	58	53	76	67	97	182
50%		66	67	70	102	95	127	121	169	310
70%		96	97	102	150	146	179	187	246	445
90%		**130**	**131**	**142**	**205**	**220**	**235**	**279**	**340**	**590**
100%		180	192	234	282	313	297	399	472	745
Relative evap. rate, n-Butyl acetate = 1.0		3.6	3.6	3.3	2.3	2.1	2.0	1.7	1.4	0.79
Typical composition, %v										
Saturates		88.7	80.7	89.2	22.1	91.3	0.01	22.1	93.8	12.6
Olefins		0.3	0.3	0.3	0.1	0.2	0.0	0.1	0.1	<0.1
Total Aromatics		11.0	19.0	10.5	77.8	8.5	99.9+	77.8	6.1	87.4
Benzene		<10 ppm	<10 ppm	<10 ppm	<50 ppm	<10 ppm	50 ppm	<50 ppm	Nil¹	Nil
Toluene/ethylbenzene		11.0	19.0	10.0	77.8	6.0	99.9+	57.3	1.1	16.4
C₈+ aromatics ex. ethylbenzene		0.0	0.0	0.5	0.0	2.5	0.0	20.4	5.0	71.0

¹Nil: < 10 ppm; a product which in normal manufacture would not contain benzene and therefore is not routinely analyzed for benzene content.

(continued)

Table 2.140: (continued)

Property	ASTM test method	Shell Xylene	Shell Cyclosol 38	Shell Sol 70	Shell Cyclosol 53	Shell TS-28B Solvent	Shell Sol 72	Shell Cyclosol 63	Shell Cyclosol 67
Specific gravity, 60/60°F	D 891/D 1298	0.865	0.857	0.747	0.874	0.851	0.768	0.887	0.857
Gravity, °API	D 287/D 1298	32.1	33.7	57.9	30.5	34.8	52.7	28.0	33.6
Pounds per gallon @ 60°F	D 1250	7.20	7.13	6.22	7.27	7.085	6.40	7.39	7.14
Color, Saybolt	D 156	+30	+30	+30	+30	+30	+30	27	21
Viscosity, cs @ 77°F	D 445	0.75	—	—	1.0	—	—	—	—
Kauri-butanol number	D 1133	95	97	27	92	78	29	89	—
Aniline point, °F	D 611/D 1012	—	—	180	—	—	183	—	—
Mixed aniline point, °F	D 611/D 1012	53	64	—	56	82	—	57	103
Flash point, closed cup, °F	D 56/D 3278	79	79	104	116	106	137	142	135
Autoignition temp., °F	D 2155	1013	—	550	968	—	532	996	—
Reid vapor pressure, psia	D 323	0.4	—	0.4	<0.4	—	—	<0.1	—
Distillation, °F	ASTM	D 850	D 86	D 86	D 86	D 86	D 86	D 86	D 86
IBP		281	268	321	320	314	366	354	352
10% recovered		—	273	325	321	319	370	360	360
30		—	274	327	—	—	372	—	371
50		—	275	329	324	323	374	364	379
70		—	276	332	—	—	377	—	389
90		—	277	338	333	330	388	398	424
Dry point		282	278	344	—	345	402	—	486
End point		—	280	347	335	360	406	410	489
Evaporation time, seconds	D 3539								
10%		69	68	145	210	195	560	615	580
30%		200	208	465	655	635	1790	1890	2110
50%		333	350	810	1110	1120	3100	3330	4150
70%		469	495	1190	1600	1680	4620	4500	7420
90%		**610**	**655**	**1640**	**2140**	**2380**	**6470**	**7120**	**20160**
100%		740	880	2250	2710	3630	8810	9260	89000
Relative evap. rate, n-Butyl acetate = 1.0		0.77	0.71	0.29	0.22	0.20	0.07	0.07	0.02
Typical composition, %v									
Saturates		<1.5	8.3	98.9	0.6	20.5	98.9	2.5	33.9
Olefins		—	<0.1	0.7	<0.1	<0.1	0.9	<0.1	<0.1
Total Aromatics		98.5+	91.6	0.4	99.4	79.5	0.2	97.5	66.1
Benzene		Nil	Nil	Nil	Nil	Nil	Nil	Nil	Nil
Toluene/ethylbenzene		~18.5	17.0	—	—	—	—	—	—
C$_8$+ aromatics ex. ethylbenzene		~80.0	74.6	0.4	99.4	79.5	0.2	97.5	66.1

Compositions are subject to variation due to feedstock makeup and should be considered as representative only.

Halogenated Hydrocarbons

FLUORINATED HYDROCARBONS

Table 3.1: 1-Chloro-1,1,3,3,3-Pentafluoro-2-Propanol *(18)*

Structural Formula:

$$CF_3-CH-CCIF_2$$
$$\underset{OH}{\underset{|}{}}$$

Empirical Formula: $C_3H_2ClF_5O$

Molecular Weight: 184.5

Boiling Point: 82°C.

Melting Point: –6°C.

Density: 1.633 g./ml. at 23°C.

Purity: 99% minimum

Description: Colorless, mobile liquid with a pungent odor.
Solubility: Soluble in water, benzene, carbon tetrachloride, methanol, hexane and petroleum ether.

Table 3.2: Chlorodifluoromethane *(11)*

"Freon"-22 $CHClF_2$

PHYSICAL PROPERTIES

Boiling point @760 mm	–40.8°C. –41.44°F @1 atm. press.
Critical density	0.523 g/cc
Critical pressure	48.7 atm. 716 psia
Critical temperature	96.0°C. 204.8°F
Chlorides	None
Freezing point	–160°C. –256.0°F @1 atm. press.
DDT, g/100 g "Freon"	1.3 (70°F)
Heat content of the liquid @70°F (21°C)	30.99 Btu/lb
Kauri-butanol number @77°F (25°C)	24.5
Latent heat of vaporization @70°F (21°C)	80.50 Btu/lb
Liquid density @70°F (21°C)	1.209 g/cc
Moisture	0.005% by wt, max.
High-boiling impurities	0.05% by vol., max.
Molecular weight	86.5

(continued)

Table 3.2: (continued)

Noncondensible gases	5. 0% by vol., max., in vapor phase
Solubility in water (g "Freon"/liter H_2O) Atm. press. (14. 7 psia)	{ 3. 0 @77°F (25°C) 1. 3 @122°F (50°C)
At the propellant vap. press.	{ 28 (152. 7 psia) @77°F (25°C) 24 (284 psia) @122°F (25°C)
Solubility of water (g H_2O/100 g "Freon")	0. 114 @70°F (21°C)
Vapor pressure @70°F (21°C)	137. 2 psi, abs.
Odor	Ethereal and similar to carbon tetrachloride
Flammability	Practically nonflammable and nonexplosive
Toxicity	Group 5-A Classification of Underwriters' Laboratories Report, MH-3134-page 13
Melting point	−256°F −160°C

Table 3.3: Dichlorodifluoromethane *(11)(17)*

"Freon"-12
"Genetron"-12 CCl_2F_2

PHYSICAL PROPERTIES

Boiling point	−29. 8°C −21. 6°F @1 atm. press.
Critical density	0. 554
Critical pressure	43. 2 atm. 582 psia
Critical temperature	198. 0°C 232. 7°F
Freezing point	−111°C −247. 0°F @1 atm. press.
Molecular weight	120. 6
Specific gravity @−30°C	1. 486
Vapor pressure @70°F	84. 8 psia
Kauri-butanol number @77°F (25°C)	18. 0
DDT, g/100 g "Freon"	0. 8 (77°F)
Solubility in water (g "Freon"/liter H_2O) Atm. press. (14. 7 psia)	{ 0. 28 @77°F (25°C), 0. 14 @122°F (50°C)
At the propellant vap. press.	{ 1. 9 (94. 4 psia) @77°F (25°C) 1. 8 (176. 2 psia) @122°F (50°C)
Solubility of water (g H_2O/100 g "Freon")	0. 0076 @70°F (21°C)

Solubility of "Freon-12" at 21. 1°C (70. 0°F) and One Atmosphere Pressure

Solvent	Wt. % "Freon-12"	Mol. Fraction "Freon-12"	Solvent	Wt. % "Freon-12"	Mol. Fraction "Freon-12"
Amyl chloride	13. 1	0. 12	Dibutyl ether	14. 1	0. 15
Benzene	9. 0	0. 060	Dibutyl oxalate	8. 9	0. 13
Bromobenzene	5. 0	0. 063	Dibutyl tartrate	6. 3	0. 12
Bromoform	1. 2	0. 025	Dichloroethyl ether	3. 9	0. 046
n-Butyl alcohol	8. 5	0. 054	Diethyl aniline	7. 2	0. 093
Butyl butyrate	13. 2	0. 14	Diethyl phthalate	4. 7	0. 082
Carbon tetrachloride	5. 2	0. 063	Dioxane	6. 9	0. 054
Chloroform	5. 5	0. 052	Ethylene dichloride	4. 7	0. 039
α-Chloronaphthalene	3. 9	0. 052	Ethylene glycol butyl ether	7. 2	0. 070
Cyclohexanone	8. 5	0. 069	Ethylene glycol ethyl ether	7. 4	0. 055
Diacetone alcohol	6. 1	0. 060			

Table 3.4: Dichlorotetrafluoroethane *(11)*

"Freon"-114 $CClF_2-CClF_2$

PHYSICAL PROPERTIES

Boiling point @ 760 mm	3. 55 °C (38. 4 °F)
Chlorides	None
Color	Clear and water-white
Critical pressure	474 psia
Critical temperature	145. 7 °C (294. 3 °F)
High-boiling impurities	Not more than 0. 05% by vol.
Flammability	Noncombustible and nonflammable
Melting point	−94 °C (−137 °F)
Moisture	Not more than 0. 0025% by wt
Noncondensible gases	Not more than 5. 0% by vol. in vapor phase
Odor	Faint and ethereal

Table 3.5: Trichlorotrifluoroethane *(17)*

Arklone P	Frigen 113TRT
Flugene 113	Genesolv D
Freon TF	Kaltron 113MDR

PHYSICAL PROPERTIES OF GENESOLV D

Chemical formula		$C_2Cl_3F_3$
Molecular weight		187.4
Boiling point, °F		117.6
°C		47.6
Freezing point, °F		−31
°C		−35
Critical temperature, °F		417
°C		214
Critical pressure, psia		495
atm		33.7
dynes/sq cm		34.14×10^6
Density of liquid at 21.1°C, g/cu cm		1.574
Approximate weight of liquid at 70°F, lb/gal.		13.1
Density of vapor at bp (compared to air = 1)		6.2
Vapor pressure at 70°F, psia		5.5
mm Hg		472
Viscosity at 68°F, centipoises		0.694
Surface tension at 68°F, dynes/cm		18.8
Evaporation rate (compared to ether = 1)		1.3
Kauri-butanol value		32
Solubility of water at 70°F, wt %		0.009
Solubility in water at 70°F, wt %		0.028
Index of refraction at 77°F		1.356
Specific heat capacity, Btu/lb • °F (cal/g•°C)		
	liquid	0.21
	saturated vapor	0.15
Latent heat of vaporization, at bp		
	Btu/lb	63.1
	cal/g	35.06
Thermal conductivity, Btu•ft/hr•sp ft•°F		
	liquid at 70°F	0.0449
	vapor at 70°F, 0.5 atm	0.0043
Dielectric strength, kv/0.1 in.		37
Relative dielectric strength (compared to nitrogen = 1)		2.6
Dielectric constant at 60 Hz		2.41
Dissipation factor at 60 Hz		0.0001
Volume resistivity, ohm.cm		10^{16}
Coefficient in air, at 70°C, 1 atm, sq cm/sec		0.069

Table 3.6: Density of Genesolv D as a Function of Temperature *(17)*

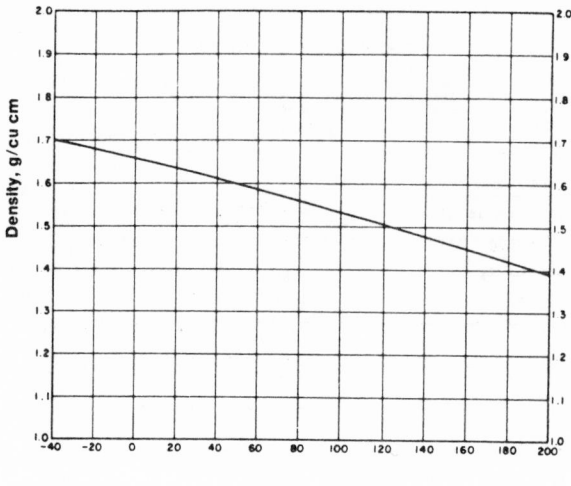

Temperature, °F

Table 3.7: Weight vs Temperature—Genesolv D Solvent *(17)*

temp, °F	weight, lb/gal	temp, °F	weight lb/gal
20	13.66	60	13.23
25	13.60	65	13.18
30	13.55	70	13.13
35	13.50	75	13.07
40	13.44	80	13.02
45	13.39	85	12.97
50	13.34	90	12.91
55	13.28		

Table 3.8: Vapor Pressure of Genesolv D Solvent as a Function of Temperature *(17)*

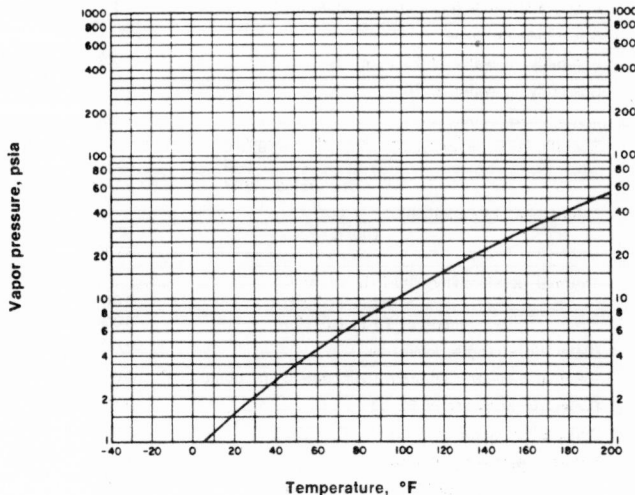

Temperature, °F

Table 3.9: Solubility of Organic Materials in Genesolv D Solvent *(17)*

Hydrocarbons, chlorinated solvents, and polar compounds of low molecular weight (alcohols, esters, ketones, ethers) are miscible with Genesolv D. However, glycols and high molecular weight alcohols and ketones are insoluble, with some exceptions. Proteinaceous materials represented by casein and gelatin are insoluble in Genesolv D. Waxes vary substantially in solubility.

SOLUBILITY OF TYPICAL ORGANIC MATERIALS
IN GENESOLV D SOLVENT AT ROOM TEMPERATURE

Alcohols		Glycols	
Ethanol	M	Ethylene glycol	I
Isopropanol	M	Propylene glycol	I
Isobutanol	M	Diethylene glycol	I
Cetyl alcohol	I	Dipropylene	M
Oleyl alcohol	I	Triethylene glycol	I
Esters		2-Ethylhexanediol-1,3	M
Ethyl acetate	M		
Propyl propionate	M		
Phenyl acetate	I	Ketones	
Phenyl benzoate	SS	Acetone	M
Dibutyl phthalate	M	Methyl ethyl ketone	M
Tricresyl phosphate	M	Methyl isobutyl ketone	M
Esters		Diacetone alcohol	M
Diethyl ether	M	Isophorone	M
1,4-Dioxane	M	Acetophenone	M
Anisole	M	Benzophenone	M
Diphenylether	M	Stearone	I

Solubility code

M = Miscible (50 g solute/100 g GENESOLV D)

SS = Slightly soluble (1-5 g solute/100 g GENESOLV D)

I = Insoluble (1 g solute/100 g GENESOLV D)

Table 3.10: Trichlorotrifluoroethane/Acetone/Nitromethane (90/9.5/0.5) Azeotrope *(17)*

Genesolv DA

Genesolv DA solvent is a ternary constant boiling mixture. Its solvent characteristics are determined by the major components, Genesolv D (trichlorotrifluoroethane) and acetone. Genesolv DA removes oils, fluxes and tars effectively and economically. Genesolv DA is effective because the acetone increases solvent power and dissolves some inorganic salts. Genesolv DA is economical because it is recoverable and stable in presence of white metals. Genesolv DA solvent retains the compatibility, penetrating power and purity of Genesolv D. The high TLV (slightly less than 1000) and nonflammability of Genesolv DA makes it a preferred solvent in many applications.

PHYSICAL PROPERTIES

Color	Clear, colorless	Specification:	
Odor	Slight ethereal odor	Assay	Wt. %
Boiling Point @ 760 mm Hg	44.4°C (112°F)	GENESOLV D	90.3 ± 0.5
		Acetone	9.4 ± 0.5
Density of Liquid @ 25°C (77°F), g/cu cm	1.42	Nitromethane	0.3 ± 0.1
Weight at 77°F, lb/gal	11.7	maximum limit of impurities:	
Viscosity @ 25°C (77.7°F), centipoises	0.55	Color, APHA, max.	10
		Combined residue, weight	0.001%
Surface Tension @ 25°C (77.7°F), dynes/cm	18.8	Water, Weight	0.05%
		Chloride, weight	0.00001%
Solvent Power (Kauri-Butanol)	46		
Flash Point	None		
Evaporation rate (Ether 1)	1.3		

Table 3.11: Weight vs Temperature—Genesolv DA Solvent *(17)*

Temp. °F	lb/gal
40	11.98
45	11.95
50	11.92
55	11.89
60	11.86
65	11.83
70	11.80
75	11.77
80	11.74
85	11.71
90	11.69

Table 3.12: Trichlorotrifluoroethane/Aliphatic Alcohols/Nitromethane (92/7/1) Azeotrope *(17)*

Genesolv DMS

Genesolv DMS solvent is an essentially constant boiling mixture comprised of aliphatic alcohols, nitromethane and Genesolv D solvent. Genesolv DMS solvent is effective, economical, and stable in presence of white metals. Genesolv DMS solvent is used to remove rosin flux residues and fingerprints from P/C board assemblies. Genesolv DMS solvent can be used in a vapor degreaser or automated equipment. Like all Genesolv D solvent constant boiling mixtures Genesolv DMS solvent is not flammable and can be recovered by distillation.

PHYSICAL PROPERTIES

GENESOLV DMS solvent is an essentially constant boiling mixture containing GENESOLV D solvent, methanol, ethanol (SDA-30), isopropanol and nitromethane. Some extra alcohol is added to the constant boiling formulation to enhance flux removal and offset the normal extraction of alcohol experienced in high humidity areas, or when removing soils. GENESOLV DMS solvent has the following properties:

Color	Clear, colorless
Odor	Slight ethereal odor
Boiling Point @ 760 mm Hg	42°C (108°F)
Latent Heat of Vaporization, btu/lb	88
Density of Liquid @ 20°C (68°F), g/cu cm	1.462
Weight at 70°F, lb/gal	12.2
Vapor Pressure @ 20°C (68°F) PSIA	8.1
Surface Tension @ 23.9°C (75°F), dynes/cm	19
Solvent Power (Kauri-Butanol)	48
Flash Point (Open Cup)	None

Specification:

Assay	Wt. %
GENESOLV D solvent (FC-113)	92.0 ± 1.0
methyl alcohol	4.0 ± 0.5
ethyl alcohol	2.0 ± 0.5
isopropyl alcohol	1.0 ± 0.2
nitromethane	1.0 ± 0.2

maximum limits of impurities, %

water, by weight	0.050
combined residue, by weight	0.0010
chloride, by weight	0.00001

Table 3.13: Trichlorotrifluoroethane/Ethanol (65/35 Blend) *(17)*

Freon T-E 35
Genesolv DE-35

Genesolv DE-35 solvent is a mixture of ethanol (SDA-30 ethyl alcohol) and trichlorotrifluoroethane which conforms to Military Specification Mil-P-46843 (MI). Genesolv DE-35 solvent dissolves solder flux residues, finger prints and water. The outstanding features of Genesolv DE-35 solvent are compatibility with most substrates, ease of handling, and nonflammability at room temperature. In addition, it can be used in extraction processes and in bench cleaning.

PHYSICAL PROPERTIES

Color	Clear, Colorless
Boiling Range	See Table 3.14
Density of Liquid at 21.1°C (70°F), g/cu cm	1.159
Weight at 21.1°C (70°F), lb/gal	9.8
Viscosity at 21.1°C (70°C), centipoises	1.8
Surface Tension at 21.1°C (70°F), dynes/cu cm	19.6
Solubility of Water at 70°F, wt %	6.2

Specification:

Assay	Wt %
GENESOLV D solvent	65.0 ± 1.0
Ethyl alcohol, SDA-30	35.0 ± 1.0

maximum limit of impurities

Color, APHA, max	10
Water, H_2O, weight %	0.050
Combined residue, weight %	0.0010
Chloride, weight %	0.00001

Table 3.14: Boiling Range of Genesolv DE-35
Solvent *(17)*

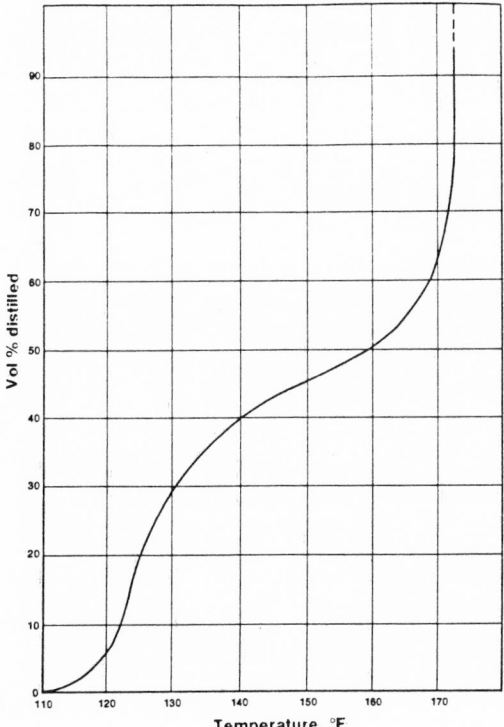

Table 3.15: Density of Genesolv DE-35
vs Temperature *(17)*

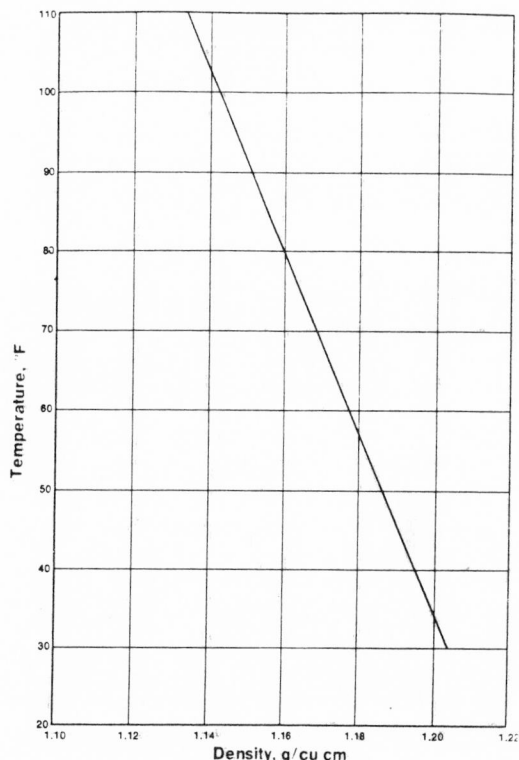

Table 3.16: Density of Various Blends of Genesolv
D/Ethanol at 70°F *(17)*

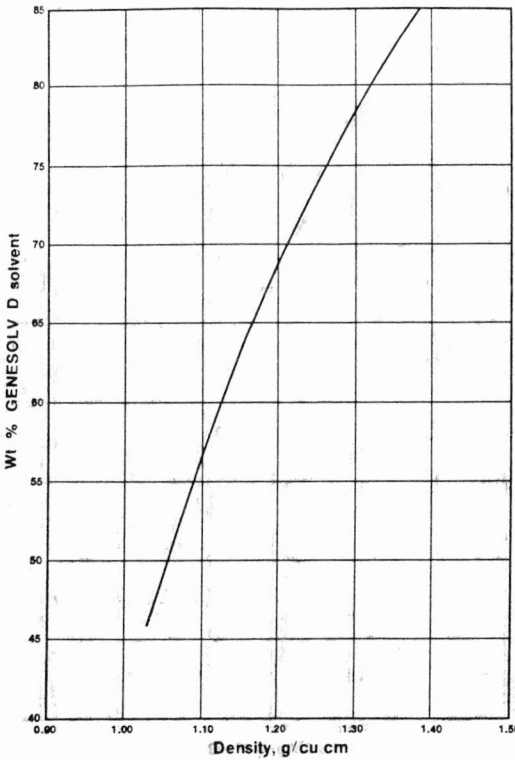

Table 3.17: Evaporation Loss vs Time for Genesolv
DE-35 *(17)*

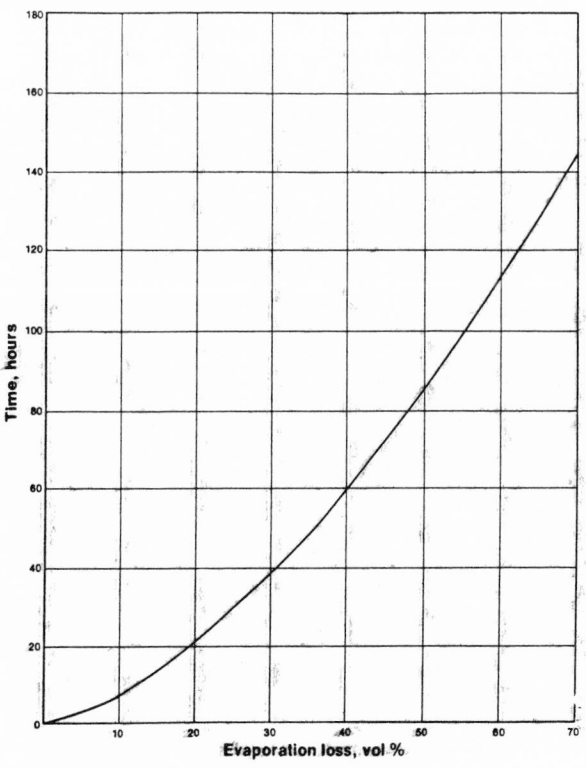

Table 3.18: Solubility of Water in Various Blends of Genesolv D/Ethanol *(17)*

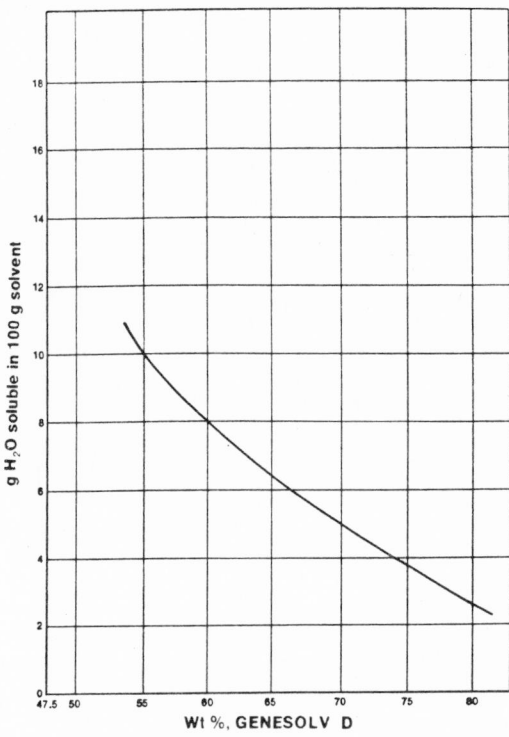

Table 3.19: Trichlorofluoroethane/Ethanol (96/4) Azeotrope *(17)*

Arklone A	Genesolv DE
Flugene 113E	Kaltron 113MDA
Frigen 113TR-E	

Genesolv DE solvent is a constant boiling mixture comprised of ethyl alcohol (SDA 30) and Genesolv D solvent, trichlorotri-fluoroethane. Genesolv DE solvent is effective to dissolve rosin flux residues, is economical, and is compatible with most electronic components. Genesolv DE solvent can be used at ambient temperatures and in vapor degreaser equipments. In addition, Genesolv DE can be recovered by distillation.

PHYSICAL PROPERTIES

Color	Clear, colorless
Odor	Slight ethereal odor
Boiling point @ 760 mm Hg	44.6°C (112.3°F)
Latent heat of vaporization at the boiling point Btu/lb	77.3
Heat capacity at 20°C (68°F) Btu/lb °F	0.27
Density of liquid @ 20°C (68°F) g/cu cm	1.49
Weight at 20°C (68°F), lb/gal	12.6
Viscosity @ 20°C (68°F) centipoises	0.7
Surface tension @ 20°C (68°F), dynes/cm	19.0
Toxicity (TLV) ppm by volume	900
Refractive index MD 20°C	1.358
Flash point	None

Specification:

Assay	Wt %
GENESOLV D	95.5 ± 0.5
Ethanol (SDA 30)	4.5 ± 0.5

maximum limits of impurities	Wt %
Water	0.02
Combined residue	0.001
Chloride	0.00001 (0.1 ppm)

Table 3.20: Weight of Genesolv DE as a Function of Temperature *(17)*

temp, °F	weight lb/gal
40	12.9
50	12.8
60	12.7
70	12.6
80	12.5
90	12.4

Table 3.21: Trichlorotrifluoroethane/Ethanol/Isopropanol/Nitromethane (93.5/3.5/2.0/1.0) Azeotrope *(17)*

Genesolv DES

GENESOLV DES solvent is a constant boiling mixture comprised of aliphatic alcohols, nitromethane and GENESOLV D.

GENESOLV DES is:
• effective,
• economical,
• stable in presence of white metals,
• nonflammable

GENESOLV DES is used to remove rosin flux residues from P/C board assemblies, as well as lubricating oils and water based emulsions from glass, metal and plastic substrates.

GENESOLV DES can be used in:
• vapor degreasers or
• automated equipment

Physical Properties

Color	Clear, colorless
Odor	Slight ethereal odor
Boiling Point @ 760 mm Hg	44.4°C (112°F)
Latent Heat of Vaporization, btu/lb	78
Density of Liquid @ 20°C (68°F), g/cu cm	1.486
Weight at 68°F, lb/gal	12.4
Viscosity @ 21.1°C (70°F), centipoises	0.70
Surface Tension @ 23.9°C (75°F), dynes/cm	19.4
Solvent Power (Kauri-Butanol)	43
Flash Point	None
Refractive Index MD 21.0°C	1.3587

Specification

Assay	wt. %
GENESOLV D	93.5 ± 1.0
Ethanol. SDA-30, anhydrous	3.5 ± 0.5
Isopropanol	2.0 ± 0.5
Nitromethane	1.0 ± 0.2

maximum limit of impurities	
Water, H_2O, weight %	0.05
Combined residue, weight %	0.0010
Chloride, weight %	0.00001

Table 3.22: Weight vs Temperature for Genesolv DES *(17)*

temp. °F (°C)	weight lb/gal
68 (20)	12.4
77 (25)	12.3
95 (35)	12.0

Table 3.23: Trichlorotrifluoroethane/Ethanol/Methyl Acetate (94/4/2) Azeotrope *(43)*

Arklone A-M

Property		Arklone A-M
Boiling Point 760 mm Hg	°C	44.5
Freezing point	°C	<−42
Vapor Pressure at 25°C	mm Hg	365
Liquid Density at 25°C	g/cc	1.453
Liquid Density at 10°C	g/cc	1.487
Surface Tension at 25°C	dyne/cm	21
Viscosity at 25°C	cp	0.72
Dielectric Constant Liquid at 25°C		3.0 est
Specific Heat of Liquid near 25°C	cal/g°C	0.28 est
Heat Capacity of Vapor at 60°C	cal/g°C	0.165 est
Latent Heat of Vaporisation at b.p.	cal/g	44.8
Solubility of water at 25°C	% w/w	0.25
Kauri Butanol Number		36
Inhalation Toxicity (TLV)	ppm	850
Flash Point		None*

*Closed cup method   est = estimated

Table 3.24: Trichlorotrifluoroethane/Isopropanol (65/35 Blend) *(17)*

Genesolv DI-35

Genesolv DI-35 solvent is a mixture of isopropyl alcohol (IPA) and trichlorotrifluoroethane. This mixture is nonflammable up to its boiling point and can be used at room temperature. Genesolv DI-35 solvent will dissolve water, rosin, oils, and some inorganic salts. Genesolv DI-35 solvent can be used effectively for drying surfaces and removing rosin solder flux residues. The outstanding features of this solvent include compatibility with most substrates, ease of handling and nonflammability at room temperature.

PHYSICAL PROPERTIES

Appearance: Genesolv DI-35 solvent is a clear, colorless liquid, having a characteristic alcoholic odor.

Boiling range: The boiling range for Genesolv DI-35 solvent is shown in Table 3.25.

Liquid density: The liquid density of Genesolv DI-35 solvent is 1.169 g/cu cm at 21.1°C (70°F). The approximate weight is 9.67 lb/gal at 21.1°C (70°F). Densities at other temperatures can be read from Table 3.26. Densities of other Genesolv D/IPA blends, at 21.1°C (70°F) are shown in Table 3.27.

Viscosity: The viscosity of Genesolv DI-35 solvent is 1.8 cp at 21.1°C (70°F). This measurement was made using an Ubbelonde viscometer.

Surface tension: The surface tension of Genesolv DI-35 solvent, at 21.1°C (70°F), is 19.6 dynes/cu cm. This measurement was made using a Du Nouy tensiometer.

Specification:

Assay	Wt. %
GENSOLV D solvent	65.0
Isopropyl alcohol	35.0
maximum limit of impurities	
Color, APHA, max	10
Acidity, mg KOH/g	0.001
Water, H_2O, weight %	0.050
Combined residue, weight %	0.0010
Chloride, weight %	0.00001

Water solubility: The water solubility of the Genesolv DI-35 blend is 6.2 g of water per 100 g of Genesolv DI-35. The number of grams of water soluble in 100 g of various blends of Genesolv D/isopropyl alcohol is shown in Table 3.29.

Table 3.25: Boiling Range of Genesolv DI-35 Solvent *(17)*

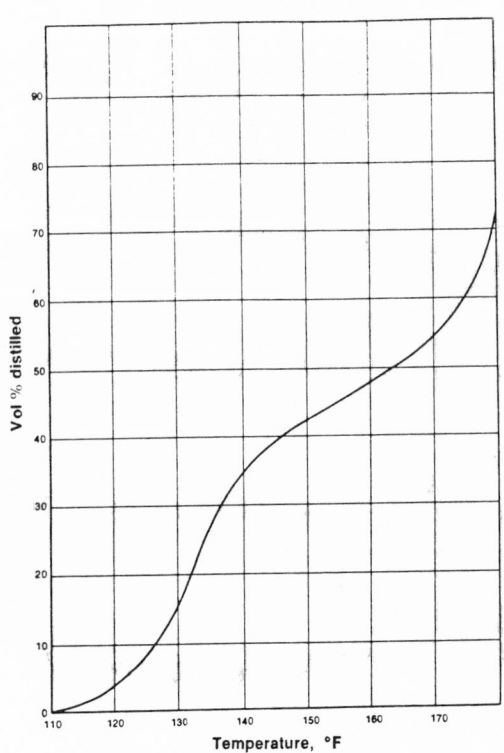

Temperature, °F

Table 3.26: Density of Genesolv DI-35 vs Temperature *(17)*

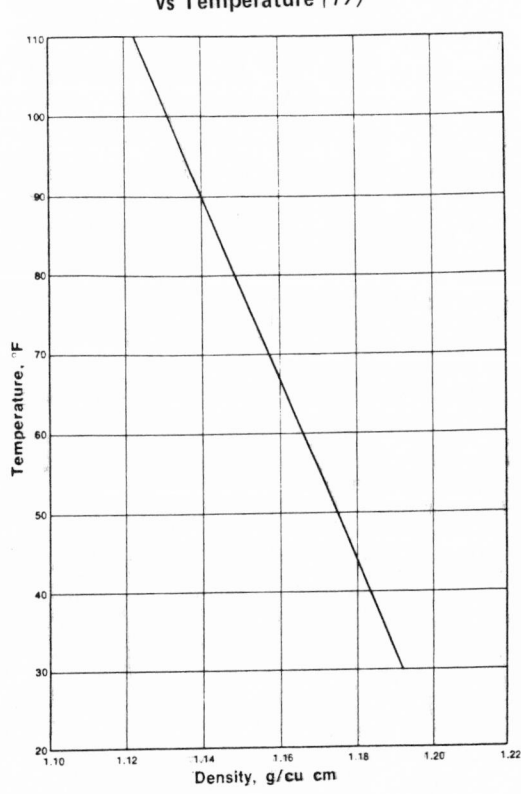

Density, g/cu cm

Table 3.27: Density of Various Blends of Genesolv D/IPA at 70°F *(17)*

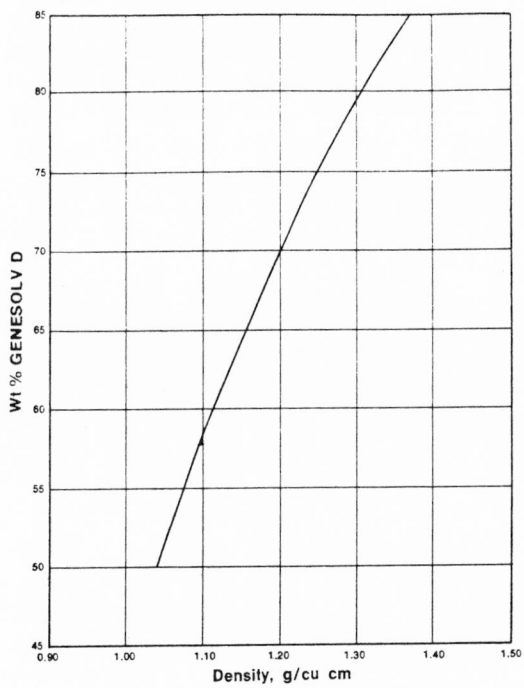

Table 3.28: Evaporation Loss vs Time for Genesolv DI-35 *(17)*

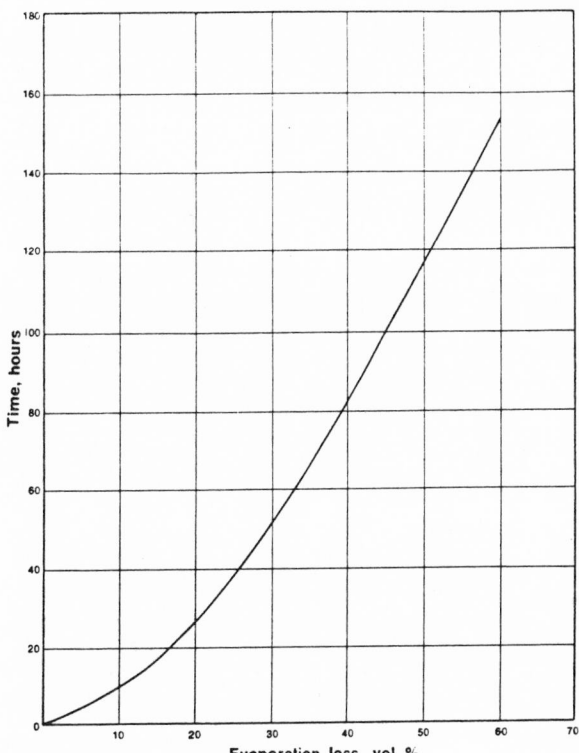

Table 3.29: Solubility of Water in Various Blends of Genesolv D/IPA *(17)*

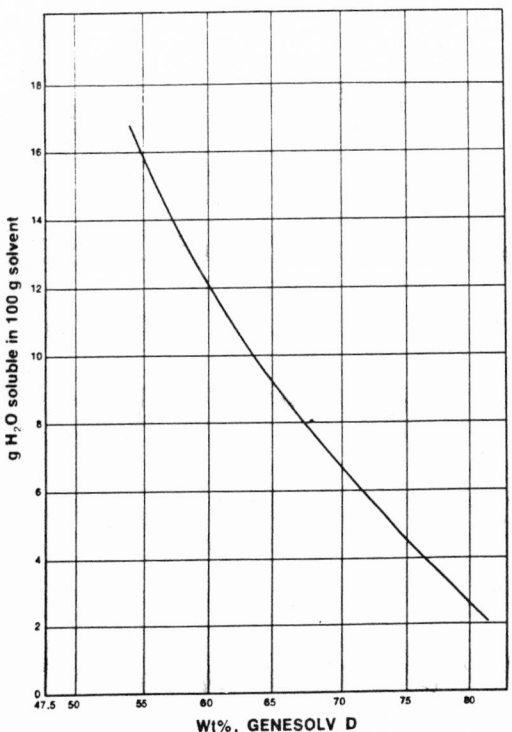

Table 3.30: Trichlorotrifluoroethane/Isopropanol (85/15 Blend) *(17)*

Genesolv DI-15

PHYSICAL PROPERTIES

Genesolv DI-15 solvent is a blend of 85 wt % Genesolv D solvent and 15 wt % isopropanol, having the following properties:

Color	Clear, colorless
Odor	Slight alcoholic odor
Boiling Range	See Table 3.31
Density of Liquid at 20°C (68°F), g/cu cm	1.36
Weight, at 20°C (68°F), lb/gal	11.37
Viscosity, at 21.1°C (70°F), centipoises	0.75
Surface Tension, at 21.1°C (70°F), dynes/cm	19.5
Solubility of Water, at 25°C (77°F), wt %	1.8
Vapor-Liquid Equilibrium	See Table 3.34
Flash Point	None*

*Genesolv DI-15 alcohol blend exhibits no Tag Open Cup flash point to boiling. Since it is a mixture, fractionation will occur upon evaporation, resulting in a solution enriched in alcohol. In time, therefore, the liquid will flash. About 80% of the Genesolv DI-15 solvent must evaporate before the residual mixture becomes flammable. The boiling point of these flammable mixtures is 147°F or above.

Specification:

Assay	Wt %
GENESOLV D solvent	85.0 ± 1.0
Isopropyl alcohol	15.0 ± 1.0

maximum limit of impurities

Color, APHA, max	10
Water, by weight %	0.050
Combined reisdue, by weight %	0.0010
Chloride, by weight %	0.00001

GENESOLV DI-15 conforms to Military Specification MIL-P-4684313 "Printed Wiring Assemblies," dated February 26, 1976.

Table 3.31: Boiling Range of Genesolv DI-15 Solvent *(17)*

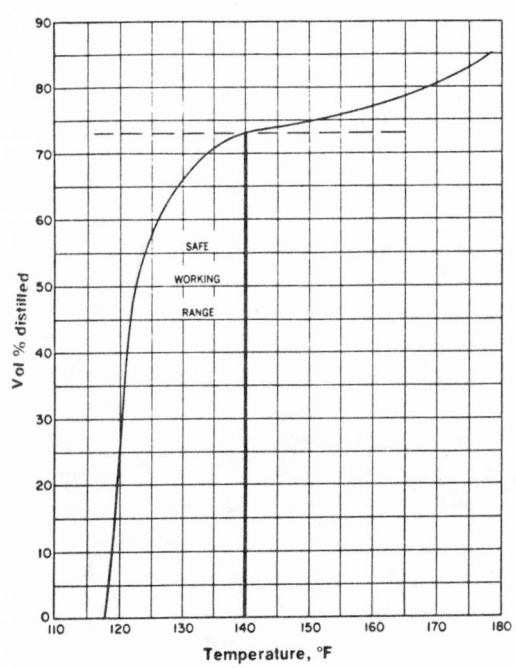

Table 3.32: Density of Genesolv DI-15 as a Function of Temperature *(17)*

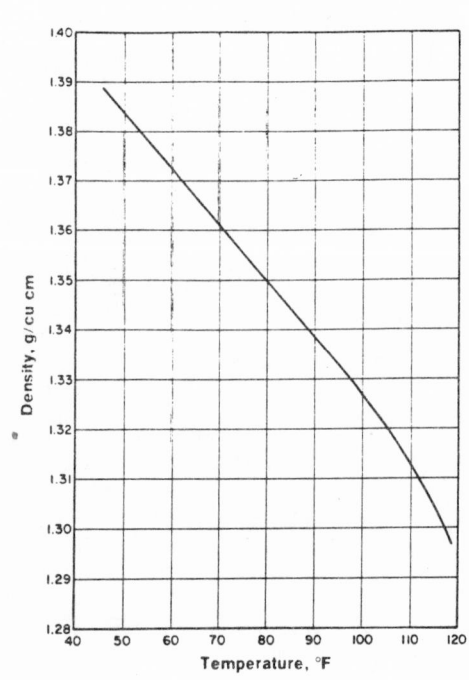

Table 3.33: Density of Genesolv D/Isopropanol Blends at Varying Concentrations of Genesolv D at 70°F *(17)*

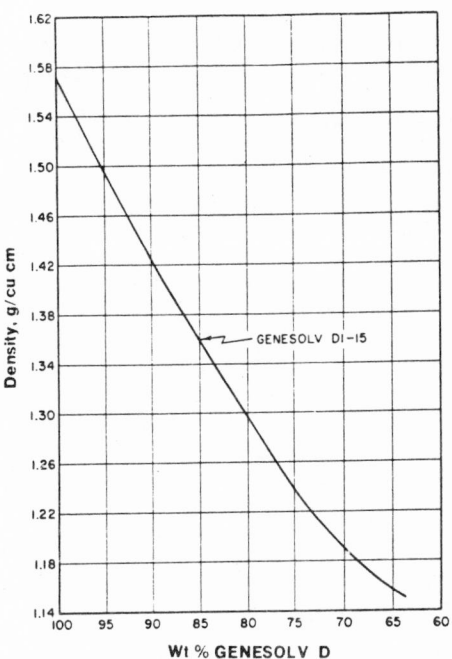

Table 3.34: Temperature-Composition Diagram of the Vapor-Liquid Equilibrium for the System Genesolv D/ Isopropanol at 750 mm Hg *(17)*

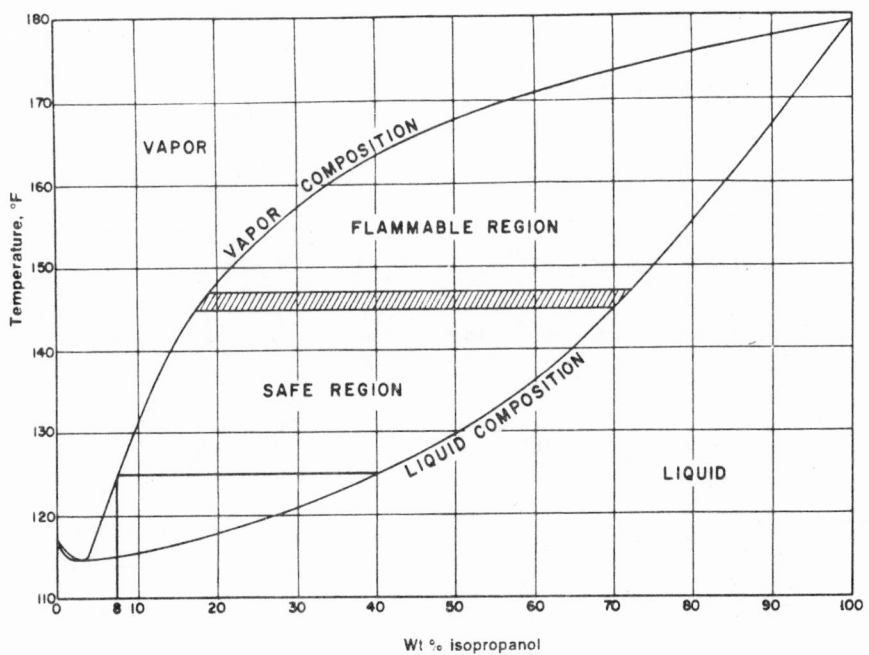

Table 3.35: Trichlorotrifluoroethane/Isopropanol (97/3) Azeotrope *(17)*

GENESOLV DI

An azeotrope of GENESOLV D fluorinated solvent and isopropanol, GENESOLV DI, is an excellent solvent for solder fluxes containing little or no activator. GENE- SOLV DI is completely safe for use in a vapor degreaser. Since it is an azeotrope, or constant boiling mixture, there is no fractionation on boiling and, therefore, no preferential concentration of one component in the vapor. Also, the solvent can be reclaimed by simple distillation.

The addition of isopropanol enhances the solvent action of GENESOLV D while retaining the outstanding features of GENESOLV D, such as compatibility with most substrates, low surface tension and high density. The latter two properties enable GENESOLV DI to penetrate into crevices and recesses to provide thorough cleaning. The advantages of GENESOLV DI are summarized below:

- Safe for use in a vapor degreaser
- Simple cleaning techniques
- Can be used with ultrasonics
- Lower cleaning costs — can be reclaimed
- Excellent compatibility
- Low surface tension
- High density
- Excellent penetration
- High purity
- Low toxicity
- Nonflammable

PHYSICAL AND CHEMICAL PROPERTIES

GENESOLV DI, an azeotrope containing GENESOLV D fluorinated solvent and isopropanol, has the following properties:

Appearance

GENESOLV DI is a clear, colorless liquid.

Odor

This solvent has a slight ethereal odor.

Boiling Point

The boiling point of GENESOLV DI azeotropic solvent, at 760 mm Hg, is 46.5°C (115.7°F).

Liquid Density

The liquid density, at 70°F, is 1.53 g/cu cm. Approximate weight is 12.77 lb/gal at 70°F.

Viscosity

Viscosity, at 70°F, is about 0.71 centipoises.

Surface Tension

Surface tension, at 70°F, is 19.2 dynes/cm. Measurement was made using a Du Noüy tensiometer.

Evaporation Rate

The evaporation rate, relative to acetone = 10, is 15. This measurement was made in accordance with ASTM D 1901-61T.

Solvent Power

Kauri-butanol solvent power value is 41.

Flash Point

This solvent exhibits no flash point.

Table 3.36: Trichlorotrifluoroethane/Methylene Chloride (51/49) Azeotrope *(17)*

Genesolv DM

Genesolv DM solvent is an azeotropic mixture comprised of Genesolv D (trichlorotrifluoroethane) and methylene chloride. It is an effective, reclaimable solvent for vapor degreasing which has low energy requirements. It is also a nonphotochemically reactive solvent. In addition Genesolv DM solvent has low boiling point, great solvency power, low diffusion rate vapors and good penetrating characteristics. Its unique properties are finding great acceptance in metal degreasing operations.

Physical Properties

Boiling Point at one atm.,	
°F	97.2
°C	36.5
Liquid Density at 68°F (20°C)	
g/cu cm	1.42
lbs/gal	11.90
lbs/cu ft	88.7
Vapor pressure at 77°F, (25°C)	
psia	9.7
Evaporation rate (ETHER=1)	0.8
Kauri-butanol value	85
Heat capacity at 70°F	
Btu (lb) (°F)	0.26
Latent heat of vaporation at the bp, btu/lb	104
Solubility of water at 70°F, wt %	0.1
Solubility in water at 70°F, wt %	0.2
Surface tension at 75°F (24°C) dynes/cm	21.4
Appearance	colorless
Odor	ethereal
Flash point	none to boiling
TLV-TWA ppm	270*

*Calculated value using American Conference of Governmental Industrial Hygienist, 1982 publication.

Table 3.37: Trichlorotrifluoroethane/Methylene Chloride/ Cyclopentane (39/51/10) Azeotrope *(17)*

Genesolv DMC

Genesolv DMC solvent is a constant boiling mixture (an azeotrope) comprised of Genesolv D solvent, methylene chloride and hydrocarbon. This constant boiling mixture is nonflammable, nonphotochemically reactive, recoverable and stable. Genesolv DMC solvent has a low boiling point, great solvency power, low diffusion rate of vapors and good penetrating characteristics. This solvent is most useful for metal degreasing and removal of heavy soils.

Physical Properties

Boiling Point at one atm., °F	97.1
°C	36.2
Liquid Density at 77°F (25°C)	
g/cu cm	1.29
lbs/gal	10.77
lbs/cu ft	80.50
Vapor pressure at 75°F, psia	9.6
Evaporation rate (ETHER=1)	0.8
Kauri-butanol value	98
Heat capacity at 70°F, btu/lb m • °F	0.26
Latent heat of vaporation at bp, btu/lb	108
Solubility of water at 70°F, wt %	0.1
Solubility in water at 70°F, wt %	0.2
Surface tension at 70°F, dynes/cm	21.5
Appearance	colorless
Odor	ethereal
Flash point	none to boiling
TLV-TWA ppm	280*

*Calculated value using American Conference of Governmental Industrial Hygienist, 1982 publication.

Table 3.38: Trichlorotrifluoroethane/Methylene Chloride/Methanol (55/42/3) Azeotrope *(17)*

Genesolv DTA

Genesolv DTA solvent, a ternary azeotrope, is a strong solvent designed especially for removing solder paste from metallic substrates, including alumina. It is also an excellent solvent for removing printing inks. Genesolv DTA solvent is nonflammable and can be used in a vapor degreaser as well as for cold cleaning. It is an ideal replacement for methyl ethyl ketone (MEK) which is flammable.

PHYSICAL PROPERTIES

GENESOLV DTA solvent is the minimum ternary azeotrope of GENESOLV D, methylene chloride and methyl alcohol, having the following properties:

Color	Clear-colorless
Odor	Slight etheral
Boiling point	34.2°C (93.5°F)
Density @ 20°C (68°F)	1.404 g/cu cm
Weight	11.74 lb./gal.
Viscosity	See figure 2
Surface tension @ 70°F	22.1 dynes/cm
Kauri-butanol solvent power	148

Specification:

Assay, GENESOLV D solvent	55.0 ± 2.0%
Methylene chloride	41.7 ± 2.0%
Methanol	3.3 + 0.7%
	− 1.0%

maximum limit of impurities

Color, APHA, max	10
Acidity, mg KOH/g	0.0005
Combined residue, %	0.005
Water, H_2O, %	0.01

Table 3.40: Viscosity of Genesolv DTA Solvent as a Function of Temperature (17)

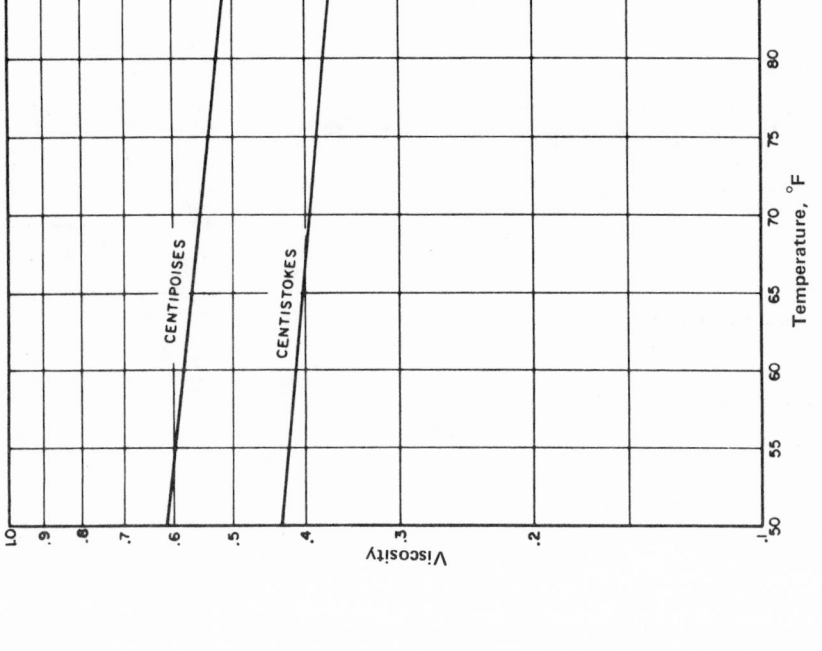

Measure was made using an Ubbelonde viscometer

Table 3.39: Density of Genesolv DTA Solvent as a Function of Temperature (17)

Table 3.41: Trichlorotrifluoroethane/Water/Surfactant Emulsion *(17)*

Genesolv DW

Genesolv DW solvent is a water-in-oil emulsion in which water is the dispersed phase and Genesolv D solvent the dispersing medium. Genesolv DW emulsion is effective in removing organic films and inorganic salts. Two of the most important applications for Genesolv DW emulsion are in the removal of fingerprints and particulate contaminants, e.g., baked-on-soil. Genesolv DW emulsion is compatible with most substrates such as plastic and metal. Some of the outstanding features of Genesolv DW emulsion are its good penetrating power, inherent regenerating ability, i.e., easily reformulated, and its ability to wet substrates covered with water-based contaminants.

PHYSICAL PROPERTIES

Appearance	Slightly hazy, pale yellow solution
Odor	Slight ethereal odor
Boiling point, at 760 mm Hg	44.4°C (112.0°F)
Liquid density at 25°C (77°F), g/cu cm	1.5
Weight at 25°C (77°F), lb/gal	12.3
Vapor pressure at 25°C (77°F), psia	7.0
Viscosity at 25°C (77°F), cp	0.94
Surface tension at 25°C (77°F), dynes/cm	17.3
Solvent power, Kauri-butanol	21
Specification:	

Assay	wt %
surfactant	2.8 ± 0.5%
water	3.5 — 4.0
GENESOLV D solvent	balance

Table 3.42: Trichloroethane/Water/Surfactant (92/6/2) Azeotrope *(43)*

Arklone W
Flugene 113AE
Frigen 113TR-N

Property		Arklone W
Boiling Point 760 mm Hg	°C	44.5
Freezing point	°C	0
Vapor Pressure at 25°C	mm Hg	362
Liquid Density at 25°C	g/cc	1.498
Liquid Density at 10°C	g/cc	1.528
Surface Tension at 25°C	dyne/cm	18
Viscosity at 25°C	cp	1 est
Dielectric Constant Liquid at 25°C		4.0
Specific Heat of Liquid near 25°C	cal/g°C	
Heat Capacity of Vapor at 60°C	cal/g°C	
Latent Heat of Vaporisation at b.p.	cal/g	40 est
Solubility of water at 25°C	% w/w	Approx. 10% total
Kauri Butanol Number		20
Inhalation Toxicity (TLV)	ppm	1000
Flash Point		None*

*Closed cup method   est = estimated

COMPARATIVE DATA

Table 3.43: "Arklone" Physical Properties *(43)*

Property		P/P-SM	A/A-S	A-M	E	W
				Grade of 'Arklone'		
Boiling Point 760 mm Hg	°C	47.6	44.5	44.5	36.5	44.5
Freezing point	°C	−35	−42	<−42	−88	0
Vapor Pressure at 25°C	mm Hg	335		365	502	362
Liquid Density at 25°C	g/cc	1.565	1.505	1.453	1.420	1.498
Liquid Density at 10°C	g/cc	1.600	1.530	1.487	1.449	1.528
Surface Tension at 25°C	dyne/cm	18	19	21	21	18
Viscosity at 25°C	cp	0.68		0.72	0.46	1 est
Dielectric Constant Liquid at 25°C		2.4	3.0 est	3.0 est	5.4	4.0
Specific Heat of Liquid near 25°C	cal/g°C	0.22	0.27 est	0.28 est	0.26	
Heat Capacity of Vapor at 60°C	cal/g°C	0.161	0.165 est	0.165 est	0.152 (20°C)	
Latent Heat of Vaporisation at b.p.	cal/g	35.1	42.9	44.8	56 est	40 est
Solubility of water at 25°C	% w/w	0.01	0.22	0.25	0.07	Approx. 10% total
Kauri Butanol Number		31	40	36	80	20
Inhalation Toxicity (TLV)	ppm	1000	1000 est	850	270 est	1000
Flash Point		None*	None*	None*	None*	None*

*Closed cup method   est = estimated

Table 3.44: "Freon" Fluorocarbon Solvents *(11)*

Properties	Pure Compound "Freon" TF[a] (100% Trichlorotrifluoroethane)	Azeotropes with "Freon" TF				Blends with "Freon" TF			Specialty Mixtures with "Freon" TF	
		"Freon" TA[b] (11% Acetone)	"Freon" TE (4% Ethanol)	"Freon" TES[b] (4% Ethanol Stabilized)	"Freon" TMC[b] (50% Methylene Chloride)	"Freon" TMS[b,c] (6% Methanol Stabilized)	"Freon" T-E 35 (35% Ethanol)	"Freon" T-P 35[b] (35% Isopropanol)	"Freon" T-WD 602[b] (6% water, 2% detergent)	"Freon" T-DF Drying Fluid
Boiling Point, °C	47.6	43.6	44.6	44.4	36.2	39.7	48.3[d]	48.9[d]	44.4[d]	47.6
°F	117.6	110.5	112.3	111.9	97.2	103.5	119.0[d]	120.0[d]	112.0[d]	117.6
D.C. Resistivity, ohm-cm	>2 x 10^15	10^8	1.8 x 10^10	−	10^9	−	10^6	2 x 10^7	−	−
Density, Liquid, 77°F lb/gal	13.06	11.73	12.56	12.48	11.85	12.33	9.75	9.60	12.47	13.0-13.5
lb/cu ft	97.7	87.78	93.96	93.40	88.65	92.21	72.92	71.79	93.27	−
g/cc	1.5649	1.406	1.505	1.496	1.420	1.477	1.168	1.150	1.494	1.55-1.50
Dielectric Constant Liquid, 75°F, at 100 Hz	2.41	5.9 (10 kHz)	3.0	−	5.2 (10 kHz)	−	18 (300 kHz)	9.2 (100 kHz)	−	−
Dissipation Factor, % at 100 Hz	<0.006	42 (10 kHz)	33.0	−	13 (10 kHz)	−	76 (300 kHz)	16.2 (100 kHz)	−	−
Freezing Point, °C	−35	−58	−42	−42	−88	−55	−78	−70	0	−
°F	−31	−72	−43	−44	−126	−66	−108	−94	32	−
Heat Capacity, Btu/(lb) (°F), 68°F Liquid	0.213	0.305	0.272	0.229	0.261	0.236	0.472	0.355	−	−
Vapor	0.152	0.210	0.189	−	0.152	−	−	−	−	−
Kauri-Butanol Number	31	51	40[e]	37	86	45	>500	>500	21	−

(continued)

Table 3.44: (continued)

Properties	Pure Compound	Azeotropes with "Freon" TF				Blends with "Freon" TF			Specialty Mixtures with "Freon" TF	
	"Freon" TF[a] (100% Trichloro-trifluoroethane)	"Freon" TA[b] (11% Acetone)	"Freon" TE (4% Ethanol)	"Freon" TES[b] (4% Ethanol Stabilized)	"Freon" TMC[b] (50% Methylene Chloride)	"Freon" TMS[b,c] (6% Methanol Stabilized)	"Freon" T-E 35 (35% Ethanol)	"Freon" T-P 35[b] (35% Isopropanol)	"Freon" T-WD 602[b] (6% water, 2% detergent)	"Freon" T-DF Drying Fluid
Latent Heat of Vaporization at B.P., Btu/lb Cal/g	63.12 35.07	85.4[e] 47.4[e]	77.2[e] 42.9[e]	76.7 42.6	104.0[e] 57.7[e]	90.7 50.4	149.0[e] 82.5[e]	119.7[e] 66.5[e]	– –	– –
Refractive Index, n_d^{20}	1.354 (77°F)	1.3583	1.3584	1.359	1.390	–	1.359	1.368	–	–
Solubility of Water, 75°F, Wt %	0.011	0.15	0.22	0.28	0.09	0.27	6.3	9.1	–	–
Surface Tension, dynes/cm, 77°F	17.3	18.7	17.7	17.2	21.4	17.4	19.6	21.0	17.3	–
Toxicity (TWA)[f] ppm by vol	1000	750[g]	925[g]	750[g]	270[g] (140[h])	475[g]	750[g]	700[g]	1000[i]	1000[i]
Vapor Pressure, Psia, 77°F 130°F	6.46 18.5	7.35 22.0	7.0 20.7	7.0 –	9.7 27.0	8.3 25.0	6.6 20.0	5.5 17.5	7.0 –	– –
Viscosity, Liquid, 77°F, centipoise	0.682	0.542	0.652	–	0.461	–	1.762	0.997	0.94	–

[a]This product is packaged for white room applications under the name "Freon" Precision Cleaning Agent.
[b]Patented Composition.
[c]"Freon" TMS is a constant boiling blend that acts like an azeotrope in a vapor degreaser.
[d]Initial.
[e]Estimated.

[f]OSHA 8 hour Time Weighted Average.
[g]Calculated from TWA Values for the individual components.
[h]Value of 1979 Intended Change for Threshold Limit Value.
[i]Based on the assumption that vapors from the mixture are only "Freon" TF.

Table 3.45: Solubility of Various Materials in "Freon" Fluorocarbon Solvents (11)

Wt % at Room Temperature

Material	"Freon" TF	"Freon" TA	"Freon" TE and TES	"Freon" TMC	"Freon" TMS	"Freon" T-E 35	"Freon" T-P 35	"Freon" T-WD 602*
Acetic Acid	M	M	–	–	–	–	–	–
Acetone	M	M	M	M	M	M	M	1.0
Benzene	M	–	–	–	–	–	–	1.2
Chloroform	M	M	M	M	M	M	M	2.0
Cottonseed Oil	M	M	M	M	M	M	M	9.7
Ethyl Acetate	M	M	M	M	M	M	M	9.8
Ethyl Alcohol	M	M	M	M	M	M	M	0.38
Ethylene Glycol	0-1.4	1.6-3.0	–	1.5-3.0	–	5-8	31	1.5
Glycerin	<0.1	–	–	–	–	6-8	7.1	1.7
n-Hexane	M	–	M	–	M	–	M	M
Isopropanol	M	–	–	–	–	M	M	0.3
Methyl Alcohol	M	M	M	M	M	M	M	1.1
Mineral Oil	M	M	M	M	M	13-14	M	23.2
Paraffin Wax, M.P. 123-127°F	6.6	3.4	–	13	–	2.2	2.8	5.0
Phenol	1.8	–	–	–	–	–	–	<0.05
Propylene Glycol	0-1.3	1.4-29	–	1.4-2.8	–	47-48	M	2.0
Silicone Oil (1000 centistokes)	M	M	M	M	M	19-24	M	1.3
Acrylic Ester Resin	30	40-45	–	50	–	–	–	–
Ester Gum 8L	10	14	–	36	–	40	63	–
Polystyrene Resin	<5	–	–	25-30	–	–	–	–

M = Miscible   *Amount which can be added before a layer separation occurs.

Table 3.46: Effect of "Freon" Fluorocarbon Solvents on Various Elastomers After Five Minutes Immersion at Room Temperature *(11)*

	Linear Swell, %							
Elastomer	"Freon" TF	"Freon" TA	"Freon" TE and TES	"Freon" TMC	"Freon" TMS	"Freon" T-E 35	"Freon" T-P 35	"Freon" T-WD 602
"Adiprene" C Urethane Rubber	0	2	*	12	†	3	0	0.6
"Adiprene" L Urethane Rubber	0	1	*	3	†	1	0	0.2
Buna N	0	1	*	9	†	1	0	−0.4
Buna S	1	2	*	11	†	1	0	0.9
Butyl	0	0	*	2	†	0	0	1.3
"Hypalon" 40 Synthetic Rubber	0	1	*	9	†	0	0	0.3
Natural Rubber	2	3	*	9	†	1	0	3.7
Neoprene W	0	1	*	6	†	1	0	−0.1
"Nordel" Hydrocarbon Rubber	1	2	*	9	†	1	0	1.5
Silicone	6	4	*	11	†	5	2	20.0
"Thiokol" FA Polysulfide	0	0	*	3	†	0	0	−0.7
"Viton" A Fluoroelastomer	0	16	*	1	†	2	0	0.4

*Assume linear swell to be equal to or less than that for "Freon" T-E 35 or "Freon" TF, whichever is greater.
† Normally used in defluxing operations at boiling temperatures. See Table 3.47.

Table 3.47: Effect of "Freon" Fluorocarbon Solvents on Various Elastomers After Five Minutes Immersion at Elevated Temperatures *(11)*

	Linear Swell, %							
Elastomer	"Freon" TF[2]	"Freon" TA[1]	"Freon" TE and TES[1]	"Freon" TMC[1]	"Freon" TMS[1]	"Freon" T-E 35	"Freon" T-P 35	"Freon" T-WD 602[1]
"Adiprene" C	–	3	2	16	2	–	–	2.2
"Adiprene" L	–	2	1	6	0	–	–	1.1
Buna N	1	2	1	14	1	–	–	0.5
Buna S	9	3	1	15	1	–	–	5.5
Butyl	23	2	1	7	1	–	–	2.3
"Hypalon" 40	5	3	0	10	0	–	–	2.0
Natural Rubber	19	4	2	14	1	–	–	13.0
Neoprene W	3	2	0	9	0	–	–	−0.2
"Nordel"	–	5	2	13	2	–	–	6.1
Silicone	36	11	5	15	4	–	–	15.0
"Thiokol" FA	1	0	0	4	0	–	–	1.2
"Viton" A	6	18	0	2	0	–	–	0.8

[1]At boiling point   [2]At 130°F, maximum swell
"Adiprene", "Hypalon", "Nordel" and "Viton" are Du Pont's registered trademarks.

Table 3.48: Effect of "Freon" Fluorocarbon Solvents on Unstressed Plastics for Five Minutes at Solvent Boiling Point (11)

Plastic	"Freon" TF	"Freon" TA	"Freon" TE and TES	"Freon" TMC	"Freon" TMS	"Freon" T-E 35*	"Freon" T-P 35*	"Freon" T-WD 602*
"Alathon"† 7050 Linear Polyethylene Resin	0	0	1	0	0	0	0**	0
"Delrin"† Acetal Resin	0	0	0	0	0	0	0	0
Epoxy Resin	0	0	0	0	0	0	0	0
Ethyl Cellulose	3	4	3	4	3	4	3	0
"Kralastic" ABS Polymer	0	1	0	3	0	0	0	0
"Lexan" Polycarbonate Resin	0	0	0	3	0	0	0	0
"Lucite"† Methylmethacrylate Resin (cast)	0	0	0	3	0	0	0	0
Polyphenylene Oxide (PPO)	0	0	0	4	0	–	0**	–
Polypropylene Resin	0	0	1	0	0	0	0	0
Polysulfone	0	0	0	3	0	–	0**	–
Polyvinyl Alcohol	0	0	2	0	2	1	0	0
Polyvinyl Chloride (unplast)	0	0	1	2	1	0	0	0
"Styron" 475 Polystyrene	0	2	3	3	3	0	0	0
"Surlyn"† A Ionomeric Resin	0	0	1	1	1	0	0	0
"Teflon"† TFE Resin	0	0	0	0	0	0	0	0
"Tenite" Polyterephthalate	0	2	0	4	0	0	0*	–
"Zytel"† 101 Nylon Resin	0	0	0	0	0	0	0	0

EFFECT KEY: 0 = No visible effect
1 = Very slight effect
2 = Compatibility should be tested
3 = Probably not suitable
4 = Disintegrated or dissolved

*At 75°F.
**At 100°F.
† Du Pont's registered trademarks.

Table 3.49: Effect of "Freon" Fluorocarbon Solvents on Various Annealed Magnet Wire Coatings (11)

(Immersed in Solvent for 5 Minutes at Boiling Point)

Coatings	"Freon" TF	"Freon" TA	"Freon" TE and TES	"Freon" TMC	"Freon" TMS	"Freon" T-E 35*	"Freon" T-P 35*	"Freon" T-WD 602*
"Acrylex" Acrylic	0	0	0	0	0	0	0	0
"Alkanex" Terephthalate Polyester	0	0	0	4	0	0	1	0
"Anavar" Isocyanate Modified Polyvinyl Formal	0	0	0	4	0	0	2	0
Enamel "G" Polyurethane	0	0	0	0	0	0	0	0
"Ensolex"/25X Solderable Acrylic	0	0	1	0	0	0	0	0
Epoxy	0	0	0	1	0	0	0	0
"Formvar" Polyvinyl Formal	0	0	0	3	0	0	0	0
"Nylaclad" Nylon-coated Polyvinyl Formal	0	0	0	0	0	0	2	0
Oleoresinous Enamel	0	0	0	4	3	0	1	0
"Pyre-ML"** Polyimide	0	0	0	0	0	0	1	0

KEY 0 = No visible effect
1 = Very slight effect
2 = Compatibility should be tested

3 = Moderate crazing or softening
4 = Severe effects

*Tested at 75°F instead of the boiling point since these solvents are not used at the boil.
**"Pyre-ML" is a Du Pont registered trademark.

Table 3.50: "Genesolv" Physical Properties *(17)*

GENESOLV solvent	Boiling point °C	°F	Liquid density, at 70°F, lb/gal	Flash point °C	TLV-TWA ppm
D	47.6	117.6	13.13	none	1000
A	23.72	74.9	12.39	none	1000
DE-15*	46.1-82.2	115-180	11.35	*	800#
DI-15*	46.1-82.2	115-180	11.36	*	650#
DE-35*	44.4-79.4	112-175	9.80	*	750#
DI-35*	46.1-82.2	115-180	9.67	none	520#
DE	44.6	112.3	12.64	none	900
DES	44.4	112	12.38	none	720#
DI	46.5	115.7	12.77	none	880#
DMS	42	107.6	12.19	none	510#
DM	36.5	97.7	11.92	none	270#
DMC	36.15	97.1	10.80	none	280#
DTA	34.2	93.5	11.72	none	270#
DA	44.4	112.1	11.80	none	940#
DS	46.5	115.7	12.60	none	830#
DS-10*	46.1-82.2	115-180	11.92	*	720#
DL	47.8	117.6	12.90	none	650#
DW	44.4	112.1	12.56	none	1000 Est

*These Genesolv blends are not flammable up to the boiling point.

\# Calculated values for the Threshold Limit Values (TLV's) Time Weighted Averages (TWA), American Conference Of Governmental Industrial Hygienist, 1979. Calculated values using OSHA standards would in some cases be higher.

CHLORINATED HYDROCARBONS

Table 3.51: Allyl Chloride *(7)*

3-Chloropropene-1 $CH_2=CHCH_2Cl$

PHYSICAL PROPERTIES

Boiling point	45°C
Fire point	4°C
Flash point	4°C
Latent heat of vaporization	84.6 cal/g
Specific gravity @25/25°C	0.933
Specific heat	0.31 cal/g/°C
Refractive index @25°C	1.412
Viscosity @25°C	0.33 centipoise
Weight per gallon @25°C	7.8 lb

Table 3.52: n-Amyl Chloride *(7)*

1-Chloropentane $CH_3CH_2CH_2CH_2CH_2Cl$

PHYSICAL PROPERTIES

Acidity as HCl	0.025% max.
Amylene	1% max.
Boiling range	105-109°C
Distillation	95% between 104.9-108.9°C
Flash point	54°F
Other hydrocarbons	None
Polychlorides content	None
Solubility in water	Insoluble
Specific gravity @20/20°C	0.885
Weight per gallon	7.38 lb

Table 3.53: Mixed Amyl Chlorides *(7)*

$C_5H_{11}Cl$

PHYSICAL PROPERTIES

Acidity as HCl	0.03% max.	Distillation range	95% between 85-109°C
Amylene and pentane content	3.0% max.	Evaporation rate @108°F:Minutes	
Boiling point (approx.)		1.30	25%
1-Chloropentane	108.2°C	1.67	50%
2-Chloropentane	96.7°C	4.30	75%
3-Chloropentane	97.3°C	6.58	100%
1-Chloro-2-methylbutane	99.9°C	Flash point (O.C.)	34°F
4-Chloro-2-methylbutane	98.8°C	Kauri-butanol value	71 cc
3-Chloro-2-methylbutane	93.0°C	Solubility in water	Negligible
2-Chloro-2-methylbutane	86.0°C	Specific gravity @20°C	0.88
		Vapor pressure @20°	42.8 mm
		Water azeotrope @77-82°C	90% $C_5H_{11}Cl$ (approx.)
		Weight per gallon	7.33 lb

Table 3.54: Benzyl Chloride (7)

α-Chlorotoluene$\qquad\qquad\qquad$$C_6H_5-CH_2-Cl$

PHYSICAL PROPERTIES

Distillation range	Not more than 2° including 179.4°C
Freezing point	-43°C
Molecular weight	126.58
Refractive index N_D^{25}	1.5365
Specific gravity @15.5°/15.5°C	1.107
Weight per gallon @15.5°C	9.23 lb

Table 3.55: n-Butyl Chloride (7)

PHYSICAL PROPERTIES

Acidity	0.01% max.
Boiling point @760 mm	78°C
Distillation range	Not less than 95% between 76.0-79.5°C
Flash point (O.C.)	20°F
Latent heat of vaporization @76.5°C	79.8 cal/g
Melting point	-123.1°C
Refractive index @20°C	1.4004
Solubility in water	Negligible
Specific gravity @20/4°C	0.884
Specific heat @20°C	0.451 cal/g
Surface tension @20°C	23.66 dynes/cm
Water content	None
Weight per gallon	7.37 lb

n-BUTYL CHLORIDE FORMS AZEOTROPES WITH:

%		B.P. °C of Azeotrope
80	Acetone	55.8
1.9	n-Butyl alcohol	77.7
57	n-Butyl nitrite	76.5
35	Ethyl acetate	76.0
20.	Ethyl alcohol	65.7
4	Isobutyl alcohol	77.7
62	Isobutyl nitrite	66.2
23	Isopropyl alcohol	70.8
29	Methyl alcohol	57.0
38	Methyl propionate	76.8
40	Methyl propyl ketone	77.0
16	Nitromethane	75.0
18	n-Propyl alcohol	74.8
38	n-Propyl formate	76.1
6.6	Water	68.1

Table 3.56: sec-Butyl Chloride (7)

FORMULA	$CH_3-\overset{\overset{\displaystyle Cl}{\displaystyle |}}{CH}-CH_2-CH_3$
PROPERTIES	**98.0% GRADE**
Composition, weight percent	
secondary-Butyl Chloride	99.5
Butenes	0.5
Purity by freezing point, mol %	
Freezing point, F	
Boiling point, F	
Distillation range, F	
Initial boiling point	151
10% Condensed	
50% Condensed	154
90% Condensed	
Dry point	156

PROPERTIES	98.0% GRADE
Specific gravity of liquid at 60/60 F	0.879
at 20/4 C	0.875
API gravity at 60 F	29.5
Density of liquid at 60 F, lbs/gal	7.32
Refractive index, 20/D	1.396
Color, Saybolt	
Acidity, distillation residue	
Nonvolatile matter, grams/100 ml	
Color Alpha	10
Flash point, approximate, F	< 80
Flammability limits, volume % in air	
Lower	
Higher	

*Literature values.

Table 3.57: Butyryl Chloride *(27)*

Butanoyl Chloride   C₃H₇COCl

Butanoyl chloride is a clear colorless liquid with a characteristic
pungent odor. It reacts with water and alcohol and is infinitely
soluble in ether. It is used for organic synthesis to introduce the
butyryl group.

PHYSICAL PROPERTIES

Molecular Weight	106.5
Freezing Point	-89°C
Boiling Point	102°C
Distillation Range	100° to 110°C
Refractive Index n20/D	1.4121
Specific Gravity, 15.5°/15.5°C	1.028
Pounds per Gallon at 15.5°C	8.56

Table 3.58: Caprylyl Chloride *(27)*

Octanoyl Chloride   CH₃(CH₂)₆COCl

Caprylyl chloride is a water-white to straw-colored liquid with a
pungent odor. It usually contains small quantities of hexanoyl and
decanoyl chlorides.

PHYSICAL PROPERTIES

Molecular Weight	162.7
Chlorine Content (typical)	21.8%
Freezing Point	<-70°C
Pour Point	<-70°C
Distillation Range[1]	183° to 212°C
Refractive Index, n20/D	1.4357
Flash Point (Cleveland open cup)	82°C
Fire Point (Cleveland open cup)	87°C
Specific Gravity, 15.5°/15.5°C	0.955
Pounds per Gallon at 15.5°C	7.96
Density Correction Factor, gm/cc/1°C	0.00085
Coefficient of Cubical Expansion at 15.5°C/1°C	0.00096

(1) Typical ASTM distillation to 90%. Decomposition occurs
beyond this point.

Table 3.59: Carbon Tetrachloride (7)

Tetrachloromethane CCl_4

PHYSICAL PROPERTIES

Acidity as HCl	None
Boiling point @760 mm	76.7°C (170.1°F)
Boiling range	Within 1°C
Coefficient of cubical expansion Av./°C, liquid	0.00127
Dielectric constant, 1000 cycle	2.24
Electrical conductivity	4×10^{-18} recip. ohm
Fire point	Nonflammable
Flash point	Nonflammable
Freezing point	-23°C
Heat of fusion	4.2 cal/kg
Heat of vaporization	46.5 cal/g
Power factor, 1000 cycle	0.057%
Purity	99.99% min.
Refractive index @20°C	1.4607
Residue	0.0010% by wt, max.
Solubility in water @20°C	0.08% by wt
Solubility of water in solvent @20°C	0.008% by wt
Specific gravity @25/4°C	1.5845
Specific heat Liquid, 25°	0.1995 cal/g/°C
76.8°C	0.2157 cal/g/°C
Specific resistivity	3.8×10^{12} ohms/cm
Thermal expansion per °C	0.127% (of liquid @158°F)
Vapor density (B.P., 760 mm)	5.37 g/liter
Vapor pressure @30°C	140 mm
Viscosity liquid @20°C	0.96 centipoise
Weight per gallon @25°C	13.22 lb

CARBON TETRACHLORIDE FORMS AZEOTROPES WITH:

	%	B.P. °C of Azeotrope
Acetone	88.5	56.4
Acetonitrile		71
Acetic acid	3	76.55
Acrylonitrile	21	66.2
Allyl alcohol	11.5	72.3
2-Butanone	71	73.8
tert-Amyl alcohol	4.5	76.6
Butyl alcohol	2.5	76.6
sec-Butyl alcohol	7.6	74.6
tert-Butyl alcohol	24	70.5
Butyl nitrite	35	74.8
1,2-Dichloroethane	21	75.6
Ethyl acetate	43	74.8
Ethyl alcohol	15.85	61.1
Ethyl nitrate	15.5	75
Formic acid	81.5	66.65
Isobutyl alcohol	5.5	75.8
Isopropyl alcohol	12	69
Methanol	20.56	55.7
Methyl propionate	25	76
Nitromethane	17	71.3
Propyl alcohol	11.5	73.1
Propyl formate	31	74.6
Water	4.1	66

Temperature-Vapor Pressure Curve for Carbon Tetrachloride

Table 3.60: Chlorinated Butane Derivatives *(73)*

Physical Properties of Intermediates and Products

Compound	B.P., °C., corr.	Press., mm.	Density d²⁵	Refractive Index, N²⁵ꝺ	Chlorine, %[1] Found	Chlorine, %[1] Calcd.
1-Chlorobutane	77.5-78.5	745		1.3995		
1,1-Dichlorobutane	114.8-115.1	752	1.0797	1.4305		
1,2-Dichlorobutane	122.9-123.3	743	1.1118	1.4425		
1,3-Dichlorobutane	133.0-133.2	744	1.1083	1.4414		
1,4-Dichlorobutane	154.1-154.2	749	1.1324	1.4522		
1,1,1-Trichlorobutane	133.1-133.3	750	1.2242	1.4483	65.76	65.88
1,1,2-Trichlorobutane	156.3-156.8	746	1.2787	1.4667	65.95	65.88
1,1,3-Trichlorobutane	153.2-153.8	750	1.2514	1.4593	65.92	65.88
1,1,4-Trichlorobutane	183.6-183.8	754	1.2967	1.4753	65.92	65.88
1,1,1,2-Tetrachlorobutane	69.1-69.4	20.0	1.3952	1.4812	72.63	72.39
1,1,1,3-Tetrachlorobutane	69.5-69.8	20.0	1.3747	1.4772	72.18	72.39
1,1,1,4-Tetrachlorobutane	86.8-87.1	20.0	1.4001	1.4858	72.81	72.39
1,1-Dichloro-1-butene	103.3-103.5	747		1.4465	56.21[2]	56.74
α-Chlorobutyraldehyde	106-108[3]	740		1.441	35.38	33.28
n-Butyryl chloride	101-101.5[3]	745		1.4098		
α-Chlorobutyryl chloride	51.5-51.7	40.0		1.4410		
β-Chlorobutyryl chloride	53.0-53.3	20.0		1.4477		
γ-Chlorobutyryl chloride	71.0-71.2	20.0		1.4597		
Ethyl α-chlorobutyrate	64.2-64.4[3]	20.0		1.4202		
Ethyl β-chlorobutyrate	69.9-70.1[3]	20.0		1.4222		
n-Propyl acetate	101-102	745		1.3823		
1-Chloropropyl acetate	48.6-48.8	20.0		1.4143		
2-Chloropropyl acetate	57.1-57.6	20.0		1.4205		
3-Chloropropyl acetate	58.4-58.8	10.0		1.4275		
n-Propyl chloroacetate	52.6-52.8	10.0		1.4233		

[1] Chlorine analysis by reaction with sodium diphenyl in dimethyl"Cellosolve" [L. M. Liggett, Anal. Chem., 26, 748(1954)].

[2] Av. of three analyses (56.17, 56.22 and 56.25%).

[3] Uncorrected.

Table 3.61: Chlorinated Hydrocarbons *(13)*

Carbon Tetrachloride	1.589*	13.22*	170	172	6.00	90.0	—	—	—	0.080*	0.013*	1.4598
1,1,1, Trichloroethane	1.319*	10.97*	162	190	6.00	100.0	—	—	—	—	—	1.4350*
Chloroform Tech	1.478*	12.31*	142		11.60	160.0	—	—	—	0.800*	0.097	1.4455
Ethylene Dichloride	1.252*	10.42*	179	186	4.46	61.6	59CC	—	—	0.810	0.150	1.4427
Methylene Chloride	1.320*	10.98*	103	104	14.50	350.0	—	—	—	1.320*	0.198*	1.4210
Monochlorobenzene	1.105*	9.19*	267	270	1.07	8.8	105	—	—	0.048	Insoluble	1.5215
Orthodichlorobenzene	1.303*	10.84*	355	362	0.15	62.0*	155	—	—	0.014	Insoluble	1.5482
Perchloroethylene	1.618*	13.46*	250	254	2.10	14.0	—	—	—	0.015*	0.010*	1.5044
Propylene Dichloride	1.159*	9.64*	204	208	3.22	43.0	63CC	—	—	—	—	1.4371*
Trichlorobenzene	1.454*	12.10*	418	427	0.06	22.0*	260	—	—	—	—	1.5690*
Trichloroethylene	1.459*	12.14*	188	190	4.46	59.0	—	—	—	0.110*	0.032*	1.4780

*Detd. at 25°C   *Detd. at 25°C   *Detd. at 100°C   *Detd. at 25/20°C   *Detd. at 25/20°C   *Detd. at 25/20°C

Table 3.62: Chlorinated Organic Solvents *(69)*

	Specific Gravity 20°/20° C	Dist. Range °F IBP	Dist. Range °F DP	Flash Pt. °F TOC
Carbon Tetrachloride	1.584	169	171	None
Chloroform	1.485	140	143	None
Ethylene Dichloride	1.255	181	183	70
Methylene Chloride	1.366	103	105	None
Monochlorobenzene	1.113	268	271	84*
Orthodichlorobenzene	1.313	355	361	170
Perchlorethylene	1.627	247	251	None
Trichlorethylene	1.455	187	190	None
1,1,1-Trichloroethane	1.316	162	190	None

*TCC

Table 3.63: Chlorinated Paraffins—"Clorafins" (28)

Typical Properties

	Clorafin 40	Clorafin 50
Color, Gardner, max	6	6
Viscosity at 25°C, poises	26-59	425-620
Chloride, % nominal	42	50
Specific gravity at 25/25°C	1.16	1.26
Bulking value at 25°C, gal/lb	0.104	0.096
Weight per gal at 25°C, lbs	9.6	10.4

COMPATIBILITY OF WAXES WITH CLORAFIN 40

The various mixtures of wax and Clorafin were melted together, observed for separation, blooming, and exudation after cooling.

Wax	Ratios of Wax:Clorafin		
	1:3	1:1	3:1
Mineral			
Amber Wax (melting point 185°F.)	C	C	C
Cardis 320	C	C	C
Ceresin, white	I	I	C
Concord Wax 407	C	C	C
Montan	C	C	C
Paraffin (melting point 130°F.)	C	C	C
Synthetic			
Acrawax C	C	C	C
Castorwax	I	I	C
Santowax M	C	C	C
Animal or Vegetable			
Beeswax	C	C	C
Candelilla	C	C	C
Carnauba	C	C	C
Japan	I	I	I
Stearic acid	I	I	C
Spermaceti	C	C	C

C—Compatible   I—Incompatible

SOME SOLVENTS WHICH ARE MISCIBLE WITH CLORAFIN

Acetone	Toluene
Methyl ethyl ketone	Xylene
Diethyl ether	Hi-Flash naphtha
Cellosolve	Carbon tetrachloride
Butyl acetate	Ethylene dichloride
Ethyl acetate	Turpentine
Mineral spirits	Nitromethane
Hexane	Nitroethane
V.M. & P. naphtha	Nitropropane

COMPATIBILITY OF CLORAFIN 40 WITH CELLULOSE DERIVATIVES

Cellulose Derivative	At a Clorafin:Cellulose Derivative Ratio			
	of 1:6	of 1:3	of 1:1	of 3:1
Ethyl cellulose, type G	C	C	SI	I
type N	C	C	C	I
Cellulose acetate, type F	C	I	I	I
type P	C	I	I	I
type W	C	I	I	I
RS® Nitrocellulose, ½-sec.	C	C	I	I
5-6-sec.	C	C	SI	I
Cellulose acetate butyrate	C	I	I	I

C—Compatible   I—Incompatible   SI—Slightly Incompatible

Table 3.64: Chlorobenzenes—Vapor Pressures *(72)*

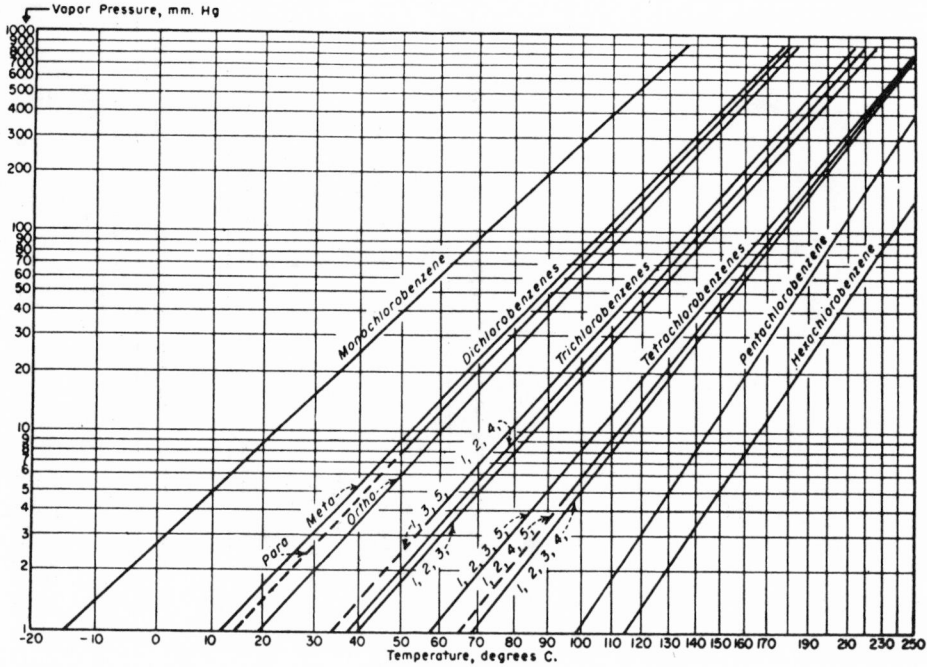

Table 3.65: Chloroform (7)

Trichloromethane   CHCl₃

PHYSICAL PROPERTIES

Acidity as HCl	0.001% by wt, max.	Viscosity @20°C	5.63 millipoises
Boiling point	61.2°C	@30°C	5.10 millipoises
Boiling range @760 mm	60.0–61.5°C	Water: no cloud @−10°C	0.021% by wt, max.
Coefficient of cubical expansion Av./°C, liquid	0.001399	Weight per gallon @25°C	12.29 lb
Color (Saybolt)	24 max.		
Dielectric constant, 1000 cycle	4.90		
Fire point	Nonflammable		
Flash point	Nonflammable		
Freezing point	63°C		

CHLOROFORM FORMS AZEOTROPES WITH:

%		B.P. °C of Azeotrope
20.5	Acetone	64.5
35	2-Bromopropane	62.2
96	2-Butanone	79.7
6.8	Ethanol	59.3
13	Ethyl formate	62.7
15	Formic acid	59.2
2.8	n-Hexane	60
4.5	Isopropanol	60.8
12.5	Methanol	53.5
23	Methyl acetate	64.8
2.8	Water	56.1

Heat of evaporation @B.P.	59.0 cal/g
Latent heat of evaporation @B.P.	106.4 Btu/lb
Refractive index @20°C	1.4467
@25°C	1.4422
Solubility in water @20°C	0.82
Solubility of water in solvent @10°C	0.06 g/water/100 g
Specific gravity 25/25°C	1.477
Specific heat Liquid, 20°	0.234 cal/g/°C
Specific resistivity	4.0 x 10⁹ ohms/cm
Thermal conductivity Liquid	0.080 Btu/hr (sq ft) (°F/ft)
Vapor density (B.P., 760 mm)	4.36 g/liter
Vapor pressure @30°C	243 mm

Table 3.66: Chloromethylene Compounds *(24)*

PRODUCT	EMPIRICAL FORMULA	MOL. WT.	PHYSICAL CONSTANTS			ASSAY (Method)	ISOMER CONTENT (Prox.)
			BOILING RANGE °C	SPECIFIC GRAVITY 25°/25°C.	REF. INDEX n_D^{25}		
COMMERCIAL							
BENZYL CHLORIDE	C_7H_7Cl	126.6	95% in 3° range incl. 179°C	1.040–1.111	1.5360–1.5370	99% min. (1)	––
para-METHYLBENZYL CHLORIDE	C_8H_9Cl	140.6	199–204°	––	1.535–1.540	98% (1)	––
METHYLBENZYL CHLORIDES	C_8H_9Cl	140.6	199–204°	1.070–1.080 (15.5°)	1.5360–1.5370	98% (2)	55% (p-) 45% (o-)
ETHYLBENZYL CHLORIDES	$C_9H_{11}Cl$	154.7	217–222°	1.046–1.047	1.5293–1.5305	99% (2)	70% (p-) 30% (o-)
ISOPROPYLBENZYL CHLORIDES	$C_{10}H_{13}Cl$	168.7	109–112° @ 15mm.	1.01–1.03	1.520–1.530	98.5% (1)	85% (p-) 15% (o-)
2,4-DIMETHYLBENZYL CHLORIDES	$C_9H_{11}Cl$	154.7	221–226°	1.050–1.065	1.5375–1.5385	98.5% (2)	86% (2,4-) 14% (2,6-)
3,4-DIMETHYLBENZYL CHLORIDES	$C_9H_{11}Cl$	154.7	225–232°	1.059–1.062	1.5370–1.5390	99% (2)	64% (3,4-) 34% (2,3-) 2% (2,4-; 2,5-; 2,6-)
DICHLOROBENZYL CHLORIDES	$C_7H_5Cl_3$	195.5	245–253°	1.410–1.418	1.5755–1.5765	94% (2)	80% (2,4-; 2,5-; 2,6-) 20% other isomers
DEVELOPMENT							
2,5-DIMETHYLBENZYL CHLORIDE	$C_9H_{11}Cl$	154.7	221–226°	1.035–1.045	1.5350–1.5360	98% (2)	––
meta-CHLOROBENZYL CHLORIDE	$C_7H_6Cl_2$	161.1	98–104° @ 15mm.	1.25–1.27	1.5532–1.5542	97.5% (1)	––
α,α'-DICHLORO-XYLENES	$C_8H_8Cl_2$	175.1	prox.133°C @ 15mm.	––	––	95% (1)	70-80% (p-) 20-30% (o-)
Bis-CHLOROMETHYL-DURENE	$C_{12}H_{16}Cl_2$	231.2	190–196° dec. (3)	––	––	98% (2)	––
RESEARCH							
ortho-METHYLBENZYL CHLORIDE	C_8H_9Cl	140.6	––	––	––	98% (1)	––
meta-METHYLBENZYL CHLORIDE	C_8H_9Cl	140.6	––	––	––	98%	––
CHLOROMETHYL-TETRALINS	$C_{11}H_{13}Cl$	180.5	135–145° @ 7mm.	––	––	98% (2)	60% (β-) 40% (α-)
α-CHLOROMETHYL-β-METHYLNAPHTHALENE	$C_{12}H_{11}Cl$	190.7	58–62° (3)	––	––	95% min. (2)	––

(1) Gas-Liquid Partition Chromatography
(2) Alcoholic potassium hydroxide hydrolysis
(3) Melting range

Table 3.67: o- and p-Chlorotoluenes *(7)*

$$C_6H_4(Cl)CH_3$$

PHYSICAL PROPERTIES

	Ortho	Para
Boiling point	159.4°C	162.5°C
Coefficient of cubical expansion @30°C		0.00092
Distillation range		
Start	158.3°C min.	
100%	165.1°C max.	
Flash point	46°C	
Freezing point	−34°C	7.5°C
Latent heat of vaporization		77 cal/g
Purity	60% approx.	40% approx.
Solubility of water @25°C		0.037 g/100 g
Solubility of water in solvent @25°C		0.014 g/100 g
Refractive index @20°C	1.5238	1.5199
Specific gravity @20/4°C	1.0817	1.0697
Surface tension @25°C		32.9 dynes/cm
Vapor pressure @100°C		132 mm
Viscosity @100°F		0.707 centistoke
		0.747 centipoise
@210°F		0.328 centistoke
		0.327 centipoise
Weight per gallon @25°C		9.1 lb

Table 3.68: p-Chlorotoluene *(7)*

$$C_6H_4(Cl)CH_3$$

PHYSICAL PROPERTIES

Acidity as acetic acid	Nil
Boiling point @760 mm Hg	162.3°C
@50 mm Hg	78.4°C
@10 mm Hg	43.8°C
Congealing point	6.8°C
Distillation range @760 mm Hg	162-166°C
Flash point (Cleveland O.C.)	140°F
Moisture content	Nil
Molecular weight	126.59
Pounds per gallon	8.85
Purity	98.0%
Refractive index N_{22}^D	1.5184
Side chain chlorine	None
Solidifies	Below 45°F
Specific gravity @25/25°C	1.067 min. - 1.071 max.
Surface tension (in air) DuNouy @25°C	34.60 dynes/cm
Vapor pressure @96.6°C	100 mm Hg

Table 3.69: Chlorotoluene *(27)*

"Halso 99" $ClC_6H_4CH_3$

Halso 99 is a solvent grade of monochlorotoluene. It is a clear, colorless liquid with the characteristic odor of ring chlorinated aromatic compounds.

PHYSICAL DATA

Molecular Weight 126.6
Kauri-Butanol Value 110
Specific Gravity, 25°/15.5°C   1.079 (9.0 lbs/gal)
Flash Point . . . . >101°F (>38°C)

Used as a solvent in many industries including paints and plastics, and as a dye carrier, sludge solvent, rubber solvent, and in metal-cleaning formulations and carbon-removal compounds.

Table 3.70: 2-Chloro-p-Xylene *(30)*

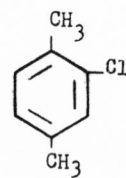

Physical Properties

Empirical Formula	C_8H_9Cl
Molecular Weight	140.61
Appearance	Colorless Liquid
Melting Point	2.0°C
Boiling Point	186°C
Density $\frac{25}{4}$	1.049
Odor	Similar to p-xylene

Approximate Solubility at 25°C

Insoluble in water and ethylene glycol

Completely miscible with acetone, cyclohexanone, benzene, ethyl acetate, n-heptane, carbon tetrachloride, methanol and diethyl ether

Table 3.71: α-Chloro-p-Xylene *(30)*

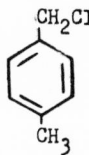

Physical Properties

Empirical Formula - - - - - - - - - - - - - - - C₈H₉Cl
Molecular Weight - - - - - - - - - - - - - - - 140.61
Appearance - - - - - - - - - - - - Colorless Liquid
Melting Point - - - - - - - - - - - - - - - - 4.5°C
Boiling Point - - - - - - - - - - - - - - - - 200°C
Vapor Pressure - - - - - - - log P = $\frac{-2762}{T}$ + 8.862

Odor - Pungent

Approximate Solubility at 25°C

Insoluble in water and ethylene glycol

Completely miscible with acetone, cyclohexanone, benzene, ethyl acetate, n-heptane, carbon tetrachloride, methanol and diethyl ether

Table 3.72: o-Dichlorobenzene *(7)*

1,2-Dichlorobenzene   C₆H₄—Cl₂

PHYSICAL PROPERTIES

	Purified	Technical
Boiling point	180.2°C	179.6°C
Boiling range (within)	3.0°C	4.0°C
Dielectric constant 1000 cycles	9.82	
Electrical conductivity @0°C	10⁻⁹ recip. ohm	
Fire point	103°C	103°C
Flash point	68°C	68°C
Freezing point	−18.3°C	−22.5°C
Heat of combustion	671.8 kg cal/mol.	
Heat of fusion	88 joules/g	
Impurities (p-dichlorobenzene, trichlorobenzene)	Not over 4%	Not over 12%
Latent heat of vaporization @B.P.		65 cal/g at.
Refractive index @22°C		1.5518
Solubility in water @25°C		Less than 0.01%
Specific gravity @20/4°C		1.3048
Specific heat		0.271 cal/g/°C
Specific resistivity		2.0 x 10⁸ ohms/cm
Weight per gallon @25°C		10.85 lb

Table 3.73: p-Dichlorobenzene *(7)*

Table 3.74: Dichlorodiisopropyl Ether *(7)*

p-DICHLOROBENZENE FORMS AZEOTROPES WITH:

%		B. P. °C of Azeotrope
20	Cineole	173.5
33.5	Cyclohexanol	153.6
37	2-Ethoxyethyl acetate	155.5
34	n-Hexyl alcohol	151.6
46	Camphene	155.0
63.5	Isoamyl ether	172.4
27	Isoamyl propionate	155.2
14	d-Limonene	174.2
50	α-Pinene	153.4
2	Phenol	156.0
43	Propyl isovalerate	154.5

$$ClCH_2CH(CH_3)-O-CH(CH_3)CH_2Cl$$

PHYSICAL PROPERTIES

Acidity as HCl	0.01% by wt, max.
Boiling point @760 mm	187.3°C
Boiling range @760 mm	Not more than 5% distills below 180°C
	Not less than 95% distills below 190°C
Color (Pt-Co scale)	25 max.
Flash point (O.C.)	185°F
Solubility in water @20°C	0.17% by wt
Solubility of water in solvent @20°C	0.11% by wt
Specific gravity @20/20°C	1.1122
Vapor pressure @20°C	0.85 mm
Weight per gallon @20°C	9.26 lb

Table 3.75: Dichloroethylene *(7)*

cis-Acetylene Dichloride

```
 H   H
 |   |
 C===C
 |   |
 Cl  Cl
```

PHYSICAL PROPERTIES
cis isomer

Acidity as HCl	0.0005% by wt, max.
Boiling point @760 mm	60.3°C
Coefficient of cubical expansion Av./°C, liquid	0.00127
Color (Saybolt)	24 max.
Flash point	6°C
Freezing point	−80.5°C
Latent heat of vaporization @B.P.	73.0 cal/g
Refractive index @15°C	1.4519
Residue on evaporation	0.007% by wt, max.
Solubility in water @25°C	0.77 g/100 g
Solubility of water in solvent @10°C	0.04 g water/100 g
Specific gravity @20/4°C	1.282
Specific heat Liquid, 20°C	0.270 cal/g/°C
Vapor density (B.P., 760 mm)	3.54 g/liter
Vapor pressure @30°C	273 mm
Viscosity liquid @20°C	0.48 centipoise
Water: no cloud @−15°C	0.001% by wt, max.
Weight per gallon @20°C	10.70 lb

trans-Acetylene Dichloride

```
 H   Cl
 |   |
 C===C
 |   |
 Cl  H
```

PHYSICAL PROPERTIES
trans isomer

Acidity as HCl	0.0005% by wt, max.
Boiling point @760 mm	48.0-48.5°C
Boiling range @760 mm	47.0-48.5°C
Coefficient of cubical expansion Av./°C, liquid	0.00136
Color (Saybolt)	24 max.
Flash point	4°C
Freezing point	50°C
Latent heat of vaporization @B.P.	73.7 cal/g
Refractive index @15°C	1.4490
Residue on evaporation	0.0007% by wt, max.
Solubility in water @25°C	0.63 g/100 g
Solubility of water in solvent @10°C	0.03 g water/100 g
Specific gravity @20/4°C	1.257
Specific heat Liquid, 20°C	0.270 cal/g/°C
Vapor density (B.P., 760 mm)	3.67 g/liter
Vapor pressure @30°C	395 mm
Viscosity liquid @20°C	0.41 centipoise
Water: no cloud @−15°C	0.004% by wt, max.
Weight per gallon @20°C	10.49 lb

Table 3.76: Dichloroethyl Ether *(7)*

2,2'-Dichloroethyl Ether

sym- or β,β'-Dichloroethyl Ether

$ClCH_2CH_2OCH_2CH_2Cl$

PHYSICAL PROPERTIES

Acidity as HCl	0.005% max.
Apparent ignition temperature in air	396°C
Boiling point	178°C
Boiling range @760 mm	Not more than 5% distills below 173°C Not less than 95% distills below 179°C
Color (500 mm tube)	Not more than 2 yellow Lovibond
Ethylene dichloride	1.0% max.
Flash point (C.C.)	55°C
Latent heat of vaporization @178°C	64.1 cal/g
Refractive index @20°C	1.457
Specific gravity @20/20°C	1.219-1.224
Specific heat @20-30°C	0.369 cal
Surface tension @25°C	41.8 dynes/ sq cm
Vapor pressure @20°C	1.2 mm
Viscosity @25°C	2.0653 centipoises
Weight per gallon @20°C	10.17 lb

Table 3.77: Dichlorohydrin *(7)*

Glycerol Dichlorohydrin

Dichloroisopropyl Alcohol

1,3-Dichloropropanol-2

α-Propenyldichlorohydrin

PHYSICAL PROPERTIES

Boiling point	(1,3-) 174°C (1,2-) 183°C
Boiling range	174-176°C (95%)
Flash point	74°C
Refractive index	1.47-1.48
Specific gravity	1.36-1.39
Vapor pressure	7 mm

$ClCH_2CH(OH)CH_2Cl$

1,2-DICHLORO-3-PROPANOL FORMS AZEOTROPES WITH:

%		B.P. °C of Azeotrope
55	o-Bromotoluene	171.6
75	Camphene	156.0
60	α-Chlorotoluene	171.0
68	Indene	160.0
60	α-Limonene	169.3
43	2-Octanol	172.5
80	α-Pinene	153.0
50	Thymene	170.8

1,3-DICHLORO-2-PROPANOL FORMS AZEOTROPES WITH:

%		B.P. °C of Azeotrope
91	Bromobenzene	155.5
39	o-Bromotoluene	170.5
32	p-Bromotoluene	172.8
62	Camphene	152.8
43	α-Chlorotoluene	168.9
85	o-Chlorotolene	158.0
78	p-Chlorotoluene	160.0
45	Cymene	165.5
55	p-Dichlorobenzene	162.2
62	2,7-Dimethyllactane	155.0
85	Dimethyl oxalate	162.0
33.5	Indene	173.5
30	Iodobenzene	173.0
10	Isoamyl butyrate	178.6
52	Isoamyl ether	165.9
43	d-Limonene	166.8
50	Mesitylene	156.0
41	p-Methylanisole	173.1
35	Methylheptenone	178.5
57	α-Phellandrene	163.0
63.5	α-Pinene	150.4
63	Pseudocumene	164.4
85	Styrene	142.5
38	α-Terpinolene	166.8
40	Thymene	166.5

Table 3.78: Dichloromethane *(22)*   •

Methylene Chloride
Methylene Dichloride

CH_2Cl_2

Methylene chloride is a clear, water-white liquid at ordinary temperatures, with a pleasant, ethereal odor. It is highly volatile and mobile. Methylene chloride is completely miscible with most organic liquids.

TYPICAL PROPERTIES

Molecular Weight	84.93
Boiling Point, °F	103.6
°C	39.8
Freezing Point, °F	– 142.1
°C	– 96.7
Flash Point (Tag open cup)	none
Ignition Temperature, °F	1224
°C	662
Specific Gravity of Vapor (air = 1.00)	2.94
Density at 20°C, pounds per gallon	11.15
Viscosity at 20°C, centipoises	0.425
Specific Heat at 20°C, cal/(g)(°C)	0.29
Vapor Pressure at 20°C, mm	348.9
Evaporation Rate at 25°C (ether = 100)	71
Heat of Vaporization, cal/g	75.3
Btu/lb	135.5
Solubility g methylene chloride/100 g water at 20°C	2.0
g water/100 g methylene chloride at 25°C	0.2
Azeotrope with Water, Boiling Point, °F	100.6
°C	38.1
Azeotropic Water Content, wt %	1.5

Specification for standard grade

Appearance	Clear, free of suspended matter
Color, APHA, maximum	10
Odor	Characteristic; no residual
Specific Gravity, 25°C/25°C	1.319 to 1.323
Acidity, ppm, maximum	5
Nonvolatile Residue, ppm, maximum	10
Free Halogen	none
Distillation Range (100%), °C	39.5 to 40.5
°F	103.1 to 104.9
Water, ppm, maximum	100

Flammability of Methylene Chloride-Oxygen-Nitrogen Mixtures

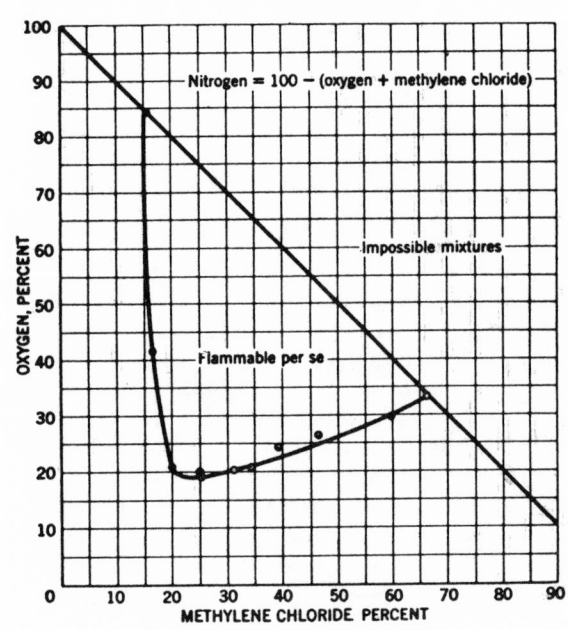

Nitrogen = 100 – (oxygen + methylene chloride)

Impossible mixtures

Flammable per se

OXYGEN, PERCENT

METHYLENE CHLORIDE, PERCENT

VAPOR PRESSURE

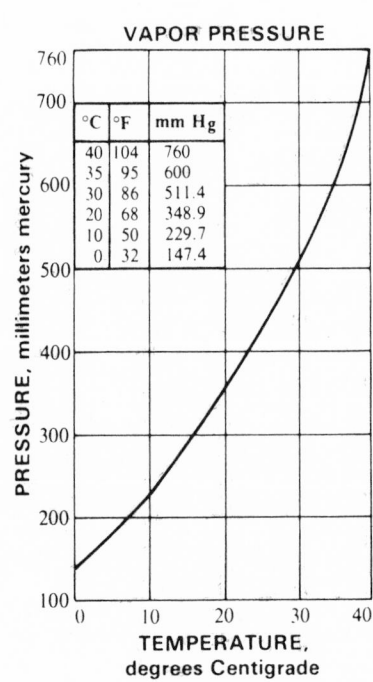

°C	°F	mm Hg
40	104	760
35	95	600
30	86	511.4
20	68	348.9
10	50	229.7
0	32	147.4

PRESSURE, millimeters mercury

TEMPERATURE, degrees Centigrade

(continued)

Table 3.78: (continued)

1,1-DICHLOROMETHANE FORMS AZEOTROPES WITH:

%		B.P. °C of Azeotrope
30	Acetone	57.6
23	Biallyl	56.5
94.8	1,3-Butadiene	-5.0
20	Chloromethyl methyl ether	54
30	Cyclopentane	38.0
55	Diethylamine	52.0
11.5	Ethanol	54.6
21	Iodomethane	39.8
8	Isopropanol	56.6
51	Pentane	35.5
23	Propylene oxide	40.6
6	tert-Butanol	57.1
1.5	Water	38.1

Table 3.79: Dichloropentanes *(7)*

$$C_5H_{10}Cl_2$$

PHYSICAL PROPERTIES

Acidity as HCl	0.025% max.
Average chlorine content	48%
Distillation	95% between 130-200°C
Evaporation rate @109°F:Minutes	
3.83	25%
8.00	50%
14.20	75%
90.00	100%
Flash point (O.C.)	97°F
Heat of vaporization	68.5 cal/g
Kauri-butanol value	67 cc
Solubility in water	Negligible
Specific gravity @20°C	1.07-1.08
Specific heat	0.369 cal/g
Surface tension @25°C	31.8 dynes/cm
Viscosity @25°C	0.016 poise
Water azeotrope @80-97°C	66% $C_5H_{10}Cl_2$ (approx.)
Water content	None
Weight per gallon	8.94 lb

Table 3.80: 2,4-Dichlorotoluene *(7)*

$$C_6H_3(Cl)_2CH_3$$

PHYSICAL PROPERTIES

Acidity as acetic acid	Nil
Boiling point @760 mm Hg	200.5°C
@50 mm Hg	113.0°C
@10 mm Hg	77.0°C
Congealing point	-13°C
Distillation Range @760 mm Hg	199-202°C
Fire point (Cleveland O.C.)	383°F
Flash point (Cleveland O.C.)	199°F
Moisture content	Nil
Molecular weight	161.04
Pounds per gallon	10.34
Purity	99.0%
Refractive index N_{22}^D	1.5480
Side chain chlorine	None
Specific gravity @25/25°C	1.247 min. -1.251 max.
Surface tension, DuNouy @25°C	38.29 dynes/cm
Vapor pressure @130°C	100 mm Hg

Table 3.81: 3,4-Dichlorotoluene *(30)*

An intermediate for the preparation of dyestuffs, pharmaceuticals, rubber chemicals and other organic chemicals; a "carrier" in the dyeing of synthetic fibers; a special solvent.

FORMULA: $C_6H_3(Cl)_2CH_3$ MOL. WT.: 161.04

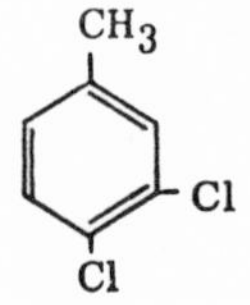

FORM: Liquid

SPECIFICATION:

    Appearance .... colorless to pale yellow liquid
    Specific Gravity, $25^\circ/25^\circ C$. .... 1.251 min. - 1.255 max.
    Distilling Range .... $207^\circ - 212^\circ C$.
    Side Chain Chlorine .... none

PROPERTIES:

    Assay 98.0%
    Congealing Point .... minus $14^\circ C$.
    Refractive Index, n/D/$22^\circ C$. 1.5490
    Flash Point, Open Cup................. $190^\circ F$. (approx.)

Table 3.82: 2,5-Dichloro-p-Xylene *(30)*

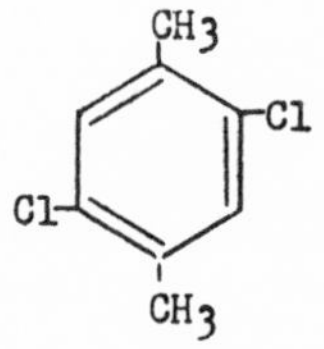

Physical Properties

    Empirical Formula- - - - - - - - - - - - - - - $C_8H_8Cl_2$
    Molecular Weight - - - - - - - - - - - - - - 175.06
    Appearance - - - - - - - - - -White Crystalline Solid
    Melting Point- - - - - - - - - - - - - - - - $71^\circ C$
    Boiling Point- - - - - - - - - - - - - - - - $222^\circ C$
    Odor - - - - - - - - - - Similar to p-dichloro-benzene

Approximate Solubility at 25°C - % by weight

    Water- - - - - - Insoluble   n-Heptane - - - - - 29
    Acetone- - - - - - 27   Carbon Tetrachloride-29
    Cyclohexanone- - - 32   Methanol- - - - - - 5
    Benzene- - - - - - 44   Diethyl Ether - - - 39
    Ethyl Acetate- - - 32   Ethylene Glycol - Insoluble

Table 3.83: α,α′-Dichloro-p-Xylene *(30)*

Physical Properties

```
Empirical Formula - - - - - - - - - - - - - - - - - - C8H8Cl2
Molecular Weight- - - - - - - - - - - - - - - - - - - 175.06
Appearance- - - - - - - - - - - - - White Crystalline Solid
Melting Point - - - - - - - - - - - - - - - - - - - - 100°C
Boiling Point - - - - - - - - - - - - - - - - - - - - 254°C
Odor- - - - - - - - - - - - - - - - - - - - - - - - - Sweet
```

Approximate Solubility (% by weight) in
Various Solvents at 25°, 50°, and 75°C

Solvent	Temperature (°C)		
	25	50	75
Acetone	22.5	44.6	--
Benzene	19.8	37.8	64.3
Carbon Tetrachloride	4.5	14.5	45.5
Cyclohexanone	27.4	43.2	66.5
Diethyl Ether	11.3	--	--
p-Dioxane	21.1	39.7	64.0
Ethyl Acetate	18.2	34.5	63.0
Ethylene Glycol	Insoluble	Insoluble	Insoluble
n-Heptane	2.0	5.3	16.6
Methanol	3.2	10.0	--
Tetrahydrofuran	30.6	46.5	--
Water	Insoluble	Insoluble	Insoluble

Table 3.84: Dimethyl 2,3,5,6-Tetrachloroterephthalate *(30)*

Physical Properties

```
Empirical Formula - - - - - - - - - - - - - - - - C10H6Cl4O4
Molecular Weight - - - - - - - - - - - - - - - - - - 331.99
Appearance - - - - - - - - - - - - White Crystalline Solid
Melting Point, °C - - - - - - - - - - - - - - - - - 154-55
Boiling Point, °C @ 750 mm/Hg - - - - - - - - - - - - 336
              °C @  20 mm/Hg - - - - - - - - - - - - 210
Odor - - - - - - - - - - - - - - - - - - - - - - - - None
```

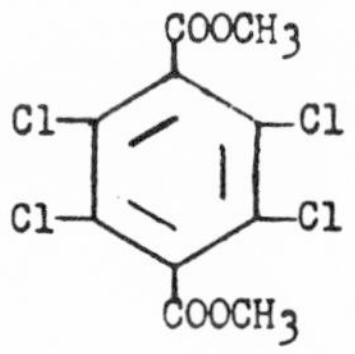

Approximate Solubility (% by weight) in
Various Solvents at 25°, 50°, and 75°C

Solvent	Temperature (°C)		
	25	50	75
Acetone	10	21	--
Benzene	16	30	48
Carbon Tetrachloride	7	14	24
Cyclohexanone	9	20	33
p-Dioxane	11	30	46
Ethyl Acetate	8	20	31
Ethyl Ether	6	--	--
Ethylene Glycol	0	0	0
n-Heptane	1	2	5
Methanol	0	1	--
Perchloroethylene	4	8	17
Tetrahydrofuran	18	31	--
Water	0	0	0

Table 3.85: Epichlorohydrin *(2)*

3-Chloropropylene-1,2-oxide $O-CH_2-CH-CH_2Cl$

Epichlorohydrin is a colorless, mobile, highly reactive liquid. It is completely miscible with many organic liquids such as acetone, carbon tetrachloride, alcohols, benzene, ethers, halogenated hydrocarbons, fixed oils, etc. It is not miscible with glycerin and water. The two reactive functional groups make it a very useful chemical intermediate. In the presence of a catalyst, its epoxy group enters into an exothermic reaction with the active hydrogen atoms of alcohols, amines, carboxylic acids, phenols, mercaptans, etc. The atom in the molecule reacts with acid salts, alkali metal phenolates, and alcoholates, amides, amines, etc.

Epichlorohydrin is used to a large extent as a raw material in the manufacture of epoxy resins. When condensed with dihydric phenols or phenolic resins, epoxy resins are obtained which range from liquids to solids. It is also used in the manufacture of ion exchange resins, adhesion resins and a large number of other chemicals.

Absolute viscosity at 20°C., cps.	1.1
$\Delta BP/\Delta P.$, at 740 to 760 mm. Hg, °C. per mm.	0.044
Boiling point, °C., 760 mm.	115.2
50 mm.	45
10 mm.	16
Freezing point, °C.	−58.1
Heat of vaporization at 1 atm., Btu/lb.	174
Molecular weight	92.53
Refractive index, n_D at 20°C.	1.4359
Solubility, % by weight at 20°C.,	
in water	5.9
water in	1.2
$\Delta SG/\Delta T.$ at 20° to 30°C.	0.00120
Specific gravity at 20/20°C.	1.1761
Vapor pressure at 20°C., mm. Hg	12.7
Flash point (open cup), °F.	105

Table 3.86: Ethyl Chloride *(22)(23)*

Monochloroethane
Muriatic Ether C_2H_5Cl

TYPICAL PROPERTIES

Molecular Weight: 64.52

Description: Ethyl chloride is a colorless mobile liquid at 1 atmosphere below 12.4°C (54°F). Above the boiling point, it is a colorless gas. Ethyl chloride has an ethereal odor and is highly volatile and flammable.

Freezing Point, °C −138.3
°F −217

Refractive Index of Vapor, n_D^{25} 1.001

Vapor Pressure, mm Hg
0°C (32°F) 464
10°C (50°F) 692
20°C (68°F) 1011

Specific Gravity of Vapor (air=1) 2.23

Solubility at 0°C,
g ethyl chloride/100 g water 0.447
g water/100 g ethyl chloride 0.07

Solubility:
Ethyl chloride is soluble in most organic solvents.

(continued)

Table 3.86: (continued)

Flash Point, Tag open cup,	°C	−43		
	°F	−45		
Explosive Limits, volume % in air		3.16 to 15		
Autoignition Temperature,	°C	519		
	°F	966		
Specific Heat at 0°C, cal/ (g) (°C) or Btu/ (lb) (°F)		0.37		
Heat of Vaporization at Boiling Point				
cal/g		92.5		
Btu/lb		165.6		
Viscosity at 10°C, cps		0.279		
Density at 20°C, pounds/gallon		7.461		

Reactivity:

At ordinary temperatures the oxidation and hydrolysis of ethyl chloride take place slowly. In the absence of air and water, it can be used with most common metals up to 200°C (392°F). Ethyl chloride burns with a green-edged flame, producing hydrogen chloride, carbon dioxide and water. It is thermally stable to 400°C (752°F); thermal splitting yields ethylene and hydrogen chloride. The reactivity of ethyl chloride as an intermediate is often based on the affinity of alkali metal atoms for its chlorine atom.

Specification and Typical Analysis:

	Specification	Typical Analysis
Purity, wt %	99.5 minimum	99.97
Color, APHA	20 maximum	<5
Appearance	clear, free of suspended matter	clear, free of suspended matter
Acidity as HCl, wt %	0.002 maximum	<0.0001
Water, wt %	0.02 maximum	0.0010
Nonvolatile Residue, wt %	0.01 maximum	<0.0001
Total Impurities, wt %	0.5 maximum	0.03
Distillation Range, °C	12 to 13	12.2 to 12.4
Specific Gravity, 0°C/4°C	0.922 to 0.925	0.922

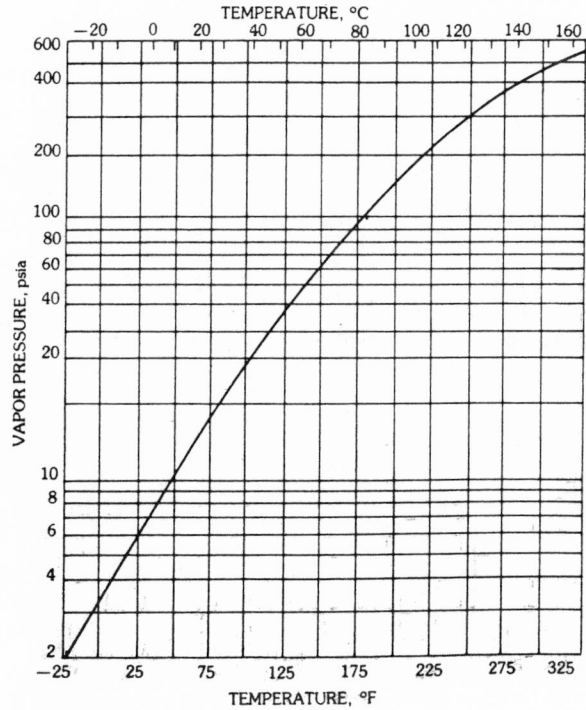

Ethyl Chloride Vapor Pressure vs Temperature

Solubility, Approximate, g/100 g Solvent at 25°C

Acetone	103
Benzene	110
n-Heptane	87
Ethanol (21°C)	48
Methanol	37
Water (20°C)	0.6

Table 3.87: Ethylene Chlorohydrin *(7)*

Glycol Chlorohydrin

2-Chlorethyl Alcohol

$ClCH_2CH_2OH$

PHYSICAL PROPERTIES

Absolute viscosity @ 20°C	3. 4 centipoises
Apparent specific gravity @ 20/20°C	1. 2040
Boiling point @ 760 mm Hg	128. 7°C
@ 50 mm Hg	60°C
@ 10 mm Hg	29°C
Coefficient of expansion @ 55°C	0. 00092
Flash point (Cleveland O. C.)	140°F
Freezing point	−62. 6°C
Molecular weight	80. 52
Pounds per gallon @ 20°C	10. 03
Solubility in water @ 20°C	Complete
Solubility of water in solvent @ 20°C	Complete
Vapor pressure @ 20°C	4. 9 mm Hg

Table 3.88: Ethylene Dichloride *(7)*

1, 2-Dichloroethane

sym-Dichloroethane

Ethylene Chloride

Dutch Oil

Elayl Chloride

$ClCH_2-CH_2Cl$

PHYSICAL PROPERTIES

Acidity as HCl	Not more than 0. 001%	Purity	Not less than 99. 0 %
Apparent ignition temperature in air	449°C	Refractive index	1. 4443
Boiling point	83. 6°C	Solubility in water @ 20°C	0. 87 % by wt
Boiling range @ 760 mm	Below 82. 5°C none Above 84. 0°C none	Solubility of water in solvent @ 20°C	0. 16% by wt
		Specific gravity @ 20/20°C	1. 2550
Coefficient of cubical expansion Av. /°C, liquid (10-30°C)	0. 00116	Specific heat	0. 31 cal/g/°C
Color (500-mm tube) (Lovibond)	Not more than 1. 0 yellow	Specific resistivity	9. 0 x 10⁶ ohms/cm
Dielectric constant @ 20°C	10. 5±0. 3	Surface tension @ 25°C	37. 5 dynes/cm
Electrical conductivity	3×10^{-8}	Thermal conductivity @ 20°C	0. 0038 cal/cm/sec/°C
Explosive limits in air	6. 2-15. 9% by vol.	Vapor density (B. P. , 760 mm)	3. 88 g/liter
Fire point	28°C	Vapor pressure @ 20°C	63 mm
Flash point (ASTM O. C.)	21°C	@ 30°C	99 mm
Freezing point	−35°C	Viscosity @ 25°C	0. 0078 poise
Heat of combustion	2720 cal/g	Water content	Not more than 0. 02%
Latent heat of evaporation @ B. P.	77. 3 cal/g	Weight per gallon @ 25°C	10. 38 lb
Nonvolatile matter	Not more than 0. 001 g/100 cc		

(continued)

Table 3.88: (continued)

Limits of Flammability of Ethylene Dichloride
in Air and Carbon Dioxide

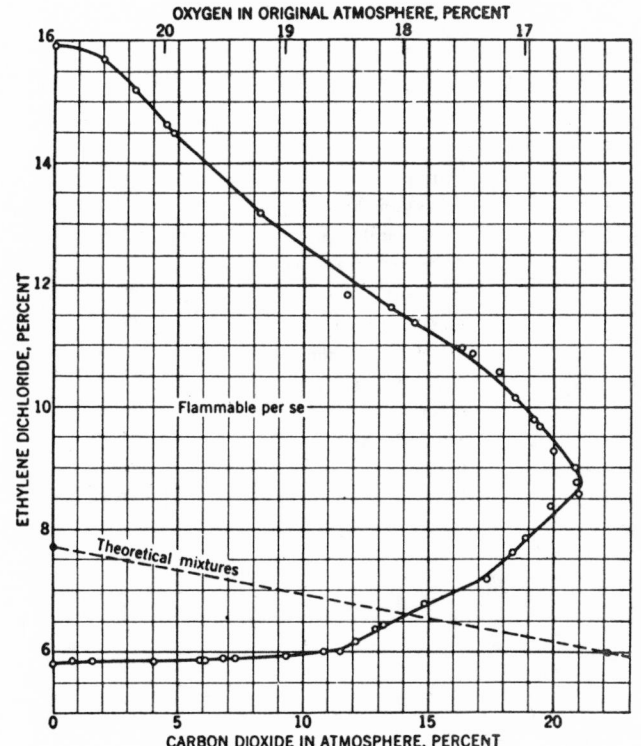

ETHYLENE DICHLORIDE FORMS AZEOTROPES WITH:

%		B.P. °C of Azeotrope
18	Allyl alcohol	79.9
6	tert-Amyl alcohol	83
79	Carbon tetrachloride	75.6
19.5	1,1-Dichloroethane	72
37	Ethanol	70.3
38	Formic acid	77.4
6.5	Isobutanol	83.5
43.5	Isopropyl alcohol	74.7
19	Propanol	80.7
10	n-Propyl formate	84.1
18	Trichloroethylene	82.9
32	Methanol	61
8.2	Water	70.5

Table 3.89: 2-Ethylhexyl Chloride (7)

$$CH_3CH_2CH_2CH_2CH(C_2H_5)CH_2Cl$$

PHYSICAL PROPERTIES

Average weight @20°C	7.33 lb/gal
Boiling point @760 mm Hg	172.9°C
Flash point (O.C.)	140°F
Molecular weight	148.67
Solubility in water @20°C	0.1% by wt
Solubility of water in solvent @20°C	0.1% by wt
Specific gravity @20/20°C	0.8833
Vapor pressure @20°C	1.3 mm Hg

Table 3.90: Glycerol α-Monochlorohydrin (7)

$$ClCH_2CHOHCH_2OH$$

PHYSICAL PROPERTIES

Boiling point	213°C
Boiling range (ASTM)	90% between 136-142°C @40 mm
Flash point (O.C.)	280°F
Refractive index @25°C	1.4781
Solubility in water	100%
Specific gravity @20/4°C	1.320
Weight per gallon	10.98 lb

Table 3.91: Hexachloroethane *(7)*

Perchloroethane

Carbon Trichloride $\qquad$ CCl_3CCl_3

Tetrachloroethylene Dichloride

PHYSICAL PROPERTIES		HEXACHLOROETHANE FORMS AZEOTROPES WITH:		
		%		B.P. °C of Azeotrope
Acidity as HCl	Less than 0.05%			
Boiling point @ 760 mm	Sublimes @ 185°C			
Latent heat of vaporization @ B.P.	46.4 cal/g 83.5 Btu/lb	34	Aniline	176.8
		12	Benzyl alcohol	182.0
Melting point in sealed tube	188.2°C	30	p-Bromotoluene	183.5
Nonvolatile matter	Less than 0.15%	25	Chloroacetic acid	171.2
Purity	98.0% min.	28	o-Cresol	181.3
Specific gravity @ 20/4°C	2.091	43	Diethyl oxalate	178.6
Specific heat @ 25°C	0.174 cal/g/°C or Btu/lb/°F	20	Diisobutyl carbonate	184.0
		55	Dimethyl malonate	176.0
Vapor density (B.P., 760 mm)	6.30 g/liter	49.5	Ethyl acetoacetate	172.5
Vapor pressure @ 30°C	2 mm	37	Isovaleric acid	172.6
Water	0.2% by wt, max.	30	Phenol	173.7
		15	Trichloroacetic acid	181.0

Table 3.92: α,α'-Hexachloro-m-Xylene *(30)*

Physical Properties

```
Empirical Formula- - - - - - - - - - - - - - - - - - - - - C8H4Cl6
Molecular Weight - - - - - - - - - - - - - - - - - - - - - 312.86
Appearance - - - - - - - - - - - - - - - - White Crystalline Solid
Melting Point- - - - - - - - - - - - - - - - - - - - - - 40-41°C
Boiling Point, at 1 mm. Hg - - - - - - - - - - - - - - -  120°C
Odor - - - - - - - - - - - - - - - - - - - - - - - - - Very Mild
```

Approximate Solubility (% by wt. at 25°C)

```
Benzene- - - - - - - - - - - - - - - - - - - - - - - - - - - 58
Carbon Tetrachloride - - - - - - - - - - - - - - - - - - - - 50
n-Heptane- - - - - - - - - - - - - - - - - - - - - - - - - - 54
Methanol - - - - - - - - - - - - - - - - - - - - - - - - - - 31
Water- - - - - - - - - - - - - - - - - - - - - - - - - - - -  0
Ethyl Acetate- - - - - - - - - - - - - - - - - - - - - - - - 84
Ethyl Ether- - - - - - - - - - - - - - - - - - - - - - - - - 80
Acetone- - - - - - - - - - - - - - - - - - - - - - - - - - - 82
Cyclohexanone- - - - - - - - - - - - - - - - - - - - - - - - 75
Ethylene Glycol- - - - - - - - - - - - - - - - - - - - - - -  0
Tetrahydrofuran- - - - - - - - - - - - - - - - - - - - - - - 88
Dioxane- - - - - - - - - - - - - - - - - - - - - - - - - - - 66
```

Table 3.93: α,α'-Hexachloro-p-Xylene *(30)*

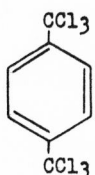

Physical Properties

Empirical Formula- - - - - - - - - - - - - - - $C_8H_4Cl_6$
Molecular Weight - - - - - - - - - - - - - - 312.85
Appearance - - - - - - - - - White Crystalline Solid
Melting Point- - - - - - - - - - - - - - - - - 110°C
Odor - None

Approximate Solubility at 25°C - % by weight

Water- - - - - -	Insoluble	n-Heptane - - - - - -	16
Acetone- - - - - -	26	Carbon Tetrachloride-	22
Cyclohexanone- - -	32	Methanol- - - - - - -	3
Benzene- - - - - -	38	Diethyl Ether - - - -	33
Ethyl Acetate- - -	27	Ethylene Glycol -	Insoluble

Table 3.94: α,α',2,3,5,6-Hexachloro-p-Xylene *(30)*

PHYSICAL PROPERTIES

Empirical Formula	$C_8H_4Cl_6$
Molecular Weight	312.86
Appearance	White crystalline solid
Melting Point, °C	179–180
Boiling Point, °C at 30 mm/Hg	220
Odor	Musty

Approximate Solubility (% by wt) in Various Solvents at 25°, 50°, and 75°C

Solvent	. . . . Temperature (°C)		
	25	50	75
Acetone	4	6	—
Benzene	12	19	35
Carbon Tetrachloride	2	5	9
Cyclohexanone	6	13	24
p-Dioxane	5	12	23
Ethyl Acetate	2	6	14
Ethyl Ether	2	—	—
Ethylene Glycol	0	0	0
n-Heptane	2	3	6
Methanol	0	0	0
Tetrahydrofuran	11	18	—
Water	0	0	0

Table 3.95: n-Hexyl Chloride *(7)*

$$C_6H_{13}Cl$$

Boiling range	133-135°C
Flash point	95°F
Specific gravity @20/20°C	0.877

Table 3.96: Isophthaloyl Chloride *(30)*

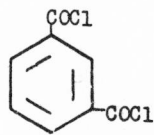

PHYSICAL PROPERTIES

Empirical Formula	$C_8H_4Cl_2O_2$
Molecular Weight	203.03
Appearance	White crystalline solid
Melting Point	$43^\circ C$
Boiling Point	$276^\circ C$
Odor	Sharp, pungent

Approximate Solubility (% by wt at $25^\circ C$)

n-Heptane	13
Benzene	73
CCl_4	62

Table 3.97: Isopropyl Chloride *(7)*

$$CH_3CHClCH_3$$

Boiling point	35.4°C
Freezing point	−117°C
Refractive index	1.3811
Solubility in water @12.5°C	0.344
Specific gravity @20/4°C	0.8590 g/100 g
Viscosity @22.5°C	0.2962 centipoise
Weight per gallon	7.5 lb

Table 3.98: Methyl Chloride *(23)*

Monochloromethane	CH_3Cl

PHYSICAL PROPERTIES

Molecular Weight	50.49
Boiling Point, 760 mm. Hg	− 23.7° C. (− 10.7° F.)
Freezing Point	− 97.6° C.
Vapor Pressure, mm., Hg. at 0° C.	1892
psia. at 0° C.	36.6
Flammable limits, (percent by volume in air)	10.7-17.4
Flash Point	−50° F calculated
Heat of Vaporization, cal./gm. at b.p.	102.45
Heat of Fusion, Cal./mole	1537
Heat of Combustion Kcal./mole	164.2
Specific Heat, cal./gm. ° C.	
Liquid at 20° C.	0.381
Vapor at 25° C. and 1.021 atmos.	0.199
Critical Temperature	143.1° C. (289.4° F.)
Critical Pressure, atmos.	65.9 (968.7 psia.)
Critical Density, gm./cc.	0.353
Refractive Index, liquid at − 23.7° C.	1.3712
Vapor at 25° C.	1.000703
Solubility, cc/100 cc. solvent at 20° C.	
Water	303
Benzene	4723
Carbon Tetrachloride	3756
Acetic Acid	3679
Ethyl Alcohol	3740

(continued)

Table 3.98: (continued)

Thermodynamic Properties of Methyl Chloride (Ideal Gas State)

T TEMP °K	C°p HEAT CAPACITY CAL /DEG / MOLE	H°T − H°298.15 HEAT CONTENT CAL /MOLE	S°T ENTROPY CAL /DEG / MOLE	F° − H°298.15 / T FREE ENERGY FUNCTION CAL /DEG / MOLE	FORMATION FROM ELEMENTS		
					HEAT Δ H°f, CAL /MOLE	FREE ENERGY Δ F°f CAL /MOLE	LOG₁₀Kp
298	9.73		55.80	55.80	-20630	-14952	10.960
300	9.76	18	55.86	55.80	-20642	-14918	10.868
400	11.50	1080	58.91	56.21	-21283	-12907	7.052
500	13.28	2332	61.70	57.04	-21825	-10750	4.699
600	14.64	3707	64.19	58.02	-22313	- 8495	3.094
700	15.92	5236	66.55	59.07	-22699	- 6144	1.918
800	17.03	6885	68.75	60.15	-23012	- 3764	1.028
900	17.76	8400	70.30	60.97	-23494	- 1111	.269
1000	18.86	10480	71.75	61.27	-23444	2086	- 455
1100	19.60*	12400	74.58	63.31	-23586	3551	- 705
1200	20.26	14400	76.32	64.32	-23673	6027	- 1.097
1300	20.82	16450	77.97	65.32	-23747	8493	- 1.427
1400	21.32	18560	79.52	66.27	-23789	10987	- 1.715
1500	21.75	20720	81.01	67.20	-23803	13502	- 1.967

*ACTUALLY IS 298.15° K.

Density of Liquid Methyl Chloride

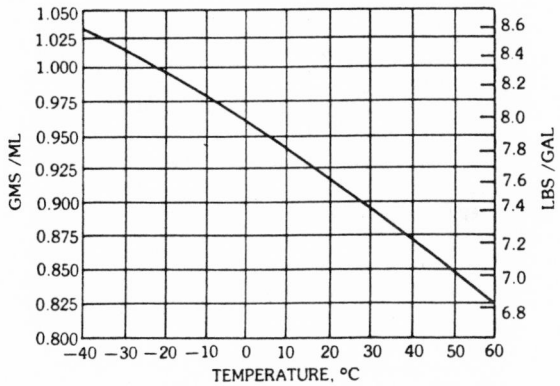

Density of Saturated Methyl Chloride Vapor

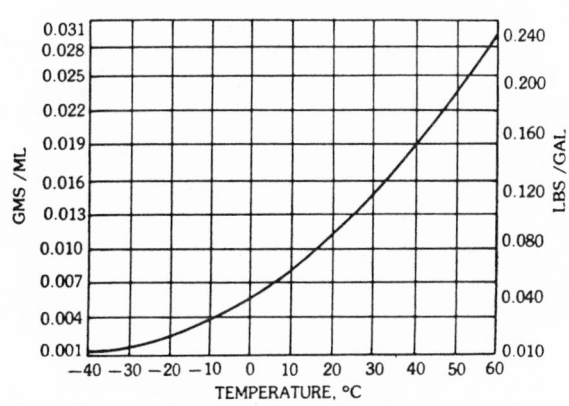

Vapor Pressure of Methyl Chloride

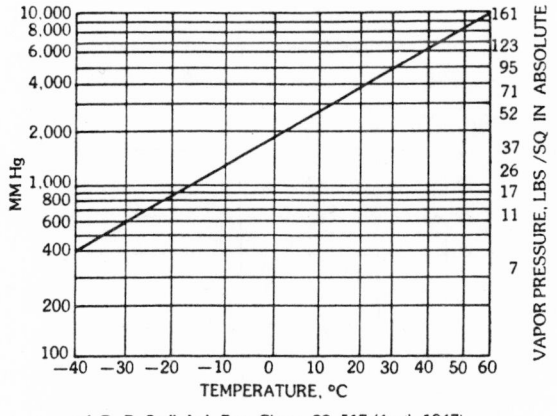

ref. D. B. Stull, Ind. Eng. Chem. *39*, 517 (April, 1947)

Viscosity of Liquid Methyl Chloride

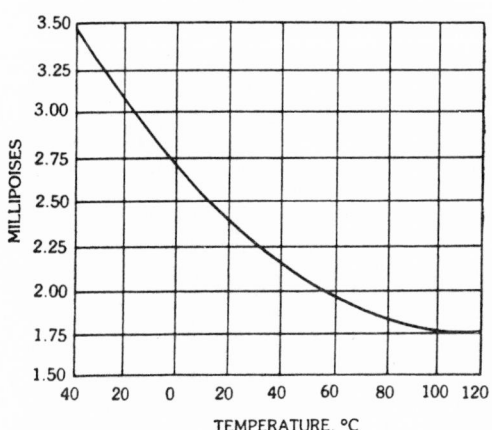

(continued)

Table 3.98: (continued)

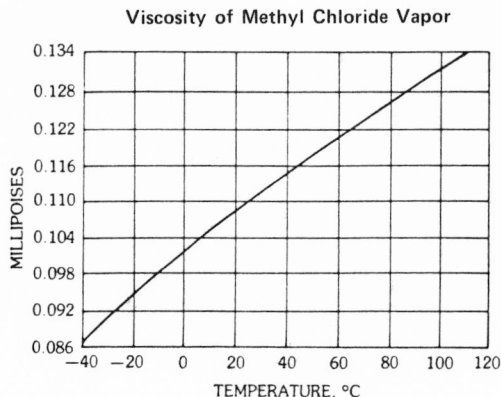

Viscosity of Methyl Chloride Vapor

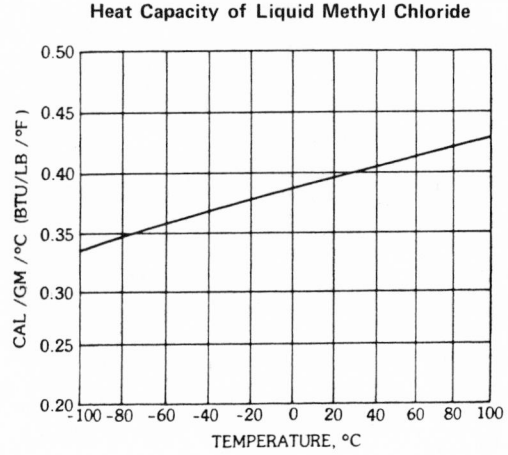

Heat Capacity of Liquid Methyl Chloride

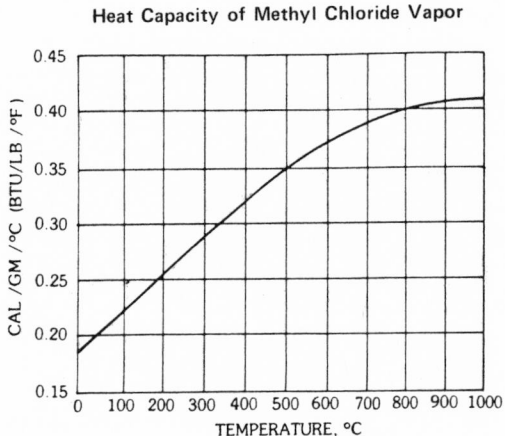

Heat Capacity of Methyl Chloride Vapor

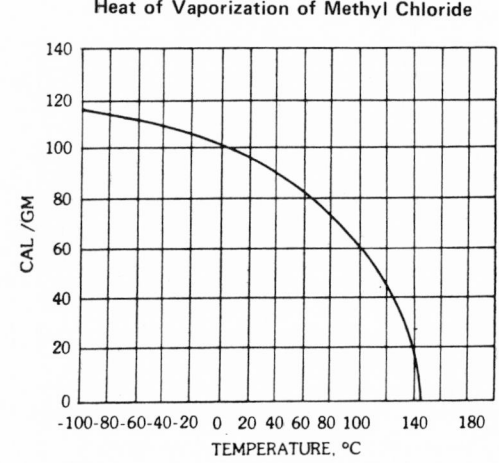

Heat of Vaporization of Methyl Chloride

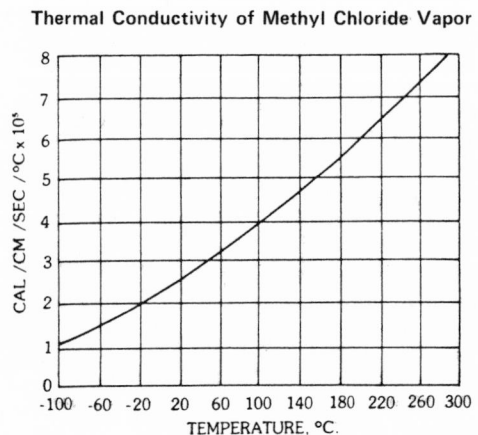

Thermal Conductivity of Methyl Chloride Vapor

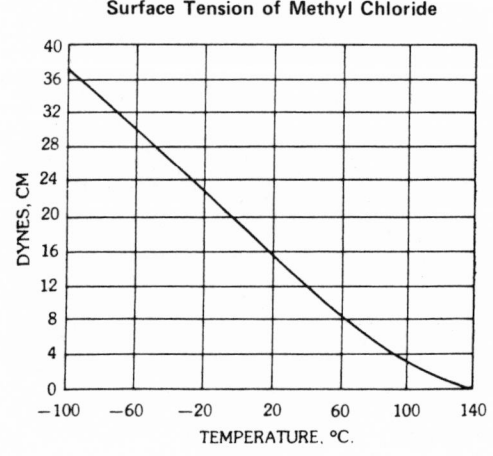

Surface Tension of Methyl Chloride

Table 3.99: Monochlorobenzene *(7)*

Chlorobenzene

Phenyl Chloride C_6H_5Cl

Monochlorobenzol

PHYSICAL PROPERTIES

Boiling point @ 760 mm	131.6°C
Dielectric constant, 1000 cycle	5.53
Electrical conductivity @ 0°C	10^{-9} recip./ohm
Fire point	36°C
Flash point	27°C
Freezing point	-45°C
Latent heat of vaporization @ B.P.	77.6 cal/g
Refractive index @ 15°C	1.52748
Solubility in water @ 30°C	0.049% by wt
Specific gravity @ 15/4°C	1.1117
Specific heat	0.30 cal/g/°C
Specific resistivity	7.8×10^9 ohms/cm
Surface tension (C.G.S. units)	33.08
Vapor pressure @ 20°C	8.71 mm
Viscosity @ 15°C	0.00844 cgs unit
Weight per gallon @ 25°C	9.19 lb

Table 3.100: Monochlorohydrin *(7)*

$ClCH_2CH(OH)CH_2OH$

PHYSICAL PROPERTIES

Boiling point	213°C
Boiling range	213-228°C (decomposes)
Specific gravity @ 18°C	1.326

Table 3.101: Pentachloroethane *(7)*

$$CHCl_2-CCl_3$$

PHYSICAL PROPERTIES

Acidity as HCl	0.001% by wt, max.
Boiling point @ 760 mm	161.9°C
Coefficient of cubical expansion Av./°C, liquid	0.0009097
Color (Saybolt)	18 max.
Explosion limits	None
Flash point	Nonflammable
Free halogen	None
Freezing point	-22.0°C
Latent heat of vaporization @ B.P.	43.6 cal/g 78.4 Btu/lb
Nonvolatile matter	0.0007% by wt, max.
Refractive index	1.5035
Solubility in water @ 25°C	0.05 g/100 g
Solubility of water in solvent @ 20°C	0.24 g water/100 g
Specific gravity @ 20.4°C	1.678
Specific heat Liquid, 20°C	0.215 cal/g/°C
Vapor density (B.P. and 760 mm)	568 g/liter
Vapor pressure @ 30°C	6 mm
Viscosity liquid @ 20°C	2.45 centipoises
Weight per gallon @ 20°C	14.00 lb

PENTACHLOROETHANE FORMS AZEOTROPES WITH:

%		B.P. °C of Azeotrope
3	Acetamide	160.5
26	Butyric acid	156.8
97	Camphene	159.3
9.9	Chloroacetic acid	158.7
36	Cyclohexanol	157.9
28	Cyclohexanone	165.4
22.5	1,3-Dichloro-2-Propanol	159.7
32	Dimethyl oxalate	157.6
65	Ethyl lactate	153.5
50	2-Furaldehyde	155.2
15	Glycol	154.5
46	Hexyl alcohol	155.8
50	Isoamyl propionate	158.7
43	Isobutyric acid	152.9
9	Isovaleric acid	160.3
56	Mesitylene	166.0
97	Methylheptenone	173.3
9.5	Phenol	160.9
89	α-Pinene	155.6

Table 3.102: Perchloroethylene *(22)*

"Perchlor"

TYPICAL PROPERTIES

Perchlorethylene is a clear, water-white liquid at ordinary temperatures. It is completely miscible with most organic liquids. The stabilized product, Perchlor, can be used with any of the common construction metals.

Chemical Names: Tetrachloroethylene; perchloroethylene

Chemical Formula: CCl_2CCl_2;

$$\begin{array}{c} Cl \\ \diagdown \\ Cl \diagup \end{array} C = C \begin{array}{c} \diagup Cl \\ \\ \diagdown Cl \end{array}$$

Molecular Weight	165.85
Boiling Point, °F	250.0
°C	121.1
Freezing Point, °F	−8.2
°C	−22.3
Pounds per Gallon at 68°F (20°C)	13.57
Kilograms per Liter at 20°C	1.63
Refractive Index, n_D^{20}	1.5053
Dielectric Constant at 1000 cps and 25°C	2.365
Specific Heat at 20°C cal/(g) (°C) or Btu/(lb) (°F)	0.205
Flash Point (Tag open cup)	None
Fire Point (Tag open cup)	None

Heat of Vaporization at 760 mm Hg. cal/g	50.1
Btu/lb	90.2
Vapor Density at 121.1°C and 760 mm Hg. g/l	5.22
lb/ft^3	0.326
Specific Gravity of Vapor (air = 1)	5.83
Vapor Pressure at 20°C. mm Hg	14.2
Evaporation Rate at 77°F (25°C) (ether = 100)	9
gal/(ft^2) (day)	0.15
Flammability	Nonflammable
Viscosity at 20°C, cps	0.88
Solubility at 25°C, g Perchlor/100 g water	0.015
g water/100 g Perchlor	0.0105
Azeotrope with Water, Boiling Point, °F	189.2
°C	87.7
Azeotropic Water Content, wt %	15.8
Permissible Exposure Limit (8-hour TWA) ppm	100

Specification and Typical Analysis, PPG Perchlor, All Grades:

	Specification	Typical Analysis
Appearance	Clear, free of suspended matter	Clear, free of suspended matter
Color, APHA	15 maximum	8
Odor	Characteristic; no residual	Characteristic; no residual
Spot Test	No spot or stain	No spot or stain
Specific Gravity, 20°C/20°C	**1.623 to 1.628**	1.624
Nonvolatile Residue, wt %	0.0025 maximum	0.0003
Free Chlorine	None	None
Moisture	No cloud at 0°C	No cloud at −5°C
Distillation Range (100%), °C	120.0 to 122.0	120.8 to 121.6
°F	248.0 to 251.6	249.4 to 250.9
pH	—	Drycleaning, 6.8 Degreasing, 8.4

Table 3.103: Propylene Chlorohydrin *(7)*

Chloroisopropyl Alcohol   CH₃CHOHCH₂Cl

Propylene chlorohydrin is a colorless liquid with a milk odor; it is freely soluble in water. It is largely used in organic syntheses, for the purpose of introducing the hydroxypropyl group.

PHYSICAL PROPERTIES

Acidity as HCl	0.02% by wt
Absolute viscosity @20°C	4.7 centipoises
Apparent specific gravity @20/20°C	1.1128
Boiling point @760 mm Hg	127.4°C
@ 50 mm Hg	59°C
@ 10 mm Hg	31°C
Coefficient of expansion @55°C	0.00097
Constant-boiling mixture @760 mm: Chlorhydrin approx. 46% Water 54%	B.P. 95.4°C
Flash point (Cleveland O.C.)	125°F
Molecular weight	94.54
Solubility in water	Miscible in all proportions
Vapor pressure @20°C	4.9 mm Hg
Weight per gallon @20°C	9.29 lb

Table 3.104: Propylene Dichloride *(7)*

1,2-Dichloropropane   CH₃CHClCH₂Cl

Propylene Chloride

PHYSICAL PROPERTIES

Acidity as HCl	0.005% max.
Boiling point	95.9°C
Boiling range @760 mm	93-99°C
Coefficient of expansion per °C	0.001108-20°C
	0.001153-55°C
Dielectric constant, 85.8 kilocycles	8.925 recip. ohms @26°
Explosive limits in air	Lower = 3.14% by vol. @25°C
	Upper = 14.5% by vol. @100°C
Flash point (ASTM O.C.)	21°C
Free halogen	None
Freezing point	-70°C
Ignition temperature in air	557°C
Latent heat of vaporization @B.P.	72.2 cal/g
Nonvolatile matter	0.005 g/100 cc, max.
Refractive index	1.4418
Solubility in water @20°C	0.26% by wt
Solubility of water in solvent @20°C	0.07% by wt
Specific gravity @20/20°C	1.157-1.163
Specific heat Liquid, 20°C	0.31 cal/g/°C or Btu/lb/°F
Surface tension @25°C	31.4 dynes/cm
Vapor density (B.P., 760 mm)	3.72 g/liter
Vapor pressure @20°C	35.8 mm
Viscosity @20°C	0.00865 poise
Weight per gallon @20°C	9.65 lb

Table 3.105: Terephthaloyl Chloride *(30)*

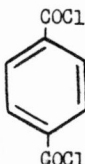

Physical Properties

```
Empirical Formula - - - - - - - - - - - - - - - - - - - - - - - - - C8H4Cl2O2
Molecular Weight - - - - - - - - - - - - - - - - - - - - - - -  203.03
Appearance - - - - - - - - - - - - - - - - - - White Crystalline Solid
Melting Point - - - - - - - - - - - - - - - - - - - - - - - - - - - 81°C
Boiling Point - - - - - - - - - - - - - - - - - - - - - - - - - 266°C
Odor - - - - - - - - - - - - - - - - - - - - - - - Sharp, Pungent
```

Approximate Solubility (% by wt. at 25°C)

```
n-Heptane - - - - - - - - - - - - - - - - - - - - - - - - - - - - - - 1
Benzene - - - - - - - - - - - - - - - - - - - - - - - - - - - - - - 37
CCl4 - - - - - - - - - - - - - - - - - - - - - - - - - - - - - - 9
```

Table 3.106: 1,1,2,2-Tetrachloroethane *(7)*

Acetylene Tetrachloride Bonoform	$CHCl_2-CHCl_2$
Acidity as HCl	0.0027% by wt, max.
Boiling point	146.5°C
Coefficient of cubical expansion Av./°C, liquid	0.000998 (15-99°C)
Fire point	Nonflammable
Flash point	Nonflammable
Free halogen	None
Freezing point	−43°C
Heat of vaporization @B.P.	55.1 cal/g
Refractive index	1.4942
Residue on evaporation	0.00062% by wt
Solubility in water @25°C	0.32% by wt
Solubility of water in solvent @20°C	0.03% by wt
Vapor pressure @30°C	9 mm
Viscosity liquid @20°C	1.7 centipoises
Water: no cloud @−10°C	0.032% by wt
Weight per gallon @25°C	13.25 lb

1,1,2,2-TETRACHLOROETHANE FORMS AZEOTROPES WITH:

%		B.P. °C of Azeotrope
45	Butyl propionate	152.5
3.8	Butyric acid	145.7
1.8	Chloroacetic acid	146.3
55	Cyclohexanone	159.1
74	2-Ethoxyethyl acetate	158.2
27	Ethyl chloroacetate	147.5
39	Ethyl orthoformate	151.5
97	2-Furaldehyde	161.6
9	Glycol	145.1
32	Isoamyl acetate	150.1
98	Isoamyl alcohol	131.3
37	Isobutyl isobutyrate	144.9
7	Isobutyric acid	144.8
15	Mesityl oxide	147.5
52	Methyl lactate	143.3
60	Propionic acid	140.4
34	Propyl butyrate	150.2
45	Styrene	143.5

Table 3.107: Tetrachloroethylene (27)

Perchloroethylene

Tetrachloroethene

$$CCl_2=CCl_2$$

PHYSICAL PROPERTIES

Acidity as HCl	0.001% by wt, max.
Boiling point	121.0°C
Boiling range @ 760 mm	120-122°C
Coefficient of cubical expansion Av./°C, liquid	0.001079 (15-90°C)
Color (Saybolt)	22
Dielectric constant, 1000 cycle	2.20
Dielectric strength, 0.1° gap	30,000 volts
Explosive limits	None
Fire point	Nonflammable
Flash point	Nonflammable
Freezing point	-22.4°C
Latent heat of vaporization @ B.P.	50.1 cal/g
Nonvolatile matter	0.0007% by wt, max.
Power factor, 1000 cycle	0.02%
Refractive index	1.5055
Residue on evaporation	0.0106% by wt, max.
Solubility in water @ 25°C	0.04% by wt
Solubility of water in solvent @ 20°C	0.02% by wt
Specific gravity @ 25/25°C	1.618
Specific heat	0.21 cal/g/°C
Specific resistivity	1.8×10^{13} ohms/cm
Vapor pressure @ 30°C	28 mm
Viscosity @ 20°C	0.90 centipoise
Water content	0.008% by wt
Weight per gallon @ 25°C	13.46 lb

TETRACHLOROETHYLENE FORMS AZEOTROPES WITH:

	%	B.P. °C of Azeotrope
Acetamide	2.6	120.5
Acetic acid	38.5	107.4
Allyl alcohol	46	93.4
1-Bromo-3-Methylbutane	52	119.3
Butanol	29	109.0
2-Chloroethanol	24.3	110.0
Diethyl carbonate	26	118.6
Epichlorohydrin	51.5	110.1
Ethanol	63	76.8
Ethyl butyrate	43	119.5
Glycol	6	119.1
Isoamyl alcohol	20	116.1
Isoamyl formate	35	117.9
Isobutanol	40	103.1
Isopropanol	70	81.7
Isobutyric acid	3	120.5
Isobutyl acetate	53	115.5
Isobutyl ether	35	119.5
Isopropyl isobutyrate	55	119.0
2-Methoxyethanol	24.5	109.7
Isobutyl nitrate	42	117.0
4-Methyl-2-Pentanone	52	113.9
Paraldehyde	32	118.8
Propanol	48	94.1
Propionic acid	8.5	119.2
Pyrrol	19.5	113.4
1,1,2-Trichloroethane	43	112.0
Triethyl borate	52	117.5

(continued)

Table 3.107: (continued)

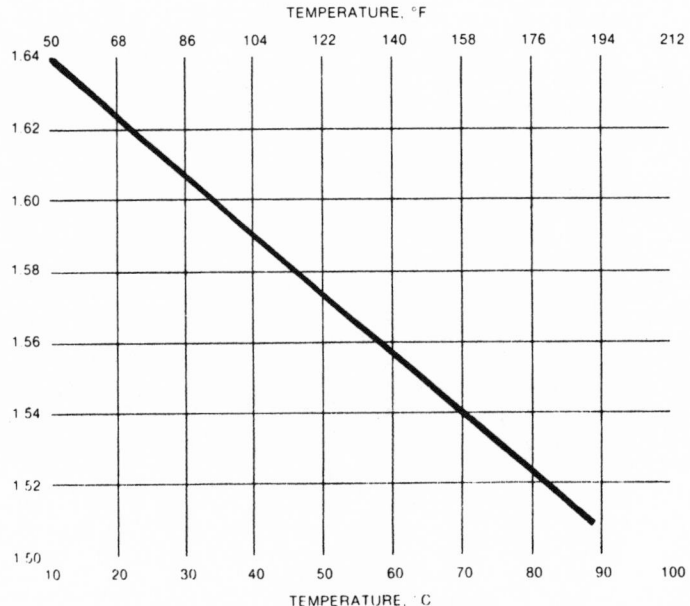

Specific Gravity vs Temperature of Hooker
Perchlorethylene

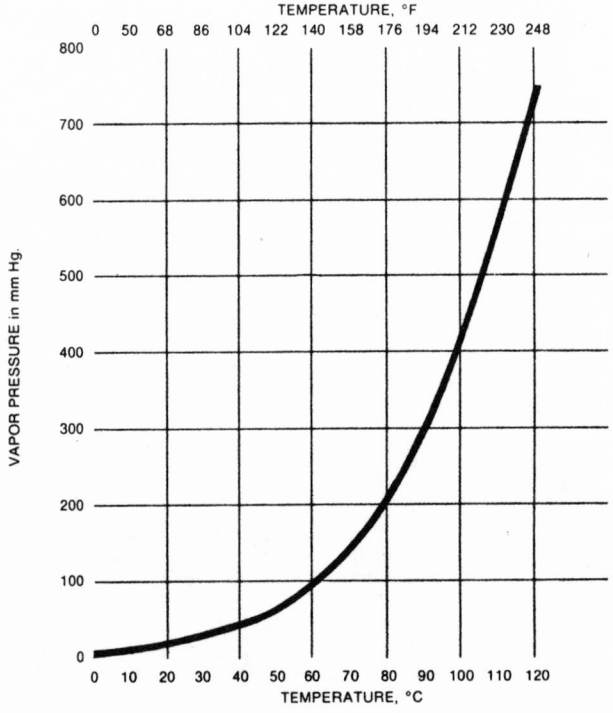

Vapor Pressure vs Temperature of Hooker
Perchlorethylene

Table 3.108: 2,3,5,6-Tetrachloroterephthaloyl Chloride *(30)*

PHYSICAL PROPERTIES

Empirical Formula	$C_8Cl_6O_2$
Molecular Weight	340.83
Appearance	White crystalline solid
Melting Point, °C	145
Boiling Point, °C	331
Odor	Odorless or (slightly sweet)

Approximate Solubility (% by wt) in Various Solvents at 25°, 50°, and 75°C

Solvent	Temperature (°C)		
	25	50	75
Acetone	5	12	—
Benzene	24	38	58
Carbon Tetrachloride	8	16	31
Cyclohexanone	8	20	40
p-Dioxane	12	26	47
Ethyl Acetate	6	14	30
Ethyl Ether	2	—	—
Ethylene Glycol	0	0	0
n-Heptane	2	6	12
Methanol	0.3	2	—
Perchloroethylene	6	13	26
Tetrahydrofuran	22	34	—
Water	0	0	0

Table 3.109: 2,3,5,6-Tetrachloro-p-Xylene *(30)*

PHYSICAL PROPERTIES

Empirical Formula	$C_8H_6Cl_4$
Molecular Weight	243.96
Appearance	White crystalline solid
Melting Point	223°C
Odor	None

Stability: No evidence of decomposition has been observed either at room temperature when allowed to stand for two or more months or on being heated to 225°C.

Approximate Solubility at 25°C (% by wt)

Water	Insoluble	n-Heptane	Insoluble
Acetone	Insoluble	Carbon Tetrachloride	1
Cyclohexanone	Insoluble	Methanol	Insoluble
Benzene	4	Diethyl Ether	Insoluble
Ethyl Acetate	Insoluble	Ethylene Glycol	Insoluble

Table 3.110: 2,3,5,6-Tetrachloro-p-Xylene-α,α'-diol *(30)*

Physical Properties

```
Empirical Formula - - - - - - - - - - - - - - - - - - - - -    C8H6Cl4O2
Molecular Weight- - - - - - - - - - - - - - - - - - - - - -    275.97
Appearance- - - - - - - - - - - - - - - - White Crystalline Solid
Melting Point, °C - - - - - - - - - - - - - - - - - - - - -    229-230
Boiling Point, °C @ 22 mm/Hg - - - - - - - - - - - - - -    253-264
Odor- - - - - - - - - - - - - - - - - - - - - - - - - - - - -    None
```

Approximate Solubility (% by weight) in
Various Solvents at 25°, 50°, and 75°C

Solvent	Temperature (°C)		
	25	50	75
Acetone	0	0.7	--
Benzene	0	0	0
Carbon Tetrachloride	0	0	0
Cyclohexanone	0	0.6	2
p-Dioxane	0	0.6	2
Ethyl Acetate	0	0.3	0.6
Ethyl Ether	0	--	--
n-Heptane	0	0	0
Methanol	0	0.8	--
Tetrahydrofuran	0	2	--

Table 3.111: Trichlorobenzenes *(7)*

	1,2,3- TRICHLOROBENZENE	1,2,4- TRICHLOROBENZENE	1,3,5- TRICHLOROBENZENE
Synonym:	vic-Trichloro- benzene	uns-Trichloro- benzene	sym-Trichloro- benzene
Physical state:	White crystals	Colorless liquid	White crystals
Boiling point: @ 760 mm	221°C (429.8°F)	213°C (415.4°F)	208.5°C (407.3°F)
Density: 25/25°C	1.69 (solid)	1.451 (liquid)	
Flash point: (Tag C.C.)	113°C (235.4°F)	110°C (230.0°F)	107°C (224.6°F)
Index ($^{20}_{D}$) of Refraction:	1.5776	1.5732	1.5662

1,3,5-TRICHLOROBENZENE FORMS AZEOTROPES WITH:

%		B.P. °C of Azeotrope
50	Camphor	210.5
43	Caproic acid	204.0
60	m-Cresol	200.5
40	p-Cresol	200.2
35	N-Ethylaniline	203.0
	Menthone	209.5
95	Phenol	181.3
	p-Toluidine	199.0

Table 3.112: 1,1,1-Trichloroethane *(7)(23)*

"Chlorothene VG" CH_3CCl_3
"Chlorothene SM"
"Aerothene MM"
Methyl Chloroform

Chlorothene VG solvent is a specially inhibited grade of 1,1,1-trichloroethane. Chlorothene SM and Aerothene MM solvents are special grades of 1,1,1-trichloroethane and methylene chloride.

Physical Properties of Chlorothene VG Solvent

NOTE: These properties are laboratory results typical of the product. but should not be confused with or regarded as specifications

Molecular Weight 133 4
Freezing Point .... - 36.9 C
Flash Point, Tag Open Cup,
　ASTM Method D-1310 .. None
Flash Point, Tag Closed Cup,
　ASTM Method D-56-70 .. None
Fire Point, Tag Open Cup,
　ASTM Method D-1310 ... None
Boiling Range at 760 mm
　Hg. IBP-DP ... 72-88 C
Refractive Index at 22 C ... 1.434

Viscosity, centipoise[1]

0°C	1.117
20	0.830
40	0.639
60	0.506

Density

0°C	1.355 g/ml
15	1.331 g/ml
20	1.323 g/ml
25	1.315 g/ml
25	10.97 lb/gal
30	1.307 g/ml
35	1.299 g/ml

Specific Gravity

60 60° F	1.333
20 20° C	1.327
25 25° C	1.320

Vapor Pressure, mm Hg[1]

30°C	157
40	238
50	344
60	490
74	760

Heat of Vaporization[2] kcal/mol

20°C	7.8
50	7.5
80	7.1

Specific Heat[1] cal/g/°C

0 C	0.238
20	0.240
40	0.250
60	0.275

Surface Tension[1], dyn/cm

0°C	27.40
10	26.26
20	25.12
30	23.98

Dielectric Strength at 22 C 0.1 in.

(ASTM D-877-67) .... 28-32 KV

Dielectric Constant at 24 C

(ASTM D-924-65)
　@ 10³ cps 10.0
　@ 10⁵ cps 7.0

Specific Resistance at 400 Hz at 24 C

(ASTM D-1169-64)   3.1 x 10⁹ ohm/cm

Thermal Conductivity Coefficients

K = Btu/(hr·ft²) • (°F/ft)

F	Liquid	K
68		0.0797
86		0.0773
104		0.0725
122		0.0700

1. Values for pure 1,1,1-trichloroethane
2. Calculated from vapor pressure data

Specifications for Chlorothene VG Solvent

Test Items	Unit	Limits
Appearance		Free of sediment and suspended matter
Water Content, Max.	PPM	100
Color, PT-CO, Max.		10
Non-volatile Residue. Max.	PPM	10
Acid Acceptance (as NAOH) Min.	%W	0.20
Purity 1.1.1-Trichloroethane, Min.	%W	94.5
Individual Halogenated Impurities. Max.	%V	0.5
Total Halogenated Impurities, Max.		1.0
Typical Properties		
Free Halogens	PPM	None
Acidity (as HCL) Max.	PPM	10
Acidity After Accelerated Oxidation		Passes MIL-T-81533A
Metals Corrosion		Passes MIL-T-81533A
Relative Density (Specific Gravity @ 25/25 C)		1.317-1.324
Distil Range MM, IBP-DP	C	72-88

THERMODYNAMIC PROPERTIES OF UNINHIBITED 1,1,1-TRICHLOROETHANE (GAS)

Temp. °C	Heat Capacity cp (cal/g. mole)	Enthalpy (cal/g. mole)	Entropy S (cal/°/mole)
25	22.6	0	77.0
27	22.6	45	77.1
77	24.3	1215	80.7
127	25.8	2465	84.1
177	27.2	3790	87.2
227	28.5	5180	90.1
277	29.6	6630	92.9
327	30.5	8130	95.5
377	31.4	9680	98.0
427	32.1	11270	100.4
477	32.8	12890	102.6
527	33.4	14545	104.7
577	34.0	16230	106.8
627	34.6	17945	108.7
677	35.1	19685	110.6

WATER CONTENT OF INHIBITED AND UNINHIBITED 1,1,1-TRICHLOROETHANE

Temperature °C	Solubility of Water in Parts per Million	
	1,1,1-Tri-chloroethane	CHLOROTHENE VG
10	150	340
12	163	368
15	183	408
20	216	480
21	224	494
22	233	510
23	243	528
24	255	546
25	266	565
30	346	680

Reflux Boiling Points and Specific Gravities of Chlorothene VG and USP Mineral Oil Solutions

Mineral Oil Volume %	Boiling Point °F at 760 mm Hg	Specific Gravity 25/25°C
0	165	1.320
5	166	1.298
10	167	1.276
15	168.4	1.253
20	170	1.231
25	171.7	1.209
30	173.6	1.187
35	175.4	1.164
40	177.8	1.142
45	180.4	1.120
50	183.9	1.098
60	191.5	1.053
70	204.8	1.009
80	231.8	0.964

SOLVENT/WATER AZEOTROPES

	Solvent/Water Azeotrope Boiling Point	Weight Percent of Azeotropic Mixture
CHLOROTHENE VG ...	149°F (65°C)	4.3 Water 95.7 Solvent
Methylene Chloride ..	100.6°F (38.1°C)	1.5 Water 98.5 Solvent
Trichloroethylene....	164°F (73.3°C)	5.4 Water 94.6 Solvent
Perchloroethylene ...	190°F (87.8°C)	15.8 Water 84.2 Solvent

(continued)

Table 3.112: (continued)

Relative Evaporation Rates:

The relative evaporation rates were determined by evaporating the solvents from equal surface areas on an analytical balance for 5 minutes. The values are compared using the Chlorothene VG solvent value as one (1). Higher values indicate faster evaporation rates.

Chlorothene VG solvent	1.0
Methylene Chloride, Technical	3.5
Trichloroethylene	0.63
Perchloroethylene	0.26
Alcohols—Methanol	0.65
—Isopropyl, Anhydrous	0.34
Acetone	1.39
Trichlorotrifluoroethane (Fluorocarbon-113 brand)	3.9
Heptane	0.40
Stoddard Solvent	0.09

Air Saturation Data Chlorothene VG Solvent

°F	°C	Vapor Pressure (in mm) of Solvent at Indicated Temperature	Pounds of Solvent Vapor per 1,000 ft³ of Air
32	0	58.7	18.9
59	15	80.0	37.1
68	20	100.0	45.6
77	25	124.0	55.6
86	30	155.0	68.3
95	35	190.0	82.4
104	40	237.0	101.1
113	45	285.0	119.7
122	50	340.0	140.5

Vapor Pressure of "Chlorothene" VG

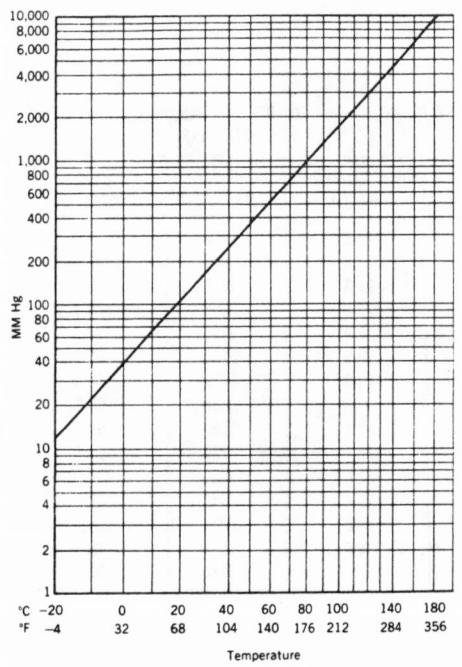

Vapor Pressure vs. Temperature of 1,1,1-Trichloroethane

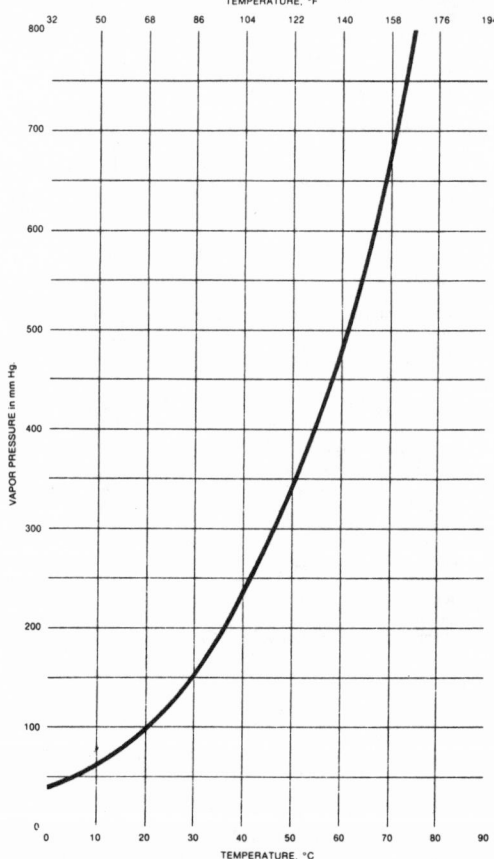

(continued)

Table 3.112: (continued)

Solubility of Various Substances in "CHLOROTHENE" VG Solvents

(Grams per 100 Grams of Solvent, at 25°C)

KEY

CM = Completely miscible
‡ = Too viscous for
   further addition.
Ins. = Insoluble

SUBSTANCE	Solubility
SOLVENTS	
acetone	CM
benzene	CM
carbon tetrachloride	CM
ethyl ether	CM
n-heptane	CM
methyl alcohol	CM
water	< 0.1
RESINS	
Abalyn rosin esterified with glycerine	100
Acryloid B-82 acrylic ester	80‡
Amberol 801-XLT phenolic resin	>100
Amberol ST-137-X phenol-formaldehyde	>100
Araldite 502 epoxy resin	100
Araldite 508 epoxy resin	100
Araldite 6010 epoxy resin	100
Araldite 7071 epoxy resin	1
Araldite 7079 epoxy resin	1
Bakelite CKR-5254 phenolic resin	<90‡
Beckacite 1001 phenolic resin	>100
Beckacite 1112 phenolic resin	>100
Cumar W-1 paracoumarone-indene	>100
D.E.N. 438 epoxy novolac resin	>100
D.E.R. 331 epoxy resin	>100
D.E.R. 332 epoxy resin	>100
D.E.R. 334 epoxy resin	>100
D.E.R. 661 epoxy resin	<1
D.E.R. 664 epoxy resin	<1
D.E.R. 667 epoxy resin	<1
Dow resin PS-3 polystyrene	>100
Epon 812 epoxy resin	>100
Epon 836 epoxy resin	< 5
Epon 1004 epoxy resin	<1
Epon 1009 epoxy resin	<1
Genepoxy 175 epoxy resin	>100
Genepoxy M-180 epoxy resin	>100
Genepoxy 190 epoxy resin	>100
Hercolyn rosin esterified with glycerine	>100
Methyl methacrylate	>100
Nevindene R5 coumarone-indene	>100
Orange shellac	Ins.
Paradene resin No. 3 Paracoumarone-indene	>100
Picco 420 ES polyindene	>100
Piccolastic A-75 polystyrene	>100
Piccolyte S-85 polyterpene	>100
Piccopale 100 hydrocarbon	>100
Polyvinyl alcohol	Ins.
Polyvinyl chloride	Ins.

SUBSTANCE	Solubility
Resin 276-V9 polyalkyl styrene	>100
Rosin wood	< 1
Saran F-120 vinylidene chloride-acrylonitrile	<1
Saran F-220 vinylidene chloride-acrylonitrile	<1
Velsicol AE9 ETO adducts	<10
Versamid 940 polyamide	<1
Vinylite AYAA vinyl acetate	>80‡
Vinylite VYHH vinyl chloride-acetate	<1
Zytel 31 nylon	Ins.
Zytel 61 nylon	Ins.
Zytel 101 nylon	Ins.
OILS & RESINS	
Alinco Z2 linseed oil	>100
Lanolin anhydrous	>100
Lard Oil	CM
Linseed oil raw	CM
Linseed oil boiled	CM
Mineral oil	CM
Motor oil	CM
OKO S-70 soybean oil	>100
WAXES	
Petrolatum white	1
Bees wax	< 5
Candelilla wax	< 1
Carnuba wax	< 1
Ceresin wax	< 1
Japan wax	< 1
Motan wax	< 1
Ozokerite white	Ins.
Paraffin 47-49°C. melt point	< 25
Paraffin 54-56°C. melt point	< 20
GUMS	
Acacia gum	Ins.
Copal gum	Ins.
Ester gum	100
Gutta percha natural gum rubber	Ins.
Kauri gum	< 1
GLYCOL DERIVATIVES	
Dowanol EEA ethylene glycol monoethyl ether acetate	CM
Dowanol EM ethylene glycol methyl ether	CM
Dowanol EB ethylene glycol n-butyl ether	CM
Dowanol PM propylene glycol methyl ether	CM

(continued)

Table 3.112: (continued)

SUBSTANCE	Solubility	SUBSTANCE	Solubility
Polyglycol E1000 polyethylene glycol	CM	Parlon S-10 chlorinated rubber	>15
Polyglycol E1450 polyethylene glycol	CM	S-1011 styrene-butadiene rubber	Ins.
Polyglycol E4000 polyethylene glycol	CM	S-1013 styrene-butadiene rubber	<15
Polyglycol E6000 polyethylene glycol	CM	Ameripol CB220 styrene-butadiene	Ins.
Polyglycol E9000 polyethylene glycol	CM	Ameripol 1006 styrene-butadiene	>15
Polyglycol P400 polypropylene glycol	CM	Ameripol 1012 styrene-butadiene	Ins.
Polyglycol P1200 polypropylene glycol	CM	Ameripol 1013 styrene-butadiene	>15
Polyglycol P2000 polypropylene glycol	CM	Butyl 268 butyl rubber	<15
Polyglycol P4000 polypropylene glycol	CM	FR-N 501 acrylonitrile	<15
		FR-N 503 acrylonitrile	<15
CELLULOSICS		FR-S 181 styrene-butadiene	Ins.
Cellulose acetate acetone soluble	Ins.	LP-2 polysulfide	>15
Cellulose triacetate	Ins.	LP-3 polysulfide	>15
Ethocel 50 cps ethyl cellulose	14‡	Vistanex LMMS polyisobutylene	>15
Ethocel 100 cps ethyl cellulose	14‡	Chemigum N-600 acrylonitrile	Ins.
Methocel 400 cps methyl cellulose	Ins.	Hycar 1022 acrylonitrile	<60
Methocel 690 cps methyl cellulose	Ins.	Paracril CV acrylonitrile	<15
Methocel 4000 cps methyl cellulose	Ins.	Diene 35NF polybutadiene	>15
(PR Grade)		Naugapol 1022 styrene-butadiene	Ins.
		Kraton 1101 styrene-butadiene	>30
RUBBERS			
Neoprene AC polychloroprene	30	**FATTY ACIDS & DERIVATIVES**	
Neoprene AD 20 polychloroprene	<15	Acrawax C amide stearate	<1
Neoprene CG polychloroprene	Ins.	Calcium stearate	<1
Neoprene FB polychloroprene	<15	Oleic acid	CM
Neoprene FC polychloroprene	>15	Potassium oleate	<1
Neoprene HC polychloroprene	>15	Sodium oleate	<1
Neoprene ILA polychloroprene	Ins.	Stearic acid	<30
Parlon S-125 chlorinated rubber	>15	**MISC.**	
		Dow Corning R5061 silicone	>100
		Dow Corning R-5581 silicone	>100
		Sealing asphalt electrical	<5
		Sulphur precipitated	Ins.

Table 3.113: 1,1,2-Trichloroethane *(7)*

beta-Trichloroethane

Vinyl Trichloride $Cl-CH_2CH-Cl_2$

Ethylene Trichloride

PHYSICAL PROPERTIES

Acidity as HCl	0.0001% by wt, max.
Boiling point @ 760 mm	113.5°C
Boiling range @ 760 mm	111.8-113.3°C (5-95%)
Fire point	Nonflammable
Flash point	Nonflammable
Free halogen	None
Freezing point	-36.7°C

(continued)

Table 3.113: (continued)

Latent heat of evaporation @ B.P.	68.7 cal/g
Nonvolatile matter	None
Refractive index	1.4711
Solubility in water @25°C	0.48 g/100 g
Solubility of water in solvent @20°C	0.03 g water/100 g
Specific gravity @20/4°C	1.441
Specific heat Liquid, 20°C	0.266 cal/g/°C
Specific resistivity	5.2×10^8 ohms/cm
Vapor density (B.P., 760 mm)	4.21 g/liter
Vapor pressure @30°C	36 mm
@90°C (194°F)	369 mm
@100°C (212°F)	505 mm
@110°C (230°F)	680 mm
@114°C (237°F)	764 mm
Water	0.007% by wt, max.
Weight per gallon @20°C	12.04 lb

Table 3.114: Trichloroethylene *(7)(27)*

1,2,2-Trichloroethylene

$$CHCl\!=\!CCl_2$$

PHYSICAL PROPERTIES

Acidity as HCl	Not more than 0.001%
Boiling point	86.7°C
Boiling range @760 mm	95% or better distills from 86.0-87.5°C
Coefficient of expansion per °C	0.00115-20°C
Color (Saybolt)	24 max.
Dielectric constant, 1000 cycle	3.27
Fire point	Nonflammable
Flash point (ASTM O.C.)	None @ B.P.
Free chlorine	None
Freezing point	−86.4°C
Latent heat of vaporization @ B.P.	57.3 cal/g
Nonvolatile matter	0.00067% by wt, max.
Power factor, 1000 cycle	2.2%
Refractive index @27°C	1.4735
Solubility in water @25°C	0.10% by wt
Solubility of water in solvent @25°C	0.02% by wt
Specific gravity @20/20°C	1.4655
Surface tension @25°C	32.0 dynes/cm
Vapor pressure @−20°C	4.5 mm
@ −9°C	9.0 mm
@ 0°C	17.4 mm
@ 20°C	56.0 mm
@ 40°C	145 mm
@ 50°C	230 mm
@ 65°C	385 mm
@ 77°C	562 mm
Viscosity @25°C	0.00550 poise
Water content	0.002% by wt
Weight per gallon @20°C	12.20 lb

TRICHLOROETHYLENE FORMS AZEOTROPES WITH:

%		B.P. °C of Azeotrope
3.8	Acetic acid	87.0
16	Allyl alcohol	81.0
7.5	tert-Amyl alcohol	86.7
2.5	Butanol	86.9
33	tert-Butanol	75.8
12	1,2-Dichloroethane	82.9
46.	Diethoxymethane	89.29
8	Isobutanol	85.4
30	Isopropanol	75.5
17	Propanol	81.8
80	Propyl formate	79.5
27	Ethanol	70.9

(continued)

Table 3.114: (continued)

Trichlorethylene—Vapor Pressure/Temperature Variation

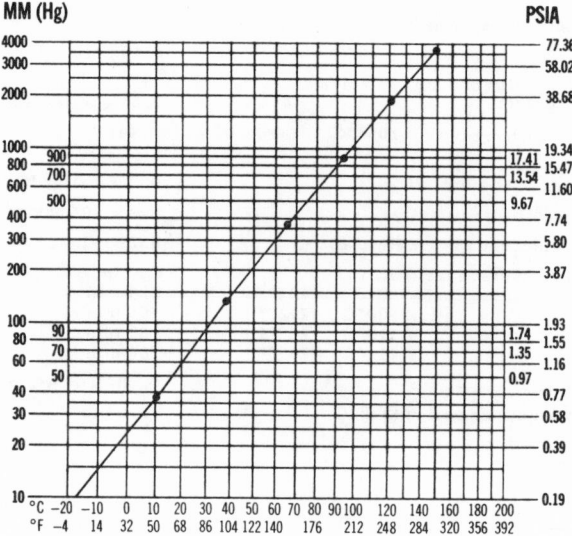

Typical Boiling Point Curve of Trichloroethylene—Mineral Oil Mixture

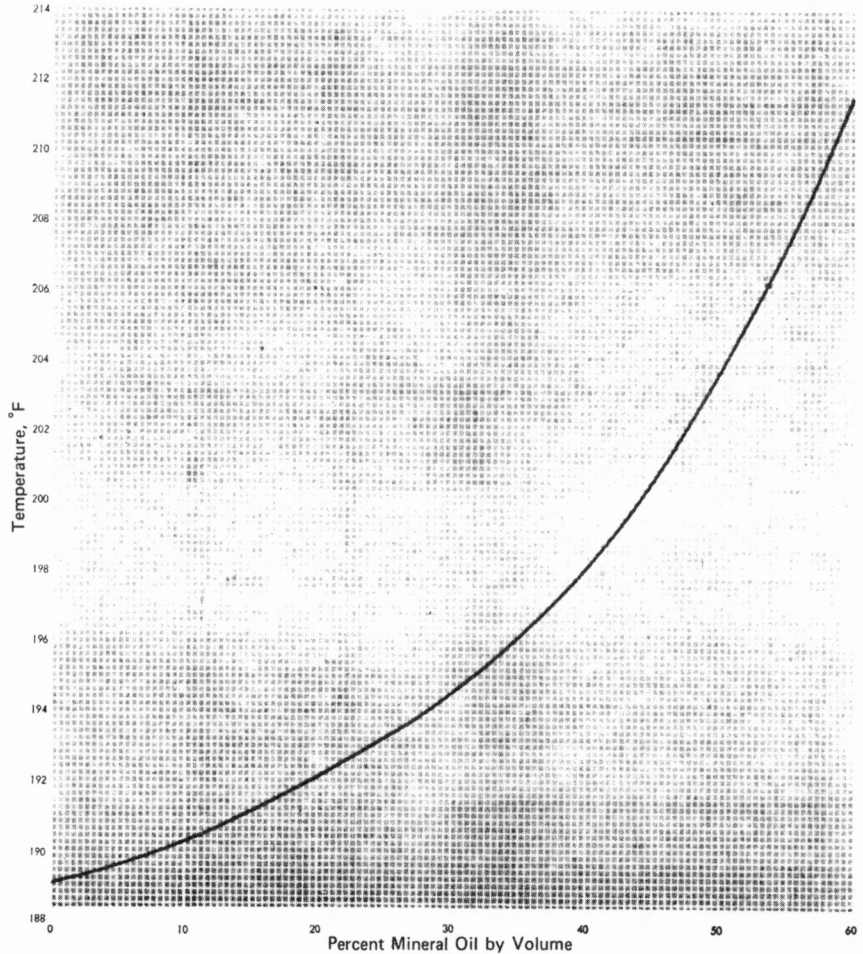

Table 3.115: Trichloropropane *(7)*

$$CH_2ClCHClCH_2Cl$$

PHYSICAL PROPERTIES

Analysis, % W	
Trichloropropane	97.0
Epichlorohydrin	1.5
Glycerol dichlorohydrin	1.5
Color, Pt-Co	15
Distillation range, IBP	150°C
90%	156.1°C
95%	156.1°C
DP	156.6°C
Flash point	165°F
Molecular weight	147.44
Refractive index	1.4832
Specific gravity @20/20°C	1.385

1,2,3-TRICHLOROPROPANE FORMS AZEOTROPES WITH

%		B.P. °C of Azeotrope
35	Camphene	152.9
30	2,7-Dimethyloctane	155.5
15	α-Pinene	150.0

Table 3.116: Triglycol Dichloride *(7)*

$$Cl(CH_2CH_2O)_2CH_2CH_2Cl$$

PHYSICAL PROPERTIES

Acidity as HCl	0.01% by wt, max.
Boiling point @760 mm	241.3°C
Boiling range @760 mm	Not more than 5% distills below 235°C Not less than 95% distills below 242°C
Color (Pt-Co scale)	25 max.
Dryness @20°C	Miscible with 19 vol. 60°Bé gasoline
Flash point (O.C.)	250°F
Solubility in water @20°C	1.89% by wt
Solubility of water in solvent @20°C	0.83% by wt
Specific gravity @20/20°C	1.1950-1.2000
Vapor pressure @20°C	0.06 mm
Weight per gallon @20°C	9.97 lb

Table 3.117: Vinyl Chloride *(7)*

Monochloroethylene $CH_2{=}CHCl$

PHYSICAL PROPERTIES

Acetaldehyde	Not more than 0.5% by wt
Boiling point @ 760 mm	−13.9°C
Boiling range @ 760 mm	Not less than 95% distills over before the temperature of the liquid reaches 10°C
Color	Water-white
Freezing point	−159.7°C
Heat of evaporization	81.6 cal/g
Residue	Not more than 0.5% by vol.
Specific gravity @ B.P.	0.97
Specific heat	0.27 cal/g/°C
Solubility in water @ 25°C	Slightly soluble
Weight per gallon @ 20°C	7.59 lb

Table 3.118: Vinylidene Chloride *(23)*

$$Cl_2C{=}CH_2$$

Molecular Weight (theoretical)	96.95
Odor	Pleasant, Sweet
Appearance	Clear liquid
Color	10-15 APHA
Solubility of monomer in H_2O at 25°C, weight %	0.021
Solubility of H_2O in monomer at 25°C, weight %	0.035
Boiling Point (760 mm Hg), °C	+ 31.56
Freezing Point, °C	−122.5

Vapor Pressure

$\log P_{mm} = 6.98200 \cdot 1104.29/(t + 237.697)$ t = temperature in °C.
Temperatures calculated at selected pressures are tabulated.

Pressure (mm Hg)	Boiling Point (°C)
760	31.56
400	14.43
200	− 1.78
100	−16.04
60	−25.49
40	−32.44
20	−43.31
10	−53.10
5	−61.94
1	−79.54

Liquid density

Temperature (°C)	Density (gm/cc)
−20	1.2902
0	1.2517
+20	1.2132

Pounds/gallon

Temperature (°C)	Pounds
−20	10.768
0	10.446
20	10.124

Index of Refraction

Temperature (°C)	
10	1.43062
15	1.42777
20	1.42468

Absolute Viscosity

Temperature (°C)	Viscosity (cps)
−20	0.4478
0	0.3939
+20	0.3302

(continued)

Table 3.118: (continued)

Flash Point (Tag closed cup), °F	0
Explosive limits in air (28°C), %	7.3 - 16.0
Auto-ignition temperature, °F	1058
Q Value =	0.22
e Value =	0.36
Latent Heat of Vaporization,	
ΔHv cal/mole at 25°C =	6,328 ± 0.3%
at Boiling Point =	6,257 ± 0.3%
Latent Heat of Fusion, ΔHm cal/mole =	1,557
Heat of Polymerization,	
ΔHp k cal/mole at 25°C =	−18.0 ± 0.9
ΔHp BTU/lb at 77°F =	334 (exothermic)
Heat of Combustion, Liquid Monomer	
ΔHc k cal/mole =	261.93 ± 0.3
Heat of Formation,	
Liquid Monomer, ΔHr k cal/mole =	−6.0 ± 0.3
Gaseous Monomer, ΔHr k cal/mole =	+0.3 ± 0.3
Heat Capacity, Liquid Monomer,	
C_p cal/mole deg at 25.15°C =	26.745
Heat Capacity, Ideal Gas State	
C_p cal/mole deg at 25.15°C =	16.04
Critical Temperature, T_c °C =	222
Critical Pressure, P_c Atmospheres =	51.3
Critical Volume, V_c cm³/mole =	219

COMPARATIVE DATA

Table 3.119: Alpha 564 and 565 *(81)*

Alpha 564 and 565 are functionally azeotropic blends of a chlorinated solvent and a polar component. Alpha 564 comprises perchloroethylene and 2-ethoxyethanol. Alpha 565 comprises 1,1,1-trichloroethane and n-propyl alcohol.

SPECIFICATIONS[1]

Property	Alpha 564	Alpha 565
Specific Gravity	1.506 ± 0.014	1.275 ± 0.010
Lb./Gal.	12.51	10.57
Residue on Evaporation (ppm max.)	25	10
pH, Typical	7.5	—
Acidity (cal. as ppm HCl)	—	10 (max.)
Acid Acceptance (cal. as Wt. % NaOH)	0.07, Min.	0.15, Min.
Appearance clarity	Clear	
color (Max. A.P.H.A.)	40	15
Boiling Point	118°C (244°F)	74°C (165°F)
Freezing Point	<0°C (<−18°F)	−34°C (−30°F)
Vapor Density at Boiling Point (lb./cu. ft)[2]	0.291	0.272
Latent Heat of Vaporization at Boiling Point (Btu/lb.)[2]	100	133
Specific Heat (Btu/lb./°F)[2]	0.240	0.268
Vapor Pressure (mm Hg)[2]	12.6	92.8
Kauri-Butanol Value[3]	90	124
Evaporation Rate (relative)[4]	0.21	0.55
Solubility of Water in Solvent (grams/100 cc)	<0.1	1.3
Toxicity (TLV)[2]	110	345
Flash Point[5]		
Tag Open Cup	None	
Tag Closed Cup	None	
Fire Point	None	

Notes:
[1]Unless specified, properties are room temperature (25°C; 77°F) and 1 atmosphere (760 mm Hg).
[2]Calculated values.
[3]Major constituent only.
[4]Relative evaporation rate calculated with a value of 1.00 assigned to F113.
[5]As per OSHA recommendations.

Table 3.120: Amsco Chlorinated Solvents *(13)*

	Specific Gravity @ 25°C	lb/gal @ 25°C	Distillation Range 760 mm Hg, °F (°C)	Relative Evaporation Rate n-BuAc = 1	Vapor Pressure mm Hg @ 20°C	Flash Point TAG CC °F	Solubility % by wt @ 20°C. in Water	of Water
Methylene chloride	1.32	10.98	103 (39.4)–104 (39.8)	14.5	390	None	1.3*	0.20*
1,1,1-Trichloroethane	1.29	10.76	162 (72.2)–190 (87.8)	6.0	134	None	0.07*	0.04*
Ethylene dichloride	1.25	10.42	179 (81.7)–186 (85.6)	4.5	66	59	0.90	0.15
Trichloroethylene	1.45	12.07	188 (86.6)–190 (87.8)	4.5	60	None	0.11*	0.032*
Perchloroethylene	1.62	13.46	250 (121.1)–254 (123.3)	2.1	14	None	0.015*	0.010*
Monochlorobenzene	1.11	9.19	267 (130.6)–270 (132.2)	1.1	9.2	105	0.048	Insoluble
Dichlorobenzene	1.30	10.84	355 (179.4)–362 (183.3)	0.2	1.2	155	0.014	Insoluble

*@25°

Table 3.121: Ashland Chlorinated Solvents *(69)*

	Specific Gravity 20° / 20°F	Dist. Range °F IBP	DP	Flash Pt. °F TOC
Carbon Tetrachloride	1.584	169	171	None
Chloroform	1.485	140	143	None
Ethylene Dichloride	1.255	181	183	70
Methylene Chloride	1.366	103	105	None
Monochlorobenzene	1.113	268	271	84*
Orthodichlorobenzene	1.313	355	361	170
Perchlorethylene	1.627	247	251	None
Trichlorethylene	1.455	187	190	None
1,1,1-Trichloroethane	1.316	162	190	None

*TCC

Table 3.122: PPG Chlorinated Solvents *(22)*

All three solvents are clear, water-white liquids. Their vapors are colorless. Their odors are not unpleasant.

	Tri-Ethane Solvent Degreasing Grade	Trichlor Degreasing Grade	Perchlor Degreasing Grade	Water
Chemical Formula	CH_3CCl_3	$CHCl:CCl_2$	$CCl_2:CCl_2$	H_2O
Molecular Weight	133.42	131.40	165.85	18.02
Boiling Point, °F	165.4	188.4	250.0	212
°C	74.1	86.9	121.1	100
Freezing Point, °F	−49	−123	−8	32
°C	−45	−86	−22	0
Specific Gravity, 20°C/20°C	1.304 25°C	1.460	1.624	1
Pounds per Gallon at 68°F (20°C)	10.84 25°C	12.2	13.6	8.3
Heat of Vaporization at 760 mm, cal/g	58	57	50	538.7
Btu/lb	104	103	90	969.7
Specific Heat at 20°C, cal/(g)(°C) or Btu/(lb)(°F)	0.258	0.225	0.205	1
Specific Gravity of Vapor (air=1)	4.6	4.5	5.8	0.6
Vapor Pressure, mm Hg at 20°C	120.0 25°C	57.8	14.2	17.5
Flash Point (Tag open cup)	None	None	None	None
Fire Point (Tag open cup)	None	None	None	None
Azeotrope with Water, Boiling Point, °F	149	163.4	189.2	NA
°C	65	73.0	87.7	
Azeotropic Water Content, weight %	4	7	15.8	NA
Kauri-Butanol Value (solvent power)	118	133	91	NA
OSHA Permissible Exposure Limit (8 hr. Time-weighted average)	350	100	100	NA

Table 3.123: Density of Chlorinated Hydrocarbon Liquids *(61)*

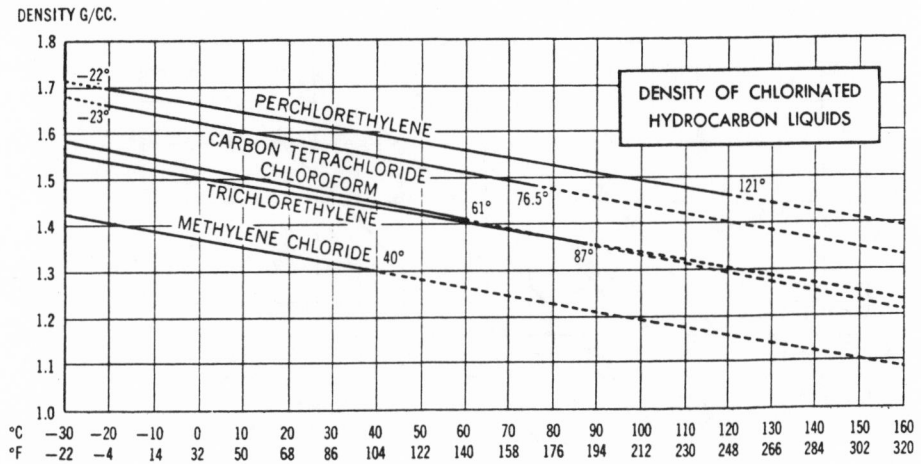

Table 3.124: Vapor Pressures of Chlorinated Hydrocarbons *(61)*

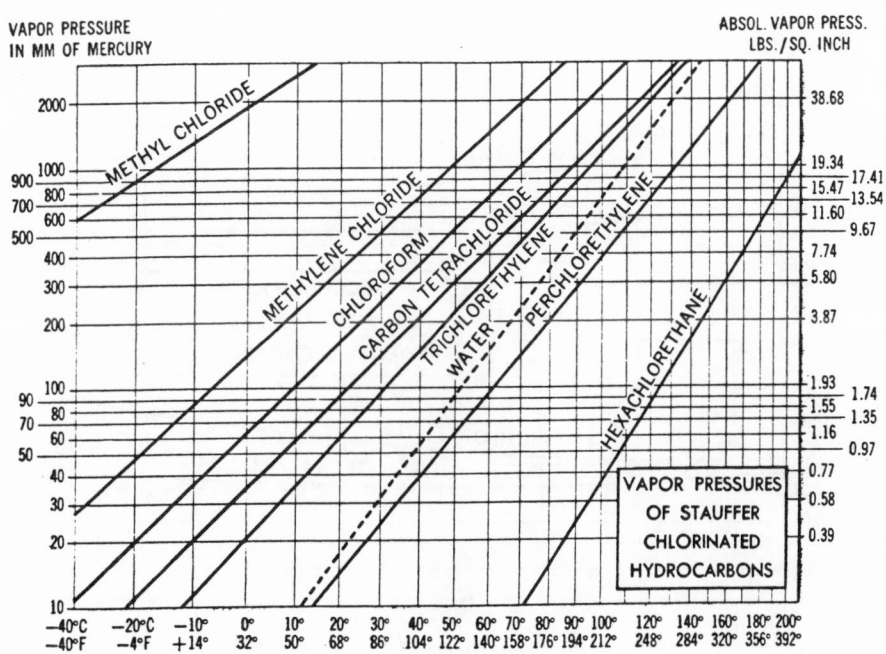

Table 3.125: Density of Saturated Vapors in Pounds per Cubic Foot *(61)*

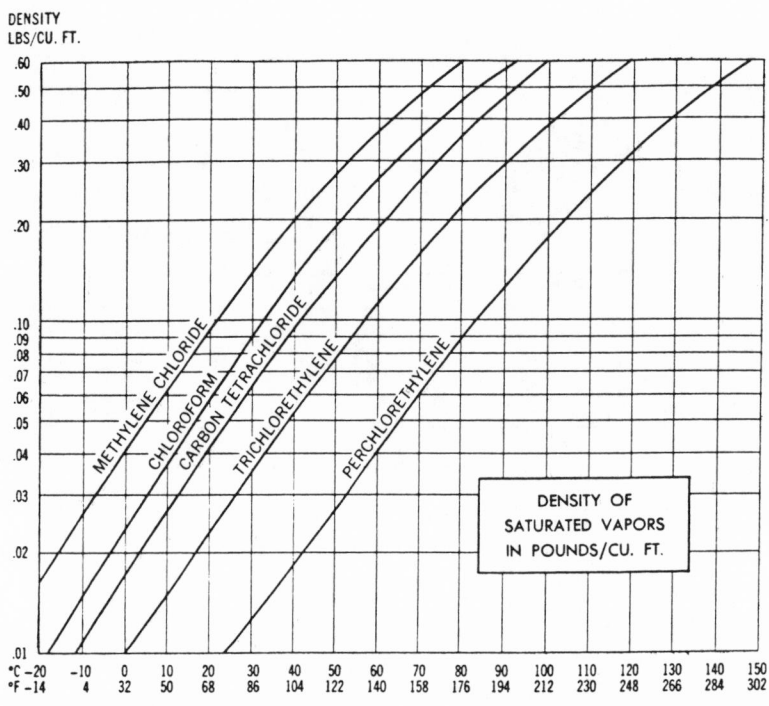

Table 3.126: Solubility of Chlorinated Hydrocarbons in Water *(61)*

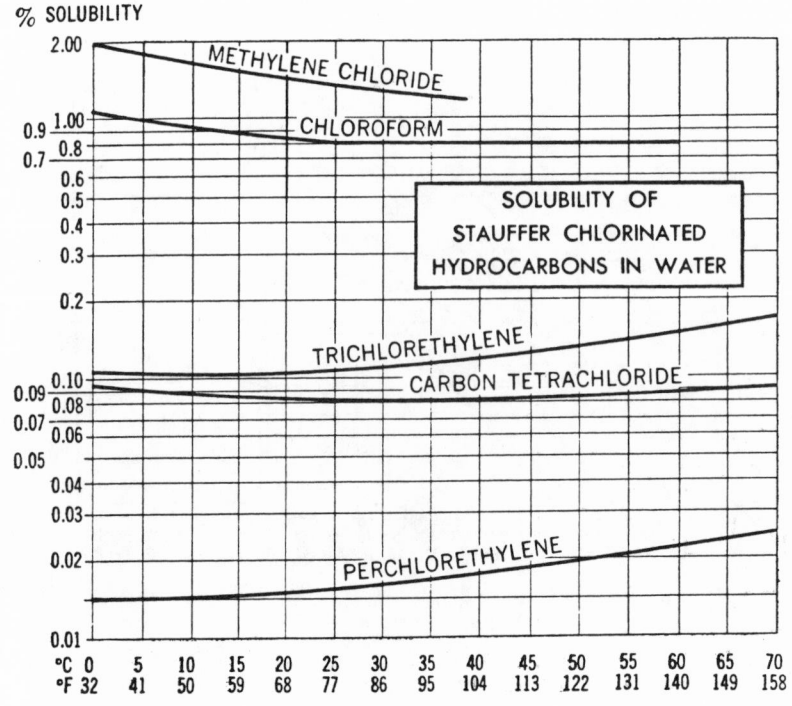

Table 3.127: Latent Heat of Vaporization of Chlorinated Hydrocarbons *(61)*

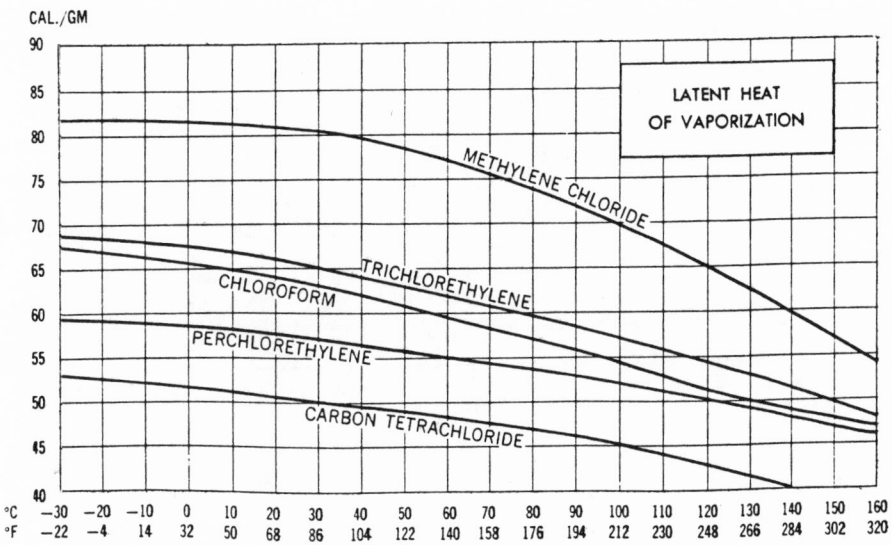

Table 3.128: Solubility of Water in Chlorinated Hydrocarbons *(61)*

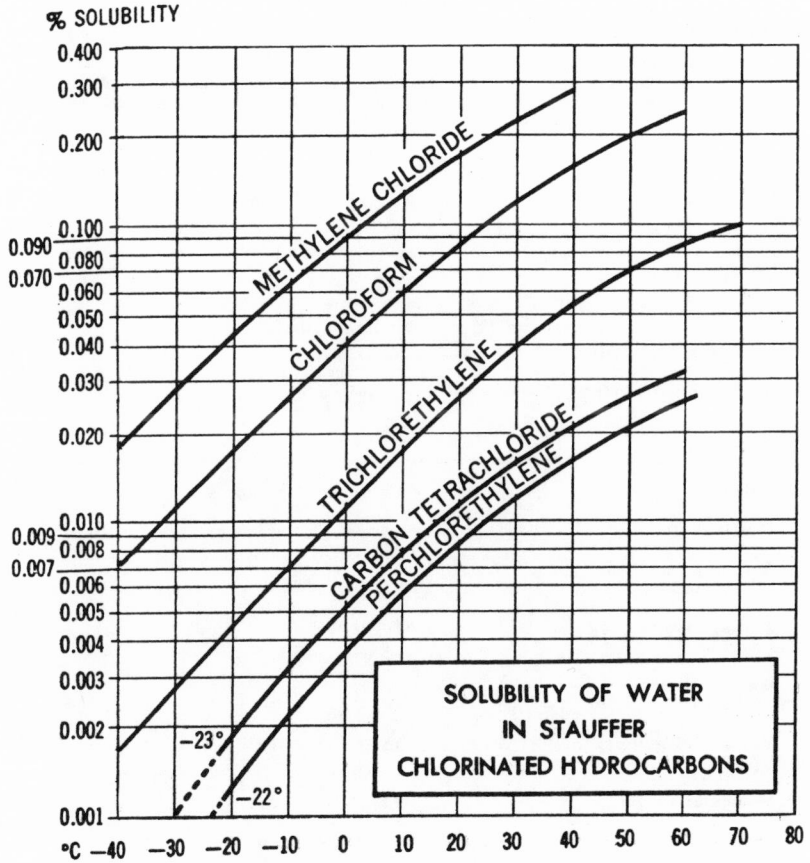

Table 3.129: Specific Heat of Chlorinated Hydrocarbon Liquids *(61)*

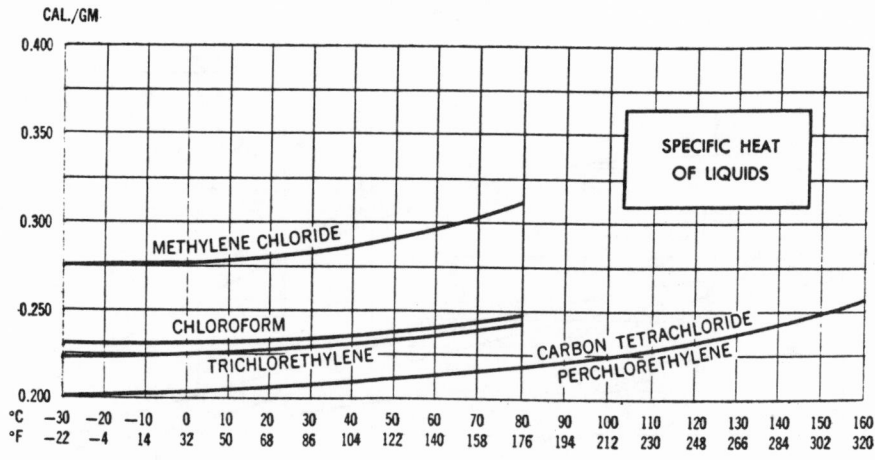

Table 3.130: Specific Heat of Chlorinated Hydrocarbon Vapors *(61)*

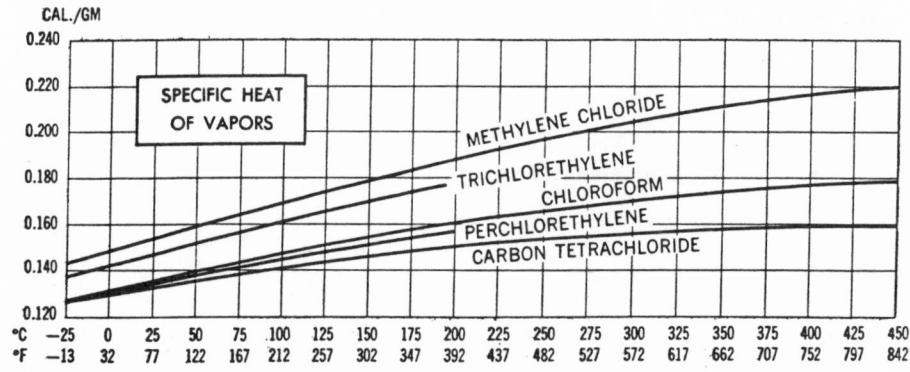

BROMINATED HYDROCARBONS

Table 3.131: Alkyl Bromides *(7)*

> I. 1,2-Dibromobutane (Butylene bromide),
> $CH_3CH_2CHBrCH_2Br$.
>
> II. 1,2-Dibromocyclohexane, $C_6H_{10}Br_2$.

Description	I	II
Color	Colorless	Colorless
Physical state	Liquid	Liquid
Boiling point @760 mm Hg	80-82°C	
Boiling range 5-95%	80-82°C	
Specific gravity @25/25°C	1.79	1.79
Pounds per gallon @25°C	15	15
Molecular weight	215.9	241.9
Solubility @25°C	Infinite in common organic solvents	
in Water g/100 g	Insoluble	Insoluble

(continued)

Table 3.131: (continued)

I. Acetylene tetrabromide, $CHBr_2CHBr_2$, is a non-flammable solvent.

II. Allyl bromide, CH_2CHCH_2Br, is a special-purpose solvent.

III. n-Amyl bromide (1-Bromopentane), $CH_3(CH_2)_4Br$.

Description	I	II	III
Color	Colorless	Colorless	Colorless
Odor	Mild, sweet	Sharp	Sweet
Physical state	Liquid	Liquid	Liquid
Boiling point @760 mm Hg	239°C	70°C	128°C
Boiling range 5-95%	239-242°C	70-71.5°C	128-130°C
Specific gravity @25/25°C	2.953	1.415	1.211
Pounds per gallon @25°C	24.57	11.8	10.0
Freezing point	0.0°C	<-20°C	<-30°C
Refractive index @25°C	1.6326 (20°C)	1.465	1.441
Flash point	None	30°F	90°F
Fire point	None	130°C	130°C
Molecular weight	345.7	121.0	151.0
Solubility @25°C in Water g/100 g	Infinite in common organic solvents	Insoluble	Insoluble
	0.07		

I. Bromodichloromethane, $CHBrCl_2$.

II. Bromotrichloromethane, $CBrCl_3$.

III. n-Butyl bromide, $CH_3(CH_2)_3Br$, is a special-purpose solvent.

Description	I	II	III
Color	Colorless	Colorless	Colorless
Odor	Mild, sweet	Sharp	Sweet
Physical state	Liquid	Liquid	Liquid
Boiling point @760 mm Hg	89.2°C	103.8°C	101.4°C
Boiling range 5-95%	89.2-90.6°C	103.8-105.1°C	101.4-103.4°C
Specific gravity @25/25°C	1.971	1.997	1.274
Pounds per gallon @25°C	16.4	16.62	10.60
Freezing point			<-30°C
Refractive index @25°C			1.437
Flash point	None	None	75°F
Fire point			130°C
Molecular weight	163.8	198.3	137.0
Solubility @25°C in Water g/100 g	Infinite in common organic solvents	<0.01	<0.01
	<1		

(continued)

Table 3.131: (continued)

I. Cetyl bromide, $CH_3(CH_2)_{15}Br$.
II. Dibromochloromethane, $CHClBr_2$.
III. Dibromodifluoromethane, CF_2Br_2.

Description	I	II	III
Color	Dark-yellow	Colorless	Colorless
Physical state	Liquid	Heavy liquid	Heavy liquid
Boiling point @760 mm Hg	186°C	118°C	23.2°C
Boiling range 5-95%	186-197°C	118-122°C	
Specific gravity @25/25°C	0.991	2.44	2.288 (15/4°C)
Pounds per gallon @25°C	8.25	20.3	19.9
Freezing point	15°C	<-20°C	-141°C
Refractive index @25°C	1.460		1.3999 (12°C)
Flash point	350°F	None	None
Molecular weight	305.3	208.3	208.3
Solubility @25°C	Infinite in common organic solvents		
in Water g/100 g	<0.1	<0.1	<0.1

I. 1,2-Dibromo-3-chloropropane, $CH_2ClCHBrCH_2Br$.
II. 1,2-Dibromo-1,1-dichloroethane, $CH_2BrCBrCl_2$.
III. Ethylene chlorobromide (1-Bromo-2-chloroethane), CH_2ClCH_2Br.

Description	I	II	III
Color	Slightly yellow	Faintly yellow	Colorless
Odor			Chloroform-like
Physical state	Liquid	Liquid	Liquid
Boiling point @760 mm Hg	200°C	176°C	106.1°C
Boiling range 5-95%			106-108°C
Specific gravity @25/25°C		2.26-2.27	1.720
Pounds per gallon @25°C		18.8	14.31
Freezing point			-18.4°C
Flash point		None	None
Fire point			None
Molecular weight	236.3	256.8	143.4
Solubility @25°C	Infinite in common organic solvents		
in Water g/100 g	Insoluble	Insoluble	0.7

(continued)

Table 3.131: (continued)

I.   Ethylene dibromide (1,2-Dibromoethane), CH₂BrCH₂Br.

II. Methyl bromide (Bromomethane), CH₃Br, is a solvent, fumigant, and extinguishing liquid.

III. n-Propyl bromide (1-Bromopropane), CH₃CH₂CH₂Br.

Description	I	II	III
Color	Colorless	Colorless	Colorless
Odor	Sweet		Penetrating
Physical state	Liquid	Liquid	Liquid
Boiling point @760 mm Hg	131.4°C	4.6°C	70°C
Boiling range 5-95%	131-133°C		70-72°C
Specific gravity @25/25°C	2.172	1.732	1.333
Pounds per gallon @25°C	18.07	14.4	11.1
Freezing point	9.3°C	-93°C	-30°C
Refractive index @25°C	1.5357		1.431
Flash point	None	None	None
Fire point	None	None	None
Molecular weight	187.9	94.9	123.0
Solubility @25°C	Infinite in common organic solvents		
in Methanol g/100 g	Soluble	∞	∞
in VM&P Naphtha	Soluble	∞	∞
in Water	0.43	0.1	Insoluble

I.   Isoamyl bromide, (CH₃)₂CHC₂H₄Br.

II. Isobutyl bromide, (CH₃)₂CHCH₂Br.

III. Isopropyl bromide, CH₃CHBrCH₃.

Description	I	II	III
Color	Colorless	Colorless	Colorless
Odor	Characteristic	Characteristic	Sweet
Physical state	Liquid	Liquid	Liquid
Boiling point @760 mm Hg	119.7°C	91°C	58.5°C
Boiling range 5-95%	119.7-121.8°C	91-93°C	58.5-60.5°C
Specific gravity @25/25°C	1.208	1.258	1.304
Pounds per gallon @25°C	10.05	10.47	10.85
Freezing point	<-40°C	<-30°C	<-90°C
Refractive index @25°C	1.440	1.433	1.422
Flash point	90°F	90°F	None
Fire point	88°F		None
Molecular weight	151.1	137.0	123.0
Solubility @25°C	Infinite in common organic solvents		
in Water g/100 g	0.1	Insoluble	Slightly soluble

(continued)

Table 3.131: (continued)

I.   Lauryl bromide, $CH_3(CH_2)_{10}CH_2Br$.
II. Methylene dibromide, CH_2Br_2.
III. Methylene chlorobromide, CH_2BrCl.

Description	I	II	III
Color	Amber	Colorless	Colorless
Physical state	Liquid	Heavy liquid	Liquid
Boiling point @760 mm Hg	137°C	95.6°C	67.8°C
Boiling range 5-95%	137-194°C	95.6-97.4°C	67.8°C
Specific gravity @25/25°C	1.021	2.485	1.930
Pounds per gallon @25°C	8.50	20.58	16.06
Freezing point	<-40°C	<-50°C	-88°C
Refractive index @25°C	1.457	1.536	1.480
Flash point	295°F	None	None
Fire point	None	None	None
Molecular weight	249.2	173.9	129.4
Solubility @25°C	Infinite in common organic solvents		
in Methanol g/100 g	11	∞	∞
in Water	0.1	Insoluble	2.4

I.   Propylene chlorobromide, $CH_3CHBrCH_2Cl$ and $CH_3CHClCH_2Br$.
II. Trimethylene bromide, $C_3H_6Br_2$.
III. Trimethylene chlorobromide, C_3H_6ClBr.

Description	I	II	III
Color	Colorless	Colorless	Colorless
Physical state	Liquid	Liquid	Liquid
Boiling point @760 mm Hg	117°C	166.5°C	143°C
Boiling range 5-95%	117-118.5°C	166.5°C	143-145°C
Specific gravity @25/25°C	1.540	1.977	1.594
Pounds per gallon @25°C	12.81	16.45	13.27
Freezing point	<-20°C	-33°C	
Refractive index @25°C	1.476	1.520	1.484
Flash point	None	None	None
Fire point	None	None	None
Molecular weight	157.5	201.9	157.5
Solubility @25°C	Infinite in common organic solvents		
in Water g/100 g	0.1	0.2(30°C)	Insoluble

(continued)

Table 3.131: (continued)

I. sec-Amyl bromide, $C_5H_{11}Br$. This is a mixture of sec-amyl bromides used as a solvent for special purposes.

II. 2-Bromo-4-tert-Butylphenol, $C_6H_3BrC_4H_9OH$, is a special-purpose solvent.

III. 2-Bromo-p-cymene (Tech.), (2-Bromo-4-isopropyl-1-methylbenzene), $C_6H_3CH_3BrC_3H_7$.

Description	I	II	III
Color	Colorless	Colorless	Straw
Odor	Characteristic	Mild	Pleasant
Physical state	Liquid	Liquid	Liquid
Boiling point @760 mm Hg	118°C	109°C	120°C
Boiling range 5-95%	118-120°C	109-129°C	120-126.5°C
Specific gravity @25/25%	1.20	1.338	1.253
Pounds per gallon @25°C		11.1	10.4
Freezing point		<-20°C	<-20°C
Refractive index @25°C		1.550	1.535
Flash point		116°C	104°C
Fire point		160°C	149°C
Molecular weight	151.0	229	213.0
Solubility @25°C in Ether	Infinite in common organic solvents		
in Methanol g/100 g	∞	∞	50
in Water	Insoluble	Insoluble	Insoluble
Viscosity @25°C centipoises		20.8	

I. Propylene dibromide (1,2-Dibromopropane), $CH_3CHBrCH_2Br$.

II. Bromoethylbenzene ("Alkazene 40"), $C_6H_4BrC_2H_5$.

III. Bromoisopropylbenzene ("Alkazene 47," Bromo-cumene), $C_6H_4BrC_3H_7$.

Description	I	II	III
Color	Colorless	Colorless	Colorless
Odor	Sweet	Characteristic	Characteristic
Physical state	Liquid	Liquid	Liquid
Boiling point @760 mm Hg	139.6°C	200.5°C	212.4°C
Boiling range 5-95%	139.6-142.6°C	200.5-202.8°C	212.4-216.2°C
Specific gravity @25/25%	1.943	1.339	1.294
Pounds per gallon @25°C	16.2	11.2	10.8
Freezing point		-65°C	<-20°C
Refractive index @25°C	1.519	1.454	1.536
Flash point	None	93°C	96°C
Fire point	None	135°C	141°C
Molecular weight	201.9	185	199
Solubility @25°C in Ether g/100 g	Infinite in common organic solvents		
	Very soluble	∞	∞
in Methanol	Soluble	∞	∞
in Water	0.2	Insoluble	Insoluble
Viscosity @25°C centipoises	1.5	1.4	1.6

(continued)

Table 3.131: (continued)

I.   sec-Butyl bromide, $C_2H_5CHBrCH_3$.

II. tert-Butyl bromide, $(CH_3)_3CBr$.

Description	I	II
Color	Colorless	Colorless
Odor	Characteristic	Characteristic
Physical state	Liquid	Liquid
Boiling point @760 mm Hg	91.4°C	73.3°C
Boiling range 5-95%	91.4-91.4°C	73.3-73.3°C
Specific gravity @25/25°C	1.257	1.215
Pounds per gallon @25°C	10.46	10.11
Freezing point	<-50°C	<-18°C
Refractive index @25°C	1.435	1.425
Flash point	70°F	None
Molecular weight	137.0	137.0
Solubility @25°C	Infinite in common organic solvents	
in Water g/100 g	<0.1	Insoluble

Table 3.132: Aryl Bromides (7)

I.   Dibromotoluene, $CH_3C_6H_3Br_2$.

II.   1-Bromonaphthalene, $C_{10}H_7Br$.

III. Bromo-m-xylene, $(CH_3)_2C_6H_3Br$.

Description	I	II	III
Color	Colorless	Faintly yellow	Colorless
Odor	Characteristic		Faint
Physical state	Liquid	Liquid	Liquid
Boiling point @760 mm Hg	243.7°C		207°C
Boiling range 5-95%	243.7-245.4°C		207-209°C
Specific gravity @25/25°C	1.836	1.48-1.49	1.355
Pounds per gallon @25°C	15.26	12.5	11.2
Refractive index @25°C		1.65-1.66	
Flash point	None		
Molecular weight	250.9	207	185.0
Solubility @25°C	Infinite in common organic solvents		
in Water g/100 g	Insoluble		Insoluble

(continued)

Table 3.132: (continued)

I. 2-Bromoethylbenzene, $C_6H_5C_2H_4Br$.
II. o-Bromotoluene, $CH_3C_6H_4Br$.
III. Bromotoluene mixture, $CH_3C_6H_4Br$.

Description	I	II	III
Color	Colorless	Colorless	Colorless
Odor	Characteristic	Characteristic	Characteristic
Physical state	Liquid	Liquid	Liquid
Boiling point @760 mm Hg		180°C	189.9°C
Boiling range 5-95%		180-182°C	189.9-183.2°C
Specific gravity @25/25°C	1.366	1.422	1.419
Pounds per gallon @25°C	11.38	11.83	11.84
Freezing point		-27°C	
Refractive index @25°C	1.558	1.552	1.551
Flash point		175°F	225°F
Fire point		None	
Molecular weight	185.1	171.0	171.0
Solubility @25°C	Infinite in common organic solvents		
in Water g/100 g	0.1	0.1	0.1
Viscosity @25°C centipoises	1.2	1.3	1.23

I. Dibromoethylbenzene, $C_6H_3C_2H_5Br_2$.
II. Bromobenzene, C_6H_5Br.
III. Bromocyclohexane, $C_6H_{11}Br$.

Description	I	II	III
Color	Colorless	Colorless	Slightly yellow
Odor		Characteristic	Characteristic
Physical state	Liquid	Liquid	Liquid
Boiling point @760 mm Hg	258°C	156°C	165.8°C
Boiling range 5-95%	258-266°C	156°C	165.8-167.3°C
Specific gravity @25/25°C	1.744	1.495	1.337
Pounds per gallon @25°C	14.51	12.44	11.13
Freezing point	-40°C	-30.6°C	
Refractive index @25°C	1.587	1.557	
Flash point	None	51°C	145°F
Fire point	None	155°C	
Molecular weight	264.0	157.0	163.1
Solubility @25°C	Infinite in common organic solvents		
in Ether g/100 g	∞	∞	71.3
in Methanol	50	∞	10.4
in VM&P naphtha	∞	71.3	∞
in Water	Insoluble	0.05	0.1
Viscosity @25°C centipoises	1.2		2.0

Table 3.133: Brominated Organic Compounds (23)

common properties (not to be considered as specifications)

PRODUCT	FORMULA	Description	Molecular weight	Freezing point °C.	Boiling point °C. 760mm. Hg	Sp. gr. 25/25°C.	Lb./gal. at 25°C.	Refractive index at 25°C.	Flash point °F.	Approx. sol. g./100g. solvent at 25°C.		
										water	methanol	ether
Acetylene Tetrabromide	CHBr₂CHBr₂	pale yellow liquid	345.7	−1.0	239 with decomposition	2.961	24.64	1.635	none	<0.1	∞	∞
Allyl Bromide	CH₂:CHCH₂Br	colorless to pale yellow liquid	121.0	below −50	70.2	1.418	11.80	1.465	30	<0.1	∞	∞
Bromobenzene	C₆H₅Br	colorless liquid	157.0	−30.6	156.0	1.495	12.44	1.557	125	0.05 (30°C.)	∞	∞
1-Bromo-4-chlorobenzene	C₆H₄BrCl	white crystals	191.5	64.5	196	—	—	—	none	<0.1	27	157
Bromochloromethane	CH₂BrCl	colorless to pale yellow liquid	129.4	−88	67.8	1.930	16.06	1.480	none	0.9	∞	∞
Bromoform	CHBr₃	colorless, heavy liquid	252.8	7.8	148	2.850	23.72	1.594	none	0.1	∞	∞
n-Butyl Bromide	C₂H₅CH₂CH₂Br	colorless liquid	136.9	−112.4	101.5	1.270	10.6	1.436	75	ins.	∞	∞
p-Dibromobenzene	C₆H₄Br₂	nearly white crystals	235.9	87.5 ¹	218.6	—	—	—	none	<0.1	8	101
Dibromodifluoromethane	CBr₂F₂	colorless, heavy liquid	242.8	22	133.7	2.408	20.0	1.496	none	—	—	—
Ethyl Bromide	C₂H₅Br	colorless liquid	109.0	−119.0	38.4	1.451	12.07	1.421	none	0.9	∞	∞
Ethylene Dibromide	CH₂BrCH₂Br	colorless liquid	187.9	9.8	131.4	2.172	18.07	1.536	none	0.27	∞	∞
Methyl Bromide	CH₃Br	colorless gas	95.0	−93	3.6	1.732 (0/0°C.)	14.45 (0°C.)	—	none	0.09 (20°C.)	44	39
Methylene Bromide	CH₂Br₂	colorless to pale yellow heavy liquid	173.9	−52	99	2.490	20.58	1.538	none	0.1	∞	∞
Tetrabromo Bisphenol A	(CH₃)₂C(C₆H₂Br₂OH)₂	off-white powder	543.9	181	—	—	—	—	above 600	<0.1	92	97
Trimethylene Chlorobromide	(CH₂)₃BrCl	colorless to pale yellow liquid	157.5	below −50	143.5	1.594	13.27	1.484	none	ins.	∞	∞
Vinyl Bromide	CH₂:CHBr	colorless liquid	106.9	−139.5	15.8	1.474²	12.4	1.435	—	—	—	—

¹ Melting point   ² Boiling range   ³ Density, gm./ml.

Table 3.134: Bromoform (7)

Tribromomethane
Formyl Tribromide
Methylene Tribromide $CHBr_3$
Methenyl Tribromide

PHYSICAL PROPERTIES

Boiling point	148.1°C	%		B.P. °C. of Azeotrope
Boiling range, 5-95%	Not more than 5.0°C			
Density @25°C	2.884	10	Butyric acid	147.6
Dielectric constant @20°C	4.5	5	Camphene	148.5
Fire point	None	5	Cyclohexanol	149.5
Flash point	None	48	Cyclohexanone	158.5
Freezing point	4.8°C	48	Formic acid	97.4
Molecular weight	252.8	6.5	Glycol	146.8
Specific gravity @25/25°C	2.847	16	Hexyl alcohol	147.7
Surface tension @20°C	41.53 dynes/cm	12	Isoamyl acetate	150.2
Weight per gallon @25°C	23.69 lb	55	Isoamyl alcohol	129.9
		65	Isobutyl butyrate	157.7
BROMOFORM FORMS AZEOTROPES WITH:		25	Isobutyl isobutyrate	151.0
		19	Isobutyric acid	145.5
%	B.P. °C of Azeotrope		Methyl lactate	152.0
		25	α-Pinene	146.5
2 Acetamide	149.0	37	Propionic acid	137.6
72 Acetic acid	118.3	10	Propyl sulfide	151.0

Table 3.135: Ethyl Bromide *(7)*

Bromoethane C_2H_5Br

PHYSICAL PROPERTIES

Specific gravity @25/25°C	1.424
Specific heat @5°C	0.22 cal/g
Surface tension @10°C	25.48 dynes/cm
@20°C	24.5 dynes/cm
@40°C	21.52 dynes/cm
Vapor pressure @-20°C	59 mm
@-10°C	101 mm
@0°C	165 mm
@10°C	257 mm
@20°C	386 mm
@30°C	564 mm
@40°C	802 mm
@50°C	1113 mm
@60°C	1512 mm
Weight per gallon @25°C	11.85 lb

ETHYL BROMIDE FORMS AZEOTROPES WITH:

%		B.P. °C of Azeotrope
3	Ethanol	37.6
65	Isoprene	32.0
1	Isopropanol	38.4
32	Isopropyl nitrite	37.7
70	2-Methylbutane	23.5
40	2-Methyl-2-Butene	35.2
50	n-Pentane	33.0
2	Acetamide	149.0
72	Acetic acid	118.3
10	Butyric acid	147.6
5	Camphene	148.5
5	Cyclohexanol	149.5
48	Cyclohexanone	158.5
48	Formic acid	97.4
6.5	Glycol	146.8
16	Hexyl alcohol	147.7
12	Isoamyl acetate	150.2

Table 3.136: Ethylene Chlorobromide *(7)*

2-Bromo-1-Chloroethane CH_2ClCH_2Br

sym-Chlorobromoethane

PHYSICAL PROPERTIES

Boiling point	106.1°C
Boiling range	5-95%, not more than 2.0°C
Fire point	Nonflammable
Flash point	Nonflammable
Freezing point	-18.4°C
Solubility in water @30°C	0.69 g/100 g solvent
@80°C	0.7 g/100 g solvent
Specific gravity @25/25°C	1.720
Weight per gallon @25°C	14.31 lb

Table 3.137: Ethylene Dibromide *(7)*

1,2-Dibromoethane CH_2BrCH_2Br

Ethylene Bromide

PHYSICAL PROPERTIES

Acidity (0.01 N NaOH/100 cc)	Less than 1 cc
Boiling point	131.4°C
Boiling range	90% within 2.0°C

(continued)

Table 3.137: (continued)

PHYSICAL PROPERTIES		ETHYLENE DIBROMIDE FORMS AZEOTROPES WITH:		

PHYSICAL PROPERTIES	
Density @10°C	2.2014
@20°C	2.1805
@30°C	2.1594
@40°C	2.1384
@60°C	2.0952
@80°C	2.0538
@100°C	2.0113
Dielectric constant, 1000 cycle	5.17
Fire point	Nonflammable
Flash point	Nonflammable
Freezing point	9.3°C
Latent heat of fusion @9.55°C	56.62 joules/g
Latent heat of vaporization @130.8°C	193.5 joules
Refractive index @25°C	1.5357
Specific gravity @25/25°C	2.172
Specific heat	0.18 cal/g/°C
Specific resistivity	2.4×10^9 ohms/cm
Surface tension @20°C	38.71 dynes/cm
Vapor pressure @-10°C	2.5 mm
@0°C	3.9 mm
@10°C	6.03 mm
@30°C	17.4 mm
@50°C	44.3 mm
@70°C	101.0 mm
@100°C	295 mm
@120°C	552 mm
@131°C	760 mm
Weight per gallon @25°C	18.07 lb

ETHYLENE DIBROMIDE FORMS AZEOTROPES WITH:

%		B.P. °C of Azeotrope
55	Acetic acid	114.4
44	Butanol	114.8
3.5	Butyric acid	131.1
45	Chlorobenzene	129.8
10	Ethylbenzene	131.1
29	Isoamyl alcohol	125.0
3.5	Glycol	130.9
88	Isoamyl formate	123.7
63	Isobutanol	106.6
6.5	Isobutyric acid	130.5
85	Mesityl oxide	129.2
44	Methyl chloroacetate	127.7
17.5	Propionic acid	127.8
33	Pyrrol	126.3
7	p-Xylene	131.3

Table 3.138: Methyl Bromide *(7)*

Monobromomethane   CH₃Br

Methyl bromide is a nonflammable gas, with a sweet odor. It is miscible with alcohols, ethers, ketones, esters, and halogenated hydrocarbons, but is only slightly soluble in water to the extent of 0.097% by weight. It is used as a low-boiling solvent, as a refrigerant, and in organic synthesis.

PHYSICAL PROPERTIES

Boiling point @760 mm	4.6°C
Critical temperature	173.3°C
Dielectric constant @100°C	0.0068
Flash point (Cleveland O.C.)	None
Freezing point	-93°C
Heat capacity	1.27 (16°C) (0.3-0.6 atm.)
Heat of combustion	184 kg cal/mol
Melting point	-84°C
Molecular weight	94.95
Purity	99.8%
Specific gravity @0/0°C	1.732
Thermal conductivity @0°C	0.574
Vapor pressure	
@-30°C	130 mm
@-10°C	380 mm
@4.6°C	760 mm
@10°C	890 mm
@20°C	1250 mm
@30°C	1800 mm
@40°C	2300 mm
Weight per gallon @0°C	14.4 lb

METHYL BROMIDE FORMS AZEOTROPES WITH:

%		B.P. °C of Azeotrope
0.55	Methanol	3.55
42.7	Butane	-4.4

Table 3.139: Methylene Chlorobromide *(7)*

Bromochloromethane

Chlorobromomethane CH_2ClBr

PHYSICAL PROPERTIES

Acidity as hydrobromic acid	0.015% max.
Boiling point @760 mm Hg	67°C
Boiling range @760 mm Hg 5-95% fraction	2° max.
Bromine content	61.76%
Chlorine content	27.40%
Dielectric constant	7.789
Distillation range @760 mm Hg	2° max.
Fire point (Cleveland O.C.)	None
Flash point (Cleveland O.C.)	None
Freezing point	Below -55°C
Melting point	-88°C
Moisture content	0.015% max.
Molal heat of vaporization @25°C @68°C	7850 cal 7173 cal
Molecular weight	129.399
Pounds per gallon @25°C	16.1
Refractive index N_{25}^{D}	1.4808
Solubility per 100 grams water @25°C	2.4 g
Specific gravity @25/4°C	1.923
Surface tension @20°C	33.32 dynes/cm
Vapor pressure @25°C	147 millimeters
Viscosity @20°C	0.3486 centistoke

Table 3.140: Propylene Dibromide *(7)*

1,2-Dibromopropane

Propylene Bromide $CH_3CHBrCH_2Br$

PHYSICAL PROPERTIES

Boiling range, 5-95%	139.6-142.6°C
Fire point	Nonflammable
Flash point	Nonflammable
Refractive index @25°C	1.519
Solubility in water @25°C @80°C	0.2% by wt 0.3% by wt
Specific gravity @25/25°C	1.943
Weight per gallon @25°C	16.2 lb

Table 3.141: Tetrabromoethane *(7)*

Acetylene Tetrabromide
1,1,2,2-Tetrabromoethane
sym-Tetrabromoethane
Muthmann's Liquid $CHBr_2CHBr_2$

PHYSICAL PROPERTIES

Boiling point @15 mm	119°C
Fire point	Nonflammable
Flash point	Nonflammable
Freezing point	-8.3°C
Melting point	0.1°C
Molecular weight	345.7
Purity	Not less than 98%
Refractive index	1.63795
Solubility in water @30°C @80°C	0.065 g/100 g solvent 0.28 g/100 g solvent
Specific gravity @25/25°C	2.953
Weight per gallon @25°C	24.57 lb

Table 3.142: Vinyl Bromide *(7)*

Molecular Weight	106.96
Bromine, Wt. %	74.71
Freezing Point, °C.	−139.5
Boiling Point, °C.:	
760 mm. Hg	15.8
100 mm. Hg	−31.7
30 mm. Hg	−52.3
Density, g./ml. 20°C.	1.493
25°C.	1.474
30°C.	1.434
Pounds per Gallon, 20°C.	12.4
Refractive Index, 25°C.	1.435
Latent Heat of Vaporization, cal./g. at 25°C.	50.5
Heat of Polymerization, Kcal./mol	18 (estimated)
dt/dp Rate of Change of BP with Pressure	0.03°C./ml. (at 25°C.)
Vapor Pressure, mm. at 25°C.	1,033

Heat of Vaporization of Vinyl Bromide

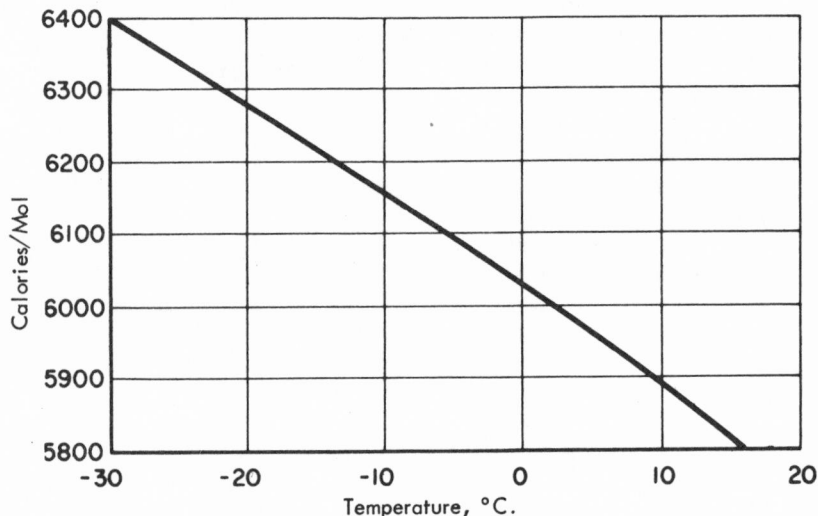

IODINATED HYDROCARBONS

Table 3.143: Iodinated Compounds *(7)*

	Description	Density	M. P.	B. P.	Refractive Index	Solubility
n-Butyl iodide CH₃CHICH₃ Mol. Wt. 184	Colorless to brown	1.617		131 °C	1.5001	Alcohol Ether Water (insol)
Isobutyl iodide CH₃CH₂CH₂I Mol. Wt. 183	Colorless to brown liquid	1.605	−93°C	120 °C	1.4960	Alcohol Ether Water (insol)
sec-Butyl iodide (CH₃)₂CHCH₂I Mol. Wt. 183	Colorless to brown liquid	1.626		119-120°C		Alcohol Ether Water (insol)
Ethyl iodide CH₃CH₂I Mol. Wt. 156	Colorless to brown liquid	1.950	−108°C		1.5168	Alcohol Ether Water (insol)

(continued)

Table 3.143: (continued)

	Description	Density	M. P.	B. P.	Refractive Index	Solubility
Methyl iodide CH₃I Mol. Wt. 142	Colorless to brown liquid	2.28	−66.5°C	42.5°C		Alcohol Ether Water (insol)
Propyl iodide (iso) CH₃CHICH₃ Mol. Wt. 170	Colorless to brown liquid	1.703	−90°C	89-90°C	1.5026	Alcohol Ether Water (insol)
Propyl iodide (n) CH₃CH₂CH₂I Mol. Wt. 170	Colorless to yellow liquid	1.747	−98°C	102-103°C	1.5051	Alcohol Ether
1-Chloro-2-iodobenzene			0-2°C			
1-Chloro-3-iodobenzene				113-114°C @14 mm		
Dichloroiodoethane ICH₂CHCl₂	Liquid	2.219		171-2°C		Water (insol)
Dichloroiodomethane Cl₂CHI	Liquid	2.403		131°C		Water (insol)
Dibromoiodomethane ICHBr₂	Solid		22.5°C			
Diiodoacetylene IC≡CI	Liquid		78-82°C	80-100°C		
Diiodobutane-1,4 I(CH₂)₄I	Liquid	2.307	5.8°C	120-5°C @12 mm		
Diiodopentane-1,5 I(CH₂)₅I	Oil	2.194	9°C	149°C @20 mm		
Diiodopropane-1,2 CH₃CH₂ICH₂I	Liquid	2.490		d		
Diiodopropane-2,2	Liquid	2.15		147.8d		
Diiodopropane-1,3	Liquid	2.561	−13°C	168-70°C @170 mm		
Iodobenzene C₆H₅I Mol. Wt. 204	Liquid	1.824	−28.5°C	188.6°C		Alcohol (sol) Ether Chloroform Water (insol)
Iodoethyl Acetate	Liquid	2.441		184°C		
o-Iodotoluene IC₆H₄CH₃	Liquid	1.698		211-2°C		Alcohol Ether Water (insol)

Table 3.144: Physical Properties and Manufacture of Miscellaneous Organic Iodine Compounds *(33)*

Compound	Formula	Formula wt.	B.p., °C.	F.p., °C.	d_4^{20}	n_D^{20}	Starting materials, method of manufacture
Ethylene iodide	CH₂ICH₂I	281.89	dec.	81-82	2.132ᵃ	—	CH₂:CH₂ plus I₂.
n-Propyl iodide	CH₃CH₂CH₂I	170.00	102.4	−101.4	1.743	1.5052	CH₃CH₂CH₂OH plus P plus I.
Isopropyl iodide	CH₃CHICH₃	170.00	89.5	−90.8	1.702	1.4992	CH₃CHOHCH₃ plus P plus I.
Allyl iodide	CH₂:CHCH₂I	167.98	103.1	−99.3	1.846	—	CH₂:CHCH₂Cl plus NaI.
Trimethylene iodide	CH₂ICH₂CH₂I	295.92	88-89₆	—	2.571	1.6470	CH₂OHCH₂CH₂OH plus P plus I.
n-Butyl iodide	CH₃CH₂CH₂CH₂I	184.03	129	−103.5	1.616	1.4998	CH₃CH₂CH₂CH₂OH plus P plus I.
sec-Butyl iodide	CH₃CHICH₂CH₃	184.03	118	−104.0	1.592	1.4991	CH₃CHOHCH₂CH₃ plus P plus I.
n-Amyl iodide	CH₃(CH₂)₃CH₂I	198.06	154.5	−85.6	1.512	1.4954	CH₃(CH₂)₃CH₂OH plus P plus I.
Isoamyl iodide	CH₃CH(CH₃)CH₂CH₂I	198.06	147.5	—	1.509	1.4939	CH₃CH(CH₃)CH₂CH₂OH plus P plus I.
2-Iodopentane	CH₃CHICH₂CH₂CH₃	198.06	141	—	1.510	1.4961	CH₃CHOHCH₂CH₂CH₃ plus P plus I.
3-Iodopentane	CH₃CH₂CHICH₂CH₃	198.06	141	—	1.511	1.4978	CH₃CH₂CHOHCH₂CH₃ plus P plus
n-Hexyl iodide	CH₃(CH₂)₄CH₂I	212.08	179.5	—	1.437	1.4926	CH₃(CH₂)₄CH₂OH plus P plus I.
n-Heptyl iodide	CH₃(CH₂)₅CH₂I	226.11	201	−48.2	1.373	1.4897	CH₃(CH₂)₅CH₂OH plus P plus I.
n-Octyl iodide	CH₃(CH₂)₆CH₂I	240.13	221	−45.9	1.330	1.4889	CH₃(CH₂)₆CH₂OH plus P plus I.
n-Decyl iodide	CH₃(CH₂)₈CH₂I	268.18	132₁₅	—	1.257	1.4827	CH₃(CH₂)₈CH₂OH plus P plus I.
Iodoacetic acid	CH₂ICOOH	185.94	dec.	—	2.2694ᵇ	—	CH₂ClCOOH plus KI.
Diiodotyrosine	HOC₆H₃I₂CH₂CH(NH₂)COOH	433.00	—	204 (dec.)	—	—	HOC₆H₄CH₂CH(NH₂)COOH plus ICl.

ᵃ Measurement made at 10°C., and 4 mm. pressure.
ᵇ Measurement made at 85°C., and 4 mm. pressure.

Nitroparaffins

Table 4.1: Angus Nitroparaffins *(34)*

Nitroparaffins	Molecular wt.	Weight per gallon, lb at 20°C	Boiling range, °C	Flash point, °F	Uses
Nitromethane NM™ CH_3NO_2	61	9.4	100-103	96	▪ Stabilizer for chlorinated solvents ▪ Chemical intermediate
Nitroethane NE™ $CH_3CH_2NO_2$	75	8.75	112-116	87	▪ Intermediate for synthesis
NiPar S-10™ 1-nitropropane $CH_3CH_2CH_2NO_2$	89	8.35	129-133	96	▪ Solvent for inks and coatings ▪ Chemical intermediate ▪ Diesel fuel additive
NiPar S-20™ 2-nitropropane $CH_3CH(NO_2)CH_3$	89	8.24	119-122	82	▪ Solvent for inks and coatings ▪ Chemical intermediate
Blends					
NITROFUEL™ NM™ with dye	61	9.4	100-103	96	▪ Fuel
NiPar S-30™ nitropropane	–	8.29	119-133	89	▪ Solvent for inks and coatings
COMSOL NM-55® nitromethane–methanol mixture	–	7.9	66	42	▪ Fuel ▪ Solvent

™Indicates a trademark of ANGUS Chemical Company

Table 4.2: Angus Nitrohydroxy Compounds *(34)*

Nitrohydroxy Compounds (Nitroparaffin derivatives)	Molecular wt.	Melting point °C	Solubility in water at 20°C, g/ml	Uses
NMP™ 2-nitro-2-methyl-1-propanol $CH_3C(CH)_3(NO_2)CH_2OH$	119	90	350	▪ Chemical intermediate ▪ Formaldehyde donor ▪ Textile reactant ▪ Reduces formaldehyde on finished cloth
NMP™-concentrate	119	–	–	
NEPD™ 2-nitro-2-ethyl-1,3-propanediol $HOCH_2C(C_2H_5)(NO_2)CH_2OH$	149	56	400	▪ Chemical intermediate ▪ Formaldehyde donor ▪ Deodorant for chemical toilets
TRIS NITRO® tris(hydroxymethyl)nitromethane $(HOCH_2)_3CNO_2$	151	172	220	▪ Chemical intermediate ▪ Registered pesticide for use as an antimicrobial agent in metalworking fluids, cooling water, oil production, drilling muds, etc. ▪ Formaldehyde donor
TRIS NITRO® 50% aqueous	151	–	misc	

Table 4.3: Angus Aminohydroxy Compounds *(34)*

Aminohydroxy Compounds (Nitrohydroxy Compound derivatives)	Neutral equivalent	Water % by wt. (max.)	Melting point °C	Solubility in water at 20°C, g/ml	Uses
AMP™ (Regular) CH₃C(CH₃)(NH₂)CH₂OH 2-amino-2-methyl-1-propanol	88.5-91	0.8	≈30	misc.	▪ Pigment dispersant ▪ Solubilizer for resins ▪ Emulsifying amine ▪ Neutralizing amine (for use in boiler water treatment) ▪ Corrosion inhibitor ▪ Catalyst
AMP-95™	93-97	5.8	-2	misc.	
AEPD® 2-amino-2-ethyl-1,3-propanediol HOCH₂C(C₂H₅)(NH₂)CH₂OH	121.5	3.8	≈38	misc.	▪ Chemical intermediate ▪ Formaldehyde scavenger ▪ Acid-salt catalyst for permanent-press resins
TRIS AMINO® tris(hydroxymethyl)-aminomethane (HOCH₂)₃CNH₂	121-122	0.5	171-172	80	▪ Chemical intermediate ▪ Buffer for diagnostics and cosmetics ▪ Resin synthesis
TRIS AMINO®, concentrate	–	≈60	–	misc.	
DMAMP™-80 2-dimethylamino-2-methyl-1-propanol (CH₃)₂NC(CH₃)₂CH₂OH	≈148	≈20	-20	misc.	▪ Solubilizer ▪ Emulsifying amine ▪ Corrosion inhibitor ▪ Urethane catalyst ▪ Synthesis
AB™ 2-amino-1-butanol CH₃CH₂CH(NH₂)CH₂OH	88-93	2.5	-2	misc.	▪ Pharmaceutical intermediate ▪ Chemical intermediate

Table 4.4: Nitromethane, Nitroethane and 1- and 2-Nitropropanes *(34)*

Specifications

	Nitromethane	Nitroethane	1-Nitropropane	2-Nitropropane
Purity, % by wt (min.)*	95	92.5	94	94
Total nitroparaffins, % by wt (min.)*	99	99	99	99
Specific gravity at 25/25°C	1.124-1.129	1.042-1.047	0.997-0.999	0.984-0.988
Acidity as acetic acid, % by wt (max.)	0.1	0.1	0.2	0.1
Water, % by wt (max.)	0.1	0.2	0.1	0.1
Color, APHA (max.)	20	20	20	20

*Determined by gas chromatograph or mass spectrometer

Typical Properties of Commercial-Grade Nitroparaffins

	Nitromethane	Nitroethane	1-Nitropropane	2-Nitropropane
Distillation range at 1 atm (90% min.), °C	100-103	112-116	129-133	119-122
Vapor density (air = 1)	2.11	2.58	3.06	3.06
Change of density with temperature, 0-50°C, g/(ml·°C)	0.0014	0.0012	0.0011	0.0011
Weight per U.S. gallon at 68°F, lb	9.4	8.75	8.35	8.24
Flash point, Tag open cup, °F	112	106	120	100
Tag closed cup, °F	96	87	96	82
Lower limit of flammability, % by vol (at °C)	7.3[33]	3.4[30]	2.2[34]	2.5[27]
Ignition temperature, °C	418	414	420	428
Evaporation rate (n-butyl acetate = 100)	139	121	88	110
Evaporation number (diethyl ether = 1)	9	11	16	10

(continued)

Table 4.4: (continued)

Physical Properties of the Nitroparaffins

	Nitromethane	Nitroethane	1-Nitropropane	2-Nitropropane
Molecular weight (calcd.)	61.041	75.068	89.095	89.095
Boiling point at 760 mmHg, °C	101.20	114.07	131.18	120.25
Vapor pressure at 25°C, mmHg	36.66	20.93	10.23	18.0
Freezing point, °C	−28.55	−89.52	−103.99	−91.32
Density at 20°C, g/ml	1.138	1.051	1.001	0.988
at 30°C, g/ml	1.124	1.039	0.991	0.977
Coefficient of expansion per °C	0.00122	0.00112	0.00101	0.00104
per °F	0.00068	0.00062	0.00056	0.00058
Refractive index, n_D, at 20°C	1.38188	1.39193	1.40160	1.39439
at 30°C	1.37738	1.38754	1.39755	1.39028
Surface tension at 20°C, dynes/cm	37.48	32.66	30.64	29.87
Viscosity at 20°C, cp	0.647	0.677	0.844	0.770
at 30°C, cp	0.576	0.602	0.740	0.677
Heat of combustion (liq.) at 25°C, kcal/mole	−169.3	−325.6	−481.9	−478.0
Heat of vaporization (liq.) at 25°C, kcal/mole	9.147	9.94	10.37	9.88
at bp, kcal/mole	8.23	9.08	9.19	8.79
Heat of formation (liq.) at 25°C, kcal/mole	−27.03	−33.9	−40.15	−43.2
Specific heat at 25°C, cal/(mole·°C)	25.33	33.10	41.96	41.87
at 25°C, cal/(g·°C)	0.415	0.441	0.471	0.470
Dielectric constant at 30°C	35.87	28.06	23.24	25.52
Dipole moment, μ, gas, Debye units	3.50	3.58	3.72	3.73
liquid, Debye units	3.17	3.19	—	—
Aqueous azeotrope, bp, °C	83.59	87.22	91.63	88.55
% NP by wt	76.4	71.0	63.5	70.6
pH of 0.01M aqueous solution	6.4	6.0	6.0	6.2
Solubility in water at 20°C, % by wt	10.5	4.6	1.5	1.7
at 25°C, % by wt	11.1	4.7	1.5	1.7
at 70°C, % by wt	19.3	6.6	2.2	2.3
Solubility of water in NP at 20°C, % by wt	1.8	0.9	0.6	0.5
at 25°C, % by wt	2.1	1.1	0.6	0.5
at 70°C, % by wt	7.6	3.0	1.7	1.6
Hydrogen bonding parameter, γ	2.5	2.5	2.5	2.5
Solubility parameter, δ	12.7	11.1	10.7	10.7

Table 4.5: Solubilities in 1- and 2-Nitropropane *(34)*

Substances Soluble in 1- AND 2-NITROPROPANE

NATURAL RESINS

Barbados Manjak
 No. 84 (SS)
Dammar (PS)
Dammar (de-waxed) (SA)
Kauri white (SA)
Pontianak (SA)
Rosin (SS)

TARS AND PITCHES

Blown asphalt (SS)
Bone pitch (SS)
Gilsonite (SS)
Petroleum asphalt (SS)
Refined coal tar (SS)
Road tar

SYNTHETIC RESINS

Aroclor 4465
Aroclor 5460
Bakelite BR-8900
Bakelite XR-9366
Bakelite XR-14590
Beckacite 1001
Beckacite 1110
Beckacite 1112
Cumar
Durez 500
Durez 525

Durez 550
Ester gum
Glyptal 2471
Paraplex 5B
Paraplex RG2
Rezyl 14
Rezyl 19
Rezyl 114 (SS)
Santolite MHP
Teglac 65
Vinylite AYAF
Vinylite AYAT

COATING MATERIALS

Acryloid C-10
Acryloid F-10
Benzyl cellulose
Cellulose acetate (SA)
Cellulose acetobutyrate
Cellulose acetopropionate
Cellulose triacetate
Chlorinated rubber
Ethyl cellulose
Hycar
Nitrocellulose
Uformite
Vinylite VYHH
Vinylite VYNS

DYES

Oil-soluble
Spirit-soluble (SS)

ORGANIC CHEMICALS

Acetic acid
Acetone
Acrylonitrile
Aniline
Benzaldehyde
Benzol
Benzoyl chloride
Butanol
Butyl acetate
Butyl butyrate
Butyl lactate
Butyl propionate
Butyric acid
Camphor
o-Chloraniline
Chloroform
Dibutylamine
Dibutyl phthalate
Diethylene glycol
Ethyl alcohol
Ethyl acetate
Ethylene chlorohydrin
Ethyl ether
a-Ethyl-*β*-propyl acrolein
Glycerol (SS)

Isobornyl acetate
Isopropyl alcohol
Lauric acid
Maleic anhydride
Naphtha, aliphatic
Naphthalene
p-Nitroaniline
Oleic acid
Phenol
p-Phenylenediamine
Phthalic anhydride
Picric acid
Pyridine
Ricinoleic acid
Salicylic acid
Terpineol
Toluol
Tributyl phosphate
Triphenyl phosphate

OILS AND FATS

Castor oil
Cocoanut oil
Kerosene
Lanolin (SS)
Linseed oil
Pine oil
Soya bean oil
Tung oil

(continued)

Table 4.5: (continued)

Substances Insoluble in 1- AND 2-NITROPROPANE

WAXES
- Beeswax
- Candelilla
- Carnauba
- Ceresine
- Montan
- Paraffin
- Spermaceti

NATURAL RESINS
- Batu
- Boea
- Congo (dark amber)
- East India
- Manila
- Shellac

TARS AND PITCHES
- Cottonseed pitch
- Mixed pitch
- Paving pitch
- Roofing pitch

COATING MATERIALS
- Casein
- Dextrin
- Egg albumen
- Gelatin
- Neoprene
- Vinylite VYNW
- Zein

DYES
- Water-soluble

ORGANIC CHEMICALS
- Adipic acid
- Aluminum stearate
- Aminobenzoic acid
- Ammonium linoleate
- Anthracene
- Citric acid
- Ethylene glycol
- Fumaric acid
- Glucose
- Glycerol bori-borate
- Glycine

- Hexamethyleneamine
- Hydroquinone
- Maleic acid
- Oxalic acid
- Pyrogallic acid
- Sebacic acid
- Sodium alginate
- Stearic acid
- Succinic acid
- Sucrose
- Tannic acid
- Tartaric acid
- Triethanolamine
- Urea
- Zinc stearate

Key: SA—soluble in the presence of alcohols. SS—soluble but less than 10 gm/100 ml. PS—only part of substance soluble.
NOTE: All unkeyed substances soluble to the extent of at least 10 gm/100 ml.

Table 4.6: Nitroparaffins *(35)*

COMPOUND	MELTING POINT	BOILING POINT	REFRACTIVE INDEX	SPECIFIC GRAVITY
	°C.	°C.		
Nitromethane	−29	101.7	1.3818	1.139 (20°/20°)
Nitroethane	−90	114	1.3916	1.052 (20°/20)
1-Nitropropane	−108	132	1.4015	1.003 (20°/20°)
2-Nitropropane	−93	120	1.3941	0.992 (20°/20°)
1-Nitrobutane		153	1.4112	0.975 (20°/20°)
2-Nitrobutane		140	1.4036	0.968 (20°/20°)
1-Nitro-2-methylpropane		140		
2-Nitro-2-methylpropane	25.5	127		
1-Nitropentane		173	1.4218	0.9475 (20°/4°)
3-Nitropentane		152–154		0.9575 (0°/4°)
1-Nitro-3-methylbutane		164	1.41806	0.9599 (20.6°/4°)
2-Nitro-2-methylbutane		150	1.4152	0.9783 (0°/4°)
1-Nitrohexane		193		0.9488 (20°/4°)
2-Nitroöctane		102–105 at 23 mm.	1.4324 (20°)	0.9224 (20°/20°)
Trinitromethane	15	45–47 at 22 mm.		
Tetranitromethane	13	126		1.650 (13°/4°)
1,1-Dinitroethane		185–186		1.3503 (23.5°/23.5°)
1,2-Dinitroethane		94–96 at 5 mm.	1.4488 (20°)	1.4597 (20°/4°)
1,1,1-Trinitroethane	56			
Hexanitroethane	142			
2,2-Dinitropropane	54	185.5		
1,4-Dinitrobutane		176–178 at 13 mm.		
1,3-Dinitro-2,2-dimethylpropane	93	140 at 15 mm.		
2,3-Dinitro-2,3-dimethylbutane	210–212			

Table 4.7: Nitroalcohols *(35)*

NITROALCOHOL	PREPARED FROM		B. P. AT 10 Mm., °C	SPECIFIC GRAVITY, d_4^{25}	REFRACTIVE INDEX, n_D^{20}	% NITROGEN	
	Nitroparaffin	Aldehyde				Theoretical	Found
2-Nitro-1-propanol	EtNO₂	HCHO	99	1.1841	1.4379	—	—
3-Nitro-2-butanol	EtNO₂	CH₃CHO	92	1.1296	1.4420	—	—
2-Nitro-3-hexanol	EtNO₂	C₃H₇CHO	108	1.0575	1.4480	9.52	9.42, 9.45
2-Nitro-1-butanol	1-PrNO₂	HCHO	105	1.1332	1.4390	—	—
3-Nitro-2-pentanol	1-PrNO₂	CH₃CHO	100	1.0818	1.4419	—	—
3-Nitro-4-heptanol	1-PrNO₂	C₃H₇CHO	115	1.0275	1.4460	8.70	8.65, 8.72
2-Methyl-2-nitro-1-propanol	2-PrNO₂	HCHO	89.5–90ᵃ	—	—	—	—
3-Methyl-3-nitro-2-butanol	2-PrNO₂	CH₃CHO	90	1.1021	1.4469	10.53	10.58, 10.72
2-Methyl-2-nitro-3-hexanol	2-PrNO₂	C₃H₇CHO	109	1.0405	1.4499	8.70	8.55, 8.69
2-Nitro-1-pentanol	1-BuNO₂	HCHO	117	1.0818	1.4405	—	—
3-Nitro-2-hexanol	1-BuNO₂	CH₃CHO	112	1.0487	1.4438	—	—
5-Nitro-4-octanol	1-BuNO₂	C₃H₇CHO	124	1.0394	1.4463	8.00	7.81, 7.94
2-Methyl-2-nitro-1-butanol	2-BuNO₂	HCHO	98	1.1047	1.4468	10.53	10.51, 10.53
3-Methyl-3-nitro-2-pentanol	2-BuNO₂	CH₃CHO	100	1.1157	1.4518	9.52	9.39, 9.42
3-Methyl-3-nitro-4-heptanol	2-BuNO₂	C₃H₇CHO	119	1.0281	1.4532	8.00	7.81, 7.93
3-Methyl-2-nitro-1-butanol	1-Iso-BuNO₂	HCHO	111	1.0886	1.4430	—	—
4-Methyl-3-nitro-2-pentanol	1-Iso-BuNO₂	CH₃CHO	96–98	1.0599	1.4477	—	—
2-Methyl-3-nitro-4-heptanol	1-Iso-BuNO₂	C₃H₇CHO	111	1.0140	1.4485	8.00	7.98, 8.11
2-Methyl-3-nitro-4-heptanol (stereoisomer)	1-Iso-BuNO₂	C₃H₇CHO	121 (53ᵃ)	—	—	8.00	7.91, 8.12
2-Methyl-2-nitro-1,3-propanediol	EtNO₂	2HCHO	149–150ᵃ	—	—	—	—
2-Ethyl-2-nitro-1,3-propanediol	1-PrNO₂	2HCHO	56ᵃ	—	—	—	—
2-Nitro-2-propyl-1,3-propanediol	1-BuNO₂	2HCHO	81–81.5ᵃ	—	—	—	—
2-Nitro-2-isopropyl-1,2-propanediol	1-Iso-BuNO₂	2HCHO	87–88ᵃ	—	—	—	—
2-Amino-2-ethyl-1,3-propanediol	—	—	—	—	—	11.76	11.56, 11.69
2-Amino-2-propyl-1,3-propanediol	—	—	—	—	—	10.52	10.25, 10.32
2-Amino-2-isopropyl-1,3-propanediol	—	—	—	—	—	10.52	10.35, 10.38

ᵃ Melting point.

Table 4.8: Azeotropes with Nitroethane *(34)*

Compound	B.p., °C	Azeotrope	
		Wt % of Compound	B.p., °C
Propyl nitrate	110.5	<79	<109.6
Propyl alcohol	97.2	68.2	94.5
Isopropyl alcohol	82.4	89.4	81.8
1-Bromobutane	101.5	75	96.0
1-Bromo-2-methylpropane	91.4	90	89.5
Butyl alcohol	117.8	45	107.7
sec-Butyl alcohol	99.5	72.4	97.2
t-Butyl alcohol	82.4	95.5	82.2
Isobutyl alcohol	108.0	60	102.5
1-Bromo-3-methylbutane	120.6	<45	<108.5
Amyl alcohol	138.2	<17	<137.8
t-Amyl alcohol	102.4	<70	<98.6
Isoamyl alcohol	131.9	22	112.0
Methylcyclopentane	72.0	<96	<71.2
4-Methyl-2-pentanone	116.0	—	<113
Ethyl butyrate	121.5	<27	<113.7
Ethyl isobutyrate	110.1	73	108.5
Isobutyl acetate	117.4	40	112.5
Isopropyl sulfide	120.5	<40	<110.9
Toluene	110.8	75	106.2
Methylcyclohexane	101.2	70	90.8
n-Heptane	98.4	72	89.2
2,5-Dimethylhexane	109.4	<38	<96.9
Acetic acid	114.2	30	112.4
Water	100.0	28.5	87.2
Ethyl alcohol	78.3	87.4	78.0

Nitroethane is reported not to form azeotropes with benzene or methanol.

Table 4.9: Azeotropes of Nitropropane at Atmospheric Pressure *(34)*

Component "A"	Boiling Point, °C	Azeotrope with 2-NP		Azeotrope with 1-NP	
		Weight % Component "A"	Boiling Point, °C	Weight % Component "A"	Boiling Point, °C
Water	100.0	29.4	88.55	36.5	91.63
Ethyl alcohol	78.4	93.6	78.28		
n-Propyl alcohol	97.2	75.1	95.97	91.2	96.95
Isopropyl alcohol	82.4	95.8	82.24		
n-Butyl alcohol	118.0	47.6	111.61	67.8	115.3
sec-Butyl alcohol	99.5	82.0	98.70	95.9	99.4
Isobutyl alcohol	108.1	66.9	105.28	84.8	105.28
n-Amyl alcohol	138.1	14.8	119.5		
Ethylene glycol monoethyl ether	135.1	14.1	119.7	36.1	128.3
n-Hexane	68.7	97	68 (2-phase)		
n-Heptane	98.4	79.2	94.5	85.8	94.6
n-Octane	125.7	53	111		
2,2,4-Trimethylpentane	99.2	79	95		
n-Nonane	150.8	26	118	36.4	126.2
Cyclohexane	80.7	90	80		
Methylcyclohexane	100.9	78	96		
Toluene	110.6	82	109		
Ethylbenzene	136.2	9	120	41.0	127.5
Perchloroethylene	121.2	57	114		
o-Xylene	143.6			15.0	130.9
Ethylene glycol monomethyl ether	124.5	34.1	115.3	58.7	121.4

The system 6.1% by weight water, 86.3% ethyl alcohol, and 7.6% 2-nitropropane is a ternary azeotrope with a boiling point of 78.2°C.

The system water--1-nitropropane--ethylbenzene is reported to form a ternary azeotrope of about equal parts of the three components.

Table 4.10: Physical Properties of Comsol NM-55 *(34)*

COMSOL NM-55 brand of solvent contains 55% by weight commercial-grade nitromethane and 45% by weight methanol. This is equivalent to a composition of about 46% by volume nitromethane and 54% by volume methanol. Other typical properties of COMSOL NM-55 are as follows:

Boiling point at reflux at 1 atm pressure	66°C (151°F)
Specific gravity at 25/25°C	0.95
Weight per gallon at 25°C (77°F)	7.9 lb
Viscosity at 25°C	0.51 cp
Surface tension at 25°C	30.1 dynes/cm^2
Flash point, Tag open cup	62°F
Tag closed cup	42°F

Table 4.11: NiPar S-30 *(34)*

Typical Properties of NiPar S-30

Distillation range at 760 mmHg (90% min.), °C	119-133
Freezing point, °C	−127
Vapor pressure at 20°C, mmHg	~10
at 25°C, mmHg	~15
Density of vapors (air = 1) (calcd.)	3.06
Specific gravity at 20/20°C	0.998
at 25/25°C	0.992
Weight per U.S. gallon at 20°C, lb	8.29
at 25°C, lb	8.26
Coefficient of expansion per °C	0.00103
Viscosity at 20°C, cp	0.80
at 30°C, cp	0.71
Flash point, Tag open cup, °F	101
Tag closed cup, °F	89
Evaporation rate, by vol (n-butyl acetate = 100)	~100
Evaporation number (diethyl ether = 1)	~11.5
Solubility of NiPar S-30 in water at 20°C, % by wt	1.7
at 70°C, % by wt	2.3
Solubility of water in NiPar S-30 at 20°C, % by wt	0.5
at 70°C, % by wt	1.7
Azeotrope with water: Boiling point at 760 mmHg, °C	~90
Water in azeotrope, % by wt	~33

Specifications for NiPar S-30

Nitropropane content, % by wt (min.)*	94
Total nitroparaffin content, % by wt (min.)*	99
Acidity as acetic acid, % by wt (max.)	0.1
Water, % by wt (max.)	0.1
Color, APHA (max.)	20

*Determined by gas chromatograph.

Table 4.13: Chloronitroparaffins (34)

The chloronitroparaffins are stable liquids miscible with most organic solvents including the lower alcohols glycols, esters, ethers, petroleum hydrocarbons, mineral oils, and vegetable oils. They are also good solvents for fats and waxes. Interesting applications of these compounds include their use as antigelling agents in certain types of rubber cement, and the use of 1,1-dichloro-1-nitroethane as an industrial fumigant.

FORMULA	1-Chloro-1-nitroethane NO_2—CH_3CHCl	1-Chloro-1-nitropropane NO_2—CH_3CH_2CHCl	2-Chloro-2-nitropropane NO_2—CH_3CClCH_3	1,1-Dichloro-1-nitroethane NO_2—CH_3CCl_2	1,1-Dichloro-1-nitropropane NO_2—$CH_3CH_2CCl_2$
Molecular Weight	109.52	123.54	123.54	143.97	157.99
Specific Gravity $\frac{20°}{20°}$C	1.258	1.209	1.193	1.405	1.314
Pounds per U.S. Gallon at 20°C	10.47	10.06	9.93	11.69	10.93
Distillation Range, °C (90%)	122.0–128.5	139.5–143.3	129.0–132.3	122.0–125.0	141.0–143.6
Flash Point, °F (Tag open cup)	133	144	135	168	151
Refractive Index, at 20°C	1.423	1.430	1.425	1.441	1.443
Solubility, at 20°C ml Solvent in 100 ml Water	<0.4	<0.8	<0.5	<0.5	<0.5
ml Water in 100 ml Solvent	<0.5	<0.4	<0.5	<0.5	<0.5

USES: As solvents in adhesives, as anti-gelling agents for highly accelerated rubber cements, as insecticides; in synthesis of diamines and other complex organic chemicals.

Table 4.12: Specific Gravities of Nitromethane-Methanol Fuel Mixtures (34)

Organic Sulfur Compounds

Table 5.1: Carbon Disulfide *(2)*

Chemical Names: Carbon Disulfide, Carbon Bisulfide
Common Names: Carbon Disulfide, Carbon Bisulfide
Formula: CS_2

PROPERTIES

Grades: Commercial or Technical, and USP
Important Physical and Chemical Properties

Physical State: Liquid.
Color: Clear, colorless liquid.
Odor: Almost odorless when pure; the commercial grade has a strong disagreeable odor, due to presence of sulfur compounds.
Specific Gravity at 20° C/4° C (68° F/39° F) (Water = 1): 1.263
Vapor Density (Air = 1): 2.63
Boiling Point (760 mm): 46.3° C (115° F)
Melting Point: −108.6° C (−163° F)
Flash Point (closed cup): −30° C (−22° F)
Explosive Limits (per cent by volume in air): 1 to 50
Ignition Temperature: 100° C (212° F)
Corrosive: Commercial grade slightly corrosive to some metals due to impurities.
Dangerously Reactive: No. However, it has an extremely low ignition temperature.
Hygroscopic: No.
Light Sensitive: Turns yellow when exposed to sunlight.

VAPOR PRESSURE OF CARBON DISULFIDE

TEMPERATURE		VAPOR PRESSURE MM MERCURY
°C	°F	
−78.2	−109	0.68
−42.6	−45	11.81
−25.35	−14	34.3
−21.5	−7	42.7
0	+32	127.0
11.54	53	211.3
19.7	67	294.3
46.3	115	760.0

Table 5.3: Vapor Pressure vs Temperature for DMSO *(36)*

Vapor Pressure-Temperature
Dimethyl Sulfoxide

DIMETHYL SULFOXIDE (DMSO)

Table 5.2: Physical Properties of DMSO *(36)*

Molecular weight	78.13
Boiling point at 760 mm Hg	189° C (372° F)
Freezing point	18.55° C (65.4° F)
Molal freezing point constant, °C/(mol)(kg)	4.07
Refractive index n_D^{25}	1.4768
Surface tension at 20° C	43.53 dynes/cm
Vapor pressure, at 25°C	0.600 mmHg
Density, g/cm³, at 25°C	1.096
Viscosity, cP, at 25°C	2.0
Specific heat at 29.5°C	0.47±0.015 cal/g/°C
Heat capacity (liq.), 25°C	0.47 cal/g/°C
Heat capacity (ideal gas), $Cp(T°K)=6.94+5.6\times10^{-2}T-0.227\times10^{-4}T^2$	
Heat of fusion	41.3 cal/g
Heat of vaporization at 70°C	11.3 kcal/mol (260 BTU/lb)
Heat of solution in water at 25°C	52 cal/g
Heat of combustion	6054 cal/g
Flash point (open cup)	95°C (203°F)
Auto ignition temperature in air	300-302°C (572-575°F)
Flammability limits in air lower (100°C)	3-3.5% by volume
upper	42-63% by volume
Coefficient of expansion	0.00088/°C
Dielectric constant, 10MHz	48.9 (20°C)
	45.5 (40°C)
Solubility parameter	13
Dipole moment, D	4.3
Conductivity, 20°C	$3\times10^{-8}(ohm^{-1}cm^{-1})$
80°C	$7\times10^{-8}(ohm^{-1}cm^{-1})$
pKa	35.1

Table 5.4: Freezing Point Curves for DMSO-Water Solutions *(36)*

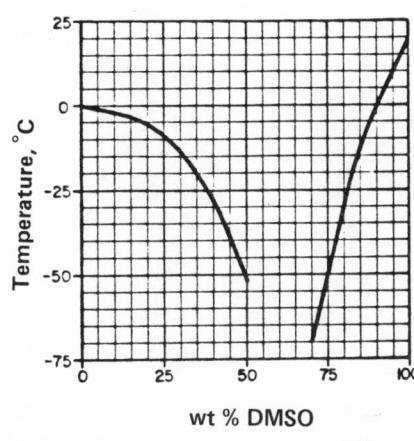
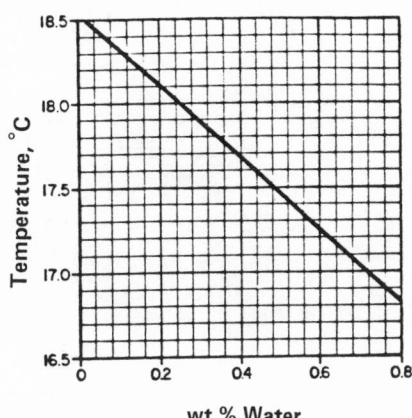

Table 5.5: Viscosity of DMSO and DMSO-Water Solutions *(36)*

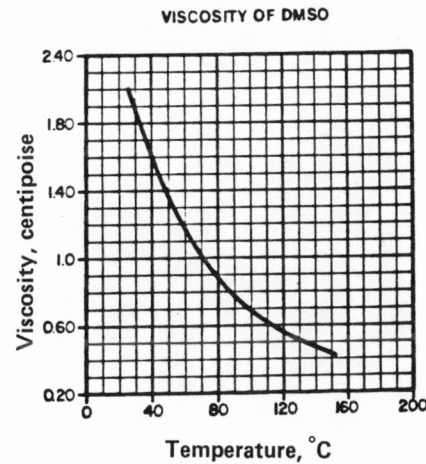
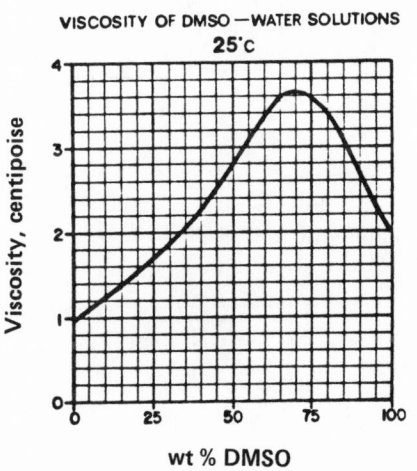

Table 5.6: Thermal Stability of DMSO *(36)*

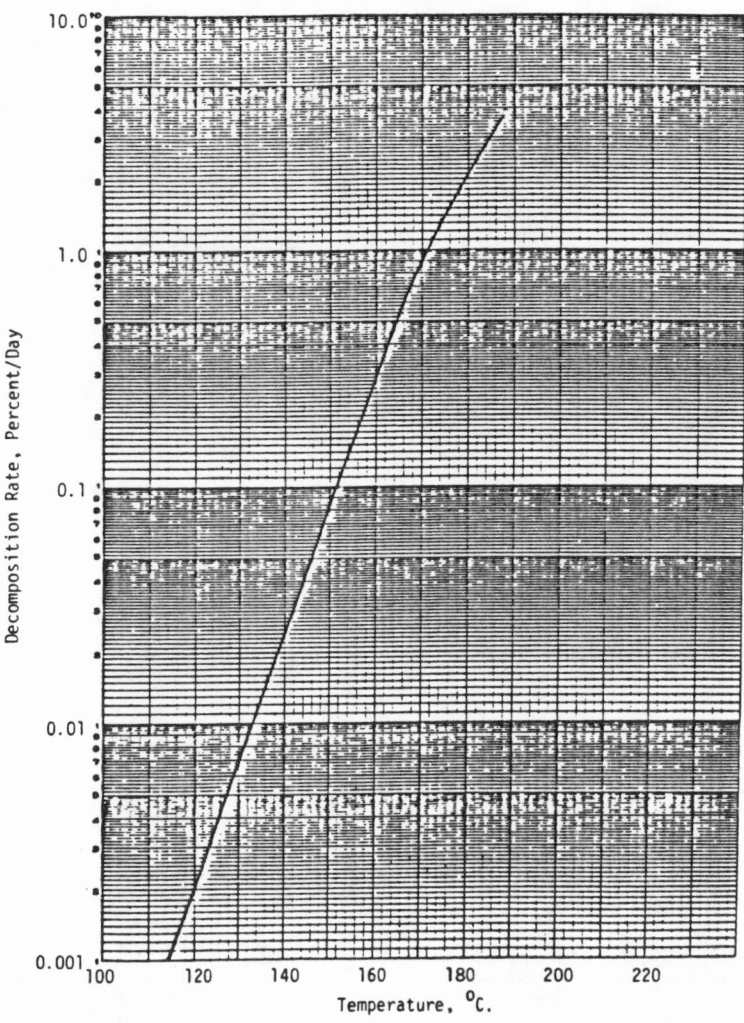

Table 5.7: Solubility of Salts in DMSO *(36)*

	Solubility Grams/100 cc DMSO			Solubility Grams/100 cc DMSO	
	25°C.	90-100°C.		25°C.	90-100°C.
Aluminum sulfate (18H$_2$O)	Insol.	5	Calcium dichromate (3H$_2$O)	50	-
Ammonium borate (3H$_2$O)	10	-	Calcium nitrate (4H$_2$O)	30	-
Ammonium carbonate (H$_2$O)	1	-	Ceric ammonium nitrate	1	-
Ammonium chloride	Insol.	10	Cobaltous chloride (6H$_2$O)	30	Misc. m.p. 86°C.
Ammonium chromate	1	-	Cupric acetate (H$_2$O)	Insol.	6
Ammonium dichromate	50	-	Cupric bromide	1	20 150°C.
Ammonium nitrate	80	-	Cupric chloride (2H$_2$O)	Insol.	27
Ammonium thiocyanate	30	-	Cuprous iodide	1	-
Barium nitrate	1	-	Ferric ammonium sulfate (12H$_2$O)	Insol.	Misc. m.p. 40°C.
Beryllium nitrate (4H$_2$O)	10	-	Ferric chloride (6H$_2$O)	30	90
Bismuth trichloride	1	-	Ferrous chloride (4H$_2$O)	30	90
Cadmium chloride	20	-	Gold chloride	5	-
Cadmium iodide	30	-	Lead chloride	10	-
Calcium chloride	Insol.	1	Lead nitrate	20	60

(continued)

Table 5.7: (continued)

	Solubility Grams/100 cc DMSO			Solubility Grams/100 cc DMSO	
	25°C.	90-100°C.		25°C.	90-100°C.
Lithium dichromate (2H$_2$O)	10	-	Silver nitrate	130	180
Lithium nitrate	10	-	Sodium dichromate (2H$_2$O)	10	-
Magnesium chloride (6H$_2$O)	1	-	Sodium iodide	30	-
Magnesium nitrate (6H$_2$O)	40	-	Sodium nitrate	20	-
Manganous chloride (4H$_2$O)	20	-	Sodium nitrite	20	-
Mercuric acetate	100	-	Sodium thiocyanate	1	-
Mercuric bromide	90	-	Stannous chloride (2H$_2$O)	40	-
Mercuric iodide	100	-	Strontium bromide (6H$_2$O)	5	-
Molybdenum bromide	1	Reacts	Strontium chloride (2H$_2$O)	10	-
Nickel chloride (6H$_2$O)	60	-	Tungsten hexachloride	5	-
Nickel nitrate (6H$_2$O)	60	-	Uranyl nitrate (6H$_2$O)	30	-
Potassium iodide	20	20	Vanadium chloride	-	1
Potassium nitrate	10	-	Zinc chloride	30	60
Potassium nitrite	2	-	Zinc nitrate (6H$_2$O)	550	-
Potassium thiocyanate	20	50			

Table 5.8: Solubility of Resins and Polymers in DMSO (36)

Material	Grams Soluble in 100 cc DMSO		
	20-30°C	90-100°C	Comments
Polyacrylics			
Orlon (du Pont)	-	20	Viscous soln.
Acrilan	> 25	-	
Verel (Eastman)	> 5		25 at 130°C with some decomposition
Creslan (Am. Cyanamid)	5		25 at 130°C
Zefran (Badische)	-	Insol.	
Polyamides			
Nylon 6	-	Insol.	40 at 150°C
Nylon 6/6	-	Insol.	25 at 150°C
Nylon 6/10	-	Insol.	40 at 150°C
Cellulose			
Cellulose triacetate	10	20	
Viscose rayon	-	< 1	
Cellophane	-	Insol.	
Carboxymethyl cellulose	-	Insol.	
Epoxies			
Epon 1001 (Shell)	50	-	
Epon 1004 (Shell)	50	-	
Epon 1007 (Shell)	50	-	
Methacrylates			
Lucite 41, 45 (du Pont)	-	< 1	
Plexiglas (Rohm & Haas)	-	< 1	
Poly Carbonates			
Lexan (General Electric)	-	> 5	
Merlon (Mobay)	-	Insol.	
Polyesters			
Dacron (du Pont)	-	> 1	Dissolves at 160°C ppts. 130°C
CX 1037 (Goodyear)	-	7	
Atlac (ICI-America)	-	50	
Silicones			
Dow Corning 803 soln.	Miscible	-	
Dow Corning 805 soln.	Miscible	-	
Dow Corning "Sylkyd 50"	Miscible	-	
Dow Corning Z6018 (flake)	70	-	
Urethanes			
Vithane (Goodyear)	-	100	

(continued)

Table 5.8: (continued)

Material	Grams Soluble in 100 cc DMSO 20-30°C	90-100°C	Comments
Vinyls - Polymers & Co-polymers			
Butvar B-76 (Monsanto)	-	20	Very viscous
Formvar 7/70 E (Monsanto)	-	42	Very viscous
Elvanol 51-05 (du Pont)	-	90	Viscous
Elvanol 52-22 (du Pont)	-	15	Viscous
Elvanol 71-24 (du Pont)	-	30	Viscous
Polyvinyl pyrrolidone (GAF)	30	> 100	
Geon 101 (PVC Goodrich)	-	10	
Vinylite VYHH (Union Carbide)	2	30	
Teslar (du Pont)	-	-	Partially sol. at 160-170°C
Vinylidenes			
Darvan (Goodrich)	5	-	Soln. cloudy and viscous
Saran film (Dow)	-	30	
Geon 200 x 20 (Goodrich)	-	20	
DNA (Goodrich)	> 5	-	25 at 130°C
Other Resinous Materials			
Melmac 405 (Am. Cyanamid)	70	-	
Neoprene	Insol.	Insol.	
Polyethylene	Insol.	Insol.	
Polystyrene	-	-	Sol. at 150°C ppts. at 130°C
Rosin (Hercules)	>100	-	
Penton (chlorinated polyether)(Hercules)	-	5	
Teflon (du Pont)	Insol.	Insol.	
Vinsol (Hercules)	50	> 100	

Table 5.9: Solubility of Miscellaneous Materials in DMSO *(36)*

Material	Solubility Grams/100 cc. DMSO 20-30°C	90-100°C	Material	Solubility Grams/100 cc. DMSO 20-30°C	90-100°C
Acetic acid	Miscible	-	Cyclohexane	4.67	-
Acetone	Miscible	-	Cyclohexylamine	Miscible	-
Acrawax	< 1	> 1	Decalin	4.5	-
Acrawax B	Insol.	4	n-Decane	0.7	-
Aniline	Miscible	-	Di-n-butylamine	11	-
Beeswax	-	< 1	o-Dichlorobenzene	Miscible	-
Benzene	Miscible	-	p-Dichlorobenzene	Very soluble	-
Benzidine	Soluble	-	Dichlorodiphenyltrichloroethane	4	100
Benzidine methane sulfonate	Insol.	-	Dicyandiamide	40	-
Bromine	Reacts	-	Dicyclohexylamine	4.5	-
Butenes	2.1	-	Diethanolamine	Miscible	-
Calcium methyl sulfonate	Soluble	-	Diethylamine	Miscible	-
Camphor	Soluble	Soluble	Diethyl ether	Miscible	-
Candelilla wax	-	< 1	bis-(2-ethylhexyl)amine	0.7	-
Carbon	Insol.	-	Diethyl sulfide	Miscible	-
Carbon disulfide	90	-	Di-isobutyl carbinol	Miscible	-
Carbon tetrachloride	Miscible	-	Di-isobutylene	3.3 (0.6% DMSO soluble in di-	
Carbowax 600	Miscible	-		isobutylene)	
Carbowax 6000	Insol.	8	Dimethyl ether	4.4	-
Carnauba wax	-	< 1	Dimethyl formamide	Miscible	-
Castor oil	Miscible	-	Dimethyl sulfide	Miscible	-
Ceresin wax	-	< 1	Dimethyl sulfone	33.9	Miscible
Chlorine	Reacts	-	Dioxane	Miscible	-
Chloroform	Miscible	-	Diphenyl	Very soluble	-
Chlorosulfonic acid	Reacts	-	Dipentene	10	-
Citric acid	> 70	-	n-Dodecane	0.38	-
Coconut oil	0.3	1.3	Dodecylbenzene (Neolene 400)	3.5	-
		Misc.-160°C	Dyes		
Cork	Softens	Softens	Burnt Sugar	Soluble	-
Cresylic acid	Miscible	-	FD&C Blue	Soluble	-
Cumene	Miscible	-	Pistachio Green B	Soluble	-

(continued)

Table 5.9: (continued)

Material	Solubility Grams/100 cc. DMSO 20-30°C	90-100°C
Ethyl benzoate	Miscible	-
Ethyl alcohol	Miscible	-
Ethyl bromide	Miscible	Reacts
Ethyl ether	Miscible	-
Ethylene dichloride	Miscible	-
Formalin (37%)	Miscible	-
Formamide	Miscible	-
Formic acid	Miscible	-
Glycerine	Miscible	-
Glycine	< 0.05	0.1
Hexane	2.9	-
Hy-Wax 120	-	< 1
Iodine	Soluble	-
Isoprene	Miscible	-
Kerosene	0.5 (0.5% DMSO soluble in kerosene)	
Lanolin, hydrated (Lanette 0)	-	11 (Gets cold)
Lauryl amide (Armid 12)	10	> 20
Lorol 5	Miscible	-
Lubricating oil	0.4	-
Methionine	0.1	0.3
Methyl borate	Miscible	-
Methyl caprate	-	Miscible
Methyl iodide	Miscible	Reacts
Methyl laurate	7	Miscible
Methyl mercaptan	40	-
N-methyl morpholine	Miscible	-
Methyl palmitate	Immiscible	Misc. 130-180°C
Methyl salicylate	Miscible	-
Methyl sulfonic acid	Miscible	-
Methylene chloride	Miscible	-
Microcrystalline wax	-	< 1
Morpholine	Miscible	-
Naphthalene	40	Miscible
Neoprene	Insol.	Insol.
Nitrobenzene	Miscible	-
Oleic acid	Miscible	-
Ouricuri wax	-	1
Oxalic acid	38	-
Paint (dried)	Softens and dissolves	
Palmitic acid	100	,
Paraffin	Insoluble	-
Paraformaldehyde	Insoluble	Slightly soluble
Pentaerythritol	5-10	30
n-Pentane	0.35	-
Pentene 1 & 2	7.1	-
Perchloric acid	Reacts violently	-

Material	Solubility Grams/100 cc. DMSO 20-30°C	90-100°C
Petroleum ether	3 (DMSO soluble 0.3-0.5% in petroleum ether)	
Phenol	Soluble	-
Phosphoric acid	Miscible	-
Phosphorus trichloride	Reacts vigorously	-
Phthalic acid	90	
Isophthalic acid	68	76
Terephthalic acid	26	33
Picric acid	Soluble	-
Pyridine	Miscible	-
Pyrogallol	50	-
Rosin	> 100	-
Rosin soap (Hercules Dresinate X)	Slightly soluble	0.9
Sevin	50	-
Shellac, white, dried	-	80
Silicon tetrachloride	Reacts vigorously	
Sodium	-	Reacts
Sorbitan sesquioleate	2.5	-
Sorbitan trioleate	-	Miscible
Sorbitol	60	> 180
Soybean oil	0.6	-
Starch, soluble	> 2	-
Stearic acid	2	Miscible
Succinic acid	30	-
Sugar (sucrose)	30	100
Sulfamic acid	40	-
Sulfur	-	< 1
Sulfuric acid	Miscible	-
Tallow	Insol.	1.9
Tallow amide, hydrogenated (Armour Armide HT)	Insol.	> 40
Tetrahydrophthalic anhydride	50	-
Thiourea	40	85
Toluene	Miscible	-
Toluene di-isocyanate	Miscible	-
Tributylamine	0.9	-
Tricresyl phosphate	Miscible	-
Triethanolamine laurylsulfate	Soluble	-
Triethanolamine	Miscible	-
Triethylamine	10	-
Trinitrotoluene	Soluble	-
Turpentine	10	-
Urea	40	110
Water	Miscible	-
Xylene	Miscible	-

Table 5.10: Solubility of Gases in DMSO at Atmospheric Pressure and 20°C *(36)*

	Grams Gas per 100 Grams Solution	Gas Volume per Volume of DMSO
Acetylene	2.99	28.1
Butadiene	4.35	-
Mixed butylenes	2.05	-
Carbon dioxide	0.5	3.0
Carbon monoxide	0.01	
Ethylene	0.32	2.8
Ethylene oxide	60.0	306.0
Freon 12	1.8	3.7
Helium	Insol.	
Hydrogen	0.00	

	Grams Gas per 100 Grams Solution	Gas Volume per Volume of DMSO
Hydrogen sulfide	0.5 (Reacts)	
Isobutylene	2.5-3.0	-
Methane	0.00	
Nitric oxide (NO)	0.00	
Nitrogen	0.00	0.06
Nitrogen dioxide (NO$_2$, N$_2$O$_4$)	Miscible (Possible reaction)	
Oxygen	0.01	
Ozone	Reacts	
Sulfur dioxide	57.4	

SULFOLENE

Table 5.11: 3-Methylsulfolene *(4)*

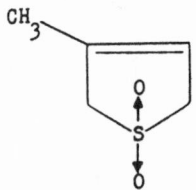

3-Methyl-2,5-Dihydrothiophene-1,1-Dioxide

Typical Properties

Molecular Weight	136
Melting Point, °C.	63
Total Sulfur, wt. %	24.3
Boiling Point	decomposes above melting point
Solubility in water at 25°C., wt. %	5.9

SULFOLANE

Table 5.12: Physical Properties of Sulfolane *(4)*

Formula	$C_4H_8SO_2$
Molecular Weight	120.17
Appearance	Clear Liquid
Specific Gravity, 30/4C	1.261
100/4C	1.2012
Density, lbs/gal, 60°F	10.5
Density, g/ml, 60°F	1.276
Boiling Point, °F	545
Freezing Point, °F (99.8+ wt. % Sulfolane)	78.8
Flash Point, °F	330
Vapor Pressure, mmHg., at 302°F	14.53
320°F	21.55
392°F	85.23
410°F	115.1
500°F	421.4
Viscosity, cp, at 30°C	10.3
50°C	6.1
100°C	2.5
150°C	1.4
200°C	1.0
Refractive Index,30°C/D	1.48
Heat of Vaporization,Btu/lb. mole, at 212°F	27,000
392°F	26,500
Heat Capacity of Liquid,Btu/lb. °F,at 212°F	0.40
392°F	0.46
Heat of Fusion, Btu/lb	4.92
Dielectric Constant (30°C)	44.0
Coefficient of Thermal Expansion at 82°F per °F	3.83×10^{-4}(1)
Thermal Conductivity (2), Btu/(hr.) (Sq. Ft.) (°F/ft.)	
50	0.107 (3)
50	0.098 (4)
100	0.114
135	0.112

(1) This value calculated from the empirical equation a = 0.000383 + 2.5 x 10^{-7} (t-82).

where a = coefficient of expansion

t = temperature, °F in the applicable range of 82°F to 400°F. Base temperature is 82°F which is the melting point of pure sulfolane.

(2) The sulfolane used in the conductivity test contained 1.5 wt. % water as determined by freezing point and Karl Fisher water analysis.

(3) Below the freezing point.

(4) The sulfolane used in this conductivity test was the same as in Note 2 except it contained 4.5 wt. % benzene + 5.0 wt. % toluene.

Table 5.13: Typical Solvent Properties of Sulfolane (4)

Solubility of Sulfolane in Various Chemical Compounds

CHEMICAL COMPOUND	TEMP., °F	GRAMS SULFOLANE PER 100 GMS OF CHEMICAL
Benzene	77°	Miscible
Cyclohexane	77°	0.4
2,3,Dimethylbutane	77°	0.3
Hexene-1	77°	1.0
Normal Hexane	77°	0.3
Perchloroethylene	76°	1.6
Toluene	77°	Miscible
Mixed Xylenes	77°	Miscible

Solubility of Various Chemical Compounds in Sulfolane

CHEMICAL COMPOUND	TEMP., °F	GRAMS CHEMICAL PER 100 GMS OF SULFOLANE
Hydrogen Chloride (gas)	77°	9.3
Ethyl Mercaptan	80°	Miscible
Methyl Mercaptan	32°	Miscible
Methyl Mercaptan	77°	21.7*
Tertiary Dodecyl Mercaptan	77°	2.0
Perchloroethylene	76°	37.5
Polystyrene	392°	0.02
Trichloroethylene	76°	Miscible

*Test performed at atmospheric pressure and approximately 34°F above the normal boiling point (42.6°F) of methyl mercaptan.

Thermal Stability of Sulfolane

Thermal stability is an important property of sulfolane because the commercial processes using this compound normally operate at elevated temperatures. For this reason tests were made to determine the relative stability of sulfolane containing 10 mg. of SO_2 per 250 ml. at various temperatures.

TEMPERATURE		MG SO_2 LIBERATED PER HOUR FROM 250 ML. SULFOLANE
°C	°F	
180	356	0.6
200	392	2.8
220	428	3.3
240	464	24.1

These tests show that sulfolane has good thermal stability up to and including 428°F but has a rather sharp decomposition rate beyond this temperature. Excessive temperatures will cause sulfolane "to crack" to a dark polymer and SO_2.

Comparative Freezing Point Depression

Impurity	Mol. % Impurity Required to Lower Freezing Point 1°F
Normal Butylbenzene	0.098
2-Phenylpentane	0.103
Sulfolene	0.228
3-Methylsulfolane	0.189
Water	0.163

(continued)

Table 5.13: (continued)

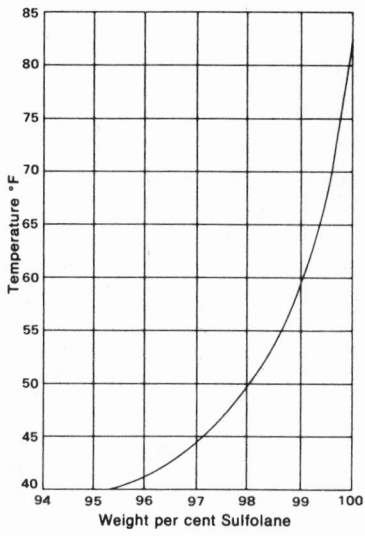

Freezing point curve for Sulfolane — water mixtures

Table 5.14: **Specific Gravity (60°/60°F) for Blends of Sulfolane-Water-Diisopropylamine (DIPA)** *(4)*

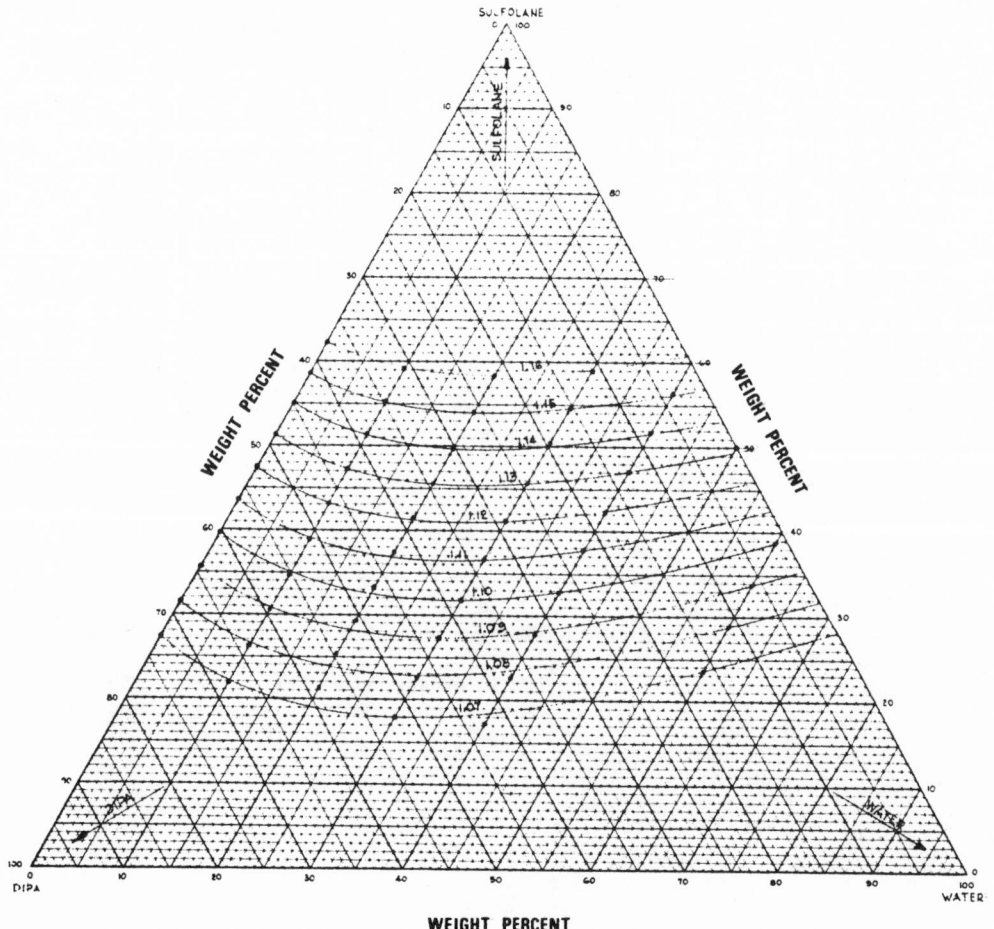

ORGANOSULFUR COMPOUNDS

Table 5.15: Pennwalt Organosulfur Compounds (37)

Fig. 1. Physical properties of Organosulfur Intermediates.

Product	Formula	Purity (1) (wt. % min.)	Color (2) (APHA max.)	Mercaptan Sulfur (3) (wt. % min.)	Distillation Range (4) IBP (°C min.)	Distillation Range (4) 95% Below (°C max.)	Molecular Weight (calc.)	Specific Gravity (15.5/15.5°C)	Average Wt./Gal. (lb.)	Flash Point (COC°F)	Coefficient of Cubical Expansion/°C
MERCAPTANS											
Methyl Mercaptan	CH_3SH	99.5	15	66.3	4.5*	7.5*	48.1	0.875	7.3	0	.00143*
Ethyl Mercaptan	C_2H_5SH	99.3	40	51.2	34.4	36.1	62.1	0.841	7.0	0	.00114
n-Propyl Mercaptan	$n\text{-}C_3H_7SH$	99.0	20	41.3	66.2	69.4	76.2	0.843	7.0	5	.00110
Isopropyl Mercaptan	$(CH_3)_2CHSH$	98.0	15	40.8	51.0	56.1	76.2	0.817	6.8	-30	.00106
n-Butyl Mercaptan	$n\text{-}C_4H_9SH$	98.0	15	34.7	96.3	100.0	90.2	0.846	7.1	55	.00103
t-Butyl Mercaptan	$t\text{-}C_4H_9SH$	98.0	15	34.5	62.1	67.8	90.2	0.804	6.7	18	.00103
n-Hexyl Mercaptan	$n\text{-}C_6H_{13}SH$	96.0	15	26.0	149.0	158.0	118.2	0.843	7.0	125	.00076
Cyclohexyl Mercaptan	$C_6H_{11}SH$	99.0	15	27.3	155.0	161.0	116.2	0.950	7.9	120	.00120
n-Octyl Mercaptan	$n\text{-}C_8H_{17}SH$	98.0	15	21.3	194.0	203.0	146.3	0.847	7.1	175	.00078
t-Octyl Mercaptan	$t\text{-}C_8H_{17}SH$	97.6	15	21.4	157.0	168.0	146.3	0.846	7.1	115	.00087
t-Nonyl Mercaptan	$t\text{-}C_9H_{19}SH$	97.0	15	19.4	186.7	201.0	160.3	0.850	7.1	150	.00083
n-Dodecyl Mercaptan	$n\text{-}C_{12}H_{25}SH$	98.0	20	15.5	269.0	285.0	202.4	0.846	7.1	190	.00074
t-Dodecyl Mercaptan	$t\text{-}C_{12}H_{25}SH$	98.5	15	15.6	227.0	248.0	202.4	0.859	7.2	205	.00072
Mixed Primary Tridecyl Mercaptans	$C_{13}H_{27}SH$	97.8	15	14.5	260.0	275.0	216.4	0.857	7.1	200	.00080
OTHER ORGANOSULFUR INTERMEDIATES											
Thiophene	$\begin{smallmatrix} S \\ HC{-}CH \\ HC{=}CH \end{smallmatrix}$	98.8	10	—	83.0	85.0	84.1	1.070	8.9	20	.00112
Tetrahydrothiophene	$\begin{smallmatrix} S \\ H_2C{-}CH_2 \\ H_2C{-}CH_2 \end{smallmatrix}$	99.0	20	—	118.0	122.0	88.2	1.002	8.4	55	.00096
Di-t-Nonyl Polysulfide (TNPS)	$(C_9H_{19})_2S_5$	—	—	—	SEE FOOTNOTE (5).		—	1.026	8.7	325	.00058

Test Methods:
(1) Determined by Vapor Phase Chromatography or Wet Chemistry.
(2) Platinum-Cobalt Color Scale.
(3) Silver-Nitrate Titration.
(4) ASTM-D-1078
(*) Literature values
(5) Di-t-Nonyl Polysulfide (TNPS) is a pale yellow, oily liquid containing a minimum of 37% sulfur. Its odor is mild. It is high in active sulfur. It is non-irritating and non-toxic.

(continued)

Table 5:15: (continued)

Fig. 2. Physical properties of selected development compounds.

Product	Formula	Purity (wt. % min.)	Color (APHA max.)	Distillation Range IBP (°C min.)	95% (°C max.)	Mercaptan Sulfur (wt. % min.)	Molecular Weight (calc.)	Specific Gravity 15.5/15.5°C	Average Wt./Gal. (lb.)	Flash Point (COC°F)	Vapor Pressure @ 100°F	Coefficient of Cubical Expansion/°C	Product
MERCAPTANS													**MERCAPTANS**
Sec-Butyl Mercaptan	$C_2H_5(CH_3)CHSH$	97.0	15	82	87	34.5	90.2	0.835	7.0	25	2.9	.00103	Sec-Butyl Mercaptan
n-Amyl Mercaptan	$n-C_5H_{11}SH$	96.0	15	120	87	29.5	104.2	0.845	7.0	80	0.4	.00090	n-Amyl Mercaptan
t-Amyl Mercaptan	$t-C_5H_{11}SH$	98.0	15	98.4	99.6	30.1	104.2	0.83	6.9	30	1.7	.00133	t-Amyl Mercaptan
n-Decyl Mercaptan	$n-C_{10}H_{21}SH$	96.0	15	235	245	17.6	174.4	0.846	7.1	215	<0.1	.00076	n-Decyl Mercaptan
Ethylcyclohexyl Dimercaptan	$C_8H_{14}(SH)_2$	98.0	15	—	—	—	176.0	1.06	8.8	250	<0.1	.0100	Ethylcyclohexyl Dimercaptan
d-Limonene Dimercaptan	$C_{10}H_{18}(SH)_2$	97.0	15	—	—	—	204.0	1.03	8.6	250	<0.1	.0007	d-Limonene Dimercaptan
n-Tetradecyl Mercaptan	$n-C_{14}H_{29}SH$	93.0	20	—	—	12.9	230.5	0.846	7.1	305	<0.1	.00069	n-Tetradecyl Mercaptan
n-Hexadecyl Mercaptan	$n-C_{16}H_{33}SH$	92.5	35	—	—	11.5	258.5	0.842[1]	7.0[1]	300	<0.1	.00067	n-Hexadecyl Mercaptan
n-Octadecyl Mercaptan	$n-C_{18}H_{37}SH$	82.0	50	—	—	9.2	286.6	0.817[2]	6.8[2]	380	<0.1	.00065	n-Octadecyl Mercaptan
SULFONIC ACIDS & CHLORIDES													**SULFONIC ACIDS & CHLORIDES**
Methane Sulfonic Acid	CH_3SO_3H	99.5	100	—	—	—	96.1	1.48	12.3	—	<0.1	.00064	Methane Sulfonic Acid
Methane Sulfonyl Chloride	CH_3SO_2Cl	99.5	20	—	—	—	114.6	1.48	12.3	230	3.0	.00115	Methane Sulfonyl Chloride

[1] Measured @ 60/15.5°C
[2] Measured @ 25/15.5°C

Table 5.16: Vapor Pressure Curves of Pennwalt Organosulfur Intermediates *(37)*

Methyl Mercaptan.

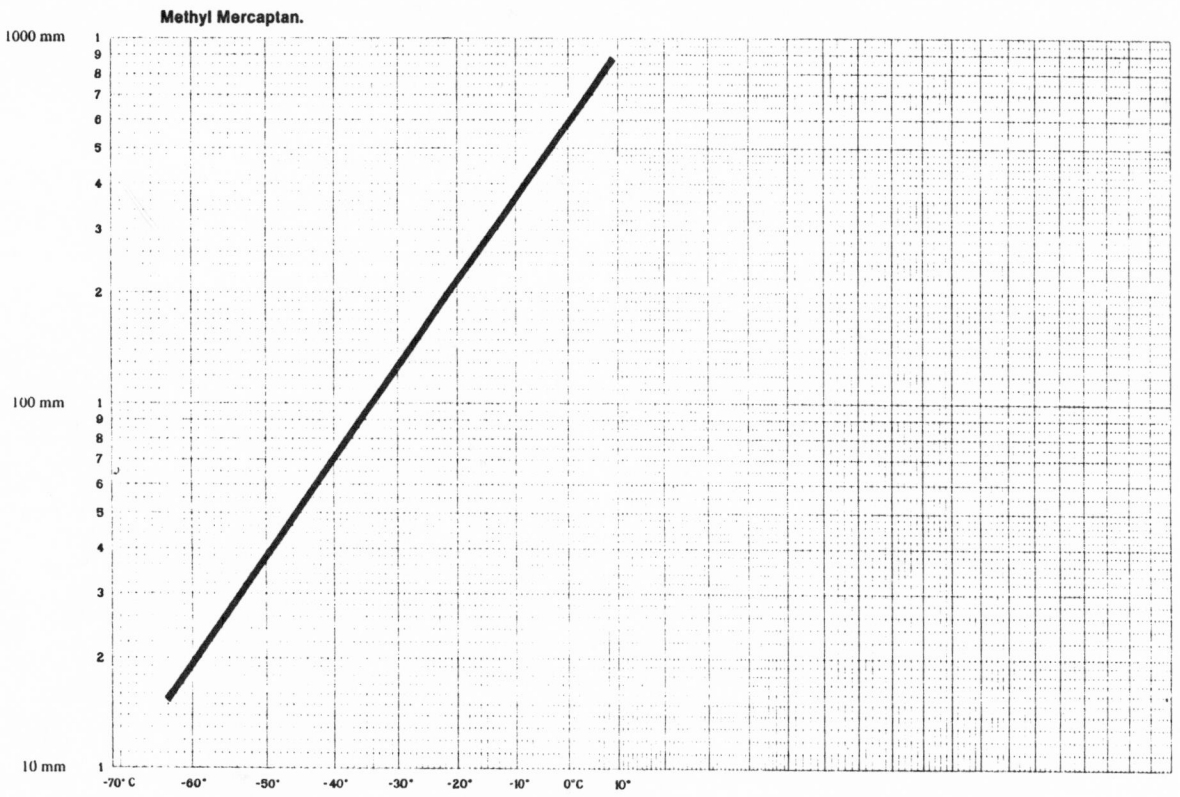

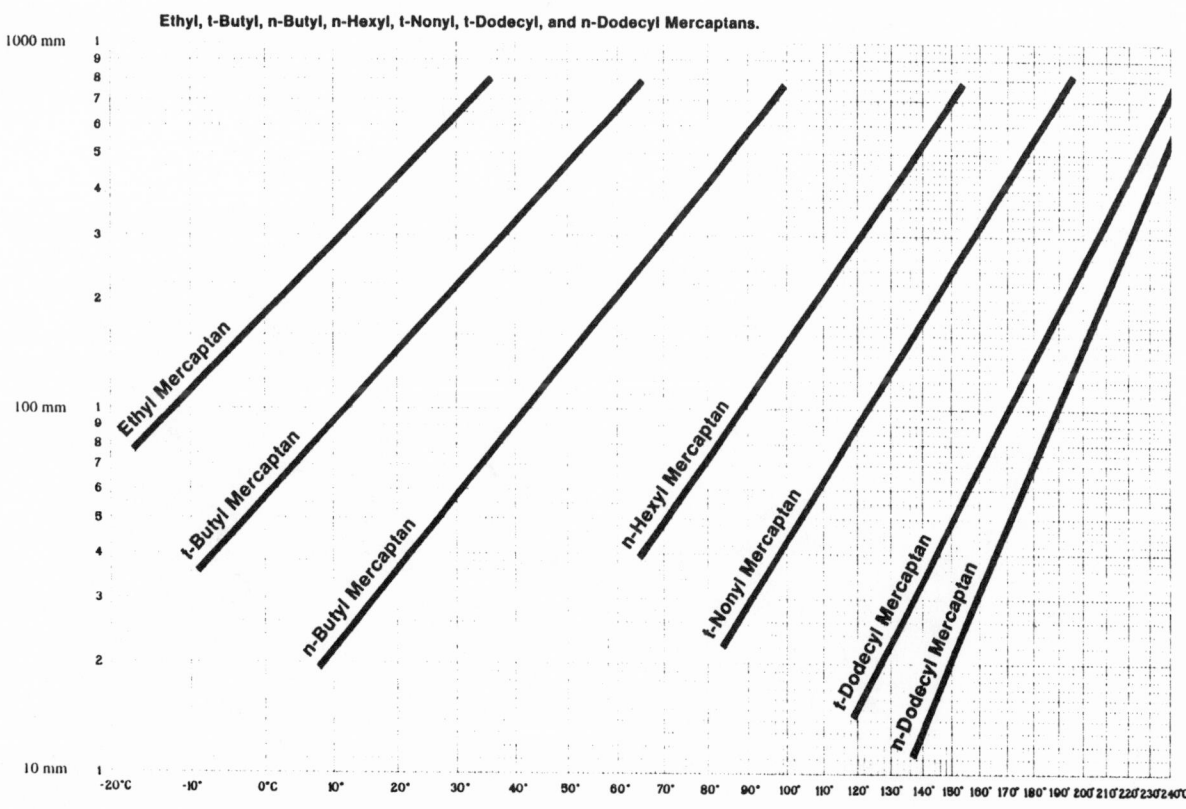

Ethyl, t-Butyl, n-Butyl, n-Hexyl, t-Nonyl, t-Dodecyl, and n-Dodecyl Mercaptans.

(continued)

Table 5.16: (continued)

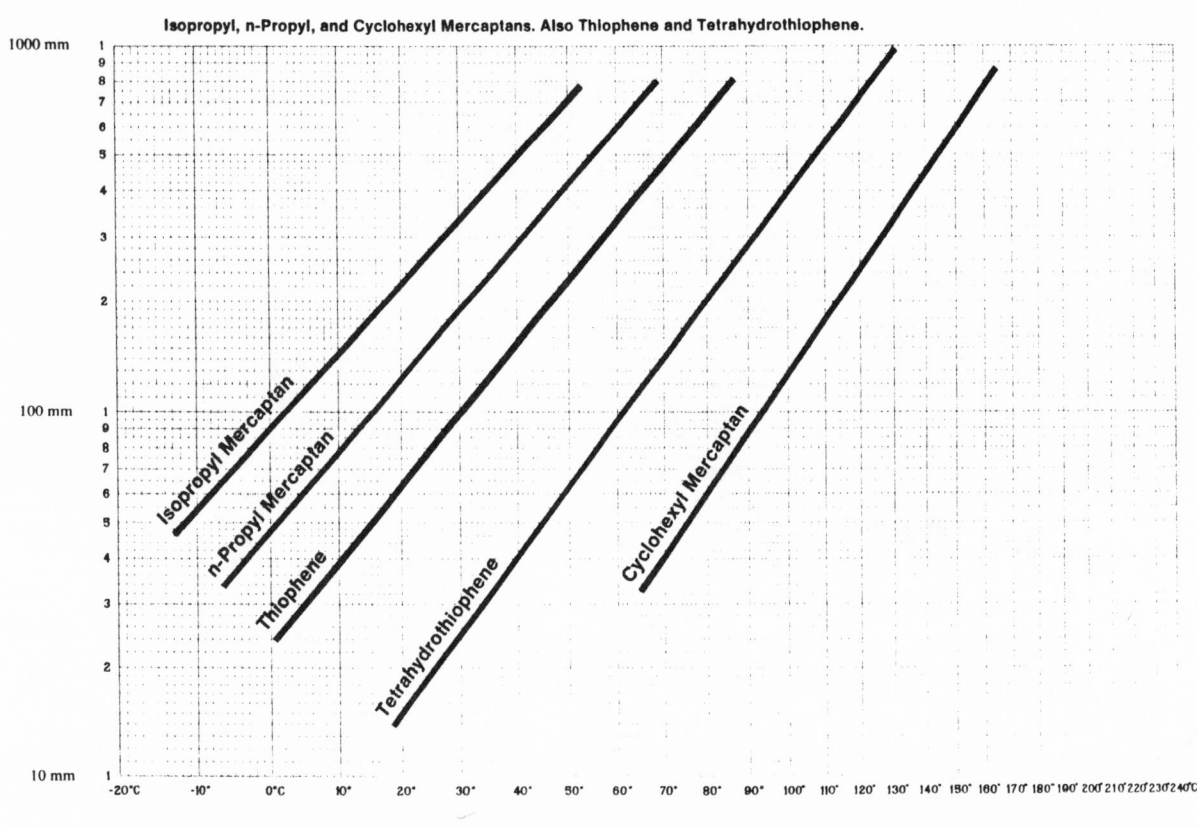

Isopropyl, n-Propyl, and Cyclohexyl Mercaptans. Also Thiophene and Tetrahydrothiophene.

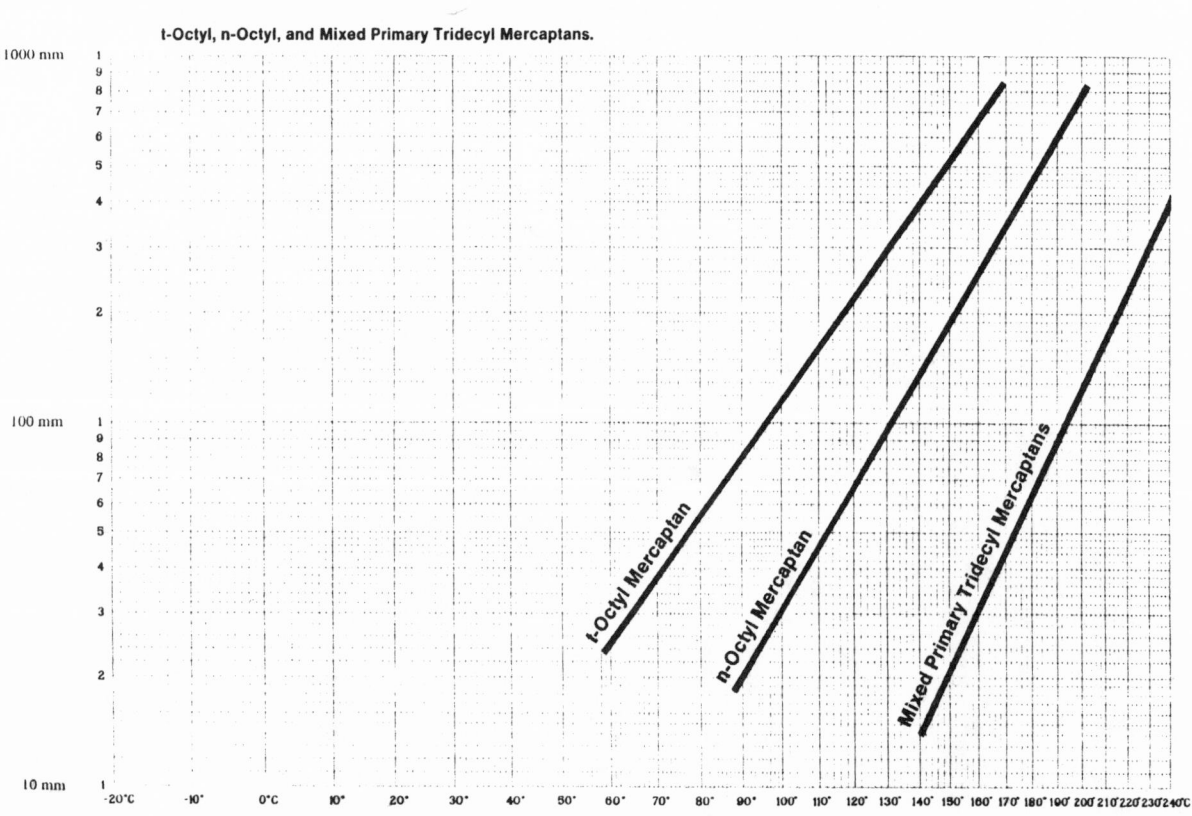

t-Octyl, n-Octyl, and Mixed Primary Tridecyl Mercaptans.

Table 5.17: Phillips Organosulfur Compounds *(4)*

PHYSICAL PROPERTIES

Product	Formula	Molecular Weight	Boiling Point[1]	Vapor Pressure[2]	Specific Gravity[3]	Density[4] of Liquid	Flash Point	Freezing Point	Sulfur Content (Wt %)	Typical Purity (Wt %)	Viscosity[5]	Coefficient of Expansion/°C
Ethyl Mercaptan	C₂H₅SH	62	95 °F / 35 °C	16.2 psia / 111.7 Kpa	0.845	7.02 lb/gal. / 0.842 Kg/dm³	−55 °F / −48 °C	−234 °F / −148 °C	51.4	99	0.293	1.14 × 10⁻³
Normal Propyl Mercaptan	C₃H₇SH	76	154 °F / 68 °C	5.1 psia / 35.2 Kpa	0.647	7.04 lb/gal. / 0.845 Kg/dm³	−5 °F / −20 °C	−172 °F / −113 °C	41.9	99	0.399	1.10 × 10⁻³
Isopropyl Mercaptan	C₃H₇SH	76	127 °F / 53 °C	8.8 psia / 60.7 Kpa	0.820	6.83 lb/gal. / 0.819 Kg/dm³	−30 °F / −34 °C	−203 °F / −130 °C	41.1	97	0.369	1.06 × 10⁻³
Normal Butyl Mercaptan	C₄H₉SH	90	209 °F / 98 °C	1.6 psia / 11.0 Kpa	0.847	7.05 lb/gal. / 0.845 Kg/dm³	35 °F / 2 °C	−176 °F / −116 °C	35.2	99	0.497	1.03 × 10⁻³
Tertiary Butyl Mercaptan	C₄H₉SH	90	148 °F / 64 °C	5.9 psia / 40.7 Kpa	0.806	6.71 lb/gal. / 0.800 Kg/dm³	−15 °F / −26 °C	34 °F / 1 °C	35.5	99	0.638	1.03 × 10⁻³
Isobutyl Mercaptan	C₄H₉SH	90	191 °F / 88 °C	2.4 psia / 16.6 Kpa	0.839	6.99 lb/gal. / 0.840 Kg/dm³	15 °F / −9 °C	−220 °F / −140 °C	35.2	97	0.506	—
Secondary Butyl Mercaptan	C₄H₉SH	90	185 °F / 85 °C	2.7 psia / 18.6 Kpa	0.834	6.94 lb/gal. / 0.832 Kg/dm³	−10 °F / −23 °C	−220 °F / −140 °C	35.0	97	0.463	1.03 × 10⁻³
Normal Hexyl Mercaptan	C₆H₁₃SH	118	307 °F / 153 °C	<1 psia / <7 Kpa	0.847	7.05 lb/gal. / 0.846 Kg/dm³	>80 °F / >27 °C	−113 °F / −81 °C	26.9	96	0.813	7.60 × 10⁻⁴
Cyclohexyl Mercaptan	C₆H₁₁SH	116	316 °F / 158 °C	<1 psia / <7 Kpa	0.953	7.94 lb/gal. / 0.950 Kg/dm³	110 °F / 43 °C	—	27.5	99	—	1.20 × 10⁻³

Note (1) °F at 14.7 psia / °C at 101.3 Kpa (2) psia at 100°F / Kpa at 37.8 (3) @ 60/60°F (4) lb gal at 60°F / Kg dm³ at 15.6°C (5) Centipoise at 68°F or 20°C

Product	Formula	Molecular Weight (Average)	Boiling Range IBP @ 5 mm Hg	Boiling Range 95%	Boiling Point @760 mm Hg	Specific[1] Gravity	Density[2] of Liquid	Flash Point	Mercaptan Sulfur (Wt %)	Typical Purity (Wt %)	Viscosity[3]	Coefficient of Expansion/°C
Normal Octyl Mercaptan	C₈H₁₇SH	146	—	—	390 °F / 199 °C	0.848	7.06 lb/gal / 0.847 Kg/dm³	156 °F / 69 °C	21.8	97	1.33	7.8 × 10⁻⁴
Tertiary Octyl Mercaptan	C₈H₁₇SH	(146)	—	—	307 °F / 153 °C	0.846	7.05 lb/gal / 0.846 Kg/dm³	105 °F / 41 °C	21.3	96	—	8.7 × 10⁻⁴
Pinanyl Mercaptan Type 2	C₁₀H₁₇SH	169	—	—	419 °F / 215 °C	0.974	8.10 lb/gal / 0.972 Kg/dm³	>190 °F / >88 °C	18.3	96	—	—
Normal Dodecyl Mercaptan	C₁₂H₂₅SH	202	—	—	531 °F / 277 °C	0.849	7.07 lb/gal / 0.848 Kg/dm³	>200 °F / >93 °C	15.6	91	2.98	7.4 × 10⁻⁴
Tertiary Nonyl Mercaptan (Sulfole™ 90)	C₉H₁₉SH	(160.3)	125 °F / 52 °C	148 °F / 65 °C	—	0.855	7.12 lb/gal / 0.854 Kg/dm³	154 °F / 68 °C	19.4	97.4[4]	—	8.3 × 10⁻⁴
Mixed Tertiary Mercaptan (Sulfole™ 100)	C₁₁H₂₁SH	(171.3)	126 °F / 52 °C	217 °F / 103 °C	—	0.855	7.12 lb/gal / 0.854 Kg/dm³	150 °F / 66 °C	18.0	96.4[4]	1.77	—
Tertiary Dodecyl Mercaptan (Sulfole™ 120)	C₁₂H₂₅SH	(199)	190 °F / 88 °C	225 °F / 107 °C	—	0.861	7.17 lb/gal / 0.860 Kg/dm³	230 °F / 110 °C	15.5	96.0[4]	2.84	7.2 × 10⁻⁴
Mixed Tertiary Mercaptan (Sulfole™ 132)	C₁₃H₂₇SH	(212)	190 °F / 88 °C	270 °F / 132 °C	—	0.864	7.21 lb/gal / 0.863 Kg/dm³	235 °F / 113 °C	13.8	91.0[4]	3.98	8.0 × 10⁻⁴
Tertiary Hexadecyl Mercaptan (Sulfole™ 160)	C₁₆H₃₃SH	(258)	253 °F / 123 °C	301 °F / 150 °C	—	0.875	7.28 lb/gal / 0.870 Kg/dm³	290 °F / 143 °C	10.5	81.5[4]	—	—

(1) @ 60 60°F (2) lb/gal at 60°F / Kg/dm³ at 15.6°C (3) Centipoise at 68°F or 20°C (4) Mercaptan Purity

Monohydric Alcohols

METHANOL

Table 6.1: Physical Properties of Methanol *(79)*

Chemical Family	Alcohol	Critical Temperature, °C (°F)	240 (464)
Chemical Formula	CH_3OH	Critical Volume, cc/g (cu ft/lb)	3.6829 (.05899)
Chemical Structure	H_3-C-OH		
Chemical Abstract Service Number	67-56-1	Density, lb/gal @ 15.6°C (60°F)	6.63
Molecular Weight	32.04	Explosive Limits, % Volume in Air,	
Synonyms	Methyl Alcohol Carbinol Wood Alcohol	Lower Upper	6.0 36.5
Auto Ignition Temperature, @ 760 mm Hg, °C (°F)	385 (725)	Flash Point, Tag Closed Cup, °C (°F)	11 (52)
Boiling Point, @ 760 mm Hg, °C (°F)	64.7 (148.4)	Heat of Formation, Liquid, @ 25°C, K cal/g mol @ 77°F, BTU/lb mol	-57.021 -102.6×10^3
Freezing Point, °C (°F)	-97.7 (-143.8)	Heat of Formation, Vapor, @ 25°C, K cal/g mol @ 77°F, BTU/lb mol	-48.08 -86.5×10^3
Coefficient of Expansion, per °C @ 20°C per °F @ 68°F	0.00119 0.00066	Heat of Fusion, @ -97°C, K cal/g mol @ -142.6°F, BTU/lb mol	16.4 29.5
Critical Compressibility	0.224		
Critical Density, g/cc (lb/cu ft)	0.272 (16.952)	Refractive Index, N_D^{20}	1.3286
Critical Pressure, Kg/cm² (PSIA)	81.12 (1153.95)	Solubility in Water, @ 20°C (68°F)	Completely Miscible

Table 6.2: Properties of Aqueous Solutions of Methanol *(31)*

METHANOL WT. %	VOL. %	FREEZING POINT °F.	BOILING POINT °F.	FLASH POINT °F. (Closed Cup)	DENSITY (g./ml.) AT VARIOUS TEMPERATURES 0°C.	10°C.	15°C.	20°C.	VISCOSITY (MILLIPOISES) AT VARIOUS TEMPERATURES 25°C.	35°C.	45°C.	55°C.
0	0	32	212	—	.9999	.9997	.9993	.9982	8.9	7.2	5.9	5.1
10	12.35	21.7	197.2	130	.9842	.9834	.9824	.9815	11.8	9.2	7.4	6.2
20	24.33	5.9	187.3	107	.9725	.9700	.9681	.9666	14.1	10.9	8.6	7.1
30	35.95	−14.6	180.0	94	.9604	.9560	.9537	.9515	15.5	11.9	9.4	7.7
40	47.11	−39.1	174.2	84	.9459	.9403	.9372	.9345	15.8	12.3	9.7	7.9
50	57.71	−65.7	169.5	76	.9287	.9221	.9185	.9156	15.7	12.2	9.7	7.9
60	67.69	−101.2	165.6	69	.9090	.9018	.8978	.8946	14.0	10.9	8.8	7.2
70	76.98	−156.1	161.6	63	.8869	.8794	.8751	.8715	12.2	9.6	7.8	6.4
80	85.50	−175.0*	157.5	58	.8634	.8551	.8505	.8469	10.1	8.1	6.7	5.6
90	93.19	−171.4	153.0	53	.8374	.8287	.8240	.8202	7.9	6.5	5.5	4.6
100	100.0	−142.6	148.3	49	.8102	.8009	.7958	.7917	5.5	4.8	4.1	3.6

WT. %	VAPOR PRESSURE (mm.Hg) AT VARIOUS TEMPERATURES 20°C.	60°C.	100°C.	140°C.	WT. % METHANOL IN VAPOR AT 760 mm.	THERMAL CONDUCTIVITY (CAL./SEC./CM.²/°C./CM.) AT VARIOUS TEMPERATURES 10°C.	40°C.	70°C.	SPECIFIC HEAT AT VARIOUS TEMPERATURES 30°C.	50°C.	80°C.	100°C.
0	17.5	149	760	2700	0	.00138	.00149	.00160	0.990	0.994	1.000	1.004
10	28.0	206	1030	3640	43.4	.00126	.00135	.00145	1.015	1.022	1.032	1.039
20	35.5	258	1260	4300	61.2	.00115	.00122	.00129	1.000	1.014	1.035	1.049
30	41.5	307	1450	4780	70.5	.00105	.00110	.00115	0.974	0.997	1.031	1.054
40	46.5	350	1600	5200	76.5	.00096	.00098	.00100	0.947	0.979	1.026	1.057
50	52.0	390	1740	5620	81.0	.00088	.00088	.00088	0.888	0.928	0.988	1.028
60	59.0	427	1880	6040	84.8	.00079	.00078	.00076	0.821	0.869	0.941	0.990
70	66.5	462	2020	6470	88.5	.00072	.00069	.00066	0.764	0.820	0.905	0.961
80	75.5	503	2190	6970	92.2	.00065	.00061	.00057	0.726	0.790	0.886	0.951
90	87.0	557	2380	7550	96.0	.00059	.00054	.00049	0.665	0.737	0.846	0.918
100	99.0	620	2600	8150	100.0	.00053	.00048	.00043	0.626	0.706	0.826	0.887

*The eutectic point or minimum freezing temperature is approximately −128.7°C. (−199.7°F.) at a composition of 82.9% Wt. methanol (87.8% Vol.). In the vicinity of the eutectic, the solutions become vitreous and direct determinations of the freezing point are difficult to make.

Table 6.3: Freezing Points of Methanol-Water Solutions *(34)*

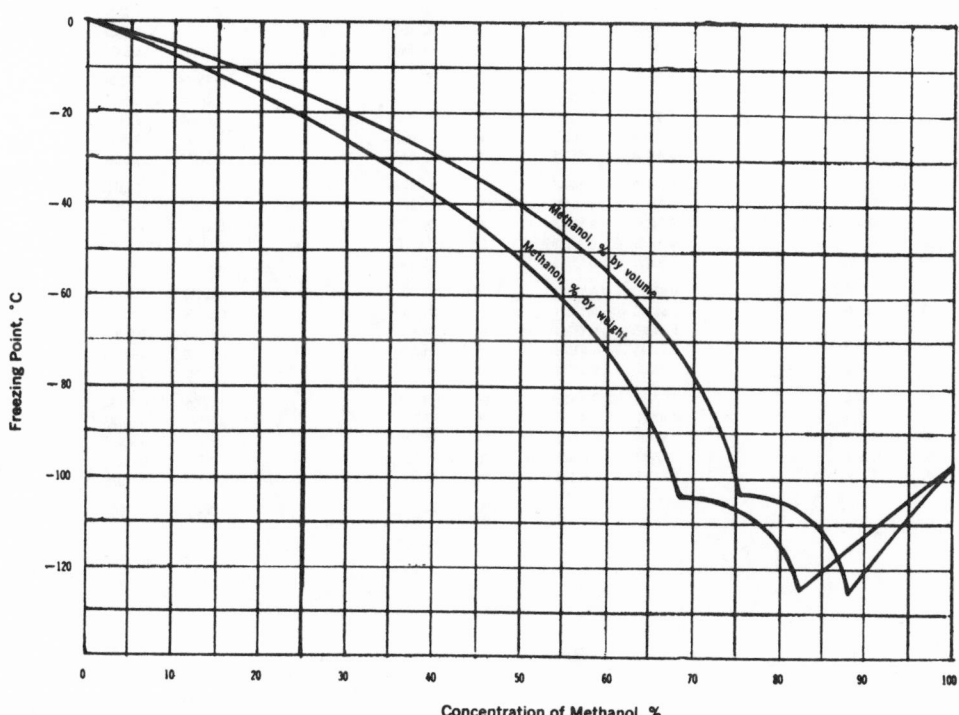

Table 6.4: Density and Specific Gravity of Methanol-Water Solutions at 15°C *(34)*

Methanol % by Weight	Methanol % by Volume	Density, 15/4°C	Specific Gravity, 15/15°C	Methanol % by Weight	Methanol % by Volume	Density, 15/4°C	Specific Gravity, 15/15°C
0	0	0.99913	1.00000	50	57.712	0.91852	0.91931
1	1.253	0.99727	0.99813	51	58.739	0.91653	0.91732
2	2.502	0.99543	0.99629	52	59.759	0.91451	0.91530
3	3.746	0.99370	0.99456	53	60.773	0.91248	0.91327
4	4.986	0.99198	0.99284	54	61.781	0.91044	0.91123
5	6.222	0.99029	0.99115	55	62.783	0.90839	0.90918
6	7.454	0.98864	0.98950	56	63.778	0.90631	0.90709
7	8.682	0.98701	0.98786	57	64.767	0.90421	0.90499
8	9.906	0.98547	0.98632	58	65.750	0.90210	0.90268
9	11.128	0.98394	0.98479	59	66.725	0.89996	0.90074
10	12.345	0.98241	0.98326	60	67.693	0.89781	0.89859
11	13.559	0.98093	0.98178	61	68.654	0.89563	0.89640
12	14.779	0.97945	0.98030	62	69.607	0.89341	0.89418
13	15.977	0.97802	0.97887	63	70.552	0.89117	0.89194
14	17.181	0.97660	0.97745	64	71.490	0.88890	0.88967
15	18.382	0.97518	0.97602	65	72.420	0.88662	0.88739
16	19.579	0.97377	0.97461	66	73.344	0.88433	0.88510
17	20.773	0.97237	0.97321	67	74.252	0.88203	0.88279
18	21.963	0.97096	0.97180	68	75.172	0.87971	0.88047
19	23.149	0.96955	0.97039	69	76.077	0.87739	0.87815
20	24.322	0.96814	0.96898	70	76.976	0.87507	0.87583
21	25.512	0.96673	0.96757	71	77.864	0.87271	0.87346
22	26.688	0.96533	0.96614	72	78.746	0.87033	0.87108
23	27.860	0.96392	0.96475	73	79.618	0.86792	0.86867
24	29.029	0.96251	0.96334	74	80.480	0.86546	0.86621
25	30.193	0.96108	0.96191	75	81.336	0.86300	0.86375
26	31.354	0.95963	0.96046	76	82.182	0.86051	0.86125
27	32.510	0.95817	0.95900	77	83.022	0.85801	0.85875
28	33.662	0.95668	0.95751	78	83.855	0.85551	0.85625
29	34.809	0.95518	0.95601	79	84.680	0.85300	0.85374
30	35.952	0.95366	0.95499	80	85.499	0.85048	0.85122
31	37.091	0.95213	0.95295	81	86.310	0.84794	0.84867
32	38.224	0.95056	0.95138	82	87.110	0.84536	0.84609
33	39.352	0.94896	0.94978	83	87.899	0.84274	0.84347
34	40.476	0.94734	0.94816	84	88.677	0.84009	0.84082
35	41.594	0.94570	0.94652	85	89.448	0.83742	0.83814
36	42.708	0.94404	0.94486	86	90.212	0.83475	0.83547
37	43.816	0.94237	0.94319	87	90.968	0.83207	0.83279
38	44.919	0.94067	0.94148	88	91.716	0.82937	0.83009
39	46.016	0.93894	0.93975	89	92.456	0.82667	0.82738
40	47.109	0.93720	0.93801	90	93.118	0.82396	0.82467
41	48.195	0.93543	0.93624	91	93.912	0.82124	0.82195
42	49.277	0.93365	0.93446	92	94.627	0.81849	0.81920
43	50.353	0.93185	0.93266	93	95.326	0.81568	0.81639
44	51.422	0.93001	0.93081	94	96.017	0.81285	0.81355
45	52.486	0.92815	0.92895	95	96.697	0.80999	0.81069
46	53.544	0.92627	0.92707	96	97.370	0.80713	0.80783
47	54.595	0.92436	0.92516	97	98.036	0.80428	0.80498
48	55.639	0.92242	0.92322	98	98.696	0.80143	0.80212
49	56.678	0.92048	0.92128	99	99.351	0.79859	0.79928
				100	100.000	0.79577	0.79646

Table 6.5: Density and Specific Gravity of Methanol-Water Solutions at 30°C (34)

Methanol % by Weight	% by Volume	Density, 30/4°C	Specific Gravity, 30/30°C	Methanol % by Weight	% by Volume	Density, 30/4°C	Specific Gravity, 30/30°C
0	0.000	0.9957	1.0000	50	58.089	0.9084	0.9123
1	1.271	0.9939	0.9982	51	59.121	0.9064	0.9103
2	2.538	0.9921	0.9964	52	60.140	0.9043	0.9082
3	3.800	0.9903	0.9946	53	61.148	0.9021	0.9060
4	5.057	0.9886	0.9929	54	62.149	0.8999	0.9038
5	6.310	0.9868	0.9911	55	63.146	0.8977	0.9016
6	7.559	0.9850	0.9893	56	64.136	0.8955	0.8994
7	8.802	0.9832	0.9874	57	65.114	0.8932	0.8971
8	10.042	0.9815	0.9857	58	66.093	0.8910	0.8948
9	11.278	0.9798	0.9840	59	67.059	0.8887	0.8925
10	12.511	0.9782	0.9824	60	68.019	0.8864	0.8902
11	13.738	0.9765	0.9807	61	68.981	0.8842	0.8880
12	14.962	0.9749	0.9791	62	69.929	0.8819	0.8857
13	16.182	0.9733	0.9775	63	70.872	0.8796	0.8834
14	17.398	0.9717	0.9759	64	71.809	0.8773	0.8811
15	18.610	0.9701	0.9743	65	72.731	0.8749	0.8787
16	19.820	0.9686	0.9728	66	73.656	0.8726	0.8764
17	21.024	0.9670	0.9712	67	74.566	0.8702	0.8740
18	22.224	0.9654	0.9696	68	75.471	0.8678	0.8715
19	23.420	0.9638	0.9680	69	76.369	0.8654	0.8691
20	24.612	0.9622	0.9664	70	77.261	0.8630	0.8667
21	25.799	0.9606	0.9647	71	78.146	0.8606	0.8643
22	26.983	0.9590	0.9631	72	79.017	0.8581	0.8618
23	28.162	0.9574	0.9615	73	79.881	0.8556	0.8593
24	29.338	0.9558	0.9599	74	80.729	0.8530	0.8567
25	30.509	0.9542	0.9583	75	81.580	0.8505	0.8542
26	31.676	0.9526	0.9567	76	82.425	0.8480	0.8517
27	32.839	0.9510	0.9551	77	83.253	0.8454	0.8491
28	33.998	0.9494	0.9535	78	84.085	0.8429	0.8465
29	35.149	0.9477	0.9518	79	84.900	0.8403	0.8439
30	36.296	0.9460	0.9501	80	85.719	0.8378	0.8414
31	37.439	0.9443	0.9484	81	86.522	0.8352	0.8388
32	38.577	0.9426	0.9467	82	87.317	0.8326	0.8362
33	39.706	0.9408	0.9449	83	88.095	0.8299	0.8335
34	40.831	0.9390	0.9431	84	88.867	0.8272	0.8308
35	41.952	0.9372	0.9412	85	89.631	0.8245	0.8281
36	43.067	0.9354	0.9394	86	90.389	0.8218	0.8253
37	44.179	0.9336	0.9376	87	91.139	0.8191	0.8226
38	45.285	0.9318	0.9358	88	91.883	0.8164	0.8199
39	46.387	0.9300	0.9340	89	92.608	0.8136	0.8171
40	47.479	0.9281	0.9321	90	93.327	0.8108	0.8143
41	48.572	0.9263	0.9303	91	94.038	0.8080	0.8115
42	49.654	0.9244	0.9284	92	94.730	0.8051	0.8086
43	50.732	0.9225	0.9265	93	95.415	0.8022	0.8057
44	51.805	0.9206	0.9246	94	96.080	0.7992	0.8027
45	52.867	0.9186	0.9226	95	96.737	0.7962	0.7996
46	53.925	0.9166	0.9206	96	97.400	0.7933	0.7967
47	54.977	0.9146	0.9185	97	98.054	0.7904	0.7938
48	56.024	0.9126	0.9165	98	98.702	0.7875	0.7909
49	57.059	0.9105	0.9144	99	99.355	0.7847	0.7881
				100	100.000	0.7819	0.7853

Table 6.6: Resultant Volume When Methanol and Water Are Mixed *(31)*

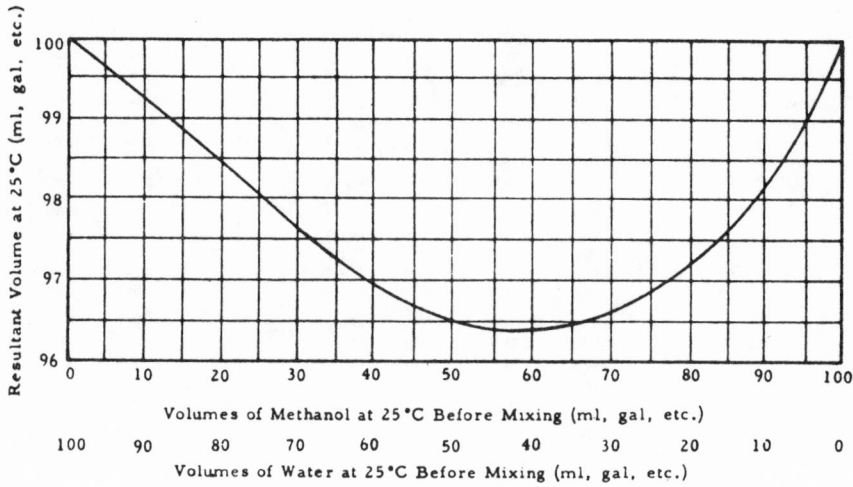

Table 6.7: Solubility of Methanol in Gasoline from 15° to 30°C *(31)*

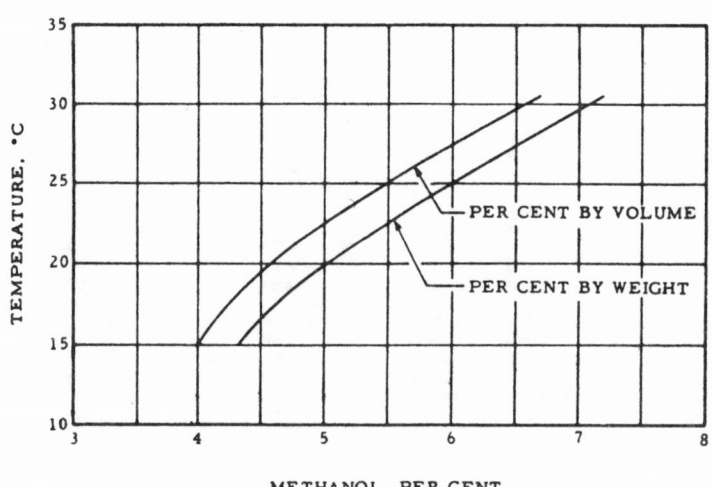

Table 6.8: Liquid Density of Methanol *(79)*

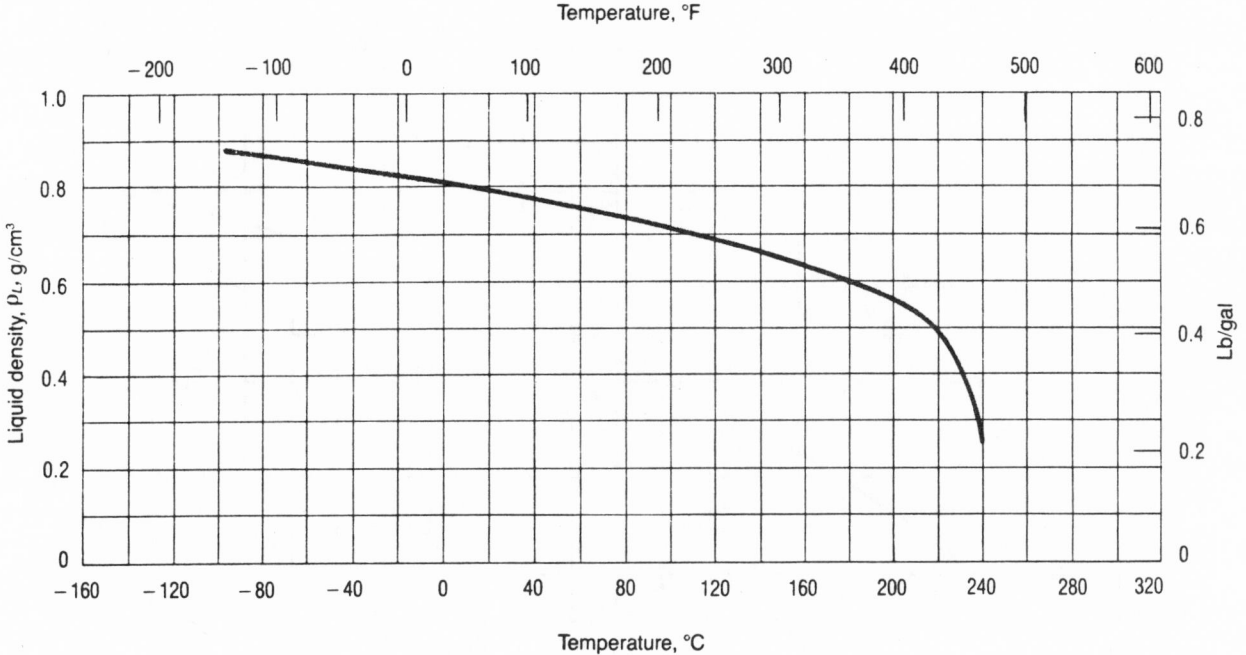

Table 6.9: Liquid Heat Capacity of Methanol *(79)*

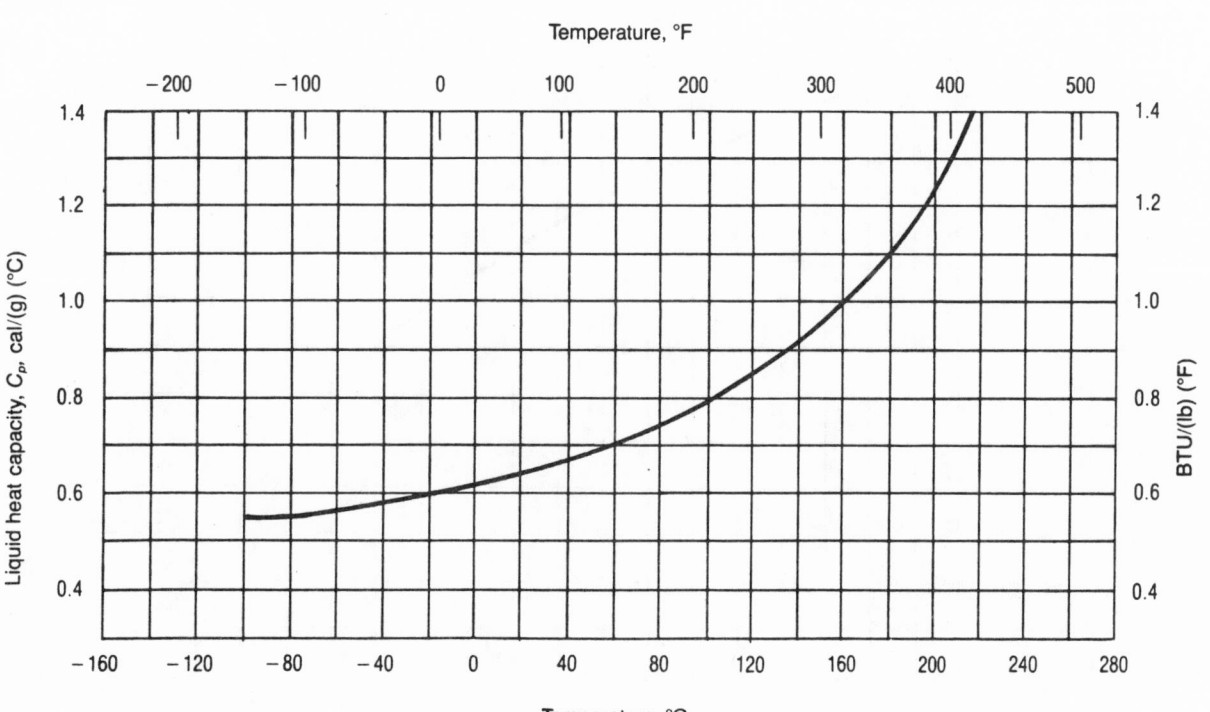

Table 6.10: Vapor Heat Capacity of Methanol *(79)*

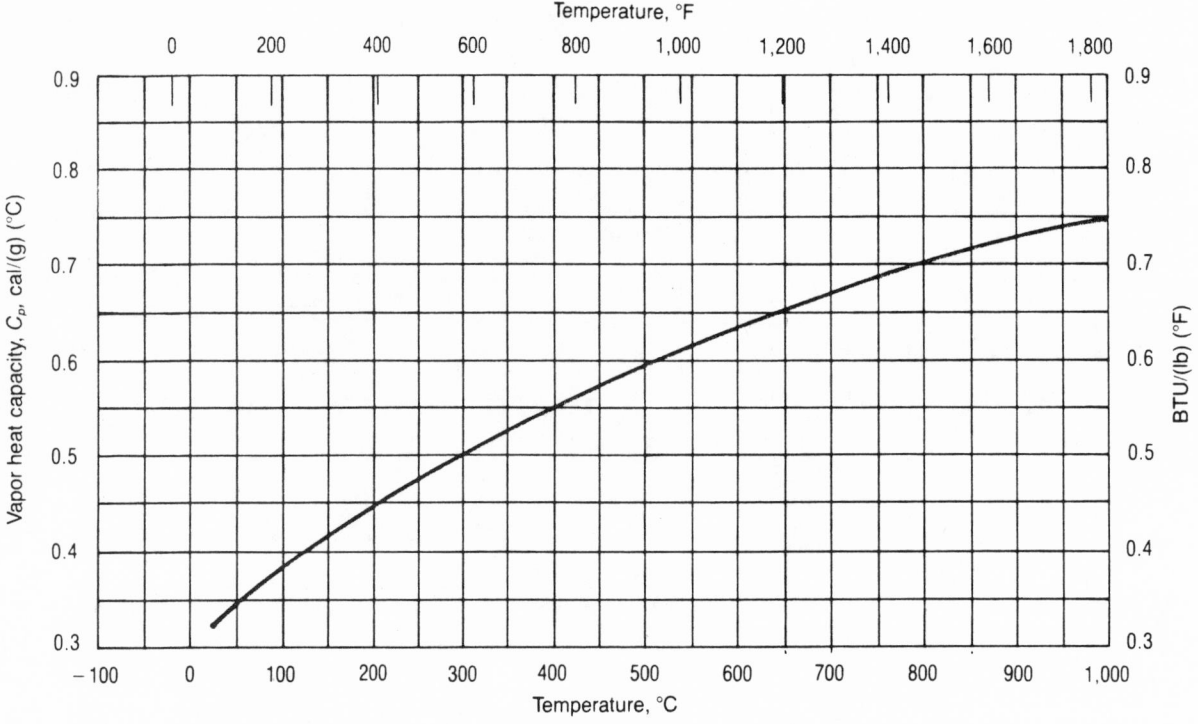

Table 6.11: Heat of Vaporization of Methanol *(79)*

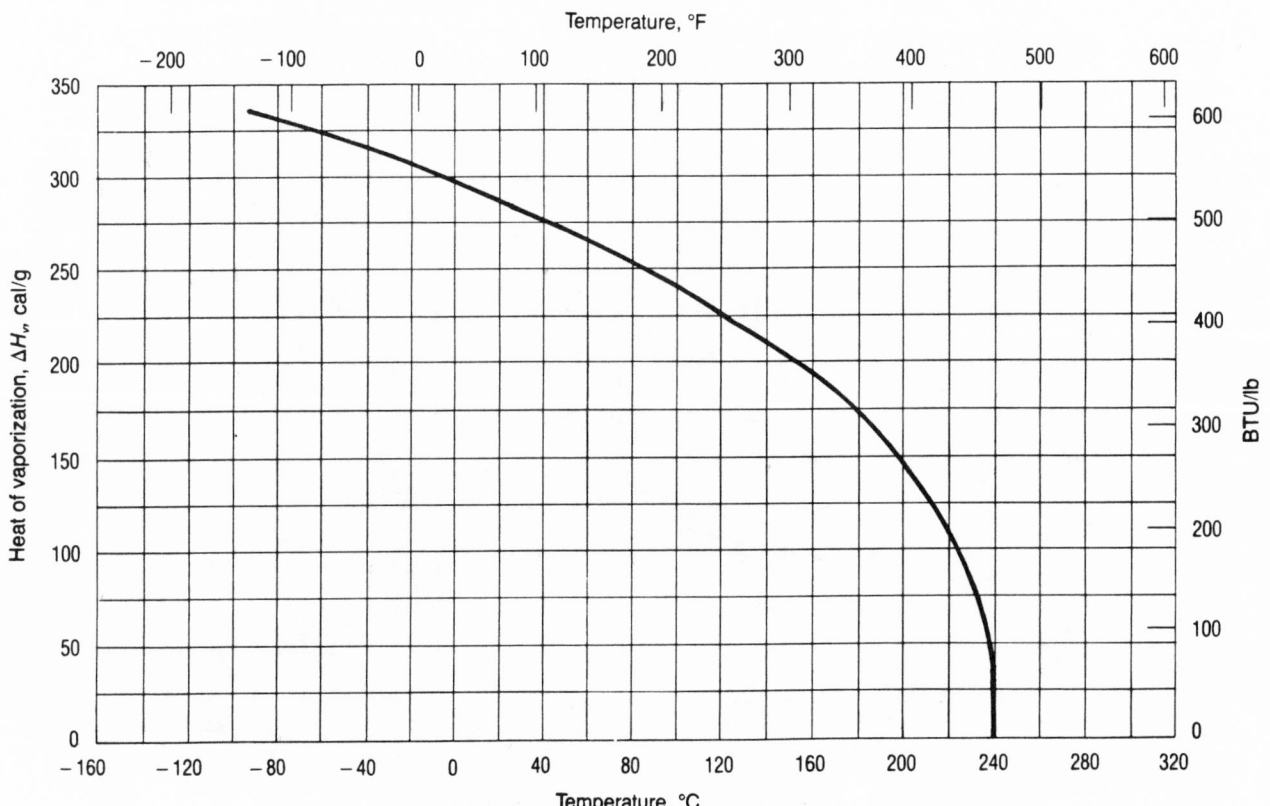

Table 6.12: Surface Tension of Methanol *(79)*

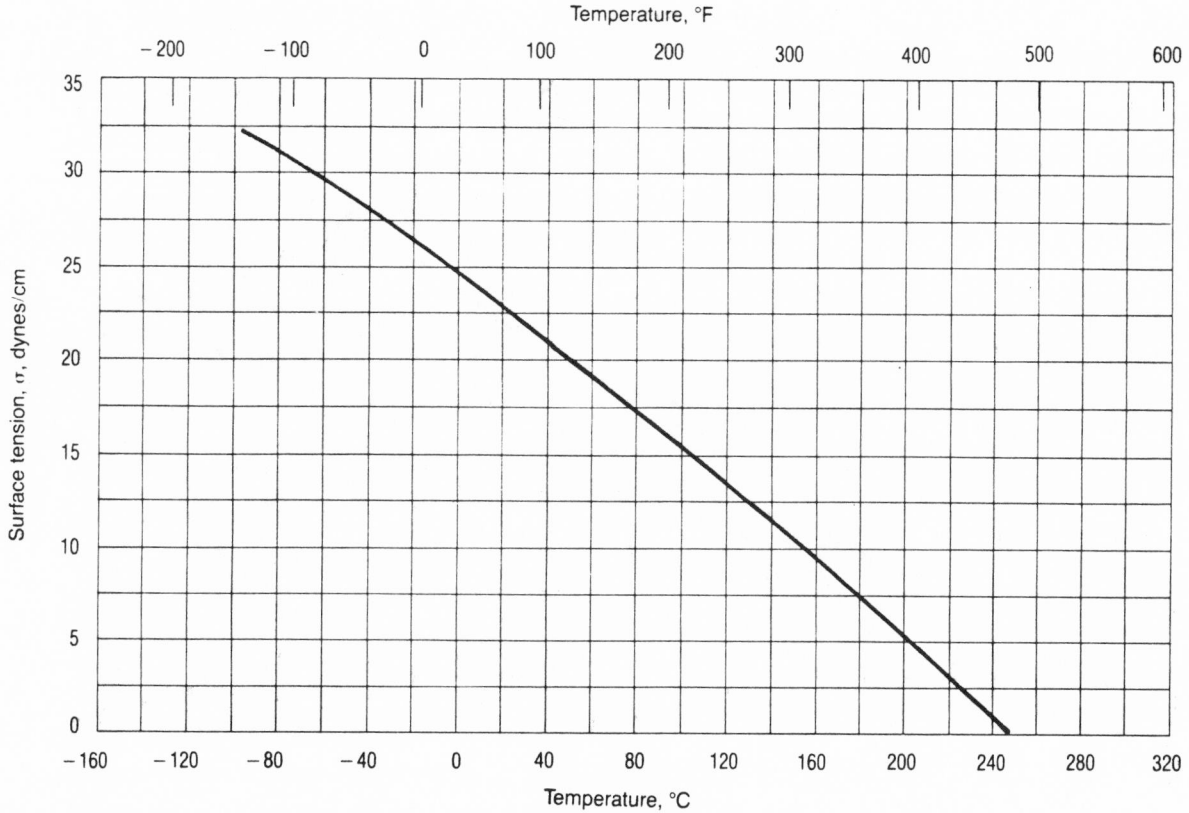

Table 6.13: Liquid Thermal Conductivity of Methanol *(79)*

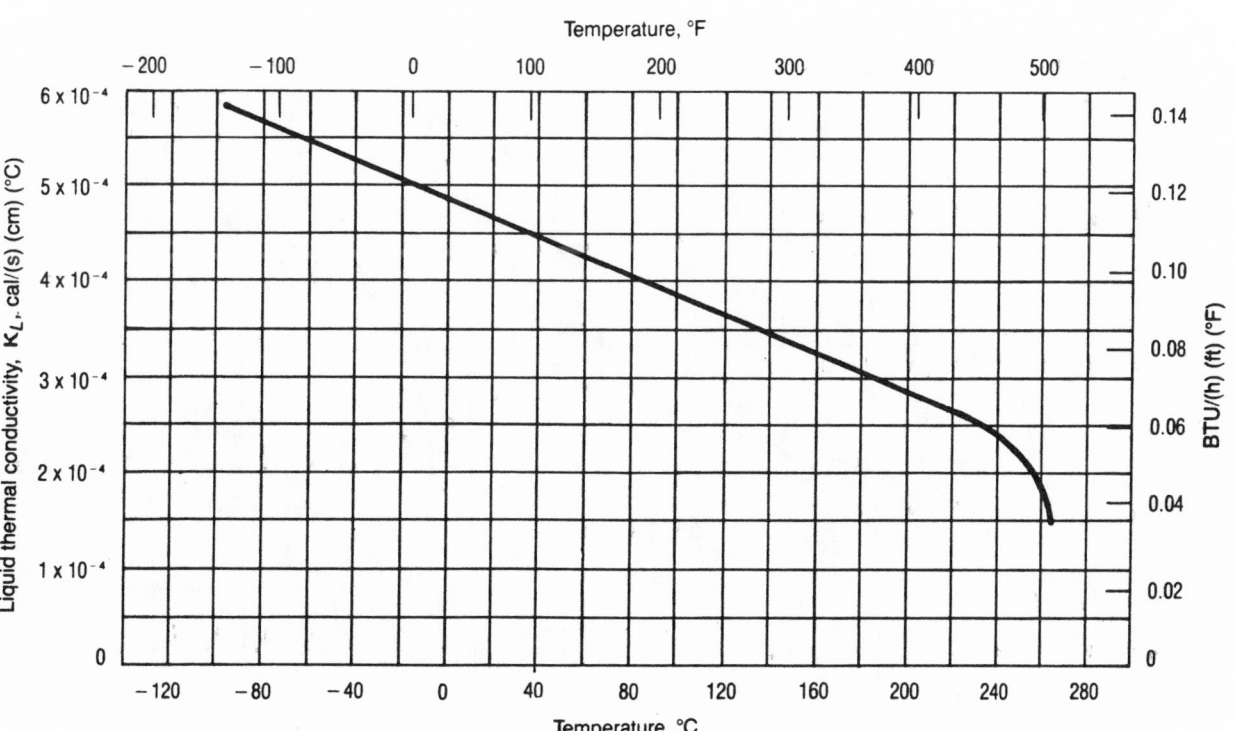

Table 6.14: Vapor Thermal Conductivity of Methanol *(79)*

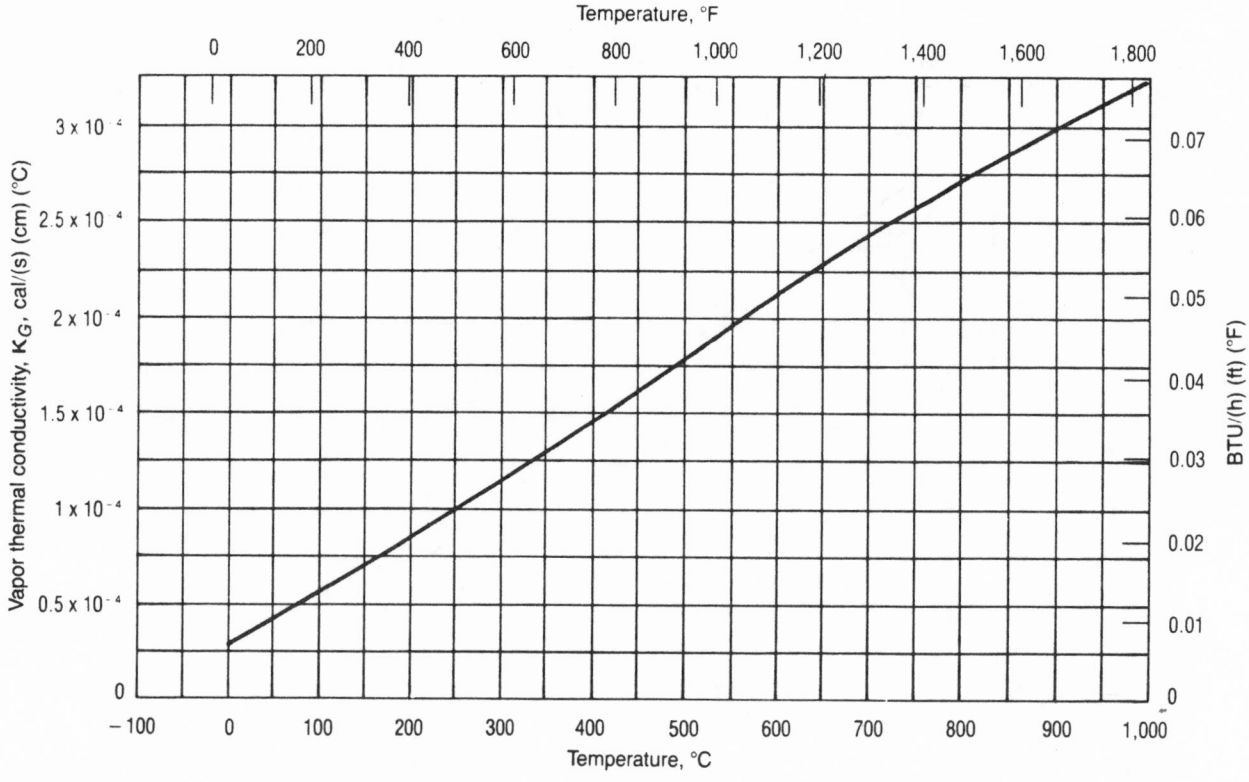

Table 6.15: Vapor Pressure of Methanol *(79)*

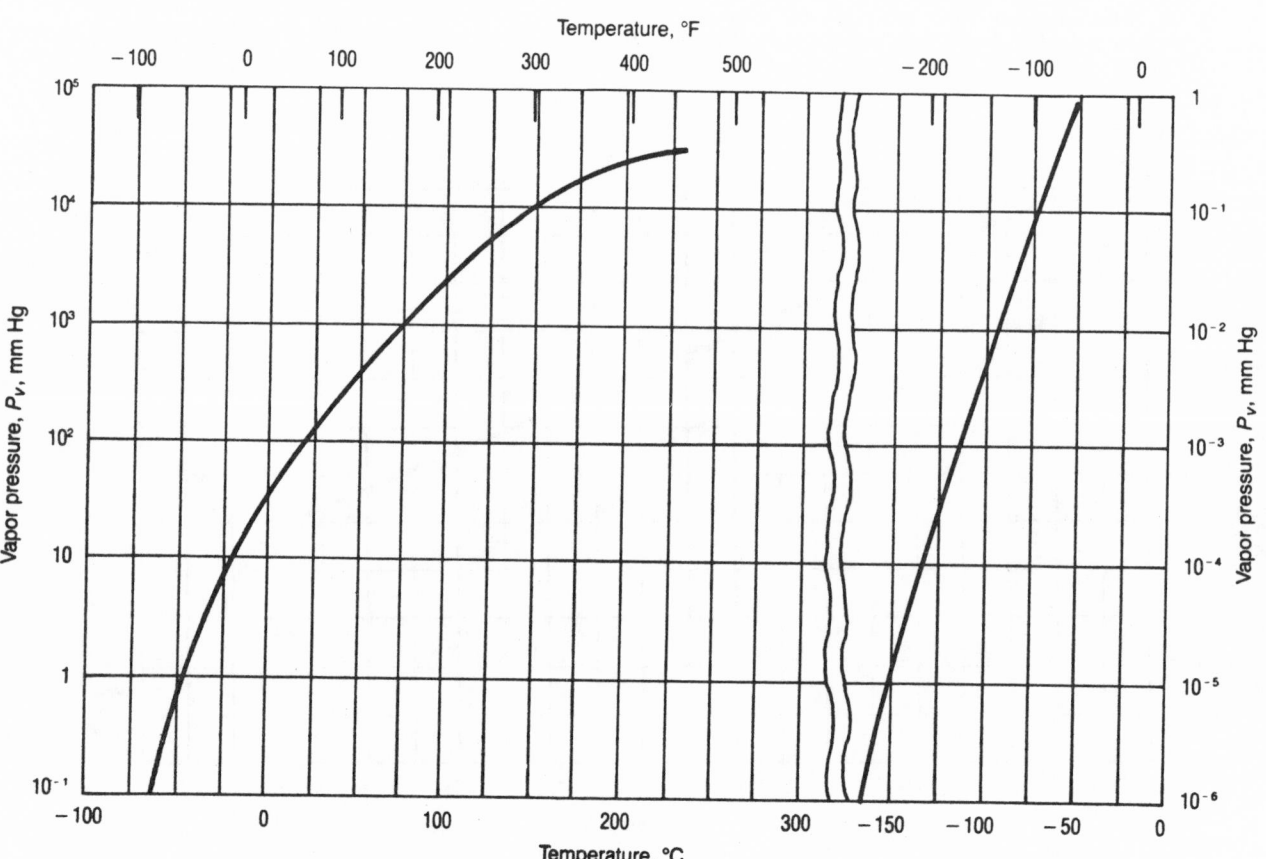

Table 6.16: Vapor Viscosity of Methanol *(79)*

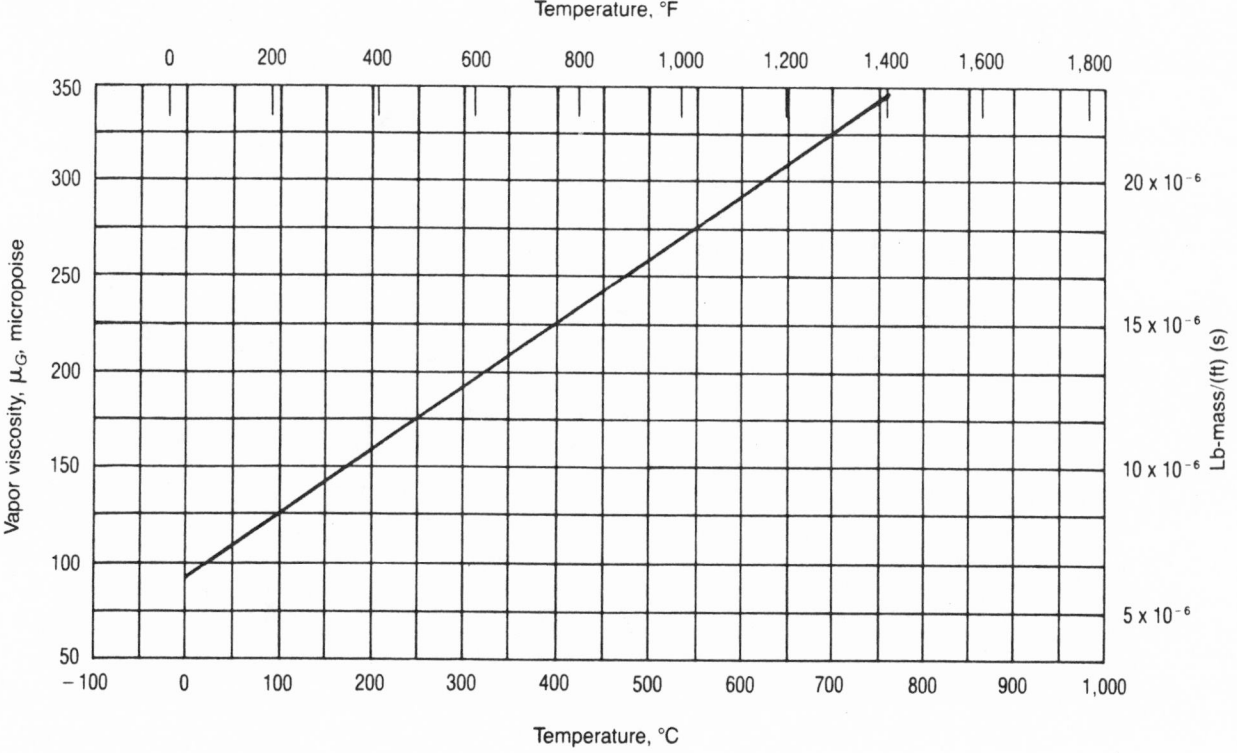

Table 6.17: Liquid Viscosity of Methanol *(79)*

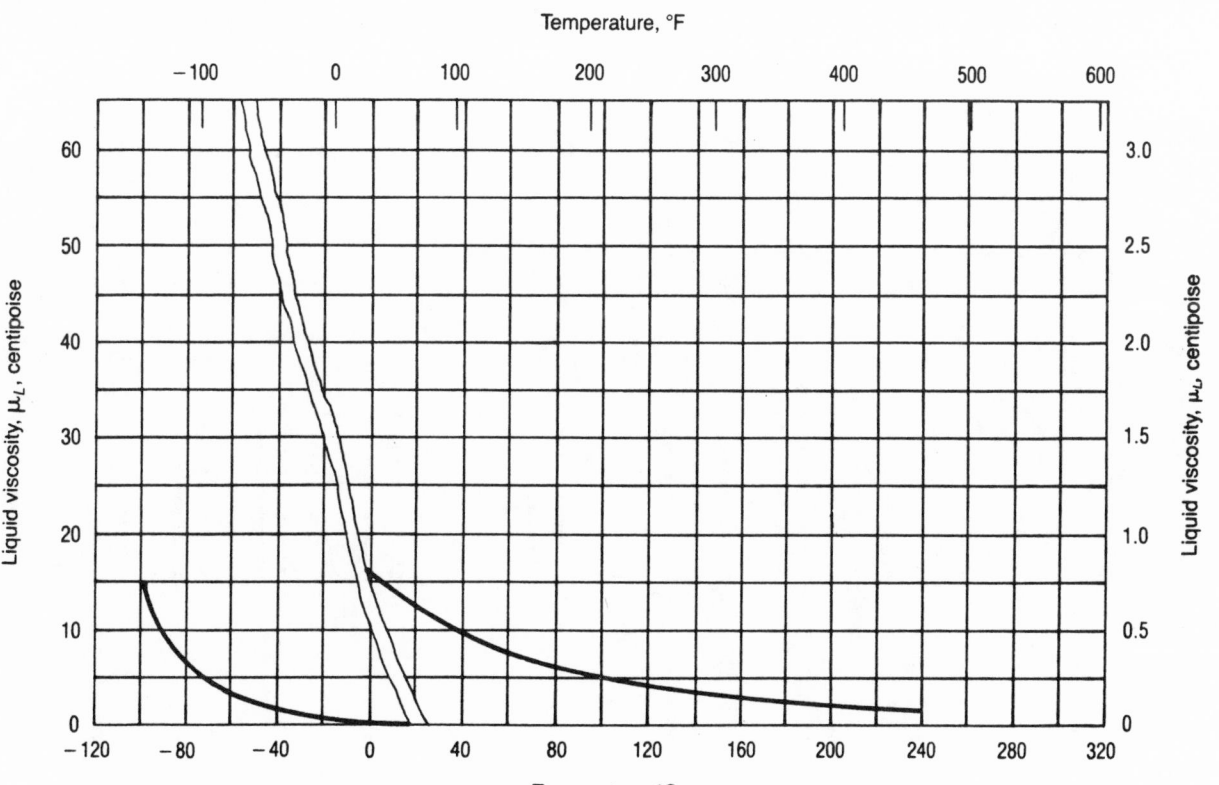

Table 6.18: Azeotropes of Methanol (31)

METHANOL FORMS BINARY AZEOTROPES WITH:

Component	B.P. of Azeotrope °C.	%
Acetone	55.5	88
Acetonitrile	63.5	81
Acrylonitrile	61.4	87
tert-Amyl methyl ether	62.3	38.7
Benzene	58.3	50
Biallyl	47.1	60.5
1-Bromobutane	63.7	77.5
2-Bromobutane	61.5	41
Bromoethane	35	58.5
1-Bromo-2-methylpropane	61.6	95
2-Bromo-2-methylpropane	55.6	58
1-Bromopropane	54.5	76
2-Bromopropane	49.0	79
cis-1-Bromopropene	48	85.5
trans-1-Bromopropene	50.8	88
2-Bromopropene	42.7	85
3-Bromopropene	54.0	89
2-Butanone	63.5	79.5
Butyl methyl ether	56.3	30
Carbon disulfide	37.7	64.6
Carbon tetrachloride	55.7	86
1-Chlorobutane	57.2	79.4
2-Chlorobutane	52.7	71.5
Chloroform	53.5	80
1-Chloro-3-methylbutane	62.0	87
Chloromethyl methyl ether	56.0	43
1-Chloro-2-methylpropane	53.1	65
2-Chloro-2-methylpropane	43.8	77
1-Chloropropane	40.6	90
2-Chloropropane	33.4	94
2-Chloropropene	22.0	97
3-Chloropropene	39.9	90
1,3-Cyclohexadiene	56.4	61.2
1,4-Cyclohexadiene	58.0	57.5
Cyclohexane	55.9	60
Cyclopentane	38.8	86
1,1-Dibromoethane	64.2	18
trans-1,2-Dibromoethylene	64.1	38
2,3-Dichloro-1,3-butadiene	61.5	50
1,1-Dichloroethane	59.0	88.5
1,2-Dichloroethane	61.0	88
cis-1,2-Dichloroethylene	51.5	68
1,2-Dichloropropane	62.9	87
2,2-Dichloropropane	55.5	47
1,2-Dichloro-1-propene	56.5	79
Diethoxymethane	63.2	75
1,2-Dimethoxyethylene	63.5	35
Dimethyl acetal	57.5	10
2,3-Dimethylbutane	45.0	75.8
2,5-Dimethylhexane	61.0	80
Ethyl acetate	77.1	40
Ethyl butyl ether	62.6	54
Ethylene dichloride	59.5	44
Ethylene sulfide	47.0	65
Ethyl formate	51.0	79
Ethyl nitrate	61.8	84
Ethyl propyl ether	55.5	43
Ethyl sulfide	61.2	76
Fluorobenzene	59.7	38
Furan	30.5	68
n-Heptane	59.1	93
Iodoethane	55.0	48.5
1-Iodopropane	63.1	83
2-Iodopropane	61.0	50
3-Iodopropene	63.5	62
Isobutyraldehyde	62.7	38
Isopropyl acetate	64.5	60
Isopropyl formate	57.2	20
Methylal	41.8	67
Methyl acetate	54.0	92.1
Methyl acrylate	62.5	80.5
Methyl borate	54.6	46
2-Methyl-2-butene	31.8	68
3-Methyl-1-butene	19.8	93
Methyl tert-butyl ether	51.6	97
Methyl carbonate	62.7	85
Methylcyclohexane	59.2	30
Methylcyclopentane	51.3	46
Methylcyclopentene	53.0	65
2-Methylfuran	51.5	77.7
Methyl isobutyrate	64.0	25
Methyl propionate	62.5	52.5
Methyl propyl ether	38.0	88
Methyl sulfide	34.5	87
n-Pentane	30.8	91
α-Pinene	64.6	9.3
Propyl ether	63.8	28
Propyl formate	61.9	49.8
Octane	63.0	28
Tetrachloroethylene	63.8	36.5
Thiophene	59.6	45
Toluene	63.8	31
1,1,1-Trichloroethane	56.0	78.3
1,1,2-Trichloroethane	64.5	3
Trichloroethylene	59.3	62
2,2,4-Trimethylpentane	59.4	47

METHANOL FORMS TERNARY AZEOTROPES WITH:

Components	%	B.P. of Azeotrope °C.
Acetone	43.5	51.1
Cyclohexane	40.5	
Acetone	5.8	53.7
Methyl acetate	76.8	
Carbon disulfide	40	33.9
Bromoethane	50	
Carbon disulfide	55	35.6
Methylal	38	
Chloroform	47	47.0
Acetone	30	
Methyl acetate	57.2	37.0
Carbon disulfide	46.5	
Methyl acetate	48.6	50.8
Cyclohexane	33.6	
Methyl acetate	27	45.0
Hexane	59	
Water	5.3	67.9
Methyl chloroacetate	13.5	

ETHYL ALCOHOL

Table 6.19: Physical Properties of Anhydrous Ethyl Alcohol *(31)*

Acidity as acetic acid	0.0015% by wt, max.	Latent heat of fusion	24.9 cal/g
Boiling point at 760 mm Hg	78.32°C	Latent heat of vaporization at 78.3°C	204.3 cal/g
dt/dp at 760 mm Hg	0.0334°C/mm Hg	MAC	1000 ppm in air
Coefficient of cubical expansion	0.00060 per 1°F	Melting point	-114.4°C
Color, Pt-Co scale	10 max.	Molecular weight	46.07
Critical pressure	63.1 atm	Non-volatile matter	Not more than 0.0025 gram when 100 ml are evaporated and heated to constant weight at 100°C to 110°C
Critical temperature	243.1°C		
Density at 25°C	0.7851 g/ml		
Dielectric constant at 20°C	25.7		
Dipole moment, $\mu \times 10^{18}$	1.70 μ		
Electrical conductivity at 25°C	1.35×10^{-9} ohm^{-1} cm^{-1}	Reducing substances	At least 25 minutes permanganate time at 15°C
Explosive range	3.28 - 19%		
Fire hazard	Dangerous when exposed to heat or flame	Refractive index at 25°C, n_D	1.3596
Flash point, Tag open cup	61°F	Specific gravity at 15.56°C (60/60°F)	0.7937
Free energy of formation, $\Delta F°$ at 25°C	-40.2 kcal/mole	Specific heat at 20°C	0.61 cal/g
Freezing point	-114.1°C	Specific tension at 25°C	22.1 dynes/cm
Heat capacity, Cp, Liquid at 25°C	0.581 cal/g	Thermal conductivity, k, at 68°F	0.105 (Btu) (ft) (sq ft) (°F)
Cp, Vapor, 90°C,1 atm	0.406 cal/(g) (°C)		
Cv, Vapor, 90°C,1 atm	0.359 cal/(g) (°C)	Toxicity	Moderately toxic by ingestion and inhalation
Heat of combustion	328 kcal/mole		
Heat of formation, Liquid, ΔH at 25°C	-64.7 kcal/mole	Vapor pressure at 20°C	44.0 mm Hg
Heat of solution in Water at 13°C	2.54 kcal/mole solute	Viscosity at 20°C	1.22 centipoises
		Weight per gallon at 20°C	6.61 lbs
Heat of solution of Water in Ethyl Alcohol, mole fraction of Water			
0.640 at 77°C	-0.018		
0.843 at 79.2°C	-0.038		

Table 6.20: Physical Properties of 95% Ethanol *(31)*

Acidity as acetic acid	0.0025 g/100 ml, max.
Color, Pt-Co scale	10 max.
Distillation range at 760 mm Hg	77°C - 80°C
Non-volatile matter	Not more than 0.0025 gram when 100 ml are evaporated and heated to constant weight at 100°C to 110°C
Permanganate time	30 minutes, min.
Reducing substances	At least 25 minutes permanganate time at 15°C
Relative evaporation rate, n-Butyl Acetate = 100	230
Specific gravity at 15.56 (60/60°F)	0.8160
Weight per gallon at 20°C	6.76 lbs

Table 6.21: Conversion Table—Volume and Weight Percent Ethyl Alcohol-Water Solutions *(31)*

% Alcohol By Volume at 60°F	% to be Converted	% Alcohol By Weight	% Alcohol By Volume at 60°F	% to be Converted	% Alcohol By Weight	% Alcohol By Volume at 60°F	% to be Converted	% Alcohol By Weight	% Alcohol By Volume at 60°F	% to be Converted	% Alcohol By Weight
1.257	1	0.795	31.555	26	21.285	58.844	51	43.428	82.121	76	68.982
2.510	2	1.593	32.719	27	22.127	59.852	52	44.374	82.967	77	70.102
3.758	3	2.392	33.879	28	22.973	60.854	53	45.326	83.805	78	71.234
5.002	4	3.194	35.033	29	23.820	61.850	54	46.283	84.636	79	72.375
6.243	5	3.998	36.181	30	24.670	62.837	55	47.245	85.459	80	73.526
7.479	6	4.804	37.323	31	25.524	63.820	56	48.214	86.275	81	74.686
8.712	7	5.612	38.459	32	26.382	64.798	57	49.187	87.083	82	75.858
9.943	8	6.422	39.590	33	27.242	65.768	58	50.167	87.885	83	77.039
11.169	9	7.234	40.716	34	28.104	66.732	59	51.154	88.678	84	78.233
12.393	10	8.047	41.832	35	28.971	67.690	60	52.147	89.464	85	79.441
13.613	11	8.862	42.944	36	29.842	68.641	61	53.146	90.240	86	80.662
14.832	12	9.679	44.050	37	30.717	69.586	62	54.152	91.008	87	81.897
16.047	13	10.497	45.149	38	31.596	70.523	63	55.165	91.766	88	83.144
17.259	14	11.317	46.242	39	32.478	71.455	64	56.184	92.517	89	84.408
18.469	15	12.138	47.328	40	33.364	72.380	65	57.208	93.254	90	85.689
19.676	16	12.961	48.407	41	34.254	73.299	66	58.241	93.982	91	86.989
20.880	17	13.786	49.480	42	35.150	74.211	67	59.279	94.700	92	88.310
22.081	18	14.612	50.545	43	36.050	75.117	68	60.325	95.407	93	89.652
23.278	19	15.440	51.605	44	36.955	76.016	69	61.379	96.103	94	91.025
24.472	20	16.269	52.658	45	37.865	76.909	70	62.441	96.787	95	92.423
25.662	21	17.100	53.705	46	38.778	77.794	71	63.511	97.459	96	93.851
26.849	22	17.933	54.746	47	39.697	78.672	72	64.588	98.117	97	95.315
28.032	23	18.768	55.780	48	40.622	79.544	73	65.674	98.759	98	96.820
29.210	24	19.604	56.808	49	41.551	80.410	74	66.768	99.386	99	98.381
30.388	25	20.443	57.830	50	42.487	81.269	75	67.870	100.000	100	100.000

Table 6.22: Index of Refraction of Ethyl Alcohol-Water Mixtures at 60°F *(31)*

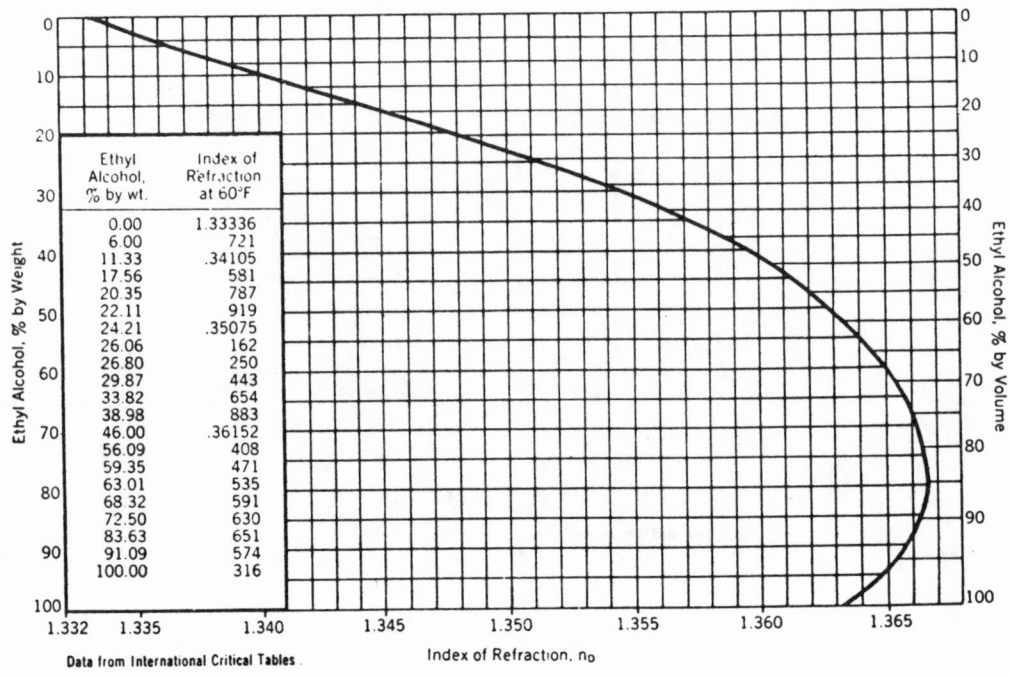

Table 6.24: Resultant Volume When Ethyl Alcohol and Water
Are Mixed *(34)*

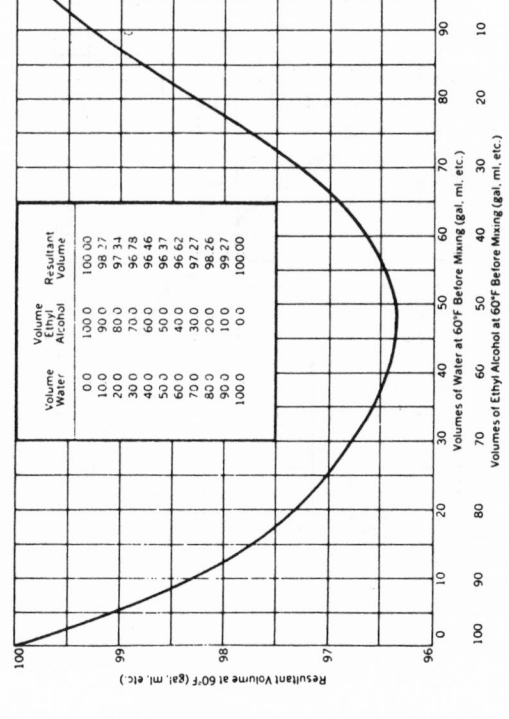

Volume Water	Volume Ethyl Alcohol	Resultant Volume
0.0	100.0	100.00
10.0	90.0	98.27
20.0	80.0	97.34
30.0	70.0	96.78
40.0	60.0	96.46
50.0	50.0	96.37
60.0	40.0	96.62
70.0	30.0	97.27
80.0	20.0	98.26
90.0	10.0	99.27
100.0	0.0	100.00

Table 6.25: Boiling Points of Ethyl Alcohol-Water Solution *(34)*

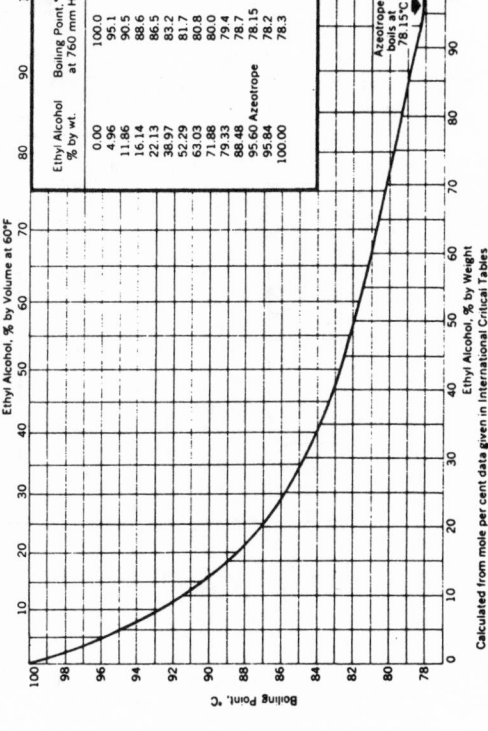

Ethyl Alcohol % by wt.	Boiling Point, °C at 760 mm Hg
0.00	100.0
4.96	95.1
11.86	90.5
16.14	88.6
22.13	86.5
38.97	83.2
52.29	81.7
63.03	80.8
71.88	80.0
79.33	79.4
88.48	78.7
95.60 Azeotrope	78.15
95.84	78.2
100.00	78.3

Calculated from mole per cent data given in International Critical Tables

Table 6.23: Heat of Solution of Ethyl Alcohol
in Water *(34)*

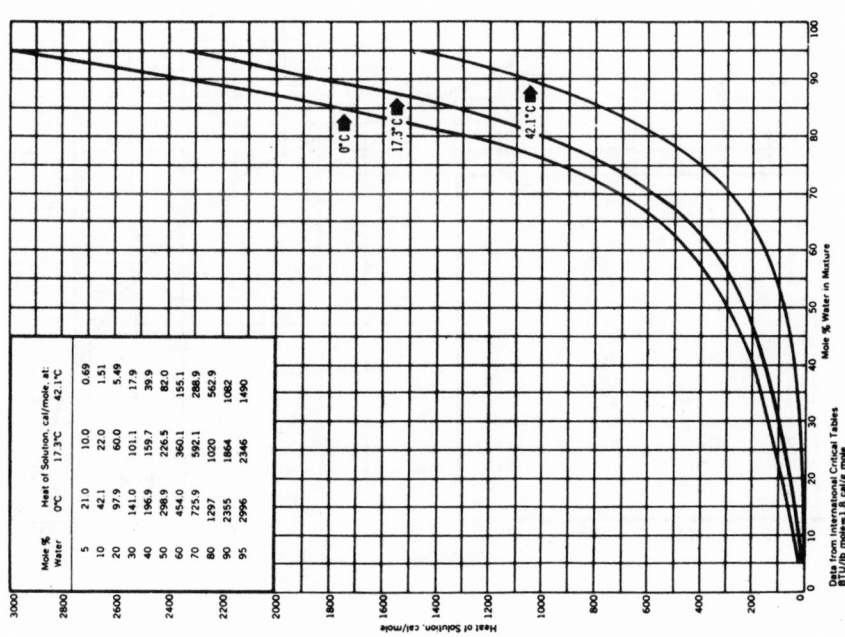

Mole % Water	Heat of Solution, cal/mole at:		
	0°C	17.3°C	42.1°C
5	21.0	10.0	0.69
10	42.1	22.0	1.51
20	97.9	60.0	5.49
30	141.0	101.1	17.9
40	196.9	159.7	39.9
50	298.9	226.5	82.0
60	454.0	360.1	155.1
70	725.9	592.1	288.9
80	1297	1020	562.9
90	2355	1864	1062
95	2996	2346	1490

Data from International Critical Tables
BTU/lb mole=1.8 cal/g mole

Table 6.26: Latent Heat of Vaporization of Ethyl Alcohol *(34)*

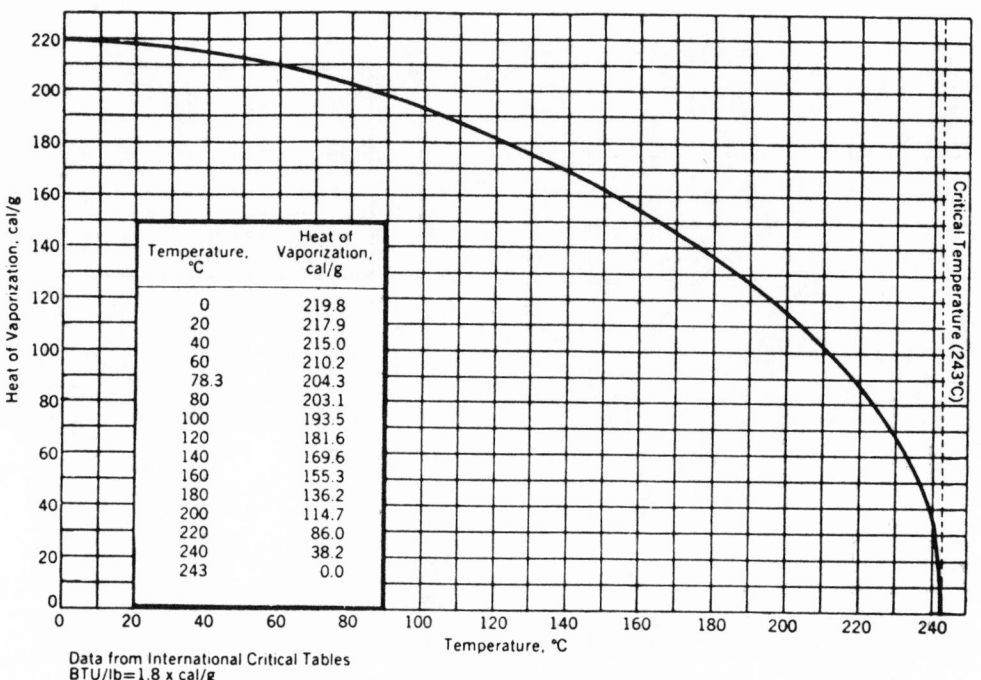

Temperature, °C	Heat of Vaporization, cal/g
0	219.8
20	217.9
40	215.0
60	210.2
78.3	204.3
80	203.1
100	193.5
120	181.6
140	169.6
160	155.3
180	136.2
200	114.7
220	86.0
240	38.2
243	0.0

Data from International Critical Tables
BTU/lb = 1.8 x cal/g

Table 6.27: Freezing Points of Ethyl Alcohol-Water Mixtures *(34)*

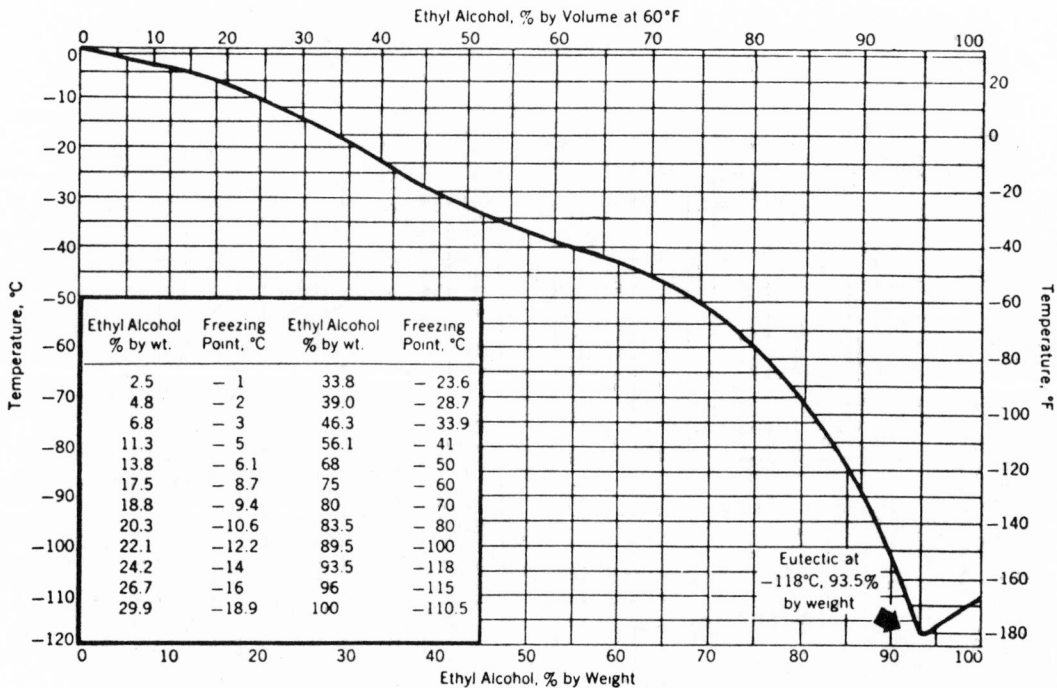

Ethyl Alcohol % by wt.	Freezing Point, °C	Ethyl Alcohol % by wt.	Freezing Point, °C
2.5	− 1	33.8	− 23.6
4.8	− 2	39.0	− 28.7
6.8	− 3	46.3	− 33.9
11.3	− 5	56.1	− 41
13.8	− 6.1	68	− 50
17.5	− 8.7	75	− 60
18.8	− 9.4	80	− 70
20.3	−10.6	83.5	− 80
22.1	−12.2	89.5	−100
24.2	−14	93.5	−118
26.7	−16	96	−115
29.9	−18.9	100	−110.5

Table 6.28: Specially Denatured Alcohol Formulas and Authorized Uses *(78)*

Formula No. and Composition	Authorized Uses
To every 100 gallons of ethyl alcohol add: 1. Four gallons of methyl alcohol and one gallon of methyl isobutyl ketone; or four gallons methyl alcohol and ⅛ avoirdupois ounce denatonium benzoate, N. F., (Bitrex).	*As a solvent:* 011. Cellulose coatings. 012. Synthetic resin coatings. 013. Shellac coatings. 014. Other natural resin coatings. 016. Other coatings. 021. Cellulose plastics. 022. Non-cellulose plastics, including resins. 031. Photographic film and emulsions. 032. Transparent sheeting. 033. Explosives. 034. Cellulose intermediates and industrial collodions. 035. Soldering flux. 036. Adhesives and binders. 041. Proprietary solvents (standard formulations). 042. Other solvents and thinners. 043. Special solvents (restricted sale). 051. Polishes. 052. Inks (including meat branding inks). 053. Stains (wood, etc.). 141. Shampoos. 142. Soap and bath preparations. 311. Cellulose compounds (dehydration). 312. Sodium hydrosulphite (dehydration). 315. Other dehydration products. 320. Petroleum products. 331. Processing pectin. 332. Processing other food products. 341. Processing crude drugs. 342. Processing glandular products, vitamins, hormones and yeasts. 343. Processing antibiotics and vaccines. 344. Processing medicinal chemicals (including alkaloids). 345. Processing blood and blood products. 349. Miscellaneous, drug processing (including manufacture of pills). 351. Processing dyes and intermediates. 352. Processing perfume materials and fixatives. 353. Processing photographic chemicals. 354. Processing rosin. 355. Processing rubber (latex). 358. Processing other chemicals. 359. Processing miscellaneous products. 410. Disinfectants, insecticides, fungicides and other biocides. 420. Embalming fluids and related products. 430. Sterilizing and preserving solutions. 440. Industrial detergents and soaps.

(continued)

Table 6.28: (continued)

Formula No. and Composition	Authorized Uses
To every 100 gallons of ethyl alcohol add: 1. Four gallons of methyl alcohol and one gallon of methyl isobutyl ketone; or four gallons methyl alcohol and ⅛ avoirdupois ounce denatonium benzoate, N. F., (Bitrex).	*As a solvent:* 450. Cleaning solutions (including household detergents). 481. Photo-engraving and rotogravure dyes and solutions. 482. Other dye solutions. 485. Miscellaneous solutions (including duplicating fluids).
To every 100 gallons of ethyl alcohol add: 2-B. One-half gallon benzene or one-half gallon rubber hydrocarbon solvent. (This formula must be used in a closed and continuous system unless it is shown that it is not practical to do so.)	*As a solvent:* 021. Cellulose plastics. 022. Non-cellulose plastics, including resins. 031. Photographic film and emulsions. 032. Transparent sheeting. 033. Explosives. 311. Cellulose compounds (dehydration). 312. Sodium hydrosulphite (dehydration). 315. Other dehydration products. 320. Petroleum products. 331. Processing pectin. 332. Processing other food products. 341. Processing crude drugs. 342. Processing glandular products, vitamins, hormones and yeasts. 343. Processing antibiotics and vaccines. 344. Processing medicinal chemicals, including alkaloids. 349. Miscellaneous drug processing (including manufacture of pills). 351. Processing dyes and intermediates. 352. Processing perfume materials and fixatives. 353. Processing photographic chemicals. 358. Processing other chemicals. 359. Processing miscellaneous products.
To every 100 gallons of ethyl alcohol add: 2-C. Thirty-three pounds, or more, of metallic sodium and either one-half gallon benzene or one-half gallon rubber hydrocarbon solvent —Continued. (This formula must be used in a closed and continuous system unless it is shown that it is not practical to do so.)	*As a solvent:* 344. Processing medicinal chemicals (including alkaloids). 358. Processing other chemicals. 359. Processing miscellaneous products.
To every 100 gallons of ethyl alcohol add: 3-A. Five gallons methyl alcohol	*As a solvent:* 021. Cellulose plastics. 022. Non-cellulose plastics, including resins. 031. Photographic film and emulsions. 032. Transparent sheeting. 033. Explosives.

(continued)

Table 6.28: (continued)

Formula No. and Composition	Authorized Uses
To every 100 gallons of ethyl alcohol add: 3-A. Five gallons methyl alcohol.	*As a solvent.* 034. Cellulose intermediates and industrial collodions. 035. Soldering flux. 036. Adhesives and binders. 051. Polishes. 052. Inks (including meat branding inks). 053. Stains (wood, etc.). 141. Shampoos. 142. Soaps and bath preparations. 311. Cellulose compounds (dehydration). 312. Sodium hydrosulphite (dehydration). 315. Other dehydration products. 320. Petroleum products. 331. Processing pectin. 332. Processing other food products. 341. Processing crude drugs. 342. Processing glandular products, vitamins, hormones and yeasts. 343. Processing antibiotics and vaccines. 344. Processing medicinal chemicals (including alkaloids). 345. Processing blood and blood products. 349. Miscellaneous (including manufacture of pills). 351. Processing dyes and intermediates. 352. Processing perfume materials and fixatives. 353. Processing photographic chemicals. 354. Processing rosin. 355. Processing rubber (latex). 358. Processing other chemicals. 359. Processing miscellaneous products. 410. Disinfectants, insecticides, fungicides, and other biocides. 420. Embalming fluids and related products. 430. Sterilizing and preserving solutions. 440. Industrial detergents and soaps. 450. Cleaning solutions (including household detergents). 470. Theater sprays, incense and room deodorants. 481. Photoengraving and rotogravure dyes and solutions. 482. Other dye solutions. 485. Miscellaneous solutions (including duplicating fluids).
To every 100 gallons of ethyl alcohol add: 3-B. One gallon of pine tar N.F.	*As a solvent:* 111. Hair and scalp preparations. 141. Shampoos. 142. Soap and bath preparations. 410. Disinfectants, insecticides, fungicides and other biocides.

(continued)

Table 6.28: (continued)

Formula No. and Composition	Authorized Uses
To every 100 gallons of ethyl alcohol add: 4. One gallon of the following solution: Five gallons of an aqueous solution containing 40 percent nicotine; and 3.6 av. ounces of methylene blue, N.F.; water sufficient to make 100 gallons.	*As a solvent:* 460. Tobacco sprays and flavors.
To every 100 gallons of ethyl alcohol add: 12-A. Five gallons of benzene.	*As a solvent:* 021. Cellulose plastics. 022. Non-cellulose plastics, including resins. 036. Adhesives and binders. 342. Processing glandular products, vitamins, hormones and yeasts. 343. Processing antibiotics and vaccines. 344. Processing medicinal chemicals (including alkaloids). 345. Processing blood and blood products. 351. Processing dyes and intermediates. 352. Processing perfume materials and fixatives. 354. Processing rosin. 358. Processing other chemicals. 359. Processing miscellaneous products. 430. Sterilizing and preserving solutions.
To every 100 gallons of ethyl alcohol add: 13-A. Ten gallons of ethyl ether.	*As a solvent:* 015. Candy glazes. 021. Cellulose plastics. 022. Non-cellulose plastics, including resins. 031. Photographic film and emulsions. 032. Transparent sheeting. 034. Cellulose intermediates and industrial collodions. 052. Inks (including meat branding inks). 241. Collodion (U.S.P. or N.F.). 331. Processing pectin. 332. Processing other food products. 342. Processing glandular products, vitamins, hormones and yeasts. 343. Processing antibiotics and vaccines. 344. Processing medicinal chemicals (including alkaloids). 345. Processing blood and blood products. 349. Miscellaneous drug processing (including manufacture of pills). 352. Processing perfume materials and fixatives. 353. Processing photographic chemicals. 358. Processing other chemicals. 359. Processing miscellaneous products. 430. Sterilizing and preserving solutions. 481. Photoengraving and rotogravure solutions and dyes.

(continued)

Table 6.28: (continued)

Formula No. and Composition	Authorized Uses
To every 100 gallons of ethyl alcohol add: 17. Five-hundredths (0.05) gallon (6.4 fluid ounces) of bone oil (Dipple's oil).	*As a solvent:* 344. Processing medicinal chemicals (including alkaloids). 358. Processing other chemicals. 359. Processing miscellaneous products.
To every 100 gallons of ethyl alcohol add: 19. One hundred gallons of ethyl ether.	*As a solvent:* 031. Photographic film and emulsions. 034. Cellulose intermediates and industrial collodions. 241. Collodion (U.S.P.).
To every 100 gallons of ethyl alcohol add: 22. Ten gallons of formaldehyde solution (U.S.P.).	*As a solvent:* 420. Embalming fluids and related products. 430. Sterilizing and preserving solutions. 470. Theater sprays, incense and room deodorants.
To every 100 gallons of ethyl alcohol add: 23-A. Ten gallons of acetone, N.F.	*As a solvent:* 011. Cellulose coatings. 012. Synthetic resin coatings. 013. Shellac coatings. 014. Other natural resin coatings. 015. Candy glazes. 016. Other coatings. 032. Transparent sheeting. 034. Cellulose intermediates and industrial collodions. 035. Soldering flux. 036. Adhesives and binders. 042. Solvents and thinners (other than proprietary solvents or special industrial solvents). 052. Inks (including meat branding inks). 053. Stains (wood, etc.). 111. Hair and scalp preparations. 112. Bay rum. 113. Lotions and creams (hand, face and body). 114. Body deodorants and deodorant creams. 141. Shampoos. 142. Soaps and bath preparations. 210. External pharmaceuticals (not U.S.P. or N.F.). 244. Antiseptic solutions (U.S.P. or N.F.). 249. Miscellaneous external pharmaceuticals (U.S.P. or N.F.). 331. Processing pectin. 332. Processing other food products. 341. Processing crude drugs. 342. Processing glandular products, vitamins, hormones and yeasts. 343. Processing antibiotics and vaccines.

(continued)

Table 6.28: (continued)

Formula No. and Composition	Authorized Uses
To every 100 gallons of ethyl alcohol add: 23-A. Ten gallons of acetone, N.F.	*As a solvent:* 344. Processing medicinal chemicals (including alkaloids). 345. Processing blood and blood products. 349. Miscellaneous drug processing (including manufacture of pills). 358. Processing other chemicals. 359. Processing miscellaneous products. 410. Disinfectants, insecticides, fungicides and other biocides. 420. Embalming fluids and related products. 430. Sterilizing and preserving solutions. 440. Industrial detergents and soaps. 450. Cleaning solutions (including household detergents). 482. Miscellaneous dye solutions. 485. Miscellaneous solutions.
To every 100 gallons of ethyl alcohol add: 23-F. Three pounds of salicylic acid, U.S.P., 1 pound resorcin, U.S.P.; and 1 gallon bergamot oil, N.F., or bay oil, N.F.	*As a solvent:* 111. Hair and scalp preparations. 210. External pharmaceuticals (not U.S.P. or N.F.).
To every 100 gallons of ethyl alcohol add: 23-H. Eight gallons of acetone, N.F. and 1.5 gallons of methyl isobutyl ketone. *Standard formula for rubbing alcohol compound:* S.D.A. No. 23-H . . . . .103.3 fl. oz. Sucrose octa-acetate . . . . .0.5 av. oz. Water q.s. .1 gallon All rubbing alcohol compounds or preparations coming under the general classification of rubbing alcohols must be manufactured with specially denatured alcohol Formula No. No. 23-H according to the above formula except that manufacturers may also add to the formula other odorous or medicinal ingredients provided they are shown in the formula submitted for approval and that the finished product contains 70 percent absolute alcohol by volume.	*As a solvent:* 111. Hair and scalp preparations. 113. Lotions and creams (hand, face, and body). 210. External pharmaceuticals (not U.S.P. or N.F.). 220. Rubbing alcohol. 410. Disinfectants, insecticides, fungicides and other biocides. 450. Cleaning solutions (including household detergents).

(continued)

Table 6.28: (continued)

Formula No. and Composition	Authorized Uses
To every 100 gallons of ethyl alcohol add:	*As a solvent:*

25. Twenty pounds of iodine, U.S.P. and 15 pounds of either potassium or sodium iodide U.S.P.

Formula for Strong Iodine Tincture N.F.:

Iodine U.S.P. ....6.50 av. oz.
Potassium iodide U.S.P. ....4.50 av. oz.
Distilled water ....6.40 fl. oz
S.D.A. No. 25 q.s. ....128.00 fl. oz.

Formula for Iodine Tincture U.S.P.:

Iodine U.S.P. ....1.0 av. oz. 11.0 gr.
Sodium iodide U.S.P. ..1.0 av. oz. 431.0 gr.
Distilled water ....65.0 fl. oz. 134.0 min.
S.D.A. No. 25 q.s. ....128.00 fl. oz.

NOTE.— N.F. and U.S.P. preparations. In preparation of N.F. and U.S.P. formulae, above, the quantities of iodine and potassium or sodium iodide referred to as separate items in the formula are exclusive of the denaturants in the specially denatured alcohol, and are the quantities that must be added in order that the finished products may comply with the official U.S.P. or N.F. preparations.

As a solvent:
230. Tinctures of iodine.
249. Miscellaneous external pharmaceuticals (U.S.P. or N.F.).

To every 100 gallons of ethyl alcohol add:

25-A. A solution composed of 20 pounds of iodine U.S.P., 15 pounds of potassium or sodium iodide U.S.P. and 15 pounds of water.

Formula for Strong Iodine Tincture N.F.:

Iodine U.S.P. ....6.50 av. oz.
Potassium iodide U.S.P. ....4.50 av. oz.
Distilled water ....4.40 fl. oz.
S.D.A. No. 25-A q.s. ....128.0 fl. oz.

Formula for Iodine Tincture U.S.P.:

Iodine U.S.P. ....1.0 av. oz. 11.0 gr.
Sodium iodide U.S.P. ....1.0 av. oz. 431 gr.
Distilled water ....64.0 fl. oz.
S.D.A. No. 25-A q.s. ....128.0 fl. oz.

NOTE.— N.F. and U.S.P. preparations. In preparation of N.F. and U.S.P. formulae, above, the quantities of iodine and potassium or sodium iodide referred to as separate items in the formula are exclusive of the denaturants in the specially denatured alcohol, and are the quantities that must be added in order that the finished products may comply with the official U.S.P. or N.F. preparations.

(1) As a solvent:
230. Tinctures of iodine.
249. Miscellaneous external pharmaceuticals (U.S.P. or N.F.).

(continued)

Table 6.28: (continued)

Formula No. and Composition	Authorized Uses
To every 100 gallons of ethyl alcohol add: 27. One gallon of rosemary oil, N.F. and 30 pounds of camphor, U.S.P. *Formula for Camphor and Soap Liniment N.F.:* Hard soap, N.F. dried and granulated or powdered ....8.0 av. oz. 5 gr. Camphor U.S.P. . . . .(small 2.0 av. oz. 280 gr. pieces) Rosemary oil N.F. ....185 min. S.D.A. No. 27 ....93.75 fl. oz. Distilled water q.s. ....128.0 fl. oz. NOTE.— N.F. preparation. In the preparation of N.F. formula, above, the quantities of soap, camphor and oil of rosemary referred to as separate items in the formula are exclusive of the denaturants in the specially denatured alcohol and are quantities that must be added in order that the finished product may comply with the official N.F. preparation.	*As a solvent:* 243. Liniments, U.S.P. or N.F.
To every 100 gallons of ethyl alcohol add: 27-A. Thirty-five pounds of camphor, U.S.P. and 1 gallon of clove oil, U.S.P.	*As a solvent:* 210. External pharmaceuticals (not U.S.P. or N.F.). 410. Disinfectants, insecticides, fungicides and other biocides.
To every 100 gallons of ethyl alcohol add: 27-B. One gallon of lavender oil, U.S.P., and 100 pounds of medicinal soft soap, U.S.P. *Formula for medicinal soft soap liniment U.S.P.:* Medicinal soft soap, U.S.P. . . ..81.0 av. oz. 240 gr. Lavender oil, U.S.P. ......2.0 fl. oz. 66 min. S.D.A. No. 27-B q.s. ....128.0 fl. oz. NOTE.— U.S.P. preparation. In the preparation of U.S.P. formula, above, the quantities of ingredients referred to as separate items in the formula are exclusive of the denaturants in the specially denatured alcohol and are necessary additions in order that the finished product may comply with the official U.S.P. formula.	*As a solvent:* 141. Shampoos. 210. External pharmaceuticals (not U.S.P. or N.F.). 243. Liniments (U.S.P. or N.F.). 410. Disinfectants, insecticides, fungicides and other biocides.
To every 100 gallons of ethyl alcohol add: 30. Ten gallons of methyl alcohol.	*As a solvent:* 011. Cellulose coatings. 012. Synthetic resin coatings. 021. Cellulose plastics. 022. Non-cellulose plastics, including resins.

(continued)

Table 6.28: (continued)

Formula No. and Composition	Authorized Uses
To every 100 gallons of ethyl alcohol add: 30. Ten gallons of methyl alcohol	*As a solvent:* 031. Photographic film and emulsions. 035. Soldering flux. 036. Adhesives and binders. 051. Polishes. 052. Inks. 053. Stains. 142. Soap and bath preparations. 331. Processing pectin. 332. Processing other food products. 341. Processing crude drugs. 342. Processing glandular products, vitamins, hormones and yeasts. 343. Processing antibiotics and vaccines. 344. Processing medicinal chemicals (including alkaloids). 345. Processing blood and blood products. 349. Miscellaneous drug processing (including manufacture of pills). 352. Processing perfume materials and fixatives. 353. Processing photographic materials. 358. Processing other chemicals. 359. Processing miscellaneous products. 410. Disinfectants, insecticides, fungicides and other biocides. 430. Sterilizing and preserving solutions. 440. Industrial detergents and soaps. 450. Cleaning solutions (including household detergents). 481. Photoengraving and rotogravure solutions and dyes. 482. Other dye solutions. 485. Miscellaneous solutions (including duplicating fluids).
To every 100 gallons of ethyl alcohol add: 31-A. One hundred pounds of glycerol, U.S.P. and 20 pounds of hard soap, N.F.	*As a solvent:* 113. Lotions and creams (hand, face and body). 131. Tooth paste and tooth powder. 141. Shampoos.
To every 100 gallons of ethyl alcohol add: 32. Five gallons of ethyl ether.	*As a solvent:* 031. Photographic film and emulsions. 034. Cellulose intermediates and industrial collodions. 052. Inks (including meat branding inks). 241. Collodion (U.S.P.). 311. Ethyl cellulose compounds (dehydration). 332. Processing miscellaneous food products. 342. Processing glandular products, vitamins, hormones and yeasts.

(continued)

Table 6.28: (continued)

Formula No. and Composition	Authorized Uses
To every 100 gallons of ethyl alcohol add: 32. Five gallons of ethyl ether.	*As a solvent:* 343. Processing antibiotics and vaccines. 430. Sterilizing and preserving solutions. 481. Photoengraving and rotogravure solutions and dyes.
To every 100 gallons of ethyl alcohol add: 33. Thirty pounds of methyl violet, U.S.P. Meat branding inks made with S.D.A. No. 33 do not meet U. S. Department of Agriculture meat inspection specifications for use in federally inspected establishments. Such inks must be made with FD&C Violet No. 1. Specially denatured alcohol Formulae No. 23-A and 32 are authorized for this purpose.	*As a solvent:* 052. Inks.
To every 100 gallons of ethyl alcohol add: 35. 29.75 gallons of ethyl acetate having an ester content of 100 percent by weight or the equivalent thereof not to exceed 35 gallons of ethyl acetate with an ester content of not less than 85 percent by weight.	*As a solvent:* 015. Candy glazes.
To every 100 gallons of ethyl alcohol add: 35-A. 4.25 gallons of ethyl acetate having an ester content of 100 percent by weight or the equivalent thereof not to exceed 5 gallons of ethyl acetate with an ester content of not less than 85 percent by weight.	*As a solvent:* 015. Candy glazes. 331. Processing pectin. 332. Processing other food products. 342. Processing glandular products, vitamins, hormones and yeasts. 343. Processing antibiotics and vaccines. 344. Processing medicinal chemicals (including alkaloids). 349. Miscellaneous drug processing (including manufacture of pills). 358. Processing miscellaneous chemicals. 359. Processing miscellaneous products.
To every 100 gallons of ethyl alcohol add: 36. Three gallons of ammonia, aqueous, 27 to 30 percent by weight: three gallons of strong ammonia solution, U.S.P.: 17.5 pounds of caustic soda, liquid grade, containing 50 percent sodium hydroxide by weight: or 12.0 pounds of caustic soda, liquid grade, containing 73 percent sodium hydroxide by weight.	*As a solvent:* 141. Shampoos. 142. Soap and bath preparations. 210. External pharmaceuticals (not U.S.P. or N.F.). 450. Cleaning solutions (including household detergents).
To every 100 gallons of ethyl alcohol add: 37. Forty-five fluid ounces of eucalyptol, U.S.P., 30 av. ounces of thymol, N.F. and 20 av. ounces of menthol, U.S.P.	*As a solvent:* 111. Hair and scalp preparations. 112. Bay rum. 113. Lotions and creams (hand, face and body).

(continued)

Table 6.28: (continued)

Formula No. and Composition	Authorized Uses
To every 100 gallons of ethyl alcohol add: 37. Forty-five fluid ounces of eucalyptol, U.S.P., 30 av. ounces of thymol, N.F. and 20 av. ounces of menthol, U.S.P.	*As a solvent:* 131. Dentifrices. 132. Mouth washes. 210. External pharmaceuticals (not U.S.P. or N.F.). 244. Antiseptic solutions (U.S.P. or N.F.). 410. Disinfectants, insecticides, fungicides and other biocides. 430. Sterilizing and preserving solutions. 470. Theater sprays, incense and room deodorants.
To every 100 gallons of ethyl alcohol add: 38-B. Ten pounds of any one or a total of 10 pounds of two or more of the oils and substances listed below: Anethole, U.S.P. Anise oil, U.S.P. Bay oil (myrcia oil), N.F. Benzaldehyde, N.F. Bergamot oil, N.F. Bitter almond oil, N.F. Camphor, U.S.P. Cedar leaf oil, U.S.P. XIII. Chlorothymol, N.F. Cinnamic aldehyde, N.F. IX. Cinnamon oil (cassia oil), U.S.P. Citronella oil, natural. Clove oil, U.S.P. Coal tar, U.S.P. Eucalyptol, U.S.P. Eucalyptus oil, N.F. Eugenol, U.S.P. Guaiacol, N.F. Lavender oil, U.S.P. Menthol, U.S.P. Mustard oil, volatile (allyl isothiocyanate), U.S.P. XII. Peppermint oil, U.S.P. Phenol, U.S.P. Phenyl salicylate (salol), N.F. Pine oil, N.F. Pine needle oil, dwarf, N.F. Rosemary oil, N.F. Safrol Sassafras oil, N.F. Spearmint oil, N.F. Spearmint oil, terpeneless. Spike lavender oil, natural. Storax, U.S.P. Thyme oil, N.F.	*As a solvent:* 111. Hair and scalp preparations. 113. Lotions and creams (hand, face and body). 114. Deodorants (body). 121. Perfumes and perfume tinctures. 122. Toilet waters and colognes. 131. Dentifrices. 132. Mouth washes. 141. Shampoos. 142. Soap and bath preparations. 210. External pharmaceuticals (not U.S.P. or N.F.). 243. Liniments, U.S.P. or N.F. 244. Antiseptic solutions, U.S.P. or N.F. 249. Miscellaneous external pharmaceuticals U.S.P. or N.F. 410. Disinfectants, insecticides, fungicides and other biocides. 430. Sterilizing and preserving solutions. 470. Theater sprays, incense and room deodorants.

(continued)

Table 6.28: (continued)

Formula No. and Composition	Authorized Uses
Thymol, N.F. Tolu balsam, U.S.P. Turpentine oil, N.F. Wintergreen oil (methyl salicylate), U.S.P. Where it is shown that none of the above single denaturants or combinations can be used in the manufacture of a particular product, application (in duplicate) may be submitted to the Director, Alcohol, Tobacco & Firearms Division, requesting permission to use another essential oil or substance having denaturing properties satisfactory to the Director. In such case the applicant shall furnish the Director with specifications, assay methods, and duplicate 8-ounce samples of the denaturant for examination.	
To every 100 gallons of ethyl alcohol add: 38-C. Ten pounds of menthol, U.S.P. and 1.25 gallons of formaldehyde solution, U.S.P.	*As a solvent:* 131. Dentifrices. 132. Mouth washes.
To every 100 gallons of ethyl alcohol add: 38-D. Two and one-half pounds of menthol, U.S.P. and .2.5 gallons of formaldehyde solution, U.S.P.	*As a solvent:* 131. Dentifrices. 132. Mouth washes.
To every 100 gallons of ethyl alcohol add: 38-F. (1) Six pounds of boric acid, U.S.P., 1⅓ pounds thymol, N.F., 1⅓ pounds chlorothymol, N.F. and 1⅓ pounds menthol, U.S.P.; or (2) Seven pounds of boric acid, U.S.P., and a total of 3 pounds of any two or more denaturing materials listed under S.D.A. No. 38-B. The denaturants selected and the amounts must be stated on Form 1479-A.	*As a solvent:* 132. Mouth washes. 210. External pharmaceuticals (not U.S.P. or N.F.). 244. Antiseptic solutions (U.S.P. or N.F.).
To every 100 gallons of ethyl alcohol add: 39 Nine pounds of sodium salicylate or salicylic acid, U.S.P., 1.25 gallons fluid extract of quassia, N.F. VII and ⅛ gallon of tertiary butyl alcohol.	*As a solvent:* 111. Hair and scalp preparations. 112. Bay rum. 113. Lotions and creams (hand, face and body). 121. Perfume and perfume tinctures. 122. Toilet waters and colognes.
To every 100 gallons of ethyl alcohol add: 39-A. Sixty av. ounces of any one of the following alkaloids or salts together with ⅛ gallon of tertiary butyl alcohol. Quinine, N.F.	*As a solvent:* 111. Hair and scalp preparations. 122. Toilet waters and colognes. 141. Shampoos.

(continued)

Table 6.28: (continued)

Formula No. and Composition	Authorized Uses

To every 100 gallons of ethyl alcohol add:
 	
39-A. Sixty av. ounces of any one of the following alkaloids or salts together with ⅛ gallon of tertiary butyl alcohol.
 	
    Quinine bisulfate, N.F.
    Quinine hydrochloride, U.S.P.
    Cinchonidine
    Cinchonidine sulfate, N.F. IX.
 	
The denaturant selected must be stated on Form 1479-A.
	As a solvent:

To every 100 gallons of ethyl alcohol add: 39-B. Two and one-half gallons of diethylphthalate and ⅛ gallon of tertiary butyl alcohol.	*As a solvent:* 111. Hair and scalp preparations. 112. Bay rum. 113. Lotions and creams (hand, face and body). 114. Deodorants (body). 121. Perfume and perfume tinctures. 122. Toilet waters and colognes. 141. Shampoos. 142. Soaps and bath preparations. 210. External pharmaceuticals (not U.S.P. or N.F.). 410. Disinfectants, insecticides, fungicides and other biocides. 450. Cleaning solutions (including household detergents). 470. Theater sprays, incense and room deodorants. 485. Miscellaneous solutions.
To every 100 gallons of ethyl alcohol add: 39-C. One gallon of diethylphthalate. Preparations manufactured with S.D.A. No.39-C must contain in each gallon of finished product not less than 2 fluid ounces of perfume material (essential oils, isolates, aromatic chemicals, etc.) satisfactory to the Director, Alcohol, Tobacco & Firearms Division.	*As a solvent:* 111. Hair and scalp preparations. 113. Lotions and creams (hand, face and body). 114. Deodorants (body). 121. Perfumes and perfume tinctures. 122. Toilet waters and colognes. 142. Soaps and bath preparations. 470. Theater sprays, incense and room deodorants. 482. Miscellaneous dye solutions (own use only).
To every 100 gallons of ethyl alcohol add: 39-D. One gallon of bay oil, N.F. and either 50 av. ounces of quinine sulfate, U.S.P., 50 av. ounces of quinine bisulfate, N.F., or 200 av. ounces of sodium salicylate, U.S.P. The denaturant selected must be stated on Form 1479-A.	*As a solvent:* 111. Hair and scalp preparations. 112. Bay rum.

(continued)

Table 6.28: (continued)

Formula No. and Composition	Authorized Uses
To every 100 gallons of ethyl alcohol add: 40. One and one-half avoirdupois ounces of brucine (alkaloid), or brucine sulfate (N.F. IX), or quassin, or one and one-half avoirdupois ounces of any combination of two or of three of those denaturants, and 1/8 gallon of tertiary butyl alcohol.	*As a solvent:* 051. Polishes. 111. Hair and scalp preparations. 112. Bay rum. 113. Lotions and creams (hand, face and body). 114. Deodorants (body). 121. Perfumes and perfume tinctures. 122. Toilet waters and colognes. 141. Shampoos. 142. Soaps and bath preparations. 210. External pharmaceuticals (not U.S.P. or N.F.). 410. Disinfectants, insecticides, fungicides, and other biocides. 450. Cleaning solutions (including household detergents). 470. Theater sprays, incense and room deodorants. 482. Miscellaneous dye solutions. 485. Miscellaneous solutions.
To every 100 gallons of ethyl alcohol add: 40-A. One pound of sucrose octa-acetate and 1/8 gallon of tertiary butyl alcohol.	*As a solvent:* 051. Polishes. 111. Hair and scalp preparations. 112. Bay rum. 113. Lotions and creams (hand, face and body). 114. Deodorants (body). 121. Perfumes and perfume tinctures. 122. Toilet waters and colognes. 141. Shampoos. 142. Soaps and bath preparations. 210. External pharmaceuticals (not U.S.P. or N.F.). 410. Disinfectants, insecticides, fungicides, and other biocides. 450. Cleaning solutions (including household detergents). 470. Theater sprays, incense and room deodorants. 482. Miscellaneous dye solutions. 485. Miscellaneous solutions.
To every 100 gallons of ethyl alcohol add: 40-B. One-sixteenth avoirdupois ounce of denatonium benzoate, N.F., (Bitrex) and 1/8 gallon of tertiary butyl alcohol.	*As a solvent:* 051. Polishes. 111. Hair and scalp preparations. 112. Bay rum. 113. Lotions and creams (hand, face and body). 114. Deodorants (body). 121. Perfumes and perfume tinctures. 122. Toilet waters and colognes. 141. Shampoos. 142. Soaps and bath preparations. 210. External pharmaceuticals (not U.S.P. or N.F.). 410. Disinfectants, insecticides, fungicides, and other biocides.

(continued)

Table 6.28: (continued)

Formula No. and Composition	Authorized Uses
To every 100 gallons of ethyl alcohol add: 40-B. One-sixteenth avoirdupois ounce of denatonium benzoate, N.F., (Bitrex) and 1/8 gallon of tertiary butyl alcohol.	*As a solvent:* 450. Cleaning solutions (including household detergents). 470. Theater sprays, incense and room deodorants. 482. Miscellaneous dye solutions. 485. Miscellaneous solutions.
To every 100 gallons of ethyl alcohol add: 40-C. Three gallons of tertiary butyl alcohol. *Conditions governing use:* This formula shall be used only in the manufacture of products which will be packaged in pressurized containers in which the liquid contents are in intimate contact with the propellant and from which the contents are not easily removable in liquid form.	*As a solvent:* 051. Polishes. 111. Hair and scalp preparations. 112. Bay rum. 113. Lotions and creams (hand, face and body). 114. Deodorants (body). 121. Perfumes and perfume tinctures. 122. Toilet waters and colognes. 141. Shampoos. 142. Soaps and bath preparations. 210. External pharmaceuticals (not U.S.P. or N.F.). 410. Disinfectants, insecticides, fungicides, and other biocides. 450. Cleaning solutions (including household detergents). 470. Theater sprays, incense and room deodorants. 482. Miscellaneous dye solutions. 485. Miscellaneous solutions.
To every 100 gallons of ethyl alcohol add: 42. (1) Eighty grams of potassium iodide, U.S.P. and 109 grams of red mercuric iodide, N.F.: or (2) Ninety-five grams of thimerosal, N.F.; or (3) Seventy-six grams of any of the following: phenyl mercuric nitrate, N.F.; phenyl mercuric chloride, N.F.; or phenyl mercuric benzoate.	*As a solvent:* 430. Sterilizing and preserving solutions.
To every 100 gallons of ethyl alcohol add: 44. Ten gallons of n-butyl alcohol.	*As a solvent:* 430. Sterilizing and preserving solutions.
To every 100 gallons of ethyl alcohol add: 45. Three hundred pounds of refined white or orange shellac.	*As a solvent:* 015. Candy glazes.
To every 100 gallons of ethyl alcohol add: 46. Twenty-five fluid ounces of phenol, U.S.P. and 4 fluid ounces of wintergreen oil (methyl salicylate), U.S.P. *Conditions governing use:* This formula may be used only by institutions and organizations which are of a semi-public character and engaged in charitable work.	*As a solvent:* 220. An antiseptic, sterilizing and bathing solution having restricted use.

Table 6.29: Properties of Specially Denatured Alcohols *(78)*

Weight of one gallon water at 60°F., 8.32823 pounds in air.

Specially Denatured Alcohol No.*	Weight Per Gallon at 60°F. (In Pounds)	Specific Gravity at 60°F. 60°F. in Vacuum	Per cent Alcohol by Volume	Per cent Alcohol by Weight	Apparent Proof of Finished Formula at 60°F.
1	6.788	0.8153	90.48	88.10	190.4
2-B	6.795	0.8161	94.53	91.95	190.0
3-A	6.785	0.8149	90.48	88.14	190.6
3-B	6.810	0.8179	94.06	91.29	189.0
4	6.823	0.8195	94.06	91.12	188.2
12-A	6.820	0.8192	90.48	87.69	188.3
13-A	6.740	0.8095	86.36	84.69	193.2
17	6.795	0.8161	94.95	92.36	190.0
19	6.468	0.7769	47.50	48.54	200.0
22	7.037	0.8451	86.36	81.12	172.9
23-A	6.788	0.8153	86.36	84.09	190.4
23-F	6.808	0.8177	93.64	90.91	189.1
23-H	6.785	0.8149	86.76	84.52	190.6
25	7.080	0.8502	94.14	87.89	169.5
25†	7.083	0.8506	94.18	87.89	169.2
25-A	7.119	0.8550	92.71	86.08	166.2
25-A†	7.117	0.8548	92.69	86.08	166.4
27	6.846	0.8222	90.71	87.58	186.7
27-A	6.867	0.8247	90.21	86.83	185.3
27-B	7.027	0.8439	84.85	79.81	173.6
30	6.785	0.8149	86.36	84.13	190.6
31-A	7.167	0.8608	85.17	78.55	162.1
32	6.769	0.8130	90.48	88.35	191.5
33	6.893	0.8279	93.10	89.27	183.5
35‡	6.956	0.8355	74.51	70.80	179.1
35-A‡	6.817	0.8187	90.48	87.73	188.6
36	6.837	0.8211	92.23	89.16	187.3
37	6.794	0.8160	94.18	91.63	190.0
38-B	6.804	0.8172	93.76	91.08	189.4
38-C	6.832	0.8206	92.59	89.58	187.6
38-D	6.863	0.8242	92.49	89.08	185.6
38-F	6.828	0.8201	94.09	91.08	187.9
39	6.867	0.8247	93.14	89.65	185.3
39-A	6.810	0.8179	94.58	91.80	189.0
39-B	6.857	0.8236	92.57	89.23	186.0
39-C	6.819	0.8189	94.06	91.17	188.5
39-D	6.819	0.8190	93.38	90.51	188.4
40	6.795	0.8161	95.20	92.28	190.0
40-A	6.798	0.8164	94.88	92.26	189.8
40-B	6.794	0.8160	94.91	92.34	190.1
40-C	6.788	0.8153	92.23	89.72	190.6
42	6.797	0.8164	94.98	92.36	189.8
44	6.790	0.8155	86.36	84.07	190.3
45	7.545	0.9061	73.18	64.11	126.5
46	6.805	0.8173	94.78	92.06	189.3

* Prepared from 190 proof alcohol.   † With Sodium Iodide.   ‡ Calculated on the basis of 85 per cent ethyl acetate.

Table 6.30: Azeotropes of Ethanol (31)

ETHYL ALCOHOL FORMS BINARY AZEOTROPES WITH:

%		B.P. of Azeotrope °C	%		B.P. of Azeotrope °C
79	tert-Amyl ethyl ether	66.6	27.3	Ethyl acrylate	77.5
67.6	Benzene	68.3	25	Ethyl propionate	78.0
57	1-Bromobutane	75.0	75	Ethyl propyl ether	61.2
67	2-Bromobutane	72.5	44	Ethyl sulfide	72.6
22.5	cis-1-Bromo-1-butene	69.6	75	Fluorobenzene	70.0
64.3	trans-1-Bromo-1-butene	72.8	51	n-Heptane	70.9
66.3	cis-2-Bromo-2-butene	72.3	79	n-Hexane	58.6
73.3	trans-2-Bromo-2-butene	69.1	30	2-Iodobutane	77.2
77.8	2-Bromo-1-butene	67.4	30	1-Iodo-2-methylpropane	77.0
18	1-Bromo-3-methylbutane	77.3	56	1-Iodopropane	75.4
59	1-Bromo-1-methylpropane	71.4	75	2-Iodopropane	70.2
85	2-Bromo-2-methylpropane	63.8	58	3-Iodopropene	75.4
83.7	1-Bromopropane	63.6	33	Isobutyl formate	77.0
88.5	2-Bromopropane	55.5	97	Isoprene	32.7
94	2-Bromopropene	46.2	47	Isopropyl acetate	76.8
89	trans-1-Bromopropene	58.7	97	Methyl acetate	56.9
91	cis-1-Bromopropene	56.4	57.6	Methyl acrylate	73.5
79	tert-Butyl ethyl ether	66.6	75	Methyl borate	63.0
91	Carbon disulfide	42.4	96.5	2-Methylbutane	26.8
84.2	Carbon tetrachloride	64.9	17	Methyl butyrate	78.0
79.7	1-Chlorobutane	65.7	55	Methyl carbonate	73.5
84.2	2-Chlorobutane	61.2	72	Methylcyclopentene	63.3
85.2	cis-1-Chloro-1-butene	57.0	75	Methylcyclopentene	60.3
79.8	trans-1-Chloro-1-butene	61.2	66	Methyl ethyl ketone	74.8
88.5	2-Chloro-1-butene	53.6	67	Methyl propionate	72.0
84.6	cis-2-Chloro-2-butene	56.8	22	Octane	77.0
93.0	Chloroform	59.4	95	Pentane	34.3
59	1-Chloro-3-methylbutane	74.8	8.8	2-Pentanone	77.7
83.7	1-Chloro-2-methylpropane	61.5	19	Perchloroethylene	78.0
94	1-Chloropropane	46.7	81	Propanediol	63.5
97.2	2-Chloropropane	36.6	15	Propyl acetate	78.2
96	trans-1-Chloropropene	36.7	56	Propyl ether	74.4
95	3-Chloropropene	44.0	55	Thiophene	70.0
66	1,3-Cyclohexadiene	60.7	32	Toluene	76.7
69.5	Cyclohexane	64.9	73	Trichloroethylene	70.9
66	Cyclohexene	66.7			
92.5	Cyclopentane	44.7			
47.2	1,1-Dichloropropane	74.7			
85.5	2,2-Dichloropropane	63.2			
41	2,5-Dimethylhexane	73.6			
58	Diethoxymethane	74.2			
69	Ethyl acetate	71.8			

ETHYL ALCOHOL FORMS TERNARY AZEOTROPES WITH:

%		%		B.P. of Azeotrope °C
11.4	Water	61	Acetal	77.8
74	Water	74.1	Benzene	64.9
7.4	Water	70	Bromodichloromethane	72.0
8	Water	65	1-Bromo-2-methylpropane	69.5
3	Water	91	cis-1-Bromopropane	54.0
4	Water	7.5	trans-Bromopropane	54.5
5	Water	83	1-Bromopropane	60.0
1	Water	95	2-Bromopropane	43.3
1.6	Water	3.4	Carbon disulfide	41.3
3.5	Water	92.5	Chloroform	55.4
4.5	Water	82.5	1-Chloro-2-methylpropane	58.6
7	Water	73	Cyclohexene	64.1
5	Water	78	1,2-Dichloroethane	66.7
2.85	Water	90.5	cis-1,2-Dichloroethylene	53.8
1.1	Water	94.5	trans-1,2-Dichloroethylene	44.4
12.8	Water	69.5	Dimethoxymethane	73.2
9	Water	82.6	Ethyl acetate	70.2
17.5	Water	20.8	Ethyl chloroacetate	81.4
5.5	Water	78.4	Trichloroethylene	67.0
9	Water	78	Triethylamine	74.7

n-PROPYL ALCOHOL

n-Propanol $CH_3CH_2CH_2OH$

n-Propyl alcohol is a colorless, volatile liquid.

Table 6.31: Physical Properties of n-Propyl Alcohol *(31)*

Acidity as acetic acid	0.003% by wt, max.
Alkalinity as ammonia	0.003% by wt, max.
Autoignition temperature	540°C
Boiling point at 760 mm	97.15°C
Coefficient of cubical expansion, 0 - 94°C	0.000956×10^{-3}
Color, APHA	5 max.
Critical density	0.2734
Critical pressure	49.9 atm
Critical temperature	263.7°C
Distillation range at 760 mm	2°C including 97.15°C
Electrical conductivity, mhos per cm at 25°C	2×10^{-8}
Explosive limits, Lower	2.6% by vol. in air
Upper	13.5% by vol. in air
Fire hazard	Dangerous when exposed to heat or flame.
Flash point (open cup)	90°F
(closed cup)	59°F
Freezing point	-127°C (-196°F)
Heat of combustion	8020 cal/g
Heat of vaporization at 97.15°C	162.6 cal/g
Limits of flammability (in air by volume)	2.5% (Lower)
Melting point	-127.0°C
Molecular weight	60.09
Non-volatile material	0.001 gm/100 ml sample, max.
Odor	Alcohol-like
Refractive index at 20°C	1.3845
Reid vapor pressure at 100°F	0.1 psi
Relative evaporation rate (butyl acetate = 1)	1.3
Specific gravity at 20/4°C	0.8036
at 20/20°C	0.8050
Specific heat at 25°C	0.586 cal/g/°C
Surface tension in air, -5°C	25.9 dynes/cm
20°C	23.8 dynes/cm
60°C	20.5 dynes/cm
Toxicity	Slight

Vapor pressure   mm Hg

°C	°F	
0	32	3.44
10	50	7.26
20	68	14.5
30	86	27.6
40	104	50.2
50	122	87.2
60	140	147.0
70	158	239.0
80	176	376.0
90	194	574.0
97.19	206.9	760.0

Viscosity, 0°C	3.8827 centipoises
20°C	2.2563 centipoises
40°C	1.4050 centipoises
90°C	0.5310 centipoises
Water content	0.2% by wt, max.
Weight per gallon at 20°C (68°F)	6.7 lbs

Table 6.32: Azeotropes of n-Propyl Alcohol *(31)*

n-PROPYL ALCOHOL FORMS BINARY AZEOTROPES WITH:

%		B.P. of Azeotrope °C
63	Acetal	92.4
83.1	Benzene	77.1
75	Biacetyl	85.0
31	1-Bromobutane	89.5
89.5	2-Bromobutane	85.3
82	n-Butyl chloride	74.8
36	Butyl formate	95.5
88.5	Carbon tetrachloride	73.1
17	Chlorobenzene	96.9
82	1-Chlorobutane	74.8
91	2-Chlorobutane	67.2
69	1-Chloro-3-methylbutane	89.4
78	1-Chloro-2-methylpropane	67.7
89	Diethoxymethane	86.2
45	Dioxane	95.3
70	Di-n-propyl ether	85.7
49	Ethyl propionate	93.4
72	Ethyl sulfide	85.5
81	Ethylene chloride	80.7
82	Fluorobenzene	80.2
96	n-Hexane	65.7
34	1-Iodobutane	96.2
47	2-Iodobutane	94.2
55	1-Iodo-2-methylpropane	93.0
60	Isobutyl formate	93.2
30	Isobutyronitrile	95.0
94.6	Methyl acrylate	70.9
65	3-Methyl-2-butanol	93.5
53	Methyl butyrate	94.4
74	Methyl isobutyrate	89.5
32	2-Pentanone	96.0
37	3-Pentanone	96.0
1.5	α-Pinene	97.1
60	Propyl acetate	94.2
91	n-Propyl bromide	69.7
90.2	Propyl formate	80.6
47.5	Toluene	92.4
28.3	Water	87.7

n-PROPYL ALCOHOL FORMS TERNARY AZEOTROPES WITH:

%		B.P. of Azeotrope °C
27.4	Water	
21	Acetaldehyde dipropylacetal	87.6
7.6	Water	
82.3	Benzene	67.0
5	Water	
84	Carbon tetrachloride	65.4
9	Water	
79	1,3-Cyclohexadiene	67.8
8.5	Water	
81.5	Cyclohexane	66.6
9	Water	
79.5	Cyclohexene	63.2
8	Water	
47.2	Dipropoxymethane	86.4
17.6	Water	
59.5	Ethoxypropoxymethane	83.8
8	Water	
72	3-Iodopropene	78.2
17.5	Water	
55.9	Nitromethane	82.3
20	Water	
60	3-Pentanone	81.2
21	Water	
59.5	Propyl acetate	82.2
25.3	Water	
16.5	Propyl chloroacetate	88.6
11.7	Water	
68.1	Propyl ether	74.8
13	Water	
82	Propyl formate	70.8
7	Water	
81	Trichloroethylene	71.6

Table 6.33: n-Propanol-Water-Benzene *(19)*

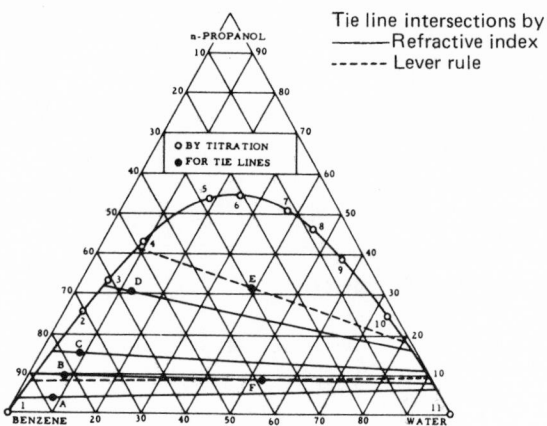

Table 6.34: n-Propanol-Water-n-Butanol *(19)*

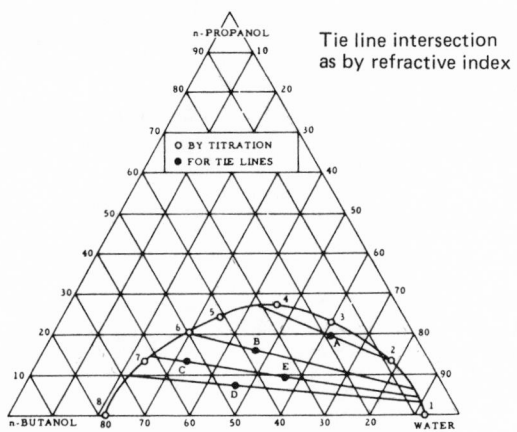

Table 6.35: n-Propanol-Water-Heptane *(19)*

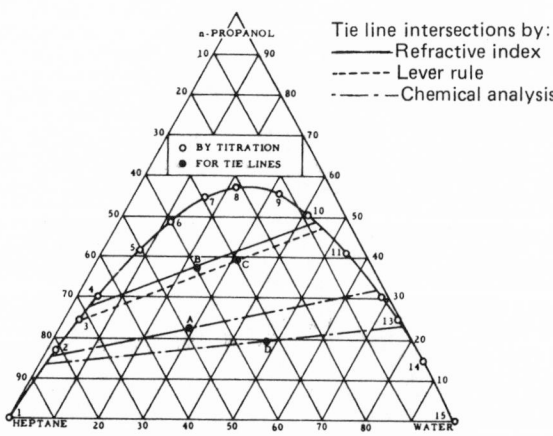

Table 6.36: n-Propanol-Water-Hexane *(19)*

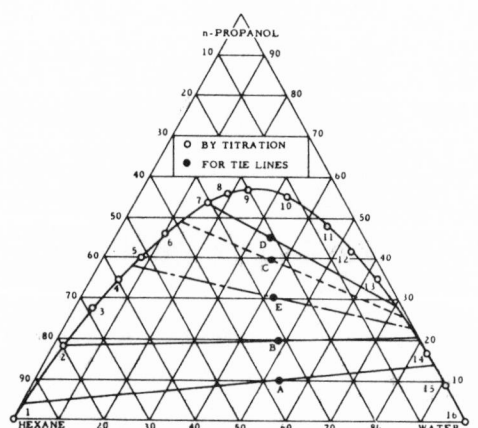

ISOPROPYL ALCOHOL

Isopropanol, Dimethyl Carbinol, 2-Propanol $(CH_3)_2CHOH$

Table 6.37: Physical Properties of Anhydrous Isopropyl Alcohol (31)

Acidity, as acetic acid	0.002% by wt, max.	Heat of fusion	21.08 cal/g
Boiling point at 760 mm	82.3°C	Heat of vaporization	287 Btu/lb
50 mm	27°C	Lower limit of flammability	
10 mm	2°C	in air	2.65% vol.
Coefficient of expansion at 55°C	0.00111	MAC	400 ppm in air
Color, Pt-Co scale	10 max.	Molecular weight	60.09
Critical pressure	53 atm.	Non-volatile matter	0.002 g/100 ml, max.
Critical temperature	234.9°C	Odor	Non-residual
Dielectric constant at 20°C	18.62	Purity	99.5% - 99.9% by vol.
40°C	16.24	Refractive index at 20°C	1.3772
80°C	11.91	Specific gravity at 20/20°C	0.7862 - 0.7867
Distillation at 760 mm	Distills entirely within 1.0°C range which includes 82.3°C	Specific heat at 20°C	0.596 cal/g/°C
		Surface tension at 25°C	20.8 dynes/cm
Fire hazard	Dangerous when exposed to heat or open flame	Toxicity	Moderate by ingestion; otherwise slight
Flash point (open cup)	70°F	Vapor pressure at 20°C	33.0 mm Hg
Freezing point	-87.8°C	Viscosity at 20°C	2.4 cps.
Heat of combustion	7970 cal/g	Water content	0.5% by wt, max.
	7942 cal/g	Weight per gallon at 20°C	6.55 lb
		60°F	6.58 lb

Table 6.38: Physical Properties of 91% Isopropyl Alcohol (31)

Acidity, as acetic acid	0.0024% by wt, max.
Color, Pt-Co scale	15 max.
Distillation at 760 mm	Distills entirely within a 1.0°C range which includes 80.4°C
Fire hazard	Dangerous when exposed to heat or open flame
Flash point (open cup)	75°F
Non-volatile matter	0.005 g/100 ml, max.
Odor	Non-residual
Permanganate time	30 minutes, minimum, at 15°C when using the Barbet end point
Purity	91.09% by volume, minimum, at 15.56°C.
Specific gravity at 20/20°C	0.8175 - 0.8185
Toxicity	Moderate by ingestion; otherwise slight
Water content	9% by volume, max.
Weight per gallon at 20°C	6.81 lb
60°F	6.84 lb

Table 6.39: Specific Gravity of Isopropyl Alcohol-Water Mixtures (9)

Specific Gravity	20/20°C % Vol.	% Wt.	Specific Gravity	20/20°C % Vol.	% Wt.	Specific Gravity	20/20°C % Vol.	% Wt.
1.000	0.0	0.0	0.9240	50.7	43.3	0.8490	80.60	74.95
0.9990	0.8	0.6	0.9230	51.2	43.7	0.8480	80.96	75.37
0.9980	1.6	1.3	0.9220	51.6	44.2	0.8470	81.32	75.79
0.9970	2.4	1.9	0.9210	52.0	44.6	0.8460	81.68	76.21
0.9960	3.2	2.6	0.9200	52.5	45.0	0.8450	82.04	76.63
0.9950	4.0	3.3	0.9190	52.9	45.5	0.8440	82.40	77.04
0.9940	4.8	3.9	0.9180	53.4	45.9	0.8430	82.76	77.45
0.9930	5.6	4.5	0.9170	53.8	46.3	0.8420	83.12	77.86
0.9920	6.5	5.2	0.9160	54.2	46.7	0.8410	83.48	78.27
0.9910	7.3	5.8	0.9150	54.7	47.2	0.8400	83.84	78.68
0.9900	8.1	6.5	0.9140	55.1	47.6	0.8390	84.20	79.09
0.9890	8.9	7.1	0.9130	55.5	48.0	0.8380	84.55	79.50
0.9880	9.8	7.8	0.9120	56.0	48.5	0.8370	84.90	79.91
0.9870	10.6	8.4	0.9110	56.4	48.9	0.8360	85.25	80.32
0.9860	11.5	9.1	0.9100	56.4	48.9	0.8350	85.60	80.73
0.9850	12.3	9.8	0.9090	57.3	49.7	0.8340	85.95	81.14
0.9840	13.2	10.5	0.9080	57.7	50.2	0.8330	86.30	81.55
0.9830	14.0	11.2	0.9070	58.1	50.6	0.8320	86.65	81.96
0.9820	14.9	11.9	0.9060	58.6	51.0	0.8310	87.00	82.37
0.9810	15.7	12.6	0.9050	59.0	51.4	0.8300	87.33	82.78
0.9800	16.6	13.3	0.9040	59.4	51.8	0.8290	87.69	83.19
0.9790	17.4	14.1	0.9030	59.8	52.3	0.8280	88.03	83.60
0.9780	18.3	14.8	0.9020	60.3	52.7	0.8270	88.36	84.01
0.9770	19.1	15.5	0.9010	60.7	53.1	0.8260	88.69	84.42
0.9760	19.9	16.2	0.9000	61.1	53.5	0.8250	89.02	84.83
0.9750	20.8	16.9	0.8990	61.5	53.9	0.8240	89.35	85.24
0.9740	21.7	17.5	0.8980	62.0	54.4	0.8230	89.68	85.65
0.9730	22.5	18.2	0.8970	62.4	54.8	0.8220	90.01	86.06
0.9720	23.4	18.8	0.8960	62.8	55.2	0.8210	90.34	86.47
0.9710	24.2	19.4	0.8950	63.2	55.6	0.8200	90.67	86.88
0.9700	25.1	20.1	0.8940	63.6	56.0	0.8190	91.00	87.29
0.9690	25.8	20.7	0.8930	64.1	56.5	0.8180	91.32	87.70
0.9680	26.6	21.3	0.8920	64.5	56.9	0.8170	91.63	88.10
0.9670	27.3	22.0	0.8910	64.9	57.3	0.8160	91.93	88.50
0.9660	28.0	22.6	0.8900	65.3	57.7	0.8150	92.23	88.90
0.9650	28.7	23.2	0.8890	65.7	58.1	0.8140	92.53	89.30
0.9640	29.4	23.8	0.8880	66.1	58.6	0.8130	92.83	89.70
0.9630	30.1	24.4	0.8870	66.5	59.0	0.8120	93.13	90.10
0.9620	30.8	25.0	0.8860	66.9	59.4	0.8110	93.43	90.50
0.9610	31.4	25.6	0.8850	67.3	59.8	0.8100	93.72	90.90
0.9600	32.1	26.2	0.8840	67.7	60.2	0.8090	94.01	91.30
0.9590	32.7	26.7	0.8830	68.0	60.7	0.8080	94.30	91.70
0.9580	33.3	27.2	0.8820	68.4	61.1	0.8070	94.58	92.10
0.9570	33.9	27.7	0.8810	68.8	61.5	0.8060	94.86	92.49
0.9560	34.5	28.2	0.8800	69.2	61.9	0.8050	95.14	92.88
0.9550	35.1	28.7	0.8790	69.6	62.3	0.8040	95.42	93.27
0.9540	35.7	29.2	0.8780	69.9	62.8	0.8030	95.69	93.66
0.9530	36.3	29.7	0.8770	70.3	63.2	0.8020	95.96	94.04
0.9520	36.8	30.3	0.8760	70.7	63.6	0.8010	96.23	94.42
0.9510	37.4	30.8	0.8750	71.1	64.0	0.8000	96.50	94.80
0.9500	38.0	31.3	0.8740	71.4	64.4	0.7990	96.77	95.18
0.9490	38.5	31.8	0.8730	71.8	64.9	0.7980	97.04	95.56
0.9480	39.0	32.3	0.8720	72.2	65.3	0.7970	97.31	95.94
0.9470	39.6	32.8	0.8710	72.6	65.7	0.7960	97.57	96.32
0.9460	40.1	33.3	0.8700	72.9	66.1	0.7950	97.83	96.70
0.9450	40.6	33.8	0.8690	73.3	66.5	0.7940	98.08	97.08
0.9440	41.1	34.3	0.8680	73.7	67.0	0.7930	98.33	97.46
0.9430	41.6	34.8	0.8670	74.0	67.4	0.7920	98.58	97.84
0.9420	42.1	35.2	0.8660	74.4	67.8	0.7910	98.83	98.22
0.9410	42.7	35.7	0.8650	74.8	68.2	0.7900	99.08	98.60
0.9400	43.2	36.1	0.8640	75.2	68.6	0.7890	99.33	98.98
0.9390	43.7	36.6	0.8630	75.5	69.1	0.7880	99.58	99.36
0.9380	44.2	37.0	0.8620	75.9	69.5	0.7870	99.83	99.74
0.9370	44.7	37.5	0.8610	76.3	69.9	0.7863	100.00	100.00
0.9360	45.2	38.0	0.8600	76.6	70.3			
0.9350	45.6	38.4	0.8590	77.00	70.75			
0.9340	46.1	38.8	0.8580	77.36	71.17			
0.9330	46.6	39.3	0.8570	77.72	71.59			
0.9320	47.1	39.7	0.8560	78.08	72.01			
0.9310	47.5	40.2	0.8550	78.44	72.43			
0.9300	48.0	40.6	0.8540	78.80	72.85			
0.9290	48.5	41.1	0.8530	79.16	73.27			
0.9280	48.9	41.5	0.8520	79.52	73.69			
0.9270	49.4	42.0	0.8510	79.88	74.11			
0.9260	49.8	42.4	0.8500	80.24	74.54			
0.9250	50.3	42.9						

Gravity-temperature coefficient of
100% alcohol 20-22 - .00086/°C

Table 6.40: Vapor-Liquid Compositions of Isopropyl Alcohol-Water Mixtures and Their Boiling Points *(9)*

Table 6.41: Refractive Index vs Composition of Isopropyl Alcohol-Water Mixtures at 25°C *(19)*

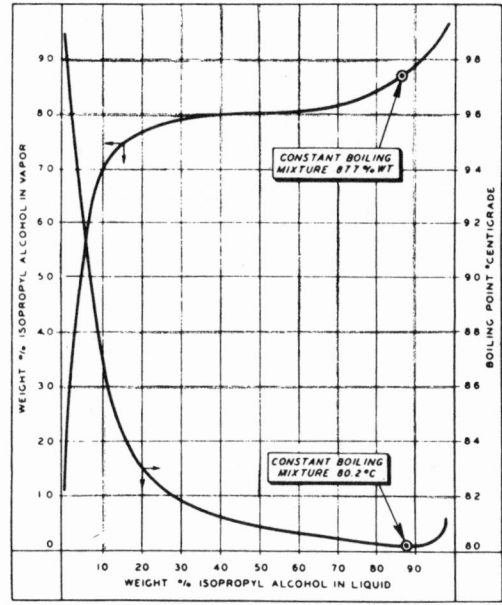

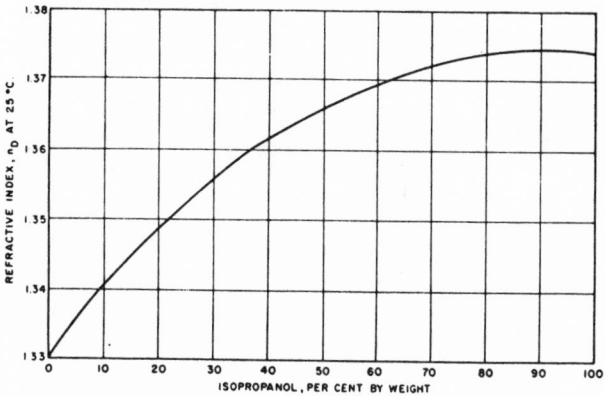

Table 6.42: Isopropyl Alcohol-Water: Kinematic Viscosity vs Composition at 25°C *(19)*

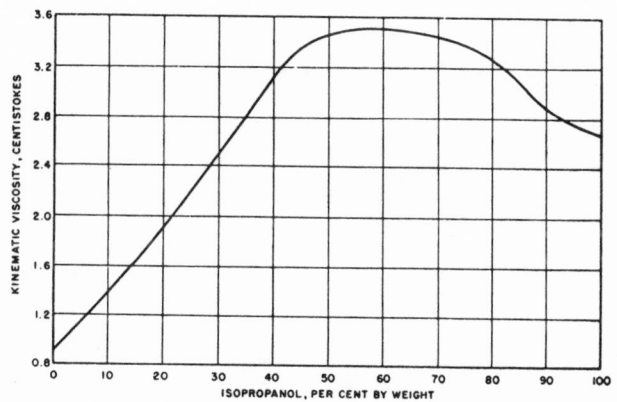

Table 6.43: Azeotropes of Isopropyl Alcohol *(31)*

ISOPROPYL ALCOHOL FORMS BINARY AZEOTROPES WITH:

%		B.P. of Azeotrope °C
56	Acrylonitrile	71.7
66.7	Benzene	71.9
66	2-Bromobutane	77.5
68	2-Butanone	77.9
40	n-Butylamine	84.7
28.1	Butyl isopropyl ether	79.0
92	Carbon disulfide	44.6
82	Carbon tetrachloride	67.0
77	1-Chlorobutane	70.8
82	2-Chlorobutane	64.0
57	1-Chloro-3-methylbutane	79.2
64	1,3-Cyclohexadiene	70.4
67	Cyclohexane	68.6
73	Cyclohexene	70.5
48	Diethoxymethane	79.6
45.5	Diisobutylene	77.8
91	2,3-Dimethylbutane	53.8
22	1,3-Dimethylcyclohexane	81.0
38	2,5-Dimethylhexane	79.0
74	Ethyl acetate	74.0
90	Ethyl propyl ether	62.0
48	Ethyl sulfide	78.0
60.8	Ethylene dichloride	72.7
70	Fluorobenzene	74.5
49.5	n-Heptane	76.4
77	Hexane	62.7
30	1-Iodo-2-methylpropane	81.5
81	Isobutyl chloride	63.8
47.7	Isopropyl acetate	80.1
83.7	Isopropyl ether	66.2
53.5	Methyl acrylate	76.0
47	Methylcyclohexane	77.6
75	Methylcyclopentane	63.3
70	Methyl ethyl ketone	77.3
35	Methyl isobutyrate	81.4
62	Methyl propionate	76.4
16	Octane	81.8
94	Pentane	35.5
48	Propyl ether	78.2
64	Propyl formate	76.9
19	Tetrachloroethylene	81.7
57	Thiophene	76.0
31	Toluene	80.6
72	Trichloroethylene	74.0
77.6	Vinyl acetate	70.8
12.2	Water	79.5

ISOPROPYL ALCOHOL FORMS TERNARY AZEOTROPES WITH:

%		B.P. of Azeotrope °C
7.5	Water	66.5
73.8	Benzene	
7.5	Water	64.8
74	Cyclohexane	
7.5	Water	61.1
71	Cyclohexene	
10	Water	72.3
9.3	Diisobutylene	
10.4	Water	73.4
67.7	Ethyl butyl ether	
7.7	Water	69.7
73.3	Ethylene dichloride	
11	Water	75.5
76	Isopropyl acetate	
4.7	Water	61.6
88	Isopropyl ether	
6	Water	78.0
32	Nitromethane	
13.1	Water	76.3
48.7	Toluene	

Table 6.44: The Effect of Isopropyl Alcohol on the Dilution Ratio of Solvents (14)

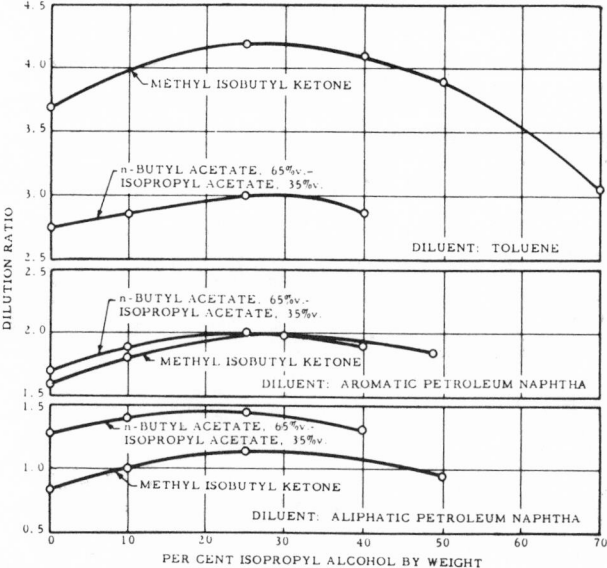

Table 6.45: Viscosity of RS ½ Sec. Nitrocellulose in Mixtures of Toluene, Isopropyl Alcohol and Methyl Isobutyl Ketone (14)

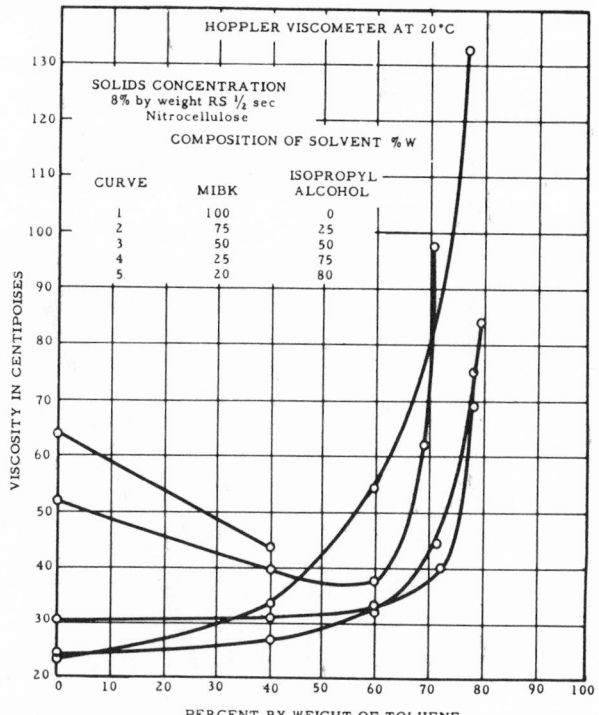

Table 6.46: Methanol-Isopropyl Alcohol: Boiling Point vs Composition at 760 mm Hg (19)

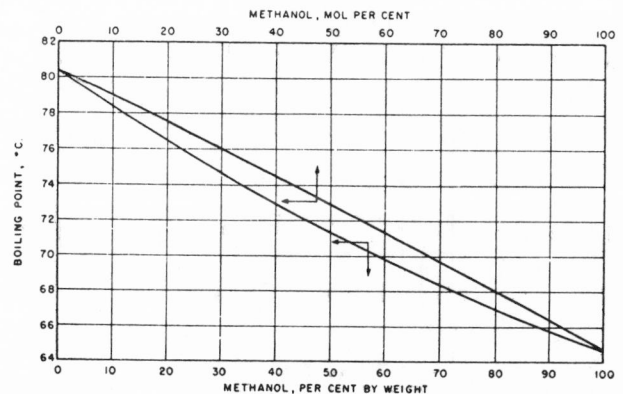

Table 6.47: Methanol-Isopropyl Alcohol: Liquid-Vapor Equilibria at Atmospheric Pressure (19)

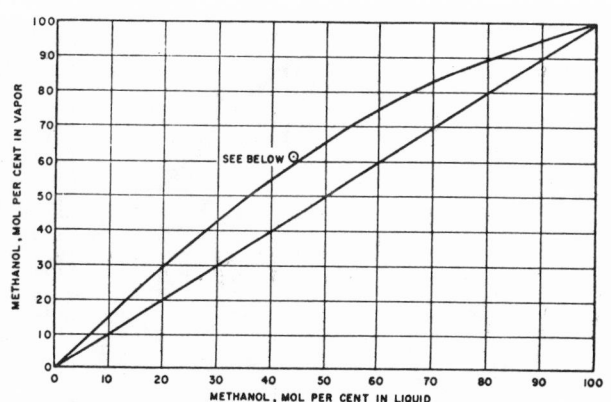

Composition at 200 mm. Hg: 43.9 mol per cent in liquid, 60.9 mol per cent in vapor. Isopropanol-Water azeotrope contains 68.3 mol per cent isopropanol.

Table 6.49: Specific Gravities of Alcohols vs Temperature (8)

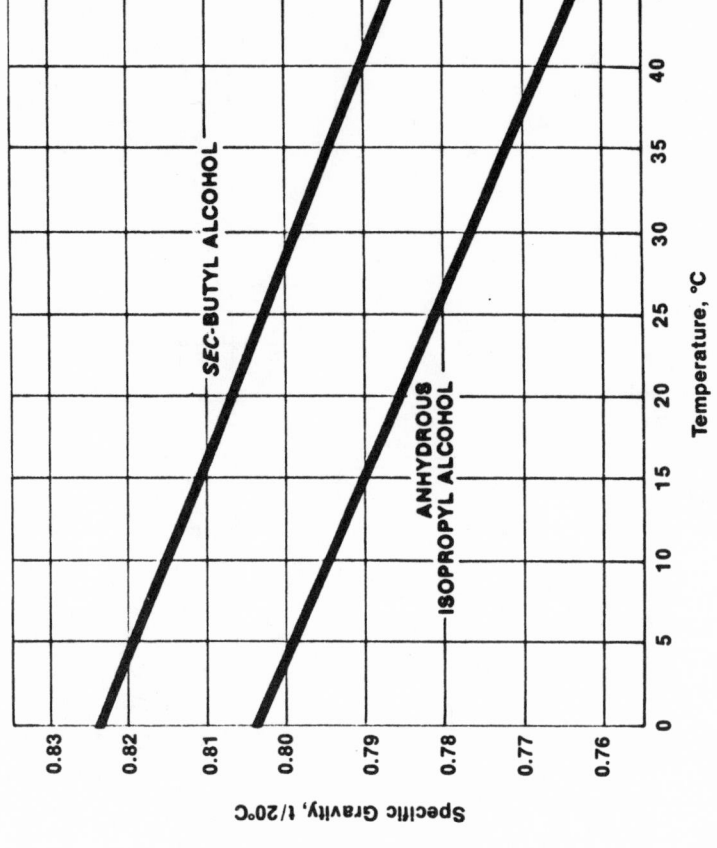

Table 6.48: Vapor Pressure of Isopropyl Alcohol (Anhydrous) and sec-Butyl Alcohol at Various Temperatures (8)

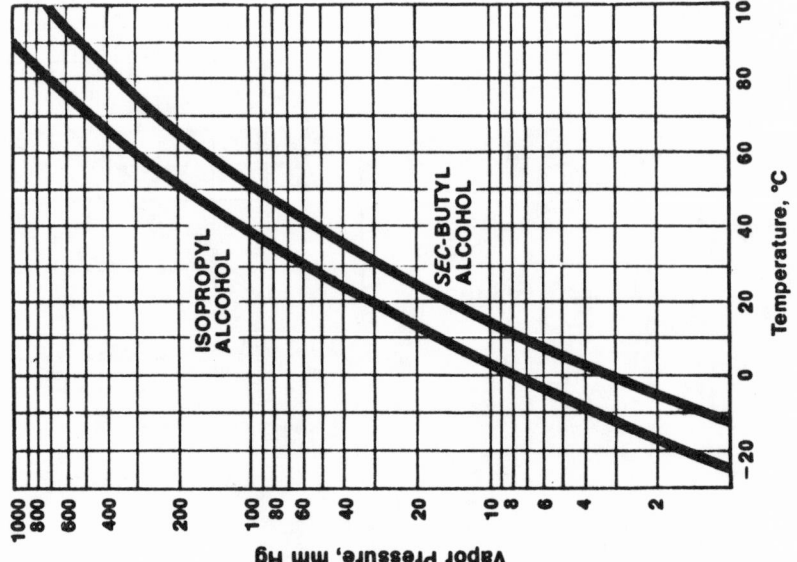

n-BUTYL ALCOHOL

n–Butanol, Butanol–1, Butyric Alcohol $CH_3CH_2CH_2CH_2OH$

Table 6.50: Physical Properties of n-Butyl Alcohol *(31)*

PHYSICAL PROPERTIES OF n-BUTYL ALCOHOL

Acidity as acetic acid	0.005% by wt, max.
Aldehydes	None
Boiling point at 760 mm	117.7°C
Chlorides	None
Coefficient of cubical expansion	
per °C	0.00093
per °F	0.00052
Color, Pt-Co	10 max.
Critical pressure	48.4 atm
Critical temperature	287°C
Dielectric constant at 25°C	16.1
Distillation range (including 117.7°C)	1.5°C max.
Electrical conductivity at 25°C	9.12×10^{-9} reciprocal ohms
Explosive limits in air, Lower	1.45% by vol.
Upper	11.25% by vol.
Fire hazard	Moderate
Flash point, Tag open cup	115°F
Freezing point	-89.0°C
Heat of combustion	8626 cal/g
Heat of fusion	29.9 cal/g
Heat of vaporization at boiling point	141.3 cal/g
Ignition temperature	367°C
Iron	None
MAC	100 ppm in air
Melting point	-89.8°C
Molecular weight	74.12 calculated
Non-volatile matter	0.005 g/100 ml, max.
Odor	Characteristic, non-residual
Refractive index at 20°C, n_D	1.3992
Relative evaporation rate, n-butyl acetate = 1	0.45
Solubility in water at 20°C	7.8% by wt
Solubility of water in n-Butanol at 20°C	20.1% by wt
Specific gravity at 20/20°C	0.8109
Specific heat of liquid at 20°C	0.563 cal/g
Sulfuric acid test (Pt-Co)	25 max.
Surface tension at 20°C	24.6 dynes/cm
Suspended matter	Substantially free
Toxicity	Moderately toxic by inhalation, ingestion and skin absorption
Vapor pressure at 20°C	4.39 mm Hg
40°C	18.6 mm Hg
60°C	59.2 mm Hg
75°C	131.3 mm Hg
100.8°C	400.0 mm Hg
Viscosity at 20°C	2.948 centipoises
Water content	0.10% by wt, max.
Weight per gallon at 20°C	6.756 lbs

Table 6.52: Solubility of Water in Butyl Alcohol at Various Temperatures (19)

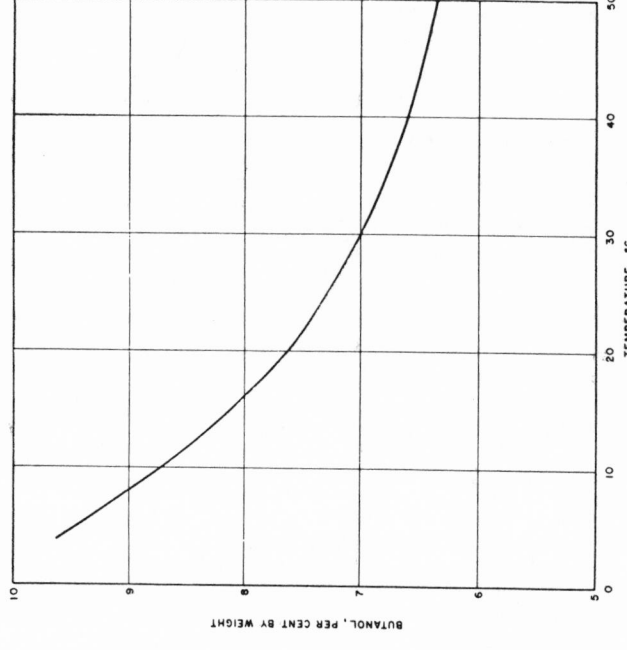

Table 6.51: Vapor Pressure of Butyl Alcohol at Various Temperatures (19)

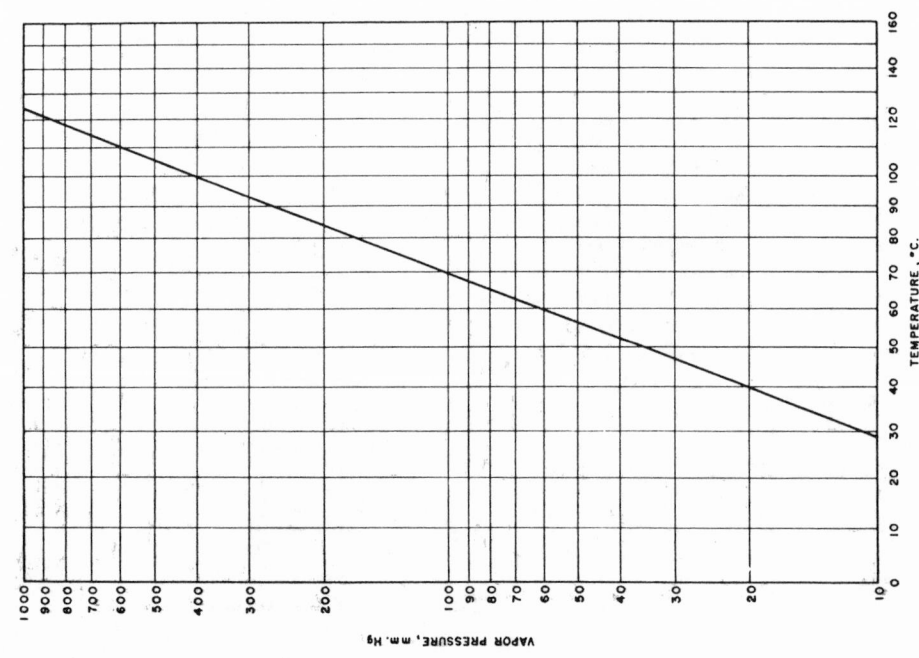

Table 6.53: Solubility of Butyl Alcohol in Water at Various Temperatures *(19)*

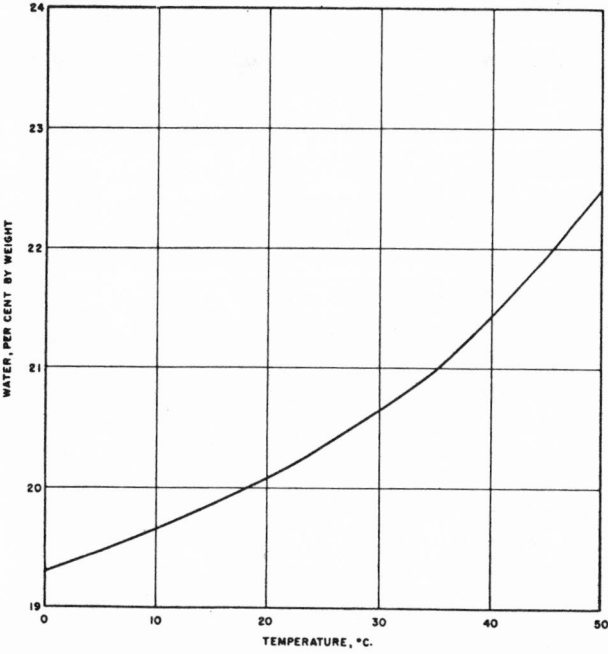

Table 6.54: Azeotropes of n-Butyl Alcohol *(31)*

n-BUTYL ALCOHOL FORMS BINARY AZEOTROPES WITH:

%		B.P. of Azeotrope °C	%		B.P. of Azeotrope °C
87	Acetal	101.0	60	Methyl isovalerate	116.3
68.5	1-Bromo-3-methylbutane	110.7	70	4-Methyl-2-pentanone	114.4
32.8	Butyl acetate	117.6	50	Octane	110.2
12	Butyl ether	117.3	48	Paraldehyde	115.8
76.4	Butyl formate	105.8	12	α-Pinene	117.4
92.2	Butyl vinyl ether	93.3	29	Pyridine	118.7
2	Camphene	117.8	21	Styrene	116.5
44	Chlorobenzene	115.3	68	Tetrachloroethylene	110.0
88	1-Chloro-3-methylbutane	97.0	73	Toluene	105.6
90	Cyclohexane	79.8	25	o-Xylene	116.8
95	Cyclohexene	82.0	32	p-Xylene	115.7
17.5	Dibutyl ether	117.6			
57	1,3-Dimethylcyclohexane	108.5			
72	2,5-Dimethylhexane	101.9			
48	Ethyl borate	113.0			
36	Ethyl butyrate	115.7			
37	Ethyl carbonate	116.5			
83	Ethyl isobutyrate	109.2			

n-BUTYL ALCOHOL FORMS TERNARY AZEOTROPES WITH:

82	Heptane	93.3	%		B.P. of Azeotrope °C
97	Hexane	67.0	37.3	Water	
18.2	2-Hexanone	116.5	35.3	Butyl acetate	89.4
20	3-Hexanone	117.2			
22	1-Iodo-3-methylbutane	117.3	41.8	Water	
31	Isoamyl formate	115.9	7.9	Butyl chloroacetate	93.1
50	Isobutyl acetate	114.5			
52	Isobutyl ether	113.5	29.3	Water	
46	Isopropyl isobutyrate	115.5	27.7	Butyl ether	91.0
55	Isopropyl sulfide	112.0			
86	Methylcyclohexane	95.3	21.3	Water	
92	Methylcyclopentane	71.8	68.7	Butyl formate	83.6
			3.1	Water	
			85.0	Carbon tetrachloride	64.7

ISOBUTYL ALCOHOL

Isobutanol, 2-Methyl Propanol-1, Isopropyl Carbinol $(CH_3)_2CHCH_2OH$

Table 6.55: Physical Properties of Isobutyl Alcohol (31)

Alkalinity	0.003% by wt, max.
Boiling point at 760 mm	107.9°C
Coefficient of cubical expansion at 10 to 30°C	0.95×10^{-3}
Color, APHA	10 max.
Critical pressure	48 atm
Critical temperature	265°C
Distillation range (including 107.9°C)	2°C max.
Electrical conductivity at 25°C	8×10^{-3} mho per cm.
Evaporation rate (n-Butyl Acetate = 1.0)	0.8
Explosive limits in air, lower limit	1.68% by volume
Fire hazard	Moderate
Flash point, Tag open cup	103°F
Heat of combustion	6382 cal/g
Heat of vaporization at boiling point	138 cal/g/mole
Ignition temperature	440°C
Melting point	-108°C
Molecular weight	74.12 calculated
Non-volatile matter	0.001 g/100 ml, max.
Refractive index at 20°C, n_D	1.3959
Solubility in water at 25°C	8.8 ml per 100 ml
Solubility of water in isobutyl alcohol at 25°C	20.0 ml per 100 ml
Specific gravity at 20/20°C	0.8034
Specific heat at 15°C	0.716 cal/g/°C
Surface tension at 20°C	22.8 dynes/cm
Toxicity	Highly toxic by inhalation or ingestion
Vapor density (Air = 1.0)	2.55
Vapor pressure at 20°C	8.8 mm
Viscosity at 20°C	6.68 centipoises
Water content	0.2% by wt, max.
Weight per gallon at 20°C	6.68 lbs

Table 6.56: Azeotropes of Isobutyl Alcohol (31)

ISOBUTYL ALCOHOL FORMS BINARY AZEOTROPES WITH:

%		B.P. of Azeotrope °C
90.7	Benzene	79.9
36.4	1-Bromo-3-methylbutane	103.4
60	Butyl formate	103.0
37	Chlorobenzene	107.1
78	1-Chloro-3-methylbutane	94.5
88	1,3-Cyclohexadiene	79.4
86	Cyclohexane	78.1
85.8	Cyclohexene	80.5
44	1,3-Dimethylcyclohexane	102.2
48	Ethyl isobutyrate	105.5
87	Ethyl propionate	98.9
91	Fluorobenzene	84.0
73	Heptane	90.8
97.5	Hexane	68.3
45	Isobutyl acetate	107.4
79.4	Isobutyl formate	97.8
93.8	Isobutyl vinyl ether	82.7
27	Isopropyl sulfide	105.8
75	Methyl butyrate	101.3
68	Methylcyclohexane	92.6
95	Methylcyclopentane	71.0
10	Methyl isovalerate	107.5
9	4-Methyl-2-pentanone	107.9
81	2-Pentanone	101.8
80	3-Pentanone	101.7
68	Pinacolone	105.5
83	Propyl acetate	101.0
90	Propyl ether	89.5
55	Toluene	101.2
73	2,2,4-Trimethylpentane	92.0

ISOBUTYL ALCOHOL FORMS TERNARY AZEOTROPES WITH:

30.4	Water	86.8
46.5	Isobutyl acetate	
33.6	Water	90.2
13.3	Isobutyl chloroacetate	
17.3	Water	80.2
76	Isobutyl formate	

Table 6.57: Relative Evaporation Rates of Various Butyl Alcohols and Acetates *(41)*

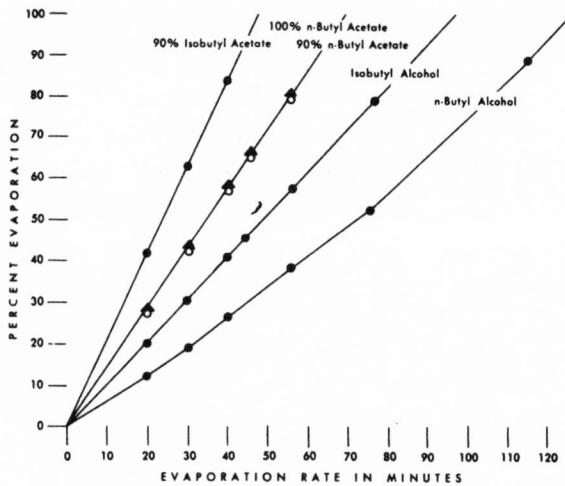

sec-BUTYL ALCOHOL

sec–Butanol, Butanol–2, Methyl Ethyl Carbinol $CH_3CHOHCH_2CH_3$

Table 6.58: Physical Properties of sec-Butyl Alcohol *(31)*

Acidity as acetic acid	0.003% by wt, max.
Boiling point at 760 mm	99.5°C
Coefficient of cubical expansion at 20°C	0.00101°C
Color, Pt-Co (Hazen)	10 max.
Critical pressure	46.9 atm
Critical temperature	265.19°C
Distillation range	98.0-101.0°C
Fire hazard	Dangerous when exposed to heat or flame
Flash Point, Tag open cup	80°F
Tag closed cup	75°F
Freezing point	-114.7°C
Heat of vaporization at 1 atm.	134.4 g cal/g
Molecular weight	74.12
Non-volatile matter	0.002 ml, max.
Purity	99.0% min.
Refractive index at 20°C n_D	1.39719
Relative evaporation rate, n-Butyl acetate = 100	120
Solubility in water at 20°C	22.5% by wt
Solubility of water in, at 20°C	60.0% by wt
Specific gravity at 20/20°C	0.8079
Specific heat at 8.5°C	0.596
Surface tension at 20°C	23.0 dynes/cm
Toxicity	Moderate
Vapor pressure at 20°C	12.1 mm
Viscosity at 20°C	3.78 cps.
Water, presence of	Miscible without turbidity with 19 vol. of n-heptane at 20°C
Weight per gallon at 20°C	6.73 lb

Table 6.59: Azeotropes of sec-Butyl Alcohol *(31)*

sec-BUTYL ALCOHOL FORMS BINARY AZEOTROPES WITH:

%		B.P. of Azeotrope °C
61	tert-Amyl ethyl ether	94.5
93	tert-Amyl methyl ether	86.0
84.6	Benzene	78.6
13.7	sec-Butyl acetate	99.6
32	Butyl formate	98.0
71	1-Chloro-3-methylbutane	91.5
79	Cyclohexene	78.7
46	2, 5-Dimethylhexene	93.0
53	Ethyl propionate	95.7
68	Ethyl sulfide	89.0
62	Heptane	89.0
92	Hexane	67.2
60	Isobutyl formate	94.7
41	Methyl butyrate	97.7
59	Methylcyclohexane	89.9
88.5	Methylcyclopentane	69.7
77	Methyl isobutyrate	92.0
42	3-Pentanone	98.0
16	Pinacolone	99.1
48	Propyl acetate	96.5
78	Propyl ether	87.0
45	Toluene	95.3

tert-BUTYL ALCOHOL

tert-Butanol, 2-Methyl Propanol-2 $(CH_3)_3COH$

Table 6.60: Physical Properties of tert-Butyl Alcohol (31)

Acidity as acetic acid	0.003% by wt, max.
Boiling point at 760 mm	82.36°C
Coefficient of cubical expansion at 26°C	0.00132°C
Color, Pt-Co (Hazen) max.	10 max.
Compressibility at 20°C, between 100-500 megabars	79.6×10^{-6} megadynes/cm
Critical pressure	46 atm
Critical temperature	234.9°C
Dielectric constant at 19°C (audio)	11.4 cgs units
Distillation range	81.5-83.0°C
Dipole moment	1.65×10^{18}
Fire hazard	Dangerous when exposed to heat or flame
Flash point, Tag open cup	60°F (approx.)
Tag closed cup	48°F (approx.)
Freezing point	25.57°C
Heat of combustion	
Liquid at constant volume	6290 cal/g
constant pressure	6302 cal/g
Vapor at constant pressure	6426 cal/g
Heat of fusion at 25.5°C	21.88 cal/g
Heat of solution at 15°C, of the solid alcohol in water	3.23 kg cal
Heat of vaporization at 1 atm.	130.6 g cal/g
Melting point	25.57°C
Molecular weight	74.12
Molecular volume, $20/V_M$	94.3 cc
Non-volatile matter	0.002 g/100 ml, max.
Purity	99.0% by wt, min.
Refractive index at 20°C, n_D	1.3841
Solubility at 20°C, in water	Complete
water in	Complete
Specific heat at 26°C	0.726 g cal/g
Specific gravity at 26/4°C	0.7793
Surface tension at 20°C	20.7 dynes
34.5°C	19.45 dynes
80°C	14.6 dynes
Toxicity	Moderate
Vapor pressure at 30°C	57.3 mm
Viscosity at 30°C	3.316 cps.
Water	Miscible without turbidity with 19 vol. of n-heptane at 20°C
Weight per gallon at 26°C	6.50 lb

Table 6.61: Azeotropes of tert-Butyl Alcohol (31)

tert-BUTYL ALCOHOL FORMS BINARY AZEOTROPES WITH:

%		B.P. of Azeotrope °C
63.4	Benzene	74.0
94	Carbon disulfide	45.7
76	Carbon tetrachloride	29.5
41	1-Chloro-3-methylbutane	81.2
61.5	1,3-Cyclohexadiene	73.4
63	Cyclohexane	71.3
60	Cyclohexene	73.2
93	Cyclopentane	48.2
65	Dibromodichloromethane	79.0
94	1,1-Dichloroethane	57.1
33	Diisobutyl alcohol	81.5
87	2,3-Dimethylbutane	55.3
10	1,3-Dimethylcyclohexane	82.2
23	2,5-Dimethylhexane	81.5
75	Ethyl acetate	76.0
38	Ethyl nitrate	78.0
30	Ethyl sulfide	79.8
69	Fluorobenzene	76.0
38	Heptane	78.0
78	Hexane	63.7
83	Isobutyl chloride	65.5
34	Methylcyclohexane	78.8
74	Methylcyclopentane	66.6
70	Methylcyclopentene	69.5

PRIMARY AMYL ALCOHOL

Primary amyl alcohol, a mixture of isomers all of which are primary alcohols, is composed of approximately 60% pentanol-1 ($CH_3CH_2CH_2CH_2CH_2OH$); 35% 2-methyl butanol-1 ($CH_3CH_2CH(CH_3)CH_2OH$); and 5% 3-methyl butanol-1 ($CH_3CH(CH_3)CH_2CH_2OH$).

Table 6.62: Physical Properties of Primary Amyl Alcohol (19)

Acidity as acetic acid	0.01% by wt, max.
Boiling point at 760 mm	133.1°C
50 mm	68°C
10 mm	39°C
Carbonyl, as C_5 aldehyde	0.20% by wt, max.
Coefficient of expansion at 20°C	0.00092 per °C
Color, Pt-Co	15, max.
Distillation at 760 mm Ibp	127.5°C
Dp	139.0°C, max.
Fire hazard	Moderate
Flash point (open cup)	118°F
Freezing point	Sets to glass below -90°C
Heat of vaporization at 133°C	242 Btu/lb
Purity, as primary amyl alcohols	98.0% by wt, min.
Refractive index at 20°C, n_D	1.4084
Solubility in water at 20°C	1.7% by wt
Solubility of water in, at 20°C	9.2% by wt
Specific gravity at 20/20°C	0.8134

PRIMARY n-AMYL ALCOHOL

Table 6.63: Physical Properties of Primary n-Amyl Alcohol (31)

Acidity (mg KOH/g)	0.06 max.
Boiling point	137.8°C
Clarity	No turbidity or suspended matter
Coefficient of expansion per °C	0.00092
Distillation, initial	Not below 134.8°C
final	Not above 140.0°C
Fire point	140°F
Fire hazard	Moderate
Flash point (open cup)	135°F
Heat of vaporization	120.6 cal/g (calculated)
Melting point	-78.5°C
Molecular weight	88.15 (calculated)
Non-volatile matter at 100°C	5.0 mg/100 ml, max.
Refractive index at 20°C	1.4099
Specific gravity at 20/20°C	0.82
Specific heat	0.712 cal/g
Toxicity	Highly toxic by inhalation and ingestion
Viscosity at 25°C	3.31 centipoises
60°C	1.33 centipoises
Weight per gallon	6.82 lbs

sec-AMYL ALCOHOL

Table 6.64: Physical Properties of sec-Amyl Alcohol *(31)*

Acidity as acetic acid	0.02% max.
Boiling point	119.3°C
Coefficient of expansion	
per 1°F	0.00053
1°C	0.00095
Distillation range	105° - 125°C
Evaporation rate at 90°F	
in minutes: 5%	2.25
25%	11.75
50%	24.25
75%	38.25
90%	50.25
95%	56.50
Flash point, Open cup	105°F
Fire hazard	Moderate
Heat of vaporization	97.8 cal/g (calculated)
Non-volatile at 100°C	0.003 g/100 cc max.
Purity	99% by wt, min.
Refractive index at 25°C, n_D	1.4041
Solubility of water in	8.2% by vol.
Specific gravity at 20°C	0.811
Toxicity	Highly toxic by ingestion and inhalation
Weight per gallon at 20°C	6.75 lbs

Table 6.65: Azeotropes of sec-Amyl Alcohol *(31)*

sec-AMYL ALCOHOL FORMS BINARY AZEOTROPES WITH:

%		B.P. of Azeotrope °C
45	Chlorobenzene	118.2
62	1,3-Dimethylcyclohexane	113.0
33	Ethylbenzene	118.0
53	Ethyl butyrate	118.5
85	Heptane	96.0
68	Isobutyl acetate	116.5
59	Isobutyl ether	115.0
82	Methylcyclohexane	98.6
80	Methyl isovalerate	115.8
44	Octane	114.8
72	Toluene	107.0
30	m-Xylene	118.3

sec-n-AMYL ALCOHOL

Table 6.66: Physical Properties of sec-n-Amyl Alcohol *(31)*

Acidity as acetic acid	0.06% max.
Boiling point	115.6°C
Coefficient of expansion	
per °C	0.00149
Distillation, 95%	Between 113.6 - 117.6°C
Fire hazard	Moderate
Flash point	100°F
Freezing point	Less than -75°C
Heat of vaporization	96.8 cal/g (calculated)
Non-volatile at 100°C	0.003 g/100 cc max.
Refractive index at 20°C	1.4098
Specific gravity at 20°C	0.82
Toxicity	Highly toxic by inhalation and ingestion
Viscosity at 25°C	4.12 centipoises
60°C	1.09 centipoises
Weight per gallon at 20°C	6.81 lbs

Table 6.67: Azeotropes of sec-n-Amyl Alcohol *(31)*

sec-n-AMYL ALCOHOL FORMS BINARY AZEOTROPES WITH:

%		B.P. of Azeotrope °C
97	Cyclohexane	80.0
80	Heptane	96.0
77	Methylcyclohexane	97.4
65	4-Methyl-2-pentanone	115.0
65	Toluene	106.0

tert-AMYL ALCOHOL, REFINED

2-Methyl Butanol-2, Dimethylethyl Carbinol, Amylene Hydrate, tert-Pentanol $(CH_3)_2COHCH_2CH_3$

Table 6.68: Physical Properties of Refined tert-Amyl Alcohol (31)

Acidity as acetic acid	0.15% max.
Boiling point	101.8°C
Clarity	No turbidity or sus-pended matter
Coefficient of expansion per °C	0.00133 (calculated)
Distillation, 95% between	98.8 - 103.8°C
Fire hazard	Dangerous when exposed to heat or flame
Flash point, Open cup	70°F
Freezing point	-11.9°C
Heat of vaporization	93.4 cal/g
Molecular weight	88.15 (calculated)
Neutralization value, mg KOH/g	0.06 max.
Non-volatile matter	0.003 g/100 cc, max.
Odor	Camphor-like
Refractive index at 20°C	1.4052
Specific gravity at 20/20°C	0.81 - 0.82
Specific heat	0.753 cal/g
Toxicity	Moderate
Viscosity at 25°C at 63°C	3.70 centipoises 0.99 centipoises
Water content	None
Water tolerance, water per 100 cc alcohol	18.0 min.
Weight per gallon	6.75 lbs

Table 6.69: Azeotropes of tert-Amyl Alcohol (31)

tert-AMYL ALCOHOL FORMS BINARY AZEOTROPES WITH:

%		B.P. of Azeotrope °C
85	Benzene	80.0
85	1,3-Cyclohexadiene	79.7
84	Cyclohexane	78.5
83	Cyclohexene	80.8
32	1,3-Dimethylcyclohexane	101.1
50	2,5-Dimethylhexane	97.0
73.5	Heptane	92.2
96	Hexane	68.3
60	Methylcyclohexane	92.0
95	Methylcyclopentane	71.5
25	Octane	101.1
80	Propyl ether	88.8
44	Toluene	100.5

ISOAMYL ALCOHOL

3-Methyl-1-Butanol, Primary Isoamyl Alcohol, Isobutyl Carbinol $(CH_3)_2CHCH_2CH_2OH$

Table 6.70: Physical Properties of Isoamyl Alcohol (31)

Acidity as acetic acid	0.01% max.
Boiling point at 760 mm	131.4°C
Coefficient of expansion	
per °C	0.00090
per °F	0.00050
Color, APHA	No. 10 max.
Critical temperature	307°C
Distillation range, below 128°C	None
above 132°C	None
Dryness	A 5 ml. sample is clearly miscible with at least 19 parts of 60 Bé gasoline at 60°F
Esters	Not more than 0.060% as amyl acetate
Fire hazard	Moderate
Flash point, Open cup	125°F
Heat of combustion	794.5 gram calories per gram
Latent heat of vaporization	105.4 gram calories per gram
MAC	100 ppm in air
Melting point	-117.2°C
Molecular weight	88.15
Non-volatile matter	0.003% max.
Odor	Alcoholic, non-residual
Refractive index at 20°C	1.4014
Solubility in water at 14°C	2.0% by wt
Specific gravity at 20/20°C	0.810 - 0.813
Specific heat at 20°C	0.544 gram calories per gram per °C
Surface tension at 20°C	23.8 dynes per cm
Toxicity	Highly toxic by ingestion and inhalation
Vapor pressure at 20°C	2.8 mm Hg
Viscosity (absolute) at 23.8°C	3.86 centipoises
Weight per gallon at 20°C	6.76 lbs approx.

Table 6.71: Azeotropes of Isoamyl Alcohol (31)

ISOAMYL ALCOHOL FORMS BINARY AZEOTROPES WITH:

%		B.P. of Azeotrope °C
15	Bromobenzene	131.7
82.5	Butyl acetate	125.9
35	Butyl ether	129.8
76	Camphene	130.9
66	Chlorobenzene	124.4
6	Cumene	131.6
73	1,3-Dimethylcyclohexane	116.6
85	2,5-Dimethylhexane	107.6
42	Ethyl isovalerate	130.5
93	Heptane	97.7
2.6	Isoamyl acetate	129.1
74.5	Isoamyl formate	123.6
88	Isoamyl vinyl ether	112.1
78	Isobutyl ether	119.8
28	Isobutyl propionate	131.2
76	Mesityl oxide	129.2
87	Methylcyclohexane	98.2
65	Octane	120.0
78	Paraldehyde	123.5
26	α-Pinene	137.7
47	Propyl isobutyrate	130.2
21	Propyl sulfide	130.5
95	2,2,4-Trimethylpentane	99.0
48	o-, m-, or p-Xylene	125-126

ISOAMYL ALCOHOL FORMS TERNARY AZEOTROPES WITH:

%		B.P. of Azeotrope °C
44.8	Water	93.6
24.0	Isoamyl acetate	
46.2	Water	95.4
6.5	Isoamyl chloroacetate	
32.4	Water	89.8
48	Isoamyl formate	

ACTIVE AMYL ALCOHOL

Table 6.72: Physical Properties of Active Amyl Alcohol *(31)*

Acidity (mg KOH per g)	0.06 max.
Boiling point	128°C
Coefficient of expansion per °C	0.00078
Distillation: 95%	Between 125 - 131°C min.
30%	Above 130°C max.
Flash point, Open cup	120°F
Freezing point	Less than -70°C
Heat of vaporization	100.0 cal/g (calculated)
Refractive index at 20°C	1.4097
Residue	0.003 g/100 cc
Specific gravity at 20/4°C	0.816
Viscosity at 20°C	5.09 centipoises
60°C	1.44 centipoises
Weight per gallon at 20°C	6.80 lbs

FUSEL OIL, REFINED

Refined fusel oil is a volatile, poisonous, oily mixture consisting largely of amyl alcohols.

Table 6.73: Physical Properties of Refined Fusel Oil *(31)*

Acidity as acetic acid	0.01% max.
Coefficient of expansion per 1°C	0.00051 - 0.0006
1°F	0.00092 - 0.0011
Color, APHA	No. 10 max.
Distillation range (ASTM)	
below 110°C	None
below 120°C	Not more than 15%
below 130°C	Not less than 60%
above 135°C	None
Dryness	A 5 ml. sample is clearly miscible with at least 19 parts of 60° Bé gasoline at 60°F
Evaporation rate at 95°F	

%	Minutes
5	3.5
25	17.0
50	36.5
75	64.75
90	90.25
95	103.5

Fire hazard	Moderate
Flash point, Open cup	123°F, approx.
Closed cup	106°F, approx.
Specific gravity at 20/20°C	0.810 - 0.815
Toxicity	Highly toxic by ingestion and inhalation
Water solubility at 25°C, 100 cc solvent dissolves	9.9 cc water
Weight per gallon at 20°C	6.76 - 6.77 lbs

METHYLAMYL ALCOHOL

Methyl Isobutyl Carbinol, 4-Methylpentanol-2, MIBC $(CH_3)_2CHCH_2CHOHCH_3$

Methylamyl alcohol is a secondary alcohol.

Table 6.74: Physical Properties of Methylamyl Alcohol *(31)*

Acidity as acetic acid	0.005% by wt, max.	Heat of vaporization at 1 atm.	98.6 g cal/g
Azeotrope with water:		MAC	25 ppm in air
boiling point, 760 mm, °C	94.3	Molecular weight	102.17
methyl amyl alcohol, %w	55.6	Non-volatile matter	0.005 g/100 ml max.
Boiling point at 760 mm	131.63 - 131.8°C	Odor	Mild and nonresidual
Coefficient of cubical expansion		Purity, minimum	97.5% by wt
at 20°C/°C	0.00103	Refractive index at 20°C, n_D	1.4081 - 1.4113
Color, Pt-Co scale	10 max.	Solubility in water at 20°C	1.7 - 1.8% by wt
Critical pressure, atm.	42.4	Solubility of water in, at 20°C	5.8 - 6.2% by wt
Critical temperature	312°C	Specific gravity at 20/20°C	0.8079 - 0.8080
Distillation range, 760 mm	130.0 - 133.0°C	Specific heat at 20°C	0.52 g cal/g-°C
Fire hazard	Moderate	Surface tension at 20°C	22.8 dynes/cm
Flash point, Tag open cup	131°F	Suspended matter	Substantially free
Tag closed cup	106°F		
Freezing point	-90°C, sets to a glass below		

2-ETHYLBUTYL ALCOHOL

Table 6.75: Physical Properties of 2-Ethylbutyl Alcohol *(31)*

Acidity as acetic acid	0.02% max.	Solubility in water at 20°C	0.43% by wt
Boiling point at 760 mm	147.0°C	Solubility of water in,	
Boiling range at 760 mm		at 20°C	4.6% by wt
below 140°C	None	Specific gravity at 20/20°C	0.8328
below 145°C	Not more than 5%	Specific heat at 25°C	0.586 cal/g
below 155°C	Not less than 95%	Surface tension at 28°C	28.05 dynes/cm
above 160°C	None	Vapor pressure at 20°C	1.2 mm
Coefficient of expansion		Viscosity at 20°C	5.63 cps.
per °C	0.000892 to 20°C	Weight per gallon at 20°C	6.93 lbs
	0.000921 to 55°C		
Dryness at 20°C	Miscible with 19 vol. of 60° Bé gasoline		
Flash point (ASTM open cup)	58°C (137°F)		
Heat of vaporization, 1 atm.	196 Btu/lb		
Molecular weight	102.17		
Non-volatile matter	0.005% max.		
Refractive index at 20°C	1.4229		

n-HEXYL ALCOHOL

n-Hexanol, Hexanol-1, Amyl Carbinol $CH_3(CH_2)_4CH_2OH$

Table 6.76: Physical Properties of n-Hexyl Alcohol *(31)*

Acidity as acetic acid	0.01% by wt	Refractive index at 20°C, n_D	1.4181
Boiling point at 760 mm	157.1°C	Solubility in water at 20°C	0.58% by wt
50 mm	89°C	Solubility of water in, at 20°C	7.2% by wt
10 mm	60°C	Specific gravity at 20/20°C	0.8203 - 0.8208
Boiling range at 760 mm	153 to 160°C	Specific heat at 16.9°C	0.544 Cal/gm/°C
Color (Pt-Co Scale)	15 max.	at 13°C	0.500 Cal/gm/°C
Fire hazard	Moderate	Surface tension at 30°C	23.6 dynes/cm
Flash point (Open cup)	165°F	Suspended matter	Substantially free
Freezing point	-44.6°C	Vapor pressure at 20°C	0.43 mm
Heat of vaporization at 1 atm.	213 Btu/lb	Viscosity (absolute) at 20°C	5.4 cps
Hydroxyl number	530 min.	Water content	Miscible without turbidity with 19 vol. of 60° API gasoline at 20°C
Iodine number	1.2 min.		
Molecular weight	102.17		
Odor	Mild	Weight per gallon at 20°C	6.83 lbs

Table 6.77: Solubility of Water in n-Hexyl Alcohol *(31)*

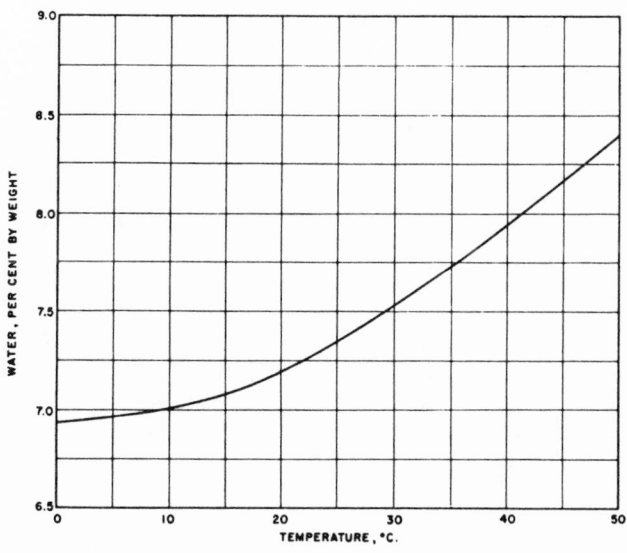

Table 6.78: Azeotropes of Hexyl Alcohol *(31)*

HEXYL ALCOHOL FORMS BINARY AZEOTROPES WITH:

%		B.P. of Azeotrope °C
63.5	Anisole	151.0
27	Benzyl methyl ether	156.7
52	Camphene	150.8
56	o-Chlorotoluene	153.5
46	p-Chlorotoluene	154.0
65	Cumeme	149.5
53	2,7-Dimethyloctane	152.5
11	Isoamyl ether	157.0
40	Isoamyl propionate	156.7
60	Isobutyl butyrate	155.0
45	Mesitylene	153.5
19	Phenetole	157.7
60	α-Pinene	150.8
55	Propylbenzene	152.5
32	Pseudocumene	156.3
77	Styrene	144.0
85	m-Xylene	138.3
82	o-Xylene	143.6
87	p-Xylene	137.0

CYCLOHEXYL ALCOHOL

Table 6.79: Physical Properties of Cyclohexyl Alcohol *(31)*

Boiling point at 760 mm	161.1°C (322°F)	Toxicity	Moderate by ingestion and inhalation
Boiling range at 760 mm, 5-95%	156-163°C		
Color, APHA	10 max.	Vapor density (air = 1.00)	3.45
Crystallization point	-10°C min.		
Dielectric constant at 25°C	15.0	Vapor pressure at 20 °C	0.8 mm
Evaporation rate at 45°C (toluene = 100)	8 approx.	70	15
Fire hazard	Moderate	80	27
Flash point (Closed cup)	145°F	100	78
(Open cup)	154°F	120	187
Freezing point	18-25.15°C	140	398
Heat of combustion, liquid	8893 cal/g	150	554
Heat of fusion	4.9 cal/g	161.1	760
Heat of vaporization	108 cal/g	Viscosity at 25 °C	49.8 centipoises
Ketone as cyclohexanone	0.5% max.	39.1°C	20.3 cps.
Phenol	0.05% max.	65.9°C	5.8 cps.
Refractive index at 20°C	1.4656	90 °C	2.45 cps.
Solubility in water at 20°C	3.6% by wt.	Water	0.5% max.
Solubility of water in at 20°C	20% by wt.	Weight per gallon at 20°C (68°F)	7.91 lbs.
Specific gravity at 20/4°C	0.9493		
Specific heat at 15-18°C	0.417 cal/gm		
Surface tension at 16.2°C	34.23 dynes/cm		

Table 6.80: Azeotropes of Cyclohexyl Alcohol *(31)*

CYCLOHEXYL ALCOHOL FORMS BINARY AZEOTROPES WITH

%		B.P. of Azeotrope °C
70	Anisole	152.5
38	Benzyl methyl ether	159.0
59	Camphene	151.9
85	Chloroacetal	155.6
62	o-Chlorotoluene	155.5
45	p-Chlorotoluene	156.5
8	Cineole	160.55
72	Cumene	150.0
28	Cymene	159.5
25	Indene	160.0
22	Isoamyl ether	158.8
37	Isoamyl propionate	157.7
80	Isobutyl butyrate	156.0
35	α-Phellandrene	158.0
60	Propylbenzene	153.8
83	Propyl isovalerate	155.1
40	Pseudocumene	158.0
35	α-Terpene	158.3
22	Thymene	159.8
95	m-Xylene	138.9
86	o-Xylene	143.0

HEPTYL ALCOHOL

Heptanol-1, Alcohol C-7 $C_7H_{15}OH$

Table 6.81: Physical Properties of Heptyl Alcohol *(31)*

Boiling point at 765 mm	175°C
Freezing point	-34.6°C
Refractive index at 20°C, n_D	1.4233
Specific gravity at 20/4°C	0.824

Table 6.82: Azeotropes of Heptyl Alcohol *(31)*

HEPTYL ALCOHOL FORMS BINARY AZEOTROPES WITH:

%		B.P. of Azeotrope °C
80	Benzyl methyl ether	167.0
90	Camphene	159.3
53	Cymene	172.5
50	Dipentene	171.7
63	Isoamyl ether	170.4
92	Isobutyl isovalerate	171.0
48	p-Methylanisole	173.0
72	Phenetole	169.0
60	α-Terpinene	169.7

2-HEPTYL ALCOHOL

Heptanol-2, Methylamyl Carbinol $CH_3(CH_2)_4CHOHCH_3$

2-Heptyl alcohol is a secondary alcohol.

Table 6.83: Physical Properties of 2-Heptyl Alcohol *(31)*

Acidity as acetic acid	0.03% by wt., max.
Boiling point at 760 mm	160.4°C
Boiling range at 760 mm, below 155°C	None
below 158°C	Not more than 5%
below 162°C	Not less than 95%
above 165°C	None
Color (Pt-Co scale)	15, max.
Dryness at 20°C	Miscible with 19 vols. 60° Bé gasoline
Fire hazard	Moderate
Flash point (Open cup)	160°F
Solubility in water at 20°C	0.35% by wt.
Solubility of water in at 20°C	5.80% by wt.
Specific gravity at 20/20°C	0.8187
Vapor pressure at 20°C	1.0 mm
Weight per gallon at 20°C	6.81 lbs.

3-HEPTYL ALCOHOL

Heptanol-3 $CH_3CH_2CH(OH)C_4H_9$

Table 6.84: Physical Properties of 3-Heptyl Alcohol *(31)*

Acidity as acetic acid	0.02% by wt.
Boiling point at 760 mm	156.2°C
Boiling range	153-158°C
Color, APHA (Pt-Co scale)	5
Fire hazard	Moderate
Flash point	140°F
Freezing point	-70°C sets to glass below
Specific gravity at 20/20°C	0.8224
Vapor pressure at 20°C	0.58 mm
Weight per gallon at 20°C	6.84 lbs.

2-ETHYLHEXYL ALCOHOL

2-Ethylhexanol, Octyl Alcohol $CH_2CH_2CH_2CH(C_2H_5)CH_2OH$

Table 6.85: Physical Properties of 2-Ethylhexyl Alcohol *(31)*

Acidity as acetic acid	0.01%, max.	Heat of vaporization, 1 atm.	167 Btu/lb
Aldehydes	None	Molecular weight	130.22
Boiling point at 760 mm	184.8°C	Refractive index at 20°C, n_D	1.4316
Boiling range, below 180°C	None	Solubility in water at 20°C	0.10% by wt
above 192°C	None	Solubility of water in, at 20°C	2.6% by wt
Coefficient of expansion per °C	0.000875 to 20°C	Specific gravity at 20/20°C	0.8339
	0.000902 to 55°C	Specific heat at 25°C	0.564 cal/gm/°C
Color, APHA	5 max.	Surface tension at 22°C	30.0 dynes/cm
Constant-boiling mixture, solvent 20% water 80%	b.p. 99.1°C	Unsaturates, as ethyl hexanol	0.2% max.
Fire hazard	Slight	Vapor pressure at 20°C	0.05 mm
Flash point, Open cup	185°F	Viscosity at 20°C	9.8 cps.
Freezing point	-70°C sets to glass below	Weight per gallon at 20°C	6.94 lbs

n-OCTYL ALCOHOL

n-Octanol, Octanol-1 $CH_3(CH_2)_6CH_2OH$

Table 6.86: Physical Properties of n-Octyl Alcohol *(31)*

Acid number	0.2 max.	Molecular weight	130.22
Boiling point at 760 mm	195°C (383°F)	Refractive index at 20°C, n_D	1.42920
Boiling range at 760 mm	194 - 197°C	Solubility in water at 25°C	0.059 g per 100 g water
Color, dichromate	0.002 max.		
Ester number	1.3 max.	Specific gravity at 20/4°C	0.827
Fire hazard	Slight	Viscosity at 20°C	8.925 centipoise
Flash point (Open cup)	195°F	Water	0.25% max.
Freezing point	-15°C (5°F)		
Heat of combustion	9690 cal/g		
Hydroxyl number	415 - 440		
Iodine number	1.3 max.		

Table 6.87: Azeotropes of n-Octyl Alcohol *(31)*

OCTYL ALCOHOL FORMS BINARY AZEOTROPES WITH:

%		B.P. of Azeotrope °C
80	N,N-Dimethyl-o-toluidine	184.8
88	Indene	182.4
85	Isoamyl isovalerate	192.6
70	Isobornyl methyl ether	191.9
80	Isobutyl carbonate	189.5
92	d-Limonene	177.5
20	Phorone	193.5
90	γ-Terpinene	182.5
93	Thymene	179.6

sec-OCTYL ALCOHOL

Table 6.88: Physical Properties of sec-Octyl Alcohol *(31)*

	85% Grade	95% Grade
Boiling range first 5%		173 - 178°C
90%	174 - 181.5°C	178 - 182.5°C
Density, lbs per gallon	6.8	6.8
Fire hazard	Moderate	Slight
Flash point	164°F	185°F
Hydroxyl number	376 - 388	408 - 414
Melting point		-38°C
Methyl hexyl ketone content	10 - 15%	Less than 5%
Molecular weight	130.23	130.23
Refractive index at 20°C	1.4244 - 1.4252	1.4258 - 1.4262
Specific gravity at 20°C	0.814 - 0.820	0.818 at 25°C
Water content	1.0 - 1.2%	0.3 - 0.5%

Table 6.89: Azeotropes of sec-Octyl Alcohol *(31)*

sec-OCTYL ALCOHOL FORMS BINARY AZEOTROPES WITH:

%		B.P. of Azeotrope °C
14	Amyl ether	179.8
50	Butylbenzene	178.2
89	Butyl isovalerate	177.4
73.5	Cineole	175.9
56	Cymene	174.0
40	Indene	176.0
28	Isoamyl butyrate	180.3
83	Isoamyl ether	172.7
55	d-Limonene	174.5
73	α-Terpinene	171.8
43	Terpinolene	179.0
48	Thymene	176.0

ISOOCTYL ALCOHOL

Table 6.90: Physical Properties of Isooctyl Alcohol *(31)*

Acidity as acetic acid	0.001% by wt	Surface tension at 20°C	29.5 dynes/cm
Carbonyl number	0.10 mg KOH/g	Vapor pressure, °C °F	mm
Coefficient of expansion per °C	0.000814	50 122	1.95
		75 167	8.4
Color (Hazen, Pt-Co)	5	100 212	30
		125 257	94
Fire hazard	Slight	150 302	250
		175 347	600
Flash point (Tag open cup)	180°F	180 356	700
Pour point	-95°F		
Purity	99.5% by wt	Viscosity, °C °F	Centistokes
		37.8 100	6.4
Refractive index at 20°C, n_D	1.4308	20.0 68	12.7
Solubility in water at 25°C	0.06 g/100g	-9.4 15	51.3
at 50°C	0.08 g/100g	-17.8 0	84.4
Solubility of water in, at 5°C	3.4 g/100g	-31.7 -25	224.2
20°C	3.8 g/100g	Water	0.02% by wt
40°C	4.1 g/100g	Weight per gallon at 60°F	6.95 lbs, approx.
Specific gravity at 20/20°C	0.832		
60/60°F	0.834		
Specific heat, 50 - 150°C	0.79 cal/g/°C		

CAPRYL ALCOHOL

2-Octanol

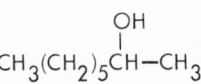

Table 6.91: Physical Properties of 85% Capryl Alcohol *(52)*

Empirical Formula	$C_8H_{18}O$
Molecular Weight	130.23

Solvent C-8

Boiling Range	174-181.5°C.
Specific Gravity	0.814 - 0.820 at 20°C.
Density, pounds per gallon	6.8
Methyl hexyl ketone content	10 - 15%
Appearance	Water-white, clear liquid
Hydroxyl Number	376 - 388
Refractive Index at 20°C.	1.4244 - 1.4252
Flash Point	164°F.
Water Content	1.0 - 1.2%
Solubility	Practically insoluble in water. Miscible with common organic solvents including paraffin hydrocarbons.

NONYL ALCOHOL

Table 6.92: Physical Properties of Nonyl Alcohol *(31)*

Aldehyde content	0.30% by wt
Boiling point at 760 mm	173.3°C
Color, Saybolt	30
Distillation (ASTM), initial	193°C
5%	196°C
50%	198°C
95%	201°C
max.	206°C
Fire hazard	Moderate
Flash point (Open cup)	80°C (176°F)
Freezing point	-65°C
Heat of vaporization (Lv),	
100°F	22,000 Btu/lb mole
300°F	19,000 Btu/lb mole
400°F	17,400 Btu/lb mole
Mixed aniline point	-15°C
Neutralization number	0.02 mg KOH/g
Refractive index at 20°C, n_D	1.4390
Solubility in water at 20°C	0.06% by wt
Solubility of water in, at 20°C	0.99% by wt
Specific gravity at 20/20°C	0.8121
Vapor pressure at 20°C	0.3 mm
Viscosity at 0°C	56.0 cps.
at 20°C	14.3 cps.
Weight per gallon at 20°C	6.75 lbs

3,5,5-TRIMETHYLHEXYL ALCOHOL

Table 6.93: Physical Properties of 3,5,5-Trimethylhexyl Alcohol *(31)*

Boiling point at 10 mm	83°C (181°F)
760 mm	194°C (381°F)
Boiling range at 760 mm,	
first drop	190°C
90%	194 ± 1°C
dry	195.5°C
Color (APHA)	25 max.
Flash point (Open cup)	200°F
Freezing point	Below -70°C
Molecular weight	144.25
Purity (by hydroxyl number)	97.5% min.
Refractive index at 25°C, n_D	1.4300
Specific gravity at 25/4°C	0.8236
Viscosity at 25°C	11.06 centipoises
Water content	0.15% max.
Weight per gallon at 25°C	6.86 lbs

DECYL ALCOHOL

Table 6.94: Physical Properties of Decyl Alcohol *(31)*

	Oxo Process	Fatty acid Process
Acidity	0.0015% by wt.	
Acid number		0.2 max.
Aldehydes, as decanal	0.20%, max.	
Boiling point at 760 mm	217.3°C	231°C (448°F)
Boiling range at 760 mm	219-221.5°C	90% between 229-233°C
Coefficient of expansion at 55°C	0.00086	
Color, Hazen Pt-Co	5	0.003 max., Dichromate
Ester, as decyl formate	Less than 0.1%	
Ester number		1.3 max.
Fire hazard	Slight	Slight
Flash point (Open cup)	225°F	220°F
Freezing point	Sets to a glass below -60°C	6.9°C (44°F)
Heat of combustion		9963 cal/g
Hydroxyl number		345-365
Iodine number		0.5 max.
Molecular weight	158.28	158.28
Pour point	-95°F	
Purity	99.7%-99.9% by wt.	
Refractive index at 20°C, n_D	1.4388-1.4390	1.43682

	Oxo Process	Fatty Acid Process
Solubility in water at 20°C	Less than 0.01% by wt.	
Solubility of water in at 20°C	2.3% by wt.	
Specific gravity at 20/20°C	0.837-0.840	0.829 at 20/4°C
Sulfur	4 ppm, max.	
Suspended matter	Substantially free	

Vapor pressure,	°C	°F	mm
	75	167	2.1
	100	212	8.4
	125	257	28.2
	150	302	82
	175	347	225
	200	392	500

Viscosity,	°C	°F	Centistokes	
	99	210	1.76	
	20	68	21	13.83 centipoises
	-9.4	15	115	
	-17.8	0	209	
	-31.7	-25	701	
	-40.0	-40	1649	
	-53.9	-65	8826	

	Oxo Process	Fatty Acid Process
Water content	0.03-0.07% by wt.	0.25%
Weight per gallon at 20°F	7.03 lbs.	
60°F	6.96 lbs. approx.	

ISODECYL ALCOHOL

Table 6.95: Physical Properties of Isodecyl Alcohol *(31)*

Acidity as acetic acid	0.002% by wt., max.
Aldehydes, as decanal	0.05% by wt., max.
Boiling point at 760 mm	220.1°C
Boiling range at 760 mm, Ibp	215°C, min.
Dp	225°C, max.
Coefficient of expansion at 55°C	0.00083
Color, (Pt-Co scale)	10, max.
Fire hazard	Slight
Flash point (Open cup)	220°F
Freezing point	Sets to a glass below -60°C
Molecular weight	158.29
Odor ·	Characteristic, non-petroleum
Purity, as decanol	98.5% by wt., min.
Refractive index at 20°C, n_D	1.4408
Solubility in water at 20°C	Less than 0.01% by wt.
Solubility of water in at 20°C	2.4% by wt.
Specific gravity at 20/20	0.8423
Sulfuric acid test (Pt-Co scale)	50, max.
Suspended matter	Substantially free
Vapor pressure at 20°C	Less than 0.01 mm
Viscosity at 20°C	18.9 cps.
Water content	0.10% by wt., max.
Weight per gallon at 20°C	7.01 lbs.

TRIDECYL ALCOHOL

Table 6.96: Physical Properties of Tridecyl Alcohol *(31)*

Acidity as acetic acid	0.002% by wt
Carbonyl number	0.7 mg KOH/g
Color, Hazen, Pt-Co	5
Distillation: initial	252°C
dry point	269°C
Fire hazard	Slight
Flash point (Tag open cup)	180°F
Hydroxyl number	278 mg KOH/g
Odor	Characteristic, non-petroleum
Pour point	-95°F
Purity	99.6% by wt
Refractive index at 20°C, n_D	1.4475
Specific gravity at 20°C	0.8454
Sulfur	2 ppm

Vapor pressure,	°C	°F	mm
	90	194	1.3
	100	212	2.2
	125	257	7.8
	150	311	24
	175	347	64
	200	401	155
	225	437	340
	250	491	685

Viscosity,	°C	°F	Centipoises
	99	210	2.61
	20	68	47.5
	- 9.4	15	382.2
	-17.8	0	808.3
	-31.7	-25	3,692
	-40.0	-40	11,081
	-53.9	-65	95,433

Water	0.10% by wt.
Weight per gallon at 60°F	7.0 lbs.

OTHER ALCOHOLS AND ALCOHOL BLENDS

Table 6.97: Alfol 4 Alcohol *(40)*

Chemicals Names: 1-butanol, normal butyl alcohol, n-propyl carbinol

Description:

ALFOL 4 Alcohol is a very high purity, linear primary butyl alcohol. It is a colorless liquid having limited water solubility but is soluble in most organic solvents. It has the characteristic odor of this molecular weight alcohol.

Properties	Specification	Typical
Total Alcohol Content, Wt. %	99.0 min	99.5
Molecular Weight Distribution (100% alcohol basis)		
$<C_4OH$	1.0 max	Tr
C_4OH (1-butanol)	98.0 min	99.8
$>C_4OH$	1.0 max	0.2
Average Molecular Weight	—	74.1
Color, APHA	10 max	0
Water, Wt. %	0.1 max	0.05
Iodine Number	0.2 max	0.05
Acidity, as Acetic Acid, Wt. %	0.01 max	0.001
Boiling Range, °C	4°C max. range	2.5°C Range (118-120°C)
Physical Appearance	—	Clear Colorless Liquid
Specific Gravity @ 60°F	—	0.814
Melting Point, °F	—	-128
Viscosity, 70°F, cSt	—	3.4
Flash Point, Pensky-Martens, °F	—	97

Table 6.98: Alfol C_6-C_{10} Alcohol Low-Range Blends *(40)*

Description: These ALFOL Alcohols consist of blends of high purity, synthetic linear primary alcohols having even-numbered carbon chains in the C_6-C_{10} range. They are clear, colorless liquids having a mild odor. These alcohols are insoluble in water and soluble in alcohol, ether and benzene. They are physically and chemically equivalent to the corresponding natural alcohols derived from coconut oil.

Properties	ALFOL 610 Alcohol Specification	Typical	ALFOL 610 AFC Specification	Typical	ALFOL 810 Alcohol Specification	Typical
Total alcohol, wt. %	98.5 min.	99.4	98.5 min.	99.4	98.5 min.	99.3
Homolog distribution, wt. % (100% alcohol basis)						
C_4OH and lower	0.5 max.	0.2	0.5 max.	Tr	—	—
C_6OH	4.0±2	4.0	4.0±2	4.5	1.0 max.	0.4
C_8OH	42.0±4	42.5	54.0±4	51.5	43.0±4	42.0
$C_{10}OH$	52.5±4	52.8	41.0±4	43.7	55.0±4	57.1
$>C_{10}OH$ and higher	1.0 max.	0.5	1.0 max.	0.3	1.0 max.	0.5
Average molecular weight	—	143	—	141	—	145
Color, APHA	10 max.	0	10 max.	0	10 max.	0
Appearance	—	Clear Colorless Liquid	—	Clear Colorless Liquid	—	Clear Colorless Liquid
Water, wt. %	0.15 max.	0.06	0.1 max.	0.04	0.15 max.	0.05

(continued)

Table 6.98: (continued)

Properties	ALFOL 610 Alcohol Specification	Typical	ALFOL 610 AFC Specification	Typical	ALFOL 810 Alcohol Specification	Typical
Acidity, as acetic acid, wt. %	0.005 max.	0.002	0.005 max.	0.002	0.005 max.	0.002
Iodine number	0.1 max.	0.02	0.1 max.	0.04	0.1 max.	0.03
Acid heat color, APHA	50 max.	10	—	10	—	—
Specific gravity, 60° F	—	0.830	—	0.830	—	0.831
Flash point, Pensky-Martens, ° F	—	182	—	182	—	188
Boiling range, ° F	—	350-460	—	350-460	—	400-460
Melting point, ° F	—	1-5	—	1-5	—	3-7
Viscosity, 100° F, cSt	—	6.4	—	6.6	—	7.4

Table 6.99: Alfol C_6-C_{18} High Purity Alcohol Homologs *(40)*

Description: The ALFOL Alcohols are high purity, synthetic linear primary alcohol homologs having even-numbered carbon chains. As such, derivatives are highly biodegradable. They are physically and chemically equivalent to the natural alcohols such as those derived from coconut oil and tallow.

Properties	ALFOL 6 Specification	Typical	ALFOL 8 Specification	Typical	ALFOL 10 Specification	Typical	ALFOL 12 Specification	Typical
Total alcohol, wt. %	98.5 min.	99.4	99.0 min.	99.5	98.5 min.	99.2	98.5 min.	99.4
Molecular weight distribution (100% alcohol basis)								
$C_6H_{13}OH$	98.0 min.	98.6	0.5 max.	Tr	—	—	—	—
$C_8H_{17}OH$	—	—	98.8 min.	99.7	—	0.1	—	—
$C_{10}H_{21}OH$	—	—	0.5 max.	0.3	98.0 min.	99.3	—	0.1
$C_{12}H_{25}OH$	—	—	—	—	—	0.6	98.5 min.	99.2
$C_{14}H_{29}OH$	—	—	—	—	—	—	—	0.7
$C_{16}H_{33}OH$	—	—	—	—	—	—	—	—
$C_{18}H_{37}OH$	—	—	—	—	—	—	—	—
$C_{20}H_{41}OH$	—	—	—	—	—	—	—	—
Color, APHA	10 max.	0	10 max.	0	10 max.	0	15 max.	0
Water, wt. %	0.15 max.	0.06	0.10 max.	0.05	0.15 max.	0.04	0.1 max.	0.04
Iodine number	0.2 max.	0.06	0.2 max.	0.05	0.2 max.	0.06	0.2 max.	0.03
Hydroxyl number	—	545	—	430	—	352	295-302	300
Molecular weight average	—	102	—	130	—	158	—	187
Carbonyl, ppm C = O	—	—	80 max.	50	100 max.	25	100 max.	64
Acidity, as acetic acid, %	0.005 max.	0.001	0.005 max.	0.001	.005 max.	.001	—	—
Specific gravity, @ 68°F.	—	0.823	—	0.826	—	0.828	—	0.831 (80°F)
Flash point, Pensky-Martens, °F.	—	130	—	180	—	235	—	265
Melting point, °F.	—	-49	—	1	—	44	—	74
Boiling range, °F.	—	313-316	—	381-385	—	448-453	—	490-498
Saponification number	—	—	—	—	—	—	—	—
Viscosity, cSt 70°F.	—	5.5	—	10.5	—	14.5	—	—
100°F.	—	3.5	—	6.0	—	9.0	—	12.3

(continued)

Table 6.99: (continued)

Properties	ALFOL 14 Specification	Typical	ALFOL 16 Specification	Typical	ALFOL 16 NF Specification	Typical	ALFOL 18 Specification	Typical	ALFOL 18 NF Specification	Typical
Total alcohol, wt. %	98.5 min.	99.0	97.5 min.	98.9	97.5 min.	98.9	98.0 min.	98.2	98.0 min.	98.2
Molecular weight distribution (100% alcohol basis)										
$C_6H_{13}OH$	—	—	—	—	—	—	—	—	—	—
$C_8H_{17}OH$	—	—	—	—	—	—	—	—	—	—
$C_{10}H_{21}OH$	—	—	—	—	—	—	—	—	—	—
$C_{12}H_{25}OH$	—	0.9	as is sample basis		as is sample basis		as is sample basis		as is sample basis	
$C_{14}H_{29}OH$	96.0 min.	98.4	3.0 max.	0.7	3.0 max.	0.7	0.5	0.2	0.5	0.2
$C_{16}H_{33}OH$	—	0.7	95.0 min.	97.4	95.0 min.	97.4	3.0 max.	0.7	3.0 max.	0.7
$C_{18}H_{37}OH$	—	—	3.0 max.	0.8	3.0 max.	0.8	95.0 min.	96.6	95.0 min.	96.6
$C_{20}H_{41}OH$	—	—	—	—	—	—	3.0 max.	0.9	3.0 max.	0.9
Color, APHA	20 max.	0	40 max.	25	40 max.	25	40 max.	20	40 max.	20
Water, wt. %	0.1 max.	0.04	0.1 max.	0.03	0.1 max.	0.03	0.1 max.	0.05	1.0 max.	0.05
Iodine number	0.4 max.	0.05	0.6 max.	0.15	0.6 max.	0.15	1.0 max.	0.58	1.0 max.	0.58
Hydroxyl number	255-264	258	218-238	227	218-238	227	200-220	200	200-220	200
Molecular weight average	—	214	—	242	—	242	—	271	—	271
Carbonyl, ppm C = O	200 max.	100	350 max.	140	350 max.	140	700 max.	600	700 max.	600
Acidity, as acetic acid, %	—	—	—	—	0.21 max.	0.01	—	—	0.21 max.	0.01
Specific gravity, @ 68°F.	—	0.815 (120°F)	—	0.813 (125°F)	—	0.813 (125°F)	—	0.811 (140°F)	—	0.811 (140°F)
Flash point, Pensky-Martens, °F.	—	290	—	300	—	300	—	355	—	355
	—	100	—	118-121	113-122	117	—	132-136	131-140	136
Melting point, °F.	—	567-573	—	626-631	—	626-631	—	662-670	—	662-670
Boiling range, °F.	—	—	1.0 max.	0.8	1.0 max.	<1	1.5 max.	0.4	1.5 max.	0.4
Saponification number	—	—	—	—	—	—	—	—	—	—
Viscosity, cSt 70°F. 100°F.	—	15	—	18 (120°F)	—	18 (120°F)	—	13.5 (160°F)	—	13.5 (160°F)

Table 6.100: Alfol C_{10}-C_{16} Alcohol Blends *(40)*

Description: These ALFOL Alcohols are blends of high purity, synthetic linear primary alcohols having even-numbered carbon chains in the C_{10}-C_{16} range. They are clear, colorless liquids and are physically and chemically equivalent to the natural alcohols such as those derived from coconut oil.

Properties	ALFOL 1012 HA Specification	Typical	ALFOL 1014 CDC Specification	Typical	ALFOL 1214 Specification	Typical	ALFOL 1214 GC Specification	Typical
Total alcohol, wt. %	98.5 min.	99.4	98.0 min.	99.0	98.0 min.	99.0	98.0 min.	99.0
Homolog distribution, wt. % (100% alcohol basis)								
C_8OH	2.0 max.	0.8	1.0 max.	0.4	0.5 max.	0.2	—	—
$C_{10}OH$	85.0±4	83.8	30.0±3	31.0	1.5 max.	0.6	1.5 max.	0.3
$C_{12}OH$	8.5±2	8.3	38.0±3	36.6	54.0±3	54.5	69.0±3	70.5
$C_{14}OH$	6.5±2	6.9	31.0±3	31.2	44.0±3	43.8	29.0±3	28.8
$C_{16}OH$	0.5 max.	0.2	2.0 max.	0.8	1.5 max.	0.9	1.5 max.	0.4
$C_{18}OH$	—	—	—	—	—	—	—	—
$C_{20}OH$	—	—	—	—	—	—	—	—

(continued)

Table 6.100: (continued)

Properties	ALFOL 1012 HA Specification	Typical	ALFOL 1014 CDC Specification	Typical	ALFOL 1214 Specification	Typical	ALFOL 1214 GC Specification	Typical
Average molecular weight	—	164	—	188	—	199	—	195
Hydroxyl number	335-348	341	295-309	302	274-287	277	280-292	285
Color, APHA	20 max.	0	30 max.	0	30 max.	5	30 max.	5
Water, wt. %	0.15 max.	0.03	0.15 max.	0.05	0.1 max.	0.05	0.1 max.	0.05
Iodine number	0.25 max.	0.04	0.3 max.	0.07	0.3 max.	0.10	0.3 max.	<0.3
Saponification number	1.0 max.	0.1	1.0 max.	0.1	0.7 max.	0.2	1.0 max.	<1.0
Specific gravity @ 72°F	—	0.834	—	0.836	—	0.838	—	0.838
Flash point, Pensky-Martens, °F	—	237	—	250	—	265	—	260
Boiling range, °F	—	425-525	—	450-545	—	518-575	—	518-575
Melting range, °F	—	35-40	—	41-45	—	70-75	—	70-75
Viscosity, cSt @ 100°F	—	10.4	—	12.5	—	14.3	—	14.3
Appearance	—	Clear White Liquid	—	Clear White Liquid	—	Clear White Liquid	—	Clear White Liquid

Properties	ALFOL 1216 Specification	Typical	ALFOL 1216 CO Specification	Typical	ALFOL 1218 DCBA Specification	Typical	ALFOL 1412 Specification	Typical
Total alcohol, wt. %	98.0 min.	99.0	98.0 min.	99.0	97.5 min.	98.9	98.0 min.	98.8
Homolog distribution, wt. % (100% alcohol basis)								
C_8OH	—	—	0.2 max.	0.1	—	—	—	0.1
$C_{10}OH$	2.0 max	0.3	1.0 max.	0.3	2.0 max.	0.7	2.0 max.	0.6
$C_{12}OH$	63.0±3	64.3	65.0 min.	67.8	39.0±3	38.9	38.0±4	37.9
$C_{14}OH$	24.0±3	24.0	Diff.	25.1	31.0±3	30.4	59.0±4	59.5
$C_{16}OH$	10.0±2	11.4	7.0±3	6.6	19.0±3	18.4	5.0 max.	1.9
$C_{18}OH$	1.0 max.	Tr	1.0 max.	0.1	11.0±2	10.9	—	—
$C_{20}OH$	—	—	—	—	2.0 max.	0.7	—	—
Average molecular weight	—	201	—	198	—	213	—	205
Hydroxyl number	275-285	276	275-295	281	250-268	260	267-281	272
Color, APHA	30 max.	5	30 max.	5	40 max.	10	30 max.	5
Water, wt. %	0.1 max.	0.05	0.1 max.	0.05	0.1 max.	0.05	0.1 max.	0.06
Iodine number	0.5 max.	0.10	0.3 max.	<0.3	0.7 max.	0.3	0.3 max.	0.1
Saponification number	1.0 max.	0.5	1.0 max.	<1.0	1.0 max.	<1.0	1.0 max.	<1.0
Specific gravity @ 72°F	—	0.840	—	0.840	—	0.840	—	0.839
Flash point, Pensky-Martens, °F	—	265	—	265	—	275	—	270
Boiling range, °F	—	520-590	—	520-590	—	525-660	—	520-575
Melting range, °F	—	64-70	—	64-70	—	68-73	—	72-75
Viscosity, cSt @ 100°F	—	14.5	—	14.5	—	15.0	—	14.4
Appearance	—	Clear White Liquid	—	Clear White Liquid	—	Clear White Liquid	—	Clear White Liquid

Table 6.101: Alfol C$_{14}$-C$_{20}$ Alcohol Blends *(40)*

Description: These ALFOL Alcohols are blends of high purity synthetic linear primary alcohols having even-numbered carbon chains in the C$_{14}$–C$_{20}$ range. They are white solids and are physically and chemically equivalent to the natural alcohols such as those derived from coconut oil and tallow.

Properties	ALFOL 1418 Specification	Typical	ALFOL 1618 Specification	Typical	ALFOL 1620 Specification	Typical	ALFOL 1620 H+ Specification	Typical
Total alcohol, wt. %	98.0 min.	99.0	97.0 min.	98.7	96.0 min.	98.5	97.0 min.	98.5
Homolog distribution, wt. % (100% alcohol basis)								
C$_{12}$OH and lower	2.5 max.	0.7	—	—	—	—	—	—
C$_{14}$OH	38.0±3	39.5	2.0 max.	1.0	3.0 max.	1.4	3.0 max	1.2
C$_{16}$OH	38.0±3	38.7	59.0±5	58.0	51.0±3	50.7	43.0±4	44.8
C$_{18}$OH	20.0±2	19.6	36.0±5	38.6	32.0±3	32.5	29.0±4	28.6
C$_{20}$OH	3.5 max.	1.5	5.0 max.	2.4	14.0±2	14.4	25.0±3	23.9
Higher than C$_{20}$	—	—	—	—	2.0 max.	1.0	3.0 max.	1.5
Average molecular weight	—	240	—	253	—	259	—	263
Hydroxyl number	229-242	231	210-223	215	202-220	208	198-216	205
Color, APHA	30 max.	10	40 max.	5	100 max.	15	100 max.	15
Water, wt. %	0.1 max.	0.03	0.1 max.	0.04	0.1 max.	0.05	0.1 max.	0.05
Iodine number	—	—	1.5 max.	0.9	3.0 max.	1.3	2.5 max.	1.3
Saponification number	—	—	1.0 max.	0.5	2.0 max.	0.6	2.0 max.	0.6
Specific gravity	—	0.819 (110°F)	—	0.820 (120°F)	—	0.820 (120°F)	—	0.820 (120°F)
Flash point, Pensky-Martens, °F	—	290	—	325	—	340	—	340
Melting range, °F	—	97-102	—	110-120	—	113-120	—	113-120
Boiling range, °F	—	592-651	—	630-665	—	630-670	—	630-670
Appearance	—	White Solid	—	White Solid	—	White Solid	—	White Solid
Viscosity, cSt.	—	14.6 (110°F)	—	13.5 (140°F)	—	13.7 (140°F)	—	13.7 (140°F)

Table 6.102: Alfol 20+ and 22+ Alcohols *(40)*

Description:

ALFOL 20+ and 22+ Alcohols are mixtures containing high molecular weight, linear primary alcohols in the C$_{20}$–C$_{28}$ range. They are off-white, nearly odorless waxy solids.

Properties	ALFOL 20+ Typical	ALFOL 22+ Typical	Properties	ALFOL 20+ Typical	ALFOL 22+ Typical
Total alcohol, wt. %	74	65-68	Iodine number	12	15
Homolog distribution, wt. % (100% alcohol basis)			Carbonyl, as C=O, %	0.3	0.4
C$_{18}$OH and lower	1.5	Tr.	Water, wt. %	0.04	0.04
C$_{20}$OH	49.0	4.6	Saponification number	6	11
C$_{22}$OH	29.4	51.3	Color, Klett (4 cm cell)	600	1,100
C$_{24}$OH	12.4	26.8	+ Flash point, Pensky-Martens, °F	390	410
C$_{26}$OH	5.0	10.9	Melting range, °F	113-129	113-136
C$_{28}$OH and higher	2.2	4.6	Appearance	Off-White Waxy Solid	Off-White Waxy Solid
Hydroxyl number	130	112			

COMPARATIVE DATA

Table 6.103: Amsco Alcohols *(13)*

Alcohols	Specific Gravity @ 20°/20°C	lb/gal @ 20°/20°C	Distillation Range 760 mm Hg °F (°C)	Relative Evaporation Rate n-BuAc = 1	Vapor Pressure mm Hg @ 20°C	Flash Point TAG CC °F	Solubility % by wt @ 20°C in Water	of Water
Methanol	0.792	6.60	147 (63.9)–149 (65.0)	3.5	96	52	∞	∞
Proprietary Solvents 190	0.815*	6.78	165 (73.9)–176 (80.0)	1.7	57	40	∞	∞
Proprietary Solvents 200	0.796*	6.64	167 (75.0)–176 (80.0)	1.7	53	41	∞	∞
Special Ind. Solvents 190	0.817*	6.80	173 (78.3)–176 (80.0)	1.7	47	50	∞	∞
Special Ind. Solvents 200	0.795*	6.65	174 (78.9)–176 (80.0)	1.7	45	54	∞	∞
Isopropanol-99	0.786	6.55	179 (81.7)–181 (82.8)	1.7	37	53	∞	∞
n-Propanol	0.805	6.71	204 (95.5)–208 (97.8)	1.1	18	74	∞	∞
sec-Butanol	0.808	6.73	208 (97.8)–214 (101.1)	0.9	16	74	20.6	30.7
Isobutanol	0.803	6.68	225 (107.2)–229 (109.0)	0.6	9.6	82	9.5	14.3
n-Butanol	0.811	6.75	241 (116.1)–246 (118.9)	0.5	5.9	95	7.9	20.8
Methyl Amyl Alcohol (MIBC)	0.808	6.72	266 (130.0)–271 (132.8)	0.3	2.6	102	1.6	6.3
Cyclohexanol	0.948	7.89	320 (160.1)–324 (162.2)	0.1	<1	154**	0.1	11.0
2-Ethylhexanol	0.834	6.94	360 (182.2)–367 (186.1)	<0.01	<1	164	0.07	2.6
Texanol***	0.950	7.90	471 (243.9)–477 (247.2)	<0.01	<1	248**	Insol.	0.9

*15.56°/15.56°C
**COC
***Eastman Chemicals

Table 6.104: Ashland Alcohols *(69)*

	Specific Gravity 20° / 20°C	Dist. Range °F IBP	Dist. Range °F DP	Flash Pt. °F TCC
Amyl Alcohol, Primary	0.8134	262	282	120
n-Butanol	0.8108	241	245	96
2-Butanol	0.8090	206	216	72
Cyclohexanol	0.948 (25°C)	320	325	140
Ethyl Alcohol, Industrially Denatured	All commercial formulas.			
2-Ethylhexanol	0.834	360	367	164
Furfuryl Alcohol	1.135	338	—	149
Isobutanol	0.8030	225	228	86
Isopropanol, 91%	0.8176	176	178	63
Isopropanol, Anhydrous	0.7861	180	182	53
Methanol	0.7924	147	149	54
Methyl Amyl Alcohol	0.8083	266	271	103
n-Propanol	0.8060	205	208	74
Tetrahydrofurfuryl Alcohol	1.054	352	—	183
TEXANOL[1]	0.950	356	360	248*

[1] Eastman.   *COC

Table 6.105: Celanese Alcohols *(42)*

	Pure Compound Molecular Weight	SPECIFICATIONS Purity wt. % min.	Color Platinum Cobalt Scale, max.	Water Content wt. % max.	Acidity as Acetic wt. % max.	PHYSICAL PROPERTIES Solubility wt. % In H₂O	H₂O In	Distillation Range °C	Normal Boiling Point 760mm Hg. °C	°F	Freezing Point °C	°F	Specific Gravity 20/20°C	Lbs./ Gal. Average 68°F	Flash Point TCC, °F
Methyl Alcohol	32.04	99.85	5	0.10	0.003	Complete	Complete	1.0 incl. 64.7	64.6	148.2	−97.8	−144.0	0.7925	6.59	54
n-Propyl Alcohol	60.09	99.8	5	0.10	0.003	Complete	Complete	1.0 incl. 97.15	97.2	207.0	−127.0	−196.6	0.8044	6.69	71
n-Butyl Alcohol	74.12	99.8	10	0.10	0.005	7.8	20.1	117.0–118.0	117.7	243.9	−89.8	−129.6	0.8109	6.75	98
I-Butyl Alcohol	74.12	99.0	5	0.1	0.003	8.0	15.0	1.5 incl. 107.9	107.9	226.2	−108.0	−162.4	0.8026	6.68	86

Table 6.106: Eastman Alcohols *(41)*

ALCOHOLS	Lb/U.S. Gal at 20°C	Color Pt-Co, Max	Sp Gr at 20°/20°C	Assay, Min Wt %	Acidity, as Acetic Acid, Max Wt %	Flash Point, TCC °F	(°C)	Boiling Range, °C	Freezing Point, °C
TEXANOL® Ester-Alcohol $(CH_3)_2CHCH(OH)C(CH_3)_2CH_2OCOCH(CH_3)_2$	7.90	20	0.950	—	0.2[a]	248	(120) (COC)	244-247	—50
Isobutyl Alcohol $(CH_3)_2CHCH_2OH$	6.68	5	0.803	—	0.005	85	(29)	106-109	—108
n-Butyl Alcohol C_4H_9OH	6.75	5	0.811	—	0.005	97	(36)	117-119	—89
2-Ethylhexanol $C_4H_9CH(C_2H_5)CH_2OH$	6.94	5	0.834	—	0.01	164	(73)	182-186	—70
n-Propyl Alcohol C_3H_7OH	6.71	5	0.805	99.0	0.001	74	(23)	96.2-98.2	—127
Isopropyl Alcohol (99%) $(CH_3)_2CHOH$	6.54	10	0.786	—	0.002	55	(13)	80.8-83.8	—88
CDA-19[b] (anhyd)	6.62[c]	15	0.793-0.796[c]	199 proof[d]	0.006	44	(7)	76-82	—
C_2H_5OH (95%)	6.78[c]	15	0.812-0.820[c]	188 proof[d]	0.006	52	(11)	76-82	—
PROPRIETARY SOLVENTS[e]									
TECSOL® 1 (anhyd)	6.66	20	0.796-0.800[c]	197 proof[d]	0.005	48	(9)	74-80	—114
C_2H_5OH (95%)	6.83[c]	20	0.816-0.820[c]	188 proof[d]	0.005	39	(4)	74-80	—114
TECSOL 3 (anhyd)	6.64[c]	20	0.794-0.797[c]	199 proof[d]	0.005	40	(4)	74-80	—114
C_2H_5OH (95%)	6.78[c]	20	0.813-0.816[c]	189 proof[d]	0.005	40	(4)	74-80	—114
SPECIAL IND SOLVENTS[e]									
TECSOL A (anhyd)	6.57[c]	15	0.793-0.798[c]	199 proof[d]	0.005	55	(13)	75-80	—114
C_2H_5OH (95%)	6.77[c]	15	0.811-0.816[c]	190 proof[d]	0.005	55	(13)	75-80	—114
TECSOL A-2 (anhyd)	6.61[c]	15	0.791-0.796[c]	199 proof[d]	0.005	55	(13)	75-80	—114
C_2H_5OH (95%)	6.77[c]	15	0.810-0.815[c]	191 proof[d]	0.005	55	(13)	75-80	—114
TECSOL B (anhyd)	6.58[c]	15	0.789-0.803[c]	199 proof[d]	0.005	53	(12)	75-80	—114
C_2H_5OH (95%)	6.77[c]	15	0.811-0.816[c]	191 proof[d]	0.005	53	(12)	75-80	—114
TECSOL C (anhyd)	6.65[c]	15	0.795-0.800[c]	198 proof[d]	0.005	53	(12)	75-80	—114
C_2H_5OH (95%)	6.80[c]	15	0.817-0.821[c]	189 proof[d]	0.005	53	(12)	75-80	—114
TECSOL D (anhyd)	6.58[c]	15	0.791-0.796[c]	200 proof[d]	0.005	53	(12)	75-80	—114
C_2H_5OH (95%)	6.74[c]	15	0.810-0.815[c]	191 proof[d]	0.005	53	(12)	75-80	—114
TECSOL D-2 (anhyd)	6.58[c]	15	0.791-0.796[c]	200 proof[d]	0.005	55	(13)	75-80	—114
C_2H_5OH (95%)	6.72[c]	15	0.810-0.815[c]	191 proof[d]	0.005	55	(13)	75-80	—114
TECSOL H (anhyd)	6.58[c]	15	0.787-0.792[c]	200 proof[d]	0.005	55	(13)	75-82	—114
C_2H_5OH (95%)	6.67[c]	15	0.798-0.803[c]	190 proof[d]	0.005	55	(13)	77-82	—114

[a] As isobutyric
[b] Completely denatured
[c] Specific gravity at 15.6°/15.6°C
[d] Apparent U.S. Internal Revenue proof
[e] The number and letter designations for the TECSOL Solvents are the same as those used for formula identification in U.S. Government regulations

Table 6.107: "Epal" Linear Primary Alcohols (82)

Description: EPAL Alcohols are even-numbered, straight-chain cuts of high purity. As carbon numbers increase, these products range from clear, water-white, mobile liquids, to white, waxy solids.

Products	Typical Composition, by wt.		Properties				
			Specific Gravity 25°C/25°C	Hydroxyl value mg. KOH/g. Typical	Distill. range, °C @ 1 atm.	Color APHA, Typical	Flash Point °F (closed cup)
EPAL 6	99.1% hexanol		0.815	549	151-160	5	142
EPAL 8	99.5% octanol		0.821	431	184-195	5	190
EPAL 10	99.4% decanol		0.826	355	226-230	5	235
EPAL 12	99.4% dodecanol		0.832	310	258-264	5	270
EPAL 14	98.4% tetradecanol	0.1% decanol 0.9% dodecanol 0.6% hexadecanol	0.822 (40°C)	262	—	5	300
EPAL 16	98.1% hexadecanol Meets all requirements of USP National Formulary	0.1% decanol 0.2% dodecanol 0.4% tetradecanol 1.2% octadecanol	0.819 (50°C)	232	—	25	347
EPAL 18	97.2% octadecanol	0.1% dodecanol 0.2% tetradecanol 1.4% hexadecanol 1.1% eicosanol	0.813 (60°C)	208	—	20	375
EPAL 108	55.0% octanol 41.0% decanol	4.0% hexanol	0.823	404	197-237	5	190
EPAL 610	42.0% octanol 53.6% decanol	4.3% hexanol 0.1% dodecanol	0.824	393	183-242	5	175
EPAL 810	45.4% octanol 53.9% decanol	0.5% hexanol 0.2% dodecanol	0.824	386	178-240	5	195
EPAL 1012	75.0% decanol 22.8% isomeric & linear dodecanol	0.2% hexanol 1.8% octanol 0.2% tetradecanol	0.828	344	230-265	10	235
EPAL 12/85	86.1 dodecanol 13.7% tetradecanol	0.1% decanol 0.1% hexadecanol	0.827	297	—	5	280
EPAL 12/70	69.5% dodecanol 29.0% tetradecanol	0.5% decanol 1.0% hexadecanol	0.833	288	—	5	275
EPAL 1214	66.3% dodecanol 26.6% tetradecanol	7.1% hexadecanol	0.831	283	233-299	5	280
EPAL 1218	48.2% dodecanol 19.7% tetradecanol 17.2% hexadecanol 14.3% octadecanol	0.3% decanol 0.3% eicosanol	0.834	266	—	5	257
EPAL 1416	62.4% tetracanol 35.9% hexadecanol	0.3 dodecanol 1.4% octadecanol	0.827 (40°C)	250	—	5	290
EPAL 1418 (waxy solid)	35.2% tetradecanol 40.2% hexadecanol 23.1% octadecanol	0.7% dodecanol 0.8% eicosanol	0.825	235	300-315	5	300
EPAL 1618 (waxy solid)	46.5% hexadecanol 50.0% octadecanol	1.5% tetradecanol 2.0% eicosanol	0.819 (50°C)	219	—	5	395
EPAL 1618T (waxy solid)	31.6% hexadecanol 65.2% octadecanol	0.2% dodecanol 1.0% tetradecanol 2.0% eicosanol	0.821 (37.5°C)	216	—	5	350
EPAL 20 + (waxy solid)	66% linear and branched alcohols. C_{18} through C_{32} 34% hydrocarbons, C_{24} through C_{40}		0.814 (50°C)	105	—	300	350

Table 6.108: Procter and Gamble Fatty Alcohols (39)

Product	Hydroxyl Value	Acid Value	Saponification Value	Iodine Value	Moisture (% KF)	P&G Acid Heat Stability (% transmittance at 450 nm)*	Color—APHA	Specific Gravity	Melting Point	Appearance	C_6	C_8	C_{10}	C_{12}	C_{14}	C_{15}	C_{16}	C_{17}	C_{18}	C_{20}	Chemical Abstract Number
	Chemical Properties							Physical Properties			Composition (GLC %)										
CO-810 Octyl/Decyl	380 406 (399)	0.1 max (0.02)	0.6 max (0.2)	0.5 max (0.01)	0.1 max (0.02)		10 max (3)	(0.823) 25°C/25°C	(−15°C)	water white mobile liquid	1.0 max (0.1)	56 64 (59)	(40)	2.0 max (0.7)	(0.1)						68603-15-6
CO-1218 Broad Range	256 270 (258)	1.0 max (0.3)	(1.2)	0.7 max (0.5)	0.2 max (0.1)		40 max (5)	(0.826) 25°C/25°C	(28°C)	white semi-solid		0.2 max (0.1)	1.0 max (0.3)	43 54 (50)	15 24 (19)		9 15 (12)		17 23 (18)	1.0 max (0.2)	67762-25-8
CO-1214 Lauryl	280 290 (285)	0.1 max (0.03)	0.7 max (0.2)	1.0 max (0.1)	0.1 max (0.04)	90 min (96)	10 max (<6)	(0.823) 35°C/25°C	(22°C)	water white mobile liquid		0.3 max (0)	1.0 max (0.3)	65.0 min (67)	21 28 (26)		4.0 8.0 (6.0)		0.5 max (0.2)		67762-41-8
CO-1670 Cetearyl	218 228 (223)	1.0 max (0.15)	2.0 max (0.4)	2.0 max (0.4)	0.1 max (0.04)		35 max (8)	(0.814) 55°C/25°C	(50°C)	waxy white solid				1.5 max (0.1)	1.5 max (0.1)		65 78 (71)		22 35 (26)		67762-27-0
CO-1695 Cetyl	220 235 (228)	0.5 max (0.04)	1.0 max (0.3)	2.0 max (0.6)	0.1 max (0.04)		35 max (6-10)	(0.814) 55°C/25°C	(49°C)	waxy white solid				2.5 max (0.1)	2.5 max (0.1)		95.0 min (96.2)		(2.3)		36653-82-4
CO-1895 Stearyl	200 215 (208)	0.5 max (0.03)	2.0 max (0.3)	2.0 max (0.6)	0.1 max (0.04)		35 max (10-15)	(0.811) 65°C/25°C	(58°C)	waxy white solid							(1.8)		95.0 min (96.9)	(0.6)	112-92-5
CO-1897 Stearyl	200 215 (208)	0.5 max (0.03)	1.0 max (0.3)	2.0 max (0.5)	0.1 max (0.03)		35 max (6-15)	(0.811) 65°C/25°C	(58°C)	waxy white solid							2.5 max (1.4)		97.5 min (98)	(0.5)	112-92-5
TA-1618 Tallow Type	204 216 (213)	1.0 max (0.1)	3.0 max (2)	1.5 max (0.5)	0.1** max (0.05)		35 max (15-25)	(0.810) 65°C/25°C	(53°C)	waxy white solid			0.5 max (0.1)	4.0 max (2.4)	23 min (28)		60 min (67)	(1 4)	1.5 max (0.4)		67762-30-5

*Coleman Jr. Spectrophotometer

**Moisture limit applies only to bulk shipments

Both Cetyl and Stearyl alcohol available as N.F. grade.

Table 6.109: Shell Chemical Alcohols *(14)*

Typical Properties of the Alcohols

	Ethyl Alcohol	Isopropyl Alcohol	Isobutyl Alcohol	Normal Butyl Alcohol	Secondary Butyl Alcohol	Tertiary Butyl Alcohol	Methyl Isobutyl Carbinol	2-Ethyl Hexanol
Molecular Weight	46.07	60.096	74.124	74.124	74.124	74.124	102.178	130.231
Specific Gravity (Apparent)								
60/60°F	0.7936	0.7893	0.8060	0.8135	0.8109	-	0.8107	0.8362
20/20°C	0.7905	0.7864	0.8033	0.8109	0.8080	-	0.8078	0.8338
25/25°C	0.7872	0.7832	0.8006	0.8082	0.8050	0.7817[a]	0.8048	0.8312
Wt. per U.S. Gallon (in air)								
60°F	6.6097	6.574	6.712	6.775	6.753	-	6.751	6.964
20°C	6.577	6.544	6.685	6.748	6.724	-	6.722	6.938
25°C	6.540	6.510	6.654	6.718	6.691	6.50[b]	6.689	6.909
Boiling Point @ 760 mm								
°C	78.32	82.33	107.89	117.73	99.50	82.57	131.8	184.8
°F	173.0	180.19	226.20	243.91	211.10	180.63	269.24	364.64
Boiling Point Change								
°C/mm @ 760 mm	0.0332	0.0325	0.0360	0.0370	0.0349	0.0334	0.0407	0.049
Vapor Pressure @ 20°C, mm	43.9	32.8	8.77	4.3	12.5	44.2	2.2	0.20
Freezing Point @ 760 mm, °C	-114.1	-88.43	-108	-89.3	-114.7	25.66	-90	<-75
Refractive Index, $n\frac{20}{D}$	1.36143	1.37720	1.3959	1.3993	1.3969	1.384[c]	1.4110	1.4328
Heat of Vaporization								
cal/g @ 760 mm	202.0	159.23	139	141.5	134.41	126.94	99.87	93
Heat of Fusion at Melting Pt.								
cal/g	25.765	21.37	-	-	-	21.88	-	-
Specific Heat (liquid)								
cal/g°C	0.574	0.541	0.581	0.564	0.540	0.725	0.52	0.564
Flash Point, Tag Open Cup								
°F, Approx.	60	60	100	110	80	60	131	185
Flash Point, Tag Closed Cup								
°F, Approx.	56	53	86	98	72	52	103	166
Autoignition Temp.								
°F. Approx.	793	750	800	650	761	892	-	-
Flammable Limits in Air								
% of Compound								
Upper	19.0	12v	10.9[d]	11.2v	9.0v	8.00v	5.5v	-
Lower	4.3	2.0v	1.7[d]	1.4v	1.7v	2.35v	1.0v	-
Solubility, % wt.								
in water @ 20°C	complete	complete	8.7	7.7	15.4	complete	1.6[e]	0.07
water in @ 20°C	complete	complete	15	20.1	65.1	complete	6.3	2.6
Azeotrope with Water								
% w compound	96	87.70	67	57.5	72.7	88.3	55.6	20
Boil Pt. @ 760 mm, °C	78.17	80.16	89.8	92.7	87.5	79.91	94.3	99.1
Viscosity, cps								
@ 15°C	-	2.859	-	-	-	-	-	-
@ 20°C	-	-	3.98	2.96	3.78	-	-	8.14
@ 25°C	1.078	2.4	3.4	2.6	2.9	4.5[f]	3.8	7.7
@ 30°C	-	-	-	-	-	3.316	-	-
Surface Tension,								
dyne/cm @ 20°C	22.27	21.35	22.8	24.6	23.0	-	22.8	-

NOTES: a) 26°C   c) (n$\frac{26}{D}$)   e) 25°C

            b) 78°F   d) 212°F   f) supercooled

Table 6.110: Union Carbide Alcohols (19)

Product	Formula	Formula Molecular Weight	Purity of Tested Sample, % by wt.	Apparent Specific Gravity, 20/20°C.	Boiling Point, °C., 760 mm.	Vapor Pressure, mm. Hg at 20°C.	Freezing Point, °C.	Solubility, % by weight at 20°C. In Water	Water In
Ethanol, 190 proof (undenatured)	C_2H_5OH	46.07	(e)	0.8038	78.2	43	−123	Complete	
Ethanol, 200 proof (undenatured)	C_2H_5OH	46.07	(c)	0.7905	78.3	44	−114.1	Complete	
1-Propanol	C_3H_7OH	60.10	(c)	0.8046	97.3	15	−126.2	Complete	
Isopropanol, 91% by vol.	$(CH_3)_2CHOH(f)$	60.10	(e)	0.8179	80.4	34	−50	Complete	
Isopropanol, 95% by vol.	$(CH_3)_2CHOH(f)$	60.10	(e)	0.8045	80.3	36	<−88.4	Complete	
Isopropanol, 99% by vol. (anhydrous)	$(CH_3)_2CHOH$	60.10	(c)	0.7861	82.3	33	−88.5	Complete	
Butanol	C_4H_9OH	74.12	(c)	0.8108	117.7	4	−89.3	7.5 25°C.	20.5 25°C.
Isobutanol	$CH_3CH(CH_3)CH_2OH$	74.12	(c)	0.8030	107.9	7	−108	10.2 25°C.	16.9 25°C.
1-Pentanol	$CH_3CH_2CH_2CH_2CH_2OH$	88.15	(c)	0.8160	137.9	1	−78.2	2.2 25°C.	7.5 25°C.
2-Methyl-1-Butanol, commercial	$C_5H_{11}OH$	88.15	(c)	0.8202	128.7	2	−70.0	3.0 25°C.	9.2 25°C.
Primary Amyl Alcohol	$C_5H_{11}OH$ (mixed isomers)	88.15	99.6	0.8134	133	2	−90(d)	1.7	9.2
2-Methylpentanol	$CH_3CH_2CH_2CHCH_3CH_2OH$	102.18	(c)	0.8254	148.0	1	.....	0.31	5.4
Methyl Amyl Alcohol [4-methyl-2-pentanol]	$(CH_3)_2CHCH_2CH(OH)CH_3$	102.18	(c)	0.8083	131.7	4	−90(d)	1.64 25°C.	6.35 25°C.
Diisobutyl Carbinol [2,6-dimethyl-4-heptanol]	$(CH_3)_2CHCH_2CH(OH)CH_2CH(CH_3)_2$	144.26	99.9	0.8121	178.1	<1	−65	0.06	0.99
2,6,8-Trimethyl-4-Nonanol	$(CH_3)_2CHCH_2CH(OH)$ $CH_2CH(CH_3)CH_2CH(CH_3)_2$	186.34	98.8	0.8193	225.2	<0.01	−60(d)	<0.02	0.60

ALLYL ALCOHOL

Table 6.111: Physical Properties of Allyl Alcohol *(31)*

Boiling point at 760 mm	96.90°C	Specific heat, Cp for liquid, 20-95°C	0.665 g cal/g -°C
Coefficient of expansion at 20°C	0.00101 per °C	Surface tension at 20°C	25.68 dynes/cm
Color (Pt-Co, Hazen)	15 max.	Toxicity	Highly toxic by inhalation and ingestion
Critical temperature	271.9°C		
Distillation range, IBP	95°C, min.	Vapor pressure at 20°C	17.3 mm
DP	98°C, min.	Viscosity at 30°C	0.01072 poises
Fire hazard	Dangerous when exposed to heat or flame	Water	0.3% by wt., max.
Flash point (Open cup)	90°F	Weight per gallon at 20°C	7.11 lbs.
(Closed cup)	72°F		
Freezing point	Becomes a glass at -190°C		
Heat of combustion (vapor)	442.4 kg cal/gm mole		
Ignition temperature in air	443°C		
in oxygen	348°C		
Latent heat of vaporization at 760 mm	9550 cal/mole (295 BTU/lb)		
MAC	5 ppm in air		
Melting point	-129°C		
Molecular weight	58.078		
Purity	98.0% by wt., min.		
Refractive index at 20°C, n_D	1.4134		
Specific gravity at 25/25°C	0.8501		

Table 6.112: Azeotropes of Allyl Alcohol *(31)*

ALLYL ALCOHOL FORMS BINARY AZEOTROPES WITH

%		B. P. of Azeotrope °C
70	Allyl ether	89.8
82.6	Benzene	76.8
70	1-Bromobutane	89.5
91	1-Bromopropane	69.4
17.5	Chlorobenzene	96.5
85	1-Chlorobutane	74.5
71	1-Chloro-3-methylbutane	88.3
93	1-Chloro-2-methylpropane	67.0
80	Cyclohexane	74.0
78.3	Cyclohexene	76.3
89	Diethoxymethane	87.0
46	Ethyl propionate	93.2
55	Ethyl sulfide	85.1
63	Heptane	84.5
95.5	Hexane	65.5
48	Isobutyl formate	93.0
64	3-Methyl-2-butanone	93.5
49	Methyl butyrate	94.7
77	Methyl carbonate	86.4
58	Methylcyclohexane	85.0
72	Methyl isobutyrate	89.8
32	Octane	93.4
30	2-Pentanone	96.0
28	3-Pentanone	96.0
48	Propyl acetate	94.6
26	Propyl alcohol	96.7
70	Propyl ether	85.7
50	Toluene	91.5

CROTYL ALCOHOL

Crotyl alcohol is a clear, stable liquid with a straight-chain, bifunctional molecular structure, $CH_3-CH=CH-CH_2OH$. A highly reactive compound, crotyl alcohol should find use in the manufacture of agricultural chemicals, plastics and polymer additives, varnish ingredients, and pharmaceuticals.

The bifunctionality or two reactive points — hydroxy group and point of unsaturation — account for the high degree of chemical reactivity of crotyl alcohol. The hydroxy group undergoes such reactions as esterification and etherification; whereas the double bond enters into polymerization and addition reactions.

Table 6.113: Physical Properties of Crotyl Alcohol *(41)*

Empirical formula	C_4H_8O
Molecular weight (theoretical)	72.10
Physical form	Clear liquid
Color, APHA, ppm.	15
Purity, by gas chromatography, %	97-98
Acidity, as crotonic acid, %	0.049
Boiling range, 760 mm., °C.	
Initial boiling point	121
Dry point	126
Specific gravity, 20°/20°C.	0.8550
Bulk density, lb./gal., 20°C.	7.12
Flash point, Tag Open Cup, °F.	113 (45°C.)
Fire point, Tag Open Cup, °F.	113 (45°C.)
Isomer concentration (approximate)	3:1 trans:cis
Viscosity, 75°F. (23.9°C.), cs.	32.7
Solubility, 25°C., wt. %	
in water	Completely miscible
water in	with water in all proportions
ethyl alcohol	miscible
acetone	miscible

METHYLBUTYNYL ALCOHOL

Methylbutynol, 2-Methyl-3-Butyn-2-ol $\qquad$ $HCCCOH(CH_3)_2$

Methylbutynyl alcohol is a tertiary acetylenic alcohol with an isoprenoid structure.

Table 6.114: Physical Properties of Methylbutynyl Alcohol *(31)*

Boiling point	104 - 105°C
Fire hazard	Dangerous when exposed to heat or flame
Flash point, Tag open cup	87.4°F
Freezing point	2.6°C
Refractive index at 20°C, n_D	1.4211
Specific gravity, 20/20°C	0.8672
Surface tension at 25°C	23.8 dynes/cm (pure) 41.7 dynes/cm (5% in water)
Vapor pressure at 20°C	12 mm
at 52°C	80 mm
Weight per gallon	7.24 lbs

METHYLPENTYNYL ALCOHOL

Table 6.115: Physical Properties of Methylpentynyl Alcohol *(31)*

Boiling point	121 - 122°C
Fire hazard	Moderate
Flash point, Tag open cup	101.3°F
Freezing point	-30.6°C
Refractive index at 20°C, n_D	1.4318
Solubility in water at 25°C	12.8 g (100 g)
Specific gravity at 20/20°C	0.8721
Surface tension at 25°C	23.8 dynes/cm (pure)
	34.1 dynes/cm (5% in water)
Vapor pressure at 20°C	4 mm
at 68°C	90 mm
Weight per gallon	7.28 lbs

HIGHER UNSATURATED ALCOHOLS

Table 6.116: Unsaturated Aliphatic Alcohols *(69)*

Systematic Name	Common Name	Empirical Formula	Mol. Wt.	Double Bonds	* Boiling Pt. oC.
9:10-Dodecenol	Lauroleyl	$C_{12}H_{23}OH$	184.31	1	157/15 mm
9:10-Tetradecenol	Myristoleyl	$C_{14}H_{27}OH$	212.36	1	
9:10-Hexadecenol	Palmitoleyl	$C_{16}H_{31}OH$	240.41	1	
9:10-Octadecenol	Oleyl	$C_{18}H_{35}OH$	268.46	1	208-209/15 mm
9:10-Eiscosenol	Gadoleyl	$C_{20}H_{39}OH$	296.51	1	
13:14-Docosenol	Erucyl	$C_{22}H_{43}OH$	324.57	1	240.5-241.5/10 mm
9:10, 12:13-Octadecadienol	Linoleyl	$C_{18}H_{33}OH$	266.45	2	148-150/1 mm
9:10, 12:13, 15:16-Octadecatrienol	Linolenyl	$C_{18}H_{31}OH$	264.43	3	
9:10, 11:12, 13:14-Octadecatrienol	Elaeostearyl	$C_{18}H_{31}OH$	264.43	3	
9:10-Octadecen-1,12-diol	Ricinoleyl	$C_{18}H_{34}(OH)_2$	284.47	1	
5:6, 8:9, 11:12, 14:15-Eicosatetraenol	Arachidonyl	$C_{20}H_{33}OH$	290.31	4	
4:5, 8:9, 12:13, 15:16, 19:20-Docosapentenol	Clupanodonyl	$C_{22}H_{35}OH$	316.0	5	

* Ralston, A. W., "Fatty Acids and Their Derivatives", p. 733.
* Hilditch, T. A., "The Chemical Constitution of Natural Fats".
* Brockelsby, H. P., "The Chemistry and Technology of Marine Oils with Particular Reference to Those of Canada". p. 90.

DIACETONE ALCOHOL

Table 6.117: Physical Properties of Diacetone Alcohol *(31)*

Acidity as acetic acid	0.01% by wt, max.	Molecular weight	116.16
Azeotrope with water:		Refractive index at 20°C, n_D	1.4232
boiling point, 760 mm	98.8°C	Relative evaporation rate	
diacetone	12.7% by wt	(n-butyl acetate = 100)	14
Boiling point, 760 mm	169.2°C	Specific gravity at 20°C	0.9406
Coefficient of expansion at 55°C	0.00100	Specific heat at 15°C	0.500 cal/gm/°C
Fire hazard	Moderate	Toxicity	Slight
Flash point, Open cup	155°F	Vapor pressure at 20°C	0.97 mm
Freezing point	-42.8°C	Viscosity at 20°C	3.2 cps
Heat of vaporization, 1 atm.	162 Btu/lb	Water at 20°C	Miscible without turbidity with 19 vol. of 60° Bé gasoline
Hydrocarbon solubility	Complete		
MAC	50 ppm in air	Weight per gallon at 20°C	7.82 lbs

2-MERCAPTOETHYL ALCOHOL

Table 6.118: Physical Properties of 2-Mercaptoethyl Alcohol *(31)*

Boiling point at 760 mm	156.9°C
50 mm	83°C
10 mm	53°C
Coefficient of expansion at 55°C	0.00080
Fire hazard	Moderate
Flash point, Open cup	170°F
Heat of vaporization	257 Btu/lb
Molecular weight	78.13
Refractive index at 20°C, n_D	1.5011
Relative evaporation rate (n-butyl acetate = 100)	13
Solubility in water at 20°C	Complete
Solubility of water in, at 20°C	Complete
Specific gravity at 20/20°C	1.1168
Vapor pressure at 20°C	1.2 mm
Viscosity (absolute) at 20°C	3.4 cps.
Toxicity	Moderate (acute local)
Weight per gallon at 20°C	9.30 lbs

2-ETHYLSULFONYLETHYL ALCOHOL

Table 6.119: Physical Properties of 2-Ethylsulfonylethyl Alcohol *(37)*

Acidity as acetic acid	0.25% max.
Boiling range at 2.5 mm	155 to 156°C
Fire hazard	Slight
Fire point	406°F
Flash point, Tag open cup	370°F
Moisture content	1.5% max.
Molecular weight	138.19
Refractive index at 26°C, n_D	1.4679
Set point	40.5 to 42.5°C
Specific gravity at 45/20°C	1.252 to 1.258 g/ml
Toxicity	Slight
Viscosity at 60°C	12.8 cps.

1,1,1-TRIFLUOROETHYL ALCOHOL

Table 6.120: Physical Properties of 1,1,1-Trifluoroethyl Alcohol *(31)*

Acidity as acetic acid	0.25% max.	Refractive index at 22°C, n_D	1.2907
Boiling point	74.05°C	Specific gravity at 22/25°C	1.387 ± 0.003
Boiling range	72 - 75°C	Surface tension at 32.5°C	20.6 dynes/cm
Fire point	108°F	Viscosity at 20°C	1.996 cps.
Fire hazard	Moderate	at 60°C	0.796 cps.
Flash point, Open cup	90°F		
Freezing point	-43.5°C		
Heat of vaporization (calculated for B.P. 74.05°C)	97.1 g cal/g		
Moisture content	2.0% max		
Molecular weight	100 (calculated)		
Purity	96% min.		

1H,1H,3H-TETRAFLUORO-1-PROPYL ALCOHOL

Table 6.121: Physical Properties of 1H,1H,3H-Tetrafluoro-1-Propyl Alcohol *(31)*

Acid number	0.82
Boiling point at 760 mm	109 - 110°C
Density at 20°C	1.4853 g/ml
Distillation range at 760 mm	90% between 99.5° and 108.5°C
Fluorine content	57.5%
Formula weight	132.06
Hydroxyl number	398
Melting point	-15°C
Moisture content	0.40%
Purity	> 95%
Refractive index at 20°C, n_D	1.3197
Surface tension at 20°C	27.6 dynes/cm

1H,1H,5H-OCTAFLUORO-1-PENTYL ALCOHOL

Table 6.122: Physical Properties of 1H,1H,5H-Octafluoro-1-Pentyl Alcohol *(31)*

Acid number	0.70
Boiling point at 760 mm	140 - 141°C
Density at 20°C	1.6647 g/ml
Distillation range, ASTM, at 760 mm	90% between 133.0° and 141.0°C
Fluorine content	65.5%
Formula weight	232.08
Hydroxyl number	224
Moisture content	0.08%
Purity	> 95%
Refractive index at 20°C, n_D	1.3190
Surface tension at 20°C	24.5 dynes/cm

BENZYL ALCOHOL

Table 6.123: Physical Properties of Benzyl Alcohol *(31)*

Acidity as benzoic acid	0.15% max.	Latent heat of evaporation at 204.25°C	111.58 gm cal/gm
Aldehyde as benzaldehyde	0.50% max.	Molecular weight	108.13
Boiling point	205.3°C	Refractive index at 20°C, n_D	1.5334-1.5397
Chlorine as benzyl chloride	0.15% max.	Solubility in water	1 part in 30 parts of water
Dielectric constant	1.66		
Distillation range, Ibp	195°C min.	Specific gravity at 25/25°C	1.044-1.058
5%	204°C	Specific heat at 15-20°C	0.5402 cal/gm
90%	207°C	Surface tension (c.g.s. units)	39.71
95%	210°C max.	Toxicity	Slight
Electrical conductivity at 25°C	18 x 10⁻⁷ recip. chms.	Vapor pressure at 30°C	0.100 mm
Fire hazard	Slight	Viscosity at 20°C	0.05582 cps.
Flash point (Open cup)	213°F	Weight per gallon at 20°C	9.78 lbs.
Freezing point	-15.3°C		
Heat of combustion	893 kg cal/mole		

Table 6.124: Azeotropes of Benzyl Alcohol (31)

%		B. P. of Azeotrope °C
28	Diethylaniline	204.2
93.5	Dimethylaniline	193.9
50	Ethylaniline	202.8
61	Ethyl isobornyl ether	201.0
89	d-Limonene	176.4
70	Methylaniline	195.8
40	Naphthalene	204.1
86	Thymene	179.0
50	Veratrole	202.5

α-METHYLBENZYL ALCOHOL

Table 6.125: Physical Properties of α-Methylbenzyl Alcohol (31)

Boiling point at 760 mm	203.9°C
Coefficient of expansion at 20°C	0.00080 per °C
Freezing point	21.4°C
Heat of vaporization at 760 mm	180 Btu/lb
Molecular weight	122.16
Optical antipodes, d-form	αD = 42°88'
1-form	αD = 10°94'
Refractive index at 20°C, n_D	1.5211
Solubility in water at 20°C	2.3%
Solubility of water in, at 20°C	5.9%
Specific gravity at 20/20°C	1.0150
Vapor pressure at 20°C	0.01 mm
Weight per gallon at 20°C	8.45 lbs

FURFURYL ALCOHOL

Table 6.126: Physical Properties of Furfuryl Alcohol (46)

General Properties

Molecular Weight	98.10
Boiling Point (at 760 mm)	
°C	170
°F	338
Freezing Point, metastable crystalline form	
°C	−29
°F	−20.2
Freezing Point, stable crystalline form	
°C	−14.63
°F	5.7
Density (at 20°C, 68°F), g/cm³	1.1285
Specific Gravity, 20/20°C	1.1351
Refractive Index	
n_D^{20}	1.4868-1.4870
n_D^{25}	1.4843-1.4845
Vapor Density (air = 1)	3.38
Vapor Pressure (at 31.8°C, 89.2°F), mm Hg	1

Thermodynamic Properties

Heat of Vaporization, cal/g	122
Heat Capacity, cal/g-°C	
liquid at −20°C	0.450
liquid at 0°C	0.472
liquid at 25°C	0.502
stable crystalline form at −40°C	0.256
stable crystalline form at −20°C	0.278
Thermal Conductivity, kcal/m-hr-°C	0.154
Heat of Combustion, kcal/gmole	
at constant volume	608.9
at constant pressure	609.2
Heat of Formation, liquid, kcal/gmole	−66.06
Heat of Fusion, stable crystalline form, cal/g	31.8
Thermal Expansion Coefficient*	
β/°C (−17.8 to 37.8°C)	8.52×10^{-4}
β/°F (0 to 100°F)	4.53×10^{-4}

$$*\beta = \frac{\rho_1^2 - \rho_2^2}{2(t_2 - t_1)\rho_1\rho_2}$$ (Note: ρ = specific gravity, t = temperature)

Fluid Properties

Viscosity (at 25°C, 77°F), cps	5
Surface Tension (at 25°C, 77°F), dynes/cm	38.2

Furfuryl Alcohol-Water Azeotrope (at 760 mm)

Boiling Point of Vapor	
°C	99
°F	210.2
Composition, wt %	
Furfuryl Alcohol	ca 9
Water	ca 91

Other Properties

Physical State	Liquid
Color	Colorless to Yellow
Odor	Mild & Characteristic
Chemical Oxygen Demand, lb/lb FA	1.75
Biochemical Oxygen Demand (5 days, 20°C), lb/lb FA	0.81
Dipole Moment, e.s.u.	1.92×10^{18}
Solubility Parameter, (cal/cm³)$^{1/2}$	12.5
Solubility in	
Water	∞
Alcohol	∞
Ether	∞

Flammability Properties of Commercial QO® Furfuryl Alcohol/FA®

Flash Point	
Tagliabue, closed cup	
°C	77
°F	170
Pensky-Martens, closed cup	
°C	83
°F	182

(Based on flash point, furfuryl alcohol is classified as a Combustible Liquid Class IIIA.*)

Flammability Limits (in dry air at 72.5 − 122°C)	
% by volume	
Lower limit	1.8
Upper limit	16.3
Ignition Temperature	
In air	
°C	391
°F	736
In oxygen	
°C	364
°F	687

*Refers to Code 29 CFR 1910.106 of Federal Regulations.

Table 6.127: Vapor Pressure of Furfuryl Alcohol (46)

Temperature		Pressure	
°C	°F	mm Hg	
31.8	89.2	1	△
40	104	1.8	○
55.5	131.9	5.5	□
56.0	132.8	5	△
60	140	6.3	○
68.0	154.4	10	△
75.5	167.9	16	□
80	176	20.3	○
81.0	177.8	20	△
95.5	203.9	44	□
95.7	204.3	40	△
100	212	53.5	○
104.0	219.2	60	△
108.5	227.3	78	□
115.9	240.6	100	△
120	248	127.4	○
129.5	265.1	194	□
133.1	271.6	200	△
140	284	271.0	○
144.0	291.2	343	□
151.8	305.3	400	△
157.0	314.6	522	□
170.0	338	760	△

□ Quaker Oats Chemicals, Inc., Research Laboratory, unpublished data.
△ D.R. Stull, *Ind. & Eng. Chem.,* 39, 517, 1947
○ G.S. Parks, private communication

Table 6.128: Pounds per Gallon of Furfuryl Alcohol at Various Temperatures (46)

T, °F	T, °C	lbs/gal	T, °F	T, °C	lbs/gal
0	−17.78	9.7380	51	10.56	9.5182
1	−17.22	9.7337	52	11.11	9.5139
2	−16.67	9.7294	53	11.67	9.5096
3	−16.11	9.7251	54	12.22	9.5053
4	−15.56	9.7208	55	12.78	9.5010
5	−15	9.7165	56	13.33	9.4967
6	−14.44	9.7122	57	13.89	9.4924
7	−13.89	9.7079	58	14.44	9.4881
8	−13.33	9.7036	59	15	9.4838
9	−12.78	9.6992	60	15.56	9.4795
10	−12.22	9.6949	61	16.11	9.4752
11	−11.67	9.6906	62	16.67	9.4708
12	−11.11	9.6863	63	17.22	9.4665
13	−10.56	9.6820	64	17.78	9.4622
14	−10	9.6777	65	18.33	9.4579
15	−9.44	9.6734	66	18.89	9.4536
16	−8.89	9.6691	67	19.44	9.4493
17	−8.33	9.6648	68	20	9.4450
18	−7.78	9.6605	69	20.56	9.4407
19	−7.22	9.6562	70	21.11	9.4364
20	−6.67	9.6518	71	21.67	9.4321
21	−6.11	9.6475	72	22.22	9.4278
22	−5.56	9.6432	73	22.78	9.4234
23	−5	9.6389	74	23.33	9.4191
24	−4.44	9.6346	75	23.89	9.4148
25	−3.89	9.6303	76	24.44	9.4105
26	−3.33	9.6260	77	25	9.4062
27	−2.78	9.6217	78	25.56	9.4019
28	−2.22	9.6174	79	26.11	9.3976
29	−1.67	9.6131	80	26.67	9.3933
30	−1.11	9.6087	81	27.22	9.3890
31	−0.56	9.6044	82	27.78	9.3847
32	0	9.6001	83	28.33	9.3803
33	0.56	9.5958	84	28.89	9.3760
34	1.11	9.5915	85	29.44	9.3717
35	1.67	9.5872	86	30	9.3674
36	2.22	9.5829	87	30.56	9.3631
37	2.78	9.5786	88	31.11	9.3588
38	3.33	9.5743	89	31.67	9.3545
39	3.89	9.5700	90	32.22	9.3502
40	4.44	9.5657	91	32.78	9.3459
41	5	9.5613	92	33.33	9.3416
42	5.56	9.5570	93	33.89	9.3373
43	6.11	9.5527	94	34.44	9.3329
44	6.67	9.5484	95	35	9.3286
45	7.22	9.5441	96	35.56	9.3243
46	7.78	9.5398	97	36.11	9.3200
47	8.33	9.5355	98	36.67	9.3157
48	8.89	9.5312	99	37.22	9.3114
49	9.44	9.5269	100	37.78	9.3071
50	10	9.5226			

Table 6.130: Vapor Pressure of Furfuryl Alcohol as a Function of Temperature *(46)*

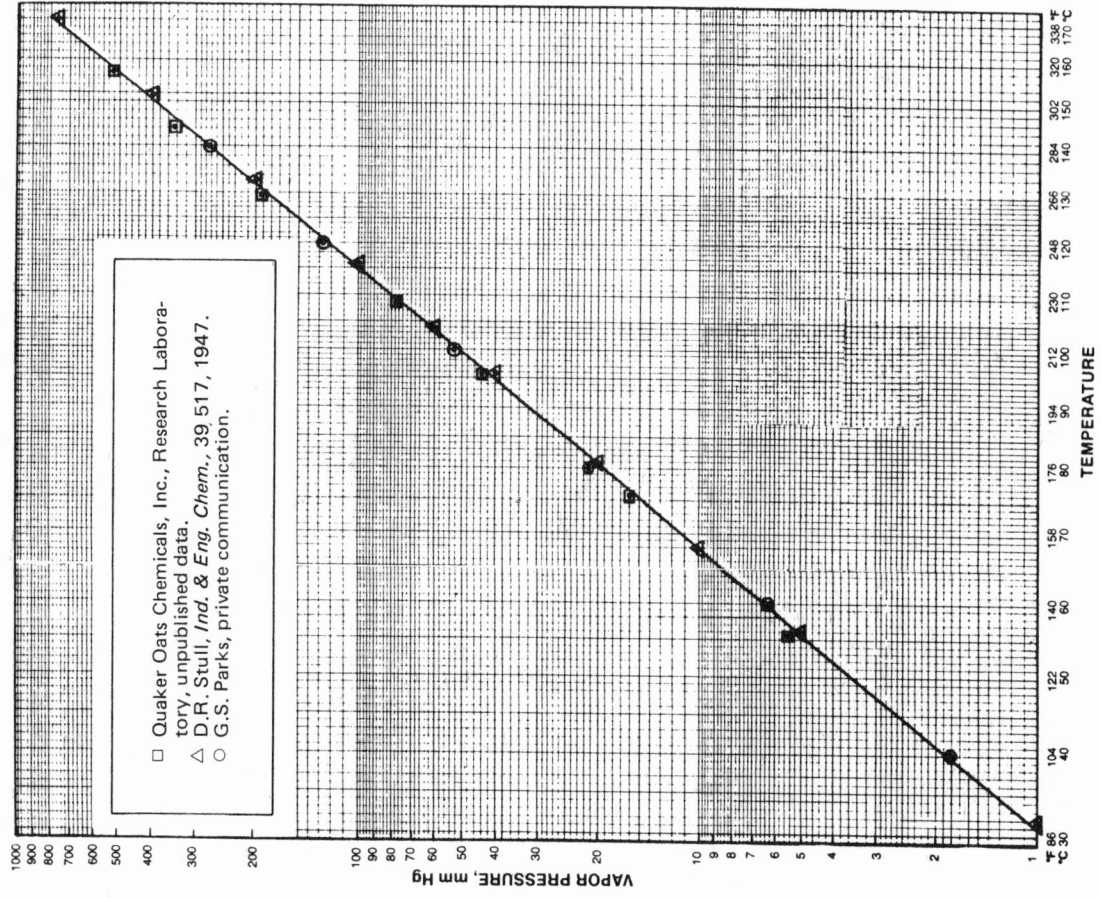

□ Quaker Oats Chemicals, Inc., Research Laboratory, unpublished data.
△ D.R. Stull, *Ind. & Eng. Chem.*, 39, 517, 1947.
○ G.S. Parks, private communication.

Table 6.129: Density of Furfuryl Alcohol-Water Solutions as a Function of Composition (at 25°C, 77°F) *(46)*

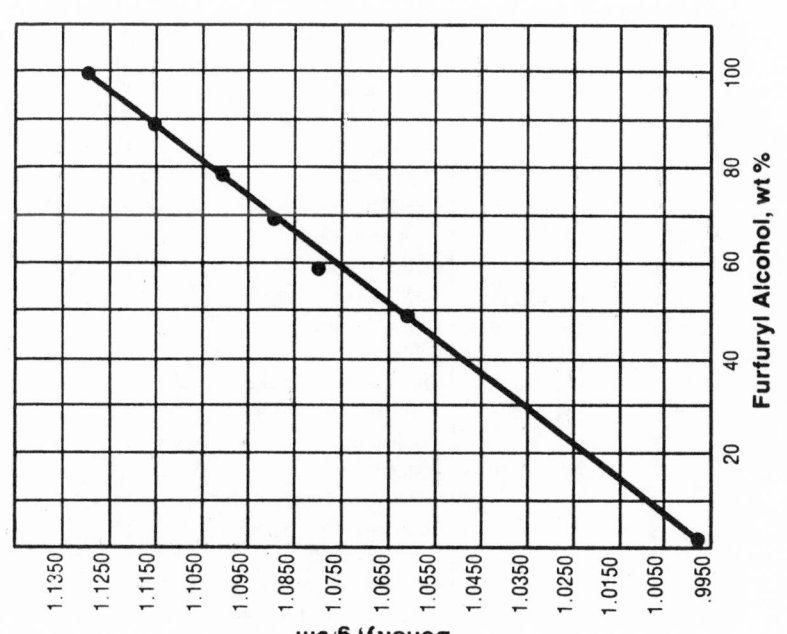

Table 6.131: Solubility of Liquid Organic Compounds in Furfuryl Alcohol (at 25°C, 77°F) (46)

Compound	5 cc/5 cc Furfuryl Alcohol	5 cc/10 cc Furfuryl Alcohol
Acid, dichloroacetic, C.P.	R	
Acid, lactic, U.S.P.	S	
Acid, valeric	S	
Alcohol, amyl	S	
Alcohol, benzyl, tech.	S	
Alcohol, ethyl	S	
Alcohol, isoamyl, tech.	S	
Alcohol, isobutyl, C.P.	S	
Alcohol, isopropyl, tech.	S	
Alcohol, propyl	S	
Aniline	S	
1,2-Butanediol	S	
Chloroform, U.S.P.	S	
Crotonaldehyde, tech.	S	
o-Dichlorobenzene	S	
Dichloroethyl ether, tech.	S	
Diethylaniline, tech.	S	
Diethyl carbonate	S	
Diethylene glycol	S	
Diethylene glycol dioleate	S	
Diethylene glycol monobutyl ether, tech.	S	
Diethyl phthalate, C.P.	S	
Diethyl sulfate, tech.	R	
N,N-Dimethylaniline, tech.	S	
Dimethyl sulfate	R	
Ether, ethyl	S	
Ether, isopropyl	S	
Ethyl acetate, tech.	S	
Ethyl acetoacetate	S	
N-Ethyl-N-benzylaniline	SS	S
Ethyl bromide	S	
Ethylchlorocarbonate	R	
Ethylene chlorohydrin	SS	SS
Ethylene dichloride	S	
Ethylene glycol monobutyl ether, tech.	S	
Glycerol, U.S.P.	S	
Methyl acetate, tech.	S	
Methyl ethyl ketone	S	
Nitrobenzene	S	
o-Nitrotoluene, tech.	S	
Oil, lard	I	I (S at 125°C)
Oil, linseed	SS	SS (S at 120°C)
Oil, neatsfoot	I	I (S at 120°C)
Oil, peanut	I	I (S at 125°C)
Oil, rapeseed	I	I (S at 120°C)
Oil, Turkey red	S	
Oil, whale	I	I (S at 125°C)
Paraldehyde, U.S.P.	S	
Pyridine, tech.	S	
1,1,2,2-Tetrachloroethane, tech.	S	
o-Toluidine, tech.	S	
Xylene	S	

Table 6.132: Solubility of Solid Organic Compounds in Furfuryl Alcohol (at 25°C, 77°F) (46)

Compound	1 g/5 cc Furfuryl Alcohol	1 g/10 cc Furfuryl Alcohol
Acid, acetylsalicylic, U.S.P.	S	
Acid, anthranilic	S	
Acid, benzoic, U.S.P.	S	
Acid, citric, U.S.P.	SS	S
Acid, monochloroacetic	S	
Acid, naphthionic, tech.	I	I (R at 115°C)
Acid, oxalic, tech.	I	R
Acid, stearic	SS	SS (S at 95°C)
Acid, sulfanilic	I	I (R at 115°C)
Acid, tannic	SS	SS (SS at 125°C)
Acid, tartaric, U.S.P.	I	I (SS at 125°C)
Acid, trichloroacetic	R	
Anthracene, tech.	I	SS (S at 110°C)
Anthraquinone	I	I (S at 130°C)
Benzidine	SS	S–R
3-Bromo-d-camphor	S	
Carbazole	I	I (S at 120°C)
Casein	I	I (I at 125°C)
Chloral hydrate, U.S.P.	S	
o-Chloronitrobenzene, tech.	S	
Dextrose	I	I (SS at 125°C)
Dianisidine, tech.	S	
p-Dichlorobenzene	S	
Diglycol stearate	SS	I (S at 100°C)
N,N-Dimethyl-para-nitrosoaniline	S	
Dinitrochlorobenzene, tech.	S	
Dinitronaphthalene	I	I (S at 120°C)
Dinitrophenol	S	
Diphenyl	S	
Diphenylamine	S	
Diphenylguanidine	S	
Hexamethylenetetramine, U.S.P.	S	
Iodoform, U.S.P.	I	I (S at 92°C)
Naphthalene	I	I (S at 92°C)
alpha-Naphthol, tech.	S	
beta-Naphthol, tech.	S	
beta-Naphthylamine, tech.	I	I (S at 92°C)
alpha-Naphthylamine hydrochloride	R (violent reaction)	
m-Nitroaniline	S	
p-Nitroaniline	S	
p-Nitrophenol, tech.	S	
p-Nitrotoluene	S	
m-Phenylenediamine	S	
Resorcinol, white, U.S.P.	S	
Saccharin, U.S.P.	I	I (SS at 125°C)
Sodium acetate	I	I (S at 115°C)
Sodium benzoate, U.S.P.	I	I (I at 125°C)
Sodium naphthionate, tech.	I	I (R at 112°C)
Sodium picramate, tech.	I	
Thiocarbanilide	I	I (S at 92°C)
2,4,6-Tribromophenol, tech.	S	
Triphenylguanidine, tech.	S	

S = Soluble   SS = Slightly soluble   I = Insoluble   R = Reacts*

*Reactions of furfuryl alcohol in the presence of acid or acid generators may be violent; use caution.

Table 6.133: Solubility of Thermoplastic Resins in Furfuryl Alcohol (at Room Temperature) *(46)*

Resin Type	Tradename (Manufacturer)	Solubility
Cellulose acetate butyrate	CAB-500-1 (Tennessee Eastman)	VS
Cellulose nitrate	RS (Hercules)	VS
Ethylcellulose	N-50 (Hercules)	VS
Methyl methacrylate	Plexiglass V(052)100 (Rohm & Haas)	S*
Methyl methacrylate	Plexiglass VM100 (Rohm & Haas)	S*
Nylon	Elvamide 8023 (DuPont)	VS
Nylon	Elvamide 8061 (DuPont)	S*
Nylon	Elvamide 8061M (DuPont)	VS*
Nylon	Elvamide 8064 (DuPont)	S*
Nylon	Elvamide 80625 (DuPont)	S
Nylon	Elvamide PB8066 (DuPont)	S
Nylon	Rilsan BMNO (Rilsan Corp.)	I
Polyethylene	Dowlex 2045 (Dow Chemical)	I
Polyethylene	Dowlex 2598TB (Dow Chemical)	I
Vinyl acetate	Bakelite AYAT (Union Carbide)	VS
Vinyl acetate-chloride	Bakelite VYHH (Union Carbide)	I
Vinyl butyral	Bakelite XYHL (Union Carbide)	VS
Vinylidene chloride	Saran F-310 (Dow Chemical)	I

S = Soluble from 1 g to 10 g per 100 g solvent
VS = Soluble 10 g or more per 100 g solvent
I = Less than 1 g per 100 g solvent
* = Slowly

Table 6.134: Effect of Time at Elevated Temperature on Certain Characteristics of Furfuryl Alcohol (Under Neutral Conditions) *(46)*

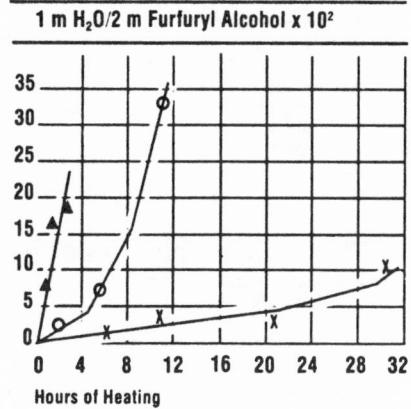

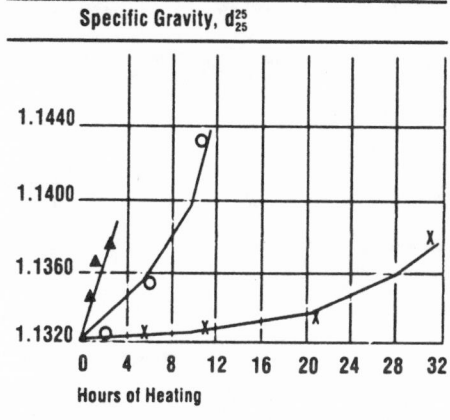

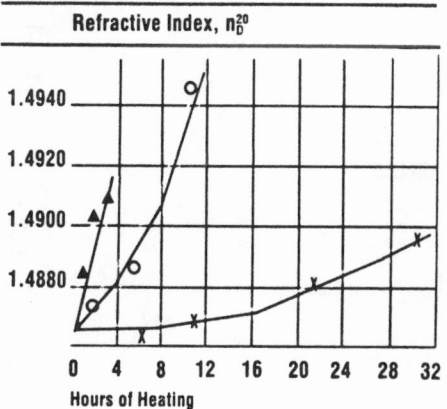

Legend:
x — At 100°C
○ — At 150°C
▲ — At 200°C

A.P. Dunlop and F.N. Peters, Jr., *Ind. & Eng. Chem.*, 34, 814 (1942).

TETRAHYDROFURFURYL ALCOHOL

Table 6.135: Physical Properties of Tetrahydrofurfuryl Alcohol *(46)*

General Properties

Molecular weight	102.13
Boiling point (760 mm), °C	178
Freezing point, °C	below −80
Refractive index (n t/D)	
at 20°C	1.4520
at 25°C	1.4499
Specific gravity	
at 20/20°C	1.0543
at 24/24°C	1.0511
at 31/31°C	1.0450
Vapor density (Air = 1)	3.522
Weight per gal., lbs. at 20°C	8.791

Thermodynamic Properties

Heat capacity, cal./g.	
20-27°C	0.424
30-37°C	0.432
40-47°C	0.445
Heat of combustion, kcal./mole	
at constant volume	708.6
at constant pressure	709.5
Thermal expansion, 20-37.8°C., $\alpha \times 10^{-3}$	0.52

Fluid Properties

Viscosity, absolute, centipoises at 20°C	6.24
Surface tension, dynes/cm. at 25°C	37

Electrical Properties

Dielectric constant at 23°C	13.6

Other Properties

Octane number	82.5
Solubility: Water, alcohol, ether, acetone, chloroform, benzene	∞
Kauri-Butanol value	71.5
Evaporation rate (*n*-butyl acetate = 100)	7
Dilution ratio (lacquer ingredients)	4.5

Flammability Properties

Flash point, Tagliabue, open cup, °F	183
Ignition temperature	
in air, minimum, °C	282
in oxygen, minimum, °C	273
Inflammability, air	
upper limit, % by vol	9.7
lower limit, % by vol	1.5

Table 6.136: Solubility of Various Substances in Tetrahydrofurfuryl Alcohol *(46)*

ACIDS:	Solubility 1 g. or 1 cc./5 cc. THFA	Solubility 1 g. or 1 cc./10 cc. THFA
Acetylsalicylic (U.S.P.)	S	
Anthranilic	S	
Benzoic	S	
Butyric	S	
Citric	SS	S
Cresylic	S	
Lactic (U.S.P.)	SS (S at 120°)	
Naphthionic	SS	S
Oxalic	SS	S
Stearic	I	(S at 100°)
Sulfanilic	I	I (I at 130°)
Tannic	I	I (SS at 130°)
Tartaric (U.S.P.)	SS (Jelly)	
Trichloroacetic	S	
Valeric	S	

ALCOHOLS:		
Benzyl alcohol	S	
Chloral hydrate	S	
Dinitrophenol	S	
Ethanol	S	
Ethylene glycol	S	
Glycerol (U.S.P.)	S	
Isobutanol	S	
Isopropanol	S	
α-Naphthol	S	
β-Naphthol	S	
Pentanol	S	
Propanol	S	

AMINES:		
Aniline	S	
Benzidine	S	
Dianisidine	SS	S
Diethyl aniline	S	
Dimethyl aniline	S	
Diphenylamine	S	
Hexamethylenetetramine (U.S.P.)	I	I (SS at 130°)
β-Naphthylamine	SS	S
m-Phenylenediamine	S	
Pyridine	S	
o-Toluidine	S	
Triphenylguanidine	SS	S
Xylidine	S	

ALDEHYDES:	Solubility 1 g. or 1 cc./5 cc. THFA	Solubility 1 g. or 1 cc./10 cc. THFA
Benzaldehyde	S	
Crotonaldehyde	S	
Paraldehyde	S	

AROMATICS:		
Anthracene	I	I (S at 100°)
Benzene	S	
Dinitronaphthalene	I	I (S at 120°)
Diphenyl	SS	S
Naphthalene	S	
p-Nitrophenol	S	
o-Nitrotoluene	S	
p-Nitrotoluene	I	S
Xylol	S	

ESTERS:		
Amyl acetate	S	
Butyl acetate	S	
Cellulose acetate	S	
Diethyl phthalate	S	
Ethyl acetate	S	
Ethyl acetoacetate	S	
Methyl acetate	S	

ETHERS:		
Dichloroethyl	S	
Diethylene glycol monobutyl	S	
Diethylene glycol monoethyl	S	
Ethyl	S	
Ethylene glycol monobutyl	S	
Ethylene glycol monoethyl	S	

HALIDES:		
Benzyl chloride	S	
Bromobenzene	S	
Bromoform	S	
Chloroform	S	
o-Dichlorobenzene	S	
p-Dichlorobenzene	SS	S
Dinitrochlorobenzene	SS	S
Ethyl bromide	S	
Ethylene chloride	S	
Iodoform (U.S.P.)	S	
o-Nitrochlorobenzene	SS	S
Tetrachloroethane	S	

KETONES:	Solubility 1 g. or 1 cc./5 cc. THFA	Solubility 1 g. or 1 cc./10 cc. THFA
Acetone	S	
Anthraquinone	I	I per (I at 130°)
Ethylmethyl ketone	S	

OILS:		
Aniline	S	
Castor	I	
Chinawood	I	
Coconut	I	
Cottonseed	I	
Lard	I	(S at 120°)
Linseed	I	(S at 120°)
Menhaden	I	(S at 120°)
Neatsfoot	I	(S at 120°)
Peanut	I	(S at 120°)
Rape-seed	I	(S at 120°)
Sperm	I	(S at 120°)
Turkey Red	S	
Whale	I	(S at 120°)

MISCELLANEOUS:		
Caffeine	I	
Camphor, Monobromo	I	
Casein	I	
Chloramine	I	SS (SS at 130°)
Dextrose	I	I (S at 100°)
Sodium acetate	SS	S
Sodium benzoate (U.S.P.)	I	SS (S at 135°)

KEY: S = soluble; **SS** = slightly soluble; **I** = insoluble

Table 6.138: Vapor Pressure of Tetrahydrofurfuryl Alcohol *(46)*

Table 6.137: Vapor-Liquid Equilibria in the Tetrahydrofurfuryl Alcohol-Water System *(46)*

Liquid Phase			Vapor Phase			Boiling Point	
	Mole Fraction			Mole Fraction			
Weight %	THFA	Water	Weight %	THFA	Water	Temperature °C	Pressure mm HG
1.0	0.0018	0.9982	1.0	0.0018	0.9982	100.0	749.5
2.6	0.0048	0.9952	2.0	0.0036	0.9964	100.4	749.5
5.7	0.0106	0.9894	2.3	0.0041	0.9959	100.5	747.8
6.1	0.0113	0.9887	3.4	0.0062	0.9938	100.6	747.8
9.7	0.0186	0.9814	4.3	0.0079	0.9921	100.6	749.5
13.5	0.0267	0.9733	4.9	0.0090	0.9910	101.0	749.2
17.2	0.0354	0.9646	5.7	0.0106	0.9894	101.0	747.8
20.3	0.0430	0.9570	6.2	0.0115	0.9885	101.0	749.0
24.6	0.0544	0.9456	7.2	0.0135	0.9865	101.0	747.8
30.2	0.0709	0.9291	7.3	0.0137	0.9863	102.0	749.5
30.8	0.0728	0.9272	8.0	0.0151	0.9849	101.0	749.0
36.9	0.0936	0.9064	9.6	0.0185	0.9815	102.0	748.9
44.8	0.125	0.875	9.7	0.0186	0.9814	102.0	752.5
44.9	0.126	0.874	10.5	0.0202	0.9798	102.0	753.7
49.3	0.146	0.854	11.3	0.0220	0.9780	102.1	755.9
53.4	0.168	0.832	11.8	0.0231	0.9769	102.5	754.1
58.9	0.202	0.798	13.3	0.0262	0.9738	103.0	754.6
63.6	0.234	0.766	14.5	0.0290	0.9710	103.0	754.9
70.4	0.296	0.704	18.5	0.0384	0.9616	104.0	755.2
77.3	0.375	0.625	18.5	0.0384	0.9616	105.5	755.2
79.3	0.403	0.597	18.9	0.0394	0.9606	106.0	747.8
81.1	0.431	0.569	20.4	0.0430	0.9570	106.0	748.4
82.3	0.451	0.549	20.3	0.0432	0.9568	107.0	741.1
84.9	0.496	0.504	24.7	0.0547	0.9453	107.0	748.4
86.1	0.522	0.478	27.4	0.0625	0.9375	108.0	745.9
87.1	0.546	0.454	40.1	0.106	0.894	107.0	741.9
92.6	0.688	0.312	65.3	0.249	0.751	119.0	745.6
95.4	0.785	0.215	87.5	0.553	0.447	139.5	745.8
98.0	0.896	0.104				148.0	750.9

TERPENES

Table 6.139: Yarmor 302 and 302W Pine Oil *(28)*

YARMOR® 302[1] pine oil is a clear, pale yellow to near water-white oily liquid with a distinct pine-like odor. Derived from terpene oils of pinewood origin, it is a blend of related compounds, mainly terpene alcohols. Yarmor 302 meets requirements of Federal Specification LLL-P-400a for Type I pine oil. It is especially indicated for manufacture of top-performance cleaners and disinfectants and for all other uses where a high-quality pine oil of uniform, highest terpene alcohol content is required.

Product Specifications

Specific gravity at 15.6/15.6°C	0.938-0.946
Total terpene alcohols, %	85 min
Moisture, %	0.5 max
Distillation range, °C	
5%	209 min
95%	225 max

Typical Properties

Specific gravity at 15.6/15.6°C	0.941
Secondary alcohols, %	16.1
Tertiary alcohols, %	75.9
Total terpene alcohols, %	92.0
Moisture, %	0.35
Refractive index at 20°C	1.481
Color, Hazen	20
Kauri-butanol value	>500
Flashpoint, TCC, °F (°C)	172 (78)
Freezing point, °C (°F)	5 (41)
Weight per gal, lbs	7.85

Outstanding Characteristics

Clear, pale color; high terpene alcohol content; piney odor; high solvent activity; excellent wetting, penetrating, and dispersing properties; high bactericidal activity when properly formulated.

Miscible with most common organic solvents. Trace solubility in water.

[1]Hercules pine oil, Yarmor 302, is registered with the Pesticide Division of the Environmental Protection Agency under EPA Registration Number 891-174.

YARMOR® 302W[1] pine oil is a clear, pale-yellow to near-water-white oily liquid with a distinct pinelike odor. Derived from terpene oils of pinewood origin, it is a blend of related compounds, predominantly terpene alcohols with minor amounts of terpene hydrocarbons. It is suitable for all uses where a general-purpose grade of pine oil is required.

[1]Hercules pine oil, Yarmor 302W, is registered with the Pesticide Division of the Environmental Protection Agency under EPA Registration Number 891-176.

Product Specifications

Specific gravity at 15.6/15.6°C	0.918-0.928
Total alcohols, %	70 min
Moisture, %	0.6 max
Distillation range, °C	
5%	190 min
95%	225 max

Typical Properties

Specific gravity at 15.6/15.6°C	0.923
Secondary alcohols, %	7.5
Tertiary alcohols %	64.1
Total alcohols, %	71.6
Monocyclic terpenes, %	28.4
Moisture, %	0.35
Distillation range, °C	
5%	198
95%	220
Refractive index at 20°C	1.480
Color, Hazen	35
Kauri-butanol value	>500
Flashpoint, TCC, °F (°C)	130 (54)
Weight per gal, lbs	7.67

Outstanding Characteristics

Clear, pale color; piney odor; high solvent activity; excellent wetting, penetrating, and dispersing properties; high bactericidal activity when properly formulated; uniform.

Miscible with most common organic solvents. Trace solubility in water.

Table 6.140: Terpineol 318 Prime *(28)*

TERPINEOL 318™ PRIME is a mixture of isomeric terpineols obtained by dehydration of terpin hydrate. It is composed predominantly of *alpha*-terpineol, with lesser amounts of *beta*- and *gamma*-terpineols. At normal temperatures, Terpineol 318 prime is a water-white, oily liquid with a hyacinthlike odor. Terpineol 318 prime is used by chemical specialties manufacturers for its odor, and by the essential-oils industry to produce perfume ingredients, particularly for soaps.

Product Specifications

	Max	Min
Color[a]	0.5	—
Specific gravity (15.6/15.6°C)	—	0.9350
Moisture, %	0.5	—
Gas chromatograph components, %		
Monocyclic hydrocarbons	0.1	—
Terpin-1-ol	2.0	—
Terpin-4-ol	1.0	—
trans-beta-terpineol	14.0	7.0
delta/cis-beta-terpineol	4.5	2.0
alpha/gamma-terpineol	90.0	80.0
High boilers	0.2	—

Typical Properties

Color[a]	0.3
Specific gravity (15.6/15.6°C)	0.9375
Moisture, %	0.3
Gas chromatograph components, %	
Monocyclic hydrocarbons	0.1
Terpin-1-ol	1.3
Terpin-4-ol	0.6
trans-beta-terpineol	10.3
delta-terpineol	0.8
cis-beta-terpineol	2.6
alpha/gamma-terpineol	84.2
High boilers	0.1
Tertiary alcohols	99.2
Freezepoint, °C (°F)	<−10 (+14)
Flashpoint, COC, °C (°F)	88 (190)
Weight/gal, lbs (kg/L)	7.8 (0.94)

Outstanding Characteristics
High purity; light color; pleasant floral odor; excellent solvent; promotes surface activity; resistant to alkalies.

Miscible with most common organic solvents; relatively insoluble in water.

[a] Hercules terpene scale.

Table 6.141: Hercules alpha-Terpineol *(28)*

HERCULES® *alpha*-TERPINEOL is a high-purity grade of the tertiary terpene alcohol *alpha*-terpineol. Derived by fractional distillation of oils extracted from pinewood, it is a water-white, oily liquid at normal temperatures, with an odor suggestive of lilacs. Its chemical nature, pleasant odor, and surface-active properties account for its usefulness to the essential-oil industry and to manufacturers of disinfectants, cleaners, and other chemical specialties.

Product Specifications

Specific gravity at 15.6/15.6°C	0.945 max
Tertiary alcohols, %	94 min
Moisture, %	0.6 max
Distillation range, °C (°F)	
5%	216 (421) min
95%	221 (430) max

Typical Properties

Specific gravity at 15.6/15.6°C	0.939
Tertiary alcohols, %	96.3
Moisture, %	0.10
Distillation range, °C (°F)	
5%	219 (426)
95%	220 (428)
Color, Hazen	20
Flashpoint, COC, °C (°F)	90 (194)
Freezing point, °C (°F)	<25 (<77)
Weight per gal, lbs (kg/L)	7.85 (0.94)

Outstanding Characteristics
High purity; light color; pleasant, distinctive odor; strong masking agent; excellent solvent; promotes surface activity; antibacterial activity when properly formulated.

Miscible with most common organic solvents. Relatively insoluble in water.

MISCELLANEOUS DATA

Table 6.142: Solubility Data for Alcohols (57)

	Cellulose Acetate (LL-1)	CAB 17% Butyryl	CAB 37% Butyryl	CAP 13-15% Propionyl	CAP 31% Propionyl	Ethyl Cellulose	Polystyrene	Methyl Methacrylate	VYHH	AYAF	XYHL	Hydrocarbons	Linseed Oil (Raw)	Rosin	Ester Gum	Shellac	Unvulcanized Rubber	Relative Evap. Rate (n-Butyl Acetate = 100)	Density (lb gal) at 20°C
Methanol	—	—	—	—	—	S	—	—	—	S	S	PS	SS	PS	—	S	—	610	6.60
Ethanol	—	—	—	—	—	S	—	—	—	S	S	S	SS	S	—	S	—	340	6.76
Isopropanol	—	—	—	—	—	PS**	—	—	—	SS	S**	S	SS	S	—	S	—	300	6.55
n-Propanol	—	—	—	—	—	S	—	—	—	SS	S**	S	S	S	—	S	—	—	6.75
n-Butyl alcohol	—	—	—	—	—	S	—	—	—	SS	S**	S	S	S	PS	S	—	45	6.68
Isobutanol	—	—	—	—	—	S	—	—	—	SS	S**	S	S	S	PS	S	—	80	
Mixed amyl alcohols	—	—	—	—	—	S	—	—	—	—	S**	S	S	S	S	S	—	—	—
3-Methoxy butanol	—	—	S	—	—	S	—	—	—	SS	S	S	S	S	S	S	—	12	7.68
Pentanol-3	—	—	S	—	—	S	—	—	—	SS	S**	S	S	S	S	S	—	54	6.84
Methyl amyl alcohol	—	—	—	—	—	S**	—	—	—	SW	S**	S	S	S	S	SS	—	33	6.72
2-Ethylbutanol	—	—	—	—	—	S**	—	—	—	SW	S**	S	S	S	S	S	—	8	6.92
n-Hexanol	—	—	—	—	—	S**	—	—	—	—	S**	S	S	S	S	SS	—	5	6.83
Heptanol-3	—	—	—	—	—	S**	—	—	—	SW	S**	S	S	S	S	SS	—	6	6.84
2-Ethylhexanol	—	—	—	—	—	S**	—	—	—	—	S**	S	S	S	S	S	—	<1	6.94
Diisobutyl carbinol	—	—	—	—	—	PS**	—	—	—	—	SW	S	S	S	S	—	—	2	6.75
Trimethyl nonyl alcohol	—	—	—	—	—	PS**	—	—	—	SW	SW	S	S	S	S	—	—	<1	6.83
Undecanol	—	—	—	—	S	S**	—	—	S	SW	G	S	S	S	S	PS	PS	<1	6.97
Tetradecanol	—	—	—	S	S	SS**	—	—	SS	—	—	S	S	S	S	—	PS	<1	6.95
Heptadecanol	—	—	SS	—	—	—	—	—	—	—	—	S	S	S	S	—	PS	<1	7.05
Trimethyl cyclohexanol	—	—	—	—	S	S	—	—	S	S	S-G	S	S	S	S	—	SW	<1	8.21***
Tetrahydropyran-2-methanol	S	—	S	—	S	S	S	—	S	—	S	—	S	S	S	S	—	3	—
2-Mercaptoethanol	S	S**	S	S	S	S	SS	S	SS	S	S	PS†	Imm	S	S	S	—	13	9.30
Phenyl methyl carbinol	SW	SW	S	SW	S	S	S	SS	S	S	S	PS†	S	S	S	S	—	<1	8.45
Diacetone alcohol	S	PS	S	S	S	S	—	S	S	S	S	S	S	S	S	PS	—	14	7.82

Cellulose Acetate Butyrate

Cellulose Acetate Propionate

Bakelite* Vinyl Resins . . . . (VYHH, AYAF, XYHL)

*UCC trademark.

**0.5 g resin to 9.5 ml solvent.

***At 55°C.

†Miscible with toluene and xylene, immiscible with Apcothinner.

Legend:

S	Soluble
PS	Partly soluble
SS	Slightly soluble
S-G	Soluble, tendency to gel
G	Gel
I	Insoluble
SW	Swelling
Imm	Immiscible

Table 6.143: Melting Points of Saturated Monohydric Alcohols *(69)*

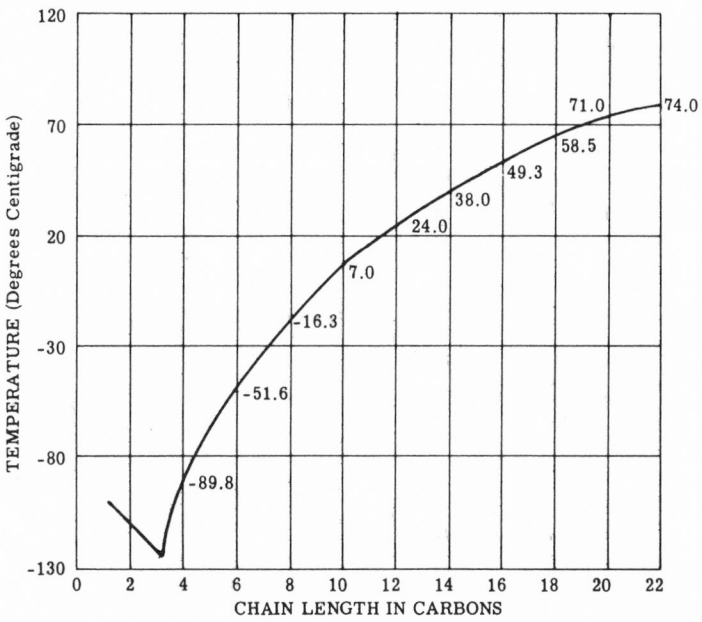

Table 6.144: Rate of Evaporation of Various Solvents at Room Temperature *(19)*

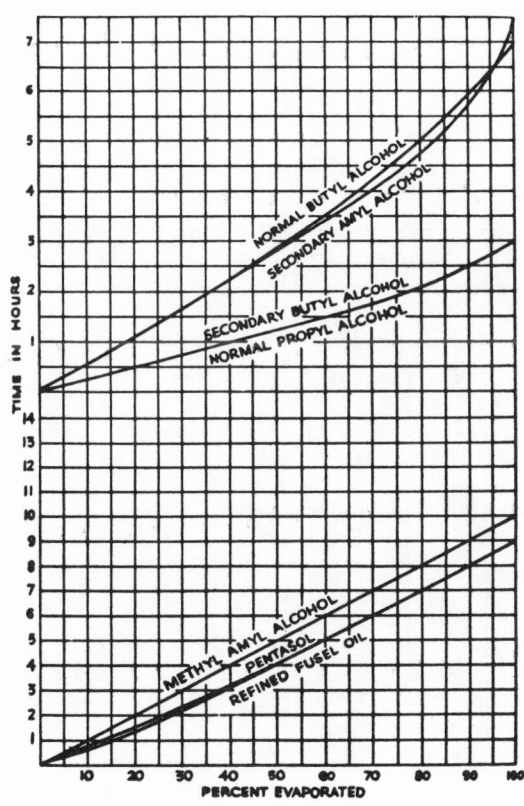

Table 6.145: Comparative Evaporation Rates of Alcohols *(19)*

(Relative Values on 5 cc Samples at 21°C. and 734.4 mm. Hg)

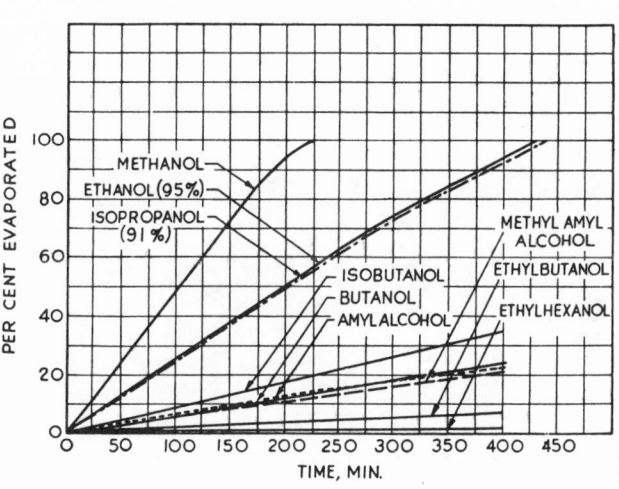

Table 6.147:   Freezing Points (Initial Crystallization) of Aqueous Solutions of Alcohols (19)

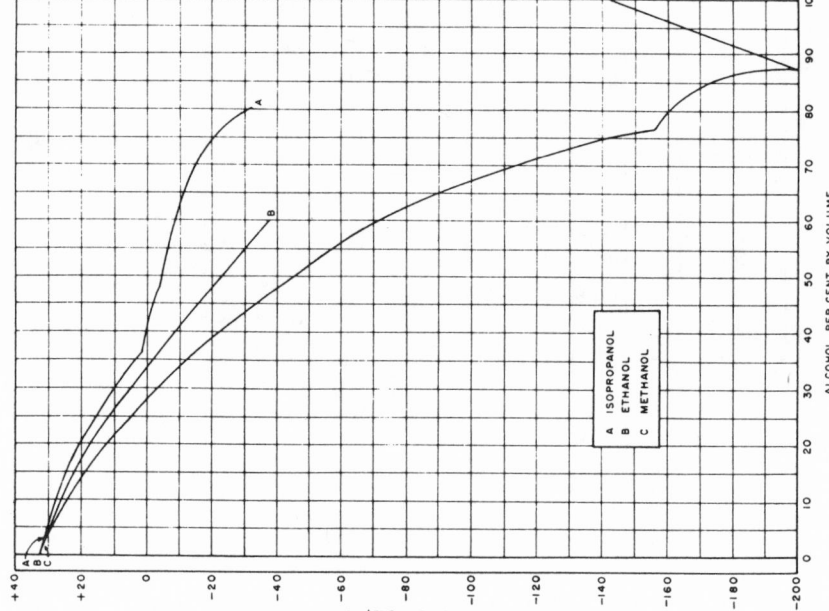

Table 6.146:   Vapor Pressure of Alcohols at Various Temperatures (19)

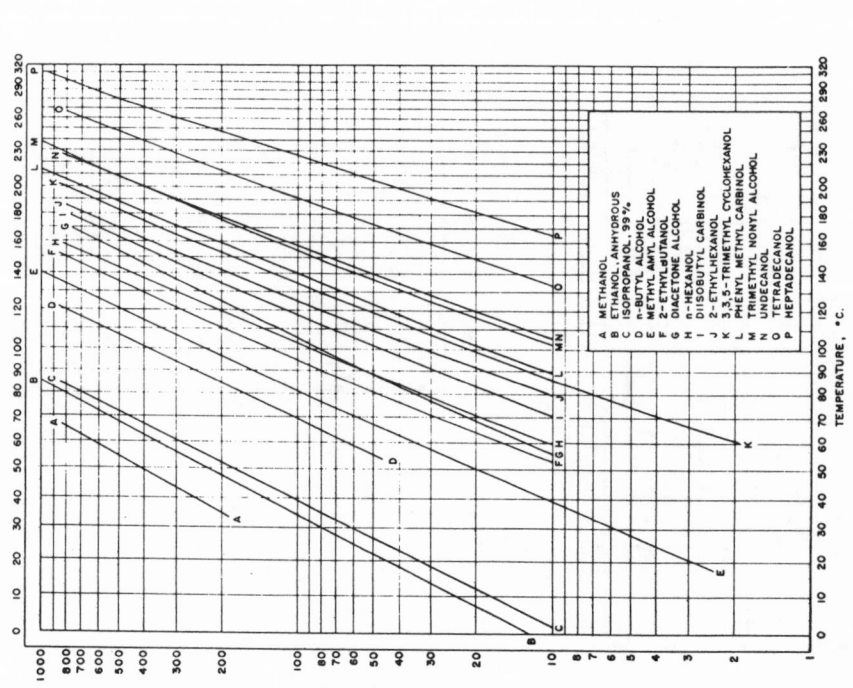

Table 6.148: Specific Gravity of Aqueous Solutions of Alcohols at 20°C *(19)*

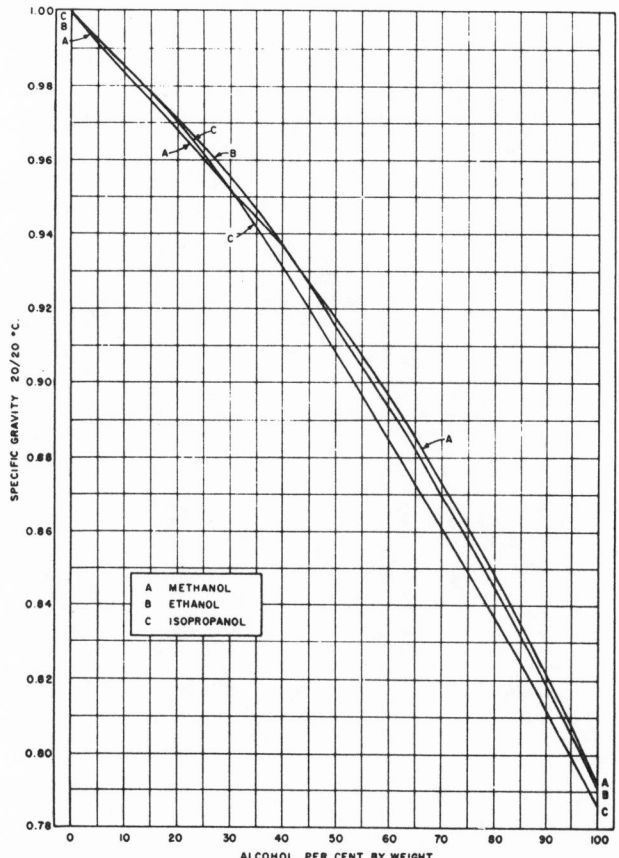

Table 6.149: Relation of Specific Gravity to Freezing Point for Solutions of Antifreeze Base Materials *(70)*

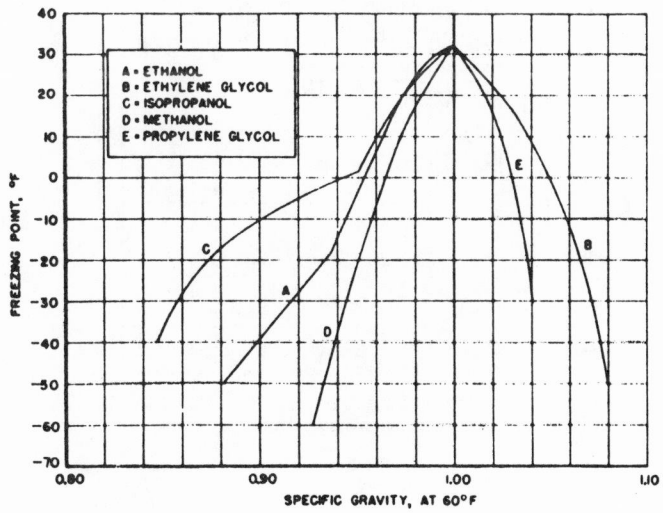

Table 6.150: Evaporation Data for Various Solvents *(14)*

Active Solvents	Rate of Evaporation (Normal Butyl Acetate = 1.00)
Acetone	7.7
Methyl Ethyl Ketone	4.6
Ethyl Acetate (85-90%)	4.6
Isopropyl Acetate 95%	3.9
Secondary Butyl Acetate	1.8
Methyl Isobutyl Ketone	1.6
Methyl Isobutyl Ketone (82.5%w.)— Methyl Isobutyl Carbinol (17.5%w.)	1.0
Normal Butyl Acetate	1.0
Mesityl Oxide	0.9
Secondary Amyl Acetate	0.8
Amyl Acetate (mixed isomers)	0.6
Methyl Amyl Acetate	0.5
"Cellosolve" Acetate	0.2
Diacetone Alcohol	0.2
Butyl "Cellosolve"	0.1

Latent Solvents	
Ethyl Alcohol (anhydrous)	1.9
Isopropyl Alcohol (anhydrous)	1.7
Ethyl* Alcohol (190 proof)	1.7
Normal Butyl Alcohol (50%v.)— Anhydrous Ethyl* Alcohol (50%v.)	0.7
Methyl Isobutyl Carbinol (30%v.)— Anhydrous Ethyl* Alcohol (70%v.)	0.7
Methyl Isobutyl Carbinol (30%v.)— Anhydrous Isopropyl Alcohol (70%v.)	0.7
Normal Propyl Alcohol	1.1
Secondary Butyl Alcohol	1.0
Normal Butyl Alcohol	0.5
Methyl Isobutyl Carbinol (60%v.)— Anhydrous Isopropyl Alcohol (40%v.)	0.5
Secondary Amyl Alcohol	0.5
Amyl Alcohol (mixed isomers)	0.3
Methyl Isobutyl Carbinol	0.3

*Proprietary grade.

Diluents	
Toluene	2.1
Xylene	0.8

Table 6.151: Viscosity of Ethyl Cellulose in Alcohol-Hydrocarbon Mixtures *(14)*

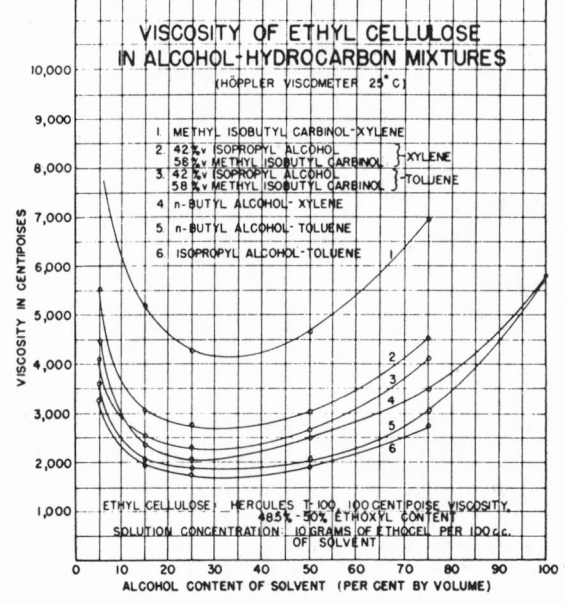

Polyhydric Alcohols

ETHYLENE GLYCOL

Glycol
1,2-Ethanediol

$HOCH_2CH_2OH$

Table 7.1: Physical Properties and Specifications of Ethylene Glycol (32)

Acidity as acetic acid	0.01% by wt., max.	Fire point, Cleveland, tag ASTM, open cup	250° F 245° F
Ash	0.005 g./100 ml., max.		
Boiling point at 760 mm. Hg	197.2-197.6° C	Flash point (open cup) ASTM, open cup	245° F 240° F
Coefficient of expansion at 20° C	0.00062/°C 0.0006375/°C	Free energy of formation at 25° C	-80.2 kcal./mole
Color, APHA	10-15 max.	Heat of combustion (const. pressure) at 20° C	-283.3 kcal./mole
Density (true) at 20° C	1.1134 g./ml.	Heat of dilution [$C_2H_4(OH)_2$ x 2 H_2O]	0.06 cal./g.
		Heat of formation at 20° C	-108.1 kcal./mole
Dielectric constant, 20° C	38.66 esu	Heat of fusion	44.7 cal./g.
Distillation at 760 mm. Hg Ibp	193° C, min.	Heat of vaporization at 760 mm. Hg	191 cal./g. 344 Btu/lb.
5 ml. 95 ml. Dp	194° C, min. 200° C, min. 205-208° C	Inorganic chlorides, as Cl	0.1 ppm, max.
		Iron	0.15 ppm, max.
Electric conductivity at 25° C	1.07 x 10^6 recip. ohms (mhos) cm.		

Molecular weight	62.07	Viscosity at 10° C (50° F) 33.6 cp. 25° C (77° F) 17.4 cp. 35° C (95° F) 12.3 cp. 60° C (140° F) 5.2 cp.
Odor	Mild	
Pour point	-75° F	
Refractive index n_D 25° C n_D 20° C	1.4306 1.4316	Water content 0.3% by wt., max. Weight per gallon at 20° C 9.28 lb.

Specific gravity (apparent), 25/25° C 20/20° C	1.1133 1.1155

Ethylene Glycol

Glycol % by Wt.	% by Vol.	Flash Point °F Cleveland, Tag	Fire Point °F Cleveland
100	100	245	250
95	94.7	260	270
90	89.4	270	280

Specific heat at 20° C at 0° C	0.561 0.544
Spontaneous ignition temperature	398.9° C 412.8° C
Sulfates	Not detectable
Surface tension at 20° C	48.4 dynes/cm.
Suspended matter	Substantially free
Vapor at 20° C (68° F) 25° C (77° F) 93° C (200° F) 132.2° C (270° F)	0.06 mm. Hg 0.12 mm. Hg 11.0 mm. Hg 75.0 mm. Hg

Table 7.2: Boiling Points of Aqueous Ethylene Glycol Solutions *(32)*

Glycol, % by Wt.	% by Vol.	Boiling Point °F	Glycol, % by Wt.	% by Vol.	Boiling Point °F	Glycol, % by Wt.	% by Vol.	Boiling Point °F
0	0.0	212	70	68.4	238	90	89.4	279
10	9.1	214	72	70.5	240	91	90.5	284
20	18.4	216	74	72.6	243	92	91.5	289
25	23.2	217	76	74.7	245	93	92.6	294
30	28.0	218	78	76.8	248	94	93.6	301
35	32.8	219	80	78.9	252	95	94.7	309
40	37.8	221	81	79.9	254	96	95.8	319
45	42.8	223	82	81.0	256	97	96.8	330
50	47.8	225	83	82.0	258	98	97.9	345
55	52.9	227	84	83.1	260	99	98.9	363
60	58.0	230	85	84.1	262	100	100	388
62	60.1	232	86	85.2	265			
64	62.2	233	87	86.2	268			
66	64.2	235	88	87.3	271			
68	66.3	236	89	88.4	275			

Table 7.3: Density of Aqueous Ethylene Glycol Solutions *(32)*

Ethylene Glycol Percentage

By Wt.	0	10	20	30	40	50	60	70	80	90	100
By Vol.	0	9.1	18.4	28.0	37.8	47.8	58.0	68.4	78.9	89.4	100
Temp. °F					Density in g./ml.						
-50							1.110	1.125	1.137		
-40							1.108	1.122	1.134		
-30						1.087	1.105	1.120	1.131		
-20						1.086	1.103	1.117	1.128	1.138	
-10					1.068	1.084	1.100	1.114	1.125	1.135	
0					1.066	1.082	1.097	1.111	1.122	1.131	
10				1.048	1.064	1.080	1.095	1.107	1.118	1.128	1.136
20			1.031	1.147	1.063	1.077	1.092	1.104	1.115	1.124	1.132
30		1.015	1.030	1.045	1.061	1.075	1.089	1.101	1.111	1.121	1.128
40	1.000	1.014	1.029	1.044	1.059	1.073	1.086	1.098	1.108	1.117	1.124
50	1.000	1.013	1.027	1.042	1.056	1.070	1.083	1.094	1.105	1.113	1.120
60	0.999	1.012	1.026	1.040	1.054	1.067	1.080	1.091	1.101	1.109	1.116
70	0.998	1.011	1.024	1.038	1.051	1.064	1.076	1.087	1.097	1.105	1.113
80	0.997	1.009	1.022	1.035	1.049	1.061	1.073	1.084	1.093	1.101	1.109
90	0.995	1.007	1.020	1.033	1.046	1.058	1.069	1.080	1.088	1.097	1.105
100	0.993	1.005	1.018	1.030	1.043	1.054	1.066	1.076	1.085	1.094	1.101
110	0.991	1.003	1.015	1.027	1.039	1.051	1.062	1.072	1.882	1.090	1.097
120	0.989	1.000	1.012	1.024	1.036	1.047	1.058	1.068	1.078	1.186	1.093
130	0.986	0.997	1.009	1.021	1.033	1.044	1.055	1.064	1.074	1.082	1.089
140	0.983	0.994	1.006	1.018	1.029	1.040	1.051	1.060	1.069	1.078	1.085
150	0.980	0.991	1.003	1.014	1.026	1.036	1.047	1.056	1.065	1.074	1.081
160	0.977	0.988	0.999	1.011	1.022	1.032	1.043	1.052	1.061	1.069	1.077
170	0.974	0.985	0.996	1.007	1.018	1.028	1.039	1.048	1.057	1.065	1.073
180	0.970	0.981	0.992	1.003	1.014	1.024	1.034	1.044	1.053	1.061	1.068
190	0.967	0.977	0.988	0.999	1.009	1.020	1.030	1.040	1.048	1.057	1.064
200	0.963	0.974	0.984	0.995	1.006	1.016	1.026	1.035	1.044	1.052	1.060
210	0.959	0.970	0.980	0.991	1.001	1.011	1.021	1.031	1.040	1.048	1.056
220	0.995	0.965	0.976	0.987	0.997	1.007	1.017	1.026	1.035	1.044	1.051
230	0.951	0.961	0.972	0.982	0.992	1.003	1.012	1.022	1.031	1.039	1.047
240	0.947	0.957	0.967	0.978	0.988	0.998	1.008	1.017	1.026	1.034	1.042
250	0.942	0.952	0.963	0.973	0.983	0.993	1.003	1.012	1.021	1.030	1.038
260	0.938	0.948	0.958	0.968	0.978	0.988	0.998	1.008	1.017	1.025	1.033
270	0.933	0.943	0.953	0.963	0.973	0.983	0.993	1.003	1.012	1.020	1.029
280	0.928	0.938	0.948	0.958	0.968	0.978	0.988	0.998	1.007	1.016	1.024
290	0.923	0.933	0.943	0.953	0.963	0.973	0.983	0.993	1.002	1.011	1.019
300	0.918	0.928	0.938	0.948	0.958	0.968	0.978	0.988	0.977	1.006	1.014
310	0.913	0.923	0.933	0.943	0.953	0.963	0.973	0.983	0.992	1.001	1.010
320	0.907	0.917	0.928	0.938	0.948	0.958	0.968	0.977	0.987	0.996	1.005
330	0.902	0.912	0.922	0.932	0.942	0.952	0.962	0.972	0.982	0.991	1.000
340	0.896	0.906	0.917	0.927	0.937	0.947	0.957	0.967	0.976	0.985	0.994
350	0.890	0.900	0.911	0.921	0.931	0.941	0.951	0.961	0.971	0.980	0.989

Table 7.5: Freezing Points of Aqueous Ethylene Glycol Solutions (11)

Ethylene Glycol		Freezing Point		Ethylene Glycol		Freezing Point	
Wt. %	Vol. %	°C	°F	Wt. %	Vol. %	°C	°F
0	0.0	0.0	32.0	40	37.8	-24	-11
2	1.8	-0.6	30.9	42	39.8	-26	-15
4	3.6	-1.3	29.7	44	41.8	-28	-18
6	5.4	-2.0	28.4	46	43.8	-31	-23
8	7.2	-2.7	27.0	48	45.8	-33	-27
10	9.1	-3.5	25.6	50	47.8	-36	-32
12	10.9	-4.4	24.0	52	49.8	-38	-37
14	12.8	-5.3	22.4	54	51.9	-41	-42
16	14.6	-6.3	20.6	56	53.9	-44	-48
18	16.5	-7.3	18.8	58	56.0	-48	-54
20	18.4	-8	17	80	78.9	-47	-52
22	20.3	-9	15	82	81.0	-43	-46
24	22.2	-11	12	84	83.1	-40	-40
26	24.1	-12	10	86	85.2	-36	-33
28	26.0	-13	8	88	87.3	-33	-27
30	28.0	-15	5	90	89.4	-29	-21
32	29.9	-17	2	92	91.5	-26	-15
34	31.9	-18	-1	94	93.6	-23	-9
36	33.8	-20	-4	96	95.8	-19	-3
38	35.8	-22	-7	98	97.9	-16	+3
				100	100.0	-13	+9

Table 7.4: Specific Gravity at 60°F of Aqueous Ethylene Glycol Solutions vs Composition (19)

Table 7.7: Vapor-Liquid Composition Curves for Aqueous Ethylene Glycol Solutions (23)

Figure: Vapor-Liquid Composition Curves for Aqueous Ethylene Glycol Solutions. Axes: TEMPERATURE, DEGREES FAHRENHEIT (vertical, 120–380 °F); MOLE FRACTION ETHYLENE GLYCOL (horizontal, 0–1.0). Curve labels: VAPOR; LIQUID 600 mm Hg; LIQUID 300 mm Hg; LIQUID 100 mm Hg.

Table 7.6: Specific Heat of Aqueous Ethylene Glycol Solutions (32)

					Ethylene Glycol Percentage						
By Wt.	0	10	20	30	40	50	60	70	80	90	100
By Vol.	0	9.1	18.4	28.0	37.8	47.8	58.0	68.4	78.9	89.4	100
Temp. °F					Specific Heat in Btu/lb. °F						
60	0.9996	0.968	0.928	0.882	0.835	0.785	0.734	0.687	0.642	0.599	0.556
70	0.9987	0.968	0.930	0.887	0.841	0.792	0.742	0.695	0.650	0.606	0.563
80	0.9982	0.969	0.933	0.892	0.847	0.799	0.750	0.703	0.658	0.613	0.570
90	0.9980	0.970	0.935	0.896	0.852	0.805	0.758	0.711	0.665	0.620	0.575
100	0.9980	0.971	0.938	0.900	0.858	0.813	0.766	0.719	0.672	0.627	0.581
110	0.9982	0.972	0.940	0.904	0.863	0.819	0.773	0.727	0.680	0.634	0.588
120	0.9985	0.973	0.942	0.907	0.868	0.825	0.780	0.734	0.687	0.640	0.594
130	0.9989	0.974	0.944	0.910	0.872	0.831	0.787	0.740	0.694	0.647	0.600
140	0.9994	0.975	0.947	0.914	0.877	0.837	0.794	0.747	0.700	0.653	0.606
150	1.0001	0.977	0.949	0.917	0.881	0.842	0.800	0.753	0.707	0.659	0.612
160	1.0008	0.978	0.951	0.921	0.886	0.847	0.805	0.759	0.713	0.666	0.619
170	1.0017	0.980	0.954	0.924	0.890	0.852	0.810	0.765	0.720	0.673	0.625
180	1.0027	0.981	0.956	0.927	0.894	0.857	0.816	0.771	0.726	0.679	0.631
190	1.0039	0.983	0.959	0.931	0.898	0.861	0.821	0.777	0.733	0.686	0.637
200	1.0052	0.985	0.961	0.934	0.902	0.866	0.826	0.783	0.739	0.692	0.644
210	1.0067	0.987	0.964	0.937	0.905	0.870	0.831	0.789	0.745	0.698	0.650
220	1.008	0.989	0.966	0.940	0.909	0.875	0.836	0.794	0.750	0.704	0.656
230	1.010	0.992	0.969	0.943	0.913	0.879	0.841	0.799	0.756	0.710	0.662
240	1.013	0.994	0.972	0.947	0.917	0.884	0.846	0.805	0.762	0.716	0.668
250	1.015	0.997	0.976	0.951	0.922	0.889	0.852	0.811	0.768	0.723	0.675
260	1.018	1.000	0.979	0.954	0.926	0.893	0.857	0.817	0.774	0.729	0.681
270	1.021	1.003	0.983	0.958	0.930	0.898	0.862	0.822	0.780	0.735	0.687
280	1.024	1.006	0.986	0.962	0.935	0.903	0.867	0.828	0.786	0.741	0.693
290	1.027	1.010	0.990	0.966	0.939	0.908	0.873	0.834	0.792	0.747	0.700
300	1.030	1.014	0.994	0.970	0.943	0.913	0.878	0.840	0.798	0.754	0.706
310	1.034	1.018	0.998	0.975	0.948	0.918	0.883	0.845	0.804	0.760	0.712
320	1.039	1.023	1.003	0.980	0.953	0.923	0.889	0.851	0.810	0.766	0.718
330	1.044	1.028	1.008	0.985	0.958	0.928	0.894	0.857	0.816	0.772	0.724
340	1.050	1.033	1.013	0.990	0.963	0.933	0.900	0.863	0.822	0.778	0.731
350	1.056	1.038	1.018	0.995	0.968	0.939	0.906	0.869	0.828	0.784	0.737

Table 7.9: Viscosity of Aqueous Ethylene Glycol Solutions (19)

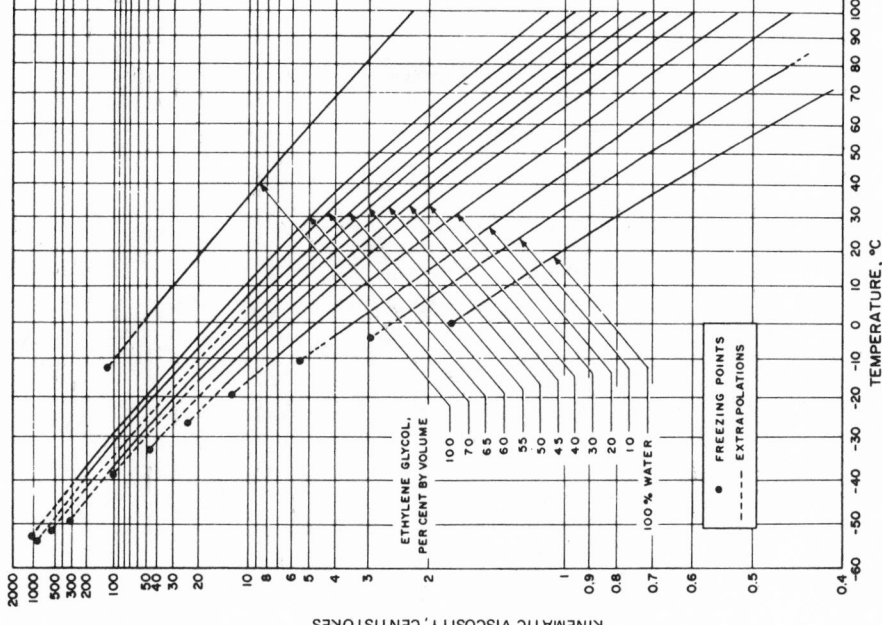

Table 7.8: Vapor Pressure of Aqueous Ethylene Glycol Solutions (11)

Ethylene Glycol Percentage

By Wt.	70	75	80	85	90	95	97	100
By Vol.	68.4	73.6	78.9	84.1	89.4	94.7	96.8	100
Temp. °F	Absolute Pressure in psi							
150	2.2	2.0	1.7	1.4	1.1	0.6	0.4	0.04
160	2.9	2.6	2.2	1.8	1.4	0.8	.5	.06
170	3.6	3.2	2.8	2.3	1.7	1.0	.7	.08
180	4.5	4.1	3.5	2.9	2.2	1.3	.8	.12
190	5.6	5.1	4.4	3.6	2.7	1.6	1.0	.16
200	7.0	6.3	5.5	4.5	3.4	2.0	1.3	0.2
210	8.5	7.7	6.7	5.5	4.1	2.4	1.6	.3
220	10.4	9.4	8.2	6.7	5.0	3.0	2.0	.4
230	12.6	11.4	9.9	8.2	6.1	3.6	2.5	.5
240	15.2	13.7	11.9	9.9	7.4	4.4	3.0	.7
250	18.1	16.4	14.3	11.8	8.9	5.3	3.7	0.9
260	21.6	19.5	17.0	14.1	10.6	6.4	4.4	1.1
270	25.5	23.0	20.1	16.7	12.6	7.6	5.3	1.4
280	30.1	27.1	23.7	19.7	14.9	9.1	6.4	1.8
290	35.2	31.8	27.9	23.2	17.6	10.8	7.6	2.3
300	41.1	37.1	32.5	27.1	20.6	12.7	9.0	2.8
310	47.7	43.1	37.8	31.5	24.0	14.9	10.6	3.5
320	55.2	49.9	43.8	36.6	27.9	17.4	12.5	4.3
330	63.5	57.5	50.5	42.2	32.3	20.2	14.6	5.2
340	72.9	66.0	58.0	48.5	37.2	23.5	17.1	6.3
350	83.3	75.5	66.4	55.6	42.7	27.1	19.8	7.6

Table 7.10: Relative Humectant Values of Aqueous Solutions of Ethylene Glycol *(17)*

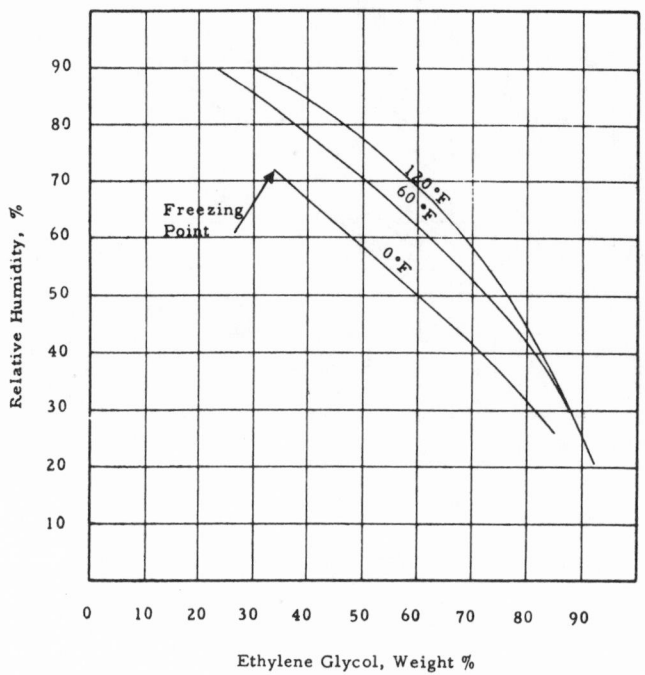

Table 7.11: Water Vapor Dew Points over Aqueous Ethylene Glycol Solutions *(23)*

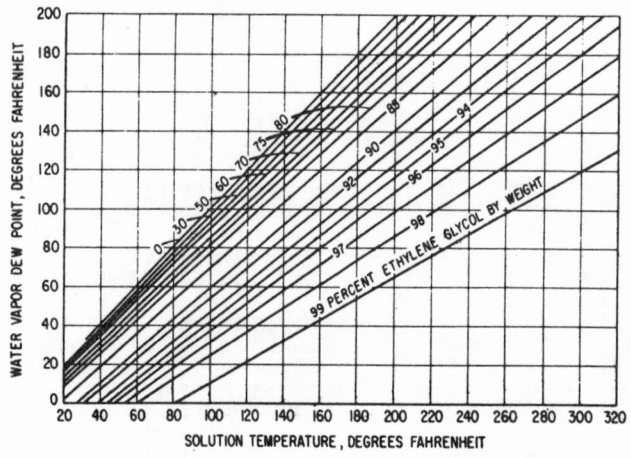

Table 7.12: Key Hygroscopicity Curve for Ethylene Glycol *(55)*

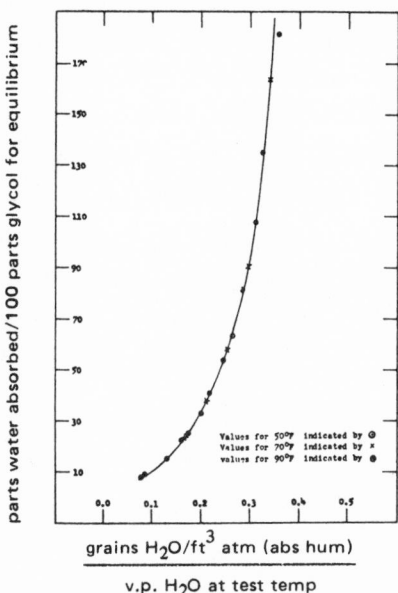

Key hygroscopicity curve for ethylene glycol showing influence of vapor pressure of water at test temperature on amounts of moisture absorbed by ethylene glycol for system equilibrium at various temperatures and various absolute humidities.

Table 7.13: Moisture Absorption of Ethylene Glycol at Various Relative Humidities *(55)*

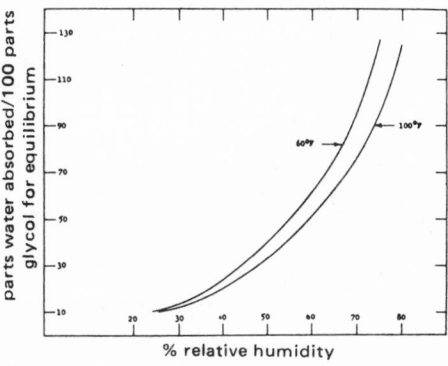

The effect of temperature on the moisture absorption of ethylene glycol for system equilibrium at various relative humidities. Values plotted were calculated from those of the key hygroscopicity curve for ethylene glycol.

Table 7.14: Moisture Absorption of Ethylene Glycol at Various Absolute Humidities *(55)*

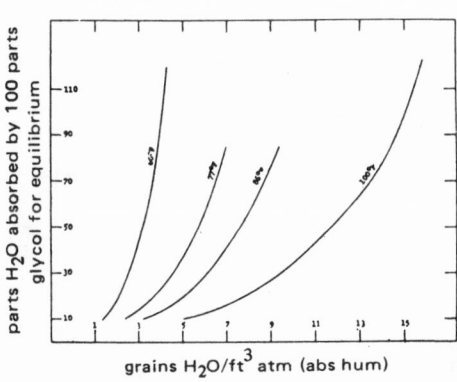

The effect of temperature on the moisture absorption of ethylene glycol for system equilibrium at various absolute humidities. Values used were calculated from those in key hydroscopicity curve for ethylene glycol.

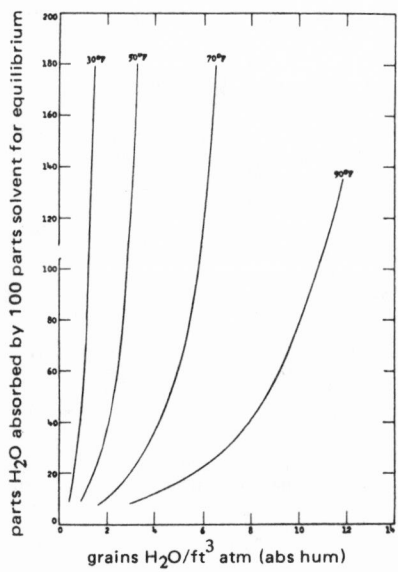

The effect of temperature on the moisture absorption of ethylene glycol for system equilibrium of various absolute humidities. Values plotted were from experimentally obtained data.

Table 7.15: Kinematic Viscosity of Anhydrous Ethylene Glycol and Trimethylene Glycol Solutions *(32)*

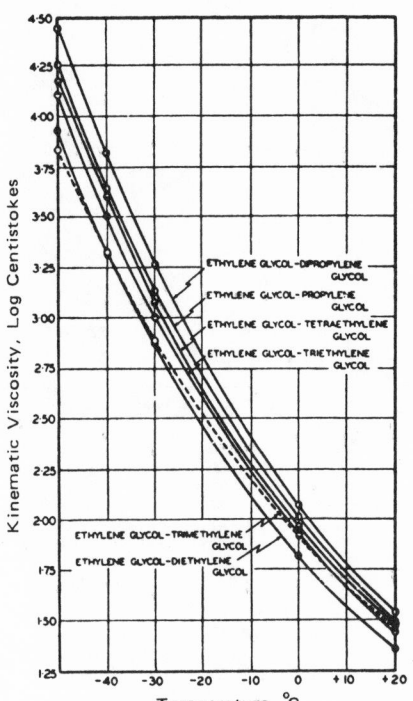

Table 7.16: Freezing Points of Anhydrous Ethylene Glycol and Trimethylene Glycol Solutions *(32)*

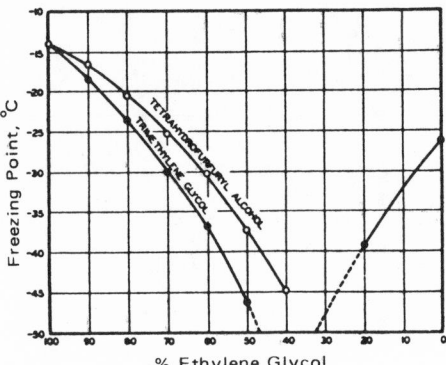

Table 7.17: Azeotropes of Ethylene Glycol *(19)*

	Components			Azeotrope					
Compound	Specific Gravity at 20/20° C	Boiling Point, °C at 760 mm.	Boiling Point, °C at 760 mm.	Composition, % by wt.			Relative Volume of Layers at 20° C	Sp. Gr. 20/20° C of Azeotrope or Layers	
				In Azeotrope	In Upper Layer	In Lower Layer			
Ethylene Glycol	1.1155	197.5	196.2	72.5				1.074	
Butyl Carbitol	0.9536	230.6		27.5					
Ethylene Glycol	1.1155	197.5	139.5	6.4	2	99	U 95	U 0.777	
Dibutyl Ether	0.7694	142.1		93.6	98	1	L 5	L 1.114	
Ethylene Glycol	1.1155	123†	92.7†				U 9.9		
Dichlorethyl Ether	1.2220	96†					L 90.1		
Ethylene Glycol	1.1155	197.5	178.0	26.1				0.959	
Diethyl Carbitol	0.9082	188.4		73.9					
Ethylene Glycol	1.1155	91‡	87‡				U 50		
Di(2-ethylhexyl) Ether)	0.8121	135‡					L 50		
Ethylene Glycol	1.1155	123†	112.8†	35.6	0.1	99.9	U 71.8	U 0.795	
Di-N-hexyl Ether	0.7942	137†		64.4	99.9	0.1	L 28.2	L 1.115	
Ethylene Glycol	1.1155	123†	120.4†	62.3	0.2	98.5	U 37.6	U 1.076	
Diphenyl Ether	1.0677#	161†		37.7	99.8	1.5	L 62.4	L 1.114	
Ethylene Glycol	1.1155	197.5	192.3	64.5	0.22	98.28	U 35.3●	U 1.068#	
Diphenyl Ether	1.0677#	257.4		35.5	99.78	1.72	L 64.7●	L 1.108#	

(continued)

Table 7.17: (continued)

Compound	Specific Gravity at 20/20° C	Boiling Point, °C at 760 mm.	Boiling Point, °C at 760 mm.	Composition, % by wt.			Relative Volume of Layers at 20° C	Sp. Gr. 20/20° C of Azeotrope or Layers
				In Azeo-trope	In Upper Layer	In Lower Layer		
Ethylene Glycol	1.1155	197.5	192	45.5				1.050
Exthoxydiglycol	0.9898	208.8		54.5				
Ethylene Glycol	1.1155	123†	114	4				1.025
Methyl Carbitol	1.0211	115†		96				
Ethylene Glycol	1.1155	157.1▲	149▲	12				1.033
Methyl Carbitol	1.0211	151.2▲		88				
Ethylene Glycol	1.1155	197.5	192	30				1.051
Methyl Carbitol	1.0211	193.6		70				

†At 50 mm. Hg
‡At 10 mm. Hg
 Heterogeneous at 20° C
#At 30/20° C
●At 30° C
▲At 200 mm. Hg

PROPYLENE GLYCOL

1,2-Propanediol $CH_3CHOHCH_2OH$

Table 7.18: Physical Properties of Propylene Glycol *(32)*

Boiling point at 10 mm. Hg	85° C
50 mm. Hg	116° C
760 mm. Hg	187.4° C
ΔBoiling point/Δpressure	0.042° C/mm.Hg
Coefficient of expansion to 20° C	0.695×10^{-3}
to 55° C	0.743×10^{-3}
Evaporation rate (n-butyl acetate—1.0)	0.01
Fire point, ASTM open cup	225° F
Flash point, Cleveland open cup	210° F
Freezing point	-60 (sets to glass below this temperature)
Heat of combustion at 25° C	5728 cal./g. 10,312 Btu/lb.
Heat of vaporization at boiling point at 1 atm.	168.9 cal./g. 304 Btu/lb.
Ignition temperature	421° C
Molecular weight, calculated	76.094
Pour point	-59.5° C
Refractive index, n_D 20° C	1.4326
Specific heat at 20° C	0.593 cal./g./°C
Specific gravity, 20/20° C	1.0381
Δ Specific gravity/Δtemperature, 0 to 40° C	0.00073/°C
Surface tension at 25° C	72.0 dynes/cm.
Vapor density (air—1.0)	2.52
Vapor pressure at 20° C	0.05 mm. Hg 0.08 mm. Hg
Viscosity at 0° C	243 cp.
20° C	56 cp.
40° C	18 cp.
Weight per gallon at 25° C	8.64 lb.

Table 7.19: Propylene Glycol Specifications *(19)*

	Standard Grade	U.S.P. Grade	Air-Treatment Grade	Special Grade
Specific gravity at 20/20° C	1.0370 to 1.0390	1.0375 to 1.0400	1.0375 to 1.0400	1.0380 to 1.0390
Distillation at 760 mm. Hg	Lbp, 185° C, max. 95 ml. 109° C, max. Dp, 194° C, max.	†	†	‡
Propylene glycol, min.	—	97.5% by wt.	97.5% by wt.	99.0% by wt.
Acidity, max.	0.005% by wt.§	0.005% by wt.§	0.005% by wt.§	0.005% by wt. #
Refractive Index at 20° C, n_D	—	—	1.4316 to 1.4335	—
Solubility	—	●	▲	—
Chlorides, max. (as Cl)	0.001% by wt.	0.001% by wt.	0.001% by wt.	0.001% by wt.
Oxidizing substances	—	—	—	none
Carbonyl groups	—	—	—	shall pass test
Sulfates	—	none	—	—
Heavy metals, max. (as Pb)	—	5 ppm	—	—
Lead, max. (as Pb)	—	—	—	0.0003% by wt. **
Arsenic, max. (As_2O_3)	—	1 ppm	—	0.001% by wt. ††
Water, max.	0.5% by wt.	0.2% by wt.	0.5% by wt.	—
Ash, max.	0.005% by wt.	0.005% by wt.	0.007% by wt.	—
Color, max. (Pt-Co Scale)	10	10	15	15
Odor	—	mild	—	mild
Suspended matter	substantially free	substantially free	substantially free	substantially free

†Shall entirely distill within a 5° C range which shall include 187.3° C.
‡Shall entirely distill within a 5° C range, and 90 ml. shall distill within a 2.2° C range.
§Calculated as acetic acid. This is equivalent to 0.047 mg. KOH per g. sample.
#Calculated as hydrochloric acid. This is equivalent to 0.077 mg. KOH per g. sample.
●Miscible in all proportions with water, acetone, and chlorform at 25° C.
▲Completely miscible in all proportions with water at 20° C.
**This is equivalent to 3 ppm.
††This is equivalent to 10 ppm.

Table 7.20: Boiling Points of Aqueous Propylene Glycol Solutions *(19)*

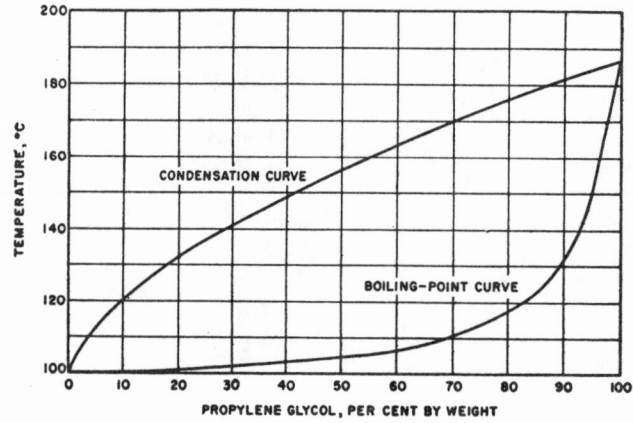

Table 7.21: Conversion Chart for Aqueous Propylene Glycol Solutions *(23)*

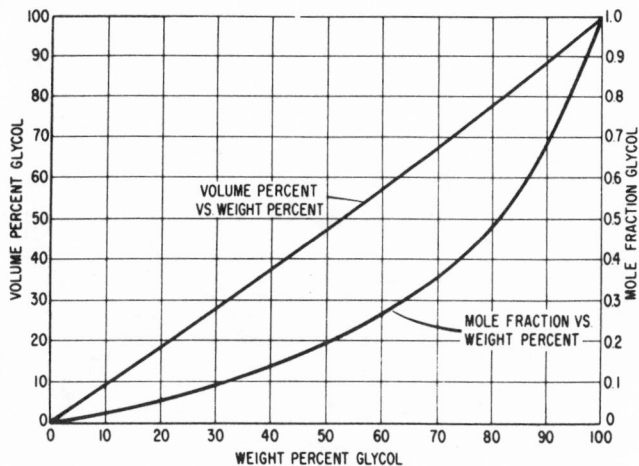

Table 7.22: Density of Aqueous Propylene Glycol Solutions (Percent by Weight) *(23)*

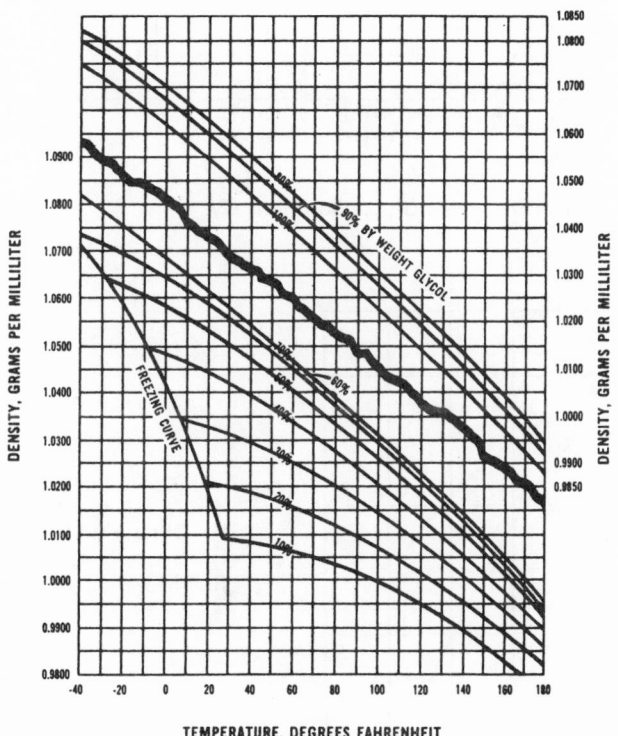

Table 7.23: Effect of Aqueous Propylene Glycol Solutions on Dew Points at Various Contact Temperatures *(19)*

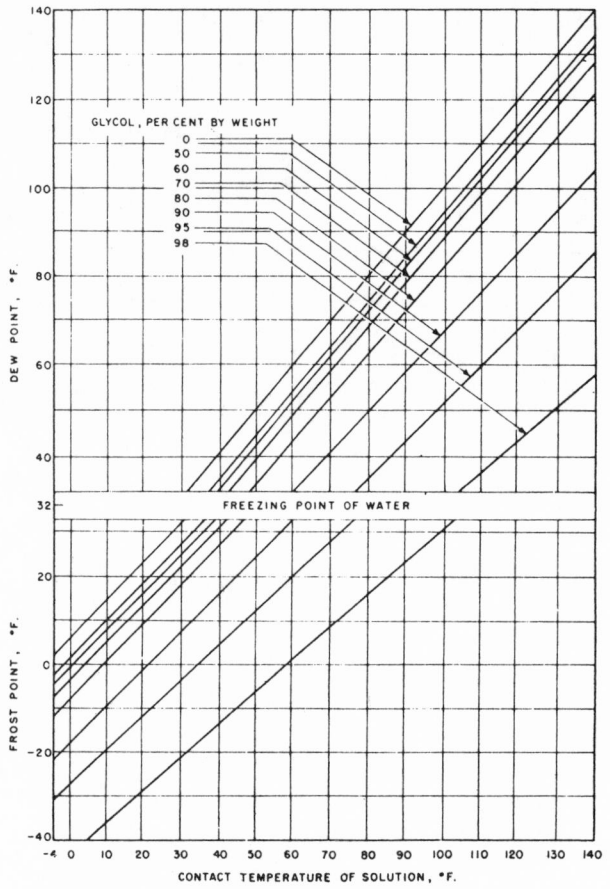

Table 7.24: Freezing Points of Aqueous Propylene Glycol Solutions *(2)*

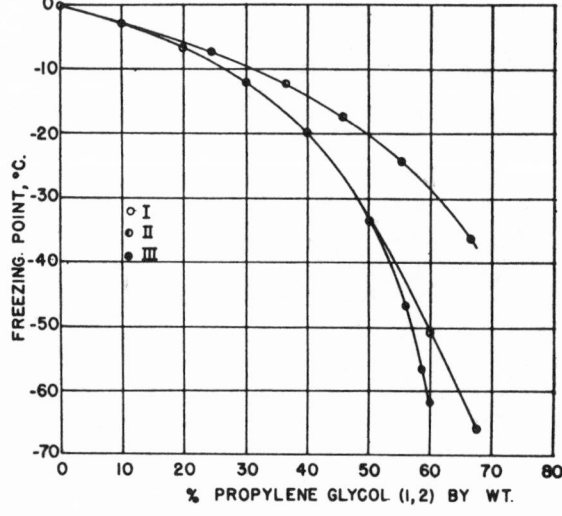

(I)   Observed;

(II) Theoretical, without hydration;

(III) Theoretical, with complete hydration.

Table 7.25: Heat of Vaporization of Propylene Glycol at Various Temperatures *(19)*

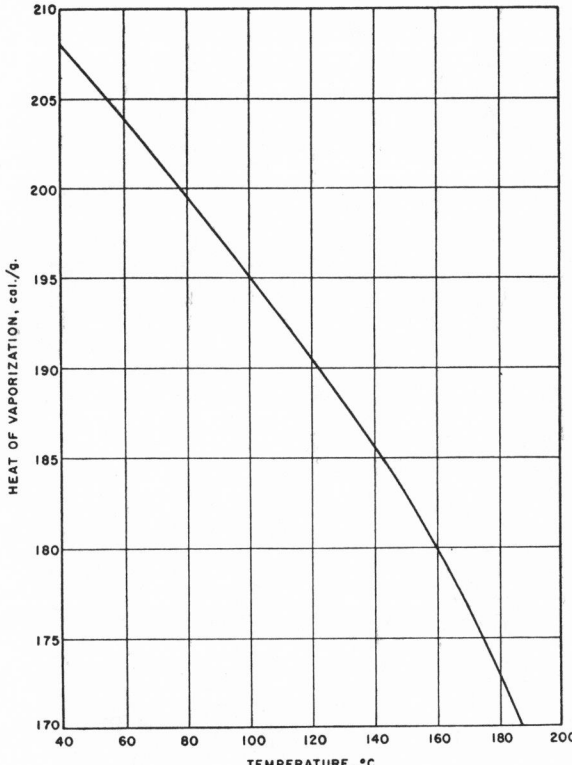

Table 7.26: Refractive Indices of Aqueous Propanediol Solutions at 20°, 30° and 40°C *(32)*

	1,2-Propanediol			1,3-Propanediol			
Glycol, %	n_D^{20}	n_D^{30}	n_D^{40}	Glycol, %	n_D^{20}	n_D^{30}	n_D^{40}
9.94	1.3435	1.3422	1.3411	10.98	1.3433	1.3430	1.3410
20.03	1.3552	1.3540	1.3522	19.96	1.3540	1.3528	1.3511
30.23	1.3670	1.3650	1.3630	30.21	1.3654	1.3640	1.3623
40.01	1.3780	1.3758	1.3732	40.34	1.3770	1.3755	1.3735
49.41	1.3887	1.3863	1.3833	49.94	1.3880	1.3861	1.3839
60.04	1.3995	1.3970	1.3940	60.32	1.3997	1.3975	1.3951
69.50	1.4082	1.4055	1.4028	70.24	1.4103	1.4080	1.4065
79.43	1.4174	1.4144	1.4111	79.87	1.4205	1.4183	1.4155
89.74	1.4252	1.4221	1.4190	89.68	1.4300	1.4276	1.4250
100	1.4324	1.4295	1.4255	100	1.4389	1.4364	1.4332

Table 7.27: Relative Humectant Values of Propylene Glycol, N.F. *(23)*

values are given as the per cent by weight of glycol in water
solutions that will be in equilibrium with air of
various temperatures and humidities

Temperature of Air	RELATIVE HUMIDITIES							
	20%	30%	40%	50%	60%	70%	80%	90%
0° F	93.0	88.0	78.0	73.7	70.0	62.5	45.0	
10° F	93.5	87.5	78.0	73.7	70.5	63.0	46.0	30.0
20° F	93.0	87.5	78.5	73.7	71.0	63.0	47.0	30.0
30° F	92.7	88.0	79.5	74.0	71.0	62.0	48.0	30.0
40° F	93.0	89.5	81.0	76.0	71.5	64.0	50.0	30.0
50° F	93.5	90.5	83.0	77.5	72.0	66.0	51.0	31.0
60° F	93.7	90.8	84.0	78.0	72.0	66.0	52.0	32.0
70° F	94.0	91.0	85.0	78.5	73.0	66.5	52.5	33.0
80° F	94.3	91.2	85.0	79.0	73.0	66.0	52.5	34.0
90° F	94.4	91.2	85.5	79.5	73.5	67.0	53.0	35.0
100° F	94.4	91.25	85.8	80.5	74.0	67.0	53.0	35.0
110° F	94.4	91.26	86.0	81.0	75.0	67.5	53.0	33.0
120° F	94.4	91.27	86.5	81.3	75.0	68.0	54.0	33.0

Table 7.28: Specific Gravity of Aqueous Propylene Glycol Solutions at Various Temperatures *(19)*

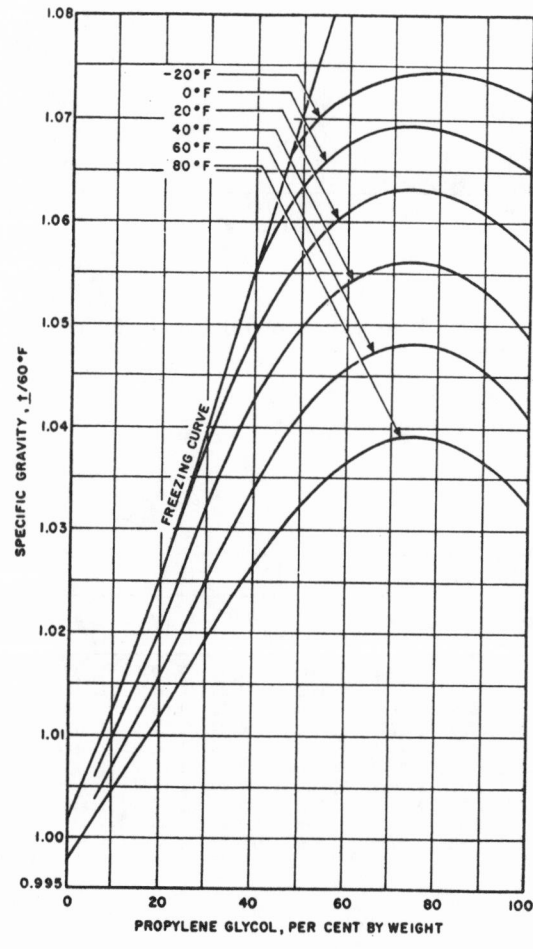

Table 7.29: Specific Heat of Aqueous Propylene Glycol Solutions *(19)*

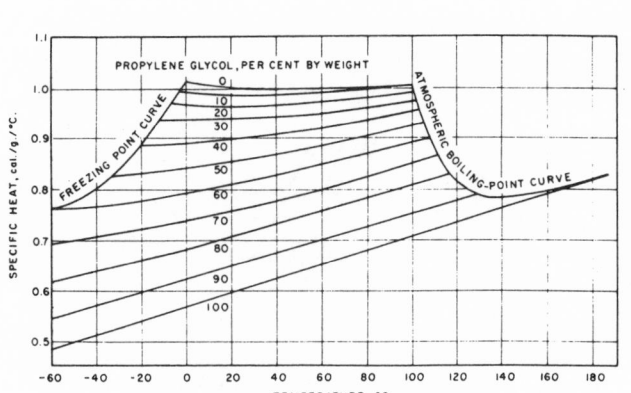

Table 7.30: Thermal Conductivity of Aqueous Propylene Glycol Solutions at Various Temperatures *(19)*

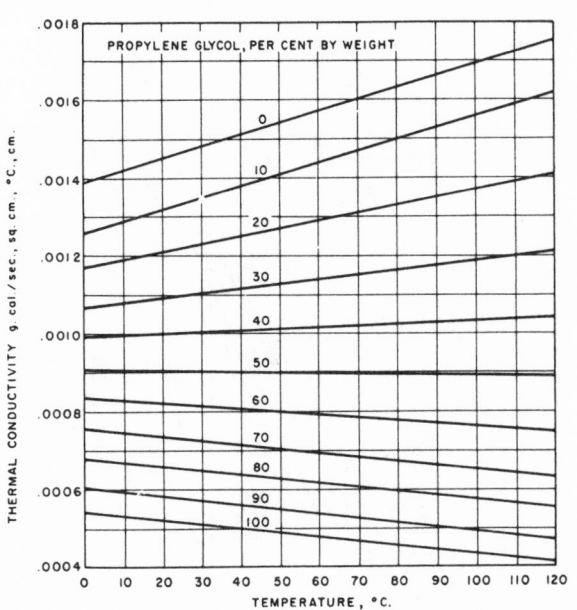

Table 7.31: Total Pressure over Aqueous Propylene Glycol Solutions Versus Temperatures *(23)*

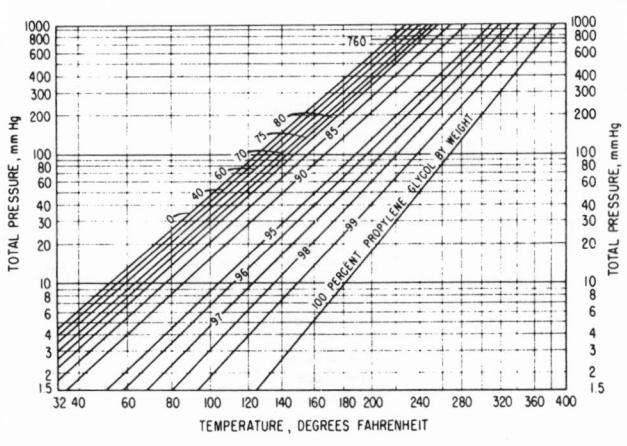

Table 7.32: Vapor-Liquid Composition Curves for Aqueous Propylene Glycol Solutions *(23)*

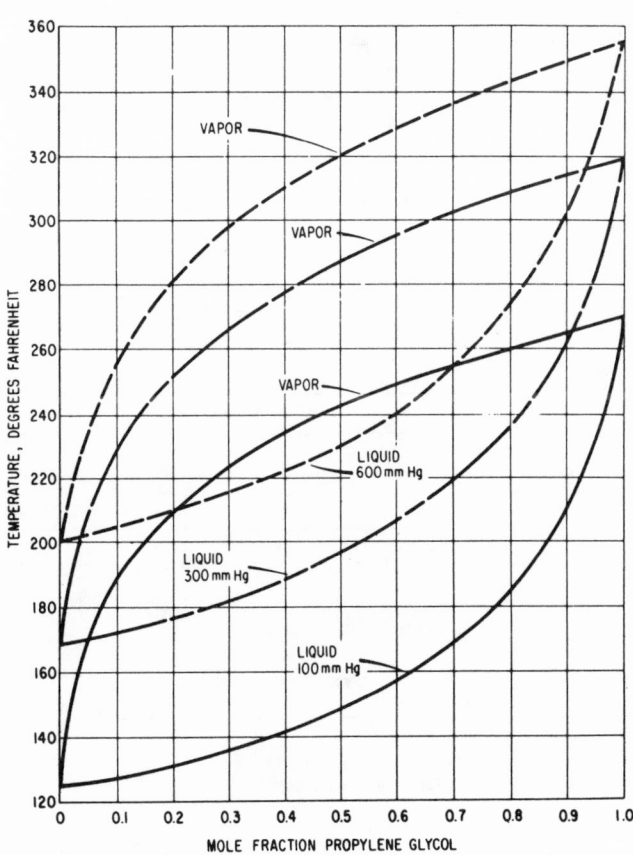

Table 7.33: Vapor Pressures of Aqueous Propylene Glycol Solutions *(19)*

Table 7.34: Viscosities of Aqueous Propylene Glycol Solutions *(23)*

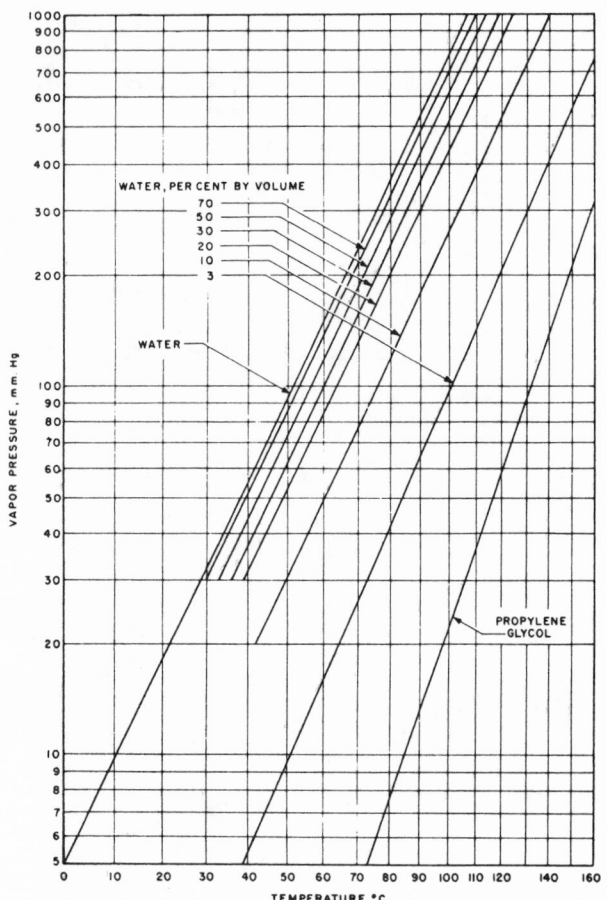

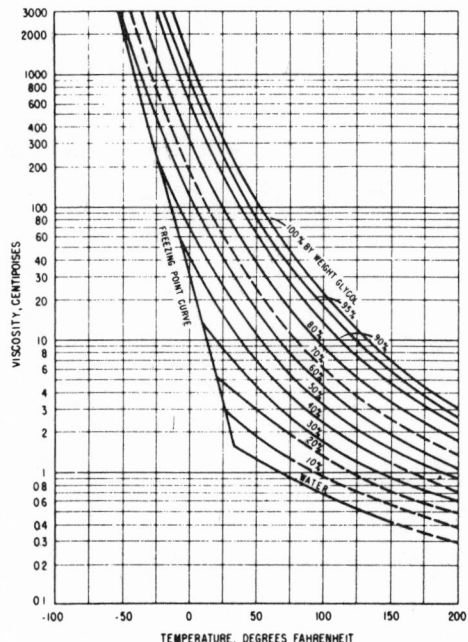

Table 7.35: Azeotropes of Propylene Glycol *(19)*

Components			Azeotrope		
Compound	Specific Gravity at 20/20° C	Boiling Point, °C at 760 mm. Hg	Boiling Point, °C at 760 mm. Hg.	Relative Volume of Layers at 20° C	Specific Gravity at 20/20° C
Propylene glycol	1.0381	187.4	136	Upper layer 93	
dibutyl ether	0.7694	142.1		Lower layer 7	
Propylene glycol	1.0381	85†	84†		‡
di-(2-ethylhexy) ether	0.8121	135†			
Propylene glycol	1.0381	187.4	108	Upper layer 98	
toluene	0.8683	110.6		Lower layer 2	

†At 10 mm. Hg.
‡Heterogeneous at 20° C.

1,3-PROPANEDIOL

Trimethylene Glycol
1,3-Dihydroxypropane
Beta-Propylene Glycol

$CH_2OHCH_2CH_2OH$

Table 7.36: Physical Properties of 1,3-Propanediol *(32)*

Boiling point at 760 mm. Hg	214° C (210-211° C)		

Freezing points of aqueous solutions, °C

10%	-2.86
20%	-6.5
30%	-11.8
40%	-18.8
50%	-27.7
60%	-40.0

Molecular weight	78.1

Refractive indices of aqueous solutions at 20, 30, and 40° C

	n_D^{20}	n_D^{30}	n_D^{40}
11.0%	1.3433	1.3430	1.3410
20.0%	1.3540	1.3528	1.3511
30.2%	1.3654	1.3640	1.3623
40.3%	1.3770	1.3755	1.3735
50.0%	1.3880	1.3861	1.3839
60.3%	1.3997	1.3975	1.3951
70.2%	1.4103	1.4080	1.4065
79.9%	1.4205	1.4183	1.4155
89.7%	1.4300	1.4276	1.4250
100.0%	1.4389	1.4364	1.4332

Specific gravity

at 20/20° C	1.0554
at 0° C	1.0625
at 214° C	0.9028

Thermal expansion of aqueous solutions between 20 and 40° C ($\alpha \times 10^3$)

20%	0.39
40%	0.47
60%	0.55
80%	0.60
100%	0.61

Isothermal contraction in volume on mixing with water between 20 and 40° C (ml. contraction per 100 ml. of initial volume)

	20° C	40° C
20%	0.37	0.29
40%	0.90	0.81
60%	1.19	1.07
80%	1.01	0.89

Table 7.37: Freezing Points of Aqueous Solutions of 1,3-Propanediol *(32)*

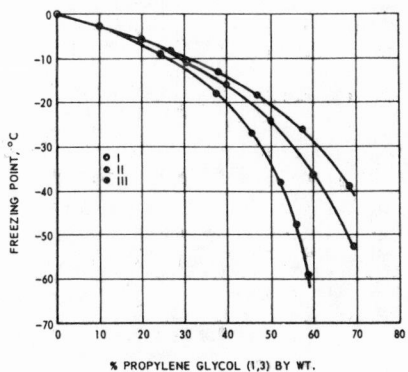

Freezing Points of Propylene Glycol (1,3)-Water Mixtures. (1) Observed; (II) Theoretical, without hydration; (III) Theoretical, with complete hydration.

Table 7.38: Specific Gravity of Aqueous Solutions of 1,3-Propanediol at 20° and 40°C *(32)*

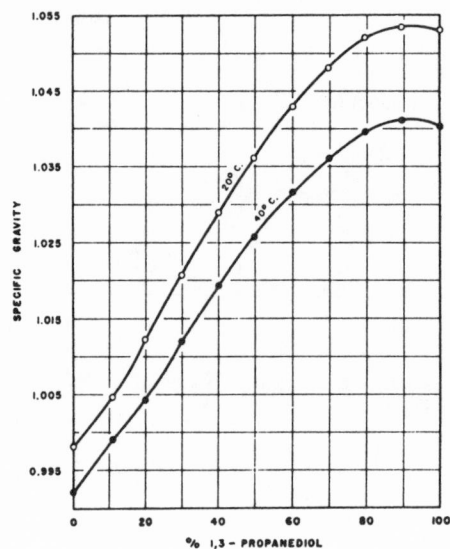

1,2-BUTANEDIOL

Table 7.39: Physical Properties of 1,2-Butanediol *(32)*

Freezing points of aqueous solutions, °C

10%	-2.6
20%	-6.0
30%	-11.0
40%	-16.5
50%	-22.4
60%	-29.0

Refractive indices of aqueous solutions at 20, 30, and 40° C	n_D^{20}	n_D^{30}	n_D^{40}
10.13%	1.3452	1.3436	1.3420
19.69%	1.3572	1.3553	1.3534
29.72%	1.3693	1.3672	1.3650
39.79%	1.3813	1.3788	1.3760
49.68%	1.3920	1.3892	1.3865
59.88%	1.4027	1.4000	1.3966
69.37%	1.4120	1.4090	1.4058
79.73%	1.4211	1.4185	1.4165
69.40%	1.4297	1.4265	1.4230
100.0%	1.4375	1.4347	1.4310

Viscosity of aqueous solutions at 20 and 40° C, in centistokes	20° C	40° C
10.125%	1.520	0.910
19.7%	2.187	1.243
29.7%	3.310	1.690
39.8%	4.802	2.311
49.7%	6.739	3.088
59.9%	9.72	4.227
36.4%	13.82	5.744
79.7%	21.37	8.372
89.4%	35.54	12.59
100.0%	68.0	21.25

Thermal expansion of aqueous solutions between 20 and 40° C ($\alpha \times 10^3$)	
20%	0.454
40%	0.654
60%	0.726
80%	0.765
100%	0.775

Isothermal concentration in volume on mixing with water between 20 and 40° C (ml. contraction per 100 ml. of initial volume)	20° C	40° C
20%	1.12	1.01
40%	1.96	1.67
60%	1.92	1.65
80%	1.27	1.10

Table 7.40: Specific Gravity of Aqueous 1,2-Butanediol Solutions at 20° and 40°C *(32)*

Table 7.41: Absolute Viscosity of Aqueous 1,2-Butanediol Solutions at 20° and 40°C *(32)*

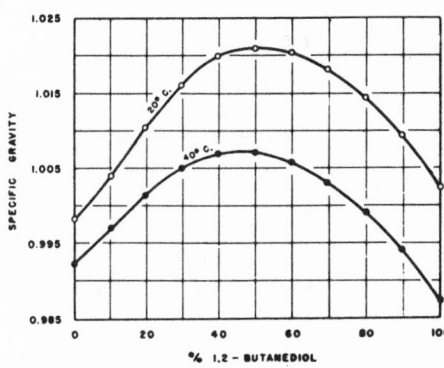

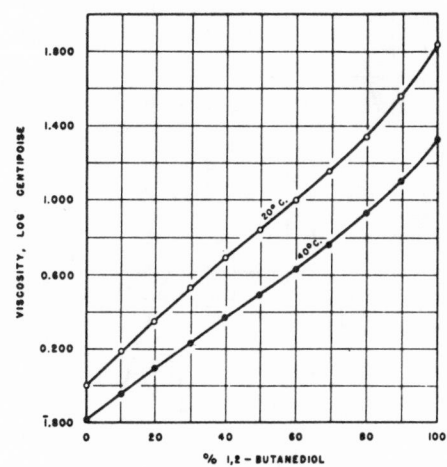

1,3-BUTANEDIOL

1,3-Butylene Glycol   CH3CH(OH)CH2CH2OH

Table 7.42: Physical Properties of 1,3-Butanediol *(32)*

Acid as acetic	0.005% by wt., max.	Refractive index at 20° C/D	1.4401
Boiling point	207.5° C	Solubility (% by weight)	
		in castor oil	18%
Color, APHA	15, max.	in ether	7%
		either in	9%
Distillation range	200–215° C	in ethyl acetate	32%
		ethyl acetate in	41%
Flash point, tag open cup	250° F	in dibutyl phthalate	2%
Freezing point	Below –50° C	Specific gravity at 20/20° C	1.0062
Heat of vaporization	155 cal./g.	Surface tension at 25° C	37.8 dynes/cm.
Hygroscopicity, weight % water absorbed in 144 hours at:		Vapor pressure at 20° C	0.06 mm. Hg
25–28° C and 81% relative humidity	38.5	Viscosity at 25° C	104 cp.
25–28° C and 47% relative humidity	12.5	at 35° C	89 cp.
25–28 C and 20% relative humidity	4.3		
		Water	0.5% by wt., max.
Molecular weight, calculated	90.12		
		Weight per gallon at 20° C	8.38 lb.
Purity	95% by wt., min.		

Table 7.43: Freezing Point of Aqueous Solutions of 1,3-Butanediol *(32)*

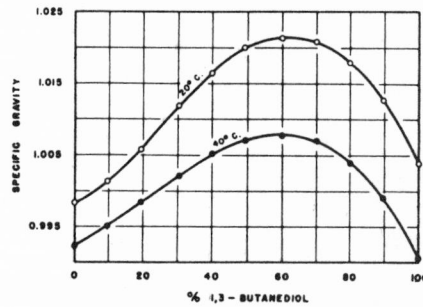

Table 7.44: Refractive Index and Freezing Point of Aqueous Solutions of 1,3-Butanediol *(32)*

Content of 1,3-Butanediol, % by Weight	n 25°C D	Freezing Point °C	Freezing Point °F
19.4	1.3561	-4	+25
39.4	1.3806	-15.5	+4
49.3	1.3922	-25	-13
58.5	1.4032	-37	-35
64.5	1.4093	-42	-44
69.0	1.4138	-51	-60
79.5	1.4237	Viscous	
89.0	1.4319	liquid	

Table 7.45: Specific Gravity of Aqueous 1,3-Butanediol Solutions at 20° and 40°C *(32)*

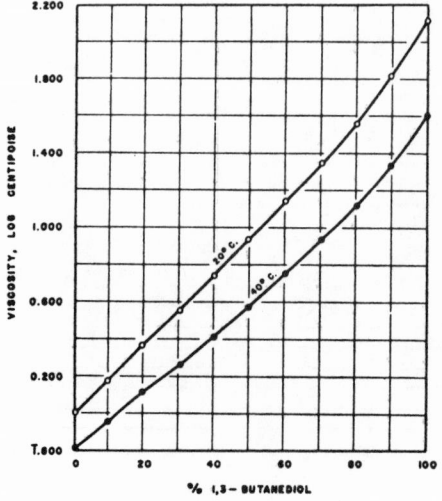

Table 7.46: Viscosity of Aqueous Solutions of 1,3-Butanediol *(32)*

Content of 1,3-Butanediol, % by Weight	Viscosity, centipoises 25.0° C	Viscosity, centipoises -171° ± C	Viscosity, centipoises -37 ± 1° C
19.4	2.1		
39.4	4.7		
49.3	6.7	95	
58.5	10.2	172	
69.0	16.7	304	
79.5	27.7	620	7,000
89.0	50.8	1,360	18,500
100.0	98.3	3,150	35,000

Table 7.47: Absolute Viscosity of Aqueous 1,3-Butanediol Solutions at 20° and 40°C *(32)*

1,4-BUTANEDIOL

Tetramethylene Glycol $HOCH_2CH_2CH_2CH_2OH$

Table 7.48: Physical Properties of 1,4-Butanediol *(32)*

Acetals (as CH_2O)	Less than 0.8%	Freezing point	20.9° C
Acidity (as HCO_2H)	Less than 1%		
Ash	0%	Refractive index, n_D^{25}	1.4446
Boiling range	221-231° C		
1-Butanol	Less than 0.5%	Solubility at 25° C (g./100 ml. solvent)	
Flash point (ASTM open cup)	More than 250° F	in water	Infinite
Free aldehyde as CH_2O	Less than 0.1%	in methanol	Infinite
Freezing point range	18-19.5° C	in ethanol	Infinite
Purity	Over 96%	in acetone	Infinite
Refractive index, n_D^{25}	1.4435-1.4445	benzene	0.3
Specific gravity, d_4^{25}	1.012-1.016	carbon tetrachloride	0.4
Unsaturation (as butendiol)	Less than 1%	chlorobenzene	0.4
Viscosity, 25° C	65-70 cp.	ethyl acetate	14.1
Water content	Less than 0.8%	ethyl ether	3.1
		petroleum ether (35-60° C)	0.9

Pure 1,4-Butanediol

Specific gravity, d_4^{25} 1.0154

Boiling point at			% Water in 1,4-Butanediol	Freezing Point (°C)	Viscosity (cp. at 25° C)
10 mm. Hg	118° C		0.0	20.0	71.5
20 mm. Hg	133° C		0.1	19.8	71.3
100 mm. Hg	170° C		0.5	19.0	70.2
200 mm. Hg	187° C		1.0	18.1	68.9
760 mm. Hg	228° C				

Table 7.49: Absolute Viscosity of Aqueous 1,4-Butanediol Solutions at 20° and 40°C *(32)*

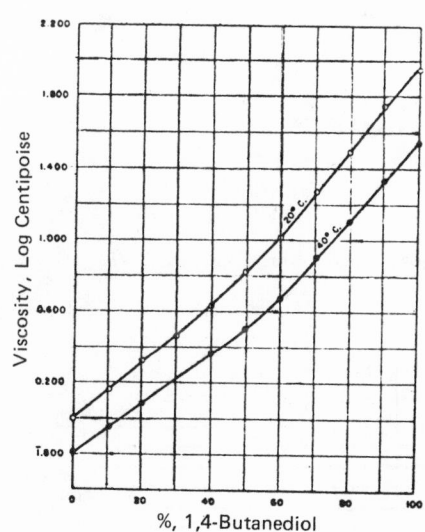

Table 7.50: Specific Gravity of Aqueous 1,4-Butanediol Solutions at 20° and 40°C *(32)*

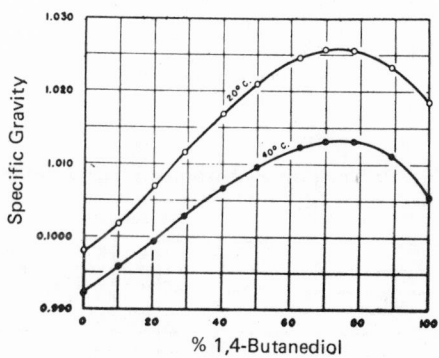

2,3-BUTANEDIOL

2,3–Butylene Glycol
2,3–Dihydroxybutane

$$CH_3CH(OH) \cdot CH(OH)CH_3$$

Table 7.51: Physical Properties of 2,3-Butanediol *(32)*

Acidity as acetic	0.005% by wt., max.
Boiling point at 760 mm. Hg	182.5° C
Color, APHA	15 max.
Density of liquid	1.048
Distillation range	175-195° C
Flash point, tag open cup	185° F
Freezing point	19° C (5% water lowers F. P. to +10° C)
Hygroscopicity (% water pickup-400 hrs.) 25° C and 50% rel. hum. 25° C and 75% rel. hum.	24 33
Molecular weight	90.12
Purity	95% by wt., min.
Refractive index, n_D^{20}	1.4377
Solubility (1% by weight) in castor oil in ether ether in in ethyl acetate ethyl acetate in in dibutyl phthalate	78% 5% 5% 14% 9% 2%
Specific gravity at 20/20° C	1.0093
Specific heat at 30° C	0.60 cal./g.
Specific tension at 25° C	36 dynes/cm.
Vapor pressure at 20° C	17 mm. Hg
Viscosity at 25° C at 35° C	121 cp. 90 cp.
Water content	0.5% by wt., max.
Weight per gallon	8.41 lb.

Table 7.52: Boiling Points of Aqueous levo-2,3-Butanediol Solutions at Atmospheric Pressure *(32)*

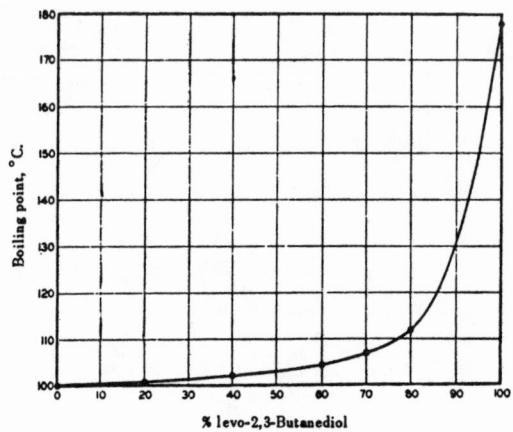

Table 7.53: Boiling Points of Aqueous levo-2,3-Butanediol-Ethanol Solutions *(32)*

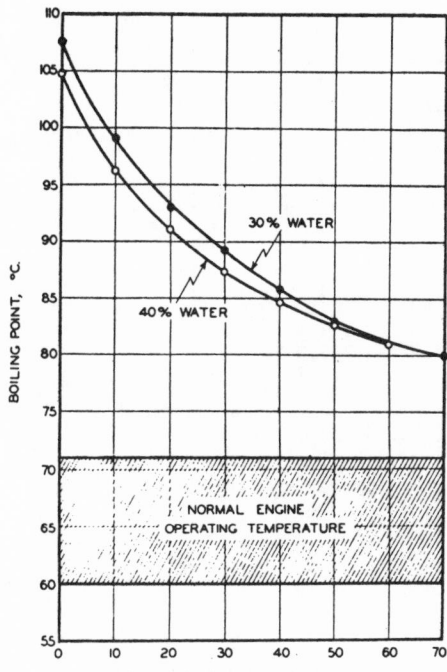

Table 7.54: Boiling Points of Aqueous levo-2,3 Butanediol-Methanol Solutions *(32)*

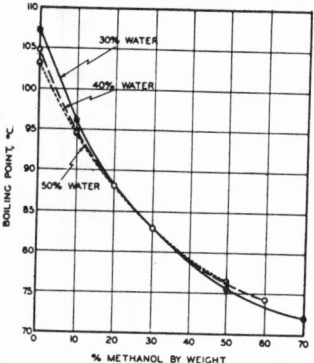

Table 7.55: Freezing Points of Aqueous levo-2,3-Butanediol Solutions *(32)*

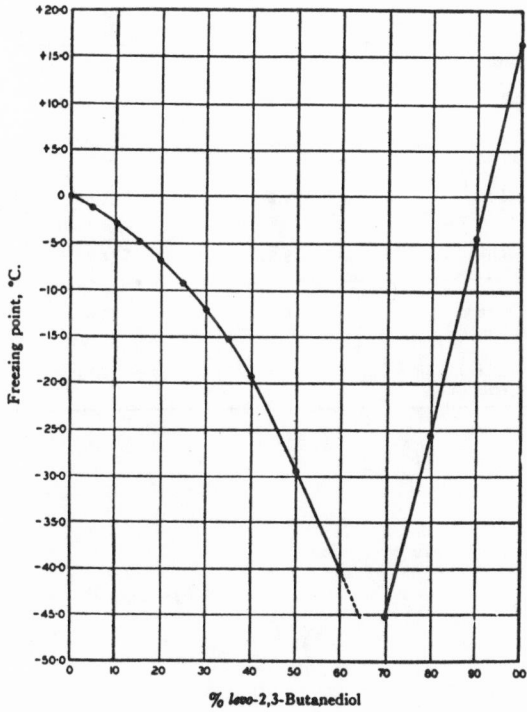

Table 7.56: Freezing Points of Aqueous meso-dextro-2,3-Butanediol Solutions *(32)*

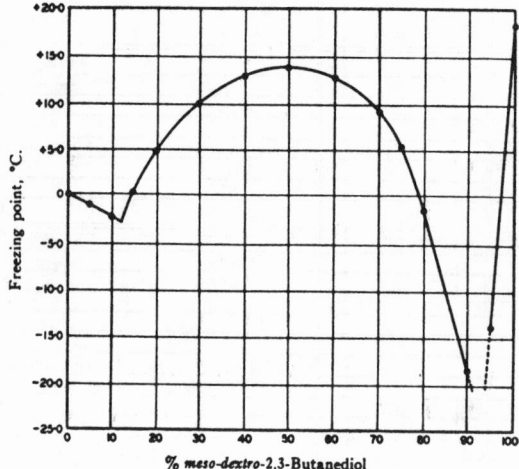

Table 7.57: Effect of meso-2,3-Butanediol on the Freezing Point of Aqueous levo-2,3-Butanediol Solutions *(32)*

Composition of Diol	40% Water	60% Water
100% levo	−40.4° C	−19.4° C
95% levo 5% meso	−37.0	−21.0
90% levo 10% meso	−28.2	−21.0
85% levo 15% meso	−18.6	−17.2
80% levo 20% meso	−14.0	−12.4
50% levo 50% meso	+1.55	+1.55

Table 7.58: Freezing Points of Aqueous levo-2,3-Butanediol-Ethanol Solutions *(32)*

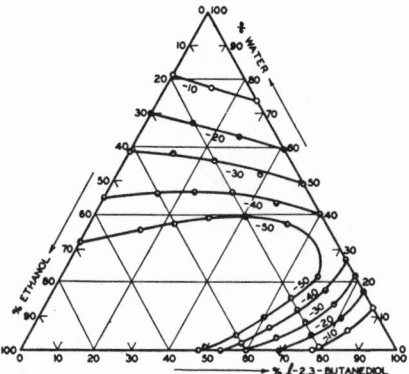

Table 7.59: Freezing Points of Aqueous levo-2,3-Butanediol-Ethylene Glycol Solutions *(32)*

Table 7.60: Freezing Points of Aqueous levo-2,3-Butanediol-Methanol Solutions *(32)*

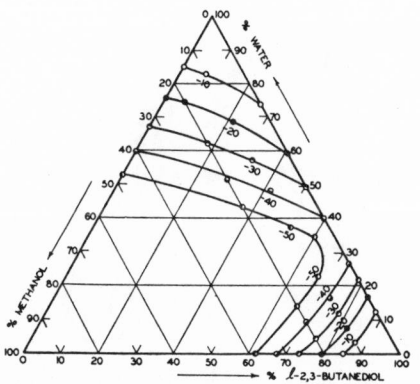

Table 7.61: Freezing Points of Aqueous levo-2,3-Butanediol-Tetrahydrofurfuryl Alcohol Solutions *(32)*

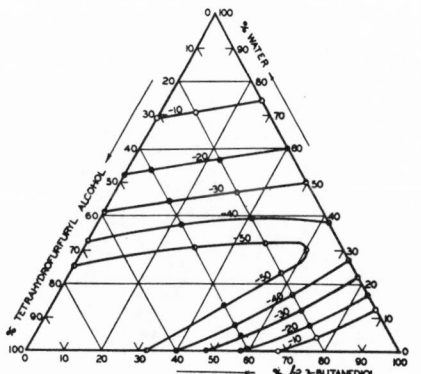

Table 7.62: Kinematic Viscosity of Aqueous levo-2,3-Butanediol Solutions, Expressed Logarithmically, as a Function of Concentration and Temperature *(32)*

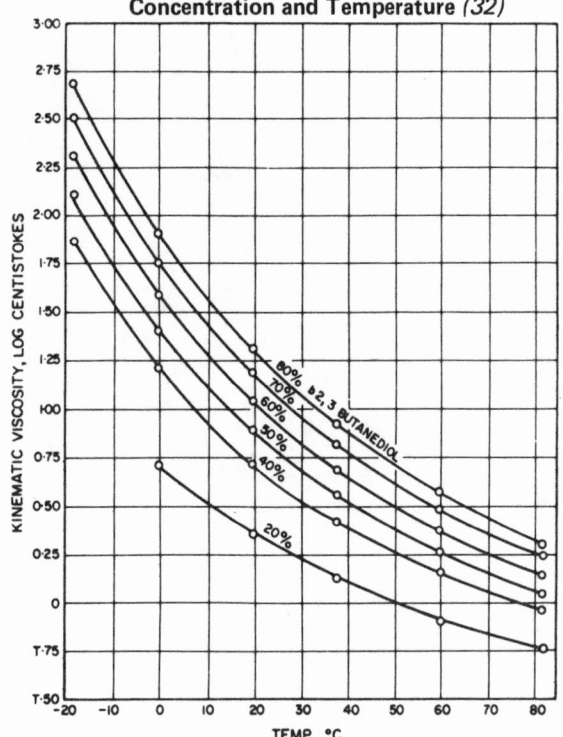

Table 7.63: Kinematic Viscosity of Aqueous levo-2,3-Butanediol Solutions in Relation to Concentration and Temperature *(32)*

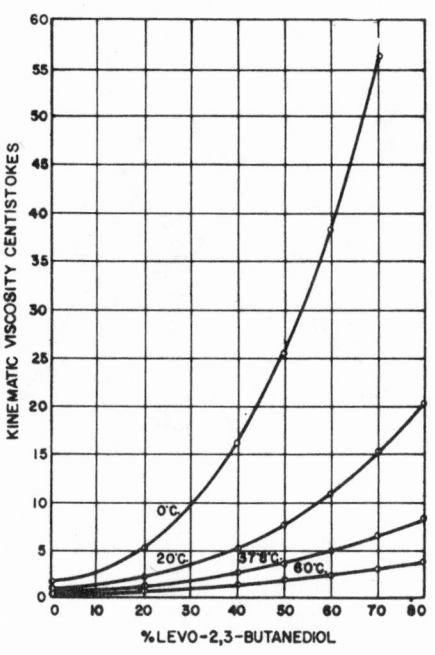

Table 7.64: Kinematic Viscosity of 60% levo-2,3-Butanediol, Glycerol and Ethylene Glycol Solutions at Low Temperatures *(32)*

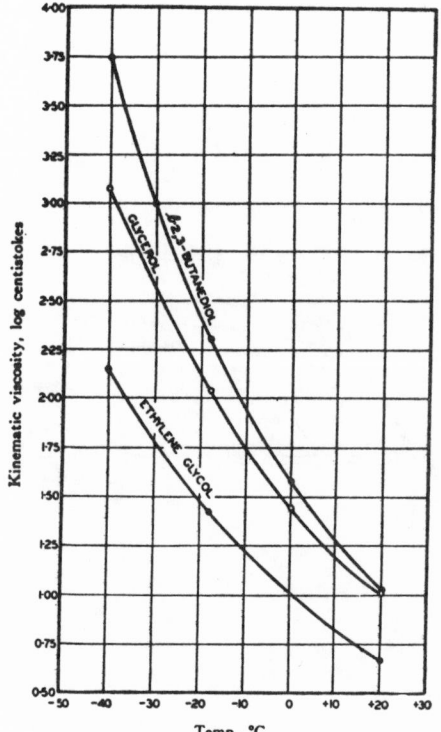

Table 7.65: Kinematic Viscosity of Aqueous levo-2,3-Butanediol-Ethanol Solutions at 20°C, Expressed in Centistokes *(32)*

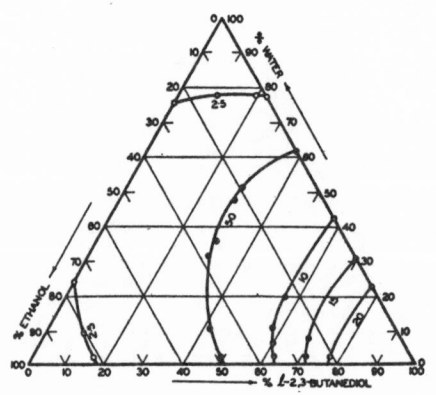

Table 7.66: Kinematic Viscosity of Aqueous levo-2,3-Butanediol - Methanol Solutions at 20°C., Expressed in Centistokes *(32)*

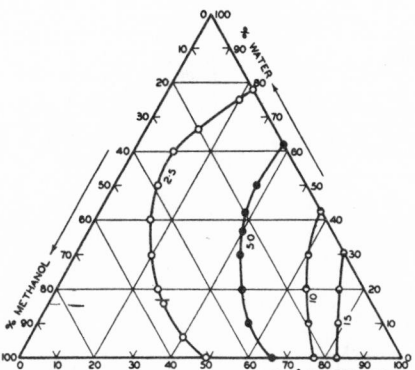

Table 7.67: Kinematic Viscosity of Aqueous levo-2,3-Butanediol - Ethylene Glycol Solutions at 20°C., Expressed in Centistokes *(32)*

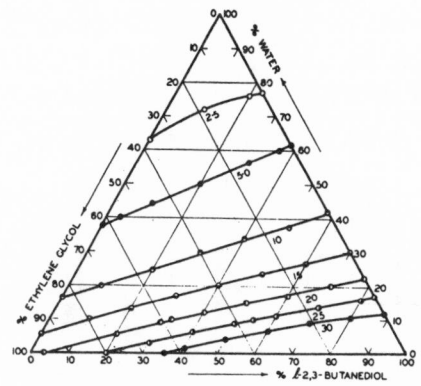

Table 7.68: Kinematic Viscosity of Aqueous levo-2,3-Butanediol - Tetrahydrofurfuryl Alcohol Solutions at 20°C., Expressed in Centistokes *(32)*

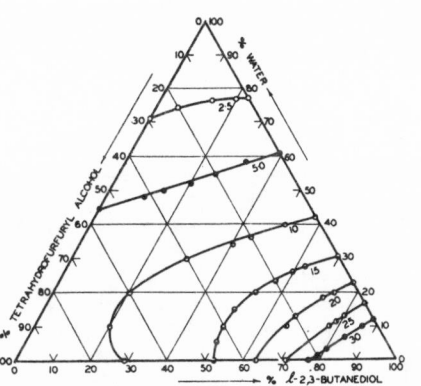

Table 7.69: Absolute Viscosity of Aqueous Solutions of Ethylene Glycol, levo-2,3-Butanediol, meso-dextro-2,3-Butanediol and Glycerol at 20°C. *(32)*

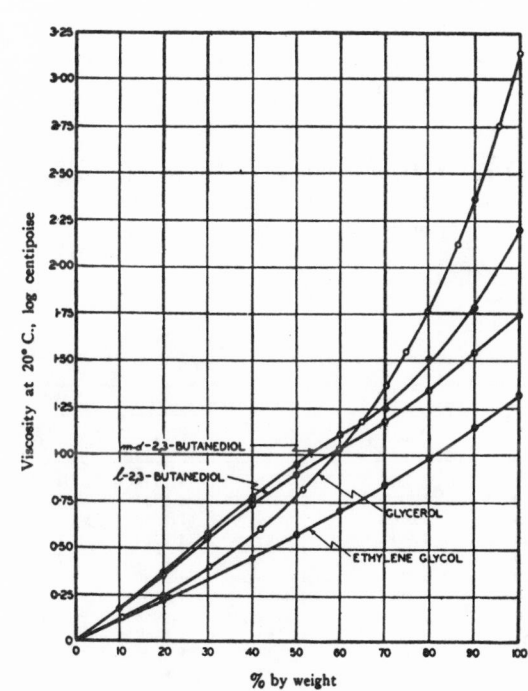

Table 7.70: Optical Rotatory Power of Aqueous levo-2,3-Butanediol Solutions at 20°C *(32)*

Table 7.71: Effects of Concentration and Temperature on the Specific Rotatory Power of Aqueous levo-2,3-Butanediol Solutions *(32)*

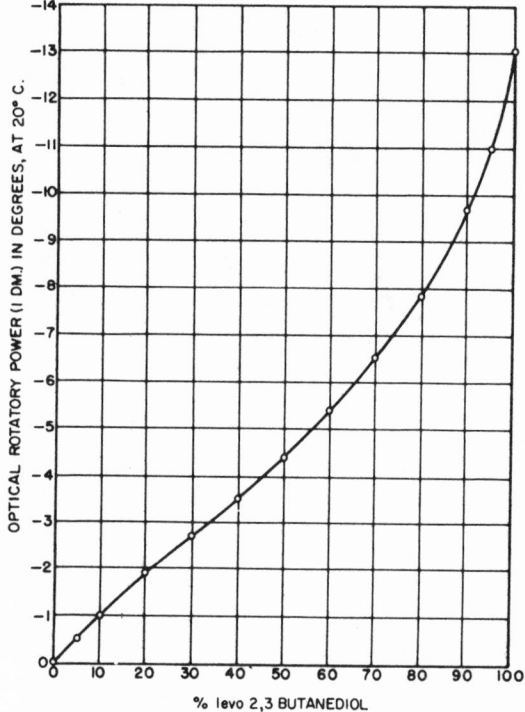

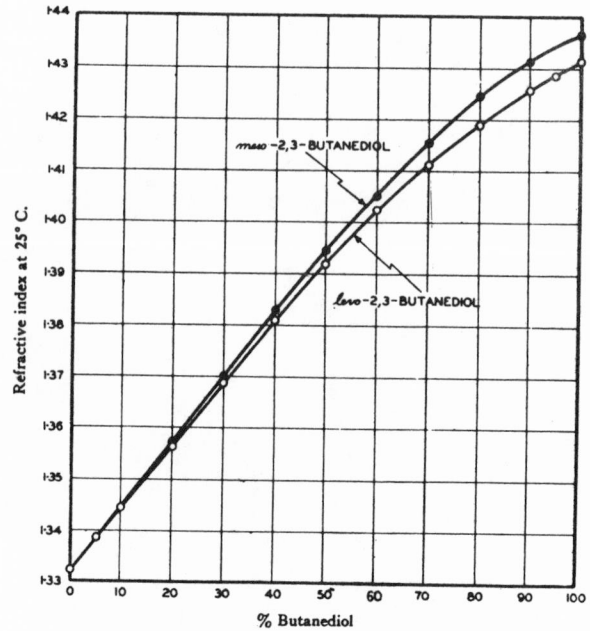

Table 7.72: Refractive Indices of Aqueous levo-2,3-Butanediol Solutions at Different Temperatures *(32)*

Table 7.73: Refractive Indices of Aqueous Solutions of meso- and levo-2,3-Butanediol at 25°C *(32)*

Diol, %	Temperature, °C			
	20	25	30	35
0	1.3330	1.3325	1.3319	1.3312
10.0	1.3450	1.3445	1.3437	1.3429
19.9	1.3574	1.3566	1.3557	1.3549
29.9	1.3700	1.3689	1.3677	1.3666
39.9	1.3820	1.3807	1.3793	1.3779
49.9	1.3930	1.3915	1.3900	1.3885
59.6	1.4027	1.4012	1.3997	1.3982
70.0	1.4115	1.4098	1.4082	1.4065
79.7	1.4197	1.4180	1.4162	1.4146
89.7	1.4264	1.4247	1.4229	1.4212
99.5	1.4322	1.4302	1.4283	1.4264

Table 7.74: Specific Gravity of Aqueous levo-2,3-Butanediol Solutions at 20°, 30° and 40°C *(32)*

Table 7.75: Specific Gravity of Aqueous meso-2,3-Butanediol Solutions at 20°, 30° and 40°C *(32)*

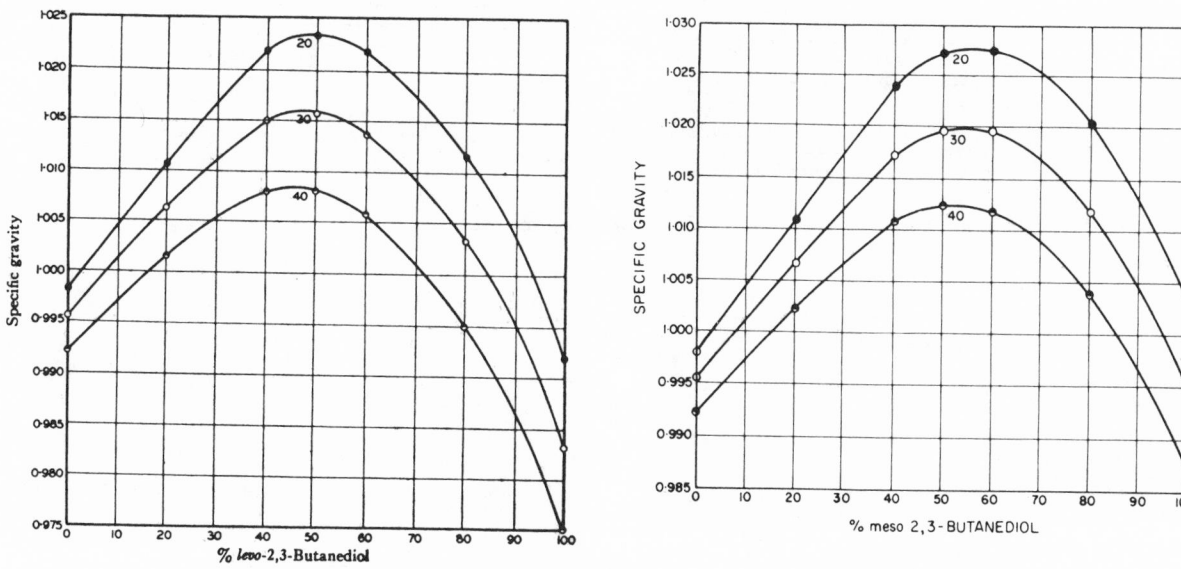

Table 7.76: Surface Tension of Aqueous Solutions of levo-2,3-Butanediol and Ethylene Glycol *(32)*

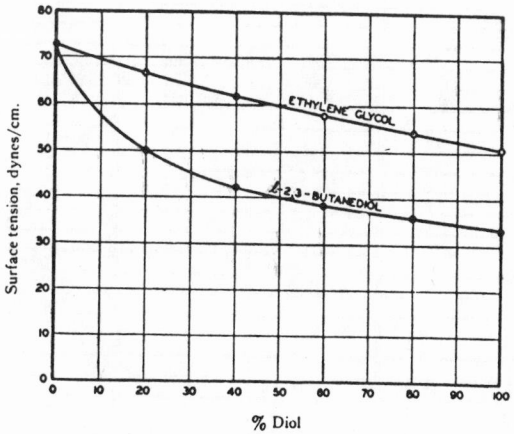

BUTANEDIOLS

Table 7.77: Refractive Indices of Aqueous Butanediol Solutions at 20°, 30° and 40°C *(32)*

1,2-Butanediol				1,3-Butanediol				1,4-Butanediol			
Glycol, %	n_D^{20}	n_D^{30}	n_D^{40}	Glycol, %	n_D^{20}	n_D^{30}	n_D^{40}	Glycol, %	n_D^{20}	n_D^{30}	n_D^{40}
10.13	1.3452	1.3436	1.3420	9.51	1.3442	1.3430	1.3417	10.51	1.3444	1.3432	1.3420
19.69	1.3572	1.3553	1.3534	19.18	1.3552	1.3548	1.3520	20.01	1.3563	1.3550	1.3532
29.72	1.3693	1.3672	1.3650	30.20	1.3688	1.3670	1.3649	30.02	1.3682	1.3671	1.3659
39.79	1.3813	1.3788	1.3760	39.94	1.3800	1.3778	1.3755	39.86	1.3802	1.3790	1.3768
49.68	1.3920	1.3892	1.3865	49.45	1.3920	1.3895	1.3870	49.70	1.3935	1.3918	1.3898
59.88	1.4027	1.4000	1.3966	60.02	1.4040	1.4012	1.3983	59.95	1.4052	1.4042	1.4020
69.37	1.4120	1.4090	1.4058	70.10	1.4145	1.4118	1.4090	70.15	1.4183	1.4167	1.4140
79.73	1.4212	1.4185	1.4165	80.20	1.4242	1.4215	1.4185	79.85	1.4283	1.4258	1.4236
89.40	1.4297	1.4265	1.4230	89.67	1.4323	1.4295	1.4264	90.10	1.4370	1.4349	1.4318
100	1.4375	1.4347	1.4310	100	1.4398	1.4370	1.4331	100	1.4451	1.4425	1.4395

Table 7.78: Kinematic Viscosity of Aqueous Butanediol Solutions at 20° and 40°C, Expressed in Centistokes *(32)*

1,2-Butanediol	Viscosity		1,3-Butanediol	Viscosity		1,4-Butanediol	Viscosity	
Glycol, %	20°C	40°C	Glycol, %	20°C	40°C	Glycol, %	20°C	40°C
10.125	1.520	0.910	9.505	1.51	0.91	10.51	1.446	0.89
19.69	2.187	1.243	19.175	2.295	1.291	20.01	2.109	1.218
29.72	3.310	1.690	30.20	3.529	1.818	30.02	2.867	1.6602
39.79	4.802	2.311	39.94	5.419	2.593	39.86	4.258	2.382
49.685	6.739	3.088	49.45	8.313	3.695	49.70	6.57	3.202
59.88	9.72	4.227	60.02	13.44	5.600	59.95	10.20	4.707
69.37	13.82	5.744	70.1	21.57	8.413	70.15	18.48	7.982
79.73	21.37	8.372	80.20	35.36	12.88	79.85	30.63	12.62
89.40	35.54	12.57	89.67	63.43	21.21	90.1	54.35	21.40
100	68.0	21.25	100	129.8	39.70	100	87.62	33.8

2-BUTENE-1,4-DIOL

$$HOCH_2CH=CHCH_2OH$$

Table 7.79: Physical Properties of 2-Butene-1,4-diol *(32)*

Physical Properties of Technical Cis-2-Butene-1,4-Diol		Purified Cis-2-Butene-1,4-Diol	
Boiling point range	232–235° C	Boiling point at 760 mm. Hg	234° C
		100 mm. Hg	177° C
Fire point (Cleveland open cup)	270° F	20 mm. Hg	140° C
		10 mm. Hg	122° C
Flash point (Cleveland open cup)	263° F	5 mm. Hg	109° C
Freezing point range	4.0–7.0° C	Freezing point	12.5° C
Molecular weight	88.1	Refractive index, n_D^{25}	1.4768–1.4773
Refractive index, n_D^{25}	1.476–1.478	Specific gravity at 25/15° C	1.070
Specific gravity at 25/15° C	1.067–1.074		
Viscosity at 68° F	22 cp.		
100° F	10.8 cp.		
210° F	2.5 cp.		

2-BUTYNE-1,4-DIOL

$$HOCH_2C \equiv CCH_2OH$$

Table 7.80: Physical Properties of 2-Butyne-1,4-diol *(32)*

Physical Properties of Commercial 2-Butyne-1,4-Diol

Acetals (as CH_2O)	Less than 0.6%
Aldehydes (as CH_2O)	Less than 0.5%
Butynediol content	35 ± 1%
Freezing point	Less than -7° C
Methanol (by distillation)	0.0%
pH	4 to 6
Propargyl alcohol	Less than 0.5%
Saponification No. (as mg. KOH/g. product)	Less than 6
Specific gravity, d_4^{25}	1.04 to 1 ˜˜
Weight per gallon	8.7 lb.

Purified 2-Butyne-1,4-Diol

Boiling point at 10 mm. Hg	140° C
100 mm. Hg	194° C
Crystal structure	
system	Orthorhombic
principal forms	Basal pinacoids and prisms with crystals flattened parallel to the basal pinacoids
Melting point	57.5° C
Refractive indices n_D^{25}	α ± 1.450 - 0.002
	β ± 1.528 - 0.002
Solubility (g./100 ml. solvent)	
in water at 0° C	121
in water at 25° C	374
in ethyl alcohol at 25° C	83
in acetone at 25° C	70
in ethyl ether at 25° C	2.6
in benzene at 25° C	0.04

1,5-PENTANEDIOL

Pentamethylene Glycol $HOCH_2CH_2CH_2CH_2CH_2OH$

Table 7.81: Physical Properties of 1,5-Pentanediol *(32)*

Boiling point at 760 mm. Hg	242.5° C
50 mm. Hg	166° C
10 mm. Hg	134° C
Coefficient of expansion at 55° C	0.00061/°C
Flash point (open cup)	265° F
Freezing point	-15.6° C
Molecular weight	104.16
Refractive index at 20° C	1.4489
Specific gravity at 20/20° C	0.9921
Surface tension at 20° C	43.2 dynes/cm.
Vapor pressure at 20° C	Less than 0.01
Viscosity at 0° C (absolute)	415 cp.
20° C	128 cp.
40° C	48 cp.
Weight per gallon at 20° C (average)	8.23 lb.

Table 7.82: Absolute Viscosity of Aqueous 1,5-Pentanediol Solutions at 20° and 40°C *(32)*

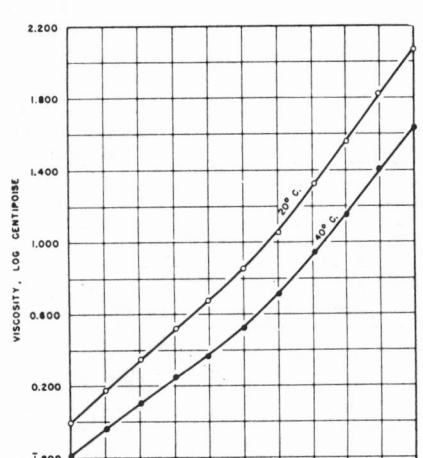

Table 7.83: Specific Gravity of Aqueous 1,5-Pentanediol Solutions at 20° and 40°C *(32)*

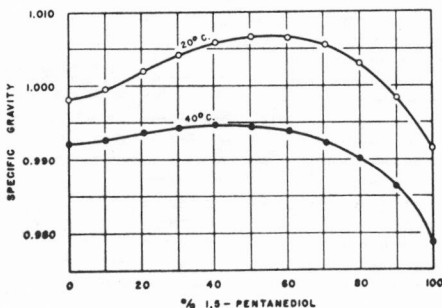

2,4-PENTANEDIOL

Amylene Glycol   CH₃CHOHCH₂CHOHCH₃

Table 7.84: Physical Properties of 2,4-Pentanediol *(32)*

Boiling point at 760 mm. Hg	199° C
Flash point, Cleveland open cup	210° F
Melting point	45° C
Molecular weight	104.15
Specific gravity (apparent), 20/20° C	0.964 (supercooled liquid)

NEOPENTYL GLYCOL

"NPG" Glycol
2,2-Dimethyl-1,3-Propanediol

$$HO-CH_2-\underset{\underset{CH_3}{|}}{\overset{\overset{CH_3}{|}}{C}}-CH_2-OH$$

Table 7.85: Physical Properties of Neopentyl Glycol *(41)*

Empirical Formula	C₅H₁₂O₂	Bulk Density, 21°C., g./cc.	1.06
Molecular Weight (calcd.)	104.15	lb./cu. ft.	66.4
Equivalent Weight (theor.)	52.08	Color, APHA, ppm., max.	25ᵃ
Acid Number	0.01	Critical Pressure, atm. (estd.)	36
Hydroxyl Number (average)	1075	Critical Temp., °K. (estd.)	653
Saponification Number	0.14	Critical Volume, cu. ft./lb. (estd.)	0.059
Acid, as acetic acid, wt. %	0.05 max.	cc./g. (estd.)	3.683
Aldehyde, as hydroxypivaldehyde, wt. %	0.70 max.	Crystal Density, 25°C., g./cc.	1.11
Ester, as neopentyl hydroxypivaldehyde,		lb./cu. ft.	69.3
wt.%	1.50 max.	Crystallization Point, °C. 128 (same as m.p.)	
Water, wt. %	1.00 max.	Effect on Metals: No corrosive effect on mild steel,	
Appearance	White, crystalline solid	galvanized steel or tinplate.	
Autoignition Temp. (ASTM D286-30), °F.	750	Slightly corrosive to aluminum.	
,°C.	399	Fire Point (Cleveland Open Cup), °F.	305
Boiling Range, °C., at 3.35 mm. Hg	93-94	°C.	151.6
25 mm. Hg	121-123	Flash Point (Cleveland Open Cup), °F.	305
760 mm. Hg	210	°C.	151.6

(continued)

Table 7.85: (continued)

Solubility

Heat Capacity, Solid, B.t.u./lb./°F. (estd.)	0.383
cal., g./g./°C. (estd.)	0.383
Heat of Combustion, B.t.u./lb. (estd.)	−12,917
cal., g./g. (estd.)	−7,176
B.t.u./lb. mole	−1,345,306
cal., g./g. mole	−747,391
Heat of Fusion, B.t.u./lb. (estd.)	90
cal., g./g. (estd.)	50

ᵃMolten

	Solubility, g./100 g. of Solvent, at		
Solvent	5°C.	15°C.	60°C.
Water	173	181	400
Acetone	23	60	439
Benzene	0.6	12	199
Cyclohexane	0.0	<1	0.4
Hexane	0.5	—	1.8
Isobutyl alcohol	87.5	—	—
Methyl ethyl ketone	25	41	>309
Methyl isobutyl ketone	7.9	14	76
Toluene	0	<1	39
Trichloroethylene	0.2	<1	117

Specific Gravity, 25°/4°C.		1.066

PENTANEDIOLS

Table 7.86: Kinematic Viscosity of Aqueous Pentanediol Solutions at 20° and 40°C Expressed in Centistokes *(32)*

1,2-Pentanediol			1,5-Pentanediol		
	Viscosity			Viscosity	
Glycol, %	20° C	40° C	Glycol, %	20° C	40° C
10.36	1.5475	0.9275	10.17	1.516	0.9210
19.97	2.264	1.258	20.09	2.246	1.277
30.18	2.88	1.538	30.42	3.300	1.795
40.13	4.06	2.08	39.82	4.735	2.331
50.02	5.73	2.82	50.04	7.08	3.350
59.96	8.02	3.742	60.12	11.30	5.250
69.97	13.03	5.725	70.45	20.9	8.842
79.85	19.85	8.138	80.20	36.22	14.46
90.05	38.20	13.62	89.75	66.25	25.70
100	68.55	20.82	100	115.65	43.58

Table 7.87: Refractive Indices of Aqueous Pentanediol Solutions at 20° and 40°C *(32)*

1,2-Pentanediol			1,5-Pentanediol		
Glycol, %	n_D^{20}	n_D^{40}	Glycol, %	n_D^{20}	n_D^{40}
10.36	1.3452	1.3430	10.17	1.3444	1.3420
19.97	1.3585		20.29		1.3543
20.64		1.3500	20.59	1.3572	
30.94	1.3705	1.3682	30.42	1.3700	1.3682
41.26	1.3830	1.3800	40.43	1.3833	1.3800
51.05	1.3930	1.3895	50.45	1.3960	1.3910
61.28	1.4050	1.3990	60.51	1.4080	1.4033
70.00	1.4120	1.4068	70.73	1.4198	1.4159
80.04	1.4223	1.4182	80.08	1.4304	1.4260
90.05	1.4320	1.4254	90.15	1.4417	1.4367
100	1.4390	1.4326	100	1.4500	1.4448

1,6-HEXANEDIOL

$$CH_2OHCH_2CH_2CH_2CH_2CH_2OH$$

Table 7.88: Physical Properties of 1,6-Hexanediol *(32)*

This glycol is very soluble in water.

Boiling point at 760 mm. Hg	243° C
Flash point, Cleveland open cup	265° F
Melting point	42° C
Molecular weight	118.17
Specific gravity (apparent)	0.958

2,5-HEXANEDIOL

$$CH_3CHOHCH_2CH_2CHOHCH_3$$

Table 7.89: Physical Properties of 2,5-Hexanediol *(32)*

This six-carbon glycol is the most viscous of the family. It is completely miscible with water.

Boiling point at 760 mm. Hg	220.8° C
Flash point, Cleveland open cup	220° F
Freezing point	Sets to a glass below -50° C
Molecular weight	118.17
Refractive index at 20° C, n_D	1.4474
Specific gravity (apparent) at 45/15.6° C	0.9617
Viscosity at 20° C	37 cp.

HEXYLENE GLYCOL

2-Methyl-2,4-Pentanediol
Methyl Amylene Glycol
Diacetone Glycol

$$CH_3COH(CH_3)CH_2CHOHCH_3$$

Table 7.90: Physical Properties and Specifications of Hexylene Glycol *(32)*

Acidity as acetic acid	0.005% by wt., max.		
		Density (in vacuo) at 0° C	0.9360 g./cc.
Boiling point at 760 mm. Hg	198.27° C	20° C	0.0216 g./cc.
	197.1° C	30° C	0.9145 g./cc.
at 50 mm. Hg	125° C		
at 10 mm. Hg	94° C	dt/dp at the boiling point	0.045° C/mm.
Color, Pt-Co (Hazen) standard	15, max.	Flash point, Cleveland open cup	210° F
			215° F
Critical properties, P_c	499 psia		
T_c	1221° R	Freezing point	Becomes semisolid at -40° C
V_c	6.78 ft./mole		without crystalline formation
			Sets to glass below -50° C
Density (in air) at 760 mm. Hg	0.928 g./cc.		
		Distillation range (ASTM D-1078)	195 to 200° C
Density in air at any temp. may be obtained from equation:	$D_t - 0.952 - 4.02 \times 10^4 t$	(95% will distill between 196° C and 199° C)	

(continued)

Table 7.90: (continued)

Latent heat of vaporization	12.3 x 13^3 cal./g.-mole 104.1 g.-cal./g. 208 Btu/lb.
Molecular weight	118.17
Pour point	-37.2° C (35° F)
Refractive dispersion, $(N_F - N_C)$ x 10^4	72.5
Refractive index, n_D^{20}	1.4276
n_D^{30}	1.4243
Specific gravity at 20/4° C	0.9216
20/20° C	0.9234
Δ Sp. Gr./Δt, 0 to 55° C	0.00097
Surface tension, 20° C	33.1 dynes/cm.
Vapor pressure, 20° C	0.05 mm. Hg
Viscosity (absolute), 20° C	34.4 cp.
Water at 20° C	Miscible without turbidity with 19 vols. of n-heptane
Weight per gallon at 20° C	7.69 lb.

Table 7.91: Freezing Points of Hexylene Glycol-Water Mixtures *(32)*

Table 7.92: Specific Gravity and Freezing Point of Hexylene Glycol-Water Mixtures *(14)*

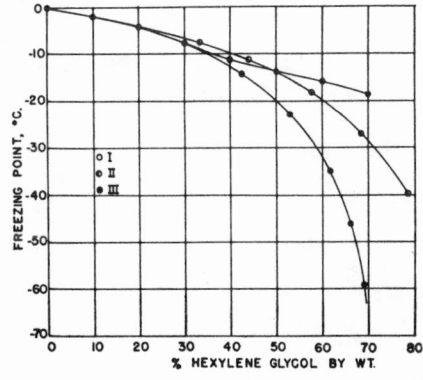

Freezing Points of Hexylene Glycol (2-Methyl-2,4-pentanediol)-Water Mixtures. (I) Observed; (II) Theoretical, without hydration; (III) Theoretical, with complete hydration.

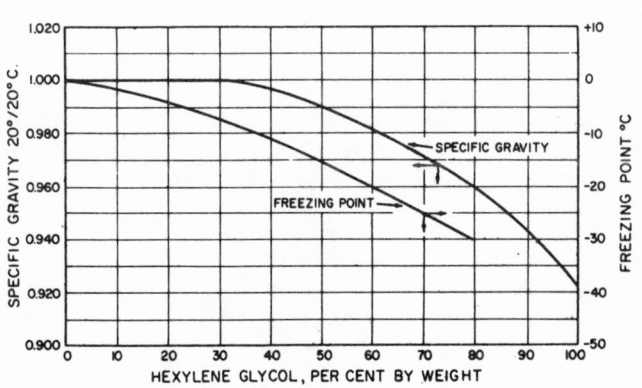

PINACOL

Pinacone
2,3-Dimethyl-2,3-Butanediol
Tetramethylethylene Glycol $HOC(CH_3)_2 \cdot C(CH_3)_2OH$

Table 7.93: Physical Properties of Pinacol *(32)*

Boiling point at 760 mm. Hg	174.4° C
Melting point	41.1° C
Molecular weight	118.17

The Hexahydrate

Melting point	45.4° C
Specific gravity, d^{15}	0.967 (supercooled liquid)

2,2-DIETHYL-1,3-PROPANEDIOL

$$HOCH_2C(C_2H_5)_2CH_2OH$$

Table 7.94: Physical Properties of 2,2-Diethyl-1,3-Propanediol *(32)*

Boiling point at 10 mm. Hg	125° C
Freezing point	61.3° C
Molecular weight	132.20
Solubility in water at 20° C	25% by wt.
Specific gravity (apparent) at 20° C	1.052

2-ETHYL-1,3-HEXANEDIOL

Ethohexadiol
Octanediol $HOCH_2CH(C_2H_5)CHOHC_3H_7$

Table 7.95: Physical Properties of 2-Ethyl-1,3-Hexanediol *(32)*

Acidity as acetic acid	0.01% by wt., max.
Boiling point at 760 mm. Hg	243.1° C
Color, Pt–Co	15, max.
Distillation range	241 to 249° C
Flash point, open cup	265° F
Freezing point	Sets to glass below –40° C
Molecular weight	146.22
Refractive index, 20° C, n_D	1.4511
Solubility in water, 20° C	4.2% by wt.
Solubility of water in, 20° C	11.7% by wt.
Specific gravity, 20/20° C	0.9422
Suspended matter	Substantially free
Vapor pressure, 20° C	Less than 0.01 mm. Hg
Viscosity, 20° C	323 cp.
Weight per gallon (average), 20° C	7.83 lb.

"TMPD" GLYCOL

$$CH_3-CH-CH-C-CH_2-OH$$

(2,2,4-trimethyl-1,3-pentanediol)

Table 7.96: Typical Properties of Eastman "TMPD" Glycol *(41)*

Typical Properties*

Empirical Formula	$C_8H_{18}O_2$
Molecular Weight	146.22
Physical Form	White solid
Density at 21°C, g/cm³ (lb/cu ft)	
Solid cake	0.897 (56)
Granulated	0.688 (43)
Specific Gravity at	
55/15°C	0.928
118/15°C	0.883
188/15°C	0.835
Assay by Gas Chromatography, wt%	96
Melting Range, °C	46 to 55
Boiling Point at 760 mm Hg, °C	
Initial	215
95%	235
Molten Color, Pt-Co scale	50
Acid as Isobutyric Acid, wt %	0.10
Flash Point, COC, °C (°F)	113 (235)
Fire Point, COC, °C (°F)	118 (245)
Autoignition Temperature, °C (°F)	346 (655)
Hygroscopicity at equilibrium at 25°C and 50% RH, wt % H_2O	0.1 to 0.2
DOT Class	Not required
DOT Label Required	None
NFPA Class	I
Solubility at 25°C, wt %, in	
Water	1.9
Methanol	75
Ethanol	75
Isopropanol	80
Ethylene Glycol	35
Propylene Glycol	50
Acetone	25
Diethyl Ether	29
Benzene	22
Gasoline (white)	4.7
Kerosene	<1
Metal Corrosion, Aluminum, mm/yr (mils/yr) at	
80°C	0.059 (2.33)
200°C	0.077 (3.05)

*Reported for information only. Eastman makes no representation that the material in any particular shipment will conform exactly to the values listed.

Table 7.97: Solubility of Water in Mixtures of "TMPD" and Hexylene Glycol *(41)*

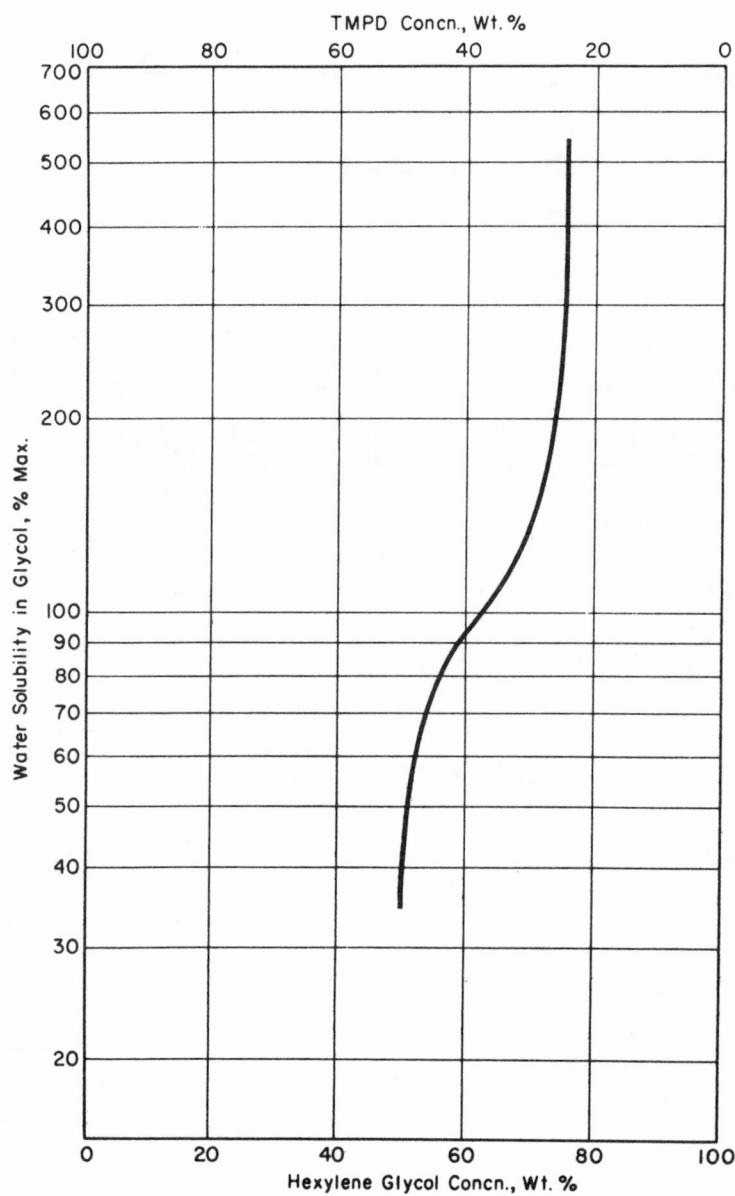

2,5-DIMETHYL-3-HEXYNE-2,5-DIOL

Dimethyl Hexynediol

$$CH_3 - \underset{\underset{OH}{|}}{\overset{\overset{CH_3}{|}}{C}} - C \equiv C - \underset{\underset{OH}{|}}{\overset{\overset{CH_3}{|}}{C}} - CH_3$$

Table 7.98: Physical Properties of 2,5-Dimethyl-3-Hexyne-2,5-diol *(32)*   IOL (32)

Boiling point	205-6° C
Freezing point	94-5° C
Surface tension at 25° C	
5% in water	41.2 dynes/cm.
0.1% in water	60.9 dynes/cm.
0.01% in water	66.9 dynes/cm.

1,4-CYCLOHEXANEDIMETHANOL

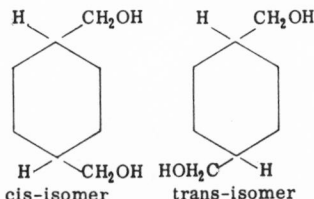

cis-isomer   trans-isomer

Table 7.99: Properties of 1,4-Cyclohexanedimethanol

(approx. 70% trans-, 30% cis-isomers)

Empirical formula	$C_8H_{16}O_2$		Specific gravity	
Molecular weight (calcd.)	144.21		Liquid:	
Equivalent weight	72.1		25°/4°C. (super-cooled)	1.026
Crystallization point, °C.	35		50°/4°C. (super-cooled)	1.010
Pour point, °C. (super-cooled)	10		100°/4°C. (molten)	0.978
Melting point, °C.	41-61		150°/4°C.	0.946
cis isomer	43		200°/4°C.	0.914
trans isomer	70		Solid:	
Boiling point, °C.			27°/4°C.	1.069
760 mm.	285		Density	
100 mm.	216			
10 mm.	160		Liquid, lb./gal.:	
1 mm.	118		70°F. (super-cooled)	8.59
cis isomer	288		100°F. (super-cooled)	8.49
trans isomer	284		200°F.	8.20
			300°F.	7.90
			400°F.	7.60
			Solid, 70°F., lb./cu. ft.	66.74

(continued)

Table 7.99: (continued)

Acid number		<0.03
Hydroxyl number		22.89
Saponification number		0.91
Refractive index, $n_D^{20°}$ (super-cooled)		1.4893
Flash point, Cleveland Open Cup, °F.		165
Fire point, Cleveland Open Cup, °F.		255
Solubility, 25°C., wt. %		
	in water	miscible
	water in	miscible
	in methanol	miscible
	in ethanol	miscible
	in diethyl ether	2.5
	in VM & P naphtha	<1
	in benzene	<1
	in acetone	56.4

Heat capacity, (estd.)

Liquid:	Temp., °C.	C_v, B.t.u./lb. °F.	C_p, B.t.u./lb. °F.
	50	0.505	0.648
	100	0.553	0.716
	150	0.609	0.794
	200	0.669	0.877

Solid: C_p, B.t.u./lb. °F. (estd.)   0.410

Thermal Conductivity (estd.)

Vapor:	Temp., °C.	k, B.t.u./hr. ft. °F.
	50	0.00602
	100	0.00772
	150	0.00960
	200	0.01229

Liquid:	Temp., °C.	k, B.t.u./hr. ft. °F.
	50	0.1118
	100	0.1229
	150	0.1280
	200	0.1311

Critical temperature, T_c, °C. (estd.)

cis isomer	457
trans isomer	451

Critical pressure, P_c, atm. (estd.)   34.85

Critical volume, V_c, cu. ft./lb. (estd.)   0.0506

Viscosity:	Temp., °F.	cs.	S.U.S.
	75	1421.6	6568
	100	478.1	2209
	125	183.4	847

p-XYLYLENE GLYCOL

ω,ω'-Dihydroxy-p-Xylene

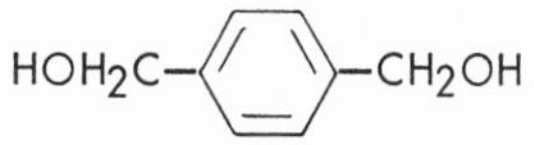

$HOH_2C-\langle\rangle-CH_2OH$

Table 7.100: Physical Properties of p-Xylylene Glycol *(32)*

Chlorine (total)	0.6% max.
Flash point (Cleveland open cup)	370° F
Molecular weight	138.16
Purity	95% min.
Set point	115–117.6° C
Specific gravity at 117° C	1.100
Toluene insolubles	0.5% max.

2-ETHYL-2-BUTYL-1,3-PROPANEDIOL

$HOCH_2C(C_2H_5)(C_4H_9)CH_2OH$

Table 7.101: Physical Properties of 2-Ethyl-2-Butyl-1,3-Propanediol *(32)*

Boiling point at 100 mm. Hg	195° C
Melting point	41.4° C
Molecular weight	160.25
Solubility in water at 20° C	0.8% by wt.
Specific gravity (apparent) at 50/20° C	0.931

3,6-DIMETHYL-4-OCTYNE-3,6-DIOL

Dimethyl Octynediol

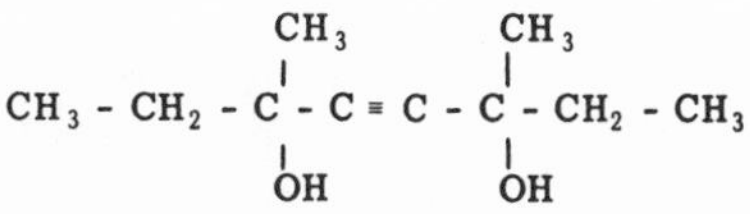

$$CH_3 - CH_2 - \underset{\underset{OH}{|}}{\overset{\overset{CH_3}{|}}{C}} - C \equiv C - \underset{\underset{OH}{|}}{\overset{\overset{CH_3}{|}}{C}} - CH_2 - CH_3$$

Table 7.102: Physical Properties of 3,6-Dimethyl-4-Octyne-3,6-diol *(32)*

Boiling point at 20 mm. Hg	135° C
Freezing point	55.6° C
Surface tension at 25° C	
5% in water	30.7 dynes/cm.
0.1% in water	55.3 dynes/cm.
0.01% in water	63.9 dynes/cm.

THIODIGLYCOL

Thiodiethylene Glycol

β, β'-Dihydroxyethyl Sulfide

$$HOCH_2CH_2-S-CH_2CH_2OH$$

Table 7.103: Physical Properties of Thiodiglycol (32)

Acidity	1.0 mg. KOH/g. sample, max.	Heat of vaporization at 1 atm.	235 Btu/lb.
Boiling point at 760 mm. Hg	283° C	Molecular weight	122.19
50 mm. Hg	194° C	Refractive index at 20° C n_D	1.5217
Δ Boiling point/Δ p	0.055° C/mm. Hg	Specific gravity	1.1847
Color (Pt-Co)	200 max.	Δ Sp. Gr./Δ t	0.00072
Coefficient of expansion at 55° C	0.00061/° C	Vapor pressure at 20° C	Less than 0.01 mm. Hg
Flash point (open cup)	320° F	Viscosity at 20° C	65.2 cp.
Freezing point	-10° C	Weight per gallin at 20° C	9.85 lb.
		at 15.56° C	9.88 lb.

Table 7.104: Vapor Pressure of Thiodiglycol at Various Temperatures (19)

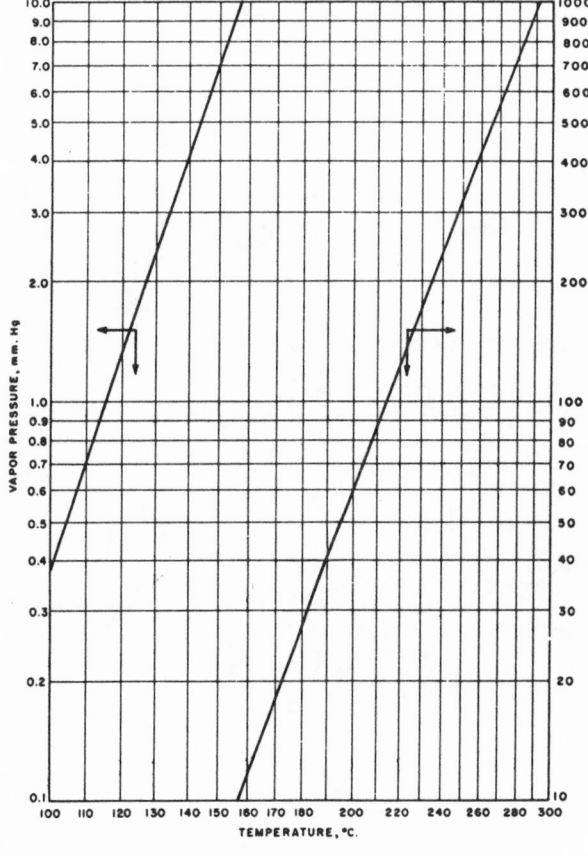

MISCELLANEOUS GLYCOLS

Table 7.105: Hydrates of Aliphatic Glycols *(32)*

Number of C Atoms	Glycol			Hydrate	
	Name	Skeletal Structural Formula	M.p. (°C)	M.p. (°C)	n in $R(OH)_2 \times nH_2O$
2	Ethylene glycol	HO-C-C-OH	-12.9	-49.6 (cong.)	2
2	Ethylene glycol	HO-C-C-OH	-12.9	-40.7	0.67
4	meso-2,3-Butanediol	HO OH C-C-C-C	34.4	16.8	6(5)†
4	±2,3-Butanediol	HO OH C-C-C-C	7.6	---	0
5	2-Methyl-2,3-butanediol	HO OH C-C-C-C C	liq.	23.5-4	6
6	Pinacol	HO OH C-C-C-C C C	41.4	41.25	1
6	Pinacol	HO OH C-C-C-C C C	41.4	46.5	6
8	2,5-Dimethyl-2,5-hexanediol	HO OH C-C-C-C-C-C C C	92	41-2	6
9	2,6-Dimethyl-2,6-heptanediol	HO OH C-C-C-C-C-C-C C C	76-77	60-61	1
10	2,7-Dimethyl-2,7-octanediol	HO OH C-C-C-C-C-C-C-C C C	92	59	2
13	2,10-Dimethyl-2,10-undecanediol	HO OH C-C-CCCCCCC-C-C C C	61	---	?
14	2,11-Dimethyl-2,11-dodecanediol	HO OH C-C-CCCCCCCC-C-C C C	67.5	---	?

†$5H_2O$ (50% H_2O) has been assigned. The maximum of the very flat freezing point curve has been found at 55% H_2O, no formula being assigned. This composition agrees excellently with $6H_2O$ which requires 54.5% H_2O.

Table 7.106: Hydrates of Cyclic Glycols *(32)*

	Glycol			Hydrate	
Number of C Atoms	Name	Skeletal Structural Formula	M.p. (°C)	M.p. (°C)	n in $R(OH)_2 \times nH_2O$
9	trans-Octa-hydroindan-8,9-diol		73-4	---	0.5 to 1.0
10	trans-Decahydro-naphthalene-9,10-diol		96	80-5	1.0 ?
10	cis-Decahydro-naphthalene-9,10-ciol		89.5	---	Unknown
10	trans-p-Menth-8(9)-ene-1,2-diol		73	60	3
10	cis-p-Menth-8(9)-ene-1,2-diol		71-2	---	0
10	cis(?)-p-Menth-1(2)-ene-3,6-diol		53-4	27	3?
10	cis-Terpin†		105	121	1
10	trans-Terpin‡		156-8	---	0
10	p-Menthane-2,5-diol		88-9	58-9	3
10	p-Menthane-1,2-diol		89	52	3 & 1
10	p-Menthane-2,8-diol (neoisodihydrocarveol hydrate)		93-4	65-75	Unknown

†Subject of crystallographic studies.
‡Stelzner's Literatur Register 1919-21 reports the formation of a hydrate and cites O. Aschan, "Bidrag till kännedon af Finlands natur ochfolk," 77, No. 1 (1918). The report appears to be without foundation.

(continued)

Table 7.106: (continued)

Number of C Atoms	Glycol Name	Skeletal Structural Formula	M.p. (°C)	Hydrate M.p. (°C)	Hydrate n in $R(OH)_2 \times nH_2O$
10	(+)-Carene-β-glycol or 3,4-Carane-diol (<u>trans</u>-2,3-dihydroxy-3,7,7-trimethyl-<u>bicyclo</u>-0,1,4-heptane)		90–91	75	1
11	Homoterpin		75–6	---	1
14	<u>iso</u> (±)-Hydrobenzoin	PhCHOHCHOHPh	121	96	Unknown
20	Dihydrodicarveol	$C_{20}H_{26}O_2$	166	100	2
29	3,3'-Dihydroxy-3,3'-diphenyl-2,2'-<u>spiro</u>-biindan		164	125–30	1 & 3
31	3,3'-Dihydroxy-3,3'-dibenzyl-2,2'-<u>spiro</u>-biindan		169	134	3
38	α-<u>s</u>-2,2'-Diphenylbenzopinacol		175	---	1
38	β-<u>s</u>-2,2'-Diphenylbenzopinacol		152–8	---	1

Table 7.107: Freezing Points of Aqueous Ethylene Glycol and Propylene Glycol Solutions *(42)*

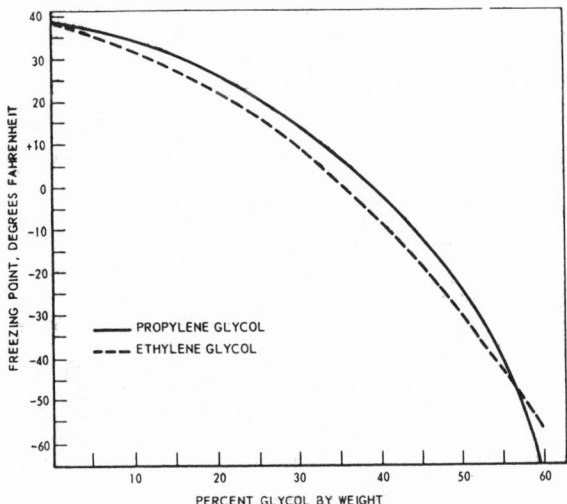

Table 7.108: Freezing Points of Various Aqueous Glycol Solutions, °C *(32)*

Glycol, %	1,2-Propane-diol	1,3-Propane-diol	1,2-Bu-tane-diol	1,3-Bu-tane-diol	levo-2,3-Bu-tane-diol	1,4-Bu-tane-diol	1,2-Pen-tane-diol	1,5-Pen-tane-diol
10	-3.12	-2.86	-2.60	-2.34	-3.1	-2.30	-2.3	-2.3
20	-7.6	-6.5	-6.0	-5.2	-7.1	-5.48	-4.8	-4.9
30	-14.0	-11.8	-11.0	-10.5	-12.4	-10.0	-6.8	-8.4
40	-22.7	-18.8	-16.5	-16.8	-19.4	-14.8	-8.4	-11.3
50	-34.5	-27.7	-22.4	-25.2	-29.6	-22.0	-10.2	-15.3
60	-48.2	-40.0	-29.0	-35.3	-40.4	-31.3	-12.6	-21.0

Table 7.109: Freezing Points of Various Aqueous Alcohols, Glycols and Glycerol *(32)*

| Solute by Weight, % | Methanol | | Ethanol | | Ethylene Glycol | | Glycerol | | levo-2,3-Butanediol | |
	F.p. Ob-served, °C	F.p. Calcu-lated, °C	F.p. Ob-served, °C	F.p. Calcu-lated, °C	F.p. Ob-served, °C	F.p. Calcu-lated, °C	F.p. Ob-served, °C	F.p. Calcu-lated, °C	F.p. Ob-served, °C	F.p. Calcu-lated, °C
10	-6.3	-6.46	-4.5	-4.49	-3.6	-3.33	-2.0	-2.25	-3.1	-2.30
20	-15.3	-14.5	-10.5	-10.1	-8.3	-8.27	-5.2	-5.05	-7.1	-5.17
30	-26.3	-24.9	-20.0	-17.3	-14.7	-12.9	-9.9	-8.67	-12.4	-8.85
40	-39.7	-38.8	-29.4	-27.0	-23.5	-20.0	-15.9	-12.0	-19.4	-14.3
50	-55.2	-58.1	-37.0	-40.4	-35.0	-30.0	-24.6	-20.2	-29.6	-20.7
60			-43.8	-60.7	-50	-45.0	-37.9	-30.3	-40.4	-31.0

Table 7.110: Compatibility of Coupling Solvents with Carbon Tetrachloride and Water *(14)*

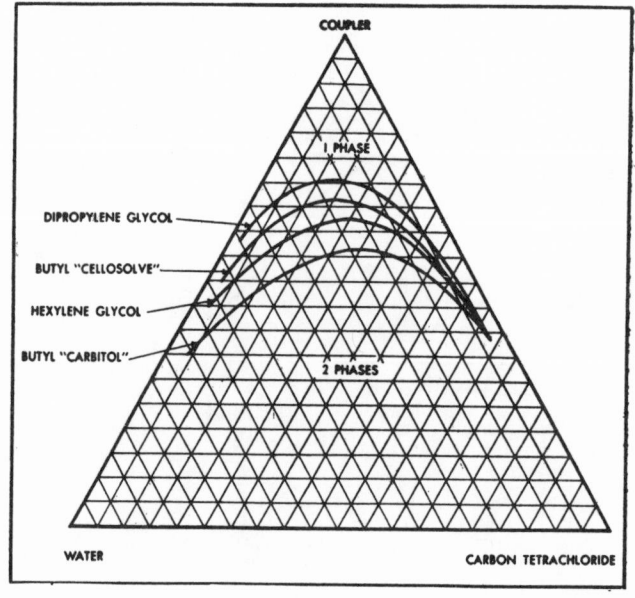

Table 7.111: Key Hygroscopicity Curve *(55)*

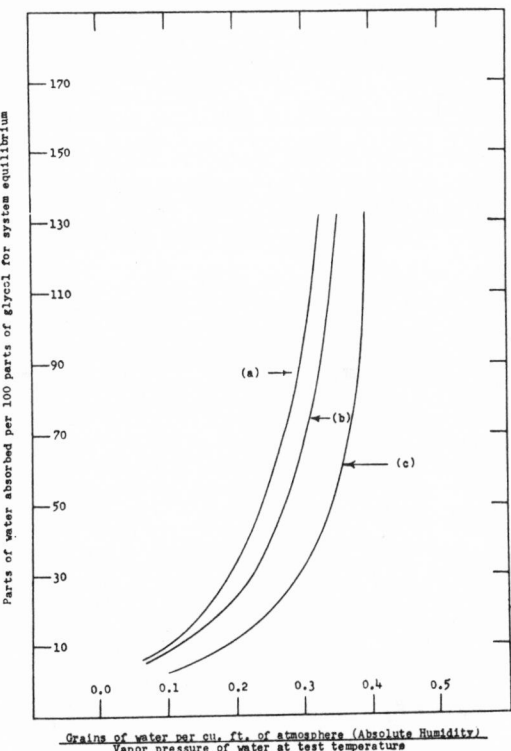

Key hygroscopicity curves for the various glycols: (a) ethylene glycol; (b) diethylene glycol; and (c) dipropylene glycol.

Table 7.112: Surface Tension of Glycol-Water Systems *(14)*

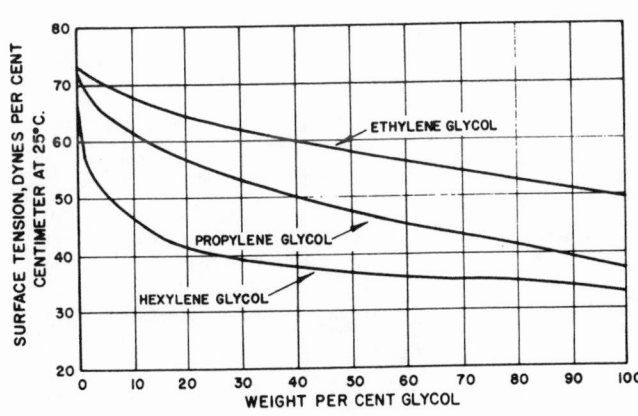

Table 7.113: Vapor Pressure of Glycols *(14)*

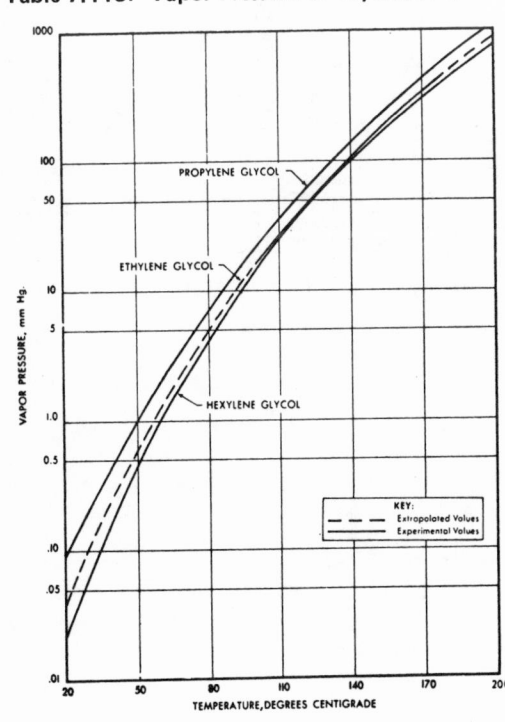

Table 7.114: Viscosity of Glycols *(32)*

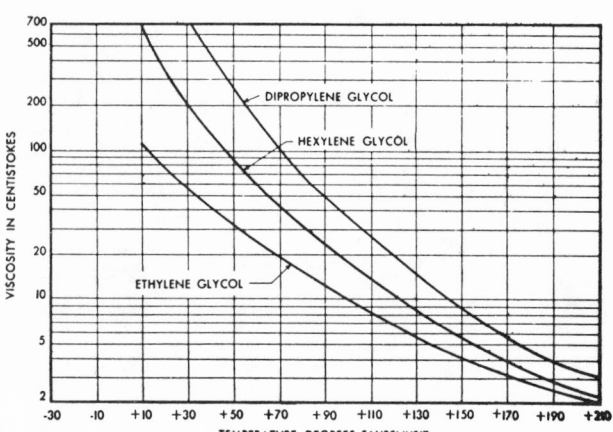

Table 7.115: Water Absorption by Glycols as a Function of Time *(14)*

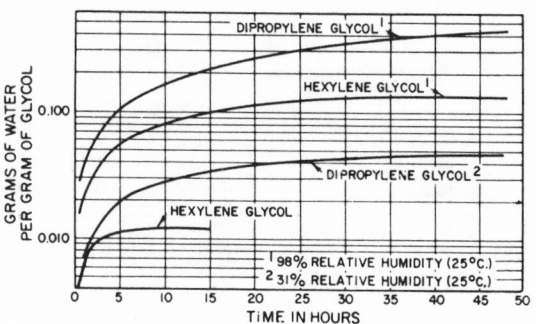

Table 7.116: Water Absorption by Glycols as a Function of Relative Humidity *(14)*

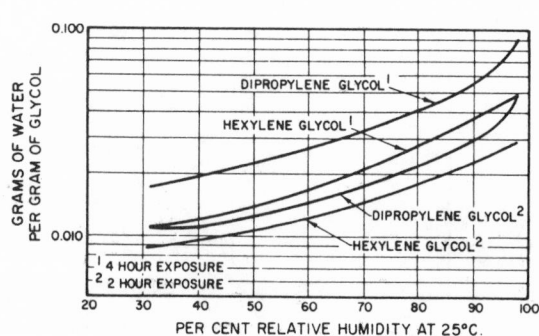

Table 7.117: Refractive Index, Specific Gravity, and Boiling Point Measurements of Various Glycols *(32)*

Compound	Refractive Index	Specific Gravity, d_4^t	Boiling Point, °C., 760 mm.
1,2-Propanediol	25°C. 1.4316 (17) 1.4313[†]	23°C. 1.0354 (17)	187 (17)
	20°C. 1.4331 (14) 1.4324[†]	20°C. 1.0364 (14) 1.0361[†]	186[†]
1,3-Propanediol	25°C. 1.4385 (17) 1.4380[†]	20°C. 1.0538 (17) 1.0529[†]	215 (17)
	21°C. 1.4394 (17) 20°C. 1.4389[†]		213.5[†]
1,2-Butanediol	20°C. 1.4378[†]	20°C. 1.0024[†]	190.5[†]
1,3-Butanediol	25°C. 1.4410 (17) 1.4391 (12) 1.4388[†]	20°C. 1.0053 (17) 1.0035 (12) 1.002 (2)	207.5 (17) 208 (2) 207[†]
	20°C. 1.4404 (2) 1.4398[†]	1.0037[†]	
1,4-Butanediol	20°C. 1.4467 (10) 1.4459 (2) 1.4460[†]	20°C. 1.0171 (10) 1.0160 (2) 1.0185[†]	230 (2) 228[†]
1,2-Pentanediol	24°C. 1.4390 (16) 25°C. 1.4380[†] 20°C. 1.4390[†]	24°C. 0.9691 (16) 20°C. 0.9723[†]	210 (16) 206[†]
1,5-Pentanediol	26°C. 1.4480 (16) 25°C. 1.4484[†] 20°C. 1.4500[†]	26°C. 0.9890 20°C. 0.9914[†]	239 (16) 238[†]

[†]Authors' observations.
[§]As cited in the fifth and earlier editions of Getman and Daniels' Outlines of Physical Chemistry, John Wiley and Sons Inc., New York, 1931.

Table 7.118: Relative Solvent Properties of Glycols *(23)*

S = Completely Soluble I = Insoluble < = Less Than > = Greater Than	Ethylene Glycol	Diethyl-ene Glycol	Tri-ethyl-ene Glycol	Tetra-ethylene Glycol	Propyl-ene Glycol	Di-propyl-ene Glycol	Tri-propyl-ene Glycol
Benzene	5.7	31.3	S	S	19.2	S	S
Carbon Tetrachloride[1]	6.2	26.2	33.6	62	23.4	S	S
Dibutyl Phthalate	0.5	10.6	16.5	S	8.1	S	S
Dichloroethyl Ether[1]	10.6	S	S	S	37.1	S	S
Diethanolamine[1]	S	S	S	S	S	S	S
DOWANOL* PM Glycol Ether[1]	S	S	S	S	S	S	S
DOWANOL* DPM Glycol Ether[1]	S	S	S	S	S	S	S
Ethyl Alcohol	S	S	S	S	S	S	S
Ethyl Ether	8.2	16.3	16.9	20	S	S	S
Methyl Alcohol	S	S	S	S	S	S	S
Methyl Isobutyl Carbinol	S	S	S	S	S	S	S
Methyl Isobutyl Ketone	12	S	S	S	S	S	S
Monochlorobenzene[1]	5.7	S	S	S	22.5	S	S
Monoethanolamine[1]	S	S	S	S	S	S	S
ortho-Dichlorobenzene[1]	4.5	48.4	S	S	19.4	S	S
Perchloroethylene[1]	0.7	10.7	15.0	19.0	14.5	S	S
Phenol[1]	S	S	S	S	S	S	S
Styrene[1]	3.4	36	S	S	15	S	S
Toluene	2.9	17.2	24.8	89	12.3	S	S
Urea	48	30	37	28	29	12	10
Castor Oil	1	<0.5	<0.5	<1	0.8	S	S
Coconut Oil	1	1	1	<1	1	1	3
Cottonseed Oil	1	1	1	<1	1	1	<1
Hydrous Wool Fat	<0.5	<0.5	<0.5	<1	<0.5	<0.5	<1
Lard Oil	1	1	1	<1	1	1	<1
Linseed Oil	1	1	1	<1	1	1.4	2.5
Oiticica Oil	<1	<1	<1	<1	<1	<1	<1
Olive Oil	1	1	1	<1	1	0.7	1.5
Pine Oil	S	S	S	S	S	S	S
Soya Bean Oil	1	1	1	<1	1	1	<1
Sperm Oil	1	1	1	<1	1	1	<1
Tall Oil	<1	<1	<1	<1	<1	S	S
Tung Oil	1	1	1	<1	1	1	<1
Turkey Red Oil	<1	<1[2]	1[2]	1[2]	<1[2]	3[2]	4[2]
Paraffin Oil	1	1	1	<1	1	1	<1
SAE No. 10 Oil	1	1	1	<1	1	1	<1
VMP Naphtha	<1	<1	<1	1	1	10	14
Animal Glue (Dry)	<0.5	<0.5	<0.5	<1	<0.5	<0.5	<1
Dextrin	<1	<1	<1	<1	<1	<1	<1
Gum Damar	<0.5	<0.5	<0.5	<1	<0.5	<0.5	<1
Kauri Gum	<0.5	<0.5	<0.5	>16[3]	<5	<5	>16[3]
Sudan III	<0.5	<0.5	<0.5	<1	<0.5	<0.5	<1
Shellac	<0.5	<0.5	<0.5	<1	<0.5	<0.5	<1

[1]Product of The Dow Chemical Company
[2]Forms stable emulsion from this concentration to 100%.
[3]Becomes too viscous to stir beyond 16%.
*Trademark of The Dow Chemical Company

Table 7.119: Effect of Various Glycols on Synthetic Rubber Samples—Results Reported as % Volume and % Weight Increase *(23)*

Glycol	GN-427T1 % Vol	GN-427T1 % Wt	GRS-53115T % Vol	GRS-53115T % Wt	FA-Thiokol % Vol	FA-Thiokol % Wt	Gum Rubber % Vol	Gum Rubber % Wt
3 Days Immersion								
Ethylene	−.2	−.2	−.1	−.5	.5	.2	.2	.2
Diethylene	−.12	−.2	−.1	−.5	.3	.0	.1	.0
Triethylene	−.1	−.1	−.1	−.5	.5	.1	.3	.1
Propylene	−.2	−.2	.1	−.3	.3	.0	.2	.0
Dipropylene	−.1	−.2	.0	−.5	.3	−.1	.1	.1
10 Days Immersion								
Ethylene	−.2	−.2	−.3	−.7	.3	.2	.3	.5
Diethylene	−.3	−.3	−.2	−.8	.3	.0	−.2	.0
Triethylene	.0	−.1	−.1	−.6	.6	.3	.0	.1
Propylene	−.1	−.1	−.3	−.7	.0	−.1	−.1	.0
Dipropylene	.1	−.2	−.1	−.6	.1	−.2	.0	.1

Table 7.120: Solubility of Cellulose Derivatives in Glycols *(23)*

Glycol	50 CPS. ST. E/C	½ Second Cellulose Nitrate	Cellulose Acetate FM 3
Ethylene	Insoluble	Swelled	Insoluble
Diethylene	Insoluble	>20% Soluble	Insoluble
Triethylene	Insoluble	>20% Soluble	Insoluble
Propylene	Insoluble	Swelled	Insoluble
Dipropylene	Insoluble	>20% Soluble	Insoluble

Table 7.121: Compatibility of Film Cast from 80/20 Toluene/Alcohol *(23)*

Glycol	50 CPS. ST. E/C			½ Second Cellulose Nitrate			Cellulose Acetate FM 3		
	Clear	Haze	Opaque	Clear	Haze	Opaque	Clear	Haze	Opaque
Ethylene	1%	3%	10%	3%	5%	10%	>10	20	—
Diethylene	1%	3%	10%	1%	5%	10%	>20	30	—
Triethylene	—	1%	3%	1%	3%	15%	>20	30	—
Propylene	1%	3%	10%	1%	3%	10%	>10	20	—
Dipropylene	20%	25%	30%	>50%	—	—	>40	50	—

Note: Table shows % glycol in film with the properties shown.

Table 7.122: Relative Humectant Values *(23)*

Temperature of Air °F	Glycol	Relative Humidities								
		10%	20%	30%	40%	50%	60%	70%	80%	90%
20 (−6.7°C)	Ethylene	97.5	93.4	89.3	85.7	82	78	72	63	48
	Diethylene	97.8	95.1	92.0	89.0	86	83	78	68	52
	Triethylene	98.5	96.8	94.0	91.1	89	83	78	66	51
	Propylene	96.8	91.4	90.0	84.6	77	73	68	55	40
	Dipropylene	98.5	97.0	95.1	92.6	89	85	79	67	51
40 (4.4°C)	Ethylene	97.3	93.2	89.1	85.4	82	76	69	60	42
	Diethylene	97.7	95.0	92.0	89.0	86	82	77	67	50
	Triethylene	98.4	96.5	93.8	91.0	88	83	77	65	51
	Propylene	97.0	92.3	90.2	85.2	78	74	68	55	40
	Dipropylene	98.4	96.9	95.0	92.5	89	85	79	67	51
60 (15.6°C)	Ethylene	97.1	93.0	88.9	85.0	81	75	66	57	37
	Diethylene	97.7	95.0	92.0	89.0	86	82	76	66	48
	Triethylene	98.2	96.2	93.6	90.8	86	82	77	65	50
	Propylene	97.1	92.9	90.4	85.8	80	74	68	55	40
	Dipropylene	98.4	96.8	94.8	92.4	89	85	79	67	51
80 (26.7°C)	Ethylene	96.8	92.8	88.6	84.7	80	73	64	55	36
	Diethylene	97.6	94.9	92.0	89.0	85	81	75	65	47
	Triethylene	98.1	96.0	93.4	90.7	85	82	76	64	50
	Propylene	97.1	93.5	90.5	86.3	81	75	68	55	40
	Dipropylene	98.3	96.7	94.7	92.3	89	85	79	67	51
100 (37.8°C)	Ethylene	96.6	92.7	88.4	84.3	79	72	63	53	35
	Diethylene	97.6	94.8	92.0	89.0	85	81	74	64	46
	Triethylene	98.0	95.7	93.2	90.6	84	82	76	64	49
	Propylene	97.2	93.9	90.6	86.6	82	75	68	55	40
	Dipropylene	98.3	96.6	94.6	92.1	89	85	79	67	51
120 (48.9°C)	Ethylene	96.4	92.5	88.2	84.0	78	71	62	51	34
	Diethylene	97.6	94.8	92.0	89.0	85	80	73	63	45
	Triethylene	97.8	95.4	93.0	90.5	83	82	75	63	49
	Propylene	97.2	94.3	90.7	86.7	83	76	68	55	40
	Dipropylene	98.2	96.5	94.5	92.0	89	85	79	67	51

Note: Values are given as the percent by weight of glycol in water solution required to maintain equilibrium in contact with air of various temperatures and humidities.

Table 7.123: Water Vapor Dew Points over Aqueous Ethylene Glycol Solutions (23)

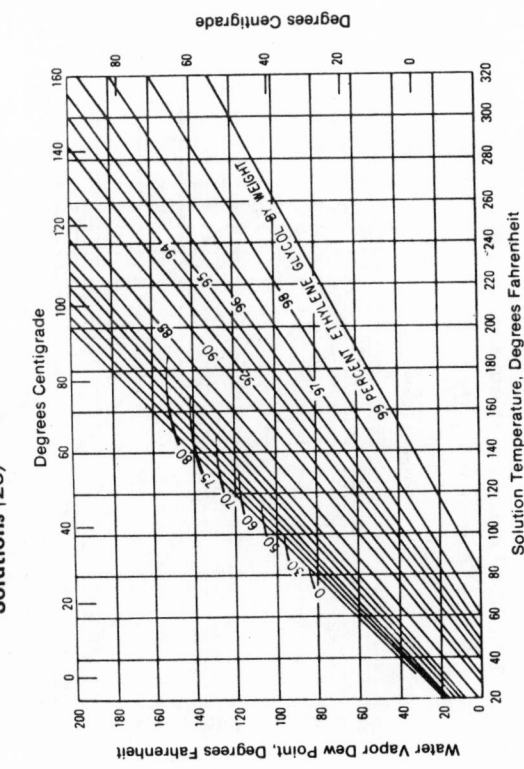

Table 7.124: Water Vapor Dew Points over Aqueous Diethylene Glycol Solutions (23)

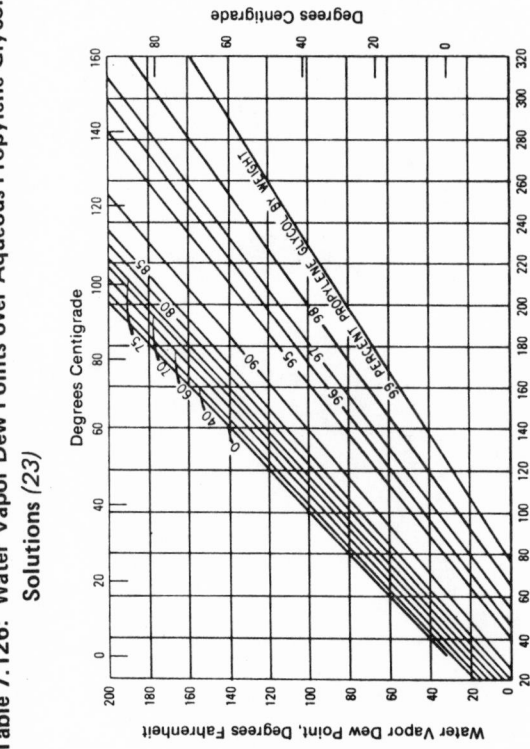

Table 7.125: Water Vapor Dew Points over Aqueous Triethylene Glycol Solutions (23)

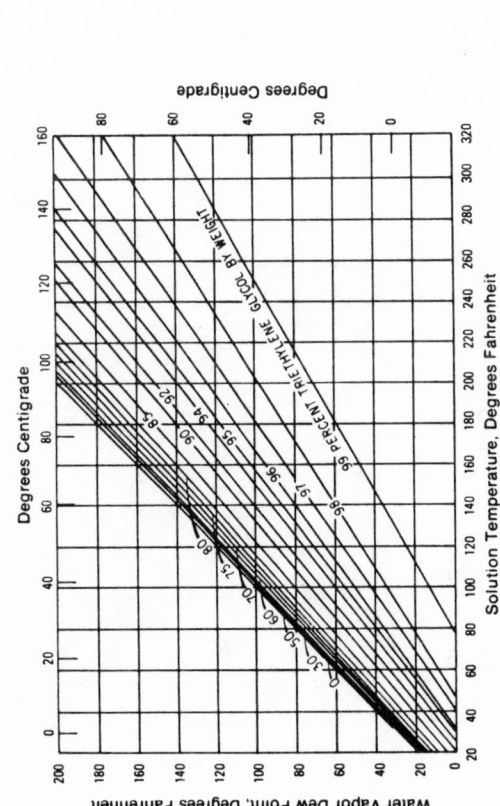

Table 7.126: Water Vapor Dew Points over Aqueous Propylene Glycol Solutions (23)

Table 7.128: Boiling Points of Glycols at 50 mm Hg *(23)*

Water	100°F (37.8°C)
Ethylene Glycol	258°F (125.5°C)
Diethylene Glycol	338°F (170°C)
Triethylene Glycol	387°F (197.2°C)
Tetraethylene Glycol	453°F (233.9°C)
Propylene Glycol	240°F (115.6°C)
Dipropylene Glycol	307°F (152.8°C)
Tripropylene Glycol	356°F (180°C)

Table 7.127: Water Vapor Dew Points over Aqueous Dipropylene Glycol Solutions *(23)*

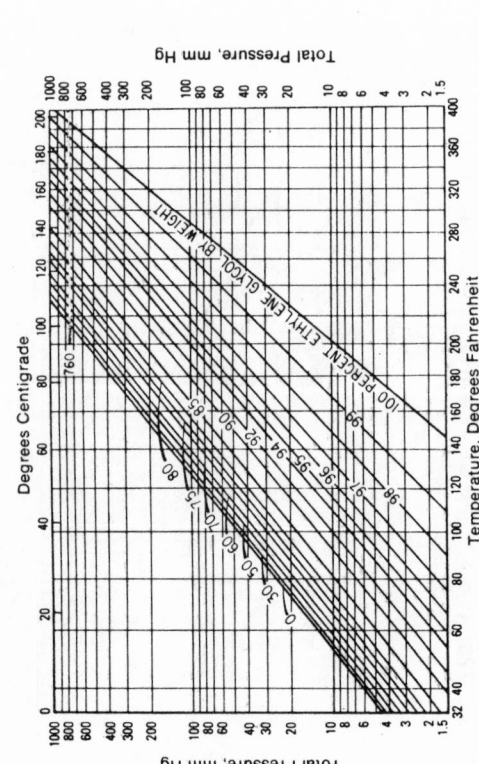

Table 7.130: Total Pressure over Aqueous Diethylene Glycol Solutions vs Temperature *(23)*

Table 7.129: Total Pressure over Aqueous Ethylene Glycol Solutions vs Temperature *(23)*

Table 7.131: Total Pressure over Aqueous Triethylene Glycol Solutions vs Temperature (23)

Table 7.132: Total Pressure over Aqueous Propylene Glycol Solutions vs Temperature (23)

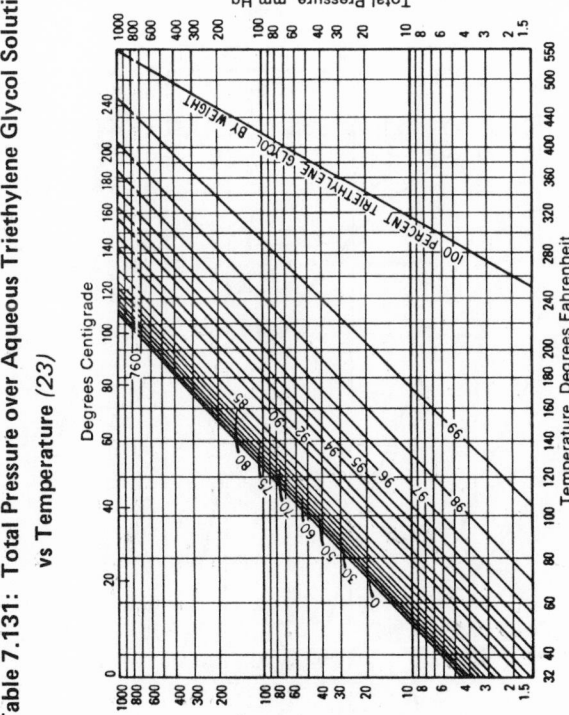

Table 7.133: Total Pressure over Aqueous Dipropylene Glycol Solutions vs Temperature (23)

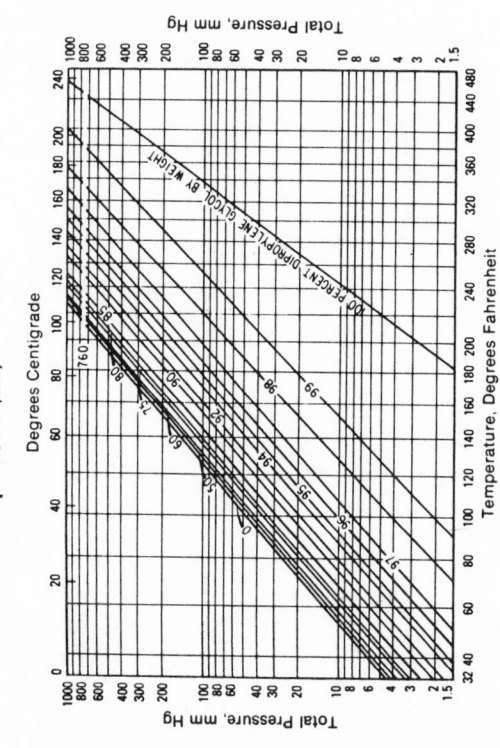

Table 7.135: Vapor-Liquid Composition Curves for Aqueous Diethylene Glycol Solutions (23)

Table 7.134: Vapor-Liquid Composition Curves for Aqueous Ethylene Glycol Solutions (23)

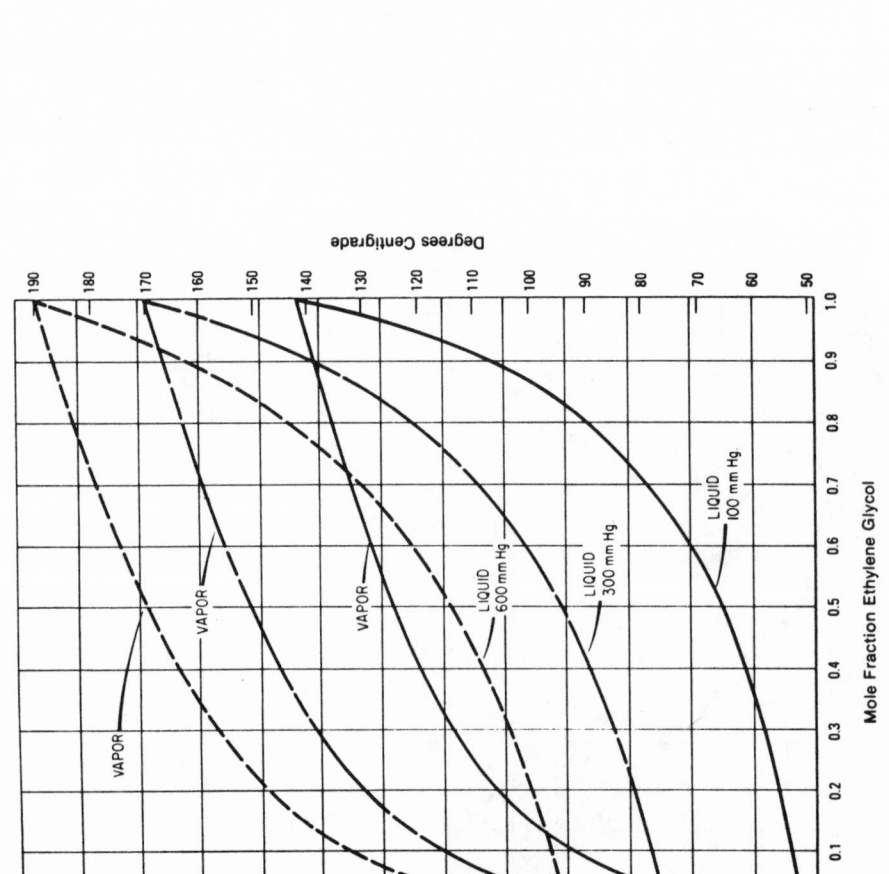

Table 7.137: **Vapor-Liquid Composition Curves for Aqueous Propylene Glycol Solutions** *(23)*

Table 7.136: **Vapor-Liquid Composition Curves for Aqueous Triethylene Glycol Solutions** *(23)*

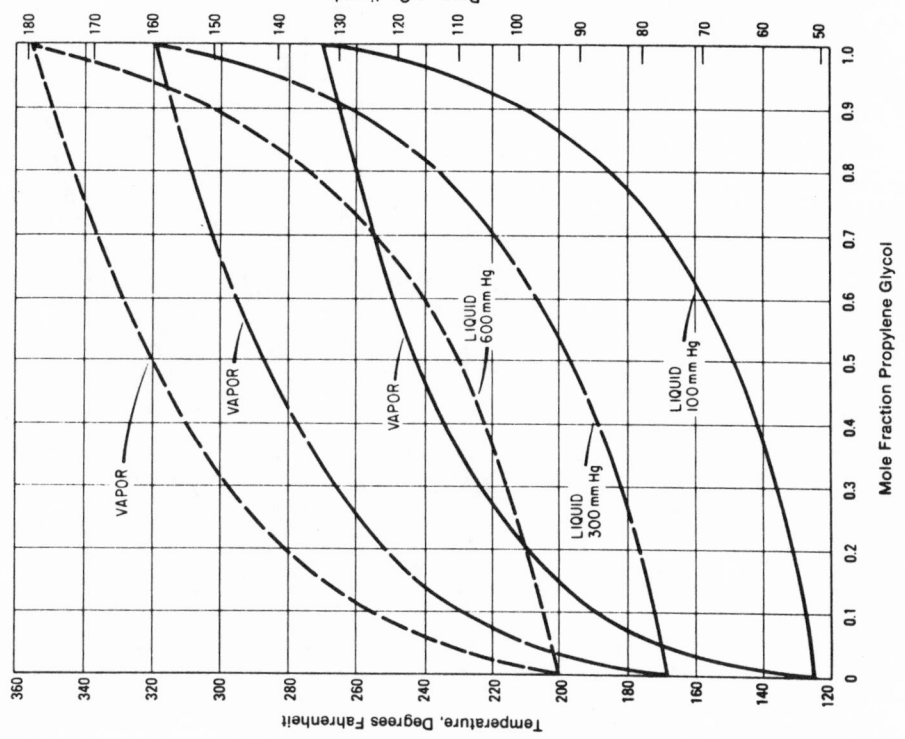

Table 7.139: Pour Points of Glycols (23)

Ethylene Glycol	<−75° F (−59°C)
Diethylene Glycol	−65° F (−54°C)
Triethylene Glycol	−73° F (−58°C)
Tetraethylene Glycol ..	−42° F (−41°C)
Propylene Glycol	−71° F (−57°C)
Dipropylene Glycol	−38° F (−39°C)
Tripropylene Glycol....	−42° F (−41°C)

Table 7.140: Viscosities of Anhydrous Glycols (23)

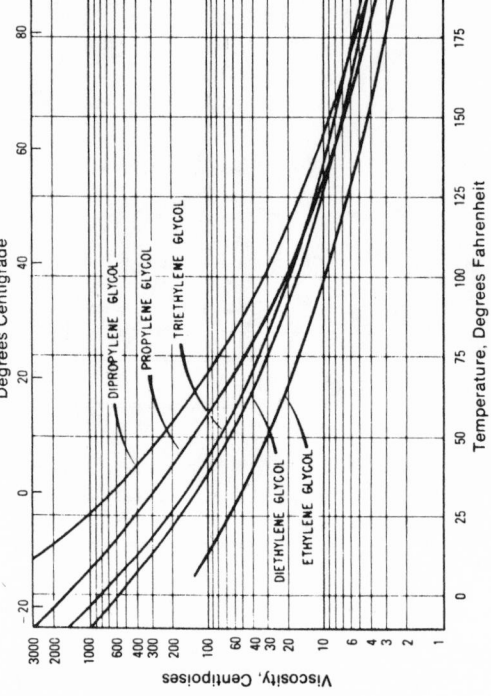

Table 7.138: Vapor-Liquid Composition Curves for Aqueous Dipropylene Glycol Solutions (23)

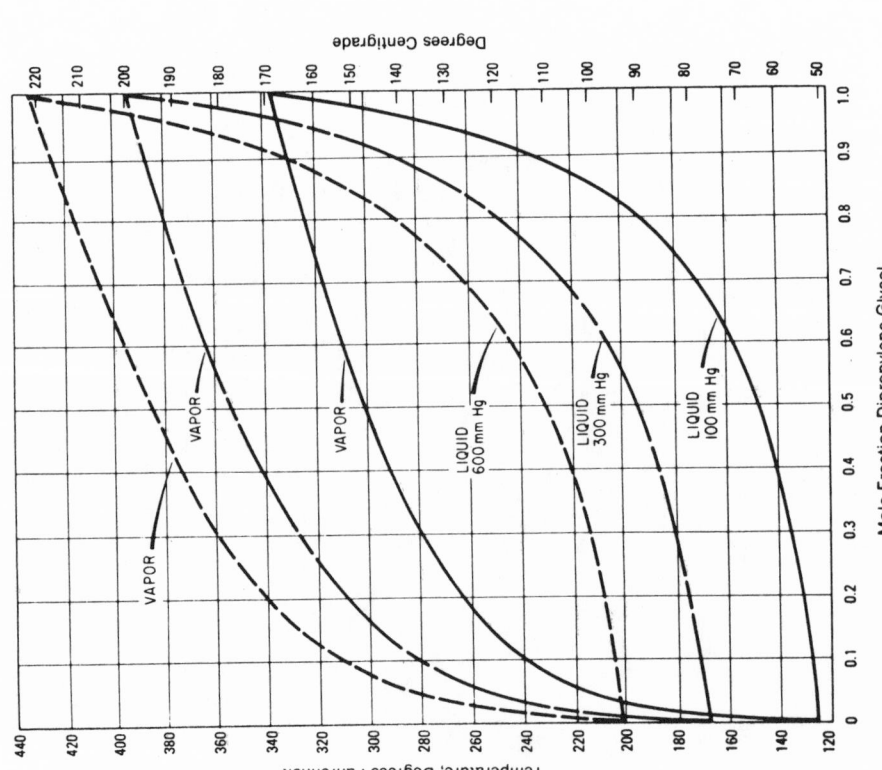

Table 7.142: Viscosities of Aqueous Diethylene Glycol Solutions *(23)*

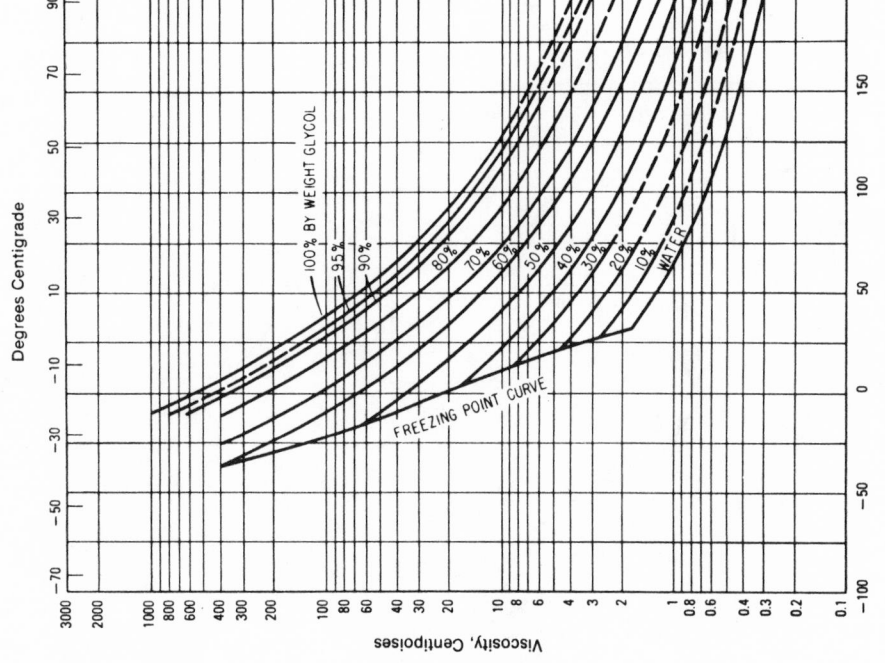

Table 7.141: Viscosities of Aqueous Ethylene Glycol Solutions *(23)*

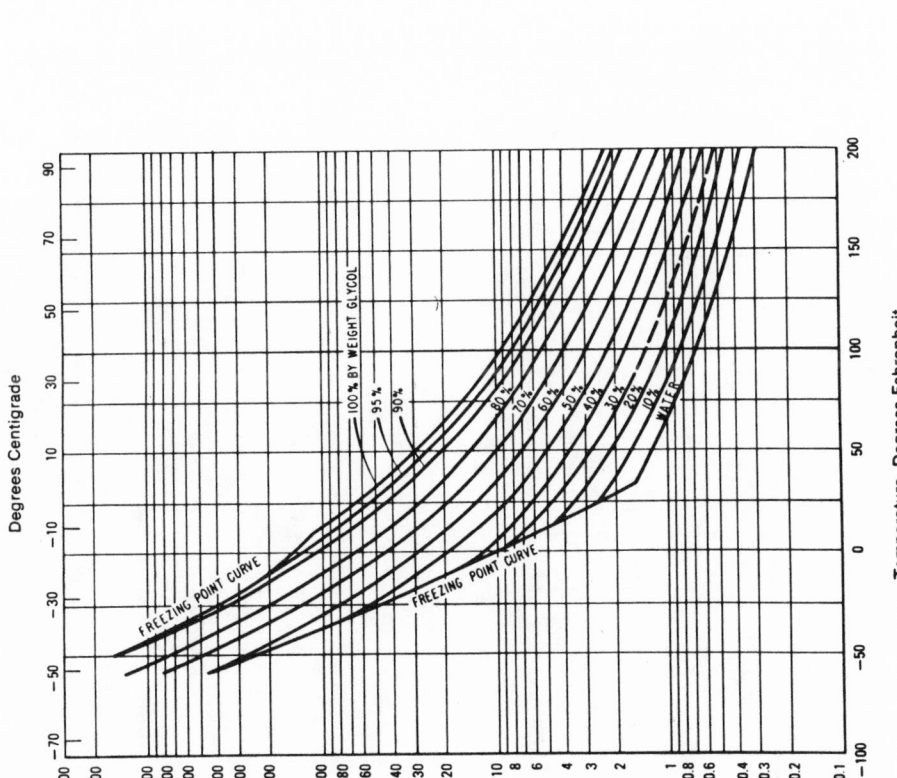

Table 7.143: Viscosities of Aqueous Triethylene Glycol Solutions *(23)*

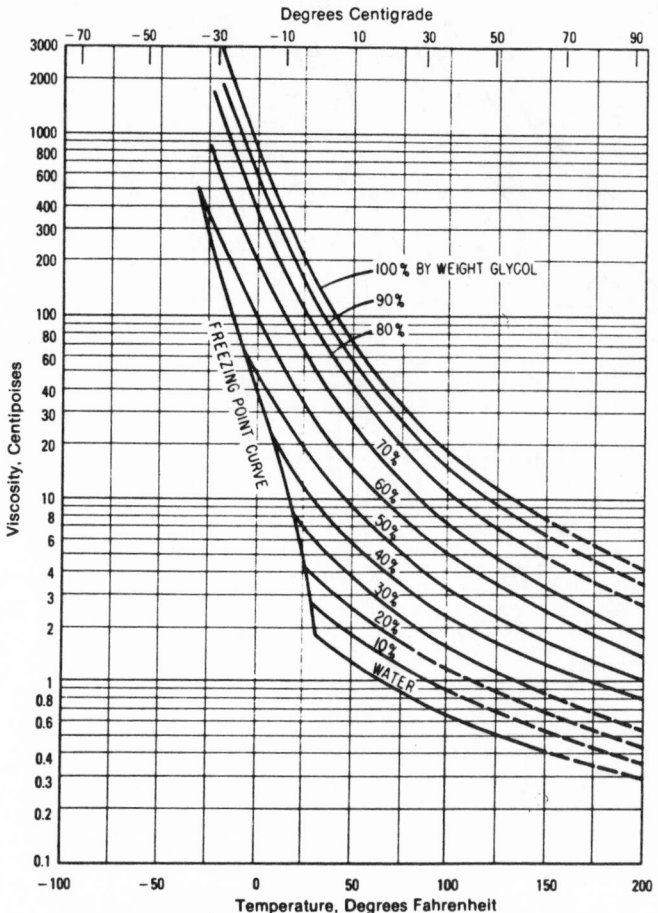

Table 7.144: Viscosities of Aqueous Tetraethylene Glycol Solutions *(23)*

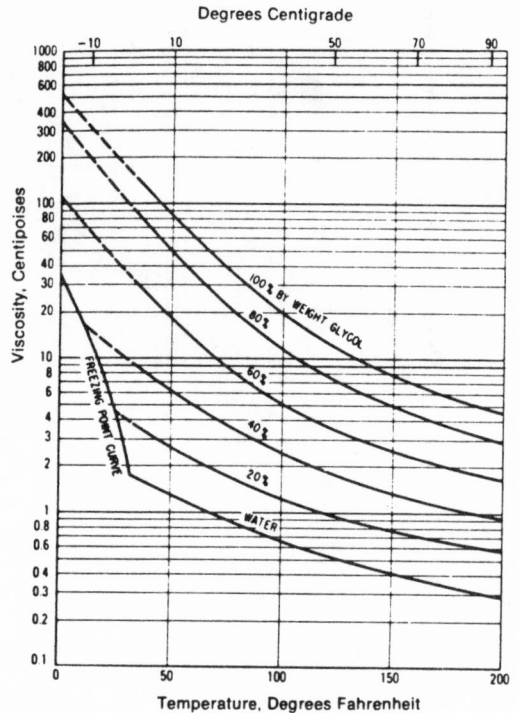

Table 7.145: Viscosities of Aqueous Propylene Glycol Solutions *(23)*

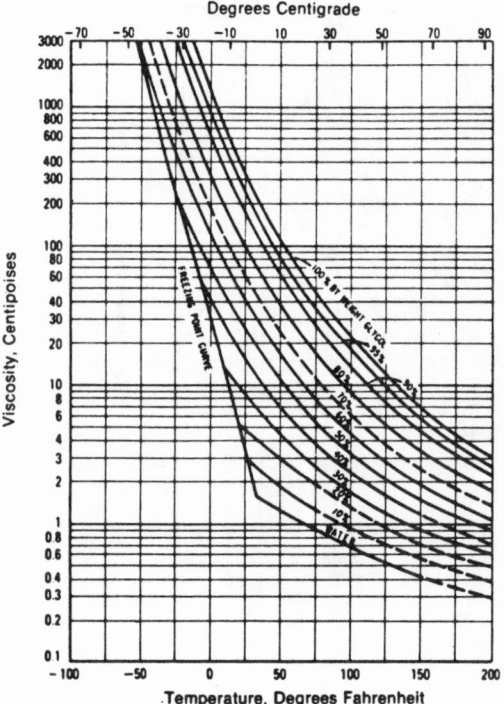

Table 7.146: Viscosities of Aqueous Dipropylene Glycol Solutions *(23)*

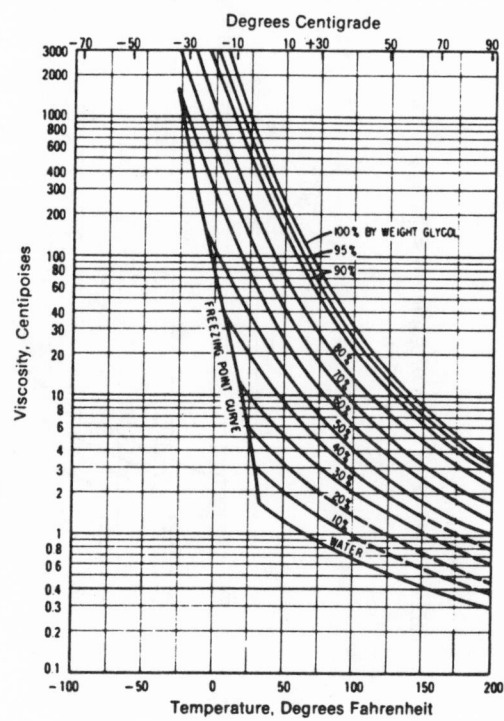

Table 7.148: Freezing Points of Aqueous Glycol Solutions (23)

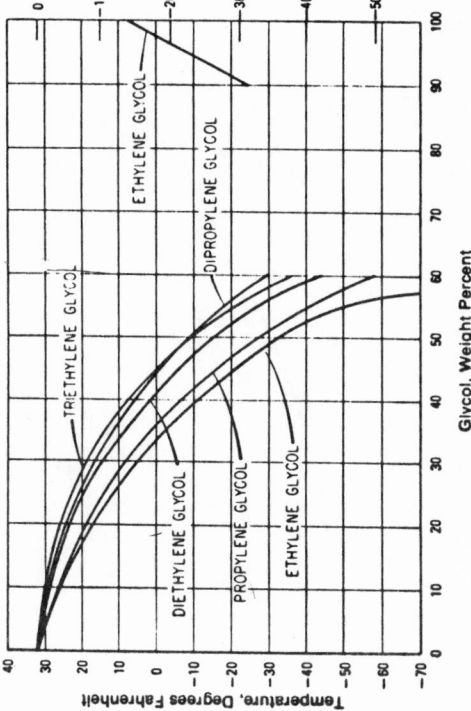

Table 7.149: Specific Heat of Anhydrous Glycols (23)

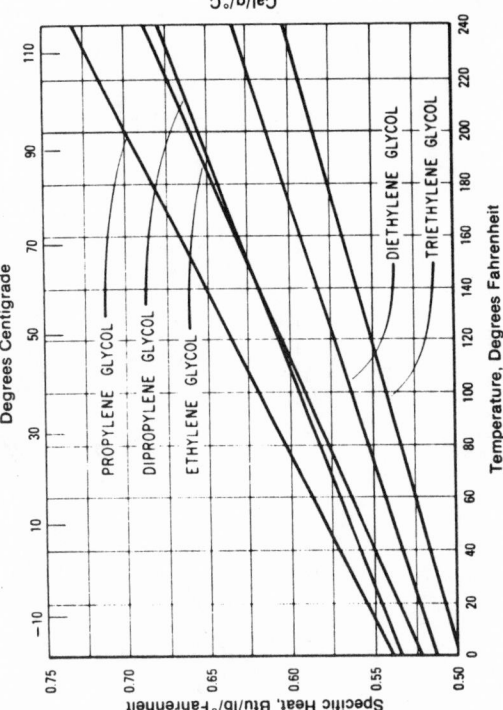

Table 7.147: Viscosities of Aqueous Tripropylene Glycol Solutions (23)

Table 7.151:   Densities of Aqueous Ethylene Glycol Solutions (% by wt) (23)

Chart axes: Pounds Per Gallon (9.5, 9.4, 9.3, 9.2, 9.1, 9.0, 8.9, 8.8, 8.7, 8.6, 8.5, 8.4, 8.3); Degrees Centigrade (−40 to 80); Density, Grams Per ml (0.9900 to 1.1500); Temperature, Degrees Fahrenheit (−40 to 180). Curves labeled 100% BY WEIGHT GLYCOL, 90%, 80%, 70%, 60%, 50%, 40%, 30%, 20%, 10%, FREEZING CURVE.

Table 7.150:   Specific Heats of Aqueous Glycol Solutions (Btu/lb/°F) (23)

Temp. °F	Glycol, % by Weight						Temp. °C
	100	80	60	40	20	10	
ETHYLENE GLYCOL							
60	.563	.660	.757	.855	.940	.976	15.6
80	.576	.673	.769	.864	.942	.977	26.7
100	.590	.685	.780	.872	.944	.978	37.8
120	.604	.697	.792	.880	.946	.979	48.9
140	.618	.710	.803	.888	.948	.980	60.0
160	.632	.722	.814	.896	.950	.981	71.1
180	.646	.735	.825	.905	.952	.982	82.2
200	.660	.748	.837	.914	.954	.982	93.3
220	.674	.761	.849	.922	.956	.983	104.4
240	.688	.774	.861	.930	.958	.984	115.5
DIETHYLENE GLYCOL							
60	.543	.631	.736	.849	.922	.949	15.6
80	.555	.645	.749	.855	.927	.954	26.7
100	.565	.659	.762	.861	.932	.960	37.8
120	.575	.672	.774	.868	.937	.965	48.9
140	.583	.686	.787	.874	.943	.970	60.0
160	.593	.700	.800	.880	.948	.975	71.1
180	.603	.714	.813	.886	.954	.980	82.2
200	.613	.728	.826	.893	.960	.985	93.3
220	.623	.742	.839	.900	.965	.990	104.4
240	.634	.756	.852	.907	.971	.995	115.5
TRIETHYLENE GLYCOL							
60	.525	.637	.749	.866	.935	.979	15.6
80	.534	.648	.758	.872	.938	.980	26.7
100	.540	.659	.768	.878	.941	.981	37.8
120	.550	.669	.777	.884	.944	.981	48.9
140	.562	.680	.787	.890	.946	.982	60.0
160	.577	.690	.796	.895	.949	.983	71.1
180	.586	.701	.806	.901	.952	.984	82.2
200	.595	.711	.815	.907	.955	.985	93.3
220	.605	.722	.825	.913	.957	.985	104.4
240		.782	.834	.919	.960	.986	115.5
PROPYLENE GLYCOL							
60	.587	.687	.795	.900	.970	.985	15.6
80	.603	.702	.808	.907	.972	.986	26.7
100	.619	.717	.821	.913	.975	.988	37.8
120	.635	.733	.833	.919	.977	.990	48.9
140	.651	.748	.846	.925	.980	.991	60.0
160	.667	.763	.857	.930	.983	.992	71.1
180	.683	.778	.871	.936	.984	.994	82.2
200	.699	.794	.882	.944	.987	.995	93.3
220	.715	.809	.895	.949	.990	.996	104.4
240	.731	.824	.907	.954	.993	.998	115.5
DIPROPYLENE GLYCOL							
60	.570	.687	.801	.900	.967	.985	15.6
80	.582	.698	.810	.905	.970	.986	26.7
100	.594	.708	.819	.910	.972	.988	37.8
120	.606	.718	.828	.915	.974	.990	48.9
140	.618	.728	.836	.920	.976	.991	60.0
160	.631	.739	.845	.924	.978	.993	71.1
180	.644	.749	.854	.929	.980	.995	82.2
200	.656	.760	.863	.934	.983	.997	93.3
220	.668	.770	.872	.939	.985	.998	104.4
240	.680	.781	.881	.944	.988	.999	115.5

Table 7.153: Densities of Aqueous Triethylene Glycol Solutions (% by wt) (23)

Table 7.152: Densities of Aqueous Diethylene Glycol Solutions (% by wt) (23)

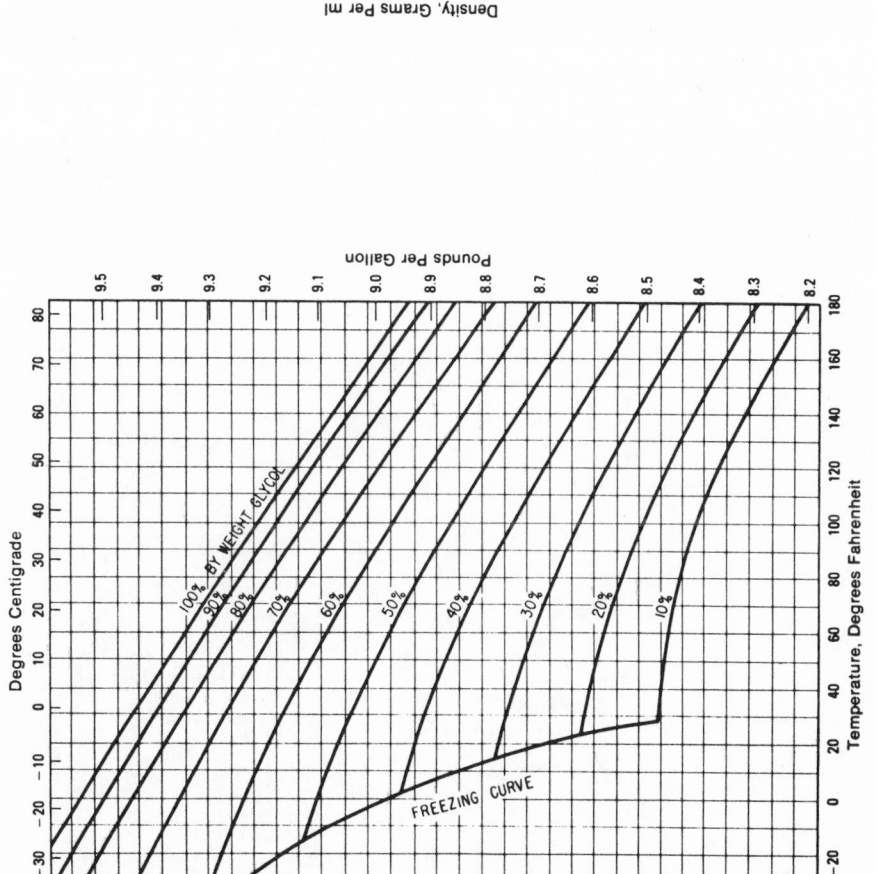

Table 7.155: Densities of Aqueous Propylene Glycol Solutions (% by wt) (23)

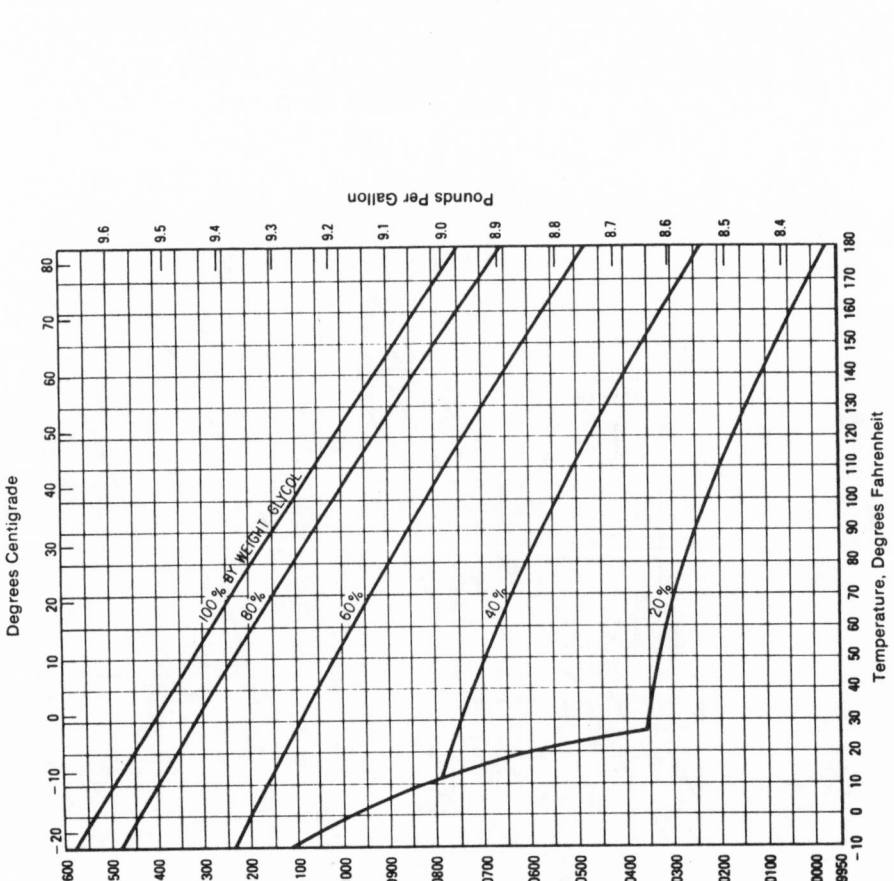

Table 7.154: Densities of Aqueous Tetraethylene Glycol Solutions (% by wt) (23)

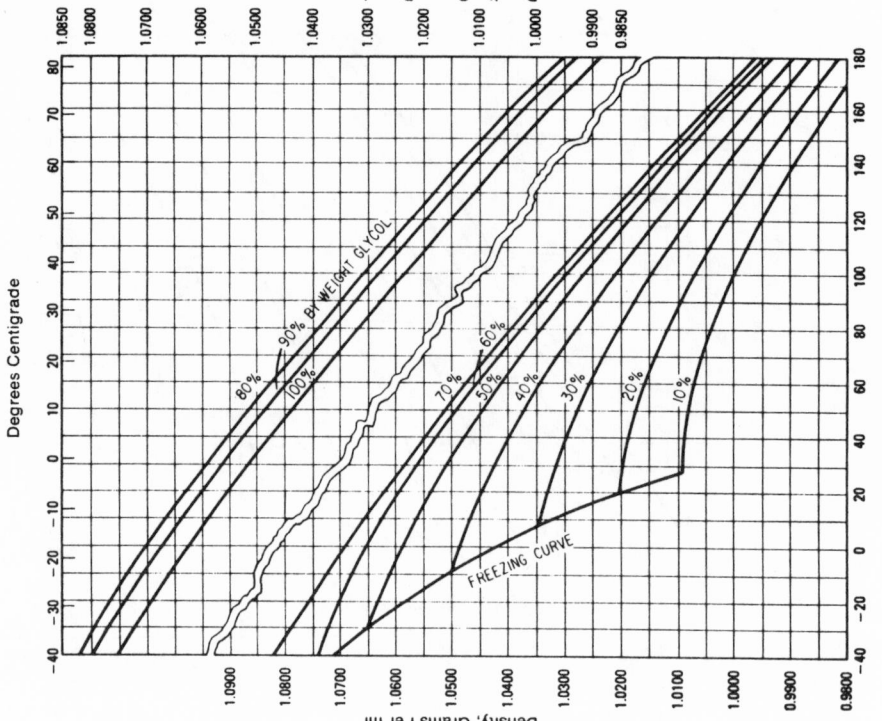

Table 7.157: Densities of Aqueous Tripropylene Glycol Solutions (% by wt) (23)

Table 7.158: Surface Tensions of Aqueous Solutions of Glycols at 77°F (23)

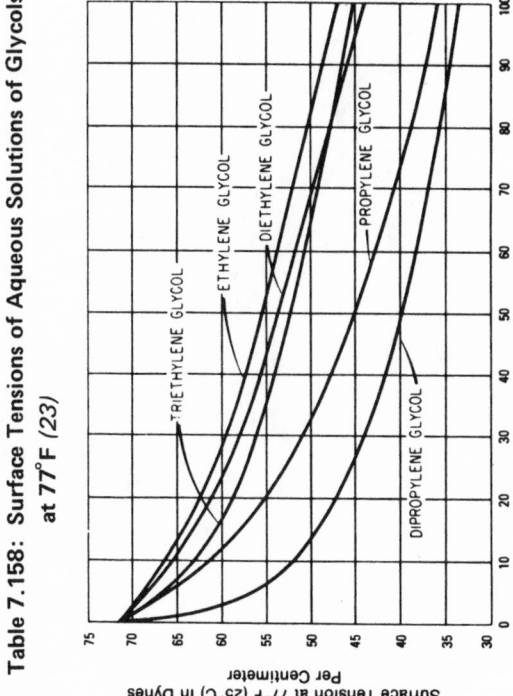

Table 7.156: Densities of Aqueous Dipropylene Glycol Solutions (% by wt) (23)

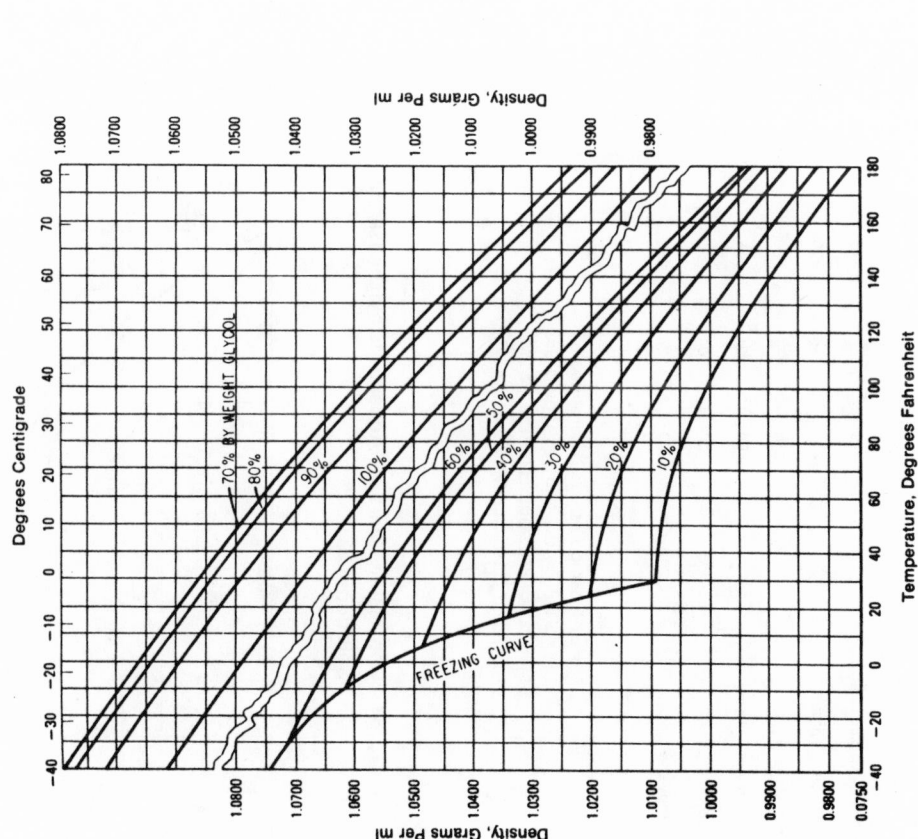

Table 7.160: Refractive Indices of Aqueous Glycol Solutions at 77°F (25°C) (23)

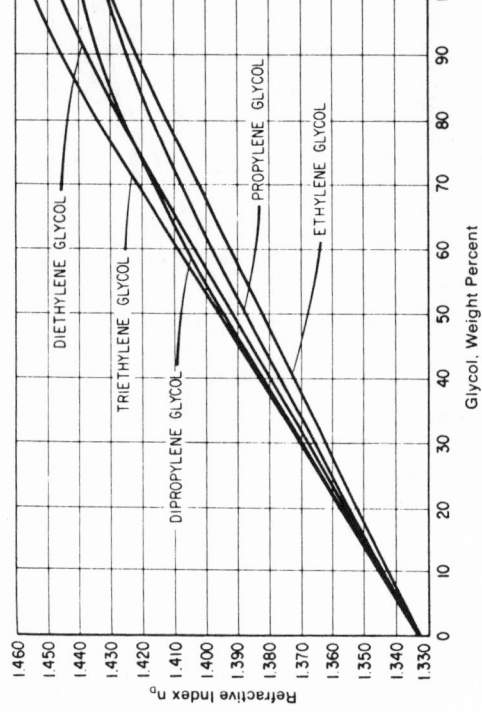

Table 7.159: Flammability of Glycols (23)

Glycol	Flash Point		Fire Point	
	°F	°C	°F	°C
Ethylene Glycol	240	116	245	119
Diethylene Glycol	255	124	290	142
Triethylene Glycol	350	177	330	166
Tetraethylene Glycol	400	204	375	191
Propylene Glycol	220	104	220	104
Dipropylene Glycol	260	127	260	127
Tripropylene Glycol	285	141	310	154

Note: Flash points are determined by the ASTM Pensky-Martens Closed Cup Method and fire points by the ASTM Cleveland Open Cup Method.

Table 7.162: Conversion Chart for Aqueous Diethylene Glycol Solutions (23)

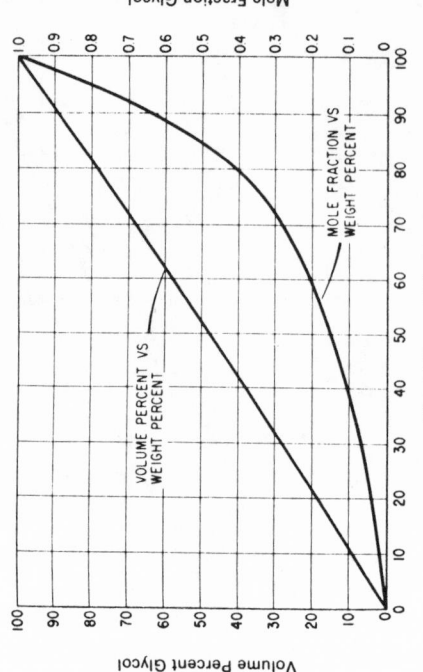

Table 7.161: Conversion Chart for Aqueous Ethylene Glycol Solutions (23)

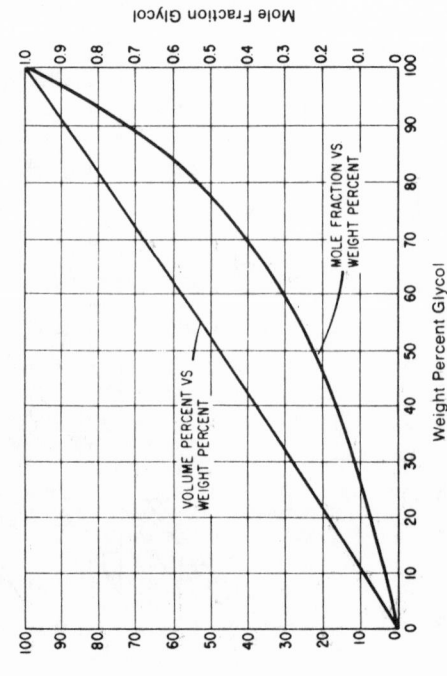

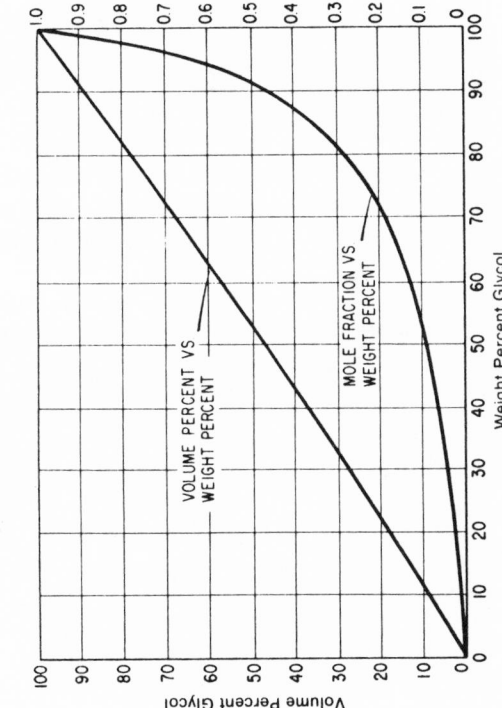

Table 7.163: Conversion Chart for Aqueous Triethylene Glycol Solutions (23)

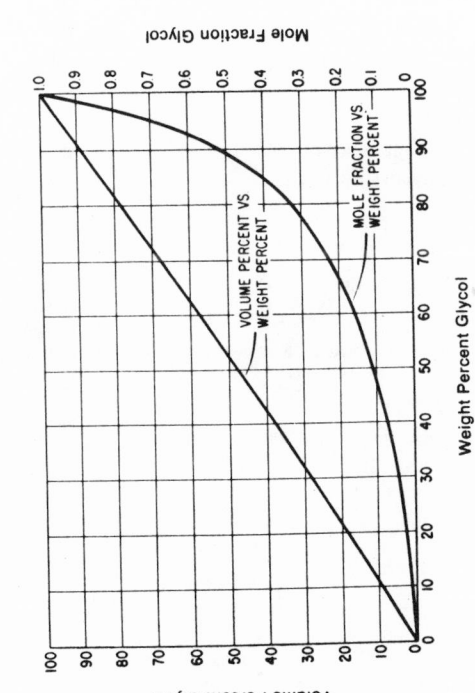

Table 7.164: Conversion Chart for Aqueous Tetraethylene Glycol Solutions (23)

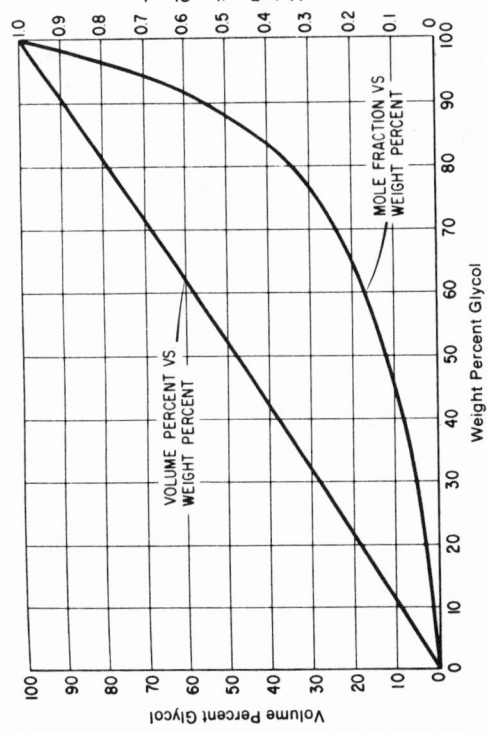

Table 7.165: Conversion Chart for Aqueous Propylene Glycol Solutions (23)

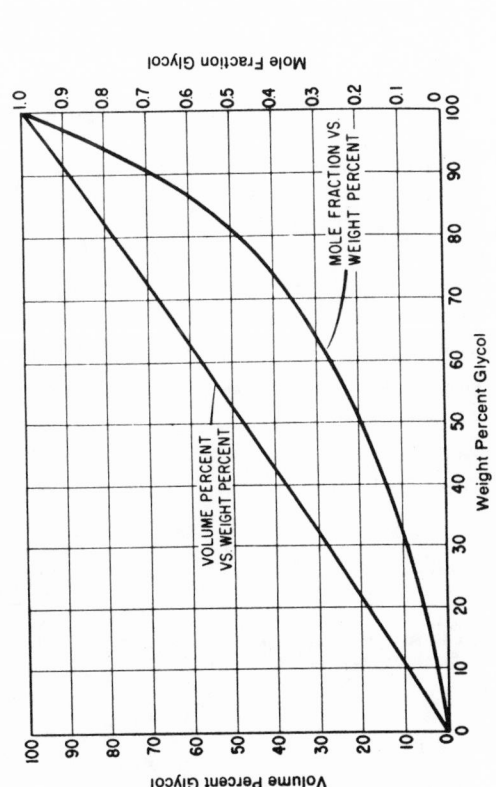

Table 7.166: Conversion Chart for Aqueous Dipropylene Glycol Solutions (23)

Table 7.167: Conversion Chart for Aqueous Tripropylene Glycol Solutions *(23)*

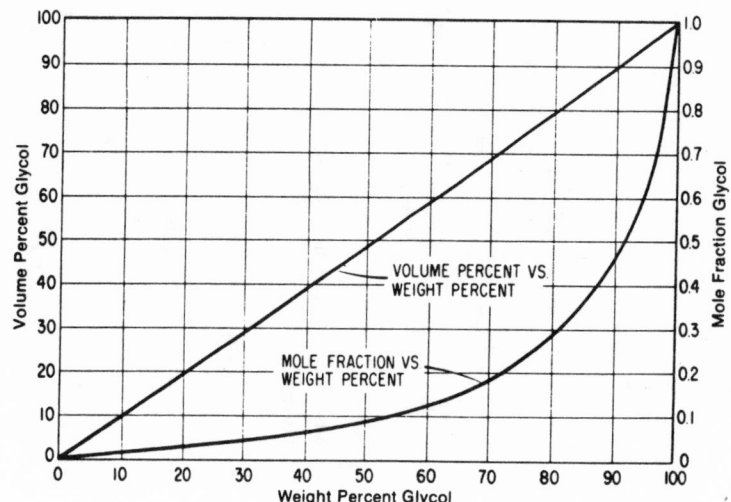

GLYCEROL (GLYCERINE)

1,2,3- Propanetriol $CH_2OH \cdot CHOH \cdot CH_2OH$

Table 7.168: Physical Properties and Specifications of Glycerol *(32)*

Acidity	Neutral to litmus	Heat of fusion	47.5 cal./g.
Ash	0.01% by wt., max.	Latent heat of vaporization at 55° C	228.7 g.-cal./g.
Auto ignition point (on glass)	804° F†	at 195° C	197.3 g.-cal./g.
Boiling point at 760 mm. Hg	290° C*	Melting point	17.9° C*
Boiling points at low pressures:		Molecular weight	92.094
at 1 mm.	125.0° C	Refractive index at 25° F	1.4722†
5 mm.	153.8° C		
10 mm.	167.2° C	Specific gravity at 25/25° C	1.262†
20 mm.	182.2° C	Specific heat at 25° C	0.577 cal./g. °C†
40 mm.	198.0° C		
Chlorine	0.0005% by wt., max.	Surface tension at 20° C	63.3 dynes/cm.
		90° C	58.6 dynes/cm.
Color, Pt-Co (Hazen) standards	20 max.	150° C	51.9 dynes/cm.
Fatty acids, mez/100 g.	1 max.	Vapor pressure at 20° C	0.0016 mm. Hg
Fire point	400° F†	200° C	42 mm. Hg
Flash point, tag open cup	350° F†	Viscosity at 25° C	945 cp.†
tag closed cup	320° F†		
Freezing point	17.9° C*	Weight per gallon at 25° C	10.50 lb.
Glycerol	99.5% by wt., min. (sp. gr. at 20° C, in air 1.2626)*		

*D. R. Stull, Ind. Engl. Chem., 39, 517 (1947).
†ACS Monograph, No. 117.

Table 7.169: Boiling Points and Specific Gravities of Aqueous Glycerol Solutions *(23)*

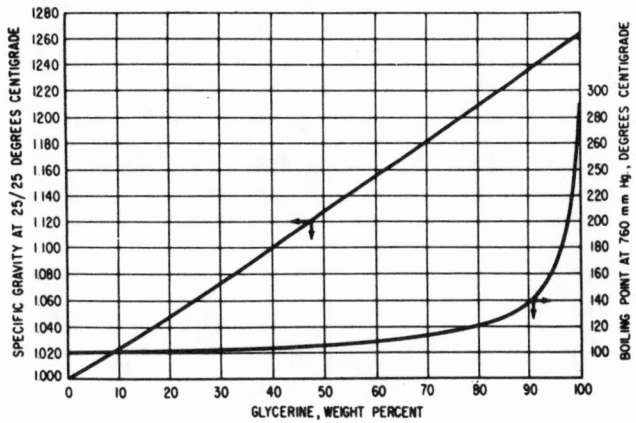

Table 7.170: Conversion Chart for Aqueous Glycerol Solutions (25°C) *(23)*

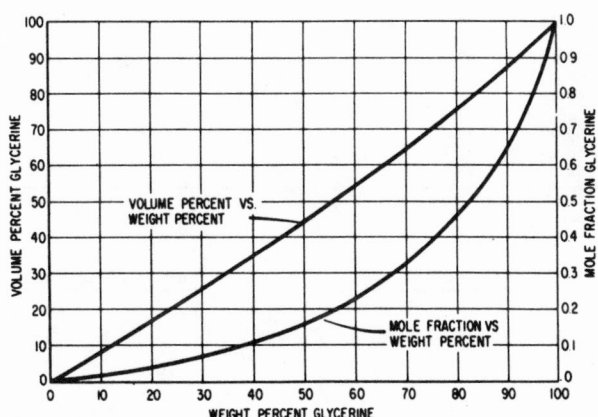

Table 7.171: Density of Glycerol Solutions *(23)*

Glycerol %	Temperature F.p.	-5°	-10°	-20°	-30°	-40°
10	-1.6°	---	---	---	---	---
20	-4.8°	---	---	---	---	---
30	-9.5°	1.0810	---	---	---	---
40	-15.4°	1.1096	1.1109	---	---	---
50	-23.0°	1.1387	1.1407	1.1450	---	---
60	-34.7°	1.1663	1.1685	1.1732	1.1787	---
66.7	-46.5°	1.1860	1.1889	1.1945	1.1985	1.2034
70	-38.5°	1.1954	1.1993	1.2038	1.2079	---
80	-20.3°	1.2210	1.2255	1.2305	---	---
90	-1.6°	---	---	---	---	---

Table 7.172: Freezing Points of Glycerol-Water Mixtures (23) (32)

Glycerol by Weight Per Cent	Water Per Cent	Freezing Points °C	Glycerol by Weight Per Cent	Water Per Cent	Freezing Points °C
0.0	100.0	0	65.0	35.0	-43.0
5.0	95.0	-0.6	65.6	34.4	-44.5
10.0	90.0	-1.6	66.0	34.0	-44.7
11.5	88.5	-2.0	66.7	33.3	-46.5
15.0	85.0	-3.1	66.1	32.9	-45.5
20.0	80.0	-4.8	67.3	32.7	-44.5
22.6	77.4	-6.0	68.0	32.0	-44.0
25.0	75.0	-7.0	70.0	30.0	-38.9
30.0	70.0	-9.5	70.9	29.1	-37.5
33.3	67.0	-11.0	75.0	25.0	-29.8
35.0	65.0	-12.2	75.4	24.6	-28.5
40.0	60.0	-15.4	79.0	21.0	-22.0
44.5	55.5	-18.5	80.0	20.0	-20.3
45.0	55.0	-18.8	84.8	15.2	-10.5
50.0	50.0	-23.0	85.0	15.0	-10.9
53.0	47.0	-26.0	90.0	10.0	-1.6
55.0	45.0	-28.2	90.3	9.7	-1.0
60.0	40.0	-34.7	95.0	5.0	7.7
60.4	39.6	-35.0	95.3	4.7	7.5
64.0	36.0	-41.5	98.2	1.8	13.5
64.7	35.3	-42.5	100.0	0.0	17.0

Freezing Points of Glycerol-Water Mixtures (°F.)

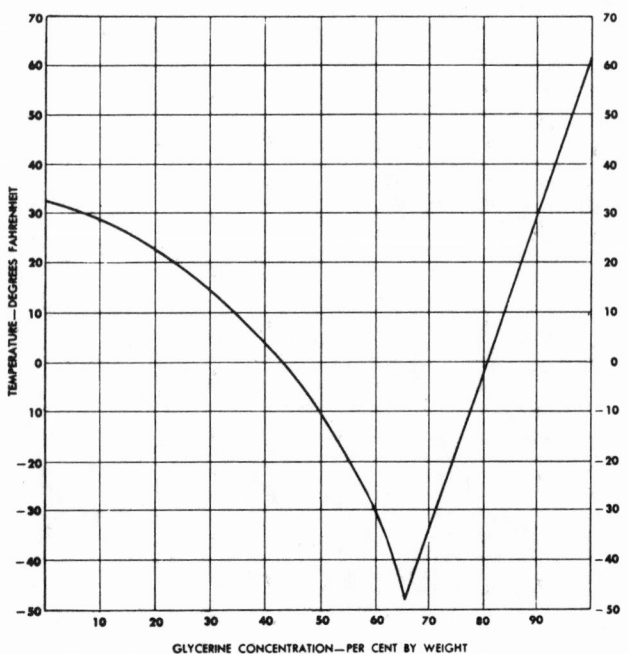

Freezing Points of Glycerol-Water Mixtures (°C.)

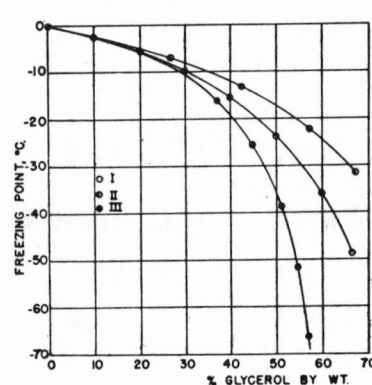

(I)   Observed
(II)   Theoretical, without hydration
(III) Theoretical, with complete hydration

Table 7.173: Hygroscopicity Curves for Glycerol
and 1,3-Butylene Glycol *(42)*

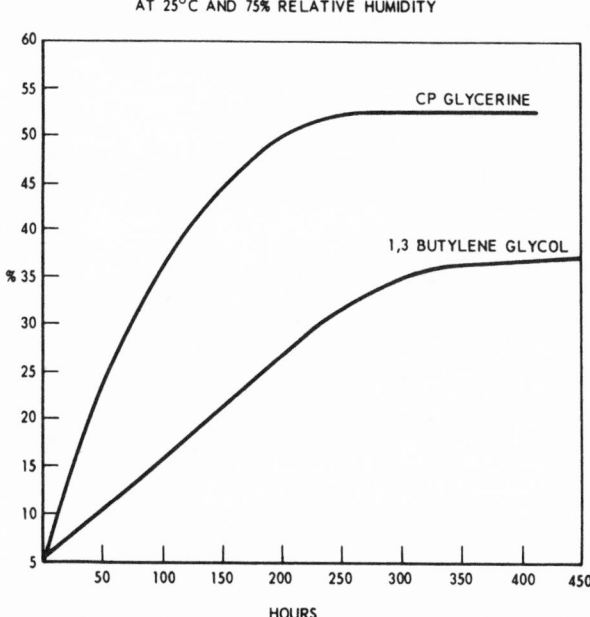

Table 7.174: Hygroscopicity Curves for Glycerol
and 2,3-Butylene Glycol *(42)*

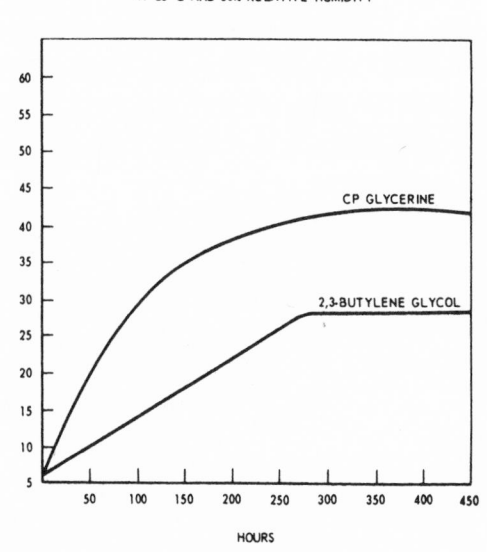

Table 7.175: Relative Humidities over Aqueous Glycerol Solutions, 20° to 100°C *(23)*

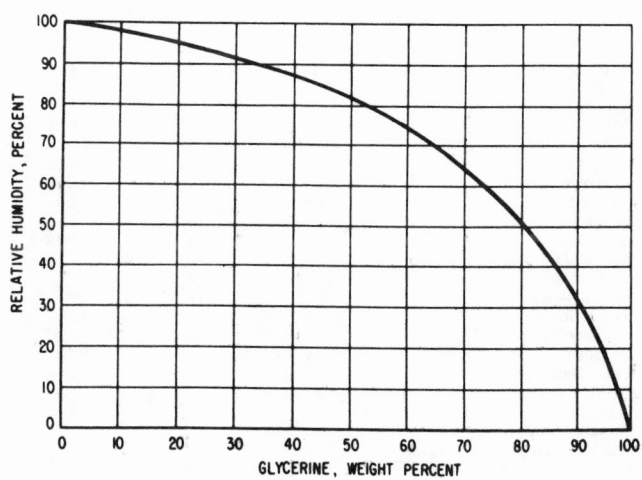

Table 7.176: Solubility of Sucrose and Dextrose in Aqueous Glycerol at 15°, 24° and 35°C *(32)*

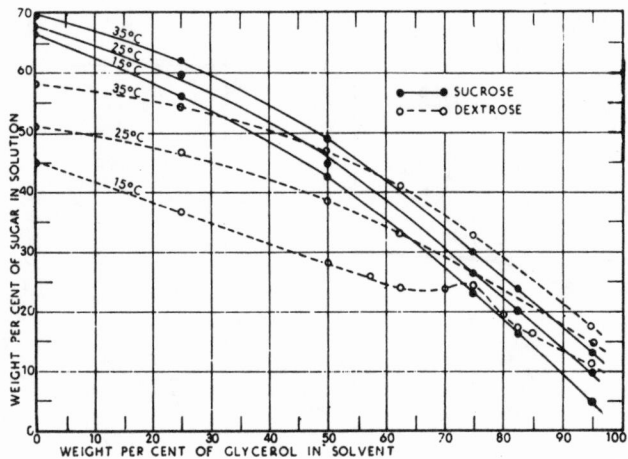

Table 7.177: Solubility of Various Compounds in Glycerol *(32)*

Substance	Glycerol Concentration % Weight	Temperature °C	Solubility in Parts per 100 Parts of Solvent
Alum	†	15	40
Ammonium carbonate	†	15	20
Ammonium chloride	†	15	20.06
Atropine	†	15	3
Benzoic acid	98.5	--	2
Boric acid	98.5	20	24.80
Calcium hydroxide	35	25	1.3
Calcium hypophosphite	99.04	20	2.5
Calcium sulfate	†	15	5.17
Codeine hydrochloride	99.04	20	11.1
Ethyl ether	99.04	20	0.65
Ferrous sulfate	†	15	25
Guaiacol	99.04	20	13.1
Iodine	†	15	2
Iodoform	95	15	0.12
Iron and potassium tartrate	†	15	8
Iron lactate	†	15	16
Morphine acetate	†	15	20
Novocaine	99.04	20	11.2
Phenacetin	99.04	20	0.47
Phenol	99.04	20	276.4
Potassium iodide	†	15	39.72
Quinine sulfate	98.5	--	1.32
Salicin	†	15	12.5
Sodium bicarbonate	†	15	8.06
Sodium carbonate (crystals)	†	15	98.3
Sodium tetraborate (borax)	†	15	60
Tannic acid	†	15	48.8
Tartar emetic	†	15	5.5
Urea	†	15	50
Zinc chloride	†	15	49.87
Zinc iodide	†	15	39.78

†Glycerol concentration not specified, probably 95 to 100 per cent.

Table 7.178: Specific Gravity and Percent Glycerol *(32)*

Glycerol	Apparent Specific Gravity				Glycerol	Apparent Specific Gravity			
	15/15° C	15.5/15.5° C	20/20° C	25/25° C		15/15° C	15.5/15.5° C	20/20° C	25/25° C
Per Cent					Per Cent				
100	1.26557	1.26532	1.26362	1.26201	50	1.12985	1.12970	1.12845	1.12720
99	1.26300	1.26275	1.26105	1.25945	49	1.12710	1.12695	1.12570	1.12450
98	1.26045	1.26020	1.25845	1.25685	48	1.12440	1.12425	1.12300	1.12185
97	1.25785	1.25760	1.25585	1.25425	47	1.12165	1.12150	1.12030	1.11915
96	1.25525	1.25500	1.25330	1.25165	46	1.11890	1.11880	1.11760	1.11650
95	1.25270	1.25245	1.25075	1.24910	45	1.11620	1.11605	1.11490	1.11380
94	1.25005	1.24980	1.24810	1.24645	44	1.11345	1.11335	1.11220	1.11115
93	1.24740	1.24715	1.24545	1.24380	43	1.11075	1.11060	1.10950	1.10845
92	1.24475	1.24450	1.24280	1.24115	42	1.10800	1.10790	1.10680	1.10575
91	1.24210	1.24185	1.24020	1.23850	41	1.10525	1.10515	1.10410	1.10310
90	1.23950	1.23920	1.23755	1.23585	40	1.10255	1.10245	1.10135	1.10040
89	1.23680	1.23655	1.23490	1.23320	39	1.09985	1.09975	1.09870	1.09775
88	1.23415	1.23390	1.23220	1.23055	38	1.09715	1.09705	1.09605	1.09510
87	1.23150	1.23120	1.22955	1.22790	37	1.09445	1.09435	1.09335	1.09245
86	1.22885	1.22855	1.22690	1.22520	36	1.09175	1.09165	1.09070	1.08980
85	1.22620	1.22590	1.22420	1.22255	35	1.08905	1.08895	1.08805	1.08715
84	1.22355	1.22325	1.22155	1.21990	34	1.08635	1.08625	1.08535	1.08455
83	1.22090	1.22055	1.21890	1.21720	33	1.08365	1.08355	1.08270	1.08190
82	1.21820	1.21790	1.21620	1.21455	32	1.08100	1.08085	1.08005	1.07925
81	1.21555	1.21525	1.21355	1.21190	31	1.07830	1.07815	1.07735	1.07660
80	1.21290	1.21260	1.21090	1.20925	30	1.07560	1.07545	1.07470	1.07395
79	1.21015	1.20985	1.20815	1.20655	29	1.07295	1.07285	1.07210	1.07135
78	1.20740	1.20710	1.20540	1.20380	28	1.07035	1.07025	1.06950	1.06880
77	1.20465	1.20440	1.20270	1.20110	27	1.06770	1.06760	1.06690	1.06625
76	1.20190	1.20165	1.19995	1.19840	26	1.06510	1.06500	1.06435	1.06370
75	1.19915	1.19890	1.19720	1.19565	25	1.06250	1.06240	1.06175	1.06115
74	1.19640	1.19615	1.19450	1.19295	24	1.05985	1.05980	1.05915	1.05860
73	1.19365	1.19340	1.19175	1.19025	23	1.05725	1.05715	1.05655	1.05605
72	1.19090	1.19070	1.18900	1.18755	22	1.05460	1.05455	1.05400	1.05350
71	1.18815	1.18795	1.18630	1.18480	21	1.05200	1.05195	1.05140	1.05095
70	1.18540	1.18520	1.18355	1.18210	20	1.04935	1.04935	1.04880	1.04840
69	1.18260	1.18240	1.18080	1.17935	19	1.04685	1.04680	1.04630	1.04590
68	1.17985	1.17965	1.17805	1.17660	18	1.04435	1.04430	1.04380	1.04345
67	1.17705	1.17685	1.17530	1.17385	17	1.04180	1.04180	1.04135	1.04100
66	1.17430	1.17410	1.17255	1.17110	16	1.03930	1.03925	1.03885	1.03850
65	1.17155	1.17130	1.16980	1.16835	15	1.03675	1.03675	1.03635	1.03605
64	1.16875	1.16855	1.16705	1.16560	14	1.03425	1.03420	1.03390	1.03360
63	1.16600	1.16575	1.16430	1.16285	13	1.03175	1.03170	1.03140	1.03110
62	1.16320	1.16300	1.16155	1.16010	12	1.02920	1.02920	1.02890	1.02865
61	1.16045	1.16020	1.15875	1.15735	11	1.02670	1.02665	1.02640	1.02620
60	1.15770	1.15745	1.15605	1.15460	10	1.02415	1.02415	1.02395	1.02370
59	1.15490	1.15465	1.15325	1.15185	9	1.02175	1.02175	1.02155	1.02135
58	1.15210	1.15190	1.15050	1.14915	8	1.01935	1.01930	1.01915	1.01900
57	1.14935	1.14910	1.14775	1.14640	7	1.01690	1.01690	1.01675	1.01660
56	1.14655	1.14635	1.14500	1.14365	6	1.01450	1.01450	1.01435	1.01425
55	1.14375	1.14355	1.14220	1.14090	5	1.01210	1.01205	1.01195	1.01185
54	1.14100	1.14080	1.13945	1.13815	4	1.00965	1.00965	1.00955	1.00950
53	1.13820	1.13800	1.13670	1.13540	3	1.00725	1.00725	1.00720	1.00710
52	1.13540	1.13525	1.13395	1.13265	2	1.00485	1.00485	1.00480	1.00475
51	1.13265	1.13245	1.13120	1.12995	1	1.00240	1.00240	1.00240	1.00235

Table 7.179: Specific Gravities of Glycerol and Glycol Mixtures *(23)*

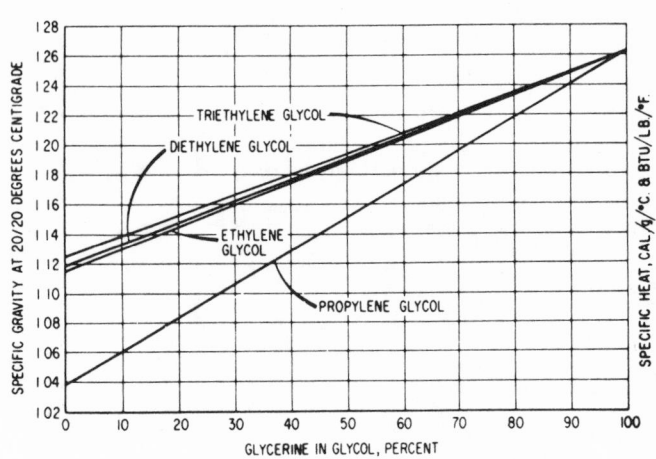

Table 7.180: Specific Heat of Glycerol *(23)*

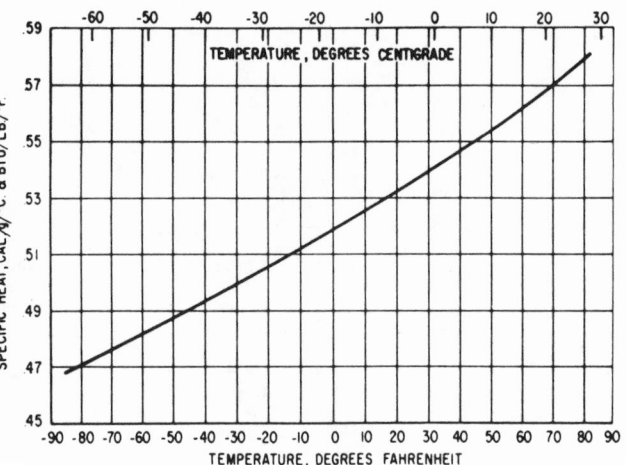

Table 7.181: Vapor Pressure of Glycerol *(23)*

Temperature, °C	V. P. mm. Hg.	Temperature, °C	V. P. mm. Hg.
120	--	210	63.8
130	1.47	220	91.9
140	2.61	230	130
150	4.48	240	181
160	7.44	250	248
170	12.0	260	334
180	18.9	270	445
190	29.0	280	586
200	43.4	290	760

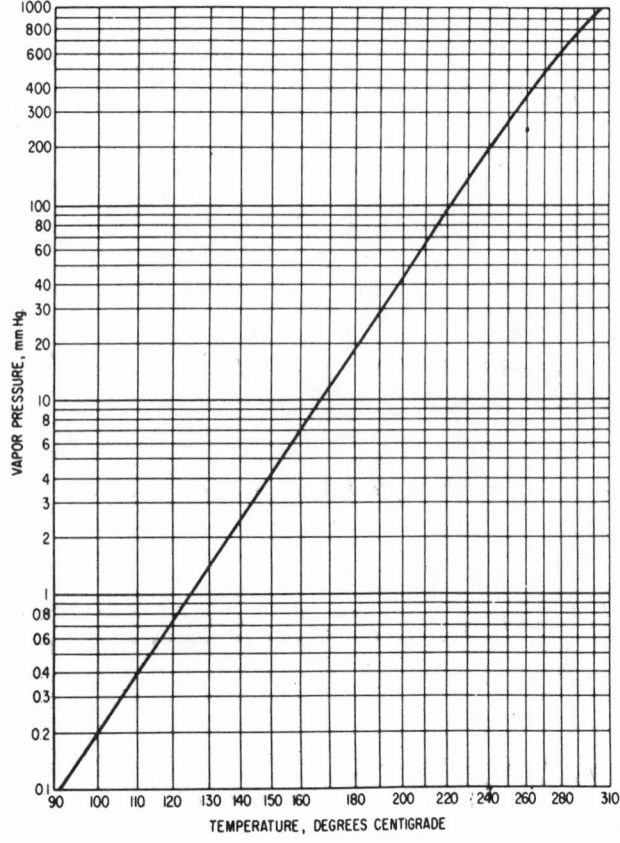

Table 7.182: Vapor Pressure of Glycerol-Water Solutions *(23)*

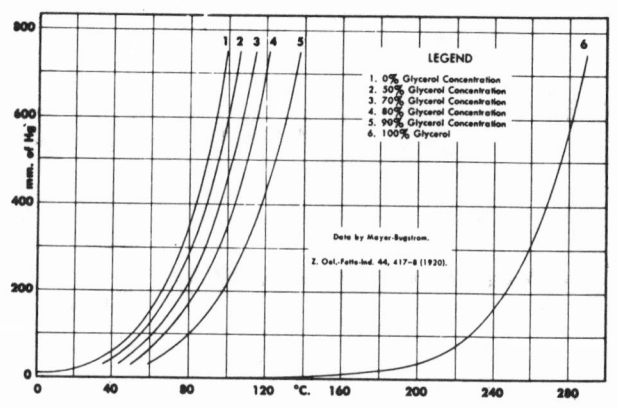

VISCOSITY, CENTIPOISES

100% GLYCERINE BY WEIGHT
90%
80%
70%
60%
50%
40%
30%
20%
10%
0%

TEMPERATURE, DEGREES CENTIGRADE

Table 7.183: Viscosity of Glycerol Solutions in Centipoises (23) (32)

Glycerol %	F.p.	-5°	-10°	-20°	-30°	-40°
10	-1.6°	---	---	---	---	---
20	-4.8°	---	---	---	---	---
30	-9.5°	6.5	---	---	---	---
40	-15.4°	10.3	14.4	---	---	---
50	-23.0°	18.8	24.4	48.1	---	---
60	-34.7°	41.6	59.1	108.0	244.0	---
66.7	-46.5°	74.7	113.0	289.0	631.0	1398.0
70	-38.5°	110.0	151.0	394.0	1046.0	---
80	-20.3°	419.0	683.0	1600.0	---	---
90	-1.6°	---	---	---	---	---

Temperature °C.

Glycerol % Wt.	0	10	20	30	40	50	60	70	80	90	100
0†	1.792	1.308	1.005	0.8007	0.6560	0.5494	0.4688	0.4061	0.3565	0.3165	0.2838
10	2.44	1.74	1.31	1.03	0.826	0.680	0.575	0.500	---	---	---
20	3.44	2.41	1.76	1.35	1.07	0.879	0.731	0.635	---	---	---
30	5.14	3.49	2.50	1.87	1.46	1.16	0.956	0.816	0.690	---	---
40	8.25	5.37	3.72	2.72	2.07	1.62	1.30	1.09	0.918	0.763	0.668
50	14.6	9.01	6.00	4.21	3.10	2.37	1.86	1.53	1.25	1.05	0.910
60	29.9	17.4	10.8	7.19	5.08	3.76	2.85	2.29	1.84	1.52	1.28
65	45.7	25.3	15.2	9.85	6.80	4.89	3.66	2.91	2.28	1.86	1.55
67	55.5	29.9	17.7	11.3	7.73	5.50	4.09	3.23	2.50	2.03	1.68
70	76.0	38.8	22.5	14.1	9.40	6.61	4.86	3.78	2.90	2.34	1.93
75	132.	65.2	35.5	21.2	13.6	9.25	6.61	5.01	3.80	3.00	2.43
80	255.	116.	60.1	33.9	20.8	13.6	9.42	6.94	5.13	4.03	3.18
85	540.	223.	109.	58.0	33.5	21.2	14.2	10.0	7.28	5.52	4.24
90	1310.	498.	219.	109.	60.0	35.5	22.5	15.5	11.0	7.93	6.00
91	1590.	592.	259.	127.	68.1	39.8	25.1	17.1	11.9	8.62	6.40
92	1950.	729.	310.	147.	78.3	44.8	28.0	19.0	13.1	9.46	6.82
93	2400.	860.	367.	172.	89.0	51.5	31.6	21.2	14.4	10.3	7.54
94	2930.	1040.	437.	202.	105.	58.4	35.4	23.6	15.8	11.2	8.19
95	3690.	1270.	523.	237.	121.	67.0	39.9	26.4	17.5	12.4	9.08
96	4600.	1580.	624.	281.	142.	77.8	45.4	29.7	19.6	13.6	10.1
97	5770.	1950.	765.	340.	166.	88.9	51.9	33.6	21.9	15.1	10.9
98	7370.	2460.	939.	409.	196.	104.	59.8	38.5	24.8	17.0	12.2
99	9420.	3090.	1150.	500.	235.	122.	69.1	43.6	27.8	19.0	13.3
100	12070.	3900.	1410.	612.	284.	142.	81.3	50.6	31.9	21.3	14.8

†Viscosity of water taken from Properties of Ordinary Water–Substances by N. E. Dorsey, New York, publisher   1940, p. 184.

COMPARATIVE DATA

Table 7.184: Emery Glycerines *(63)*

						SPECIFICATIONS					
	Glycerol %, min.	Specific Gravity 25/25°C min.	Color APHA max.	Residue on Ignition PPM, max.	Chloride PPM max.	Sulfate PPM max.	Arsenic PPM max.	Heavy Metals PPM max.	Chlorinated Compounds PPM max.	Fatty Acids and Esters[2]	Readily Carbon- izable
Emery 912 96% CP/USP Glycerine	96.0	1.2517	20[1]	100	10	20	1.5	5	30	1.0	—[3]
Emery 915 99% High Gravity Glycerine	99.0	1.2595	150	—	100	—	—	—	—	—	—
Emery 916 99.5% CP/USP Glycerine	99.5	1.2607	20[1]	100	10	20	1.5	5	30	1.0	—[3]
Emery 918 99.7% CP/USP Ultra Glycerine	99.7	1.2612	10	100	5	20	1.5	5	30	0.18	—[3]

[1]Meets USP specification which is equivalent to 20 APHA.
[2]Ml. 0.5N NaOH per 50 g of glycerine.
[3]Lighter than matching H fluid.

Table 7.185: "Kemstrene" Glycerines *(26)*

Product	Description (CTFA adopted name)	Specific gravity 25/25° Min	Color	% Ash Max	% Chloride Max	% Sulfate Max
Kemstrene 99.5%	99.5% USP *(Glycerine)*	1.26073	Not darker than FeCl3 standard**	0.01	0.001	0.002
Kemstrene 99.0%	99.0% USP *(Glycerine)*	1.25945	Not darker than FeCl3 standard**	0.01	0.001	0.002
Kemstrene 96.0%	96.0% USP *(Glycerine)*	1.25165	Not darker than FeCl3 standard**	0.01	0.001	0.002
Kemstrene High Gravity	High Gravity* *(Glycerine)*	1.2587	Not darker than standard	0.10	0.01	—

Product	% Arsenic Max	% Heavy metals Max	Readily carbonizable substances	% Chlorinated compounds Max	Acrolein, glucose and ammonium compounds	Fatty acids and esters
Kemstrene 99.5%	0.00015	0.0005	Not darker than matching fluid H	0.003	Not yellow, no ammonia odor	1.0 ml of 0.5N NaOH Max
Kemstrene 99.0%	0.00015	0.0005	Not darker than matching fluid H	0.003	Not yellow, no ammonia odor	1.0 ml of 0.5N NaOH Max
Kemstrene 96.0%	0.00015	0.0005	Not darker than matching fluid H	0.003	Not yellow, no ammonia odor	1.0 ml of 0.5N NaOH Max
Kemstrene High Gravity	—	—	Silver test—no precipitate formed in 10 minutes	—	Saponification equivalent 0.05% Max	See Footnote†

*As per Federal Specification O-G-49 lb.
**All tests run per *U.S. Pharmacopeia XX*, Page 353, effective July 1, 1980.
†Acidity/Alkalinity measures 0.6 ml of 0.5N HCl or 0.5 NaOH.

Table 7.186: Procter and Gamble Glycerines *(39)*

Superol, Moon, Star brands available in Kosher grade.

	SUPEROL™ GLYCERINE—U.S.P. FOOD GRADE	MOON™ GLYCERINE—U.S.P. FOOD GRADE	STAR™ GLYCERINE—U.S.P. FOOD GRADE	HIGH GRAVITY GLYCERINE Conforms to or surpasses the requirements of Federal Specification O-G-491 (current revision) and of numerous consumer specifications.	
Glycerol (Bosart & Snoddy tables)	99.5% minimum	99% minimum	96% minimum	Glycerol (Bosart & Snoddy tables)	99% minimum
Specific Gravity, by pycnometer: at 25°/25°C (77° / 77°F)	1.2608 minimum	1.2595 minimum	1.2517 minimum	Specific Gravity, by pycnometer: at 25°/25°C (77° / 77°F)	1.2595 minimum
Color, APHA Pt-Co (Hazen) scale	12 maximum	12 maximum	12 maximum	Color, APHA Pt-Co (Hazen) scale	150 maximum
Residue on ignition	0.007% (70 ppm) maximum	0.007% (70 ppm) maximum	0.007% (70 ppm) maximum	Acidity or Alkalinity	0.1 ml. N NaOH or HCl maximum
Chlorides (as chlorine)	0.0006% (6 ppm) maximum	0.0006% (6 ppm) maximum	0.0006% (6 ppm) maximum	Ash	0.05% maximum
Sulfates	0.002% (20 ppm) maximum	0.002% (20 ppm) maximum	0.002% (20 ppm) maximum	Total Chlorides (as Cl)	30 ppm (0.003%) maximum
Arsenic (as As₁)	0.00015% (1.5 ppm) maximum	0.00015% (1.5 ppm) maximum	0.00015% (1.5 ppm) maximum	Saponification equivalent (as Na₂O)	0.05% maximum
Heavy Metals (as Pb)	0.0005% (5 ppm) maximum	0.0005% (5 ppm) maximum	0.0005% (5 ppm) maximum	Silver nitrate precipitation	No flocculent precipitate in 10 minutes
Chlorinated Compounds (as Cl)	0.003% (30 ppm) maximum	0.003% (30 ppm) maximum	0.003% (30 ppm) maximum	Test procedures are as set forth in Federal Specification O-G-491 (current revision).	
Acrolein, Glucose and Ammonium Compounds	None—No yellowing or odor of ammonia	None—No yellowing or odor of ammonia	None—No yellowing or odor of ammonia		
Fatty Acids and Esters	Not more than 0.3 ml. N/2 NaOH is absorbed by 50 g of glycerine, which is equivalent to 0.009% as Na₂O, maximum.	Not more than 0.4 ml. N/2 NaOH is absorbed by 50 g of glycerine, which is equivalent to 0.012% as Na₂O, maximum.	Not more than 0.4 ml. N/2 NaOH is absorbed by 50 g of glycerine, which is equivalent to 0.012% as Na₂O, maximum.	Chemical Abstract Number: 56-81-5, for all P&G Glycerine.	

1,2,4-BUTANETRIOL

$$HOCH_2CHOHCH_2CH_2OH$$

Table 7.187: Physical Properties of 1,2,4-Butanetriol *(32)*

Boiling point at 760 mm. Hg	312° C*
0.17 mm. Hg	116° C
Fire point, Cleveland open cup	393° F
Flash point	343° F
Freezing point	Supercools (resistance to crystallization)
Refractive index at 25° C, n_D	1.473
Specific gravity, d/4	1.182
Viscosity at 25° C	1038 cs. (kinematic) 1227 cp.
Weight per gallon at 25° C	9.86 lb.

Purified 1,2,4-Butanetriol

Fire point, Cleveland open	387° F
Flash point, Cleveland open cup	332° F
Heat of combustion	555 kcal./mole
Heat of formation	165.1 kcal./mole (liquid) 157 kcal./mole (gas)
Heat of vaporization	14.0 kcal./mole
Specific gravity, d/4	1.184

*Decomposes before reaching boiling point at atmospheric pressure. This is an extrapolated value.

1,2,6-HEXANETRIOL

$$HOCH_2CHOH(CH_2)_3CH_2OH$$

Table 7.188: Physical Properties of 1,2,6-Hexanetriol *(32)*

Boiling point at 5 mm. Hg	178° C	Specific gravity at 20/20° C	1.1063
Coefficient of expansion at 20° C	0.00054/°C	Δ Sp. Gr./Δ t at 10 to 40° C	0.00059/°C
Flash point, open cup	375° F	Vapor pressure at 20° C	Less than 0.01 mm. Hg
Freezing point	-32.8° C (freezes under controlled conditions; usually sets to glass at below -20° C)	Viscosity at 20° C	2584 cp.
		Weight per gallon at 20° C	9.19 lb.
Molecular weight	134.17	Δ lb./gal./Δ t	0.00499° C
Refractive index	1.4771		

Table 7.189: Freezing Points of 1,2,6-Hexanetriol-Water Mixtures *(32)*

(I)   Observed
(II)   Theoretical, without hydration
(III) Theoretical, with complete hydration

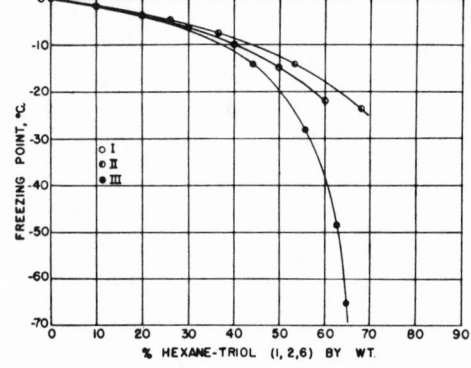

Table 7.190: Vapor Pressure of 1,2,6-Hexanetriol *(19)*

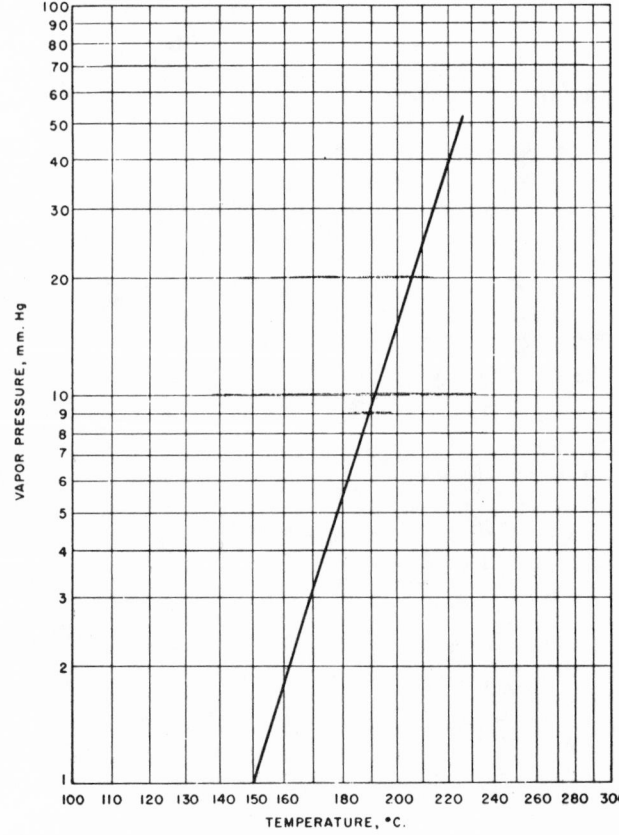

Table 7.191: Solubility of 1,2,6-Hexanetriol in Organic Solvents *(32)*

4cc. solvent and 1cc. triol at 20°C.

Acetone	M	Ethyl Acetate	I
Benzene	I	Ethyl Alcohol (Absolute)	M
Butanol	M	Ethyl Ether	I
Butyl Acetate	I	Heptane	I
Butyl CELLOSOLVE	M	Isophorone	M
Castor Oil	I	Methyl Isobutyl Ketone	I
CELLOSOLVE Acetate	I	Mineral Oil	I
CELLOSOLVE Solvent	M	Pine Oil	M
Diacetone Alcohol	M	Toluene	I
Dibutyl Phthalate	I	Trichlorethylene	I
Dichlorethyl Ether	I		

M = Miscible   I = Immiscible

Table 7.192: Compatibility of 1,2,6-Hexanetriol *(32)*

4 parts material to 1 part triol

Animal Glue	C	Gelatin	PC
Beeswax	I	Nitrocellulose	I
Carnauba Wax No. 3	I	Paraffin Wax	I
Casein	C	Rosin	I
Ester Gum C	I	Shellac	PC
Ethyl Cellulose	I	Zein	C

C = Compatible   I = Incompatible   PC = Partly Compatible

Table 7.193: Viscosities and Freezing Points of 1,2,6-Hexanetriol *(32)*

1,2,6-Hexanetriol, % by wt. in H_2O	Viscosity, in cps. at 100°F.	Freezing Point, °C.
10	0.977	−2.5
20	1.37	−4.5
30	2.01	−7.0
50	5.06	−15.5

TRIMETHYLOLPROPANE

2,2-Dihydroxymethyl-1-Butanol

Ethyl Trimethylolmethane $C_2H_5C(CH_2OH)_3$

TMP

Table 7.194: Physical Properties of Trimethylolpropane *(32)*

Acidity as formic acid	0.002% by wt., max.
Ash	0.01% by wt., min.
Boiling point at 5 mm. Hg abs.	160° C
50 mm. Hg abs.	210° C
760 mm. Hg (extrapolated)	295° C
Bulk density (free-flowing)	35.5 lb/ft^3
Color of 10% aqueous solution	5 Pt-Co units, max.
Combining weight	44.72
Fire point, Cleveland open cup	380° F
Flash point, Cleveland open cup	355° F
Freezing point	59° C
Hydroxyl content	37.5% by wt., min.
Hygroscopicity (water absorbed in 68 hrs.):	
at 27° C and 18 to 26% RH	0.00% by wt.
at 25° C and 29 to 44% RH	0.06% by wt.
at 27° C and 70 to 80% RH	0.23% by wt.
Melting point range	57 to 59° C
Molecular weight	134.18
Phthalic color, Gardner	1 max.
Water content as packaged	0.05% by wt., max.

PENTAERYTHRITOL

Tetramethylolmethane

PE

$$HOH_2C - \underset{\underset{CH_2OH}{|}}{\overset{\overset{CH_2OH}{|}}{C}} - CH_2OH$$

Table 7.195: Physical Properties of Pentaerythritol *(32)*

Ash	0.01% by wt., max.
Bulk density	40 lb./ft.^{3}
Dipentaerythritol (combined)	0.3%
Hydroxyl content	47.0% min. (technical)
	49.5% (pure)
Melting point (capillary final)	240° C
	250° C initial (pure)
Melting point range	185-245° C (technical)
Moisture	0.40% by wt. (technical)
	0.10% by wt. (pure)
Molecular weight	136.1
Monopentaerythritol	88.0% by wt. (technical)
	97.0% by wt. (pure)
Nonvolatile	99.50% min.
Specific gravity at 25/4° C	1.38

SORBITOL

d–Sorbitol
Sorbit
Sorbol
d–Glucitol

$CH_2OH(CHOH)_4CH_2OH$

Table 7.196: Physical Properties of Sorbitol *(38)*

Density at –5° C	1.472
Heat of combustion	3994 cal./gm.
Negative heat of solution	–26.5 cal./gm.
Molecular weight	182.17
Melting point, metastable form	93° C
stable form	97.7° C
Refractive index at 25° C, in 10% aqueous solution	1.3477
Rotation, $\frac{25}{D}$	–0.985° C

Table 7.197: Boiling Point of Sorbitol Solutions *(38)*

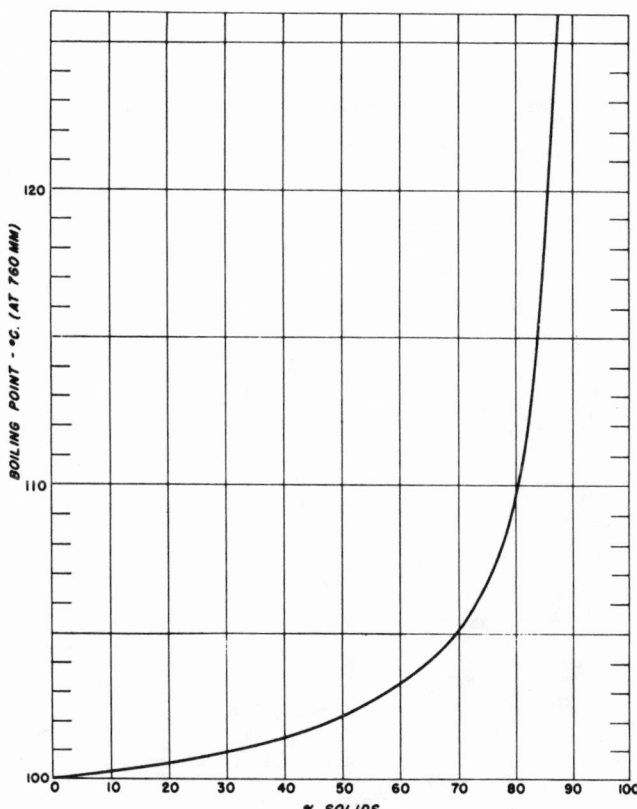

Table 7.198: Hydrogenolysis of Sorbitol and Glycerol at a Hydrogen Pressure of 2,000 psi *(32)*

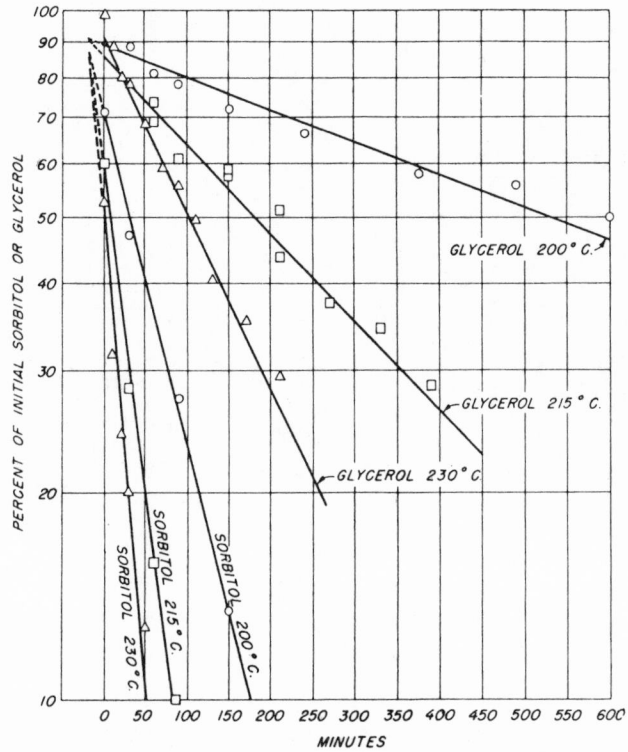

Table 7.199: Hydrogenolysis of Sorbitol at 215°C and a Hydrogen Pressure of 2,000 psi *(32)*

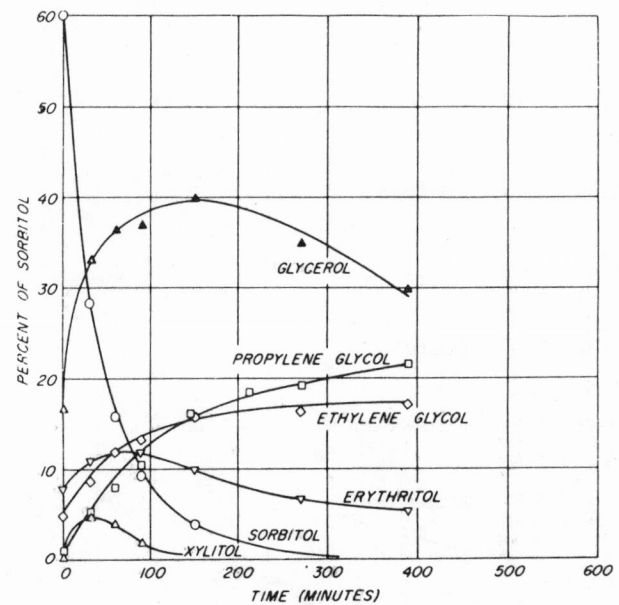

Table 7.200: Phase Diagram of Sorbitol Solubility in Hydroalcoholic Liquids at 25°C *(38)*

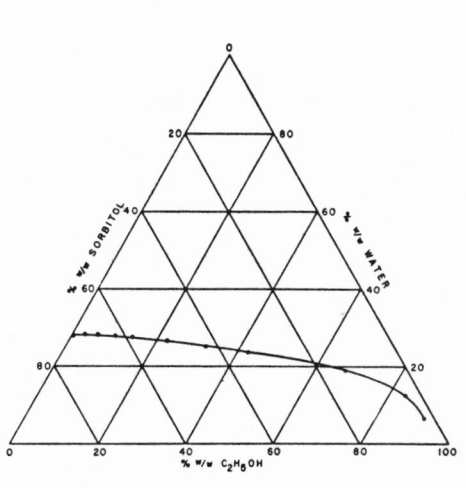

Table 7.201: Solubility of Sorbitol in Hydroalcoholic Liquids at 25°C *(38)*

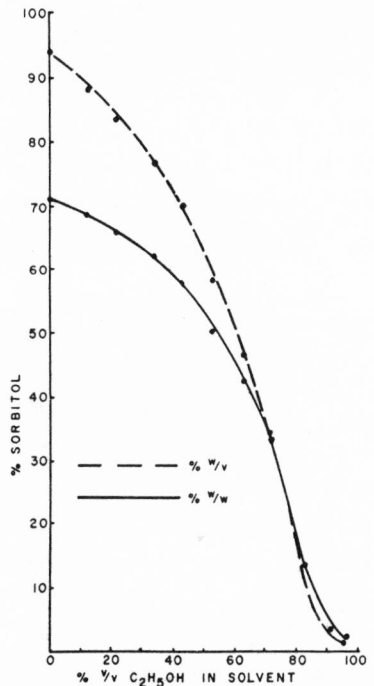

Table 7.202: Viscosity Curve for Pure d-Sorbitol Solutions of Various Concentrations *(38)*

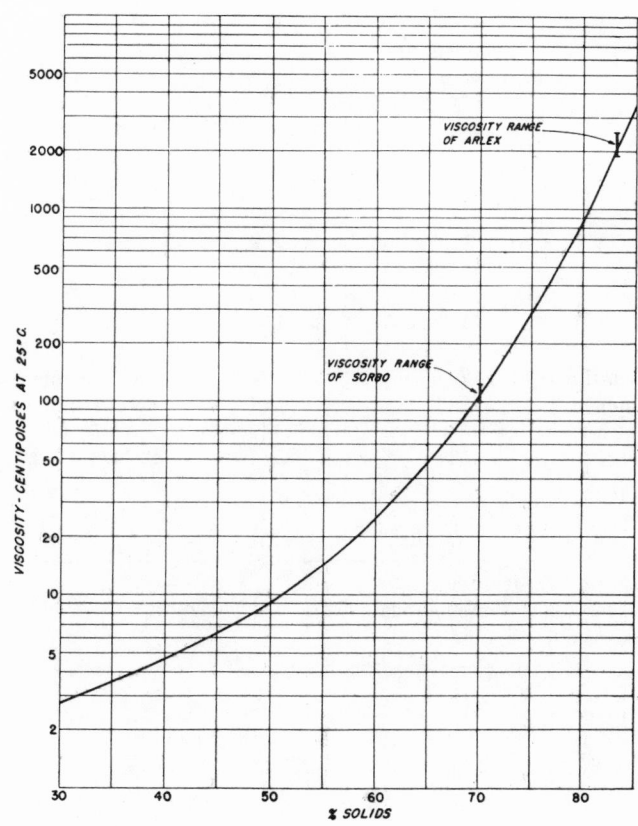

SUGAR ALCOHOLS

Table 7.203: Physical Properties of the Sugar Alcohols *(38)*

Sugar alcohol	Melting point, °C	Optical activity in H_2O, $[\alpha]_D^{20-25}$	Solubility, g/100 g H_2O^c	Heat of combustion, constant volume, kcal/mole
tetritols				
erythritol	120	meso	61.5	499.9 (94)
D-threitol	88.5–90	+4.3	very soluble	
L-threitol	88.5–90	−4.3		
D,L-threitol	69–70			
pentitols				
ribitol	102	meso	very soluble	
xylitol	61–61.5 (meta-stable) 93–94.5 (stable)	meso	179	
D-arabitol	103	+131[a]	very soluble	
L-arabitol	102–103	−130[a]		611.7 (124)
hexitols				
allitol	155	meso	very soluble	
dulcitol	189	meso	3.2 (15° C)	720.3 (94)
sorbitol (D-glucitol)	90.4–91.8 (meta-stable) 96.7–97.7 (stable)	−1.98	235	723.5 (6)
L-glucitol	89–91	+1.7		
D,L-glucitol	135–137			
D-mannitol	166	−0.2	21.3	722.1 (6)
L-mannitol	162–163			
D,L-mannitol	168			
D-talitol	88–89	+3.2	very soluble	
L-talitol	87–88	−2.9		
D,L-talitol	95–96			
D-iditol	73.5	+3.5		
L-iditol	75.7–76.7	−3.5		
heptitols				
glycero-gulo-heptitol	129	meso	very soluble	
D-*glycero*-D-*ido*-heptitol	129	+0.7	very soluble	
perseitol	187	−1.1	7.4 (18° C)	835.8 (124)
volemitol	153	+2.15	22.2 (14° C)	
octitol				
D-*erythro*-D-*galacto*-octitol	169-170	−11[b]		

[a] In aqueous molybdic acid (46).
[b] In 5% aqueous ammonium molybdate (27).
[c] At 25°C unless otherwise indicated.

MISCELLANEOUS POLYHYDRIC ALCOHOLS

Table 7.204: Hydrates of Polyhydric Alcohols *(32)*

Alcohol				Hydrate	
Number of C Atoms	Name	Skeletal Structural Formula	M.p. (°C)	M.p. (°C)	n in $R(OH)_n \cdot nH_2O$
A. Trihydric Alcohols					
6	α (or cis)-Phloro-glucitol		185	115	2
9	4(1,2-Dihydroxy-n-propyl)-cyclo-hexanol		63	31	3
10	p-Menthane-1,4,8-triol		110–112	96	1
10	p-Menthane-1,2,4-triol		129	115	---
10	Glycol (a dihydroxyether?)	$C_{10}H_{18}O_3$	103–105	---	1
13	2(2,3-Dihydroxy-n-propyl)-2-hydroxy camphane		---	---	---
B. Tetrahydric Alcohols					
6	cycloHexane-1,2,4,5-tetrol		---	195	1
6	cycloHexane-1,2,4,5-tetrol		242	---	2
8	A dimethylether of an inositol	$C_6H_6(OH)_4(OCH_3)_2$	230	--	3
10	trans(?)-p-Menthane-1,2,6,8-tetrol		156	100–105	2

(continued)

Table 7.204: (continued)

Number of C Atoms	Name	Skeletal Structural Formula	M.p. (°C)	M.p. (°C)	n in R(OH)$_m$ nH$_2$O
		Alcohol		**Hydrate**	
10	p-Menthane-1,2, 4,8-tetrol	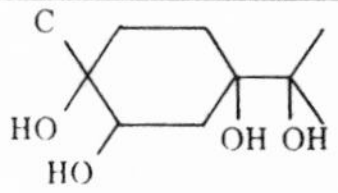	149	100	1
10	p-Menthane-1,2, 3,4-tetrol	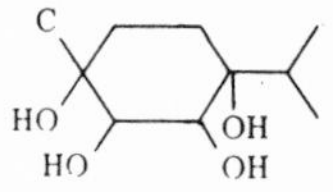	130	---	1
38	2,2'-Dihydroxy-6,6'-bis (α-hydroxybenz-hydryl)-diphenyl	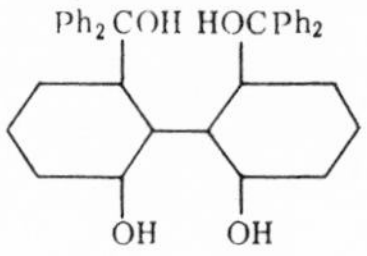	308	141-145	2

C. Pentahydric Alcohols

6	Viburnitol (cyclohexane-2,3,5/4,6-pentol)	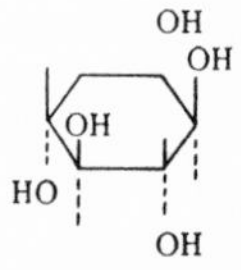	181	---	1
6	Inositol bromo-hydrin	C_6H_6 (OH)$_5$ Br	170-5	---	1
6	Inositol chloro-hydrin	C_6H_6 (OH)$_5$ Cl	180-5	---	2
6	Scyllitol chloro-hydrin	C_6H_6 (OH)$_5$ Cl	---	---	2
7	1-Methylene-cyclohexane-2,4,6/3,5-pentol	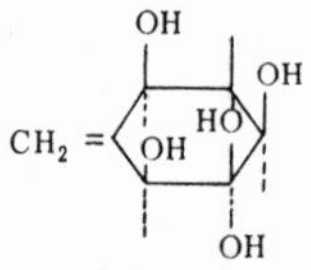 $CH_2 =$	205	---	2

D. Hexahydric Alcohols

6	(+)-Sorbitol	HOH_2C (CHOH)$_4$ CH_2OH	111	55 / 75	1 / 0.5
6	meso-Inositol (1,2,3,5/4,6-cyclohexane-hexol)	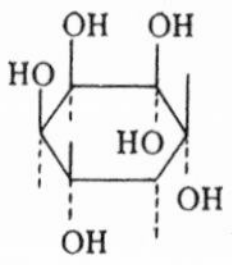	225	---	2
6	d- and l-Inositols (active) (1,3,4/2,5,6-cyclohexane-hexol)	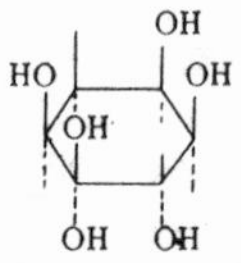	248	---	2

Phenols

Table 8.1: Phenol (2)

Carbolic Acid C_6H_5OH

PHYSICAL PROPERTIES OF PHENOL

Boiling point	181.6°C
Distillation range	95% distills within a range of 1.5°C
Flash point (Open cup)	175°F
Freezing point	Not less than 40°C
MAC	5 ppm in air
Odor	Characteristic
Purity	98%, min.
Solidifying point	Not less than 40.7°C
Solubility in water, above 68°C	In all proportions
at 20°C	8.3%
Specific gravity at 41/4°C	1.058
Toxicity	Highly toxic

PHENOL FORMS BINARY AZEOTROPES WITH

%		B.P. of Azeotrope °C.	%		B.P. of Azeotrope °C.
92.2	Acetophenone	202.0	28	Heptyl alcohol	185.0
22	Amyl ether	180.2	55	Indene	173.2
58	Aniline	186.2	85	Isoamyl ether	172.2
49	Benzaldehyde	185.5	74	Isobutyl carbonate	192.5
55	Benzylamine	196.8	17	Isopropyl lactate	184.8
57	m-Bromotoluene	175.7	79	Mesitylene	163.5
60	o-Bromotoluene	174.4	20	2-Methylcyclohexanol	183.1
37	2-Butoxyethanol	186.4	77	Methyl fumarate	194.9
54	Butylbenzene	175.0	33	Methylheptenone	184.6
30	Butyl isovalerate	184.0	32	2-Octanone	184.5
78	Camphene	156.1	87	n-Octyl alcohol	195.4
97	o-Chlorotoluene	159.0	50	sec-Octyl alcohol	184.5
28	Cineole	182.9	65	α-Phellandrene	165.0
13	Cyclohexanol	183.0	82	Phorone	198.8
28	Cyclohexanone	184.5	29	Pinacol	185.5
65	Decane	168.0	81	α-Pinene	152.8
59	Ethyl oxalate	189.5	75	Pseudocumene	166.0
75	Fenchone	196.2	55	Terpinene	171.5
60	Glycol diacetate	189.9	60	Thymene	172.3

Table 8.2: Octylphenol and Nonylphenol *(52)*

TYPICAL PROPERTIES AND SPECIFICATIONS

	OCTYLPHENOL		NONYLPHENOL	
	Typical Properties	Specifications	Typical Properties	Specifications
Appearance	White flakes	White to light pink flakes	Clear, straw-colored liquid, no foreign matter	Clear, straw-colored liquid, no foreign matter
Color (APHA)	40 (molten)	100 max.	35	50 max.
Sp. Gr., 25°C.	--	--	0.9483	--
60°C.	--	--	0.9228	--
80°C.	--	--	0.9093	--
85°C.	0.9199	--	--	--
100°C.	0.9081	--	--	--
120°C.	0.8938	--	--	--
Density, 25°C., lbs./gal.	7.84	--	7.85	--
Congealing Point, °C.	77.5	75-80	5.5-6.5 (pour point)	--
Viscosity				
Centipoises, 25°C.	--	--	1690[a]	--
60°C.	--	--	260[b]	--
80°C.	22.5	--	90[b]	--
83°C.	17	--	--	--
100°C.	--	--	40[b]	--
120°C.	8	--	--	--
Hydroxyl Number	268	255-275	250	245-255
Distillation Range				
°C., at 10 mm.	90%--150-175	--	--	--
°C., at 760 mm.	--	--	First drop - 298	First drop - 295 min.
	--	--	5% - 300	5% - 298 min.
	90%--278-302	--	95% - 318	95% - 325 max.
n_D^{25}	--	--	1.508	--
Water, % (K.F.)	--	--	0.03	0.05 max.
Specific Heat, B.T.U./lb./°F.	0.4	--	--	--

a   Brookfield Spindle #1 at 12 rpm.
b   Brookfield Spindle #2 at 12 rpm.

Table 8.3: Physical Properties of Phenolic Compounds *(52)*

Formula	Compound	Melting Point[a] °F	Melting Point[a] °C	Boiling Point[a] °C 20 mm.	Boiling Point[a] °C 100 mm.	Boiling Point[a] °C 760 mm.	Refractive Index, n_D^{20}
C_6H_6O	Phenol	105.8	41.0	85	119.5	182	
C_7H_8O	2-Methylphenol (o-cresol)	86	30.0	90	125.5	190.8	1.5453
	3-Methylphenol (m-cresol)	52.7	11.5	101	138	202.2	1.5438
	4-Methylphenol (p-cresol)	94.5	34.7	101	138	202.1	1.5359
$C_8H_{10}O$	2,3-Dimethylphenol (2,3-xylenol)	167	75.0	112	150	218	1.5420
	2,4-Dimethylphenol (2,4-xylenol)	80.6	27	105	143	210	
	2,5-Dimethylphenol (2,5-xylenol)	166.1	74.5	105	143	210	
	2,6-Dimethylphenol (2,6-xylenol''	120.2	49.0	107[b]	145	212	
	3,4-Dimethylphenol (3,4-xylenol)	144.5	62.5	122	160	225	
	3,5-Dimethylphenol (3,5-xylenol)	154.4	68.0	117.5	155	219.5	
			63.2				
	2-Ethylphenol	<0	<-18	101.5	138.5	207	
	3-Ethylphenol	24.8	-4.0	114.5	151	214	
	4-Ethylphenol	116.6	47.0	115	153	219	1.5239[m]
$C_9H_{12}O$	2-Propylphenol			122[b]	158	220	
	3-Propylphenol	78.8	26.0	127	164	228	
	4-Propylphenol	71.6	22.0	128[b]	166	232.6	
$C_{10}H_{14}O$	2-tert-Butylphenol			114	153	221	
	3-tert-Butylphenol	105.8	41	132.5	172	240	
	4-tert-Butylphenol	212	100	130.5	170	237	
	2-sec-Butylphenol	69.8	21				
	4-sec-Butylphenol	143.6	62				1.5150
	2-n-Butylphenol	<-4	-20	123	163.5	235	1.496[n]
	3-n-Butylphenol			138	179	248	
	4-n-Butylphenol	71.6	22	138	179	248	1.5165[o]
$C_{11}H_{16}O$	2-Methyl-x-tert-butylphenol	80.6	27	118	159	231	
	2-Methyl-y-tert-butylphenol	80.6	27	132	174	247	
	3-Methyl-x-tert-butylphenol	70.3	21.3	129	171	244	
	4-Methyl-2-tert-butylphenol	130.1	51.7	126.5	167	237	
	4-Methyl-2-sec-butylphenol			129	169	237	
	4-tert-Amylphenol	199.4	93	138.5[d]		262.5	
	4-n-Amylphenol	73.4	23	135[e]		250.5	1.5272[m]
$C_{12}H_{18}O$	2-Methyl-4-tert-amylphenol			153[b]	198	273	
	3-Methyl-4-tert-amyphenol			153[b]	198	273	
	4-Methyl-2-tert-amylphenol	80.6	27	137.5	180	252	
	2,3-Dimethyl-x-tert-butylphenol	126.9	52.5	139[b]	181	252	
	2,3-Dimethyl-y-tert-butylphenol			145[b]	188	259	
	2,4-Dimethyl-6-tert-butylphenol	72.1	22.3	131	174	249	1.5183[a]
	2,5-Dimethyl-4-tert-butylphenol	160.2	71.2	151	193.5	264	1.5311[a]
	2,6-Dimethyl-4-tert-butylphenol	183.6	82.4	135	176.5	248	
	3,4-Dimethyl-6-tert-butylphenol	114.8	46.0	144.7	187		1.5222[a]
	2-Ethyl-x-tert-butylphenol			129[b]	170		
	2-Ethyl-4-tert-butylphenol			141[b]	184	257	
	3-Ethyl-x-tert-butylphenol			142[b]	185	257	
	4-Ethyl-2-tert-butylphenol	73.4	23	137	179	250	
$C_{14}H_{22}O$	2,4-Di-tert-butylphenol	133.7	56.5	146	190	263.5	
	2,6-Di-tert-butylphenol	102.2	39				
	4-Diisobutylphenol[c]	183.2	84	163	207	279	
$C_{15}H_{18}O$	2,4,6-Triallylphenol			167.5	215	294	
$C_{15}H_{24}O$	2-Methyl-4,6-di-tert-butylphenol	123.8	51	149.5	194	269	
	3-Methyl-4,6-di-tert-butylphenol	143.8	62.1	167	211	282	
	4-Methyl-2,6-di-tert-butylphenol	158	70	147	191	265	
	2-Methyl-4-diisobutylphenol[c]	121.1	49.5	95-105f			
	3-Methyl-4-diisobutylphenol[c]	121.6	49.8	163.5[b]	206	275	
	4-Methyl-2-diisobutylphenol[c]	115.2	46.2	158[b]	200	269	
$C_{16}H_{26}O$	2,3-Dimethyl-4,6-di-tert-butylphenol	185.9	85.5	174[b]	217	284	
	2-Ethyl-4,6-di-tert-butylphenol	86	30	156.5[b]	201	275	
	3-Ethyl-4,6-di-tert-butylphenol	177	80.5	174	218	289	
	4-Ethyl-2,6-di-tert-butylphenol	111	44.0	154	198	272	
$C_{17}H_{28}O$	4-Methyl-2,6-di-tert-amylphenol			165	210	283	1.4950[a]
$C_{18}H_{30}O$	2,4,6-Tri-tert-butylphenol	267.8	131	158	203	278	

For notes, see end of Table.

(continued)

Table 8.3: (continued)

(Only compounds for which data are available are listed.)

Formula	Compound	Solid	Specific Gravity[a], d_4^t			Viscosity, Centistokes[a]		
			80°C (176°F)	120°C (248°F)	160°C (320°F)	80°C (176°F)	120°C (248°F)	160°C (320°F)
C_6H_6O	Phenol		1.020	0.985	0.949	1.51	0.828	0.540
C_7H_8O	2-Methylphenol (o-cresol)		0.994	0.960	0.924	1.47	0.784	0.515
	3-Methylphenol (m-cresol)		0.986	0.954	0.921	1.76	0.890	0.570
	4-Methylphenol (p-cresol)		0.986	0.954	0.921	2.00	0.992	0.620
$C_8H_{10}O$	2,5-Dimethylphenol (2,5-xylenol)	1.063[g]	0.965	0.932	0.899	1.61	0.825	0.528
	3,4-Dimethylphenol (3,4-xylenol)	1.064[h]	0.983	0.952	0.920	3.05	1.270	0.737
	3,5-Dimethylphenol (3,5-xylenol)	1.008[h]	0.968	0.935	0.902	2.50	1.075	0.635
	2-Ethylphenol		1.037[g]					
	3-Ethylphenol		1.025[g]					
	4-Ethylphenol		1.011[g]					
$C_9H_{12}O$	2-Propylphenol		1.015[j]					
	4-Propylphenol		1.009[g]					
$C_{10}H_{14}O$	4-tert-Butylphenol		0.908[k]					
	4-sec-Butylphenol		0.969[l]					
	2-n-Butylphenol		0.975[g]					
	3-n-Butylphenol		0.974[g]					
	4-n-Butylphenol		0.978					
$C_{11}H_{16}O$	2-Methyl-x-tert-butylphenol		0.924	0.894	0.864	1.60	0.906	0.610
	2-Methyl-y-tert-butylphenol		0.929	0.899	0.869	4.40	1.580	0.865
	3-Methyl-x-tert-butylphenol		0.922	0.892	0.862	2.12	1.030	0.642
	4-Methyl-2-tert-butylphenol		0.922	0.892	0.862	2.55	1.170	0.713
	4-tert-Amylphenol		0.962[g]					
	4-n-Amylphenol		0.960[g]					
$C_{12}H_{18}O$	2,4-Dimethyl-6-tert-butylphenol		0.917	0.888	0.859	2.10	1.060	0.670
	2,5-Dimethyl-4-tert-butylphenol	1.001[i]	0.939	0.911	0.883	8.30	2.280	1.130
	2,6-Dimethyl-4-tert-butylphenol	0.959[i]	0.916	0.883	0.851	2.72	1.320	0.820
	3,4-Dimethyl-6-tert-butylphenol	0.973[i]	0.920	0.892	0.863	3.50	1.370	0.782
$C_{15}H_{24}O$	2-Methyl-4,6-di-tert-butylphenol	0.940[i]	0.891	0.862	0.833	4.75	1.900	1.080
	3-Methyl-4,6-di-tert-butylphenol		0.912	0.882	0.853	9.90	2.870	1.420
	4-Methyl-2,6-di-tert-butylphenol	1.048[g]	0.899	0.870	0.841	3.47	1.540	0.920
	3-Methyl-4-diisobutylphenol[c]		0.904	0.876	0.847	5.00	1.970	1.125
	4-Methyl-2-diisobutylphenol[c]	0.930[i]	0.904	0.876	0.847	5.70	2.130	1.158
$C_{17}H_{28}O$	4-Methyl-2,6-di-tert-amylphenol		0.931[l]					
$C_{18}H_{30}O$	2,4,6-Tri-tert-butylphenol	0.864[i]						

Notes

[a] Experimental data.
[b] Experimental data incomplete; these data obtained from "family" curve plot using a single experimental point.
[c] As used here, "diisobutyl" refers to the alkyl group 1,1,3,3-tetramethylbutyl, which results from the alkalylation of phenol and of cresols with diisobutylene. Diisobutylene is a mixture of two isomers, 2,4,4-trimethyl-1-pentene and 2,4,4-trimethyl-2-pentene which are assumed to give identical alkylation products.

[d] 15 mm.	[g] d_4^{20}	[j] d_4^0	[m] n_D^{25}
[e] 10 mm.	[h] $d_4^{27.6}$	[k] d_4^{114}	[n] n_D^{15}
[f] 1 mm.	[i] $d_4^{26.7}$	[l] d_4^{25}	[o] n_D^{22}

Table 8.4: Vapor Pressures of Phenolic Compounds *(52)*

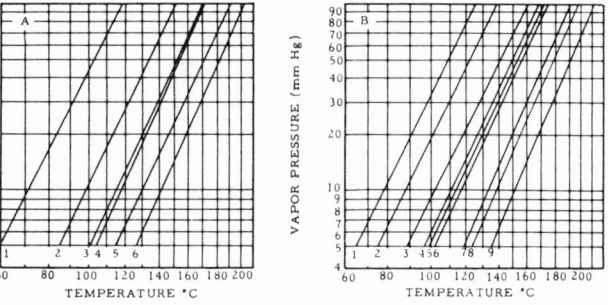

Vapor Pressure Data

A. PHENOL AND tert-BUTYLPHENOLS

C_6H_6O	Phenol	1
$C_{10}H_{14}O$	2-tert-Butylphenol	2
	3-tert-Butylphenol	4
	4-tert-Butylphenol	3
$C_{14}H_{22}O$	2,4-Di-tert-butylphenol	5
$C_{18}H_{30}O$	2,4,6-Tri-tert-butylphenol	6

B. CRESOLS AND tert-BUTYLCRESOLS

C_7H_8O	2-Methylphenol	1
	3-Methylphenol	2
	4-Methylphenol	2
$C_{11}H_{16}O$	2-Methyl-x-tert-butylphenol	3
	2-Methyl-y-tert-butylphenol	6
	3-Methyl-x-tert-butylphenol	5
	4-Methyl-2-tert-butylphenol	4
$C_{15}H_{24}O$	2-Methyl-4,6-di-tert-butylphenol	8
	3-Methyl-4,6-di-tert-butylphenol	9
	4-Methyl-2,6-di-tert-butylphenol	7

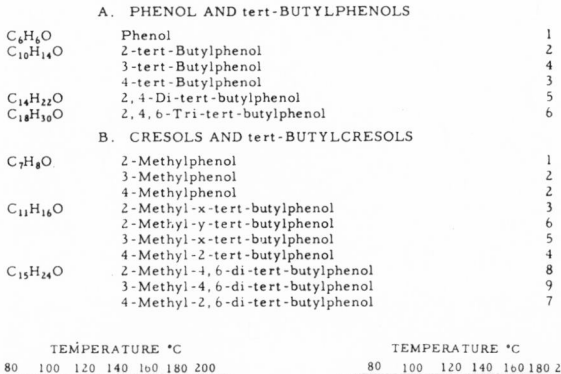

Vapor Pressure Data

C. ETHYLPHENOLS AND tert-BUTYLETHYLPHENOLS

$C_8H_{10}O$	2-Ethylphenol	1
	3-Ethylphenol	2
	4-Ethylphenol	3
$C_{12}H_{18}O$	2-Ethyl-x-tert-butylphenol	4
	2-Ethyl-y-tert-butylphenol	6
	3-Ethyl-x-tert-butylphenol	7
	4-Ethyl-2-tert-butylphenol	5
$C_{16}H_{26}O$	2-Ethyl-4,6-di-tert-butylphenol	9
	3-Ethyl-4,6-di-tert-butylphenol	10
	4-Ethyl-2,6-di-tert-butylphenol	8

D. XYLENOLS AND tert-BUTYLXYLENOLS

$C_8H_{10}O$	2,3-Dimethylphenol	3
	2,4-Dimethylphenol	1
	2,5-Dimethylphenol	1
	2,6-Dimethylphenol	2
	3,4-Dimethylphenol	5
	3,5-Dimethylphenol	4
$C_{12}H_{18}O$	2,3-Dimethyl-x-tert-butylphenol	8
	2,3-Dimethyl-y-tert-butylphenol	9
	2,4-Dimethyl-6-tert-butylphenol	6
	2,5-Dimethyl-4-tert-butylphenol	10
	2,6-Dimethyl-4-tert-butylphenol	7
	3,4-Dimethyl-6-tert-butylphenol	9
	2,3-Dimethyl-4,6-di-tert-butylphenol	11

Table 8.5: Norpetro Phenol and Substituted Phenols *(45)*

PRODUCT	DESCRIPTION	TYPICAL COMPOSITION	Water % Max.	Neutral Oil Max.	Total Sulphur Max.	USES
98% PHENOL M.P. 39°C	Phenol is a highly refined crystalline material with a minimum melting point of 39° centigrade.	Phenol 98%, Ortho Cresol 2%	0.4	0.1	0.02 (less than 0.01 mercap. sul.)	Phenol and blends of phenol are used in the manufacture of chemical intermediates for resins; herbicides; pharmaceuticals; pentachlorophenol; picric acids; and germicides. It is also used in selective solvents for manufacturing lubricating oils.
ORTHO CRESOL M.P. 30°C	Ortho cresol is a highly refined crystalline material with a minimum melting point of 30° centigrade. HIGHER OR LOWER PURITIES ALSO AVAILABLE.	Ortho Cresol 98%, Phenol 1%, Meta-Para Cresol and 2, 6- Xylenol 1%	0.2	0.1	0.02 (less than 0.01 mercap. sul.)	Ortho cresol is used as a chemical intermediate for agricultural and pharmaceutical chemicals. It is also used as a disinfectant and in the manufacture of phenolic resins to control the reaction rate and final hardening time of the resin.
META-PARA CRESOL No. 60	Meta-Para No. 60 is available in TCP grade which is guaranteed to contain less than 1% ortho cresol and Resin grade which contains between 1% and 3% ortho cresol.	Ortho Cresol 0%, Meta Cresol 60%, Para Cresol 26%, Xylenols 8%, Ortho Ethyl Phenol 6%	0.2	0.1	0.02 (less than 0.01 mercap. sul.)	Meta-Para cresol is used in plastics, resins, TCP, gasoline additives, chemical intermediates and functional fluids.
XYLENOL No. 100	Xylenol No. 100 is a high purity fraction of 2, 4, and 2, 5 xylenol. (Ortho Xylenols)	2, 4 – 2, 5-Xylenol 93%, Meta-Para Cresol 2%, Other Xylenols 5%	0.2	0.1	0.02 (less than 0.01 mercap. sul.)	Xylenol No. 100 is used in phosphate esters, alkylated xylenols, antioxidants, and as a chemical intermediate in pharmaceuticals, insecticides and fungicides.
XYLENOL No. 200	Xylenol No. 200 is a concentrated fraction of 3, 5 xylenol. (Meta Xylenol)	3,5 Xylenol 50%, 2,4 – 2,5 Xylenols 5%, 2,3 and Meta- 3,4 Xylenol 12%, Para Ethyl Phenol 25%, C_9 + Phenol 8%	0.2	0.1	0.02 (less than 0.01 mercap. sul.)	Xylenol No. 200 is used in resins, phosphate esters chemical intermediates, and functional fluids.
CRESYLIC ACID No. 300	Cresylic Acid No. 300 is a blend of cresols and xylenols.	Ortho Cresol 2%, Meta-Para Cresol 45%, Xylenols 40%, C_9 + Phenols 13%	0.5	0.1	0.02 (less than 0.01 mercap. sul)	Cresylic Acid No. 300 is used as a functional fluid, engine cleaning compound, disinfectant, metal degreaser, flotation reagent and wire enamel solvent.
CRESYLIC ACID No. 400	Cresylic Acid No. 400 is a blend of low and high boiling xylenols.	2,4-2,5 Xylenols 35%, 3,4-3,5 Xylenols and Ethyl Phenols 50%, C_9 + Phenols 15%	0.5	0.1	0.02 (less than 0.01 mercap. sul.)	Cresylic Acid No. 400 is used as a functional fluid, engine cleaning compound, disinfectant, metal degreaser, flotation reagent and wire enamel solvent.
CRESYLIC ACID No. 500	Cresylic Acid No. 500 is a blend of high boiling xylenols and alkylated phenols.	3,4-3,5 Xylenol and Ethyl Phenols 80%, C_9 + Phenols 20%	0.5	0.1	0.02 (less than 0.01 mercap. sul.)	Cresylic Acid No. 500 is used as a functional fluid engine cleaning compound, disinfectant, metal degreaser, flotation reagent and wire enamel solvent.
CRESYLIC ACID No. 600	Cresylic Acid No. 600 is a blend of higher molecular weight alkylated phenols.	3,4 Xylenol 40%, C_9 + Phenols 45%, 3,5 Xylenol 15%	0.5	0.1	0.02 (less than 0.01 mercap. sul.)	Cresylic Acid No. 600 is used as a functional fluid, engine cleaning compound, disinfectant, metal degreaser, flotation reagent and wire enamel solvent.
CRESYLIC ACID No. 650	Cresylic Acid No. 650 is a black mixture of high molecular weight alkylated phenols and tars.	3,4 Xylenol 35%, C_9 + Phenols 45%, 3,5 Xylenol 15%, Tars 5%	0.5	0.1	0.05	Cresylic Acid No. 650 is used mainly as a flotation agent.
CRESYLIC ACID No. 700	Cresylic Acid No. 700 is a broad range blend of phenols, cresols, and xylenols.	Phenol 15%, Cresols 40%, Xylenols 35%, C_9 + Phenols 10%	1.0	0.1	0.05	Cresylic Acid No. 700 is used as a functional fluid, engine cleaning compound, disinfectant, metal degreaser, flotation reagent, wire enamel solvent, and in resin and plastic manufacture.

Table 8.6: Pitt-Consol Substituted Phenols *(40)*

o-Cresol 120

Description:

Pitt-Consol Ortho Cresol 120 is a high quality synthetic ortho cresol. It is free of sulfur compounds and tar bases. The high purity of the synthetic cresol ensures the production of derivatives of exceptionally high quality.

Properties	Specification	Typical
Specific Gravity (@60°F (solid)	—	1.050
@175°F	—	0.994
@320°F	—	0.924
Density, lbs/gal @70°F	—	8.75
Color, Barrett*	C-1 max.	—
Water, Wt. %*	0.1 max.	—
Freeze Point, °C	30.5 min.	—
Viscosity, cs @175°F	—	1.470
@250°F	—	0.784
@320°F	—	0.515
Homolog Distribution, Wt. %		
Ortho Cresol	—	99.2
Phenol	—	0.1
2, 6-Xylenols	—	0.7
Flash Point, °F, Pensky-Martens	—	178

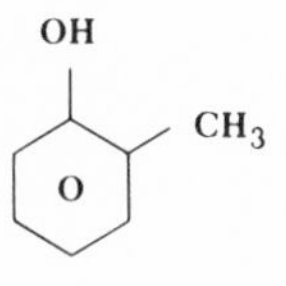

Cresol 174

Description:

Pitt-Consol Cresol 174 is a synthetic cresol blend. It is used in metal degreasing, carbon removal, paint stripping, and other solvent-type applications.

Properties	Specification	Typical
Specific Gravity @60°F	—	1.054
Density, lb/gal @70°F	—	8.77
Color, Barrett*	C-5 max.	—
Water, Wt. %*	0.2 max.	—
Freeze Point, °F	—	20
Homolog Distribution, Wt. %		
Phenol	27-32	29
Ortho Cresol	50-55	52
2,6-Xylenols	5 max.	4
Meta/Para Cresol	13-18	15
Flash Point, °F, Pensky-Martens	—	180

Xylenol 235

Description:

Pitt-Consol Xylenol 235 is a synthetic 2,6-xylenol. It is an attractive intermediate for the manufacture of polyphenylene oxide and modified phenolic resins and phosphate esters used in the production of hydraulic fluids. It also lends itself to halogenation and aminolysis reactions.

Properties	Specification	Typical
Specific Gravity (@ 60°F	—	1.018
Density, lb/gal (@ 70°F	—	8.49
Color, Barrett*	C-2 max.	B-5
Water, Wt. %*	0.2 max.	0.1
Freeze Point, °C	—	40
Homolog Distribution, Wt. %		
Ortho Cresol	—	0.5
2,6-Xylenols	88 min.	92
2,4/2,5 Xylenols	1.0 max.	0.5
Meta/Para Cresol	—	7
Flash Point, °F, Pensky-Martens	—	190

(continued)

Table 8.6: (continued)

Xylenol 410

Description:

Pitt-Consol Xylenol 410 is a synthetic xylenol blend used in the production of high-viscosity phosphate esters. It is also used as a feedstock for hindered phenol antioxidant and specialty modified phenolic resin manufacture.

Properties	Specification	Typical
Specific Gravity @ 60°F	—	1.03
Density, lb/gal @ 70°F	—	8.58
Color, Barrett*	C-2 max.	—
Water, Wt. %*	0.1 max.	—
Homolog Distribution, Wt. %		
2,4/2,5-Xylenols	80 min.	—
2,6-Xylenol	2 max.	1
2,4-Xylenol	—	42
2,5-Xylenol	—	43
2,3-Xylenol Group	—	14
Trimethylphenols	5 max.	—
Flash Point, °F, Pensky-Martens	—	188

Cresylic Acid 514

Description:

Pitt-Consol Cresylic Acid 514 is a synthetic cresylic acid blend used primarily as a wire enamel solvent. Other applications where it can be used include metal degreasing, carbon and rubber removal, alkyd paint strippers, odor masking agents, ore flotation agents, and specialty resins.

Properties	Specification	Typical
Specific Gravity @ 60°F	—	1.02
Density, lb/gal @ 70°F	—	8.50
Color, Barrett*	C-3 max.	C-2
Water, Wt. %*	0.2	—
Freeze Point, °F*	Below 0	-40
Corrosion, ASTM D-489*	Must Pass	—
Homolog Distribution, Wt. %		
Phenol	2 max.	1
Ortho Cresol	31.5-36.5	33
2,6-Xylenols	32.5-37.5	34
2,4/2,5 Xylenols	—	12
2,3-Xylenols + TMP's	—	16
Xylenols + TMP's	25.0 min.	—
Meta/Para Cresol	8.0 max.	4
Flash Point, °F, Pensky-Martens	—	210

Cresylic Acid 530

Description:

Pitt-Consol Cresylic Acid 530 is a synthetic cresylic acid blend used as a wire enamel solvent. It also is used in metal degreasing and carbon removal applications.

Properties	Specification	Typical
Specific Gravity @ 60°F	—	1.02
Density, lb/gal @ 70°F	—	8.48
Color, Barrett*	C-3 max.	C-2
Water, Wt. %*	0.3 max.	0.10
Freeze Point, °F	—	10
Corrosion, ASTM D-849	Must Pass	—
Homolog Distribution, Wt. %		
Phenol	2.0 max.	1
Ortho Cresol	22.0-27.0	24
2,6-Xylenols	39.5-44.5	42
Meta/Para Cresol	10.0 max.	5
2,4/2,5-Xylenols	—	15
2,3-Xylenols + TMP's	—	13
Xylenols + TMP's	25.0 min.	28
Flash Point, °F, Pensky-Martens	—	196

(continued)

Table 8.6: (continued)

Cresylic Acid 560

Description:

Pitt-Consol Cresylic Acid 560 is a synthetic cresylic acid used in disinfectants and ore flotation agents.

Properties	Specification	Typical
Specific Gravity @60°F	—	1.02
Density, lbs/gal @70°F	—	8.50
Color, Barrett*	C-3	C-2
Water, Wt. %*	0.2	0.1
Freeze Point, °F*	0 max.	—
Flash Point, °F, Pensky-Martens	—	200

Aldehydes

FURFURAL

Furfuraldehyde
Furol
Pyromucic Aldehyde

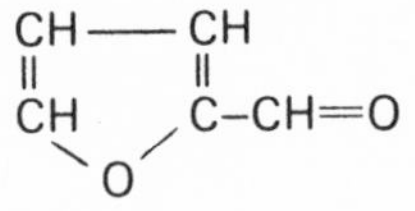

$$
\begin{array}{c}
\text{CH}\!-\!\!-\!\!\text{CH} \\
\| \qquad \| \\
\text{CH} \quad \text{C}\!-\!\text{CH}\!\!=\!\!\text{O} \\
\diagdown\;\diagup \\
\text{O}
\end{array}
$$

Table 9.1: Properties of Pure Furfural *(46)*

Furfural (2-furaldehyde), C_4H_3OCHO, is a liquid aldehyde with a pungent almond-like odor. Colorless when freshly distilled, it darkens on contact with air. Industrial furfural is light yellow to brown in color.

General

Molecular weight	96.08
Boiling point (at 760 mm), °C (°F)	161.7 (323.06)
Freezing point, °C (°F)	−36.5 (−33.7)
Refractive index (n t/D)	
at 20° C (68° F)	1.5261
at 25° C (77° F)	1.5235
Density (d t/4)	
at 20° C (68° F)	1.1598
at 25° C (77° F)	1.1545
Vapor pressure	See Table 9.7
Vapor density (air=1)	3.3

Thermodynamic properties

Heat of vaporization, ΔH_v g cal/g mole	11,614.6
Specific heat (liquid), cal/g/deg	
14 to 80° C (57.2 to 176° F)	0.401
20 to 100° C (68 to 212° F)	0.416
Thermal conductivity,	
Btu/(hr) (ft²) (°F/ft) at 100° F	0.1525
cal/(sec) (cm²) (°C/cm) at 38° C	6.3×10^{-4}
Heat of combustion (liquid), $\Delta H_{298.2}$ kcal/mole	−560.3

Fluid properties

Viscosity, cps, at 0° C (32° F)	2.48
at 25° C (77° F)	1.49
at 38° C (100.4° F)	1.35
at 54° C (129.2° F)	1.09
at 99° C (210.2° F)	0.68
Surface tension, dynes/cm	
at 0° C (32° F)	43.5
at 29.9° C (85.9° F)	40.7
at 30.0° C (86° F)	41.1
Vapor diffusion coefficient, cm²/sec	
at 17° C (62.6° F)	0.076
at 25° C (77° F)	0.087
at 50° C (122° F)	0.107

(continued)

Table 9.1: (continued)

Electrical properties

Dielectric constant

at 1° C (33.8° F)	46.9
at 20° C (68° F)	41.9
at 25° C (77° F)	38
at 50° C (122° F)	34.9

Specific conductivity, mho

Minimum	0.26×10^{-5}
Maximum	0.37×10^{-5}

Other properties

Critical pressure, psia	798
kg/cm²	56.1
Critical temperature, ° C (° F)	397 (746.6)
Molar volume, 25° C, ml/mole	83.19
Molecular association	1.11
Solubility in	
water, wt. % at 20° C (68° F)	8.3
alcohol	∞
ether	∞

Note: Furfural is miscible with most common organic solvents except saturated aliphatic hydrocarbons.

Flammability properties

Explosive limits (% by vol.)

Lower limit (at 125° C [257° F] and 740 mm Hg)	2.1

Flash point

Tag closed cup, °C (°F)	61.7 (143)
Pensky-Martens, °C (°F)	61.7 (143)

(Based on flash point, furfural is classified as Class III A.*)

Ignition temperature, °C (°F)	393 (739)

Note:

Furfural has a high order of thermal stability in the absence of oxygen. At temperatures as high as 230° C (446° F), exposure for many hours is required to produce detectable changes in the physical properties of furfural, with the exception of color (29).

*Refers to Code of Federal Regulations: 29CFR 1910.106.

Table 9.2: Typical Properties and Specifications of Furfural *(2)*

Acidity, as acetic	Technical 0.3%	Refractive index at 68°F	1.5261
	Refined 0.1%	Solubility in water at 20°C	8.3%
Boiling point	158–162°C	Specific gravity at 20/20°C	1.161
Density at 60°F	1.164		1.158–1.160 Technical
100°	1.140		1.59–1.161 Refined
150°	1.110		
175°	1.095	Surface tension	49 dynes/cm.
200°	1.080	Vapor pressure at 60°F	0.035 lbs./sq. in. abs.
250°	1.049	100°	0.130 lbs./sq. in. abs.
300°	1.019	150°	0.540 lbs./sq. in. abs.
Distillation range (Engler)		175°	0.950 lbs./sq. in. abs.
1%, °F (min.)	300	200°	1.650 lbs./sq. in. abs.
End point, °F (max.)	335	250°	4.40 lbs./sq. in. abs.
Recovery, % (min.)	98.5	300°	11.50 lbs./sq. in. abs.
Residue, % (max.)	0.9	350°	22.50 lbs./sq. in. abs.
Loss, % (max.)	0.9	400°	43.5 lbs./sq. in. abs.
Explosive limit, lower	2.1% at 257°F	450°	77.0 lbs./sq. in. abs.
Flash Point (Cleveland Open Cup)	131–5°F	Viscosity at 100°F	1.35 centipoises
Freezing point	−34°F	130°	1.09 centipoises
Heat of Vaporization	107.51 cal./g	210°	0.68 centipoises
Purity	98.5% Technical	Weight per gallon (20°C)	9 lbs.
	99.0–99.5% Refined		

Table 9.3: Solubility of Various Substances in Furfural *(46)*

Acetone	S	Isobutyl	S
Acids:		n-Octyl	S
Abietic (technical)	9.4	Amyl acetate	M
Acetic	S	Benzene	S
Benzoic	14.8	Butyl acetate	M
Butyric (technical)	S	Carbon tetrachloride	S
Cinnamic	4.1	Castor oil	M
Citric	3.6	Chinawood oil	M
Formic	S	Chloroform	S
Lactic	S	Diethylene glycol monobutyl ether	M
Maleic	R	Diethylene glycol monoethyl ether	M
Naphthenic acids (practical)	S	Diethyl phthalate	M
Oleic (U.S.P.)	S	Ethyl acetate	S
Oxalic	4.8	Ethylene glycol	S
Oxalic (anhydrous)	3.6	Ethylene glycol monobutyl ether	M
Palmitic (technical)	1.6	Ethylene glycol monoethyl ether	M
Phthalic	17.6	Ferric chloride	0.55
Propionic (technical)	S	Ferric chloride hexahydrate	20.0
Salicylic	11.0	Hydrogen cyanide	M
Sebacic (mp 132-133° C		Linseed oil	M
[269.6-271.4° F])	0.8	Nitrobenzene	M
Stearic (U.S.P.)	2.1	Nitrotoluene	M
Succinic	3.0	Paraldehyde	M
Tartaric	10.9	Pyridine	S
Alcohols:		Quinoline	S
Amyl	M	Toluene	S
n-Butyl	S	Xylol	M
Ethylene glycol	S	Zinc chloride	20.6
Glycerol	2.1-2.8		

S=infinitely soluble
M=miscible in equal volume at room
    temperature
R=reaction

Table 9.4: Solubility of Selected Thermoplastic Resins in Furfural *(46)*

(At 23°C [73.4° F])

RESIN TYPE	MANUFACTURER	SOLVENT ACTION	RESIN TYPE	MANUFACTURER	SOLVENT ACTION
Nitrocellulose	Hercules (RS)	VS	PVC	Goodrich (Geon® 222)	SH
Ethylcellulose	Hercules (N-50)	VS	Nylon	Du Pont (Zytel® 31)	1 (B)
Cellulose acetate butyrate	Eastman	VS	Nylon	Du Pont (Elvamide)	1 (B)
Polyvinyl butyral	Union Carbide (Bakelite®)	S;VSH	Polyethylene	Du Pont (Alathon®)	1 (B)
Vinyl acetate	Union Carbide (Bakelite®)	SH (B)	Acrylic	Du Pont (Lucite® 140)	SH (B)
Vinyl acetate chloride	Union Carbide (Bakelite®)	SH (B)	Acrylic	Du Pont (Lucite® 130)	VS
PVC	Uniroyal (Marvinol® VR-10)	1	Polystyrene	Dow (PS-3)	SH

S=Soluble from 1 g to 10 g per 100 g solvent
VS=Soluble 10 g or more per 100 g solvent
H=Temperature, 70-75° C (158-167° F); time
    one hour
B=Cloudy
1=Less than 1 g per 100 g solvent

Table 9.5: Specific Gravity and Pounds per Gallon of Furfural *(46)*

(Change per °C: Sp. Gr. −0.00110; lbs./gal. −0.00917)

TEMPERATURE °F	°C	SP. GR.[1]	LBS./GAL.	TEMPERATURE °F	°C	SP. GR.[1]	LBS./GAL.
122.0	50	1.127	9.403	57.2	14	1.167	9.733
118.4	48	1.129	9.421	53.6	12	1.169	9.752
114.8	46	1.131	9.440	50.0	10	1.171	9.770
111.2	44	1.134	9.458	46.4	8	1.173	9.788
107.6	42	1.136	9.476	42.8	6	1.175	9.807
104.0	40	1.138	9.494	39.2	4	1.178	9.825
100.4	38	1.140	9.502	35.6	2	1.180	9.833
96.8	36	1.142	9.531	32.0	0	1.182	9.861
93.2	34	1.145	9.549	28.4	− 2	1.184	9.879
89.6	32	1.147	9.568	24.8	− 4	1.186	9.898
86.0	30	1.149	9.586	21.2	− 6	1.189	9.916
82.4	28	1.151	9.604	17.6	− 8	1.191	9.935
78.8	26	1.153	9.623	14.0	−10	1.193	9.953
75.2	24	1.156	9.631	10.4	−12	1.195	9.971
71.6	22	1.158	9.660	6.8	−14	1.197	9.990
68.0	20	1.160	9.678	3.2	−16	1.200	10.008
64.4	18	1.162	9.696	−1.6	−18	1.202	10.027
60.8	16	1.164	9.715	−4.0	−20	1.204	10.045

[1] Referred to water at 4°C.

Table 9.6: Composition/Density of Furfural-Water Solutions[1] *(46)*

FURFURAL (% BY WEIGHT)	DENSITY $\frac{t°}{4}$ 20° C	25° C	FURFURAL (% BY WEIGHT)	DENSITY $\frac{t°}{4}$ 20° C	25° C
0	0.9982	0.9971	4.6	1.0068	1.0054
0.2	0.9986	0.9974	4.8	1.0072	1.0058
0.4	0.9990	0.9978	5.0	1.0075	1.0062
0.6	0.9993	0.9982	5.2	1.0079	1.0065
0.8	0.9997	0.9985	5.4	1.0083	1.0069
1.0	1.0001	0.9989	5.6	1.0086	1.0073
1.2	1.0005	0.9993	5.8	1.0090	1.0076
1.4	1.0008	0.9996	6.0	1.0094	1.0080
1.6	1.0012	1.0000	6.2	1.0098	1.0084
1.8	1.0016	1.0003	6.4	1.0101	1.0087
2.0	1.0020	1.0007	6.6	1.0105	1.0091
2.2	1.0023	1.0011	6.8	1.0109	1.0094
2.4	1.0027	1.0014	7.0	1.0113	1.0098
2.6	1.0031	1.0018	7.2	1.0116	1.0102
2.8	1.0034	1.0022	7.4	1.0120	1.0105
3.0	1.0038	1.0025	7.6	1.0124	1.0109
3.2	1.0042	1.0029	7.8	1.0127	1.0113
3.4	1.0046	1.0033	8.0	1.0131	1.0116
3.6	1.0049	1.0036	8.2	1.0135	1.0120
3.8	1.0053	1.0040	8.3[2]	1.0137	1.0122
4.0	1.0057	1.0044	8.4	—	1.0124
4.2	1.0060	1.0047	8.6[3]	—	1.0127
4.4	1.0064	1.0051			

[1] Mains, G.H., **Chem. & Met. Eng.**, 26,779 (1922).
[2] Saturated solution of furfural in water at 20°C (68°F).
[3] Saturated solution of furfural in water at 25°C (77°F).

Table 9.8: Solution Temperature of Furfural-Water System *(46)*

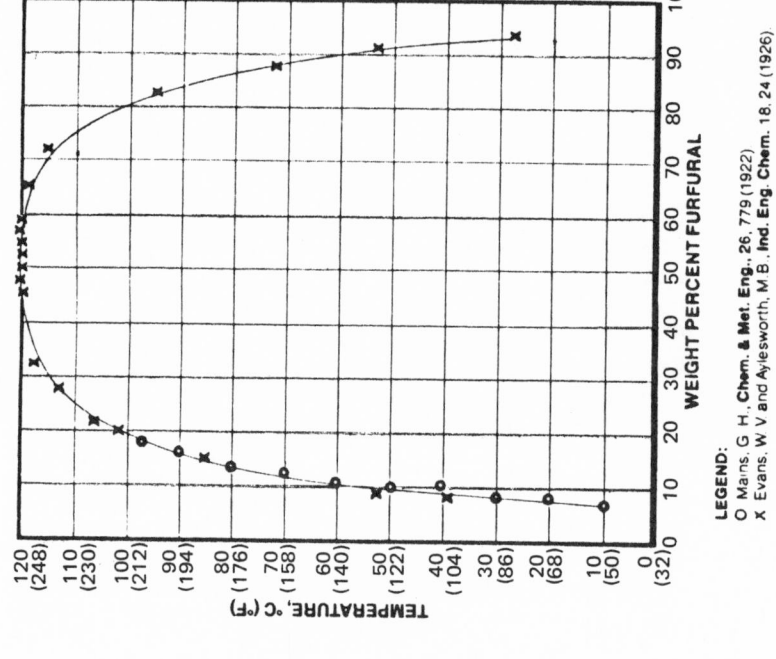

Table 9.7: Vapor Pressure of Furfural *(46)*

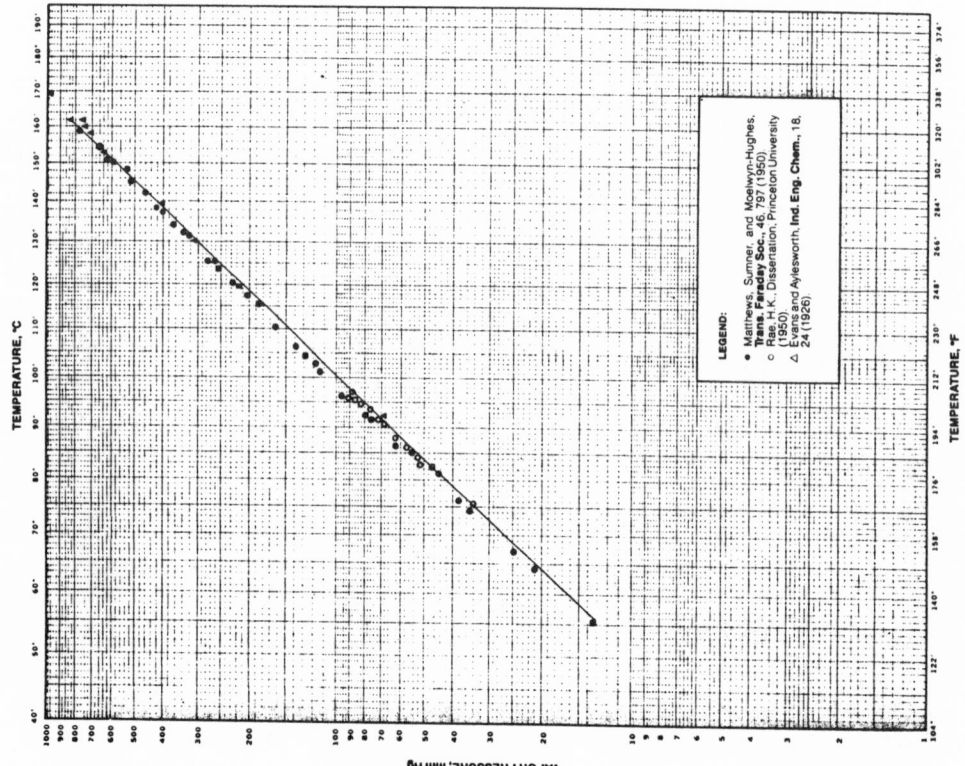

Table 9.10: Vapor-Liquid Equilibria in the Furfural-Water System[1] (46)

(PRESSURE = 760 mm Hg)

% FURFURAL BY WEIGHT		BOILING POINT	
COMPOSITION OF LIQUID	COMPOSITION OF VAPOR	°C	°F
0.2	1.5	99.90	211.8
0.4	3.0	99.82	211.68
0.6	4.4	99.74	211.53
0.8	5.8	99.67	211.41
1.0	7.0	99.60	211.28
1.5	10.0	99.42	210.96
2.0	12.7	99.25	210.65
2.5	15.0	99.11	210.40
3.0	17.1	98.99	210.18
3.5	19.0	98.87	209.97
4.0	20.7	98.76	209.77
4.5	22.2	98.66	209.59
5.0	23.6	98.58	209.44
5.5	24.8	98.50	209.30
6.0	25.8	98.43	209.17
6.5	26.8	98.37	209.07
7.0	27.7	98.31	208.96
7.5	28.5	98.26	208.87
8.0	29.2	98.21	208.78
8.3[2]	29.6	98.19	208.74
8.5	29.9	98.17	208.71
9.0	30.5	98.13	208.63
10.0	31.7	98.07	208.53
11.0	32.6	98.02	208.44
12.0	33.3	97.98	208.36
13.0	33.9	97.95	208.31
14.0	34.4	97.93	208.27
15.0	34.7	97.92	208.26
16.0	34.8	97.91	208.24
17.0	34.9	97.91	208.24
18.0	35.0	97.90	208.22
18.4[3]	35.0	97.90	208.22
18.4-84.1[4]	35.0	97.90	208.22

[1]Mains, G.H., Chem. & Met. Eng., 26, 779 (1922).
[2]Saturated solution of furfural in water at 20° C (68° F).
[3]Saturated solution of furfural in water at the boiling point.
[4]Range over which both furfural and water layers are present.

Table 9.9: Temperature-Composition Diagram of Furfural-Water System* (46)

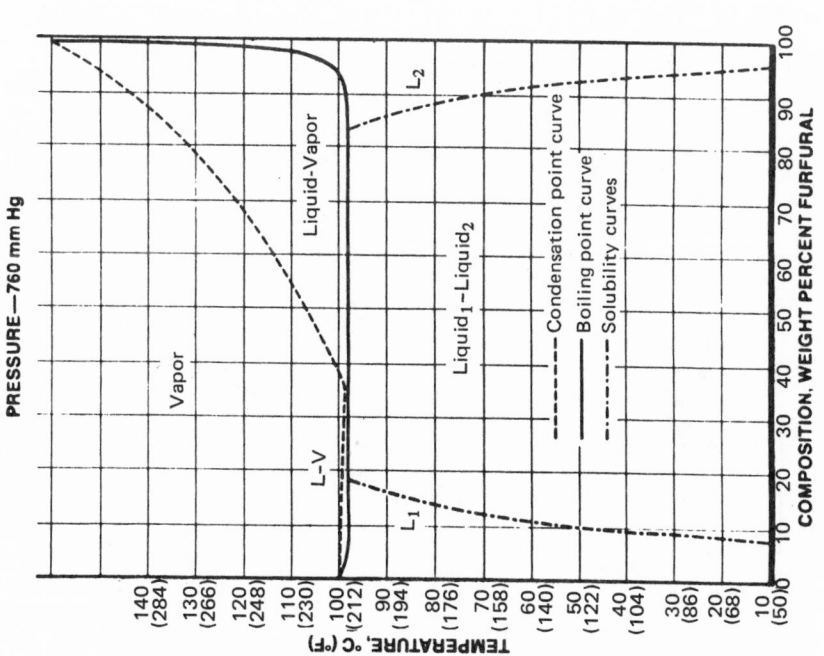

*Mains, G.H., Chem. & Met. Eng., 26, 779(1922)

Table 9.11: Vapor-Liquid Composition of Furfural-Water System *(46)*

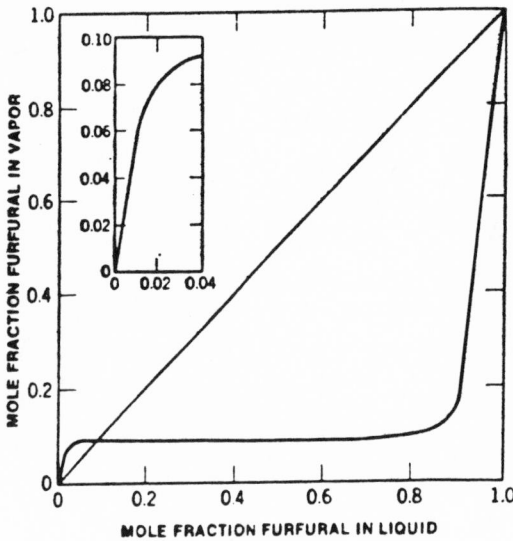

TEREPHTHALALDEHYDE

Table 9.12: Physical Properties of Terephthalaldehyde *(30)*

Empirical Formula – – – – – – – – – – – – – – – – – – – $C_8H_6O_2$
Molecular Weight– – – – – – – – – – – – – – – – – – 134.12
Appearance– – – – – – – – – – – – White, Crystalline Solid
Melting Point, °C – – – – – – – – – – – – – – – – – 114
Boiling Point, °C @ 15-16 mm/Hg – – – – – – – – – 151-154
　　　　　　　°C @ Atmospheric Pressure – – – – – – 248
Odor– – – – – – – – – – – – – – – – – – – Pleasant, Aldehyde

Approximate Solubility (% by weight) in
Various Solvents at 25°, 50°, and 75°C

Solvent	Temperature (°C)		
	25	50	75
Acetone	15	26	--
Benzene	6	9	--
Carbon Tetrachloride	0	1	4
Cyclohexanone	9	19	40
p-Dioxane	14	27	47
Ethyl Acetate	8	15	--
Ethyl Ether	2	--	--
Ethylene Glycol	5	34	--
n-Heptane	0	0	4
Methanol	46	64	--
Perchloroethylene	0	1	3
Tetrahydrofuran	14	24	--
Water	0	0.3	0.6

OTHER ALDEHYDES

Table 9.13: Vapor Pressures of Various Aldehydes *(19)*

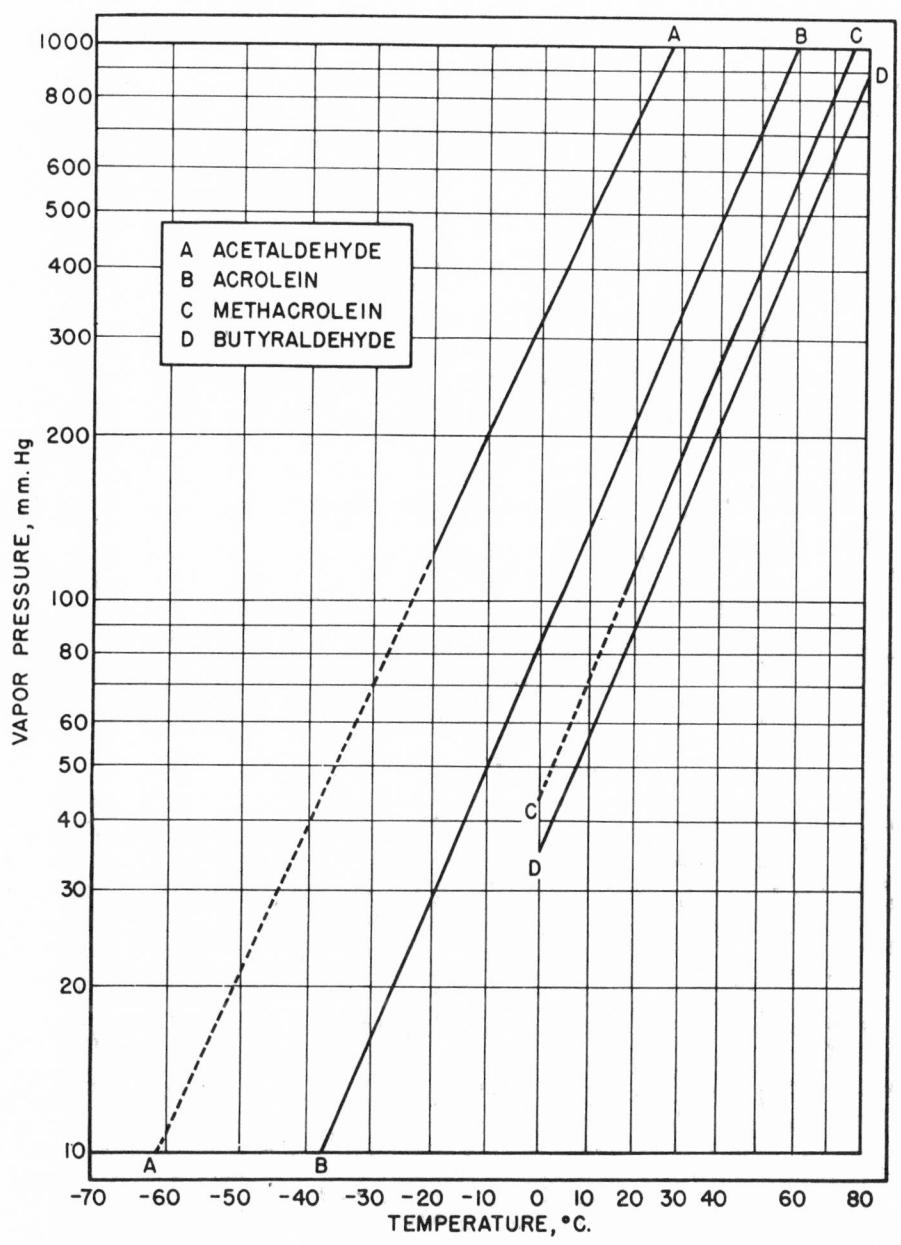

Table 9.14: Physical Properties of Various Aldehydes (19)

Product	Formula	Formula Molecular Weight	Purity of Tested Sample, % by wt.	Apparent Specific Gravity, 20/20°C.	Boiling Point, °C., 760 mm.	Vapor Pressure, mm. Hg at 20°C.	Freezing Point, °C.	Solubility, % by weight at 20°C.		Pounds Per Gal. at 20°C.	Flash Point, °F. (a)
								In Water	Water In		
Formaldehyde, 37% (uninhibited)	HCHO	30.03	(c)	0.816(g)	−19.1	3284	−117	Complete		9.24(h)	None
Paraldehyde	[CH₃CHO]₃	132.16	(c)	0.9961	124	26	12.6	10.5	1.1	8.27	96
Propionaldehyde	C₂H₅CHO	58.08	(c)	0.7982	48.0	258	−80	22	35	6.72	<0
Butyraldehyde	C₃H₇CHO	72.11	(c)	0.8028	74.8	88.5	−96.4	7.1 25°c.	3.0 25°c.	6.69	15
Isobutyraldehyde	CH₃CH(CH₃)CHO	72.11	(c)	0.7905	64.1	138	.....	6.5	2.9	6.58	13
Valeraldehyde	C₄H₉CHO	86.13	(c)	0.8109	103.0	26	−91	1.35	1.35	6.75	54
2-Methylpentaldehyde	C₃H₇CH(CH₃)CHO	100.16	(c)	0.8102	118	14	−100(d)	0.42	0.83	6.74	72
2,3-Dimethyl Pentaldehyde	C₂H₅CH(CH₃)CH(CH₃)CHO	114.19	(c)	0.8293	140.5	5	−110	0.21	0.60	6.91	94
Acrolein	CH₂:CHCHO	56.06	99	0.8427	53	220	−87.0	20.6	6.8	7.02(i)	<0(i)
Tetrahydrobenzaldehyde	CH₂:CHCHCH₂CH₂CH₂CHCHO	110.16	99.8	0.9721	165	2	−100(d)	0.5	1.0	8.08	135
UCAR Glyoxal 40 (aq. sol.)	OHCCHO	58.04	(e)	1.2798	.....	.....	−15	Complete	Complete	10.65	None
UCAR Glyoxal LV	.....	.....	(e)	1.2851	.....	.....	−15	Complete	Complete	10.69	None
Glutaraldehyde, 25% aq. sol.	OHCC₃H₆CHO	100.12	(e)	1.062	.....	17	−7.0	Complete	Complete	8.83	None
Glutaraldehyde, 50% aq. sol.	OHCC₃H₆CHO	100.12	(e)	1.124	.....	17	−14.0	Complete	Complete	9.38	None

(a) All flash points were determined by either ASTM method D 1310 using Tag open cup or ASTM method D 92 using Cleveland open cup.

(c) 99+ mol per cent material.

(d) Sets to glass below this temperature.

(e) Typical commercial material.

(f) Made from anhydrous isopropanol diluted with demineralized water.

(g) True density at −19°C.

(h) 37% Solution.

(i) Inhibited material.

Ethers

Table 10.1: Dimethyl Ether (34)

Methyl Ether   CH3—O—CH3

Physical Properties

Molecular weight (calc.) 46.07
Boiling point at 760 mm . . . . −24.9°C
Vapor pressure at 20°C. . . . . 5.24 atm
Freezing point . . . . −141.5°C
Density at 20°C. . . . . 0.661 g/ml
Vapor density (air = 1.0). 1.59
Critical pressure . . . . 52.5 atm
    temperature . . . . 128.8°C
    density . . . . 0.2714 g/ml
Heat of combustion, gas . . . . 347.6 kcal/mole
Heat of formation, gas . . . . −44.3 kcal/mole
Heat of melting . . . . 25.621 cal/g
Heat of vaporization at −24.8°C . . . 111.64 cal/g
Free energy of formation, 25°C . . . . −27.3 kcal/mole
Entropy at 25°C. . . . . 63.72 cal/°C—mole
Specific heat at −27.68°C . . . . . . 0.5351 cal/g
Surface tension, liquid-vapor interface,
    at −40°C. . . . . 21 dynes/cm
    −20 18
    −10 16

Viscosity of gas at 0°C, η x 10⁷ 825
               20 855
Dielectric constant at 25°C . . . . 5.02 c.s.u.
Flash point, Tag closed cup. . . . . . . −42°F
Autoignition temperature . . . . 662°F
Explosive limits, % by vol. in air 3.45-26.7%
Solubility* in water at 24°C . . . . . . 35.3% by wt.
Solubility* of water in methyl ether
    at 24°C. . . . . 7.0% by wt.
Solubility in gasoline (unleaded)
    at −40°C. . . . . 64% by wt.
     0 19
     25 7
Solubility at 25°C in:
    carbon tetrachloride at 782 mm . . . . 16.33 mole %
    acetone. . . . . 762 . . . 11.83
    benzene . . . . 761 . . . 15.29
    chlorobenzene. . . . 795 . . . 18.55
    methyl acetate. . . . 704 . . . 11.17

* At about 5 atm.

SOME PHYSICAL AND THERMODYNAMIC PROPERTIES OF DIMETHYL ETHER AT VARIOUS TEMPERATURES

Tem-perature °C	Vapor pressure atm.	Density		Dielec-tric constant	Heat of vapori-zation kcal/kg	Enthalpy		Entropy	
		liquid g/ml	vapor g/ml			liquid kcal/kg	vapor kcal/kg	liquid cal/(g)(°K)	vapor cal/(g)(°K)
− 40	0.392	—	—	—	116.13	77.58	193.71	0.9109	1.4090
− 30	0.741	—	—	—	113.17	83.08	196.25	0.9342	1.3996
− 20	1.35	0.7174	0.0027	—	110.12	88.64	198.76	0.9568	1.3918
− 10	1.97	.7040	.0039	—	106.95	94.23	201.23	0.9787	1.3851
0	2.80	.6905	.0055	—	103.64	100.00	203.64	1.0000	1.3794
10	3.86	.6759	.0076	—	100.17	105.79	205.96	1.0206	1.3744
20	5.24	.6610	.0104	5.15	96.44	111.75	208.19	1.0410	1.3700
30	7.00	.6455	.0142	4.90	92.64	117.60	210.24	1.0604	1.3660
40	9.06	.6292	.0188	4.67	88.48	123.63	212.11	1.0795	1.3620
50	11.6	.6116	.0241	4.41	—	—	—	—	—
60	14.7	.5932	.0306	4.18	—	—	—	—	—
70	18.4	.5735	.0385	3.93	—	—	—	—	—
80	22.7	.5517	.0484	3.70	—	—	—	—	—
90	27.4	.5257	.0623	3.48	—	—	—	—	—
100	33.0	.4950	.0810	3.25	—	—	—	—	—
110	39.5	.4575	.1060	3.00	—	—	—	—	—
120	46.6	.4040	.1465	—	—	—	—	—	—

(continued)

Table 10.1: (continued)

Some Properties of (CH₃)₂O·BF₃

Molecular weight (calc.) . . 113.89
Melting point . . . . . . −12°C
Boiling point . . . . . . 128°C
Density at 20°C . . . . . 1.241 g/ml
Vapor pressure at 30°C . . 6.1 torr
           at 70°C . . 52.7 torr
Surface tension at 20.5°C . 33.03 dynes/cm
Dissociation constant . . . $\log K = (-2983)/T + 7.228$

Binary Azeotropes Containing Dimethyl Ether

Component A	Azeotrope Boiling point, °C	Component A, % by wt
Boron trifluoride	127	60
Hydrogen chloride	−2	38
Ammonia at 1 atm	−37	42.5
at 11 atm	25	56
Sulfur dioxide at 1 atm . . .	0	65
at 56.1 atm . .	6.6	60
at 77.1 atm . .	12.1	60
at 108.7 atm . .	26.7	60
Dichlorodifluoromethane at 3 atm	0	90

Solubility of Methyl Ether at Various Pressures
Temperature = 25°C

Carbon tetrachloride		Acetone		Benzene		Chlorobenzene		Methyl acetate	
p,mm	Methyl ether, Mole %	p,mm	Methyl ether, Mole %	p,mm	Methyl ether, Mole %	p,mm	Methyl ether, Mole %	p,mm	Methyl ether, Mole %
112.4	0.000	229.2	0.0	93.7	0.0	11.6	0.0	213.4	0.0
237.6	3.0	311.7	1.79	196.9	2.30	120.4	6.21	293.2	1.75
360.1	5.96	403.1	3.78	372.6	6.32	310.5	7.20	440.6	5.08
464.8	8.52	548.2	7.01	503.0	9.32	423.3	9.74	576.0	8.17
612.8	12.17	650.8	9.33	634.8	12.29	550.8	12.78	704.4	11.17
782.4	16.33	762.3	11.83	761.4	15.29	795.3	18.55	812.3	13.65
932.7	19.93	939.1	15.77	913.0	18.84	957.9	22.14	923.5	16.27
1072.9	23.30	1075.0	18.93	1006.7	21.00	1072.1	24.71	1039.7	19.50

Table 10.2: Chlorodimethyl Ether *(2)*

$$CH_3-O-CH_2Cl$$

Chlorodimethyl ether is a colorless liquid which decomposes in water and in hot ethyl alcohol. It is soluble in acetone, carbon disulfide and concentrated hydrochloric acid.

Physical Properties

Boiling Point (760 mm. Hg), °C.	59
Dipole Moment	
In Carbon Tetrachloride D	1.88
In Benzene D	1.82 - 1.85
Melting Point, °C.	−103.5
Molecular Weight	80.52
Purity	90% min.
Refractive Index n_D^{20}	1.39737
Specific Gravity D_4^{20}	1.0703

Table 10.3: Chloromethyl Ethyl Ether *(2)*

$$ClCH_2-O-CH_2-CH_3$$

This ether is a colorless liquid which is an irritant to the mucous membranes. It is used as a raw material in organic syntheses.

Physical Properties

Assay (chlorine)	App. 98%
Boiling Range, 760 mm. Hg, °C.	79 - 83
Density D_4^{20}	1.03 - 1.05
Refractive Index n_D^{20}	1.40 - 1.41

Table 10.4: Ethyl Ether *(1)(19)(23)(49)*

Ether Ethyl oxide Sulfuric ether	$C_2H_5-O-C_2H_5$

Typical Properties and Specifications

Apparent ignition temperature in air	190°C.
Boiling point at 760 mm.	34.5°C.
Coefficient of expansion	0.00164 per 1°C.
Constant-boiling mixtures (% by wt.)	
Ethyl ether 99% Carbon disulfide 1.0%	B.P. at 760 mm. 34.5°C.
Ethyl ether 44.5% Methyl formate 55.5%	B.P. at 760 mm. 28.2°C.
Ethyl ether 98.9% Water 1.1%	B.P. at 760 mm. 34.1°C.
Electrical conductivity at 25°C.	4×10^{-13} recip. ohm
Explosive limits	2.34 - 6.15%
Flash point	-40°F.
Freezing point	-116.2°C.
Heat of combustion	651 Cal./mol
Heat of vaporization	83.96 cal./g at B.P.
Refractive index at 17°C.	1.3542
Specific gravity at 20/20°C.	0.7146
Specific heat at 30°C.	0.5476 cal./g.
Surface tension at 20°C.	17.0 dynes/sq. cm.
Solubility in water at 20°C.	6.9% by wt.
Solubility of water in solvent at 20°C.	1.3% by wt.
Viscosity at 20°C.	0.00233 poise
Vapor pressure at 20°C.	442.0 mm. Hg
Weight per gallon at 20°C.	5.95 lbs.
Weight per gallon at 17°C.	5.3542 lbs.
Acidity (as acetic)	0.002% by wt., max.

Table 10.5: Flammability of Ethyl Ether-Oxygen-Helium Mixture *(1)*

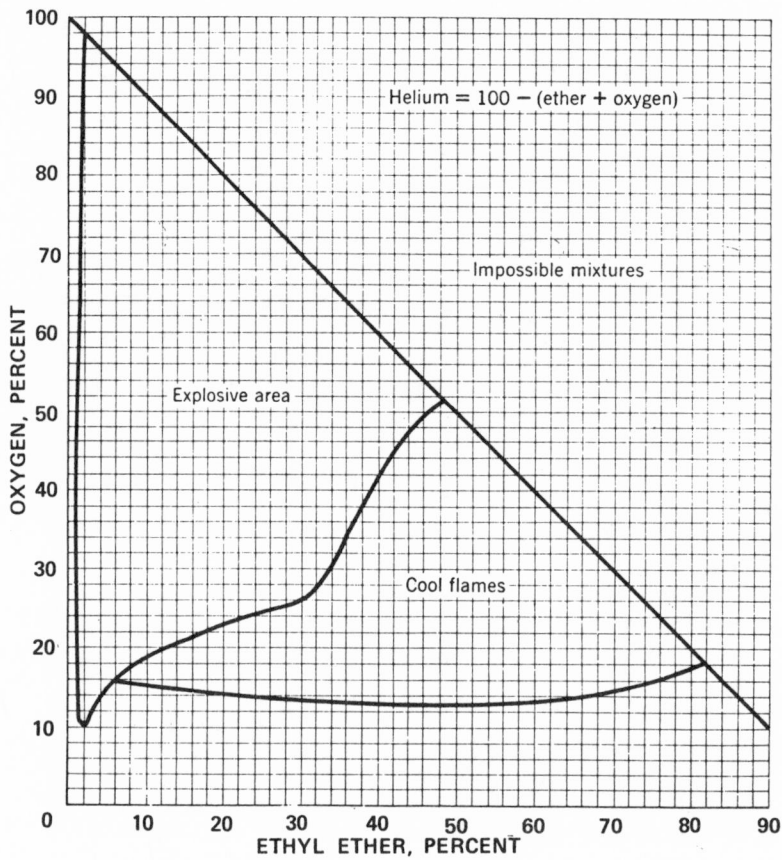

Table 10.6: Dichloroethyl Ether *(2)*

2,2'-Dichlorethyl Ether
β,β'-Dichlorodiethyl Ether

$$Cl-CH_2-CH_2-O-CH_2-CH_2-Cl$$

Acidity (as hydrochloric)	0.005% by wt. max.	Heat of vaporization at 178°C.	64.1 cal./g.
Apparent ignition temperature in air	396°C.	Refractive index at 20°C.	1.457
Boiling point at 760 mm.	178.5°C.	Specific gravity at 20/20°C.	1.2220
Boiling range at 760 mm.	170–180°C.	Specific heat (at 20–30°C.)	0.369 cal.
Coefficient of expansion at 20°C.	0.00097 (per °C.)	Surface tension at 25°C.	41.8 dynes/sq. cm.
Coefficient of expansion at 55°C.	0.00100 (per °C.)	Solubility in water at 20°C.	1.1% by wt.
Constant boiling mixture (% by wt.)		Solubility of water in dichlorethyl	
Dichlorethyl ether	34.4	ether at 20°C.	0.28% by wt.
Water	65.6	Viscosity at 25°C.	2.0653 centipoises
B.P. at 760 mm.	97.7°C.	Vapor pressure at 20°C.	0.7 mm. Hg
Flash Point (ASTM, open cup)	79°C.	Water content	0.10% by wt. max.
Flash Point (ASTM, closed cup)	55°C.	Weight per gal. at 20°C.	10.17 lbs.

ISOPROPYL ETHER

Table 10.7: Properties of Pure Isopropyl Ether *(14)*

Molecular Formula	$C_6H_{14}O$	Specific Heat (Liquid), cal/g°C (at 20°C)	0.506
Molecular Weight	102.172	Thermal Conductivity (Vapor at 100°C)	
Boiling Point, °C	68.5	cal/(sec) (cm²) (°C/cm)	0.0000483
Boiling Point Change, °C/mm at 760 mm	0.042	Viscosity, cps at -20°C	0.545
Freezing Point, °C	-85.5	0°C	0.419
Density at 20°C, g/ml (in vacuo)	0.7235	20°C	0.333
at 60°F, lb/US gal (in air)	6.07	50°C	0.255
Specific Gravity, 20/20°C (in air)	0.7244	Surface Tension (6), 25°C, dynes/cm	17.28
Coefficient of Expansion (1) at 20°C, per °C	0.00143	Dielectric Constant (7), 85.8 kHz, 25°C	4.449
Refractive Index, n0/D	1.3784		
n20/D	1.36820		
n30/D	1.36301		

Other Properties of Commercial IPE

Critical Temperature, °C	288	Autoignition Temp., °F	830
Critical Pressure, atm	27.5	Flash Point (8), Tag Open Cup, °F (approx.)	+15
Critical Volume, cc/g	3.80	Tag Closed Cup, °F (approx.)	-18
Heat of Vaporization (2,3), 760mm, cal/g	68.16	Flammable Limits of Vapor with Air	
Heat of Fusion at Melting Point (4), cal/g	25.79	% vol. of Compound, Upper	21
Heat of Formation (5) (vapor at 25°C)		Lower	1.4
k-cal/mole	-77	Relative Evaporation Rate at 25°C and 0%	
Free Energy of Formation (4) (vapor at		R.H.; Shell Thin Film Evaporometer	
25°C) k-cal/mole	-31	(n-BuOAc = 1.0)	8.04

References

1. Calculated from density measurements as $\dfrac{.7242 - .7139}{.7190 \times 10}$

2. Calculated via Clapeyron Equation Z = 0.95.
3. Fife & Reid, Ind. Eng. Chem. *22*, 513 (1930)
4. Parks, et. al., J. Am. Chem. Soc. *55*, 2735 (1933)
5. Kharasch, M. S., J. Research, Nat'l Bur. Stds. *2*, 359 (1929)
6. Vogel, A. I., J. Chem. Soc. Part I, 616 (1948)
7. Kirk-Othmer, "Ency. of Chem. Tech." *5*, 870 (1950)
8. Petroleum Engineer, June 1945, 219.

Table 10.8: Vapor Pressure of Isopropyl Ether[1] *(14)*

t°C	mm Hg	t°C	mm Hg	t°C	mm Hg
-20	13.4	15	94.4	50	406.6
-15	18.4	20	119.4	55	485.8
-10	24.9	25	149.5	60	576.7
-5	33.3	30	185.6	65	680.3
0	44.0	35	228.4	70	797.8
5	57.3	40	278.9	75	930.1
10	74.0	45	338.0	80	1078.7

[1]Log VP mm Hg = 23.16817 - 2382.7/T - 5.2545Log T
T = 273.15 + t°C

Table 10.9: Isopropyl Ether-Water Solubility *(14)*

t°C	%wt		t°C	%wt	
	IPE in H_2O	H_2O in IPE		IPE in H_2O	H_2O in IPE
-10	—	0.41	50	0.73	0.82
0	—	0.43	60	0.73	0.93
10	1.43	0.47	70	0.76	1.06
20	1.07	0.53	80	0.83	1.20
30	0.88	0.62	90	0.92	1.34
40	0.78	0.72	100	1.04	1.49

Table 10.10: Mutual Solubility for the System: Isopropyl Ether-Isopropyl Alcohol-Water at 25°C, % wt *(14)*

IPE	H_2O	IPA	Sp. Gr.25/4° C
99.5	0.5		0.7210
93.4	1.1	5.5	0.7274
89.0	1.5	9.5	0.7326
84.4	2.2	13.4	0.7380
79.9	3.2	16.9	0.7427
74.4	4.6	21.0	0.7490
72.8	4.7	22.5	0.7509
70.3	5.2	24.5	0.7547
68.7	5.8	25.5	0.7564
65.3	6.7	28.0	0.7605
64.0	7.1	28.9	0.7620
61.5	7.8	30.7	0.7641
58.3	8.9	32.8	0.7698
56.4	9.6	34.0	0.7726
50.8	11.6	37.6	0.7812
47.6	13.0	39.4	0.7864
42.6	15.5	41.9	0.7958
38.6	17.8	43.6	0.8029
35.7	19.7	44.6	0.8091
31.5	23.0	45.5	0.8189
28.3	26.0	45.7	0.8275
24.8	29.7	45.5	0.8379
22.6	32.4	45.0	0.8450
18.9	37.6	43.5	0.8590
16.3	41.9	41.8	0.8707
14.5	45.0	40.5	0.8789
12.6	48.4	39.0	0.8884
12.2	49.0	38.8	0.8897
10.6	52.1	37.3	0.8982
8.6	55.6	35.8	0.9084
6.6	60.2	33.2	0.9200
5.9	61.8	32.3	0.9245
5.2	63.6	31.2	0.9293
4.7	65.0	30.3	0.9334
3.4	69.6	27.0	0.9437
2.2	74.8	23.0	0.9568
1.6	78.3	20.1	0.9634
1.3	83.3	15.4	0.9716
1.2	89.4	9.4	0.9796
1.0	93.8	5.2	0.9864
0.9	99.1		0.9928

Reference: Frere, F. J., Ind. Eng. Chem. *41*, 2365 (1949)

Table 10.11: Conjugate Solutions in the System: Isopropyl Ether-Isopropyl Alcohol-Water at 25°C, % wt *(14)*

Upper Layer			Tie Line[1]	Lower Layer		
IPE	H₂O	IPA		IPE	H₂O	IPA
96.4	0.8	2.8	1	1.0	92.1	6.9
93.0	1.2	5.8	2	1.2	89.0	9.8
90.1	1.5	8.4	3	1.2	86.9	11.9
86.1	2.0	11.9	4	1.2	85.4	13.4
82.8	2.7	14.5	5	1.2	83.8	15.0
76.3	4.0	19.7	6	1.3	82.4	16.3
72.1	5.0	22.9	7	1.4	81.6	17.0
64.2	7.0	28.8	8	1.4	80.3	18.3
54.2	10.2	35.6	9	1.7	77.8	20.5
45.4	14.0	40.6	10	2.3	74.1	23.6
36.9	18.7	44.4	11	2.9	71.2	25.9
31.5	23.0	45.5	12	3.7	68.3	28.0
25.3	29.0	45.7	13	4.7	64.6	30.7
	Estimated plait point[2]			9.8	53.3	36.9

[1] See Table 10.12
[2] Point at which two layers converge into one phase.

Reference: Frere, F.J., *Ind. Eng. Chem.* 41, 2365 (1949).

Table 10.12: Miscibility of Isopropyl Ether-Isopropyl Alcohol-Water at 25°C *(14)*

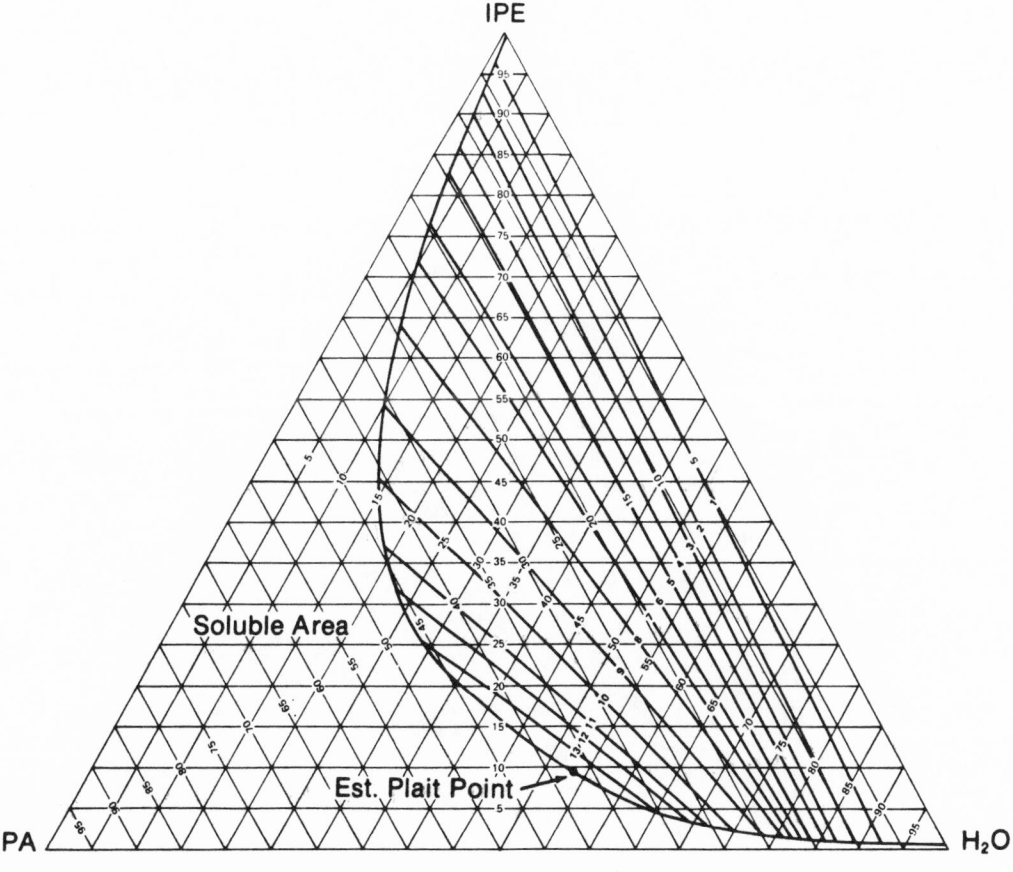

Table 10.13: Azeotropic Information—Isopropyl Ether *(14)*

Ternary Azeotrope: IPA — IPE — Water
(Boiling Point 61.7°C)

Component	Azeotrope	Upper Layer	Lower Layer
IPA (Isopropyl Alc.)	6.0	5.8	10.0
IPE	89.0	93.1	1.0
Water	5.0	1.1	89.0
%w	100	95.6	4.4

Other Azeotropic Information

Binary Azeotropes

B.P. °C	Other Components	%wt Other Component
61.	Boron Triflouride	40
62.2	Water	4.5
70.5	Chloroform	36
<67.5	Propronitrile	> 4
74.0	2, 2-Dichloropropane	60
54.2	Acetone	61
66.2	Isopropyl Alcohol	16.3
66.0	1-Propanethiol	65
>69.0	1-chloro-2-methylpropane	—
<68.0	Methylocyclopentane	< 20
67.5	Hexane	47

Ternary Azeotropes

B.P. °C	Components and %wt
66	H_2O, 7.0%; Ethyl alcohol 14.7%; IPE 78.3%
Min B.P.	H_2O — %; Acetone — %; IPE — %
Nonazeotrope	H_2O — Sec. Butyl alcohol — IPE

IPE does not form azeotropes with

Trichloroethylene   2-Bromo-2-methylpropane
1,1-Dichloroethane   1-Chlorobutane
1,2-Dichloroethane   Ethyl sulfide
2-Chloroethanol   Diethoxymethane
Iodoethane   Benzene
2-Bromopropane   Hexyl Alcohol
Thiophene

Table 10.14: Vapor Pressure of Isopropyl Ether at Various Temperatures *(8)*

Table 10.15: Specific Gravity of Isopropyl Ether vs Temperature *(8)*

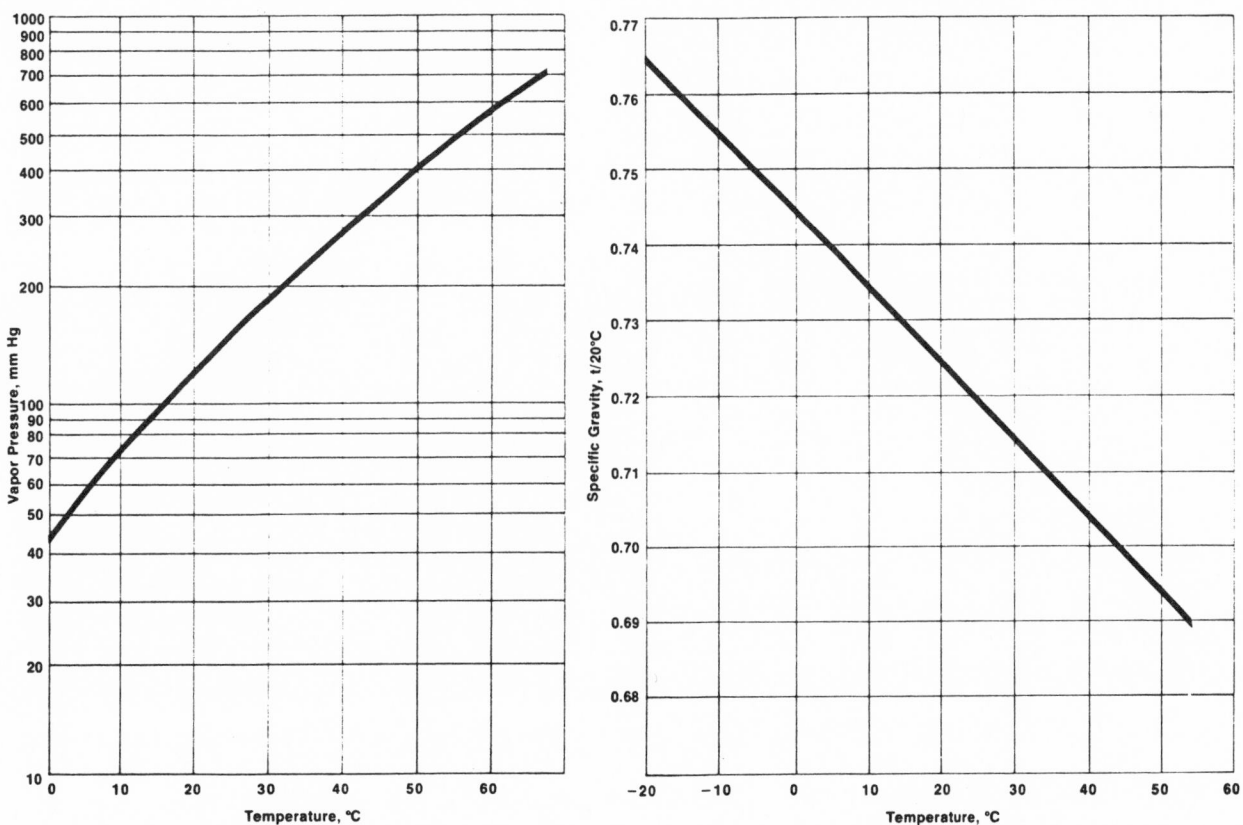

Table 10.16: Mutual Solubility and Specific Gravity of Isopropyl Ether, Water and Isopropyl Alcohol at 25°C *(2)*

Isopropyl Ether	Water	Isopropyl Alcohol	SG d$\frac{25°C.}{4}$
99.5	0.5	- -	0.7210
93.4	1.1	5.5	0.7274
89.0	1.5	9.5	0.7326
84.4	2.2	13.4	0.7380
79.9	3.2	16.9	0.7427
74.4	4.6	21.0	0.7490
72.8	4.7	22.5	0.7509
70.3	5.2	24.5	0.7547
68.7	5.8	25.5	0.7564
65.3	6.7	28.0	0.7605
64.0	7.1	28.9	0.7620
61.5	7.8	30.7	0.7641
58.3	8.9	32.8	0.7698

(continued)

Table 10.16: (continued)

Isopropyl Ether	Water	Isopropyl Alcohol	SG d$^{25°C.}_4$
56.4	9.6	34.0	0.7726
50.8	11.6	37.6	0.7812
47.6	13.0	39.4	0.7864
42.6	15.5	41.9	0.7958
38.6	17.8	43.6	0.8029
35.7	19.7	44.6	0.8091
31.5	23.0	45.5	0.8189
28.3	26.0	45.7	0.8275
24.8	29.7	45.5	0.8379
22.6	32.4	45.0	0.8450
18.9	37.6	43.5	0.8590
16.3	41.9	41.8	0.8707
14.5	45.0	40.5	0.8789
12.6	48.4	39.0	0.8884
12.2	49.0	38.8	0.8897
10.6	52.1	37.3	0.8982
8.6	55.6	35.8	0.9084
6.6	60.2	33.2	0.9200
5.9	61.8	32.3	0.9245
5.2	63.6	31.2	0.9293
4.7	65.0	30.3	0.9334
3.4	69.6	27.0	0.9437
2.2	74.8	23.0	0.9568
1.6	78.3	20.1	0.9634
1.3	83.3	15.4	0.9716
1.2	89.4	9.4	0.9796
1.0	93.8	5.2	0.9864
0.9	99.1	- -	0.9928

Table 10.17: n-Butyl Ether *(2)*

$$C_4H_9OC_4H_9$$

n-Butyl ether is a colorless, stable liquid, soluble in water. Having two butyl groups, this ether is an excellent solvent for many natural and synthetic resins, gums, oils, fats, organic acids, esters, and alkaloids. Beeswax and carnauba wax have limited solubility in butyl ether at room temperature, but become quite soluble at higher temperatures. n-Butyl ether will not dissolve cellulose acetate, benzyl cellulose, or cellulose nitrate, but when it is mixed with ethyl or butyl alcohol it becomes a solvent for ethylcellulose. Butyl ether is used as a reaction medium in organic synthesis and in the extraction and purification of essential oils, organic acids, waxes and resins.

Typical Properties and Specifications

Boiling point at 760 mm	142.4°C
50	63
10	28
Color	Water-white
Flash point	30.6°C
Heat of vaporization	68.8 cal./g
Freezing point	Approx. −96°C
Specific gravity at 20/20°C	0.769-0.771
Refractive index at 20°C	1.3992
Surface tension at 20°C	22.9 dynes per sq cm
Solubility in water at 20°C	0.03%
Solubility of water in solvent at 20°C	0.19%
Vapor pressure at 20°C	4.8 mm Hg
Weight per gallon at 20°C	6.4 lbs
Acidity (as butyric)	0.05% by wt., max.
Distillation range	137-143°C
Water content	0.10% by wt., max.

Table 10.18: Diamyl Ether *(2)*

$$C_5H_{11}OC_5H_{11}$$

Commercial diamyl ether consists principally of di-n-amyl ether and di-isoamyl ether, with small percentages of isomeric amyl ethers and diamylene. It is a colorless to light yellow liquid which is quite stable. It is insoluble in water but soluble in methanol, ethyl ether, ethyl acetate, acetone, aliphatic and aromatic hydrocarbons, fixed oils, oleic and hot stearic acids, hot paraffin and carnauba waxes, the latter two solidifying when cooled. Unlike the lower aliphatic ethers, it will not dissolve nitrocellulose when admixed with ethanol. However, a mixture of diamyl ether and 20% ethanol will dissolve ethylcellulose.

Typical Properties and Specifications

Dielectric constant	3.14
Flash point (open cup)	146°F
Heat of vaporization	65.9 cal./g
	(calc'd)
Specific gravity at 20/20°C	0.78–0.80
Specific heat	0.513 cal/g
Refractive index at 20°C	1.4198
Surface tension at 20°C	24.8 dynes/sq cm
Freezing point	Below −75°C
Vapor pressure at 20°C	0.67 mm
Water azeotrope at 96–98°C	41% amyl ether (approx.)
Weight per gallon at 20°C	6.61 lbs
Acidity (mg. KOH per g)	0.4, max.
Distillation	
Initial boiling point	Not below 170°C
Not less than 95%	Below 200
Final boiling point	Not above 210
Water content	0.2% by wt., max.

Table 10.19: n-Hexyl Ether *(2)*

$$C_6H_{13}OC_6H_{13}$$

n-Hexyl ether is a colorless, stable liquid with a mild odor. It is less volatile than the lower members of the aliphatic ether group and its solubility in water is very slight. It is miscible with most organic solvents and can replace butyl ether for many similar applications. It is used as a solvent medium in chemical reactions and is a foam breaker for certain processes.

Boiling point at 760 mm.	226.2°C.
Boiling point at 50 mm.	136°C.
Boiling point at 10 mm.	100°C.
Flash Point	170°F.
Specific gravity at 20/20°C.	0.7942
Solubility in water at 20°C.	0.01% by wt.
Solubility of water in solvent at 20°C.	0.12% by wt.
Vapor pressure at 20°C.	0.07 mm. Hg
Weight per gallon at 20°C.	6.61 lbs.
Acidity (as acetic)	0.01% by wt., max.
Distillation range at 760 mm.	205 – 235°C.
Color (A.P.H.A.)	15 max.
Water content	0.10% by wt.

Table 10.20: Solubility Data for Various Ethers *(19)*

Substance	Ethyl	Isopropyl	Butyl	Dioxane	Dichlor-ethyl	Dichlor-isopropyl
"Bakelite" vinyl resin AYAF	I	I	SW-G	S	S	S
"Bakelite" vinyl resin VYHH	I	I	I	S	S	SW
Cellulose nitrate (dry)	SA	SA	I	SA	SA	I
Cellulose acetate	I	I	I	S	SA	I
Buna S	—	—	—	—	—	—
Neoprene GN	—	—	—	—	—	—
Carnauba wax	—	SS	—	SW	SW	—
Paraffin wax	—	S	—	SW	S	—
Beeswax	—	SW	—	SW	SW	S
Rosin	S	S	—	S	I	S
Dewaxed dammar	S	SS	S	S	I	S
Zein	—	—	—	—	—	—
Soluble starch	—	—	—	—	—	—
Gelatin	—	—	—	—	—	—
Hydrocarbons	S	S	S	S	S	S
Linseed oil (raw)	S	S	S	S	S	S
Shellac	I	I	SSW	S	I	SSW
Kauri gum	I	I	S	S	I	SW
Ester gum	S	S	S	S	S	S
Unvulcanized rubber	S	S	S-G	SS	I	SS-G

Types of Ethers as Solvents

S: soluble	SS: slightly soluble	I: insoluble
G: tendency to gel	SA: soluble with alcohol	W: when warm

Table 10.21: Comparative Evaporation Rates of Various Ethers *(19)*

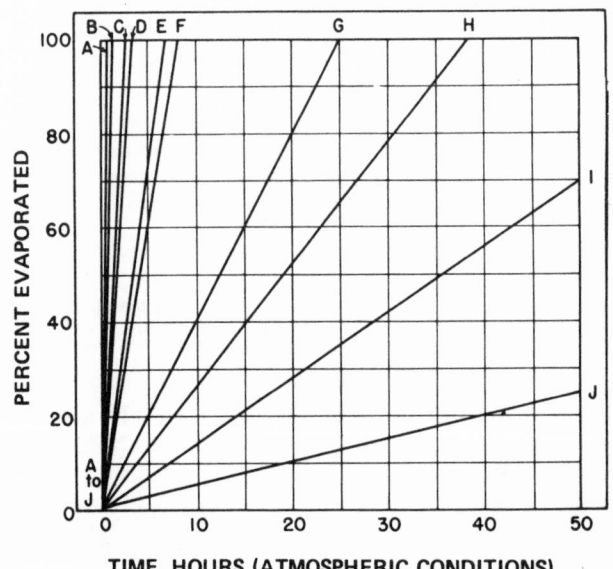

(A) Ethyl ether		(F) Butyl acetate (90%)
(B) Acetone		(G) "Cellosolve" solvent
(C) Isopropyl ether		(H) "Cellosolve" acetate
(D) Ethyl acetate		(I) Dichlorethyl ether
(E) Dioxane		(J) Diethyl "Carbitol"

Table 10.22: Specific Gravities of Various Ethers *(19)*

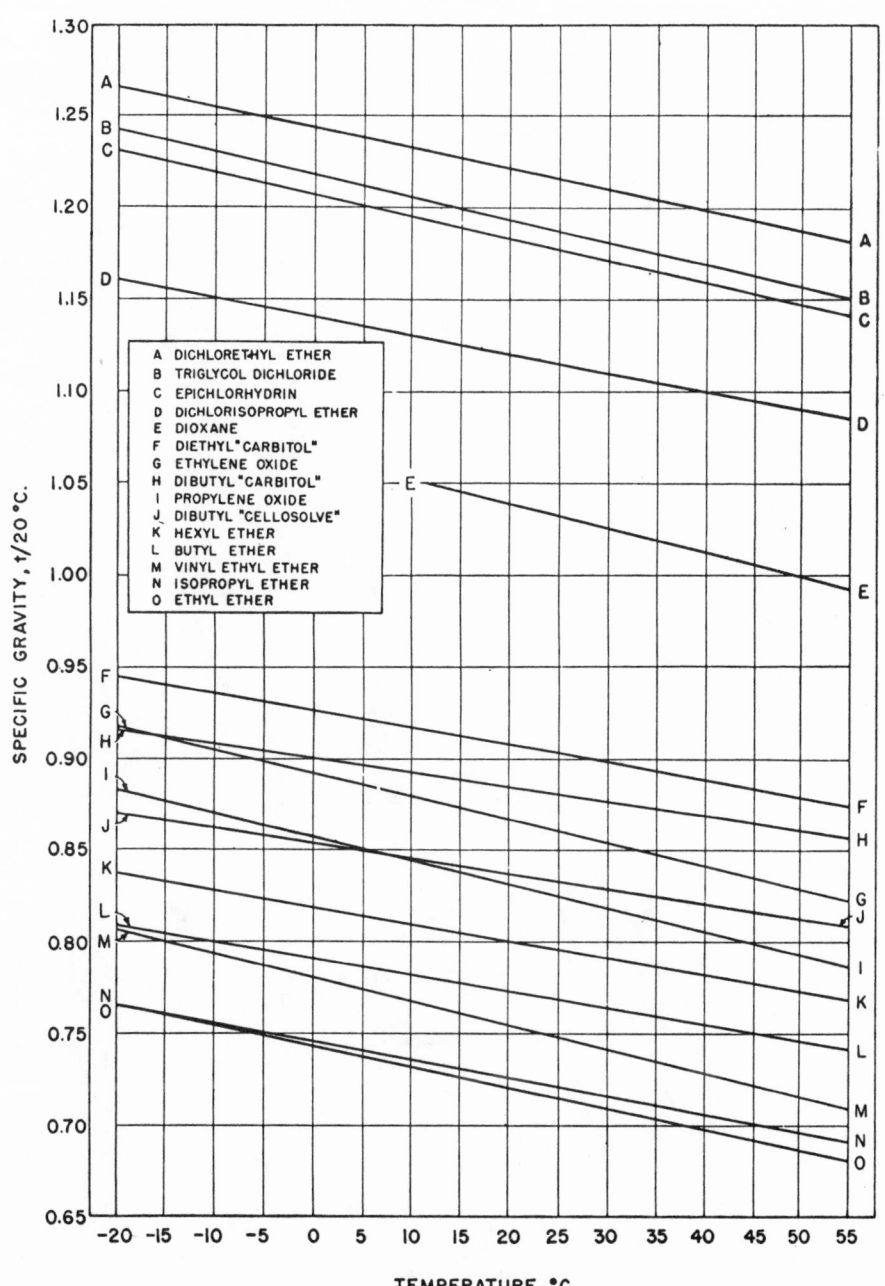

A DICHLORETHYL ETHER
B TRIGLYCOL DICHLORIDE
C EPICHLORHYDRIN
D DICHLORISOPROPYL ETHER
E DIOXANE
F DIETHYL "CARBITOL"
G ETHYLENE OXIDE
H DIBUTYL "CARBITOL"
I PROPYLENE OXIDE
J DIBUTYL "CELLOSOLVE"
K HEXYL ETHER
L BUTYL ETHER
M VINYL ETHYL ETHER
N ISOPROPYL ETHER
O ETHYL ETHER

SPECIFIC GRAVITY, t/20 °C.

TEMPERATURE, °C.

Table 10.23: Vapor Pressures of Various Ethers *(19)*

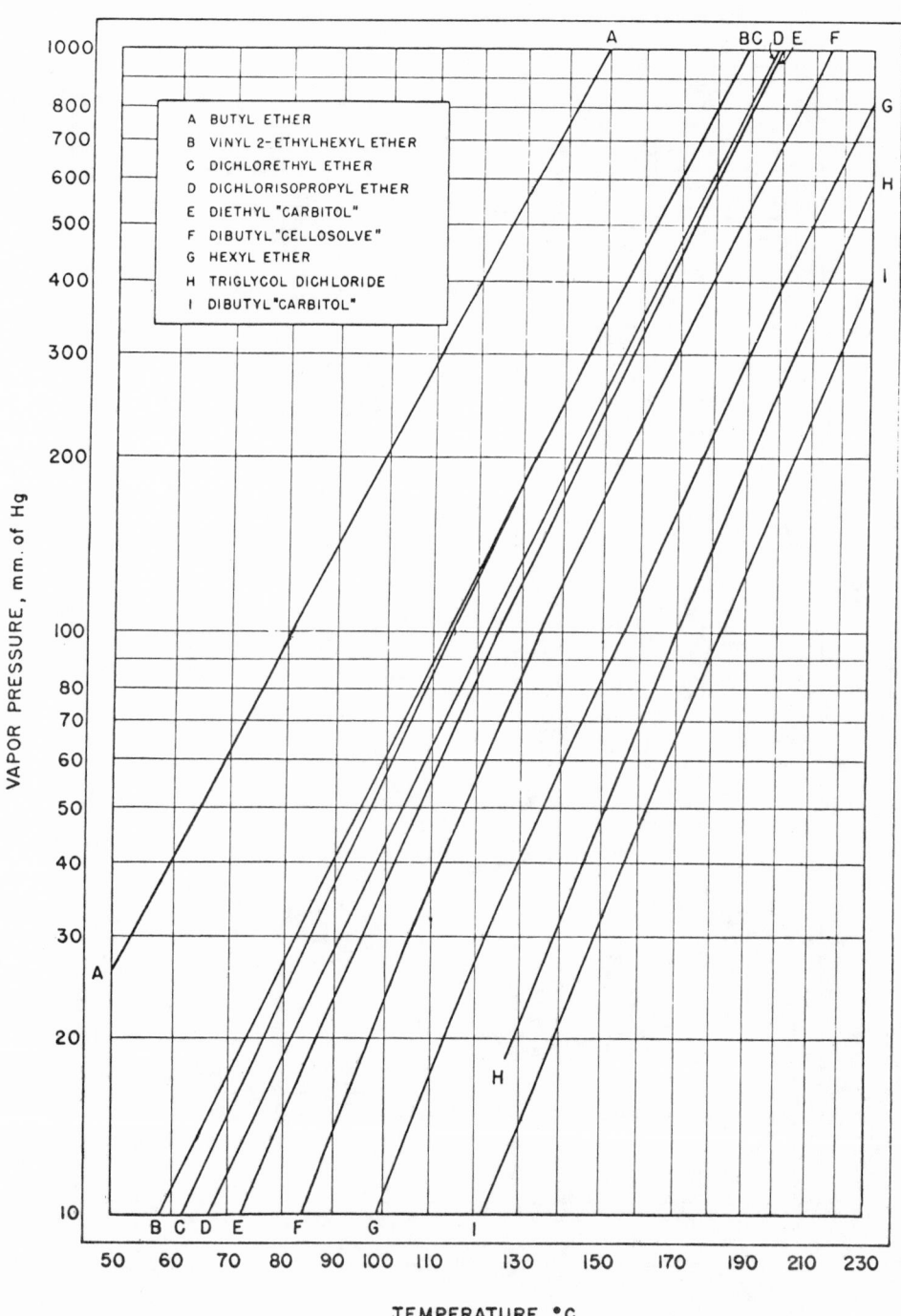

(continued)

Table 10.23: (continued)

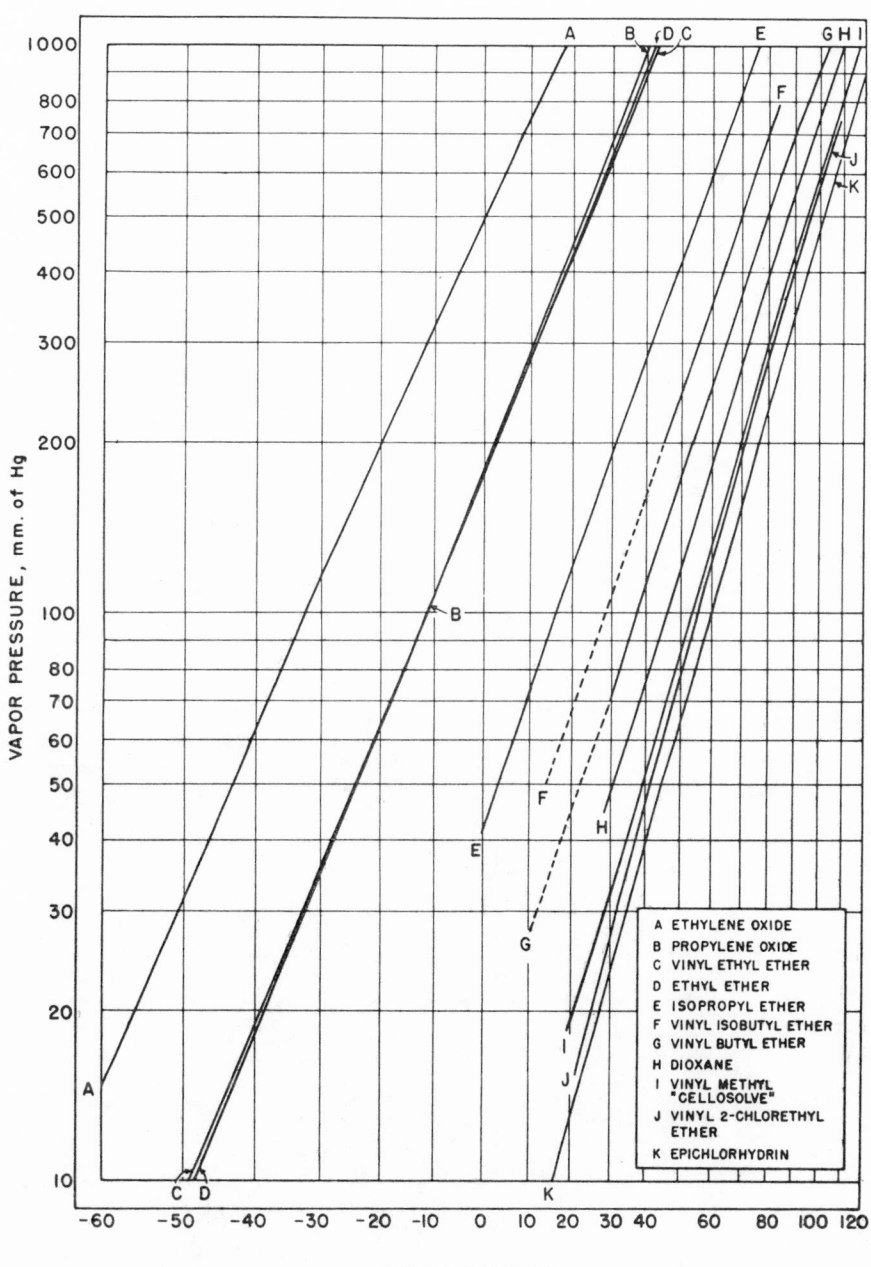

A ETHYLENE OXIDE
B PROPYLENE OXIDE
C VINYL ETHYL ETHER
D ETHYL ETHER
E ISOPROPYL ETHER
F VINYL ISOBUTYL ETHER
G VINYL BUTYL ETHER
H DIOXANE
I VINYL METHYL "CELLOSOLVE"
J VINYL 2-CHLORETHYL ETHER
K EPICHLORHYDRIN

Table 10.24: Ethylene Oxide *(2)*

Epoxyethane
Dimethylene Oxide

CH_2-CH_2 over O

Acidity (as acetic acid), % by wt.	0.005 (max.)
Boiling point, °C.	
760 mm.	10.4
50 mm.	-44
10 mm.	-66
Δ bp/Δ p, °C./mm. Hg	0.033
Coefficient of expansion at 55°C.	0.00177
Flash point (open cup), °F.	below 0
Freezing point, °C.	-112.5
Heat of vaporization (Btu/lb. at 1 atm.)	245
Molecular weight	44.05
Refractive index (n_D at 7°C.)	1.3597
Solubility, % by wt. at 20°C.	
in water	infinite
water in	infinite
Specific gravity, 20/20°C.	0.8711
ΔSG/Δ T	0.00140
Specific heat at 20°C.	0.8763
lb./gal. at 60°F.	7.30
Vapor pressure, mm. Hg at 20°C.	1120
Viscosity (absolute) in centipoises, 0°C.	0.3

Table 10.25: Enthalpy and Entropy of Ethylene Oxide *(19)*

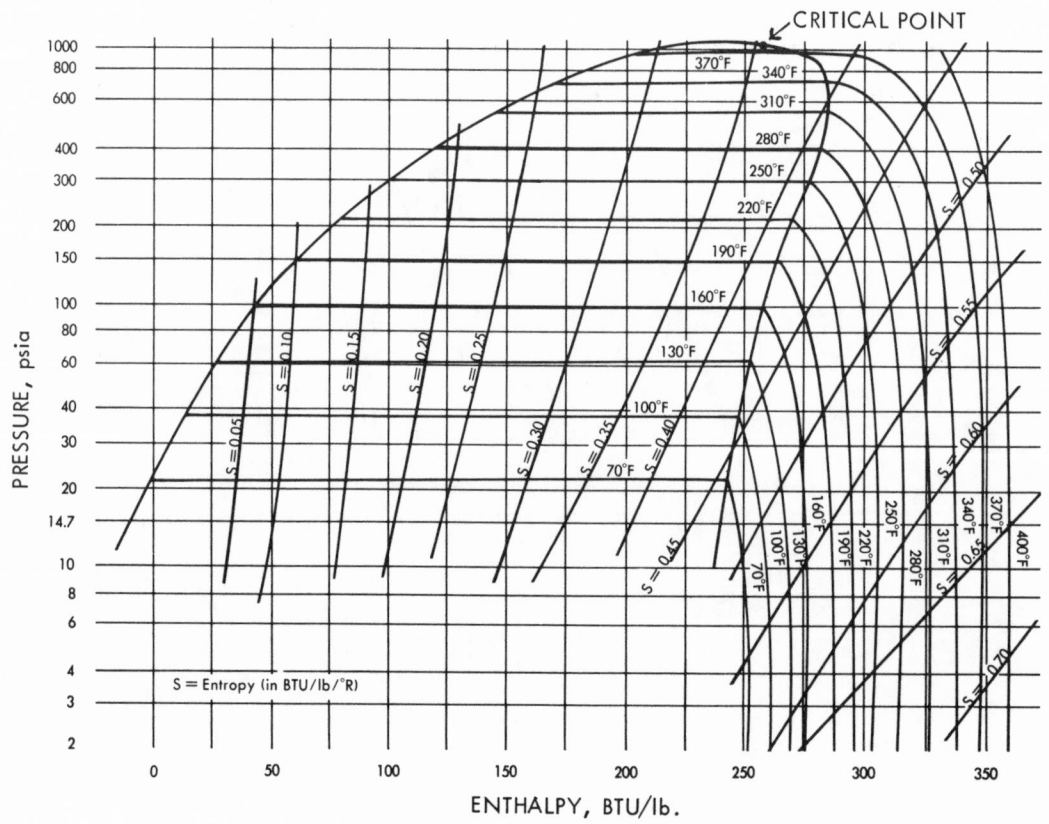

Table 10.26: Propylene Oxide *(2)*

1,2-Epoxypropane

$$CH_3-CH-CH_2$$
$$\diagdown O \diagup$$

Propylene oxide is soluble in water and miscible with most organic solvents. It is found to be an excellent low-boiling solvent for cellulose acetate, nitrocellulose, adhesive compositions and vinyl chloride-acetate resins. It is also a solvent for hydrocarbons, gums and shellac. Some of its uses are as a solvent and stabilizer in DDT aerosol-type insecticides, and as a fumigant and food preservative. Since it is an acid acceptor, it is also used as a stabilizer for vinyl chloride resins and other chlorinated systems.

Acidity (as acetic acid), % by wt. (max.)	0.01
Boiling point, °C.:	
760 mm. Hg	34.0
50 mm. Hg	-26
10 mm. Hg	-52
Δ BP/Δ P., °C./mm. Hg	0.037
Coefficient of expansion at 55°C.	0.00157
Distillation at 760 mm., °C.:	
Initial BP/min.	33.0
DP, max.	37.0
Flash point (open cup), °F.	-35
Freezing point, °C.	-104.4
Heat of vaporization (Btu/lb. at 1 atm.)	160
Molecular weight	58.08
Refractive index (n_D at 20°C.)	1.3657
Solubility, % by wt. at 20°C.:	
in water	40.5
water in	12.8
Specific gravity, 20/20°C.	0.8304
SG/T.	0.00125
Specific heat at 15°C.	0.465
lb./gal. at 60°F.	6.96
Vapor pressure, mm. Hg at 20°C.	449
Viscosity (absolute) in centipoises, 20°C.	0.4

Table 10.27: Freezing Points of Solutions of Ethylene Oxide and Propylene Oxide *(19)*

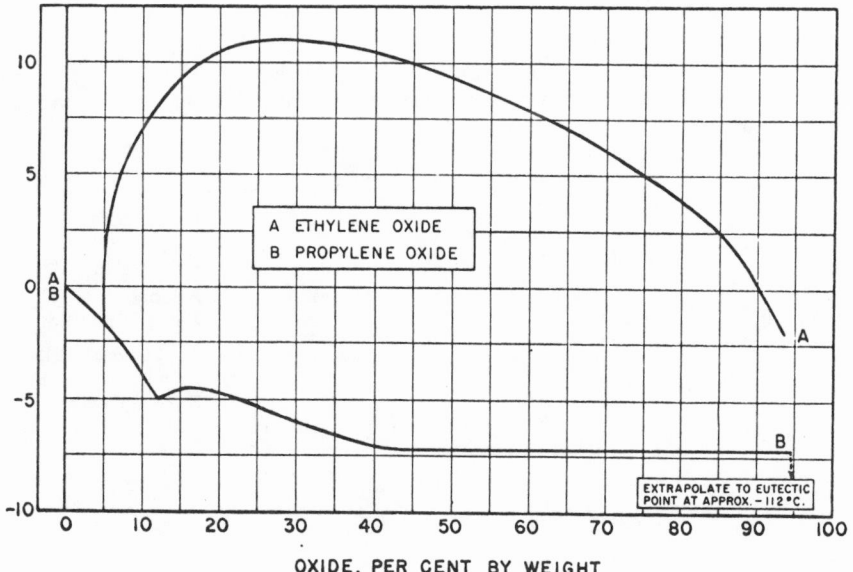

Table 10.28: 1,2-Butylene Oxide *(2)*

1,2-Epoxybutane

$$CH_3CH_2CHCH_2$$
$$\underset{O}{\diagdown\diagup}$$

1,2-Butylene oxide is a colorless mobile liquid. This low boiling liquid has but limited water solubility, yet is miscible with most common organic solvents. It undergoes the usual reactions of epoxides with compounds having labile hydrogen atoms. Some of these are acids, amines, ammonia, alcohols, phenols, polyols, thiols, etc. Butylene oxide can be polymerized or copolymerized with other alkylene oxides to yield polyethers. The resulting polymers are less water soluble than the polymers made from ethylene and propylene oxide, of equivalent chain length.

Boiling point, °C. at 760 mm.	63.2	Surface tension at 20°C., dynes/cm.	23.9
Coefficient of expansion at 20°C.	0.00132	Vapor pressure, mm. Hg at 20°C.	141
Freezing point, °C.	-150	Viscosity (absolute) in centipoises:	
Heat of combusion (Btu/lb. at 25°C.)	14,665	0°C.	0.54
Heat of vaporization (Btu/lb. at 1 atm.)		20°C.	0.41
and 63.2°C.	181	40°C.	0.33
Molecular weight	72.11		
Refractive index (n_D at 20°C.)	1.3840		
Solubility, % by wt. at 20°C.:			
in water	5.91		
water in	2.65		

Table 10.29: "Cardolite" NC-513 *(71)*

"Cardolite" NC-513 is a reactive mutual solvent for epoxy coatings. It is a clear amber liquid, which is a flexibilizer and reactive diluent for epoxy resins. This low viscosity fluid having one epoxy group per molecule combines chemically in any cured epoxy formulation.

Typical Properties

Viscosity	55 cps @ 25°C.
Epoxide Equivalent	480
Specific Gravity	~0.97 @ 25°C.
Odor	Slight
Flash Point (Pensky-Martens closed cup)	370°F.
Storage Life	Extensive
Volatile material @ 105°C.	0.2%
Gardner Color	13

"Cardolite" NC-513 is compatible with a wide variety of organic solvents, resins, and reactive diluents and is notable for its miscibility with aliphatic hydrocarbons in all proportions. It is compatible with the following typical solvents:

Aliphatic Hydrocarbons	Heptane	Esters	"Cellosolve" Acetate
	Hexane		Ethyl Acetate
	Mineral Oil		Ethyl "Carbitol"
	Mineral Spirits		Ethyl "Cellosolve"
	VM&P Naphtha		
		Ketones	Acetone
Aromatic Hydrocarbons	Toluene		Methyl Ethyl Ketone
	Xylene		
		Reactive Diluents	Allyl Glycidyl Ether
			Phenyl Glycidyl Ether
			Styrene Oxide

Table 10.30: 1,4-Dioxane *(2)*

1,4-Dioxan
1,4-Diethylene Oxide
Dioxyethylene Ether
Diethylene Ether
Diethylene Dioxide

1,4-Dioxane is a colorless, stable liquid with a faint, pleasant odor. Although it has been known as far back as 1863, it was not until 1929 that is became commercially available. It is chemically a di-ether obtained by the loss of water from two molecules of ethylene glycol. It is completely soluble in water, as well as most organic solvents. It is freely soluble in mineral, vegetable, blown and heat-bodied oils, and oil soluble dyes. Most waxes are more readily soluble in dioxane when heated and examples of these are beeswax, carnauba, montan, paraffin, gilsonite, and Japan wax.

Acidity (as acetic)	0.010% by wt, max
Boiling point 760 mm	101.3°C
Distillation range at 760 mm	95–103°C
Coefficient of expansion	0.001030 (per °C) to 20°C
	0.001070 (per °C) to 55°C
Electrical conductivity at 25°C	2×10^{-8} recip. ohms
Flash point	65°F
Freezing point	11.7°C
Heat of combustion	581 kg cal/mol
Heat of fusion	33.8 cal/g
Heat of vaporization	98.6 cal/g
Refractive index at 20°C	1.4221
Specific gravity at 20/20°C	1.0356
	1.0353
Specific heat at 20°C	0.420 cal/g
Surface tension at 25°C	36.9 dynes/cm
Solubility in water at 20°C	Complete
Solubility of water in dioxane at 20°C	Complete
Viscosity at 25°C	0.0120 poise
Vapor pressure at 20°C	29.0 mm Hg
Water content at 20°C	Miscible without turbidity with 19 vol. 60° Be gasoline
Weight per gal at 20°C	8.61 lbs

Table 10.31: Trioxane *(2)*

Cyclic Trimeric Polymer of Formaldehyde

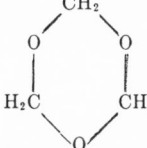

Trioxane is a most unusual chemical. It is an excellent solvent for many classes of materials. Concentrated aqueous solutions of trioxane have solvent properties which are not possessed by trioxane itself. Molten trioxane dissolves numerous organic compounds, such as naphthalene, urea, camphor, dichlorobenzene, etc. It is stable in alkaline or neutral solutions, yet it is depolymerized to formaldehyde by small amounts of strong acid or acid-forming materials, and the rate of depolymerization can be readily controlled.

Properties

Colorless, crystalline compound		Solubility:	
Molecular weight	90.05	Water	Readily soluble
Odor	Mild, pleasant	Alcohols	Readily soluble
Melting point	61°C	Ketones	Readily soluble
Boiling point	115°C	Ethers	Readily soluble
Vapor Pressure:		Esters	Readily soluble
25°C	13 mm	Chlorinated hydrocarbons	Readily soluble
86°C	283 mm	Aromatic hydrocarbons	Readily soluble
114.5°C	759 mm	Vegetable oils	Readily soluble
129°C	1212 mm	Naphthalene	Readily soluble
Flash point	45°C	Phenol	Readily soluble
Density (molten) at 65°C	1.170	Petroleum ether	Slightly soluble

Table 10.32: Vinyl Methyl Ether *(2)*

$$CH_2{=}CHOCH_3$$

Vinyl methyl ether is a gas at ordinary temperature and pressure. When condensed it is a colorless, mobile liquid having a vapor pressure at 760 mm. at 5.5°C. It is miscible with most organic solvents, but only slightly soluble in water or polyhydroxy organic compounds such as glycols. In volatility and flammability it resembles liquefied petroleum gases.

Boiling point, 760 mm.	5.5°C.	Solubility of water in ether at 25°C.	0.51% by wt.
Flash point (Cleveland open cup)	-69°F. (-56°C.)	Specific gravity at 5.7/4°C.	0.7694
Freezing point	-122°C.	Specific gravity at 20/4°C.	0.7511
Molecular weight	58.08	Vapor pressure at 25°C.	1550 mm. Hg
Odor	Sweet, pleasant	Vapor pressure at 70°F.	28 psi abs.
Refractive Index	1.3947	Weight per gallon at 25°C.	6.17 lbs.
Solubility in water at 25°C.	0.82% by wt.		

Table 10.33: Vinyl Ethyl Ether *(2)*

$$CH_2{=}CHOCH_2CH_3$$

Boiling point, °C.:		Molecular weight	72.10
760 mm.	35.5	Refractive index (n_D at 20°C.)	1.3774
50 mm.	-24	Solubility, % by wt. at 20°C.:	
10 mm.	-49	in water	0.9
Δ BP/Δ P., °C./mm. Hg	0.038	water in	0.2
Coefficient of expansion at 55°C.	0.00165	Specific gravity, 20/20°C.	0.7541
Flash point (open cup), °F.	below 0	Δ SG/Δ T.	0.00117
Freezing point, °C.	-115.3	Vapor pressure, mm. Hg at 20°C.	428
Heat of vaporization (Btu/lb. at 1 atm.)	161	Viscosity (abs.) in centipoises, at 20°C.	0.2

Table 10.34: Vinyl 2-Chloroethyl Ether *(2)*

$$CH_2{=}CHOCH_2CH_2Cl$$

Boiling point, °C.:		Solubility, % by wt. at 20°C.:	
760 mm.	109.1	in water	0.6
50 mm.	39	water in	0.4
10 mm.	10	Specific gravity, 20/20°C.	1.0498
Δ BP/Δ P., °C./mm. Hg	0.044	Δ SG/Δ T.	0.00123
Coefficient of expansion at 55°C.	0.00118	Vapor pressure, mm. Hg at 20°C.	18.6
Flash point (open cup), °F.	90	Viscosity (abs.) in centipoises:	
Freezing point, °C.	-69.7	0°C.	1.1
Heat of vaporization (Btu/lb. at 1 atm.)	154	20°C.	0.8
Molecular weight	106.55	40°C.	0.6
Refractive index (n_D at 20°C.)	1.4381		

Table 10.35: Vinyl Butyl Ether *(2)*

$$CH_2{=}CHOC_4H_9$$

Boiling point, °C.:		Molecular weight	100.16
760 mm.	94.2	Refractive index (n_D at 20°C.)	1.4007
50 mm.	24	Solubility, % by wt. at 20°C.:	
10 mm.	-4	in water	0.30
Δ BP/Δ P., °C./mm. Hg	0.044	water in	0.09
Coefficient of expansion at 55°C.	0.00133	Specific gravity, 20/20°C.	0.7803
Flash point (open cup), °F.	15	Δ SG/Δ T.	0.00100
Freezing point, °C.	-112.7	Vapor pressure, mm. Hg at 20°C.	40.4
Heat of vaporization (Btu/lb. at 1 atm.)	137	Viscosity (abs.) in centipoises, 20°C.	0.5

Table 10.36: Vinyl Isobutyl Ether *(2)*

$$CH_2{=}CHOCH_2CH(CH_3)_2$$

Boiling point, °C.:	
760 mm.	83.4
50 mm.	17
10 mm.	-7
Δ BP/Δ P., °C./mm. Hg	0.045
Coefficient of expansion at 55°C.	0.00140
Flash point (open cup), °F.	15
Freezing point, °C.	-132.3
Heat of vaporization (Btu/lb. at 1 atm.)	144
Molecular weight	100.16

Refractive index (n_D at 20°C.)	1.3961
Solubility, % by wt. at 20°C.:	
in water	0.2
water in	0.08
Specific gravity, 20/20°C.	0.7706
Δ SG/Δ T.	0.00104
Specific heat at 15°C.	0.512
Vapor pressure, mm. Hg at 20°C.	59.5
Viscosity (abs.) in centipoises, at 20°C.	0.4

Table 10.37: Vinyl 2-Ethylhexyl Ether *(2)*

$$CH_2{=}CHOCH_2CH(C_2H_5)C_4H_9$$

Boiling point, °C.:	
760 mm.	177.7
50 mm.	95
10 mm.	62
Δ BP/Δ P., °C./mm. Hg	0.053
Coefficient of expansion at 55°C.	0.00107
Flash point (open cup), °F.	135
Freezing point, °C.	100*
Heat of vaporization (Btu/lb. at 1 atm.)	129
Molecular weight	156.26

Refractive index (n_D at 20°C.)	1.4273
Solubility, % by wt. at 20°C.:	
in water	0.01
water in	0.05
Specific gravity, 20/20°C.	0.8102
Δ SG/Δ T.	0.00084
Vapor pressure, mm. Hg at 20°C.	0.60
Viscosity (abs.) in centipoises:	
0°C.	1.5
20°C.	1.0

*Sets to a glass below this temperature

Table 10.38: Typical Properties of the Vinyl Ethers *(49)*

Vinyl Ether	Melting Point °C	Boiling Point Temp. °C	at mm Hg Pressure	Flammability °F
Methyl	-122°	5-6°	760	-69° (a)
Isopropyl	-140°	55-56°	760	-
Isobutyl	-112°	25°	77	20° (a)
		83°	760	
2-Ethylhexyl	-100°	62-64°	18	-
		178°	760	
Isooctyl	-80°	80°	25	140° (a)
		175-6°	760	
Decyl	-41°	60-98°	5	185° (a)
Cetyl	16°	142°	1	325° (a)
		173°	5	
Octadecyl	28°	124-168°	2	350° (a)
		147-187°	5	
Dimethylaminoethyl	-	42-44°	30	110° (b)

(a) Flash point - open cup method.   (b) Fire point - ASTM D-92.

(continued)

Table 10.38: (continued)

Vinyl Ether	Refractive Index	Specific Gravity	Pounds per Gallon @ 25°C	Typical Vinyl Ether Content
Methyl	$1.3947 \frac{25}{D}$	$0.7694 \frac{5.7}{4}$	6.17	99%
Isopropyl	$1.3849 \frac{20}{D}$	$0.753 \frac{20}{4}$	6.28	98%
Isobutyl	$1.3965 \frac{20}{D}$	$0.768 \frac{20}{4}$	6.40	98%
2-Ethylhexyl	$1.4273 \frac{20}{D}$	$0.810 \frac{20}{20}$	6.74	95%
Isooctyl	$1.4256 \frac{25}{D}$	$0.802 \frac{20}{4}$	6.66	98%
Decyl	$1.4278 \frac{25}{D}$	$0.812 \frac{20}{4}$	6.75	98%
Cetyl	$1.4444 \frac{25}{D}$	$0.822 \frac{27}{15}$	6.85	97%
Octadecyl	$1.4440 \frac{30}{D}$	0.80 cast solid 0.821$\frac{30}{4}$ liquid 6.84		95%
Dimethylaminoethyl	$1.4225 \frac{25}{D}$	$0.830 \frac{20}{20}$	6.85	99%

Table 10.39: Phenyl Methyl Ether *(2)*

Anisole

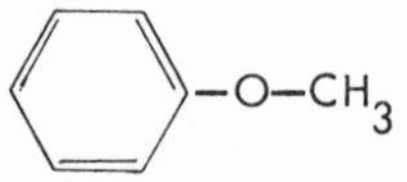

Anisole is a high-boiling, mobile, straw-colored liquid with excellent thermal stability. It is immiscible in water and glycols but completely miscible with most common solvents. It is useful as a solvent for many organic compounds and it has unusual solvency for asphalts and pitches.

Boiling point at 760 mm. Hg, °C.	153.8
Flash point (Cleveland open cup), °F.	125
Heat of combustion, kcal./g. mol	905.2
Heat of vaporization at boiling point, cal./g. mol	8.8
Refractive index (n_D at 20°C.)	1.5165
Molecular weight	108.13
Specific gravity, 18°/4°C.	0.996
Specific heat:	
24°C.	0.422
31.6°C.	0.462
Vapor pressure, mm. Hg:	
40°C.	8.4
60°C.	25
80°C.	63
100°C.	140
120°C.	275

Table 10.40: Dibenzyl Ether *(2)*

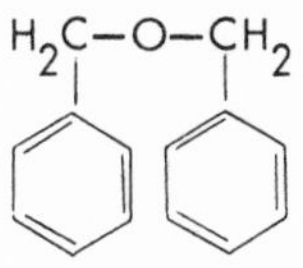

$$H_2C-O-CH_2$$

Dibenzyl ether is a clear, almost colorless liquid. It is miscible with alcohols and ethers, but insoluble in water. Dibenzyl ether is used as special solvent and delustering agent for textiles.

Boiling point, °C.:		Flash point (open cup) °F.	275
760 mm.	298	Melting point, °C.	5 (approx.)
15 mm.	165–168	Molecular weight	198.3
Distilling range, °C.:		Specific gravity, 25°/25°C.	1.041–1.045
5%	220 (min.)	Solidifying temperature, °F.	below 45
50%	300		
dry point	308		

Table 10.41: Diphenyl Oxide *(43)*

Diphenyl Ether

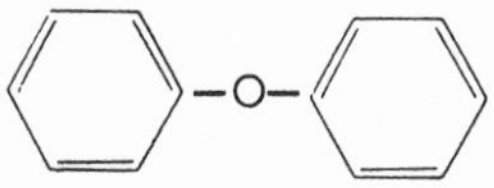

Diphenyl oxide is a practically colorless crystalline solid with a strong geranium-like odor. It is almost completely insoluble in water, but dissolves in most of the common organic solvents. Its high thermal stability at temperatures as high as 350° to 400°C. together with its noncorrosiveness and general chemical inertness make it eminently suitable as a component of high-boiling heat transfer media.

TYPICAL PROPERTIES

Molecular weight	170
Diphenyl oxide	>99% by weight
Crystallising point	26·6°C
Distillation range at 760 mm Hg :	
initial boiling point	253·0°C
5% by volume	254·0°C
95% by volume	256·0°C
final boiling point	260·0°C
Flash point (Pensky Martens closed cup)	240°F (115°C)
Ash	0·001% by weight
Acidity (as hydrochloric acid)	0·001% by weight
Water content	0·02% by weight
Phenol content	0·02% by weight

PHYSICAL PROPERTIES

The following are values for pure diphenyl oxide :

Density at 30/4°C	1·066 g/ml
Latent heat of fusion	22·9 cal/g
Specific heat at 30°C	0·40
Vapour pressure at 25°C	0·02 mm Hg
Viscosity at 25°C	3·86 cP
Refractive index n_D^{25}	1·57870

Table 10.42: Miscellaneous Alkyl Aryl Ethers *(2)*

These ethers are generally high-boiling, water insoluble liquids of pleasant odor, miscible with a variety of organic solvents and commercial oils, fats, waxes and resins.

Physical Properties

	Formula	Boiling Range °C	Flash Point °F	Sp. Gravity 20/20	Molecular Wt.
Methyl Phenyl Ether (Anisole)	$CH_3OC_6H_5$	150–160	120	0.993	
n-Butyl Phenyl Ether	$C_4H_9OC_6H_5$	202–212	180	0.929	
Amyl Phenyl Ether	$C_5H_{11}OC_6H_5$	214–229	185	0.924	164.1
p-tert-Amylphenyl Methyl Ether	$C_5H_{11}C_6H_4OCH_3$	239–243	210	0.942	
p-tert-Amylphenyl-*n*-Amyl Ether	$C_5H_{11}C_6H_4OC_5H_{11}$	285–295	260	0.905	234.2
Amyl Benzyl Ether	$C_5H_{11}OCH_2C_6H_5$	224–239	175	0.912	
Amyl Tolyl Ether	$C_5H_{11}OC_6H_4CH_3$	240–264	195	0.916	
Amyl beta Naphthyl Ether	$C_5H_{11}OC_{10}H_7$	320–350	310	1.01	
Amyl Xylyl Ether	$C_5H_{11}OC_6H_3(CH_3)_2$	250–263	205	0.907	

Table 10.43: Furan *(11)*

Furan is a cyclic dienic ether stabilized by benzene-like resonance. Because of its conjugated unsaturation and heterocyclic atom, furan will undergo many types of reactions. It is, therefore, of interest as a chemical intermediate for pharmaceuticals, insecticides and fine chemicals. The heterocyclic oxygen atom in a ring with conjugated unsaturation gives furan a combination of ether, aromatic and olefinic characteristics. This polyfunctionality permits it to undergo a variety of reactions. Compared to benzene, the furan ring has greater reactivity, and is more susceptible to cleavage, thus resembling the vinyl ethers. Like the vinyl ethers, the furan ring is cleaved by aqueous acids. This reaction is accompanied by resinification.

PHYSICAL PROPERTIES

Physical State	Liquid
Color	Colorless
Odor	Characteristic ethereal
Specific Gravity at 20°/4°C.	0.937
Freezing Point	-85.61°C. (-122.10°F.)
Vapor Density	0.170 lb./cu. ft.
Boiling Point (760 mm.)	31.3°C. (88.45°F.)
Flash Point (Tag. closed cup)	-32°F.
Refractive Index n20/D	1.4214
Molecular Weight	68.07
Flammability or Explosive Limits	2.3-14.3 vol. % in air
Heat of Vaporization at 31.2°C.	95.5 cal./gram
Heat of Combustion at constant vol.	500.1 kg.-cal./gram-mole
Critical Temperature	214°C.
Heat of Formation at 25°C.	-14.9 kcal./mole
Solubility in:	

Water (wt. % at 25°C.) 1
Most organic solvents ∞

Table 10.44: 2-Methylfuran *(2)*

Sylvan

2-Methylfuran is a cyclic diene possessing ether-like properties. It is highly reactive with many inorganic and organic compounds yielding a variety of new derivatives which await exploration for the development of commercial applications.

(continued)

Table 10.44: (continued)

Appearance	Colorless, mobile liquid
Odor	Ether-like
Molecular weight	82.098
Boiling point at 760 mm	
	62–64°C (144–47°F)
Freezing point	—88°C (−126.4°F)
Specific gravity, 20°C./4°C.	0.915
Index of refraction, N20/D	1.434
Flash point	−30°C (−22°F)
Vapor pressure at 15°C. (59°F)	110.5 mm
20°C. (68°F)	139 mm
25°C. (77°F)	174 mm
30°C. (96°F)	216 mm
Solubility in water at 25°C	Less than 0.3 gm/100 gm.

Table 10.45: Tetrahydrofuran *(46)*

THF

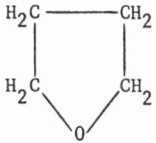

The most valuable characteristic of tetrahydrofuran is its high solvent power for a wide variety of synthetic and natural materials. This characteristic is particularly evident in the case of polyvinyl chlorides, vinyl chloride copolymers, and vinylidene chloride copolymers which dissolve readily in THF, even at room temperature.

Physical Properties of Tetrahydrofuran

Physical state	Liquid	
Color	Water-white	
Odor	Characteristic ethereal	
Specific gravity, 20°C	0.888	
Weight per gal. (20°C), lbs.	7.41	
Freezing point, °C	—108.5	
°F	—163.0	
Vapor density (air = 1)	2.5	
Boiling point (760 mm Hg) °C	66	
°F	151	
Flash point	°C	°F
Cleveland open cup	−20	−4
closed cup	−14.5	6
Refractive index n20/D	1.4073	
Molecular weight	72.1	
Critical temperature, T_c, °C	268	
Critical pressure, P_c, psia	753	
Critical density, d_c, g/cc	0.322	
Explosive limits		
(% by volume in air, 25°C)	1.8 to 11.8	
Ignition temperature, °C	321	
°F	610	
Corrosive	Noncorrosive at normal atmospheric temperatures	
Solubility in water, alcohol, ether	Miscible in all proportions	
Specific heat, 68°F	0.469 Btu/lb./°F	
122°F	0.496 Btu/lb./°F	
Heat of vaporization, 151°F	176 Btu/lb.	
Heat of combustion	598 Kg-cal./mole	
Dipole moment	1.70	
Dielectric constant, 25°C	7.4	
Electrical conductivity	4.5×10^{-5} to 9.3×10^{-4} mho/cm	
Rate of evaporation[a]	800 (Between acetone and methyl ethyl ketone)	
Solubility parameter[b]	9.1	
Hydrogen-bonding capacity	5.3	
Threshold limit value, ppm[c]	200	

[a] Relative to butyl acetate (—100)

[b] Calculated using H. Burrell's empirical corrections (Offic. Dig. Feder. Soc. Paint Technol. 27, 726-758 [1955]).

[c] Values in parts by volume per million parts of air, recommended by the American Conference of Governmental Industrial Hygienists (1970).

(continued)

Table 10.45: (continued)

VAPOR PRESSURE OF THF

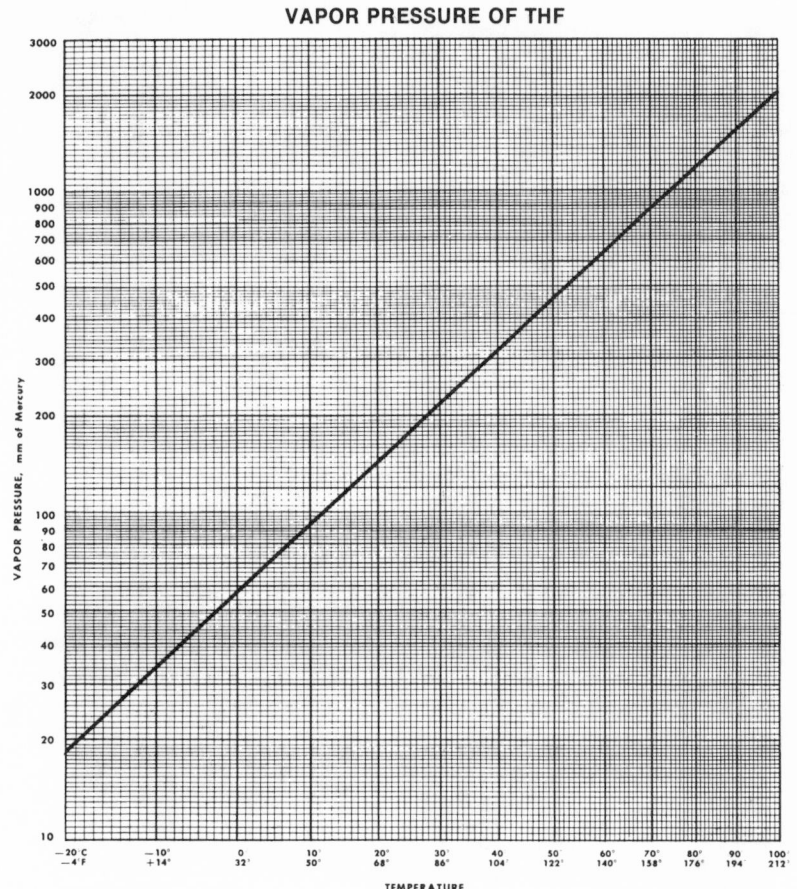

THF-SOLUBLE PLASTICS, RESINS, AND ELASTOMERS

ACRYLIC RESINS

Methyl methacrylate polymers
Ethyl, butyl, and other methacrylate polymers
Acrylic polymers and copolymers

ALKYD AND AMINO RESINS

Alkyd resins
Urea formaldehyde resins (uncured)
Phenol formaldehyde resins (uncured)

CELLULOSICS

Cellulose acetate
Cellulose acetate butyrate
Cellulose acetate stearate
Ethyl cellulose
Nitrocellulose

MISCELLANEOUS RESINS

Acrylonitrile-butadiene-styrene copolymers
Styrene-acrylonitrile copolymers
Chlorinated polyethylene
Polycarbonates
Polysulfones
Epoxy (uncured)
Silicones (uncured)
Polyesters (low-mol.-wt.)
Polyamides (low-mol.-wt)
Polystyrene
Styrene-butadiene copolymers (some)

ELASTOMERS

Butadiene-acrylonitrile copolymers (some)
Chlorinated rubbers
Chlorosulfonated polyethylenes
Polysulfides
Polyurethanes (uncured)
Rubber (natural, unvulcanized)
Chloroprene elastomers

VINYL RESINS

Polyvinyl acetate
Polyvinyl butyrate
Polyvinyl butyrals
Polyvinyl chloride
Vinyl chloride copolymers
Vinylidene chloride copolymers
Vinyl acetate/ethylene (some)

NATURAL RESINS

Congo ester
Coumarone-indene
Raw dammar
Ester gum
Manila copal
Pentaerythritol ester gum
Rosin
Shellac (many)

COMPARATIVE SOLUBILITIES OF SYNTHETIC RESINS IN THF AND MEK

Resin	Solubility, %* THF	Solubility, %* MEK
"Bakelite" QYNV[i] polyvinyl chloride	13	<3
PVC-71[a] dispersion resin	14	<5
"Marvinol" 10[h] polyvinyl chloride	14	<5
"Vygen" 120[e] polyvinyl chloride	15	<5
"Exon" 654[d] polyvinyl chloride	15	—
Hi-Temp "Geon" vinyl (CPVC)[f]	16	<5
"Geon" 121[f] polyvinyl chloride	16	<5
"Opalon" 410[g] PVC plastisol	16	—
"Estane" 5701[f] F-1 polyurethane	17	<3
"Geon" 101[f] polyvinyl chloride	17	<6
"PVC Pearls" 2250[c] resin	18	<3
"Geon" 103EP[f] polyvinyl chloride	18	<8
"Vygen" 110[e] polyvinyl chloride	18	—
"PVC Pearls" 2200[c] resin	20	—
"Bakelite" VYNS[i] vinyl copolymer	27	17
"Bakelite" VMCH[i] vinyl copolymer	35	—
"Bakelite" VYHH[i] vinyl copolymer	35	22
"Bakelite" VAGH[i] vinyl copolymer	40	—
"Geon" 106[f] polyvinyl chloride	40	—
Saran F-242[b] vinylidene chloride resin	44	<3

*wt. % resin in solution at 2500 cp (25°C)
[a]Diamond Shamrock Chemical Co.
[b]Dow Chemical Co.
[c]Escambia Chemical Corp.
[d]Firestone Plastics Co.
[e]General Tire & Rubber Co., Chemical/Plastics
[f]B. F. Goodrich Chemical Co.
[g]Monsanto Co., Plastic Products & Resins Div.
[h]Uniroyal Chemical, Div. Uniroyal, Inc.
[i]Union Carbide Corp., Chemicals & Plastics

COMPARATIVE DIFFUSION RATES OF THF AND MEK

Film	Average Diffusion Rate* THF	Average Diffusion Rate* MEK
Unplasticized PVC†	0.223	0.113
Dioctyl phthalate-plasticized PVC†	0.350	0.191
Saran wrap	0.208	0.053

*Measured in g. solvent/hr./unit area at 25° ± 2°C.
†"Geon" 121 dispersion resin (B. F. Goodrich Chemical Co.)

(continued)

Table 10.45: (continued)

DENSITY OF THF

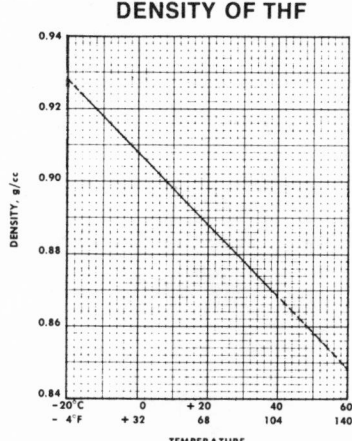

Table 10.46: 2,3-Dihydropyran *(2)*

$$
\begin{array}{c}
CH_2 \\
CH_2 \quad\quad CH \\
CH_2 \quad\quad CH \\
O
\end{array}
$$

Physical Properties

	Colorless liquid with characteristic odor
Molecular weight	84.11
Boiling point	85–86°C
Sp., g. $\frac{20°C}{4°C}$	0.923

Table 10.47: Tetrahydropyran *(2)*

Pentamethylene Oxide

$$
\begin{array}{c}
{}^{(5)}H_2C \\
{}^{(6)}H_2C \quad\quad CH_2{}^{(4)} \\
O \quad\quad CH_2{}^{(3)} \\
{}^{(1)} \quad {}^{(2)}CH_2
\end{array}
$$

Tetrahydropyran reacts with chlorine to form mono-, di-, tri- and tetrachlorotetrahydropyrans. Reaction with acid chlorides yields omega-haloamyl esters. Conversion to dihalides such as 1,5-dibromopentane and 1,5-dichloropentane can be effected. Ammonia and aliphatic and aromatic amines yield piperidine and substituted piperidines. It is used as a solvent for resins, plastics and rubbers. Lacquers can be made by dissolving certain organic film-forming substances in tetrahydropyran. Solutions of high solids content at a working viscosity can be obtained. A solution of nitrocellulose in tetrahydropyran gives clear, nonblushing films. Tetrahydropyran is miscible with water, drying oils and most common organic solvents. The ether-like structure and ability of tetrahydropyran to dissolve a wide range of nonresinous materials suggest its use as a reaction medium for chemical processes such as Grignard reactions.

Appearance	Colorless, mobile, liquid
Odor	Ether-like
Molecular weight	86.13
Boiling point	88°C at 760 mm
Specific gravity, 20°C/4°C	0.8814
Index of refraction, N20/D	1.420
Flash point	−4°F

Solubility—Miscible with water; less soluble in hot than cold water. Miscible with alcohol, ether, and most common organic solvents.

Table 10.48: Tetrahydropyran-2-Methanol *(19)*

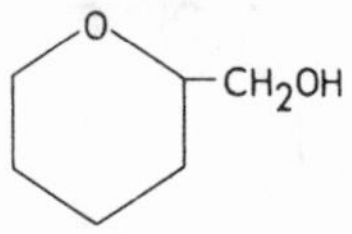

PHYSICAL PROPERTIES	Determined on specially purified sample

Molecular Weight	116.16
Apparent Specific Gravity at 20/20°C.	1.0272
Δ Sp.Gr./Δ t., 10 to 40°C.	0.00083 per °C.
True Density at 20°C.	1.0254 g. per ml.
Boiling Point	
at 760 mm. Hg	186.8°C.
at 300 mm. Hg	154°C.
at 10 mm. Hg	72°C.
Vapor Pressure at 20°C.	< 0.1 mm. Hg
Δ b.p./Δ p., 750 to 770 mm. Hg	0.051°C. per mm. Hg.
Absolute Viscosity	
at 0°C.	29.3 cps.
at 20°C.	11.0 cps.
at 40°C.	5.4 cps.
Surface Tension at 25°C.	34.1 dynes per cm.
Freezing Point	–70°C. (a)
Heat of Vaporization at 1 atm.	164 Btu per lb.
at 300 mm. Hg	173 Btu per lb.
Refractive Index, n_D 20°C.	1.4581
Δ n_D/Δ t., 20 to 40°C.	0.00043 per °C.
Solubility	
In Water at 20°C.	Complete
Water In at 20°C.	Complete
Solubility in Organic Solvents at 25°C.	
acetone, benzene, ethyl ether, heptane,	
methanol, carbon tetrachloride	Complete
Flash Point, Cleveland open cup (ASTM Method D92)	200°F. (b)

(a) Sets to a glass below this temperature
(b) Commercial material

Table 10.49: Terpinyl Methyl Ether *(2)*

Terposol No. 3

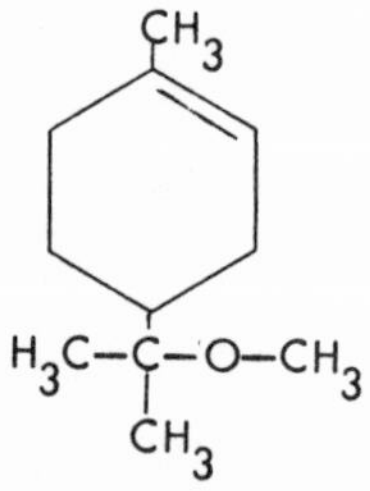

This terpine ether known as terpinyl methyl ether is a light, colored liquid with a pleasant odor, which contains some impurities. It is a strong solvent for resins and is used in alkyd enamels to the extent of 2 per cent to which it imparts flow.

(continued)

Table 10.49: (continued)

Aniline point	Below −20°C
Color (Lovibond 500 Amber Series Glasses)	1.0
Distillation range (ASTM)	
5%	215.0°C
50%	216.5°C
90%	217.5°C
95%	218.°C
Flash point (Cleveland open cup)	178°F
Freezing point	Below −10°C
Kauri-Butanol solvency value	Approx. 500
Moisture	0.10%
Refractive index at 20°C	1.4712
Specific gravity at 15.5/15.5°C	0.9192
Viscosity at 25°C (Ubbelohde)	31.8 cp

Glycol Ethers

Table 11.1: "Amsco-Solvs" (13)

Amsco-Solv	Specific Gravity @ 20°/20°C	lb/gal @ 20°/20°C	Distillation Range 760 mm Hg °F (°C)	Relative Evaporation Rate n-BuAc = 1	Vapor Pressure mm Hg @ 20°C	Flash Point TAG CC °F	Solubility % by wt @ 20°C in Water	of Water
EM	0.966	8.04	253 (122.8)–257 (125.0)	0.5	6.1	103	∞	∞
EE	0.931	7.74	273 (134.0)–277 (136.0)	0.3	3.8	108	∞	∞
EP	0.913	7.59	301 (149.5)–308 (153.5)	0.2	1.3*	120	∞	∞
EB	0.902	7.51	336 (168.8)–343 (172.5)	0.1	0.6	140	∞	∞
DM	1.021	8.51	379 (192.8)–389 (198.3)	<0.01	<0.1	188	∞	∞
DE	0.990	8.25	388 (197.8)–399 (203.9)	<0.01	0.05	195	∞	∞
DB	0.954	7.94	446 (230.0)–455 (235.0)	<0.01	<0.1	214	∞	∞

*@20%.

Note: EM is ethylene glycol methyl ether   DM is diethylene glycol methyl ether
EE is ethylene glycol ethyl ether   DE is diethylene glycol ethyl ether
EP is ethylene glycol propyl ether   DB is diethylene glycol butyl ether
EB is ethylene glycol butyl ether

Table 11.2: "Ansuls" (15)

Physical Properties

	E-121	E-222	E-141	E-142
CAS Registry Number	110-71-4	629-14-1	111-96-6	1002-67-1
Empirical Formula	$C_4H_{10}O_2$	$C_6H_{14}O_2$	$C_6H_{14}O_3$	$C_7H_{16}O_3$
Chemical Formula	$CH_3OCH_2\text{-}$ CH_2OCH_3	$C_2H_5OCH_2\text{-}$ $CH_2OC_2H_5$	$CH_3(OCH_2\text{-}$ $CH_2)_2OCH_3$	$C_2H_5(OCH_2\text{-}$ $CH_2)_2OCH_3$
Molecular Weight	90.12	118.18	134.17	148.20
Specific Gravity, 20/20°C	0.8683	0.8417	0.9451	0.9245
Refractive Index, n 20°/D	1.3792	1.3922	1.4078	1.4094
Boiling Point, °C. 760 mm Hg	85	121	162	176
Freezing Point, °C	−69	−74	−64	−72
Vapor Pressure, mm Hg, 20°C	48	10	3	<1
Viscosity, Absolute, cp, 20°C	1.1	0.7	2.0	1.2
Surface Tension, dynes/cm, 20°C	22.9		27.0	24.7[a]
Specific Heat, Cal/gm/°C	0.438	0.519	0.403	0.5118
Heat of Fusion, Cal/gm	33.3		24.2	
Heat of Vaporization, Kcal/mole	6.7	8.6	9.9	10.4
Heat of Combustion, Kcal/mole	602	922	902	1042
Heat of Formation, Kcal/mole	118	123	143	166
Dielectric Constant	5.50		5.79	
Volume Resistivity, megohm-cm	21		15	
Flash point, °C, open cup	1	35	70	82
Autoignition Temperature. °F	379	208	370	

(continued)

Table 11.2: (continued)

	E-121	E-222	E-141	E-142
Solubility in				
a) Water	Miscible	~ 20% by wt.	Miscible	Miscible
b) Ethanol	Miscible	Miscible	Miscible	Miscible
c) Acetone	Miscible	Miscible	Miscible	Miscible
d) Benzene	Miscible	Miscible	Miscible	Miscible
e) Ether	Miscible	Miscible	Miscible	Miscible
f) Octane	Miscible	Miscible	Miscible	Miscible
g) water in	Miscible	~ 4% by wt.	Miscible	Miscible
Weight per gallon. lb., 20°C	7.24	7.02	7.88	7.70
Color	Colorless	Colorless	Colorless	Colorless
Odor	Ethereal	Mild	Faint, Pleasant	Mild
pH	Neutral	Neutral	Neutral	Neutral
Volatility Hoffman Dish Method (n-Butyl Acetate Standard as 100)	499	110[c]	36	28[c]

	E-242	E-444	E-161	E-262	E-181	E-282
CAS Registry Number	112-36-7	112-73-2	112-49-2	4499-99-4	143-24-8	4353-28-0
Empirical Formula	$C_8H_{18}O_3$	$C_{12}H_{26}O_3$	$C_8H_{18}O_4$	$C_{10}H_{22}O_4$	$C_{10}H_{22}O_5$	$C_{12}H_{26}O_5$
Chemical Formula	$C_2H_5(OCH_2\cdot CH_2)_2OC_2H_5$	$C_4H_9(OCH_2\cdot CH_2)_2OC_4H_9$	$CH_3(OCH_2\cdot CH_2)_3OCH_3$	$C_2H_5(OCH_2\cdot CH_2)_3OC_2H_5$	$CH_3(OCH_2\cdot CH_2)_4OCH_3$	$C_2H_5(OCH_2\cdot CH_2)_4OC_2H_5$
Molecular Weight	162.23	218.34	178.22	206.28	222.28	250.340
Specific Gravity, 20/20°C	0.9082	0.8853	0.9862	0.9566	1.0130	0.9695
Refractive Index, n 20°/D	1.4115	1.4233	1.4233	1.4269	1.4322	1.4295
Boiling Point, °C, 760 mm Hg	189	256	216	246	275	153[d]
Freezing Point, °C	− 44	− 60	− 45	− 24	− 27	
Vapor Pressure, mm Hg, 20°C	<1	<0.01	0.9	<0.01	<0.01	<0.01
Viscosity, Absolute, cp. 20°C	1.4	2.4	3.8	2.7	4.1	
Surface Tension, dynes/cm, 20°C	27.2	27.0[a]	29.4	24.7	31.1	
Specific Heat, Cal/gm/°C		0.495	0.424	0.504	0.427	
Heat of Fusion, Cal/gm			20.2		27.3	
Heat of Vaporization, Kcal/mole	10.5	12.0	14.3	12.1	18.7	
Heat of Combustion, Kcal/mole	1199	1823	1191	1475	1480	1764
Heat of Formation, Kcal/mole	152	175	179	188	217	226
Dielectric Constant					9.16	
Volume Resistivity, megohm-cm			33		27	
Flash point, °C, open cup	91	118	111		141	
Autoignition Temperature, °F	420	397	375		510	
Solubility in						
a) Water	Miscible[b]	0.3% by wt.[a]	Miscible	Miscible	Miscible	Miscible
b) Ethanol	Miscible	Miscible	Miscible	Miscible	Miscible	Miscible
c) Acetone	Miscible	Miscible	Miscible	Miscible	Miscible	Miscible
d) Benzene	Miscible	Miscible	Miscible	Miscible	Miscible	Miscible
e) Ether	Miscible	Miscible	Miscible	Miscible	Miscible	Miscible
f) Octane	Miscible	Miscible	Miscible	Miscible	Miscible	Miscible
g) water in	Miscible[b]	1.4% by wt.[a]	Miscible	Miscible	Miscible	Miscible
Weight per gallon, lb., 20°C	7.56	7.40	8.24	8.00	8.43	8.10
Color	Colorless	Colorless	Colorless	Colorless	Colorless	Colorless
Odor	Mild	Very Mild	Mild	Mild	Very Mild	Mild
pH	Neutral	Neutral	Neutral	Neutral	Neutral	Neutral
Volatility Hoffman Dish Method (n-Butyl Acetate Standard as 100)	21[c]	1.0[c]	14	2[c]	0.2	<0.1[c]

(a) at 25°C.
(b) at room temperature; relatively immiscible at higher temperatures.
(c) estimated from values of similar compounds.
(d) at 10 mm Hg pressure.

Table 11.3: "Arcosolvs"[1] (79)

ARCO Chemical Company Nomenclature	Chemical Name	Molecular Weight	Boiling Point, °C (760 mm Hg)	Flash Point[2] °F	Specific Gravity at 25°C	Evaporation Rate BuAc = 100	Vapor Pressure at 25°C mm Hg	lb/gal at 25°C	Viscosity (cs) at 25°C	Surface Tension dynes/cm at 25°C	Freeze Point °F	Specific Heat cal/g/°C at 25°C	Heat of Vaporization cal/g (760 mm Hg)	Thermal Conductivity K x 10⁴ (cal/cm²sec°C/cm) at 60°C	Electro-conductivity (μmhos)	Dilution Ratio Toluene	Dilution Ratio Naphtha	Solubility Parameter[3]	Hydrogen Bonding Parameter[3]	Dipole Moment[4] (debye)	Solubility in Water ml/100 ml	Dow	Union Carbide	Eastman
ARCOSOLV™ PM	Propylene Glycol Methyl Ether	90.1	120.1	89	.919	66	10.9	7.65	1.8	27.7	-139	.57	102.0	3.78	.45	5.2	.9	10.4	14.6	1.67	∞	DOWANOL PM	—	—
ARCOSOLV™ DPM	Dipropylene Glycol Methyl Ether	148.2	188.3	167	.951	2	.40	7.91	3.6	28.8	-112	.54	73.1	3.63	.16	4.4	.8	9.6	13.3	1.97	∞	DOWANOL DPM	—	—
ARCOSOLV™ TPM	Tripropylene Glycol Methyl Ether	206.3	242.4	237	.965	< 1	.02	8.03	5.8	30.0	-110	.51	60.2	3.54	.03	3.0	.7	9.1	10.9	2.18	∞	DOWANOL TPM	—	—
—	Ethylene Glycol Methyl Ether	76.1	124.6	105	.963	47	9.6	8.01	1.6	30.8	-121	.53	123.9	4.19	1.84	5.3	Insol.	10.8	15.3	1.34	∞	DOWANOL EM	METHYL CELLOSOLVE	EKTASOLVE EM
—	Ethylene Glycol Ethyl Ether	90.1	135.5	109	.928	32	5.5	7.73	2.1	28.2	-148	.56	107.5	3.78	2.50	6.6	1.1	9.9	15.5	1.69	∞	DOWANOL EE	CELLOSOLVE	EKTASOLVE EE
—	Ethylene Glycol Butyl Ether	118.2	171.1	150	.900	6	.90	7.49	3.2	27.4	-125	.56	88.4	3.55	.17	5.2	2.2	8.9	15.5	1.80	∞	DOWANOL EB	BUTYL CELLOSOLVE	EKTASOLVE EB
—	Diethylene Glycol Methyl Ether	120.1	194.0	192	1.021	< 1	.25	8.47	3.4	34.8	-92	.54	92.7	4.10	.78	4.6	Insol.	9.6	13.4	1.91	∞	DOWANOL DM	METHYL CARBITOL	EKTASOLVE DM
—	Diethylene Glycol Ethyl Ether	134.2	202.0	196	.988	< 1	.27	8.24	3.8	31.8	-105	.55	84.5	4.09	.52	6.4	.6	9.6	13.5	1.96	∞	DOWANOL DE	CARBITOL	EKTASOLVE DE
—	Diethylene Glycol Butyl Ether	162.2	230.0	222	.952	< 1	.06	7.94	5.2	30.0	-90	.54	74.3	3.91	.19	6.5	1.9	8.9	13.5	2.05	∞	DOWANOL DB	BUTYL CARBITOL	EKTASOLVE DB

[1]Values should not be regarded as specifications, maxima or minima.
[2]Flash points below 200°F by Tag Closed Cup.
Flash points above 200°F by Pensky-Martens Closed Cup.
[3]For a discussion of solubility parameters, see H. Burrel, *Interchemical Review*,
Vol. 14, No. 1.
[4]For a discussion of hydrogen bonding parameters, see A. Orr, *Journal of Paint Technology*,
Vol. 47, No. 607, p. 45.

Table 11.4: "Carbitols" and "Cellosolves" (19)

	Molecular Weight	Apparent Specific Gravity at 20/20° C	ΔSp. Gr./Δt., 10-40° C, per °C	Boiling Point, °C, mm Hg 760	50	10	Δb.p./Δp., 750 to 770 mm Hg, °C per mm Hg	Vapor Pressure at 20° C, mm Hg	Relative Evaporation Rate (f) (Bu Ac = 100)	Freezing Point, °C	Absolute Viscosity, cP at 20°C	Solubility at 20° C, % by wt. In Water	Water In	Solubility % by wt. In Heptane (c)	Refractive Index, n_D^{20}	Heat of Vaporization, Btu/lb at 1 Atm.	300 mm Hg	Flash Point, Closed Cup, °F
Methyl CELLOSOLVE	76.10	0.9663	0.00092	124.5	56	27	0.043	6	44	−85.1	1.7	Complete	Complete	11.2	1.4021	223	235	103
CELLOSOLVE Solvent	90.12	0.9311	0.00090	135.6	64	35	0.0422	4	29	−90(a)	2.5	Complete	Complete	0.7	1.4077	191	202	108
Butyl CELLOSOLVE	118.18	0.9022	0.00083	171.2	94	62	0.048	0.6	5	−70.4	6.4	Complete	Complete	Complete	1.4198	158	167	140
Methyl CARBITOL	120.16	1.023	0.00088	194.0	115	82	0.049	0.1	<1	−85(c)	3.9	Complete	Complete	1.5	1.4263	170	180	188
CARBITOL Solvent, Low Gravity	134.18	0.9910	0.00089	202.7	120	87	0.051	0.08	<1	−78(c)	4.5	Complete	Complete	0.7	1.4273	151	159	182
CARBITOL Solvent PM-600(g)		1.027	0.00089	195.0				0.05	<1	−70(a)	6.9(b)	Complete	Complete	0.7	1.4297			204
Butyl CARBITOL	162.23	0.9563	0.00084	230.6	144	109	0.053	0.01	<1	−68.1	6.5	Complete	Complete	Complete	1.4316	132	141	214
Methoxytriglycol	164.20	1.053	0.00088	249.0	162	126	0.053	<0.01	<0.1	−38.2	7.27	Complete	Complete	1.5	1.4381	141	150	238
Ethoxytriglycol	178.23	1.0250	0.00088	255.9	167	130	0.055	<0.01	<0.1	−18.7	7.8	Complete	Complete	2	1.4376	129	137	255
Butoxytriglycol	206.28	1.0021	0.00082	(dec)	188	148	—	<0.01	<0.1	−47.6	10.9	Complete	Complete	Complete	1.4394	176(e)		>250

FOOTNOTES:

(a) Sets to a glass below this temperature.
(b) 25° C.
(c) At 25° C, glycol ethers listed in this table are completely soluble in acetone, benzene, ethyl ether, methanol, and carbon tetrachloride.
(d) Estimate.
(e) At 190° C and 50 mm Hg.
(f) Evaporation rate of glycol ether at 20° C referenced to n-butyl acetate at 25° C.
(g) A mixture of CARBITOL Solvent, Low Gravity and ethylene glycol.

Table 11.5: "Dowanols" (23)

DOWANOL	CHEMICAL NAME	STRUCTURAL FORMULA	Mol. Wt.	Boiling Pt., °C	Flash Point (TCC)°F	Evap. Rate Bu Ac. = 1.0	Sp. Grav. 25/25°C	Lbs/Gal 25°C	Visc. Centistokes 25°C	Vapor Pressure at 25°C (mm Hg)	Surface Tension (dynes/cm)	DIL. RATIO Toluene	DIL. RATIO Naptha	Sol. Para.	Hydrogen Bonding	Dipole Moment (Debye)	Solubility in Water ml/100 ml
PM	Propylene Glycol Methyl Ether	$CH_3OCH_2CHOHCH_3$	90.1	120.1	94	0.71	0.919	7.65	1.86	12.5	27.7	5.2	0.9	10.2	14.6	1.67	∞
DPM	Dipropylene Glycol Methyl Ether	$CH_3O[CH_2CH(CH_3)O]_2H$	148.2	188.3	175	0.03	0.951	7.91	3.57	0.55	28.8	4.2	0.8	9.4	12.0	1.97	∞
DPM-SG	Dipropylene Glyco Methyl Ether Special Grade	$CH_3O[CH_2CH(CH_3)O]_2H$	148.2	188.3	175	0.03	0.951	7.91	3.57	0.55	28.8	4.2	0.8	9.4	12.1	1.97	∞
TPM	Tripropylene Glycol Methyl Ether	$CH_3O[CH_2CH(CH_3)O]_3H$	206.3	242.4	240	<0.01	0.965	8.03	5.80	0.02	30.0	3.1	0.7	8.7	10.9	2.18	∞
PiBT	Propylene Glycol Isobutyl Ether and higher homologs	$(CH_3)_2C_2H_3OCH_2CHOHCH_3$		172	138	0.09	0.883	7.33	4.01	1.3	25.1	2.3	1.5	8.6	14.8	1.97	2.9
EM	Ethylene Glycol Methyl Ether	$CH_3OC_2H_4OH$	76.1	124.6	105	0.56	0.963	8.01	1.60	9.5	30.8	5.3	Immis.	11.4	15.3	1.34	∞
EE	Ethylene Glycol Ethyl Ether	$C_2H_5OC_2H_4OH$	90.1	135.5	109	0.35	0.928	7.73	2.00	5.5	28.2	6.6	1.1	10.6	15.5	1.69	∞
EB	Ethylene Glycol Butyl Ether	$C_4H_9OC_2H_4OH$	118.2	171.1	150	0.06	0.900	7.49	3.15	0.88	27.4	5.2	2.2	9.8	15.5	1.80	∞
DM	Diethylene Glycol Methyl Ether	$CH_3OC_2H_4OC_2H_4OH$	120.1	194.1	192	0.02	1.021	8.47	3.42	0.25	34.8	4.6	Immis.	10.7	14.7	1.91	∞
DE	Diethylene Glycol Ethyl Ether	$C_2H_5OC_2H_4OC_2H_4OH$	134.2	202.0	196	0.01	0.988	8.24	3.85	0.26	31.8	6.4	0.6	10.2	14.8	1.96	∞
DE-SG	Diethylene Glycol Ethyl Ether Special Grade			190-205	198	0.01	1.025	8.52	6.9		33.5			11.4	16.6	2.05	∞
DB	Diethylene Glycol Butyl Ether	$C_4H_9OC_2H_4OC_2H_4OH$	162.2	230.0	222	<0.01	0.952	7.94	5.17	0.06	30.0	6.5	1.9	9.5	14.8	2.05	∞
EPh	Ethylene Glycol Phenyl Ether	$C_6H_5OC_2H_4OH$	138.0	245.0	260	<0.01	1.104	9.18	20.5	0.03	42.0	Insol.	Insol.	11.4	16.6	1.67	2.3
PPh	Propylene Glycol Phenyl Ether	$C_6H_5OC_3H_6OH$	152.2	242.7	260	<0.01	1.063	8.84	23.2		38.1	Insol.	Insol.	10.5	15.6	1.70	1.1
CX	Propylene Based Glycol Ether Blend			179	147	0.08	0.917	7.62	3.8	0.6	26.4			9.0	14.1	1.97	12.0
DALPAD A	Aromatic Based Glycol Ether		138.0	245.0	260	<0.01	1.104	9.18	20.5	0.03	42.0	Insol.	Insol.	11.4	16.6	1.67	2.3

Table 11.6: "Ektasolves" (10)

GLYCOL ETHERS	Lb/U.S. Gal at 20°C	Color Pt-Co, Max	Sp Gr at 20°/20°C	Assay, Min Wt %	Acidity, as Acetic Acid, Max Wt %	Flash Point, TCC °F	Flash Point, TCC (°C)	Boiling Range, °C	Freezing Point, °C
Ektasolve® EM $CH_3OC_2H_4OH$	8.04	10	0.965	—	0.01	102	(39)	123.5-125	—85
Ektasolve EE $C_2H_5OC_2H_4OH$	7.75	10	0.931	—	0.005	110	(43)	134-136	—94
Ektasolve EP $C_3H_7OC_2H_4OH$	7.59	10	0.913	—	0.01	120	(49)	149.5-153.5	—
Ektasolve EB $C_4H_9OC_2H_4OH$	7.51	10	0.902	—	0.01	143	(62)	169-172.5	—75
Ektasolve DM $CH_3(OC_2H_4)_2OH$	8.51	10	1.023	—	0.01	191	(88)	191-198	—84
Ektasolve DE $C_2H_5(OC_2H_4)_2OH$	8.25	10	0.990	—	0.01	195	(91)	198-204	—90
Ektasolve DB $C_4H_9(OC_2H_4)_2OH$	7.94	10	0.955	—	0.01	232	(111) (COC)	227-235	—76

Table 11.7: "Jeffersols" *(48)*

Methyl Ethers
JEFFERSOL EM (ethylene glycol mono-
methyl ether) $CH_3OCH_2CH_2OH$
Mol. Wt. 76.09
JEFFERSOL DM (diethylene glycol mono-
methyl ether) $CH_3OCH_2CH_2OCH_2CH_2OH$
Mol. Wt. 120.15

Ethyl Ethers
JEFFERSOL EE (ethylene glycol monoethyl
ether) $CH_3CH_2OCH_2CH_2OH$
Mol. Wt. 90.12
JEFFERSOL DE (diethylene glycol monoethyl
ether) $CH_3CH_2OCH_2CH_2OCH_2CH_2OH$
Mol. Wt. 134.17

Butyl Ethers
JEFFERSOL EB (ethylene glycol monobutyl
ether) $CH_3CH_2CH_2CH_2OCH_2CH_2OH$
Mol. Wt. 118.17
JEFFERSOL DB (diethylene glycol monobutyl
ether)
$CH_3CH_2CH_2CH_2OCH_2CH_2OCH_2CH_2OH$
Mol. Wt. 162.22

SPECIFICATIONS		JEFFERSOL®					
		EM	DM	EE	DE	EB	DB
Acidity as acetic acid, wt. %	max.	0.01	0.01	0.01	0.01	0.01	0.01
Appearance		Clear, substantially free of suspended matter					
Boiling range, ASTM, °C							
IBP	min.	123.5	188	134	198	169	220
DP	max.	125.5	198	136	205	173	235
Color, Pt-Co	max.	10	10	10	10	10	15
Odor, room temperature		Mild and not objectionable					
Specific gravity, 20/20°C	min.	0.964	1.021	0.929	0.990	0.901	0.953
	max.	0.967	1.025	0.933	0.994	0.904	0.958
Water, wt. %	max.	0.2	0.2	0.1	0.1	0.2	0.2
SELECT PROPERTIES							
Boiling point, 760 mm., °C		124.5	194.2	135.1	201.9	171.2	230.4
Flash point, °F		115[1]	200[2]	118[1]	210[2]	165[1]	225[2]
		106[3]	194[3]	110[3]	194[3]	141[3]	230[3]
Freezing point, °C		−85	—	—	—	—	−68.1
Viscosity, cp., 20°C		1.7	3.9	2.0	3.85	6.4	6.49
Weight, lbs./gal., 20°C.,		8.03	8.51	7.75	8.24	7.5	7.93

[1]TOC   [2]COC   [3]TCC

Table 11.8: "Poly-Solvs" (66)

	EM	EE	EB	DM	DE (High Gravity)	DE (Low Gravity)	DB	TM	TE	TB	MPM	DPM	TPM
Typical Properties													
Molecular weight	76.09	90.12	118.17	120.15	—	134.17	162.22	164.20	178.23	206.28	90.12	148.20	206.28
Specific gravity @ 20°/20°C	0.966	0.931	0.903	1.021	1.0253	0.989	0.955	1.048	1.020	0.988	0.923	0.954	0.969
Boiling point, °C													
@ 760 mm	124	135	171	194	195	202	230	249	256	decomposes	121	187	242
@ 50 mm	55	64	94	115	105	121	145	152	158	188	51	110	162
@ 10 mm	27	35	61	82	80	87	109	126	130	148	23	80	122
Flash point													
TCC, °C	41	45	63	87	—	85	—	—	—	—	36	76	—
TCC, °F	106	113	145	188	—	185	—	—	—	—	97	169	—
COC, °C	—	—	—	—	96	—	116	118	124	143	—	—	121
COC, °F	—	—	—	—	205	—	240	245	255	290	—	—	250
Freezing point, °C	-85	-70	-70	-85	-75	-76	-68	-55	-21	-41	-96	-83	-78
Refractive index, n_D													
@ 20°C	1.4021	1.4076	1.4193	1.4263	1.4297	1.4273	1.4316	1.4381	1.4376	1.4394	1.4036	—	—
@ 25°C	—	—	—	—	—	—	—	—	—	—	—	1.419	1.428
Vapor pressure @ 20°C, mm Hg	6.2	3.8	0.6	0.2	0.05	0.1	0.02	<0.01	<0.01	<0.01	8.4	0.4	0.02
Viscosity @ 20°C, cp	1.7	2.1	3.4	3.9	6.9	4.5	6.5	7.5	7.8	10.9	1.9	3.9	6.1
Solubility parameter	10.8	9.9	8.9	9.6	11.5	9.6	8.9	10.8	10.2	9.4	10.4	9.6	9.1
Weight @ 20°C, lb/gal	8.05	7.76	7.52	8.51	8.53	8.24	7.95	8.74	8.50	8.23	7.68	7.94	8.06
Net contents, 55 gal drum, lb	440	420	410	470	470	450	440	460	450	450	420	440	440
Specifications													
Specific gravity @20°/20°C	0.964-0.967	0.929-0.931	9.900-0.904	1.024-1.031	1.024-1.030	0.989-0.993	0.953-0.956	1.037-1.055	1.020-1.035	0.980-0.995	0.922-0.925	0.953-0.957	0.964-0.967*
Color, APHA, max	10	10	10	15	10	10	10	50	50	50	10	15	15
Acidity, as acetic acid, max, %	0.01	0.005	0.01	0.01	0.01	0.01	0.01	0.01	0.01	0.01	0.01	0.01	0.01
Distillation range @ 760 mm Hg													
IBP, min, °C	123.5	134	167	188	190	198	220	220	225	258	119	180	236
5%, min, °C	—	—	—	—	—	—	224	230	235	270	—	—	—
95%, max, °C	—	—	—	—	200	—	232	—	—	—	—	—	—
DP, max, °C	125.5	136	173	198	205	205	235	—	—	—	125	196	251
Water, max, %	0.05	0.1	0.15	0.1	0.2	0.1	0.15	0.15	0.1	0.15	0.1	0.1	—
Suspended matter	Substantially free												

* @ 25°/25°C.

Table 11.9: Shell Glycol Ethers *(14)*

	Butyl Dioxitol Butyl Oxitol Dioxitol High Gravity Dioxitol Low Gravity		Methyl Dioxitol Methyl Oxitol Oxitol		
	Butyl Dioxitol Glycol Ether[1]	Butyl Oxitol Glycol Ether[2]	Dioxitol- Low Gravity Glycol Ether[3]	Dioxitol- High Gravity Glycol Ether[3]	Oxitol Glycol Ether[4]
Molecular weight	162.22	118.17	134.18	—	90.12
Boiling point					
°C	230.4	171.2	201.9	190–205	135.1
°F	446.7	340.2	395.4	374–401	275.2
Freezing point, °C*	–68.1	–75	–76	—	–100
Flash point, °F ± 5°					
(Tag Closed Cup)	220	138	192	201	110
Vapor pressure, mm Hg 20°C*	0.02	0.6	0.1	0.1	3.8
Refractive index, n 20*/D	1.4316	1.4193	1.4273	1.4286	1.4076
Surface tension					
dynes/cm 20°C*	30.0†††	27.3	35.5	—	27.9
Coefficient of expression					
@ 20°C	0.00085	0.00092	0.00090	0.0009	0.00097
Average lb/gal					
@ 25°C	7.95	7.49	8.21	8.51	7.72
@ 60°F	8.01	7.55	8.28	8.58	7.79
Average specific gravity					
@ 25°/25°C	0.956	0.901	0.988	1.024	0.928
@ 60°/60°F	0.962	0.907	0.994	1.030	0.935
Evaporation characteristics					
Seconds to 90% evap.**	150,390	6,750	27,800	36,300	1,213
Relative rate nBuOAc = 1.0	<0.01	0.07	0.02	0.01	0.38
Viscosities, cp @ 25°C					
Neat compound	5.3	2.9	4.0	7.0	1.9
8 g N.C. solution***	215	107	135	320	59
Blush resistance % R.H.					
@ 80°F***	85	96	<50 §	—	67 §
Dilution ratio†					
Toluene	3.9	3.3	4.8	2.0	4.9
Aliphatic naphtha††	1.9	1.8	Imm.	Imm.	1.1
Solubility of pure compound					
@ 20°C, % by wt					
In water	complete	complete	complete	complete	complete
Water in	complete	complete	complete	complete	complete
Physical chemical parameters					
Solubility parameter	8.9	8.9	9.7	11.0	9.9
Fractional polarity	0.028	0.048	0.043	0.045	0.086
Hydrogen bonding					
Index	0	0	0	3.4	0
Characteristics	DA §§	DA	DA	DA	DA

[1] $C_4H_9O(C_2H_4O)_2H$
[2] $C_4H_9OC_2H_4OH$
[3] $C_2H_5O(C_2H_4O)_2H$
[4] $C_2H_5OC_2H_4OH$

*Determined on pure material.

**Shell thin film evaporometer, 25°C and 0% R.H.

***8 g R.S. ½" N.C. (dry)/100 ml solvent.

†At final concentration of 8 g R.S. ½" N.C. (dry)/100 ml combined solvent and diluent.

††Tolu-Sol 17 or similar.

†††25°C.

§Nitrocellulose blush.

§§Donor-Acceptor.

METHYLAL

Dimethoxymethane

Methylene Dimethyl Ether

$$CH_3OCH_2OCH_3$$

Methylal is a low-boiling solvent, stable in the presence of alkalis and mild acids, and to high temperatures and pressures. It differs from other ethers in that it forms only minute amounts of peroxides. It will dissolve such synthetic resins as nitrocellulose, cellulose acetate and propionate, ethyl cellulose, vinyl, "Epons" and polystyrene, and also many of the natural gums and waxes. Methylal as a latent solvent is activated by the addition of esters, ketones or alcohols. Its evaporation rate, twice that of acetone, places this ether in a class with such solvents as acetone, methyl acetate and ethyl acetate in resin formulations.

Table 11.10: Physical Properties of Methylal *(2)*

Acidity (as acetic acid), % by wt. (max.)	0.1
Aldehydes, % by wt. (max.)	0.1
Appearance	water-white
Boiling point at 760 mm. Hg, °C.	42.3
Boiling range, °C.	42.0 to 43.5
Flash point (Cleveland open cup), °F.	0
Freezing range, °C.	-104.8
Heat of combustion (Btu/lb.) at 20°C.	10.97
Refractive index (n_D at 25°C.)	1.35335
Melting point, °C.	-104.8
Methylal content, % (min.)	97
Molecular weight	76.1
Specific gravity, 20/20°C.	0.8601
Surface tension at 25°, dynes/cm.	21.1
Vapor pressure at 20°C., mm. Hg	330
Viscosity at 20°C., centipoises	0.325

ETHYLENE GLYCOLS

Table 11.11: Ethylene Glycol Monomethyl Ether *(2)*

Amsco-Solv EM
Dowanol EM
Ektasolve EM
Jeffersol EM
Methyl Cellosolve
Methyl Oxitol
Poly-Solv EM

$$HO-CH_2CH_2-O-CH_3$$

Ethylene glycol monomethyl ether is a colorless, limpid liquid of mild odor. It is miscible with water and with aliphatic and aromatic hydrocarbons. It is a solvent for essential oils, lignin, dammar, elemi, ester gum, kauri, mastic, rosin, sandarac, shellac, Zanzibar, nitrocellulose, cellulose acetate, alcohol-soluble dyes and many synthetic resins. Its solvency for cellulose esters is augmented when a ketone or a halogenated hydrocarbon is added. The uses for methyl "Cellosolve" are as a solvent in quick-drying varnishes and enamels, in conjunction with aliphatic, aromatic and halogenated hydrocarbons, alcohols and ketones; in solvent mixtures and thinners for lacquers and dopes; in the manufacture of synthetic resin plasticizers and as a penetrating and leveling agent in dyeing processes, especially in the dyeing of leather, animal and vegetable fibers. Other uses are as a fixative in perfumes and as a solvent in odorless nail-polish lacquers. "Dowanol EM" should not be added to nitrocellulose lacquers containing coumarone resins or ester gum because it will cause incompatibility between these substances.

(continued)

Table 11.11: (continued)

Acidity (as acetic acid) % by wt. (max.)	0.01
Boiling point at 760 mm. Hg, °C.	124.2
Color (APHA, max.)	15
Coefficient of expansion at:	
20°C.	0.00095
55°C.	0.00099
Flash point (Cleveland open cup), °F.	115
Freezing range, °C.	–85.1
Heat of vaporization (Btu/lb.)	239
Refractive index (n_D at 25°C.)	1.4021
Molecular weight	76.09
Specific gravity, 20/20°C.	0.9663
Specific heat (average) cal./°C.	0.534
Surface tension at 25°C., dynes/cm.	30.8
Solubility:	
in water at 20°C.	complete
water in at 20°C.	complete
Vapor pressure at 20°C., mm. Hg	9.7
Viscosity:	
at 25°F., centipoises	1.53
at 60°F., centipoises	0.85
Weight per gal. at 20°C., lb.	8.01

Table 11.12: Ethylene Glycol Dimethyl Ether *(15)*

"Ansul" E–121

1,2-Dimethoxyethane

$$CH_3OCH_2CH_2OCH_3$$

Ethylene glycol dimethyl ether is a clear water white neutral liquid of ethereal odor. It is miscible in all proportions with water and many hydrocarbons. This unusual combination of properties suggests many uses both in the laboratory and the plant. As a general solvent this ether shows a remarkable versatility, since the degree of solvent power may be greatly altered by proper dilution with water or with any of the common organic solvents. The boiling point is low enough to allow the solvent to be readily removed from finished preparations, and is at the same time high enough for convenient handling.

Acidity (as acetic acid), % by wt. (max.)	0.015
Appearance	white–white liquid
Boiling point, °C.:	
630 mm. Hg	70 to 80
760 mm. Hg	85.2
Flash point (open cup), °F.	40
Freezing point, °C.	–68
Molecular weight	90.1
Refractive index (n_D at 20°C.)	1.3792
Specific gravity, 25/25°C.	0.8683
Surface tension at 25°C., dynes/cm.	48.0
Viscosity (absolute) in centistokes, 20°C.	1.1
Weight per gal. at 20°C., lb.	7.23

Table 11.13: Ethylene Glycol Monoethyl Ether *(2)*

Amsco-Solv EE
Cellosolve Solvent
Dowanol EE
Ektasolv EE
Jeffersol EE
Oxitol
Poly-Solv EE

$HOCH_2CH_2OC_2H_5$

This colorless liquid has a mild and agreeable odor and combines a low evaporation rate with a strong solvent action. It is miscible in all proportions with acetone, benzene, carbon tetrachloride, ethyl ether, methanol and water. It has a powerful solvent action on nitrocellulose and alkyd resins and an extremely high dilution ratio with coal-tar hydrocarbons. This solvent will tolerate 4.9 times its own volume of toluene before the mixture will cease to dissolve nitrocellulose, while butyl acetate will tolerate only 2.9 times its volume.

Acidity (as acetic acid), % by wt. (max.)	0.01	Specific gravity, 25/25°C.	0.9311
Appearance	water-white	Specific heat (average) cal./°C.	0.53
Boiling point at 760 mm. Hg, °C.	134.7	Surface tension at 25°C., dynes/cm.	28.2
Fire point (open cup), °F.	115	Vapor pressure at 25°C., mm. Hg	5.3
Flash point (Cleveland open cut), °F.	115	Viscosity:	
Freezing range, °C.	–59	at 25°C., centipoises	1.84
Refractive index (n_D at 20°C.)	1.4076	at 60°C., centipoises	0.94
Molecular weight	90.1	Weight per gal. at 20°C., lb.	7.72

Table 11.14: Ethylene Glycol Diethyl Ether *(2)*

Ansul E-222

$C_2H_5OCH_2CH_2OC_2H_5$

Boiling point at 760 mm. Hg, °C.	121.4	Specific gravity, 20/20°C.	0.8417
Flash point (Cleveland open cup), °F.	80	Vapor pressure at 20°C., mm. Hg	9.4
Freezing range, °C.	–74	Viscosity at 20°C., centipoises	0.65
Refractive index (n_D at 25°C.)	1.3922	Weight per gal. at 20°C., lb.	7.01
Molecular weight	118.2		

Table 11.15: Ethylene Glycol Monobutyl Ether *(2)*

Amsco-Solv EB
Butyl Cellosolve
Butyl Oxitol
Dowanol EB
Ektasolve EB
Jeffersol EB
Poly-Solv EB

$HOCH_2CH_2OC_4H_9$

Ethylene glycol monobutyl ether is colorless liquid, miscible in all proportions with many ketones, ethers, alcohols, aromatic paraffin and halogenated hydrocarbons. More specifically, it mixes in all proportions with acetone, benzene, carbon tetrachloride, ethyl ether, n-heptane and water. Because of its excellent solvency, low evaporation rate and high dilution ratios, it is used as a solvent in the manufacture and formulation of lacquers, enamels, inks and varnishes, employing such resins as alkyd, phenolic, nitrocellulose, maleic modified, styrene and epoxy. In lacquers butyl "Cellosolve" imparts a slow evaporation rate, strengthens blush resistance, heightens gloss, improves flow-out and helps prevent orange peel. Hot spray lacquers usually contain about 10% of "Dowanol" EB based on the solvent-diluent weight.

(continued)

Table 11.15: (continued)

Acidity (as acetic acid), % by wt. (max.)	0.01	Molecular weight	118.2
Appearance	water–white	Specific gravity, 25/25°C.	0.899
Boiling point at 760 mm. Hg, °C.	170.6	Specific heat (average), cal./°C.	0.54
Fire point (open cup), °F.	165	Surface tension at 25°C., dynes/cm.	27.4
Flash point (Cleveland open cup), °F.	165	Vapor pressure at 75°C., mm. Hg	0.88
Freezing range, °C.	–40	Viscosity at 25°C., centistokes	2.83
Heat of vaporization, Btu/lb.	88.4	Weight per gal. at 20°C., lb.	7.48
Refractive index (n$_D$ at 25°C.)	1.417		

Table 11.16: Water-Solubility of Ethylene Glycol n-Butyl Ether *(23)*

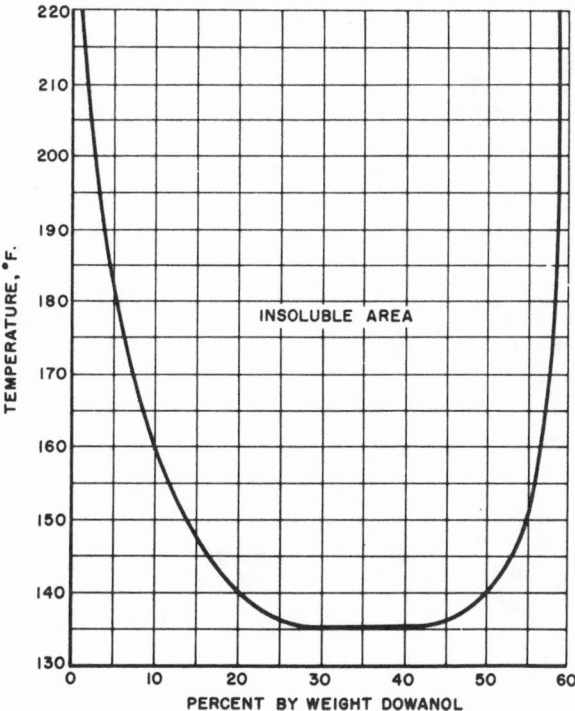

Table 11.17: Ethylene Glycol Monoisobutyl Ether *(41)*

$$CH_3-CH-CH_2-O-CH_2-CH_2OH$$
$$|$$
$$CH_3$$

Ethylene glycol monoisobutyl ether is a high boiling ether solvent for alkyd phenolic, malic, epoxy, alcohol-soluble butyrate, and ethyl cellulose nitrate resins.

Specifications

Color (Pt-Co Scale), ppm, max	10	Water, wt %, max	0.20
Specific Gravity, 20°/20°C	0.891-0.894	Appearance	Free from insoluble matter or haze
Boiling Range, 760 mm, °C			
Initial boiling point, min	158.0	Odor	Mild, characteristic, nonresidual
Dry point, max	162.0		
Acidity, as acetic acid, wt %, max	0.01		

(continued)

Table 11.17: (continued)

Typical Properties

Molecular Weight ($C_6H_{14}O_2$)	118.17
Evaporation Rate (n-butyl acetate = 1)	0.1
Weight/Vol, 20°C,	
lb./gal. (U.S.)	7.40
kg/liter	0.89
lb./gal. (Imperial)	8.88
Solubility, 20°C, wt %	
In water	Complete
Water in	Complete
Dilution Ratio, toluene	3.1
VM & P naphtha	1.6
Refractive Index, 20°C	1.4168
Vapor Pressure, 163°C, mm Hg	752
Flash Point, Tag Closed Cup, °F (°C)	136 (58)
Tag Open Cup, °F (°C)	145 (63)
Fire Point, °F (°C)	146 (63)
Flammable Limits in Air, % by volume	
Lower	1.21
Upper	9.4
Autoignition Temperature (ASTM D-2155), °F (°C)	540 (282)
NFPA Classification 30:	Flammable Liquid, Class II
ICC Labels Required	None
Bureau of Explosives Classification	Nonhazardous Liquid

Table 11.18: Ethylene Glycol Dibutyl Ether (2)

$$C_4H_9OCH_2CH_2OC_4H_9$$

This glycol diether is a colorless liquid. It is completely miscible with acetone, ethyl alcohol, ethyl acetate, isopropyl ether, heptane, ethylene dichloride and castor oil. Because of its being a good solvent for metallic reagents, it is particularly suitable for the Grignard type of reaction. It is also a solvent for inorganic halides and chlorosilanes, and is therefore used in silicone rubber formulations and in the extraction of aliphatic acids from dilute aqueous solutions.

Acidity (as acetic acid), % by wt.	0.01	Heat of vaporization (Btu/lb. at 1 atm.)	118
Boiling point, °C.		Molecular weight	174.28
760 mm.	203.6	Refractive index (n_D at 20°C.)	1.4131
50 mm.	117	Solubility, % by wt. at 20°C.	
10 mm.	83	in water	0.2
ΔBP/ΔP, °C./mm. Hg	0.056	water in	0.6
Coefficient of expansion at 55°C.	0.00105	Specific gravity, 20/20°C.	0.8374
Distillation at 760 mm., °C.		Δ SG/Δ T.	0.00085
Initial BP, min.	195	Specific heat at 20°C.	0.480
DP, max.	208	lb./gal. at 60°F.	7.0
Flash point (open cut), °F.	185	Vapor pressure, mm. Hg at 20°C.	0.09
Freezing point, °C.	-69.1		

Table 11.19: Ethylene Glycol Monophenyl Ether *(23)*

Dowanol EPh

$$HO-CH_2CH_2-O-\langle\text{phenyl ring}\rangle$$

Ethylene glycol monophenyl ether is a colorless, high-boiling, nonhygroscopic, water-immiscible liquid with a faint rose-like odor.

Molecular Weight	138.2	Specific Heat, (cal/g/°C) @ 25°C	0.52
Freezing Point, °F	51	Surface Tension, (dynes/cm)	
Boiling Point, 760 mm Hg, °C	245	25°C	42
10 mm Hg, °C	131.3	75°C	38
Vapor Pressure @ 20°C, mm Hg	0.03	Heat of Vaporization, (cal/g) @ 760 mm Hg	90.2
Specific Gravity @ 25/25°C	1.104	Thermal Conductivity, K x 10^4 (cal/cm^2 sec °C/cm) @ 60°C	3.86
Viscosity, centistokes, 25°C	20.5	Pounds/Gallon @ 25°C	9.20
60°C	4.3	Phenol content, max. %	0.5
Flash Point, °F (TCC)	260		

Table 11.20: Ethylene Glycol Monobenzyl Ether *(2)*

$$HOCH_2CH_2OCH_2C_6H_5$$

Ethylene glycol monobenzyl ether is a water-white liquid which will dissolve a large number of organic substances among which are oils, fats, greases, some vinyl resins, dewaxed dammar, rosin, ester gum, etc. It is used principally as a high boiling solvent in lacquers, inks, and textile dyeing.

Acidity (as acetic acid), % by wt. (max.)	0.010
Boiling point at 760 mm. Hg, °C.	255.9
Distillation range at 760 mm. Hg, °C.	248 to 260
Flash point, °F.	265
Specific gravity at 20/20°C.	1.0670 to 1.0720
Solubility in water at 20°C., % by wt.	0.4
Solubility of water in benzyl "Cellosolve" 20°C., % by wt.	18
Vapor pressure at 20°C., mm. Hg	0.02
Weight per gal. at 20°C., lb.	8.90

Table 11.21: Terpinyl Ethylene Glycol Ether *(2)*

Terpinyl ethylene glycol ether is a light-colored liquid used in enamels, inks, paints, and varnish.

Aniline point	Below –20°C	Flash point (Cleveland open cup)	284°F
Color (Lovibond 500 Amber Series Glasses)	3.0	Freezing point	Below –10°C
Distillation range (ASTM) 5%	248.0°C	Kauri butane solvency value	Infinite
10%	252.0°C	Moisture	0.05%
30%	295.0°C	Refractive index at 20°C	1.4786
50%	263.5°C	Specific gravity at 15.5/15.5°C	0.9813
70%	268.0°C	Viscosity at 25°C (Ubbelohde)	44.6 cp
90%	278.0°C		
95%	284.0°C		

Table 11.22: Ethylene Glycol Butylphenyl Ethers *(2)*

I.	– Ethylene Glycol p-sec-Butylphenyl Ether	$C_{12}H_{18}O_2$
II.	– Ethylene Glycol p-tert-Butylphenyl Ether	$C_{12}H_{18}O_2$

	I	II
Boiling point, °F.	313 to 322/10	311 to 350/10
Flash point (Cleveland open cup), °F.	300	315
Freezing range, °F.	below –4	54
Fire point, °F.	320	325
Molecular weight	194.3	194.3
Specific gravity, 77/77°F.	1.008	1.016
Solubility (g./100 grams water)	0.1	0.1
Viscosity at 77°F., centistokes	64.6	120.7
140°F.	8.8	11.6

Table 11.23: Diethylene Glycol *(2)*

Diglycol	$HOCH_2CH_2OCH_2CH_2OH$

This hygroscopic glycol is a clear colorless, odorless and stable liquid. It is also slightly viscous, noncorrosive and non-volatile. Because of its ether and alcohol group, diethylene glycol exhibits chemical properties characteristic of both primary alcohols and ethers. Its boiling point is considerably higher than that of ethylene glycol, and its solvent is greater. Diethylene glycol is miscible with water, ethers, lower aliphatic alcohols, aldehydes and ketones and is partially soluble in benzene, carbon tetrachloride, monobenzene, orthodichlorobenzene and toluene. It dissolves many dyes, resins, oils, nitrocellulose and many organic substances. Because of its solvent power, low volatility and hygroscopicity, it is used in textile lubricants, cutting oils, dry cleaning soap, printing inks, steam-set inks, and nongrain wood stains. In the textile industry diethylene glycol is used as a conditioning agent for wool, rayon, and cotton. As a solvent for dyes it makes a valuable assistant in dyeing and printing. The high hygroscopicity of diethylene glycol makes it an efficient softening agent for tobacco, paper, synthetic sponges, glues and casein. Diethylene glycol is especially useful in the dehydration of natural gas. A mixture of diethylene glycol and monoethanolamine will remove moisture, hydrogen sulfide and carbon dioxide from natural gas.

Acidity (as acetic acid), % by wt. (max.)	0.02	Ignition temperature in air, °C.	351
Boiling point at 760 mm. Hg, °C.	244.5	Spontaneous ignition temperature	412.8
Coefficient of expansion at 20°C.	0.000635/°C.	Molecular weight	106.12
Density (true) at 20°C., g./°C.	1.1161	Refractive index (n_D at 20°C.)	1.4475
Distillation range at 760 mm. Hg		Specific gravity at 20/20°C.	1.1185
Below 320°C.	none	Specific heat at 20°C., cal./g./°C.	0.5509
Below 240°C.	not over 20%	Surface tension at 25°C., dynes/cm.	48.5
Below 250°C.	not less than 85%	Vapor pressure, mm. Hg	
Below 270°C.	not less than 95%	20°C.	0.015
Electrical conductivity (reciprocal ohms)	0.586×10^{-6}	130°C.	8.0
Fire point, °C.	146 (approx.)	180°C.	96.0
Flash point (Cleveland open cup), °C.	143.3 (approx.)	Viscosity (absolute), in centipoises	
Freezing point, °C.	–8	15°C.	50.0
Heat of combustion (constant pressure) at 20°C.,		20°C.	38.0
kcal/mol	567	25°C.	30.0
Heat of formation (constant pressure), kcal/mol	148.42	Water, % by wt. (max.)	0.30
Heat of vaporization at 760 mm. Hg and		Weight per gal. at 20°C., lb.	9.308
244.5°C.			
Btu/lb.	270		
cal./g.	150		

Table 11.24: Diethylene Glycol Monomethyl Ether *(2)*

Amsco-Solv DM
Dowanol DM
Ektasolve DM
Jeffersol DM
Methyl Carbitol
Methyl Dioxitol
Poly-Solv DM

$$HOCH_2CH_2OCH_2CH_2OCH_3$$

Diethylene glycol monomethyl ether is a colorless, stable hygroscopic liquid with an agreeable odor. It is completely miscible with water, ketones, alcohol, ethers, aromatic hydrocarbons and halogenated hydrocarbons. More specifically, it is miscible with acetone, benzene, carbon tetrachloride, ethyl ether, methanol and water. It is a solvent for dyes, oils, fats, waxes, many natural and synthetic resins, nitrocellulose and cellulose acetate. It is used as a high-boiling solvent in such formulations as printing inks and pastes, stamp pad inks, textile dye pastes, lacquers, and synthetic resin coatings. Its presence in lacquers eases brushability and flow-out, and minimizes lifting of undercoats.

Acidity (as acetic acid), % by wt. (max.)	0.02	Specific gravity, 25/25°C.	1.018
Appearance	water-white	Specific heat (average), cal./°C.	0.54
Boiling point at 760 mm. Hg, °C.	194.1	Surface tension at 25°C., dynes/cm.	34.8
Fire point (open cup), °F.	200	Vapor pressure at 75°C., mm. Hg	0.18
Flash point (Cleveland open cup), °F.	200	Viscosity at 20°C., centistokes	3.87
Freezing range, °C.	-50	25°C.	3.47
Heat of vaporization, Btu/lb.	163	60°C.	1.64
Refractive index (n_D at 20°C.)	1.424	Weight per gal. at 20°C., lb.	8.47
Molecular weight	120.2		

Table 11.25: Diethylene Glycol Dimethyl Ether *(2)*

Ansul E-141

$$CH_3OCH_2CH_2OCH_2CH_2OCH_3$$

Diethylene glycol dimethyl ether is a clear, water-white neutral liquid of faint, pleasant odor. This ether may be used as a solvent for alkali metal hydrides for use in such reactions as reduction, alkylation and condensation. It may also be used as a lacquer solvent.

Acidity (as acetic acid), % by wt. (max.)	0.015	Specific gravity, 20/20°C.	0.9451
Appearance	water-white	Surface tension at 25°C., dynes/cm.	27.0
Flash point (Cleveland open cup), °F.	168	Vapor pressure at 100°C., mm. Hg	3.0
Freezing range, °C.	-68	Viscosity at 20°C., centistokes	2.0
Refractive index (n_D at 20°C.)	1.40778	Weight per gal. at 20°C., lb.	7.87
Molecular weight	134.2		

Table 11.26: Diethylene Glycol Monoethyl Ether *(2)*

Amsco-Solv DE
Carbitol Solvent Low Gravity
Dioxitol Low Gravity
Dowanol DE
Ektasolve DE
Jeffersol DE
Poly-Solv Low Gravity

$$HOCH_2CH_2OCH_2CH_2OC_2H_5$$

Diethylene glycol monoethyl ether is a colorless, stable, hygroscopic liquid of a mild, pleasant odor. It is completely miscible with water, alcohols, ethers, ketones, aromatic and aliphatic hydrocarbons, and halogenated hydrocarbons. Owing to the fact that it contains an ether-alcohol-hydrocarbon group in the molecule, it has the power to dissolve a wide variety of substances such as oils, fats, waxes, dyes, camphor and natural resins like copal, kauri, mastic, rosin, sandarac, shellac, as well as several types of synthetic resins. It is used as a solvent in synthetic resin coating compositions, and in lacquers, where high-boiling solvents are desired.

(continued)

Table 11.26: (continued)

Acidity (as acetic acid), % by wt. (max.)	0.02	Molecular weight	134.2
Boiling point at 760 mm. Hg, °C.	202.0	Specific gravity, 25/25°C.	0.986
Fire point (open cup), °F.	210	Specific heat (average), cal./°C.	0.54
Flash point (Cleveland open cup), °F.	205	Surface tension at 25°C., dynes/cm.	31.8
Freezing range, °C.	–55	Vapor pressure at 75°C., mm. Hg	0.13
Heat of vaporization, Btu/lb.	84.5	Viscosity at 20°C., centistokes	3.78
Refractive index (n$_D$ at 25°C.)	1.425	Weight per gal. at 20°C., lb.	8.20

Table 11.27: Diethylene Glycol Monoethyl Ether (Special Grade) *(23)*

Carbitol Solvent
Dioxitol High Gravity
Dowanol DE-SG
Jeffersol DE-75
Poly-Solv DE

Boiling point, °C	190–205	Surface tension (dynes/cm)	33.5
Flash point (TCC), °F	198	Solvent constants	
Evaporation rate Bu Ac = 1.0	0.01	Solubility parameter	11.4
Specific gravity 25°/25°C	1.025	Hydrogen bonding	16.6
Pounds per gallon 25°C	8.52	Dipole moment (Debye)	2.05
Viscosity, cs 25°C	6.9	Solubility in water ml/100 ml	∞

Table 11.28: Diethylene Glycol Diethyl Ether *(2)*

Ansul E-242 $C_2H_5OCH_2CH_2OCH_2CH_2OC_2H_5$

Diethylene glycol diethyl ether is a colorless liquid. It is completely miscible with many alcohols, ketones, ethers, halogenated hydrocarbons, etc. At room temperature this glycol diether is miscible with water. It is used as a mutual solvent for oils and water, in the dyeing of textiles and leather and in the extraction of uranium ores.

Acidity (as acetic acid), % by wt.	0.02	Heat of vaporization (Btu/lb. at 1 atm.)	130
Boiling point, °C.:		Molecular weight	162.22
760 mm. Hg	188.4	Refractive index (n$_D$ at 20°C.)	1.4115
50 mm. Hg	107	Solubility, % by wt. at 20°C.	
10 mm. Hg	72	in water	infinite
Δ BP/Δ P., °C./mm. Hg	0.049	water in	infinite
Coefficient of expansion at 55°C.	0.00111	Specific gravity, 20/20°C.	0.9082
Distillation at 760 mm. Hg, °C.		Δ SG/Δ T.	0.00097
Initial BP, min.	180	Specific heat at 15°C.	0.503
BP, max.	190	Weight/gal. at 60°F., lb.	7.60
Flash point (open cup), °F.	180	Vapor pressure, mm. Hg at 20°C.	0.38
Freezing point, °C.	–44.3		

Table 11.29: Diethylene Glycol Divinyl Ether *(2)*

1,5-Bis(Vinyloxy)-3-Oxapentane $CH_2{=}CHOCH_2CH_2OCH_2CH_2OCH{=}CH_2$

This vinyl ether is monomeric in character and is used as a chemical intermediate or as a crosslinking agent. Addition of isocyanic acid produces secondary diisocyanates. Divinyl ethers hydrolyze to the glycol and acetaldehyde. Chlorine or bromine add to the double bonds. Reaction with an alcohol in the presence of water produces a diacetal. Polymerization of divinyl ether of diethylene glycol with acidic catalysts produce crosslinked gels. Unsaturated polyesters, crosslinked with styrene, have been made noncorrosive to metals through use of divinyl ethers to reduce hydroxyl and acid numbers.

Boiling point, °C.:		Physical form	colorless liquid
10 mm. Hg	81–82	Purity	95%
12 mm. Hg	85	Refractive index	1.4441–1.4452
50 mm. Hg	115–116	Stabilizer	0.01%
Density, 29°C.	0.975		

Table 11.30: Diethylene Glycol Monobutyl Ether *(2)*

Amsco-Solv DB
Butyl Carbitol
Butyl Dioxitol
Dowanol DB
Ektasolve DB
Jeffersol DB
Poly-Solv DB

$$HOCH_2CH_2OCH_2CH_2OC_4H_9$$

Diethylene glycol monobutyl ether is a colorless, high-boiling liquid. It is miscible in proportions with water, alcohol (methanol), ketones (acetone), ethers (ethyl ether), aromatic hydrocarbons (benzene), paraffinic hydrocarbons (n-heptane), and halogenated hydrocarbons (carbon tetrachloride). As it is an ether-alcohol type compound it possesses solvent action for many substances such as oils, dyes, gums, and natural and synthetic resins. It is used as a high-boiling solvent in nitrocellulose lacquers and other synthetic coatings, baking lacquers, flash-dry printing inks, and dye baths.

Acidity (as acetic acid), % by wt. (max.)	0.01	Molecular weight	162.2
Appearance	water-white	Specific gravity, 25/25°C.	0.952
Boiling point at 760 mm. Hg, °C.	230	Specific heat (average), cal./°C.	0.54
Fire point (open cup), °F.	240	Surface tension at 25°C., dynes/cm.	30.0
Flash point (Cleveland open cup), °F.	230	Vapor pressure at 25°C., mm. Hg	0.023
Freezing range, °C.	-40	Viscosity at 25°C., centistokes	4.92
Heat of vaporization, Btu/lb.	74.3	Weight per gal. at 20°C., lb.	7.92
Refractive index (n_D at 25°C.)	1.430		

Table 11.31: Diethylene Glycol Monoisobutyl Ether *(41)*

$$(CH_3)_2CHCH_2OCH_2CH_2OCH_2CH_2OH$$

Specifications

Color (Pt-Co Scale), ppm, maximum	10	Specific gravity, 20°/20°C	0.945-0.949
Acidity, as acetic acid, weight percent, maximum	0.01	Water, weight percent, maximum	0.10
Boiling Range, 760 millimeters, °C		Appearance	Free from
Initial boiling point, minimum	217.0		insoluble
Dry point, maximum	225.0		matter or haze

Table 11.32: Diethylene Glycol Dibutyl Ether *(2)*

Ansul E-444

$$C_4H_9OCH_2CH_2OCH_2CH_2OC_4H_9$$

Ansul E-444 is a colorless liquid. It is completely miscible with many ethers, monohydric alcohols, esters, ketones and halogenated hydrocarbons. It is used to extract aliphatic acids for dilute aqueous solution.

Acidity (as acetic acid), % by wt. (max.)	0.01	Freezing point, °C.	-60.2
Boiling point, °C.		Heat of vaporization (Btu/lb. at 1 atm.)	110
760 mm. Hg	254.6	Molecular weight	218.33
50 mm. Hg	162.0	Refractive index (n_D at 20°C.)	1.4233
10 mm. Hg	122.0	Solubility, % by wt. at 20°C.	
$\Delta BP/\Delta P.$, °C./mm. Hg	0.056	in water	0.3
Coefficient of expansion at 55°C.	0.00097	water in	1.4
Distillation at 760 mm. Hg, °C.		Specific gravity, 20/20°C.	0.8853
Initial BP, min.	230	$\Delta SG/\Delta T.$	0.00084
5 ml., min.	248	Specific heat at 13°C.	0.430
DP, max.	258	Weight per gal. at 60°F., lb.	7.39
Flash point (open cup), °F.	245	Vapor pressure, mm. Hg at 20°C.	0.01

Table 11.33: Triethylene Glycol *(2)*

Triglycol

$$CH_2OCH_2 \cdot CH_2OH$$
$$|$$
$$CH_2OCH_2 \cdot CH_2OH$$

Triethylene glycol is a clear, colorless, viscous, stable liquid with a slightly sweetish odor. Because it has two ether and two hydroxyl groups its chemical properties are closely related to ethers and primary alcohols. It is a good solvent for gums, resins, nitrocellulose, steam-set printing inks and wood stains. With a low vapor pressure and a high boiling point, its uses and properties are similar to those of ethylene glycol and diethylene glycol. Because it is an efficient hygroscopic agent it serves as a liquid desiccant for removing water from natural gas. It is also used in air conditioning systems designed to dehumidify air.

Acidity (as acetic acid), % by wt.	0.01
Boiling point at 760 mm. Hg, °C.	287.4
Coefficient of expansion at 20°C.	0.00069
Density (true) at 20°C., g./°C.	1.1242
Fire point, °C. (approx.)	173.9
Flash point (Cleveland open cup), °C. (approx.)	330.0
Freezing point, °C.	–72
Heat of combustion (constant pressure) at 20°C., kcal/mol	850
Heat for formation (constant pressure), kcal/mol	192.9
Heat of vaporization at 760 mm. Hg and 244.5°C., cal./g.	179
Spontaneous ignition temperature, °F.	206
Surface tension at 25°C., dynes/cm.	45.2
Vapor pressure, mm. Hg:	
20°C.	0.01
101°C.	0.4
159°C.	10
202°C.	60
Water	miscible in all proportions
Weight per gal. at 20°C., lb.	7.37

Table 11.34: Triethylene Glycol Dimethyl Ether *(2)*

Ansul E-161

$$CH_3OCH_2CH_2OCH_2CH_2OCH_2CH_2OCH_3$$

Triethylene glycol dimethyl ether is a clear, colorless and odorless liquid. It is miscible in all proportions with water and many hydrocarbons. It may be used as a lacquer solvent. It serves as a solvent for alkali metal hydrides for such reactions as reduction, alkylation and condensation. This ether is also used as a solvent for the extraction of acetic acid.

Acidity (as acetic acid), % by wt. (max.)	0.015
Appearance	water-white
Boiling point at 760 mm. Hg, °C.	216.0
Flash point (Cleveland open cup), °F.	232
Heat of combustion at 25°C., Btu/lb.	–47
Refractive index (n_D at 20°C.)	1.4233
Molecular weight	178.3
Specific gravity, 20/20°C.	0.9871
Vapor pressure at 20°C., mm. Hg	0.9
Viscosity at 20°C., centistokes	3.8
Weight per gal. at 20°C., lb.	8.24

Table 11.35: Tetraethylene Glycol *(2)*

Tetraethylene glycol is a high-boiling, clear liquid of low volatility. It is completely miscible with water and a wide variety of organic solvents. For certain aliphatic hydrocarbons, it has a very slight affinity. Tetraethylene glycol is used as a coupling agent for blending water-soluble and water-insoluble compounds in such formulations as lubricants, glues, cork and textile products, etc.

Acidity (as acetic acid), % by wt. (max.)	0.01	Refractive index (n_D at 25°C.)	1.457
Ash, % by wt. (max.)	0.01	Specific gravity, 20/20°C.	1.125-7
Boiling point:		Specific heat at 77°F. (25°C.)	0.52
760 mm. Hg, °F.	586.0	Surface tension at 25°C., dynes/cm.	45
760 mm. Hg, °C.	307.8	Vapor pressure at 25°C., mm. Hg	0.01
Color, Pt-Co scale (max.)	200	Viscosity (absolute) in centistokes:	
Fire point, (Cleveland open cup), °F.	375	25°C.	39.9
Flash point (Cleveland open cup), °F.	365	60°C.	10.2
Freezing range, °F.	22	Water, % by wt.	0.20
Molecular weight	194.2	Weight per gal. at 25°C., lb.	9.34

Table 11.36: Tetraethylene Glycol Dimethyl Ether *(2)*

Ansul E-181 $CH_3(OCH_2CH_2)_4OCH_3$

Tetraethylene glycol dimethyl ether is a clear, colorless, odorless neutral liquid. It is miscible in all proportions with water, hydrocarbons and many organic solvents. This ether may be used as a solvent for alkali metal hydrides for such reactions as reduction, alkylation and condensation. It is also found to be useful, due to its lack of active hydrogen, as a reaction medium for the sodium condensation and Grignard reactions. It is a good selective solvent and as such is applied to purifying gases such as acetylene, sulfur dioxide, etc. It has been successfully used as an additive to hydrocarbon diesel fuels to reduce the auto-ignition temperature of the fuel for cold starts. When exposed to air, peroxide is formed.

Absolute viscosity at 20°C., centipoises	4.05	Molecular weight	222.28
Average weight per gal. at 20°C., lb.	8.43	Refractive index (n_D at 20°C.)	1.4336
Boiling point, °C. at:		Solubility, % by wt.:	
760 mm. Hg	275.3	in water at 20°C.	complete
100 mm. Hg	214	water in at 20°C.	complete
50 mm. Hg	184	Specific heat, cal./g.	0.46
10 mm. Hg	146	Specific gravity, 20/20°C.	1.0132
Flash point (tag, closed cup), °F.	285	Vapor pressure at 20°C., mm. Hg	0.01
Freezing point, °C.	-29.7		

PROPYLENE GLYCOLS

Table 11.37: Propylene Glycol Monomethyl Ether *(2)*

Arcosolv PM
Dowanol PM
Poly-Solv MPM $CH_3CHOHCH_2OCH_3$
Propasol Solvent M

Propylene glycol monomethyl ether is a colorless high-boiling liquid. It is miscible in all proportions with acetone, benzene, carbon tetrachloride, ethyl ether, petroleum ether and water. It shares in common the general characteristics of the "Dowanols".

Acidity (as acetic acid), % by wt. (max.)	0.02	Specific gravity, 20/20°C.	0.919
Appearance	water-white	Specific heat (average), cal./°C.	0.58
Boiling point, 760 mm. Hg, °C.	120.1	Surface tension at 25°C., dynes/cm.	27.7
Fire point (open cup), °F.	100	Vapor pressure at 20°C., mm. Hg	10.9
Flash point (Cleveland open cup), °F.	100	Viscosity at 20°C., centistokes	1.75
Freezing range, °F.	-142	75°C.	0.70
Refractive index (n_D at 20°C.)	1.4021	Weight per gal. at 20°C., lb.	7.65
Molecular weight	90.1		

Table 11.38: Propylene Glycol Monophenyl Ether *(2)*

Dowanol PPh

Typical Physical Properties

Molecular Weight	152.2
Boiling Point, °C 760 mm Hg	242.7
Boiling Point, °C, 10 mm Hg	115.9
Freezing Point, °F	55
Specific Gravity, 25/25°C	1.063
Pounds/Gallon at 25°C	8.80
Viscosity, cs, 25°C	23.2
Flash Point, °F (TCC)	260
Specific Heat, 25°C, cal/gm/°C	0.52
Surface Tension, 25°C, dynes/cm	38.1
Refractive Index, 25°C	1.522
Solubility in Water, 25°C, g/100g	1.1
Vapor Pressure, 25°C, mm Hg	<0.1
Color, APHA	<25

Table 11.39: Propylene, Dipropylene and Tripropylene Glycol Ethyl Ethers *(2)*

I. Propylene Glycol Ethyl Ether		$C_5H_{12}O_2$
II. Dipropylene Glycol Ethyl Ether		$C_8H_{18}O_3$
III. Tripropylene Glycol Ethyl Ether		$C_{11}H_{24}O_4$

	I.	II.	III.
Boiling point, °F.:			
760 mm. Hg	270	388	486
10 mm. Hg	87	179	250
Fire point (open cup), °F.	110	200	270
Flash point (Cleveland open cup), °F.	110	195	270
Refractive index, 77°F.	1.405	1.419	1.427
Molecular weight	104.1	162.2	220.3
Specific gravity:			
167/77°F.	0.845	0.884	0.904
77/77°F.	0.895	0.930	0.948
Specific heat (average), 77°F.	0.65	0.59	0.58
Surface tension, dynes/cm.			
77°F.	25.9	27.7	28.2
167°F.	20.2	22.2	23.6
Vapor pressure, at 77°F. mm. Hg	7.2	0.3	0.02
Viscosity in centistokes: 167°F.	0.72	1.06	1.56
77°F.	1.88	3.34	5.45
22°F.	16.9	55.0	133

Table 11.40: Propylene, Dipropylene and Tripropylene Glycol Isopropyl Ethers *(2)*

I. Propylene Glycol Isopropyl Ether	$C_6H_{14}O_2$	
II. Dipropylene Glycol Isopropyl Ether	$C_9H_{20}O_3$	
III. Tripropylene Glycol Isopropyl Ether	$C_{12}H_{26}O_4$	

	I.	II.	III.
Boiling point, °F., 760 mm. Hg	118.2	176.2	234.8
Fire point (open cup), °F.	125	205	265
Flash point (Cleveland open cup), °F.	120	195	255
Refractive index, 77°F.	1.407	1.421	1.428
Molecular weight	118.2	176.2	234.8
Specific gravity:			
167/77°F.	0.831	0.878	0.900
77/77°F.	0.879	0.931	0.942
Specific heat (average), 77°F.	0.62	0.58	0.52
Surface tension, dynes/cm.:			
77°F.	23.9	25.9	27.4
167°F.	19.6	21.6	22.8
Vapor pressure at 77°F., mm Hg	4.5	0.2	0.01
Viscosity in centistokes:			
267°F.	0.75	1.29	2.00
77°F.	2.27	440	8.18
22°F.	24.2	100	295

Table 11.41: Propylene, Dipropylene and Tripropylene Glycol n-Butyl Ethers *(2)*

I. Propylene Glycol n-Butyl Ether	$C_7H_{16}O_2$	
II. Dipropylene Glycol n-Butyl Ether	$C_{10}H_{22}O_3$	
III. Tripropylene Glycol n-Butyl Ether	$C_{13}H_{28}O_4$	

	I.	II.	III.
Boiling point, °F.:			
760 mm. Hg	340	444	529
10 mm. Hg	139	221	288
Fire point (open cup), °F.	160	245	300
Flash point (Cleveland open cup), °F.	160	235	275
Refractive index, 77°F.	1.415	1.425	1.430
Molecular weight	132.2	190.3	248.4
Specific gravity:			
167/77°F.	0.832	0.870	0.892
77/77°F.	0.878	0.913	0.934
Specific heat (average), 77°F.	0.63	0.59	0.58
Surface tension, dynes/cm.:			
77°F.	26.5	28.2	28.8
167°F.	21.9	20.2	24.2
Vapor pressure at 77°F., mm. Hg	1.2	0.06	0.005
Viscosity in centistokes:			
167°F.	0.97	1.33	1.88
77°F.	2.90	4.61	7.04
-22°F.	39.7	93	197

Table 11.42: Propylene Glycol Isobutyl Ether and Higher Homologs *(23)*

Dowanol PiBT

Boiling point, °C	172	Dilution ratio	
Flash point (TCC)°F	138	Toluene	2.3
Evaporation rate BuAc = 1.0	0.09	Naphtha	1.5
Specific gravity 25°/25°C	0.883	Solvent constants	
Pounds per gallon 25°C	7.33	Solubility parameter	8.6
Viscosity, cs 25°C	4.01	Hydrogen bonding	14.8
Vapor pressure @ 25°C (mm Hg)	1.3	Dipole moment (Debye)	1.97
Surface tension (dynes/cm)	25.1	Solubility in water ml/100 ml	2.9

Table 11.43: Propylene Glycol Substituted Phenyl Ethers *(2)*

I.   Propylene Glycol p–Cyclohexylphenyl Ether $C_{15}H_{22}O_2$

II. Propylene Glycol p–Chlorophenyl Ether $C_9H_{11}ClO_2$

III. Propylene Glycol 2,2–Dichlorophenyl Ether $C_9H_{10}Cl_2O_2$

	I.	II.	III.
Boiling point, °F.	--	296–301/10	311–329/10
Flash point (Cleveland open cup), °F.	--	290	335
Freezing range, °F.	174–180 *	81	below 10
Fire point, °F.	--	325	390
Refractive index, 77°F.	--	1.536	1.548
Molecular weight	234.3	186.6	221.1
Solubility (g./100 grams of solvent at 77°F.):			
Acetone	80	--	--
Benzene	60	--	--
Carbon tetrachloride	45	--	--
Ether	27	--	--
Methanol	134	--	--
V.M.P. Naphtha	--	23	--
Water	0.1	0.1	0.1
Viscosity in centistokes:			
77°F.	--	67.1	116.0
140°F.	--	7.3	10.2

* Melting point

Table 11.44: Propylene and Dipropylene Glycol Phenyl Ethers *(2)*

I.   Dipropylene Glycol Phenyl Ether $C_{12}H_{18}O_3$

II. Propylene Glycol o–Chlorophenyl Ether $C_9H_{11}ClO_2$

III. Propylene Glycol p–sec–Butylphenyl Ether $C_{13}H_{20}O_2$

	I.	II.	III.
Boiling point, °F.	301–325/10	279–293/10	306–326/10
Flash point (Cleveland open cup), °F.	315	270	295
Freezing range, °F.	below –13	below –4	below –4
Fire point, °F.	320	305	315
Molecular weight	210.2	186.6	208.3
Specific gravity, 77/77°F.	--	1.199	0.990
Solubility (g./100 grams of solvent at 25°C.):			
V.M.P. Naphtha	--	30	--
Water	0.1	0.1	0.1
Viscosity in centistokes:			
77°F.	26.3	40.4	50.8
140°F.	5.2	6.0	6.4

Table 11.45: Propylene Based Glycol Ether Blend *(23)*

Dowanol CX

Boiling point, °C	179
Flash point (TCC)°F	147
Evaporation rate Bu Ac = 1.0	0.08
Specific gravity 25°/25°C	0.917
Pounds per gallon 25°C	7.62
Viscosity, cs 25°C	3.8
Vapor pressure @ 25°C (mm Hg)	0.6
Surface tension (dynes/cm)	26.4
Solvent constants	
Solubility parameters	9.0
Hydrogen bonding	14.1
Dipole moment (Debye)	1.97
Solubility in water ml/100 ml	12.0

Table 11.46: Dipropylene Glycol *(2)*

$$O \Big\langle \begin{matrix} CH_2CHOHCH_3 \\ CH_2CHOHCH_3 \end{matrix}$$

Dipropylene glycol is a clear, colorless, viscous liquid. It is miscible with water in all proportions and the presence of the ether linkage makes it more compatible with hydrocarbon-like substances than propylene glycol. It is also a partial solvent for the both low and high cellulose acetate. Dipropylene glycol is a solvent for castor oil, and therefore serves as a component of hydraulic brake fluid formulations.

Acidity (acetic acid), % by wt. (max.)	0.01	Solubility, g./100 g. of solvent at 25°C.:	
Boiling point at 760 mm. Hg, °C.	231.8	in water	complete
Color	water-white	in methanol	complete
Flash point (open cup), °F.	280	Specific gravity, 20/20°C.	1.0252
Freezing range, °C.	none	Specific heat at 25°C.	0.55
Heat of vaporization (Btu/lb. at 1 atm.)	308	Vapor pressure, mm. Hg at 25°C.	0.01
Molecular weight	134.17	Viscosity (absolute) in centipoises:	
Refractive index (n_D at 20°C.)	1.4440	20°C.	107

Table 11.47: Dipropylene Glycol Monomethyl Ether *(2)*

Arcosolv DPM
Dowanol DPM
Poly-Solv DPM
Propasol Solvent DM

$$HOC_3H_6OC_3H_6OCH_3$$

"Dowanol" DPM has a mild, pleasant odor. Because of its structure it is completely miscible with water and a wide variety of organic substances, and has the combined solubility characteristics of an alcohol, an ether and a hydrocarbon. It is used in formulations of brake fluids, lacquers, paints, varnishes, dye and ink solvents, wood stains, textile processes, dry cleaning soaps and cleaning compounds.

Acidity (as acetic acid), % by wt. (max.)	0.02	Specific gravity, 20/20°C.	0.951
Appearance	water-white	Specific heat (average), cal./°C.	0.54
Boiling point at 760 mm. Hg, °C.	188.3	Surface tension at 25°C. dynes/cm.	28.8
Fire point (open cup), °F.	185	Vapor pressure at 20°C., mm. Hg	0.4
Flash point (Cleveland open cup), °F.	185	Viscosity in centistokes: 25°C.	3.33
Freezing range, °C.	-117	75°C.	1.07
Refractive index (n_D at 20°C.)	1.419	Weight per gal. at 20°C., lb.	7.91
Molecular weight	148.2		

Table 11.48: Tripropylene Glycol *(2)*

Tripropylene glycol is a water-white liquid. One of its unique features is its combination of water-solubility and good solubility for many organic compounds. Because of high boiling point and low volatility it is used in the formulation of textile soaps, lubricants, cutting oils, and similar applications.

Boiling point:		Specific heat at 77°F. (25°C.)	0.51
760 mm. Hg, °F.	513.0	Surface tension at 25°C., dynes/cm.	34
760 mm. Hg, °C.	267.2	Vapor pressure at 25°C., mm. Hg	0.01
Fire point (Cleveland open cup), °F.	310	Viscosity (absolute) in centistokes:	
Flash point (Cleveland open cup), °F.	285	25°C.	55.1
Freezing range	supercools	60°C.	9.80
Molecular weight	192.3	Weight per gal at 25°C., lb.	8.51
Refractive index (n_D at 25°C.)	1.442		

Table 11.49: Tripropylene Glycol Monomethyl Ether *(2)*

Arcosolv TPM
Dowanol TPM $HOC_3H_6OC_3H_6OC_3H_6OCH_3$
Poly-Solv TPM

Tripropylene glycol monomethyl ether is a colorless liquid possessing a mild, pleasant odor. It is completely miscible with a wide variety of organic products and water. This solubility for a wide range of organic products is due to the presence of the hydroxyl, ether and alkyl group in the molecule. It is used in the manufacture of cosmetics, liquid soaps, cleaning formulation, printing and writing inks, dyeing formulations, wood stains and in lacquers, paints and varnish formulations.

Acidity (as acetic acid), % by wt. (max.)	0.02	Specific gravity, 25/25°C.	0.967
Appearance	water-white	Specific heat (average), cal./°C.	0.51
Boiling point at 760 mm. Hg, °C.	242.4	Surface tension at 25°C., dynes/cm.	30.0
Fire point (open cup), °F.	270	Vapor pressure at 75°C., mm. Hg	0.022
Flash point (Cleveland open cup), °F.	260	Viscosity in centistokes:	
Freezing range, °C.	-42	25°C.	6.16
Refractive index (n_D at 25°C.)	1.428	75°C.	1.67
Molecular weight	206.3	Weight per gal. at 20°C., lb.	8.05

Table 11.50: Aromatic Based Glycol Ether *(23)*

Dalpad A

Molecular weight	138.0
Boiling point, °C	245.0
Flash point (TCC)°F	260
Evaporation rate Bu Ac = 1.0	<0.01
Specific gravity 25°/25° C	1.104
Pounds per gallon, 25°C	9.18
Viscosity, cs, 25°C	20.5
Vapor pressure @ 25°C (mm Hg)	0.03
Surface tension (dynes/cm)	42.0
Dilution ratio	
Toluene	Insoluble
Naphtha	Insoluble
Solvent constants	
Solubility parameter	11.4
Hydrogen bonding	16.6
Dipole moment (Debye)	1.67
Solubility in water ml/100 ml	2.3

BUTYLENE GLYCOLS

Table 11.51: 1,3-Butylene Glycol Methyl Ether (2)

Methyl Ether of 1,3-Butylene Glycol

3-Methoxyl Butanol

$$CH_3OCHCH_2CH_2OH$$
$$|$$
$$CH_3$$

1,3-Butylene glycol methyl ether is a colorless liquid with mild odor, combining the solvent characteristics of an ether linkage and a hydroxyl group. This high-boiling solvent is completely soluble in water and is a good solvent for cellulose derivatives, oils, resins, fats and waxes. It will dissolve nitrocellulose, ethyl cellulose, melamine and urea formaldehyde resins, maleic modified resins, phenolic and alkyd resins, cellulose acetate butyrate, ester gum, linseed oil, polyvinyl butyral and shellac.

Recommended use for this solvent is as a diluent in hydraulic brake fluids. Its low rubber swell and good stability are added properties of value for this application. It is of interest in nitrocellulose lacquer formulations for its good solvency, slow evaporation rate and high hydrocarbon dilution, imparting good flow and gloss, retarding blush and preventing orange peel in the finished product. It is also useful in brushing, dipping and hot spray lacquers as well as "high-low" lacquer thinner formulations. In addition, it is applied as a film-forming and a freeze-thaw stabilizer in PVA emulsions.

1,3-Butylene glycol is a mutual solvent for soluble cutting and textile oils and for specialty dry cleaning soaps. Other solvent applications are in textile and leather dyes, printing inks, and wood stains, in insecticide and herbicide concentrate and as an extractant and crystallizing agent for pharmaceuticals.

Acidity (as acetic acid), % by wt. (max.)	0.01	Molecular weight	104.2
Boiling point at 760 mm. Hg, °C.	161.1	Specific gravity, 20/20°C.	0.9229
Color (APHA, max.)	15	Solubility in water	complete
Flash point (Cleveland open cup), °F.	163	Vapor pressure at 20°C., mm. Hg	0.9
Freezing range, °C.	-85	Viscosity at 20°C., centipoises	3.7
Heat of vaporization, cal./g.	116.7	Weight per gal. at 20°C., lb.	7.68
Refractive index (n_D at 20°C.)	1.4160		

Table 11.52: Divinyl Ether of Butanediol (2)

1,4-Di(Vinyloxy)Butane
DVBD

$$CH_2{=}CHOCH_2CH_2CH_2CH_2OCH{=}CH_2$$

This monomer ether is used as a crosslinking agent and a chemical intermediate. DVBD polymerizes at 0°C. in the presence of $BF_3 \cdot (C_2H_5)_2O$ complex to form a brittle polymer. Rubbery copolymers have been produced with methyl vinyl ether and vinyl acetate using DVBD as a crosslinking agent. Transparent oils have resulted under certain reaction conditions when DVBD is homopolymerized. Unsaturated polyesters, crosslinked with styrene, have been made noncorrosive to metals through the use of DVBD to reduce hydroxyl and acid numbers. Addition of isocyanic acid to divinyl ethers produces secondary diisocyanates.

Boiling point, °C.:		Flash point	145°F.
10 mm. Hg	62–64	Physical form	liquid
20 mm. Hg	70–71	Purity	95%
100 mm. Hg	108–109	Refractive index	1.4376
Density, 25°C.	0.896	Stabilizer	0.01%

TRIGLYCOLS

Table 11.53: Methoxytriglycol *(19)*

Poly-Solv TM

Molecular weight	164.20	Freezing point, °C	–38.2
Apparent specific gravity,		Absolute viscosity, cp	
@ 20°/20°C	1.053	@ 20°C	7.27
ΔSpecific gravity/Δt.,		Solubility	
10°–40°C, per °C	0.00088	@ 20°C in water	complete
Boiling point, °C		% by wt water in	complete
@ 760 mm Hg	249.0	Solubility, % by wt in	
@ 50 mm Hg	162	heptane**	1.5
@ 10 mm Hg	126	Refractive index, n_D^{20}	1.4381
Δb.p./Δp., 750 to 770		Heat of vaporization, Btu/lb	
mm Hg,°C per mm Hg	0.053	@ 1 atm	141
Vapor pressure at 20°C,		@ 300 mm Hg	150
mm Hg	<0.01	Flash point, closed cup, °F	238
Relative evaporation rate*			
(Bu Ac = 100)	<0.1		

*Evaporation rate of glycol ether at 20°C referenced to n-butyl acetate at 25°C.
**A 25°C the glycol ether is completely soluble in acetone, benzene, ethyl ether, methanol, and carbon tetrachloride.

Table 11.54: Ethoxytriglycol *(19)*

Poly-Solv TE

Molecular weight	178.23	Freezing point, °C	–18.7
Apparent specific gravity,		Absolute viscosity, cp	
@ 20°/20°C	1.0250	@ 20°C	7.8
ΔSpecific gravity/Δt.,		Solubility	
10°–40°C, per °C	0.00088	@ 20°C in water	complete
Boiling point, °C		% by wt water in	complete
@ 760 mm Hg	255.9	Solubility, % by wt in	
@ 50 mm Hg	167	heptane**	2
@ 10 mm Hg	130	Refractive index, n_D^{20}	1.4376
Δb.p./Δp., 750 to 770		Heat of vaporization, Btu/lb	
mm Hg,°C per mm Hg	0.055	@ 1 atm	129
Vapor pressure at 20°C,		@ 300 mm Hg	137
mm Hg	<0.01	Flash point, closed cup, °F	255
Relative evaporation rate*			
(Bu Ac = 100)	<0.1		

*Evaporation rate of glycol ether at 20°C referenced to n-butyl acetate at 25°C.
**At 25°C the glycol ether is completely soluble in acetone, benzene, ethyl ether, methanol, and carbon tetrachloride.

Table 11.55: Butoxytriglycol *(19)*

Poly-Solv TB

Molecular weight	206.28	Freezing point, °C	–47.6
Apparent specific gravity,		Absolute viscosity, cp	
@ 20°/20°C	1.0021	@ 20°C	10.9
ΔSpecific gravity/Δt.,		Solubility	
10°–40°C, per °C	0.00082	@ 20°C in water	complete
Boiling point, °C		% by wt water in	complete
@ 760 mm Hg	(dec)	Solubility, % by wt in	
@ 50 mm Hg	188	heptane**	complete
@ 10 mm Hg	148	Refractive index, n_D^{20}	1.4394
Δb.p./Δp., 750 to 770		Heat of vaporization, Btu/lb	
mm Hg, °C per mm Hg	–	@ 1 atm	176***
Vapor pressure at 20°C,		@ 300 mm Hg	–
mm Hg	<0.01	Flash point, closed cup, °F	>250
Relative evaporation rate*			
(Bu Ac = 100)	<0.1		

*Evaporation rate of glycol ether at 20°C referenced to n-butyl acetate at 25°C.
**At 25°C the glycol ether is completely soluble in acetone, benzene, ethyl ether, methanol, and carbon tetrachloride.
***At 190°C and 50 mm Hg.

MISCELLANEOUS GLYCOLS

Table 11.56: Freezing Points of Aqueous Glycol Solutions *(19)*

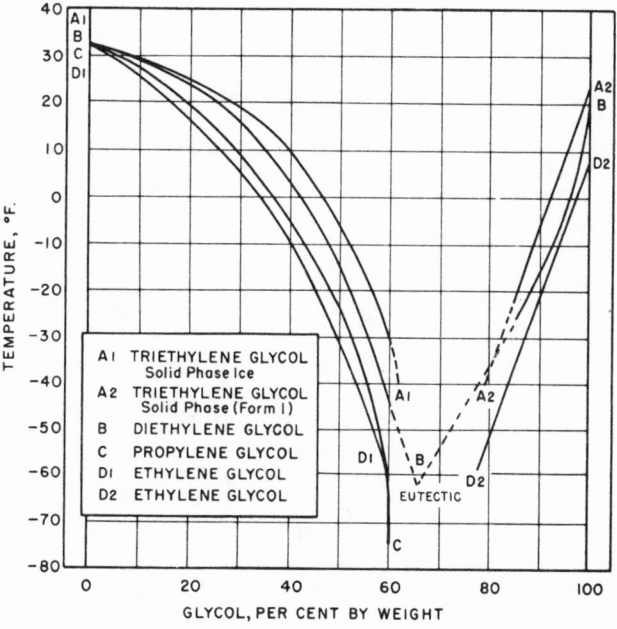

Table 11.57: Viscosities of Anhydrous Glycols *(23)*

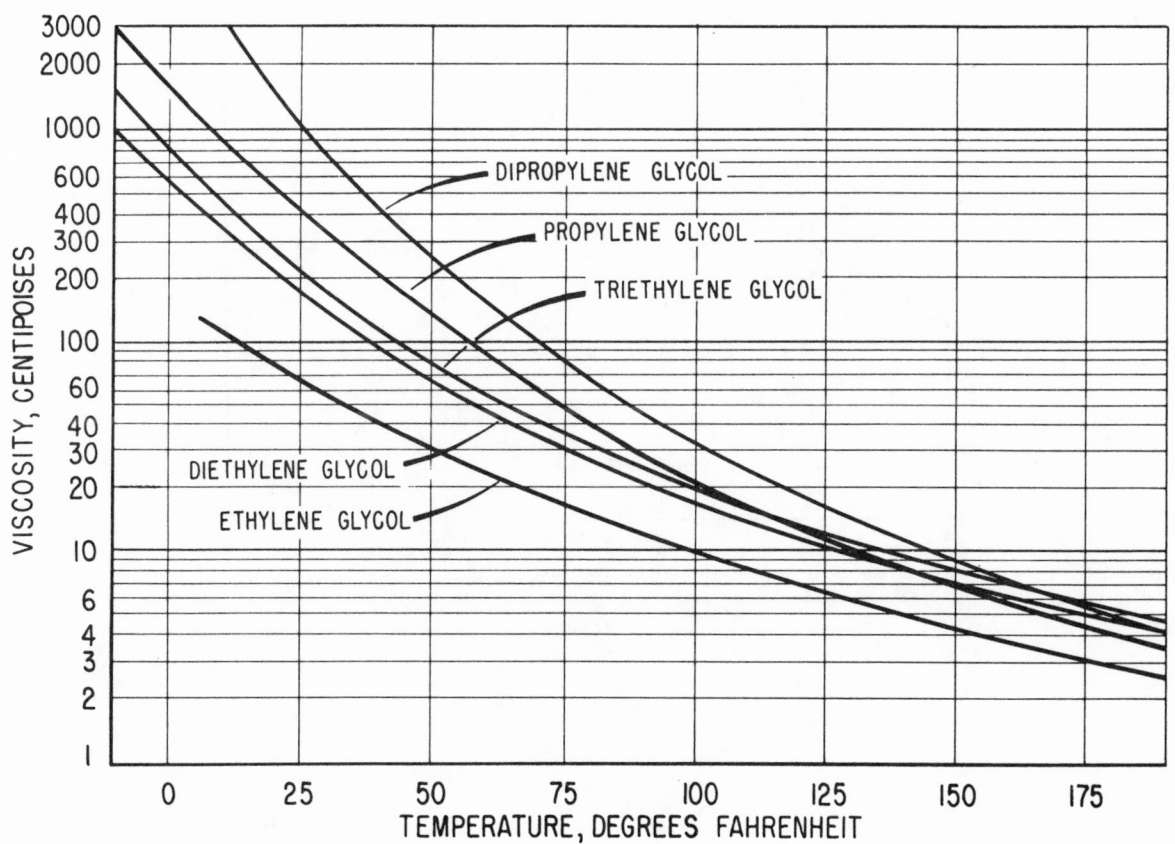

Table 11.58: Comparative Hygroscopicities of Glycols at 70° to 80°F *(19)*

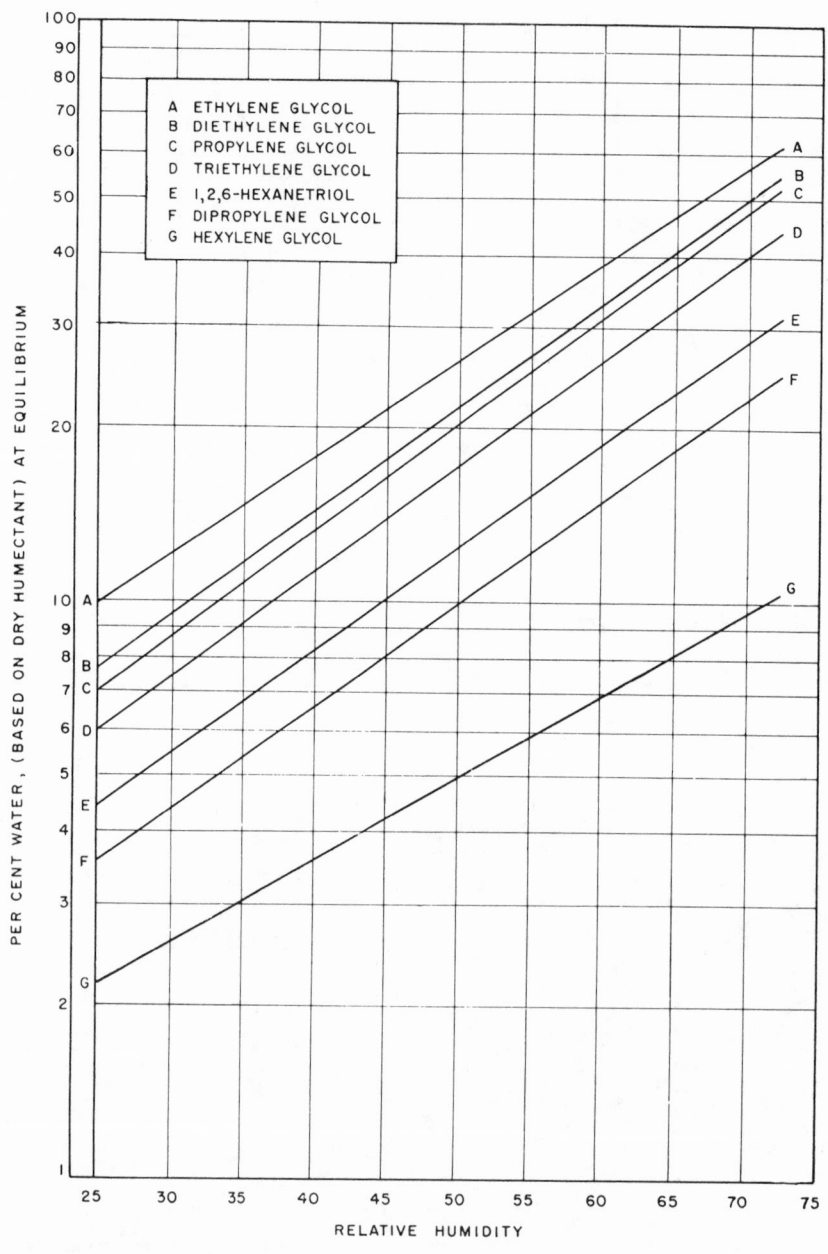

Table 11.60: Vapor Pressures of "Dowanols" (23)

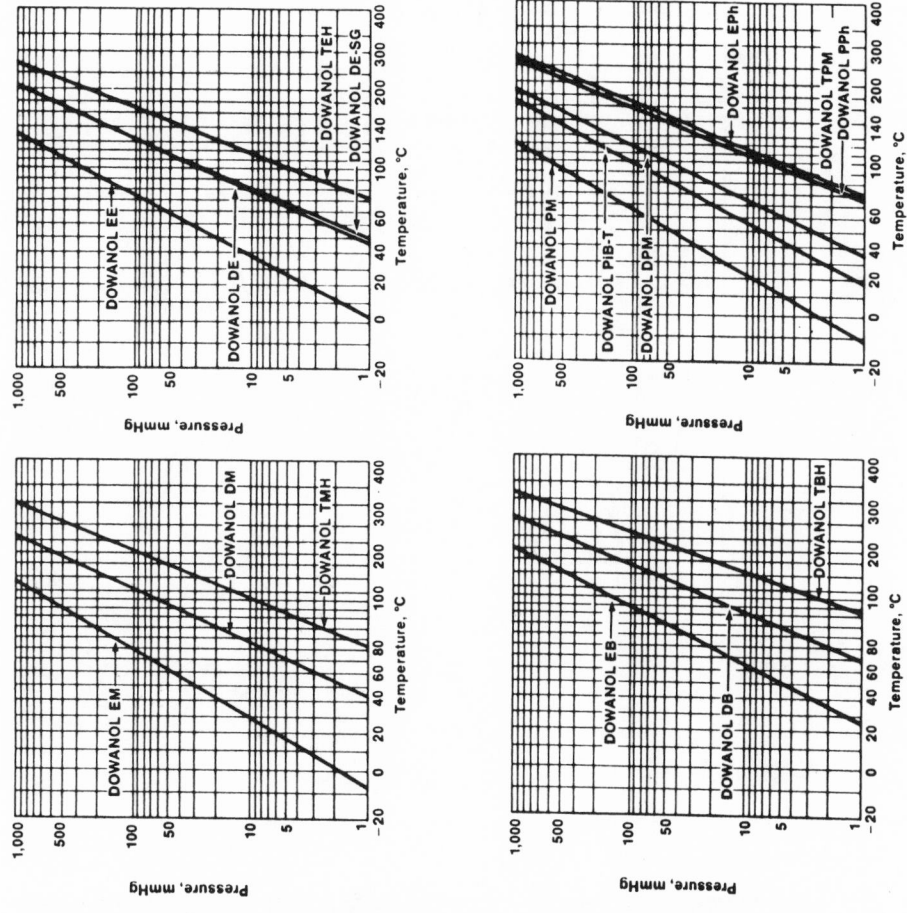

Table 11.59: Vapor Pressure vs Temperature of "Carbitols" and "Cellosolves" (19)

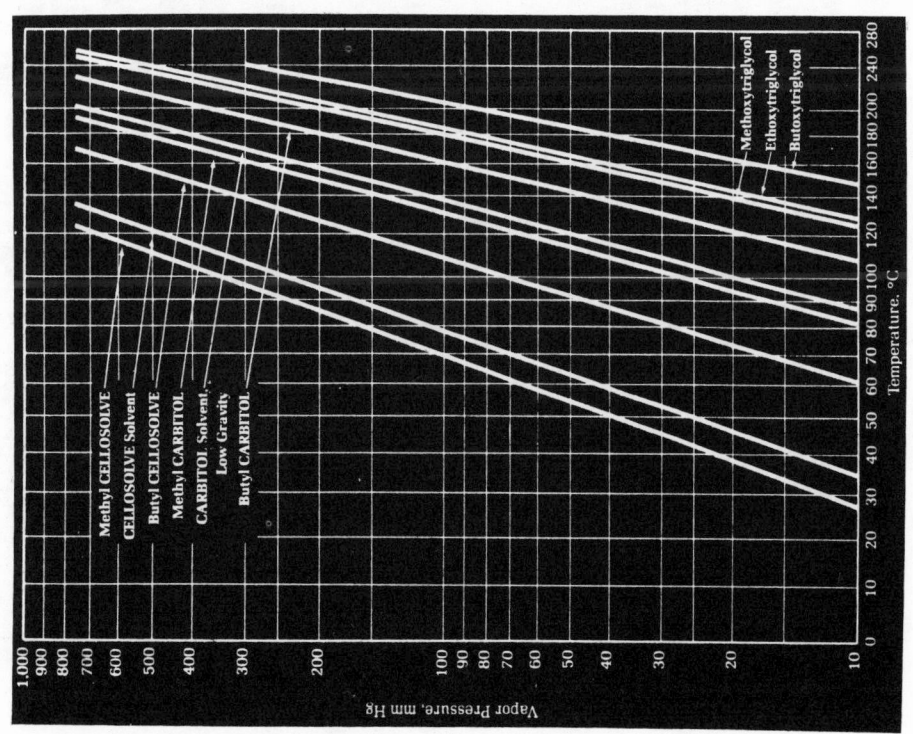

Table 11.62: Specific Gravity vs Temperature of "Carbitols" and "Cellosolves" (19)

Table 11.61: Vapor Pressures of "Ansuls" (15)

Table 11.64: Specific Gravity vs Temperature of Triglycols *(19)*

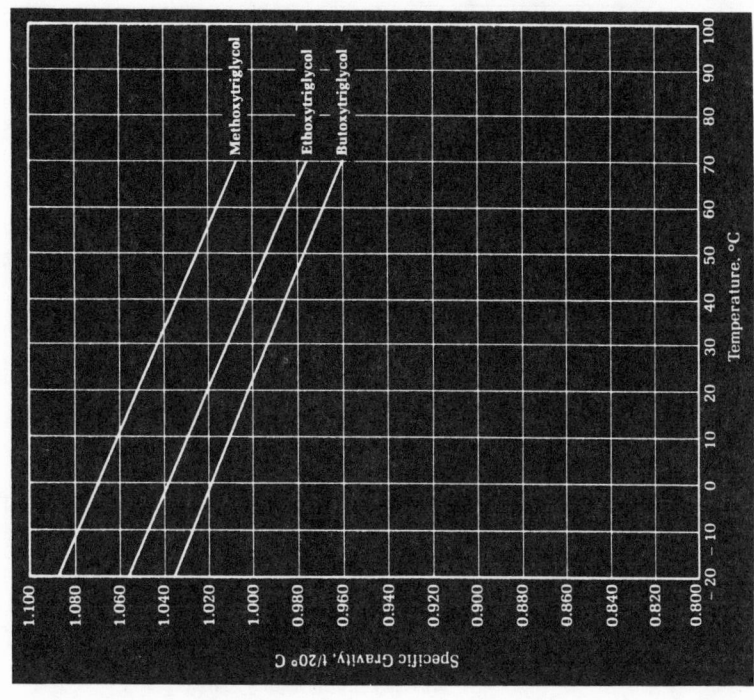

Table 11.63: Densities of "Dowanols" *(23)*

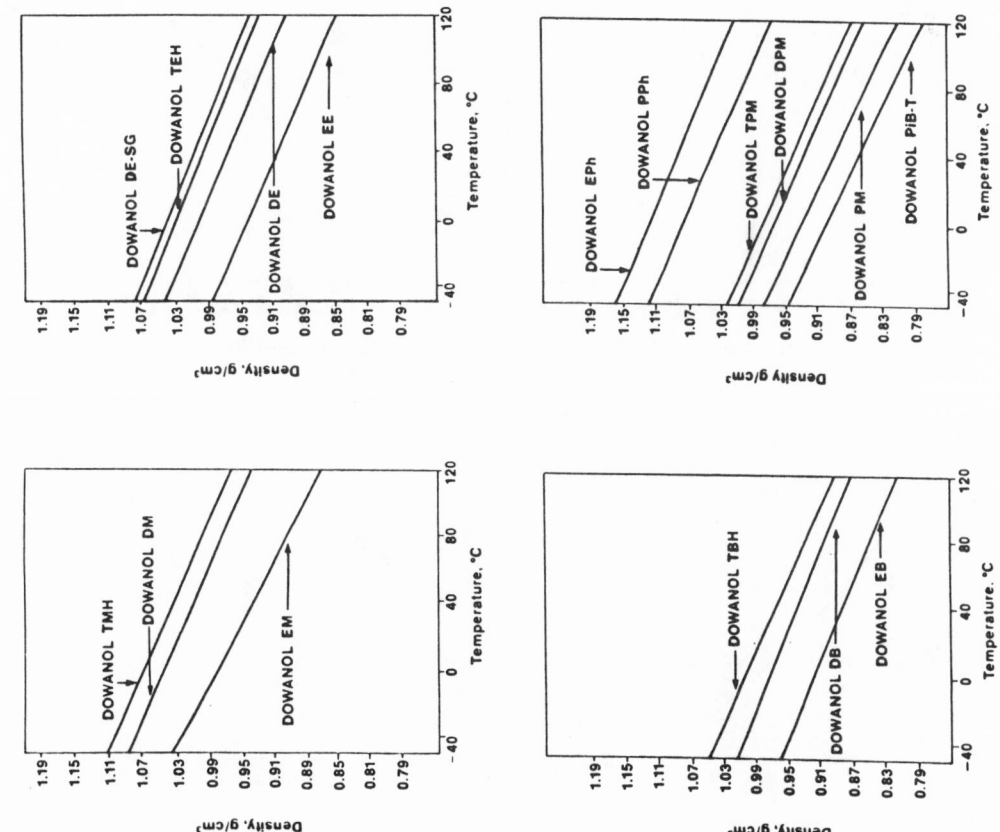

Table 11.66: Surface Tension of Aqueous Solutions of "Carbitols" and "Cellosolves" (19)

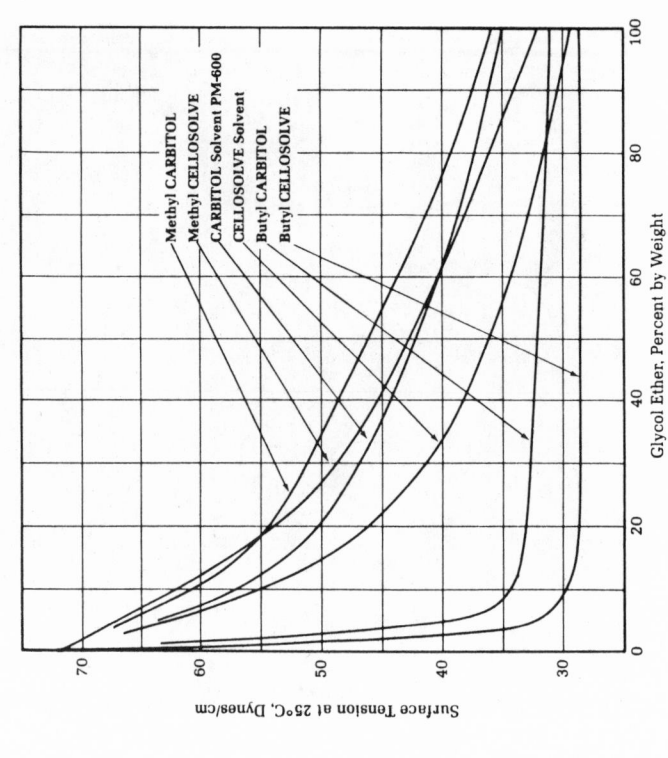

Table 11.65: Densities of Aqueous Solutions of "Dowanols" (23)

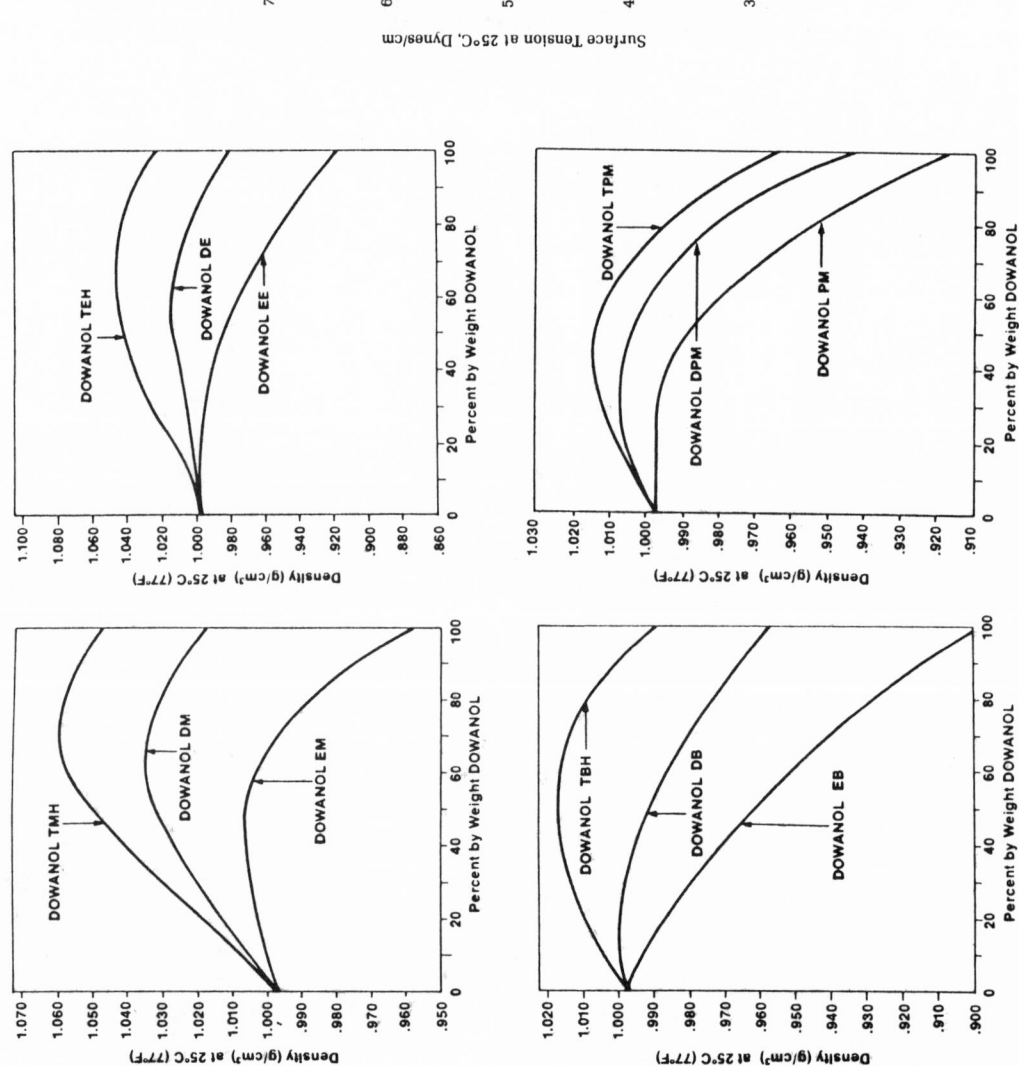

Table 11.68: Freezing Points of Aqueous Solutions of "Carbitol" and "Cellosolve" Solvents *(19)*

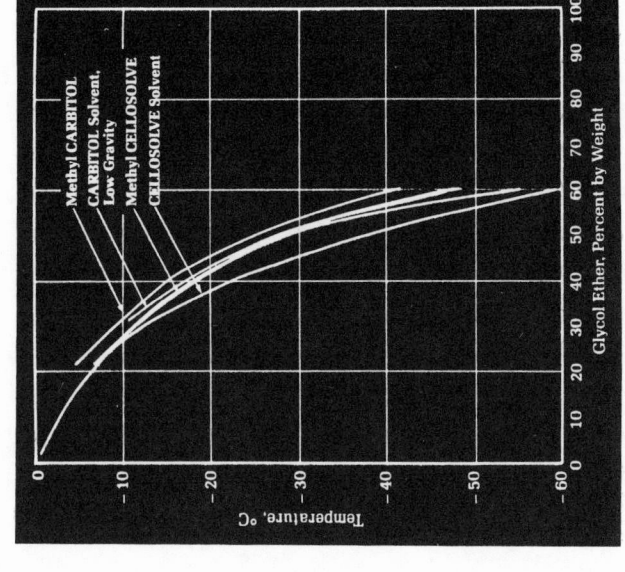

Table 11.67: Surface Tension of Aqueous Solutions of "Dowanols" *(23)*

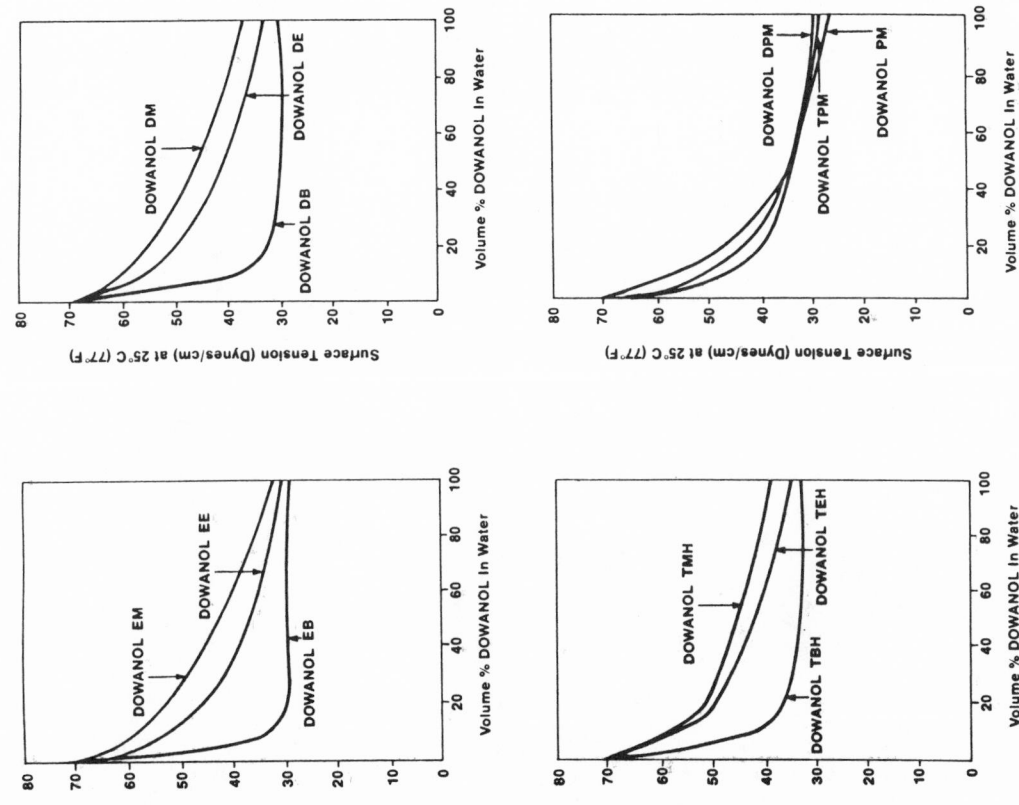

Table 11.70: Viscosity of Aqueous Solutions of "Carbitol" and "Cellosolve" Solvents *(19)*

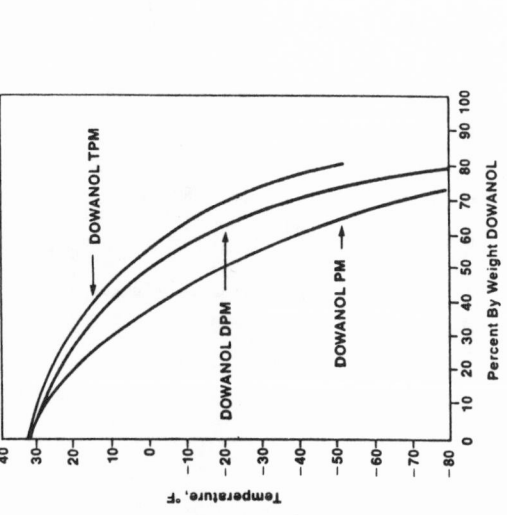

Table 11.69: Freezing Points of Aqueous Solutions of "Dowanols" *(23)*

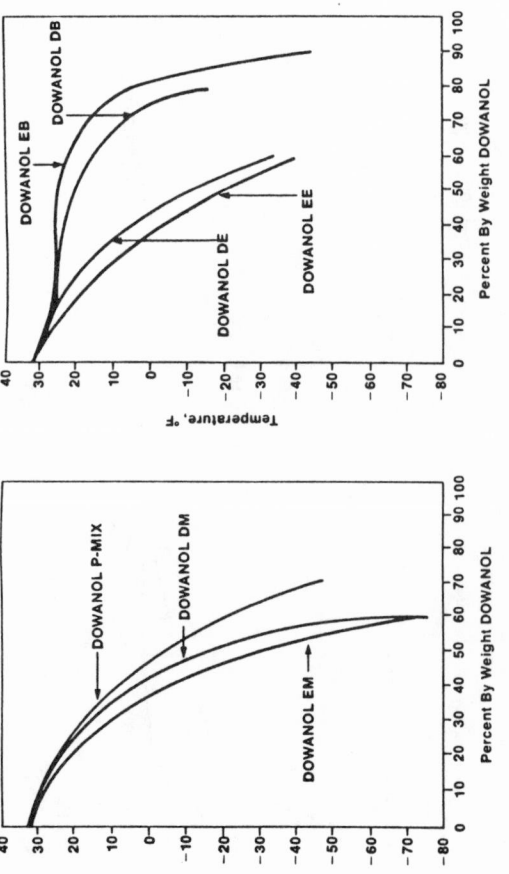

Table 11.72: Liquid-Vapor Equilibria of "Carbitols" and "Cellosolves" and Water at 760 mm Hg (19)

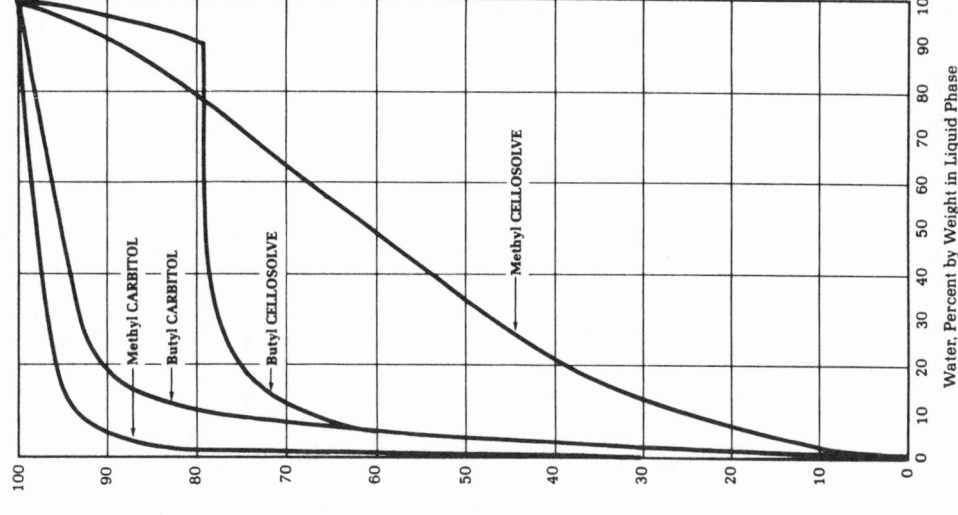

Table 11.71: Boiling Point of Aqueous Solutions of "Cellosolves" (19)

Table 11.74: Relative Evaporation of "Cellosolves" (19)

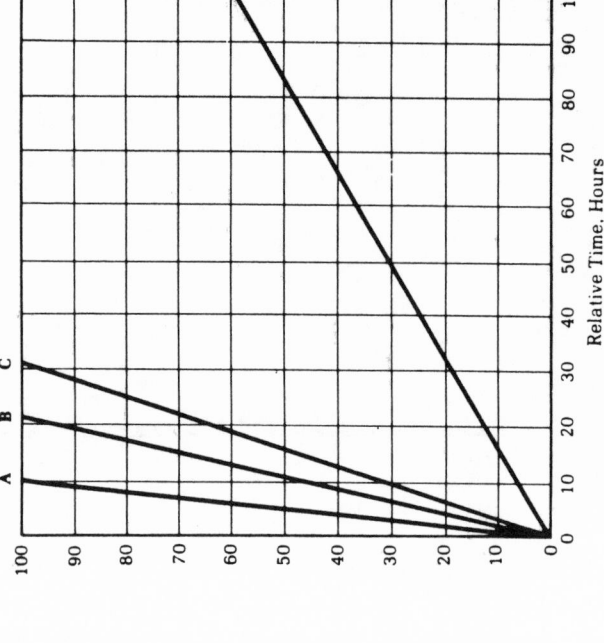

CHART KEY
A. Butyl Acetate
B. Methyl CELLOSOLVE
C. CELLOSOLVE Solvent
D. Butyl CELLOSOLVE

Table 11.73: Vapor-Liquid Equilibrium of "Dowanols" and Water
at 760 mm Hg (23)

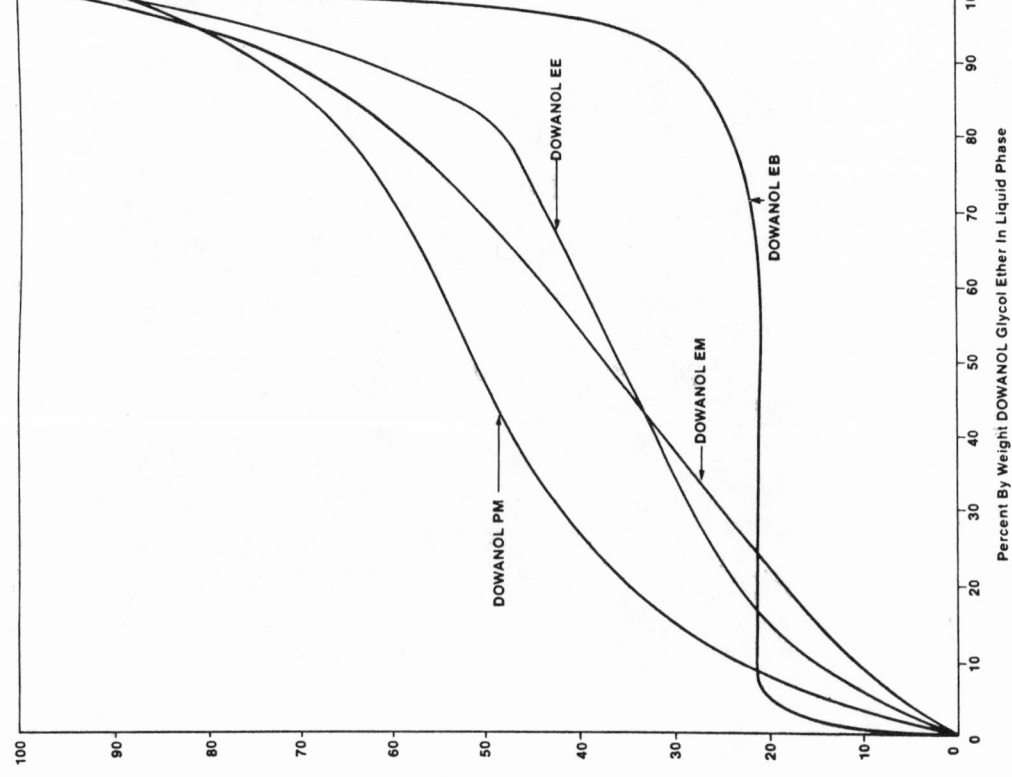

Table 11.75: Evaporation Rates of "Dowanols" (Butyl Acetate = 100) * *(23)*

PM	70	35%PM/65%DPM	10
EM	50	DPM	2
80%PM/20%DPM	50	TPM	<1
EE	20	DM	<1
65%PM/35%DPM	20	DE	<1
PiB-T	15	DB	<1
55%PM/45%DPM	15	EPh	<1
EB	10	PPh	<1

ASTM 3539

* Chemists use the evaporation rate of butyl acetate as the standard for determining evaporation rates of solvents. Butyl acetate has an arbitrary value of 100. All solvents evaporating faster than butyl acetate have a number higher than 100. Those evaporating more slowly have evaporation rates lower than 100. All glycol ethers evaporate more slowly than butyl acetate.

Table 11.76: Constant Boiling Mixtures of "Carbitols" and "Cellosolves" *(19)*

Solvent	Specific Gravity, at 20/20° C	Boiling Point, °C at 760 mm Hg	Boiling Point, °C at 760 mm Hg	Composition % by Wt, at 20° C in Azeotrope	in Upper Layer	in Lower Layer	Relative Volume of Layers at 20° C	Sp. Gr. 20/20° C of Azeotrope Layer
Butyl CARBITOL	0.9536	230.6	196.2	27.5	—	—	—	1.074
Ethylene Glycol	1.1155	197.6		72.5	—	—	—	
Butyl CELLOSOLVE	0.9022	171.2	98.8[a]	20.8	57	10	—	0.989[b]
Water	1.0000	100.0		79.2	43	90	—	
CELLOSOLVE Solvent	0.9311	135.6	125.8	35.7	—	—	—	0.896
Butyl Acetate	0.8826	126.0		64.3	—	—	—	
CELLOSOLVE Solvent	0.9311	135.6	110.0	10.0	—	—	—	0.874
Toluene	0.8683	110.6		90.0	—	—	—	
CELLOSOLVE Solvent	0.9311	135.6	99.4	28.8	—	—	—	1.003
Water	1.0000	100.0		71.2	—	—	—	
Methyl CARBITOL	1.0230	194.0	192	70	—	—	—	1.051
Ethylene Glycol	1.1155	197.6		30	—	—	—	
Methyl CELLOSOLVE	0.9663	124.5	105.9	25	—	—	—	0.887
Toluene	0.8683	110.6		75	—	—	—	

(a) Heterogeneous at the boiling point.
(b) Homogeneous at 20° C.

Table 11.78: Refractive Indices of Aqueous Solutions of "Dowanols" (23)

Table 11.77: Weight Loss of "Dowanols" by Evaporation at 77°F and 15% Relative Humidity (23)

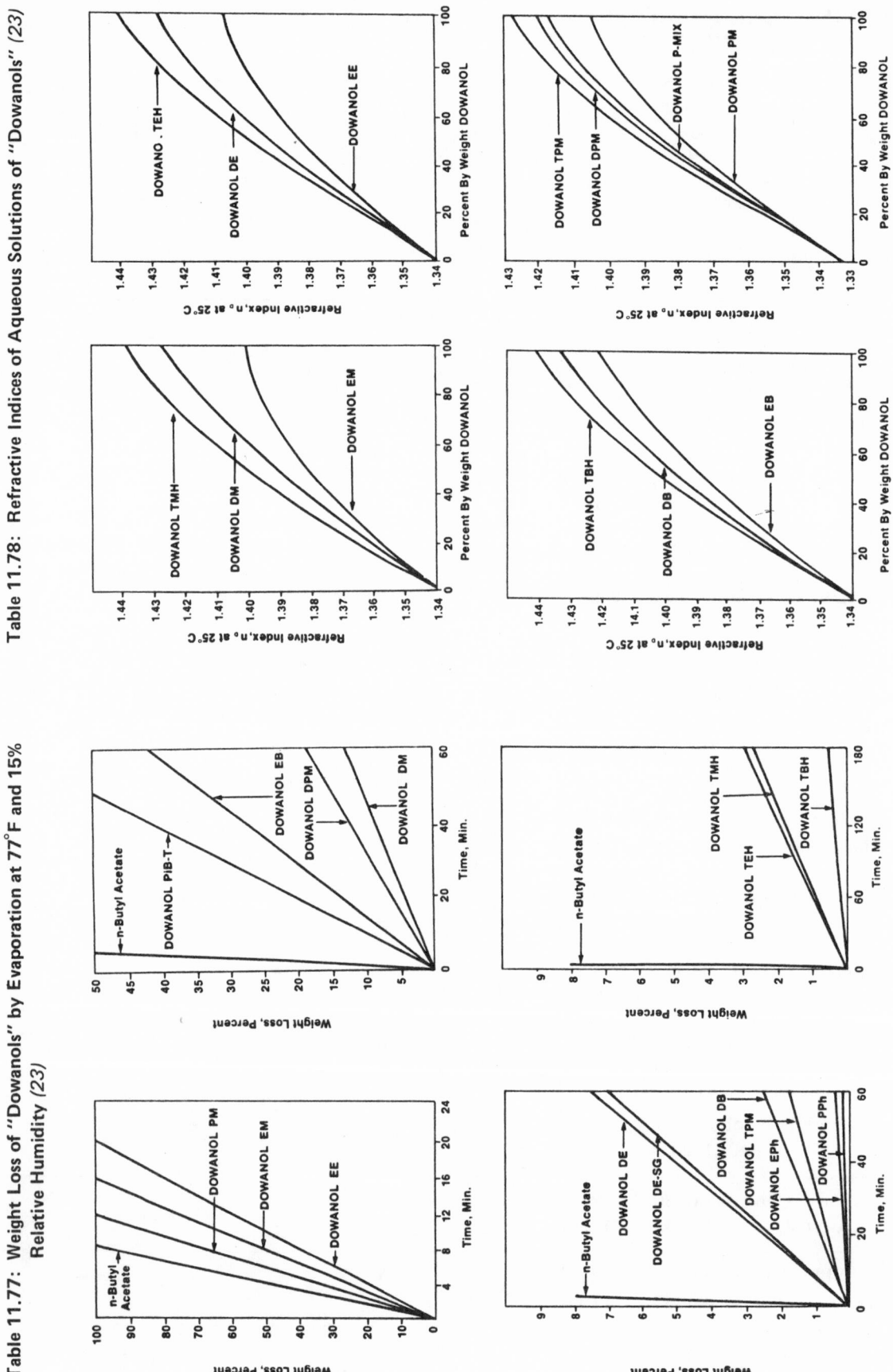

Table 11.79: Hygroscopicity of "Dowanols" at 21°C (70°F) and 77% Relative Humidity (23)

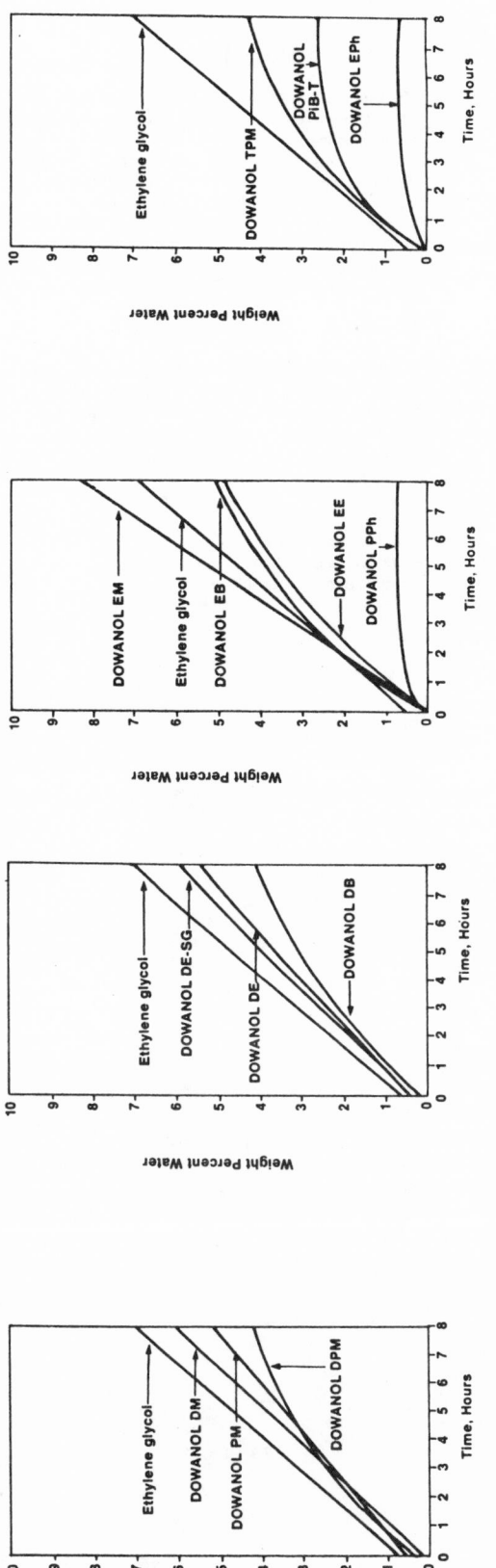

Table 11.80: Rubber Swell Properties of "Dowanols" (23)

| | Natural Rubber Swell[1] | | Synthetic Rubber Swell[2] | | | | | | | | | | | |
| | Average % Dimension Change | Average % Volume Change | Average % Dimension Change in Length | | | Average % Dimension Change in Width | | | Average % Dimension Change in Thickness | | | Average % Volume Change | | |
			Buna (GR-S)	Butyl	Neoprene	Buna (GR-S)	Butyl	Neoprene	Buna (GR-S)	Butyl	Neoprene	Buna (GR-S)	Butyl	Neoprene
DOWANOL EM	2.56	8.13	2.73	1.10	6.42	4.65	3.18	8.66	24.96	22.34	41.78	9.27	3.24	31.73
DOWANOL EE	6.95	22.50	8.78	4.90	10.32	10.73	7.35	12.16	30.33	27.10	50.33	25.79	17.02	44.18
DOWANOL EB	20.80	92.02	20.42	15.43	24.34	26.47	16.32	24.12	48.93	33.45	59.94	58.36	42.88	104.86
DOWANOL DM	2.12	7.32	1.39	-0.58	7.18	4.00	3.31	8.86	22.83	18.86	47.72	7.36	-0.16	35.40
DOWANOL DE	3.92	13.30	3.31	1.22	10.81	6.55	3.97	12.04	25.77	19.81	38.85	11.14	3.30	43.65
DOWANOL DB	10.83	31.73	10.60	7.26	27.93	13.90	9.61	28.51	33.95	24.89	69.18	37.27	21.75	126.17
DOWANOL PM	5.56	20.97	7.61	5.70	9.75	7.15	3.14	10.50	6.15	4.72	41.76	23.97	14.72	37.41
DOWANOL DPM	7.46	26.24	12.35	9.00	20.43	11.39	6.64	21.61	9.74	10.45	59.09	40.26	29.11	82.07
DOWANOL TPM	8.61	29.35	12.38	7.03	21.70	12.55	6.43	22.37	11.03	7.21	57.86	43.75	22.92	86.55
DOWANOL P-Mix	6.56	23.31	11.00	7.86	15.89	10.50	6.43	18.27	9.23	9.21	58.73	36.13	23.93	63.96

[1] Tests made in manner specified for hydraulic fluids by SAE (Lockheed Wagner FC-666-XO brake fluid cups, 120 hours at 158°F).
[2] Tests were carried out using cured rubber strips measuring approximately 2 x 1 x 0.11 inch, 120 hours at 158°F.

Table 11.81: Solubilities of Liquids in "Dowanols" *(23)*

The liquids listed below are miscible with the Dowanol glycol ether products except as noted below.

Acetaldehyde
Acetic Acid (Glacial)
Acetic Anhydride
Acetone
* Acetylene Tetrabromide
Acrylonitrile
* ALKAZENE 42 ar-Di-
   bromoethylbenzene
* Allyl Alcohol
Amyl Alcohol
* tert-Amyl Alcohol
Amyl Acetate
Aniline
Benzaldehyde
Benzene
Benzyl Alcohol
n-Butyl Acetate
n-Butyl Alcohol
* n-Butyl Lactate
* Butyl Oleate
n-Butyraldehyde
Carbon Bisulfide
Carbon Tetrachloride
Castor Oil

Chloroform
CHLOROTHENE® solvent
   (Methyl Chloroform)
Cyclohexanol
* Cyclohexanone
Cyclohexene
* Dehydrated Castor Oil
   9-11 Acids
* Diacetone Alcohol
* Dibutoxy Ethyl Phthalate
Dibutyl Phthalate
* Dibutyl Sebacate
o-Dichlorobenzene
Dichloroethyl Ether
* Dichloroisopropyl Ether
Diethanolamine
* Didecyl Phthalate
Diethyl Ether
Diethylene Glycol
* Di-2-Ethylhexyl Phthalate
* Di-2-Ethylhexyl Sebacate
* Diisoctyl Phthalate
* Dimethoxy Ethyl Phthalate
p-Dioxane

Diphenyl Oxide
Dipropylene Glycol
Ethanol (95%)
Ethyl Acetate
* Ethyl Benzene
* Ethyl Bromide
* Ethylene Chlorohydrin
Ethylene Dibromide
Ethylene Dichloride
Ethylene Glycol
* Ethylidene Dichloride
* Ethyl Lactate
Furfural
Isopropanol
Isopropylbenzene (Cumene)
* Isopropyl Chloride
Lactic Acid 85%
Methanol
* Methyl Cyclohexanol
Methyl Ethyl Ketone
Methyl Isobutyl Ketone
Methyl Salicylate
Methylene Bromide
* Methylene Chlorobromide

Methylene Chloride
* αMethyl Styrene
Monochlorobenzene
Monoethanolamine
Monoisopropanolamine
Morpholine
Nitrobenzene
* Nitroethane
Nitromethane
Octyl Alcohol
Oleic Acid
Paraldehyde
* Pentachlorodiphenyl Oxid
Perchloroethylene
* Phenethyl Acetate
* Phenethyl Alcohol
* Phenetole
Phenyl Acetate
Phosphoric Acid (Conc.)
Pine Oil
* Piperidine
* Polyethylene Glycol 400
Polyethylene Glycol 600
* Polypropylene Glycol 400

* Polypropylene Glycol 750
Polypropylene Glycol 1200
n-Propanol
Propylene Dichloride
Propylene Glycol
* Pyridine
* Ricinoleic Acid
* Styrene N-99
* Styrene Oxide
Tetrachloroethane
Tetrahydrofurfuryl Alcohol
Toluene
Trichloroethylene
* Tricresyl Phosphate
Triethanolamine
Triethylene Glycol
* Trimethylene Chlorobromide
Tripropylene Glycol
Tall Oil
Xylene

* The solubility of these products has not been determined for DOWANOL PiB-T or DOWANOL EPh.

Table 11.82: Solubility Limits of Various Liquids in "Dowanols" (ml/100 ml) *(23)*

Compound	EM	EE	EB	DM	DE	DE-SG	DB	TMH	TEH	TBH	PM	DPM	TPM	PiB-T	EPh	PPh
n-Butyl Stearate	25	∞	∞	8	∝	7	∝	—	—	—	∞	∝	∝	∝	∝	∝
Cottonseed Oil	Ins.	∞	∞	Ins.	14	Ins.	∝	—	—	—	∞	∝	∝	∝	6	∝
Cyclohexane	∞	∞	∞	64	∝	45	∝	—	—	—	∞	∝	∝	∝	33	∝
Diamylnaphthalene	∞	∞	∞	∞	∝	28.5	∝	—	—	—	∞	∞	∝	—	—	—
Di-2-Ethylhexyl Adipate	∞	∞	∞	∞	∝	18.5	∝	—	—	—	∞	∞	∝	—	—	—
Fish Oil	1	∞	∞	1.5	∝	Ins.	∞	—	—	—	∞	∞	∝	—	—	—
Formaldehyde (40%)	∞	∞	∞	∞	∝	∝	∝	·∝	∝	∝	∞	∝	∝	∝	∞	21
Formamide (37-38% Stabilized)	∞	∞	∞	∞	∞	∞	∝	—	—	—	∞	∞	∞	9	∞	12
Gasoline	60	∞	∞	22	∝	20	∞	∞	∞	∞	∞	∞	∞	∞	10	∞
n-Heptane	47	∞	∞	19	∝	20	∞	23	19	∞	∞	∞	∞	∞	8	46
Hexane	58.5	∞	∞	21	∝	21	∞	25	28	93	∞	∞	∞	∞	12	70
Hydrochloric Acid (conc.)	∞	∞	∞	∞	∞	∝	∝	∞	∞	∝	∝	∝	∝	∝	∞	5.2
Kerosene	<0.5	∞	∞	3	13.5	1.5	∞	—	—	—	∞	∞	∝	∝	0.5	∞
Linseed Oil (Boiled)	<0.4	∝[1]	∞	<0.4	∝[2]	Ins.	∞	Ins.	Ins.	∞	∝[3]	∞	∝	∝	∞	∞
Lemon Oil	∞	∞	∞	∞	∝	88	∝	—	—	—	∝	∞	∝	—	—	—
Methyl Cyclohexane	∞	∝	∞	46	∝	31	∞	—	—	—	∞	∞	∝	—	—	—
Oiticica Oil	∝[4]	∝	∞	∝[5]	∝	Ins.	∝	—	—	—	∞	∞	∝	—	—	—
Olive Oil	Ins.	∞	∞	Ins.	9	Ins.	∝	Ins.	Ins.	∞	∞	∝	∝	∞	∞	∞
Peanut Oil	Ins.	∞	∞	Ins.	7	Ins.	∞	Ins.	Ins.	∞	∞	∞	∝	—	—	—
Safflower Oil	Ins.	∝[6]	∞	Ins.	Ins.	Ins.	∝	Ins.	Ins.	∞	∞	∝	—	—	—	—
Soy Bean Oil	Ins.	∞	∞	Ins.	∝	∝	∝	Ins.	Ins.	∞	∞	∞	—	—	—	—
Turpentine	∞	∞	∞	55	∝	44	∝	28	∞	∞	∞	∞	∝	—	—	—
Tung Oil	Ins.	Ins.	∞	Ins.	Ins.	Ins.	∞	Ins.	Ins.	∞	∝[7]	∝[8]	∞	—	—	—
Water	∞	∞	∞	∞	∞	∞	∞	∞	∞	∞	∞	∞	∞	2.9	2.3	1.1

[1] Above 64 Mls. solute; smaller quantities give hazy solutions.
[2] Above 76 Mls. solute; smaller quantities give hazy solutions.
[3] Above 48 Mls. solute; smaller quantities give hazy solutions.
[4] Above 38 Mls. solute; smaller quantities give hazy solutions.
[5] Above 80 Mls. solute; smaller quantities give hazy solutions.
[6] Above 46 Mls. solute; smaller quantities give hazy solutions.
[7] Above 86 Mls. solute; smaller quantities give hazy solutions.
[8] Above 20 Mls. solute; smaller quantities give hazy solutions.

Solubility data refer only to room temperature.

Table 11.83: Solubility of Various Soaps in "Dowanols" (g/100 g) *(23)*

Compound	Dowanol Glycol Ether Products											
	EM	EE	EB	DM	DE	DB	PM	DPM	TPM	PiB-T	EPh	PPh
Monoethanolamine Laurate	38	33	37	15	22	21	19	4	3	8	3	7
Monoethanolamine Oleate	>100	>100	>100	>100	>100	>100	>100	>100	>100	>100	>100	>100
Monoethanolamine Stearate	<1	2	2	<1	<1	<1	3	1	<1	4	<1	2
Diethanolamine Laurate	>100	>100	>100	>100	>100	>100	>100	97	28	>100	>100	>100
Diethanolamine Oleate	>100	>100	>100	>100	>100	>100	>100	>100	>100	>100	>100	>100
Diethanolamine Stearate	9	18	26	7	12	14	28	7	4	21	38	66
Triethanolamine Laurate	>100	>100	90	67	72	58	>100	22	18	45	80	68
Triethanolamine Oleate	>100	>100	>100	>100	>100	>100	>100	>100	>100	>100	>100	>100
Triethanolamine Stearate	1	11	13	<1	5	5	15	6	5	13	27	21
Monoisopropanolamine Laurate	>100	>100	>100	>100	>100	>100	>100	37	15	>100	>100	>100
Monoisopropanolamine Oleate	>100	>100	>100	>100	>100	>100	>100	>100	>100	>100	>100	>100
Monoisopropanolamine Stearate	5	5	5	<1	3	4	3	2	1	13	<1	11
Monoethanolamine Tall Oil	>100	>100	>100	>100	>100	>100	>100	>100	>100	>100	>100	>100
Triethanolamine Tall Oil	>100	>100	>100	>100	>100	>100	>100	>100	>100	>100	>100	>100
Mixed Isopropanolamine Tall Oil	>100	>100	>100	>100	>100	>100	>100	>100	>100	>100	>100	>100
Potassium Oleate	>100	>100	>100	>100	>100	>100	>100	>100	>100	>100	—	—
Sodium Oleate	1	1	1	1	1	1	<1	<1	<1	<1	—	—

The solubilities of the various soaps in the Dowanol glycol ether products were determined by the following method: The various substances were added by weight to 25 g of Dowanol glycol ether. The samples were then shaken mechanically for 15 hours. All solubility studies were carried out at room temperature. Solubility was determined on basis of a true solution. Solubility in all cases is reported as grams dissolved in 100 g Dowanol.

Table 11.84: Solubility of Resins in "Dowanols" *(23)*

	Grams of Resin Soluble in 100 g Solvent, 25°C										
	EM	EE	EB	DM	DE	DB	PM	DPM	TPM	EPh	PPh
Acrylic Ester											
Acryloid[1] B-72	65	27	4			31	49	55	33	—	—
Acryloid[1] B-82	68	80	8	>70	60	36	65	56	40	50	—
Chlorinated Rubber											
Parlon[2] S-5	>90	>90	>90	>90	>90	>90	<1	>90	>90	<1	<1
Parlon[2] S-20	<1	>70	>70	>70	>70	>70	<1	>70	>70	<1	<1
Parlon[2] S-300	<1	>35	>35	>35	>35	>35	<1	>35	>35	<1	<1
Nitrocellulose											
RS½ sec	59	51	63	—	—	49	57	41	41	—	—
Polyamide											
Versamide[3] 940	<1	<1	<1	<1	<1	<1	<1	<1	<1	<1	<1
Rosin Ester											
Abalyn[2]	>100	>100	>100	>100	>100	>100	>100	>100	>100	> 100	>100
Rosin (Wood)	>100	>100	>100	>100	>100	>100	>100	>100	>100	<75	—
Shellac Orange	96	88	69	—	—	57	31	22	9	—	—
Ester Gum	1	34	36	1	38	36	>100	>100	85	—	—

1. Trademark of Rohm & Haas Company
2. Trademark of Hercules, Incorporated
3. Trademark of General Mills Chemicals, Incorporated

The solubilities of the various resins in DOWANOL were determined by the following method: The various substances were added by weight to 25 grams of DOWANOL. The samples were then shaken for 15 hours.

All solubility studies were carried out at room temperature except where otherwise indicated. Solubility was determined on the basis of a true solution. In all cases, solubility is reported as grams dissolved in 100 grams of DOWANOL.

Table 11.85: Solubility of Film Formers in "Dowanols" *(23)*

Film Former	PM	EE	EB	DM	DE	DB	PM	DPM	TPM
				Appearance of 10% Solution of Film Former in Dowanol Glycol Ether					
ETHOCEL 20 cps Med .... ethylcellulose	HG	HG	HG	HG	HG	HG	PS	PS	HG
ETHOCEL 10 cps Std.	H;PS	CS	CS	HG	HS	CS	HG	HG	CS
METHOCEL 60HG hydroxypropyl .... methylcellulose	HG	HG	PS	HG	PS	PS	HG	I	I
Cellulose Acetate Butyrate	CS	CS	SG	CS	CS	SG	SG	SG	S;PS
Cellulose Acetate ..	HG	I	I	HG	I	I	I	I	I
Polyvinyl Alcohol	I	I	I	I	I	I	I	I	I
Nitrocellulose RS ½ sec	CS	CS	CS	CS	CS	CS	CS	CS	CS

Legend: **HG** = Hazy Gel
　　　　　PS = Partly Soluble
　　　　　CS = Clear Solution
　　　　　　I = Insoluble
　　　　　SG = Swollen Gel
　　　　　HS = Hazy Solution
　　　　　　H = Hazy
　　　　　　S = Swollen

Table 11.86: Azeotropic Data for "Ansuls" *(15)*

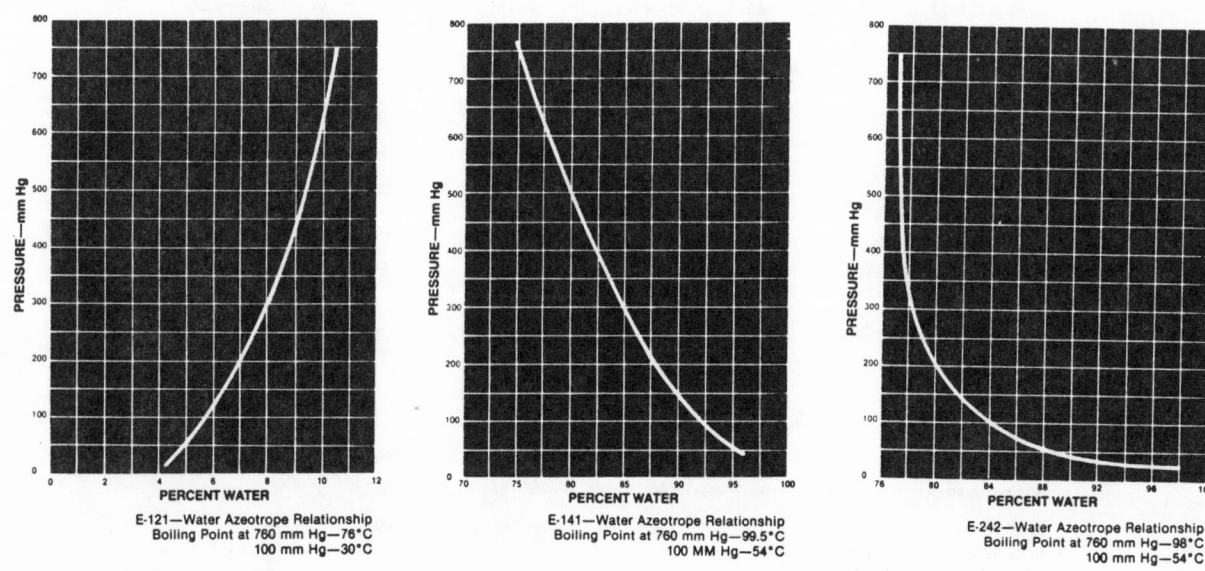

E-121—Water Azeotrope Relationship
Boiling Point at 760 mm Hg—76°C
100 mm Hg—30°C

E-141—Water Azeotrope Relationship
Boiling Point at 760 mm Hg—99.5°C
100 MM Hg—54°C

E-242—Water Azeotrope Relationship
Boiling Point at 760 mm Hg—98°C
100 mm Hg—54°C

E-161, E-181 and E-444 do not form water azeotropes

POLYETHYLENE GLYCOLS

Polyethylene glycols are water-soluble, nonvolatile unctuous liquids and solids. These polymers have the general formula $HO-CH_2-(CH_2-O-CH_2)_n-CH_2-OH$. The number following these compounds represents their average molecular weight, such as "Carbowax" polyethylene glycol 200, 300, 400, 600, 1000, 1500, 1540, 4000, 6000 and 20,000. These compounds range from oily to viscous liquids to waxlike solids and tough solids. Polymers with a molecular weight of above 1,000 are solids, those between 200 and 600 are liquids and molecular weights between 600 and 1,000 are more or less unctuous. When taken as a group, the freezing or melting point range, flash point, viscosity and specific gravity increase as the molecular weight increases. With these increases, such properties as vapor pressure, hygroscopicity, and water and organic solvent solubility decrease.

Table 11.87: Ashland Polyethylene Glycols *(69)*

	Specific Gravity 20°/20°C	Mol. Wt.	Freeze Range °C	Flash Pt. °C TOC	Visc. CPS at 212°F
Polyethylene Glycol 200	1.126	190-210	Supercools	340	4.3
Polyethylene Glycol 300	1.125	285-315	−15 to −8	385	5.8
Polyethylene Glycol 400	1.128	380-420	4 to 8	435	7.3
Polyethylene Glycol 600	1.128	570-630	20 to 25	475	10.5
Polyethylene Glycol 1000	1.101	950-1050	36.5 to 39.5	510	17.4
Polyethylene Glycol 1500	1.151	500-600	38 to 41	430	13-18
Polyethylene Glycol 1540	1.091	1300-1600	43 to 46	510	25-32
Polyethylene Glycol 4000	1.204	3000-3700	53 to 56	515	75-85
Polyethylene Glycol 6000	—	6000-7500	60 to 63	520	700-900

Table 11.88: "Carbowax" Polyethylene Glycols *(19)*

Product	Formula Molecular Weight/Range	Apparent Specific Gravity, 20/20°C.	Freezing Range, °C.	Water Solubility at 20°C., % by wt.	Viscosity, Centistokes at 210°F.	Comparative Hygroscopicity (Glycerine =100)	Surface Tension, 25°C., Dynes/Cm.	Refractive Index, n_D20°C.
CARBOWAX Polyethylene Glycol 200	190-210	1.1266	(e)	Complete	4.3	70	44.5	1.459
CARBOWAX Polyethylene Glycol 300	285-315	1.127	−15 to −8	Complete	5.8	60	44.5	1.463
CARBOWAX Polyethylene Glycol 400	380-420	1.1281	4 to 8	Complete	7.3	55	44.5	1.465
CARBOWAX Polyethylene Glycol 540 Blend (g)	500-600	1.151(h)	38 to 41	73	13-18	35	(f)	(f)
CARBOWAX Polyethylene Glycol 600	570-630	1.1279	20 to 25	Complete	10.5	40	44.5	1.467
CARBOWAX Polyethylene Glycol 1000	950-1050	1.101 (55/20°)	37 to 40	70, approx.	17.4	35	(f)	(f)
CARBOWAX Polyethylene Glycol 1540	1300-1600	1.0910(h)	43 to 46	70	25-32	30	(f)	(f)
CARBOWAX Polyethylene Glycol 4000 (flake and powder)	3000-3700	1.204, approx.	54 to 58	62	75-110	Low	(f)	(f)
CARBOWAX Polyethylene Glycol 6000 (flake and powder)	7000-9000	1.207, approx.	60 to 63	50, approx.	700-900	Low	(f)	(f)
CARBOWAX Polyethylene Glycol 14,000 (flake)	12,500-15,000	1.2024(h)	61 to 67	50, approx.	2700-4800	Low	(f)	(f)
CARBOWAX Methoxy Polyethylene Glycol 350	335-365	1.094	−5 to 10	Complete	4.1	....	40	1.455
CARBOWAX Methoxy Polyethylene Glycol 550	525-575	1.089 (40/20°)	15 to 25	Complete	7.5	....	37.5₄₀°C.	1.455₄₀°C.
CARBOWAX Methoxy Polyethylene Glycol 750	715-785	1.094 (40/20°)	27 to 32	Complete	10.5	....	40.7₄₀°C.	1.459₄₀°C.
CARBOWAX Methoxy Polyethylene Glycol 2000	1900, approx.	....	51.9, approx.	....	54.6	....	(f)	(f)
CARBOWAX Methoxy Polyethylene Glycol 5000	5000, approx.	....	59.2, approx.	....	613	....	(f)	(f)
Polyethylene Glycol Compound 20M	15,000, approx.	1.207	50 to 55(i)	50, approx.	14,500	....	52(j)	(f)

(e) Sets to glass below −65°C.

(f) Solid at 25°C.

(g) A blend of approximately equal parts of CARBOWAX Polyethylene Glycols 300 and 1540.

(h) Density, g./ml. at 20°C.

(i) Softening point.

(j) 50% by wt. aqueous solution.

Table 11.89: Dow Polyethylene Glycols *(23)*

Polyethylene Glycol Comparison Sheet

Polyglycol	Viscosity cstks. at 210° F.	Average Molecular Wt.	Freezing Range, °C.	Water Solubility	Flash Point, °F. COC	Solubility in Various Solvents[*]						Spec. Gravity 25/25° C.
						Acetone	Methanol	CCl₄	Mineral Oil	n-Heptane	Benzene	
E200	4.4	200	Super cools	Complete	360	>100	>100	2	<1	<1	>100	1.124
E300	5.7	300	−15 to −6	Complete	415	>100	>100	5	<1	<1	>100	1.125
E400	7.4	400	+4 to +8	Complete	460	>100	>100	>100	<1	<1	>100	1.125
E600	11.0	600	+20 to +25	Complete	480	>100	>100	>100	<1	<1	>100	1.126
E500M	16	(Mixture) 550	38 - 41	Complete	430	>100	>100	60	<1	<0.1	>100	1.200
E1000	18.5	1000	36 - 39	Complete	490	>100	>100	>100	<1	<1	>100	1.117
E1450	29	1450	43 - 46	Complete	490	60	>100	8	<1	<0.1	64	1.210
E2000	47	2000	47 - 50	Complete	510	4	67	<0.1	<1	<0.1	45	1.211
E4000	180	4500	54 - 57	Complete	515	<0.1	28	<0.1	<1	<0.1	32	1.212
E6000	580	7500	56 - 59	Complete	515	<0.1	10	<0.1	<1	<0.1	12	1.212
E9000	1120	9500	60 - 64	Complete	520	<0.1	<0.1	<0.1	<1	<0.1	1	1.212
E20.000	5960	20,000	54 - 58	Complete	550	<0.1	12	<0.1	<1	<0.1	18	1.215

[*]gms 100 gms at 77°F.

Competitive Products

DOW	CARBIDE	ANTARA	OLIN MATHIESON	ALLIED	JEFFERSON	WYANDOTTE
Polyglycol E200 Polyglycol E300 Polyglycol E400 Polyglycol E600	Carbowax[§] P.E.G. 200 Carbowax P.E.G. 300 Carbowax P.E.G. 400 Carbowax P.E.G. 600	Gafanol[*] E200 Gafanol E300 Gafanol E400 Gafanol E600	Poly-G[*] 200 Poly-G 300 Poly-G 400 Poly-G 600	P.E.G. 200 P.E.G. 300 P.E.G. 400 P.E.G. 600	P.E.G. 200 P.E.G. 300 P.E.G. 400 P.E.G. 600	Pluracol[*] E200 Pluracol E300 Pluracol E400 Pluracol E600
Polyglycol E1000 *Polyglycol E500M Polyglycol E1450 Polyglycol E2000 Polyglycol E4000 Polyglycol E6000 Polyglycol E9000 Polyglycol E20,000	Carbowax P.E.G. 1000 *Carbowax P.E.G. 1500 Carbowax P.E.G. 1540 Carbowax P.E.G. 4000 Carbowax P.E.G. 6000 Carbowax P.E.G. 20M		Poly-G 1000 *Poly-G B1530 Poly-G 1500	P.E.G. 1000 P.E.G. 1450 P.E.G. 4000	P.E.G. 1000 P.E.G. 550B P.E.G. 1450 P.E.G. 4000 P.E.G. 6000	Pluracol E1000 Pluracol E Blend 500 Pluracol E1500 Pluracol E2000 Pluracol E4000 Pluracol E6000

P.E.G.—Polyethylene Glycol.
Mixtures of a solid and liquid Polyethylene Glycol
E300 + E1450 (50-50).

Table 11.90: "Jeffox" Polyethylene Glycols *(48)*

$$HOCH_2CH_2(OCH_2CH_2)_nOH$$

DESCRIPTION
JEFFOX polyethylene glycols of average molecular weight 200, 300, 400 and 600 are viscous, water-white, hygroscopic, completely water-soluble liquids with low vapor pressure.

SPECIFICATIONS[1]

	JEFFOX			
PEG-	200	300	400	600
Ash, wt. % ...max.	0.05	0.01	0.01	0.01
Average molecular weight..... min.	190	285	380	570
max.	210	315	420	630
Color, Pt-Co.. max.	18	18	18	18
pH, 5% aqueous solution, 25°C ...... min.	4.5	4.5	4.5	4.5
max.	7.5	7.5	7.5	7.5
Water solubility, 25% aqueous ...	Free from haze or turbidity			

NOTE: Liquid polyethylene glycols of all molecular weights are clear and substantially free of foreign matter.

SELECT PROPERTIES

	JEFFOX			
PEG-	200	300	400	600
Flash point, COC, °F	340	385	435	475
PMCC, °F	328	350	390	410
Viscosity, SUS, 210°F......	40.3	45.4	52.7	61.7
Weight, lbs./gal., 20°C	9.38	9.38	9.39	9.34

[1] Products meeting the *National Formulary* specifications are also available.

Table 11.91: "Pluracol" Liquid Polyethylene Glycols *(47)*

Product	Average Molecular Weight	Form	Viscosity at 99°C. CS	Specific Gravity at 25°C.	Cloud Point (1% aqueous sol.) °C.	Flash Pt. °C.	Pour Point. °C.
E200	200	Liquid	4.4	1.120	>100	182	—
E300	300	Liquid	5.9	1.124	>100	210	-13
E400	400	Liquid	7.4	1.124	>100	238	5
E600	600	Liquid	10.8	1.125	>100	249	20
E1000	1000	CS	17.5	1.160	>100	255	38c
E1450	1450	CS	28.5	1.200	>100	255	45c
E2000	2000	CS	43.5	1.210	>100	>260	52c
E4000	4000	Flake/CS	134	1.230	>100	>260	59c
E4500	4500	Flake/CS	170	1.230	>100	>260	60c
E8000	8000	Prill/CS	750	1.240	>100	>260	61c

NF Grades Available: E400, E600, E1450 Prill, E4000 Prill, E8000 Prill

c Melting point.

Table 11.92: Compatibilities of Polyethylene Glycols *(19)*

	Polyethylene Glycol 400	1500	4000		Polyethylene Glycol 400	1500	4000
Nitrocellulose	C	C	PC	Rosin	C	PC	PC
Ethyl cellulose	I	I	I	Gum arabic	I	I	I
Methyl cellulose	PC	I	I	Raw castor oil	I	I	I
Bakelite BR3360	C	C	C	Tung oil	I	I	I
Bakelite BR8900	C	C	PC	Mineral oil	I	I	I
Vinylite AYAF	C	C	PC	Olive oil	I	I	I
Shellac	PC	PC	I	Pine oil	C	PC	I
Carnauba wax (No. 3)	I	i	I	Casein	C	C	PC
Paraffin wax	I	I	I	Zein	C	C	PC
Beeswax	I	I	I	Chlorinated starch	C	C	C
Ester gum	I	I	I	Gelatin	I	I	I

C = Compatible or soluble   PC = Partly compatible   I = Incompatible

Table 11.93: Comparative Hygroscopicity of Polyethylene Glycols *(23)*

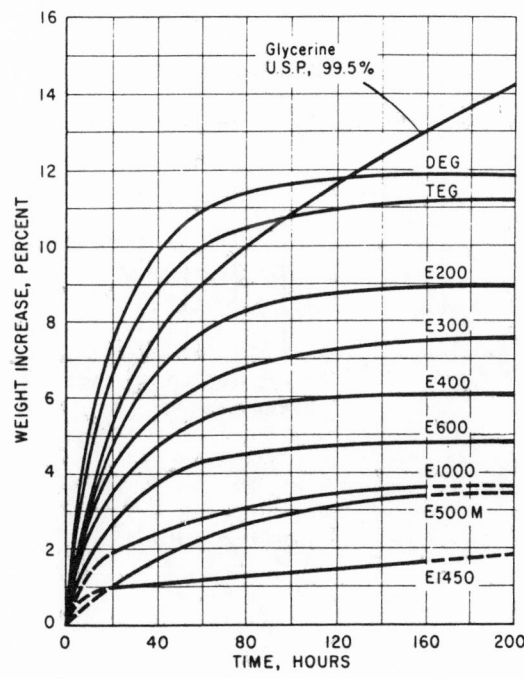

Table 11.94: Kinematic Viscosity of Polyethylene Glycols *(19)*

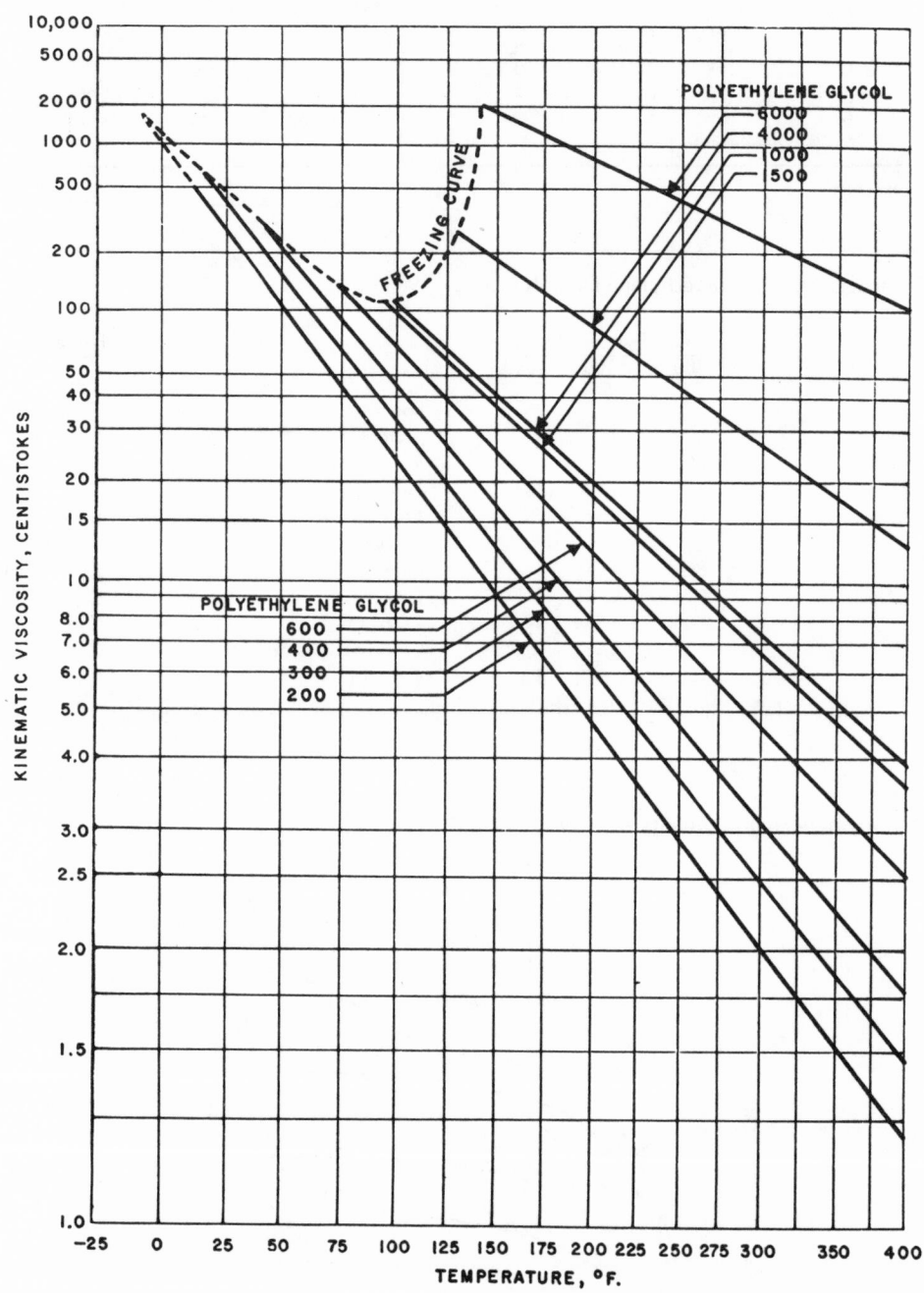

Table 11.95: Phase Diagrams of Aqueous Solutions of Polyethylene Glycols *(19)*

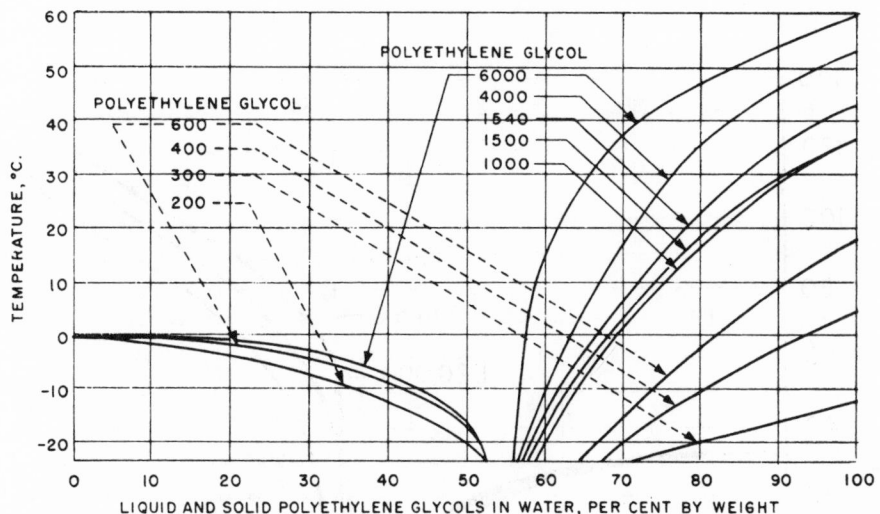

Table 11.96: Freezing Ranges of Polyethylene Glycols Plotted Against Their Average Molecular Weights *(19)*

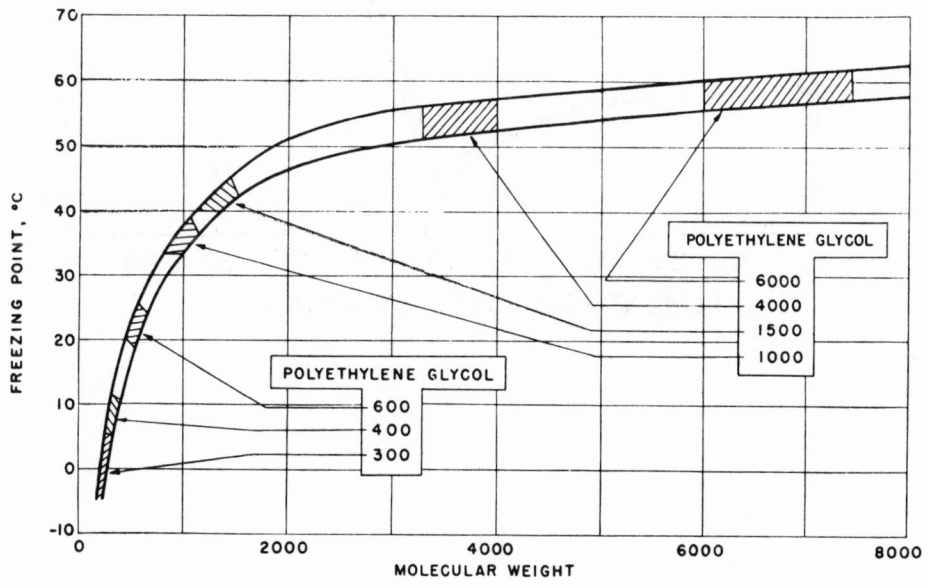

Table 11.97: Freezing Points of Aqueous Polyethylene Glycol *(19)*

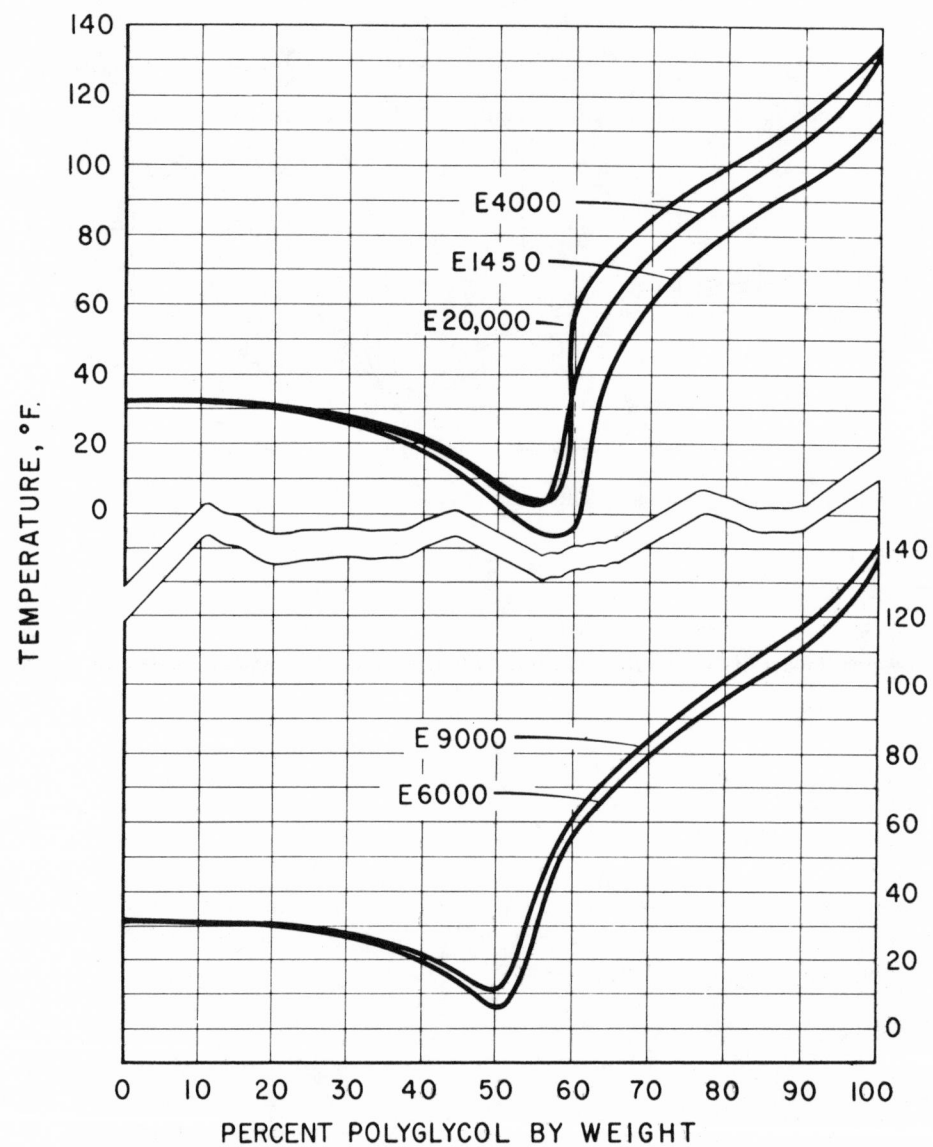

Table 11.98: Liquids Insoluble or Partly Soluble in the Liquid Polyethylene Glycols at 75°F *(23)*

	Approx. Solubility, vol. %					Approx. Solubility, vol. %			
	E200	E300	E400	E600		E200	E300	E400	E600
n–Butyl stearate	I	I	I	I	Isopropylbenzene	I	25	35	S
Butyraldehyde	I	I	I	I	Isopropylchloride	25	55	S	S
Carbon disulfide	10	10	10	25	Kerosene	I	I	I	I
Carbon tetrachloride	40	45	S	S	Lard oil	I	I	I	S
Castor oil	I	I	I	I	Lemon oil	I	I	I	I
Cod liver oil	I	I	I	I	Methyl laurate	I	I	I	I
Cottonseed oil	I	I	I	I	alpha–Methylstyrene	35	S	S	S
Cyclohexane	I	I	I	I	Olive oil	I	2	10	30
Decahydronaphthalene	I	I	I	I	Orange oil	I	I	I	I
Diamylnaphthalene	I	I	I	I	Pentachlorodiphenyl oxide	I	S	S	S
Dibutyl sebacate	I	I	I	I	Perchloroethylene	I	I	10	25
Diethyl benzene	I	I	10	25	Ricinoleic acid	I	I	I	I
Diethyl ether	25	25	25	25	Soya oil	I	I	I	I
Diisopropylbenzene	I	I	I	I	Sperm oil	I	I	I	I
Dodecyl alcohol	I	I	I	I	Tetrahydronaphthalene	10	25	45	S
Ethylbenzene	10	35	75	S	Tributyl aconitate	I	I	I	10
Ethylcyclohexane	I	I	I	I	Triethylbenzene	I	I	I	I
Gasoline	I	I	I	I	Xylene	10	35	65	S

S = soluble in all proportions   I = insoluble

Table 11.99: Solubilities of Pharmaceutical Materials in Polyethylene Glycol 400 at 75°C *(19)*

Organics			
Acetanilide	B	Salol	A
Acetophenetidin	B	Sulfadiazine	C
Acetylsalicylic acid	B	Sulfamerazine	C
Aloin	A	Sulfanilamide	B
Antipyrine	B	Sulfathiazole	B
Barbital	C	Sucrose	F
Benzocaine	A	Tannic acid	A
Benzoic acid	B	Terpin hydrate	C
Benzyl alcohol	A	Thymol	A
Caffeine	D	Urea	C
Camphor	B	Vanillin	B
Cetyl alcohol	D	Zinc sulfocarbolate	D
Chloral hydrate	A	**Flavoring Oils**	
Chlorobutanol	B	Anise oil	A
Chlorothymol	A	Benzaldehyde	A
Citric acid	A	Cinnamon oil	A
Diethoxin	E	Clove oil	A
Ethyl carbamate	A	Lemon oil	D
Gum arabic	G	Methyl salicylate	A
Hexamethylene tetramine	E	Peppermint oil	D
Menthol	B	Sweet orange oil	D
Paraldehyde	A	**Inorganics**	
Phenobarbital	B	Arsenic trioxide	G
Phenol	A	Boric acid	E
Piperazine *	B	Cupric oxide	G
Quinine	B	Ferric oxide	G
Resorcinol	A	Zinc oxide	G

* Solubility of piperazine at 20°C. is 11.5% by weight; 40°C., 18.5%; 80°C., 50.5%.

Descriptive Term	Quantity of PE Glycol 400 to Dissolve 1 Part Solute	Descriptive Term	Quantity of PE Glycol 400 to Dissolve 1 Part Solute
A Very Soluble	Less than one part	E Slightly Soluble	From 100 to 1000 parts
B Freely Soluble	From 1 to 10 parts	F Very Slightly Soluble	From 1000 to 10,000 parts
C Soluble	From 10 to 30 parts	G Practically Insoluble	More than 10,000 parts
D Sparingly Soluble	From 30 to 100 parts		

Table 11.100: Solubilities in Polyethylene Glycols 400, 1500 and 4000 at Room Temperature *(19)*

Solvents	Polyethylene Glycol			Gums and Waxes	Polyethylene Glycol		
	400	1500	4000		400	1500	4000
Castor oil, raw	I	I	I	Beeswax	I	I	I
Ethanol, 200 proof	C	<I	<I	Casein	C	C	PC
Ethyl acetate	C	15	<I	Ethyl cellulose	I	I	I
Ethyl ether	I	I	I	Gelatin	I	I	I
Heptane	I	0.5	0.01	Gum arabic	I	I	I
Isopropyl ether	I	I	I	Methyl cellulose	PC	I	I
Mineral oil	I	I	I	Paraffin wax	I	I	I
Olive oil	I	I	I	Shellac	PC	PC	I
Pine oil	C	PC	I	Zein	C	C	PC
Toluene	C	13	<I				
Water	C	69	60				

C = Compatible or soluble   PC = Partly compatible   I = Insoluble

Where numbers appear they indicate approximate percent by weight.

Table 11.101: Solubilities of Various Liquids in Polyethylene Glycols 400, 1500 and 4000 *(19)*

(Approximate percent by weight)

Liquids	Polyethylene Glycol 400		Polyethylene Glycol 1500		Polyethylene Glycol 4000	
	20°C.	50°C.	20°C.	50°C.	20°C.	50°C.
Water	C	C	68.8	97	60.3	84
Methanol	C	C	48	96	35	C
Ethanol (200 proof)	C	C	I	C	I	C
Acetone	C	C	20	C	I	99
Dichlorethyl ether	C	C	44	C	25	85
Trichlorethylene	C	C	50	90	30	80
Cellosolve solvent	C	C	I	C	I	88
Butyl Cellosolve	C	C	I	C	I	52
Carbitol solvent	C	C	2	C	I	63
Butyl Carbitol	C	C	I	C	I	64
Ethyl acetate	C	C	15	C	I	93
Dimethyl phthalate	C	C	30	90	13	74
Dibutyl phthalate	C	C	I	C	I	55
Ethyl ether	I	Ia	I	I	I	I
Isopropyl ether	I	I	I	I	I	I
Toluene	C	C	13	C	I	C
Heptane	I	I	0.50	0.01	0.01	0.01

I   Insoluble
Ia   Insoluble at boiling point
C   Greater than 100 cc of solvent

Table 11.102: Solubility of Various Solids in Liquid Polyethylene Glycols E200, E300 and E400 *(23)*

Solid	Approximate Solubility (g./100 g. Solvent) at 77°F.		
	E200	E300	E400
Alkyd resin plasticizer ("Paraplex" G-50)*	<2	<2	<2
Beeswax	<1	<1	<1
Butylated urea-formaldehyde resin ("Uformite" F240)*	<1	<1	<1
Carboxymethylcellulose	<1	<1	<1
Carnauba wax	<1	<1	<1
Casein	<1	<1	<1
Dibasic acid-modified rosin resin ("Teglac" 127)**	<60***	20	20
Dibasic acid-modified rosin resin ("Teglac" 152)**	<1	<1	<1
Dibasic acid-modified rosin resin ("Teglac" 161)**	<1	<1	<1
Gelatin	<1	<1	<1
Glue (animal)	<1	<1	<1
Long oil oxidizing alkyd resin ("Rezyl" 869)**	<1	<1	<1
Medium oil oxidizing alkyl resin ("Rezyl" 412)**	<1	<1	<1
Melamine-formaldehyde resin ("Melmac" 405)**	17	17	<1
Melamine-urea-formaldehyde resin ("Melurac" 301)**	<1	<1	<1
Methyl cellulose (400 cp.)	I	I	I
Paraffin	I	I	I
Rosin-modified maleic acid resin ("Amberol" 820)****	60	17	17
Shellac	I	I	I
Short oil oxidizing alkyd resin ("Rezyl" 387)**	I	I	I
Starch	I	I	I
Urea-formaldehyde resin ("Urac" 110)**	I	I	I

*	Resinous Products and Chemical Company	I = insoluble
**	American Cyanamid Company	
***	Mixture too viscous to continue addition of salts	
****	Rohm & Haas Company	

Table 11.103: Liquids Miscible in All Proportions with Liquid Polyethylene Glycols E200, E300, E400 and E600 at 75°F *(23)*

Acetaldehyde	o-Chlorophenol	Ethylidene dichloride*	Nitromethane
Acetic acid (glacial)	o-Cresol	Formamide	1-Nitropropane
Acetic anhydride	Cyclohexanol*	Furfural	2-Nitropropane
Acetone	Cyclohexanone	Glycerine*	Octyl alcohol
Acetylene tetrabromide*	Diacetone alcohol	Hydrochloric acid (conc.)*	Paraldehyde
Acrylonitrile	Dichloroacetic acid*	Isophorone	Phenetole*
Allyl alcohol*	o-Dichlorobenzene*	Isopropanol (99%)	Phenyl acetate
Allyl bromide*	Dichloroethyl ether*	Isopropyl bromide*	Phenyl ethyl acetate
Amyl acetate	Dichloroisopropyl ether*	Lactic acid (85%)	Phenyl ethyl alcohol
Amyl alcohol	Diethanolamine*	Mesityl oxide	4-Phenyl-m-dioxane
tert-Amyl alcohol	Diethylene glycol*	Methanol	Phosphoric acid (85%)
Aniline*	1,4-Dioxane	Methyl chloroform*	Piperidine
Benzaldehyde	Diphenyl oxide*	4-Methylcyclohexanol	n-Propanol
Benzene	Dipropylene glycol*	Methylene bromide*	Propylene dibromide*
Benzyl alcohol	Ethanol (95%)	Methylene chloride*	Propylene dichloride*
Bromobenzene*	Ethanolamine*	Methylene chlorobromide*	Pyridine
Bromoform*	Ethyl acetate	Methyl ethyl ketone	Styrene oxide*
n-Butyl acetate	Ethyl bromide*	Methyl formate	Tetrahydrofurfuryl alcohol
n-Butyl bromide*	Ethyl chloroacetate	Methyl isobutyl carbinol	Triacetin
n-Butyl phosphate	Ethyl lactate	Methyl isobutyl ketone	Trimethylene bromide*
n-Butyl stearate	Ethylene chlorohydrin	Methyl salicylate*	Trimethylene chlorobromide*
o-Chloroaniline	Ethylene dibromide*	Morpholine	Tripropylene glycol*
Chlorobenzene*	Ethylene dichloride	Nitrobenzene	Water
Chloroform*	Ethylene glycol*	Nitroethane	

* Products of Dow Chemical Company

POLYPROPYLENE GLYCOLS

Table 11.104: Dow Polypropylene Glycols *(23)*

Polyglycol Data Sheet

PRODUCT	VISCOSITY CENTISTOKES 100° F.	AVERAGE MOLECULAR WEIGHT	POUR POINT, °F.	FLASH POINT, °F. COC	SPECIFIC GRAVITY, 25/25°C.	SOLUBILITY IN VARIOUS SOLVENTS AT 25°C.					
						WATER	ACETONE	METHANOL	CARBON TETRACHLORIDE	n-HEXANE	BENZENE
Polyglycol P400	33	400	−49	390	1.007	S	S	S	S	S	S
Polyglycol P1200	91	1200	−40	460	1.003	P	S	S	S	S	S
Polyglycol P2000	167	2000	−31	445	1.002	I	S	S	S	S	S
Polyglycol P4000	537	4000	−20	445	1.001	I	S	S	S	S	S
Polyglycol 112-2	400	4500	−18	485	1.028	I	S	S	S	S	P	S
Polyglycol 112-3	230	3000	−22	457	1.019	I	S	S	S	S	S
Polyglycol 15-200	200	2600	−40	470	1.063	S	S	S	S	S	S
AMBIFLO' H149 Lubricant	27.0 (a)	1700	−33	440	1.044	S	S	S	S	I	S
AMBIFLO H438 Lubricant	73.0 (a)	2500	−25	440	1.055	S	S	S	S	I	S
AMBIFLO H813 Lubricant	125 (a)	2800	−15	440	1.060	S	S	S	S	I	S
AMBIFLO H1000 Lubricant	165 (a)	3500	−10	440	1.062	S	S	S	S	I	S
AMBIFLO L317 Lubricant	54 (a)	2500	−20	440	0.997	I	S	S	S	S	S

(a) 210°F.

S —soluble
I —insoluble
P —partially soluble

Competitive Products

DOW NAME	CARBIDE NAME
Polyglycol P400	Propylene Glycol 425
Polyglycol P1200	Propylene Glycol 1025
Polyglycol P2000	Propylene Glycol 2025
Polyglycol P4000	None
Polyglycol 112-2	None
Polyglycol 112-3	None
Polyglycol 15-200	None
AMBIFLO' H149 Lubricant	UCON' 50HB-660 Lubricant
AMBIFLO H438 Lubricant	UCON 50HB-2000 Lubricant
AMBIFLO H813 Lubricant	UCON 50HB-3520 Lubricant
AMBIFLO H1000 Lubricant	UCON 50HB-5100 Lubricant
AMBIFLO L317 Lubricant	UCON LB-1715 Lubricant

The above comparison is based on the similarity of the viscosity and melting points of Polyglycols as offered in the manufacturers' literature.

Table 11.105: "Jeffox" Polypropylene Glycols *(48)*

$$\underset{\text{Mol. Wt. 400 to 2,000 (avg.)}}{\overset{\overset{\displaystyle CH_3}{|}\qquad\overset{\displaystyle CH_3}{|}}{HOCHCH_2O(CH_2CHO)_n\text{—}H}}$$

DESCRIPTION

JEFFOX polypropylene glycols are essentially water-white, nonvolatile, viscous liquids with a faint ether-like odor. In general, their viscosity increases and their water solubility decreases with increasing molecular weight. Polypropylene glycols are substantially less water-soluble, less hygroscopic, and have lower pour points but are more oil-soluble than polyethylene glycols of similar molecular weights. They are better solvents for organic materials, both natural and synthetic, than are the corresponding polyethylene glycols. Typical specifications for two molecular weight polypropylene glycols are shown.

SPECIFICATIONS

		JEFFOX PPG- 400	2000
Acid number, mg. KOH/g.	max.	—	0.1
Appearance		Clear and substantially free of foreign matter.	
Ash, wt. %	max.	—	0.005
Color, Pt-Co	max.	75	50
Hydroxyl number, mg. KOH/g.	min.	250	54.5
	max.	280	57.5
pH; 10:1 methanol-water solvent	min.	5.0	5.0
	max.	7.0	7.0
Total unsaturation, meq./g.	max.	—	0.04
Viscosity, SUS at 100°F	min.	150	—
	max.	175	—
Water, wt. %	max.	0.3	0.1

Table 11.106: "Pluracol" P Polypropylene Glycols *(47)*

The "Pluracol" P's are a series of polypropylene glycols formed by adding propylene oxide to a propylene glycol nucleus. These materials are virtually water-white, nonvolatile liquids with a slight ether-like odor. With increasing molecular weight, their water-solubility decreases; "Pluracol" P410 being completely water-soluble, P710 moderately soluble, while P1010 and P2010 are practically insoluble at room temperature. Members of the "Pluracol" series are designated by their approximate molecular weight. The letter "P" indicates a polypropylene glycol. ("E" would indicate a polyethylene glycol.) Thus, "Pluracol" P410 is a polypropylene glycol having an approximate molecular weight of 400.

Product	Average Molecular Weight	Form	Brookfield Viscosity, cps[a]	Specific Gravity at 25°C.	Cloud Point (1% aqueous sol.) °C.	Flash Pt. °C.	Pour Point, °C.
P410	425	Liquid	72	1.005	>100	>205	-37
P710	775	Liquid	120	1.004	43	>205	-37
P1010	1050	Liquid	140	1.006	34	>205	-37
P2010	2000	Liquid	340	1.002	Ins.	>205	-37
P4010	3900	Liquid	—	1.000	Ins.	>205	-29

[a] Measured for liquids at 25°C. for pastes at 60°C. and for solids at 77°C.

Table 11.107: Compatibility of Various Materials with Polypropylene Glycols *(19)*

Material [a]	Polypropylene Glycol 425	1025	2025
Beeswax [b]	C	C	C
Carnauba wax no. 3 [b]	C	C	C
Casein	I	I	I
Castor oil	PC	PC	PC
Ethyl cellulose	I	I	I
Gum arabic	I	I	I
Mineral oil	PC	PC	PC
Nitrocellulose	C	C	C
Paraffin	PC	PC	PC
Pine oil	PC	PC	PC
Rosin	C	C	C
Shellac	I	I	I
Bakelite vinyl resin AYAF	I	I	I

(a)   The compatibility of typical batches of polypropylene glycols with various materials were determined in a ratio of 20 to 1.

(b)   When tested in a ratio of five parts wax to one part glycol carnauba wax and beeswax were found to be compatible, while paraffin wax was incompatible.

C = compatible   PC = partly compatible   I = incompatible

Table 11.108: Solubility of Liquids in Polypropylene Glycols at 77°F *(23)*

Liquids	Approximate Solubility, volume % P400	P1200	P2000
Diethanolamine	S	I	I
Diethylene glycol	S	10	10
Ethylene glycol	S	8	I
Glycerine	I	I	I
Oleic acid	I	S	S
Polyglycol E200	S	S	9
Polyglycol E400	S	S	I
Polyglycol E600	S	S	I
Propylene glycol	S	S	10
Sperm oil	20	S	S
Triethanolamine	S	I	I
Triethylene glycol	S	S	9

S = soluble in all proportions   I = insoluble

Table 11.109: Comparative Hygroscopicity of Polypropylene Glycols at 50% Relative Humidity and 70°F *(23)*

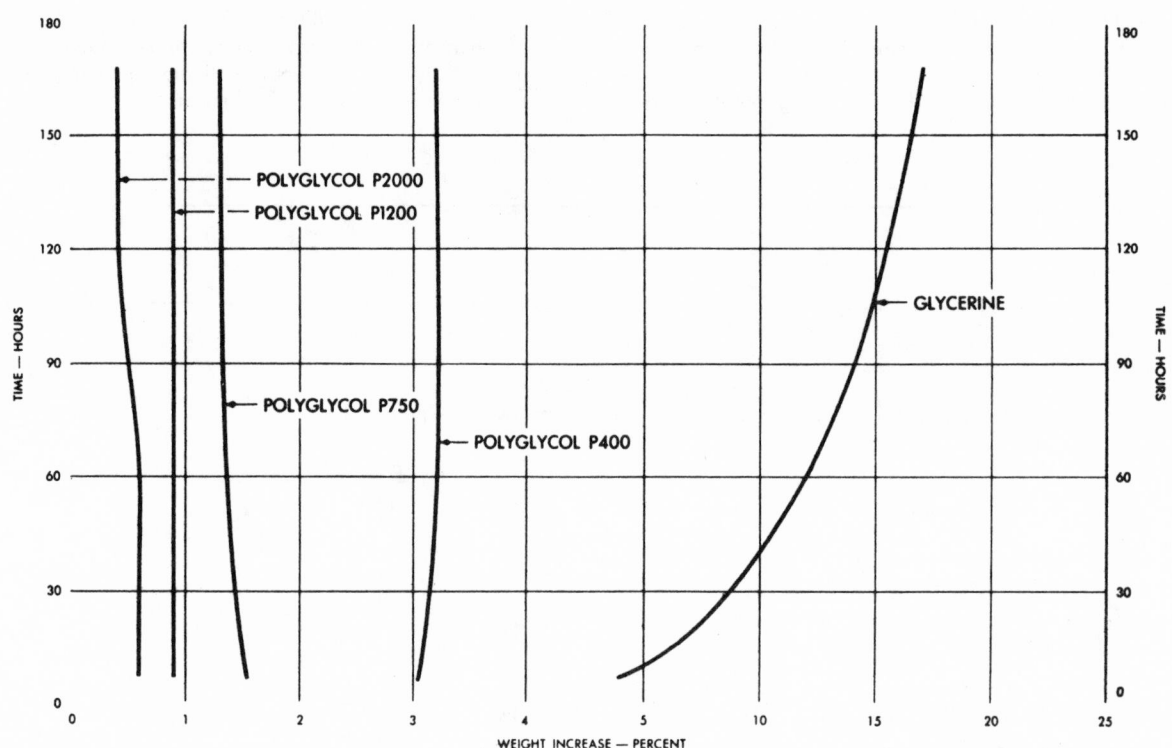

Table 11.110: Solubility of Aliphatic Hydrocarbons in Polypropylene Glycols *(23)*

	Approximate Solubility, vol. %					
	P400 77° to 120°F.		P1200 77° to 120°F.		P2000 77° to 120°F.	
Hexane	S	S	S	S	S	S
VM and P naphtha	S	S	S	S	S	S
No. 2 fuel oil	S	S	S	S	S	S
Mineral spirits	35	S	S	S	S	S
Hi-flash naphtha	30	S	S	S	S	S
SAE 20 lube oil	15	20	25	S	30	S
Light paraffin oil	5	10	20	40	25	S
Heavy mineral oil	2	10	5	15	10	20

Note: The solubility of aliphatic hydrocarbons in polyglycols P400, P750, P1200 and P2000 diminishes with an increase in the chain length of the hydrocarbon.

S = soluble in all proportions

Table 11.111: Solubility of Various Solids in Polypropylene Glycols at 77°F *(23)*

Solid Solid	Approximate solubility, g./100 g. of solvent		
	P400	P1200	P2000
Alkyd resin plasticizer ("Paraplex" G-50)	S	S	N
Beeswax	I	I	I
Butylated urea-formaldehyde resin ("Uformite" F240)	S	S	S
Carboxymethylcellulose	I	I	I
Carnauba wax	I	I	I
Casein	I	I	I
Dibasic acid-modified rosin resin ("Teglac" 127)	20	10	30
Dibasic acid-modified rosin resin ("Teglac" 152)	I	I	I
Dibasic acid-modified rosin resin ("Teglac" 161)	I	I	I
Gelatin	I	I	I
Glue	I	I	I
Long oil oxidizing alkyd resin ("Rezyl" 869)	200	200	200
Medium oil oxidizing alkyd resin ("Rezyl" 412)	120	30	35
Melamine-formaldehyde resin ("Melmac" 405)	I	I	I
Melamine-urea-formaldehyde resin ("Melurac" 301)	I	I	I
Methylcellulose	I	I	I
Nitrocellulose	I	I	N
Paraffin	I	I	I
Rosin-modified maleic acid resin ("Amberol" 820)	10	15	30
Shellac	S	I	I
Short oil oxidizing alkyd resin ("Rezyl" 387)	I	I	I
Starch	I	I	I

I = insoluble   S = soluble   N = not tested

POLYEPICHLOROHYDRINS

These viscous polyglycols are designated as the 166 series, and the number which follows represents the average molecular weight, such as 166-450, 166-900 and 166-1150. Their physical properties are influenced by the ether linkages, terminal hydroxyl groups and the chloromethyl side chains. Some of the other glycols are the polystyrene glycols, the trihydroxy polypropylene having primary terminal hydroxyl groups and those made by reacting ethylene and propylene oxide with glycerol.

"PLURONIC" POLYOLS

The "Pluronic" polyol compounds contain water-soluble polyoxyethylene groups attached to both ends of the water-insoluble polyoxypropylene chain. These block polymers are in fact prepared by adding either propylene oxide or polyethylene groups to the two hydroxyl groups of a propylene glycol nucleus. With this procedure compounds can be made which contain both hydrophilic and hydrophobic groups. The size of the hydrophilic portion of the molecule can be controlled to consist of from 10 to 80% of the final molecule and their molecular weights will vary from 800 to over 16,000. The simplified structure is represented as:

$$HO(CH_2-CH_2-O)_a (CH-CH_2-O)_b (CH_2-CH_2-O)_c H$$
$$\underset{CH_3}{|}$$

The "Pluronic" polyols range in physical form from mobile liquids to solids and they vary from being water-soluble to almost water-insoluble.

Table 11.112: Physical Properties of the "Pluronic" Polyols *(47)*

Product	Average Molecular Weight	Form	Brookfield Viscosity, cps[s]	Specific Gravity [a]/25°C.	Cloud Point (1% aqueous sol.)°C	Surface Tension, dynes/cm. 0.1% 25°C	Pour Point, °C	Dynamic Foam Height.[b] mm	HLB Value. 25°C.
L10	3200	Liquid	660	1.04	32	40.6	-5	90	14.0
L31	1100	Liquid	175	1.02	37	46.9	-32	18	4.5
L35	1900	Liquid	375	1.06	77	48.8	7	145	18.5
F38	4700	Prill/CS	260	1.07	>100	52.2	48[c]	>600	30.5
L42	1630	Liquid	280	1.03	37	46.5	-26	10	8.0
L43	1850	Liquid	310	1.04	42	47.3	-1	18	12.0
L44	2200	Liquid	440	1.05	67	45.3	16	360	16.0
L61	2000	Liquid	325	1.01	24	Ins.	-29	10	3.0
L62*	2500	Liquid	450	1.03	32	42.8	-4	35	7.0
L62D	2360	Liquid	400	1.03	35	43.0	-1	15	7.0
L63	2650	Liquid	490	1.04	34	43.3	10	120	11.0
L64	2900	Liquid	850	1.05	61	43.2	16	>600	15.0
P65	3400	Paste	180	1.06	82	46.3	27	>600	17.0
F68*	8400	Prill/CS	1000	1.06	>100	50.3	52[c]	>600	29.0
L72	2750	Liquid	510	1.03	25	39.0	-7	20	6.5
P75	4150	Paste	250	1.06	82	42.8	27	>600	16.5
F77	6600	Prill/CS	480	1.04	>100	47.0	48[c]	>600	24.5
L81	2750	Liquid	475	1.02	20	Ins.	-37	10	2.0
P84	4200	Paste	285	1.03	74	42.0	18	>600	14.0
P85	4600	Paste	310	1.04	85	42.5	29	>600	16.0
F87	7700	Prill/CS	700	1.04	>100	44.0	49[c]	>600	24.0
F88	11,400	Prill/CS	2300	1.06	>100	48.5	54[c]	>600	28.0
L92	3650	Liquid	700	1.03	26	35.9	7	25	5.5
F98	13,000	Prill/CS	2700	1.06	>100	43.0	58[c]	>600	27.5
L101	3800	Liquid	800	1.02	15	Ins.	-23	10	1.0
P103	4950	Paste	285	1.04	86	34.4	21	>600	9.0
P104	5900	Paste	390	1.04	83	33.1	32	>600	13.0
P105	6500	Paste	750	1.05	92	39.1	35	>600	15.0
F108	14,600	Prill/CS	2800	1.06	>100	41.2	57[c]	>600	27.0
L121	4400	Liquid	1200	1.01	14	33.0[d]	5	5	0.5
L122	5000	Liquid	290[e]	1.03	19	33.0	20	15	4.0
P123	5750	Paste	350	1.01	90	34.1	31	360	8.0
F127	12,600	Prill/CS	3100	1.05	>100	40.6	56[c]	>600	22.0
10R5	1950	Liquid	440	1.058	69	50.9	15	45	11.6
10R8	4550	Prill/CS	400	1.062	98	54.1	46[c]	120	16.5
12R3	1800	Liquid	340	1.040	53	42.7	<-20	<5	5.0
17R1	1900	Liquid	300	1.018	32	33.0	-27	5	2.5
17R2	2150	Liquid	450	1.030	35	41.9	-25	5	4.1
17R4	2650	Liquid	600	1.048	46	44.1	18	20	6.7
17R8	7000	Prill/CS	—	1.064	81	47.3	53[c]	45	13.4
22R4	3350	Liquid	950	1.048	40	—	24	—	6.3
25R1	2700	Liquid	460	1.017	27	36.3	-27	<5	2.1
25R2	3100	Liquid	680	1.028	29	37.5	-5	<5	3.5
25R4	3600	Liquid	1110	1.048	40	40.9	25	70	6.0
25R5	4250	Paste	—	1.036	44	43.5	30	160	9.0
25R8	8550	Prill/CS	2600	1.062	45	46.1	54[c]	50	12.1
31R1	3250	Liquid	660	1.018	25	34.1	-25	<5	1.7
31R2	3300	Liquid	850	1.029	30	38.9	9	<5	2.9
31R4	4150	Paste	300	1.028	31	41.2	24	80	6.0

[a] Measured for liquids at 25°C, for pastes at 60°C, and for solids at 77°C;   [b] 0.1% solution at 400 ml/minute at 49°C;   [c] Melting point;
[d] Not completely soluble;   [e] Measured at 60°C.
CS=Cast Solid   * Low foaming grade also available

GLYCERINE ETHERS

Table 11.113: Glyceryl α-Monomethyl Ether *(2)*

α–Monomethyl Ether of Glycerine $CH_3OCH_2CHOHCH_2OH$

Glyceryl α-monomethyl ether is a colorless liquid, soluble in benzene, ethyl alcohol, glycerol and water but insoluble in gasoline and carbon tetrachloride. It is a solvent for rosin, and when mixed with butyl acetate is compatible with nitrocellulose. It may be used as a selective solvent and in the manufacture of alkyd resins.

Boiling range at 745 mm	90% between 215–220°C
Refractive index, n $\frac{25°}{D}$	1.442
Specific gravity at 25/25°C	1.1147
Weight per gal	9.29 lb

Table 11.114: Glyceryl α,γ-Dimethyl Ether *(2)*

α,γ-Dimethyl Ether of Glycerine $H_3COCH_2CHOHCH_2OCH_3$

Glycerine α,γ-dimethyl ether is a water-white liquid soluble in benzene, gasoline, carbon tetrachloride, ethyl alcohol, water and glycerine, but insoluble in linseed oil and other fixed oils. It is a solvent for rosin, cellulose acetate and when mixed with butyl acetate is compatible with nitrocellulose. It has use as a solvent and plasticizer.

Boiling range at 736 mm.	90% between 164–170°C
Specific gravity at 25/25°C	1.003
Weight per gal	8.36 lbs

Table 11.115: Glyceryl α-Mono-n-Butyl Ether *(2)*

α-Mono-n-Butyl Ether of Glycerine $C_4H_9OCH_2CHOHCH_2OH$

α-Mono-n-butyl ether of glycerine is a colorless liquid, soluble in benzene, gasoline, ethyl alcohol and carbon tetrachloride, but only slightly soluble in water and glycerol. It is a solvent for rosin and ester gum and may be used in the preparation of varnishes made with these substances.

Boiling range at 18 mm.	90% between 133–137°C
Refractive index, n $\frac{25°}{D}$	1.434
Specific gravity at 25/25°C	0.945
Weight per gal	7.87 lbs

Table 11.116: Glyceryl α-Monoisoamyl Ether *(2)*

α-Monoisoamyl Ether of Glyceryl $C_5H_{11}OCH_2CHOHCH_2OH$

α-Monoisoamyl ether of glyceryl is a colorless liquid which generally contains small amounts of other amyl isomers. It is soluble in benzene, ethyl alcohol, halogenated hydrocarbons, carbon tetrachloride, gasoline, linseed oil, and other fixed oils and, in certain amounts, soluble in glycerol and water. It is a solvent for rosin and, when mixed with butyl acetate, is compatible with nitrocellulose. It may be used as a solvent in the preparation of alkyd resins and in the synthesis of ester derivatives.

Boiling range at 745 mm	90% between 252–260°C
Refractive index, N $\frac{25°}{D}$	1.442
Specific gravity at 25/25°C	0.987
Weight per gal	8.22 lbs

Table 11.117: Glyceryl α,γ-Diisoamyl Ether *(2)*

α,γ-Diisoamyl Ether of Glycerine $C_5H_{11}OCH_2CHOHCH_2OC_5H_{11}$

Glyceryl α,γ-diisoamyl ether is a water-white liquid which may contain small quantities of other amyl isomers. It is soluble in ethyl alcohol, benzene, gasoline, carbon tetrachloride and linseed oil, but insoluble in water and glycerol. It is a solvent for ester gum and rosin and has use as a solvent and plasticizer.

Boiling range at 10 mm	90% between 147–153°C
Refractive index, n $\frac{25°}{D}$	1.432
Specific gravity at 25/25°C	0.903
Weight per gal	7.52 lbs

Table 11.118: Miscellaneous Glycerine Ethers (2)

Glycerine ethers range widely from low-boiling liquids to high-boiling solids. The solubility varies equally from complete water miscibility to complete water insolubility. The following lists these glyceryl ethers with their density and boiling points. At the present time, these ethers have limited commercial application.

Glyceryl-Ether	$\underline{d}$	b.p. (or m.p.) °C.
α-Isoamyl	0.987^{15}_{15}	$137\text{-}9_{27}$
		$251\text{-}2_{758}$
α, γ—di-Isoamyl	0.903^{15}_{15}	$147\text{-}53_{10}$
		269
α-Benzyl	1.196^{15}_{15}	$124\text{-}6_2$
α-n—Butyl	0.945^{15}_{15}	$133\text{-}7_{18}$
Cresyl		
α-Ethyl	1.063	$231\text{-}2_{762}$
α, γ-di-Ethyl	0.920_{21}	190
tri-Ethyl	0.886^{25}_{4}	$103\text{-}5_{60}$
		181_{760}
Epiethylin	0.94_{12}	128-9
Glycidol	1.1143_{25}	41_1
α-Methyl	1.1147^{25}_{25}	110_{12}
		221_{754}
β-Methyl		
α, γ-di-Methyl	1.003^{15}_{15}	$69.5\text{-}70.5_{15}$
		$164\text{-}70_{738}$
tri-Methyl	0.937^{25}_{4}	148_{766}
Epimethylin	1.002_4	$113\text{-}4_{770}$
mono-α-Naphthyl		m.p. 91-2
mono-β-Naphthyl		m.p. 109-10
α-Phenyl		$185\text{-}7_{15}$
		$150\text{-}5_4$
		m.p. 53-4
α, γ-di-Phenyl		287-8
		m.p. 80-1
α-o-Cl-Phenyl		m.p. 56
α-p-Cl-Phenyl		m.p. 76
mono-2, 4-di-Nitrophenyl		m.p. 83
Epiphenylin	1.08^{25}_{4}	$115\text{-}6_{3\text{-}4}$
α-Propyl	1.074^{15}_{4}	$118\text{-}22_{15}$
α, γ-di-Isopropyl	0.915_{15}	112-3
α, γ-di-n-Propyl		215-7
mono-p-Tolyl		m.p. 73-4

Ketones

ACETONE

Dimethyl Ketone, Methylacetyl, Propanone-2 $CH_3-CO-CH_3$

Acetone is a colorless, limpid, mobile, hygroscopic, flammable liquid having a mint-like odor.

Table 12.1: Physical Properties of Acetone (41)

Typical Properties

Molecular Weight	58.08	Boiling Range, 760 mm, °C	
Color (Pt-Co Scale), max	5	Initial Boiling Point, min	55.1
Weight/Vol, 20°C,		Dry Point, max	57.1
lb/gal (U. S.)	6.59	Freezing Point, °F (°C)	−138 (−95)
kg/litre	0.79	Flash Point, Tag Closed Cup, °F (°C)	−4 (−20)
lb/gal (Imperial)	7.91	Tag Open Cup, °F (°C)	−2 (−19)
Solubility, 20°C, wt %		Fire Point, °F (°C)	−2 (−19)
In water	Complete	Flammable Limits in Air, % by volume	
Water in	Complete	Lower	2.6
Evaporation Rate (n-butyl acetate = 1)	7.7	Upper	12.8
Dilution Ratio, toluene	4.6	Autoignition Temperature (ASTM D-2155),	
VM & P naphtha	0.55	°F (°C)	1000 (538)
Refractive Index, 20°C	1.3589	NFPA Classification 30	IB
Vapor Pressure, 20°C, mm Hg	180	DOT Classification	Flammable Liquid
Specific Gravity 20°/20°C	0.792	DOT Labels Required	Flammable Liquid

Table 12.2: Low Temperature Characteristics of Aqueous Solutions of Acetone (19)

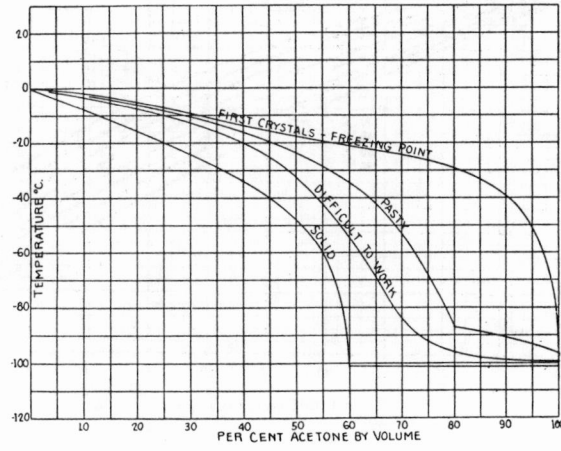

Table 12.3: Solubility of Various Materials in Acetone *(44)*

SOLUBILITY OF SHELLACS IN ACETONE

TYPE	PER CENT SOLUBLE AT BOILING POINT OF ACETONE
Superfine orange shellac	98.8
Superfine shellac	92.8
T. N. shellac, No. 1	95.6
T. N. shellac, No. 2	98.8
A. C. garnet	81.3
Refus lac	63.0

PERCENTAGE OF ACETONE-INSOLUBLE MATTER IN VARIOUS RESINS

TYPE	PER CENT INSOLUBLE IN ACETONE
Kauri, pale	8.90
Kauri, brown	38.70
Kauri, bush	20.70
Rosin	Soluble
Burgundy pitch	Soluble
Stockholm tar	Soluble
Mastic	9.50
Sandarac	Soluble
Madagascar copal, fused	84.80

SOLUBILITY OF FATS, OILS AND GREASES IN ACETONE

TYPE	MISCIBILITY A 25°C.	PER CENT SOLUBLE AT 25°C.	SOLUBILITY AT BOILING POINT OF ACETONE
Chinawood oil	M		
Coconut oil (refined and bleached)	M		
Corn oil (raw)	M		
Cottonseed oil (refined and bleached)	M		
Cottonseed oil (hydrogenated, Crisco)	M	100.0	M
Cottonseed oil (hydrogenated)		32.0	M
Cottonseed oil (stearin)	M		
Cottonseed oil (summer)	M		
Cottonseed oil (winter)	M		
Fish oil (herring, raw)	M		
Fish oil (hydrogenated)		35.8	M
Fish oil (menhaden, raw)	m	99.8	M
Grease, brown	m	96.4	99.8
Grease, garbage	m	99.6	99.7
Grease, white	m	97.3	M
Linseed oil, raw	M		

M—miscible in all proportions.
m—part soluble at 25°C is miscible in all proportions.

SOLUBILITY OF COPAL RESINS IN ACETONE

TYPE	PER CENT SOLUBLE AT BOILING POINT OF ACETONE	SOLUBILITY OF PART SOLUBLE AT 25°C
Congo	40.8	M¹
Manila, soft	96.6	M
Elemi	100.0	M
Yacca	96.6	M
Sandarac	97.0	M¹
Sierra Leone	55.5	M¹
Borneo pontianac	93.5	M¹
Batavia dammar	88.7	M¹
Red, accrodites	95.2	M

M—miscible in all proportions.
M¹—miscible in concentrated solutions with separation on dilution.

SOLUBILITY OF WATER GUMS IN ACETONE

TYPE	PER CENT SOLUBLE AT BOILING POINT OF ACETONE
Arabic gum	11.9
Indian gum	16.7
Senegal gum	12.0
Tragacanth, Allepa	9.2
Tragacanth, Persian	8.0
Tragacanth, Turkey	7.2

SOLUBILITY OF ASPHALTS AND BITUMENS IN ACETONE

TYPE	PER CENT SOLUBLE AT BOILING POINT OF ACETONE
Alberite	5.8
Asphalt, blown, from mid-continental petroleum	56.4
Bermudez pitch, refined	62.4
Coal-tar pitch, refined	70.4
Fatty acid pitch, soft grade	62.4
Fatty acid pitch, medium grade	54.3
Gilsonite selex	25.0
Grahamite	1.6
Mexican petroleum asphalt, steam-distilled, medium grade	44.2
Mexican petroleum asphalt, steam-distilled, soft grade	64.3
Petroleum asphalt, steam-distilled, California, medium grade	81.0
Residual oil from Gulf Coast	61.0
Residual oil from steam distillation of mid-continental petroleum asphalt	97.2
Syrian asphalt	5.9
Trinidad pitch, refined	42.0

Table 12.4: Specific Gravity of Aqueous Solutions of Acetone at Different Temperatures *(19)*

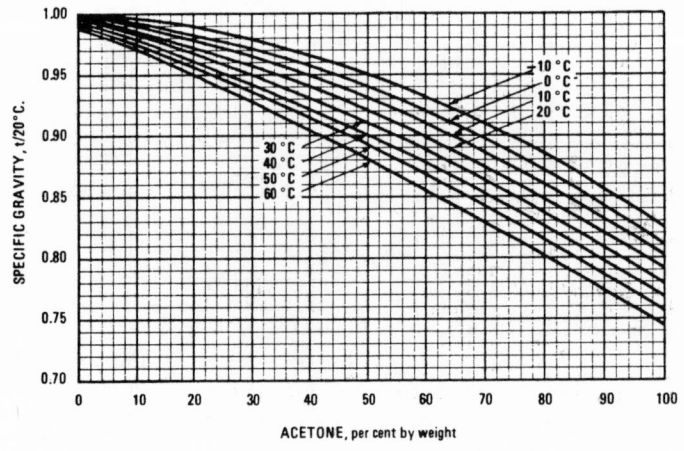

Table 12.5: Surface Tension of Aqueous Solutions of Acetone at 25°C *(19)*

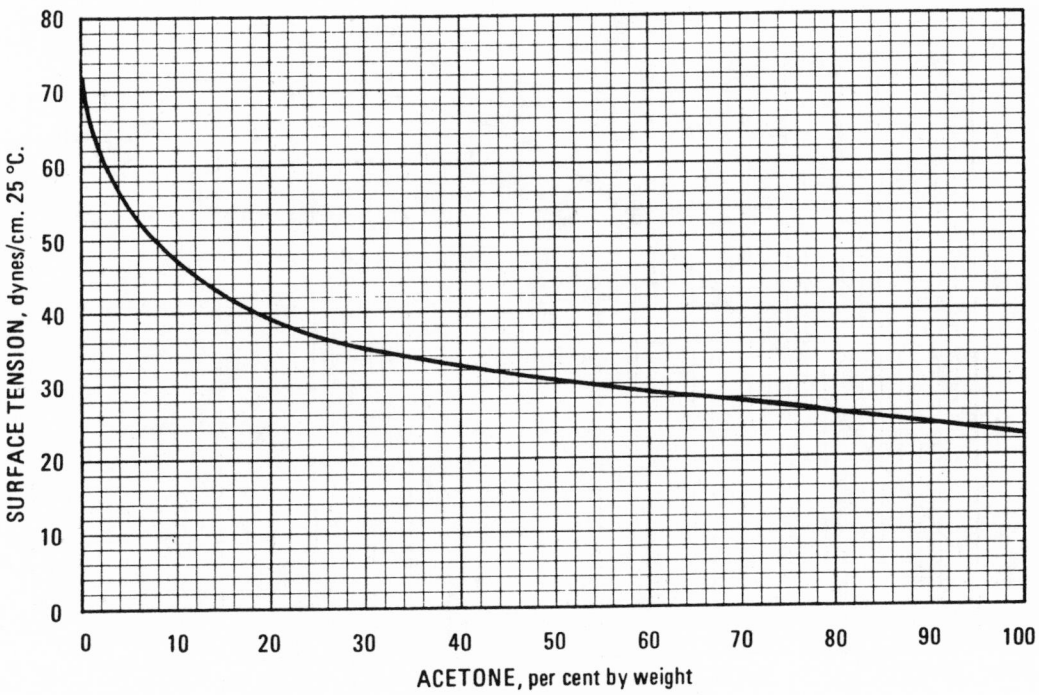

Table 12.6: Viscosity of Aqueous Acetone Solutions at 25°C *(19)*

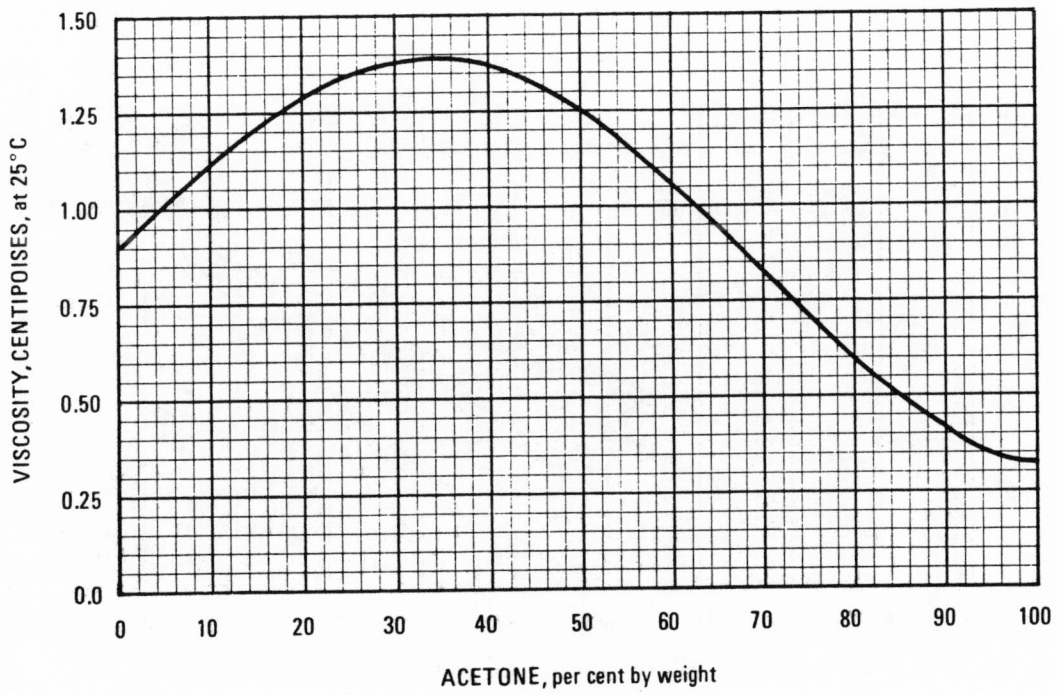

Table 12.7: Refractive Index of Aqueous Solutions of Acetone at 25°C *(19)*

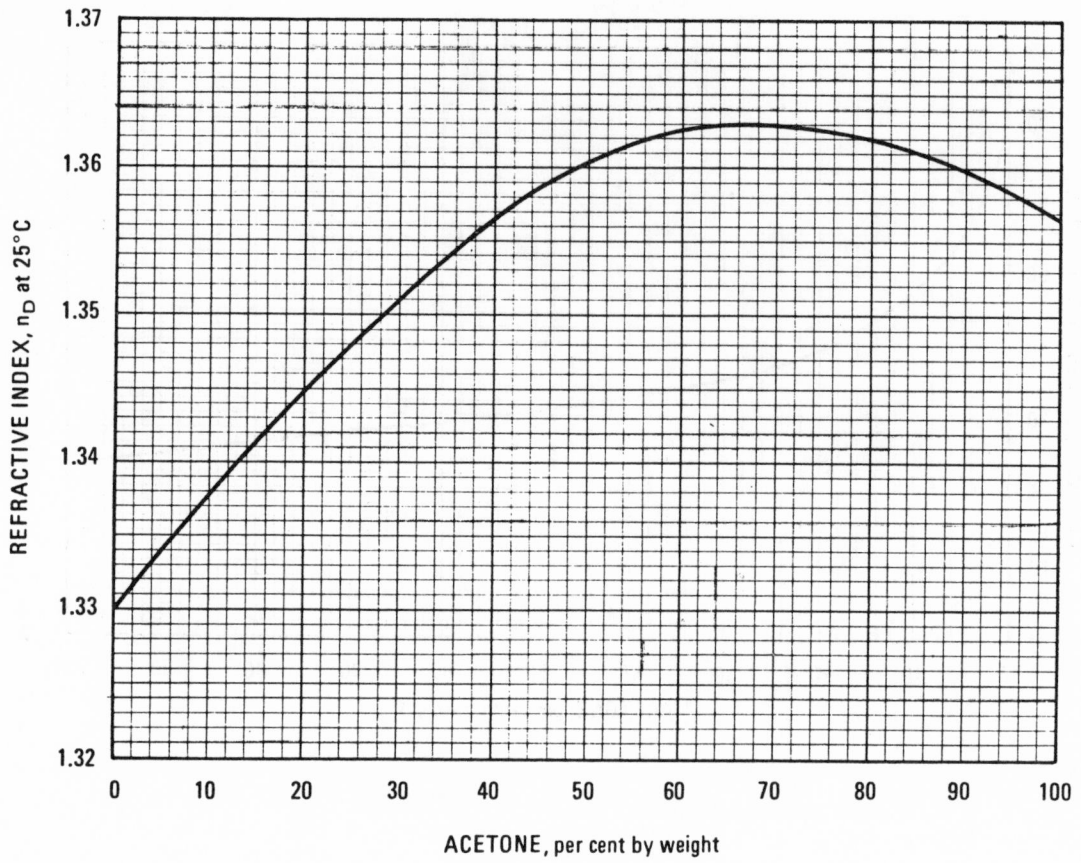

Table 12.8: Liquid-Vapor Equilibria for Aqueous Solutions of Acetone at Different Pressures *(19)*

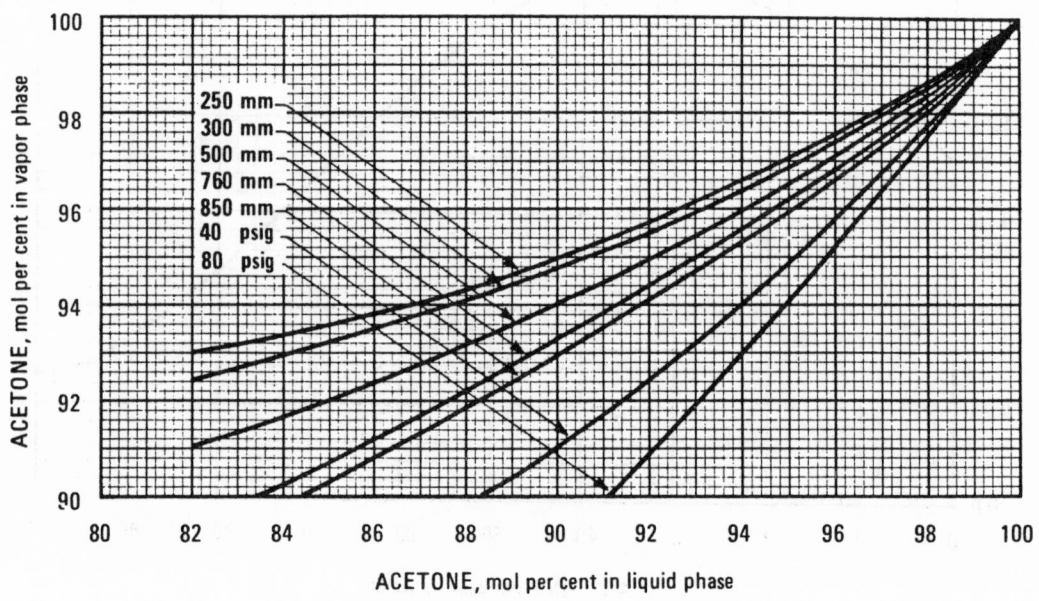

Table 12.9: Freezing Point of Aqueous Solutions of Acetone *(19)*

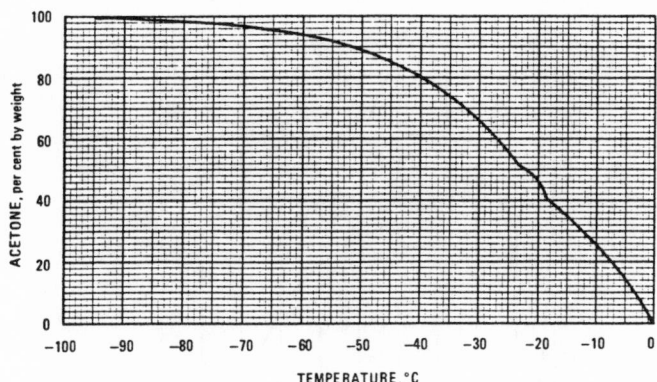

METHYL ETHYL KETONE

MEK, Butanone-2, Ethyl Methyl Ketone $\qquad$ $CH_3—CO—C_2H_5$

Methyl ethyl ketone is a colorless, stable, mobile, flammable liquid with an odor like acetone.

Table 12.10: Physical Properties of Methyl Ethyl Ketone *(2)*

Azeotropic Mixtures

	% by wt.		% by wt.	B.P. (°C)
Methyl ethyl ketone	37.5	Benzene	62.5	78.4
	73	*tert*-Butyl alcohol	27	77.5
	84.7	Carbon disulfide	15.3	45.9
	29	Carbon tetrachloride	71	73.8
	40	1,3-Cyclohexadiene	60	73.0
	40	Cyclohexane	60	72.0
	12	Ethyl acetate	82	77.0
	60	Ethyl alcohol	40	74.8
	20	Ethyl sulfide	80	77.5
	70	Isopropyl alcohol	30	77.5
	52	Methyl propionate	48	79.3
	55	Propyl formate	45	79.5
	75	Propyl mercaptan	25	55.5
	45	Thiophene	55	76

Ternary Mixtures

						B.P. (°C)
(1) Methyl ethyl ketone	22.2	Water	3.0	CCl_4	74.8	65.7
(2)	17.8		8.9	C_6H_6	73.8	68.9
Upper layer of (2)	19.0		0.4		80.6	
Lower layer of (2)	3.5		96.4		0.1	

Typical Properties and Specifications

Boiling point at 760 mm	79.6°C
Coefficient of expansion	0.00076 per °F
Electrical Conductivity	1.0×10^{-7} ohms at 25°C
Explosive limits	1.97%—10.2%
Flash point (Tag Closed Cup)	25°F
Freezing point	−86.4°C
Heat of combustion	582 Cal./mole
Latent heat of Vaporization at 20°C	106.0 cal./g
Refractive Index, N 20/D	1.3788
Solubility of water in solvent at 20°C	10% by wt.
Specific gravity at 20/20°C	0.805–0.807
Specific heat	0.55 cal./g
Surface tension	
0°C 26.9 dynes/sq cm	
20 24.6	
40 22.3	
75 18.4	
Viscosity at 15°C	0.00423 poise
Weight per gallon at 20°C	6.72 lbs.
Acidity (as acetic)	0.0025 by wt. (max.)
Distillation range (ASTM)	70°–80.5°C
Non-volatile matter	3 mg. per 100 ml. (max.)
Purity	99%

Table 12.11: Methyl Ethyl Ketone and Water *(14)*

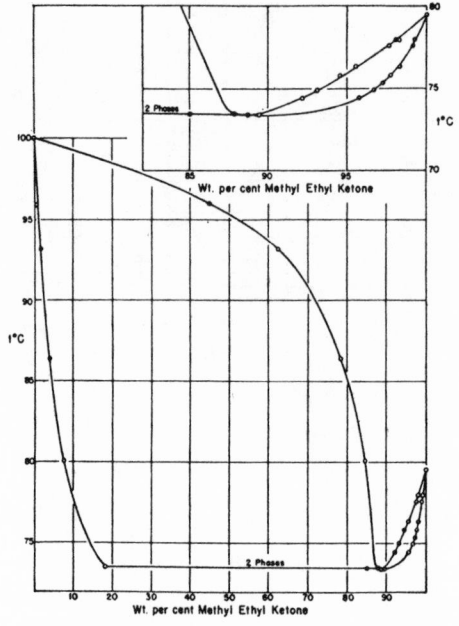

Vapor and Liquid Compositions at the Boiling Point

Pressure = I Atm.

METHYL n-PROPYL KETONE

Pentanone-2 $CH_3 \cdot CO \cdot CH_2 \cdot C_2H_5$

Commercial methyl n-propyl ketone, produced synthetically by dehydrogenation of the corresponding alcohol, consists of a mixture of methyl n-propyl and diethyl ketones in the approximate ratio of 3 to 1, and contains at least 97% of these ketones, the balance being secondary amyl alcohol. It is a colorless liquid, soluble in alcohol and ether but only very slightly soluble in water.

Table 12.12: Properties of Methyl n-Propyl Ketone *(41)*

Typical Properties

Molecular Weight ($C_5H_{10}O$)	86.13	Specific Gravity at 20°/20°C	0.807
Branched-Chain Ketones, wt % (max)	10	Boiling Range at 760 mm, °C	
Color (Pt-Co Scale), max	15	Initial Boiling Point, min	101
Evaporation Rate (n-butyl acetate = 1)	2.3	Dry Point, max	105
Weight/Vol at 20°C		Freezing Point, °F (°C)	-122 (-86)
lb/gal (U.S.)	6.72	Flash Point, Tag Closed Cup, °F (°C)	46 (8)
kg/L	0.81	Tag Open Cup, °F (°C)	50 (10)
lb/gal (Imperial)	8.06	Fire Point, °F (°C)	50 (10)
Solubility at 20°C, wt %		Flammable Limits in Air, % by volume	
In water	3.1	Lower, at 94°F (34°C)	1.56
Water in	4.2	Upper, at 144°F (62°C)	8.7
Dilution Ratio, toluene	3.9	Autoignition Temperature (ASTM D 2155), °F (°C)	840 (449)
VM & P naphtha	1.0	NFPA Classification 30	IB
Refractive Index at 20°C	1.3904	DOT Classification	Flammable Liquid
Vapor Pressure at 20°C, mm Hg	27.8	DOT Labels Required	Flammable Liquid

(continued)

Table 12.12: (continued)

Comparison of Solvent Power
MPK vs Other Solvents

Solvent	Solution Viscosity at 25°C, cP (mPa · s)			
	RS ½-Sec Nitrocellulose 8%	CAB-381-0.5[a] 10%	*Elvacite* 2010 Acrylic[b] 10%	VYNS[c] 10%
Ethyl Acetate[a]	23	28.3	5.8	–
Isopropyl Acetate[a]	25	31.0	6.6	–
MEK	14	18.8	3.6	22.8
MPK[a]	16	20.8	4.5	31.5
MIBK[a]	23	27.0	5.8	59.2

[a]*an Eastman product*
[b]*product of Du Pont Company*
[c]*product of Union Carbide Corporation*

METHYL n-BUTYL KETONE

Hexanone-2 $CH_3 \cdot CO \cdot C_4H_9$

Methyl n-butyl ketone is a colorless liquid, freely soluble in alcohol and ether but very slightly soluble in water.

Table 12.13: Properties of Methyl n-Butyl Ketone *(41)*

Molecular Weight ($C_6H_{12}O$)	100.16	Water, wt %	0.05
Melting Point, °C	-56.9	Branched-Chain Ketones, max, wt %	5
Boiling Point, °C, 760 mm	127	Refractive Index, 20°C	1.3969
Evaporation Rate (n-butyl acetate = 1)	1.0	Flash Point (Tag Closed Cup), °F (°C)	77 (25)
Weight/Vol, at 20°C		(Tag Open Cup), °F (°C)	83 (28)
lb/gal. (U.S.)	6.75	Fire Point, °F (°C)	86 (30)
kg/liter	0.81	Flammable Limits in Air, % by volume	
lb/gal. (Imperial)	8.10	Lower	1.3
Solubility, 20°C, wt %		Upper	8.0
In water	1.4	Autoignition Temperature (ASTM D-2155),	
Water in	2.1	°F (°C)	795 (424)
Dilution Ratio, toluene	4.0	NFPA Classification 30:	Flammable Liquid, Class IC
VM & P naphtha	1.1		
Color (Pt-Co Scale), ppm	5	ICC Labels Required	None
Acidity, as acetic acid, wt %	0.01	Bureau of Explosives Classification	Nonhazardous Liquid

Several of the solvent characteristics of Methyl n-Butyl Ketone are listed in the following table. Similar values for other solvents are included for comparison.

Eastman Solvent	Evap. Rate	Blush Res., % R.H. @ 80°F (27°C)	Solution Viscosity, 25°C, cp		
			RS. ½-Sec Cellulose Nitrate[a] 10 Wt %	Exon[b] 470 20 Wt %	Elvacite[c] 2010 20 Wt %
Methyl n-Butyl Ketone	1.0	80	28	24	65
Methyl Isobutyl Ketone	1.6	78	30	24	64
Isobutyl Acetate	1.4	80	49	38	83
n-Butyl Acetate	1.0	83	46	33	77

[a]*product of Hercules, Inc.*   [c]*product of E. I. du Pont de Nemours Co., Inc.*
[b]*product of Firestone Plastics Co.*

Table 12.14: Solubility of Dry Half-Second R.S. Nitrocellulose in a System of Methyl Butyl Ketone-sec-Butanol-Toluene *(2)*

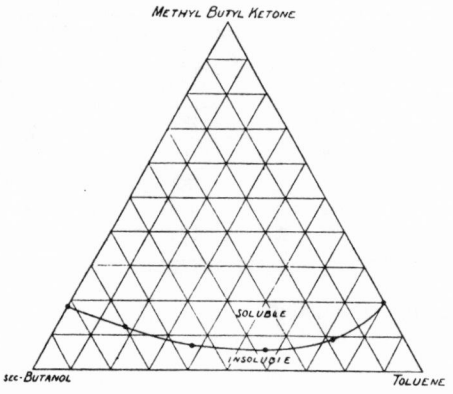

METHYL ISOBUTYL KETONE

Hexone, 4-Methylpentanone-2, 2-Methyl-4-Pentanone $CH_3 \cdot CO \cdot C_4H_9$

Although first prepared in 1849, methyl isobutyl ketone was not made synthetically and on a large scale until the last decades. It is a stable, colorless liquid classified as a medium boiler. It is miscible with most organic solvents and with mineral and vegetable oils. When compared with butyl acetate its rate of evaporation is somewhat faster so that it can either replace esters or be combined with them. Its rate of evaporation is somewhat faster than that of butyl acetate. It is used in the vinyl type resins for coatings where it helps to prevent gelling and lowers viscosity, in nitrocellulose lacquer manufacture, in extraction processes and in chemical synthesis. It may be used in dewaxing oils.

Table 12.15: Properties of Methyl Isobutyl Ketone *(41)*

Typical Properties

Molecular Weight ($C_6H_{12}O$)	100.16
Color (Pt-Co Scale), max	10
Weight/Vol at 20°C,	
lb/gal (U.S.)	6.67
kg/L	0.80
lb/gal (Imperial)	8.00
Solubility at 20°C, wt %	
In water	2.0
Water in	1.0
Evaporation Rate (n-butyl acetate = 1)	1.6
Dilution Ratio, toluene	3.5
VM & P naphtha	1.0
Refractive Index at 20°C	1.3958
Vapor Pressure at 20°C, mm Hg	15
Specific Gravity at 20°/20°C	0.802
Boiling Range at 760 mm, °C	
Initial Boiling Point, min	114
Dry Point, max	117
Freezing Point, °F (°C)	-119 (-84)
Flash Point, Tag Closed Cup, °F (°C)	60 (16)
Tag Open Cup, °F (°C)	68 (20)
Fire Point, °F (°C)	70 (21)
Flammable Limits in Air, % by volume	
Lower, at 200°F (93°C)	1.22
Upper, at 200°F (93°C)	7.96
Autoignition Temperature (ASTM D-2155), °F (°C)	840 (449)
NFPA Classification 30	IB
DOT Classification	Flammable Liquid
DOT Labels Required	Flammable Liquid

(continued)

Table 12.15: (continued)

Several of these solvent characteristics of MIBK are listed in the following table. Similar values for other solvents are included for comparison.

Eastman Solvent	Evap. Rate	Blush Res., % R.H. @ 80°F (27°C)	Solution Viscosity at 25°C, cP (mPa•s)		
			RS ½-Sec Cellulose Nitrate[a] 10 Wt %	FPC 470 Resin[b] 20 Wt/%	Elvacite 2010 Resin[c] 20 Wt/%
Methyl Isobutyl Ketone	1.6	78	30	24	64
Isobutyl Acetate	1.4	80	49	38	83
n-Butyl Acetate	1.0	83	46	33	77

[a]*Product of Hercules Incorporated*
[b]*Product of Firestone Plastics Company*
[c]*Product of Du Pont Company*

Table 12.16: Solubility of Miscellaneous Materials in Methyl Isobutyl Ketone at 20° to 25°C *(2)*

Soluble, Over 5% by Weight Concentration

Acid
  Oleic (Technical Red Oil)

Oils
  Castor, Refined Raw
  Cottonseed, Raw
  China Wood
  Coconut, Crude
  Fish, Processed
  Linseed, Pure Raw
  Mineral, 70/80 viscosity
  Pine
  Soybean, 2–3 viscosity

Gums
  Elemi
  Kauri (Pale Bold)
  Mastic
  Pontianak

Resins, Natural
  Dammar (dewaxed)
    Batavia
    Singapore
    Light Rosin
  Sandarac

RESINS, SYNTHETIC

Trade Name	Type
Amberlac 80-X	Modified drying type phthalic alkyd
Amberol 801	Rosin modified maleic alkyd
Arochem 519	Modified maleic
Aroclor 1260	Chlorinated diphenyl
Bakelite BR-254	Non-heat-hardening 100% para-phenylphenol resin
No. 1 Solid Beckosol	Phenolic modified drying type alkyd
Beckosol 1313	Drying type alkyd
Beetle 227-8	Unmodified urea-formaldehyde
Cellolyn 102	Modified rosin ester
Ester gum	Rosin ester
Ethyl methacrylate	Acrylic ester
Glyptal 2477	Non-drying type alkyd
Melmac 245-8	Unmodified melamine-formaldehyde
Neville R-21 (soft)	Unmodified coumarone-indene
Nevillite 1	Naphthene polymers
Nitrocellulose	Cellulose ester
Parlon X (20 cps.)	Chlorinated rubber
Phenac 608	Modified phenolic
Santolite K	Alkyl-arylsulfonamide-formaldehyde
Saran F-120	Vinylidenechloride-acrylonitrilecopolymers
Staybelite	Hydrogenated rosin ester
Teglac Z-152	Rosin modified maleic alkyd
Vinylite AYAF	Polyvinyl acetate
Vinylite VMCH	Maleic modified vinyl chloride-vinyl acetate copolymers
Vinylite VYHH	Vinyl chloride-vinyl acetate copolymers

METHYL n-AMYL KETONE

Heptanone-2 $CH_3(CH_2)_4 \cdot CO \cdot CH_3$

This ketone is a colorless, stable liquid, miscible with most lacquer solvents and only very slightly soluble in water. It is used as a high-boiling solvent for nitrocellulose and is particularly applicable in vinyl resin finishes, where its slow rate of evaporation prevents quick drying, improves the flow and gives blush resistance; also used with some effect in insecticidal preparations.

Table 12.17: Properties of Methyl n-Amyl Ketone *(41)*

Typical Properties

Molecular Weight ($C_7H_{14}O$)	114.19	Specific Gravity at 20°/20°C	0.817
Branched-Chain Ketones. wt % max	2.0	Boiling Range at 760 mm. °C	
Color (Pt-Co Scale), max	10	Initial Boiling Point. min	149
Evaporation Rate (n-butyl acetate = 1)	0.4	Dry Point, max	153.5
Weight/Vol. at 20°C		Freezing Point, °F (°C)	−27 (−33)
lb/gal (U. S.)	6.80	Flash Point, Tag Closed Cup, °F (°C)	102 (39)
kg/L	0.81	Tag Open Cup. °F (°C)	114 (46)
lb/gal (Imperial)	8.16	Fire Point. °F (°C)	115 (46)
Solubility at 20°C, wt %		Flammable Limits in Air, % by volume	
In water	0.46	Lower, at 150°F (66°C)	1.11
Water in	1.31	Upper, at 250°F (121°C)	7.9
Dilution Ratio, toluene	3.9	Autoignition Temperature (ASTM D 2155), °F (°C)	740 (393)
VM & P naphtha	1.2	NFPA Classification 30	II
Refractive Index at 20°C	1.4085	DOT Classification	Combustible Liquid
Vapor Pressure at 20°C, mm Hg	2.14	DOT Labels Required	None

COMPARISON OF PROPERTIES OF HIGH-BOILING SOLVENTS

Solvent	Evap. Rate	Blush Res., % R.H. @ 80°F (27°C)	Solution Viscosity at 25°C, cP		
			RS ½-Sec Cellulose Nitrate[a] 10 Wt %	CAB-381-0.5[b] 10 Wt %	VMCH Copolymer[c] 20 Wt %
Methyl n-Amyl Ketone[b]	0.4	93	40	37	158
Methyl Isoamyl Ketone[b]	0.5	89	42	37	164
Isobutyl Isobutyrate[b]	0.4	92	128	Insol	Gel
Ethyl Amyl Ketone	0.3	94	69	Insol	320
Diisobutyl Ketone[b]	0.2	95	143	Insol	Gel
Ektasolve® EE Acetate[b]	0.2	94	113	89	1040

[a] *product of Hercules Incorporated*   [b] *an Eastman product*   [c] *product of Union Carbide Corporation*

METHYL ISOAMYL KETONE

MIAK $CH_3{-}CO{-}C_5H_{11}$

MIAK is a retarder solvent, having an evaporation rate of 0.5, but it also possesses exceptional solvent power for most film-formers. In lacquers, the low evaporation rate of MIAK promotes good flow and leveling properties; whereas the high solvency provides low viscosities or permits a higher nonvolatile content.

Table 12.18: Properties of Methyl Isoamyl Ketone *(41)*

Typical Properties

Molecular Weight ($C_7H_{14}O$)	114.19	Boiling Range, 760 mm, °C	
Color (Pt-Co Scale), max	10	Initial Boiling Point, min	141
Weight/Vol, 20°C,		Dry Point, max	148
lb/gal (U. S.)	6.76	Freezing Point, °F (°C)	-101 (-74)
kg/litre	0.81	Flash Point, Tag Closed Cup, °F (°C)	96 (36)
lb/gal (Imperial)	8.14	Tag Open Cup, °F (°C)	106 (41)
Solubility, 20°C, wt %		Fire Point, °F (°C)	107 (42)
In water	0.5	Flammable Limits in Air, % by volume	
Water in	1.2	Lower, at 200°F (93°C)	1.05
Evaporation Rate (n-butyl acetate = 1)	0.5	Upper, at 200°F (93°C)	8.2
Dilution Ratio, toluene	4.1	Autoignition Temperature (ASTM D-2155), °F (°C)	795 (425)
VM & P naphtha	1.2	NFPA Classification 30	IC
Refractive Index, 20°C	1.4069	DOT Classification	Flammable Liquid
Vapor Pressure, 20°C, mm Hg	4.5	DOT Labels Required	Flammable Liquid
Specific Gravity, 20°/20°C	0.814		

(continued)

Table 12.18: (continued)

Solvent	Evap. Rate	Blush Res., % R. H. @ 80°F (27°C)	Solution Viscosity, 25°C, cP		
			RS ½-Sec Cellulose Nitrate[a] 10 Wt/%	FPC 470 Resin[b] 20 Wt/%	Elvacite 2010 Resin[c] 20 Wt/%
Methyl Amyl Acetate	0.5	92	128	Insol	Insol
Methyl Isoamyl Ketone	0.5	89	42	34	68
Isobutyl Isobutyrate	0.4	92	128	Insol	Insol
Ektasolve® EE Acetate	0.2	94	113	Insol	284

[a] product of Hercules Incorporated
[b] product of Firestone Plastics Company
[c] product of Du Pont Company

Table 12.19: Properties of Methyl Isoamyl Ketone vs Other Solvents (41)

Solvent	Evaporation Rate	Dilution Ratio (Toluene)	Blush Resistance, % R.H. at 80°F.	Specific Gravity, 20/20°C.	Flash Point, Tag Open Cup, °F.	Boiling Range, 760 mm., °C.
Methyl isobutyl ketone	1.6	3.6	78	0.8018	73	114-117
Isobutyl acetate	1.4	2.7	78	0.8728	90	114-118
n-Butyl acetate	1.0	2.7	82	0.8109	100	116-118
Amyl acetate	0.6	2.4	92	0.862	93	100-150
MIAK	**0.50**	**4.1**	**92**	**0.813**	**110**	**141-148**
Methyl amyl acetate	0.5	1.7	92	0.8595	110	143-150
2-Ethoxyethanol	0.3	4.9	65	0.9311	130	132-136
4-Methoxy-4-methyl-pentanone-2	0.3	3.1	91	0.904	141	147-163
Ethyl amyl ketone	0.2	2.2	94	0.822	135	156-162
2-Ethoxyethyl acetate	0.2	2.5	91	0.9748	150	145-165
4-Methoxy-4-methyl-pentanol-2	0.2	4.7	93	0.890	140	164-169
Cyclohexanone	0.2	5.8	92	0.945	129	153-160
2-Butoxyethanol	0.06	3.33	96	0.9019	165	166-173
Isophorone	0.03	6.2	97	0.9229	205	205-220

Table 12.20: Butyrate-Acrylic Wood Lacquer—Substituting Methyl Isoamyl Ketone for 2-Ethoxyethyl Acetate (41)

Ingredients	Part A Wt. %		Part B Wt. %
Half-Second Butyrate	8.5		8.5
Acryloid B-66 resin (40%)[1]	21.3		21.3
Santicizer 160 plasticizer[2]	3.0		3.0
Dow-Corning 510 (1000 cs.) fluid[3]	0.01		0.01
Eastman Inhibitor DOBP[4]	0.09		0.09
Toluene	26.3		36.3
Isobutyl acetate	13.6		13.6
Isobutyl alcohol	13.6		3.6
Methyl ethyl ketone	6.8		6.8
MIAK	6.8	—	6.8
2-Ethoxyethyl acetate	—	6.8	—
	100	100	100
Solids, %	20.12	20.12	20.12
Viscosity, cp.	45	50	42
Wt./gal., lb.	7.45	7.51	7.50
Flow out	excellent	excellent	excellent

[1] Product of Rohm and Haas Company   [3] Product of Dow Corning Corporation
[2] Product of Monsanto Chemical Company   [4] 2-Hydroxy-4-dodecyloxy benzophenone

METHYL HEXYL KETONE

Octanone-2 $CH_3(CH_2)_5CO \cdot CH_3$

A colorless liquid with a characteristic odor, methyl hexyl ketone is used as a solvent for vinyl compounds and dyes, and has been found particularly suitable in dispersing dyes in light petroleum oils for newsprint inks.

Table 12.21: Properties of Methyl Hexyl Ketone *(2)*

Purity	95%, min.
Specific gravity at 20°C.	0.81-0.83
Weight per gallon at 20°C.	6-8 lbs.

METHYL HEPTYL KETONE

MHK
5-Methyl-2-Octanone

$$CH_3\underset{\substack{\| \\ O}}{C}CH_2CH_2\underset{\substack{| \\ CH_3}}{C}HCH_2CH_2CH_3$$

Methyl heptyl ketone, a high-boiling, active solvent, imparts desirable drying characteristics in many high-temperature baked coatings.

Table 12.22: Properties of Methyl Heptyl Ketone *(41)*

Molecular Weight ($C_9H_{18}O$), calcd	142.24	Color (Pt-Co Scale), ppm	5-25
Melting Point, °C	−9	Acidity, as acetic acid, wt %	0.018
Boiling Range, °C, 760 mm	183-195	Water, wt %	0.01-0.05
Evaporation Rate (n-butyl acetate = 1)	0.08	Flash Point (Tag Closed Cup), °F (°C)	140 (60)
Weight/Vol, at 20°C		(Tag Open Cup), °F (°C)	160 (71)
lb/gal (U.S.)	6.87	Fire Point, °F (°C)	168 (76)
kg/liter	0.83	Flammable Limits in Air, % by volume	
lb/gal (Imperial)	8.59	Lower (at 180°F)	0.9
Solubility, 20°C, wt %		Upper (at 313°F)	5.9
In water	0.5	Autoignition Temperature (ASTM D-2155), °F (°C)	680 (360)
Water in	0.95	NFPA Classification 30:	Combustible Liquid,
Dilution Ratio, toluene	3.0		Class IIIA
VM & P naphtha	1.0	ICC Labels Required	None
Refractive Index, 20°C	1.422	Bureau of Explosives Classification	Nonhazardous Liquid

Solvent	Evap Rate	Dilution Ratio		Blush Res, % RH @ 80°F(27°C)	Sp Gr 20°/20°C	Lb/gal @20°C
		Toluene	VM & P Naphtha			
MAK	0.4	3.9	1.2	93	0.815	6.80
EKTASOLVE® *EB* Solvent*	0.1	3.4	2.1	96	0.902	7.51
MHK	0.08	3.0	1.0	97	0.827	6.87
Isophorone	0.03	6.2	1.2	97	0.922	7.68

*EKTASOLVE EB (ethylene glycol monobutyl ether) is an Eastman product.

"PENToXONE" SOLVENT

4-Methoxy-4-Methyl-2-Pentanone

PENToXONE solvent, a keto-ether, is a colorless, high boiling, slow evaporating liquid with strong solvent power and pleasant odor. It is available in high purity and is essentially anhydrous. Its slow evaporation rate and true solvency for a large variety of resinous materials, including vinyls, nitrocellulose and acrylics, makes it a valuable solvent for many surface coating systems.

Table 12.23: Properties of "Pentoxone" (14)

Purity, percent weight, minimum	98.5
Apparent specific gravity, 25/25°C	0.899-0.909
20/20°C	0.903-0.913
Color, Pt-Co, maximum	15
Distillation range, °C	157-162
Acidity (as acetic acid, CO_2 free basis), percent weight, maximum	0.01
Water, percent weight, maximum	0.1
Methanol, percent weight, maximum	0.2

Table 12.24: Viscosity of Various Resins in "Pentoxone" (41)

Resin	Viscosity, Cps., 25°C. in PENToXONE® Solvent	Resin	Viscosity, Cps., 25°C. in PENToXONE® Solvent
Nitrocellulose		Melamines	
R.S. ½-Second Grade, Eight Grams per 100 Mls	67	Cymel 248-8	11
S.S. ½-Second Grade, Eight Grams per 100 Mls	38	MM-55	14
		Resimene 872	16
Acrylics		Maleics	
Acryloid A-21, 15 Per Cent Weight	22	Cellolyn 102	9
Acryloid B-44, 15 Per Cent Weight	20	Amberol 801	17
Acryloid B-66, 15 Per Cent Weight	15		
Acryloid AT-50, 30 Per Cent Weight	63	Alkyds	
		Aroplaz 2480	24
Phenolics		Rezyl 412	14
Bakelite BKS-2600	46	Cellolyn 502	19
Amberol F-7	19	Cellolyn 582	12
Methylon 75108	5	Duraplex ND 77B	31
		Aroplaz 6006	23
Ureas		Cycopol 102	62
Beetle 227-8	26	Other	
Uformite F-240	7	Half-Second Butyrate, 10 Per Cent Weight	75
Beckamine P-196	8	Parlon P	630
		Buton 200	31
		EPON® 1002, 50 Per Cent Weight	240
		EPON® 1007	107

(1) *Commercially available resin (solids or in solution) reduced to 30 per cent weight solids, except as indicated. Viscosities of resulting solutions measured in absolute units with capillary tube viscometers.*

ETHYLBUTYL KETONE

Heptanone-3 $C_2H_5COCH_2CH_2CH_2CH_3$

Ethylbutyl ketone is a stable, high-boiling solvent of special value in lacquers and synthetic resin coatings. Its evaporation rate in relation to those of comparable solvents is indicated in the following tabulation:

Solvent	Hours
Methyl isobutyl ketone	4.5
Butyl acetate	8
Ethylbutyl ketone	14
Amyl acetate	16
Methylamyl acetate	17
Methylamyl ketone	20
"Cellosolve" acetate	38
Diisobutyl ketone	44

The unusual combination of good solvent power with medium evaporation rate makes ethylbutyl ketone generally useful for coating solutions having adequate flow without unduly long drying time. It bakes out of films somewhat faster than other comparable ketones.

Table 12.25: Properties of Ethylbutyl Ketone *(2)*

Boiling point	147.8°C.
Freezing point	-36.7°C.
Coefficient of expansion at 20°C.	0.00107
Flash point	125°F.
Solubility in water at 20°C.	0.43% by wt.
Solubility of water in at 20°C.	0.78% by wt.
Refractive index at 20°C.	1.4085
Specific gravity at 20/20°C.	0.8197

ETHYL AMYL KETONE

EAK, 5-Methyl-3-Heptanone

$$CH_3CH_2 - \overset{\displaystyle O}{\overset{\displaystyle \|}{C}} - CH_2\overset{\displaystyle CH_3}{\overset{\displaystyle |}{CH}}CH_2CH_3$$

Ethyl amyl ketone, a high boiling ketone, is a colorless, stable liquid with a mild pleasant odor. It is compatible with alcohols, ethers, other ketones and organic liquids, and in addition, exhibits low water miscibility. Ethyl amyl ketone's high solvency for cellulose esters, vinyl polymers and copolymers, synthetic and natural protective coating resins, coupled with its slow evaporation rate, high blush resistance and good diluent tolerance makes it a valued surface coating raw material.

Table 12.26: Properties of Ethyl Amyl Ketone *(14)*

Apparent specific gravity, 20/20°C	0.820-0.824	Acidity (as acetic acid), % w, Max	0.01
25/25°C	0.816-0.820	Water, % w, Max	0.15
Color, Pt-Co, Max	25	Alcohol (as ethyl amyl carbinol), % w, Max	0.50
Distillation range, °C	156-162		

DI-n-PROPYL KETONE

Heptanone-4, Butyrone, Amyl Ketone $(CH_3CH_2CH_2)_2CO$

Di-n-propyl ketone is a colorless, stable liquid having a pleasant odor. It is miscible with many organic solvents, and dissolves a wide variety of materials, some of which are crude rubber, nitrocellulose, raw and blown oils, many natural and synthetic resins like dewaxed dammar, manila, rosin, ester gum, and waxes.

Table 12.27: Properties of Di-n-Propyl Ketone *(2)*

Boiling point	143.7°C.
Coefficient of expansion	0.001073 (per °C.) to 20°C.
	0.001115 (per °C.) to 55°C.
Dilution ratio ("Kemsolene")	0.8
(Toluene)	3.1
Flash point (ASTM Open Cup)	49°C.
Freezing point	-32.1°C.
Heat of combustion	1051 cal./mol
Latent heat of vaporization	75.8 cal./g.
Solubility in water at 20°C.	0.53% by wt.
Solubility of water in solvent at 20°C.	1.27% by wt.
Specific gravity at 20/20°C.	0.8162
Refractive index at 20°C.	1.4068
Specific heat at 25°C.	0.553 cal./g.
Surface tension at 25°C.	25.2 dynes/sq. cm.
Vapor pressure at 20°C.	5.2 mm. Hg
Viscosity at 20°C.	0.0074 poise
Weight per gallon at 20°C.	6.79 lbs.

DIISOBUTYL KETONE

Valerone $(C_4H_9)_2CO$

A water-white, stable liquid, miscible with most organic liquids, diisobutyl ketone has good solvency for cellulose acetate, nitrocellulose, vinyl resins, waxes, gums, natural and synthetic resins, and crude rubber. It is used principally as a high-boiler in nitrocellulose lacquers and vinyl resin coatings, where its slow evaporation rate is advantageous.

Table 12.28: Properties of Diisobutyl Ketone *(41)*

Typical Properties

Molecular Weight ($C_9H_{18}O$)	142.23	Boiling Range, 760 mm, °C	
Color (Pt-Co Scale), max	20	Initial Boiling Point, min	163
Evaporation Rate (n-butyl acetate = 1)	0.2	Dry Point, max	173
Weight/Vol, 20°C,		Freezing Point, °F (°C)	−43 (−42)
lb/gal (U.S.)	6.76	Flash Point, Tag Closed Cup, °F (°C)	120 (49)
kg/liter	0.81	Tag Open Cup, °F (°C)	131 (55)
lb/gal (Imperial)	8.11	Fire Point, °F (°C)	137 (58)
Solubility, 20°C, wt %		Flammable Limits in Air, % by volume	
In water	0.05	Lower, at 200°F (93°C)	0.81
Water in	0.75	Upper, at 200°F (93°C)	7.1
Dilution Ratio, toluene	1.5	Autoignition Temperature (ASTM D-2155), °F (°C)	745 (396)
VM & P naphtha	0.8	NFPA Classification 30	II
Refractive Index, 20°C	1.4230	DOT Classification	Combustible Liquid
Vapor Pressure, 20°C, mm Hg	1.7	DOT Labels Required	None
Specific Gravity, 20°/20°C	0.807-0.814		

(continued)

Table 12.28: (continued)

Solvent	Evap. Rate	Blush Res., % R. H. @ 80°F (27°C)	Solution Viscosity, 25°C, cp		
			RS. ½-Sec Cellulose Nitrate, 10 Wt/%	*FPC* 470 Resin[a], 20 Wt %	*Elvacite* 2010 Resin[b], 20 Wt %
Methyl Isoamyl Ketone	0.5	89	50	34	68
Methyl Amyl Acetate	0.5	92	128	Insol	Insol
Isobutyl Isobutyrate	0.4	92	128	Insol	Insol
Diisobutyl Ketone	0.2	95	160	75	Insol

[a]*product of Firestone Plastics Company*   [b]*product of E. I. du Pont de Nemours Co., Inc.*

CYCLOHEXANONE

"Sextone", "Anon", Pimelin Ketone, Keto Hexamethylene

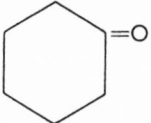

Cyclohexanone is a colorless to pale yellow, stable liquid with an odor suggestive of peppermint. It is made by the dehydrogenation of cyclohexanol. It is miscible in all proportions with most solvents, especially the common lacquer solvents and diluents, hydrogenated and chlorinated hydrocarbons, phenols, pyridine, and turpentine. It is a good solvent for cellulose ethers, esters, basic dyes, latex, fats, blown oils, waxes, crude rubber, and such gums and resins as ester gum, alkyds, vinyls, coumarone, 100% and modified phenol resins, cyclohexanone resins and many natural resins. It forms constant-boiling mixtures with camphor, tetrachloroethane, and trichloropropane. It has a very high dilution ratio as compared with the coal-tar hydrocarbons, a fact which accounts for its excellence as a solvent, especially in the lacquer industry.

Its low rate of evaporation and strong solvent powers impart blush resistance, good flow and working qualities to lacquers and give films that are clear, smooth and glossy and show good adhesion. It is also used in spraying and brushing lacquers and as a medium boiler. It is particularly effective for blending nitrocellulose with spirit-soluble and hydrocarbon-soluble resins and oils. Its solvency for basic dyes makes it applicable in wood stains. Other uses are in the air-drying and stoving type of synthetic resins, in plastics and molding powders, in paint and varnish removers, in spot and stain removers, in metal-degreasing preparations, in polishes, printing inks, as a leveling agent in dyeing, in delustering cellulose acetate, insecticides and pharmaceuticals.

Table 12.29: Properties of Cyclohexanone *(2)*

Boiling point	155.6°C.
Color	Water-white to pale yellow
Dielectric constant at 25°C.	18.2
Evaporation rate, approximate (toluene = 100)	20
Flash point (open cup)	130°F.
Freezing point	-45°C.
Solubility in water at 20°C.	8.7%
Specific gravity at 20°C.	0.944 – 0.950
Specific heat 15° to 18°C.	0.433 cal./g.
Refractive index	1.443 – 1.451
Viscosity (SUV at 100°F.)	33
Weight per gallon at 20°C.	7.9 lbs.
Acidity	Neutral
Distillation range	95% within 151° – 157°C.
Purity	98 – 100%
Residue	0.02%
Water content	0.2% max.

Table 12.30: Resin Solubility in Cyclohexanone *(19)*

Resin	Manufacturer	Viscosity at 25°C., cps. (1)	Toluene Dilution (2)	Heptane Dilution (2)
Acrylic				
"Acryloid" B-82	Rohm & Haas	32	>50	2
"Elvacite" 2010	DuPont	54	>50	7
Cellulosics				
Cellulose Acetate AB-141-95 (14% acetyl)	Eastman	9200	28	6.5
Cellulose Acetate Butyrate EAB-171-2 (17% butyryl)	Eastman	892	34	9.5
Cellulose Acetate Butyrate EAB-381-20 (37% butyryl)	Eastman	5060	>50	14
Ethyl Cellulose (N-22, 24 sec.)	Hercules	1408	>50	23
Half Second Butyrate AB-H	Eastman	242	>50	17
"Hercose" "C" Type A	Hercules	806	38	8.5
Nitrocellulose (RS 1/2 sec.)	Hercules	218	>50	10.5
Styrene				
Polystyrene	--	96	>50	24
SMA 4000A	Sinclair	19	>50	14.5
Vinyl				
BAKELITE Vinyl Resin AYAF	UCC	74	>50	7
BAKELITE Vinyl Resin VYHH	UCC	68	>50	14.5
BAKELITE Vinyl Resin XYHL	UCC	(3)	–	–
"Saran" F-120 (1000 cps.)	Dow	484	21	6.5
Epoxy				
BAKELITE Epoxy Resin EKR 2002	UCC	21	>50	9.5
Urethane				
"Estane" 5701F1	Goodrich	282	14	4.5
"Estane" 5707F1	Goodrich	388	6	2
Rosin-Ester				
"Amberol" 801 LT	Rohm & Haas	14	>50	19
"Cellolyn" 104	Hercules	20	>50	<1
Melamine-Formaldehyde				
"Cymel" 300	Am. Cyanamid	16	>50	>50
Alkyd				
"Beckasol" # 7	Reichhold	28	>50	47
"Beckasol" #31	Reichhold	25	>50	>50
Rubber				
"Parlon" S-20 (18 cps.)	Hercules	46	>50	22
"Pliolite" S-5	Goodyear	56	>50	30
Phenolic				
BAKELITE Phenolic Resin BKR 2620	UCC	23	16	5.5
Phenoxy				
BAKELITE Phenoxy Resin PKHH	UCC	Insoluble	–	–

(1) 10 grams resin, 90 grams cyclohexanone
(2) 10 grams of 10% resin solution, titrated with diluent (in ml.)
(3) Partially soluble

METHYL CYCLOHEXANONE

Methyl "Anon", "Sextone" B

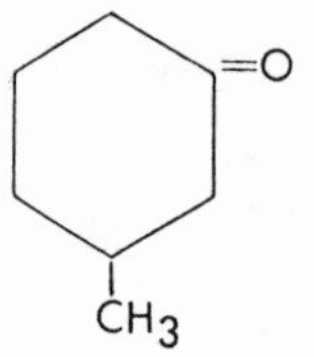

Meta 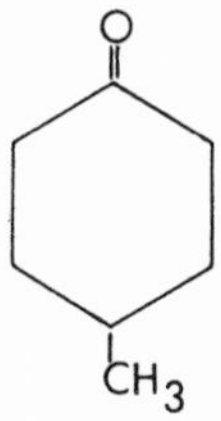   Para

Methyl cyclohexanone is a water-white to pale yellow liquid with an acetone-like odor. It is a mixture of two isomeric cyclic ketones made by the dehydrogenation of methyl cyclohexanol. It closely resembles cyclohexanone in its physical properties, miscibility, tolerance for non-solvents and solvent action. It differs from cyclohexanone in its somewhat slower evaporation rate and lower dilution ratios with aromatic hydrocarbons. Methyl cyclohexanone is especially suitable for phenolic and alkyd resins, crude rubber, nitrocellulose, ester gum and kauri. It is also an excellent agent for blending pyroxylin with resins, oils and rubber in lacquers. It is used in crystalizing lacquers, where its low evaporation rate retards evaporation sufficiently to permit crystal growth. It is also used in slow-setting varnish removers and in rubber cements.

Table 12.31: Properties of Methyl Cyclohexanone (2)

Boiling point	169.0° - 170.5°C.
Evaporation rate (approximate) (toluene = 100)	20
Flash point	53°C.
Freezing point	-70°C.
Refractive index at 25°C.	1.442 - 1.446
Solubility in water at 20°C.	2 - 3%
Specific gravity at 25/4°C.	0.910 - 0.914
Viscosity (SUV at 100°F.)	33
Weight per gallon	7.6 lbs.
Distillation range	165.0° - 172°C.
	95% distills within 3.0°
Purity	98 - 100%
Residue	None
Water content	0.2%, max.

METHYL ACETONE

Methyl Ketone

Methyl acetone is a clear, colorless, flammable, volatile liquid, obtained from the product of the destructive distillation of wood. Although it varies in composition it is generally composed of acetone 35 to 60%, methanol 20 to 40%, and methyl acetate 20 to 30%.

DIACETONE ALCOHOL

Diacetone
4-Hydroxy-4-Methylpentanone-2
"Pyranton A"
Diacetonyl Alcohol

$$CH_3-\overset{\overset{\displaystyle O}{\|}}{C}-CH_2-\overset{\overset{\displaystyle OH}{|}}{\underset{\underset{\displaystyle CH_3}{|}}{C}}-CH_3$$

Diacetone alcohol is a flammable liquid that is colorless when pure, becoming yellow on aging; it has a mint-like odor. Made by the condensation of acetone, the commercial product contains up to 15% of acetone. For this reason the technical product is superior in its solvent power to the acetone-free grade. It is miscible with most organic liquids, as well as with water. It is a good solvent for cellulose acetate, nitrocellulose, cellulose acetobutyrate, cellulose acetopropionate, hydrocarbons, oils, fats, resins, gums and dyes. It has only limited solvency for dammar gum, polyvinyl acetate and the petroleum resins. A high-boiling solvent, diacetone alcohol also exhibits the desirable properties of reducing the viscosities of organic solutions of high solids content, and of minimizing temperature effects on viscosities. In most respects it is quite similar to acetone with the exception of a very much slower rate of evaporation

It is used in cellulose ester lacquers, particularly of the brushing type, where it produces brilliant gloss and hard film and where its lack of odor is desirable. It is used in lacquer thinners, dopes, wood stains, wood preservatives and printing pastes; in coating compositions for paper and textiles; in making artificial silk and leather; in imitation gold leaf; in celluloid cements; as a preservative for animal tissue; in metal-cleaning compounds; in the manufacture of photographic film; and in hydraulic brake fluids, where it is usually mixed with an equal volume of castor oil

Diacetone alcohol is available in two grades: technical, containing up to 15% acetone, and acetone-free.

Table 12.32: Physical Properties of Acetone-Free Diacetone Alcohol (2)

Boiling point at 760 mm.	167.9°C.
Coefficient of expansion (Cubical)	0.000533 per °F.
Color	Water-white to light straw
Flash point (open cup)	144°F.
Heat of combustion	8,601 cal./g.
Melting point	-47°C.
Specific gravity at 20/20°C.	0.937 - 0.946
Refractive index at 20°C.	1.4235
Viscosity (Saybolt)	
113 seconds at	-12°C.
674 seconds at	-30°C.
1,980 seconds at	-48°C.
Weight per gallon at 20°C.	7.83 lbs.
Acidity (as acetic)	0.05%
Distillation range at 760 mm.	
Below 135°C.	None
Below 158°C.	Not more than 5%
Above 170°C.	None
Nonvolatile matter	0.005% by wt. (max.)

ACETONYL ACETONE

Hexanedione-2,5 $(CH_3 \cdot CO \cdot CH_2)_2$

Acetonyl acetone, a diketone, is a water-white liquid with an agreeable odor. It is completely soluble in water, almost entirely soluble in such substances as toluene, kauri gum and rosin, and only partly soluble in raw linseed oil, shellac, dewaxed dammar and ester gum. It has been suggested as an intermediate in the manufacture of rubber accelerators, dyes, inhibitors, insecticides, and pharmaceuticals and for the preparation of derivatives of thiophene, furan and pyrrole. It may also be employed in tanning hides and skins.

Table 12.33: Properties of Acetonyl Acetone *(2)*

Boiling point	191.4°C.
Dilution ratio (xylene)	1.8
Flash point	158°F.
Specific gravity at 20/20°C.	0.9710 – 0.9760
Solubility in water at 20°C.	Complete
Vapor pressure at 20°C.	0.5 mm. Hg
Weight per gallon at 20°C.	8.10 lbs.
Acidity (as acetic)	0.020% by wt., max.
Boiling range at 760 mm.	185° to 195°C.
Purity	98.0% by wt., min.
Water	Miscible with 19 vol. 60° Bé gasoline at 20°C.

MESITYL OXIDE

4–Methyl–3–Pentenone–2
Isopropylidone Acetone
Methyl Isobutenyl Ketone

$$(CH_3)_2C{=}CH \cdot CO \cdot CH_3$$

Mesityl oxide is an unsaturated, medium–boiling ketone made by the dehydration of diacetone alcohol. It is a colorless to straw–yellow, oily liquid with a peppermint–like odor. It will darken and form a solid residue on exposure and aging. It is miscible with most organic liquids and it is a good solvent for such substances as nitrocellulose, ethylcellulose, low–viscosity cellulose acetate, polyvinyl chloride, vinyl resins, hydrocarbons, raw linseed oil, kauri gum, rosin, ester gum and synthetic rubber. It will only partly dissolve shellac and dewaxed dammar.

Mesityl oxide is used in lacquers and thinners where its presence in the solution lowers the viscosity and gives it both a high tolerance for hydrocarbons and resistance to humidity. Its excellent solvent power for gums and resins is especially applicable in vinyl–type resins, where it produces films that are tough, glossy and have good flow; its presence permits use of larger proportions of aromatic hydrocarbon diluents.

Table 12.34: Properties of Mesityl Oxide *(2)*

Boiling point at 760 mm.	129.5°C.
Coefficient of expansion	0.000599 per °F.
Color	Straw–yellow
Dielectric constant at 20°C.	15.4
Flash point (Tag closed cup)	83°F.
Heat of combustion	846.7 Cal. per mol
Heat of vaporization	85.9 cal./g.
Melting point	–59°C.
Solubility in water at 25°C.	3.4% by vol.
Solubility of water in solvent at 20°C.	3.4% by wt.
Specific gravity at 20/20°C.	0.853 – 0.856
Specific heat (21 – 121°C.)	0.521 cal./g.
Refractive index at 20°C.	1.4456
Vapor pressure at 20°C.	8.0 mm. Hg
30°C.	14.3 mm. Hg
40°C.	24.5 mm. Hg
Viscosity at 25°C.	8.79 millipoises
Weight per gallon at 20°C.	7.12 lbs.
Acidity (as acetic)	0.05%, max.
Distillation range (ASTM)	Below 120°C. None
	Above 135°C. None
	More than 95% distills over below 131°C.
Purity	95% by wt., min.
Water	Miscible without turbidity with 19 vols. of 60° Bé gasoline at 20°C. (approx. 0.20% by wt.)

ISOPHORONE

3,5,5-Trimethylcyclohexene-2-one-1

Isophorone is a stable, colorless, volatile liquid with a mild odor. It is only slightly soluble in water, but miscible with most lacquer solvents. It is an excellent solvent for many types of cellulose esters, cellulose ethers, oils, fats, gums and resins, both natural and synthetic. It is the most powerful solvent for nitrocellulose and "Vinylite" resins. Isophorone has one of the highest aromatic hydrocarbon dilution ratios for nitrocellulose—5.7 for toluene and 5.1 for xylene. It will dissolve 30% of "Vinylite" resin without gelling. At ordinary temperatures solutions can be made of 1/2 second RS nitrocellulose containing 45% solids. Isophorone is used in the manufacture of coatings, inks, stencil pastes and as a thinner in synthetic resin finishes.

Table 12.35: Properties of Isophorone *(2)*

Boiling point at 760 mm.	215.2°C.
Dilution ratios	
Toluene	5.7
Xylene	5.1
"Troluoil"	1.0
Mineral spirits	0.7
Flash point (open cup)	205°F.
Freezing point	-8.1°C.
Solubility in water at 20°C.	1.2% by wt.
Solubility of water in solvent at 20°C.	3.8% by wt.
Specific gravity at 20/20°C.	0.9200 - 0.9250
Vapor pressure at 20°C.	0.25 mm. Hg
Weight per gallon at 20°C.	7.68 lbs.
Acidity (as acetic)	0.02% by wt., max.
Distillation range at 760 mm.	205° - 220°C.
Color	Not darker than 0.05 g. $K_2Cr_2O_7$ per l. of water
Purity	98.0% by wt., min.
Water content	Miscible with 19 vol. 60° Bé gasoline at 20°C.

FENCHONE

Fenchone is a liquid ketone closely resembling camphor.

Table 12.36: Properties of Fenchone *(2)*

			Distillation Range (Calculated from 50:50 Min. Spirits)	
Boiling point	191.0°C.			
Dilution ratio:				
with coal-tar naphtha	1.3 final conc. 8.0			
with hi-flash naphtha	1.2 final conc. 8.2		5%	193.0°C.
Kauri-butanol	All proportions in 50%		10%	193.4°C.
	sol. with mineral		20%	193.8°C.
	spirits 131		40%	194.2°C.
Optical activity	+7.4		60%	194.5°C.
Refractive index at 20°C.	1.4625		80%	195.4°C.
Specific gravity at 15.5°C.	0.9457		90%	196.0°C.
Aniline point (—)	54°C.		95%	197.5°C.

BETA-PROPIOLACTONE

"BPL"

$$CH_2—CH_2—C=O$$

difunctionality

y---CH$_2$—CH$_2$—C=O
|
OH

HOCH$_2$—CH$_2$—C=O
|
y

Beta-Propionic Acid Derivatives   Hydracrylic Acid Derivatives

Table 12.37: Physical Properties of Beta-Propiolactone *(42)*

Physical state	Liquid
Color	Colorless
Odor	Pungent, acrylic
Boiling point at 10 mm Hg, deg C	51
100 mm Hg	100.0
400 mm Hg	139.7
760 mm Hg	162.3
Melting point, deg C	—33.4
Refractive index n_D^{20}	1.4131
Specific gravity, 20/20 C	1.1490
Pounds per gallon at 20 C	9.56
Flash point, Tag open cup, deg F	165

Solubility: BPL is miscible at room temperature with most organic solvents such as ether, alcohol (reacts), benzene, acetone, and acetic acid. Solubility in water at 25 C is 37 per cent by volume, with moderately fast hydrolysis to hydroxypropionic (hydracrylic) acid.

GAMMA-BUTYROLACTONE

"BLO"

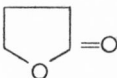

Gamma-butyrolactone is a powerful solvent and undergoes many reactions that make it of considerable interest in synthesis. It is a colorless hygroscopic liquid over a wide temperature range. It is soluble in acetone, benzene, carbon tetrachloride, ethyl ether, methanol, monochlorobenzene and water in all proportions.

Table 12.38: Properties of Gamma-Butyrolactone *(49)*

PHYSICAL PROPERTIES

Appearance ....clear liquid	Specify gravity (d$_4^{25}$) 1.124	Specific heat (25°C) ....0.40 cal/g/°C
Color (APHA) 40	Flash point, open cup ....98°C (208°F)	(60°C) ....0.45 cal/g/°C
Purity ....98.0% min.	Fire point ....99°C (210°F)	Dielectric constant (20°C) ...39
Moisture ....0.3% max.	pH (10% aqueous solution) ..4.5	Critical pressure ....500 psi
Free acid, as	Refractive index (n$_D^{25}$) 1.435	(35 kg/cm²)
hydroxybutyric ....0.1% max.	Heat of vaporization,	Critical temperature ....436°C
Molecular weight 86	Clausius-Clapeyron	Solubility: soluble in acetone, benzene,
Boiling point ....204°C	(calc.) ....133 cal/g	carbon tetrachloride, ethyl ether, methanol,
Freezing point ....—44°C	Heat of solution ....598 cal/mol	chlorobenzene, and water in all proportions.
Viscosity (25°C) ....1.7 cp	Heat of combustion ....492 kcal/mol	

(continued)

Table 12.38: (continued)

Vapor Pressure of Butyrolactone

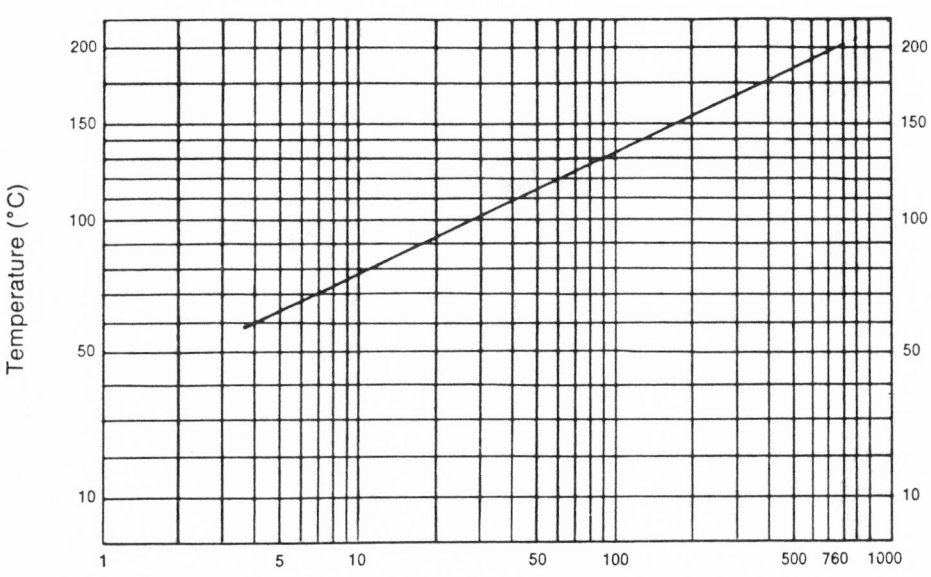

Vapor Pressure (mm Hg)

Table I. Percentage of Butyrolactone Hydrolyzed under Acid Conditions as Function of Time, Temperature, and Concentration

Concentration (%)		Time (hours)							
		1		3		5		24	
BLO	Dilute HCl	Room Temperature	65°C	Room Temperature	65°C	Room Temperature	65°C	Room Temperature	65°C
99	1	—	—	—	0.34	—	0.54	—	0.56
98	2	—	0.32	0.24	0.97	0.42	1.18	0.97	1.21
95	5	—	0.73	0.49	2.35	0.68	2.53	1.94	3.11
90	10	—	1.23	0.59	4.43	0.99	4.87	3.14	5.95
80	20	0.28	2.17	1.02	8.15	1.62	9.15	5.52	10.95
50	50	0.92	6.48	2.57	15.98	4.07	17.92	12.10	18.41

Table II. Percentage of Butyrolactone Hydrolyzed at pH 7 as Function of Time and Concentration*

Concentration (%)[†]		Time (hours)		
BLO	H_2O	8	24	48
80	20	—	0.33	1.7
50	50	1.7	11.1	17.4

*Tests were conducted at 65°C. No observable hydrolysis was detected at room temperature.
†At concentrations of up to 10 per cent water, no hydrolysis was observed in 48 hours.

Table III. Bunsen Coefficients of Butyrolactone (cc gas/cc solvent converted to STP)

Gas	25°C	45°C	75°C
Hydrogen	0.12	0	0
Carbon Monoxide	0.09	0.044	0
Carbon Dioxide	3.6	2.7	1.1
Methyl Acetylene	37.8	12.5	10.8
Acetylene	11.8	8	1.45
Vinyl Acetylene	145.1 (27°C)	33.1	23.1 (73°C)

COMPARATIVE DATA

Table 12.39: Amsco Ketones (13)

Ketones	Specific Gravity @ 20°/20°C	lb/gal @ 20°/20°C	Distillation Range 760 mm Hg °F (°C)	Relative Evaporation Rate n-BuAc = 1	Vapor Pressure mm Hg @ 20°C	Flash Point TAG CC °F	Solubility % by wt @ 20°C in Water	of Water
Acetone	0.791	6.59	132 (55.6)–134 (56.7)	7.7	186	0	∞	∞
Methyl Ethyl Ketone	0.806	6.71	174 (78.9)–176 (80.0)	4.6	70	24	26.8	11.8
Methyl Propyl Ketone	0.807	6.72	214 (101)–221 (105)	2.3	27.8	46	3.1	4.2
Methyl Isobutyl Ketone	0.802	6.67	237 (113.9)–243 (117.2)	1.6	15	60	1.9	1.6
Methyl Isoamyl Ketone	0.814	6.76	286 (141.1)–298 (147.7)	0.7	6	96	0.54	1.3
Diacetone Alcohol	0.941	7.82	275 (135.0)–342 (172.2)	0.2	<1	126	∞	∞
Methyl n-Amyl Ketone	0.817	6.80	300 (148.9)–304 (151.1)	0.4	4	102	0.46	1.31
Cyclohexanone	0.948	7.89	308 (153.3)–317 (158.3)	0.2	2	111	2.5	8.0
Diisobutyl Ketone	0.808	6.72	325 (162.8)–343 (172.8)	0.2	1.7	118	0.05	0.75
Isophorone	0.923	7.67	412 (211.1)–426 (218.9)	0.03	<1	179	1.2	4.3

Table 12.40: Ashland Ketones (69)

	Specific Gravity 20°/20°C	Dist. Range °F IBP	DP	Flash Pt. °F TCC
Acetone	0.7907	131	133	−4
Cyclohexanone	0.944	309	315	112
Diacetone Alcohol	0.9406	293	342	120
Diisobutyl Ketone	0.8076	325	343	120
Isophorone	0.923	410	424	180
Methyl Amyl Ketone	0.818	297	309	102
Methyl Ethyl Ketone	0.8006	172	176	24
Methyl Isoamyl Ketone	0.8164	286	298	96
Methyl Isobutyl Ketone	0.8024	237	243	61
Methyl Propyl Ketone	0.807	214	221	46

Table 12.41: Shell Ketones (14)

Typical properties of the compounds

	Acetone	Methyl ethyl ketone	Methyl isobutyl ketone	Mesityl oxide	Diacetone alcohol
Molecular weight	58.080	72.108	100.162	98.146	116.162
Specific gravity (apparent)					
60/60° F	0.7967	0.8105	0.8055	0.8590	0.9441
20/20° C	0.7925	0.8065	0.8022	0.8556	0.9409
25/25° C	0.7879	0.8023	0.7986	0.8520	0.9374
Weight per U.S. gallon (in air)					
60° F	6.636	6.750	6.709	7.154	7.863
20° C	6.595	6.711	6.676	7.120	7.830
25° C	6.549	6.668	6.638	7.082	7.792
Boiling point @ 760 mm					
°C	56.13	79.64	116.2	129.76	169.2
°F	133.03	175.26	241.16	265.57	336.6
Boiling point change					
°C/mm @ 760 mm	0.0385	0.04	0.046	0.0468	0.075
Vapor pressure at 20° C, mm	185.95	70.21	14.96	7.93	0.81
Freezing point @ 760 mm, °C	−94.897	−86.37	−83.5	−52.85	−44.
Refractive index n $\frac{20}{D}$	1.35900	1.37880	1.3957	1.44575	1.4234
Heat of vaporization					
cal/g @ 760 mm	122.09	105.95	82.50	85.6	90.0
Heat of fusion at melting point					
cal/g	23.53	24.86			
Specific heat (liquid)					
cal/g °C @ 25° C	0.51	0.51	0.53	0.52	0.62
Flash point, tag open cup, °F approx.	15.	20.	79.	98.	135.
tag closed cup, °F approx.	−15.	23.	60.	73.	126.
Flammable limits in air					
% of compound, upper	11.0v	11.5v	7.5v		
lower	3.0v	1.81v	1.4v		
Solubility, % wt.					
in water, @ 20° C	complete	27.1	2.04	3.1	complete
water in, @ 20° C	complete	12.5	2.41	3.1	complete
Azeotrope with water,					
%w compound	none	88.73	75.7	65.2	12.7
Boil pt. @ 760 mm. °C		73.41	87.93	91.8	98.8
Viscosity, cps					
@ 20° C			0.583	0.639	
@ 25° C	0.3075	0.41	0.55	0.62	2.9
@ 30° C		0.365			
Surface tension					
dyne/cm 20° C	22.32	24.6	23.64	22.9	28.9

Table 12.42: Union Carbide Ketones (19)

Ketone	Chemical Formula	Molecular Weight	Apparent Specific Gravity at 20/20°C.	Δ Sp. gr. Δt 10-40°C.	Refractive Index, n_D^{20}	Boiling Point, °C. 760 mm	300 mm	10 mm	Relative Evaporation Rate (BuAc=100)	Vapor Pressure at 20°C, mm. Hg	Δt Δp 750-770 mm	Freezing Point, °C.	Solubility, % by weight at 20°C. In Water	Water In	Absolute Viscosity at 20°C, cP	Heat of Vaporization, Btu/lb 1 atm	300 mm	Status under Rule 66 (a)	Flash Point, °F. (b)
Acetone	CH_3COCH_3	58.08	0.7905	0.00111	1.3590	56.1	31	-32	1160	186	0.039	-94.7	Complete		0.33	219	228	N-PCR	0
Methyl Ethyl Ketone (MEK, 2-Butanone)	$CH_3COC_2H_5$	72.10	0.8061	0.00104	1.3788	79.6	53	-12	570	70	0.043	-86.3	26.8	11.8	0.43	187	202	N-PCR	24
Methyl Isobutyl Ketone (MIBK, Hexone, 4-Methyl-2-Pentanone)	$CH_3COCH_2CH(CH_3)_2$	100.16	0.8020	0.00094	1.3957	116.2	87	13	165	15	0.046	-84	1.9	1.6	0.6	147	156	PCR-20%	61
Diisobutyl Ketone	$(CH_3)_2CHCH_2COCH_2CH(CH_3)_2$	142.24	0.8076	0.00082	1.4127	169.4	136	53	18	1.7	0.051	-41.5	0.05	0.75	1.0	119	127	PCR-20%	118
Isobutyl Heptyl Ketone (2,6,8-Trimethyl-4-Nonanone)	$(CH_3)_2CHCH_2CH_2COCH_2CH(CH_3)[CH_2CH(CH_3)_2]$	184.32	0.8180	0.00077	1.4257	218.2	183	92	1.4	<0.1	0.055	-75(c)	<0.01	0.2	1.9	104	111	PCR-20%	189
Mesityl Oxide	$(CH_3)_2C{:}CHCOCH_3$	98.15	0.8598	0.00082	1.4434	129.8	99	23	94	8	0.047	-46.4	3.1	3.4	0.60	157	166	PCR-5%	97
Diacetone Alcohol	$(CH_3)_2C(OH)CH_2COCH_3$	116.16	0.9406	0.00091	1.4226	169.2	137	56	14	<1	0.050	-42.8	Complete	4.5	3.2	154	162	PCR-20%	117
2,4-Pentanedione (Acetyl Acetone)	$CH_3COCH_2COCH_3$	100.12	0.9753	0.00102	1.4504	140.4	108	27	75	7	0.051	-23.5	16.6		0.58	152	159	N-PCR	99
Isophorone	$COCH{:}C(CH_3)CH_2C(CH_3)_2CH_2$ $C_9H_{14}O$	138.21	0.9229	0.00078	1.4781	215.3	179	86	3	<1	0.057	-8.1	1.2	4.3	2.6	135	143	PCR-5%	179
Cyclohexanone (Cyclohexyl Ketone, Ketohexamethylene, Pimelic Ketone)	$C_6H_{10}O$	98.15	0.9482	0.00090	1.4502	155.8	125	46	23	2	0.048	-31.2	2.5	8.0	2.2	174	184	N-PCR	111
Acetophenone (Phenyl Methyl Ketone)	$C_6H_5COCH_3$	120.15	1.0296	0.00084	1.5342	201.7	167	79	3	0.3	0.054	19.7	0.55	1.65	1.8	157	166	PCR-8%	180
Propiophenone (Phenyl Ethyl Ketone)	$C_6H_5COC_2H_5$	134.17	1.012	–	–	218.0	–	–	1.1	0.2	–	18.2	0.2	1.0	–	–	–	–	185

(a) N-PCR, Non-Photochemically Reactive; PCR-20%, PCR-8%, and PCR-5%, Photochemically Reactive – Volume % without requiring emission control.

(b) Determined with commercial material.

All flash points determined by closed cup in accordance with DOT regulations.

(c) Sets to glass below this temperature.

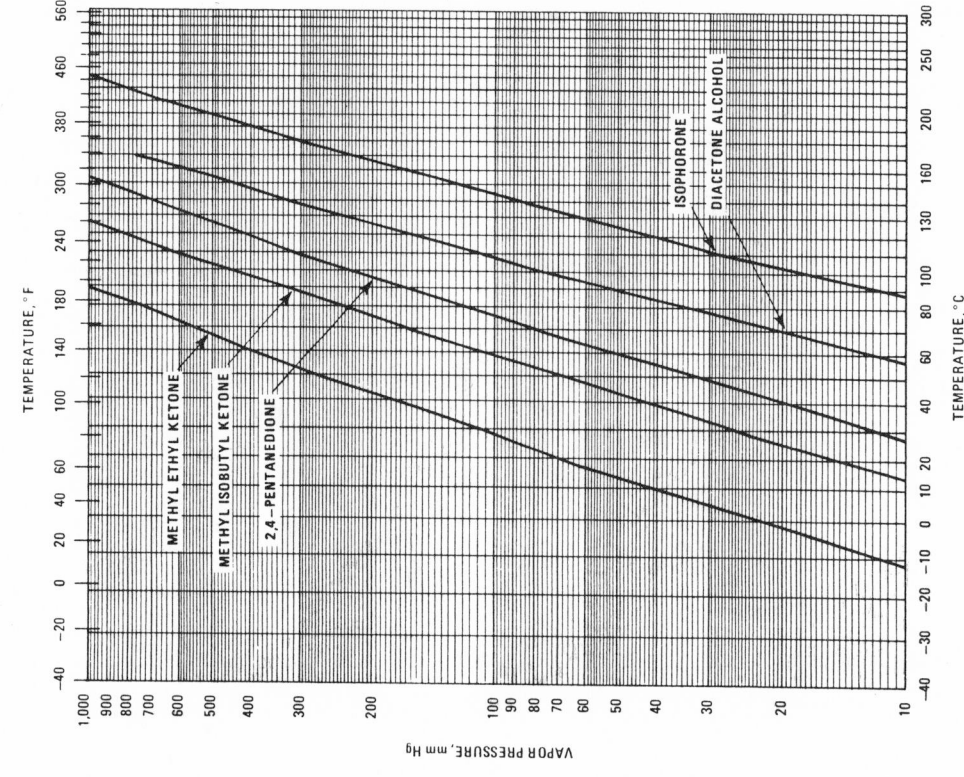

Table 12.43: Vapor Pressure of Various Ketones at Different Temperatures (19)

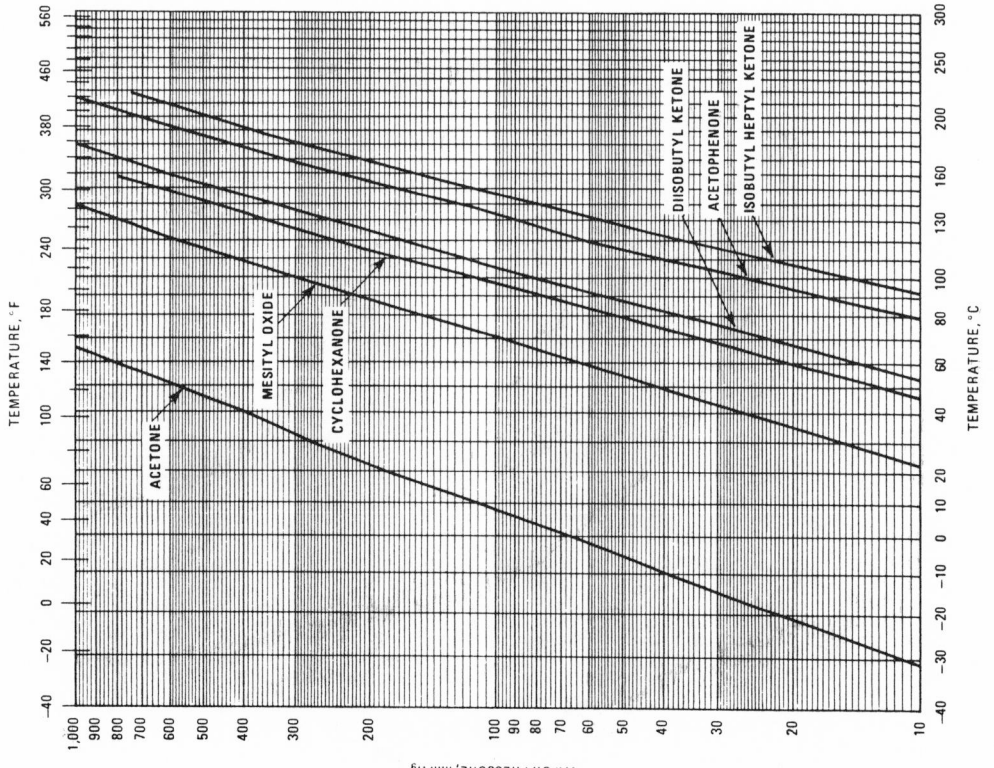

Table 12.44: Specific Gravities of Ketones *(19)*

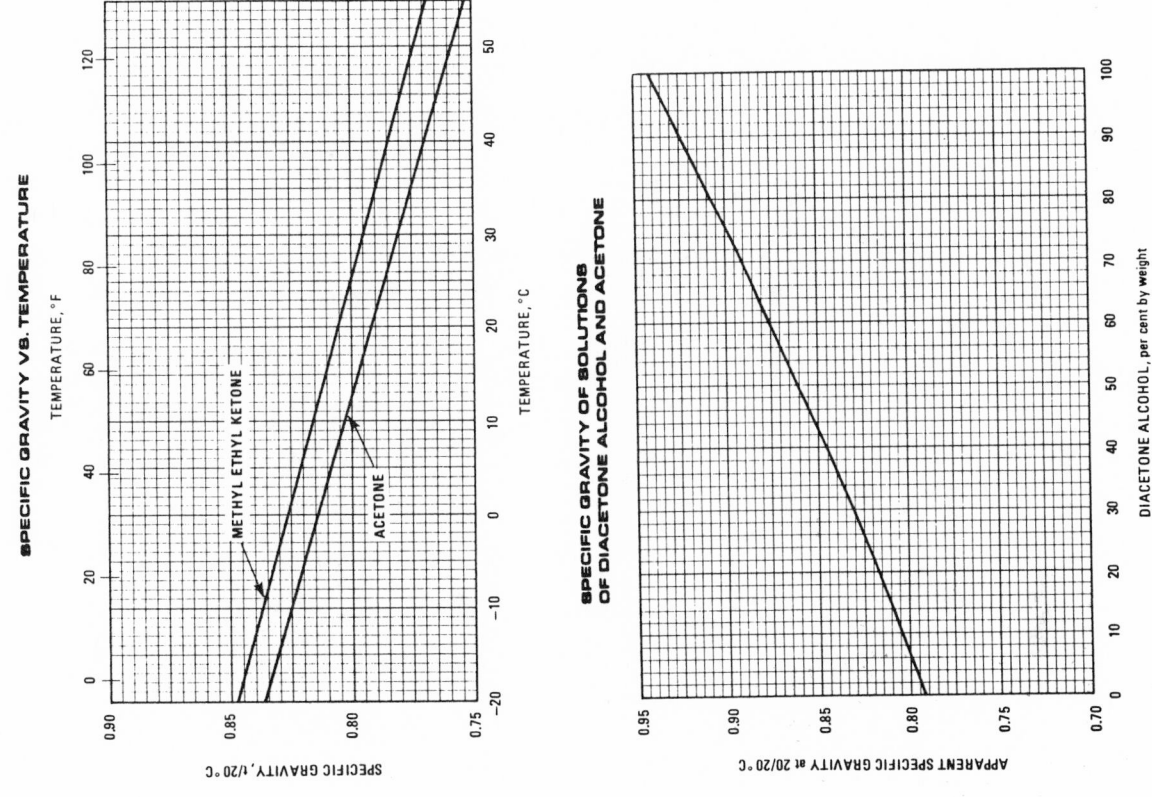

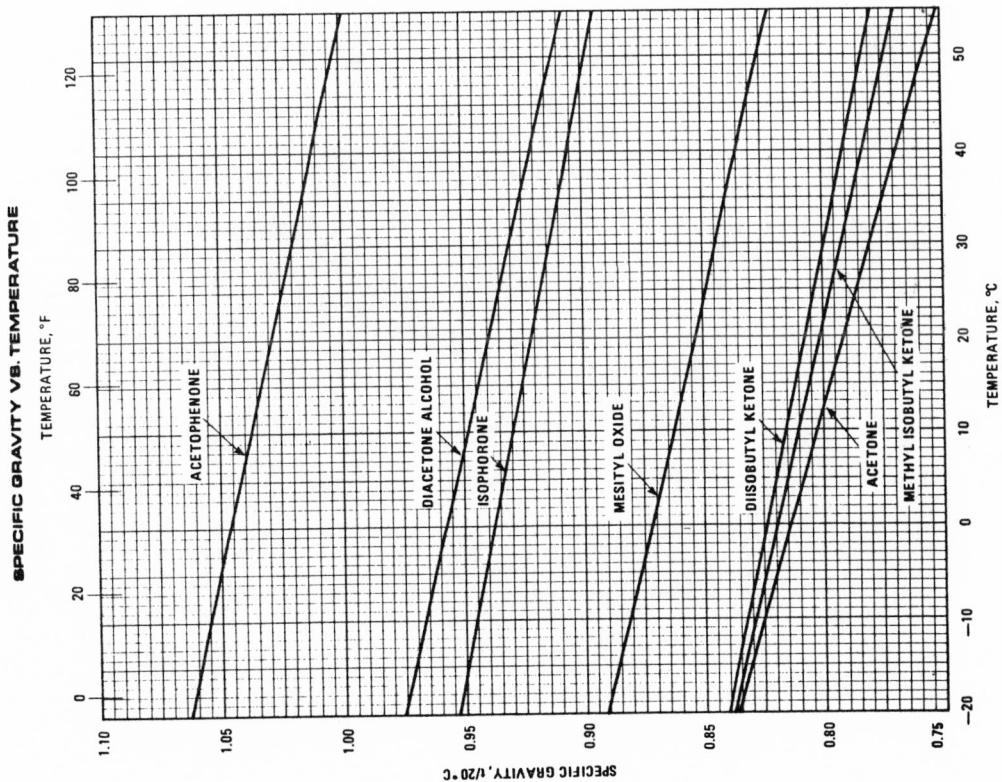

Table 12.45: Solubility of Ketones in Water *(19)*

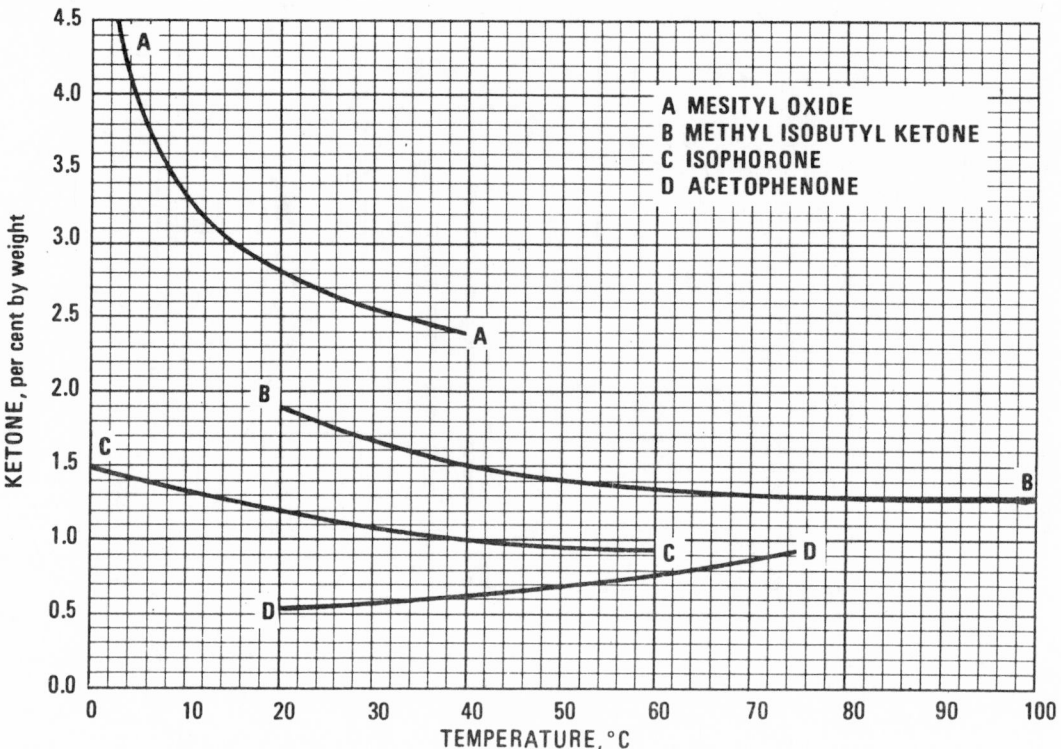

Table 12.46: Solubility of Water in Ketones *(19)*

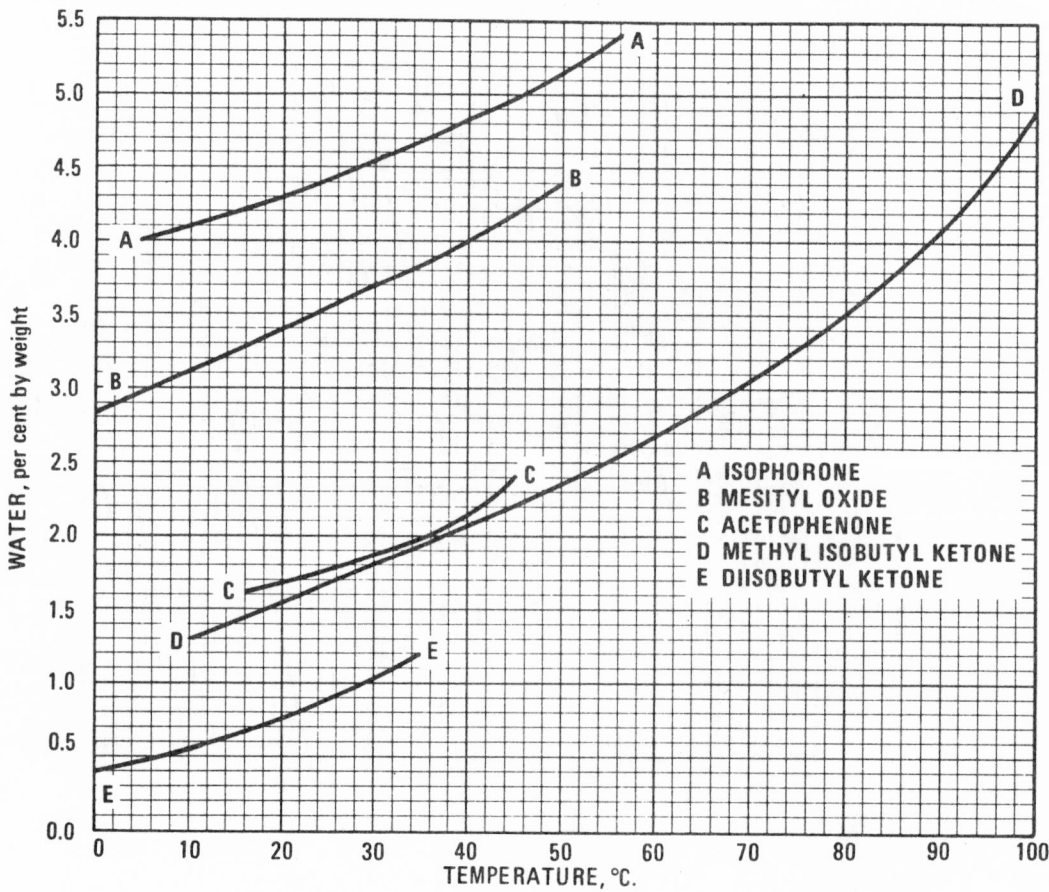

Table 12.47: Resin Solubilities *(19)*

Ketone	Cellulose Acetate 41% Acetyl	Cellulose Acetate Butyrate 17% Butyryl	Cellulose Acetate Butyrate 37% Butyryl	Cellulose Acetate Propionate 13-15% Propionyl	Cellulose Acetate Propionate 31% Propionyl	Ethyl Cellulose 47-49% Ethoxyl	Poly-Styrene	Poly (Methyl meth-acrylate)	BAKELITE Vinyl Resins VYHH Vinyl Chloride Acetate Copoly-mer	BAKELITE Vinyl Resins AYAF Vinyl Acetate	BAKELITE Vinyl Resins XYHL Vinyl Butyral
Acetone	S	S	S	S	S	S	PS	S	S	S	SW
Methyl Ethyl Ketone	S	S	S	S	S	S	S	S	S	S	S
Methyl Isobutyl Ketone	I	I	S	I	S	S	S	PS	S	S	PS
Diisobutyl Ketone	I	I	I	I	I	SW	SW	I	S-G	PS	SW
Isobutyl Heptyl Ketone	I	I	I	I	I	SW	I	I	I	I	I
Mesityl Oxide	PS	SW	S	S-G	S	S	S	SI S	S	S	G
Diacetone Alcohol	S	PS	S	S	S	S	I	S	S	S	S
2,4-Pentanedione	S	S	S	S	S	SW	S	S	S	S	G
Isophorone	S	S	S	S	S	S	S	PS	S	S	S
Cyclohexanone	S	S	S	S	S	S	S	S	S	S	S
Acetophenone	PS	PS-G	S	S	S	S	S	S	S	S	S

CODE:

S — Soluble SW — Soluble Warm I — Insoluble
PS — Partially Soluble G — Gels

Table 12.48: Relative Evaporation of Ketones—Fast to Intermediate Evaporating Liquids *(19)*

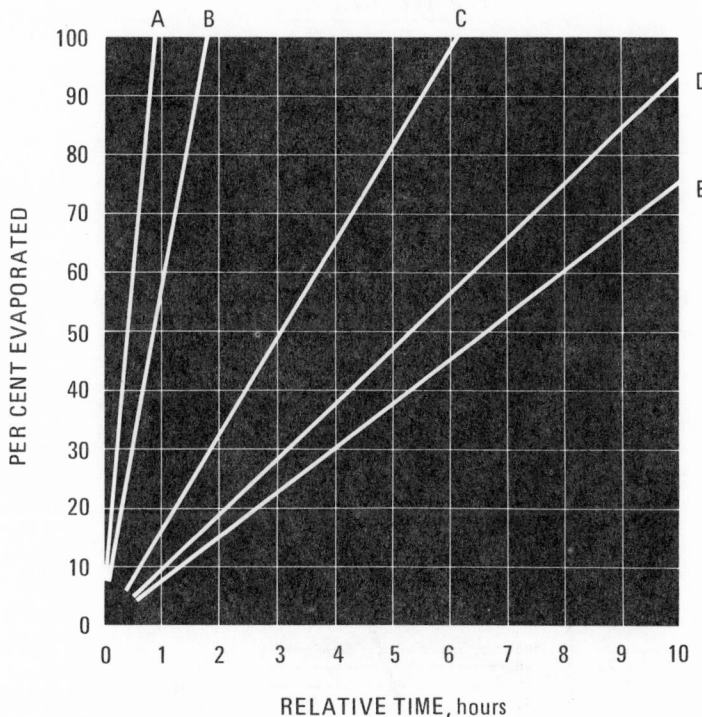

CHART KEY

A Acetone
B Methyl Ethyl Ketone
C Methyl Isobutyl Ketone
D Mesityl Oxide
E 2,4—Pentanedione

Table 12.49: Relative Evaporation of Ketones—Intermediate to Slow Evaporating Liquids *(19)*

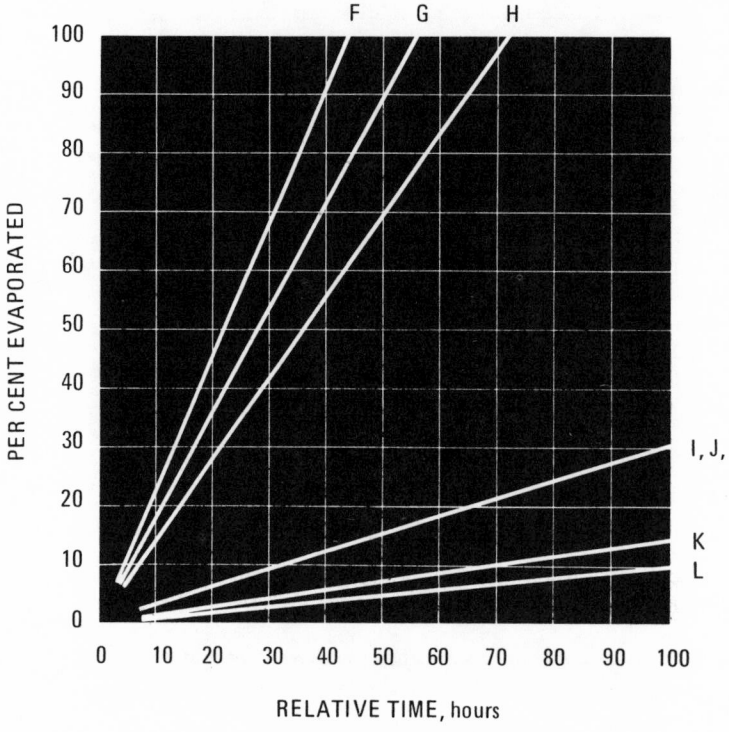

CHART KEY

F Cyclohexanone   J Isophorone
G Diisobutyl Ketone K Isobutyl Heptyl Ketone
H Diacetone Alcohol L Propiophenone
I Acetophenone

Table 12.50: Viscosity vs Concentration of Chlorinated Rubber (Hercules Parlon S-20) in MEK *(8)*

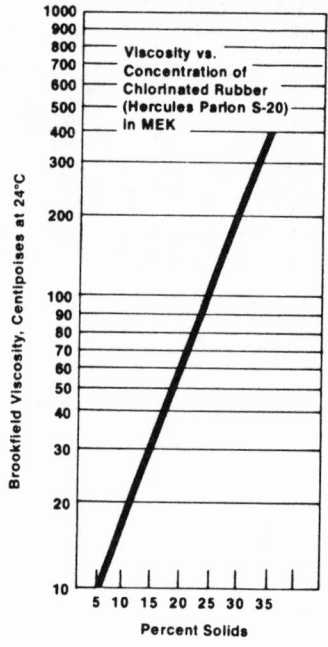

Table 12.51: Effect of Diluent Type on Viscosity of VAGH System at 18% Weight Solids *(14)*

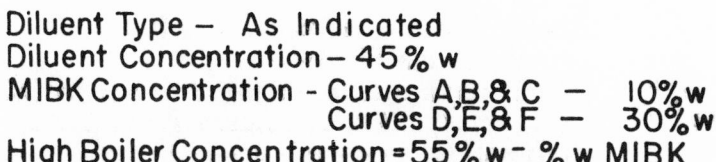

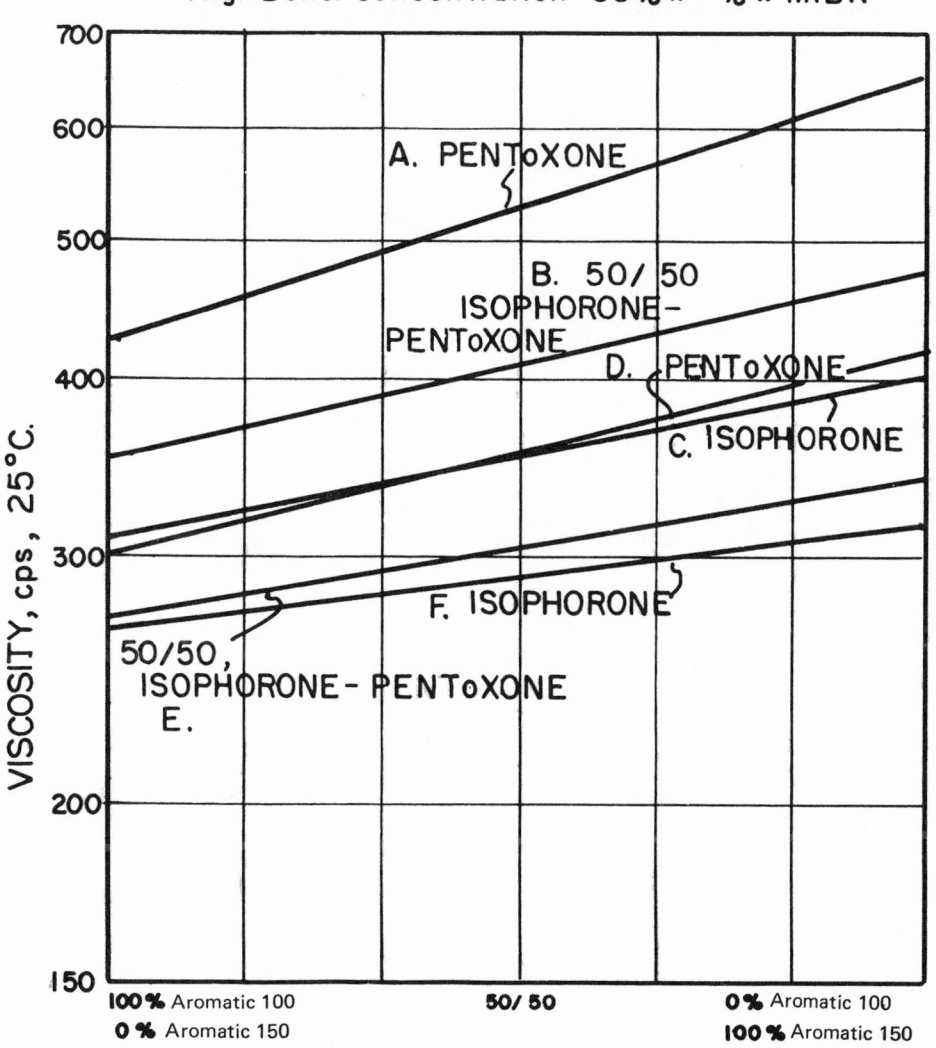

Table 12.52: Effect of Diluent Type on Viscosity of 50/50 VMCH/VYHH Combination at 20% Weight Solids *(14)*

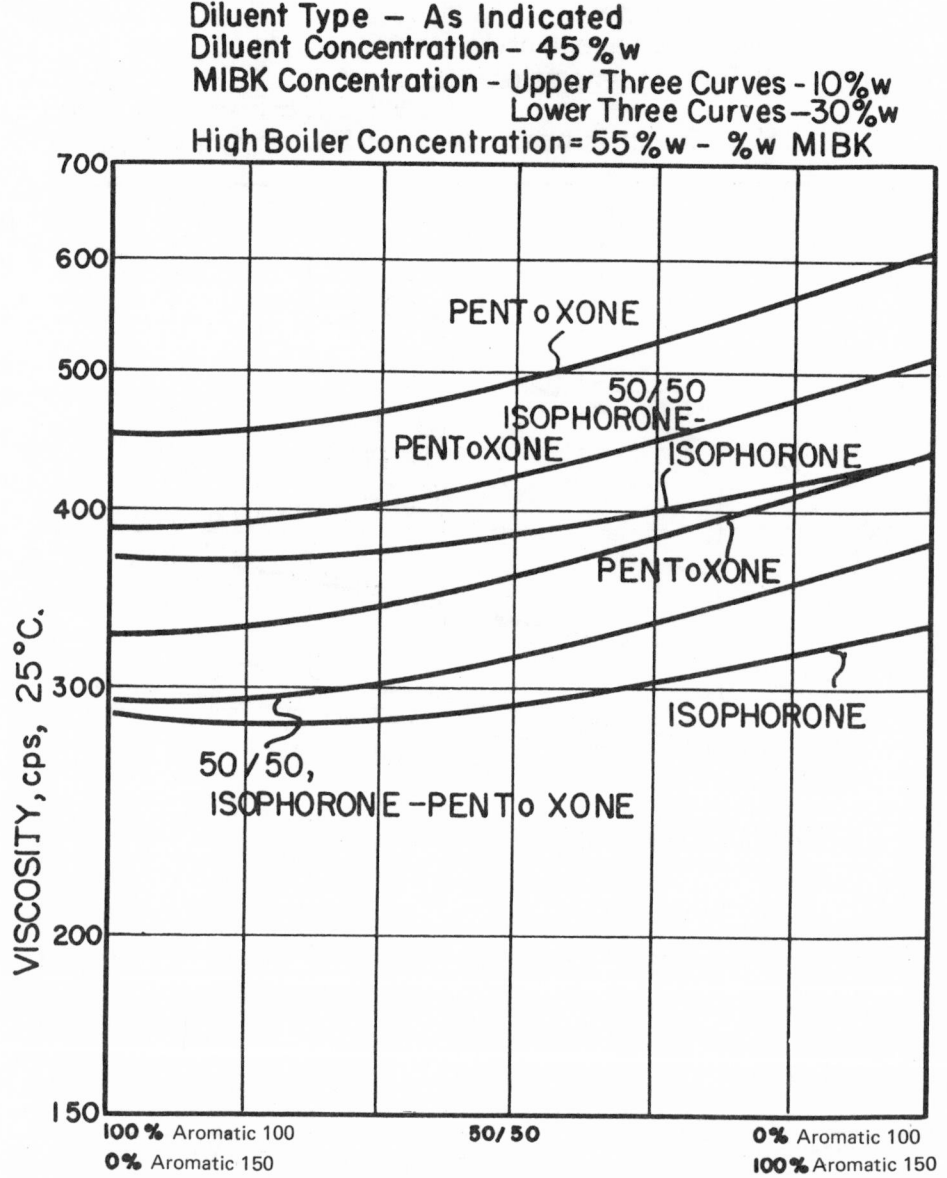

Acids

Vinegar Acid
Methanecarboxylic Acid
Ethanoic Acid

$$CH_3COOH$$

Acetic acid is a colorless liquid with a pungent odor; it is made synthetically from acetylene or by the oxidation of alcohol. It is soluble in water, alcohols, ethyl ether, and other organic solvents. It is used as a precipitant for albumen, casein, and rubber latex. It is also employed in the manufacture of leather, cordage, linoleum, acetate solvents, acetyl derivatives, dyes, matches, printing inks, and polishes, and as an assistant in dyeing processes.

Specifications (Glacial Acetic)

	Standard + Laundry Special	U.S.P. XII	C.P.
Acetaldehyde	0.05% (max.)		
Acidity, as acetic acid	99.5% (min.)	99.5% (min.)	99.8% (min.)
Color	Water-white	Water-white	Water-white
Formic acid	0.2% (max.)	—	0.00% (max.)
Freezing point	15.6°C (min.)	15.6°C (min.)	16.24°C (min.)
Non-volatile matter	—	0.0265% (max.)	0.0008% (max.)
Water content	0.5% (max.)	0.5% (max.)	0.2% (max.)
Weight per gallon at 20°C	8.74 lbs.	8.74 lbs.	8.74 lbs.

Table 13.2: Viscosity of Acetic Acid and Acetic Anhydride Mixtures at 15°C and 76.5°C *(19)*

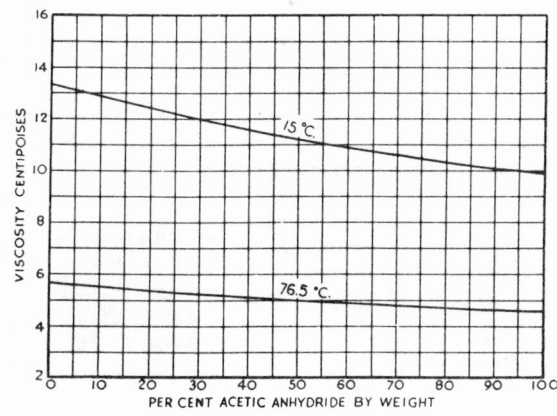

501

Table 13.3: Butyric Acid *(2)*

Ethylacetic Acid
Butanoic Acid
Propylformic Acid

$$CH_3CH_2CH_2COOH$$

Butyric acid is a water–white liquid having a characteristically pronounced and highly disagreeable odor. It is soluble in most organic solvents and completely soluble in water. The importance of butyric acid is found in its butyrate, made with alcohols; these compounds are used as flavors because of their pleasant fruity odors. Other uses are in the manufacture of flavor esters, plastics, drugs, in leather tanning and for deliming hides.

Typical Properties and Specifications

Boiling point at 760 mm	163.5°C	Solubility in water at 20°C	Complete
Coefficient of expansion at 20°C	0.001026°C	Solubility of water in solvent at 20°C	Complete
at 55	0.001064	Specific gravity at 20/20°C	0.9595
Color	Water-white	Specific heat	0.514 (20–100°C)
Critical temperature	355°C	Refractive index at 19°C	1.3980
Critical density	0.302	Surface tension at 20°C	26.8 dynes/sq cm
Dissociation constant at 25°C	1.48×10^{-5} recip. ohm	Vapor pressure at 20°C	0.84 mm Hg
Electrical conductivity at 25°C	0.00039×10^{-4} recip. ohm	Viscosity at 25°C	0.01529 poise
Flash point (ASTM open cup)	170°F	Weight per gallon at 20°C	7.985 lbs.
Heat of combustion	5905 cal. (15)/g	Chlorides	None
Heat of fusion	20.1 cal. (15)/g	Distillation range at 760 mm	160–165°C
Heat of vaporization	1.59 cal./g	Purity	99.0% by wt., min.
Melting point	−5.7°C		

Table 13.4: Viscosity of Aqueous Butyric Acid Solution at 25°C *(19)*

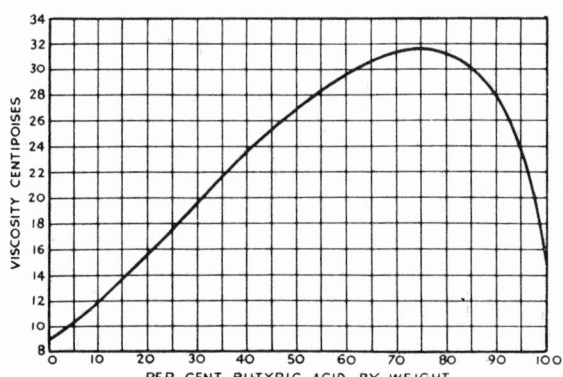

Table 13.5: Butyric Anhydride *(2)*

$$(C_3H_7CO)_2O$$

Butyric anhydride is a water–white liquid which hydrolyzes to butyric acid in the presence of water. Like butyric acid, it is used in making butyrates, flavors, drugs and tanning agents.

Boiling point at 760 mm.	199.5°C.	Weight per gallon at 20°C.	8.1 lbs.
Color	Water-white	Distillation range at 760 mm.	
Flash point	190°F.	Below 190°C.	None
Melting point	-75°C.	Above 200°C.	None
Specific gravity at 20/20°C.	0.965-0.970	Below 195°C.	Not more than 10%
Vapor pressure at 20°C.	0.37 mm. Hg	Purity	85% by wt., min.

Table 13.6: Solubility of Water in Caproic Acid at Various Temperatures *(19)*

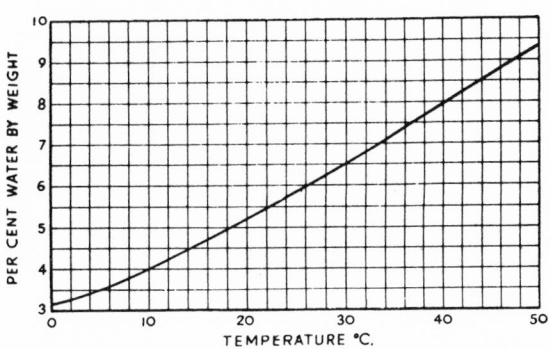

Table 13.7: 2-Ethylbutyric Acid *(2)*

Diethyl Acetic Acid
2-Ethylbutanoic Acid

$(C_2H_5)_2CHCOOH$

2-Ethylbutyric acid is a water-white liquid, similar to butyric acid in most of its properties, except that its odor is less strong and it is not as soluble in water. Its halogenated derivatives are finding use in the manufacture of drugs. Its esters with higher glycols are outstanding vinyl resin plasticizers.

Boiling point at 760 mm.	194°C.	Vapor pressure at 20°C.	0.14 mm. Hg
Flash point	210°F.	Weight per gallon at 20°C.	7.68 lbs.
Solubility in water at 20°C.	0.22% by wt.	Distillation range at 760 mm.	185°-200°C.
Solubility of water in solvent at 20°C.	3.3% by wt.	Purity	90% by wt., min.
Specific gravity at 20/20°C.	0.9225		

Table 13.8: 2-Ethylhexoic Acid *(2)*

Octoic Acid
2-Ethylhexanoic Acid

$CH_3(CH_2)_3CH(C_2H_5)COOH$

This acid possesses a mild odor and a high boiling point. It is important for its metallic esters, the properties of which suggest usefulness as varnish driers. These metallic salts are stable, mild-odored, light-colored compounds, and are soluble in hydrocarbons. The glycol esters of this acid are excellent vinyl resin plasticizers.

Boiling point at 760 mm.	226.9°C.	Vapor pressure at 20°C.	0.03 mm. Hg
Flash point	260°F.	Weight per gallon at 20°C.	7.55 lbs.
Solubility in water at 20°C.	0.25% by wt.	Distillation range at 760 mm.	220°-230°C.
Solubility of water in solvent at 20°C.	1.2% by wt.		(90% distills within this range
Specific gravity at 20/20°C.	0.9077	Purity	95% by wt. min.

Table 13.9: Solubility of Water in Ethylhexoic Acid, Ethylbutyraldehyde and Ethylpropylacrolein *(19)*

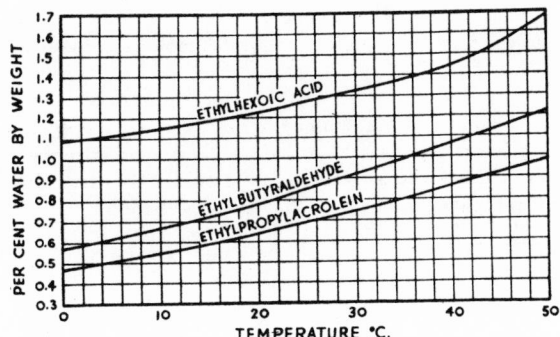

Table 13.10: Gluconic Acid, Technical—50% Solution *(25)*

Gluconic Acid, Technical—
50%, Pfizer is produced as a
50% (by weight) aqueous solution. It is colorless to pale yellow, and has a slight acetous
odor. Aqueous solutions of the
acid are an equilibrium mixture of gluconic acid and the
gamma- and delta-lactones.
Gluconic acid is a mild organic
acid, has a very low order of
toxicity, and is noncorrosive
and nonirritating.

Formula	$C_6H_{12}O_7$
M. W.	196.16
Specific Rotation $\left[\alpha\right]_D^{20°}$	−6.7°
Specific Gravity	
50% Solution	1.24
Viscosity	
50% Solution	
40°F.	32 cps.
60°F.	18 cps.
80°F.	11 cps.
100°F.	7 cps.
pH, Approximate	
1% by wt.	3.2
5% by wt.	2.8
50% by wt.	2.2

```
   COOH
    |
   HCOH
    |
   HOCH
    |
   HCOH
    |
   HCOH
    |
   CH₂OH
```

Table 13.11: Glucono-Delta-Lactone *(25)*

Glucono-Delta-Lactone, Pfizer
is an inner ester of gluconic
acid. The compound is an odorless, free-flowing white powder,
stable in air. Solutions of glucono-delta-lactone, like gluconic acid itself, have a very
low order of toxicity and are
relatively noncorrosive and
nonirritating.

Formula	$C_6H_{10}O_6$
M. W.	178.14
Melting Point	153°C.
Specific Rotation $\left[\alpha\right]_D^{20°}$	+61.7°
Solubility, 25°C.	
Ethyl Alcohol	
Water—	1.0 gm./100 ml.
	Dissolves to form equilibrium solution of gluconic acid and gamma and delta lactones.

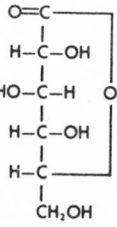

Table 13.12: Lactic Acid *(2)*

α-Hydroxypropionic Acid $CH_3CHOH \cdot COOH$

Lactic acid, which is among the oldest known organic acids, is obtained from sour
milk by the reduction of hexose sugars or by the interaction of acetaldehyde and
carbon monoxide. It is miscible with water and many organic reagents. Since it
has an asymmetrical carbon atom, lactic acid exists in two optical isomeric forms.
Peckham states that "the nomenclature used to designate the isomeric forms was,
until recently, very confusing. The form of the acid commonly known as sarco-
lactic, the form occurring in blood, has (+) rotation but the l configuration. It
is therefore correctly designated as l(+) lactic acid and its mirror image as d(−)
lactic acid. The salts of the l(+) form are levorotatory and the salts of the d(−)
form are dextrorotatory. Because of the low optical rotatory power of the free
acids, rotation of the pure acid or its simple salts is not a suitable criterion for
establishing the optical form of the acids, or the percentage composition in case
of a mixture".

Commercial lactic acid has been determined to be a mixture of α-hydroxypropionic
acid, lactyllactic acid, and water. When dilute lactic acid is concentrated, two
molecules of lactic acid unite to form lactyllactic acid and water. The lactyllactic
acid splits off from the water as shown on the following page.

(continued)

Table 13.12: (continued)

$$2CH_2CHOH \cdot COOH \rightleftarrows CH_2 \cdot CHOH \cdot COOCH(CH_3) \cdot COOH + H_2O$$

$$\downarrow\uparrow$$

$$\underline{CH(CH_3)OCOCH(CH_3)OCO}$$

Polylactyllactic acids may also be formed by loss of water between the carboxyl and the alcohol groups, thus:

$$CH_3CHOHCOCH_3 \cdot CH \cdot COCH_3 \cdot CHCOOH$$

dilactyllactic acid

The conditions which affect the production of a lactic acid solution from lactyllactic acid are temperature, concentration and age of solution.

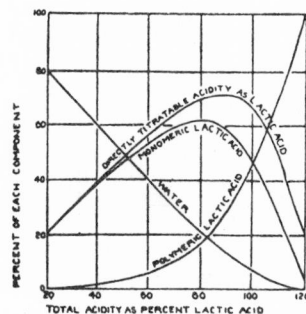

A graph showing the Composition of Aqueous Lactic Acid Systems at Equilibrium and at Progress States of Dehydration.

Table 13.13: Vapor Pressure of Organic Acids and Anhydrides at Various Temperatures *(19)*

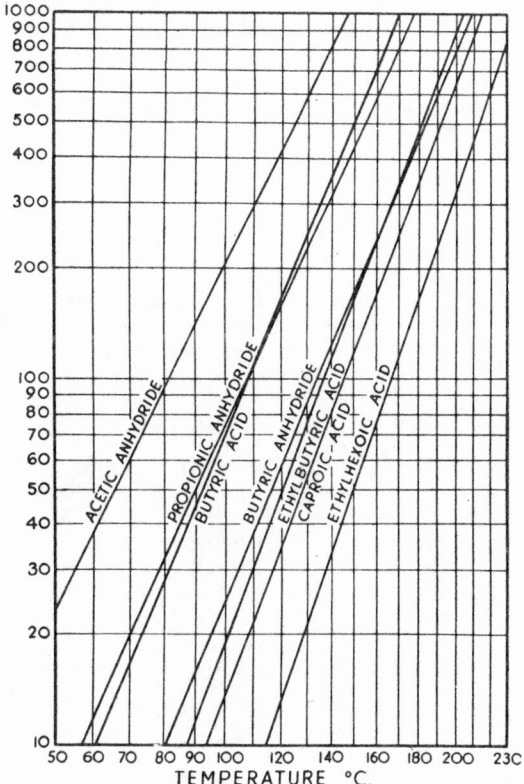

Table 13.14: Fatty Acid Composition of Various Fats and Oils (26)

Fat / Oil	Butyric	Caproic	Caprylic	Capric	Lauric	Lauroleic	Myristic	Myristoleic	Pentadecanoic	Palmitic	Palmitoleic	Margaric	Stearic	Oleic	Linoleic	Linolenic	Ricinoleic	Dihydroxystearic	Licanic	Eleostearic	Arachidic	Eicosenoic	Eicosapolyenoic	Behenic	Erucic (Docosenoic)	Docosapolyenoic	Lignoceric	Tetracosenoic	Tetracosapolyenoic	Iodine Value	Saponification Value	Titer °C
Babassu	3.0		6.0	5.5	45.0		16.5			7.0			3.0	14.5	1.5															12–18	247–251	22–26
Butterfat	3.0	1.0	1.5	3.0	3.5		12.0	1.5	1.0	28.0	3.0		13.0	28.5	1.0															25–42	210–235	33–38
Castor										1.5			0.5	4.0	4.0	0.5	87.5	0.5			0.5									81–91	177–187	1–3
Cocoa Butter					1.5		0.5			25.0			34.5	36.5	3.0						1.0									35–40	190–200	45–50
Coconut		0.5	7.0	6.0	48.0		19.0			9.0			3.0	6.0	1.5															8–12	250–260	20–24
Corn							1.0			12.0			2.0	25.0	60.0	0.5														118–128	186–194	14–20
Cottonseed							1.0			23.0	0.5		2.5	16.5	57.0						0.5									103–113	189–199	30–37
Herring							7.0			12.0	10.0		0.5	13.0	8.0							25.5			24.0					125–145	180–193	23–27
Lard							1.5			25.0	3.0		13.0	45.5	10.5	1.0														58–68	192–202	34–43
Linseed										6.0			3.5	20.0	14.5	56.0														180–195	188–196	19–21
Menhaden							9.0	0.5	0.5	19.0	16.0	0.5	5.5	24.5							0.5	15.0				9.0			0.5	140–180	188–196	31–33
Mustard Seed (Montana)										3.0			6.5	22.0	22.5	15.5					0.5	11.5			18.0					114–128	176–184	6–10
Neatsfoot						0.5	1.0	0.5		20.5	6.0		4.5	56.5	9.5	0.5						0.5								65–75	190–199	20–30
Oiticica										6.0			5.0	6.0	5.0			2.0	76.0											140–160	188–196	17–21
Olive							1.0			13.0	1.0		2.5	74.0	9.0	0.5														79–89	188–196	17–26
Palm							1.0			43.5			4.5	40.0	11.0															45–55	195–205	40–47
Palm Kernel			3.5	3.5	48.5		16.5			8.5			2.5	14.5	2.0															14–24	245–255	20–28
Peanut (Southwest)										11.0			2.5	50.0	30.5	1.0					1.0			3.0			1.0			93–98	188–196	28–34
Peanut (West Coast)										12.5			3.0	38.5	38.0	1.5					1.5			3.5			1.5			96–101	188–196	26–32
Perilla										8.0				16.0	14.0	62.0														193–208	188–197	12–19
Rapeseed (High-Erucic)					0.5					4.0			0.5	12.5	14.5	16.5						10.5			51.5					97–104	169–179	11–15
Rapeseed (Montana)										3.0			1.5	32.0	19.0	10.0					0.5	10.5			23.5		1.0			104–110	170–180	11–15
Rice Bran							0.5			17.0			2.5	45.5	32.0	1.0					0.5			0.5						92–109	184–195	26–28
Safflower										6.5			2.5	11.5	79.0	0.5														138–145	186–195	15–18
Sardine							6.0			11.5	12.0		2.5	11.5	11.5							25.5				19.5				160–190	188–196	28–32
Sesame										9.0			6.0	41.5	43.0	0.5														108–113	188–196	20–25
Soybean										11.0			4.0	21.0	55.5	8.5														125–135	188–196	20–22
Sperm-Body Fatty Acids				3.0			5.0	4.0		8.0	15.0		2.0	37.0								19.0			1.0					76–88	140–144	8–14
Sperm-Head Fatty Acids				3.0	16.0	4.0	14.0	14.0		8.0	15.0			17.0								7.0								55–70	140–144	12–18
Sunflower							0.5			6.5			4.0	17.0	72.5															128–138	186–196	16–20
Tall Oil (Distilled Fatty Acids)										1.0	0.5		1.5	50.5	46.5						0.5									128–133	186–196	1–8
Tallow							3.5	1.0	0.5	25.5	4.0	2.5	19.5	41.0	2.5						0.5									40–50	192–202	40–46
Tung Oil										4.0			1.0	8.5	3.5	4.0				80.0										160–175	189–195	34–42
Whale							8.0	2.0		17.0	13.0		2.0	39.0		3.0						12.5				6.5				110–140	185–195	22–24

Fatty Acid	Number of Carbon Atoms	Number of Double Bonds	Molecular Weight	Neutralization Value	Iodine Value	Boiling Point °C @ 5 mm Hg	Melting Point °C
Butyric	4	0	88	636	0	50.0	−8.0
Caproic	6	0	116	483	0	86.5	−3.4
Caprylic	8	0	144	389	0	113.5	16.7
Capric	10	0	172	325	0	137.0	31.6
Lauric	12	0	200	280	0	158.0	44.2
Lauroleic	12	1	198	282	128	—	—
Myristic	14	0	228	245	0	178.0	53.9
Myristoleic	14	1	226	247	112	—	—
Pentadecanoic	15	0	242	231	0	187.0	52.3
Palmitic	16	0	256	218	0	197.0	63.1
Palmitoleic	16	1	254	220	99	—	0.5
Margaric	17	0	270	207	0	206.0	61.3
Stearic	18	0	284	197	0	214.0	69.6
Oleic	18	1	282	198	89	209.0	13.4
Linoleic	18	2	280	200	181	—	−5.0
Linolenic	18	3	278	201	273	—	<−10.0
Ricinoleic	18	1	298	187	85	—	5.0
Dihydroxystearic	18	0	316	177	0	—	141.0
Licanic	18	3	292	191	260	—	74.0–75.0
Eleostearic	18	3	278	201	273	—	—
Arachidic	20	0	312	179	0	233.0	75.3
Eicosenoic	20	1	310	180	81	—	—
Eicosapolyenoic	20	2–5	—	—	—	—	—
Behenic	22	0	340	164	0	247.0	79.9
Erucic (Docosenoic)	22	1	338	165	74	—	34.7
Docosapolyenoic	22	2–5	—	—	—	—	—
Lignoceric	24	0	368	152	0	255.0	84.2
Tetracosenoic	24	1	366	153	69	—	—
Tetracosapolyenoic	24	2–5	—	—	—	—	—

Table 13.15: "Crodym" Dimer and "Cromon" Monomer Acids *(80)*

PRODUCT	DESCRIPTION	SPECIFICATIONS		TYPICAL PROPERTIES								TYPICAL COMPOSITION		
		ACID NUMBER ASTM D 1980	GARDNER COLOR ASTM D 1544-68 MAXIMUM	ACID NUMBER ASTM D 1980	UNSAPON-IFIABLES ASTM D 803 %	GARDNER COLOR ASTM D 1544-68	SAPON. NUMBER ASTM D 803	VISCOSITY GARDNER-HOLDT @ 25°C	FLASH POINT ASTM D 92 C.O.C. °F	DENSITY @ 25°C G/CC	DENSITY @ 25°C LB/GAL	TRIMER ACIDS SYLVACHEM GEL PERMEATION CHROMATOGRAPHY METHOD APRX. %	DIMER ACIDS APRX. %	MONOMER ACIDS APRX. %
DIMER ACIDS														
CRODYM T-18	A dimerized fatty acid produced by selective reaction of high quality tall oil fatty acid distillates mainly composed of C_{36}-C_{54} di and tri carboxylic acids.	185-195	8	190	0.5	7	193	Z4-Z5	>500	0.95	7.9	18	81	<1
CRODYM T-22	C_{36}-C_{54} di and tri carboxylic acids similar to CRODYM T-18 with higher trimer content.	185-195	9	190	0.5	8	193	Z4-Z5	>500	0.95	7.9	20	77	3
CRODYM MX	A dimerized fatty acid produced from quality fatty acid distillates with higher concentrations of monomer acid.	185-195	12	188	0.5	10	191	Z4	>500	0.95	7.9	16	80	4
CRODYM MXB	A mixture of dimerized and monomeric fatty acids produced from the dimerization process.	185-195	12	187	1.0	10	—	Z2	>500	0.95	7.9	14	78	8
MONOMER ACIDS														
CROMON D-1	The monomeric volatile fatty acid component removed by vacuum distillation from a dimerized tall oil fatty acid composite. A mixture of branched and straight chain fatty acids which may be saturated or mono-unsaturated. Typical Titer is 25-35°C with an Iodine Number of 75-95.	170-185	6	178	4.5	4	188	—	370	0.92	7.6	0	5.0	90
CROMON D-1R	A purified form of CROMON D-1 with a Typical Titer of 25-35°C and an Iodine Number of 75-95.	170-185	4	174	8.0	3	180	—	370	0.92	7.6	0	1.5	90

Table 13.16: "Emery" Fatty and Dibasic Acids (63)

Dimer, Trimer and Polybasic Acids

| | SPECIFICATIONS | | | | COMPARATIVE TYPICAL COMPOSITION[1] | | | | | |
| | Acid Value | Sap. Value | Color 1963 Gardner, max. | Unsap. max.,% | Short-Path Methyl Ester Distillation | | | High Pressure Liquid Chromatography | | |
					Mono	Di	Poly	Mono	Di	Poly
Empol 1010 Dimer Acid (polymer grade)	195-201	196-203	1	nil[2]	0	97	3	4	94	2
Empol 1014 Dimer Acid	194-198	197-201	5	0.5[2]	1	95	4	4	91	5
Empol 1016 Dimer Acid	190-198	195-204	6	0.5	1	80	19	6	76	18
Empol 1018 Dimer Acid	188-196	192-198	8	1	Tr	83	17	6	79	15
Empol 1022 Dimer Acid	189-197	191-199	8	1	3	75	22	9	77	14
Empol 1024 Dimer Acid	189-197	191-199	8	1	1	75	20	8	77	15
Empol 1040 Trimer Acid	175-192	192-200	—	—	—	7	93	2	18	80
Empol 1041 Trimer Acid	161-181	180-195	11	—	—	18	82	3	35	62
Empol 1052 Polybasic Acid	250-265[2]	258-269[2]	dark	—	—	—	—	3[3]	34[3]	63[3]

[1]Short path methyl ester fractionation measures the relative molecular size of the various components of these acids. High pressure liquid chromatography (HPLC) separates components according to their functionality.

[2]Not a specification.

[3]By thin layer chromatography. This method determines composition according to functionality

Coconut Fatty Acids

| | SPECIFICATIONS | | | | | | TYPICAL COMPOSITION[1] | | | | | | | |
	Titer, °C	Iodine Value	Color % Trans. 440/550 nm., min.	Color Gardner 1963, max.	Acid Value	Sap. Value	Caprylic $C_8H_{16}O_2$	Capric $C_{10}H_{20}O_2$	Lauric $C_{12}H_{24}O_2$	Myristic $C_{14}H_{28}O_2$	Palmitic $C_{16}H_{32}O_2$	Stearic $C_{18}H_{36}O_2$	Oleic $C_{18}H_{34}O_2$	Linoleic $C_{18}H_{32}O_2$
Emery 621 Coconut Fatty Acid	23-27	8-16	30/80	5[2]	258-268	258-268	4	5	48	20	10	2	10	1
Emery 622 Coconut Fatty Acid	22-26	5-10	65/96	2[2]	268-276	268-276	7	6	48	19	9	2	8	1
Emery 625 Partially Hydrogenated Coconut Fatty Acid	23-26	2-5	85/98	1	266-274	266-274	7	6	49	19	9	7	3	
Emery 626 Low IV Ultra Coconut Fatty Acid	23-26	1.0 max.	85/99	1	270-276	270-276	7	6	51	18	10	7	1	
Emery 627 Low IV, Stripped, Ultra Coconut Fatty Acid	28-32	1.0 max.	80/98	1	252-258	252-258		1	55	22	11	10	1	

[1]By GLC analysis, AOCS Ce 1-62.

[2]Typical property.

Food Grade[1] Fatty Acids

| | SPECIFICATIONS | | | | | | TYPICAL COMPOSITION[2] | | | | | | | | | |
| | | | | | | | Saturated Acids | | | | | Unsaturated Acids | | | | |
	Titer, °C	Iodine Value max.	Color % Trans. 440/550 nm., min.	Acid Value	Sap. Value	Unsap. %, max.	Myristic $C_{14}H_{28}O_2$	Pentadecanoic $C_{15}H_{30}O_2$	Palmitic $C_{16}H_{32}O_2$	Margaric $C_{17}H_{34}O_2$	Stearic $C_{18}H_{36}O_2$	Myristoleic $C_{14}H_{26}O_2$	Palmitoleic $C_{16}H_{30}O_2$	Oleic $C_{18}H_{34}O_2$	Linoleic $C_{18}H_{32}O_2$	Linolenic $C_{18}H_{30}O_2$
Emersol 6320 DP Stearic Acid	53.9-54.7	3.5-5.0	88/99	205-210	206-211	0.5	2.5	0.5	50	1	40			6		
Emersol 6332 USP/NF TP Stearic Acid[3,4]	54.7-55.7	0.5	93/99	205-211	207-212	0.5	1.5	0.5	50	1	47					
Emersol 6349 Stearic Acid	59.0-60.5	0.5	88/99	203-206	204-207	0.5	3	0.5	26.5	1	69			Tr		
Emersol 6351 Stearic Acid	65-68	1.0	84/98	196-201	197-202	—	1	Tr	7.5	2.5	88			1		
Emersol 6313 USP/NF Low-titer Oleic Acid[4]	6 max.	88-93	75/98	201-204	201-207	0.5	3	Tr	5	1	Tr	3	6	75	6	1
Emersol 6321 USP/NF Low-titer White Oleic[4]	6 max.	87-92	85/99	201-204	201-207	0.5	3	Tr	5	1	Tr	3	6	75	6	1
Emersol 6333 USP/NF LL Oleic Acid[4,5]	8-10	86-91	85/99	200-204	201-207	0.5	3	Tr	6.5	1	1.5	3	5.5	73.5	5.5	0.5

[1]Meet the requirements of Federal Food Additive Regulation Section 21CFR 172.860.

[2]By GLC analysis, AOCS Ce 1-62.

[3]Available in powdered form, Sieve Test: 20 mesh-99%, 40 mesh-75%, 60 mesh-30%, 80 mesh-15%, 100 mesh-10%; color, 90 min. @ 440 nm. NOTE: Stearic acid powder may present an explosion hazard if the particles are suspended in air in certain concentrations and ignited; thus special care should be taken in the handling of this product.

[4]USP XX Revision/National Formulary XV Edition.

[5]LL (low-linoleic content) oleic: polyunsaturate 6% max.

(continued)

Table 13.16: (continued)

Isostearic Acids

	SPECIFICATIONS					
	Titer, °C, max.	Iodine Value	Color % Trans. 440/550 nm., min.	Acid Value	Sap. Value	Unsap. %, max.
Emersol 871 Isostearic Acid	10	12 max.	30/85	175 min.	180 min.	6.0[1]
Emersol 875 Isostearic Acid	10	3 max.	85/98	187-197	195-202	3.0

[1]Typical property.

Oleic and Linoleic Acids

| | **SPECIFICATIONS** | | | | | | **TYPICAL COMPOSITION[1]** | | | | | | | | | | |
| | | | | | | | **Saturated Acids** | | | | | | **Unsaturated Acids** | | | | |
	Titer, °C	Iodine Value	Color % Trans. 440/550 nm., min.	Acid Value	Sap. Value	Unsap. %, max.	Lauric $C_{12}H_{24}O_2$	Myristic $C_{14}H_{28}O_2$	Pentadecanoic $C_{15}H_{30}O_2$	Palmitic $C_{16}H_{32}O_2$	Margaric $C_{17}H_{34}O_2$	Stearic $C_{18}H_{36}O_2$	Myristoleic $C_{14}H_{26}O_2$	Palmitoleic $C_{16}H_{30}O_2$	Oleic $C_{18}H_{34}O_2$	Linoleic $C_{18}H_{32}O_2$	Linolenic $C_{18}H_{30}O_2$
Emersol 210 Oleic Acid	8-11	89-93	2/30	199-204	201-206	1.5	Tr	3	Tr	5	1	1	4	6	71	8	1
Emersol 213 USP/NF Low-titer Oleic Acid[2,3]	5 max.	88-95	50/86	199-204	201-206	1.5	Tr	3	Tr	5	1	Tr	3	6	73	8	1
Emersol 221 USP/NF Low-titer White Oleic Acid[2,3]	5 max.	88-95	71/99	199-204	201-206	1	Tr	3	Tr	4	1	Tr	3	7	73	8	1
Emersol 233 LL Oleic Acid[4]	6 max.	86-90	78/99	200-204	202-206	0.5	Tr	3	Tr	4	1	Tr	3	11	74	4	Tr
Emersol 315 Linoleic Acid	5 max.	145-160	60/95	195-201	197-203	1		0.5	Tr	3.5	Tr	0.5	Tr	Tr	19.5	65.5	10.5

[1]By GLC analysis, AOCS Ce 1-62.
[2]Corresponding food grade products available.
[3]For external use only USP XX Revision/NF XV Edition.
[4]LL (low-linoleic content) oleic: polyunsaturates 5% max.

Short-Chain Acids

| | **SPECIFICATIONS** | | | | | **TYPICAL COMPOSITION[1]** | | | | | |
	Titer, °C	Iodine Value max.	Color % Trans. 440/550 nm., min.	Acid Value	Sap. Value	Caproic $C_6H_{12}O_2$	Caprylic $C_8H_{16}O_2$	Capric $C_{10}H_{20}O_2$	Lauric $C_{12}H_{24}O_2$	Myristic $C_{14}H_{28}O_2$	Palmitic $C_{16}H_{32}O_2$
Emery 657 Caprylic Acid	14-16	0.2	88/99	385-390	386-391	Tr	99	1			
Emery 658 Caprylic-Capric Acid	1-6	0.3	88/99	359-366	361-368	3	56	40	1		
Emery 659 Capric Acid	28-31	0.2	88/99	322-326	323-329		1	97	2		
Emery 650 Lauric Acid	33-35	0.4	85/97	268-272	268-272				71	28	1
Emery 651 Lauric Acid	41-43	0.2	90/98	276-280	277-281		Tr	1	96	3	
Emery 652 Lauric Acid	43 min.	0.2	90/98	277-281	278-282		Tr	0.3	99	0.7	
Emery 655 Myristic Acid	52.0-53.5	0.5	90/99	243-246	244-247				1	97	2

[1]By GLC analysis, AOCS Ce 1-62.

(continued)

Table 13.16: (continued)

Stearic and Palmitic Acids

| | SPECIFICATIONS | | | | | TYPICAL COMPOSITION[1] | | | | | | | |
| | | | | | | Saturated Acids | | | | | Unsaturated Acids | | |
	Titer °C	Iodine Value	Color % Trans. 440/550 nm., min.	Acid Value	Sap. Value	Myristic $C_{14}H_{28}O_2$	Pentadecanoic $C_{15}H_{30}O_2$	Palmitic $C_{16}H_{32}O_2$	Margaric $C_{17}H_{34}O_2$	Stearic $C_{18}H_{36}O_2$	Palmitoleic $C_{16}H_{30}O_2$	Oleic $C_{18}H_{34}O_2$	Linoleic $C_{18}H_{32}O_2$
Emersol 110 Stearic Acid	52.8-53.5	8-12	60/94	205-210	206-211	2.5	0.5	50	2	35		9	1
Emersol 120 Stearic Acid[2]	53.7-54.7	5-7	88/99	205-210	206-211	2.5	1	50	2.5	39	Tr	5	Tr
Emersol 132 USP/NF Lily Stearic Acid[2,3,4]	54.5-55.5	0.5 max.	93/99	205-210	207-211	2.5	0.5	50	1.5	45.5			
Emersol 140 Palmitic Acid[5]	53.4-55.5	2 max.	93/99	209-214	210-215	1.5	0.5	74.5	0.5	23		Tr	
Emersol 143 Palmitic Acid	58-61	1 max.	93/99	215-223	215-223	Tr	0.5	91	4.5	4			
Emersol 150 Stearic Acid[6]	63.9-65.0	1 max.	93/99	197-202	198-203	2	1	11	2	83		1	
Emersol 153 USP/NF Stearic Acid[4]	67-69	1 max.	80/97	196-199	197-200			~5		95			
Emery 400 Stearic Acid	52 min.	9.5 max.	1/40	197-209	195-211[7]								
Emery 404 Stearic Acid	53.5-54.5	6-9	1/50	197-209	197-211								
Emery 410 Stearic Acid	56.1-60.0	7 max.	40/86	195-209	197-210	3	0.5	30	2	58	2.5	4	
Emery 420 Stearic Acid	57.2-63.0	2 max.	71/97	200-207	201-208	4	0.5	29	1.5	65	Tr	Tr	
Emery 422 Stearic Acid	55.8-60.0	1 max.	90/99	203-209	204-210	3	Tr	41	1	55			

[1]By GLC analysis, AOCS Ce 1-62.

[2]Corresponding food grade products available.

[3]Also available in powdered grades, Sieve Test: 20 mesh-99%, 40 mesh-75%, 60 mesh -30%, 80 mesh-15%, 100 mesh-10%; color 90 min. @ 440 nm. NOTE: Stearic Acid powder may present an explosion hazard if the particles are suspended in air in certain concentrations and ignited; thus special care should be taken in the handling of this product.

[4]For external use only. USP XX Revision/NF XV Edition.

[5]67% minimum palmitic content.

[6]80% minimum stearic content.

[7]Not a specification.

Table 13.17: "Empol" Dimer and Trimer Acids (63)

Specifications	Empol 1010[1] Dimer Acid	Empol 1014 Dimer Acid	Empol 1016 Dimer Acid	Empol 1018 Dimer Acid	Empol 1022 Dimer Acid	Empol 1024 Dimer Acid	Empol 1040 Trimer Acid	Empol 1041 Trimer Acid	Empol 1052 Polybasic Acid
Acid value	191-197	194-198	190-198	188-198	189-197	189-197	175-192	161-181	240-260[4]
Saponification value	193-200	197-201	194-200	192-198	191-199	191-199	192-200	180-195	245-260
Color, 1963 Gardner, max.	1	5	6	8	9	9	dark[2]	11	dark
Unsaponifiables, %, max.	—	—	0.5	1.0	1.0	1.0	—	—	1.0
Monomeric content, %, max.[3]	—	1.5	1.0	1.0	2-5	1.0	—	—	—
Typical Characteristics									
Refractive index, 25°C	1.475	1.485	1.479	1.470	1.474	1.476	1.495	1.480	
Specific gravity, 25/20°C	0.936	0.947	0.961	0.950	0.948	0.947	0.975	0.941	
Specific gravity, 100/20°C	0.897	0.910	0.904	0.911	0.910	0.911	0.930	0.898	
Density, lbs./gal., 25°C	7.8	7.9	8.0	7.8	7.9	7.9	8.1	7.8	
Pour point, °F	10	35	35	35	35	35	55	20	
Flash point, °F (COC)	585	530	530	565	530	530	595	625	
Fire point, °F (COC)	660	610	600	600	600	600	680	695	
Viscosity, Gardner-Holt, 25°C	Z-3	Z-4	Z-3	Z-4	Z-4	Z-5	Z-6	Z-6	
Viscosity, cSt. @ 210°F	72	85	75	90	89	94	320	185	
100°F	1800	2500	1800	2700	2650	2900	21,500	6700	
25°C	5200	5600	5300	7500	6200	9500	60,000	18,000	
Unsaponifiables, %	0.1	0.3	0.3	0.2	0.3	0.3	0.3	0.3	
Surface tension, dynes/cm, 25°C	31.1	31.8	31.9	33.4	29.2	32.9	—	—	

[1]Tentative specification.

[2]Not a specification, typical value = 14.

[3]By distillation at 270°C under 2.0 mm torr pressure.

[4]The value given is the hydrous acid value. The anhydrous acid value (212-222) differs because some of the acid groups exist in the anhydride form. This must be taken into consideration when testing.

Table 13.18: "Industrene" and "Hystrene" Fatty and Dibasic Acids (26)

Coconut oil acids.

Product	Description (CTFA adopted name)	Titer °C	Iodine value	Acid value	Sap value	% Unsap Max	% Trans. 440/550 nm, Min	Lovibond 5¼" cell	C6	C8	C10	C12	C14	C16	C18	C18:1	C18:2
							Color		Saturated							Unsaturated	
Industrene 365	Caprylic/Capric (Mixture Caprylic/Capric Acid)	5 Max	1.0 Max	355-369	355-374	1.0	66/93	10.0Y-1.0R	1	55	42	2					
Industrene 325	Distilled Coconut (Coconut Acid)	22-27	6-15	270-280	270-281	0.5	65/90	10.0Y-1.0R		10	8	51	18	8	2	2	1
Industrene 328	Stripped Coconut (Coconut Acid)	27-30	5-14	253-260	253-261	0.5	73/95	4.5Y-0.6R		1	1	55	24	12.5	1.5	4	1
Industrene 223	Hydrogenated Crude Coconut (Hydrogenated Coconut Acid)	23-26	3.0 Max	270-280	270-281	2.0	72/96	5.0Y-0.5R		10	8	51	18	8	3	2	
Hystrene 5012	Hydrogenated Stripped Coconut (Hydrogenated Coconut Acid)	24-33	2.0 Max	250-264	250-266	0.5	74/96	4.0Y-0.4R		1	1	55	24	12.5	6.5		
Hystrene 9512	95% Lauric (Lauric Acid)	41-43	0.5 Max	276-281	276-282	0.25	78/97	3.0Y-0.3R			2	95	3				
Hystrene 9014	90% Myristic (Myristic Acid)	50-52	0.5 Max	240-243	240-244	0.3	78/97	3.0Y-0.3R				4	90	6			

Stearic acids.

Product	Description (CTFA adopted name)	Titer °C	Iodine value Max	Acid value	Sap value	% Unsap Max	% Trans. 440/550 nm, Min	Lovibond 5¼" cell	C14	C15	C16	C17	C18	C18:1
							Color		Saturated					Unsat.
Industrene 4518	Single Pressed Grade (Stearic Acid)	53-55	8-11	204-211	204-212	1.0	82/94	2.0Y-0.2R	4	0.5	50	2	37	6.5
Industrene 4516	45% Palmitic (Palmitic Acid)	54-56	4 Max	205-211	205-212	1.0	88/98	1.0Y-0.1R	3	1	42	2	50	2
Industrene 5016	Double Pressed Grade (Stearic Acid)	53-56	4-7	207-210	207-211	0.5	92/98	1.0Y-0.1R	3	0.5	50	1.5	42	3
Hystrene 5016	Triple Pressed Grade (Stearic Acid)	54.5-56.5	0.5 Max	206-210	206-211	0.2	92/98	1.0Y-0.1R	1	0.5	52	2.5	44	
Hystrene 8016	80% Palmitic (Palmitic Acid)	55-59	0.5 Max	213-216	213-218	0.2	92/98	1.0Y-0.1R	4	1	80	2	13	
Hystrene 9016	90% Palmitic (Palmitic Acid)	58.0-60.5	0.5 Max	216-220	216-221	0.2	92/98	1.0Y-0.1R	1		92		7	
Industrene 9018	90% Stearic (Stearic Acid)	64-68	2 Max	196-201	196-202	0.5	65/93	10.0Y-1.0R			10		90	
Hystrene 9718	92% Stearic (Stearic Acid)	66.5-68.0	0.8 Max	196-201	196-202	0.3	82/96	2.0Y-0.2R			5		95	

Marine oil acids.

Product	Description (CTFA adopted name)	Titer °C	Iodine value Max	Acid value	Sap value	% Unsap Max	% Trans. 440/550 nm, Min	Lovibond 5¼" cell	C14	C16	C18	C20	C22	C18:1
							Color		Saturated					Unsat.
Hystrene 1522	Hydrogenated Fish Fatty Acid (Hydrogenated Menhaden Acid)	45-49	25	199-206	199-208	2.0	45/87	3 Max*	8	32	29		8	4
Hystrene 3022	Hydrogenated Fish Fatty Acid (Hydrogenated Menhaden Acid)	50-54	5	190-198	190-199	3.0	60/90	15.0Y-1.5R			27			
Hystrene 5522	Hydrogenated Fish Fatty Acid (Hydrogenated Menhaden Acid)	58-66	5	177-187	177-188	2.0	60/90	15.0Y-1.5R	1	14	30	52		3
Hystrene 7022	70% Arachidic & Behenic (Behenic Acid)	63-66	3.5	170-180	170-181	2.0	60/90	15.0Y-1.5R		1	29	70		3
Hystrene 9022	90% Arachidic & Behenic (Behenic Acid)	67-71	3	165-175	165-176	1.5	60/90	15.0Y-1.5R			10	88		2

*Gardner, 1963

(continued)

Table 13.18: (continued)

Tallow acids.

Product	Description (CTFA adopted name)	Titer °C	Iodine value	Acid value	Sap value	% Unsap Max	% Trans. 440/550 nm, Min	Color Other*	C8	C10	C12	C14	C15	C16	C17	C18	C14:1	C16:1	C18:1	C18:2
									Saturated								Unsaturated			
Hystrene 1835	Tallow/Coconut Blend (Mixture Tallow/Coconut Acid)	40 Max	36-42	211-219	211-220	0.5	82/96	2.0Y-0.2R Lovibond	1.5	2	10	5		20		18.5		3	36	3.5
Hystrene 7018	70% Stearic (Stearic Acid)	61.0-62.5	0.5 Max	200-205	200-206	0.2	88/98	1.0Y-0.1R Lovibond				2	0.5	30	2.5	65				
Industrene 7018	70% Stearic (Stearic Acid)	58-62	1.5 Max	200-207	200-208	0.5	82/96	2.0Y-0.2R Lovibond				3	0.5	29	2	65			0.5	
Industrene B	Hydrogenated Mixed (Hydrogenated Tallow Acid)	57-63	3.0 Max	199-207	199-208	1.0	40/86	3 Gardner				2	0.5	30	2	63.5			2	
Industrene R	Hydrogenated Rubber Grade (Stearic Acid)	52-64	15 Max	193-213	193-214	3.0	3/24	10 Gardner												
Industrene 145	Tallow Fatty Acids (Tallow Acid)	44-50	38-45	195 Min	195 Min	1.0	80/96	3.0Y-0.3R Lovibond				2.5		26	1.5	24	1	3.5	38.5	3
Industrene 143	Distilled Animal (Tallow Acid)	39-43	50-65	202-206	202-207	1.5	19/81	5 Gardner				2.5	1	27	2.5	18	1	5	39	4

*Lovibond, 5¼" cell; Gardner, 1963.

Oleic acids.

Product	Description (CTFA adopted name)	Titer °C	Iodine value	Acid value	Sap value	% Unsap Max	% Trans. 440/550 nm, Min	Color Other	C14	C15	C16	C17	C18	C14:1	C16:1	C18:1	C18:2	C18:3
									Saturated					Unsaturated				
Industrene 105	Oleic Acid; Red Oil (Oleic Acid)	11-14	135-145	195-204	195-205	1.5	10/60	6 Gardner*	3	2	6	1	2		2	69	10	1
Industrene 205	White Oleic (Oleic Acid)	11-14	85-95	195-204	195-205	2.0	66/93	10.0Y-1.0R Lovibond†	3	2	6	1	2		1	70	10	1

*Gardner, 1963.   †Lovibond, 5¼" cell

Vegetable oil acids.

Product	Description (CTFA adopted name)	Titer °C	Iodine value	Acid value	Sap value	% Unsap Max	% Trans. 440/550 nm, Min	Color Gardner, 1963	C16	C18	C18:1	C18:2	C18:3
									Saturated		Unsaturated		
Industrene 225	Distilled Soya Alkyd (Soya Acid)	25 Max	135-145	195-201	195-202	2.0	55/85	3	7	4.5	25.5	56	7
Industrene 226	Distilled Soya Alkyd (Soya Acid)	26 Max	125-135	195-203	195-204	2.0	55/85	3	12	4.5	23.5	54	6
Industrene M	Monomer	35 Max	80-105	175-190	175-200	5.0	19/81	5					

(continued)

Table 13.18: (continued)

Food grade acids.

Product	Description (CTFA Adopted Name)	Iodine value	Titer °C	Sap value	Acid value	% Unsap Max	% Trans. 440/550 nm, Min	Color Lovibond 5¼" cell	Saturated C14	C15	C16	C17	C18	Unsaturated C16:1	C14:1	C16:1	C18:1	C18:2	C18:3
Hystrene 5016 NF FG*	Triple Pressed Stearic Acid (Stearic Acid)	0.5 Max	54.5-56.5	206-211	206-210	0.2	92/98	1.0Y-0.1R	1	0.5	52	2.5	44						
Industrene 7018 FG	70% Stearic Acid (Hydrogenated Tallow Acid)	1.5 Max	58-62	200-208	200-207	0.5	82/96	2.0Y-0.2R	3	0.5	29	2	65			0.5			
Hystrene 7018 FG	70% Stearic Acid (Stearic Acid)	0.5 Max	61.0-62.5	200-206	200-205	0.2	88/98	1.0Y-0.1R	1	0.5	29	2	68.5						
Industrene 8718 FG	87% Stearic Acid (Stearic Acid)	2.0 Max	64.5-67.5	196-202	196-201	1.5	60/94	2.0Y-0.2R			12		87						
Hystrene 9718 NF FG*	92% Stearic Acid (Stearic Acid)	0.8 Max	66.5-68.0	196-202	196-201	0.3	82/96	2.0Y-0.2R	1	0.5	6	2.5	90						
Industrene 226 FG	Distilled Soya Acid (Soya Acid)	125-135	26 Max	195-204	195-203	2.0	55/85	25.0Y-2.5R			12		4.5				23.5	54	6

*Also available as NF (National Formulary) non-food grade for external use only.

Dimer acids.

Product description (CTFA adopted name)	Specification Acid value	Sap value	Color Gardner, 1963 Max	Neutral equivalent	Viscosity cSt, @25°C	% Unsap	Typical Composition % Monomer	% Dimer	% Trimer
Hystrene 3695 95% Dimer Acid (Dimer Acid)	190-196	190-202	6	286-295	7,500	0.5	1	95	4
Hystrene 3680 80% Dimer Acid (Dimer Acid)	190-197	190-199	8	285-295	8,500	1.0	1.5	83	15.5
Hystrene 3675 75% Dimer Acid (Dimer Acid)	189-197	189-199	9	285-297	9,000	1.0	1	83	16
Hystrene 3675C 75% Dimer Acid 3% Monomer (Dimer Acid)	189-197	189-199	9	285-297	7,500	1.0	3.5	82	14.5
Hystrene 5460 60% Trimer Acid (Trimer Acid)	170-190	170-202	>18	295-330	30,000	1.0	Trace	40	60

Table 13.19: "Neo-Fat" Fatty Acids (59)

Specifications columns (Titer, Iodine Value, Acid Value, Color Lovibond, Moisture, Heat Stability†, Saponification Value, Unsaponifiable) and Approximate Chemical Composition (Gas Chromatography).

Group	NEO-FAT®	Product Name	Titer °C Min	Titer °C Max	Iodine Min	Iodine Max	Acid Min	Acid Max	Color Lovibond Max	Moisture Max	Heat Stability† Max	Sapon Min	Sapon Max	Unsap Max	C-6	C-8	C-10	C-12	C-14	C-16	C-17	C-18	C-20	C-16′	C-18′	C-18″	C-18‴
SHORT CHAIN SATURATED ACIDS	8	Commercially Pure Caprylic	8.0	12.0		0.7	387	392	1.0 R- 5 Y 5-1/4″	0.2	4.0 R-40 Y	387	394	0.2	5.0	92.0	3.0										
	8-S	98% Min. Commercially Pure Caprylic	15.0			0.5	385	390	0.5 R- 3 Y 5-1/4″	0.2	2.5 R-15Y	385	392	0.2	0.4	99.2	0.4										
	10	Commercially Pure Capric	29.0	32.0		0.5	323	329	0.8 R- 3 Y 5-1/4″	0.2	2.5 R-10 Y	323	331	0.2		1.0	97.0	2.0									
	12	Commercially Pure Lauric	41.5	44.0		0.5	278	282	0.5 R- 3 Y 5-1/4″	0.2	1.5 R-10 Y	278	284	0.2			0.3	99.0	0.7								
	12-43	99% Commercially Pure Lauric	43.0			0.5	278	282	0.5 R- 2 Y 5-1/4″	0.2	1.0 R- 5 Y	278	284	0.2			1.5	97.5	1.0								
	14	Commercially Pure Myristic	52.0			0.5	244	249	0.5 R- 2 Y 5-1/4″	0.2	1.5 R- 8 Y	244	251	0.2													
	255	Stripped Coco	27.5	29.5	8.0	13.0	252	258	1.0 R- 7 Y 5-1/4″	0.3		252	260	0.5			1.0	55.0	22.0	11.0		3.0		6.0	6.0	2.0	
	265	Distilled Coco	23.0	26.0		10.0	265	275	1.0 R- 7 Y 5-1/4″	0.3		265	277	0.5	5.0	6.0	6.0	52.0	19.0	9.0		2.0		6.0	6.0	1.0	
	16	Commercially Pure Palmitic	59.0	61.0		0.5	216	220	0.5 R- 2 Y 5-1/4″	0.2	1.0 R- 6 Y	216	221	0.3				1.0	0.5	92.5	1.0	5.0					
	16-S	97% Commercially Pure Palmitic	61.6			0.5	216	220	0.5 R-2.0 Y 5-1/4″	0.2	1.0 R- 6 Y	216	221	0.3					0.1	98.0	0.2	1.7					
	16-54	Eutectic Palmitic-Stearic	53.0	55.0		0.5	211	213	0.5 R- 2 Y 5-1/4″	0.2	1.5 R- 7 Y	211	214	0.4				2.0	0.5	66.0	1.5	30.0					
	16-56	80% Commercially Pure Palmitic	56.0	58.0		1.0	214	218	0.6 R-3.5 Y 5-1/4″	0.2	2.0 R-12 Y	214	219	0.4				1.5	0.7	80.5	2.0	15.3					
LONG CHAIN SATURATED ACIDS	18	Commercially Pure Stearic	65.5	68.0		1.0	195	200	1.0 R- 5 Y 5-1/4″	0.2	3.5 R-25 Y	197	201	0.5						7.0	2.5	90.0	0.5	trace			
	18-S	Special C. P. Stearic	65.5			0.5	195	200	0.5 R-1.5 Y 5-1/4″	0.2	1.5 R-5.0 Y	197	201	0.5						7.0	2.5	90.0	0.5				
	18-53	Single Pressed Stearic	53.3	54.2	5.0	10.0	207	210	2.0 R-15 Y 5-1/4″	0.5	3.0 R-20 Y xxx	207	211	0.8				2.5	0.5	52.0	2.0	38.0		5.0			
	18-54	Double Pressed Stearic	54.0	54.6	4.5	7.0	208	211	0.5 R- 2 Y 5-1/4″	0.5	1.0 R- 7 Y	208	212	0.5				2.5	0.5	52.0	2.0	39.0		4.0			
	18-55	Triple Pressed Stearic	55.0	56.0		0.5	208	211	1.0 R- 5 Y 5-1/4″	0.5	2.5 R-15Y	208	212	0.5				2.0	0.5	52.0	2.0	43.5					
	18-58	Hydrogenated Tallow Acid	57.0	61.0		1.0	201	206	8.0 R-40 Y 1″	0.5		201	207	0.5				2.0	0.5	29.0	1.5	66.0	1.0				
	18-59	Rubber Grade Stearic	55.0	62.0		9.0	195	208	1.0 R- 5 Y 5-1/4″	0.5	3.5 R-25 Y	195	209	2.0				9.0	0.5	21.0	1.5	60.0	1.0	5.0			
	18-61	Stearic-Palmitic	60.0	64.0		1.0	198	205	1.0 R- 8 Y 5-1/4″	0.4		198	206	0.5				4.0	0.5	21.0	1.5	72.0	1.0	trace			
UNSATURATED ACIDS—OLEIC	90-04*	Low Poly-unsaturated Oleic Acid‡		7.0	84.0		200	204	1.3 R- 9 Y 5-1/4″	0.4	xx	200	205	1.0				3.0	0.5	3.0	1.0	trace	1.5	6.5	79.5	4.0	1.0
	92-04*	5°C Max Titer Crystallized White Oleic		5.0		95.0	200	204	1.0 R- 7 Y 1″	0.4	xx	200	205	1.0				3.0	0.5	3.0	1.0	trace	1.5	6.5	77.0	6.5	1.0
	94-04*	5°C Max Titer Crystallized Red Oil		5.0		95.0	199	204	1.0 R- 7 Y 1″	0.4	xx	199	205	1.5				3.5	0.5	3.0	1.0	trace	1.5	6.5	76.0	6.5	1.0
	94-10*	8-11°C Titer Crystallized Red Oil	8.0	11.0		95.0	199	204	1.5 R-10 Y 1″	0.4	xx	199	205	1.5				3.5	0.5	4.0	1.0	2.0	1.5	6.5	73.0	6.5	1.0
UNSATURATED ACIDS—OTHER	65	Distilled Animal Acid	40.0	44.0	49.0	60.0	201	206	1.5 R-10 Y 1″	0.5		201	207	1.0				3.0	0.5	29.0	1.0	18.0	1.0	3.5	40.0	3.5	0.5

* Ester Number 1 Maximum
xxx Peroxide Index Max 1
xx Flash Point Min 300°F
† 2 Hours at 200°C; 5¼″ Lovibond Cell.
‡ Polyunsaturates Max 5%

(continued)

Table 13.19: (continued)

PROPERTIES OF FATTY ACIDS—SATURATED, UNSATURATED, AND SUBSTITUTED

	Systematic name	Number of C atoms	Viscosity centipoises (T C)	Index of refraction (T C)
Saturated				
Capric	Decanoic	10	2.88 (70)	1.4130 (80)
Undecylic	Undecanoic	11	7.30 (50)	1.4164 (80)
Lauric	Dodecanoic	12	4.43 (70)	1.4191 (80)
Myristic	Tetradecanoic	14	5.83 (70)	1.4236 (80)
Pentadecylic	Pentadecanoic	15		1.4254 (80)
Palmitic	Hexadecanoic	16	7.8 (70)	1.4272 (80)
Margaric	Heptadecanoic	17		1.4287 (80)
Stearic	Octadecanoic	18	9.87 (70)	1.4299 (80)
Arachidic	Eicosanoic	20		1.4250 (100)
Behenic	Docosanoic	22		1.4270 (100)
Unsaturated				
Palmitoleic	9-Hexadecenoic	16		1.4103 (70)
Oleic	9-Octadecenoic	18	9.41 (60)	1.4582 (20)
Erucic	13-Docosenoic	22		1.4758 (20)
Linoleic	9,12-Octadecadienoic	18		1.4699 (20)
Linolenic	9,12,15-Octadecatrienoic	18		1.480 (20)
Eleostearic	9,11,13-Octadecatrienoic	18		1.5112 (50)
Substituted				
Ricinoleic	12-Hydroxy-9-octadecenoic	18		1.4716 (20)
Vernolic	12-Epoxy-9-octadecenoic	18		1.4628 (40)

SATURATED FATTY ACIDS ($C_nH_{2n}O_2$)

Common Name	Geneva Nomenclature	Chemical Formula	Molecular Weight	Acid Value	Melting Point °C
Acetic	n-Ethanoic	CH_3COOH	60.05	934.26	16.6
Butyric	n-Butanoic	C_3H_7COOH	88.10	636.79	-7.9
Caproic	n-Hexanoic	$C_5H_{11}COOH$	116.15	483.00	-3.4
Caprylic	n-Octanoic	$C_7H_{15}COOH$	144.21	389.05	16.7
Capric	n-Decanoic	$C_9H_{19}COOH$	172.26	325.69	31.6
Lauric	n-Dodecanoic	$C_{11}H_{23}COOH$	200.31	280.08	44.2
Myristic	n-Tetradecanoic	$C_{13}H_{27}COOH$	228.36	245.68	53.9
Palmitic	n-Hexadecanoic	$C_{15}H_{31}COOH$	256.42	218.80	63.1
Stearic	n-Octadecanoic	$C_{17}H_{35}COOH$	284.47	197.23	69.6
Arachidic	n-Eicosanoic	$C_{19}H_{39}COOH$	312.52	179.52	75.3
Behenic	n-Docosanoic	$C_{21}H_{43}COOH$	340.57	164.73	79.9
Lignoceric	n-Tetracosanoic	$C_{23}H_{47}COOH$	368.62	152.20	84.2

UNSATURATED FATTY ACIDS ($C_nH_{2n-x}O_2$, where x is an integer from 1 to 5)

Common Name	Geneva Nomenclature	Chemical Formula	Acid Value	Iodine Value	Melting Point °C	Molecular Weight
ONE DOUBLE BOND						
Myristoleic	9-Tetradecenoic	$CH_3(CH_2)_3CH=CH(CH_2)_7COOH$	247.87	112	-4.0	226.35
Palmitoleic	9-Hexadecenoic	$CH_3(CH_2)_5CH=CH(CH_2)_7COOH$	220.53	100	0.5	254.40
Oleic	cis-9-Octadecenoic	$CH_3(CH_2)_7CH=CH(CH_2)_7COOH$	198.63	90	13.4	282.45
Elaidic	trans-9-Octadecenoic	$CH_3(CH_2)_7CH=CH(CH_2)_7COOH$	198.63	90	46.5	282.45
Erucic	cis-13-Docosenoic	$CH_3(CH_2)_7CH=CH(CH_2)_{11}COOH$	165.72	75	34.7	338.57
TWO DOUBLE BONDS						
Linoleic	9,12-Octadecadienoic	$CH_3(CH_2)_4CH=CHCH_2CH=CH(CH_2)_7COOH$	200.06	181	-5.0	280.44
THREE DOUBLE BONDS						
Linolenic	9,12,15-Octadecatrienoic	$CH_3CH_2CH=CHCH_2CH=CHCH_2CH=CH(CH_2)_7COOH$	201.51	273	-10.5	278.42
αEleostearic	9,11,13-Octadecatrienoic	$CH_3(CH_2)_3CH=CHCH=CHCH=CH(CH_2)_7COOH$	201.51	273	49.0	278.42

Table 13.20: "P & G" Fatty Acids (39)

Composition (GLC%) (Typical Unless Otherwise Designated)

Type / Grade	Saponification Value	Acid Value	Iodine Value	Moisture (%, KF)	Unsaponifiables (%)	Color—Lov 5¼" Yellow/Red	Color—Gardner (1963)	Color—% Transmittance 440nm/550nm	Avg. Molecular Weight	Titer (°C)	C6 Caproic	C8 Caprylic	C10 Capric	C12 Lauric	C14 Myristic	C14 (1=) Myristoleic	C15 Pentadecanoic	C16 Palmitic	C16 (1=) Palmitoleic	C16 (2=) Hexadecadienoic	C17 Margaric	C18 Stearic	C18 (1=) Oleic	C18 (2=) Linoleic	C18 (3=) Linolenic	Chemical Abstract Number
Coconut-Type																										
C-108	266 274 (271)	(272)	5 max (4)	0.5 max (0.1)	0.5 max (0.1)	1.0 max R (0.3)	(<1)	85/95 min (90/96)	(207)	(25)	0.3	7.3	6.5	50.7	18.9			8.6				7.2	0.5			67701-05-7
C-109	266 274 (271)	(273)	5 10 (7)	0.5 max (0.1)	0.5 max (0.1)	1.0 max R (0.6)		(86/95)	(207)	(24)	0.4	7.1	6.4	48.5	18.0			8.9				4.4	5.9	0.4		67701-05-7
C-110	266 274 (270)	(272)	12 max (9)	0.5 max (0.1)	0.5 max (0.1)	3 max (2)		(80/94)	(208)	(23)	0.4	6.8	6.0	48.5	18.0			9.4				3.1	6.9	0.9		67701-05-7
C-130	250 265 (255)	(260)	24 max (18)	0.5 max (0.1)	1.0 max (0.4)	7 max (3)			(220)	(25)	1.0	7.0	5.5	43.0	16.4			10.5				6.0	9.2	1.3	0.1	67701-05-7
Short Chain																										
C-810	358 368 (364)	370 max (365)	0.5 max (0.2)	0.2 max (0.04)		3·0.8 max R (1/0.1)			(154)		6.0 max (4.0)	53 60 (56)	34 42 (39)	2.0 max (0.3)												67762-36-1
C-895	380 394 (383)	395 max (384)	0.2 max (0.1)	0.2 max (0.04)		3·0.8 max R (1/0.1)			(146)	(14)	(0.5)	95.0 min	(2.6)	0.5 max												124-07-02
C-1095	320 330 (322)	331 max (323)	0.5 max (0.3)	0.2 max (0.04)		3·0.8 max R (1/0.2)			(174)	(30)		(2.0)	95.0 min	(1.0)												334-48-5
Tallow-Type																										
T-11	200 min (203)	200 208 (205)	38 45 (42)	0.3 max (<0.1)	1.5 max (0.4)	1.5 max R (1.0)		(82/95)	(276)	44 50 (46)				0.2	3.3	0.5	0.7	25.3	2.0		2.2	25.0	40.4	0.4		67701-06-8
T-18	200 208 (204)	(206)	40 55 (50)	0.5 max (<0.1)	1.5 max (0.2)	3.0 max R (2.5)			(275)	(42)				0.4	2.5	1.0	0.4	26.3	3.8	0.4	1.5	17.3	43.0	3.1	0.3	67701-06-8
T-20	200 210 (205)	(207)	40 55 (52)	0.5 max (0.1)	1.5 max (0.3)	7 max (3)			(274)	40 47 (42)			0.1	1.2	3.2	0.4	0.4	24.8	2.5	0.4	1.0	18.8	44.3	2.6	0.4	67701-06-8
T-22	200 210 (205)	(207)	45 70 (63)	0.5 max (0.1)	1.5 max (0.3)	7 max (4)			(274)	36 42 (38)				0.9	2.4	0.3	0.5	22.7	1.5		2.9	17.3	43.2	7.3	0.9	67701-06-8
Soya-Type																										
S-210	197 203 (200)	197 205 (203)	133 min (137)	0.5 max (0.1)	0.5 max (0.2)	2 max (1)		(83/97)	(281)	21 25 (22)				0.2	0.2			11.8				4.1	23.7	53.3	6.7	67701-08-0
S-230	197 203 (201)	197 205 (203)	120 min (125)	0.5 max (0.1)	1.5 max (0.4)	4 max (3)			(279)	32 max (26)				0.3	0.7			14.1	0.8			6.2	27.3	44.8	5.8	67701-08-0
Linseed-Type																										
L-310	197 min (200)	197 208 (202)	180 min (189)	0.5 max (0.1)	0.5 max (0.3)	4 max (3)			(280)	22 max (16)				0.2	0.1			6.0	0.1		0.1	4.2	22.2	17.9	49.2	67701-08-0

Physical Properties · Chemical Properties

(continued)

Table 13.20: (continued)

Tallow/Coconut Type Blends	Saponification Value	Acid Value	Iodine Value	Moisture (%, KF)	Unsaponifiables (%)	Color–Lov. 5¼" Yellow/Red	Color–Gardner (1963)	Color–% Transmittance 440nm/550nm	Average Molecular Weight	Titer (°C)	C6 Caproic	C8 Caprylic	C10 Capric	C12 Lauric	C14 Myristic	C14 (1=) Myristoleic	C15 Pentadecanoic	C16 Palmitic	C16 (1=) Palmitoleic	C16 (2=) Hexadecadienoic	C17 Margaric	C18 Stearic	C18 (1=) Oleic	C18 (2=) Linoleic	C18 (3=) Linolenic	Chemical Abstract Number
TC-1005	(218)	214 218 (216)	35 42 (39)	0.2 max (0.1)	0.5 max (0.2)			74/91 min (85/96)	(260)	(39)		1 2	1.3	9	6	0.5	0.5	22.5	3			1 5	17	35	2	<0.5	67701-06-8
TC-1010T	(218)	214 218 (216)	40 48 (42)	0.5 max (0.1)	0.5 max (0.1)			76/93 min (85/96)	(260)	(39)		1 2	1.3	9	6	0.5	0.5	22.5	3			1 0	16	36	2	<0.5	67701-06-8
TC-1010	(220)	217 221 (218)	(42)	0.5 max (0.1)	0.5 max (0.1)			74/91 min (82/94)	(257)	(38)		1 2	1.3	9	6	0.5	0.5	22.5	3			1.5	16	35	3	0.5	67701-06-8

Physical Properties — Chemical Properties — Composition (GLC%) (Typical Unless Otherwise Designated)

Table 13.21: "Sylvachem" Fatty Acids (80)

TALL OIL FATTY ACIDS

	SPECIFICATIONS			
	Acid Number	Rosin Acids %	Unsaps %	Gardner Color
Sylfat 496	196 min	1.1 max	1.2 max	3 max
Sylfat 96	196 min	1.3 max	1.3 max	4 max
Sylfat 96 E	196 min	1.3 max	1.3 max	4 max
Sylfat 95	193 min	2.7 max	1.9 max	4 max
Sylfat 94	193 min	6.0 max	3.0 max	4 max
Sylfat 93	190 min	6.0 max	3.0 max	8 max

Sylfat tall oil based fatty acids provide a range of properties to suit many application needs. Sylfat 496 is an extremely high quality industrial grade, exhibiting lightest color, low rosin acid and unsaponifiable content. Sylfat 96E is a high quality fatty acid offered as epoxy grade material. Sylfats 93 and 94 are fatty acids which are higher in rosin content but quite suitable for general tall oil fatty acid applications.

DISTILLED TALL OIL

	Acid Number	Rosin Acids %	Unsaps %	Gardner Color
Sylvatal 40	185 min	26-32	2.5 max	10 max
Sylvatal 40 DD	185 min	26-32	2.4 max	5 max

Sylvatal 40 — A consistently produced distilled tall oil having lower rosin and unsaponifiable composition than crude tall oil. Used in pine oil type disinfectants, elastomer emulsion polymerizations, lubricants and oil field applications.

Sylvatal 40 DD — A redistilled form of Sylvatal 40 which has improved color.

SPECIALTY FATTY ACIDS

Croplas MM — The ester obtained from the reaction of methyl alcohol and an oleic type fatty acid. A fluid, stable liquid which contains no polyunsaturation. Recommended in the preparation of grinding oils, cutting oils, textile oils, and special lubricants; also useful as a plasticizer for rubber and various resins.

SPECIFICATIONS	
Iodine value	60-100
Acid number	5 max
Saponification number	165-180
Color, Gardner	13 max
Titer, °C	7 max

Sylvacote D — A paper coating lubricant composed of a 50% aqueous dispersion of calcium stearate. Designed for use in coated paper applications where appearance and printability are important.

% Solids	50*
pH	8.5-10*
Viscosity (cps @ 25C)	150-200*

Sylvacote K — A proprietary modified fatty acid recommended for improved anti-yellowing and better dispersability of titanium dioxide in pigmented plastics (mainly polyolefin based).

Color, Gardner	5*
Acid number	180-190*

* Typical

Table 13.22: "Union Carbide" Acids (19)

	Molecular Weight	Apparent Specific Gravity at 20/20°C.	ΔSp. Gr./Δt. at 10 to 40°C., per °C.	Boiling Point, °C. 760 mm.	Boiling Point, °C. 300 mm.	Boiling Point, °C. 10 mm.	Δb.p./Δp. at 750 to 770 mm., °C. per mm.	Vapor Pressure at 20°C., mm. Hg	Freezing Point, °C.	Solubility at 20°C., % by wt. In Water	Solubility at 20°C., % by wt. Water In	Viscosity at 20°C., cps.	Refractive Index, n_D^{20}	Surface Tension at 20°C., dynes per cm.	Ionization Constant at 25°C.	Heat of Vaporization at 1 atm., Btu./lb.	Flash Point, °F. (a)
Formic Acid	46.03	1.2225	0.00124	100.6	71.9	3.8	0.044	33	8.4	Complete		1.78	1.3714	37.0	–	209	138
Acetic Acid	60.05	1.0512	0.00114	117.9	89.9	18	0.043	11(b)	16.7	Complete		1.23	1.3719	26.9	1.7×10^{-5}	175	112
Propionic Acid	74.08	0.9952	0.00110	140.8	113	41	0.042	2.4	–20.7	Complete		1.1	1.3865	26.2	1.3×10^{-5}	249	134
Valeric Acid	102.13	0.9406	0.00090	185.5	157	80	0.045	<0.1	–34	2.4	13.0	2.2	1.4080	26.1	1.5×10^{-5}	210	205
iso-Pentanoic Acid(c)	102.13	0.9393	0.00091	180.3	111	80	0.045	0.19	–48	3.2(d)	11.1(d)	2.2	1.4076	–	1.7×10^{-5}	–	205
2-Ethylhexoic Acid	144.21	0.9077	0.00081	227	196	116	0.048	<0.1	–118.4	0.1	1.4	7.7	1.4252	–	1.5×10^{-5}	163	260
Acrylic Acid	72.06	1.0472	0.00110	141.3	113	40	0.044	3	12.3	Complete		1.3	1.4215	27.6(d)	5.5×10^{-5}	–	122

FOOTNOTES: (a) All flash points below 175°F. were determined by ASTM method D 1310 using Tag open cup; all flash points above 175°F. were determined by ASTM method D 92 using Cleveland open cup.
(b) Sublimes.
(c) Data obtained on commercial material.
(d) At 25°C.

(continued)

Table 13.22: (continued)

VAPOR PRESSURES AT VARIOUS TEMPERATURES

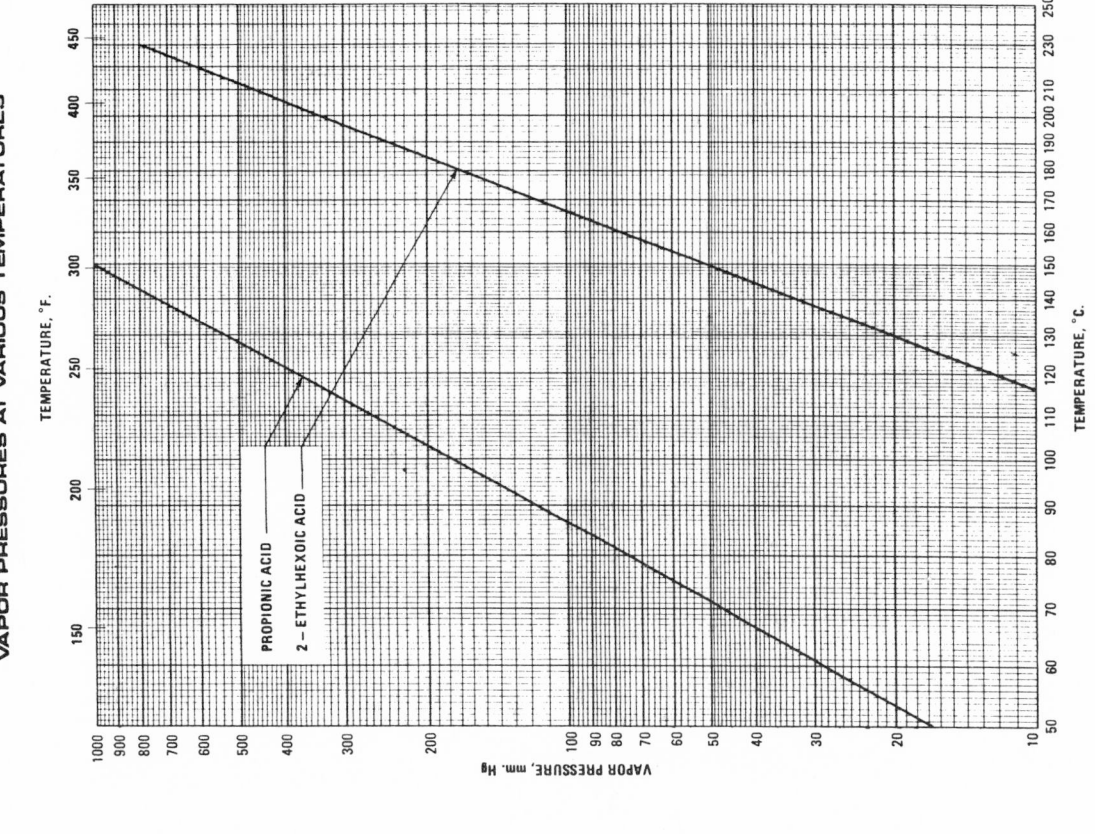

VAPOR PRESSURES AT VARIOUS TEMPERATURES

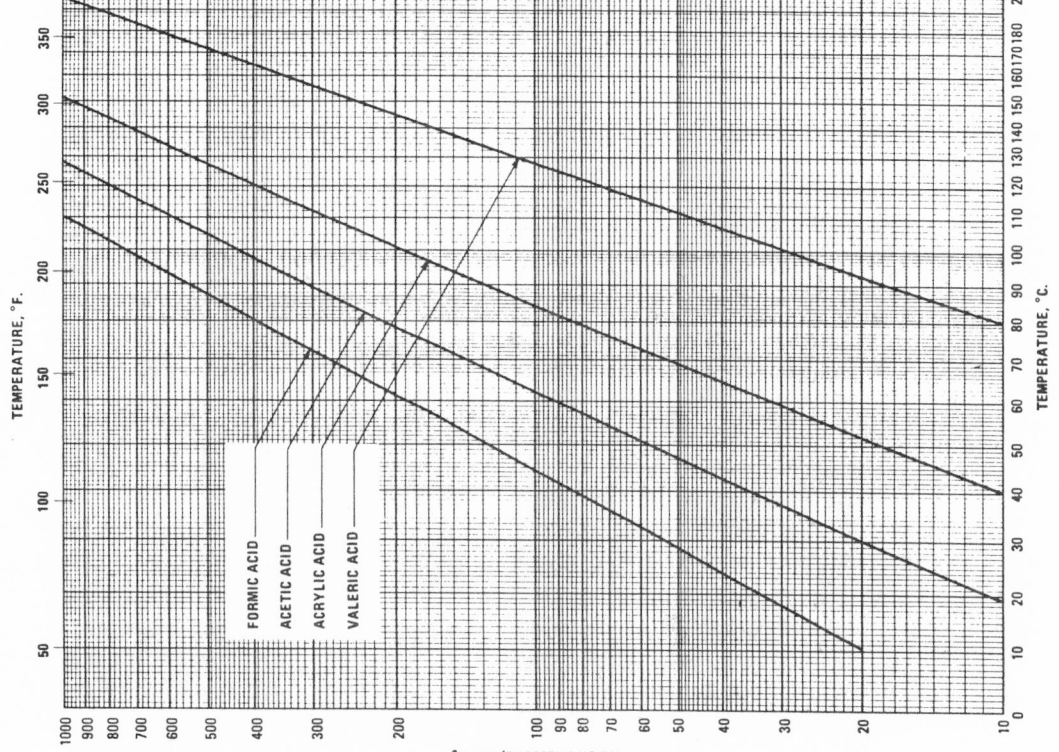

Amines

ALKYL AMINES

Table 14.1: Monomethylamine *(2)*

$$CH_3NH_2$$

Monomethylamine is a colorless, flammable gas with a strong ammoniacal odor; it is sold commercially as a 30% by weight aqueous solution. It is soluble in ethyl alcohol, ethyl ether, and many other organic substances, as well as in water. It is used in the tanning industry, in the manufacture of dyestuffs, in many synthetic products, and in the treatment of cellulose acetate rayon for dyeing.

Typical Properties and Specifications

Boiling point (75 mm.)	–6°C.
Flash point (30% solution)	0.3°C.
Melting point	–92.5°C.
Specific gravity at –10.8°/15°C.	0.699
Solubility in water	Very soluble
Weight per gallon (30% solution) at 68°F.	7.7 lbs.
Ammonia	Less than 0.2% by weight of solution
Concentration	3 to 3.5% by weight in water gas
Formaldehyde	Less than 0.3% by weight of solution
Purity	Not less than 98 mol % of total amines

Table 14.2: Dimethylamine *(2)*

$$(CH_3)_2NH$$

Dimethylamine is a colorless gas with a strong ammoniacal odor. The commercial product is an aqueous solution containing 25% by weight of dimethylamine. It is soluble in ethyl alcohol, ethyl ether, water, and many organic solvents. It is used as a dehairing agent in the tanning industry, in the manufacture of antioxidants, dyes, flotation agents, gasoline stabilizers, pharmaceuticals, rubber accelerators, emulsifiers, and cleaning compounds.

(continued)

Table 14.2: (continued)

<u>Typical Properties and Specifications</u>

Boiling point (764 mm.)	7.2 to 7.3°C.
Description	Gas at ordinary temperature
Flash point (25% solution)	Approx. 6.25°C.
Melting point	-96°C.
Specific gravity at -6°C.	0.6865
Solubility in water	Soluble
Weight per gallon (25% solution, 68°F.)	Approx. 7.8 lbs.
Ammonia	Not more than 1% by wt. of sol.
Concentration	25 to 25.5% by wt. in water
Formaldehyde	Not more than 0.5% by wt. of sol.
Purity	Not less than 98 mol % of total amines

Table 14.3: Trimethylamine *(2)*

$$(CH_3)_3N$$

Trimethylamine is a colorless, flammable, easily condensible gas with a pungent, ammoniacal odor. The commercial product is an aqueous solution containing 25% by weight of trimethylamine. It is very soluble in water and is used as a warning agent in bottled gas, as an insect attractant, and in organic synthesis.

<u>Typical Properties and Specifications</u>

Boiling point	Approx. 3.5°C.
Critical temperature	161°C.
Critical pressure	41 atm.
Decomposition temperature	800 to 1300°C.
Dielectric constant at 4°C.	2.9
Electrical conductivity	2.2×10^{-12} reciprical ohms at -33.5°C.
Heat of combustion	578.6 kg. cal. per mol
Ionization constant at 25°C.	6.5×10^{-5} for solutions 0.001 N to 0.06N
Heat of evaporization at BP	95.6 cal. per g.
Melting point	-124°C.
Specific gravity at -5°C.	0.662
Surface tension at -4°C.	17.4 dynes per cm.
Solubility in water at 19°C.	1 liter of aqueous saturated solution contains 410 g. of amine
Absolute viscosity at -33.5°C.	3.208 millipoises
Ammonia	Not more than 0.2% by wt. of solution
Formaldehyde	Not more than 0.3% by wt. of solution
Purity	Not less than 98 mol %

Table 14.4: Aqueous and Anhydrous Methylamines (52)

Product	Formula	Formula Weight	Freezing Point °C.	Boiling Point, °C. 760 mm.	Density at 20°C. g/ml	Density at 20°C. lb/gal	Vapor Pressure at 20°C. psia.	Critical Temp. °C.	Explosive Limits %	Flash Point °F. Tag. Open Cup
Anhydrous Monomethylamine	CH_3NH_2	31.06	-93.5	-6.3	0.624	5.55	41.7	156.9	4.95-20.75	-
40% Aqueous Monomethylamine		31.06	-39	42.0	0.901	7.50	5.4	-	4.95-20.75	10-40
50% Aqueous Monomethylamine*		31.06	-44	33.0	0.874	7.29	9.0	-	4.95-20.75	-
Anhydrous Dimethylamine	$(CH_3)_2NH$	45.08	-92.2	6.9	0.653	5.48	25.7	164.6	2.80-14.40	-
40% Aqueous Dimethylamine		45.08	-38	40.0	0.898	7.46	3.7	-	2.80-14.40	-4-16
60% Aqueous Dimethylamine		45.08	-60	-	0.833	6.96	8.7	-	2.80-14.40	<-4
Anhydrous Trimethylamine	$(CH_3)_3N$	59.11	-117.1	2.9	0.634	5.30	27.7	160.1	2.00-11.60	-
25% Aqueous Trimethylamine		59.11	6	38.0	0.933	7.76	6.9	-	2.00-11.60	25-50

Table 14.5: Freezing Points of Aqueous Methylamine Solutions *(34)*

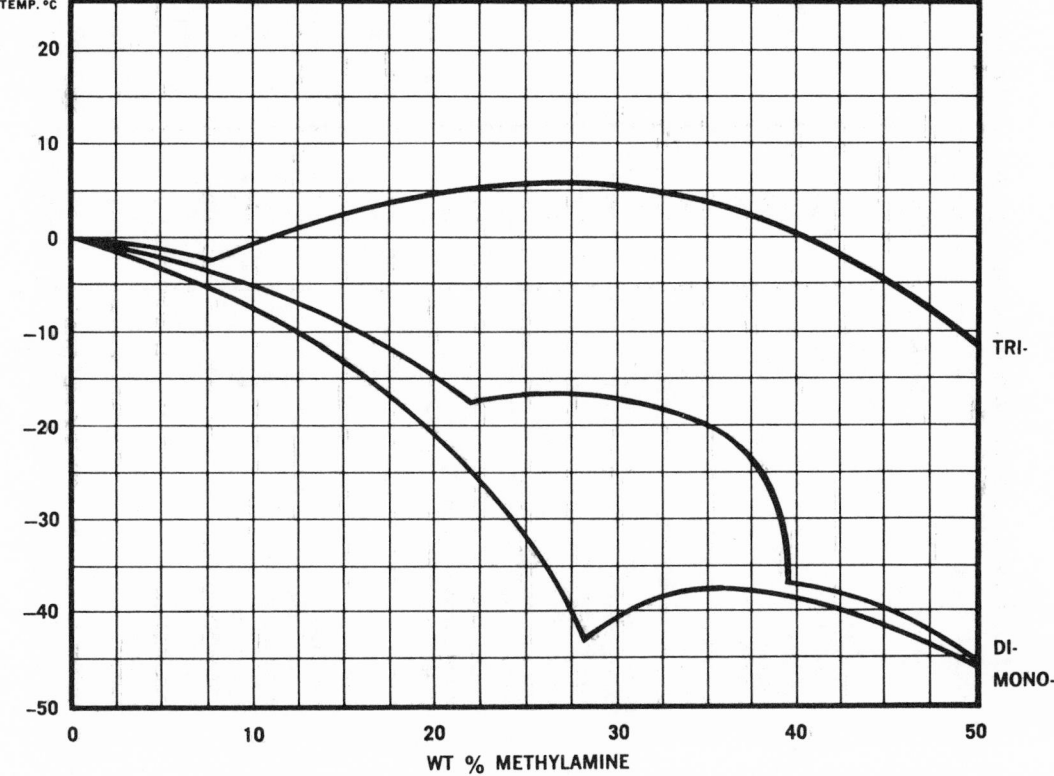

Table 14.6: Binary Azeotropes of Methylamines *(34)*

A = Monomethylamine B — Component as follows:	Azeotrope b.p., °C		Wt. % A
Trimethylamine at 760 mm	− 6.5		70
60 psig	36		85
210 psig	75		90-92
1-3 Butadiene	− 9.5		41.4
1-Butene	−13		22.2
cis-2-Butene	− 9.6		47.5
trans-2-Butene	−10.4		48.5
1-Butene-3-yne	− 6.8		97.5
Isoprene		Minimum B.P.	
2-Methylpropane	−19.9		25.5
2-Methylpropene	−14.3		32
Butane	−14.0		37.6
Amylenes		Minimum B.P.	
A = Dimethylamine B — Component as follows:			
Trimethylamine at 760 mm	3		26
107 psig	73		72
Ammonia		Non-azeotrope	
3-Methyl-1-Butene		Non-azeotrope	
A = Trimethylamine B — Component as follows:			
Dimethyl ether		Non-azeotrope	
1-Butene		Non-azeotrope	
2-Methylpropene		Non-azeotrope	
n-Butane		Non-azeotrope	
2-Methylpropane		Non-azeotrope	
Acetic acid	149		20
Ammonia	−34		27
Boron trifluoride	230		47
Formic acid	179		75

Table 14.7: Solubility Data for Methylamines (34)

solubility of some inorganic salts in methylamines

SALT	MONO	DI	TRI
I₂ [b]	v	v	s
KAg(CN)₂	—	s	—
KCN	—	s	—
KI	s	s	i
KNO₃	s	—	—
K₂PtCl₄	s	s	—
K₂PtCl₆	v	—	—
KSCN	v	m	—
LiCl	v	As	—
Mg(NO₃)₂	—	s	—
NaBr	s	s	—
NaClO₃	m	s	s
NaNO₃	v	s	—
NiSO₄	i	—	—
P (red)	s	—	i
P (yellow) [c]	m	m	—
PbBr₂	—	As	As
PbI₂ [d]	As	As	m
Pb(SCN)₂	Av	—	—
PtI₂	—	s	—
S	v	v	s
SbI₃	Ai	—	—
SnI₄	s	m	m
SrI₂	v	Av	—
Sr(NO₃)₂	i	—	—
TlNO₃	m	—	s
UrO₂(C₂H₃O₂)₂ [e]	Ai	—	—
UrO₂(NO₃)₂	v	v	i
ZnS	s	v	m

SALT	MONO	DI	TRI
AgI	v	v	Ai
AgNO₃	v	v	—
Ag₂SO₃	s	—	—
Ag₂SO₄	Ai	—	—
BaBr₂	—	—	—
BaI₂	Av	Av	As
Ba(NO₃)₂	s	s	s
Ba(SCN)₂	v	m	i
BiBr₃	—	v	—
BiCl₃	Am	—	—
BiI₃	Av	v	s
Bi₂S₃	s	s	—
Br₂	v	Rv	—
CaC₂	i	—	—
CaI₂	—	v	—
Ca(NO₃)₂	m	i	i
CdBr₂	—	Am	—
Cd(CN)₂	—	m	—
CdI₂	—	v	m
Cr₂(SO₄)₃	—	—	—
CuCl	R	—	—
CuHAsO₃	s	—	—
CuS	s	—	—
CuSCN	v	—	—
CuSO₄	i	—	—
FeI₂	—	m	—
Fe₂(SO₄)₃	Ai	—	—
Hg(CN)₂	v	v	—
HgI₂	v	v	m
Hg(SCN)₂	s	—	s

LEGEND: v = very soluble;   m = moderately soluble;   s = slightly soluble;
i = insoluble;   A = formation of an aminate;   R = marked reaction

a) Bromine reacts with dimethylamine with the evolution of heat to form a very soluble crystalline compound. With methylamine, the reaction is much more violent and a black residue is formed in addition to a soluble crystalline product.

b) Iodine is extremely soluble in mono- and dimethylamine, but not in trimethylamine. On standing, the deep color of the solution fades in color. Iodine is only slightly soluble in trimethylamine but, after some weeks, colorless crystals separate from this solution.

c) Yellow phosphorus is soluble in monomethylamine, forming almost colorless solutions, but on standing the red form, which is only slightly soluble, separates.

d) PbI₂ turns white on contact with the amines. By heating the tubes very gently, the original yellow color returns, indicating that the amine of crystallization has been removed. On cooling, the PbI₂ again turns white.

e) A solution of UrO₂(C₂H₃O₂)₂ gelatinizes on standing for some days.

solubility of methylamines in organic liquids

Volumes of gas per 1 cc of liquid Pressure = 1 atmosphere; temperature = 20°C

SOLVENT	MONO	DI	TRI
Aniline	271 cc	520 cc	300 cc
Anisole	89	252	185
Benzyl alcohol	314	528	322
i-Butanol	298	598	405
n-Butanol	303	504	379
Cedrene	34	106	86
o-Chloronaphthalene	52	174	130
Cymene	48	182	177
Decahydronaphthalene	24	116	156
Diacetone alcohol	420	457	345
Dibenzylether	115	154	120
o-Dichlorobenzene	64	252	240
Diethanolamine	313	497	74
Diethylaniline	60	180	134
Diethylene glycol mono-ethyl ether	336	588	216
Dimethylaniline	64	230	149
Dimethylcyclohexylamine	67	187	187
Dimethylformamide	132	298	78
Ethanol	440	727	600
Ethylene glycol	630	860	369
Furfuryl alcohol	413	679	410
Methanol	654	992	573
Methylcyclohexanol	219	439	256
Monoethanolamine	216	379	48
Monoethylaniline	113	324	228
Monomethylaniline	197	406	223
Morpholine	255	580	138
Nitrobenzene	88	226	154
o-Nitrotoluene	86	221	149
Pinene	34	156	176
n-Propanol	339	600	439
Quinoline	92	212	255
Tetrahydronaphthalene	40	170	151
o-Toluidine	88	430	242
Triethylene glycol	316	488	164
Trimethylene glycol	480	722	307

Table 14.8: Monoethylamine *(2)*

$$C_2H_5NH_2$$

Monoethylamine is a water-white liquid which is commercially available as a 70% aqueous solution. It is soluble in ethyl alcohol, methyl alcohol, the paraffin hydrocarbons, aromatic and aliphatic hydrocarbons, ethyl ether, ethyl acetate, acetone, fixed oils, mineral oil, oleic and stearic acids. It is also soluble in hot paraffin and carnauba waxes, which solidify when cooled.

Typical Properties and Specifications
(Anhydrous grade)

Distillation range	15–18°C
Flash point (open cup)	Below 20°F
Purity	97–99%
Specific gravity at 15/15°C	0.689
Weight per gallon	5.70 lbs

(70% Solution)

Boiling point	16.6°C
Color	Water-white
Critical temperature	183.2°C
Dissociation constant at 25°C.	5.6×10^{-4}
Heat of combustion	Gas 9157 cal./g.
	Liquid 9058 cal./g.
Heat of vaporization at 15°C	14.57 cal./g.
Heat of solution in water at 19°C	6330 cal. per mol of solute at infinite dilution
Melting point	−80.6°C
Purity	At least 70%
Specific gravity at 20°/20°C	0.79–0.80
Weight per gallon (20°C)	6.63 lbs.

Table 14.9: Diethylamine *(2)*

$$(C_2H_5)_2NH$$

Diethylamine is a water-white liquid with an ammoniacal odor. It is soluble in water, ethyl alcohol, paraffin hydrocarbons, aromatic and aliphatic hydrocarbons, ethyl ether, ethyl acetate, acetone, fixed oils, mineral oil, oleic and stearic acids. It dissolves hot paraffin and carnauba waxes, which solidify when cooled. It is used as a selective solvent for the removal of impurities from oils, fats, and waxes where its property of hydrating in aqueous solution is utilized; also used in the manufacture of rubber chemicals, textile emulsions, dyes, flotation agents, resins, polymerization inhibitors, gum inhibitors, drugs, and insecticides.

Typical Properties and Specifications

Boiling point	56.0°C
Critical density	0.246 g./cc.
Critical pressure	36.2 atm
Critical temperature	223.5°C
Dissociation constant	1.26×10^{-3}
Flash point (open cup)	Below 0°F
Heat of combustion	
Gas	9995 cal./g.
Liquid	9882 cal./g.
Heat of Vaporization at 58°C	91.03 cal./g.
Heat of solution in water at room temperature	8220 cal./mol of solute at infinite dilution
Melting point	−50.0°C
Specific gravity at 20/20°C	0.71
20/4°C	0.711
Specific heat of liquid at 22.5°C	0.516 cal./g.
Refractive Index at 17.6°C	1.3873
Viscosity at 25°C	0.346 centipoise
Weight per gallon (20°C)	5.89 lbs.
Distillation range	
Initial boiling point	Not below 53°C
Final boiling point	Not above 59.5°
Purity	At least 98%
Water insoluble	None

Table 14.10: Triethylamine *(2)*

$$(C_2H_5)_3N$$

Triethylamine is a colorless liquid, freely soluble in water at temperatures below 18°C., soluble in ethyl alcohol, methyl alcohol, ethyl ether, ethyl acetate, aliphatic and aromatic hydrocarbons, acetone, fixed oils, mineral oil, oleic and stearic acids, and in hot carnauba and paraffin waxes, the latter two solidifying when cooled. It is used in the manufacture of corrosion inhibitors, emulsifying agents, dyestuffs, and insecticides.

Typical Properties and Specifications

Boiling point	89.5°C
Critical solution temperature (in water)	18°C
Dissociation constant	6.4×10^{-4}
Flash point (open cup)	20°F
Heat of combustion	10,248 cal./g.
Heat of solution in water	10,040 cal./mol of solute at infinite dilution
Melting point	−114.8°C
Specific gravity at 20/20°C	0.730
Refractive index at 20°C	1.4003
Weight per gallon (20°C)	6.1 lbs.
Color	Water-white
Distillation range	
Initial boiling point	Not below 85°C
Final boiling point	Not above 91°C
Purity	98.5%, min.
Water insoluble (20–30°C)	None

Table 14.11: n-Propylamine *(2)*

1–Aminopropane $CH_3CH_2CH_2NH_2$

n-Propylamine is a colorless liquid soluble in water, methyl and ethyl alcohols, ethyl ether, ethyl acetate, acetone, aromatic, aliphatic and paraffin hydrocarbons, fixed oils, mineral oil, oleic and stearic acids, and hot carnauba and paraffin waxes, the latter two solidifying when cooled.

Boiling point (760 mm)	49–50°C
Distillation range	46–51°C
Flash point	Below 20°F
Melting point	−83°C
Purity	95–99% (min.)
Refractive index at 20°C	1.3910
Specific Gravity at 20°C	0.718
Weight per gallon (at 20°C)	5.99 lbs.

Table 14.12: Di-n-Propylamine *(2)*

$$(C_2H_5CH_2)_2NH$$

Di-n-propylamine is a colorless liquid, soluble in ethyl alcohol, methyl alcohol, ethyl ether, ethyl acetate, acetone, paraffin hydrocarbons, aliphatic and aromatic hydrocarbons, fixed oils, mineral oil, oleic and stearic acids. It dissolves hot paraffin and carnauba waxes which solidify when cooled. It is partly soluble in water.

Boiling point	110–1°C
Flash point	45°F
Melting point	−39.6°C
Purity	97% min.
Specific gravity at 20°C	0.74
Refractive index at 20°C	1.4063
Weight per gallon at 20°C	6.18 lbs.

Table 14.13: Mutual Solubility of Di-n-Propylamine and Water at Various Temperatures *(2)*

WEIGHT % AMINE	TEMP., °C.	WEIGHT % AMINE	TEMP., °C.
1.96	52.6	47.54	−1.5
2.42	44.1	60.40	4.2
2.91	36.1	64.06	8.0
5.86	12.2	73.33	17.5
9.33	−0.6	78.69	24.7
12.27	−2.2	82.15	31.2
15.28	−3.5	85.83	39.0
25.21	−4.5[a]	89.26	49.0
33.69	−4.8	93.25	74.8
44.68	−2.9		

[a] Upon cooling to −5.0° the first blue opalescence was noted at −4.9°.

Table 14.14: Solubility Curve at 25° for the System Di-n-Propylamine-Water-Ethanol *(29)*

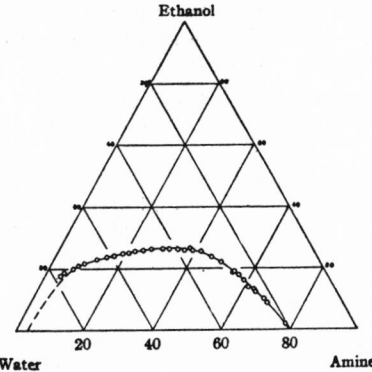

Table 14.15: Isopropylamine *(2)*

2-Aminopropane $\qquad\qquad$ $CH_3CH(NH_2)CH_3$

This water-white, primary aliphatic amine is available commercially in an anhydrous form. It is soluble in water, methyl and ethyl alcohols, ethyl ether, ethyl acetate, aromatic and aliphatic hydrocarbons, acetone, mineral oil, fixed oils, oleic and stearic acids. It is soluble in hot paraffin and carnauba waxes, which solidify on cooling. It is potentially useful as an intermediate in such manufactured products as dyestuffs, surface-active agents, textile specialties, pharmaceuticals, bactericides, insecticides, and cleaning compounds. It is also used as a dehairing agent in the leather industry.

Boiling point	31.9°C
Flash point	<20°F
Melting point	−101.2°C
Specific gravity at 25/4°C	0.686
Vapor pressure at 15°C	385 mm
pH of 0.1 N aqueous solution	11.57
Boiling range	31–35°C
Color	Water-white

Table 14.16: Di-Isopropylamine *(2)*

$$[(CH_3)_2CH]_2NH$$

Di-isopropylamine is a water-white liquid with an amine odor. It is soluble in methyl alcohol, ethyl ether, ethyl acetate, acetone, aromatic and aliphatic hydrocarbons, mineral oil, fixed oils, oleic and stearic acids, and only partly soluble in water. It dissolves hot paraffin and carnauba waxes which solidify on cooling.

Boiling range	81–85°C
Flash point	20°F
Specific gravity at 20/20°C	0.726

Table 14.17: n-Butylamine *(2)*

1-Aminobutane $CH_3CH_2CH_2CH_2NH_2$

n-Butylamine is a colorless liquid with an ammoniacal odor. It is miscible with water, ethyl alcohol, ethyl ether, paraffin hydrocarbons, and many organic solvents, and dissolves a wide range of materials. The butylamine salts and soaps are usually soluble in hydrocarbons. It behaves in many ways like monoamylamine, but will not produce a constant-boiling mixture with water. This compound is used in the manufacture of specialty soaps, emulsifying agents, desizing agents for textiles, rubber chemicals, flotation agents, corrosion inhibitors, dyestuffs, insecticides, and pharmaceuticals.

Boiling point (760 mm)	77.8°C
Flash point	45°C
Heat of combustion	710 kg. cal. per mol
Melting point	−50.5°C
Specific gravity at 20/20°C	0.7385
Solubility in water	Complete
Solubility of water in solvent	Complete
Refractive index at 20°C	1.4044
Viscosity at 25°C	0.68 centipoise
Vapor pressure at 20°C	0.01 mm
Weight per gallon at 20°C	6.15 lbs.
Acid insoluble	1.0% max.
Distillation range	
Initial boiling point	Not below 73.0°C
Not less than 95%	Below 82.0°C
Final boiling point	Not above 86.0°C
Purity	94% min.

Table 14.18: n-Dibutylamine *(2)*

$$(C_4H_9)_2NH$$

n-Dibutylamine is a water-white liquid with an ammoniacal odor. It is miscible with a large number and variety of organic solvents but its solubility in water is limited. It is soluble in ethyl and methyl alcohols, ethyl ether, ethyl acetate, acetone, aliphatic, and aromatic hydrocarbons, fixed oils, mineral oil, oleic and stearic acids. While it dissolves hot paraffin and carnauba waxes, these solidify on cooling. It is used in organic synthesis where its derivatives are used as flotation reagents, dyestuffs, rubber vulcanization accelerators and corrosion inhibitors.

Boiling point	161°C
Flash point (open cup)	135°F
Specific gravity at 20/20°C	0.76
20/4°C	0.767
Refractive index at 20°C	1.4186
Vapor pressure at 20°C	2.5 mm
Weight per gallon at 20°C	6.33 lbs.
Acid insoluble	0.6% max.
Distillation range	
Initial boiling point	Not below 153°C
Not less than 95%	Below 163°C
Final boiling point	Not above 172°C
Purity	98% min.

Table 14.19: n-Tributylamine *(2)*

$$(C_4H_9)_3N$$

n-Tributylamine is a water-white to light yellow liquid with an ammoniacal odor. It is soluble in ethyl alcohol, methyl alcohol, aliphatic and aromatic hydrocarbons, ethyl acetate, acetone, fixed oils, mineral oil, oleic and stearic acids, hot carnauba and paraffin waxes, the latter two solidifying when cooled, and most organic solvents. It is almost insoluble in water. Its sulfuric acid salts are water-soluble. It is used in the manufacture of corrosion inhibitors, emulsifying agents, dyestuffs, and insecticides.

Boiling point	214°C
Coefficient of expansion per °C	0.00105
Flash point (open cup)	187°F
Melting point	−70°C
Specific gravity at 20/20°C	0.78
20/4°C	0.778
60/15°C	0.755
Refractive index at 20°C	1.431
Surface tension at 20°C	24.9 dynes/cm.
Viscosity at 25°C	1.35 centipoise
at 60°C	0.73 centipoise
Weight per gallon at 20°C	6.5 lbs.
Acid insoluble	0.25% max.
Distillation range	
Initial boiling point	Not below 203°C
Not less than 95%	Below 216°C
Final boiling point	Not above 219°C
Purity	99% min.

Table 14.20: Isobutylamine *(2)*

$$(CH_3)_2CHCH_2NH_2$$

Isobutylamine is a colorless liquid soluble in water, methyl and ethyl alcohols, ethyl ether, ethyl acetate, acetone, aromatic and aliphatic hydrocarbons, fixed oils, mineral oil, oleic and stearic acids, and hot paraffin and carnauba waxes, the latter two solidifying when cooled.

Boiling point	68–9°C
Flash point	Below 20°F
Melting point	−85°C
Specific gravity at 20°C	0.731
Refractive index at 20°C	1.3985
Weight per gallon at 20°C	6.10 lbs.
Distillation range	66–69°C
Purity	99% min.

Table 14.21: Diisobutylamine *(2)*

$$(C_4H_9)_2NH$$

Diisobutylamine is a colorless liquid soluble in ethyl alcohol, methyl alcohol, aromatic and aliphatic hydrocarbons, ethyl ether, ethyl acetate, acetone, fixed oils, mineral oil, oleic and stearic acids. It is insoluble in water and while it dissolves hot paraffin and carnauba waxes, these solidify on cooling.

Flash point	85°F
Melting point	−70°C
Specific gravity at 20°C	0.75
Refractive index at 20°C	1.4124
Weight per gallon at 20°C	6.22 lbs.
Distillation range	136–140°C
Purity	97% min.

Table 14.22: sec-Butylamine *(2)*

$$CH_3CHNH_2C_2H_5$$

sec-Butylamine is a water-white liquid with a characteristic amine odor. It is soluble in water, methyl and ethyl alcohols, ethyl ether, ethyl acetate, acetone, aromatic and aliphatic hydrocarbons, mineral oil, fixed oils, stearic and oleic acids. It dissolves hot paraffin and carnauba waxes but these solidify on cooling.

Boiling point (772 mm)	66°C
Flash point	20°F
Melting point	−104°C
Specific gravity at 20/20°C	0.725
Boiling range	62–69°C

Table 14.23: Mono-n-Butyl Diamylamine *(2)*

$$C_4H_9N(C_5H_{11})_2$$

This amine is a light straw colored liquid with an amine odor. It is soluble in acetone, ethyl ether, ethyl acetate, aromatic and aliphatic hydrocarbons, fixed oils, mineral oil, and oleic and stearic acids. It is insoluble in water, methyl alcohol, and although soluble in hot paraffin and carnauba waxes, these solidify on cooling.

Flash point	200°F
Specific gravity at 20/20°C	0.788
Weight per gallon at 20°C	6.56 lbs.
Boiling range	229–241°C

Table 14.24: n-Amylamine *(2)*

$$CH_3(CH_2)_4NH_2$$

n-Amylamine is a colorless liquid with an ammoniacal odor. Commercially it is a mixture of the following isomers, although a pure product is available.

	B.P.°C.
tert-Amylamine	82
sec-Isoamylamine	87
2-Aminopentane	89
3-Aminopentane	90
Active amylamine	94
sec-Amylamine	95
n-Amylamine	104

It is miscible with water, ethyl and methyl alcohols, ethyl ether, ethyl acetate, acetone, aromatic and aliphatic hydrocarbons, fixed oils, mineral oil, pyridine, oleic and hot stearic acids, hot paraffin and hot carnauba waxes, the latter two solidifying when cooled. It dissolves a varied range of materials which are usually dissolved with difficulty by other organic solvents. It is used as a corrosion inhibitor and as a base for emulsifiers which are soluble in vegetable and mineral oils. It is also employed in textile lubrication, and as a raw material in the manufacture of dyestuffs, emulsifying agents, antioxidants, desizing agents for textiles and pharmaceuticals.

Boiling point	102–104°C	Viscosity at 20°C	0.01018 poise
Coefficient of expansion at 20–60°C	0.00116	Vapor pressure at 26°C	35 mm
Constant-boiling mixture		Weight per gallon at 20°C	6.41 lbs.
n-Amylamine	79–82.5% 82–85°C B.P.	Color	Water-white
Water	21–17.5%	Distillation range	
Flash point (open cup)	65°F	Initial boiling point	Not below 84°C
Heat of vaporization	108 cal./g.	Not less than 95%	Below 100°C
Melting point	−55°C	Final boiling point	Not above 110°C
Specific gravity at 20/20°C	0.76–0.78	Purity	90% min.
Specific heat at 60°F	0.65 cal./g.	Water dilution	20:1 min.
Refractive index at 19°C	1.4068		
Surface tension at 13°C	24.4 dynes/cm.		

Table 14.25: sec-Amylamine *(2)*

$$CH_3CH(NH_2)C_3H_7$$

sec-Amylamine is a colorless liquid with an amine odor. It is soluble in water, methyl and ethyl alcohols, aromatic and aliphatic hydrocarbons, ethyl ether, ethyl acetate, acetone, fixed oils, oleic and stearic acids. It dissolves hot paraffin and carnauba waxes which solidify when cooled.

Boiling point (760 mm)	91–92°C
Flash point	20°F
Specific gravity at 20°C	0.739
Refractive index at 20°C	1.4047
Weight per gallon at 20°C	6.15 lbs.
Distillation range	89–92°C
Purity	95–99% min.

Table 14.26: Diamylamine (Mixed Isomers) *(2)*

$$(C_5H_{11})_2NH$$

Diamylamine is a colorless to straw-colored liquid with an ammoniacal odor, which is composed of a mixture of amyl isomers. It is soluble in ethyl alcohol, methyl alcohol, ethyl ethers, ethyl acetate, acetone, aromatic and aliphatic hydrocarbons, fixed oils, mineral oil, oleic and stearic acids. It is insoluble in water and while soluble in hot paraffin and carnauba waxes, these solidify on cooling. It is a solvent for oils, resins, and some cellulose esters. Introduction of the amyl group imparts oil solubility to otherwise oil-insoluble substances. It is used as a corrosion inhibitor, and in chemical synthesis.

Coefficient of expansion at 20–60°C	0.00102
Flash point (open cup)	158°F
Heat of vaporization	83 cal./g.
Specific gravity at 20/20°C	0.77–0.78
Specific heat at 60°F	0.54 cal./g.
Refractive index at 20°C	1.4259
Surface tension at 13°C	24.4 dynes/cm.
Vapor pressure at 26°C	9 mm
Viscosity at 20°C	0.01264 poise
Weight per gallon at 20°C	6.45 lbs.
Acid insoluble	0.5% min.
Distillation range	
Initial boiling point	Not below 175°C
Not less than 95%	Below 202°C
Final boiling point	Not above 218°C
Purity	99% min.
Sulfur	0.06% min.

Table 14.27: Triamylamine (Mixed Isomers) *(2)*

$$(C_5H_{11})_3N$$

Triamylamine is a water-white to light yellow, stable liquid which is strongly basic in reaction. It is soluble in ethyl alcohol, ethyl ether, ethyl acetate, acetone, aromatic and aliphatic hydrocarbons, fixed oils, mineral oil, oleic and stearic acids, and in hot paraffin and carnauba waxes, the latter two solidifying when cooled. It is insoluble in water and methyl alcohol. It is an excellent corrosion inhibitor of steel in a 0.13% solution in normal sulfuric acid. It is used in the manufacture of emulsifying agents, dyestuffs, and insecticides.

Coefficient of expansion	0.00091
Flash point (open cup)	215°F
Heat of vaporization	79 cal./g.
Specific gravity at 20/20°C	0.79–0.80
Specific heat at 60°F	0.51 cal./g.
Refractive index at 18°C	1.4374
Surface tension at 13°C	24.4 dynes/cm.
Viscosity at 20°C	0.02421 poise
Vapor pressure at 26°C	7 mm
Weight per gallon at 20°C	6.60 lbs.
Acid insolubles	1.0% max.
Distillation range	
Initial boiling point	Not below 234°C
Not less than 50%	Above 244°C
Not less than 95%	Below 256°C
Final boiling point	Not above 260°C
Purity	99% min.

Table 14.28: sec-Hexylamine *(2)*

$$CH_3CHNH_2C_4H_9$$

sec-Hexylamine is a colorless liquid with an amine odor and soluble in water, ethyl alcohol, and paraffin hydrocarbons.

Flash point	55°F
Specific gravity at 20°C	0.746
Weight per gallon	6.22 lbs.
Distillation range	107–110°C
Purity	95–99%

Table 14.29: 2-Ethylbutylamine *(2)*

Hexylamine $(C_2H_5)_2CHCH_2NH_2$

2-Ethylbutylamine is a water-white liquid with an amine odor. It is soluble in methyl and ethyl alcohols, ethyl ether, ethyl acetate, acetone, aromatic and aliphatic hydrocarbons, fixed oils, mineral oil, oleic and stearic acids. It dissolves in water and while soluble in hot paraffin and carnauba waxes, these solidify on cooling.

Flash point	70°F
Specific gravity at 20/20°C	0.776
Boiling range	121–125°C

Table 14.30: n-Heptylamine *(2)*

$$C_7H_{15}NH_2$$

n-Heptylamine is a water-white liquid with an amine odor. It is insoluble in water but soluble in ethyl and methyl alcohols, ethyl ether, ethyl acetate, acetone, aromatic and aliphatic hydrocarbons, fixed oils, mineral oil, oleic and stearic acids. It dissolves hot paraffin and carnauba waxes, which solidify when cooled.

Flash point	130°C
Melting point	−23°C
Specific gravity at 20/20°C	0.779
Boiling range	150–160°C

Table 14.31: 2-Ethylhexylamine *(2)*

Octylamine $C_4H_9CH(C_2H_5)CH_2NH_2$

Octylamine is a water-white liquid with an amine odor. It is insoluble in water but soluble in methyl alcohol, ethyl ether, ethyl acetate, acetone, aromatic and aliphatic hydrocarbons, fixed oils, mineral oil, oleic and stearic acids. It dissolves hot paraffin and carnauba waxes, which solidify on cooling.

Flash point	135°F
Specific gravity at 20/20°C	0.792
Boiling range	165–169°C

Table 14.32: Di-2-Ethylhexylamine *(2)*

Dioctylamine $[C_4H_9CH(C_2H_5)CH_2]_2NH$

Dioctylamine is a colorless liquid with a faintly amine odor. It is soluble in methyl alcohol, ethyl ether, ethyl acetate, acetone, aromatic and aliphatic hydrocarbons, fixed oils, mineral oil, oleic and stearic acids. It is insoluble in water and while soluble in hot paraffin and carnauba waxes, these solidify when cooled. Among the large number of substances it will dissolve are natural and synthetic resins.

Boiling point (760 mm)	281.1°C	Solubility of water in solvent at 20°C	0.17% by wt.
Flash point	270°C	Vapor pressure at 20°C	0.01 mm
Specific gravity at 20/20°C	0.8062	Weight per gallon at 20°C	6.71 lbs.
Solubility in water at 20°C	0.02% by wt.	Boiling range	269–280°C

Table 14.33: Cyclohexylamine *(2)*

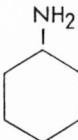

Cyclohexylamine is a colorless, caustic liquid with a fishy, amine odor. It has been known since 1893, but not until 1936 was it made in commercial quantities in the United States. It is produced by the catalytic hydrogenation of aniline. It is a strong base, even stronger than ammonia or the ethanol-amines. It is miscible with water and most of the common organic solvents, among which are the alcohols, ethers, ketones, esters, aliphatic and aromatic hydrocarbons, both pure and chlorinated. It is used as a solvent and as a corrosion inhibitor. Either alone or as a soap, it is employed as a wetting-out, cleansing, washing, emulsifying or dispersing agent in the textile industry. It may be used to absorb acidic gases, as a preservative for dyes, as an insecticide, and in the printing and dyeing of textile products.

Typical Properties and Specifications

Boiling point (760 mm)	134.5°C
Freezing point	−17.7°C
Fire point	30°C
Flash point	Below 0°C
Specific gravity at 25/25°C	0.8647
Refractive index at 25°C	1.4565
Weight per gallon at 20°C	7.206 lbs.
Azeotropic mixture	
Cyclohexylamine	44.2% by wt. B.P. (760 mm) 96.4°C
Water	55.8% by wt.
Distillation range	132.0–137.5°C

Table 14.34: Dicyclohexylamine *(2)*

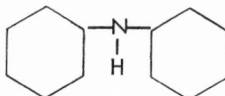

Dicyclohexylamine is a clear, colorless and strongly basic liquid with a faint odor. It is miscible with most organic solvents but only slightly soluble in water. Unlike cyclohexylamine, it does not form an azeotropic mixture with water. It is more toxic than cyclohexylamine when absorbed through the skin and when large amounts are absorbed, death may result.

Dicyclohexylamine soaps are good emulsifying agents. This solvent may be used to absorb acidic gases, to preserve rubber latex, to plasticize casein, to neutralize plant and insect poisons, and as a solvent for dyes in the textile printing and dyeing industry.

Typical Properties and Specifications

Boiling point (760 mm)	255.8°C
Freezing point	−0.1°C
Fire point	160°C
Flash point	100°C
Specific gravity at 25/25°C	0.9104
Refractive index at 25°C	1.4823
Weight per gallon at 20°C	7.59 lbs.
Boiling range	252.0–258.0°C
Purity	98%, min.

Table 14.35: t-Alkyl Primary Amines (52)

	t-Butylamine	t-Octylamine	t-Nonylamine	Primene 81-R	Primene JM-T
Formula	CH3-C(CH3)(CH3)-NH2	CH3-C(CH3)(CH3)-CH2-C(CH3)(CH3)-NH2	Principally t-C9H19NH2 and t-C10H21NH2	Principally t-C12H25NH2 to t-C14H29NH2	Principally t-C18H37NH2 to t-C21H43NH2
Molecular Weight	73	129	Principally 143-157	Principally 185-213	Principally 269-311
Neutral Equivalent	73	129	142	191	315
Boiling Range (760 mm.)	44-50°C.	137-143°C.	5-95% at 160-174°C.	5-90% at 221-238°C.	5-70% at 265-305°C.
Specific Gravity, 25°C.	0.690	0.771	0.789	0.813	0.840
Refractive Index, 25°C.	1.375	1.423	1.428	1.423	1.456
Water, %	0.09	0.30	0.28	0.40	0.10
Color	15 (APHA)	5 (APHA)	27 (APHA)	42 (APHA)	11 (Varnish Scale)
Flash point (Tag., open cup)	Below room temperature	95°F.	120°F.	205°F.	265°F.
Solubility	Miscible in all proportions with water and alcohol. Soluble in common organic solvents.	Insoluble in water. Soluble in common organic solvents. Excellent solubility in petroleum hydrocarbons.			

Table 14.36: "Alamine" Fatty Tertiary Amines (58)

Product*	Description	Amine Value	Percent Tertiary Amine, Min.	Percent Water Max.	Color Max.
Alamine 304	Tertiary trilauryl amine	98–116	90.0	1.0	500 APHA
Alamine 336	Tertiary tricaprylyl amine	136–149	95.0	0.2	500 APHA
Alamine 310	Tertiary triisodecyl amine		90	1.0	500 APHA
Alamine 308	Tertiary triisooctyl amine		90	1.0	500 APHA
Alamine 300	Tertiary tri-n-octyl amine		95	1.0	500 APHA

*Alamine is a registered trademark of Henkel Corp.

Table 14.37: "Aliquat" Fatty Quaternary Ammonium Chloride (58)

Product	Aliquat 336*
Description	Methyl tricaprylyl ammonium chloride
Percent solids	88% minimum

*Aliquat is a registered trademark of Henkel Corp.

Table 14.38: "Kemamine" Fatty Quaternary Ammonium Chlorides (26)

Product	Description (CTFA adopted name)	% Active Min	% Amine Max	% Amine HCl Max	Color Gardner, 1963 Max	Average molecular weight	pH of 5% solution Max
Kemamine Q-2802C	Dimethyl Di-80%-Behenyl (Dibehenyl Dimonium Chloride)	75	2.0	2.0	4	690	9
Kemamine BQ-2802C	Dimethyl 80%-Behenyl Benzyl (Behenalkonium Chloride)	75	2.0	2.0	4	475	9
Kemamine Q-9702C	Dimethyl Di-Hydrogenated Tallow (Quaternium-18)	75	1.5	0.5	2	575	9
Kemamine BQ-9742C	Dimethyl Tallow Benzyl (Tallow Alkonium Chloride)	75	1.5	0.5	6	420	9

Product	% Ash Max	Typical carbon chain composition									
		Saturated								Unsaturated	
		C_{10}	C_{12}	C_{14}	C_{16}	C_{18}	C_{20}	C_{22}	C_{24}	$C_{18:1}$	$C_{18:2}$
Kemamine Q-2802C	0.2				2	5	10	80	3		
Kemamine BQ-2802C	0.2				2	5	10	80	3		
Kemamine Q-9702C	0.2			4	29	67					
Kemamine BQ-9742C	0.2			4	29	25				38	4

Table 14.39: High Molecular Weight Aliphatic Amines (59)

N-alkyl Chain	Carbon Chain Length	Primary																Secondary		Diamines			
		Armeen 8D	Armeen 10D	Armeen 12	Armeen 12D	Armeen 14D	Armeen 16D	Armeen HT	Armeen HTD	Armeen 18	Armeen 18D	Armeen T	Armeen TD	Armeen S	Armeen SD	Armeen C	Armeen CD	Armeen 2C	Armeen 2HT	Duomeen C	Duomeen CD	Duomeen S	Duomeen T
Hexyl	6	3	—	—	—	—	—	—	—	—	—	—	—	—	—	—	—	—	—	—	—	—	—
Octyl	8	90	4	—	—	—	—	—	—	—	—	—	—	—	—	8	8	8	—	8	8	—	—
Decyl	10	7	90	2	2	—	—	—	—	—	—	—	—	—	—	9	9	9	—	9	9	—	—
Dodecyl	12	—	6	95	95	4	—	—	—	—	—	—	—	—	—	47	47	47	—	47	47	—	—
Tetradecyl	14	—	—	3	3	92	—	2	2	—	—	2	2	—	—	18	18	18	—	18	18	—	2
Hexadecyl	16	—	—	—	—	4	92	24	24	6	6	24	24	20	20	8	8	8	24	5	5	20	24
Octadecyl	18	—	—	—	—	—	7	71	71	90	90	28	28	17	17	5	5	10	75	5	5	17	28
Octadecenyl	18	—	—	—	—	—	1	3	3	4	4	46	46	26	26	5	5	5	1	5	5	26	46
Octadecadienyl	18	—	—	—	—	—	—	—	—	—	—	—	—	37	37	—	—	—	—	—	—	46	—
Mol. combining weight		135	166	213	195	227	250	300	275	300	280	298	274	297	275	223	208	450	530	321	310	402	400
Percent Primary Amine		90	—	82	94	92	95	85	95	85	95	85	95	86	95	85	95	—	—	40	44	40	40
Percent Secondary Amine		—	—	—	—	—	—	—	—	—	—	—	—	—	—	—	—	85	85	—	—	—	—
Approx. Melting Pt. °C		−13	8	24	24	29	38	57	55	55	55	46	41	31	22	24	21	46	68	22	20	40	46
Color—FAC		3	3	9	3	3	3	11	3	11	3	11	3	19	7	11	3	9	5	19	11	13	19
Grade: D—Distilled, T—Technical		D	D	T	D	D	D	T	D	T	D	T	D	T	D	T	D	D	D	T	D	T	T

N-alkyl Chain	Carbon Chain Length	Dimethyl Tertiary Amine										Dialkyl Tertiary Amines		
		Armeen DM16	Armeen DM16D	Armeen DM18	Armeen DM18D	Armeen DM C	Armeen DM CD	Armeen DM S	Armeen DM SD	Armeen DM HT	Armeen DM HTD	Armeen M2HT	Armeen M2C	Armeen M2S
Hexyl	6	—	—	—	—	—	—	—	—	—	—	—	—	—
Octyl	8	—	—	—	—	8	8	—	—	—	—	—	8	—
Decyl	10	—	—	—	—	9	9	—	—	—	—	—	9	—
Dodecyl	12	—	—	—	—	47	47	—	—	—	—	—	47	—
Tetradecyl	14	—	—	—	—	18	18	—	—	2	2	2	18	—
Hexadecyl	16	92	92	6	6	8	8	20	20	24	24	24	8	20
Octadecyl	18	7	7	90	90	5	5	17	17	71	71	71	5	17
Octadecenyl	18	1	1	4	4	5	5	26	26	3	3	3	5	26
Octadecadienyl	18	—	—	—	—	—	—	37	37	—	—	—	—	37
Mol. weight—theoretical		271	271	295	295	224	224	289	289	289	289	522	389	520
Mol. combining weight		338	295	369	321	280	244	361	314	361	314	564	436	594
Percent Tertiary Amine		80	92	80	92	80	92	80	92	80	92	—	—	—
Approx. Melting Pt. °C		15	10	23	20	−10	−15	0	−8	17	15	28	−5	9
Color—Gardner—1933		5	1	5	1	5	1	10	1	5	1	—	—	—
Grade: D—Distilled, T—Technical		T	D	T	D	T	D	T	D	T	D	T	T	T

Table 14.40: Solubilities of Pure Dodecyl- and Octadecyl-Trimethylammonium Chlorides in Grams per 100 Grams of Solvent (59)

Salt—Solvent		−10°	0°	10°	20°	30°	40°	45°	50°	55°	56.5°	60°	65°	70°	72°	80°	84°	86°
Dodecyl	Methanol	83.1	113.8	145.8	180	226.6	—	—	—	—	—	—	—	—	—	—	—	—
	Acetone	—	—	—	—	2.88	9.76	—	41.75	91.9	110.6	—	—	—	—	—	—	—
	Acetonitrile	—	4.8	10.9	18.2	32.8	81.2	—	—	—	—	—	—	—	—	—	—	—
	Carbon tetrachloride	—	—	—	—	1.21	34.2	102	gel	—	—	—	—	—	—	—	—	—
	Insoluble in ethyl acetate, benzene, n-hexane or cyclohexane at 95°.																	
Octadecyl	Methanol	5.7	15.4	32.5	71.6	112.8	168	—	252	—	—	—	—	—	—	—	—	—
	Ethanol (93.5%)	3.7	9.3	25.6	43.1	82.9	132	—	210	—	—	—	—	—	—	—	—	—
	Acetone	—	—	—	—	—	—	—	0.50	0.71	0.76	—	—	—	—	—	—	—
	Acetonitrile	—	—	—	0.7	1.8	3.2	—	5.1	—	—	9.9	—	32.7	—	78.6	—	—
	Carbon tetrachloride	—	—	—	—	—	—	—	—	0.40	—	—	5.04	—	36.2	—	—	—
	Chloroform	13.6	25	40.8	56	73.5	100	—	—	—	—	—	—	—	—	—	—	—
	Ethyl acetate	—	—	—	—	—	—	—	—	—	—	—	—	—	—	—	1.22	5.04
	Benzene	—	—	—	—	—	—	—	—	0.5	—	3.1	19	46	—	—	—	—
	Insoluble in n-hexane or cyclohexane at 95°C.																	

Table 14.41: Solubilities of Organic Compounds in Aliphatic Amines at 25° ± 5°C *(1)(2)*

Abbreviation		Approx. Sol'y Range per 100 cc. of Solvent *Grams*
ins.	Insoluble or extremely slightly soluble	(0+ to 10)
ss	Slightly soluble	(10+ to 40)
s	Moderately soluble	(40+ to 70)
vs	Very soluble	(70+ to 100)
vs+	More than vs	(100+ and above)
es	Extremely soluble	
misc	Miscible (in methylamine column only)	
∞	Miscible in all proportions	
n	Not soluble to an appreciably greater extent in hot solvent	
x	More soluble in heated amine, crystallizes on cooling	
m	More soluble in heated amine (in some cases because of chemical reaction)	
p	Separates into two liquid phases	
r	Solute reacts chemically with solvent. Reaction is rapid enough to be apparent. All acidic substances react more or less rapidly with amines. (The letter *r* has been omitted in these cases)	
a	Swells	
Numerals	Numerals appearing in diethylamine column indicate number of grams of solute (or of its reaction product with diethylamine) per 100 cc. of solution	

#	Compound	C_2H_5OH	$C_2H_4{>}O$	CH_3NH_2	$C_2H_5{>}NH$	$(C_2H_5)_3N$	$(n\text{-}C_3H_7)_2NH$	$n\text{-}C_4H_9NH_2$	$(n\text{-}C_4H_9)_2NH$	$(n\text{-}C_4H_9)_3N$	$iso\text{-}C_5H_{11}NH_2$	$C_6H_5CH_2NH_2$	$NH_3(-33°C.)$
1	Acenaphthene	ssx	..	..	15	sx	..	sx	s	ssx	s	s	ins
2	Acetaldehyde	..	..	misc		ins		s	ins	ins	s	s	vs
3	Acetamide	s	..	vs	ssx	ins		s	ss	ssx	sx	s	..
4	Acetanilide	vs	ss	..	11	ssx		vs	ss	ssx	sx	s	..
5	Acetic acid	..	..	vs				..	..	..	..	..	..
6	Acetoacetic ester	∞		∞	∞	..		..	∞	..	..	..	∞
7	Acetone	∞	∞	∞		..		∞	∞	..	..	..	..
8	Acetophenone	∞		..		..		∞	∞	..	..	..	s
9	p-Acetophenylene diamine	..	..	..	ins	..		ssx	ssx	.	ssx	ssx	..
10	p-Acetotoluide	ss	s	..	ss	..		..	..	ssx	ssx	..	ss
11	Acetylene	s	..	s		..		..	..	..	..	..	..
12	Acetylene tetrabromide	s	..	..	r	..		..	..	..	..	..	s
13	Acetylsalicylic acid	vs	ss	..	s	s	s	vs	s	ss	vs	s	..
14	Agar-agar	..	..	..	..	..		..	..	..	insn	..	ins
15	Alanine	ss	ins	..	..	..		..	..	..	insn	..	..
16	Aldol	∞	s	..	∞	..		∞	..	..	..	..	..
17	Alizarin	..	..	s	..	..		..	..	..	..	..	..
18	Allyl alcohol	∞	∞	..	∞	..		..	..	..	..	..	..
19	1-Aminoanthraquinone	ss	ss	..	ss	ss	ss	ssm	ss	ssx	ssm	ss	..
20	p-Aminobenzoic acid	s	.	..	ins	ins	isn	ss	ins	insn	ss	ssx	..
21	m-Aminophenol	s	s	..	vs+	ss		s	s	es	vs	s	..
22	o-Aminophenol	ss	ss	..	vs	ins	s	vs+	sx	ssn	s	s	vs
23	Aminosulfonic (sulfamic) acid	..	..	..	..	..		ssn	..	ssx	..	..	vs
24	Ammonium benzoate	..	..	..	ss	..		vsr	..	vsr	..	..	..
25	Ammonium citrate	..	..	..	ss	..		..	..	..	..	..	..
26	Amyl alcohol (iso)	∞	∞	∞	∞	..		∞	..	..	∞	..	..
27	n-Amyl formate	∞	∞	..	∞	..		..	..	∞	..	..	ss
28	Anethole	∞	∞	..	∞	..		..	..	..	..	..	..
29	Anhydroformaldehydeaniline	ss	ss	..	ssn	ssx	ins	ss	ss	ssx	ssx	ssx	..
30	Aniline blue	..	..	..	ss	ins		ss	ins	ins	ss	ss	..
31	Anthracene	ss	ss	..	ss	ssx		ssx	ss	ssx	ssx	ssx	ins
32	Anthranilic acid	s	s	..	ss	ins	ssx	s	ss	ss	s	s	..
33	Anthraquinone	ss	ss	ss	..	..		ssx	..	ssx	ss	..	ins
34	Atoxyl	..	..	..	ins	ins		ss	ins	ins	ss	ins	..
35	Azobenzene	ss	s	vs	vs	ss	vsx	vs	vs	s	vs	s	s
36	Azoxybenzene	s	s	..	es	..		..	..	..	es	..	s
37	Beeswax	..	..	..	s	s		s	s	ssm	s	ssm	..
38	Benzalacetophenone	ss	vs	..	vs	..		es	..	ssx	..	..	..
39	Benzaldehyde	∞	∞	misc	∞	..		..	..	..	..	..	es
40	Benzamide	s	vss	..	ss	ins	s	s	ss	ins	s	s	vs
41	Benzene	∞	∞	..	∞	..		∞	∞	∞	..	∞	s
42	Benzidine	s	ss	..	ss	ss		s	ssx	ins	sx	..	..
43	Benzil	vs	vs	vs	ss	ss	sx	es	ss	ss	es	es	s
44	Benzoic acid	vs	vs	vs	ssx	ssx		vs	s	ssp	s	sx	vs+
45	Benzoic sulfinide (Saccharin)	ss	ins	r	..	..		..	..	..	ss	..	..
46	Benzoin	s	..	..	..	..		vs	..	..	..	..	..
47	Benzophenone	s	s	vs	es	es		es	vs+	sx	es	es	s
48	Benzyl acetate	∞	∞	..	es	..		..	∞	..	..	∞	∞
49	Benzyl alcohol	∞	∞	..	..	..		..	..	..	..	..	..
50	Borneol	vs	vs	..	es	..		..	..	..	..	..	s
51	o-Bromoacetanilide	..	..	vs	..	..		..	..	..	..	..	..
52	p-Bromoaniline	vs	vs	..	esx	vs+	esx	vs+	vs	ssx	vs	vs+	..
53	Bromcamphor	s	vs	..	vs+x	s		vs+x	vs+	ssx	vs	vs	..
54	Bromocresol green	..	..	..	ss	ins		s	ssn	ins	s	s	..
55	1-Bromonaphthalene	∞	∞	..	∞	..		..	..	∞	..	..	..
56	p-Bromonitrobenzene	s	s	..	s	..		..	..	..	..	∞	..
57	o-Bromotoluene	s	s	..	∞	..		..	∞	∞	..	..	..
58	p-Bromotoluene	∞	s	..	es	..		..	es	..	..	..	..
59	n-Butyl alcohol	∞	..	..	∞	..		..	..	..	∞	∞	∞
60	tert-Butyl alcohol	..	..	∞	..	..		..	..	..	..	..	..
61	n-Butyl xyanide	∞	s	..	∞	..		∞	..	..	∞	..	s
62	n-Butyl ether	∞	∞	..	∞	..		∞	∞	∞	∞	..	..
63	n-Butyl formate	∞	∞	..	∞	..		..	∞	..	..	..	..
64	n-Butyl stearate	..	..	..	∞	..		∞	∞	∞	..	rss	..
65	Caffeine	ss	ss	..	ss	..		ssx	..	..	ssx	..	ss
66	Calcium acetate	ins	..	..	insn	..		..	..	..	..	..	..
67	n-Calcium butyrate	..	..	..	..	..		insn	..	..	..	..	..
68	Calcium formate	ins	..	..	..	..		..	..	..	insn	..	..
69	d-Camphor	es	vs	..	es	es		es	vs	vsx	es	es	vs
70	Carbon disulfide	∞	∞	..	r	..		esr	..	∞	..	..	r

(continued)

Table 14.41: (continued)

		C_2H_5OH	$C_6H_5 \atop C_6H_4 > O$	CH_3NH_2	$C_6H_5 \atop C_6H_4 > NH$	$(C_6H_5)_2N$	$(n\text{-}C_3H_7)_3NH$	$n\text{-}C_4H_9NH_2$	$(n\text{-}C_4H_9)_2NH$	$(n\text{-}C_4H_9)_3N$	$iso\text{-}C_5H_{11}NH_2$	$C_6H_5CH_2NH_2$	$NH_3(-33°\,C.)$
71	Casein	..	..	..	ins	ins	..	ins	ins	ins	ins	ins	..
72	Cellulose	ins	ins	..	ins	ins	..	ins	ins	ins	ins	ins	ins
73	Cellulose acetate	..	..	..	ss	ins	ins	es	ins	ins	s	insa	..
74	Cellulose nitrate	..	..	..	s	ins	..	vs	ss	ins	vs	s	es
75	Cerulein	..	..	vs	..	..	..	..	..	..	..	..	s
76	Cetyl alcohol	..	..	vs	..	..	..	..	..	..	..	..	..
77	Chloroacetic acid (mono)	s	s	vsx	..	..	..	es	..	s	..	..	ins
78	p-Chlorodiphenyl	s	s	..	srx	vsxp	..	sx	ssxr	ssp	vs	ssrm	vs
79	Chloroform	s	∞	∞	vs+	vs+x	..	vs	vs	sx	vs	vs	s
80	Cholesterol	ssx	s	..	∞	..	vs+	..	..	..	..	..	sr
81	Chromotropic salt	..	..	..	..	..	..	insn	..	..	vs+	..	ins
82	Cinchonine	..	ss	..	..	..	..	ssn	..	..	..	..	..
83	Cinnamic acid	ss	vs	..	s	ss	..	ssn	ss	ssx	ssx	ssx	ins
84	Coconut oil	s	vs	..	∞	∞	..	sx	ss	ssx	s	ssx	s
85	Copal	..	..	..	..	..	..	∞	∞	∞	∞	insn	ss
86	Crystal violet	..	..	..	ins	ins	..	vs	ins	ins	s	s	s
87	Cyclohexanol	s	s	..	∞	..	..	..	..	..	..	..	..
88	o-Dianisidine (bianisidine)	s	vs	..	ss	ins	..	sx	ss	ins	s	s	..
89	Diazoaminobenzene	s	vs	..	vs	..	..	..	..	..	es	..	vs
90	p-Dibromobenzene	s	..	..	40	vs+x	..	vsx	vs	sx	vs	vs+	ss
91	2,3-Dibromopropyl alcohol	vs	∞	..	..	..	..	..	..	..	..	..	..
92	Dichloramine-T	..	..	..	vsr	sr	..	vsr	ssr	r	vsr	vs+r	r
93	p-Dichlorobenzene	es	vs	..	53	esx	..	es	ves	vs+x	esx	es	ss
94	Dichlorogallein	..	..	s	..	..	..	..	..	..	..	..	s
95	Dichlorohydrin	∞	∞	..	∞	..	..	..	..	..	..	..	..
96	2,3-Dihydroxyquinoxaline	..	..	..	..	..	..	insn	..	..	..	..	s
97	Dimethylaminoazobenzene	x	..	..	sx	ssx	ssx	s	s	s	s	s	..
98	p-Dimethylaminobenzaldehyde	s	s	..	ss	ins	..	sx	..	..	..	..	..
99	Dimethylaniline	s	..	..	∞	..	..	..	..	..	..	..	ss
100	Dimethylethylcarbinol	s	s	..	∞	..	..	..	..	∞	..	∞	ss
101	Dimethylglyoxime	vs	vs	..	ssx	s	ssx	vs	ss	ssx	vs	s	..
102	2,6-Dimethylquinoline	vs	vs	..	vs	..	..	..	..	..	..	..	..
103	2,2-Dinaphthylamine	..	..	..	ss	ssx	..	ssx	ssx	ssx	ssx	ssx	..
104	2,4-Dinitroaniline	ss	..	vs	ss	ss	..	sx	s	ssx	ssx	s	..
105	m-Dinitrobenzene	ssx	..	vs	9.3	ss	..	s	ss	ssx	s	s	vs
106	3,5-Dinitrobenzoic acid	vs	ss	..	ss	ss	ssx	s	ss	ssx	vs	s	..
107	4,4'-Dinitrodiphenyl	ss	vs	..	ss	ss	..	ssx	ssx	ssx	ssx	ssx	..
108	2,4-Dinitro-1-naphthol-7-sulfonic acid	..	..	..	ss	ss	..	..	..	..	..	..	..
109	2,4-Dinitrophenol	ss	vs	..	ss	..	..	sx	..	ssn	vs	..	vs
110	2,4-Dinitrotoluene	ss	..	vs	sx	..	..	sx	ssx	ssx	ss	s	ss
111	Diphenyl	ss	s	..	41	s	..	vs	s	sx	vs	vs	ss
112	Diphenylamine	vs	vs	..	es	vs	es	es	vs	sx	es	es	..
113	Diphenylbenzamide	..	..	..	ss	..	..	..	..	ssx	..	..	..
114	Diphenylguanidine	..	..	..	ss	ins	..	vs	s	ssx	vs	s	..
115	Diphenyl ketoxime	..	..	vs	..	..	..	..	..	..	..	..	..
116	Diphenyl sulfone (phenyl sulfone)	ss	s	..	ssx	ssx	..	ssx	ss	ssx	ssx	sx	..
117	Diphenylurea (sym.)	s	s	..	ss	ins	..	ss	ss	ssx	s	ssx	s
118	4,4'-Dipyridyl (bipyridine)	vs	..	..	..	..	..	vsx	..	sx	..	..	vs
119	Di-p-tolylselenide	..	..	..	..	..	..	es	..	..	..	..	..
120	Eosin	..	..	..	ss	ins	..	vs	ss	ins	vs	s	s
121	Ethyl alcohol	∞	∞	..	..	..	..	..	∞	..	..	∞	..
122	Ethyl carbonate	s	..	..	ssr	..	..	∞	..	..	..	∞	sr
123	Ethyl cyanoacetate	∞	∞	..	∞	..	..	..	..	..	..	..	..
124	Ethylene dibromide	s	∞	spr	∞	..	..	∞	..	..	∞	∞	..
125	Ethylene glycol	∞	ss	..	∞	..	∞	..	..	..	∞	..	..
126	Ethyl iodide	s	s	..	∞	..	..	..	..	∞	..	esr	s
127	Ethyl malonate	∞	∞	..	∞	..	..	..	∞	..	∞	..	..
128	Ethyl oxalate	s	..	..	∞	..	..	..	∞	..	..	ssr	..
129	Ethyl sulfate	..	..	..	r	..	..	..	rs	..	..	..	..
130	Fluorene	ss	vs	..	13	ssx	sx	s	s	ssx	s	s	ins
131	Fluorescein	vs	ss	..	ss	ins	..	s	ins	ins	s	s	s
132	Galactose	ss	..	vs	..	..	..	..	..	..	..	..	..
133	Gallein	..	..	ss	..	..	..	..	..	..	..	..	..
134	Gallic acid	s	ss	..	insn	insn	..	s	insx	insn	s	ssm	s
135	Gelatin	..	..	..	ins	ins	..	insn	ins	ins	ins	ins	..
136	Glucose	ss	ins	vs	ss	ins	ssn	vs+	ins	ins	s	s	vs
137	β-Glucose pentaacetate	ss	ss	..	..	..	..	es	..	..	..	..	..
138	Glycerol	∞	ins	vs	..	..	..	..	..	..	..	..	..
139	Guaiacol	∞	∞	..	∞	..	..	..	..	..	..	..	..
140	Guanidine nitrate	..	..	..	ss	..	..	..	..	..	..	..	vs
141	Gum arabic	..	..	..	insn	ins	..	ins	ins	ins	ssn	ss?	..
142	H acid	ss	ss	..	..	..	..	insn	..	..	..	..	..
143	Hemoglobin	..	..	..	ins	..	..	insn	..	..	..	..	ss
144	Hexaethylbenzene	s	vs	..	..	..	..	ssx	..	..	ssx	..	..
145	Hexamethylenetetramine	ss	ins	..	ssn	..	..	ssn	..	..	..	..	..
146	Hippuric acid	ss	vss	vs	ssx	..	..	..	..	..	vs	..	vs
147	Hydrazine sulfate	ins	ins	vs	ins	ins	..	..	..	..	..	..	insr
148	Hydroquinone	vs	vs	vs	35	ss	s	s	s	ss	s	ssx	s
149	Hydroxylamine hydrochloride	s	ins	vs	..	ins	..	ss	ss	ss	ins	..	ss
150	Indigotin	ins	ins	vs	ins	ins	..	s	s	s	ins	ss	ss

(continued)

Table 14.41: (continued)

#		C_2H_5OH	$\frac{C_2H_5}{C_2H_5}{>}O$	CH_3NH_2	$\frac{C_2H_5}{C_2H_5}{>}NH$	$(C_2H_5)_3N$	$(n\text{-}C_3H_7)_2NH$	$n\text{-}C_4H_9NH_2$	$(n\text{-}C_4H_9)_2NH$	$(n\text{-}C_4H_9)_3N$	$iso\text{-}C_5H_{11}NH_2$	$C_6H_5CH_2NH_2$	$NH_3(-33°\,C.)$
151	Indole	vs	vs	..	es	..	..	..	vs	..	..	..	vs
152	Isatin	s	ss	..	..	..	..	vs	..	..	..	..	s
153	Isoquinoline	∞	∞	..	..	..	..	..	..	..	..	..	vs
154	Lactose	ss	ins	..	ss	ss	..	ss	ins	ins	es	ins	vs
155	Lanolin	..	..	..	es	es	..	es	es	es	es	es	..
156	Linseed oil, raw	..	..	..	∞	∞	..	∞	∞	∞	∞	∞	..
157	Lysol	..	..	..	∞	..	..	..	..	..	..	..	..
158	Maleic acid	s	ss	..	ins	ins	..	ssn	ins	ins	sex	insm	..
159	Malic acid	vs	ss	..	ins	ins	..	ssn	ins	ins	ssn	ssx	ss
160	Malonic acid	s	ss	..	ins	insn	insx	ss	ins	ss?	ss	ins	ins
161	Mannite	ss	ins	..	ss	ins	..	ss	ins	ins	ssn	ins	ss
162	d-Menthol	vs	vs	..	es	es	es	es	es	es	es	es	..
163	Mercuric acetate	ss	..	..	insr	ins	s	vs	ss	ss	ins	ins	s
164	Mercuric cyanide	s	..	vs	..	..	..	es	..	ssn	..	..	..
165	Mercury diphenyl	ss	..	..	ss	ins	ss	ss	..	ins	ss	ins	..
166	Mercury di-p-tolyl	..	..	..	..	..	..	ssx	..	ssx	ssx	..	..
167	Methanol	∞	∞	∞	..	..	..	..	..	∞	..	..	..
168	meso-Methyl acridine	vs	vs	..	..	..	..	..	..	..	es	..	..
169	Methylene aminoacetonitrile	..	..	..	ss	ss	..	ss	ins	ssx	ss	ins	..
170	Methylene dianiline	..	..	..	insn	ss	..	ss	ss	insx	ss	ss	..
171	Methyl orange	ss	ins	ss	ins	ins	ins	ssn	ins	ins	ss	ss	..
172	Mucic acid	ins	..	..	ins	ins	insn	ins	ins	ins	ss	ins	s
173	Naphthalene	ss	vs	..	27	sx	..	vsx	s	sx	vs	sx	ss
174	Naphthionic acid	ss	ins	..	ss	ins	..	s	ins	ins	s	ss	..
175	1-Naphthol	vs	vs	vs	..	..	..	..	..	..	..	..	..
176	2-Naphthol	vs	vs	vs	es	vs+	..	es	es	vs	es	es	ss
177	Naphthol yellow	..	..	..	ins	ins	..	ss	ins	ins	ins	ins	..
178	Night blue	..	..	..	s	s	..	s	s	s	s	s	..
179	p-Nitroacetanilide	s	ss	..	ss	..	..	..	..	..	..	..	..
180	m-Nitroaniline	s	ss	..	sx	ssx	..	ss	ssx	ssx	s	s	s
181	p-Nitroaniline red	..	..	..	ss	ss	..	s	ssx	ssx	s	s	..
182	m-Nitrobenzaldehyde	vs	vs	vs	..	..	..	..	..	..	..	..	..
183	Nitrobenzene	∞	∞	∞	..	..	..	∞	..	∞	..	..	vs
184	m-Nitrobenzenesulfonamide	..	..	vs	..	..	..	..	..	..	..	..	..
185	o-Nitrobenzoic acid	s	s	vs	..	..	..	..	..	..	..	..	..
186	p-Nitrobenzoic acid	ss	ss	..	ssx	insn	..	vsx	ss	ss	vsx	sx	ss
187	p-Nitrobromobenzene	ss	..	..	ss	ss	ssx	sx	ss	ssx	s	sx	..
188	o-Nitrophenol	vs	vs	vs	..	..	..	..	..	..	..	..	..
189	p-Nitrophenol	vs	s	..	13	ss	ssx	vs	ss	ssm	vs	s	s
190	p-Nitrophenylhydrazine	..	..	..	..	..	..	vs	..	ssx	..	..	..
191	o-Nitrotoluene	∞	∞	vs	..	..	..	..	..	..	..	..	..
192	p-Nitrotoluene	s	s	..	vs+x	..	..	..	..	sx	vsx	..	..
193	3-Nitro-4-toluidine	vs	..	..	sx	..	..	vsx	..	ssx	..	..	vs
194	Nitrourea	vs	vs	..	ins	..	..	..	..	..	..	..	ins
195	Oleic acid	∞	∞	..	∞	∞	..	∞	∞	∞	∞	∞	..
196	Olive oil	..	..	..	∞	∞	..	∞	∞	∞	∞	∞	..
197	Orange IV	..	..	..	ss	ins	..	ss	ss	ins	ssn	ss	ins
198	Oxalic acid.2H₂O	s	ss	..	ins	ins	..	insn	insn	ins	ins	ins	..
199	Oxanilide	ss	ss	..	0.5	..	..	..	..	..	..	..	..
200	Palmitic acid	ss	..	vs	..	..	..	..	..	..	..	..	..
201	Paraffin	..	..	..	ss	sx	..	s	sm	sx	s	ssx	ins
202	Paraffin oil	..	..	..	∞	..	..	..	∞	..	∞	..	..
203	n-Pentane	∞	..	∞	..	..	..	..	..	..	..	..	..
204	2-Pentene	..	..	..	..	..	..	..	..	..	..	..	..
205	Penacetin	ss	ss	..	ss	ins	ssx	s	ss	ssx	s	ss	..
206	Phenanthrene	ss	ss	..	s	..	..	..	..	..	..	vs	vs
207	Phenol	∞	∞	vs	es	..	..	..	..	es	..	..	vs
208	Phenolphthalein	s	∞	vs	s	ss	..	s	ss	ss	s	s	..
209	Phenylacetic acid	es	vs	vs	s	ss	..	s	ss	ss	s	s	..
210	meso-Phenyl acridine	ss	s	..	..	..	..	..	..	..	ssx	..	..
211	Phenylazo-1-naphthylamine	..	..	..	..	..	..	..	..	ssx	s	..	..
212	o-Phenylenediamine	vs	vs	..	..	..	..	sx	..	..	sx	..	vs
213	Phenylglucosazone	ss	vs	..	..	..	..	sx	..	..	vs	..	vs
214	Phenylhydrazine	∞	∞	∞	..	..	..	..	..	..	..	..	..
215	Phenylmercuric bromide	..	..	..	..	..	..	ssx	..	ssx	ssx	..	..
216	2-Phenylquinoline	vs	vs	..	vs	ins	ins	vs	..	..	s	..	..
217	Phenyl-p-tolyl sulfone	..	..	..	6.3	ins	ins	ss	ss	ssx	ssx	ss	..
218	Phenyl urea	vs	vs	..	ss	ss	ssn	s	ss	ssx	vs	s	s
219	Phthalic acid	s	ss	ins	ins	ins	..	ins	ins	insx	..	ssx	ins
220	Phthalic anhydride	s	ss	ins	sm	ss	..	s	ss	ins	s	ssm	..
221	Phthalimide	s	ss	..	s	ss	ss	s	sx	ssm	s	s	vs
222	Picric acid	s	ss	vs	ssp	ss	vs	vs+	s	ssx	vs	s	es
223	Potassium amide	r	ins	..	ins	..	..	insn	..	..	..	..	vs
224	Potassium ethyl sulfate	..	..	..	ins	..	..	..	..	..	..	..	vs
225	Potassium quinaldine	r	vs	..	..	..	..	vs	..	..	..	..	es
226	Potassium triphenylmethyl	r	s	..	..	..	..	s	..	..	..	..	..
227	Pyridine	∞	∞	∞	..	..	..	..	..	..	..	..	..
228	Pyrogallol	vs+	vs+	vs	ssm	ss	..	s	s	ss	s	sm	es
229	Pyrrole	vs	vs	..	∞	..	..	..	..	..	..	..	..
230	Quinaldine	vs	vs	..	∞	..	..	∞	..	∞	∞	..	s
231	Quinaldine picrate	ss	..	..	es	..	..	es	vsx	ss	..	..	..
232	Quinine	es	vs+	..	s	ssm	s	vs	ssm	vs	vs	s	insol
233	Quinoline	s	..	..	..	..	..	∞	∞	∞	∞	∞	∞
234	Quinoline methiodide	..	..	..	..	..	..	vsx	..	..	vs	..	vs
235	Quinoline yellow (water soluble)	..	..	..	ins	ins	..	ss	ss	ins	ssn	ss	..
236	Quinone	vs	vs	sr	var	..	..	..	..	..	..	..	..
237	Resorcinol	es	vs	vs	..	..	..	vs	..	..	s	..	es
238	Rosaniline	s	ins	vs	ins	ins	..	ss	ins	ins	ss	ss	ins
239	Rosin	vs	vs	..	s	ss	..	vs	vs	ins	s	s	ins
240	Rosolic acid	ss	s	..	ss	ss	..	s	ss	ss	s	s	ss

Table 14.42: Vapor Pressure of Various Amines *(37)*

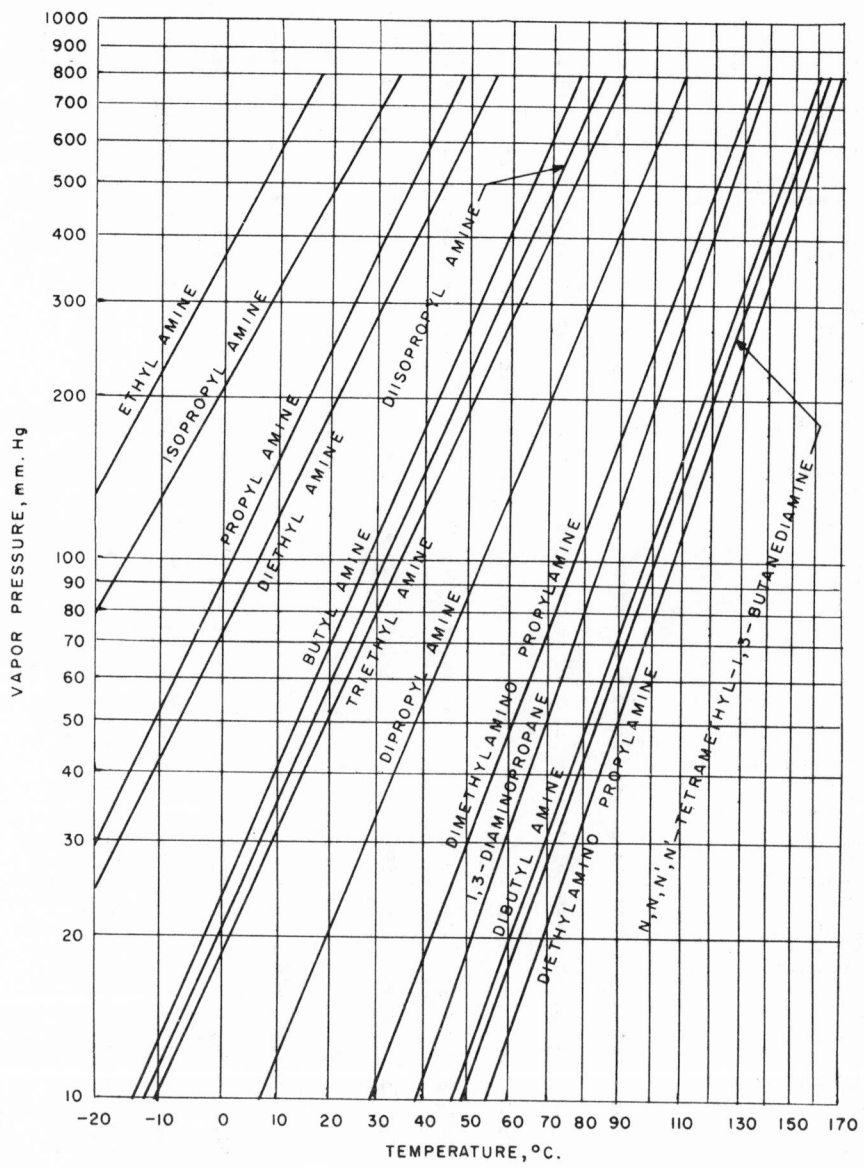

Table 14.43: Vapor Pressure of Sharples Amines (37)

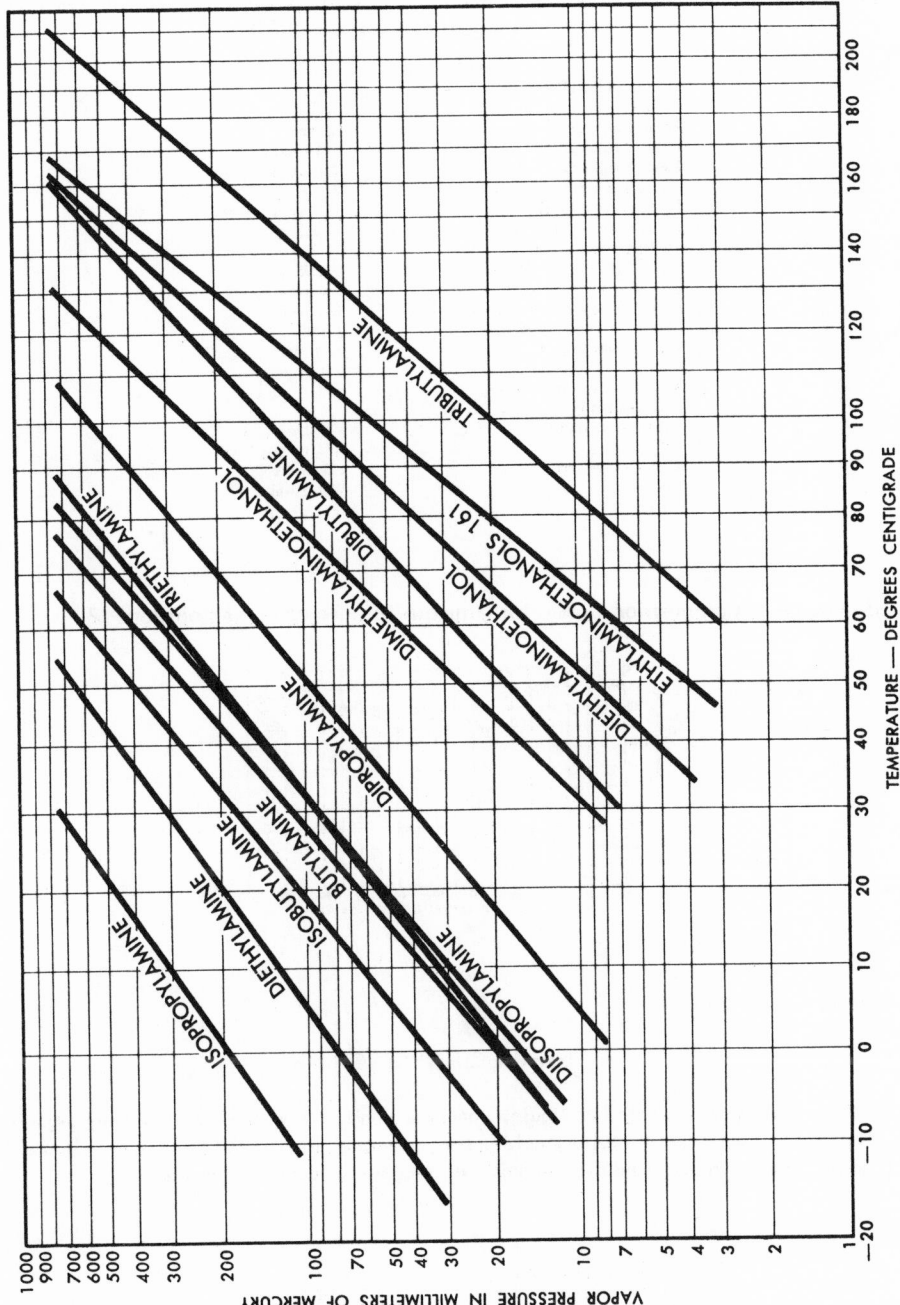

ALKYLENE DIAMINES

Table 14.44: Ethylene Diamine *(2)*

$$NH_2CH_2CH_2NH_2$$

Ethylenediamine is a water-white, hygroscopic liquid with a strong ammoniacal odor. The commercial product is a 78% solution of ethylenediamine by weight. It is used in the synthesis of organic rubber accelerators, insecticides, textile processing chemicals, emulsifiers, plastics and pharmaceuticals. It is also used as a corrosion inhibitor.

78% solution

Boiling point (760 mm)	117.2°C
Dielectric constant at 18°C	16.0
Flash point (open cup)	110°F
Heat of combustion	425.6 cal./mol
Heat of solution at 15°C	7.6 cal./mol
Ionization constant at 25°C	7.1×10^{-5}
Latent heat of evaporization	167 cal./g.
Latent heat of fusion (0°C)	77 cal./g.
Melting point	11.0°C
Specific gravity at 20/20°C	0.8995
Solubility in water at 20°C	Complete
Solubility of water in solvent at 20°C	Complete
Refractive index at 26°C	1.4540
Vapor pressure at 20°C	10.7 mm
Viscosity at 25°C	0.0154 poise
Weight per gallon at 20°C	7.49
Constant-boiling mixture	
Ethylenediamine	80% by wt. B.P.°C
Water	20% by wt. 118.5
Boiling range (760 mm)	115 to 122°C
Color	Water-white
Purity	78% by wt.

Table 14.45: Boiling Point Composition Curves for Aqueous Ethylenediamine Solutions *(2)*

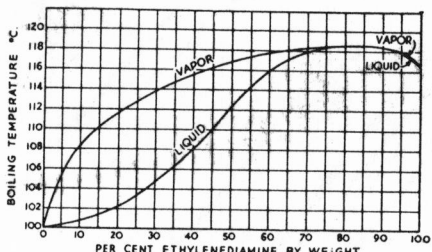

Table 14.46: Diethylenetriamine *(2)*

$$NH_2CH_2CH_2NHCH_2CH_2NH_2$$

Diethylenetriamine is a colorless liquid, completely miscible with water and many organic solvents. It is a solvent for sulfur, acid gases, numerous natural resins and dyes. It is also used in organic synthesis and as a saponification agent for acidic materials.

Boiling point (760 mm)	207.1°C
Flash point (open cup)	215°F
Specific gravity at 20/20°C	0.9542
Vapor pressure at 20°C	0.03 mm
Weight per gallon at 20°C	7.94 lbs.

Table 14.47: Boiling Point Composition Curves for Aqueous Diethylenetriamine Solutions *(2)*

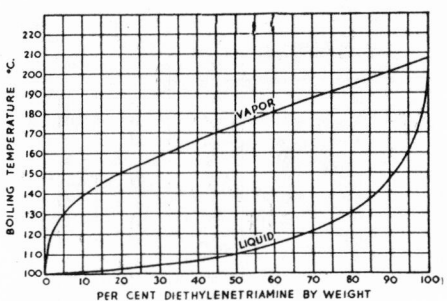

Table 14.48: Tetraethylenepentamine *(2)*

$$NH_2(CH_2CH_2NH)_3CH_2CH_2NH_2$$

Tetraethylenepentamine is a viscous, hygroscopic and high-boiling liquid. It is miscible with water and many organic solvents, and is a solvent for dyes, resins, sulfur, and acid gases. It also forms soaps with fatty acids and it is employed in the synthesis of emulsifiers, plastics, and in rubber reclaiming.

Boiling point (760 mm)	333°C
Flash point	325°F
Specific gravity at 20°C	0.998
Vapor pressure at 20°C	0.01 mm
Weight per gallon at 20°C	8.31 lbs.
Boiling range (760 mm)	320°–360°C

Table 14.49: Propylenediamine *(2)*

1,2-Diaminopropane $CH_3CH(NH_2)CH_2NH_2$

Propylenediamine is a water-white liquid with an ammoniacal odor. It is miscible with water and many organic solvents, among them being benzene and naphtha. It does not form a constant-boiling mixture with water. It is a solvent for such substances as cellulose nitrate, castor oil, shellac, pine oil, copal gum, rosin, and dyes.

It behaves much like ethylenediamine but it is considered superior in solvent power. It is used in the manufacture of gasoline additives.

Boiling point (760 mm)	119.7°C
Flash point (open cup)	120°F
Specific gravity at 20/20°C	0.8732
Solubility in water at 20°C	Complete
Solubility of water in solvent at 20°C	Complete
Vapor pressure at 20°C	9.4 mm
Weight per gallon at 20°C	7.27 lbs.
Boiling range (760 mm)	112–122°C
Purity	80% by wt. min.

Table 14.50: Solvent Properties of Alkylene Diamines *(2)*

	ETHYLENE DIAMINE	PROPYLENE DIAMINE	TRIETHY-LENE TETRAMINE	MORPHO-LINE	MORPHO-LINE ETHANOL	MORPHOLINE ETHYL ETHER
Water	M	M	M	M	M	M
Alcohol	M	M	M	M	M	M
Glycols	M	M	M	M	M	M
Glycol ethers	M	M	M	M	M	M
Acetone	M	M	M	M	M	M
Methyl butyl ketone	S	S	S	M	M	M
Ethyl ether	S	S	S	M	M	M
Butyl ether	SS	S	SS	M	M	M
Naphtha	S	S	SS	S	I	M
Benzene	M	M	S	M	M	M
Turpentine	I	I	I	M	M	M
Pine oil	M	M	M	M	M	M
Paraffin oil	I	I	I	I	I	M
Castor oil	M	M	M	M	M	M
Linseed oil	I	I	I	M	S	M
Paraffin wax	SH	SH	SH	SH	SH	SS
Beeswax	I	I	SH	I	I	SS
Shellac	S	S	S	S	S	S
Rosin	S	S	S	S	SS	S
Ester gum	SS	SS	SS	S	S	S
Dammar gum	I	I	I	PS	PS	S
Copal gum	S	S	S	S	S	S
Sulfur	VS	VS	S	SS	SS	SS
Vinylite A	G	G	G	S	G	SS
Vinylite N	S	S	S	S	S	S
Vinylite 0200	G	G	G	S	G	S
Cellulose acetate	G	G	G	S	S	I
Cellulose nitrate	S	S	S	S	S	S
Benzyl cellulose	SS	SS	SS	S	S	S
Water-sol. dye	S	S	SS	I	I	I
Alcohol-sol. dye	S	S	S	S	S	S
Oil-sol. dye	S	S	S	S	S	S
Satd. brine	M	M	M	M	M	S

M = miscible in all proportions   I = sol. to less than 1%
S = sol. to over 5%   SH = sol. hot
SS = sol. from 1 to 5%   VS = very sol.
PS = sol. in part   G = gels.

Table 14.51: Vapor Pressures of Alkylene Diamines and Other Amines *(19)*

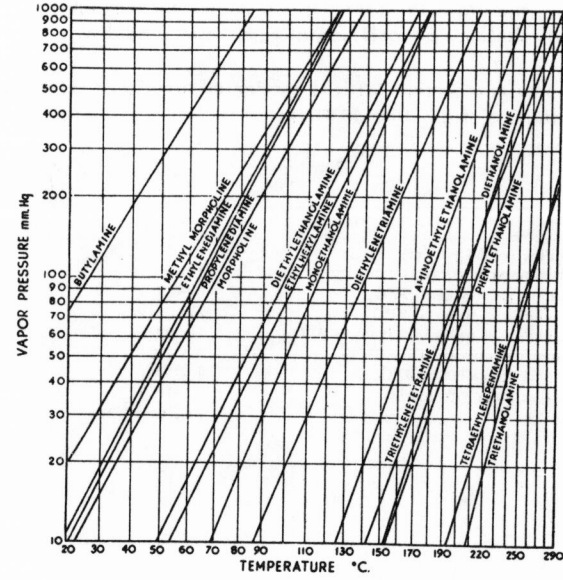

Table 14.52: Density of Ethylenediamine Solutions *(23)*

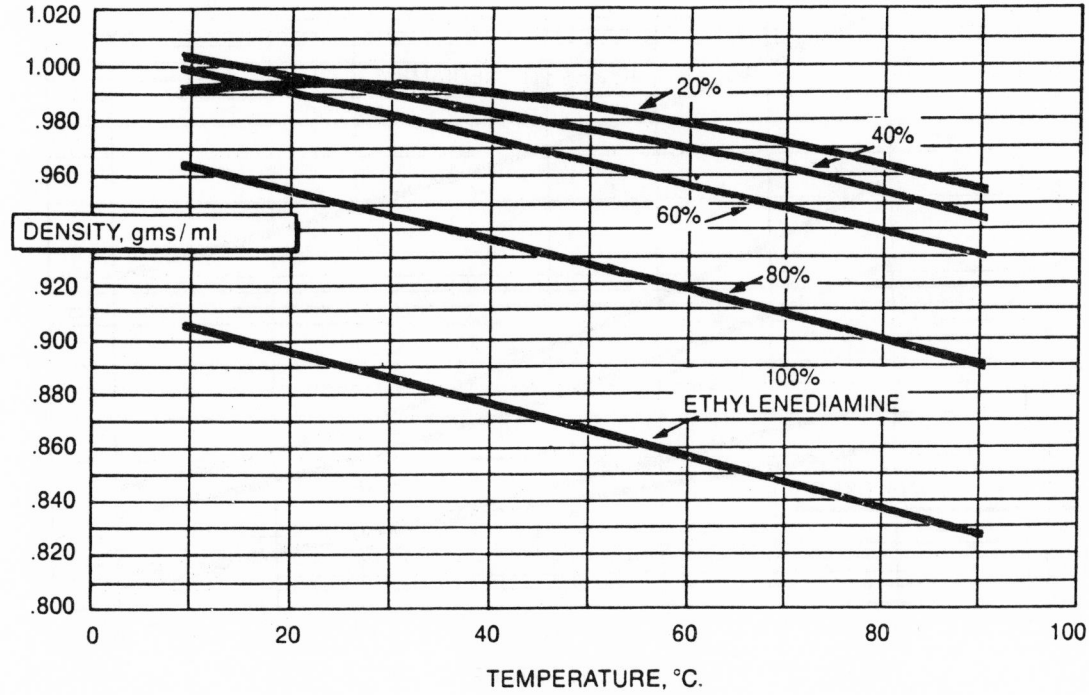

Table 14.53: Density of Higher Ethylene Amines *(23)*

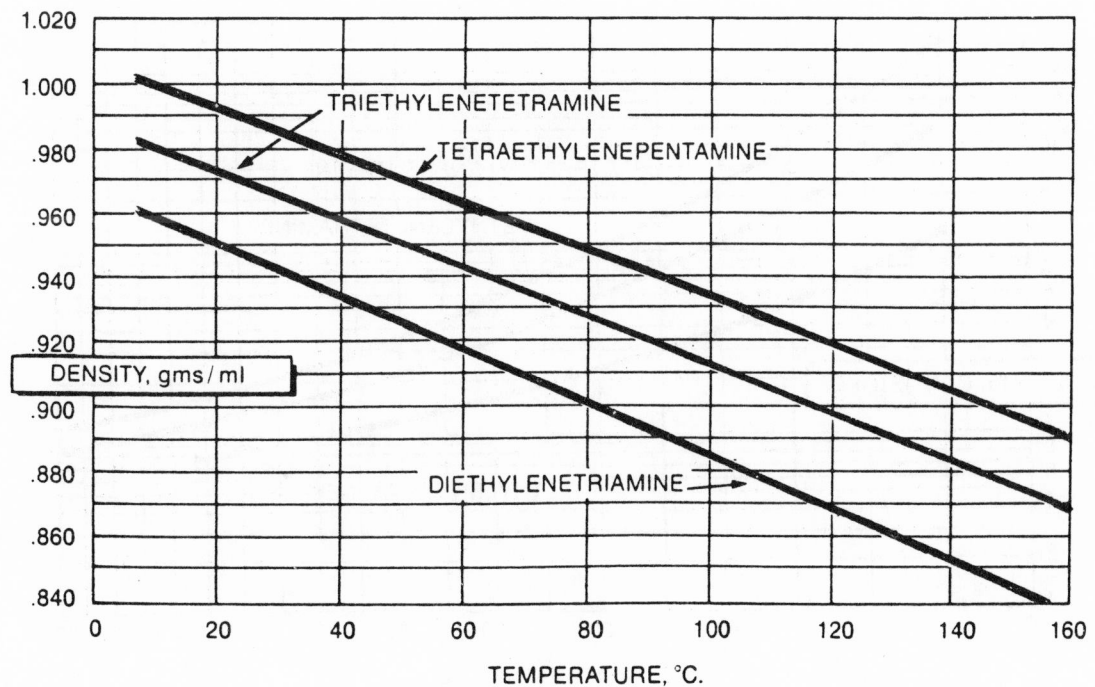

Table 14.54: Viscosity of Ethylenediamine Solutions *(23)*

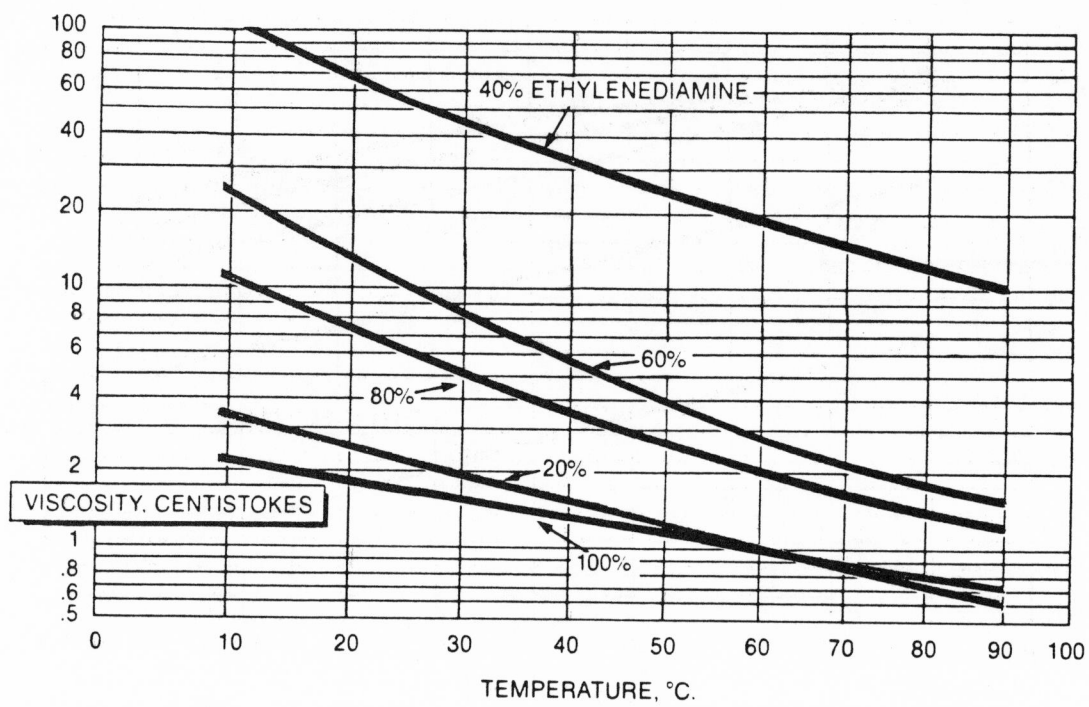

Table 14.55: Viscosity of Higher Ethylene Amines *(23)*

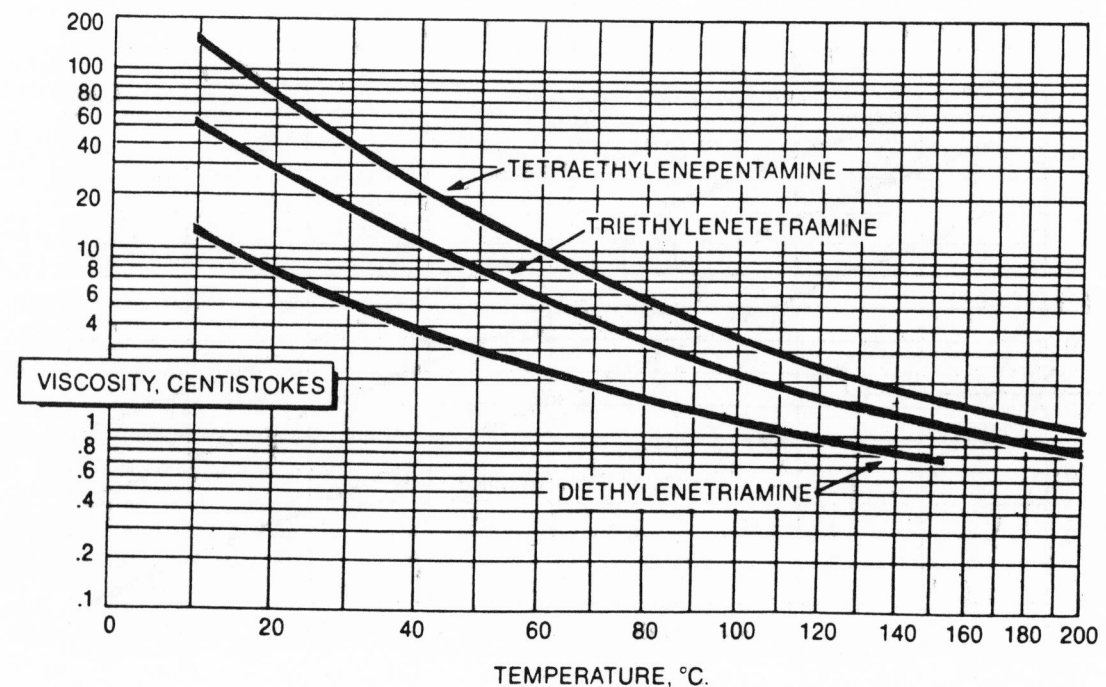

COMPARATIVE DATA

Table 14.56: "ADMA" and "DAMA" Amines *(82)*

ADMA Products Carbon No. Distribution, wt. %	Typical Properties										
	Single Cuts						Example Blends				
	ADMA C8	ADMA C10	ADMA 2	ADMA 4	ADMA 6	ADMA 8	ADMA 24	ADMA 246-621	ADMA 246-451	ADMA WC	ADMA C20 +
C8 Octyl	99									7.0	
C10 Decyl	1	99								6.5	
C12 Dodecyl		1	97	2			65	66	40	53	
C14 Tetradecyl			3	97	1		35	26	50	19	
C16 Hexadecyl				1	97	4		8	10	8.5	
C18 Octadecyl					2	92				6.0	3
C20 Eicosyl						4				<0.5	47
C22 Docosyl											35
C24 Tetracosyl											10
C26 + Hexacosyl											5
Tertiary Amine, wt. %	98	98	98	98	98	98	98	98	98	98	—
Primary & Secondary Amine, wt. %	<0.1	<0.1	<0.1	<0.1	<0.1	<0.1	<0.1	<0.1	<0.1	<0.1	—
Water, wt. %	<0.1	<0.1	<0.1	<0.1	<0.1	<0.1	<0.1	<0.1	<0.1	<0.1	—
Color, APHA	10	10	10	10	10	5	10	10	10	10	—
Amine Value, mg. KOH/g	352	300	258	229	206	187	247	245	237	247	—
Specific Gravity (25°C/25°C)	0.765	0.778	0.778	0.794	0.800	0.807	0.791	0.791	0.792	0.789	—

(ADMA C20 + column is marked "Developmental")

DAMA Products Carbon No. Distribution. wt. %	Typical Properties	
	DAMA 10	DAMA 810
C8-8 Dioctyl	—	25
C8-10 Octyldecyl	1	50
C10-10 Didecyl	99	25
Tertiary Amine, wt. %	97	97
Primary & Secondary Amine, wt. %	<0.1	<0.1
Water, wt. %	<0.1	<0.1
Color, APHA	25	25
Amine Value mg. KOH/g	177	196
Specific Gravity (25°C/25°C)	0.807	0.801

Table 14.57: Amine CS-1135 *(34)*

<u>Typical Properties of AMINE CS-1135</u>

Amine component	78% by wt
Neutral equivalent as a base	120-126
Color, APHA	100 (max)
Flash point, Tag closed cup	120°F
Freezing point	below -20°C
Specific gravity at 25/25°C	0.98-0.99
Viscosity at 25°C	~7.5 cp
pH	10.5-11.5
Weight per U.S. gallon	8.2 lb

Table 14.58: Amsco Amines *(13)*

	Specific Gravity @ 20°/20°C	Pounds per Gallon @ 20°C	Boiling Point @ 760 mm Hg °F (°C)	Vapor Pressure mm Hg @ 20°C	Flash Point TAG CC °F	Solidifi- cation Point °F	Solubility % by wt @ 20°C in Water	of Water
Alkylamines								
Isopropylamine	0.687	5.73	90 (32.4)	478	<0	-139	—	—
Ethylamine-70	0.797	6.69	102 (39.1)	393	<0	-130	—	—
Diethylamine	0.708	5.88	132 (55.5)	194	<0	-58	—	—
Diisopropylamine	0.717	5.97	183 (83.9)	60	12	-94	—	—
Triethylamine	0.729	6.06	193 (89.5)	54	10	-174	5.5	4.6
Alkyleneamines								
Ethylenediamine	0.909	7.52	243 (117)	10	105	52	—	—
Diethylenetriamine	0.954	7.92	404 (206.9)	<1	210	-38	—	—
Alkanolamines								
Monoethanolamine	1.018	8.47	339 (170.4)	0.4	200	51	—	—
Diethanolamine	1.092 (@ 30°/20°C)	9.09 (@ 30°C)	515 (268.4)**	<0.01	280***	82	95.4	—
Triethanolamine	1.126	9.37	636 (335.4)**	<0.01	382***	64	—	—
Hydroxyamine								
AMP-95* (2-amino-2-methyl-1-propanol	0.942 (@ 25°/25°C)	7.78	329 (165)	59 (@ 100°C)	182	28		
Amide								
Dimethyl formamide	0.945 (@ 25°/4°C)	7.86	304 (151.1) 308 (153.3)	5	153	-78	—	—

*IMC Corp.
**With decomposition.
***Open cup.

Table 14.59: "Armeen" Primary Amines (59)

		Primary Amines RNH_2							
	PRODUCT	ARMEEN® 12D	ARMEEN® C	ARMEEN® CD	ARMEEN® 16D	ARMEEN® 18	ARMEEN® 18D	ARMEEN® HT	ARMEEN® HTD
SPECIFICATIONS	Chemical Name	dodecylamine	cocoamine	cocoamine	hexadecylamine	octadecylamine	octadecylamine	(hydrogenated-tallow)amine	(hydrogenated-tallow)amine
	Major Component	$C_{12}H_{25}NH_2$	$C_{12}H_{25}NH_2$	$C_{12}H_{25}NH_2$	$C_{16}H_{33}NH_2$	$C_{18}H_{37}NH_2$	$C_{18}H_{37}NH_2$	$C_{18}H_{37}NH_2$	$C_{18}H_{37}NH_2$
	Primary Amine, %, min	98.0	95.0	98.0	98.0	95.0	97.0	95.0	98.0
	Secondary Amine, %, max *	2.0 (ASA)	}5.0	2.0 (ASA)	2.0 (ASA)	5.0 (ASA)	3.0 (ASA)	}5.0	2.0 (ASA)
	Tertiary Amine, %, max	—	—	—	—	—	—	—	—
	Color, Gardner, max	2	5	2	1	4	2	9	2
	Iodine Value	2.0 max	12.0 max	12.0 max	2.0 max	3.0 max	4.0 max	5.0 max	5.0 max
	Combining Weight, max	189	210	204	246	280	278	275	269
	Amine Value, min	297	267	275	228	200	202	204	209
	Moisture, %, max	1.0	0.5	1.0	0.5	—	0.5	—	1.0
PHYSICAL PROPERTIES	Physical Form @ 25°C	liquid	liquid	liquid	solid	solid	solid	solid	solid
	Melting Range, °F, ASTM	79 to 86	54 to 59	57 to 63	100 to 118	122 to 133	122 to 133	118 to 133	118 to 133
	Pour Point, °F, ASTM	80	45	55	100	115	110	110	100
	Cloud Point, °F, ASTM	90	58	60	105	132	125	115	110
	Flash Point, °F, COC	235	240	230	285	320	300	320	315
	Fire Point, °F, COC	250	270	250	325	365	350	365	360
	Specific Gravity @ 25/4°C	0.801	0.805	0.804	—	—	—	—	—
	@ 38/4°C	0.792	0.796	0.795	0.789	0.792	0.791	0.795	0.794
	@ 60/4°C	0.777	0.781	0.780	0.782	0.785	0.784	0.791	0.790
	@ 70/4°C	0.769	0.774	0.773	—	—	—	—	—
	Viscosity, SSU, @ 25°C	42.2	44.2	43.0	—	—	—	—	—
	@ 35°C	39.2	41.4	40.1	—	—	—	—	—
	@ 45°C	36.4	37.8	37.0	37.5	45.6	43.7	47.5	44.1
	@ 55°C	33.6	34.5	33.8	36.9	43.1	42.2	41.4	38.8
	@ 65°C	—	33.6	32.8	36.6	40.4	39.9	38.8	36.3
	@ 75°C	—	—	—	—	—	—	—	—
CARBON CHAIN DISTRIBUTION	SATURATED								
	Hexyl, C_6	—	—	—	—	—	—	—	—
	Octyl, C_8	—	6	6	—	—	—	—	—
	Decyl, C_{10}	1	7	7	—	—	—	—	—
	Dodecyl, C_{12}	95	51	51	—	—	—	—	—
	Tetradecyl, C_{14}	3	19	19	TR	—	—	3.5	3.5
	Pentadecyl, C_{15}	—	—	—	0.5	—	—	0.5	0.5
	Hexadecyl, C_6	1	9	9	91	9	9	31	31
	Heptadecyl, C_{17}	—	—	—	1.5	2	2	1	1
	Octadecyl, C_8	—	2	2	7	87	87	61	61
	UNSATURATED**								
	Tetradecenyl, C_{14}'	—	—	—	—	—	—	—	—
	Hexadecenyl, C_{16}'	—	—	—	—	—	—	—	—
	Octadecenyl, C_{18}'	—	}6	}6	—	}2	}2	}3	}3
	Octadecadienyl, C_{18}''	—			—				

The suffix D indicates products of distilled grade.

*ASA = Apparent Secondary Amine: bracket indicates combined secondary and tertiary amine content.

**Brackets indicate combined unsaturated C_{18} content.

NOTE: Former nomenclature for fatty acid derived alkyl chains indicated was: hexyl = caproyl (C_6); octyl = capryl (C_8); decyl = capryl (C_{10}); dodecyl = lauryl (C_{12}); tetradecyl = myristyl (C_{14}); hexadecyl = palmityl (C_{16}); heptadecyl = margaryl (C_{17}); octadecyl = stearyl (C_{18}); tetradecenyl = myristoleyl (C_{14}'); hexadecenyl = palmitoleyl (C_{16}'); octadecenyl = oleyl (C_{18}') and octadecadienyl = linoleyl (C_{18}'').
Carbon Chain Distribution determined by gas chromatography.

(continued)

Table 14.59: (continued)

PRODUCT	Primary Amines RNH_2						
	ARMEEN® T	ARMEEN® TM-97	ARMEEN® TD	ARMEEN® O	ARMEEN® OL	ARMEEN® OD	ARMEEN® SD
SPECIFICATIONS							
Chemical Name	tallowamine	tallowamine	tallowamine	oleylamine	oleylamine	oleylamine	soyaamine
Major Component	$C_{18}H_{35}NH_2$	$C_{18}H_{35}NH_2$	$C_{18}H_{35}NH_2$	$C_{18}H_{35}NH_2$	$C_{18}H_{35}NH_2$	$C_{18}H_{35}NH_2$	$C_{18}H_{35}NH_2$
Primary Amine, %, min	95.0	97.0	98.0	95.0	95.0	98.0	98.0
Secondary Amine, %, max*	}5.0	3.0	2.0 (ASA)	5.0	}5.0	2.0	2.0 (ASA)
Tertiary Amine, %, max	—	—	—	—	—	0.5	—
Color, Gardner, max	11	3	2	8	8	2	3
Iodine Value	38.0–54.0	38.0–54.0	38.0–54.0	70.0 min	85.0 min	70.0 min	70.0 min
Combining Weight, max	275	270	268	278	278	271	270
Amine Value, min	204	208	209	202	202	207	208
Moisture, %, max	1.0	1.0	1.0	1.0	1.0	1.0	1.0
PHYSICAL PROPERTIES							
Physical Form @ 25°C	paste	paste	paste	paste/liquid	liquid	paste/liquid	paste
Melting Range, °F, ASTM	93 to 104	93 to 104	90 to 100	—	—	—	81 to 86
Pour Point, °F, ASTM	70	70	70	—	60 max	—	70
Cloud Point, °F, ASTM	100	100	102	—	70 max	—	85
Flash Point, °F, COC	315	315	305	320	310	310	306
Fire Point, °F, COC	355	355	345	360	350	350	345
Specific Gravity @ 25/4°C	—	—	—	—	—	—	—
@ 38/4°C	0.813	0.813	0.812	0.820	0.819	0.819	0.810
@ 60/4°C	0.798	0.798	0.797	0.804	0.803	0.803	0.795
@ 70/4°C	0.791	0.791	0.790	0.796	0.795	0.795	0.789
Viscosity, SSU, @ 25°C	—	—	—	57.0	56.6	56.6	—
@ 35°C	47.0	47.0	45.2	47.6	46.4	46.4	46.2
@ 45°C	44.4	44.4	42.4	42.4	42.3	42.3	43.3
@ 55°C	40.6	40.6	40.0	40.4	40.4	40.4	38.5
@ 65°C	38.8	38.8	38.3	—	—	—	—
@ 75°C	37.8	37.8	37.4	—	—	—	—
CARBON CHAIN DISTRIBUTION							
SATURATED							
Hexyl, C_6	—	—	—	—	—	—	
Octyl, C_8	—	—	—	—	—	—	
Decyl, C_{10}	—	—	—	—	—	—	
Dodecyl, C_{12}	—	—	—	0.5	0.5	0.5	
Tetradecyl, C_{14}	3	3	3	1.5	1.5	1.5	
Pentadecyl, C_{15}	0.5	0.5	0.5	—	—	—	
Hexadecyl, C_6	29	29	29	4	4	4	
Heptadecyl, C_{17}	1	1	1	0.5	0.5	0.5	
Octadecyl, C_8	20	20	20	14	8	14	
UNSATURATED**							
Tetradecenyl, $C_{14'}$	0.5	0.5	0.5	0.5	0.5	0.5	
Hexadecenyl, $C_{16'}$	2	2	2	4	4	4	
Octadecenyl, $C_{18'}$	}44	}44	}44	70	74	70	
Octadecadienyl, $C_{18''}$				5	7	5	

The suffix D indicates products of distilled grade.

*ASA = Apparent Secondary Amine; bracket indicates combined secondary and tertiary amine content.

**Brackets indicate combined unsaturated C_{18} content.

NOTE: Former nomenclature for fatty acid derived alkyl chains indicated was: hexyl (C_6); octyl = capryl (C_8); decyl = capryl (C_{10}); dodecyl = lauryl (C_{12}); tetradecyl = myristyl (C_{14}); hexadecyl = palmityl (C_{16}); heptadecyl = margaryl (C_{17}); octadecyl = stearyl (C_{18}); tetradecenyl = myristoleyl ($C_{14'}$); hexadecenyl = palmitoleyl ($C_{16'}$); octadecenyl = oleyl ($C_{18'}$) and octadecadienyl = linoleyl ($C_{18''}$). Carbon Chain Distribution determined by gas chromatography.

Table 14.60: Ashland Amines *(69)*

	Specific Gravity 20° / 20°C	Dist. Range °F IBP	Dist. Range °F DP	Flash Pt. °F TOC		Specific Gravity 20° / 20°C	Dist. Range °F IBP	Dist. Range °F DP	Flash Pt. °F TOC
Diethanolamine	1.0919 (30°C)	Decomposes		280	Monoethanolamine	1.0179	331	345	200
Diethylamine	0.7079	129	140	below 0	Monoisopropanolamine	0.9619	316	329	165
Diethylenetriamine	0.955	383	419	215	Morpholine	1.0020	259	266	100
Diisopropanolamine	0.992 (40°C)	486 (5 mm)	—	200	Triethanolamine	1.1258	675	685	375
Diisopropylamine	0.718	181	184	21	Triethylamine	0.729	185	196	20
Ethylamine, 70%	0.802	102	—	below 0	Triethylenetetramine	0.980	500 (5 mm)	554	290
Ethylenediamine, 99%	0.900	239	246	104	Triisopropanolamine	1.010 (40°C)	581 (dec.)	—	320
Isopropylamine	0.689	86	93	below 0					

Table 14.61: Dow Ethylene Amines *(23)*

Property	Ethylenediamine EDA	Diethylenetriamine DETA	Triethylenetetramine TETA[a]	Tetraethylenepentamine TEPA[a]	Ethylene Amine E-100[a] PEHA + Heavier Ethylene Amines
Formula	$H_2NCH_2CH_2NH_2$	$H_2N(CH_2CH_2\overset{H}{N})_2H$	$H_2N(CH_2CH_2\overset{H}{N})_3H^b$	$H_2N(CH_2CH_2\overset{H}{N})_4H^b$	$H_2N(CH_2CH_2\overset{H}{N})_nH^b$ $N>4$
Molecular Weight	60.1	103.2	146.2[b]	189.3[b]	309[c]
Boiling Point, °C	117	206.7	IBP-DP[d] 260-290	IBP-DP @ 5 Torr[d] 155-210	—
Freezing Point, °C	11	−35	−39	<-40	<-40
Viscosity, cst, 25°C	1.56	6.00	20.0	52.5	370
Vapor Pressure, mm Hg, 20°C	9.3[e]	0.15[e]	<0.01	<0.01	<0.01
Specific Gravity, 20/20°C	0.893-0.906	0.950-0.954	0.979-0.984	0.992-0.994	1.00—1.03
Heat of Vaporization Btu/lb, 760 mm	289	214	160	144	—
Refractive Index, 25°C	1.455	1.483	1.496	1.503	—
Pounds/gallon, 25°C	7.45	7.91	8.13	8.26	8.46
% Nitrogen Content	46.5	40.3	36.9	35.5	32.8
Amine Value	1860	1610	1450	1350	1170
pK$_b$	$_pK_1 = 3.83$ $_pK_2 = 6.56$	$_pK_1 = 3.89$ $_pK_2 = 4.60$	$_pK_1 = 3.84$ $_pK_2 = 4.47$	—	—
pH, 25% aqueous solution	12.8	12.5	12.4	12.0	12.0
Heat of Formation, $\Delta H°_f$, 298°K kcal/mol	−15.1	−19.2	−23.3	−27.5	—
Electrical Conductivity Mhos/cm, 25°C	—	4.5×10^{-7}	4.4×10^{-8}	1.3×10^{-8}	—
Thermal Conductivity cal/cm-sec-°C $\times 10^{-5}$, 30°C	56.8	50.1	46.3	44.5	—
Flash Point °F	104 TCC[f]	208 PMCC[g]	245 PMCC	310 PMCC	330 PMCC
Fire Point °F	100 COC[h]	215 COC	305 COC	365 COC	—
Flammability Limits (by vol. of air)	2.5-12.0%[i]	2.0-6.7%[j]	—	—	—

[a] Linear, branched, and cyclic isomeric mixtures
[b] Linear isomer
[c] Weight average molecular weight
[d] Distillation boiling point range, initial boiling point to dry point.
[e] Extrapolated values
[f] Tag closed cup
[g] Pensky-Marten closed cup
[h] Cleveland open cup
[i] 100°C
[j] 150°C

Table 14.62: "Duomeen" Aliphatic Diamines *(59)*

TYPICAL PROPERTIES

Duomeen®	Physical Form @ 25°C	Approximate Melting Range, °F	Specific Gravity @ 40/25°C	Specific Gravity @ 60/25°C	Specific Gravity @ 70/25°C	Specific Gravity @ 80/25°C	Viscosity @ 25°C, cp	Viscosity @ 30°C, cp	Viscosity @ 35°C, cp	Typical Molecular Weight	Flash Point, °F, COC
C	Liquid	54-59	∼0.836	0.810	0.803	0.796	10	—	—	276	295
C Special	Liquid	80	∼0.836	—	—	—	—	—	—	276	295
T	Paste	80-88	—	0.821	0.813	0.806	880	70	10	350	370
T Special	Paste	111-118	—	0.821	0.813	0.806	880	70	10	330	370
O	Liquid	46-54	0.831	0.817	0.810	0.804	19	17	14	350	363
S	Pasty liquid	48-68	0.840	0.826	0.820	0.814	70	40	10	350	∼350

SPECIFICATIONS

Duomeen®	Chemical Name	Iodine Value	Apparent Diamine Activity, % (min)	Primary Amine, % (min)	Secondary Amine, % (min)	Color, Gardner (max)	Moisture, % (max)	Combining Weight (max)	Amine Number (min)
C	N-coco-1,3-diaminopropane	15 max	89	43	43	12	1	145	387
C Special*	N-coco-1,3-diaminopropane	5-15	95	45	45	7	1	139	404
T	N-tallow-1,3-diaminopropane	32-50	89	43	43	12	1	180	312
T Special	N-tallow-1,3-diaminopropane	30 min	95	45	45	12	1	168	334
O	N-oleyl-1,3-diaminopropane	60 min	89	43	43	11	1	180	312
S	N-soya-1,3-diaminopropane	60 min	89	43	43	12	1	180	312

*Prepared from distilled feedstock.

Table 14.63: "Jeffamine" Polyoxypropyleneamines *(48)*

SELECT PROPERTIES	JEFFAMINE Diamines D-230	JEFFAMINE Diamines D-400	D-2000	JEFFAMINE Triamine T-403
Total acetylatables, meq./g.	8.75	5.17	1.01	6.75
Total amines, meq./g.	8.45	4.99	0.96	6.45
Primary amines, meq./g.	8.30	4.93	0.95	6.16
Color, Pt-Co	30	50	100	10
Flash point COC, °F	256	347	460	380
PMCC, °F	256	330	395	340
pH, 5% aqueous	11.7	11.6	10.5	11.6
Specific gravity, 20/20°C	0.9480	0.9702	0.9964	0.9812
Viscosity, 20°C, cs.	14.4	30	344	97
Water, wt. %	0.10	0.13	0.10	0.08

Table 14.64: "Kemamine" Amines *(26)*

Primary amines.

Product	Description (CTFA adopted name)	% Primary Min	Total amine value Min	Color Gardner, 1963 Max	% H2O Max	Iodine value	Typical carbon chain composition						
							Saturated			Unsaturated			
							C14	C16	C18	C18:1	C18:2	C18:3	C20:1
Kemamine P-970	Technical Hydrogenated Tallow (*Hydrogenated Tallow Amine*)	93	205	3	0.5	3 Max	4	29	67				
Kemamine P-970D	Distilled Hydrogenated Tallow (*Hydrogenated Tallow Amine*)	97	210	1	0.5	3 Max	4	29	67				
Kemamine P-974D	Distilled Tallow (*Tallow Amine*)	97	210	1	0.5	38 Min	4	29	25	38	4		
Kemamine P-989D	Distilled Oleyl (*Oleamine*)	97	205	1	0.5	70 Min	4	14	10	65	7		
Kemamine P-990D	Distilled Stearyl (*Stearamine*)	97	205	1	0.5	2 Max		10	90				
Kemamine P-999	Technical Vegetable	93	200	5	0.5	85 Min		15	6	55	15	5	4

1,3-Propylenediamines.

Product	Description	% Diamine	Amine values				Color Gardner, 1963 Max	% H2O Max	Iodine value
			1°	2°	3°	Total			
Kemamine D-190	N-90% Arachidyl-Behenyl	85 Min	150 Typ	120 Typ	10 Typ	275 Min	9	0.5	5.0 Max
Kemamine D-999	N-Vegetable	85 Min	150 Typ	140 Typ	10 Typ	300 Min	8	0.5	70 Min

Tertiary amines.

Product	Description (CTFA adopted name)	% Tertiary Min	Total amine value, Min	Color Gardner, 1963 Max	% H2O Max	Typical carbon chain composition							
						Saturated						Unsaturated	
						C14	C16	C18	C20	C22	C24	C18:1	C18:2
Kemamine T-2801	Methyl Dibehenyl (*Dibehenyl Methylamine*)	95	85	2	0.5		2	5	10	80	3		
Kemamine T-2802D	Distilled Dimethyl Behenyl (*Dimethyl Behenamine*)	95	150	1	0.5		2	5	10	80	3		
Kemamine T-9701	Methyl Di-Hydrogenated Tallow	95	103	3	0.5	4	29	67					
Kemamine T-9742D	Distilled Dimethyl Tallow (*Dimethyl Tallow Amine*)	95	180	1	0.5	4	29	25				38	4
Kemamine T-9902D	Distilled Dimethyl Stearyl (*Dimethyl Stearamine*)	95	180	1	0.5		10	90					
Kemamine T-9992D	Distilled Dimethyl Veg. (*Dimethyl Veg. Amine*)	95	180	2	0.5		18	7				55	20

Table 14.65: Pennwalt Amines (37)

Typical Properties

Name	Formula	Molecular Weight	Distillation Range, °C	Amine Content % Minimum	Specific Gravity @ 20°C/20°C	Refractive Index @ 20°C	Viscosity Centipoises	Vapor Pressure mm Hg @20°C	Heat of Vaporization (cal/gram)	Heat of Combustion kcal/mol	Flash Point TAG Closed Cup °F	Average Weight Lbs/gal
Monoethylamine*	$C_2H_5NH_2$	45.1	16.6(b.p.)	99.5	0.70-0.71 (0°/20°C)	1.3688	0.437@-33.5°C	950	145.9	413	-52	5.9
Diethylamine	$(C_2H_5)_2NH$	73.1	54.0-58.0	99.5	0.70-0.71	1.3873	0.32@20°C	180	91	731	-6	5.9
Triethylamine	$(C_2H_5)_3N$	101.2	85.0-91.0	99.5	0.72-0.73	1.3994	0.42@0°C	52	74.8	1042	22	6.0
n-Propylamine	$C_3H_7NH_2$	59.1	45.0-52.0	99.5	0.71-0.72	1.3861	0.353@25°C	250	110.3	558.3	-22	6.0
Di-n-propylamine	$(C_3H_7)_2NH$	101.2	105.0-112.0	99.5	0.73-0.74	1.4024	0.54@20°C	22	75.7	1055	53	6.2
Tri-n-propylamine	$(C_3H_7)_3N$	143.3	150.0-158.0	98.5	0.75-0.76	1.4148	0.89@0°C	2.9	61.6	1589	84	6.3
Monoisopropylamine**	$(CH_3)_2CH\ NH_2$	59.1	30.5-34.5	99.5	0.68-0.70	1.3722	0.36@25°C	470	108.4	557	-29	5.8
Diisopropylamine	$[(CH_3)_2CH]_2NH$	101.2	84 (b.p.)	99.5	0.71-0.72	1.3906	0.40@20°C	50	72.2	1113	16	6.0
n-Butylamine	$C_4H_9NH_2$	73.1	75.0-81.0	99.5	0.73-0.74	1.401	0.67@-25°C	68	99.1	710	25	6.2
Di-n-butylamine	$(C_4H_9)_2NH$	129.2	154.5-162.0	99.5	0.75-0.77	1.4160	0.89@25°C	4	69.9	1350	108	6.3
Tri-n-butylamine	$(C_4H_9)_3N$	185.4	200.0-216.0	98.5	0.77-0.78	1.4277	1.33@25°C	2.6	55.3	1949	156	6.5
Sec.-butylamine	$CH_3CH(NH_2)CH_2CH_3$	73.1	63.0-70.0	98.5	0.72-0.73	1.395	0.44@25°C	190	99.5	715	18	6.0
Monoamylamine	$C_5H_{11}NH_2$	87.2	87.0-110.0	98.0	0.74-0.76	1.4100	0.67@25°C	32	99.1	867	52	6.3
Diamylamine	$(C_5H_{11})_2NH$	157.3	185.0-205.0	98.5	0.76-0.78	1.4235	1.26@25°C	3	83.0	1660	152	6.4
Triamylamine	$(C_5H_{11})_3N$	227.2	235.0-265.0	98.0	0.77-0.79	1.4350	2.42@25°C	1	79.0	2459	196	6.6
Ethylbutylamine	$(C_2H_5)NH(C_4H_9)$	101.2	113 (b.p.)	99.0	0.73-0.74	1.4090	-	21	79.1	719	56	6.1
Cyclohexylamine	$C_6H_{11}NH_2$	99.2	134.0 (b.p.)	98.0	0.86-0.87	1.456	-	95	87.7	996	88	7.2

* Also available as a 70% aqueous solution
** Also available as a 50% aqueous solution

(continued)

Table 14.65: (continued)

Typical Properties of Alkyl Alkanol Amines

Pennwalt Code Number	Name	Molecular Formula	Viscosity @ 20°C Centipoise	Molecular Weight Calculated	Distillation Range °C (°F)	Specific Gravity 20/20°C	Vapor Pressure at 20°C mm Hg	Freezing Point °C	Flash Point TAG Closed Cup °F	Ionization Constant Kb x 10	Heat of Vaporization cal/gram	Average Weight lbs./gal.
02178	Methylethanolamine/MEA/ Methylaminoethanol/MAE	$CH_3-N-CH_2CH_2OH$ H	13	75.1	152-162 (306-324)	0.93-0.94	0.5	-4.5	160	6.3		7.8
02146	Methyldiethanolamine/ MDEA	$CH_3N(CH_2CH_2OH)_2$ $C_5H_{13}NO_2$	101	119.2	240-255 (464-291)	1.035-1.050	0.7 @ 80	-21	240		123.9	8.7
00272	Dimethylaminoethanol/DMAE/ Dimethylethanolamine/DMEA	$(CH_3)_2NCH_2CH_2OH$ $C_4H_{11}NO$	3.6	89.1	130.5-136.5 267-278	0.883-0.888	4.0	-59	104	0.63	94.8	7.4
01111	Ethylaminoethanol/EAE/ Ethylethanolamine/EEA	$CH_3CH_2NHCH_2CH_2OH$ $C_4H_{11}NO$	12.4 @ 25°C	89.1	160-170 (320-338)	0.91-0.92	<1	-9	162	96	103.0	7.5
00161	Ethylaminoethanols (Mixed)/ EAE and EDEA	$CH_3CH_2NHCH_2CH_2OH$ and $CH_3CH_2N(CH_2CH_2OH)_2$ $C_4H_{11}NO$ & $C_6H_{15}NO_2$	60	89.1 & 133.2 109-119 combining wts.	104-265 (219-509)	0.96-0.98	<1		180	4.0		8.1
00129	Diethylaminoethanol/DEAE Diethylethanolamine/DEEA	$(CH_3CH_2)_2NCH_2CH_2OH$ $C_6H_{15}NO$	3.5	117.2	158-163.5 (316-326)	0.88-0.89	1	-56	125	5.0	77.9	7.4
00287	Diisopropylaminoethanol/DIPAE Diisopropylethanolamine/DIPEA	$[(CH_3)_2CH]_2NCH_2CH_2OH$ $C_8H_{17}NO$	25	145.2	188-192 (370-378)	0.873-0.878	<1	-39	153		74.4	7.3
00270	Isopropylaminoethanols (Mixed)/ IPAE and IPDEA	$(CH_3)_2CHNHCH_2CH_2OH$ and $(CH_3)_2CHN(CH_2CH_2OH)_2$ $C_5H_{13}NO$ & $C_7H_{17}NO_2$	50.8	103.2 & 147.2	110-265 (230-509)	0.91-0.94	2	-50 (glassy)	168	158		7.8
01029	Butyldiethanolamine	$CH_3CH_2CH_2CH_2N(CH_2CH_2OH)_2$ $C_8H_{19}NO_2$	8	161.2	265-287 (509-549)	0.96-0.97	1 @ 25°C		295			8.1
00156	Dibutylaminoethanol/DBAE/ Dibutylethanolamine/DBEA	$(C_4H_9)_2NCH_2CH_2OH$ $C_{10}H_{23}NO$	6.5 @ 25°C	173.3	222-234 (432-453)	0.855-0.865	0.1	-75 (glassy)	195	5.0		7.2
02445	Dimethylamino-2-propanol/DMA-2-P/ N,N-Dimethylisopropanolamine/DMIPA anhydrous*	$(CH_3)_2NCH_2CHOHCH_3$ $C_5H_{13}NO$	1.5	103.2	121-127 (250-261)	0.845-0.855	8	-85	82 (77%-99°) (70%-104°)	190	60.9	7.1
01480	Diethylaminoethoxyethanol	$(C_2H_5)_2N(C_2H_4O)CH_2CH_2OH$ $C_8H_{15}NO(C_2H_4O)_x$	11.1			0.950-0.970	<1	<-60	201			8.0
01566	N-Propylaminoethanol	$CH_3CH_2NHCH_2CH_2OH$ $C_5H_{13}NO$	13.5	103.2	173-182 (343-360)	0.895-0.905	4 @ 75°C	-10	190			7.5

* also available in 77% and 70% aqueous concentrations

Table 14.66: Union Carbide Amines (19)

Product	Formula	Formula Molecular Weight	Purity of Tested Sample, % by wt.	Apparent Specific Gravity, 20/20°C.	Boiling Point, °C., 760 mm.	Vapor Pressure, mm. Hg at 20°C.	Freezing Point, °C.	Solubility, % by weight at 20°C. In Water	Solubility Water In	Pounds Per Gal. at 20°C.	Flash Point, °F. (a)
ALKYLAMINES											
Ethylamine, 70%	$C_2H_5NH_2$	45.09	(c)	0.797	39.1	393	....	Complete		6.69	<0
Ethylamine, anhydrous	$C_2H_5NH_2$	45.09	100	0.6837	16.6	873	-81.0	Complete		5.69	<0
Diethylamine	$(C_2H_5)_2NH$	73.14	99.4	0.7079	55.5	194	-49.8	Complete		5.88	<0
Triethylamine	$(C_2H_5)_3N$	101.19	100	0.7290	89.5	54	-114.7	5.5soc.	4.6soc.	6.06	17
Isopropylamine	$(CH_3)_2CHNH_2$	59.11	(e)	0.6873	32.4	478	-95.2	Complete		5.73	<0
Diisopropylamine	$[(CH_3)_2CH]_2NH$	101.19	100	0.7173	83.9	60	-96.3	Complete (k)		5.97	8
ALKYLENEAMINES											
Ethylenediamine, 99%, HP	$H_2NC_2H_4NH_2$	60.10	99	0.9089	117	10	11	Complete		7.52	105
Diethylenetriamine, HP	$(H_2NC_2H_4)_2NH$	103.17	99	0.9542	206.9	<1	-39	Complete		7.92	210
Triethylenetetramine (c)	$H_2N[C_2H_4NH]_3H$ and isomers	146.24	(c)	0.9818	277.4	<0.01	-35	Complete		8.17	275
Tetraethylenepentamine (c)	$H_2N[C_2H_4NH]_4H$ and isomers	189.30	(c)	0.9980	340.	<0.01	-35	Complete		8.33	280
Polyamine HPA #2	(n)	260.app.	99	1.0-1.10	Decomposes	<1	(m)	Complete		8.63	310
Dimethylaminopropylamine	$(CH_3)_2NC_3H_6NH_2$	102.18	99	0.8183	134.9	5	-100(d)	Complete		6.80	94
N,N,N',N'-Tetramethyl-1,3-Butanediamine	$CH_3CHN(CH_3)_2C_2H_4N(CH_3)_2$	144.26	99	0.8020	165.1	2	-100(d)	Complete		6.67	113
ALKANOLAMINES											
Monoethanolamine	$HOC_2H_4NH_2$	61.08	99.9	1.0179	170.8	<1	10.5	Complete		8.47	185
Diethanolamine	$(HOC_2H_4)_2NH$	105.14	99	1.0919 (30/20°)	268.4 (q)	<0.01	28.0	96.4	.....	9.09soc.	336P
Triethanolamine, 99%	$(HOC_2H_4)_3N$	149.19	99	1.1258(p)	335.4	<0.01	21.6(f)	Complete		9.37	407P
N-Methyl Ethanolamine	$CH_3NHC_2H_4OH$	75.11	99	0.9414	159.6	1	-4.5	Complete		7.83	165C
N,N-Dimethyl Ethanolamine	$(CH_3)_2NC_2H_4OH$	89.14	99.7	0.8879	134.6	4	-59.0	Complete		7.39	103
N,N-Diethyl Ethanolamine	$(C_2H_5)_2NC_2H_4OH$	117.19	99	0.8851	162.1	1	....	Complete		7.36	120
N-Aminoethylethanolamine, HP	$H_2NC_2H_4NHC_2H_4OH$	104.15	99.7	1.0304	243.8	<0.01	(l)	Complete		8.58	265
N-Methyl Diethanolamine	$CH_3N(C_2H_4OH)_2$	119.16	99	1.0418	247.3	<0.1	-21.0	Complete		8.67	276P
Polyglycolamine H-163	$HO[C_2H_4O]_2C_3H_6NH_2$	163.22	99.9	1.0548	Decomposes	<0.01	14.1	Complete		8.79	331P
Polyglycoldiamine H-221	$(H_2NC_3H_6OC_2H_4)_2O$	220.32	98	1.0075	Decomposes	<0.01	-33.1	Complete		8.41	261P
PIPERAZINES											
Piperazine (q)	$HNCH_2CH_2NHCH_2CH_2$	....	(c)	1.005 (60/20°)	146.0	11.5 25c.	49.2	Complete		8.29 65c.	None(r)
N-Aminoethyl Piperazine, HP	$H_2NC_2H_4NCH_2CH_2NHCH_2CH_2$	129.21	99.7	0.9837	222.0	<0.1	-17.6	Complete		8.20	215

(a) Unless otherwise specified with a letter-suffix, flash point was determined by ASTM method D 56 using Tag closed cup. Flash point suffixes indicate the following : "C" = determined by ASTM method D 92 using Cleveland open cup; "T" = determined by ASTM method D 1310 using Tag open cup; "P" = determined by ASTM method D 93 using Pensky-Martens closed cup; "S" = determined by ASTM method D 3278 using Seta-Flash closed cup.

(c) Typical commercial material.

(d) Sets to glass below this temperature.

(e) 99+ mol per cent material.

(f) Supercools easily.

(k) Not miscible above 27°C. and 30% amine by weight.

(l) Pour point, -19° F.

(m) Pour point, 15-18°C.

(n) Residue mixture ($C_{10}N_6$ and higher mole weight amines).

(p) Supercooled liquid.

(q) Technical Grade, 60% water solution.

(r) None by closed cup: 172°F. by Tag open cup.

ALKANOL AMINES

The most important members of this group from a commercial standpoint are monoethanolamine, diethanolamine, and triethanolamine. Also available in commercial quantities are the aminohydroxy derivatives of nitroparaffins, which are 2-amino-1-butanol, 2-amino-2-methyl-1-propanol, 2-amino-2-methyl-1,3-propanediol, 2-amino-2-ethyl-1,3-propanediol, and tris(hydroxymethyl)aminomethane.

These compounds are used as emulsifiers for cosmetic lotions and creams, mineral oil and paraffin wax emulsions, textiles, leather dressings, cleaning compounds, polishes, and "soluble oils." They are also used in the manufacture of pharmaceuticals, surface-active and wetting agents, vulcanization accelerators, photographic developers, dyestuffs, and resins. Having the property of absorbing acidic gases, such as H_2S and CO_2 in cold aqueous solutions and releasing them when hot, these compounds suggest usefulness in gas recovery and purification. They also form the basis for chemical synthesis.

Table 14.67: Monoethanolamine *(19)*

2(Hydroxyethyl)amine
2-Aminoethanol
Colamine

$H_2NCH_2CH_2OH$

Monoethanolamine is a somewhat viscous hygroscopic liquid with an ammoniacal odor. It is miscible with water and many organic solvents. Its molecule contains both a hydroxyl and an amine group, thus producing derivatives that have characteristics of both types of compounds. It is used as a softener and conditioning agent, and in the recovery and extraction of carbon dioxide and hydrogen sulfide from industrial gases. Its soaps with fatty acids are excellent emulsifiers for waxes. It is also utilized as an intermediate in the manufacture of rubber accelerators and dyestuffs.

Typical Properties and Specifications

Boiling point	172.2°C
Coefficient of expansion at 20°C	0.000770 (per°C)
Dissociation constant at 20°C	5×10^{-5}
Equivalent weight	61 to 63
Flash point (open cup)	93°C (200°F)
Heat of evaporization at B.P.	199 cal/g
Refractive index at 20°C	1.4539
Specific gravity at 20 (20°C	1.0180
Specific heat at 30°C	0.665 cal/g
Surface tension at 20°C	51 dynes/cm
Viscosity at 20°C	3.40 poises
Vapor pressure at 20°C	0.67 mm Hg
Weight per gallon at 20°C	8.472 lbs
Boiling range at 760 mm Not less than 90% over between 165 and 173°C	
Color	Water-white
pH 25% Solution at 25°C	12.1
Solubility in water	Complete

Table 14.68: Boiling Point Composition Curves for Aqueous Monoethanolamine Solutions *(19)*

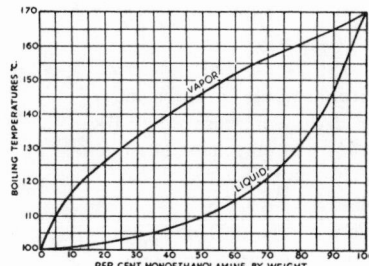

Table 14.69: Viscosity of Monoethanolamine at Various Temperatures *(19)*

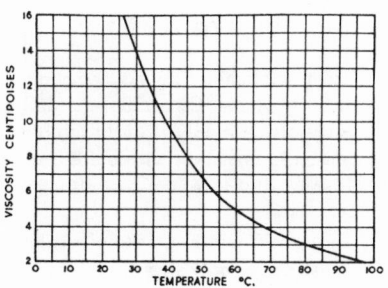

Table 14.70: Diethanolamine *(2)*

Di-2-Hydroxyethylamine   (HOCH₂CH₂)₂NH

When pure, diethanolamine is a crystalline, white solid which has a melting point of 28°C.,
or just above room temperature. The commercial material has a mild, ammoniacal odor.
Like other ethanolamines, diethanolamine enters into reactions characteristic of both amines
and alcohols; its most important property is its ability to combine directly with acids and
acidic gases. At normal temperatures, its aqueous solutions have a strong affinity for hydrogen
sulfide and carbon dioxide; and at higher temperatures, this affinity decreases, with expulsion
of the gases.

Diethanolamine finds wide use as an absorbent for acidic gases; especially for the removal,
recovery, and concentration of carbon dioxide from flue and other waste gases as well as
from hydrogen gas produced by cracking methane. Many industrial processes require pure
hydrogen free of acidic gases. Diethanolamine is used to remove troublesome hydrogen sulfide from sour natural gas in transmission lines and natural gasoline plants. It is also used as
a softening, moistening, and emulsifying agent; and in the synthesis of organic compounds by
esterification of its hydroxyl groups.

It is an excellent agent for neutralizing the acidity which is developed by the high percentage
of clays used in rubber compounding, and thus reduces the curing time considerably. It is also
used in the production of powerful synthetic detergents and in certain synthetic resins.

> Color and properties: Faintly colored, viscous liquid.
> Constants: Sp.gr. 1.0985 at 20°C./20°C.; b. p. (760 mm.) 268.0°C.; vapor pressure
> < 0.01 mm. (20°C.); flash point 280°F.; wt. 91. lbs./gal. (20°C.). *Typical speci-*
> *fications:* Sp.gr. 1.088 to 1.095 at 30°C./20°C.; water not more than 1.5%;
> monoethanolamine not more than 2%; diethanolamine not less than 95%;
> triethanolamine not more than 2%; color (100-mm. tube) not more than 3
> yellow and 1 red Lovibond; equivalent wt. 104 to 108; average wt. 9.08 lbs./gal.
> (30°C.).
> Miscible with water and most organic solvents.

Table 14.71: Viscosity of Diethanolamine at Various Temperatures *(19)*

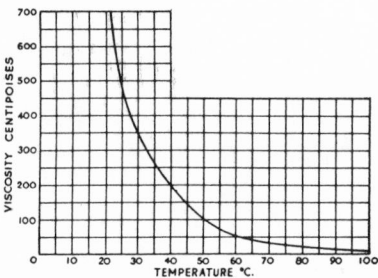

Table 14.72: Triethanolamine *(2)*

Tri-2-Hydroxyethylamine   (HOCH₂CH₂)₃N

Triethanolamine is a viscous and very hygroscopic liquid with a slight ammoniacal odor.
It boils at 244°C. at 50 mm. (360°C. at 760 mm.) and is entirely soluble in water and
alcohols, but only slightly soluble in hydrocarbons. It is a mild, organic base which like
ammonia combines with acids and acidic materials. The alkalinity of pure triethanolamine
is somewhat less than that of ammonia, its pH being 11.2 in 25% solution.

Three commercial grades of triethanolamine are available: 98%, "regular", and "SP."
These differ only slightly in physical and chemical properties from the pure compound. The
most significant variation is in equivalent weights. Pure triethanolamine has an equivalent
weight of 149; "regular," 140; and "SP," about 130. This variation is due to increasing
amounts of mono- and diethanolamine present in the respective commercial grades.   (continued)

Table 14.72: (continued)

With free fatty acids, triethanolamine forms soaps in direct molecular proportions. Triethanolamine oleate is a semi-liquid soap capable of forming solutions of marked detergent properties in water or in organic solvents such as gasoline. In water, triethanolamine oleate is soluble in all proportions; in gasoline, more than 2% soap is necessary to effect solution. The stearate is a hard, white product which finds use in cosmetic preparations. Only the 98% or regular grades should be used in cosmetic products. These soaps are practically neutral, their pH being approximately 8, and are thus free from irritating effect upon the skin or from injurious effect on fabrics. Very stable water emulsions of almost any oil, fat, or wax can in general be prepared with these soaps. The usual requirements for emulsification are between 2 and 4% triethanolamine and 5 to 15% oleic or stearic acid, each based on the weight of the oil to be emulsified. Triethanolamine emulsions are distinguished by their small particle size, non-corrosiveness, non-volatility, ease of preparation, and wide flexibility in formulation with fear of separation.

A small percentage of triethanolamine assists in the penetration of liquids into porous materials. Because of its pronounced hygroscopicity, it is employed as an economical softening agent, humectant, and plasticizing agent for such products as textiles, glues, leather coatings, as a penetrating agent in impregnating wood, paper, and cellulose products. Also, an ingredient of adhesives, rubber mixtures, and lacquers.

> Viscous, pale yellow liquid intermediate in properties between alcohol and ammonia; slightly ammoniacal odor; excellent penetrating properties; forms soaps with fatty acids; hygroscopic. Commercial product contains 70–75% triethanolamine, 20–25% diethanolamine, 0–5% monoethanolamine. Soluble in water, alcohol and chloroform. Sp.gr. 1.1204–1.1284; b.p. 360°C; vapor pressure < 0.01 mm (20°C); flash point 355°F; wt./gal. 9.4 lbs. (20°C); coefficient of expansion 0.00048 (20°C); freezing point 21.2°C; viscosity 0.10 poise (20°C).
> *Typical specifications:* Sp. gr. 1.1240–1.1300 (20/20°C); water not more than 1.0%; purity not more than 2.5% monoethanolamine, not more than 15% diethanolamine, not less than 80% triethanolamine; equivalent wt. 140–145; color (500-mm. tube) not more than 7 yellow and 2 red Lovibond; average wt./gal. 9.40 lbs. (20°C).

Table 14.73: Viscosity of Triethanolamine at Various Temperatures *(19)*

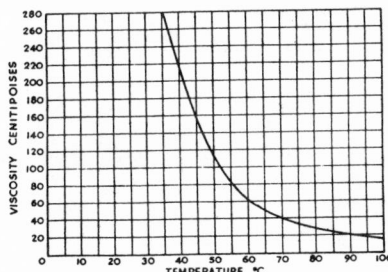

Table 14.74: Specific Heats of Aqueous Triethanolamine Solutions at 21°C *(19)*

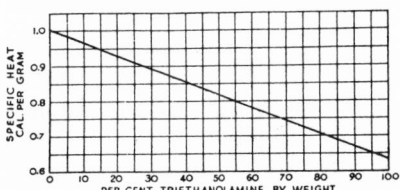

Table 14.75: Surface Tension of Aqueous Ethanolamine Solutions at 20°C *(19)*

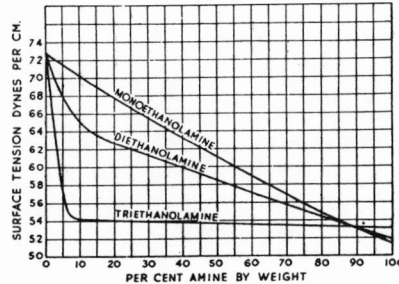

Table 14.76: Viscosity of Aqueous Ethanolamine Solutions at 20°C *(19)*

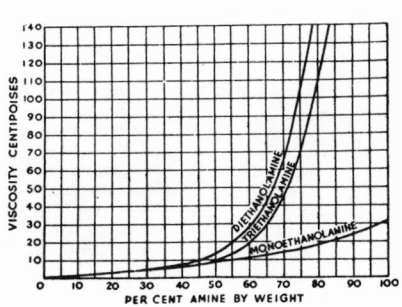

Table 14.77: Isopropanolamines (Mixed) *(2)*

The mixed isopropanolamines are available as a liquid mixture of mono-, di-, and triisopropanolamine.

Uses: The isopropanolamine soaps may be employed in all uses now found for the ethanolamine soaps. Their excellent hydrocarbon solubility and color stability make them of special interest in soluble oils, dry cleaning soaps, cosmetics, and pharmaceutical preparations. Vinyl acetate resin emulsions of the oil-in-water type for coating fabrics and leather have excellent stability when prepared by stirring 80 parts by weight of "Vinylite" resin AYAF (30% solution in toluene) and 1 part oleic acid, into 20 parts of water containing 0.6 to 0.8 parts of mixed isopropanolamine.

Kerosene solubilized with 4% by weight of mixed isopropanolamine and 15% by weight of oleic acid produces stable emulsions with water upon mechanical agitation. Stable water emulsions of chlorinated hydrocarbons or naphtha may be prepared by a similar procedure. The addition of about 2% by weight of mixed isopropanolamine has been found to improve the penetration of starch glues into heavily sized envelope stock.

Purity	
Monoisopropanolamine	14± 2% by wt.
Diisopropanolamine	43± 4% by wt.
Triisopropanolamine	43+ 4% by wt.
Specific gravity at 20/20°C	1.0040–1.0100

Table 14.78: Triisopropanolamine *(2)*

Tri-2-Hydroxyisopropylamine $(CH_3CHOHCH_2)_3N$

This compound is a white, crystalline solid, completely soluble in water. It is used as a reactant in pharmaceutical syntheses. It is important in the oral treatment of syphilis. Combined with sodium bismuthate and propylene glycol, it produces a bismuth compound stable enough to withstand chemical action of the digestive system. Triisopropanolamine can be used for the preparation of cosmetic creams, "soluble" oils, and emulsions—where the good color stability of its soaps is of interest. Formulas containing lanolin may vary in color stability. Triisopropanolamine is especially suggested for "soluble" white paraffin oils for the rayon industry, where good color and low free fatty acid content are desirable.

Boiling point (760 mm)	305.4°C
Flash point	305°F
Latent heat of vaporization	45°C
Melting point	46°C
pH 25% Solution at 25°C	10.7
Equivalent weight	188–192
Specific gravity at 50/20°C	0.9996
Solubility in water at 20°C	Complete
Solubility of water in amine	Complete
Vapor pressure at 20°C	0.01 mm Hg
Weight per gallon at 50°C	8.32 lbs.

Table 14.79: 2-Amino-2-Methyl-1-Propanol *(2)*

$$CH_3C(CH_3)NH_2CH_2OH$$

This is a water-white, syrupy, alkaline liquid, with a faint ammoniacal odor. It is soluble in water and many organic solvents. It forms soaps with higher fatty acids and these are useful as emulsifying agents in textile and leather materials, water-emulsion paints, and self-polishing waxes.

Boiling point (760 mm)	165°C
Melting point	25°C
Specificity gravity	0.934
pH (0.1 *M* solution at 20°C	11.27
Solubility in 100 cc water	Complete
Vapor pressure at 20°C	1.0 mm
Flash point (Tag. open cup)	153°F
Refractive index at 20°C	1.449
Weight per Gallon at 68°F	7.77 lbs

Table 14.80: 2-Amino-2-Methyl-1,3-Propanediol (2)

$$CH_2OHC(CH_3)NH_2CH_2OH$$

Boiling point at 10 mm	151°C
Melting point	109–111°C
pH (0.1 M solution) at 20°C	10.78
Solubility in 100 cc water	250 grams

Table 14.81: 2-Amino-2-Ethyl-1,3-Propanediol (2)

$$CH_2OHC(C_2H_5)NH_2CH_2OH$$

Boiling point at 10 mm	153°C
Flash point (Tag. open cup)	166°F
Melting point	37.5 to 38.5°C
pH of 0.17 M aqueous solution at 20°C	10.8
Solubility in water at 20°C	Complete
Specific gravity at 20/20°C	1.099
Refractive index at 20°C	1.490
Weight per gallon at 68°F	9.15 lbs.

Table 14.82: 2-Amino-1-Butanol (2)

$$\overset{\displaystyle NH_2}{\underset{}{CH_3CH_2CHCH_2OH}}$$

Boiling point at 760 mm	178°C
Flash point (Tag. open cup)	164°F
Melting point	−2°C
pH of 0.1M aqueous solution at 20°C	11.1
Specific gravity at 20°/20°C	0.944
Solubility in water at 20°C	Completely Miscible
Vapor pressure at 20°C (est)	0.5 mm
Refractive index at 20°C	1.453

Table 14.83: Tris(Hydroxymethyl)Aminomethane (2)

$$\overset{\displaystyle NH_2}{\underset{\displaystyle CH_2OH}{CH_2OH - CCH_2OH}}$$

Boiling point at 10 mm	219 to 220°C
Melting point	171 to 172°C
pH of 0.1M aqueous solution at 20°C	10.4
Solubility in water at 20°C	80 grams per 100 ml

Table 14.84: 2-Aminoethylethanolamine (2)

Hydroxyethyl Ethylenediamine

$$NH_2CH_2CH_2NHCH_2CH_2OH$$

This compound is a hygroscopic liquid with a mild ammoniacal odor; it is completely soluble in water. It is used in the manufacture of dyes, pharmaceuticals, textile specialties, flotation agents, resins, insecticides, and rubber products.

Boiling point at 760 mm	243.7°C
Flash point	275°C
Specific gravity at 20/20°C	1.0304
Solubility in water	Complete
Solubility of water in solvent	Complete
Vapor pressure at 20°C	0.02 mm Hg
Weight per gallon at 20°C	8.58 lbs.
Boiling range at 760 mm	232–250°C
Purity	99%, min

Table 14.85: 1-Diethylamino-2,3-Propanediol (2)

$$(C_2H_5)_2NCH_2CHOHCH_2OH$$

This alkylol amine is a water-white to light-straw liquid with a faintly ammoniacal odor. It is soluble in water, methyl alcohol, ethyl ether, ethyl, acetate, acetone, aromatic hydrocarbons, fixed oils, oleic and hot stearic acids, and hot carnauba wax, the latter solidifying when cooled. It is insoluble in mineral oil and paraffin wax.

Boiling range	233–235°C
Flash point	210°F
Specific gravity at 20/20°C	0.973

Table 14.86: Aminohydroxy Compounds *(34)*

Product Specifications

	2-Amino-2-methyl-1-propanol		2-Amino-2-ethyl-1,3-propanediol	Tris(hydroxymethyl)aminomethane	
	AMP Regular	AMP-95	AEPD	TRIS AMINO Crystals	TRIS AMINO 40% Concentrate
Neutral equivalent	88.5-91	93-97	121.5*	121-122	—
Water, % by wt (max.)	0.8	5.8	3.8	0.5	—
Melting point, °C (min.)	—	—	—	160	—
Color (max.)	20 APHA	20 APHA	2 Gardner	—	5 Gardner
Color of 20% aqueous solution (max.)	—	—	—	40 APHA	—
Distillation range, °C	156-177	—	—	—	—
Nonvolatile matter, % by wt (max.)	0.005	0.005	—	—	—
Amine assay by titration, calc. as % TRIS AMINO...	—	—	—	—	40 ± 2

*Anhydrous basis (max.)

Physical Properties of Purified Materials

Formula	2-Amino-2-methyl-1-propanol CH_3CCH_2OH with NH_2 and CH_3	2-Amino-2-ethyl-1,3-propanediol $HOCH_2CCH_2OH$ with NH_2 and C_2H_5	Tris(hydroxymethyl)-aminomethane $HOCH_2CCH_2OH$ with NH_2 and CH_2OH
Molecular weight (calcd.)	89.14	119.17	121.14
Boiling point at 760 mmHg, °C	165	—	—
Boiling point at 10 mmHg, °C	—	152-153	219-220
Melting point, °C	30-31	37.5-38.5	171-172
Specific gravity at 40/40°C	0.928	1.101	—
pH of 0.1M aqueous solution at 20°C	11.3	10.8	10.4
Solubility in water at 20°C, g/100 ml	miscible	miscible	80
Weight per gallon at 20°C, lb	7.78	9.15	—
pKa at 25°C	9.72	8.80	8.03

Additional Properties of AMP

	AMP Regular	AMP-95
Viscosity at 10°C, cp	—	561
25°C, cp	—	147
30°C, cp	102	—
50°C, cp	24	—
70°C, cp	9	—
90°C, cp	4	—
Vapor pressure at 100°C, mmHg	59	—
150°C, mmHg	457	—
Specific gravity at 25/25°C	—	0.942
Coefficient of expansion per °C	0.00095	0.00096
Refractive index, n_D at 20°C	1.449	—
Heat of vaporization at 110°C, kcal/mole	13.2	—
130°C, kcal/mole	12.5	—
150°C, kcal/mole	12.3	—
165°C, kcal/mole	12.1	—
Heat of dissociation at 25°C, kcal/mole	12.9	—

Table 14.87: 2-Dimethylamino-2-Methyl-1-Propanol *(34)*

DMAMP-80

$$CH_3-\underset{\underset{N(CH_3)_2}{|}}{\overset{\overset{CH_3}{|}}{C}}-CH_2OH$$

Specifications

DMAMP, % by wt......................78-82
  (as titratable amine)
Color, APHA.......................100 max.
Water, % by wt........................18-22

Typical Properties

	DMAMP-80
Neutral equivalent....................	~148
Specific gravity at 25/25°C...........................	0.95
Weight per gallon at 25°C.............................	7.9 lb
Flash point, Tag open cup.......................	150°F
Tag closed cup.....................	153°F
Freezing point.........................	−20°C
Boiling point at 760 mmHg.........................	~98°C
Viscosity at 25°C, Gardner...........................	A–A₂
pH of 0.1 N aqueous solution...........................	11.6
APHA color (max.)....................	100

ALKYLALKANOL AMINES

This group of compounds, also referred to as alkylaminoethanols, have less odor than most alkylamines and possess both water and oil solubility. The solubility degree of each is determined by the number of alkyl or hydroxyl groups present in the molecule. A larger number of hydroxyl groups gives greater water solubility, whereas a predominance of alkyl groups gives greater oil solubility. The derivatives of these compounds are of particular interest. They form soaps with fatty acids which may be employed as emulsifying, penetrating, and wetting agents, and these uses can also be applied to the ester and acid amide derivatives. These amines also serve as intermediates in the manufacture of drugs and dyes.

Table 14.88: Properties of Various Alkylalkanol Amines (2)

Dimethylethanolamine $[(CH_3)_2NCH_2CH_2OH]$. Dimethylethanolamine is a water-white liquid with an amine odor. It resembles diethylethanolamine in chemical behavior and it is used as an intermediate in the synthesis of corrosion inhibitors, dyes, pharmaceuticals, and textile auxiliaries.

Boiling point	133°C
Equivalent weight	89
Specific Gravity at 20/20°C	0.887
Refractive index	1.4300

Diethylethanolamine (Diethylaminoethanol, $(C_2H_5)_2NC_2H_4OH$). Diethylaminoethanol is a water-white, hygroscopic liquid which behaves chemically like the tertiary amines and alcohols. It is soluble in water, ethyl alcohol, methyl alcohol, ethyl ether, ethyl acetate, acetone, aromatic hydrocarbons, fixed oils, mineral oil, oleic acid, hot stearic acid, and hot paraffin and carnauba waxes, the last two solidifying when cooled. It is used in the manufacture of certain pharmaceuticals, such as procaine and "atabrine". It forms amine soaps with higher fatty acids, which are oil-soluble and useful as emulsifiers and textile lubricants. Its mild alkalinity makes it applicable as a neutralizing agent and a corrosion inhibitor.

Di-n-butylethanolamine (Di-*n*-Butylaminoethanol, $(C_4H_9)_2NCH_2CH_2OH$). This alkylolamine is a water-white liquid with a faintly amine odor. It is insoluble in water but soluble in ethyl alcohol, methyl alcohol, ethyl ether, ethyl acetate, aromatic hydrocarbons, fixed oils, mineral oil, oleic and hot stearic acids, and hot paraffin and carnauba waxes, the latter two solidifying when cooled. It is slightly soluble in paraffinic hydrocarbons.

n-Butyl diethanolamine $[C_4H_9N(CH_2CH_2OH)_2]$. This alkylol amine is a light straw-colored liquid with a faintly amine odor. It is soluble in water, ethyl alcohol, methyl alcohol, ethyl ether, ethyl acetate, acetone, aromatic hydrocarbons, castor oil, oleic and hot stearic acids, and hot carnauba wax, the latter solidifying when cooled. It is insoluble in linseed and cottonseed oils, mineral oil, and paraffin wax.

n-Butyl monoethanolamine $(C_4H_9NHCH_2CH_2OH)$. This alkylol amine is a water-white liquid with a faintly amine odor. It is soluble in water, ethyl alcohol, methyl alcohol, ethyl ether, ethyl acetate, acetone, aromatic hydrocarbons, fixed oils, oleic and stearic acids, and hot carnauba and paraffin waxes, the latter two solidifying when cooled. It is only slightly soluble in paraffinic hydrocarbons.

Ethyl diethanolamine $[C_2H_5N(CH_2CH_2OH)_2]$. Ethyl diethanolamine is a water-white liquid with an amine odor and soluble in water, ethyl alcohol, methyl alcohol, acetone, aromatic hydrocarbons, some fixed oils, oleic and hot stearic acids. It is insoluble in linseed and cottonseed oils, mineral oil, paraffin and carnauba waxes. It is only slightly soluble in paraffinic hydrocarbons.

(continued)

Table 14.88: (continued)

Ethyl monoethanolamine ($C_2H_5NHCH_2CH_2OH$). This alkylolamine is a colorless liquid, with an amine odor. It is soluble in water, ethyl alcohol, methyl alcohol, ethyl ether, ethyl acetate, acetone, aromatic hydrocarbons, fixed oils, mineral oil, oleic and stearic acids, and hot carnauba and paraffin waxes, the latter two solidifying when cooled. It is only slightly soluble in paraffinic hydrocarbons.

Tetraethanolammonium hydroxide [$N(C_2H_4OH)_4OH$]. This solvent is a white, crystalline, strongly basic solid, completely miscible with water. The commercial product is an aqueous methanol solution in which 40 to 41 per cent of this solvent is present. Although aqueous solutions of this solvent are stable at ordinary temperatures, it will decompose when heated to weakly basic tertiary amines. This property is utilized in processes where it is desired to destroy a strongly alkaline substance which is no longer needed. Tetraethanolammonium hydroxide is also a good solvent for certain types of dyes.

Color	White
Description	Crystalline solid
Melting point	123°C
Solubility in water	Complete
Solubility of water in solvent	Complete
Vapor pressure at 20°C	0.01 mm

	Diethylamino-ethanol	Ethyl Monoethanolamine	Ethyl Diethanolamine
Color	Water-white	Water-white	Water-white
Specific gravity at 20°/20°C.	0.88–0.89	0.92	1.02
Minimum amine content	99.5%	98.5%	98.5%
Initial boiling point	158°C.	161°C.	245°C.
Final boiling point	163°C.	174.5°C.	260°C.
Flash point	135°F.	160°F.	255°F.
Solidification point	< −70°C.	−8.8°C.	−50°C.
Refractive index at 20°C.	1.440	1.444	1.466
Viscosity at 25°C (centipoise)	4.05	12.40	53
Viscosity at 60°C (centipoise)	1.50	3.22	11.2
Coefficient of expansion per °C.	0.0012	0.00091	0.00080
Theoretical molecular weight	117.19	89.14	133.19
Average weight per gallon	7.36 lbs.	7.66 lbs.	8.5 lbs.

	Di-n-Butyl-aminoethanol	n-Butyl-Monoethanolamine	n-Butyl Diethanolamine
Color	Water-white	Water-white	Water-white to Light Straw
Specific gravity at 20°/20°C.	0.860	0.89	0.97
Minimum amine content	98.5%	96.0%	95.0%
Initial boiling point	222°C.	192°C.	262°C.
Final boiling point	234°C.	215°C.	290°C.
Flash point	200°F.	170°F.	245°F.
Solidification point	< −70°C.	−3.5°C.	< −70°C.
Refractive index at 20°C.	1.444	1.444	1.462
Viscosity at 25°C (centipoise)	6.50	17.4	55
Viscosity at 60°C (centipoise)	1.94	4.02	10.6
Coefficient of expansion per °C.	0.00114	0.0010	0.00077
Theoretical molecular weight	173.29	117.19	161.24
Average weight per gallon	7.16 lbs.	7.44 lbs.	7.25 lbs.

GLYCOL ETHER AMINES

Table 14.89: Properties of Various Glycol Ether Amines (47)

Amine	Methoxyethyl	Dimethoxyethyl	Ethoxyethyl	Diethoxyethyl	Methoxyisopropyl
Formula	$CH_3OC_2H_4NH_2$	$(CH_3OC_2H_4)_2NH$	$C_2H_5OC_2H_4NH_2$	$(C_2H_5OC_2H_4)_2NH$	$CH_3OCH_2\overset{\underset{\|}{CH_3}}{CH}\ NH_2$
Molecular Weight	75	133	89	161	89
Boiling Point, °C.	91	172	107	194	98
Vapor Pressure, mm. at 20°C.	--	1.0	--	0.5	--
n_D at 25°C.	1.4058	1.4190	1.4086	1.4205	1.4038
Specific Gravity	0.89	0.91	0.85	0.88	0.84
pK_b	4.62	5.49	7.74	5.53	4.60
Flash Point, °F.	60	155	70	185	60

Formula	$(C_4H_9)_2NH$	$(CH_3OCH_2CH_2)_2NH$	$(HOCH_2CH_2)_2NH$
Molecular Weight	129	133	105
Boiling Point, °C.	160	172	270
Vapor Pressure, (20°C.)	1.9	1.0	<0.01
Freezing Point, °C.	-62	<-40	28.0
Specific Gravity	0.76	0.91	1.09
pK_b	2.7	5.5	5.2
% Solubility in H_2O at 25°C.	0.47	∞	∞

Formula	$C_4H_9CH(C_2H_5)CH_2NH_2$	$CH_3OCH_2CH(CH_3)O\text{-}CH_2CH(CH_3)NH_2$
Molecular Weight	129	147
Boiling Point, °C.	169	175
Vapor Pressure, (20°C.)	1.2	1.0
Freezing Point, °C.	--	--
Specific Gravity	0.79	0.85
pK_b	3.2	4.8
% Solubility in H_2O at 25°C.	0.25	∞

ARYL AMINES

Table 14.90: Aniline *(2)*

Aminobenzene
Phenylamine

$C_6H_5NH_2$

Aniline is a colorless to straw-colored, toxic, highly refractive, oily liquid having a characteristic odor. It is soluble in ethyl alcohol, ethyl ether, carbon tetrachloride, and only slightly soluble in water. It is used in the production of such materials as indigo, aniline black, tetranitraniline, acetanilide, explosives, dyes, rubber chemicals, and pharmaceuticals.

Boiling point	184.2°C
Flash point	70°C
Freezing point	−6.3°C min.
Melting point	−6.2°C
Specific gravity at 25/25°C	1.021
Solubility in water at 25°C	3.8
80°C	6.0
Weight per gallon at 25°C	8.50 lbs.
Boiling range	1.5°C
	95% distills within 1.0°C
Nitrobenzene	None

Table 14.91: Dimethylaniline *(2)*

$C_6H_5N(CH_3)_2$

Dimethylaniline is a pale yellow, highly refractive, toxic, oily liquid with a pungent odor. It is soluble in ethyl alcohol, ethyl ether, and carbon tetrachloride, but only very slightly soluble in water. It is used in the making of dyes and the explosive tetranitroaniline ("Tetryl").

Boiling point	192.9°C
Freezing point	1.5°C
Specific gravity at 25/25°C	0.956
Weight per gallon at 25°C	7.95 lbs.
Boiling range within	2.0°C
	95% distills within 1.0°C
Color	Yellow to amber
Monomethylaniline	0.5% max.

Table 14.92: Diethylaniline *(2)*

$C_6H_5N(C_2H_5)_2$

Diethylaniline is a light-yellow, oily, toxic liquid with a pungent odor. It is soluble in ethyl alcohol, ethyl ether, and carbon tetrachloride, but only very slightly soluble in water. It is used in the preparation of dyes, pharmaceuticals, and other organic compounds.

Boiling range	216.3°C
Freezing point	−34.4°C
Specific gravity at 25/25°C	0.933
Weight per gallon at 25°C	7.76 lbs.
Aniline	None
Boiling range	5–95% within 2.5°C
	Boiling range includes 216.3°C
Color	Light yellow
Monoethylaniline	0.2% max.
Purity	99.8%, min.
Water	No visible separation

Table 14.93: N-Mono-n-Butyl Aniline (2)

$$C_6H_5NHC_4H_9$$

This secondary amine is a light straw to amber-colored liquid with an aniline odor. It is insoluble in water but soluble in ethyl alcohol, methyl alcohol, ethyl ether, ethyl acetate, acetone, aromatic and aliphatic hydrocarbons, fixed oils, mineral oil, oleic acid, and hot stearic acid. It is also soluble in hot paraffin and carnauba waxes which solidify upon cooling.

Flash point	225°F
Specific gravity at 20°C	0.93
Refractive index at 20°C	1.5351
Weight per gallon at 20°C	7.71 lbs.
Boiling range	234–242°C
Purity	95%, min.

Table 14.94: N,N-Di-n-Butyl Aniline (2)

$$C_6H_5N(C_4H_9)_2$$

N,N-Di-n-Butyl aniline is a light-straw colored liquid with a faintly aniline odor and is soluble in ethyl alcohol, aromatic hydrocarbons, ethyl ether, ethyl acetate, acetone, fixed oils, mineral oil, oleic acid and hot stearic acid. It is insoluble in water, methyl alcohol, and while soluble in hot paraffin and carnauba waxes, these solidify when cooled.

Flash point	230°F.
Specific gravity at 20°C	0.904
Refractive index at 20°C	1.5197
Weight per gallon at 20°C	7.53 lbs.
Boiling range	267–275°C
Purity	95%, min.

Table 14.95: n-Monoamyl Aniline (Mixed Isomers) (2)

$$C_6H_5NHC_5H_{11}$$

n-Monoamyl aniline is a mixture of isomers. It is a light-straw colored liquid with a faintly aniline odor. It is insoluble in water but soluble in ethyl alcohol, methyl alcohol, ethyl ether, ethyl acetate, acetone, aromatic and aliphatic hydrocarbons, fixed oils, mineral oil, oleic acid, and hot stearic acid. It is also soluble in hot paraffin and carnauba waxes which solidify on cooling.

Flash point	225°F
Specific gravity at 20°C	0.92
Refractive index at 20°C	1.5285
Weight per gallon at 20°C	7.64 lbs.
Boiling range	245–260°C
Purity	95%, min.

Table 14.96: p-tert-Amyl Aniline (2)

$$C_5H_{11}C_6H_4NH_2$$

p-tert-Amyl aniline, an aryl amine, is a straw to deep red-colored liquid with a faintly aromatic odor. It is soluble in methyl alcohol, ethyl ether, ethyl acetate, acetone, aromatic and aliphatic hydrocarbons, fixed oils, mineral oil, oleic acid and hot stearic acid. It is insoluble in water and while it dissolves hot carnauba and paraffin waxes, these solidify on cooling.

Flash point	215°F
Specific gravity at 20/20°C	0.948
Boiling range	253–259°C

Table 14.97: Di-tert-Amyl Aniline *(2)*

$$(C_5H_{11})_2C_6H_3NH_2$$

Di-tert-amyl aniline, an aryl amine, is a red-colored, almost odorless liquid, soluble in ethyl ether, ethyl acetate, acetone, aromatic and aliphatic hydrocarbons, fixed oils, mineral oil, oleic acid and hot stearic acid. It is insoluble in water and methyl alcohol, and dissolves hot carnauba and paraffin waxes, which solidify when cooled.

Flash point	265°F
Specific gravity at 20/20°C	0.923
Boiling range	289–321°C

Table 14.98: N,N-Diamyl Aniline (Mixed Isomers) *(2)*

$$C_6H_5N(C_5H_{11})_2$$

N,N-diamyl aniline is a dark-amber liquid with a faintly aniline odor. It is insoluble in methyl alcohol and water, but soluble in ethyl ether, ethyl acetate, acetone, aromatic and aliphatic hydrocarbons, fixed oils, mineral oil, oleic acid, and hot stearic acid. It is also soluble in hot carnauba and paraffin waxes, which solidify on cooling.

Flash point	260°F
Specific gravity at 20/20°C	0.898
Boiling range	276–292°C

Table 14.99: Diethylbenzylamine *(2)*

$$C_6H_5CH_2N(C_2H_5)_2$$

Diethylbenzylamine is a colorless liquid with an almond-like odor. It is soluble in ethyl and methyl alcohols, ethyl ether, ethyl acetate, acetone, aromatic hydrocarbons, fixed oils, mineral oil, oleic acid and hot stearic acid. It is insoluble in water and while soluble in hot paraffin and carnauba waxes, these solidify on cooling.

Flash point	170°F
Specific gravity at 20°C	0.890
Refractive index at 20°C	1.5002
Weight per gallon at 20°C	7.41 lbs.
Boiling range	207–215°C
Purity	97%, min.

Table 14.100: N-(n-Butyl)-α-Naphthylamine *(2)*

$$C_{10}H_7NHC_4H_9$$

This solvent is a dark-red liquid with a faintly amine odor. It is a solvent for methyl alcohol, ethyl ether, ethyl acetate, acetone, aromatic and aliphatic hydrocarbons, fixed oils, mineral oil, oleic and hot stearic acid. It is insoluble in water, and soluble in hot paraffin and carnauba waxes which solidify on cooling.

Flash point	295°F
Specific gravity at 20/20°C	1.012
Boiling range	318–325°C

IMINES

Table 14.101: Ethylene Imine *(23)*

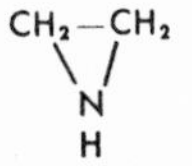

$$CH_2 - CH_2$$
$$\diagdown \; N \; \diagup$$
$$H$$

Ethylene imine is a colorless mobile liquid having an amine-like odor, with the above structure.

Molecular Weight	43.07
Density, gm./ml. —10° C.	0.865
0° C.	0.856
10° C.	0.846
25° C.	0.832
Boiling Point, °C.	57
Freezing Point, °C.	—78
Index of Refraction, n_D at 25° C.	1.4123
Viscosity, cps. at 25° C.	0.418
Surface Tension, dynes/cm.	32.8
Flash Point, °F.	12
Heat of Formation, Kcal./mole	21.95
Heat of Vaporization, Kcal./mole at 20° C.	7.9
Dissociation Constant	7.8×10^{-7}

Table 14.102: Propylene Imine *(41)*

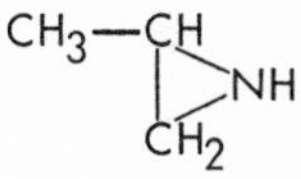

$$CH_3 - CH$$
$$\diagdown \; NH$$
$$CH_2$$

Molecular Weight	57.09
Density, g./ml.	
25°C.	0.8017
35°C.	0.7908
45°C.	0.7811
$\triangle d / \triangle t$, g./ml. per °C. at 25°C.	0.0011
Boiling Point, °C. at 760 mm. Hg	66.0
$\triangle B.P. / \triangle p$, °C. per mm. Hg at 760 mm. Hg	0.038
Vapor Pressure at 25°C., mm. Hg	140
Refractive Index, N_D at 25°C.	1.4084
Absolute Viscosity, centipoises at 25°C.	0.491
pKa at 25°C.	8.18
Heat of Vaporization at 66°C. and 1 atm, cal. per g.	139
Integral Heat of Solution of P.I. in water, 5 wt. % P.I. final conc., kcal./mole	4.5
Heat of Vaporization at 66°C. and 1 atm, BTU per lb.	250

Solubility:

Water	Soluble
Polar organic solvents	in all
Pentane	Proportions

AMIDES

Table 14.103: Formamide *(11)*

$$HCONH_2$$

Formamide is a water-white to light yellow, hygroscopic liquid. It is miscible in all proportions with the lower alcohols and glycols, but is insoluble in hydrocarbons, chlorinated solvents and ethers. Its high dielectric constant is an indication of its high ionizing power.

Formamide dissolves many metal chlorides, iodides, nitrates, phosphates, and some carbonates and is less soluble in sulfates and oxides. Proteins, saccharides, and polyvinyl alcohol dissolve or soften in formamide. Cellulose will swell in formamide as it does in water. Formamide is a nonaqueous solvent for electrolytes due to its ionizing solvent action on numerous inorganic salts.

AVERAGE ANALYSIS

Formamide, %	98.5
Methanol, %	1.0
Color (as shipped), APHA	7

PHYSICAL PROPERTIES

Molecular weight	45.04
Boiling point (760mm), °C	210
°F	410
Freezing point, °C	2.6
°F	36.7
Specific gravity, 25°/4°C (77°/39°F)	1.1339
Density, lb/gal, 60°F (15.6°C)	9.5
Refractive index, N_D^{20}	1.4481
Surface tension, dynes/cm, 20°C (68°F)	58.35
Viscosity, cp, 20°C (68°F)	3.76
Dielectric constant, 20°C (68°F)	84
Solubility parameter	19.2
Hydrogen bonding index	>16.2
Specific conductance, ohm^{-1}, 25°C (77°F)	18.9×10^{-5}
Specific heat of liquid, cal/g, 19°C	0.55
Btu/lb, 66°F	0.99
Latent heat of vaporization, cal/g, 210°C	400
Btu/lb, 410°F	720
Flash point (TOC), °F	310
°C	154.4

Table 14.104: Dimethylformamide *(11)*

DMF $(CH_3)_2NCH=O$

DMF, dimethylformanide, is a uniquely versatile and powerful solvent with the following general properties:

Appearance	Colorless, mobile liquid	Flash point, T.O.C., °C	67 (153°F)
Molecular weight	73.09	Ignition temperature, °C	445 (833°F)
Boiling point, 760mm, °C	153 (307°F)	Flammability limits in air	
Freezing point, °C	−61 (−78°F)	lower	2.2 vol %
Specific gravity 0°/4°C	0.9683	upper	15.2 vol %
25°/4°C	0.9445	Dielectric constant, 25°C	36.71
Density, lbs/gal, 20°C	7.92	Dipole moment, 20°C	3.82 Debye Units
Refractive index, $N_D 25°C$	1.4269	Hygroscopicity, 30°C (300 hrs @ 50% RH)	34% gain
Vapor pressure, 25°C	3.7mm	Relative evaporation rate (butyl acetate = 100)	17
Viscosity, 25°C	0.802 cp	Solubility parameter	12.1
Surface tension, 25°C	35.2 dynes/cm	Ionization constant (@ 20°C)	10^{-18}
Specific heat (liquid, 20°C)	0.49 Btu/lb/°F	*Azeotropes:*	
Heat of vaporization	248 Btu/lb	DMF (18.7 wt %), p-xylene (81.3%)	135.1°C at 760mm
Heat of combustion	457.5 kg cal/gm mol	DMF (69 wt %), formic acid (31 wt %)	162.4°C at 760mm
	11,280 Btu/lb	DMF (7 wt %), tetrachlorethylene (93.0 wt %)	117.5°C at 730mm
Thermal conductivity (at 23.5°C)	440 cal/sec cm°C		

(continued)

Table 14.104: (continued)

Evaporation Rate

Atmospheric conditions . . .

% Evaporation	Time, Hours
0	0
20	8
40	16
60	26
80	34
100	44

Heat of Mixing DMF-Water . . .

Temp 30°C . . .

Wt % DMF in Aqueous Solution	Btu/lb DMF
5	89
10	82.5
15	75
20	73.5
25	70
30	66
35	64

Flash and Fire Point of DMF Water Solutions . . .

Composition, DMF-H₂O Mixture—DMF % by Wt	Flash Point °F	Fire Point °F
100	145	150
90	165	170
70	215	230
65	210	225
60	none	none
50	none	none

Table 14.105: Surface Tension and Density of DMF-Water Mixtures *(11)*

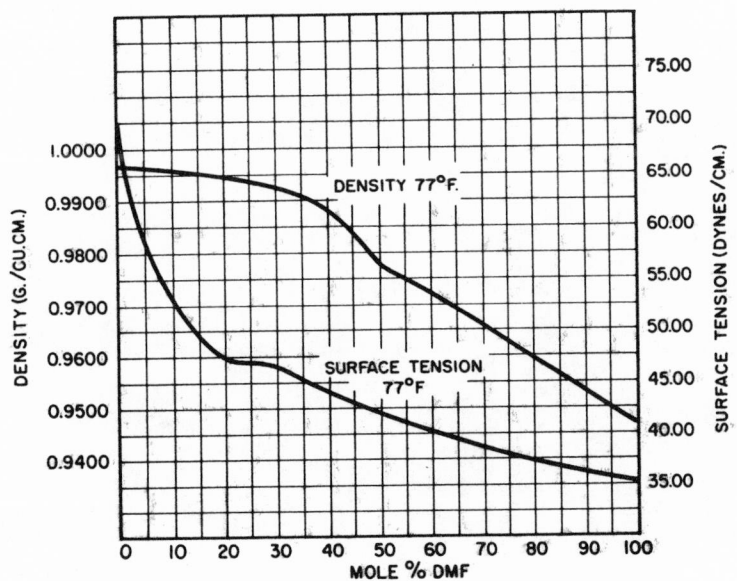

Table 14.106: Semi-Quantitative Solubilities of Inorganic Materials in DMF at 25°C (11)

Salt	Solubility g/100g DMF
AgBr	0.03
AgCl	0.01
AgI	0.04
AlCl$_3$	reaction
Al(NO$_3$)$_3$ · 9H$_2$O	20
Be(NO$_3$)$_2$ · 3H$_2$O	20
CaCl$_2$	Approx. 0.5
CaF$_2$	0.05
Ca(NO$_3$)$_2$ · 4H$_2$O	20
CaSO$_4$ · 2H$_2$O	1.2[93]
Cd(NO$_3$)$_2$	20
Co(C$_2$H$_3$O$_2$)$_2$ · 4H$_2$O	20
CoCl$_2$ · 6H$_2$O	20
Co(NO$_3$)$_2$ · 6H$_2$O	20
CoSO$_4$ · 7H$_2$O	slight
CrCl$_3$ · 6H$_2$O	40
Cu(C$_2$H$_3$O$_2$)$_2$	slight
CuCl$_2$ · 2H$_2$O	15
Cu(NO$_3$)$_2$ · 3H$_2$O	20
CuSO$_4$	1.8
FeCl$_3$	20
Fe(NO$_3$)$_3$ · 6H$_2$O	20
FeSO$_4$ · 7H$_2$O	slight
HgCl$_2$	25
I$_2$	25
KBH$_4$	1.2
KC$_2$H$_3$O$_2$	0.09
KCl	0.05
KCN	0.22
KCNO	0.12
KCNS	18.2
K$_3$Fe(CN)$_6$	0.05
KI	25
KMnO$_4$	reaction
KNO$_2$	0.7[92]*
KNO$_3$	1.5
KOH	0.1
LiBH$_4$	3.5[94]
LiCl	11.40

Salt	Solubility g/100g DMF
LiH	0.7
MgCl$_2$	moderate
MgSO$_4$	0.13[95]
MnCl$_2$ · 4H$_2$O	15
NaB(OCH$_3$)$_4$	77.8
NaBH$_4$	25.5[94]
NaCHO$_2$	0.03
NaC$_2$H$_3$O$_2$ · 3H$_2$O	1.5
NaCl	0.05
NaCNO	0.76
NaCNS	0.05
Na$_2$CO$_3$	29.2
Na$_2$Cr$_2$O$_7$ · 2H$_2$O	0.05
Na$_2$Fe(CN)$_5$(NO) · 2H$_2$O	20
Na$_2$HPO$_4$	25
NaI	0.05
NaIO$_3$	14.4
NaNO$_2$	0.05
NaNO$_3$	2.0*
NaPO$_3$	15.4
Na$_2$S$_2$O$_3$	0.05
NH$_4$Br	0.08
NH$_4$C$_2$H$_3$O$_2$	12.7[95]
NH$_4$Cl	0.1
NH$_4$CNS	0.1
(NH$_4$)$_2$CO$_3$	15.2
NH$_4$NO$_3$	0.04
NiCl$_2$ · 6H$_2$O	55.1
Ni(NO$_3$)$_2$ · 6H$_2$O	5
Pb(C$_2$H$_3$O$_2$)$_2$ · 3H$_2$O	20
Pb(HCO$_2$)$_2$	1.5
PbO	0.1
PbS	0.3
PbSO$_4$	0.15

*Urea increases solubility:
4.7g NaNO$_2$ with 4.6g urea
8.1g NaNO$_2$ with 9.3g urea
3.7g KNO$_2$ with 7.5g urea

Category	Material	Solubility
Cellulosic	Cellulose nitrate	S
	Cellulose acetate	S
	Cellulose acetate butyrate	S
	Cellulose acetate propionate	PS
	Cellulose triacetate	S
	Ethyl cellulose	S
	Cyanoethylated cellulose	–
	Chlorinated Polyether ("Penton"—Hercules Powder)	S
Nylon (polyamides)	Types 6/6, 6, 6/10	–
	Type 8 (Belding-Corticelli Industries, B.C.I.)	PS
Polyethylene		–
Polypropylene		S
Polycarbonate	("Lexan"—General Electric Company)	S
Fluorocarbons	Polytetrafluoroethylene ("Teflon"*)	S
Styrene	Polystyrene	S
	Styrene-acrylonitrile copolymer (Tyril 767—Dow)	S
Vinyl Polymers and Copolymers	Polyvinyl chloride	–
	Polyvinyl chloride-acetate	S
	Polyvinyl alcohol	S
	Polyvinyl butyral ("Butacite"*—DuPont)	S
	Vinylidene chloride/vinyl chloride copolymer (Geon 200 x 20)	–
	Vinylidene chloride/vinyl chloride copolymer (Saran B-115)	S
	Polyvinyl acetate	S
	Polyvinyl formal	PS
	Polyvinyl fluoride	S
Polyesters	Saturated ("Mylar"*)	S
	Alkyd	–
Phenolic	Phenol-formaldehyde pure resin	S
	Ester gum modified phenol-formaldehyde	S
	Urea formaldehyde	PS
	Thiourea formaldehyde	S
Coumarone	Coumarone-indene	S
Natural	Garnet shellac	PS
	Orange shellac	S
	Ester gum	S
	Kauri gum	–
	Manila copal	S
	Esterified congo copal	S
	Wood resin	–
	Damar	PS
	Soft albino asphalt	PS
Epoxy (cured)		S
Polyurethanes		S

(continued)

Table 14.106: (continued)

Polymer Solvent

The principal use of DMF is as a solvent in the spinning of acrylic and polyurethane fibres. This is a specialised outlet but illustrates the solvent power of DMF for polymers of high molecular weight.

Various polymers which are soluble in DMF together with some which are insoluble are shown in the following lists:

Soluble
polyacrylonitrile
polyurethanes
polymethylmethacrylate
cellulose acetate
cellulose nitrate
cellulose acetate butyrate
ethylcellulose
cyanoethylated cellulose
polystyrene
polyvinyl chloride
polyvinyl alcohol
polyvinyl acetate
alkyds
phenol-formaldehyde resins
coumarone-indene resins
shellac
ester gum
kauri gum

Insoluble
polyethylene
polypropylene
polytetrafluoroethylene
saturated polyesters
urea-formaldehyde resins
natural rubber
butyl rubber
styrene-butadiene rubber
nylon 66, 6, and 610

Not only does DMF allow many polymers of sparing solubility to be brought into solution at economical concentrations, but when used either alone or as a booster solvent it yields solutions with lower viscosities and higher solids content than can be obtained with other solvents. It is therefore suggested as an attractive solvent for use in the formulation of protective coatings and films, adhesives, and printing inks.

Reaction Solvent and Catalyst

The use of DMF alone or as a component of a solvent system confers a number of advantages, the relative importance of which depends upon the particular application, but the following may be specially noted:

(*a*) its high solvent power can increase the effective concentration of one of the reacting species;

(*b*) it has a high dielectric constant;

Table 14.107: Dimethylacetamide *(11)*

DMAC

PHYSICAL PROPERTIES OF DIMETHYLACETAMIDE:

Formula.....................................$CH_3\overset{\overset{O}{\|}}{C}N(CH_3)_2$
Molecular weight............................87.12
Boiling point...............................165.5°C
Vapor pressure at 25°C......................1.3 mm
Freezing point..............................—20°C
Specific gravity (15.5°C)...................0.9448
Pounds per gallon (15.5°C)..................7.87
Viscosity (25°).............................0.92 cp
Refractive index 1.4356
Dielectric constant.........................37.8
Dipole moment (in dioxane)..................3.79
Flash point (Tag Open Cup)..................70°C
Thermal conductivity (22.2°C)...............416 x 10^{-6}cal/sec/cm/°C
Specific heat (liquid, 0 to 87°C)...........0.485 BTU/lb°F
Heat of vaporization at 165°C (cal'd).......10,360 cal/g mol
Heat of combustion (20°C)...................608 k cal/g mol

Flammability limits in air at 740 mm Hg and 160°C

Lower...............................2.0% by vol
Upper..............................11.5% by vol

(continued)

Table 14.107: (continued)

Vapor pressure:—

Temperature, °C	Pressure, mm Hg
25	1.3
40	3.4
70	18
90	46
110	108
130	230
150	460
165.5	758

Azeotrope—

DMAC (77.2 wt %)—acetic acid (21.1 wt %)—170.8°C at 760 mm

Solubility—Completely miscible with water, ethers, esters, ketones, and aromatic compounds. Unsaturated aliphatics are highly soluble, but saturated aliphatics have limited solubility.

	Solubility at 25°C g/100 g DMAC
Iso-octane	33
Di-isobutylene	Compl Misc
N-hexane	" "
N-Heptane	31
Cyclohexane	Compl Misc
Cyclohexene	" "
Kerosene	16

Table 14.108: Viscosities of Resins in DMAC *(11)*

Viscosities of Surface Coating Resins in DMAC

	Viscosity at 25°C, 15 wt % Solution—cps
Acrylic Resins	
Acryloid®[a] A-21[b]	23
A-107[b]	23
B-72[b]	20
Lucite®[c] 44	26
45	38
46	30
Expoxy Resins	
Epon®[d] 1001	5
1002	6
1004	7
1007	10
Cellulosic Resins	
Half-Sec. Butyrate	555
EAB-500-1[e]	950
EAB-171-2[e]	1275
Urea-Formaldehyde Resins	
Uformite®[a] F-222[f]	20
Melamine-Formaldehyde Resin	
MM-55	18

Viscosities of Vinyl Resins in DMAC

	Viscosity at 25°C, 15 wt % resin in 50/50 Solvent/ Toluene, cps
VYHH[g]	52
VAGH[g]	53
VMCH[g]	48
Geon®[h] 121	230
Geon®[h] 101	3800

Viscosities of Nitrocellulose Solutions in DMAC

	Viscosity at 25°C, 8 wt % resin in Solvent, cps
HB-14 Nitrocellulose[c]	18

(a)—Rohm and Haas Company
(b)—Reduced to 5 wt % solids with DMAC
(c)—E. I. du Pont de Nemours & Co. (Inc.)
(d)—Shell Chemical Corporation
(e)—Eastman Chemical Products, Inc.
(f)—Reduced to 30 wt % solids with DMAC
(g)—Union Carbide Chemicals Company
(h)—B. F. Goodrich Chemicals Company

Table 14.109: 1-Formylpiperidine *(56)*

1-Formylpiperidine, an amide solvent, is a stable, highly ordered, dipolar aprotic liquid having a high boiling point and wide liquid range (-30.6° to 222°C), making this a favorable solvent for nonvolatile applications in gas absorption processes, ink and dye systems, and plastics modifiers and stabilizers. 1-Formylpiperidine is a strong solvent for both polar and nonpolar compounds. It is unusual for its solubility in both water and hexane. It is miscible with acyclic alkanes (C_6 and below), cycloalkanes, alcohols, esters, ketones, aldehydes, amines, carboxylic acids and hydrides, amides, alkyl halides beyond C_{11}, stearates, ethers, olefins, nitriles, nitro compounds, heterocyclics, aromatics, organophosphorus compounds, alkynes, organotin compounds, organosilicates and inorganic acids. The high solubility of many polymers in 1-formylpiperidine is of particular significance. 1-Formylpiperidine is a reactive solvent reacting at the carbonyl center and at the amide nitrogen.

Molecular Formula	$C_6H_{11}NO$
Molecular Weight	113.16
Density	1.02 g/ml (8.51 lb/gal)
Index of Refraction (25.0°C)	1.4823
Freezing Point	-30.6°C
Boiling Point	222°C
Vapor Pressure	
25°C	0.1 mm Hg (0.002 lb/in^2, 0.0001 atm)
100°C	14 mm Hg (0.27 lb/in^2, 0.02 atm)
Heat of Vaporization (ΔHvap)	16.5 kcal/mol (262 Btu/lb)
Heat of Fusion (ΔH$_f$)	2.2 kcal/mol (35 Btu/lb)
Heat of Sublimation (ΔHsub)	18.7 kcal/mol (297 Btu/lb)
Entropy of Vaporization (ΔSvap)	33.4 cal/°/mol
Molar Freezing Point Depression Constant	5.7°C/mol
Corrosion of Metals at 222°C	
Mild Steel	0.3×10^{-4} in/yr
Brass	3.2×10^{-4} in/yr
Copper	5.2×10^{-4} in/yr
Hydroscopicity (23°C, 100% RH,	0.16% weight gain/hour
exposed surface area/volume = 1.67 cm^{-1})	(nearly linear to 100 hr)

Solubilities of Gases (g/100 g, 23°C, 1 atm)

Ammonia	1.1
1,3-Butadiene	12.9
1-Butene	6.8
Carbon Dioxide	0.8
Methyl Chloride	12.9
Ethane	0.2
Methane	less than 0.1

Solubilities of Inorganic Solids (g/100 g, 23°C)

Aluminum Chloride	Reacts
Potassium Acetate	0.51
Potassium Cyanide	0.06
Potassium Iodide	18.9
Potassium Permanganate	Reacts
Sodium Chloride	0.08
Ammonium Bromide	2.4
Sodium Iodide	17.1
Sodium Hydroxide	0.02

Toxicity

LD$_{50}$ in rats and mice (C5)	~1,100 mg/kg of body weight

(continued)

Table 14.109: (continued)

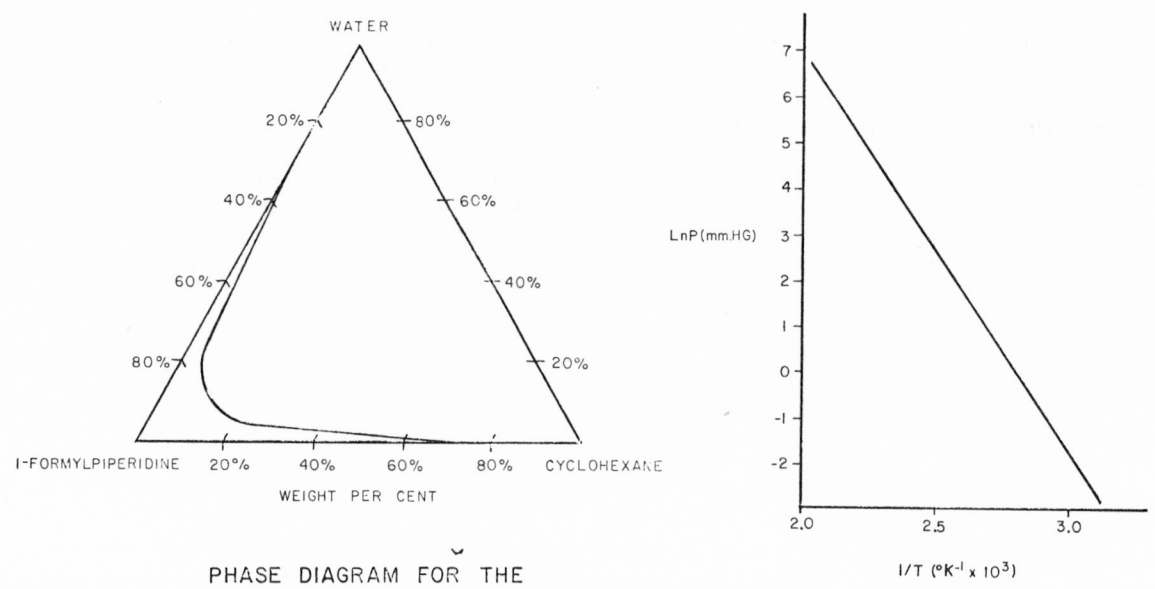

PHASE DIAGRAM FOR THE

TERNARY SYSTEM

I-FORMYLPIPERIDINE, WATER, CYCLOHEXANE

at 25°C, I ATMOSPHERE

CLAUSIUS-CLAPEYRON PLOT

VAPOR PRESSURE I-FORMYLPIPERIDINE

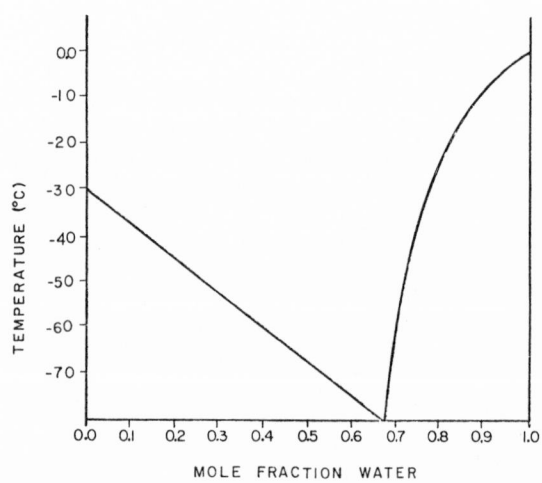

FREEZING POINT-COMPOSITION DIAGRAM FOR

THE I-FORMYLPIPERIDINE, WATER SYSTEM

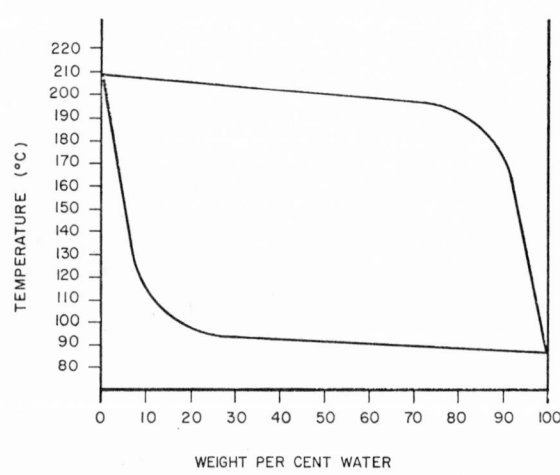

LIQUID-VAPOR COMPOSITION DIAGRAM FOR THE

I-FORMYLPIPERIDINE, WATER SYSTEM

NITRILES

Table 14.110: Acetonitrile *(54)*

Methyl Cyanide CH_3CN

PHYSICAL PROPERTIES

Molecular Weight - - - - - - - - - - - - - -	41.05
Boiling Point at 760 mm. Hg. - - - - - - - -	81.6°C.
50 mm. Hg. - - - - - - - -	13°C.
10 mm. Hg. - - - - - - -, - -	-15°C.
Vapor Pressure at 20°C. - - - - - - - - - -	72.8 mm. Hg.
dt/dp	0.042°C./mm Hg.
Specific Gravity at 20/20°C. - - - - - - - -	0.7868
Freezing Point - - - - - - - - - - - - - -	-45.7°C.
True Density at 20°C. - - - - - - - - - - -	0.7857 gm/ml.
Coefficient of Expansion at 20°C. - - - - - -	0.00137/°C.
Solubility in water at 20°C. - - - - - - - -	Complete
Solubility of water in at 20°C. - - - - - - -	Complete
Absolute Viscosity at 0°C. - - - - - - - - - -	0.43 cps.
20°C. - - - - - - - - - -	0.35 cps.
40°C. - - - - - - - - -	0.30 cps.
Refractive Index at 20°C., n/D - - - - - - -	1.3441
Flash Point (Open Cup) - - - - - - - - - -	42°F.
Specific Heat (21° to 76°C.) - - - - - - - - -	0.541 B.t.u./lb./°F.
Heat of Fusion at -46°C. - - - - - - - - - - -	94 B.t.u./lb.
Heat of Combustion at 25°C. (liquid) - - - -	13,350 B.t.u./lb.
Heat of Vaporization at 80°C. - - - - - - - -	313 B.t.u./lb.
Dielectric Constant at 0°C. - - - - - - - - -	42.0
20°C. - - - - - - - - -	38.8
81.6°C. - - - - - - - - -	26.2
Specific Conductance, OHM^{-1} - - - - - - - -	$5-9 \times 10^{-8}$(25°C.)
Surface Tension at 20°C. - - - - - - - - -	29.3 dynes/cm. air
Average Weight per gal. at 20°C. - - - - - -	6.52 lb.
Evaporation Rate (Butyl Acetate = 100) - - -	579
Nitrocellulose Dilution Ration - Toluene - -	2.0
Critical Constants - - - - - - - - - - - - - -	t_c° = 274.7°C.
	p_c = 47.7 atm.
	dc - 0.237 gm/c.c.

Enthalpy Data for Vapors ---	t°F. - - - - -	H° Btu/16-mole (ideal vapor)
	32	0.0
	60	339.0
	77	551.1
	100	841.8
	200	2181
	300	3640
	400	5222

Dipole Moment - - - - - - - - - - - - - - - - -	3.2 Debye Units

Table 14.111: Acetonitrile Vapor Pressure *(54)*

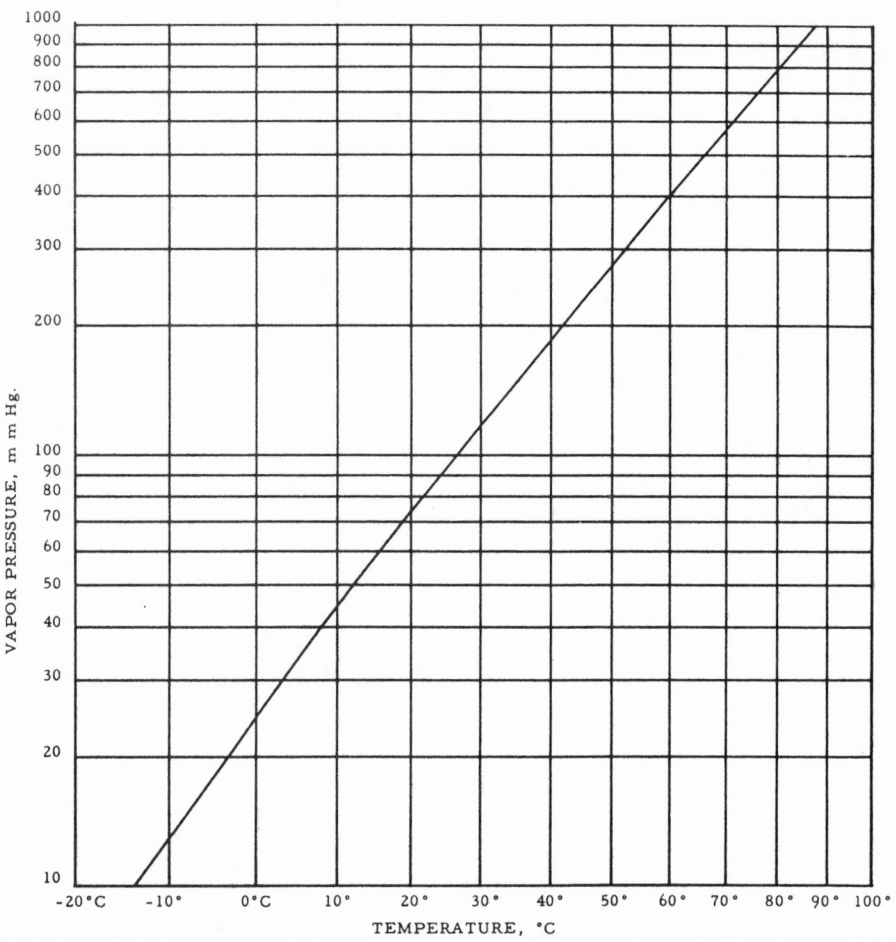

Table 14.112: Approximate Solubilities of Various Materials in Acetonitrile *(54)*

Compound	Solubility at room temp., gms./100ml.	Compound	Solubility at room temp., gms./100ml.
Acids		**Aryl Compounds**	
Formic Acid	M.*	Pyridine	M.
Acetic Acid	M.	Nitrobenzene	M.
Crotonic Acid	50	Aniline	M.
Levulinic Acid	M.	Xylene	M.
Oleic Acid	6.0	Benzene Sulfonic Acid	insoluble
		Phenol	3.3
Alcohols		**Acyl Chlorides**	
Methanol	M.	Acetyl Chloride	M.
CELLOSOLVE Solvent	M.	**Esters**	
Pentaerythritol	insoluble	Castor Oil	6.4
		Dibutyl Phthalate	M.
Aldehydes		Diglycol Stearate	M.
Formaldehyde	M.	**Ethers**	
Acetaldehyde	M.	n-Butyl Ether	M.
		Dichlorethyl Ether	M.
Amines		**Ketones**	
Di-n-Butyl Amine	M.	Acetone	M.
Triethanolamine	insoluble	Methyl Isobutyl Ketone	M.
		Nitroparaffins	
Anhydrides		Nitromethane	M.
Acetic Anhydride	M.	Nitroethane	M.
		Nitropropane	M.

M.*=Equal weights of acetonitrile and other material are miscible at room temperature.

Table 14.113: *n-Butyronitrile (19)*

$$CH_3CH_2CH_2C\equiv N$$

n-Butyronitrile is a clear, colorless liquid which is slightly soluble in water and completely miscible with common organic solvents. The product undergoes reactions typical of the aliphatic nitriles.

Physical Properties

Molecular weight	69.10
Specific gravity, 20/20°C	0.7920
Boiling point, 760 mm	117.5°C
50 mm	43°C
10 mm	13°C
Vapor pressure, 20°C	15 mm
Freezing point	-111.90°C
Solubility, in water, 20°C	3.5% by wt.
water in, 20°C	2.5% by wt.
Viscosity, 0°C	0.8 cps.
20°C	0.6 cps.
40°C	0.5 cps.
Refractive index, $n\frac{20}{D}$	1.3841
Weight per gallon, 20°C	6.60 lbs.
Flash point	79°F

Shipping Data

Net Container Contents:	
1-gallon tin can	6.5 lbs.
5-gallon iron drum	30 lbs.
55-gallon iron drum	360 lbs.

Typical Analysis of Current Production

Specific gravity	0.7907
Distillation, IBP	115.7°C
50 ml	117.6°C
-DP	118.4°C
n-Butyronitrile	98.9% by wt.
Water	0.09% by wt.
Alkalinity	0.22 meq/gm
Color	5 Pt-Co

Table 14.114: Lactonitrile *(53)*

Appearance:

Colorless liquid.

Molecular Weight (theory):

71. 08

Freezing Point

-34 ± 1°C.

Boiling Point:

Temperature °C	Pressure mm. Hg.
74	10
86	20
103	50
112 (decomposition)	75

Lactonitrile has a strong tendency to supercool making the determination of the freezing point difficult; therefore, the freezing-point method is not recommended for the determination of purity.

The boiling point is not recommended as a method for identifying lactonitrile because of the serious decomposition occurring during distillation, even when lactonitrile is stabilized with iodine or acid.

(continued)

Table 14.114: (continued)

Surface Tension

35. 8 dynes cm. $^{-1}$ at 30°C

Viscosity

0. 0201 poises (30°C) compared to

0. 008 poises for H_2O

Flash Point

76. 7°C (170°F) (Tagliabue closed cup)

Refractive Index

n_D^{25} = 1. 4027, $n_D^{18.4}$ = 1. 4058

Density

d_{25} = 0. 9834

d_t = 1. 0048 - 0. 8438t x 10^{-3}

$-0.518t^2$ x 10^{-6}
where t = 0 to 60°C.

Heat of Combustion

42. 1 Kcal. mole^{-1}

Heat of Formation

-33. 2 Kcal. mole^{-1}

Free Energy of Formation

8. 2 Kcal. mole^{-1} (estimated)

HETEROCYCLIC COMPOUNDS

Table 14.115: Pyrrole *(49)*

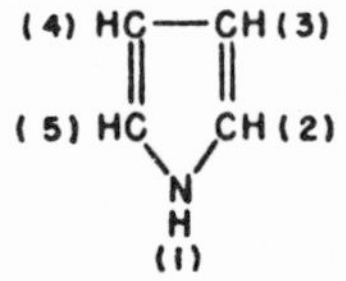

Appearance . . . . Colorless liquid, darkens on standing	Freezing Point*. -24°C. (-11°F.)
Odor . . . . Mild, nonirritating	Specific Gravity, 20/4°C.* . . ; 0.968
Molecular Weight 67.09	Index of Refraction, n20/D*. 1.5095
Boiling Point* . . . . 129°C. (264°F.) at 760 mm.	Flash Point (Tag closed cup). . . . . .39°C. (102°F.)
	Modified Reid Vapor Pressure*. . . . . .0.25 p.s.i. (±0.05)

*Determined on purified pyrrole

Table 14.116: N-Methyl-2-Pyrrolidone *(49)*

"M-PYROL"

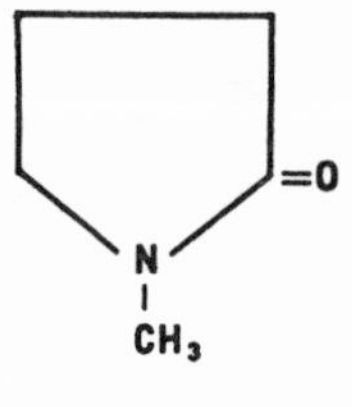

Empirical formula	C_5H_9NO
Molecular weight	99.1
Purity (N-methyl-2-pyrrolidone, area % VPC)	99.5% min
Physical form	liquid with mild amine-like odor
Color (APHA)	50
Moisture content	0.05% max
Freezing point	-24.4°C (-11.9°F)

(continued)

Table 14.116: (continued)

Boiling point	
@ 760 mm	202°C (395°F)
@ 162 mm	150°C (302°F)
@ 24 mm	100°C (212°F)
Vapor pressure, 20°C	0.29 mm
Viscosity, 25°C	1.65 cp
Refractive index, n_D^{25}	1.469
Specific gravity, d_4^{25}	1.027
@ 75°C	0.987
@ 100°C	0.969
Interfacial surface tension, 25°C	40.7 dynes/cm
Flash point (open cup)	95°C (204°F)
Explosive limits	0.058 g/ℓ lower limit
(Factory Mutual Research Corp.	2.18% vapor in air 360°F (182°C)
Apparatus)	0.323 g/ℓ upper limit
	12.24% vapor in air 370°F (188°C)
Dipole moment	4.09 ± 0.04 Debye
Dielectric constant, 25°C	32.2
Heat of combustion	719 Kcal/mol
Specific heat	0.40 Kcal/kg at 20°C
Heat of vaporization	127.3 Kcal/kg (230 Btu/lb)
Autogenous ignition temperature	287° ± 3°C (549° ± 5°F)
(ASTM Method D-2155)	
Miscibility with other solvents	completely miscible with water and most organic solvents including alcohols, ethers, ketones, aromatic and chlorinated hydrocarbons and vegetable oils
Solubility parameter, δ	11.0 (see Table 14.119)

Table 14.117: Compounds Soluble in "M-Pyrol" *(49)*

Inorganic Compounds

The following inorganic compounds give clear solutions in concentrations of 10% or more.

aluminum chloride hexahydrate
ammonium molybdate
ammonium sulfide
cobalt chloride
lead acetate
lead chloride
lead nitrite
potassium permanganate
potassium thiocyanate
sulfur
zinc chloride

Salts which exhibit very limited solubility include:
cupric sulfate
ferrous sulfate
sodium bromide
sodium nitrite

Table 14.118: Resins Soluble in "M-Pyrol" *(49)*

Test Substance	% Soluble *
Acrylonitrile/Vinyl Chloride Copolymer	>10
"Adiprene" B, urethane rubber (Du Pont)	10 [a]
Black "Tervan" Wax (Esso Standard Oil)	>10
"Cardolite" Resin (Irvington Varnish)	>10
Cellulose Triacetate	>10
"Chemigum," butadiene/acrylonitrile copolymer (Goodyear)	**
"Epi Rez" 510, epoxy resin (Jones Dabney)	—
"Epon" 1000, 1004, 1007 (Shell Chemical)	—
"Estane" 5740X1, 5740X2, polyurethane (Goodrich)	>10
Ethyl Cellulose N-100 (Hercules Powder)	25
"Formvar," polyvinylformal resin (Shawinigan)	5
"Genthane" S, polyurethane (General Tire and Rubber)	>10
"Geon" 101, 102, polyvinyl chloride (Goodrich)	10
"Hycar" OR-type, butadiene/acrylonitrile copolymer (Goodrich)	**
"Kynar" polyvinylidene fluoride resin (Pennsalt)	—
"Lexan," polycarbonate (General Electric)	10 [e]
"Mekon" 20 Wax (Warwick Wax)	>10
"Mylar" polyester film (Du Pont)	>10
Nylon	>10
"Pentalyn" M Resin, pentaerythritol ester of rosin (Hercules Powder)	50
Polyacrylonitrile	
specific viscosity 2.1 [b]	24
specific viscosity 3.1 [b]	18
specific viscosity 8.7 [b]	10
specific viscosity 31.9 [b]	5
Polymethyl-α-chloracrylate	>10
Polymethyl Methacrylate	>10
Polybutene	—
Polystyrene	25
Polyester-type Polyurethane Rubber (Mobay)	10 [c]
Polyvinyl Chloride	>10
Polyvinylpyrrolidone	>10
"Gantrez" AN Resin, methyl vinyl ether/maleic anhydride copolymer (GAF)	>10
Pyromellitic Anhydride	>10
"Vinosol" Ester Gum, neutral glycol ester of resin derived from pine wood (Hercules Powder)	10
"Sumatra" Crude, paraffinic resin (Magnus Chemical) [f]	—
"Vinac" polyvinyl acetate (Colton)	10
"Vinylite" VYHH, VMCH, and VYNS, vinyl chloride/vinyl acetate copolymers (Union Carbide)	>10
"Vinylite" NYGL, vinyl resin (Union Carbide)	25
"Vinylite" VYNW, vinyl chloride resin (Union Carbide)	15
"Zytel," nylon molding resin (Du Pont)	25 [d]

* >10 indicates test on 10 g resin in 100 g M-PYROL solvent; other figures represent limit of solubility for pourable solution; —indicates qualitative test only.
** The economical slab form as well as the crepe or crumb type is soluble.
a Solubility 10% at room temperature; 25% at 80°C.
b Specific viscosity based on 1 g polymer in 100 ml M-PYROL solvent at room temperature.
c Solubility at 80°C.
d Solubility at 200°C.
e 24 hours at 25°C or 1 hour at 60°C; gels on prolonged standing.
f 70° to 80°C (insoluble at R.T.).

Note: The following resins were insoluble in methylpyrrolidone at room temperature under the test conditions: "Delrin" polyacetal resin (Du Pont); "Epi Rez" 510, epoxy resin (Jones Dabney) crosslinked with "Versamid" 125, polyamide (General Mills); "Epolene" N, polyethylene (Eastman Chemical); "Multrathane" MA-40 and MB-40, polyurethane (Mobay Chemical); and "Teflon" fluorocarbon resin (Du Pont, sheet stock).

Table 14.119: Solubility Parameters of Resins in Moderately Hydrogen-Bonded Solvents Compared with the Solubility Parameter of "M-Pyrol" (49)

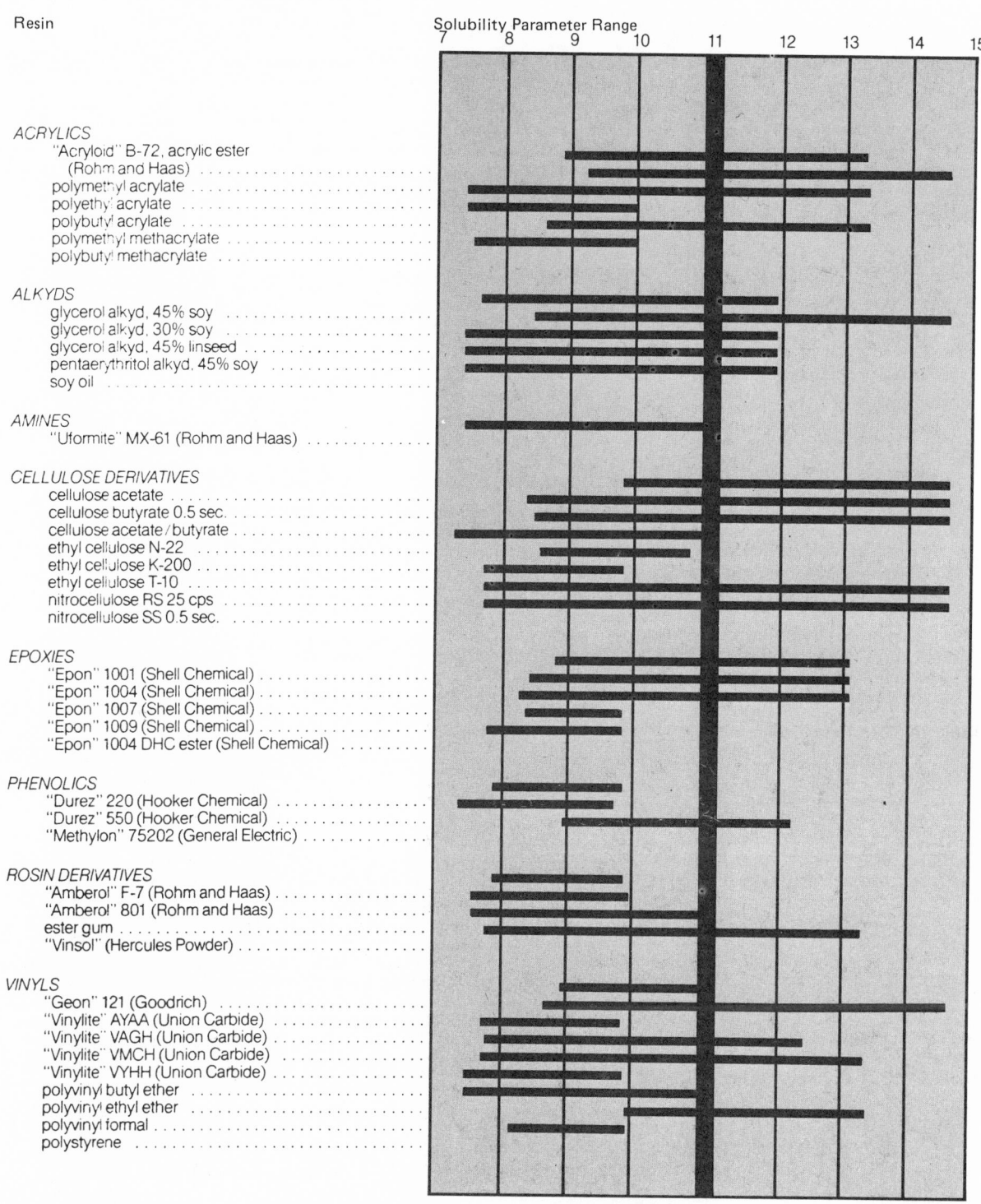

Resin — Solubility Parameter Range (7, 8, 9, 10, 11, 12, 13, 14, 15)

ACRYLICS
- "Acryloid" B-72, acrylic ester (Rohm and Haas)
- polymethyl acrylate
- polyethyl acrylate
- polybutyl acrylate
- polymethyl methacrylate
- polybutyl methacrylate

ALKYDS
- glycerol alkyd, 45% soy
- glycerol alkyd, 30% soy
- glycerol alkyd, 45% linseed
- pentaerythritol alkyd, 45% soy
- soy oil

AMINES
- "Uformite" MX-61 (Rohm and Haas)

CELLULOSE DERIVATIVES
- cellulose acetate
- cellulose butyrate 0.5 sec.
- cellulose acetate/butyrate
- ethyl cellulose N-22
- ethyl cellulose K-200
- ethyl cellulose T-10
- nitrocellulose RS 25 cps
- nitrocellulose SS 0.5 sec.

EPOXIES
- "Epon" 1001 (Shell Chemical)
- "Epon" 1004 (Shell Chemical)
- "Epon" 1007 (Shell Chemical)
- "Epon" 1009 (Shell Chemical)
- "Epon" 1004 DHC ester (Shell Chemical)

PHENOLICS
- "Durez" 220 (Hooker Chemical)
- "Durez" 550 (Hooker Chemical)
- "Methylon" 75202 (General Electric)

ROSIN DERIVATIVES
- "Amberol" F-7 (Rohm and Haas)
- "Amberol" 801 (Rohm and Haas)
- ester gum
- "Vinsol" (Hercules Powder)

VINYLS
- "Geon" 121 (Goodrich)
- "Vinylite" AYAA (Union Carbide)
- "Vinylite" VAGH (Union Carbide)
- "Vinylite" VMCH (Union Carbide)
- "Vinylite" VYHH (Union Carbide)
- polyvinyl butyl ether
- polyvinyl ethyl ether
- polyvinyl formal
- polystyrene

NOTE: The value for M-PYROL solvent δ = 11.0 is based on a total heat of vaporization of 12,200 cal/mol, reference temperature 65°C.

The solubility parameters for resins are based on figures published by Burrell (1957) and Hughes and Britt (1961).

Polystyrene seems to be a notable exception. It is known to be soluble in M-PYROL solvent yet the reported solubility parameter range is only 9.0 ± 0.9 (Burrell 1957).

Table 14.120: Pyridine *(2)*

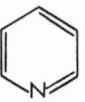

Pyridine is a liquid miscible with water, alcohol, ether, benzene and many organic liquids. It is an excellent solvent for organic materials and will dissolve many metallic salts giving comparatively stable compounds (without substitution). It is used in the preparation of water-proofing chemicals, rubber accelerators, and pharmaceuticals. It is also used as an extractant and in distilling and purifying operations. The less pure grade is used as a denaturant for industrial alcohol.

Boiling point	115°C
Melting point	−42°C
Specific gravity at 25/4°C	0.978

Table 14.121: Alpha-Picoline *(2)*

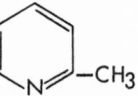

2-Methyl Pyridine

Alpha-picoline is a liquid which is very soluble in water, forming a constant-boiling mixture with it. It is also soluble in ethyl alcohol and ethyl ether. It may be used in the manufacture of alkaloids, pharmaceuticals, antioxidants, and rubber accelerators.

Boiling point	128°C
Melting point	−69.9°C
Specific gravity at 15/4°C	0.950
Distillation Range	Completely within 2°C

Table 14.122: Beta-Picoline *(2)*

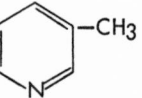

3-Methyl Pyridine

Beta-picoline is similar to the alpha compound. It is soluble in water with which it forms a constant-boiling mixture; it is also soluble in ethyl alcohol and ethyl ether. Suggested uses for it are in the manufacture of alkaloids, pharmaceuticals and rubber accelerators. It is also a starting material for the production of nicotinic acid and nicotinic acid amide.

Boiling point	143.5°C
Melting point	−18.3°C
Purity	95%, min.
Specific gravity at 15/4°C	0.961

Table 14.123: Gamma-Picoline *(2)*

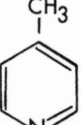

4-Methyl Pyridine

The solubility and uses for this solvent are similar to those of the alpha and beta compounds.

Boiling point	143.1°C
Melting point	+3.8°C
Purity	95%, min.
Specific gravity at 15/4°C	0.957

Table 14.124: 2,4-Lutidine *(2)*

2,4-Dimethyl Pyridine

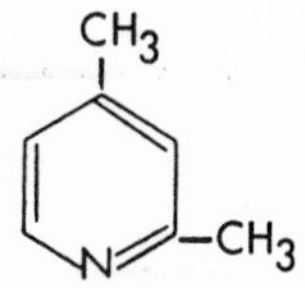

2,4-Lutidine is a liquid, very soluble in alcohols, ketone, ethers, hydrocarbons, and most organic solvents, but only 15% soluble in water. It is recommended for use in the synthesis of drugs, dyes, and other chemicals.

Boiling point	158.3°C
Distillation	90% distills within 2°C
Freezing point	Below −60°C
Specific gravity at 25/4°C	0.927

Table 14.125: 2,6-Lutidine *(2)*

2,6-Dimethyl Pyridine

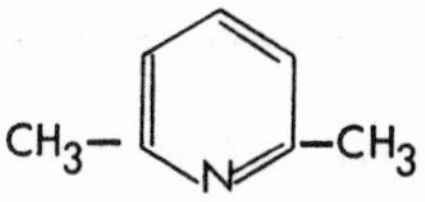

2,6-Lutidine is a liquid, very soluble in water, alcohols, ethers, ketones, hydrocarbons, and most organic solvents. It is recommended for use in the manufacture of resins, dyes, drugs, insecticides, rubbers, and organic chemicals.

Boiling point	142.9°C
Freezing point	−6.0°C
Purity	95%, min.
Specific gravity at 25/4°C	0.928

Table 14.126: Quinoline *(2)*

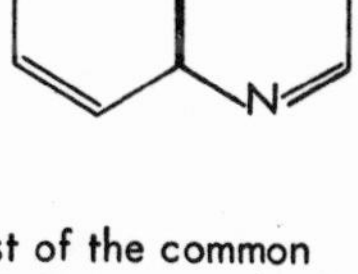

Quinoline is a liquid, soluble in alcohol, ether, carbon disulfide, and in most of the common organic solvents, but only partially soluble in water. It is a solvent for cellulose esters and ethers when used with other solvents. It is used in the manufacture of dyes, photographic sensitizers, nicotinic acid, and drugs. It is also used as an extraction agent and in organic synthesis.

Boiling point	237.7°C
Melting point	−19.5°C
Specific gravity at 20/4°C	1.095
Distillation range	95% distills within 2°C

Table 14.127: 2-Methyl-5-Vinyl Pyridine *(4)*

FORMULA	CH₂ = CH — C, CH / CH, HC, C — CH₃, N
PROPERTIES	
Purity, mol percent (water-free basis)	95.1	94.0 min *
Boiling point, at 50 mm Hg, F	212	
at 160 mm Hg, F	358	
Freezing point, C (water-free basis)	−14.16	−15.14 min
Water content, weight percent	0.20	0.5 max
Refractive index at 25 C	1.541	
Specific gravity of liquid at 60/60 F	0.962	
Density of liquid at 60 F, lbs/gal	8.01	
Color, Gardner	1	2 max
Appearance	Clear	
Polymer content (Hexane Dilution)	Negative	Negative min
Flash point, F (TOC)	165	
Inhibitor content, weight percent (Tertiary Butyl Catechol)	0.1	0.05 min − 0.15 max

Table 14.128: 2-Methylpiperazine *(56)*

	2-methylpiperazine Anhydrous	77% Aqueous Solution
Molecular Weight	100.16	ca. 130 g./mole
Boiling Point, °C.	155.6	--
Melting Point, °C.	65-66	--
Specific Gravity, 25/25°C.	--	0.9958
Flash Point, Open Cup, °C.	73	76.6
Shipping Weight, lbs./gal.	6	8.3
Solubility, % at 25°C.		
Water	78	
95% Ethanol	66	
Acetone	37	
Benzene	32	
Heptane	6	

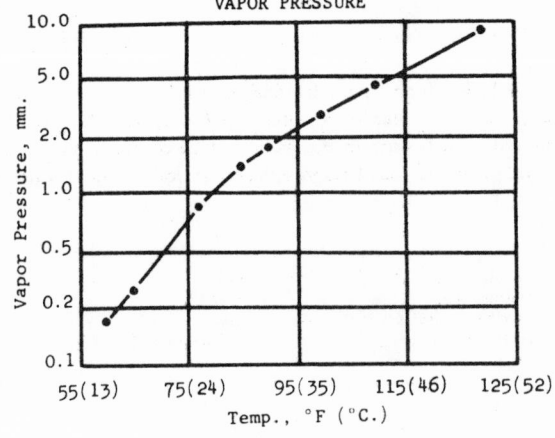

VAPOR PRESSURE

Table 14.129: **1,2,4-Trimethylpiperazine** *(47)*

TYPICAL PHYSICAL PROPERTIES

Form	Liquid
Viscosity at 25°C.	1.037 cps.
pH, 1% Aqueous Solution	10.3
Boiling Point (746 mm)	149° - 151°C.
Freezing Point	<-50°F.
Specific Gravity 25/25°C.	0.851
Refractive Index at 25°C.	1.4480
Fire Point, Cleveland Open Cup	125°F.
Pour Point	<-50°F.

SOLUBILITY

Soluble in water, acetone, methanol and benzene.

AVAILABILITY

1,2,4-trimethylpiperazine is available in semi-commercial quantities.

Table 14.130: **"DHP-MP"** *(47)*

1,4-bis(2-Hydroxylpropyl)-2-Methylpiperazine

TYPICAL PHYSICAL PROPERTIES

Form	Liquid
Viscosity, 25°C.	752 cps.
Boiling Point (3 mm.)	145°C.
Pour Point	10°F.
Specific Gravity 25/25°C.	1.001
Refractive Index, 25°C.	1.4803
Flash Point, Open Cup	300°F.
Color	Light Yellow
Molecular Weight	216
pH, 1% Aqueous Solution	10.0
Analysis, based on tertiary nitrogen content	97%

SOLUBILITY

Miscible in all proportions with water, acetone, ethanol, benzene, heptane, and carbon tetrachloride.

Table 14.131: Morpholine *(48)*

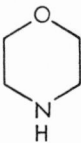

Tetrahydro-p-Oxazine

Molecular weight: 87.12

This commercially important secondary amine is a water-white, mobile liquid having an ammoniacal odor. It is very soluble in water and forms a stable solution which exhibits a constant composition during evaporation and distillation as well as preserving a constant alkalinity. The ring structure of this solvent, as well as its ether and amine groups, gives it unique solvent power for a greater than usual variety of organic substances, among which are resins, dyes, waxes, shellac, and casein.

It is used in permanent wave solutions for its mild alkalinity; in soaps which are emulsifying agents for paper coatings; in rubless polishes, lacquers, paints, insecticides, etc. It imparts water-resistance after drying. Its water-soluble salts have high phenol coefficients. Morpholine may also be used in photographic developing.

Color, Pt-Co scale	15 max.
Boiling range, °C	
IBP	125.0 min.
DP	132.0 max.
Purity, wt.%	98.0 min.
Specific gravity, 20/20°C	0.999 - 1.004
Suspended matter	Substantially free
Freezing point, °C	-4.9
Boiling point, °C	128.9
Flash point, (TOC) °C	38
°F	100
Density, g./cc. at 20°C	0.9994
Refractive index, n_D^{20}	1.4545
Surface tension, dynes/cm. at 20°C	37.5
Viscosity, centipoises at 20°C	2.23
Conductivity, mho/cm. x 10^{10}	6
pK_b	9.61
Dielectric constant	7.33
Dipole moment, Debyes	1.58
Molar polarization, P_∞ in benzene	75.3
Heat capacity, cal./mol./deg. at 25°C	41.6
Heat of vaporization, cal./mol. (45-129°C)	9510

Table 14.132: Boiling Point Composition Curves for Aqueous Morpholine Solutions *(19)*

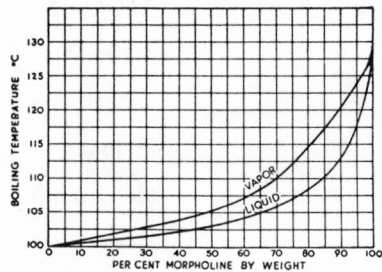

Table 14.133: pH of Aqueous Morpholine Solutions at 25°C *(19)*

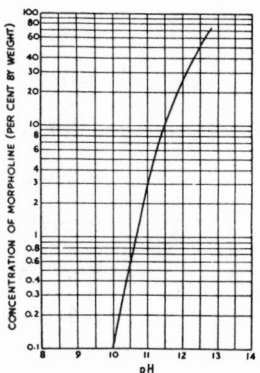

Table 14.134: Viscosity of Aqueous Morpholine Solutions at 20°C *(19)*

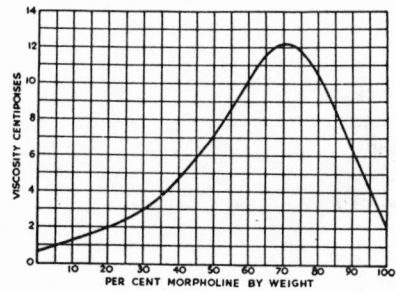

Table 14.135: Solubility of Various Substances in Morpholine *(48)*

Substance	g. Solute in 100 g. Morpholine at 25°C	Substance	g. Solute in 100 g. Morpholine at 25°C
Acetone	∞	2-Hexanone	∞
Beeswax	< 1	Linseed oil	∞
Benzene	∞	Methanol	∞
Benzyl cellulose	> 5	Methylamine	33
Butyl ether	∞	Methylcyclohexanol	∞
Castor oil	∞	Naphtha	> 5
Cellulose acetate	> 5	Paraffin oil	< 1
Cellulose nitrate	> 5	Paraffin wax (hot)	> 5
Copal gum	> 5	Pine oil	∞
Dimethylamine	109	Resin	> 5
Ester gum	> 5	Shellac	> 5
Ethanol	∞	Sulfur	< 5
2-Ethylbutanol	∞	Trimethylamine	34
Ethylene glycol	∞	Turpentine	∞
Ethyl ether	∞	Polyvinyl acetate	> 5
Glycol ether	∞	Polyvinyl butyral	> 5
		Polyvinyl chloride	> 5

Table 14.136: N-Ethyl Morpholine *(2)*

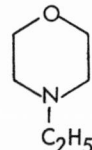

This cyclic tertiary amine is a water-white liquid miscible with water. It may be used as a solvent for oils, dyes and resins, and as an intermediate in the synthesis of rubber accelerators, emulsifying agents, drugs, and dyes.

Boiling Point . . . . 138°C.

Specific Gravity at 20/20°C. . . . 0.916

Table 14.137: N-Phenyl Morpholine *(2)*

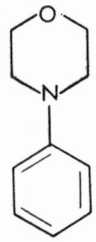

Boiling Point at 760 mm. . . . . . . 268°C.

Melting Point . . . . 57°C.

Esters

FORMATES

Table 15.1: Methyl Formate (2)

$$HCOOCH_3$$

Methyl formate is a colorless flammable liquid with a pleasant ethereal odor. It will dissolve cellulose ethers and esters but will dissolve them more readily when mixed with other solvent esters or the less volatile halogenated hydrocarbons. Mork is of the opinion that methyl formate is a more powerful solvent for cellulose derivatives than acetone. It is more powerful than the other volatile formic esters when used as a fumigant and larvicide in the treatment of tobacco, dried fruit, cereals, etc.

Acidity	Neutral to methyl orange (methyl formate hydrolyzes in presence of water)
Boiling point	31.8° C
Color	Water-white
Distillation range	Below 31.5°C None
	Above 35.0°C None
Electrical conductivity at 25°C	3.6×10^{-8} reciprocal ohms
Flash point	-32°C
Melting point	-99.8°C
Odor initial	Pleasant, ethereal
Odor residual	Non-residual
Purity	95% to 100% ester, by wt
Refractive index at 20°C	1.3431
Solubility in water at 20°C	30% by vol
Solubility of water in solvent at 25° C	24% by vol
Specific gravity at 20/20°C	0.950 to 0.980
Vapor pressure	
0°C	195.0 mm of Mercury
10°C	309.4 mm of Mercury
16°C	400.0 mm of Mercury
20°C	476.4 mm of Mercury
25.8°C	600.0 mm of Mercury
30°C	707.9 mm of Mercury

Table 15.2: Ethyl Formate (2)

Formosol $\qquad$ $HCOOC_2H_5$

Ethyl formate is a water-white, highly volatile and unstable liquid with a pleasant odor resembling peach kernels. It is partly soluble in water and miscible with benzene. It is a powerful solvent for cellulose nitrate and acetate, yielding solutions of unusual low viscosity which have a tendency to chill. It is an important fumigant and larvicide for the treatment of tobacco, cereals, dried fruit and similar products. It is used as a chemical intermediate in the manufacture of such medicinals as sulfadiazine, thiamin (Vitamin B₁), and perfumes and synthetic flavors.

(continued)

Table 15.2: **(continued)**

Acidity	Neutral to methyl orange (it hydrolyzes in the presence of water)
Boiling point	54.3°C
Color	Water-white
Distillation range	51°–55°C
Electrical conductivity at 25°C	Less than 1.45×10^{-8} recip ohms
Flash point	−19°C
Freezing point	−80.5°C
Purity	95% to 100%
Refractive index at 20°C	1.3604
Specific gravity at 20/20°C	0.900 to 0.930
Solubility in water at 20°C	10% by vol
Solubility of water in solvent at 20°C	17% by vol
Vapor pressure at 20.6°C	200 mm of Hg
at 30.2°C	300 mm of Hg
Weight per gal at 68°F	7.61 lbs

Table 15.3: Butyl Formate *(2)*

$$HCOOCH_2CH_2CH_2CH_3$$

Butyl formate is a colorless liquid, miscible with alcohols, ethers, oils, hydrocarbons and so forth. It will dissolve cellulose nitrate, some types of cellulose acetate, and many cellulose ethers. Butyl formate will also dissolve many natural and synthetic resins such as copals, dammar, elemi, mastic, shellac, cumar resins, ester gum and alkyds in the presence of ethyl alcohol. It is used as an intermediate and in perfumes.

Acidity	0.02% max.
Ester content	85% min.
Boiling range	96°–110°C.
Specific gravity	0.885–0.9108

Table 15.4: Amyl Formate *(2)*

$$HCOOC_5H_{11}$$

Commercial amyl formate is an anhydrous, colorless liquid composed of a mixture of isomeric amyl formates with the isoamyl formate in predominance. This mixture is miscible with oils, hydrocarbons, alcohols, ketones and so forth. It is a solvent for cellulose esters, "Cumar", copal, gum esters, etc. It is able, when mixed with an alcohol, to dissolve shellac and alkyd resin. It is a less odoriferous and more energetic solvent than amyl acetate. It also has both a lower boiling point and a greater speed of evaporation. n-Butyl acetate and amyl formate have similar volatility and have substantially the same solvent power which permit free interchange of these only as far as these properties allow.

Acidity	0.05% max.
Boiling point	130.4°C.
Boiling range	110°–130°C.
Flash point	80°F.
Specific gravity	0.880–0.885

Table 15.5: Chlorothiolformates *(61)*

chlorothiolformate	form	color	b.p. °C	MW	d_4^{30}	N_d^{30}	flash pt. °F	fire pt. °F
methyl	liquid	colorless to slightly yellow	110 (atm); 53.0 (90mm)	110.6	1.276	1.4844	100	100
ethyl	liquid	colorless to slightly yellow	132.5 (atm); 70.0 (90mm)	124.6	1.189	1.4763	125	125
n-propyl	liquid	colorless to slightly yellow	155 (atm with slight decomp.) 89.5 (90mm)	138.6	1.134	1.4758	145	145
iso-propyl	liquid	colorless to slightly yellow	142 (atm with slight decomp.) 80.5 (90mm)	138.6	1.119	1.4684	130	140
n-butyl	liquid	colorless to slightly yellow	177.5 (atm., slight decomp.) 109.5 (90mm)	152.6	1.098	1.4737	165	165
t-butyl	liquid	colorless to slightly yellow	42.0 (10mm)	152.6	1.081	1.4691	115	125
n-octyl	liquid	colorless to slightly yellow	124 (10mm)	208.8	1.015	1.4713	260	280
phenyl	liquid	yellow	100-101 (10mm); m.p. − 14°C	172.6	1.269	1.5786	240	270
benzyl	liquid	colorless to slightly yellow	80.0 (0.13mm)	186.5	1.237	1.5711	245	265
p-chlorophenyl	liquid	yellow	126-126.5 (10mm); m.p. + 20.5°C	207.1	1.395	1.5961	290	300

ACETATES

Table 15.6: Methyl Acetate *(2)*

$$CH_3COOCH_3$$

Methyl acetate is a water-white flammable, readily hydrolyzable liquid, with a fragrant odor. This low-boiling solvent was first prepared in 1835 by reacting acetic acid and methanol. It is miscible with most organic solvents and will completely dissolve cellulose nitrate and acetate, ethyl cellulose, resins such as ester gum, rosin, "Cumar", elemi, phenolics, and oils such as corn, linseed, castor, neatsfoot, chinawood and cottonseed. It will only partially dissolve shellac, manila, dammar, pontianac, Beckacites and alkyds. In many respects, methyl acetate resembles acetone as a solvent, particularly as to its boiling point, solvent power and miscibility, but its tendency to hydrolyze to methanol and acetic acid, in the presence of water, limits its wider use in the industries. Methyl acetate is usually admixed with higher boiling solvents. It is used in lacquers, paints, varnishes, enamels, perfumes, dyes, dopes, plastics, and synthetic finishes as well as a substitute for acetone.

Acidity (as acetic)	0.005%, max	Freezing point	−98.1°C
Boiling point	56.9°C	Heat of combustion	5371 cal/g
Distillation range	55-58°C	Heat of vaporization	104.4 cal/g
Coefficient of expansion (per °C) at 20°C	0.001390	Non-volatile matter	0.005 gram per 100 cc, max
Color	Water-white	Refractive index at 20°C	1.3593
Critical temperature	233.7°C	Solubility in water at 20°C	24% by wt
Critical pressure	46–3 atm	Solubility of water in solvent at 20°C	8% by wt
Dielectric constant at 20°C	7.3 ± 0.2	Specific gravity at 20/20°C	0.9353
Dilution ratios		Surface tension at 20°C	24.6 dynes/cm
Toluene	2.9	Vapor pressure at 20°C	173 mm Hg
Petroleum naphtha	0.9	Viscosity at 20°C	0.00381 poises
Electrical conductivity at 25°C	3.4×10^{-6} mho	Weight per gal at 20°C	7.783 lbs
Flash point (A.S.T.M. Open Cup)	−15°C		

Table 15.7: Ethyl Acetate *(2)*

Acetic Ether $CH_3COOC_2H_5$

Ethyl acetate is a water-white, flammable liquid with a pleasant, fruity odor. The 85 to 88 per cent grade of ethyl acetate suitably denatured is generally used for commercial purposes but 95 and 99 percent grades are also available. It is miscible with most organic solvents such as alcohols, ketones, esters, aromatic, aliphatic and halogenated hydrocarbons. It dissolves such materials as nitrocellulose, camphor, oils, fats, waxes, gums and natural and synthetic resins. It will tolerate fairly large amounts of lacquer diluents and like methyl acetate it not only has a wide range of solubilities but it possesses the unique property of dissolving nitrocellulose, cellulose acetate and cellulose ethers yielding solutions of low viscosity. Its solvent power for cellulose derivatives is much improved, however, by adding a small quantity of alcohol.

(85 to 88%)

Acidity (as acetic acid)	0.01% by wt, max
Blush resistance at 90°F (10% $\frac{1}{4}$ sec. R.S. nitrocellulose solution)	Clear 45% Relative humidity
	Bluish 50%
Coefficient of cubical expansion (ordinary temperatures)	0.00073 per °F
	0.00132 per °C
Color	Water-white
Critical temperature	250.1°C
Critical pressure	37.8 atmospheres
Dilution ratio	
Toluol	3.5
Petroleum naphtha	1.1
Distillation range	Below 70°C None
	Below 72°C Not more than 10%
	Above 80°C None
Dryness	Miscible without turbidity with 20 volumes 60° Bé gasoline at 20°C
Electrical conductivity at 25°C	Less than 1 × 10⁻⁹ reciprocal ohms
Evap. rate at 95°F (in min.)	
5%	$\frac{1}{4}$
25%	$1\frac{1}{2}$
50%	$3\frac{3}{4}$
75%	$6\frac{1}{4}$
90%	$9\frac{1}{2}$
95%	$11\frac{1}{4}$
Flash point	−5°C (23°F)
Freezing point	−82.4°C
Non-volatile matter	0.003 gram per 100 cc, max
Refractive index at 20°C	1.3725
Solubility in water at 25°C	9.7% by vol
Solubility of water in solvent at 25°C	9.8% by vol
Specific gravity at 20/20°C	0.883 to 0.888
Viscosity at 20°C	4.546 millipoises
Weight per gal at 68°F	7.36 lbs

(95 to 98%)

Acidity (as acetic)	0.01% by wt, max
Blush resistance at 90°F (10% $\frac{1}{4}$ sec. R.S. nitrocellulose solution)	Clear 50% Relative humidity
	Blush 55%
Coefficient of expansion per 1°F	0.00074
per 1°C	0.00133
Color	Water-white
Dilution ratio	
Toluol	3.2
Petroleum hydrocarbon	1.0
Distillation range	74 to 80°C
Dryness at 20°C	Miscible without turbidity with 20 vol 60° Bé gasoline

(95 to 98%)

Evap. rate at 95°F (in min.)	
5%	$\frac{1}{4}$
25%	$1\frac{1}{2}$
50%	$3\frac{1}{2}$
75%	6
90%	$7\frac{1}{4}$
95%	$8\frac{1}{4}$
Flash point	26°F (approximate)
Non-volatile matter	0.003 gm per 100 cc, max
Solubility of water in solvent at 25°C	4% by vol
Specific gravity at 20/20°C	0.895 to 0.900
Viscosity (10% $\frac{1}{4}$ sec. nitrocellulose solution)	33 centipoises
Water	No turbidity when mixed with 19 volumes of 60° Bé gasoline at 20°C
Weight per gal at 20°C	7.47 lbs

(99 to 100%) acetic ether grade

Acidity (as acetic)	0.01% by wt, max
Blush resistance at 90°F (10% $\frac{1}{4}$ sec. R.S. nitrocellulose solution)	Clear 55% Relative humidity
	Blush 60%
Coefficient of expansion per 1°F	0.00074
per 1°C	0.000133
Color	Water-white
Dilution ratio	
Toluol	3.0
Petroleum naphtha	1.0
Distillation range	75 to 80°C
Electrical conductivity at 25°C	3.2 × 10⁻⁷ recip ohms
Evap. rate at 95°F (in min.)	
5%	$\frac{1}{4}$
25%	$1\frac{1}{2}$
50%	$3\frac{1}{4}$
75%	$6\frac{1}{2}$
90%	$9\frac{1}{4}$
95%	$10\frac{1}{4}$
Explosive limits	2.26–11.4%
Flash point	0.5°C
Freezing point	−82.4°C
Heat of combustion	538 kg. cal/mole
Heat of vaporization at 0°C	102 cal/gm
at 80°C	102.9 cal/gm
Non-volatile matter	0.003 gram when 100 cc, max

Table 15.8: n-Propylacetate *(41)*

$CH_3COOC_3H_7$

The properties of n-propyl acetate are, approximately, intermediate between those of ethyl and n-butyl acetates. It is miscible with alcohols, ketones, esters, oils and hydrocarbons and is a good solvent for nitrocellulose and a wide range of cellulose derivatives, especially when it is admixed with the aromatic hydrocarbons or the lower aliphatic alcohols. It will also dissolve natural and synthetic resins like elemi, "Cumar" resins, ester gum, manila, mastic, rosin and sandarac. It is used principally as a low-boiling component in nitrocellulose lacquer formulations.

(continued)

Table 15.8: (continued)

Molecular Weight (Theoretical)	102.08	Boiling Range, 760 mm, °C	
Color (Pt-Co Scale), max	15	Initial Boiling Point, min	99
Weight/Vol, 20°C,		Dry Point, max	103
lb/gal (U.S.)	7.39	Freezing Point, °F (°C)	-131 (-93)
kg/liter	0.89	Flash Point, Tag Closed Cup, °F (°C)	55 (13)
lb/gal (Imperial)	8.87	Tag Open Cup, °F (°C)	58 (14)
Solubility, 20°C, wt %		Fire Point, °F (°C)	70 (21)
In water	2.3	Flammable Limits in Air, % by volume	
Water in	2.6	Lower, at 100°F (38°C)	1.71
Evaporation Rate (n-butyl acetate = 1)	2.3	Upper, at 200°F (93°C)	7.95
Dilution Ratio, toluene	3.2	Autoignition Temperature (ASTM D-2155), °F (°C)	855 (457)
VM & P naphtha	1.5	NFPA Classification 30:	IB
Refractive Index, 20°C	1.3844	ICC Labels Required	Red
Vapor Pressure, 20°C, mm Hg	23	Bureau of Explosives Classification	Flammable Liquid
Specific Gravity, 20°/20°C	0.885		

Table 15.9: Isopropyl Acetate (2)

$$CH_3COOCH(CH_3)_2$$

Isopropyl acetate is a water-white pleasant-odored liquid with properties intermediate between ethyl and butyl acetates. It is miscible with most of the common organic solvents such as alcohols, ketones, esters, oils, hydrocarbons, etc., and it is a solvent for nitrocellulose, cellulose acetate (of low viscosity) and a wide range of oils, fats, waxes, gums and natural and synthetic resins. Like n-propyl acetate, its solvent power for cellulose esters is increased when lower aliphatic alcohols are added. It is largely used in the lacquer industry where its slow evaporation rate and blush resistance are of importance. It is also used in the manufacture of plastics, artificial leather, dopes, films, cements, and in the recovery of acetic acid from aqueous solutions.

Acidity (as acetic)	0.02% by wt, max	Flash point (Tag Closed Cup)	39°F
Boiling point at 760 mm	88.6°C	Freezing point	−73.4°C
Coefficient of expansion per °F	0.000727	Heat of vaporization	79.4 cal/gm
Color	Water-white	Non-volatile matter	2 mg per 100 cc, max
Dilution ratios		Refractive index, N 20/D	1.3772
Toluene	2.7	Solubility of water in solvent	3.2% by wt
V.M. and P. naphtha	0.92	Specific gravity at 20/20°C	0.866 to 0.871
Distillation range	84.5–90°C	Specific heat at 15–25°C	0.521 cal/gm
Electrical conductivity at 25°C	5.7 × 10⁻⁷ recip ohms	Surface tension at 25°C	24.5 dynes/cm
Evaporation rate at 95°F (in minutes)		Vapor pressure at 10°C	26.2 mm Hg
5%	½	20°C	45.7 mm Hg
25%	1¾	30°C	76.1 mm Hg
50%	4¼	40°C	121.8 mm Hg
75%	7¼	Viscosity at 20°C	0.00525 poises
90%	10¼	Weight per gallon at 20°C	7.23 lbs
95%	11½		

Constant Boiling Mixtures

				% by Wt	B.P. –C
Isopropyl Acetate	47.7	Isopropanol		52.3	80.1
Isopropyl Acetate	89.4	Water		10.6	76.6

Table 15.10: n-Butyl Acetate (2)

$$CH_3COOCH_2CH_2CH_2CH_3$$

This ester is a water-white liquid with a characteristic fruity odor which is less pronounced than the odor of amyl acetate. It is miscible with alcohols, ketones, esters and most of the common organic solvents. It is a solvent for nitrocellulose and cellulose ethers, especially when previously mixed with active or latent solvents. It will dissolve oils, fats, waxes, metallic resinates, camphor, "Cumar" resins, dammar, ester gum, elemi, kauri, manila, mastic, pontianac, rosin, sandarac, chlorinated rubber, and such synthetic resins as the vinyls, polystyrene, and acrylates. In combination with 20 per cent of butyl alcohol, butyl acetate will dissolve the less highly polymerized alkyd resins and shellac. Owing to its power of imparting low viscosity, gum compatibility, and good working qualities, it is classed among the best medium boiling solvents for nitrocellulose. Its volatility meets the demands of a lacquer solvent because it is sufficiently high to leave the film readily and at the same time low enough to prevent blushing. When combined with butyl alcohol it will prevent

(continued)

Table 15.10: (continued)

gum-blush, cotton-blush and chilling. Its largest use is as a solvent in the manufacture of nitrocellulose lacquers for protective coatings, artificial leather and coated paper, plastics, polishes, safety glass, in perfumes, and flavoring materials.

<u>88-92%</u>

Acidity (as acetic)	0.01% by wt, max
Boiling point at 760 mm Hg	126.5°C
Coefficient of cubical expansion	
(ordinary temperatures)	0.00067 per°F
	0.00121 per°C
Color	Water-white
Dielectric constant at 20°C	5.0
Dilution ratio	
Toluol	2.9
Petroleum naphtha	1.4
Distillation range	Below 115°C None
	Below 120°C Not more than 8%
	Above 130°C Not more than 5%
	Above 135°C None
Evaporation rate at 95°F (in minutes)	
5%	$1\frac{1}{4}$
25%	$6\frac{1}{2}$
50%	$13\frac{1}{2}$
75%	$22\frac{1}{4}$
90%	$3\frac{1}{4}$
95%	$34\frac{1}{4}$
Flash point	28°C
Heat of vaporization	73.8 calories per gm
Freezing point	−76.8°C
Non-volatile matter	0.005 gram per 100 cc, max
Residue	None
Refractive index at 20°C	1.3947
Solubility in water at 25°C	0.5% by vol
Solubility of water in solvent at 25°C	1.6% by vol
Specific gravity at 20/20°C	0.872 to 0.880
Vapor pressure at 0° C	3.0 mm Hg
25°C	15.0 mm Hg
50°C	45.0 mm Hg
Viscosity at 25°C	0.671 centipoises
Weight per gallon at 20°C	7.28 lbs

<u>98-100%</u>

Acidity (as acetic)	0.01% by wt, max
Boiling point	126.5°C
Coefficient of expansion per 1°F	0006
per 1°C	0.0011
Color (A.P.H.A.)	10 max
Dilution ratio	
Toluol	3.05
Petroleum naphtha	1.40
Distillation range	123°-128°C
Electrical conductivity at 25°C	13×10^{-9} recip ohms
Flash point	82°F approx
Fractionation: I.P.	114.3°
10%	116.0
25%	116.9
50%	117.4
75%	117.6
90%	117.6
E.P.	118.1
Heat of vaporization	73.8 cal/gm
Non-volatile matter	Negligible
Refractive index at 20°C	1.3951
Solubility in water at 25°C	0.78% by wt
Solubility of water in solvent at 25°C	2.88% by wt
Specific gravity at 20/20°C	0.879 to 0.883
Specific heat at 21-27°C	0.505 cal/gm
Surface tension at 27°C	27.6 dynes/cm
Vapor pressure at 20°C	9.0 mm Hg
Viscosity at 25°C	0.00693 poises
Weight per gallon	7.76 lbs

Constant Boiling Mixtures

	% by Wt		% by Wt	B.P. °C
n-Butyl acetate	54.0	n-Butanol	46.0	118.0
n-Butyl acetate	60.0	n-Propanol	40.0	94.2
n-Butyl acetate	48.0	Isopropanol	52.0	80.1
n-Butyl acetate	71.3	Water	28.7	90.2

Ternary Mixtures

		B.P.
n-Butyl acetate	35.3%	
n-Butanol	27.4%	89.4°C
Water	37.3%	

Table 15.11: sec-Butyl Acetate *(2)*

$$CH_3COOCH(CH_3)CH_2CH_3$$

sec-Butyl acetate is a colorless, flammable liquid with a fruity odor. It is miscible with castor and linseed oils and hydrocarbons, and will dissolve nitrocellulose, cumarone, elemi, ester gum, kauri, mastic, manila, pontianac, asphalt and tar. It has only partial solubility for dammar and shellac. Its solvency is very similar to n-butyl acetate but it has a lower boiling range, less blush resistance and will evaporate with greater rapidity. For this reason, to replace n-butyl acetate it is necessary that this solvent should be mixed with slower evaporating solvents that can make up for its quicker rate of evaporation. It is largely used in the manufacture of nitrocellulose lacquers and similar types of coatings used in airplane dopes, artificial leather, celluloid products, coated paper, patent leather, and textile sizing and printing compounds.

Acidity (as acetic)	0.03% by wt, max
Blush resistance at 60°F	Clear 75% Relative humidity
	Blush 80%
Coefficient of expansion per 1°F	0.00063
per 1°C	0.00113
Color	Water-white
Distillation range:	
Below 104°C	None
Below 111°C	Not more than 10%
Below 114°C	Not more than 60%
Below 118°C	Not less than 90%
Above 130°C	None

(continued)

Table 15.11: (continued)

Evaporation rate at 95°F (in minutes)		Non-volatile matter	0.005 gm per 100 cc, max
5%	½	Residue	None
25%	3¾	Refractive index, N 25.3/D	1.3866
50%	8½	Specific gravity at 20/20°C	0.862-0.866
75%	13½	Solubility in water	0.74% by wt
90%	16¾	Solubility of water in solvent at 25°C	2.1% by wt
95%	18		
Flash point	66°F	Weight per gal at 20°C	7.19 lbs (approx.)

Table 15.12: Isobutyl Acetate (41)

$$CH_3COOCH_2CH(CH_3)_2$$

Isobutyl acetate is a medium-boiling solvent, colorless and with a mild, fruity ester odor. The commercial grade has an ester content of 88 to 92 percent, the balance being substantially isobutyl alcohol. The solvent power of this ester is similar to the normal and secondary acetates. It is miscible with most organic solvents and will dissolve a large number of oils, waxes and natural and synthetic resins. With the limitation set by Rule 66 on the use of branched chain ketones and aromatic solvents, isobutyl acetate is an economical replacement for MIBK and by having a similar evaporation rate, can be formulated in toluene replacements.

Molecular Weight ($C_6H_{12}O_2$)	116.2	Boiling Range, 760 mm, °C	
Color (Pt-Co Scale), max	10	Initial Boiling Point, min	112
Weight/Vol, 20°C,		Dry Point, max	119
lb/gal (U.S.)	7.25	Freezing Point, °F (°C)	-146 (-99)
kg/liter	0.87	Flash Point, Tag Closed Cup, °F (°C)	69 (20)
lb/gal (Imperial)	8.70	Tag Open Cup, °F (°C)	75 (24)
Solubility, 20°C, wt %		Fire Point, °F (°C)	88 (31)
In water	0.7	Flammable Limits in Air, % by volume	
Water in	1.6	Lower, at 200°F (93°C)	1.27
Evaporation Rate (n-butyl acetate = 1)	1.4	Upper, at 200°F (93°C)	7.5
Dilution Ratio, toluene	2.7	Autoignition Temperature (ASTM D-2155), °F (°C)	800 (427)
VM & P naphtha	1.1	NFPA Classification 30:	1B
Refractive Index, 20°C	1.3997	DOT Labels Required	Red
Vapor Pressure, 20°C, mm Hg	12.5	DOT Classification	Flammable Liquid
Specific Gravity, 20°/20°C	0.870		

Table 15.13: Amyl Acetate (2)

Banana Oil
Amyl Acetic Ether
Isoamyl Acetate

$$CH_3COOC_5H_{11}$$

Amyl acetate is a colorless, flammable liquid with an odor resembling bananas or pears. Its exceptional solvent power places it among the first solvents in the nitrocellulose lacquer industry. The amyl acetates are made by the acetylation of fusel oil or synthetic amyl alcohols. Amyl acetate is miscible with oils, hydrocarbons, alcohols, ethers and esters, and will dissolve such substances as camphor, elemi, ester gums, copal ester, copals, dammar, kauri, rosin, sandarac, tannins, waxes, zanzibar and "Cumar" resins, and when it is mixed with alcohol it will dissolve some alkyd resins. It is a good solvent for cellulose esters and ethers, the solvency of which is increased when combined with ethyl alcohol. Amyl acetate is used extensively as a solvent in nitrocellulose lacquers, both for its solvency and its power to impart blush resistance, good flow, gloss and toughness. It is also used in making smokeless powder, artificial leather and pearls, airplane dopes, waterproofing compositions, varnishes, dry cleaning compounds, bronzing liquid, films, celluloid, rayon, linoleum, oilcloth, fruit flavors, soft drinks, food preparations, confectionery, perfumes, soap solvent, and in photoengraving.

	85-88% technical grade
Acidity (as acetic)	0.03% by wt, max
Blush resistance at 90°F (10% ½ sec R.S. nitrocellulose solution)	Clear 80% Relative humidity Blush 85%
Coefficient of expansion per 1°F	0.00066
per 1°C	0.00119
Color	Water-white
Dilution ratio	
Toluol	2.7
Petroleum naphtha	1.4

(continued)

Table 15.13: (continued)

85–88% technical grade

Distillation range
Below 110°C	Not more than 15%
Below 120°C	Not more than 30%
Below 130°C	Not more than 55%
Below 140°C	Not less than 80%
Above 150°C	None
Dryness	Miscible without turbidity with 20 vols 60° Bé gasoline at 20°C

Evaporation rate at 95°F (in minutes)
5%	1
25%	6
50%	15½
75%	28½
90%	41½
95%	47½
Flash point	63°F
Non-volatile matter	0.005 gm per 100 cc, max
Solubility of water in solvent at 25°C	2.4% by vol
Specific gravity at 20/20°C	0.857–0.865
Viscosity (10% ½ sec. R.S. nitrocellulose solution)	43 centipoises
Weight per gal at 20°C	7.17 lbs

(High Test 85–88%)

Acidity (as acetic)	Not more than 0.03%
Blush resistance at 90°F (10% ½ sec. R.S. nitrocellulose solution)	Clear 85% Relative humidity Blush 90%
Coefficient of expansion per 1°F	0.00066
per 1°C	0.00119
Color	Water-white

Dilution ratio
Toluol	2.5
Petroleum naphtha	1.4

Distillation range
Below 110°C	None
Below 120°C	Not more than 10%
Below 130°C	Not more than 40%
Below 140°C	Not less than 60%
Below 150°C	None

Evaporation rate at 95°F (in minutes)
5%	1½
25%	8½
50%	18½
75%	30
90%	45
95%	51¾
Flash point	84°F

(High Test 85–88%)

Non-volatile matter	Not more than 0.005 gm per 100 cc
Odor	Mild
Odor residual	Non-residual
Purity	Ester content as amyl acetate 85–88%
Solubility of water in solvent at 25°C	100 cc solvent dissolves 1.8 cc water
Specific gravity at 20/20°C	0.859–0.863
Viscosity (10% ½ sec. R.S. nitrocellulose solution)	61 centipoises
Weight per gal at 20°C	7.17 lbs

90–95% grade

Acidity (as acetic)	0.035% by wt, max
Blush resistance at 90°F (10% ½ sec. R.S. nitrocellulose solution)	Clear 85% Relative humidity Blush 90%
Coefficient of expansion per 1°F	0.00061
per 1°C	0.00110
Color	Water-white

Dilution ratio
Toluol	2.6
Petroleum naphtha	1.4

Evaporation rate at 95°F (in minutes)
5%	1¼
25%	7
50%	15½
75%	25¾
90%	35
95%	40
Flash point	79°F

Distillation range
Below 110°C	None
Below 120°C	Not more than 10%
Below 130°C	Not more than 70%
Below 140°C	Not less than 90%
Above 150°C	None
Non-volatile matter	0.005 gram per 100 cc, max
Residue	None
Solubility in water at 25°C	2% by vol
Solubility of water in solvent at 25°C	2% by vol
Specific gravity at 20/20°C	0.868 to 0.872
Viscosity (10% ½ sec. R.S. nitrocellulose solution)	60 centipoises
Weight per gal at 20°C	7.22 lbs

Table 15.14: sec-Amyl Acetate *(2)*

$$CH_3COOC_5H_{11}$$

This medium boiling solvent is a water-white liquid made by acetylation from the secondary amyl alcohols (pentanol-2 and pentanol-3). Although its solvent power is not equal to that of amyl acetate because it has a lower tolerance for diluents, as well as a less desirable odor, it has many similar properties and it is used as a nitrocellulose and ethyl cellulose solvent.

Acidity (as acetic)	0.03% by wt, max
Blush resistance at 90°F (10% ½ sec. R.S. nitrocellulose solution)	Clear 85% Relative humidity Blush 90%
Coefficient of Expansion per 1°F	0.00060
per 1°C	0.00108
Color	Water-white

Dilution ratio
Toluol	2.1
Petroleum naphtha	1.1

(continued)

Table 15.14: (continued)

Distillation range	
Below 123°C	None
Below 126°C	Not more than 10%
Below 132°C	Not more than 60%
Below 140°C	Not less than 90%
Above 145°C	None
Evaporation rate at 95°F (in minutes)	
5%	1½
25%	7½
50%	15½
75%	24½
90%	30½
95%	33½
Flash point	89° F
Non-volatile matter	0.005 gm per 100 cc, max
Residue	None
Specific gravity at 20/20°C	0.862-0.866
Solubility of water in solvent at 25°C	0.8% by vol
Viscosity (10% ½ sec. R.S. nitro-cellulose solution)	75 centipoises
Weight per gal at 20°C	7.19 lbs

Table 15.15: Pentacetate *(2)*

Pentacetate made from synthetic amyl alcohol is a mixture of five isomeric amyl acetates with some free amyl alcohol. It is soluble in methanol, ethyl ether, ethyl acetate, fixed oils, acetone, oleic acid, hot stearic acid, and aromatic and aliphatic hydrocarbons. It is soluble in hot paraffin and carnauba waxes but these congeal on cooling. It is insoluble in water. The solvent power of this mixture being similar to that of amyl acetate, pentacetate finds its most important use in the manufacture of nitrocellulose lacquers. It is also used as an extractant in the production of penicillin. It also finds use in various types of poison bait.

Some of the esters to be found in Pentacetate are:

	B.P.
$CH_3CH_2CH_2CH_2CH_2OOCCH_3$	148°C
$(CH_3)_2CHCH_2CH_2OOCCH_3$	142°C
$CH_3CH_2CH(CH_3)CH_2OOCCH_3$	142°C
$CH_3CH(OOCCH_3)CH_2CH_2CH_3$	134°C
$CH_3CH_2CH(OOCCH_3)CH_2CH_3$	132°C

Acidity (as acetic)	0.03% by wt max
Coefficient of expansion at 10 to 35°C	0.00110 per °C
Color	Water-white
Distillation range	
100%	Above 126°C
95%	Above 130°C
75%	Above 135°C
25%	Above 140°C
End point	Not above 150°C
Dilution ratio	
Toluol	2.1
Petroleum naphtha	1.3
Evaporation rate at 114.8°F	
Minutes 3.92	25%
Minutes 7.92	50%
Minutes 12.50	75%
Minutes 20.00	100%
Flash point (Open Cup)	107°F
Heat of vaporization	68.5 cal/gm
Non-volatile	0.005 gm/100 cc max
Refractive index at 20°C	1.4013
Specific gravity at 20/20°C	0.860-0.870
Solubility in water	1% by vol
Solubility of water in solvent	1.5% by vol
Viscosity at +40°C	0.683 centipoises
−40°C	3.464 centipoises
Water azeotropic mixture at 92-95°C	67% Pentacetate (approx)
	33% water by vol
Water content	None
Weight per gal	7.21 lbs

Table 15.16: Methyl Amyl Acetate *(2)*

Methyl Isobutyl Carbinol Acetate $CH_3COOCH(CH_3)CH_2CH(CH_3)_2$

Methyl amyl acetate is a colorless liquid with a mild and pleasant odor. This medium boiling solvent is used in nitro-cellulose lacquer fabrication producing such advantages as blush resistance, reduction of "orange peel" in the lacquer film, and no swelling of oilbase undercoats.

Acidity (as acetic)	0.02% by wt, max	Color	Water-white
Boiling range at 760 mm	Below 140°C None	Dryness	Miscible with 19 vol 60° Bé gasoline at 20°C
	Above 150°C None	Purity	95% by wt, max
	Not more than 5% distills below 143°C	Specific gravity at 20/20°C	0.855 to 0.860
	Not less than 95% distills below 148°C	Weight per gal at 20°C	7.14 lbs

Table 15.17: 2-Ethyl Butyl Acetate *(2)*

2-Ethyl butyl acetate is a colorless liquid having a mild odor. It is a solvent for nitrocellulose, gums and resins, and is employed as a high-boiling solvent in lacquers.

Acidity (as acetic)	0.01% by wt, max	Boiling range	155°–164°C
Color	Water-white	Purity	90% min
Specific gravity	0.875 to 0.881 at $\frac{20°C}{20°C}$	Dryness	Miscible with 19 vols Bé gasoline at 20°C
		Average weight	7.33 lbs/gal (20°C)

Table 15.18: 2-Ethylhexyl Acetate *(41)*

Ethylhexyl acetate is a water-white, stable liquid. It will dissolve nitrocellulose and many of the natural and synthetic resins. It is used in slow-evaporating preparations such as brushing and dipping lacquers, mist coatings, baking finishes and lacquer emulsions.

Molecular Weight ($C_{10}H_{20}O_2$)	172.26	Fire Point, °F (°C)	187 (86)
Color (Pt-Co Scale), max	15	Flammable Limits in Air, % by volume	
Weight/Vol, 20°C,		Lower, at 200°F (93°C)	0.76
lb/gal (U.S.)	7.26	Upper, at 300°F (149°C)	8.14
kg/liter	0.87	Autoignition Temperature (ASTM D-2155), °F (°C)	515 (268)
lb/gal (Imperial)	8.71	NFPA Classification 30	IIIA
Solubility, 20°C, wt %		DOT Labels Required	None
In water	0.03	DOT Classification	Nonhazardous Liquid
Water in	0.55	Color (Pt-Co Scale), ppm, max	15
Evaporation Rate (n-butyl acetate = 1)	0.03	Specific Gravity, 20°/20°C	0.870 - 0.875
Dilution Ratio, toluene	1.4	Acidity, as acetic acid, wt %, max	0.02
VM & P naphtha	0.9	Boiling Range, 760 mm, °C	
Refractive Index, 20°C	1.4103	Initial boiling point, min	192.0
Vapor Pressure, 20°C, mm Hg	0.4	Dry point, max	205.0
Specific Gravity, 20°/20°C	0.872	Ester Content, wt %, min	95.0
Boiling Range, 760 mm, °C		Water, wt %, max	0.2
Initial Boiling Point, min	192.0	Odor	Mild
Dry Point, max	205.0		
Freezing Point, °F (°C)	−135 (−93)		
Flash Point, Tag Closed Cup, °F (°C)	160 (71)		
Tag Open Cup, °F (°C)	175 (79)		

Table 15.19: Cyclohexyl Acetate *(2)*

Hexalin Acetate
Hexahydrophenyl Acetate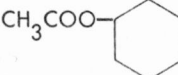
Adronol Acetate

Cyclohexyl acetate is a colorless, water-insoluble liquid with an odor resembling that of amyl acetate. It is miscible in all proportions with most of the lacquer solvents and diluents, with halogenated and hydrogenated hydrocarbons, and will completely dissolve waxed dammar and unrun congo copal. It is a good solvent for cellulose ethers and nitrocellulose and has powerful solvency for basic dyes, blown oils, raw rubber, metallic soaps, driers, shellac, bitumens, and a wide range of natural and synthetic resins and gums. It is used in spraying and brushing lacquers imparting blush resistance and good flow.

(continued)

Table 15.19: (continued)

Boiling range	160–180°C
Color	Water-white
Flash point	64°C
Freezing point	−65°C
Purity	88–95% min
Residue	None
Refractive index	1.435–1.445
Specific gravity at 20°C	0.968–0.972
Water	None
Viscosity (S.U.V. at 100°F)	32

Table 15.20: Methyl Cyclohexanyl Acetate *(2)*

Sextate
Methyl Hexalin Acetate
Hexahydrocresol Acetate
Hexahydromethylphenol Acetate

Methyl cyclohexanyl acetate is a colorless high-boiling liquid having an ester-like odor. Its miscibility and solvent action are quite similar to those of cyclohexanyl acetate but it is slower acting. It is a solvent for nitrocellulose, basic dyes, rubber, bitumens, oils, fats and waxes, and for such resins as dammar, elemi, manila, mastic, rosin, ester gum, phenolic and vinyl resins. It will dissolve, in a lesser degree, shellac, kauri and cellulose acetate. It is used as a high-boiling solvent in nitrocellulose lacquers for both spraying and brushing purposes. Its solvency and slow rate of evaporation impart resistance to blushing and good working qualities and produce films that are smooth, homogeneous and glossy. Its dilution ratio with various diluents are as follows:

Xylene	2.5
Toluene	2.5
Benzene	2.0
White spirits	1.5

Acidity	4.04% (max.)
Ester content	80–90%
Boiling range	175°–190°C.
Flash point	66°–69°C.
Specific gravity	0.95

Table 15.21: Ethylene Glycol Monoacetate *(2)*

$$CH_2OH$$
$$|$$
$$CH_2OOCCH_3$$

Ethylene glycol monoacetate is a colorless, odorless liquid and is structurally a primary alcohol and an ester. It is made by combining a dihydric alcohol and a monocarboxylic acid. It will mix completely with water and many of the lacquer solvents. Ethylene glycol monoacetate will dissolve cellulose esters and ethers and many of the resins.

Boiling point	181°C.
Specific gravity	1.109 (20°C.)
Flash point	102°C.

Table 15.22: Ethylene Glycol Diacetate *(2)*

$$CH_2OOCCH_3$$
$$|$$
$$CH_2OOCCH_3$$

Glycol diacetate is a colorless liquid having a faint odor resembling that of ethyl acetate. It will dissolve a wide range of cellulose esters, camphor, dammar, ester gum, elemi, mastic, rosin and sandarac. When it is mixed with active solvents its range of solubility is increased for a wide variety of cellulose esters and ethers and for natural and synthetic resins.

(continued)

Table 15.22: (continued)

Molecular Weight (C$_6$H$_{10}$O$_4$)	146.15	Boiling Range, 760 mm, °C	
Color (Pt-Co Scale), max	15	Initial Boiling Point, min	187.0
Weight/Vol, 20°C,		Dry Point, max	193.0
lb/gal (U.S.)	9.21	Freezing Point, °F (°C)	-43 (-42)
kg/liter	1.11	Flash Point, Tag Closed Cup, °F (°C)	191 (88)
lb/gal (Imperial)	11.04	Cleveland Open Cup, °F (°C)	210 (99)
Solubility, 20°C, wt %		Fire Point, °F (°C)	210 (99)
In water	16.4	Flammable Limits in Air, % by volume	
Water in	7.6	Lower, at 275°F (135°C)	1.6
Evaporation Rate (n-butyl acetate = 1)	0.02	Upper, at 310°F (154°C)	8.4
Dilution Ratio, toluene	1.4	Autoignition Temperature (ASTM D-2155), °F (°C)	900 (482)
Refractive Index, 20°C	1.4159	NFPA Classification 30:	IIIA
Vapor Pressure, 20°C, mm Hg	0.2	DOT Labels Required	None
Specific Gravity, 20°/20°C	1.107	DOT Classification	Nonhazardous Liquid

Table 15.23: Ethylene Glycol Monomethyl Ether Acetate *(41)*

$$CH_3OCH_2CH_2OOCCH_3$$

This is a colorless and odorless liquid. It is a solvent for cellulose acetate and it is used in nitrocellulose lacquers to prevent blush. It is also used in dips and cements, and in nonflammable celluloid.

Molecular Weight (C$_5$H$_{10}$O$_3$)	118.13	Autoignition Temperature (ASTM D-2155),	
Evaporation Rate (n-butyl acetate = 1)	0.2	°F (°C)	740 (392)
Weight/Vol, 20°C,		NFPA Code 30, Class	II
lb/gal (U.S.)	8.38	DOT Labels Required	None
kg/liter	1.01	DOT Classification	Nonhazardous liquid
lb/gal (Imperial)	10.06	Color (Pt-Co Scale), max	15
Solubility, 20°C, wt %		Acidity, as acetic acid, wt%, max	0.02
In water	Complete	Boiling Range, 760 mm, °C	
Water in	Complete	Initial boiling point, min	140.0
Dilution Ratio, toluene	2.3	Dry point, max	147.0
VM & P naphtha	0.6	Specific Gravity, 20°/20°C	1.003-1.008
Refractive Index, 20°C	1.4025	Ester Content, wt%, min	98.0
Vapor Pressure, 20°C, mm Hg	2.0	Water, wt%, max	0.1
Freezing Point, °F (°C)	-85 (-65)	Odor	Mild, nonresidual
Flash Point, Tag Closed Cup, °F (°C)	120 (49)	Appearance	Free from insoluble matter or haze
Tag Open Cup, °F (°C)	125 (52)		
Fire Point, °F (°C)	125 (52)		
Flammable Limits in Air, % by volume			
Lower, at 200°F (93°C)	1.52		
Upper, at 200°F (93°C)	12.3		

Table 15.24: Ethylene Glycol Monoethyl Ether Acetate *(41)*

Amsco-Solv EE Acetate
Cellosolve Acetate
Ektasolve EE Acetate

$$CH_2OC_2H_5$$
$$CH_2OOCCH_3$$

This is a water-white liquid with a mild, characteristic odor. It is widely used as a solvent in nitrocellulose lacquers where it imparts gloss, flow, and prevents blush.

Typical Properties

Molecular Weight (C$_6$H$_{12}$O$_3$)	132.16	Boiling Range at 760 mm, °C	
Color (Pt-Co Scale), max	15	Initial Boiling Point, min	150
Weight/Vol at 20°C,		Dry Point, max	160
lb/gal (U.S.)	8.11	Freezing Point, °F (°C)	-78 (-61)
kg/L	0.98	Flash Point, Tag Closed Cup, °F (°C)	130 (54)
lb/gal (Imperial)	9.73	Tag Open Cup, °F (°C)	139 (59)
Solubility at 20°C, wt %		Fire Point, °F (°C)	144 (62)
In water	23.8	Flammable Limits in Air, % by volume	
Water in	6.5	Lower, at 200°F (93°C)	1.24
Evaporation Rate (n-butyl acetate = 1)	0.2	Upper, at 275°F (135°C)	12.7
Dilution Ratio, toluene	2.5	Autoignition Temperature (ASTM D 2155), °F (°C)	720 (382)
VM & P naphtha	0.9	NFPA Classification 30	II
Refractive Index at 20°C	1.4030	DOT Classification	Combustible Liquid
Vapor Pressure at 20°C, mm Hg	1.7	DOT Labels Required	None
Specific Gravity at 20/20°C	0.973		

Table 15.25: Ethylene Glycol Monobutyl Ether Acetate *(41)*

Amsco EB Acetate
Butyl Cellosolve Acetate
Ektasolve EB Acetate

$$CH_3-C-O-CH_2-CH_2-O-C_4H_9,$$
$$\overset{\|}{O}$$

This is a very high boiling glycol ether ester solvent particularly useful as a coalescing aid for latex paints. With its limited water solubility and its general solvent properties, it is found useful in multicolor lacquers and lacquer emulsions.

Typical Properties

Molecular Weight ($C_8H_{16}O_3$)	160.21
Color (Pt-Co Scale), max	15
Evaporation Rate (n-butyl acetate = 1)	0.03
Weight/Vol, 20°C,	
lb/gal (U.S.)	7.84
kg/litre	0.94
lb/gal (Imperial)	9.42
Solubility, 20°C, wt %	
In water	1.1
Water in	1.6
Dilution Ratio, toluene	1.8
VM & P naphtha	1.2
Refractive Index, 20°C	1.4200
Vapor Pressure, 20°C, mm Hg	0.29
Specific Gravity, 20°/20°C	0.942
Boiling Range, 760 mm, °C	
Initial Boiling Point, min	186
Dry Point, max	194
Freezing Point, °F (°C)	−83 (−64)
Flash Point, Tag Closed Cup, °F (°C)	160 (71)
Tag Open Cup, °F (°C)	177 (81)
Fire Point, °F (°C)	180 (82)
Flammable Limits in Air, % by volume	
Lower, at 200°F (93°C)	0.88
Upper, at 275°F (135°C)	8.54
Autoignition Temperature (ASTM D-2155), °F (°C)	645 (340)
NFPA Classification 30	IIIA
DOT Classification	Combustible Liquid
DOT Labels Required	None

Table 15.26: Diethylene Glycol Monoethyl Ether Acetate *(41)*

Amsco-Solv DE Acetate
Carbitol Acetate
Ektasolve DE Acetate

$$CH_3-C-O-CH_2-CH_2-O-CH_2-CH_2-O-C_2H_5$$
$$\overset{\|}{O}$$

This is primarily used as a coalescing aid in latex paints. Both its solvency and slow evaporation rate are effective in producing slow drying characteristic brushing lacquers.

Typical Properties

Molecular Weight ($C_8H_{16}O$)	176.21
Color (Pt-Co Scale), max	15
Evaporation Rate (n-butyl acetate = 1)	< 0.01
Weight/Vol, 20°C,	
lb/gal (U.S.)	8.41
kg/litre	1.01
lb/gal (Imperial)	10.09
Solubility, 20°C, wt %	
In water	Complete
Water in	Complete
Dilution Ratio, toluene	2.2
VM & P naphtha	0.6
Refractive Index, 20°C	1.4230
Vapor Pressure, 20°C, mm Hg	0.05
Specific Gravity, 20°/20°C	1.011
Boiling Range, 760 mm, °C	
Initial Boiling Point, min	214
Dry Point, max	221
Freezing Point, °F (°C)	−13 (−25)
Flash Point, Cleveland Open Cup, °F (°C)	225 (107)
Fire Point, °F (°C)	230 (110)
Flammable Limits in Air, % by volume	
Lower, at 275°F (135°C)	0.98
Upper, at 365°F (185°C)	19.4
Autoignition Temperature (ASTM D-2155), °F (°C)	680 (360)
NFPA Classification 30	IIIB
DOT Classification	Nonhazardous Liquid
DOT Labels Required	None

Table 15.27: Diethylene Glycol Monobutyl Ether Acetate *(41)*

Amsco-Solv DB Acetate
Butyl Carbitol Acetate
Ektasolve DB Acetate

$$CH_3-C-O-CH_2-CH_2-O-CH_2-CH_2-O-C_4H_9$$
$$\overset{\|}{O}$$

This very high-boiling glycol ester is used primarily as a solvent in printing inks and high-bake enamels, and as a coalescing aid in latex paints. The very slow evaporation rate and the limited water solubility of this solvent are especially applicable in silk screen inks and as a component in polystyrene coatings for decals. Also it is a selective solvent in the separation of alcohols and ketones by distillation.

Molecular Weight (theoretical)	204.27	Solubility, 20°C, wt %,	
Weight/Vol, 20°C, lb/gal (U.S.)	8.16	In water	6.5
kg/liter	0.98	Water in	3.7
lb/gal (Imperial)	9.79	Color (Pt-Co Scale), ppm, max	15
Evaporation Rate (n-butyl acetate = 1)	<0.01	Acidity, as acetic acid, wt %, max	0.03
Dilution Ratio, toluene	1.8	Boiling Range, 760 mm, °C	
VM & P naphtha	0.9	Initial boiling point, min	235.0
Flash Point (Cleveland open cup), °F	240	Dry point, max	250.0
°C	116	Specific Gravity, 20°/20°C	0.975-0.985
Freezing Point, °F	−26	Ester Content, wt %, min	97.0
°C	−32	Water, wt %, max	0.2
Vapor Pressure, 20°C, mm Hg	0.04		

Table 15.28: Propylene Glycol Monomethyl Ether Acetate *(41)*

Ektasolve PM Acetate

Typical Properties

Molecular Weight ($C_6H_{12}O_3$)	132.2	Refractive Index at 20°C	1.40
Color (Pt-Co Scale)	15	Vapor Pressure at 20°C, mm Hg	3.7
Evaporation Rate (n-butyl acetate = 1)	0.39	Specific Gravity at 20°/20°C	0.97
Weight/Vol at 20°C,		Boiling Range at 760 mm, °C	
lb/gal (U.S.)	8.06	Initial Boiling Point, min.	140
kg/L	0.97	Dry Point, max.	150
lb/gal (Imperial)	9.68	Flash Point by Setaflash, °C (°F)	45 (114)
Solubility at 20°C, wt %		Flammable Limits in Air, % by volume	
In water	20	Lower at 78°C (173°F)	1.3
Water in	5.9	Upper at 139°C (283°F)	13.1
Dilution Ratio,		Autoignition Temperature (ASTM D 2155), °C (°F)	354 (670)
Toluene	2.6	DOT Classification	Combustible Liquid
VM & P naphtha	0.8	DOT Labels Required	None

Table 15.29: Amsco Acetate Esters *(13)*

	Specific Gravity @ 20°/20°C	Pounds per Gallon @ 20°/20°C	Distillation Range 760 mm Hg °F (°C)	Relative Evaporation Rate n-BuAc = 1	Vapor Pressure 1 mm Hg @ 20°C	Flash Point TAG CC °F	Solubility % by wt @ 20°C in Water	of Water
Methyl Acetate	0.904	7.52	127 (52.8)–138 (58.9)	11.8	206	14	24.5	8.2
Ethyl Acetate 85/88	0.884	7.36	162 (72.2)–172 (77.8)	4.2	114	27	7.4	3.1
Ethyl Acetate 99	0.902	7.51	169 (76.1)–172 (77.8)	4.1	82	24	7.4	3.3
Isopropyl Acetate	0.874	7.26	185 (85.0)–193 (89.4)	3.1	58	35	2.9	1.8
n-Propyl Acetate	0.888	7.39	210 (98.8)–217 (102.8)	2.3	34	55	2.3	2.6
sec-Butyl Acetate	0.858	7.14	219 (103.9)–239 (115.0)	1.9	30	60	0.8	3.6
Isobutyl Acetate	0.873	7.24	233 (111.7)–243 (118.9)	1.5	20	70	0.7	1.6
n-Butyl Acetate	0.883	7.34	252 (122.2)–262 (127.8)	1.0	14	84	0.7	1.6
Isobutyl Isobutyrate (IBIB)	0.855	7.13	291 (114)–304 (151)	0.4	5.5	101	<0.1	<0.2

Table 15.30: Ashland Acetate Esters (69)

	Specific Gravity 20° / 20°C	Dist. Range °F IBP	Dist. Range °F DP	Flash Pt. °F TCC
Amyl Acetate, Primary	0.8752	284	302	101
n-Butyl Acetate	0.8750	244	262	81
Ethyl Acetate 85-88%	0.8860	163	176	21
Ethyl Acetate 99%	0.902	169	172	26
2-Ethylhexyl Acetate	0.873	378	401	160
Glycol Ether DB Acetate	0.980	455	482	240*
Glycol Ether DE Acetate	1.010	417	430	235*
Glycol Ether EB Acetate	0.9424	367	381	160
Glycol Ether EE Acetate	0.9724	302	320	130
Glycol Ether EM Acetate	1.005	284	297	120
Isobutyl Acetate	0.8693	234	246	63
Isobutyl Isobutyrate	0.855	291	304	101
Isopropyl Acetate 99%	0.8730	187	194	47
n-Propyl Acetate	0.8880	210	217	55

*COC

Table 15.31: Union Carbide Esters (19)

Physical Properties
Determined on Purified Samples

	Molecular Weight	Apparent Specific Gravity at 20/20°C	ΔSp Gr/Δt at 10 to 40°C, per °C	Boiling Point, °C, mm Hg 760	Boiling Point, °C, mm Hg 300	Boiling Point, °C, mm Hg 10	Δbp/Δp at 750 to 770 mm Hg, °C per mm Hg	Vapor Pressure at 20°C, mm Hg
Ethyl Acetate	88.11	0.9018	0.00121	77.2	51	-15	0.041	76
n-Propyl Acetate	102.13	0.8883	0.00116	101.6	74	5	0.043	25
Isopropyl Acetate	102.13	0.8737	0.00115	88.5	62	- 3	0.041	42
Butyl Acetate	116.16	0.8826	0.00102	126.0	97	23	0.044	8
Isobutyl Acetate	116.16	0.8728	0.00105	117.3	89	16	0.045	13
Primary Amyl Acetate, mixed isomers	130.19	0.8757	0.00097	146	——	36	0.048	4
Methyl CELLOSOLVE Acetate	118.14	1.0067	0.00110	145	116	41	0.046	2
CELLOSOLVE Acetate	132.16	0.9748	0.00109	156.3	126	49	0.046	2
Butyl CELLOSOLVE Acetate	160.21	0.9424	0.00098	192.3	159	76	0.051	< 1
CARBITOL Acetate	176.21	1.0114	0.00102	217.4	184	98	0.052	< 0.1
Butyl CARBITOL Acetate	204.27	0.9810	0.00095	246.7	211	119	0.056	< 0.01
Glycol Diacetate	146.14	1.1063	0.00113	190.9	159	78	0.049	< 1
Triethylene Glycol Diacetate	234.25	1.1173	0.00096	285.7	——	164	0.052[f]	< 0.01
Glyceryl Triacetate	218.21	1.1604	0.00109	258.0	225	140	0.051	< 0.01

	Relative Evaporation Rate (Butyl Acetate = 100)[a]	Freezing Point, °C	Absolute Viscosity at 20°C, cP	Solubility at 20°C, % by wt, In Water	Solubility at 20°C, % by wt, Water In	Refractive Index, 20 n_D	Heat of Vaporization, Btu/lb at 760 mm	Heat of Vaporization, Btu/lb at 300 mm	Flash Point, °F[b]
Ethyl Acetate	615	- 83.6	0.45	8.7	3.3	1.3719	155	163	30
n-Propyl Acetate	275	- 95.0	0.58	2	——	1.3847	144	152	58
Isopropyl Acetate	500	- 69.3	0.60	2.6[d]	1.9[d]	1.3791	142	151	42
Butyl Acetate	100	- 73.5	0.73	0.68	1.2	1.3944	135	143	84
Isobutyl Acetate	145	- 97.1	0.70	0.63	1.02	1.3900	131	139	62
Primary Amyl Acetate, mixed isomers	42	-100[e]	0.9	0.2	0.9	1.4013	211	——	106
Methyl CELLOSOLVE Acetate	31	- 65.1	1.1	Complete		1.4019	147	156	121
CELLOSOLVE Acetate	21	- 61.7	1.3	22.9	6.5	1.4058	133	141	126
Butyl CELLOSOLVE Acetate	3	- 63.5	1.8	1.5	1.7	1.4138	118	126	165
CARBITOL Acetate	< 1	- 25	2.8	Complete		1.4213	117	125	211
Butyl CARBITOL Acetate	< 1	- 32.2	3.6	6.5	3.7	1.4262	106	113	221
Glycol Diacetate	——	- 41.5	2.9	16.4	7.0	1.4159	134	142	185
Triethylene Glycol Diacetate	——	- 50[e]	11.6	Complete		1.4389	——	——	296[c]
Glyceryl Triacetate	——	- 37[e]	22	5.8[d]	3.6[d]	1.4312	114	123	305

(a) Determined on commercial grade material.
(b) Determined on commercial grade material. Flash points were determined by ASTM Method D56, using Tag closed cup.
(c) Flash point determined by ASTM Method D93, using Pensky-Martens closed cup.

(d) At 25°C.
(e) Sets to glass below this temperature.
(f) At 740 to 760 mm Hg.

Table 15.32: Vapor Pressures of Esters vs Temperature (19)

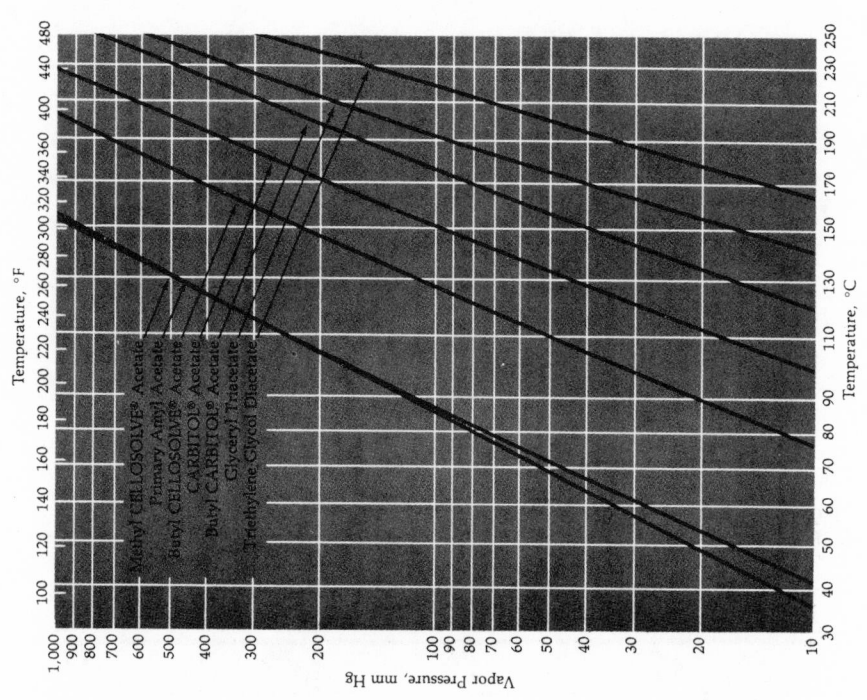

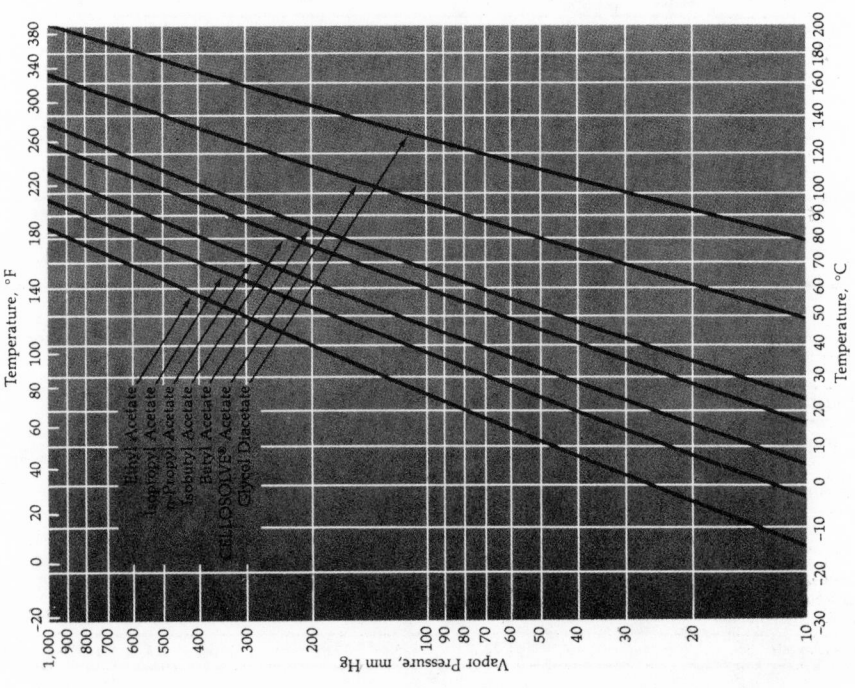

Table 15.33: Solubilities of Water in Esters *(19)*

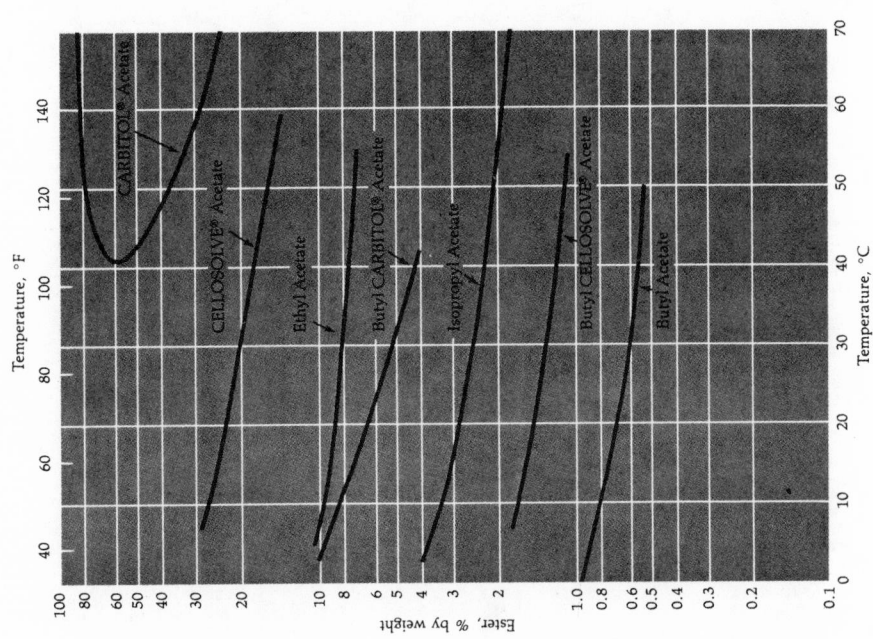

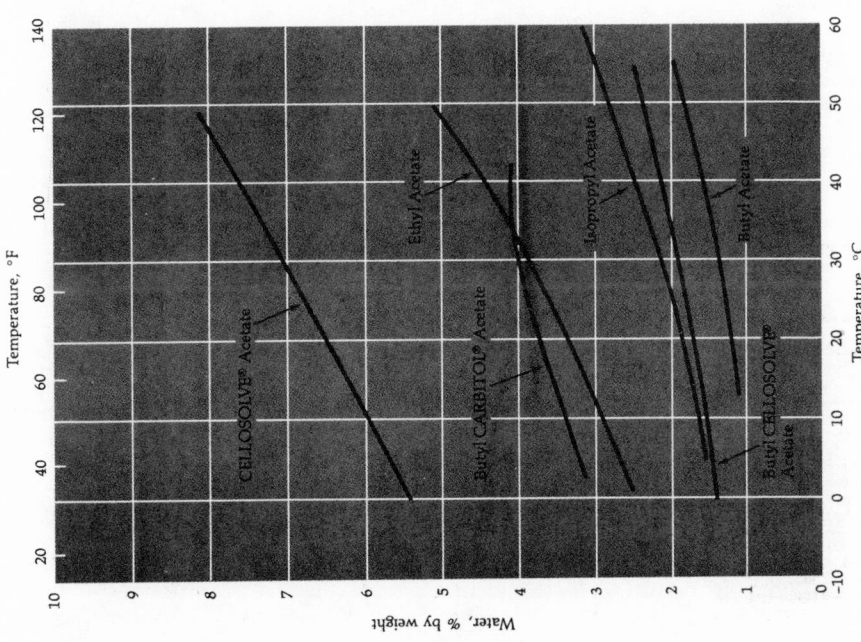

Table 15.34: Constant Boiling Mixtures (19)

Mixtures	Components Apparent Specific Gravity at 20/20°C	Components Boiling Point at 760 mm, °C	Azeotrope Boiling Point at 760 mm, °C	Azeotrope Composition, % by Weight in Azeotrope	Azeotrope Composition in Upper Layer	Azeotrope Composition in Lower Layer	Relative Volume of Layers at 20°C	Sp Gr at 20/20°C of Azeotrope or Layers
Butyl Acetate	0.8826	126.0	117.6	32.8				0.832
Butanol	0.8108	117.7		67.2				
Butyl Acetate	0.8826	126.0		63	86	1	U75.5	U0.874
Butanol	0.8108	117.7	90.7	8	11	2	L24.5	L0.997
Water	1.0000	100.0		29	3	97		
Butyl Acetate	0.8826	126.0	90.7	72.9	98.8	0.68	U75.8	U0.882
Water	1.0000	100.0		27.1	1.2	99.32	L24.2	L0.998
Butyl CELLOSOLVE Acetate	0.9424	192.3	98.8	28.1	98.4	1.1	U71.0	U0.943
Water	1.0000	100.0		71.9	1.6	98.9	L29.2	L0.999
CELLOSOLVE Acetate	0.9748	156.3	97.5	44.4	93.4	22.9	U31.2	U0.983
Water	1.0000	100.0		55.6	6.6	77.1	L68.8	L1.012
Ethyl Acetate	0.9018	77.2	71.8	69.0				0.863
Ethanol	0.7905	78.3		31.0				
Ethyl Acetate	0.9018	77.2		82.6				Heter
Ethanol	0.7905	78.3	70.2	8.4				0.901(a)
Water	1.0000	100.0		9.0				
Ethyl Acetate	0.9018	77.2	70.4	91.9	96.7	8.7	U95.0	U0.907
Water	1.0000	100.0		8.1	3.3	91.3	L 5.0	L0.999
Glycol Diacetate	1.1063	190.9	99.7	15.4				1.024
Water	1.0000	100.0		84.6				
Isobutyl Acetate	0.8728	117.3	88.4	78.0	98.98	0.63	U81	U0.874
Water	1.0000	100.0		22.0	1.02	99.37	L19	L1.000
Isopropyl Acetate	0.8737	88.5	80.1	47.4				0.822
Isopropanol	0.7861	82.3		52.6				
Isopropyl Acetate	0.8737	88.5		76	81.4	2.9	U94	U0.870
Isopropanol	0.7861	82.3	75.5	13	13.0	11.5	L 6	L0.981
Water	1.0000	100.0		11	5.6	85.6		
Isopropyl Acetate	0.8737	88.5	75.9	88.9	98.2	2.9	U91.4	U0.870
Water	1.0000	100.0		11.1	1.8	97.1	L 8.6	L0.995
Methyl CELLOSOLVE Acetate	1.0067	145	97.1	48.2				1.03
Water	1.0000	100.0		51.8				
Primary Amyl Acetate	0.8757	146.0	94	63.8	99.13	0.18	U67.5	U0.877
Water	1.0000	100.0		36.2	0.87	99.82	L32.5	L1.000
n-Propyl Acetate	0.8883	101.6	94	37				0.833
Propanol	0.8046	97.2		63				
n-Propyl Acetate	0.8883	101.6		59.5				
Propanol	0.8046	97.2	82.2	19.5				
Water	1.0000	100.0		21.0				
n-Propyl Acetate	0.8883	101.6	82.4	86				
Water	1.0000	100.0		14				

FOOTNOTE: (a) At 25/20°C

Table 15.35: General Solvent Properties of Esters (19)

Solvent	Pounds Per Gallon at 20°C	Relative Evaporation Rate (Butyl Acetate = 100)	Nitrocellulose Dilution Ratios Toluene	Nitrocellulose Dilution Ratios Naphtha	Nitrocellulose Dilution Ratios Xylene	8 Per Cent Solution of R.S. ½ s Nitrocellulose Blush Resistance % RH at 80°F	8 Per Cent Solution of R.S. ½ s Nitrocellulose Viscosity at 25°C, cP
Ethyl Acetate (85-90%)	7.37	615	3.3	1.1	— —	37	16
Isopropyl Acetate (99%)	7.26	500	2.7	1.1	— —	69	20
Propyl Acetate	7.39	275	3.0	1.2	— —	76	22
Isobutyl Acetate	7.24	145	2.7	1.1	— —	80	29
Butyl Acetate	7.34	100	2.9	1.3	2.7	83	30
Primary Amyl Acetate	7.29	42	2.3	1.3	— —	91	39
Methyl CELLOSOLVE Acetate	8.37	31	2.3	0.6	1.9	80	65
CELLOSOLVE Acetate	8.12	21	2.5	0.9	2.3	94	67
Butyl CELLOSOLVE Acetate	7.84	3	1.8	1.2	— —	96+	84
CARBITOL Acetate	8.41	<1	2.2	0.6	1.9	92	— —
Butyl CARBITOL Acetate	8.16	<1	— —	— —	1.8	96+	220
REFERENCE SOLVENTS							
Acetone	6.60	1160	4.5	0.7	— —	<35	9
CELLOSOLVE Solvent	7.74	32	4.9	1.1	4.3	59	73
Methyl Isobutyl Ketone	6.67	165	3.6	1.0	3.2	78	19

Table 15.36: Relative Evaporation of Solvents *(19)*

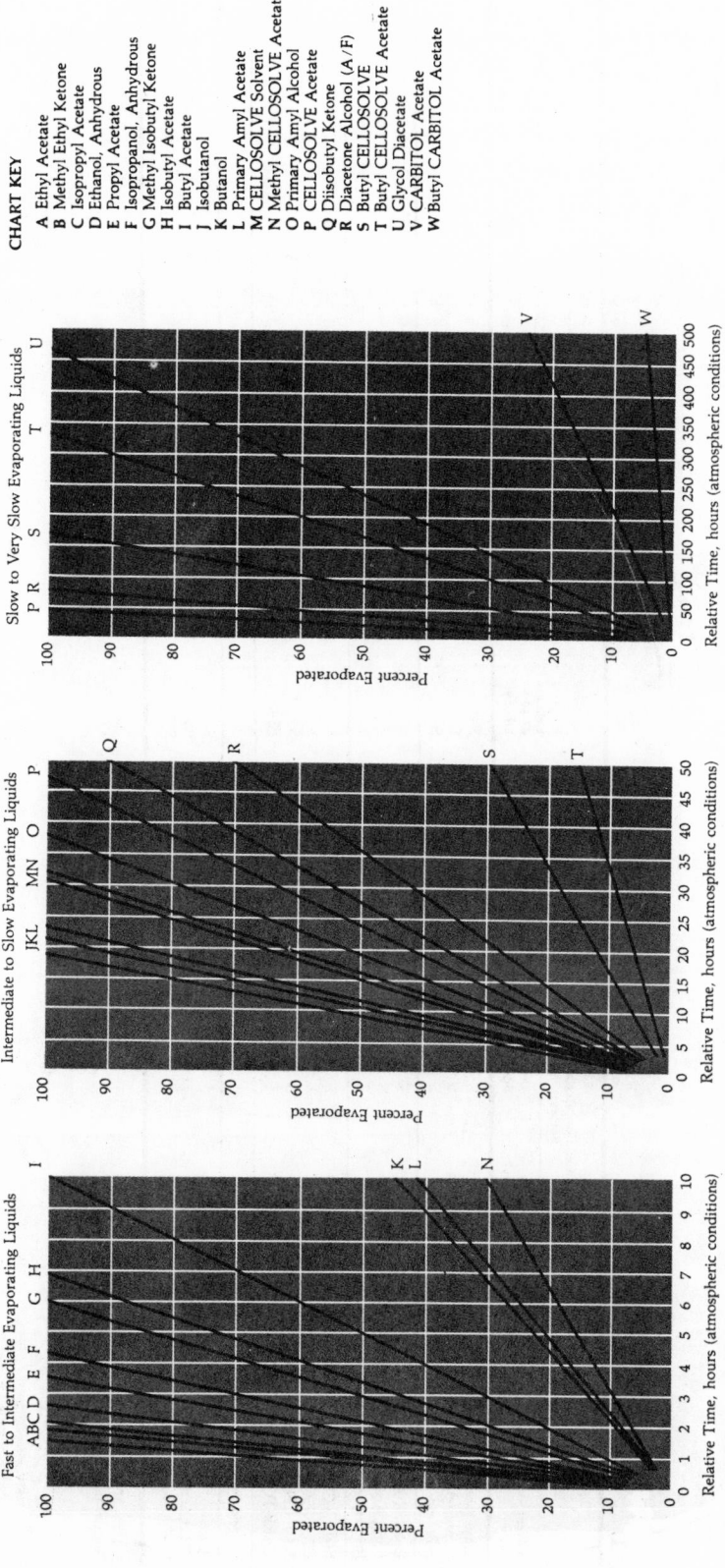

CHART KEY

A Ethyl Acetate
B Methyl Ethyl Ketone
C Isopropyl Acetate
D Ethanol, Anhydrous
E Propyl Acetate
F Isopropanol, Anhydrous
G Methyl Isobutyl Ketone
H Isobutyl Acetate
I Butyl Acetate
J Isobutanol
K Butanol
L Primary Amyl Acetate
M CELLOSOLVE Solvent
N Methyl CELLOSOLVE Acetate
O Primary Amyl Alcohol
P CELLOSOLVE Acetate
Q Disobutyl Ketone
R Diacetone Alcohol (A/F)
S Butyl CELLOSOLVE
T Butyl CELLOSOLVE Acetate
U Glycol Diacetate
V CARBITOL Acetate
W Butyl CARBITOL Acetate

Table 15.37: Resin Solubilities (19)

| Solvent | Cellulose Acetate 55-56% Acetyl | Cellulose Acetate Butyrate | | Cellulose Acetate Propionate | | Ethyl Cellulose 47-49% Ethoxyl | Poly-styrene | Methyl Meth-acrylate | UCAR Solution Vinyl | | |
		17% Butryl	37% Butryl	13-15% Propionyl	31% Propionyl				VYHH Vinyl Chloride Acetate Copolymer	AYAF Vinyl Acetate	XYHL Vinyl Butyral
Ethyl Acetate (85-90%)	Sl.S	S	S	S	S	S	S	S	S-G	S	S
Isopropyl Acetate	I	PS*	S	Sl.S-G*	S	S	S	S	S-G	S	S
n-Propyl Acetate	I	PS*	S	Sl.S-G*	S	S	S	S	S	S	SW
Isobutyl Acetate	I	I	S	I	S	S	S	PS	S	S	G
Butyl Acetate	I	SW	S	I	S	S	S	S	S	S	SW
Primary Amyl Acetate	I	I	S	I	I	S	S	SW	S	S	SW
Methyl CELLOSOLVE Acetate	S	S	S	S	S	S	S	S	S	S	G
CELLOSOLVE Acetate	Sl.S	PS	S	S	S	S	S	S	S	S	G
Butyl CELLOSOLVE Acetate	I	I	S	I	Sl.S	S	S	I	Sl.S	S	I
CARBITOL Acetate	S	SW*	S	PS-G	S	PS-G	S	S	S-G	S	SW
Reference Solvents											
Acetone	S	S	S	S	S	S	PS	S	S	S	SW
CELLOSOLVE Solvent	I	SW	PS	SW	I	PS	I	I	I	PS	S
Methyl Isobutyl Ketone	I	I	S	I	S	S	S	PS	S	S	PS

FOOTNOTES:
Concentration = 0.5 g resin to 4.5 ml solvent
   * = 0.5 g resin to 9.5 ml solvent
   S = Soluble
   I = Insoluble
   G = Gel
   SW = Swelling
   Sl.S = Slightly soluble
   PS = Partly soluble
   S-G = Soluble, tendency to gel
   PS-G = Partly soluble, tendency to gel
   Sl.S-G = Slightly soluble, tendency to gel

PROPIONATES

$$CH_3CH_2COOR$$

The propionic esters are very similar to the acetic esters in physical and chemical properties with the difference that the former have a higher boiling point, lower evaporation rate and a lesser power of solubility. They are miscible with many of the lacquer solvents and diluents and possess a distinctive but not a disagreeable odor. The consumption of these esters for solvent purposes is relatively small compared to the highly developed acetate esters.

Table 15.38: Methyl Propionate (2)

$$CH_3CH_2COOCH_3$$

Methyl propionate has been advocated as a solvent for cellulose derivatives. When it is admixed with other propionates (such as ethyl, propyl, butyl and amyl) the mixture will dissolve cellulose ethers and esters.

Boiling point	91°C.
Specific gravity	0.937 (4°C.)

Table 15.39: Ethyl Propionate (2)

Propionic Ether
Propionic Ester

$$CH_3CH_2COOCH_2CH_3$$

Ethyl propionate is a colorless liquid with an odor resembling that of pineapples. It is a solvent for cellulose ethers and esters and for a variety of natural and synthetic resins. It is used principally as an ingredient in soft drinks and fruit syrups.

Acidity (as propionic)	0.02% by wt, max	Evaporation rate	Slower than ethyl acetate
Distillation range	90 to 100% between 80 and 120°C	Purity	85 to 90%
Color	Water-white	Residue	None
Toluene dilution ratio	2.5–3.0	Specific gravity at 15½°C	0.876–0.886
Dryness	No turbidity with 19 volumes gasoline	Weight per gal	7.35 lbs

Table 15.40: n-Butyl Propionate

$$CH_3CH_2COOC_4H_9$$

n-Butyl propionate is a water-white liquid with an apple-like odor. It is miscible with most of the lacquer solvents and diluents and with oils but not miscible with water. It is a solvent for nitrocellulose and many of the natural and synthetic resins. When an active solvent is added to it, butyl propionate will dissolve many of the cellulose esters and ethers. It It may be used as a solvent in lacquer fabrication where it imparts gloss, adhesion and prevents blushing. It is also used to replace butyl and amyl acetate when lower volatility and slower evaporation are desired.

Acidity (as propionic)	0.35% by wt, max	Evaporation rate at 95°F (in minutes)	
Blush resistance at 90°F (10% ½ sec. R.S. nitrocellulose solution)	Clear 85% Relative humidity Blush 90%	5%	2¼
		25%	12
		50%	24½
		75%	41½
Distillation range	120–175°C	90%	56½
Coefficient of expansion		95%	63½
per 1°F	0.00060	Flash point	63°F
per 1°C	0.001	Non-volatile matter	0.005 gm/100 cc, max
Color	Water-white	Purity	90–92%
Dilution ratio		Residue	None
Toluol	2.1	Solubility of water in solvent at 25°C	1.2% by vol
Petroleum naphtha	1.2	Specific gravity at 20/20°C	0.868–0.872
Distillation range:		Viscosity (10% ½ sec. R.S. nitrocellulose solution)	5.9 centipoises
Below 120°C	None		
Below 140°C	Not more than 50%	Weight per gal at 20°C	7.24 lbs
Below 150°C	Not less than 85%		
Above 160°C	None		
Dryness at 20°C	Miscible without turbidity with 20 volumes 60° Bé gasoline		

Table 15.41: Amyl Propionate *(2)*

$$CH_3CH_2COOC_5H_{11}$$

Amyl propionate is a colorless, volatile liquid with an apple-like odor. It will dissolve cumar resins, elemi, ester gum, mastic, copal, kauri, sandarac, and nitrocellulose and it is miscible with most lacquer solvents and oils. It has a slow solvent action upon cellulose ethers thus acting as a latent solvent and this latency can be overcome when acetone or ethyl alcohol is added to it. It has similar solvent properties to amyl acetate but is not as rapid and its solutions are more viscous, it has a slower rate of evaporation, and it has a more agreeable odor. It is used as a desirable high-boiling lacquer solvent imparting gloss, blush resistance and a reduction in "orange peel" effect. It is also used in flavoring and perfumery.

Acidity (as propionic)	0.030% by wt, max
Blush resistance at 90°F (10% ½ sec. R.S. nitrocellulose solution)	Clear 90% Relative humidity Blush 95%
Coefficient of expansion	
per 1°F	0.00060
per 1°C	0.00108
Color	Water-white
Dilution ratio	
Toluol	1.4
Petroleum naphtha	0.7
Distillation range:	
At or below 110°C	None
At or below 150°C	Not more than 40%
At or below 170°C	Not less than 90%
Above 175° C	None
Non-volatile matter	Not more than 0.005 gms per 100 cc
Residue	None
Solubility of water in solvent at 25°C	0.3% by vol
Specific gravity at 20/20°C	0.869-0.873
Viscosity (10% ½ sec. R.S. nitrocellulose solution)	106 centipoises
Weight per gal at 20°C	7.25 lbs

BUTYRATES

Butyrates do not find extensive use in the solvent industry because of their relatively unpleasant odor and higher price.

Table 15.42: Methyl Butyrate *(2)*

$$CH_3CH_2CH_2COOCH_3$$

Methyl butyrate is a solvent for ethyl cellulose and when it is mixed with active solvents it will dissolve nitrocellulose.

Boiling point	102°C.
Specific gravity	0.898 (20°C.)

Table 15.43: Ethyl Butyrate *(2)*

$$CH_3CH_2CH_2COOC_2H_5$$

Ethyl butyrate is a nontoxic liquid having an odor suggestive of pineapples. Its solvent properties lie between those of ethyl acetate and n-butyl acetate, and when mixed with other solvents it will dissolve cellulose esters and ethers, and many of the natural and synthetic resins. It is used in flavors.

Boiling point	120°C.
Flash point	23°C.
Specific gravity	0.879 (20°C.)

Table 15.44: n-Butyl Butyrate *(2)*

$$CH_3CH_2CH_2COOC_4H_9$$

Butyl butyrate is a water-white, neutral liquid with an apple-like odor. The commercial grade is composed of a mixture of the isomeric esters. It is a solvent for nitrocellulose, "Cumar" resins, dammar, ester gum, elemi, shellac, and metallic resinates.

Acidity (as butyric)	0.02% by wt, max
Boiling point	156.9° C
Distillation range	95–100% between 140–170°C
Critical temperature	338°C
Toluene dilution ratio	1.8–2.0
Dryness	No turbidity with 19 vol 60° Bé gasoline
	Complete, standing at least 19 vols gasoline without turbidity
Flash point	51°C
Purity	90–95%
Refractive index	1.4035
Residue	None
Specific gravity at 20/20°C	0.8717
Specific heat at 20°C	0.458
Surface tension at 157°C	12.0
Vapor pressure at 20°C	113 mm Hg
Viscosity at 25°C	0.84 centipoises
Weight per gal	7.25 lbs

Table 15.45: Ethyl Hydroxy-Isobutyrate *(2)*

Ethyl Oxybutyrate

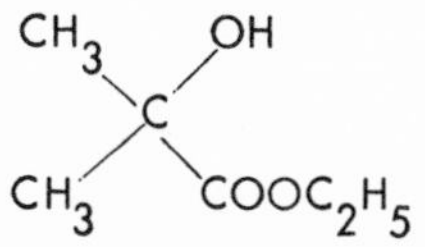

Ethyl hydroxy-isobutyrate is a water-white, stable liquid of a mild odor. It is a solvent for cellulose nitrate and acetate and when mixed with other solvents it will also dissolve cellulose ethers. Its solvent action is somewhat comparable with that of ethyl lactate, differing in the following aspects:

Its solvent action is slower and requires the presence of an active solvent to accentuate it.
Its solutions of nitrocellulose are more viscous.
Its tolerance for hydrocarbons is about the same as far as it concerns nitrocellulose and is lower in the presence of acetyl cellulose.
Its volatility is higher.

Ester content	96–100%
Boiling range	142°–146°C.
Specific gravity	0.978–0.986 (20°C.)

Table 15.46: Isobutyl Isobutyrate *(41)*

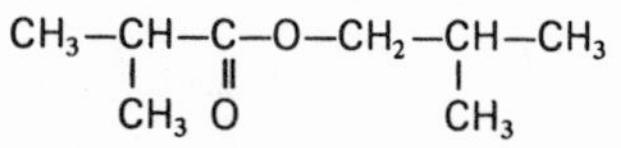

$$\begin{array}{ccccc} CH_3-CH-C-O-CH_2-CH-CH_3 \\ \quad\;\; | \quad\; || \qquad\qquad\; | \\ \quad\;\; CH_3 \;\; O \qquad\qquad CH_3 \end{array}$$

Isobutyl isobutyrate is a slow evaporating solvent with blush resistance, good flow and leveling which are favorable properties in formulating cellulose nitrate. Its solvent activity is equivalent to methyl amyl acetate and is therefore used as a direct substitute in many formulations.

(continued)

Table 15.46: (continued)

Molecular Weight (C$_8$H$_{16}$O$_2$)	144.22	Boiling Range, 760 mm, °C	
Color (Pt-Co Scale), max	15	Initial Boiling Point, min	144
Weight/Vol, 20°C,		Dry Point, max	151
lb/gal (U.S.)	7.13	Freezing Point, °F (°C)	-112 (-80)
kg/liter	0.86	Flash Point, Tag Closed Cup, °F (°C)	101 (38)
lb/gal (Imperial)	8.56	Tag Open Cup, °F (°C)	111 (44)
Solubility, 20°C, wt %		Fire Point, °F (°C)	115 (46)
In water	<0.1	Flammable Limits in Air, % by volume	
Water in	<0.2	Lower, at 200°F (93°C)	0.96
Evaporation Rate (n-butyl acetate = 1)	0.4	Upper, at 200°F (93°C)	7.59
Dilution Ratio, toluene	1.5	Autoignition Temperature (ASTM D-2155), °F (°C)	810 (432)
VM & P naphtha	0.8	NFPA Classification 30:	II
Refractive Index, 20°C	1.3913	DOT Labels Required	None
Vapor Pressure, 20°C, mm Hg	3.2	DOT Classification	Nonhazardous Liquid
Specific Gravity, 20°/20°C	0.855		

Table 15.47: 2,2,4-Trimethylpentanediol-1,3-Monoisobutyrate *(41)*

"Texanol" Ester Alcohol

This ester alcohol alcohol solvent is used as a coalescing agent in latex paints and water-base ink formulations, in polyvinyl acetate homopolymer and copolymer latices, polyvinyl acetate-acrylic copolymers, 100% acrylic, ethylene-vinyl acetate, and butadiene-styrene latices, where it also exhibits improved pigment application uniformity and stability, increased scrub resistance, and lower film forming temperatures.

Molecular Weight (C$_{12}$H$_{24}$O$_3$)	216.3	Flash Point, Cleveland Open Cup, °F	248
		°C	120
Evaporation Rate (n-butyl acetate = 1)	<0.01	Fire Point, °F	270
		°C	132
Weight/Vol, 20°C,		Flammable Limits in Air, % by volume	
lb/gal (U.S.)	7.90	Lower, at 300°F	0.62
kg/liter	0.95	Upper, at 393°F	4.24
lb/gal (Imperial)	9.48	Autoignition Temperature (ASTM D-2155), °F	740
Solubility, 20°C, wt %		°C	393
In water	Insol		
Water in	0.9	NFPA Classification 30	Combustible Liquid, Class IIIB
Refractive Index, 20°C	1.4423		
Boiling Range, 125 mm Hg, °C		ICC Labels Required	None
Initial boiling point	180.0	Bureau of Explosives Classification	Nonhazardous Liquid
Dry point	182.0		
Freezing Point, °F	−58	Color (Pt-Co Scale), ppm, max	20
°C	−50		
		Specific Gravity, 20°/20°C	0.945 - 0.955
Pour Point, °F	−71		
°C	−57	Acidity, as isobutyric acid, wt %, max	0.2
Vapor Pressure, 87°C, mm Hg	1	Carbonyl, as C=O, wt %, max	0.4

HIGHER FATTY ACID ESTERS

Table 15.48: Mixture of Dimethyl Adipate and Dimethyl Glutarate *(11)*

$$H_3COOC(CH_2)_xCOOCH_3 \quad x = 3\text{-}4$$

This mixture of dibasic esters is used as a high boiling solvent and as an intermediate.

DIESTER CONTENT, WT. % MINIMUM	99	WATER CONTENT, WT. % MAXIMUM	0.5
DIMETHYL ADIPATE, WT. %	30-45	AVERAGE MOLECULAR WEIGHT	165
DIMETHYL GLUTARATE, WT. %	55-70	SPECIFIC GRAVITY	1.082 - 1.090 @ 25/25°C
DIMETHYL SUCCINATE, WT. % MAX.	3	DISTILLATION RANGE, °C	210 -225
SOLUBILITY PARAMETERS (HANSEN SYSTEM)		EVAPORATION RATE BuAc = 100	<1
POLAR BONDING	3.29	VISCOSITY @ 25°C, CENTIPOISE	2.38
HYDROGEN BONDING	4.02	FREEZING POINT	-13°C (APPROX.)
NON-POLAR BONDING	7.03	FLASH POINT	219°F CLOSED CUP
SOLUBILITY PARAMETER	8.75		

Table 15.49: Mixture of Dimethyl Adipate, Dimethyl Glutarate and Dimethyl Succinate *(11)*

$$H_3COOC(CH_2)_xCOOCH_3 \quad x = 2\text{-}4$$

This dibasic ester mixture is used as a high boiling solvent in industrial and automotive coatings.

DIESTER CONTENT, WT. % MINIMUM	99
DIMETHYL ADIPATE, WT. %	20-30
DIMETHYL GLUTARATE, WT. %	40-60
DIMETHYL SUCCINATE, WT. %	20-30
SOLUBILITY PARAMETERS (HANSEN SYSTEM)	
POLAR BONDING	3.4
HYDROGEN BONDING	4.1
NON-POLAR BONDING	8.5
SOLUBILITY PARAMETER	10.1
WATER CONTENT, WT. % MAXIMUM	0.5
AVERAGE MOLECULAR WEIGHT	160
SPECIFIC GRAVITY	1.082 - 1.090 @ 25/25°C
DISTILLATION RANGE, °C	196 - 225
EVAPORATION RATE BuAc = 100	<1
VISCOSITY @ 25°C, CENTIPOISE	2.39
FREEZING POINT	-20°C (APPROX.)
FLASH POINT	212°F TAG CLOSED

Table 15.50: Emery Methyl Esters (63)

| | SPECIFICATIONS | | | | | | TYPICAL COMPOSITION[1] | | | | | | | | | | | | |
| | | | | | | | Saturated Esters | | | | | | | | Unsaturated Esters | | | | |
Product	Acid Value max.	Sap. Value	Iodine Value max. (range)	Color % Trans. 440/550 nm., min.	Unsap. %, max.	Typical[2] Melting Point, °C	Caproate C_6	Caprylate C_8	Caprate C_{10}	Laurate C_{12}	Myristate C_{14}	Palmitate C_{16}	Margarate C_{17}	Stearate C_{18}	Myristoleate C_{14}	Palmitoleate C_{16}	Oleate C_{18}	Linoleate C_{18}	Linolenate C_{18}
Emery 2209 Methyl Caprylate-Caprate	0.5	330-336	0.4	80/95[2]	0.2	-30	3	55	40	2									
Emery 2296 Methyl Laurate 96	0.5	260-264	0.5	92/99	0.2	5		2	96	2									
Emery 2290 Methyl Laurate 90[3]	0.5	258-262	0.5	92/99[2]	0.2	2		2	90	8									
Emery 2270 Methyl Laurate 70	0.5	251-255	0.2	92/99[2]	0.2	1		1	70	28	1								
Emery 2214 Methyl Myristate 95[3]	1	230-234	0.6	92/99[2]	0.2	17			3	95	2								
Emery 2216 Methyl Palmitate 95	0.2	206-210	0.2	92/99[2]	0.5	27			2	95	3								
Emery 2218 Methyl Stearate 95	0.5	188-192	1	71/98[2]	0.5[2]	36			4	95		1							
Emery 2219 Methyl Oleate	4	188-192	(68-76)	71/98	—	18			4	24			58	14					
Emery 2301 Methyl Oleate	4	190-200	(84-90)	23/76	0.2[2]	-16		3	4	1	2	6	76	7	1				
Emery 2253 Methyl Coconate	0.5	250-260	(7-11)	71/98[2]	0.2	4	8	7	48	17	9	2			7	2			
Emery 2254 Stripped Methyl Coconate	1	237-247	(5-10)	90/99	—	—		Tr	54	22	11	3			8	2			
Emery 2203 Methyl Tallowate	1	190-200	(47-53)	—	1	—			3	28	24			42	3				
Emery 2204 Hydro-genated Methyl Tallowate	1	192-198	1	80/95	1	—			3	28	69								

[1]By GLC analysis, AOCS Ce 1-62

[2]Not a specification.

[3]Cosmetic grade with premium quality also available.

Table 15.51: "Kemester" Methyl Esters (39)

Product	Description (CTFA adopted name)	Color Gardner, 1963 Max	Iodine value	Sap value	% Unsap Max	Acid value Max	Saturated							Unsaturated				
							C14	C15	C16	C17	C18	C20	C22	C14:1	C16:1	C18:1	C18:2	C18:3
Kemester 105	Methyl Oleate (*Methyl Oleate*)	2	80–90	194–203	1.0	4.0	3		6	1	2			2	6	69	10	1
Kemester 205	Methyl Oleate (*Methyl Oleate*)	1	80–90	194–203	1.0	4.0	3		6	1	2			1	6	70	10	1
Kemester 115	Methyl Oleate	4	90 Max	185–205	1.0	4.0												
Kemester 213	Methyl Veg By-Product	3	130–145	194–204	1.0	10.0			15		5					30	30	20
Kemester 226	Methyl Soyate	4	125–135	190–200	1.0	7.0			12		4.5					23.5	54	6
Kemester 143	Methyl Ester of Tallow Fatty Acid	8	60 Max	195 Min	2.0	4.0	2.5	1	27	2.5	18			1	5	39	4	
Kemester 9018	Methyl Stearate	1	1.5 Max	185–192	1.0	4.0			10		90							
Kemester 9022	Methyl Behenate (*Methyl Behenate*)	2	1.0 Max	150–160	1.0	20					3	9	88					

Table 15.52: Procter and Gamble Methyl Esters *(39)*

| | Chemical Properties | | | | Physical Properties | | | Composition (GLC %) | | | | | | | | | | |
	Saponification Value	Acid Value	Iodine Value	Moisture (%, KF)	Specific Gravity 25/25°C	Melting Point (°C)	% Transmittance (460 nm)	C_6	C_8	C_{10}	C_{12}	C_{14}	C_{16}	C_{18} Total	C_{18} (0=)	C_{18} (1=)	C_{18} (2=)	Chemical Abstract Number
CE-618	(255)	0.5 max. (0.3)	14 max. (8)	0.1 max. (0.04)	(0.864)	(−4)	90 min. (94)	(0.5)	7 9.5 (8.7)	(5)	44.0 49.9 (48.9)	(19.4)	5.5 10 (8.2)	(9.3)	(2)	(6.3)	(1)	67762-37-2
CE-810	(331)	0.5 max. (0.1)	0.5 max. (0.3)	0.15 max. (0.1)	(0.870)	(−29)	85 min. (96)	6.0 max. (5)	51 59 (55)	34 42 (39)	2.0 max. (0.8)							67762-39-4
CE-1095	(302)	0.5 min. (0.2)	0.6 max. (0.3)	0.15 max. (0.07)	(0.874)	(−14)	95 min. (98)		(3.1)	95.0 min. (96.5)	(.04)							110-42-9
CE-1218	(238)	1.0 max. (0.3)	(9.6)	0.1 max. (0.03)	(.866)	(10)	(90)				52 57 (55.4)	19 24 (22.2)	8 12 (10.1)	9 15 (11.5)	(2.8)	(7.1)	(1.6)	67762-26-9
CE-1270	(254)	0.5 max. (0.2)	0.8 max. (0.09)	0.05 max. (0.03)	(0.877)	(0)	96 min. (97)		(0)	1.0 max. (0.5)	70.5 74.5 (72.8)	24 29 (26.3)	1.0 max. (0.4)					67762-40-7
CE-1290	(259)	0.5 max. (0.1)	0.8 max. (0.1)	0.1 max. (0.03)	(0.867)	(5)	95 min. (98)		(0)	1 max. (0.4)	90 94 (92.5)	6 9 (7)	0.8 max. (0.01)					67762-40-7
CE-1295	(261)	0.3 max. (0.1)	0.1 max. (0.07)	0.05 max. (0.02)	(0.866)	(6)	95 min. (98)		0.3 max. (0)	2.5 max. (0.5)	95 min. (98)	2.5 max. (1.5)	0.5 max. (0)					111-82-0

Table 15.53: Stepan Esters (77)

ALCOHOL ESTERS

SPECIFICATIONS	Acid Value (Max.)	Iodine Value (Max.)	Color APHA (Max.)	Specific Gravity @ 25/20°C.	Form @ 25°C.	Flash Point °F. COC	Freezing Point °C.	Viscosity @ 25°C. CPS	Refractive Index @ 25°C.	Mid-Boiling Point @ 4 mm
Isopropyl Myristate	1.0	1.0	20	0.849-0.855	Liquid	305	−3	4.8	1.433	160°C.
Isopropyl Palmitate	1.0	1.0	20	0.849-0.855	Liquid	325	13	6.7	1.437	170°C.
Kessco 639	1.0	1.0	20	0.849-0.855	Liquid	305	7	5.9	1.436	170°C.
Butyl Stearate Cosmetic	1.0	0.5	30	0.850-0.860	Liquid	370	19	7.0	1.442	200°C.
Isobutyl Stearate	1.0	1.0	40	0.849-0.855	Liquid	360	15	8.5	1.441	200°C.
Isocetyl Stearate	1.0	5.0	35	0.849-0.855	Liquid	450	0	32.0	1.452	—
Octyl Palmitate	3.0	1.0	200	0.853-0.859	Liquid	395	0	11.3	1.4453	—
Octyl Isononanoate	3.0	1.0	35	0.854-0.858	Liquid	260	−34	4.3	1.434	—
Octyl Oxystearate	1.0	5.0	Yellow	0.889-0.895	Liquid	425	12	84.2	1.4565	—

GLYCEROL ESTERS

SPECIFICATIONS	Acid Value (Max.)	Iodine Value (Max.)	Specific Gravity 25/20°C.	Color	HLB Value	Form @ 25°C.	Flash Point °F. COC	Melting Point, °C.	Viscosity @ 25°C., cps
Glycerol Monostearate Pure	3.0	0.5	—	White	3.8	Flakes	410	56.5-58.5	—
Glycerol Monostearate 860 Food Grade	3.0	2.0	—	White	3.8	Flakes	450	58.5-61.5	—
Glycerol Distearate	5.0	1.0	—	White	2.4	Waxy Flake	470	55-60	—
Glycerol Monostearate S.E.	20.0	0.5	—	White to Cream	—	Flakes	400	56.5-59.5	—
Glycerol Monostearate 24 S.E.	20.0	3.0	—	White to Cream	—	Flakes	372	56-60	—
Glycerol Monostearate S.E., Acid Stable	3.0	1.0	—	White to Cream	11.2	Flakes	460	54-58	—
Glycerol Monooleate	5.0	77.0	0.945-0.953	Yellow	3.8	Liquid	435	<20	204
Glycerol Dioleate	5.0	82.0	0.923-0.929	Yellow-Amber	2.9	Liquid	520	0	90
Glycerol Monolaurate (Kessco 675)	5.0	1.0	—	White	4.9	Solid	425	53.9	—
Glycerol Dilaurate	5.0	2.0	—	White	4.0	Solid	480	30.0	—

(continued)

Table 15.53: (continued)

SPECIFICATIONS	Acid Value Max.	Iodine No. Max.
Ethylene Glycol Monostearate Pure	2.0	0.5
Ethylene Glycol Monostearate 70	2.0	0.5
Ethylene Glycol Distearate	15.0	0.5
Ethylene Glycol Amido Stearate	5.0	0.5
Diethylene Glycol Monostearate	5.0	0.5
Diethylene Glycol Distearate	10.0	0.5
Diglycol Stearate S.E.	103.0	7.0
Diglycol Stearate Neutral	103.0	7.0
Propylene Glycol Monostearate Pure	3.0	5.0
Propylene Glycol Distearate	10.0	1.0
Propylene Glycol Monostearate 8615	20.0	3.0
Propylene Glycol Monolaurate E	3.5	1.0
Propylene Glycol Monostearate 534	3.0	2.0

GLYCOL ESTERS

TYPICAL PROPERTIES	Form @ 25°C.	Flash Pt. °F. COC	Melting Point °C.	HLB Value
Ethylene Glycol Monostearate Pure	Flakes	390	56-60	2.9
Ethylene Glycol Monostearate 70	Flakes	370	52-56	2.9
Ethylene Glycol Distearate	Flakes	390	60-63	1.5
Ethylene Glycol Amido Stearate	Flakes	360	56.5-58.5	—
Diethylene Glycol Monostearate	Flakes	395	44.5-47.5	4.3
Diethylene Glycol Distearate	Flakes	360	42-48	2.8
Diglycol Stearate S.E.	Flakes	345	48-53	—
Diglycol Stearate Neutral	Flakes	365	42-48	2.9
Propylene Glycol Monostearate Pure	Flakes	390	33.5-38.5	3.4
Propylene Glycol Distearate	Flakes	430	36-38	2.2
Propylene Glycol Monostearate 8615	Flakes	379	57-62	—
Propylene Glycol Monolaurate E	Oily Liquid	370	10	3.2
Propylene Glycol Monostearate 534	Flakes	390	34.5-39.5	2.9

POLYETHYLENE GLYCOL ESTERS

Product Name	HLB	Form	MP/FP °C.	Acid Value	Saponification Value	Color	Specific Gravity At 25°C (a) or 65°C (b)	Cloud Point 1% in Water	Water	Propylene Glycol	Isopropyl Alcohol	Isopropyl Myristate	Mineral Oil 55 vis.	70 vis.	330 vis.
PEG 200 Monolaurate	9.3	Liquid	<5	5	132-142	Lt. Yellow	0.9833(a)	15°	D	I	S	S	PS	I	I
PEG 200 Dilaurate	5.9	Liquid	<9	10	176-186	Lt. Yellow	0.9520(a)	<5°	D	I	S	S	S	S	S
PEG 300 Monolaurate	11.4	Liquid	<8	5	104-114	Lt. Yellow	1.0100(a)	46°	D	I	S	PS	I	I	I
PEG 300 Dilaurate	7.9	Liquid	<13	10	148-158	Lt. Yellow	0.9703(a)	<5°	D	I	S	S	S	S	I
PEG 400 Monolaurate	13.0	Liquid	12	5	86-96	Lt. Yellow	1.0242(a)	68°	S	I	S	PS	I	I	I
PEG 400 Dilaurate	9.7	Liquid	18	10	127-137	Lt. Yellow	0.9884(a)	29°	D	I	S	S	I	S	I
PEG 600 Monolaurate	14.6	Liquid	23	5	64-74	Lt. Yellow	1.0505(a)	68°	S	I	S	I	I	I	I
PEG 600 Dilaurate	11.7	Soft Solid	24	10	102-112	Cream	0.9820(b)	30°	D	I	S	I	I	I	I
PEG 1000 Monolaurate	16.6	Soft Solid	40	5	41-51	Cream	1.035	88°	S	SH	S	I	I	I	I
PEG 1000 Dilaurate	14.2	Soft Solid	38	10	68-78	Cream	1.015	60°	S	PS	S	S	I	I	I
PEG 1540 Monolaurate	17.5	Wax	46	5	26-36	Cream	1.060 (b)	>100°	S	SH	SH	I	I	I	I
PEG 1540 Dilaurate	15.8	Wax	42	10	48-56	Cream	1.040 (b)	73°	S	SH	SH	PS	I	I	I
PEG 4000 Monolaurate	19.0	Wax	55	5	9-18	Cream	1.075 (b)	>100°	S	SH	SH	I	I	I	I
PEG 4000 Dilaurate	18.1	Wax	52	5	20-30	Cream	1.065 (b)	83°	S	SH	SH	I	I	I	I
PEG 6000 Monolaurate	19.3	Wax	61	5	7-13	Cream	—	>100°	S	SH	SH	I	I	I	I
PEG 6000 Dilaurate	18.7	Wax	57	9	12-20	Cream	1.077 (b)	86°	S	SH	SH	I	I	I	I
PEG 200 Monostearate	8.1	Solid	31	5.0	120-129	Wt. to Cream	0.9360	<5°	DH	I	S	S	I	I	I
PEG 200 Distearate	4.8	Solid	34	10.0	153-162	Wt. to Cream	0.9060	<5°	DH	I	S	S	S	S	S
PEG 300 Monostearate	10.3	Solid	28	5.0	97-105	Wt. to Cream	0.9660	<5°	DH	I	S	S	I	I	I
PEG 300 Distearate	6.9	Solid	32	10.0	130-139	Wt. to Cream	—	<5°	DH	I	S	S	S	S	S
PEG 400 Monostearate	11.7	Solid	32	5.0	83-92	Wt. to Cream	0.9780	<5°	DH	I	S	S	I	I	I
PEG 400 Distearate	8.5	Solid	36	10	115-124	Wt. to Cream	0.9390	<5°	DH	I	S	S	S	S	S
PEG 600 Monostearate	13.5	Solid	37	5.0	61-70	Wt. to Cream	1.0000	55°	S	PS	S	PS	I	PS	I
PEG 600 Distearate	10.7	Solid	39	10.0	93-102	Wt. to Cream	0.9670	<5°	DH	I	S	I	I	S	I
PEG 1000 Monostearate	15.7	Wax	41	5.0	40-48	Cream	1.030	91°	S	S	S	I	I	I	I
PEG 1000 Distearate	13.3	Wax	40	10.0	65-74	Cream	1.005	51°	S	S	S	I	I	I	I
PEG 1540 Monostearate	16.9	Wax	47	5.0	27-36	Cream	1.050	>100°	S	S	S	I	I	I	I
PEG 1540 Distearate	14.6	Wax	45	10.0	49-58	Cream	1.015	78°	S	S	S	PS	I	I	I
PEG 4000 Monostearate	18.7	Wax	56	5.0	10-18	Cream	1.075	>100°	S	S	S	I	I	I	I
PEG 4000 Distearate	17.6	Wax	51	5.0	19-27	Cream	1.060	91°	S	S	S	I	I	I	I
PEG 6000 Monostearate	19.1	Wax	61	5.0	7-13	Cream	1.080	>100°	S	S	S	I	I	I	I
PEG 6000 Distearate	18.4	Wax	55	9.0	14-20	Cream	1.075	83°	PS	S	S	I	I	I	I
PEG 200 Monooleate	8.2	Liquid	<-15	5	115-124	Lt. Amber	0.9742(a)	<5°	D	I	S	S	S-1%	I	I
PEG 200 Dioleate	5.0	Liquid	<-15	10	148-158	Lt. Amber	0.9405(a)	<5°	D	I	S	S	S	S	S
PEG 300 Monooleate	10.2	Liquid	<-5	5	94-102	Lt. Amber	0.998 (a)	<5°	D	I	S	PS	I	I	I
PEG 300 Dioleate	6.9	Liquid	<-5	10	128-137	Lt. Amber	0.9609(a)	<5°	D	I	S	S	S	S	I
PEG 400 Monooleate	11.6	Liquid	<10	5	80-89	Lt. Amber	1.0135(a)	65°	D	I	S	PS	I	I	I
PEG 400 Dioleate	8.3	Liquid	<7	10	113-122	Lt. Amber	0.977 (a)	<5°	D	I	S	S	S	S	I
PEG 600 Monooleate	13.6	Liquid	23	5	60-69	Lt. Amber	1.0381(a)	80°	S	I	S	I	I	I	I
PEG 600 Dioleate	10.6	Liquid	19	10	92-102	Lt. Amber	1.0038(a)	10°	D	I	S	S	I	I	I
PEG 1000 Monooleate	15.9	Soft Solid	39	5	40-49	Cream	1.035 (b)	98°	S	SH	S	I	I	I	I
PEG 1000 Dioleate	13.2	Soft Solid	37	10	64-74	Cream	1.005 (b)	47°	S	I	S	PS	I	I	I
PEG 1540 Monooleate	17.0	Wax	45	5	28-37	Cream	1.050 (b)	>100°	S	S	SH	I	I	I	I
PEG 1540 Dioleate	14.9	Wax	44	10	45-55	Cream	1.025 (b)	82°	S	S	SH	I	I	I	I
PEG 4000 Monooleate	18.7	Wax	55	5	10-18	Cream	1.075 (b)	>100°	S	S	SH	I	I	I	I
PEG 4000 Dioleate	17.7	Wax	49	5	19-27	Cream	1.060 (b)	88°	S	S	SH	I	I	I	I
PEG 6000 Monooleate	19.1	Wax	59	5	7-13	Cream	1.085 (b)	>100°	S	S	SH	I	I	I	I
PEG 6000 Dioleate	18.4	Wax	56	9	13-21	Cream	1.070 (b)	87°	S	S	SH	I	I	I	I

ACRYLATES AND METHACRYLATES

Table 15.54: Acrylate Esters *(52)*

Acrylate Esters	Formula	Formula Weight	Boiling Range, °C. 5-95% 760 mm.	Density (25°/15.6°C.) g/ml.	Density (25°/15.6°C.) lb/gal	Refractive Index at 25°C.	Flash Point °F. Open Cup
Methyl Acrylate	$CH_2=CHCOOCH_3$	86	79.8-80.3	0.950	7.92	1.4003	50 (Tag.)
Ethyl Acrylate	$CH_2=CHCOOCH_2CH_3$	100	98.8-99.8	0.917	7.65	1.4034	50 (Tag.)
Butyl Acrylate	$CH_2=CHCOOCH_2CH_2CH_2CH_3$	128	145.7-148.0	0.894	7.46	1.4160	120 (Cleveland)
2-Ethylhexyl Acrylate	$CH_2=CHCOOCH_2CHCH_2CH_2CH_2CH_3$	184	214.8-218.0	0.880	7.35	1.4332	195 (Cleveland)
	C_2H_5						

Table 15.55: Methacrylate Esters *(52)*

Methacrylate Esters	Formula	Formula Weight	Boiling Range, °C. 760 mm. 5-95%	Density (25°/15.6°C.) g/ml.	Density (25°/15.6°C.) lb/gal	Refractive Index at 25°C.	Flash Point °F. Open Cup
Methyl Methacrylate	$CH_2=C-COOCH_3$ with CH_3	100	100.3-100.8	0.939	7.83	1.4120	55 Tag.
Ethyl Methacrylate	$CH_2=C-COOCH_2CH_3$ with CH_3	114	118.0-118.8	0.909	7.58	1.4116	70 Tag.
Butyl Methacrylate	$CH_2=C-COOCH_2CH_2CH_2CH_3$ with CH_3	142	163.5-170.5	0.889	7.40	1.4220	150 Cleveland
Isobutyl Methacrylate	$CH_2=C-COOCH_2CH(CH_3)_2$ with CH_3	142	155	0.882	7.35	1.4172	120 Tag.
Lauryl Methacrylate	$CH_2=C-COO(CH_2)_nCH_3$ with CH_3; n < 11 2%, n = 11 66%, n = 13 26%, n > 13 3%	262	272.0-344.0	0.868	7.23	1.4440	270 Cleveland
Stearyl Methacrylate	$CH_2=C-COO(CH_2)_nCH_3$ with CH_3; n < 15 3%, n = 15 31%, n = 17 62%	332	310.0-370.0	0.864	7.20	1.4502	>300 Cleveland
1,3-Butylene Dimethacrylate	$CH_2=C-COOCH_2CH_2CHOOCC=CH_2$ with CH_3 CH_3 CH_3	226	110 (3mm.)	1.011	8.42	1.4502	>150 Tag. closed

OXALATES

Table 15.56: Diethyl Oxalate *(2)*

Ethyl Ethanedioate $(COOC_2H_5)_2$

Diethyl oxalate is a water-white liquid with a mild odor. It is used as a slow-evaporating nitrocellulose solvent, in special lacquers for fixing rare salts on the cathode of radio tubes and in organic synthesis.

Acidity (as oxalic)	0.05% by wt, max
Blush resistance at 60° F (10% ½ sec. R.S. nitrocellulose solution)	Clear 90% Relative humidity Blush
Coefficient of expansion	
per 1°F	0.00056
per 1°C	0.00101
Color	Water-white
Dilution ratio	
Toluol	3.5
Petroleum naphtha	0.7
Distillation range	
Below 180°C	None
Below 182°C	Not more than 10%
Below 188°C	Not less than 90%
Above 190°C	None
Dryness at 20°C	Miscible without turbidity with 20 volumes 60° Bé gasoline
Flash point (Open Cup)	168°F
Non-volatile matter	0.005 gm per 100 cc, max
Odor	Mild, non-residual
Purity	99% min
Specific gravity at 20/20°C	1.075–1.079
Water solubility at 25°C	10 cc solvent dissolves 1.5 cc water
Viscosity (10% ½ sec. R.S. nitrocellulose solution)	380 centipoises
Weight per gal at 20°C	8.96 lbs (approx)

Table 15.57: Dibutyl Oxalate

$C_4H_9OOCCOOC_4H_9$

Dibutyl oxalate is a high-boiling, water-white liquid with a mild odor and having a tendency to hydrolyze and split off oxalic acid. It is miscible with most alcohols, ketones, oils and hydrocarbons, and is a solvent for benzyl abietate, cellulose esters and ethers, "Cumar" resins, ester gum, copal ester, "Glyptal" resins and mastic. It is used in nitrocellulose lacquers as a plasticizing solvent for the purpose of fixing rare earth salts on cathode elements, and in organic synthesis.

Acidity (as oxalic)	0.05% by wt, max
Blush resistant at 90°F (1.0% R.S. ½ sec. nitrocellulose solution)	Clear 90 Relative humidity Blush
Coefficient of expansion	
per 1°F	0.00053
per 1°C	0.00095
Color	Water-white
Dilution ratio	
Toluol	2.3
Petroleum naphtha	1.0
Distillation range:	
Below 240°C	Not more than 5%
Below 248°C	Not less than 90%
Above 255°C	None
Dryness at 20°C	Miscible without turbidity with 20 volumes 60° Bé gasoline
Freezing point	−30.0°C
Non-volatile matter	0.005 gm/100 cc, max
Solubility of water in solvent at 25°C	0.5% by vol
Specific gravity at 20/20°C	0.989–0.993
Viscosity (10% ½ sec. nitrocellulose solution)	800 centipoises
Weight per gal at 20°C	8.24 lbs

Table 15.58: Diamyl Oxalate *(2)*

$$C_5H_{11}OOCCOOC_5H_{11}$$

Diamyl oxalate is a colorless, oily liquid miscible with most lacquer solvents, oils and hydrocarbons. It is a solvent for ester gum, copal ester, "Cumar" resins, alkyd resins, mastic, nitrocellulose and shellac. It is used as a plasticizer and in paint and varnish removers. Like other oxalates, it has a tendency to hydrolyze.

Boiling point	265°C.
Flash point	116°C.
Specific gravity	0.97

LACTATES

Studies made by Eder and Kutter on commercial lactic acid have found it to be a mixture of alphahydroxypropionic acid, lactyllactic acid and water. Watson has found, that when a dilute lactic acid solution is concentrated, a molecule of water is split off from two molecules of lactic acid thus forming lactyllactic acid.

$$2CH_3 \cdot CHOH \cdot COOH \rightleftarrows CH_3 \cdot CHOH \cdot COOCH(CH_3) \cdot COOH + H_2O$$

$$\Updownarrow$$

$$CH(CH_3)OCOCH(CH_3)OCO + H_2O$$

By further concentration, one molecule of lactyllactic acid loses a molecule of water forming a lactide.

$$CH_3CHOHCO \cdot CH_3 \cdot CH \cdot CO \qquad CH_3 \cdot CH \cdot COOH$$

Chain polylactyllactic acids are also formed by the loss of a molecule of water between the alcohol and carboxyl groups. These reactions are reversible when diluted with water. The composition of lactic acid solutions are dependent upon age, temperature, concentration and the method of preparation. Because of the two active hydroxyl groups present in the lactic acid molecule, there are three types of esters.

Table 15.59: Methyl Lactate *(2)*

$$CH_3CH(OH)COOCH_3$$

Methyl lactate is a water-white liquid, completely miscible with water and most organic liquids. It is a solvent for nitrocellulose, cellulose acetate, cellulose acetobutyrate and cellulose acetopropionate. It is used in the manufacture of lacquers and dopes where it contributes high tolerance for diluents, good flow and blush resistance.

Acidity (as lactic)	0.15% by wt, max
Boiling point	144.8°C
Color	Water-white
Distillation range:	
Below 115°C	None
Between 141°C and 145°C	Not less than 60%
Above 155°C	None
Flash point	51.7°C
Heat of combustion	4778 calories per gram
Freezing point	Approx 66°C
Non-volatile matter	0.01 gram per 100 cc, max
Purity	95% min
Refractive index at 20°C	1.4131
Specific gravity at 20/20°C	1.087 to 1.097
Water at 20°C	No turbidity when mixed with 19 volumes of 60° Bé gasoline
Weight per gal at 68°F	9.09 lbs

Table 15.60: Ethyl Lactate *(2)*

$$CH_3CH(OH)COOC_2H_5$$

Ethyl lactate is a colorless and almost odorless liquid, which, upon evaporation, will sometimes develop a disagreeable odor. This is owing to the lactides, or inner anhydrides, contained in the lactic acid made by fermentation. It is miscible with water, alcohols, ketones, esters, hydrocarbons and oils. Ethyl lactate will dissolve cellulose acetate and nitrate and many of the ethers of cellulose. It is also a solvent for basic dyes, alkyd resins, kauri, manila, pontianac, rosin, shellac and vinyl resins. Ethyl lactate has high solvent power and equally high tolerance for nonsolvents and diluents. These exceptional properties are accounted for by the existence of both an alcohol and an ester group in its molecule.

Its rate of evaporation is slow but this is desirable for brushing lacquers. The presence of ethyl lactate in a solvent mixture imparts good working qualities and good flow, and permits the application of a thin coat on almost any surface. The resulting films are smooth and uniform, although at times the film will remain soft for a longer period than is anticipated. Its solvent action is slower than that of butyl or amyl acetate and the resulting solution has a high viscosity. However, it will tolerate two or three times as much nonsolvent or diluent. In fact, a solution of pyroxylin in ethyl lactate will tolerate the addition of 25 percent water without precipitation. As far as water tolerance is concerned it has no rival in the field of solvents. Ethyl lactate is useful as a lacquer solvent for cellulose nitrate, acetate and ethers. It is used in the preparation of stencil sheets, incandescent mantle lacquers and in laminated glass.

Physical Properties and Specifications

Acidity (as lactic)	0.08%, max
Color	Water-white
Distillation range:	
Below 102°C	None
Below 139°C	Not more than 10%
Below 155°C	Not less than 90%
Above 173°C	None
Dryness	Miscible without turbidity with 20 vols 60° Bé gasoline at 20°C
Non-volatile matter	0.005 g/100 cc, max
Odor	Mild, non-residual
Purity	96% min
Specific gravity at $\frac{20°C}{20°C}$	1.020–1.036
Blush resistance at 90°F (10% ½-sec. R.S. nitrocellulose sol.)	Clear 80% Relative humidity Blush 85%
Coefficient of expansion	0.00058/1°F 0.00104/1°C
Dilution ratio	5.5 with toluene 0.8 with petroleum naphtha
Evaporation rate at 95°F	
Per cent	5 25 50 75 90 95
Minutes	4 23 47¼ 73 92¼ 101
Flash point	129°F. (approx)
Viscosity (10% ½-sec. R.S. nitrocellulose solution)	195 centipoises
Water solubility	Soluble in all proportions (25°C)

Table 15.61: Butyl Lactate *(2)*

$$CH_3CH(OH)COOC_4H_9$$

Butyl lactate is a colorless liquid having a mild odor. The commercial grade contains condensation products and its physical and chemical properties will vary. It is miscible with many of the lacquer solvents, diluents and oils. It will dissolve such substances as cellulose esters, "Cumar" resins, ester gum, copal ester, alkyd resins, mastic and shellac. It has a high tolerance for nonsolvents and it evaporates slowly. Its presence in a solvent mixture will impart brilliance, gloss, adhesion, flexibility and tenacity to the film. It is used as a solvent in lacquers, in stencil manufacture and in lithographic and printing inks. It is also used as an anti-skinning agent, as an intermediate, and in perfumes.

Purity	95% ester by wt, min
Specific gravity at $\frac{20°C}{20°C}$	0.974 to 0.984
Acidity (as lactic)	0.15% max
Water	No turbidity when mixed with 19 vols of 60° Bé gasoline at 20°C

(continued)

Table 15.61: (continued)

Non-volatile matter	0.01 g per 100 cc, max
Color	Water-white
Distillation range	
Below 140°C	None
Between 155°C and 195°C	Not less than 60 per cent
Between 187°C and 189°C	Not less than 90 per cent
Dry point	Not above 200°C
Molecular weight	146.11
Odor	Mild (No residual odor)
Flash point	71°C (159.8°F)
Freezing point	−43°C
Weight per gallon	8.15 lbs. (68°F)
Solubility in water	3.4% by vol (25°C)
Solubility of water	
in butyl lactate	13.0% by vol (25°C)
Refractive index	1.42162 (20°C)
Vapor pressure	0.4 mm Hg (20°C)
Heat of vaporization	77.4 cal/g (20°C)

Table 15.62: Amyl Lactate *(2)*

$$CH_3CH(OH)COOC_5H_{11}$$

Amyl lactate is a colorless to pale yellow nontoxic liquid with an odor like that of brandy. Its composition varies containing lacticides among other things. It is miscible with alcohols, ketones, esters, hydrocarbons, oils, and so forth. It is a solvent for cellulose ethers, "Cumar" resins, copal esters, mastic, nitrocellulose and shellac, and will dissolve alkyd resins when combined with alcohol. It is used as a plasticizer for cellulose derivatives.

Acidity (as lactic)	0.05% by wt., max.
Color	Water-white
Distillation range at 20 mm.	100% between 75°–150°C.
Flash point	175°F.
Purity	At least 95%, min.
Specific gravity at 20°C.	0.954–0.966
Weight per gal.	7.99 lbs.

Table 15.63: Physical Properties of Lactates *(2)*

	B.P.		SP. GR.	REFRACTIVE INDEX	SAPONIFICATION VALUE			B.P.		SP. GR.	REFRACTIVE INDEX	SAPONIFICATION VALUE	
	°C	Mm.			Calcd.	Found		°C	Mm.			Calcd.	Found
	Lactic Esters							Acetoxypropionate Esters					
Methyl	144.8	760	d^{19} 1.0898	n$_D^{18}$ 1.4132^b			Methyl	171.5	760	d$^{20}_4$ 1.088	n$_D^{18}$ 1.4111		
Ethyl	154.5	760	d^{19} 1.0308	n$_D^{19}$ 1.4121^b			Ethyl	177	733	d^{17} 1.0458	n$_D^{18}$ 1.4085^b		
n-Propyl	86	40	d$^{20}_{20}$ 0.996	n$_D^{18}$ 1.4167^b			n-Propyl	195–6	766	d$^{25}_{25}$ 1.0163	n$_D^{18}$ 1.4123		
Isopropyl	166–8	760	d$^{20}_{20}$ 0.998	n$_D^{18}$ 1.4082^b			Isopropyl	182–3	765	d$^{25}_{25}$ 0.9920	n$_D^{18}$ 1.4058		
n-Butyl	185	760	d$^{20}_{20}$ 0.973	n$_D^{18}$ 1.4214^b			n-Butyl	213–4	767	d$^{20}_4$ 1.0001	n$_D^{18}$ 1.4147		
Isobutyl	96	40	d$^{20}_{20}$ 0.971	n$_D^{18}$ 1.4183^b			Isobutyl	205	763	d$^{25}_{25}$ 0.9952	n$_D^{18}$ 1.4140		
sec-Butyl^c	180	760	d$^{20}_{20}$ 0.974				n-Amyl	226–7	763	d$^{25}_{25}$ 0.9822	n$_D^{18}$ 1.4199		
n-Amyl	112	40	d$^{20}_{20}$ 0.952	n$_D^{18}$ 1.4254^b			Isoamyl	221–2	763	d$^{25}_{25}$ 0.9838	n$_D^{18}$ 1.4190		
Isoamyl	82	7	d$^{25}_{25}$ 0.9614	n$_D^{18}$ 1.4240	350	353	n-Hexyl	135	17	d$^{25}_{25}$ 0.9770	n$_D^{18}$ 1.4232	519	519
n-Hexyl	75	2	d$^{25}_{25}$ 0.9533	n$_D^{18}$ 1.4290	322	322	2-Ethyl butyl	127	14	d$^{25}_{25}$ 0.9822	n$_D^{18}$ 1.4245	519	522
2-Ethyl butyl	104	12	d$^{25}_{25}$ 0.9615	n$_D^{18}$ 1.4307	322	321	2-Ethyl hexyl	145	13	d$^{25}_{25}$ 0.9629	n$_D^{18}$ 1.4298	460	462
2-Ethyl hexyl	112	3.6	d$^{25}_{25}$ 0.9405	n$_D^{18}$ 1.4358	277	278	Lauryl	165	4	d$^{25}_{25}$ 0.9304	n$_D^{18}$ 1.4373	373	370
Lauryl	150–3	4	d$^{25}_{25}$ 0.9108	n$_D^{18}$ .14433	217	212	Phenyl ethyl	139	4	d$^{25}_{25}$ 1.0983	n$_D^{18}$ 1.4896	475	476
Phenyl ethyl	124	4	d$^{25}_{25}$ 1.0979	n$_D^{18}$ 1.5073	289	293	Acetoxyethyl (glycol mono-lactate diacetate)	145	10	d$^{25}_{25}$ 1.1489	n$_D^{18}$ 1.4297		
Glycol^d	140	10	d$^{25}_{25}$ 1.1967	n$_D^{18}$ 1.4452	419	413	Benzyl	145.8	4	d$^{20}_4$ 1.1227	n$_D^{18}$ 1.4874		
Glycerol^d	175–80^e	2					Glycerol monolactate triacetate^e						
Benzyl	134	4	d$^{20}_4$ 1.1355	n$_D^{18}$.^{b}1.5049									
Stearyl	180^e	2											

* Where no reference is given, the properties were determined by the authors.
^{b} Properties not given in the reference but determined by the authors.
^{c} Compounds not prepared by authors.
^{d} Monolactate.
^{e} Decomposed.

ALCOHOL ESTERS

Table 15.64: Kessco Alcohol Esters *(59)*

Kessco® Alcohol Ester	SPECIFICATIONS				TYPICAL PROPERTIES						
	Maximum Acid Value or % Acidity	Iodine Value	Color, APHA	Specific Gravity @ 25/20°C	Form @ 25°C	Freezing Point, °C	Viscosity @ 25°C, cp	Refractive Index @ 25°C	Mid-Boiling Point, °C @ 4mm Hg	Pounds Per Gallon @ 25°C	Flash Point, °F, COC
		max	max								
Butoxyethyl oleate	0.2% (as acetic)	70	250	0.881-0.889	Liquid	−37	10	1.449	220	7.4	395
Butoxyethyl stearate	0.1% (as acetic)	5.0	60	0.874-0.880	Liquid	12	11	1.444	218	7.3	380
Butyl oleate	0.1% (as acetic)	−	100	0.861-0.869	Liquid	<−15	8	1.450	185	7.2	380
Butyl stearate cosmetic	1.0	0.5	30	0.850-0.860	Liquid	19	7	1.442	200	7.1	370
Butyl stearate distilled	0.5% (as stearic)	6.0	60	0.853-0.859	Liquid	18	7	1.442	200	7.1	370
Isobutyl palmitate	0.5% (as palmitic)	1.0	60	0.850-0.856	Liquid	13	6.7	1.437	170	7.1	325
Isobutyl stearate	0.5% (as stearic)	1.0	35	0.849-0.855	Liquid	15	8.5	1.441	200	7.1	360
Isocetyl stearate	3.0	5.0	200	0.853-0.859	Liquid	0	32	1.452	−	7.1	450
Isopropyl myristate	1.0	1.0	20	0.849-0.855	Liquid	−3	4.8	1.433	160	7.1	305
Isopropyl palmitate	1.0	1.0	20	0.849-0.855	Liquid	13	6.7	1.437	170	7.1	325
Kessco 639 (IPP/IPM)	1.0	1.0	30	0.849-0.855	Liquid	7	5.9	1.436	170	7.1	305
Octyl isononoate	1.0	1.0	20	0.853-0.859	Liquid	<−50	4.3	1.434	150	7.2	265

GLYCOL ESTERS

Table 15.65: Kessco Glycol Esters *(59)*

Kessco® Glycol Ester	SPECIFICATIONS		TYPICAL PROPERTIES		
	Acid Value	Iodine Value	Form @ 25°C	Melting Point, °C	Flash Point, °F, COC
	max	max			
Diethylene glycol monostearate	5.0	0.5	Flakes	44.5-47.5	395
Diglycol stearate S.E.	103	7.0	Flakes	48-53	345
Diglycol stearate neutral	103	7.0	Flakes	42-48	360
Ethylene glycol distearate	15.0	0.5	Flakes	60-63	390
Ethylene glycol monostearate	2.0	0.5	Flakes	56-60	390
Ethylene glycol monostearate 70	2.0	0.5	Flakes	52-56	370
Propylene glycol monostearate pure	3.0	0.5	Flakes	33.5-38.5	390

Table 15.66: Kessco Polyethylene Glycol Esters *(59)*

Kessco® Polyethylene Glycol Ester	Saponification Value min	Saponification Value max	Acid Value max	Iodine Value max	Color	Form	Freezing Point, °C	pH @ 20°C (3% Disp.)	Pounds/Gallon @ 25°C	HLB Value	Flash Point, °F, COC	Fire Point, °F	Specific Gravity (a) @ 25°C (b) @ 65°C
PEG 200 monolaurate	132	142	5.0	9.5	Light yellow	Liquid	< 5	4.5	8.2	9.8	385	420	0.9833(a)
PEG 200 dilaurate	176	186	10.0	12.5	Light yellow	Liquid	< 9	4.5	7.9	5.9	460	510	0.9520(a)
PEG 300 monolaurate	104	114	5.0	7.5	Light yellow	Liquid	< 8	4.5	8.4	11.4	445	500	1.0100(a)
PEG 300 dilaurate	148	158	10.0	10.0	Light yellow	Liquid	<13	4.5	8.1	9.8	475	550	0.9703(a)
PEG 400 monolaurate	86	96	5.0	6.5	Light yellow	Liquid	12	4.5	8.6	13.1	475	535	1.0242(a)
PEG 400 dilaurate	127	137	10.0	9.5	Light yellow	Liquid	18	4.5	8.3	9.8	480	555	0.9884(a)
PEG 600 monolaurate	64	74	5.0	5.0	Light yellow	Liquid	23	4.5	8.8	14.9	475	555	1.0505(a)
PEG 600 dilaurate	102	112	10.0	7.5	Cream	Soft liquid	24	4.5	–	12.2	465	530	0.9820(b)
PEG 1000 monolaurate	41	51	5.0	3.5	Cream	Soft liquid	40	4.5	–	16.5	490	505	1.035(b)
PEG 1000 dilaurate	68	78	10.0	5.5	Cream	Soft liquid	38	4.5	–	14.5	475	495	1.015(b)
PEG 1540 monolaurate	26	36	5.0	2.5	Cream	Wax	46	4.5	–	17.6	445	520	1.060(b)
PEG 1540 dilaurate	48	56	10.0	4.0	Cream	Wax	42	4.5	–	15.7	450	515	1.040(b)
PEG 4000 monolaurate	9	18	5.0	1.0	Cream	Wax	55	4.5	–	18.8	515	565	1.075(b)
PEG 4000 dilaurate	20	30	5.0	2.0	Cream	Wax	52	4.5	–	17.6	495	515	1.065(b)
PEG 6000 monolaurate	7	13	5.0	1.0	Cream	Wax	61	4.5	–	19.2	–	–	–
PEG 6000 dilaurate	12	20	9.0	1.5	Cream	Wax	57	4.5	–	18.7	435	495	1.077(b)

Kessco® Polyethylene Glycol Ester	Saponification Value min	Saponification Value max	Acid Value max	Iodine Value max	Color	Form	Melting or Freezing Point, °C	pH @ 25°C (3% Disp.)	Pounds/Gallon @ 25°C	HLB Value	Flash Point, °F, COC	Fire Point, °F	Specific Gravity (a) @ 25°C (b) @ 65°C
PEG 200 monooleate	115	124	5.0	56.0	Light amber	Liquid	<-15	4.5	8.1	8.0	395	445	0.9742(a)
PEG 200 dioleate	148	158	10.0	70.0	Light amber	Liquid	<-15	4.5	7.9	6.0	545	625	0.9405(a)
PEG 300 monooleate	94	102	5.0	45.0	Light amber	Liquid	<-5	4.5	8.3	9.6	450	–	–
PEG 300 dioleate	128	137	10.0	61.0	Light amber	Liquid	<-5	4.5	8.0	7.2	510	585	0.9609(a)
PEG 400 monooleate	80	89	5.0	39.0	Light amber	Liquid	< 10	4.5	8.4	11.4	510	585	1.0135(a)
PEG 400 dioleate	113	122	10.0	55.0	Light amber	Liquid	< 7	4.5	8.1	8.5	520	635	0.9770(a)
PEG 600 monooleate	60	69	5.0	30.0	Light amber	Liquid	23	4.5	8.7	13.5	525	620	1.0381(a)
PEG 600 dioleate	92	102	10.0	45.0	Light amber	Liquid	19	4.5	8.3	10.5	495	615	1.0038(a)
PEG 1000 monooleate	40	49	5.0	21.0	Cream	Soft solid	39	4.5	–	15.4	515	595	1.035(b)
PEG 1000 dioleate	64	74	10.0	33.0	Cream	Soft solid	37	4.5	–	13.1	505	615	1.005(b)
PEG 1540 monooleate	28	37	5.0	14.5	Cream	Wax	45	4.5	–	17.0	520	545	1.050(b)
PEG 1540 dioleate	45	55	10.0	24.5	Cream	Wax	44	4.5	–	15.0	480	560	1.025(b)
PEG 4000 monooleate	10	18	5.0	6.0	Cream	Wax	55	4.5	–	18.3	495	515	1.075(b)
PEG 4000 dioleate	19	27	5.0	11.0	Cream	Wax	49	4.5	–	17.8	500	520	1.060(b)
PEG 6000 monooleate	7	13	5.0	4.3	Cream	Wax	59	4.5	–	19.0	470	495	1.085(b)
PEG 6000 dioleate	13	21	9.0	7.5	Cream	Wax	56	4.5	–	18.3	500	525	1.070(b)

(continued)

Table 15.66: (continued)

Kessco® Polyethylene Glycol Ester	Saponification Value		Acid Value	Iodine Value	Color	Form	Melting Point, °C	pH @ 25°C (3% Disp.)	HLB Value	Flash Point, °F, COC	Fire Point, °F	Specific Gravity @ 65°C
	min	max	max	max								
PEG 200 monostearate	120	129	5.0	0.5	White to cream	Soft solid	31	5.0	7.9	410	450	0.9360
PEG 200 distearate	153	162	10.0	0.5	White to cream	Soft solid	34	5.0	5.0	475	525	0.9060
PEG 300 monostearate	97	105	5.0	0.5	White to cream	Soft solid	28	5.0	9.7	475	520	0.9660
PEG 300 distearate	130	139	10.0	0.5	White to cream	Soft solid	32	5.0	6.5	–	–	–
PEG 400 monostearate	83	92	5.0	0.5	White to cream	Soft solid	32	5.0	11.6	480	525	0.9780
PEG 400 distearate	115	124	10.0	0.5	White to cream	Soft solid	36	5.0	8.0	500	545	0.9390
PEG 600 monostearate	61	70	5.0	0.25	White to cream	Soft solid	37	5.0	13.6	480	550	1.0000
PEG 600 distearate	93	102	10.0	0.5	White to cream	Soft solid	39	5.0	10.6	490	530	0.9670
PEG 1000 monostearate	40	48	5.0	0.25	Cream	Wax	41	5.0	15.6	475	495	1.030
PEG 1000 distearate	65	74	10.0	0.25	Cream	Wax	40	5.0	12.3	485	510	1.005
PEG 1540 monostearate	27	36	5.0	0.25	Cream	Wax	47	5.0	17.3	495	540	1.050
PEG 1540 distearate	49	58	10.0	0.25	Cream	Wax	45	5.0	14.8	490	540	1.015
PEG 4000 monostearate	10	18	5.0	0.1	Cream	Wax	56	5.0	18.6	465	520	1.075
PEG 4000 distearate	19	27	5.0	0.2	Cream	Wax	51	5.0	17.3	515	545	1.060
PEG 6000 monostearate	7	13	5.0	0.1	Cream	Wax	61	5.0	18.8	480	525	1.080
PEG 6000 distearate	14	20	9.0	0.1	Cream	Wax	55	5.0	18.4	475	525	1.075

GLYCERYL ESTERS

Table 15.67: Kessco Glycerol Esters *(59)*

Kessco® Glycerol Ester	Acid Value or % Acidity	Iodine Value	Specific Gravity @ 25/20°C	Color	Form @ 25°C	Freezing Point or Melting Point, °C*	Viscosity @ 25°C, cP	Pounds per Gallon @ 25°C	Flash Point, °F, COC
	max	max							
Glycerol monolaurate	5.0	1.0	–	White	Solid	53.9	–	–	425
Glycerol monostearate pure	3.0	0.5	–	White	Flakes	(56.5-58.5)	–	–	410
Glycerol monostearate DH-1	5.0	6.0	–	White	Flakes	(55-58)	–	–	410
Glycerol monostearate 860	3.0	2.0	–	White	Flakes	(58.5-61.5)	–	–	450
Glycerol monostearate, SE, acid stable	3.0	–	–	White to cream	Flakes	54-58	–	–	–
Glycerol monooleate	5.0	77	0.945-0.953	Yellow	Liquid & Solids	< 20	204	7.9	435
Glycerol dioleate	5.0	82	0.923-0.929	Yellow to amber	Liquid	0	90	7.7	520
Acetin (glycerol monoacetate)	0.5% (as acetic)	–	1.188-1.198	75 APHA max	Liquid	–30	66.8	9.9	305
Diacetin (glycerol diacetate)	0.5% (as acetic)	–	1.173-1.183	–	Liquid	~ –35	36	9.8	295
Triacetin (glycerol triacetate)	0.005% (as acetic)	–	1.152-1.158	50 APHA max	Liquid	–50 Gel	16	9.6	290

*Melting points in parentheses are specification ranges; figures below 20°C are freezing points.

Table 15.68: Polyglyceryl Esters (68)

	Gardner Color	Hydroxyl Value	Iodine Value	Acid Value	Saponification Value	Form
Decaglycerol Monolaurate	8	690	3	2	60	Liquid
Decaglycerol Distearate	8	420	3	5	95	Paste
Decaglycerol Dioleate	8	420	25	5	95	Liquid
Decaglycerol Tristearate	10	300	3	6	120	Solid
Decaglycerol Tetraoleate	8	220	55	6	135	Liquid
Decaglycerol Octaoleate	7	80	65	7	165	Liquid
Decaglycerol Decastearate	8	25	3	7	170	Solid
Decaglycerol Decaoleate	7	35	70	7	175	Liquid
Decaglycerol Decalinoleate	7	35	125	7	165	Liquid
Triglycerol Monostearate	10	325	3	4	130	Solid
Triglycerol Monooleate	8	320	55	3	135	Liquid

SOLUBILITIES

Product	Water	Alcohol SD 40	Neobee M5	Peanut Oil	Isopropyl Myristate
Decaglycerol Monolaurate	S	Sppt	I	I	I
Decaglycerol Distearate	D	Sppt	D	D	Dg
Decaglycerol Dioleate	D	Sppt	D	D	D
Decaglycerol Tristearate	D	D	Dg	Dg	Dg
Decaglycerol Tetraoleate	D	S	S	S	S
Decaglycerol Octaoleate	I	I	S	S	S
Decaglycerol Decastearate	I	I	SH	SH	SH
Decaglycerol Decaoleate	I	I	MP	MP	MP
Decaglycerol Decalinoleate	I	I	MP	MP	MP
Triglycerol Monostearate	D	D	D	D	D
Triglycerol Monooleate	D	S	D	S	S

Product	Mineral Oil Visc 50/60	Visc 125/135	Visc 435/355	Propylene Glycol	Glycerol	Acetone	Benzene
Decaglycerol Monolaurate	I	I	I	S	S	I	I
Decaglycerol Distearate	I	D	D	Dg	Dg	I	S
Decaglycerol Dioleate	I	I	I	D	S	S	S
Decaglycerol Tristearate	PS	PS	Dg	D	Dg	I	S
Decaglycerol Tetraoleate	S	S	S	I	S	S	S
Decaglycerol Octaoleate	S	S	S	I	S	S	S
Decaglycerol Decastearate	SH	PD	PD	I	Dg	I	S
Decaglycerol Decaoleate	MP	MP	MP	PS	S	Sppt	S
Decaglycerol Decalinoleate	MP	MP	MP	I	S	Sppt	S
Triglycerol Monostearate	D	D	Dg	D	Dg	D	S
Triglycerol Monooleate	S	S	S	I	S	S	S

S = Soluble, up to 10% or more
Sppt = Soluble, a precipitate forms on standing
PS = Partially soluble
SH = Soluble hot
MP = Miscible in all proportions
D = Dispersible
PD = Partially dispersible
Dg = Dispersible, gels at higher concentrations
I = Insoluble

BENZOATES

Table 15.69: Polyol Benzoates *(65)*

TYPICAL PROPERTIES

	Neopentyl Glycol Dibenzoate	Triethylene Glycol Dibenzoate	Glyceryl Tribenzoate	Trimethylol-ethane Tribenzoate	Penta-erythritol Tetrabenzoate
Empirical Formula	$C_{19}H_{20}O_4$	$C_{20}H_{22}O_6$	$C_{24}H_{21}O_6$	$C_{26}H_{24}O_6$	$C_{33}H_{28}O_8$
Molecular Weight (Theory)	312	358	404	432	552
Melt Point °C. (approx.)	49	47	71	81	95
Form and Appearance	White Crystalline Solid (Cast)				
Color (Molten) APHA (approx. max.)	200	200	200	200	200
Specific Gravity (Solid 30°C.)	1.2154	1.2715	1.2619	1.2419	1.2801
Refractive Index (Molten 50°C.)	1.5298	1.5252	1.5584*	1.5538*	1.5715*

*Supercooled

CARBONATES

Table 15.70: Diethyl Carbonate *(2)*

Ethyl Carbonate
"Diatol" (contains 90% diethyl carbonate)

$$(C_2H_5)_2CO_3$$

Diethyl carbonate is a stable, colorless liquid with a mild and pleasant odor. It is miscible with alcohols, ketones, esters, aromatic hydrocarbons, some aliphatic solvents, and dissolves many of the natural and synthetic resins, nitro-cellulose and cellulose ethers. Its solvent power is increased by adding either an active solvent, such as acetone or an ester, or a latent solvent such as alcohol. It has a rather low tolerance for hydrocarbons. It is used in lacquer formulation, in special lacquers for coating the cathode of radio tubes with rare earths, in organic synthesis and in the manufacture of phenobarbital.

Acidity (as carbonic)	0.02% by wt, max	Evaporation rate at 95°F (in minutes)	
Blush resistance at 90°F (10% ½ sec. R.S. nitrocellulose solution)	Clear 85% Relative humidity Blush 90%	5%	1¼
		25%	7¼
		50%	14½
		75%	24
Boiling point	180°C	90%	31¼
Coefficient of expansion		95%	34¼
per 1°F	0.00066	Flash point	89°F
per 1°C	0.00119	Freezing point	48.2°C
Color	Water-white	Non-volatile matter	Not more than 0.005 gm per 100 cc
Dilution ratio		Purity	98–100%
Toluol	0.6	Solubility in water	69% by wt
Petroleum Naphtha	0.4	Solubility of water in solvent at 25°C	1.4% by vol
Distillation range:			
Below 120°C	None	Specific gravity at 20/20°C	0.973–0.977
Below 128°C	Not less than 90%	Vapor pressure at 103°C	54 mm Hg
Above 130°C	None	Weight per gal at 20°C	8.11 lbs (approx)
Dryness at 20°C	Miscible without turbidity with 20 vols 60° Bé gasoline		

PHTHALATES

Table 15.71: Dibutyl Phthalate *(34)*

Dibutyl phthalate is a colorless, odorless, nonvolatile liquid, stable to light and heat. It is miscible with the common organic solvents, a good solvent for both aliphatic and aromatic diluents, and is compatible with the usual components of protective coatings.

CSC Specifications

Purity, Ester Content by Weight	99-100%
Specific Gravity, 20°/20°C	1.047-1.049
Acidity as Phthalic Acid, maximum	0.01%
Water (naphtha test)	Pass test
Color, maximum	20 APHA
Odor	None

Molecular Weight (calc.)	278.35
Boiling Point at 760 torr	340°C
Vapor Pressure: 198.2°C	10 torr
287.0°C	200 torr
313.5°C	400 torr
Freezing Point	—35°C
Density at 25°C	1.0426 g/ml
Weight per U.S. Gallon at 68°F	8.72 lb
Coefficient of Expansion,°F	0.00048
°C	0.00086
Refractive Index, n_D at 25°C	1.4914

Physical Properties

Viscosity at 25°C	15.8 cp
Heat of Vaporization at bp	18.93 kcal/mole
Specific Heat at ordinary temperatures	0.43 cal/g (approx.)
Electrical Conductivity at 30°C	4.2 x 10⁻⁸ mho/cm
Dielectric Constant at 30°C	6.436
Flash Point, Cleveland Open Cup	340°F
Solubility in Water at 20°C	<0.01% by weight
Solubility of Water in Dibutyl Phthalate at 20°C	0.46% by weight

PHOSPHATES

Table 15.72: Tributyl Phosphate *(34)*

CSC Specifications

Specific gravity at 20/20°C	0.977-0.979
Acidity as phosphoric acid (max.)	0.01% by wt
Water by Karl Fischer test (max.)	0.3%
Butanol (max.)	0.2%
Color (max.)	15 APHA
Appearance	Clear and free from suspended matter

Physical Properties

Molecular weight (calc.)	266.31
Boiling point at 760 torr	289°C (dec)
at 27 torr	177–178°C
at 10 torr	150°C
Freezing point	<—80°C
Density at 25°C	0.973 g/ml
Weight per U.S. gallon at 68°F	8.19 lb.
Coefficient of expansion	0.00093 per °C
	0.00052 per °F
Refractive index, n_D at 25°C	1.422
Viscosity at 25°C	3.41 centipoise
at 85°F	38.6 Saybolt sec
Specific heat	0.41 cal/(g)(°C)
Surface tension at 25°C	26.7 dynes/cm
Latent heat of vaporization	55.1 cal/g
Dielectric constant at 30°C	7.96
Flash point, Cleveland open cup	295°F

Solvency Characteristics

The solubility of pure Tributyl Phosphate in water varies inversely with temperature: at 4°C it is 1 g liter; at 25°C, 0.42 g liter; at 50°C, 0.28 g liter. The solubility of water in Tributyl Phosphate at 25°C is 64 g liter. The specific gravity at 25 4°C of water-saturated Tributyl Phosphate is 0.976.

Tributyl Phosphate is miscible with many organic liquids, including the following:

Acetic acid	Chloroform	Monobutylamine
Acetone	Decalin	Nitriles
Alcohols	Esters	Nitroparaffins
Benzene	Ether	Petroleum
Carbon tetrachloride	Kerosene	naphtha
	Linseed oil	Toluene
Castor oil	Mineral oil	Xylene

The following organic materials are soluble in Tributyl Phosphate to the extent of 10 grams or more per 100 ml:

Benzoic acid	Naphthalene	Orange shellac
2,4–D acid	β-Naphthol	Picric acid
Ester gum	Nitrocellulose	Tannic acid
Ethylene glycol		

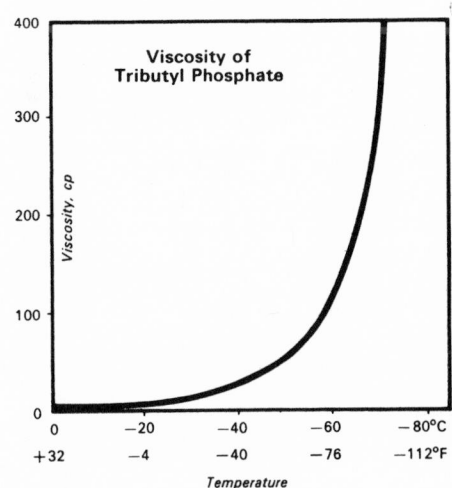

Viscosity of Tributyl Phosphate

PHOSPHITES

Table 15.73: Dialkyl Hydrogen Phosphites *(64)*

PHYSICAL PROPERTIES

Property	Dimethyl Hydrogen Phosphite	Diethyl Hydrogen Phosphite	Dibutyl Hydrogen Phosphite	Bis(2-ethylhexyl) Hydrogen Phosphite
Formula	$(CH_3O)_2P(O)H$	$(C_2H_5O)_2P(O)H$	$(C_4H_9O)_2P(O)H$	$(C_8H_{17}O)_2P(O)H$
Molecular Weight	110.1	138.1	194.2	306.4
Appearance	colorless liquid	colorless liquid	colorless liquid	colorless liquid
Odor	mild, characteristic	mild, pleasant	mild, pleasant	mild, suggestive of octyl alcohol
Boiling Point	72-3°C/25 mm	65-66°C/6 mm	118-9°C/7 mm	163-4°C/3 mm
Specific Gravity 20°/4°	1.200	1.079	0.995	0.937
Index of Refraction, n_D^{20}	1.4016	1.4073	1.4238	1.4423
Flash Point, Cleveland, open cup	205°F	195°F	250°F	330°F
Fire Point, Cleveland, open cup	220°F	220°F	300°F	400°F
Viscosity (Centistokes) 77°F	1.08	1.21	2.36	6.54
100°F	0.92	1.03	1.90	4.72
210°F	0.51	0.56	0.89	1.59
Toxicity (single dose oral LD_{50}, rats), mg/kg	3,050	1,000	3,900	11,900
Acidity	neutral	neutral	neutral	neutral
Solubility in water:	sol., hydrolyzes	sol., hydrolyzes	sl. sol., slowly hydrolyzes	insol., very slowly hydrolyzes

in other solvents: miscible with alcohol, ether, acetone and most common organic solvents.

Table 15.74: Trialkyl Phosphites *(64)*

PHYSICAL PROPERTIES

Property	Trimethyl Phosphite	Triethyl Phosphite	Tris(2-chloroethyl) Phosphite	Triisopropyl Phosphite	Tributyl Phosphite	Triisooctyl Phosphite	Tris(2-ethylhexyl) Phosphite
Formula	$(CH_3O)_3P$	$(C_2H_5O)_3P$	$(ClC_2H_4O)_3P$	$(C_3H_7O)_3P$	$(C_4H_9O)_3P$	$(C_8H_{17}O)_3P$	$(C_8H_{17}O)_3P$
Molecular Weight	124.08	166.2	269.5	208.2	250.3	418.6	418.6
Appearance	colorless liquid	colorless liquid	colorless liquid	colorless liquid	colorless liquid	colorless liquid	colorless liquid
Odor	penetrating	sweet, characteristic	mild, characteristic	sweet, characteristic	mild, not unpleasant	mild, not unpleasant	mild, not unpleasant
Boiling Point	111–112°C	65°C/24 mm	119°C/0.15 mm	94–6°C/50 mm	118–25°C/7 mm	161–4°C/0.3 mm	163–4°C/0.3 mm
Specific Gravity 20°/4°	1.046	0.969	1.353	0.914	0.925	0.891	0.902
Index of Refraction, n_D^{20}	1.4076	1.4131	1.4878	1.4101	1.4327	1.4498	1.4494
Flash Point, Cleveland, open cup	100°F	130°F	375°F	165°F	250°F	385°F	365°F
Fire Point, Cleveland, open cup	100°F	160°F	410°F	195°F	275°F	410°F	400°F
Viscosity (Centistokes) 77°F	0.58	0.74	5.22	1.18	2.08	9.49	8.35
100°F	0.52	0.65	4.11	0.99	1.65	6.85	5.86
210°F	0.32	0.40	1.45	0.57	0.86	2.24	1.90
Toxicity (single dose oral LD_{50}, rats), mg/kg	2,000	3,160	250	2,300	3,000	9,200	-----
Solubility in water:	insol. but reacts with	sl. sol., hydrolyzes	insol., slowly hydrolyzes	insol., slowly hydrolyzes	insol., slowly hydrolyzes	insol., very slowly hydrolyzes	insol., very slowly hydrolyzes

In other solvents: miscible with alcohol, acetone, benzene, ether, heptane, carbon tetrachloride, and most of the common organic solvents.

Table 15.75: Tertiary Phosphites (27)

Name	Formula	Molecular Weight	Oral Toxicity LD50 Microliters per Kg of Rat. Wt.	Color and Form	Phosphorus Content % P	Melting Point °C	Boiling Point °C	Refractive Index n 25/D	Specific Gravity 25/15°C	Flash Point COC°F	Viscosity Centistokes 68°F	100°F	210°F	Vapor Pressure
Trimethyl Phosphite	$(CH_3O)_3P$	124	2000	Water-white liquid	24.97	<−78	111±1	1.404	1.045	130	—	0.51	0.30	10 mm at 12°C / 100 mm at 55°C
Triethyl Phosphite	$(C_2H_5O)_3P$	166	3160	Water-white liquid	18.67	—	154±1	1.413	0.954	—	8.5	5.03	—	—
Tri(2-ethylhexyl) Phosphite	$(C_8H_{17}O)_3P$	418	—	Straw-colored liquid	7.45	glass at low temp.	—	1.451	0.897	340	—	—	—	—
Tridecyl Phosphite (iso)	$(C_{10}H_{21}O)_3P$	502	>10000	Water-white liquid	6.17	< 0	180 at 0.1 mm	1.454	0.886	455	—	11.24	2.90	—
Trilauryl Phosphite	$(C_{12}H_{25}O)_3P$	586	> 3160	Water-white liquid	5.29	<10	—	1.456	0.866	—	—	—	—	—
Trioctadecyl Phosphite	$(C_{18}H_{37}O)_3P$	838	>10000	White waxy solid	3.70	45-47	—	—	0.940*	—	—	—	—	—
Trilauryl Trithiophosphite	$(C_{12}H_{25}S)_3P$	634	>10000	Light straw-colored liquid	4.89	20	—	1.502	0.915	430	—	24.7	5.7	0.01 mm at 200°C
Triphenyl Phosphite	(structure)	310	Approx. 2800	Water-white to pale yellow liquid	10.0	22-25	155-160 at 0.1 mm	1.589	1.184	425	—	8.34	2.07	—
Diphenyldecyl Phosphite (iso)	P·OC₁₀H₂₁ (structure)	374	>10000	Water-white liquid	8.28	18	—	1.516	1.024	425	—	7.82	2.26	—
Phenyldidecyl Phosphite (iso)	O·P (OC₁₀H₂₁)₂ (structure)	438	>10000	Water-white liquid	7.07	<0	—	1.478	0.940	425	—	8.95	2.42	—
"Pentite" — [tetra(diphenyl phosphito) pentaerythritol]	(structure)	989	1500	White waxy solid	12.4	30-60	—	—	1.240	—	—	—	—	—
"Dipentite" — [diphenyl pentaerythritol diphosphite]	(structure)	380	5000	White solid	16.3	70-80	190-200 at 0.1 mm	—	—	—	—	—	—	—
Phenylneopentyl Phosphite	(structure)	226	1780	Water-white liquid	13.70	19	138-140 at 10 mm	1.517	1.135	—	—	—	—	—
1620 Polymeric Phosphite	Bisphenol A-Pentaerythritol Phosphite	Av 1100	>3160	White solid	16.1±0.2	100-110	—	—	—	—	—	—	—	—

*Density

Table 15.76: Organophosphites *(27)*

PHYSICAL DATA

	Triphenyl	Diphenyldecyl	Phenyldidecyl	Tridecyl
Phosphorus Content	10%	8.28%	7.07%	6.17%
Melting Point	22°–25°C.	18°C.	< 0°C.	< 0°C.
Boiling Point at 0.1 mm.	155°–160°C.	—	—	180°C.
Refractive Index, n25/D	1.589	1.5160	1.4785	1.4560
Flash Point (Cleveland open cup)	425°F.	425°F.	425°F.	455°F.
Fire Point (Cleveland open cup)	470°F.	455°F.	470°F.	485°F.
Specific Gravity, 25°/15.5°C.	1.184	1.023	0.940	0.891
Specific Gravity Correction Factor, per 1°C.	0.00085	0.00077	0.00073	0.00066
Pounds per Gallon at 25°C.	9.86	8.520	7.829	7.421
Viscosity in Centistokes:				
at 100°F.	8.34	7.82	8.95	11.24
at 210°F.	2.07	2.26	2.42	2.90

SILICATES

Table 15.77: Ethyl Silicate *(2)*

Silicon Tetraethyl Ester
Ortho–Silicic Acid Ethyl Ester

$$Si(OC_2H_5)_4$$

Ethyl silicate is a water-white liquid, soluble in alcohol. It hydrolyzes in water to an adhesive form of silicic acid and alcohol. It is used in lacquers and paint as a pigment binder giving films that are resistant to fire and chemicals and are weatherproof. A less pure, higher silica ester, Ethyl Silicate 40, is also available commercially.

Specifications—Tetraethyl Silicate	
Acidity (as HCl)	0.05% by wt, max
Available silica (as SiO₂)	28.8%
Boiling point	165.5°C
Boiling range at 760 mm	
Below 160°C	Not more than 5%
Below 170°C	Not less than 95%
Color	Water-white
Purity	97%, min
Specific gravity at 20/20°C	0.933 to 0.938
Weight per gal at 20°C	7.78 lbs

Specifications—Ethyl Silicate	
Acidity, maximum acidity (as HCl)	0.1%
Available silica (as SiO₂)	34–42%
Boiling point	165°C
Boiling range at 760 mm	
Below 80°C	None
Below 110°C	Not more than 5%
Color	Light brown
Odor initial	Mild
Specific gravity at 20/20°C	1.050 to 1.070
Weight per gal at 20°C	8.82 lbs

SOLVENTS AND PLASTICIZERS

Table 15.78: Miscellaneous Esters *(67)*

ADIPATES	Specific Gravity 25°C/25°C	Viscosity Centipoises 20°C	Pour Point °C	Flash Point °F	"PHR" for Standard Modulus	USES
Dimethyl Adipate	1.058	14	-16	285	—	Adipates are primary plasticizers used with a
Diisobutyl Adipate	0.950	19	0	320	—	wide variety of resins. Excellent permanence
Diisooctyl Adipate	0.926	18	-60	400	44	and low temperature properties are imparted
Di-2-ethylhexyl Adipate	0.925	13	-60	380	42	to adipate plasticized polyvinyl chloride compounds. In addition, plastisols prepared with
Dinonyl Adipate	0.917	37	-56	425	46	adipates give exceptional low initial and stable
Diisodecyl Adipate	0.916	26	-66	440	49	long-term viscosity. Adipates have good heat
Dibutoxyethyl Adipate	0.995	12.5	-30	375	44	and light stability and impart a soft drape to
n-Octyl n-Decyl Adipate	0.916	16	0	400	45	vinyl film.
Isooctyl Isodecyl Adipate	0.922	20	-35	400	47	
610 Adipate	0.919	19	3	400	42	
810 Adipate	0.914	22	5	410	43	

(continued)

Table 15.78: (continued)

FUMARATES	Specific Gravity 25°C/25°C	Viscosity Centipoises 20°C	Pour Point °C	Flash Point °F	Saponi-fication Number	USES
Dibutyl Fumarate	0.984	8	-12.5	280	492	Fumarates are monomers used for copolymerization with other monomers such as vinyl acetate, vinyl chloride, styrene and acrylonitrile.
Diisooctyl Fumarate	0.940	19	-50	380	330	
Di-2-ethylhexyl Fumarate	0.932	19	-57	380	330	

MALEATES	Specific Gravity 25°C/25°C	Viscosity Centipoises 20°C	Pour Point °C	Flash Point °F	Saponi-fication Number	USES
Diethyl Maleate	1.066	3.6	-15	200	651	Maleates are monomers used to react and form copolymers with vinyl acetate, vinylidene chloride, styrene and other monomers. They are also useful as raw materials in the preparation of anionic surfactants of the sulfosuccinate type.
Dibutyl Maleate	0.993	5	-45	275	492	
Dimethylamyl Maleate	0.941	12	-40	290	394	
Diisooctyl Maleate	0.942	18	-50	355	330	
Di-2-ethylhexyl Maleate	0.939	17	-50	355	330	
Ditridecyl Maleate	0.917	88	-45	395	233	

PHTHALATES	Specific Gravity 25°C/25°C	Viscosity Centipoises 20°C	Pour Point °C	Flash Point °F	"PHR" for Standard Modulus	USES
Dimethyl Phthalate	1.187	22	5.5	320	—	Dimethyl and Diethyl Phthalates are used in adhesives and to plasticize cellulose acetate.
Diethyl Phthalate	1.118	21	-4	325	—	
Diisobutyl Phthalate	1.038	30	-45	345	—	Diisobutyl, Dibutyl and Dicyclohexyl Phthalates are used in adhesives, lacquers, emulsions, heat sealing paper coatings and for cellophane.
Dibutyl Phthalate	1.045	20	-30	360	—	
Dicyclohexyl Phthalate	1.148	Solid	*63.5	405	65	
Diisohexyl Phthalate	1.005	50	-33	380	51	Diisohexyl, Butyl Benzyl, Butyl Octyl and Butyl Decyl Phthalates are primary plasticizers for vinyl chloride resins. These plasticizers are efficient, fast fluxing and high solvating.
Butyl Benzyl Phthalate	1.119	65	-45	425	48	
Butyl Octyl Phthalate	0.996	40	-44	370	50	
Butyl Decyl Phthalate	0.994	55	-40	395	54	
Diisooctyl Phthalate	0.982	83	-45	425	55	The Octyl and Decyl Phthalates are the backbone of vinyl formulations. These general purpose plasticizers afford optimum performance at low cost in film and sheeting, extrusions and plastisols.
Di-2-ethylhexyl Phthalate	0.983	82	-46	420	54	
Dicapryl Phthalate	0.969	71	-41	394	55	
Diisodecyl Phthalate	0.966	113	-39	450	58	
Isooctyl Isodecyl Phthalate	0.972	100	-44	415	55	
n-Octyl n-Decyl Phthalate	0.972	45	-45	445	53	These straight chain plasticizers provide general purpose plasticizers for vinyls exhibiting good low temperature flexibility.
610 Phthalate	0.972	43	-35	435	47	
810 Phthalate	0.968	48	-40	440	50	
Isodecyl Tridecyl Phthalate	0.958	200	-37	465	61	These plasticizers are finding wide acceptance in high temperature wire compounds.
Ditridecyl Phthalate	0.951	334	-35	480	64	
Dibutoxyethyl Phthalate	1.062	42	-48	408	52	These specialty plasticizers are used as processing aids and show compatibility with many resins.
Hatcol 59	1.013	Solid	*56	405	65	

SEBACATES	Specific Gravity 25°C/25°C	Viscosity Centipoises 20°C	Pour Point °C	Flash Point °F	"PHR" for Standard Modulus	USES
Dimethyl Sebacate	†0.983	Solid	*26	320	—	DMS is used as a chemical intermediate.
Dibutyl Sebacate	0.934	8	0	350	38	FDA approved plasticizer for food wraps.
Diisooctyl Sebacate	0.915	24	-40	440	45	These are excellent low temperature vinyl plasticizers and synthetic lubricants.
Di-2-ethylhexyl Sebacate	0.913	27	-44	440	45	

† 30°C/20°C   *Freezing Point

TRIMELLITATES	Specific Gravity 25°C/25°C	Viscosity Centipoises 20°C	Pour Point °C	Flash Point °F	"PHR" for Standard Modulus	USES
Triisooctyl Trimellitate	0.990	260	-50	500	57	These are used in formulating high temperature vinyl wire compounds.
n-Octyl n-Decyl Trimellitate	0.978	115	-56	510	58	
Tri-2-ethylhexyl Trimellitate	0.987	300	-45	500	56	

Appendix—Comparative Data for Various Solvents

Physical Properties of Some Selected Solvents (10)

Name	Mol. Wt.	Density	Temp.	Refract. Index	Melting Point	Pressure	Boiling Point	Dielec. Constant	List No.
WATER	18.02	0.9971	25.0	1.3329	0.0	760.	100.0	78.54	1
METHANOL	32.04	0.7866	25.0	1.3265	-97.7	0.	64.7	32.70	2
ACETONITRILE	41.05	0.7766	25.0	1.3416	-43.8	0.	81.6	37.50	3
ETHYLENIMINE	43.07	0.8320	25.0	1.4123	-78.0	0.	57.0	18.30	4
ACETALDEHYDE	44.05	0.7780	20.0	1.3311	-123.0	0.	20.4	21.10	5
FORMAMIDE	45.04	1.1334	20.0	1.4475	2.6	0.	210.5	109.00	6
FORMIC ACID	46.03	1.2141	25.0	1.3694	8.3	0.	100.6	58.50	7
ETHANOL	46.07	0.7850	25.0	1.3594	-114.1	0.	78.3	24.55	8
ACRYLONITRILE	53.06	0.8004	25.0	1.3888	-83.6	0.	77.3	33.00	9
1,2-BUTADIENE	54.09	0.6760	1.0	1.4205	-136.3	0.	10.9	0.0	10
2-BUTYNE	54.09	0.6910	20.0	1.3921	-32.3	760.	27.0	0.0	11
PROPIONITRILE	55.08	0.7818	20.0	1.3681	-92.8	0.	97.4	27.20	12
PROPIONITRILE	55.08	0.7768	25.0	1.3636	-92.8	0.	97.4	27.20	13
ACROLEIN	56.06	0.8389	20.0	1.4017	-87.0	0.	52.7	0.0	14
PROPARGYL ALCOHOL	56.06	0.9450	25.0	1.4300	-51.8	0.	113.6	24.50	15
ALLYLAMINE	57.10	0.7629	20.0	1.4205	-88.2	0.	53.3	0.0	16
ACETONE	58.05	0.7900	20.0	1.3587	-94.7	0.	56.3	20.70	17
ACETONE	58.08	0.7844	25.0	1.3560	-94.7	0.	56.3	20.70	18
ALLYL ALCOHOL	58.08	0.8540	20.0	1.4135	-129.0	0.	97.0	0.0	19
2-PROPEN-1-OL	58.08	0.8421	30.0	1.4090	-129.0	0.	97.1	21.60	20
PROPIONALDEHYDE	58.08	0.7912	25.0	1.3593	-80.0	0.	48.0	18.50	21
PROPYLENE OXIDE	58.08	0.8287	20.0	1.3660	-111.9	0.	33.9	0.0	22
N-ME FORMAMIDE	59.07	0.9988	25.0	1.4300	-3.8	0.	182.5	182.40	23
ISOPROPYLAMINE	59.11	0.6821	25.0	1.3711	-95.2	0.	32.4	5.45	24
N-PROPYLAMINE	59.11	0.7173	20.0	1.3882	-83.0	0.	48.5	5.31	25
ACETIC ACID	60.05	1.0492	20.0	1.3719	16.7	0.	117.9	6.15	26
METHYL FORMATE	60.05	0.9742	20.0	1.3433	-99.0	0.	31.5	8.50	27
ETHYLENEDIAMINE	60.10	0.8859	30.0	1.4513	11.3	0.	117.3	12.90	28
PROPANOL-1	60.10	0.8038	20.0	1.3856	-126.2	0.	97.2	20.33	29
PROPANOL-1	60.10	0.7998	25.0	1.3837	-126.2	0.	97.2	20.33	30
PROPANOL-2	60.10	0.7854	20.0	1.3772	-88.0	0.	82.3	19.92	31
NITROMETHANE	61.04	1.1312	25.0	1.3796	-28.5	0.	101.2	35.87	32
2-AMINOETHANOL	61.08	1.0116	25.0	1.4521	10.5	0.	170.0	37.72	33
1,2-ETHANEDIOL	62.07	1.1135	20.0	1.4318	-13.2	0.	197.3	37.70	34
ETHANETHIOL	62.13	0.8391	20.0	1.4311	-144.4	760.	35.0	0.0	35
METHYLSULFIDE	63.13	0.8423	25.0	1.4323	-98.3	0.	37.3	6.20	36
CHLOROETHANE	64.52	0.9039	20.0	1.3790	-136.4	0.	12.3	9.45	37
3-BUTENENITRILE	67.09	0.8329	20.0	1.4060	-84.0	760.	119.0	0.0	38
TRANS-CROTONONITRILE	67.09	0.8239	20.0	1.4225	-51.5	760.	120.5	0.0	39
METHYLACRYLONITRILE	67.09	0.8001	20.0	1.4007	-35.8	0.	90.3	0.0	40
PYRROLE	67.09	0.9699	21.0	1.5002	-23.4	0.	129.8	8.13	41
FURAN	68.08	0.9378	20.0	1.4214	0.0	0.	31.4	2.94	42
1,3-PENTADIENE	68.11	0.6830	0.0	1.4280	0.0	0.	41.8	0.0	43
ISOPRENE	68.13	0.6810	20.0	1.4219	-146.0	760.	34.0	2.10	44
1,2-PENTADIENE	68.13	0.6926	20.0	1.4209	-137.3	760.	44.9	0.0	45
1,4-PENTADIENE	68.13	0.6608	20.0	1.3888	-148.3	760.	0.0	0.0	46
2,3-PENTADIENE	68.13	0.6950	20.0	1.4284	-125.7	760.	48.3	0.0	47
BUTYRONITRILE	69.11	0.7954	15.0	1.3860	-111.9	0.	117.9	20.30	48
BUTYRONITRILE	69.11	0.7865	25.0	1.3820	-111.9	0.	117.9	20.30	49
ISOBUTYRONITRILE	69.11	0.7656	25.0	1.3712	-75.0	0.	103.9	20.40	50
PROPYNOIC ACID	70.05	1.1380	20.0	1.4306	18.0	0.	144.0	0.0	51
CROTONALDEHYDE	70.09	0.8516	20.0	1.4373	-76.5	0.	104.1	0.0	52
CYCLOPENTANE	70.13	0.7454	0.0	1.4065	-93.8	0.	49.3	1.96	53
1-PENTENE	70.13	0.6405	20.0	1.3715	-165.2	0.	30.0	2.02	54
2-PENTENE	70.13	0.6545	20.0	1.3798	-138.0	0.	36.7	0.0	55
CIS-2-PENTENE	70.13	0.6556	20.0	1.3830	-151.4	0.	36.9	0.0	56
TRANS-2-PENTENE	70.13	0.6482	20.0	1.3793	-140.2	0.	36.4	0.0	57
2-METHYL-1-BUTENE	70.14	0.6504	20.0	1.3378	-137.6	760.	31.2	2.20	58
2-METHYL-2-BUTENE	70.14	0.6623	20.0	1.3874	-133.8	760.	38.6	0.0	59
2-CYANOETHANOL	71.08	1.0404	25.0	0.0	-46.0	0.	220.0	0.0	60
3-HYDROXY PROPIONITRILE	71.08	1.0588	20.0	1.4240	0.0	760.	230.0	0.0	61
LACTONITRILE	71.08	0.9877	20.0	1.4058	-40.0	0.	183.0	0.0	62
PYRROLIDINE	71.12	0.8520	22.0	1.4270	0.0	0.	88.7	0.0	63
ACRYLIC ACID	72.06	1.0511	20.0	1.4224	13.5	0.	141.2	0.0	64
PROPIOLACTONE	72.06	1.1460	20.0	1.4131	-33.4	0.	155.0	0.0	65
ALLYL METHYL ETHER	72.11	0.7610	25.0	1.3786	0.0	0.	41.5	0.0	66
2-BUTANONE	72.11	0.8049	20.0	1.3788	-86.7	0.	79.6	18.51	67
2-BUTANONE	72.11	0.7997	25.0	1.3764	-86.7	0.	79.6	18.52	68
2-BUTENEOL-1 (CIS)	72.11	0.8540	20.0	1.4342	-89.4	760.	123.6	0.0	69
1,2-BUTYLENE OXIDE	72.11	0.8297	20.0	1.3840	-150.0	0.	63.2	0.0	70
BUTYRALDEHYDE	72.11	0.8016	20.0	1.3791	-96.4	0.	74.8	13.40	71
ISO-BUTYRALDEHYDE	72.11	0.7891	20.0	1.3727	-65.0	0.	64.1	0.0	72
ETHYLVINYL ETHER	72.11	0.7531	20.0	1.3754	-115.8	0.	35.7	0.0	73
METHALLYL ALCOHOL	72.11	0.8574	19.0	1.4255	0.0	0.	114.5	0.0	74
TETRAHYDROFURAN	72.11	0.8892	20.0	1.4050	-108.5	0.	66.0	7.58	75
TETRAHYDROFURAN	72.11	0.8811	25.0	1.4050	-108.5	0.	66.0	7.58	76
1,2-EPOXY-2-ME PROPANE	72.12	0.8650	20.0	1.3712	0.0	760.	52.0	0.0	77
2,2-DIMETHYL PROPANE	72.15	0.6135	20.0	1.3476	-20.0	0.	9.5	0.0	78
2-METHYL BUTANE	72.15	0.6197	0.0	1.3537	-159.9	0.	27.9	1.84	79
PENTANE	72.15	0.6262	20.0	1.3579	-129.7	0.	36.1	0.0	80
NN-DIMETHYLFORMAMIDE	73.10	0.9440	25.0	1.4282	-60.4	760.	153.0	36.71	81
METHYL ISOTHIOCYANATE	73.12	1.0691	37.0	1.5258	36.0	758.	119.0	0.0	82
METHYL THIOCYANATE	73.12	1.0678	25.0	1.4669	-5.1	757.	132.9	0.0	83
SEC-BUTYL AMINE (D)	73.14	0.7240	20.0	1.3440	-104.5	0.	63.0	0.0	84
SEC-BUTYL AMINE (DL)	73.14	0.7271	17.0	1.3950	-72.0	772.	67.0	0.0	85

Note: Missing data is indicated by 0, 0., or 0.0.

(continued)

Name	Mol. Wt.	Density	Temp.	Refract. Index	Melting Point	Pressure	Boiling Point	Dielec. Constant	List No.
N-BUTYLAMINE	73.14	0.7346	25.0	1.3987	-49.1	0.	77.4	4.88	86
SEC-BUTYLAMINE	73.14	0.7246	20.0	1.3934	0.0	0.	62.5	0.0	87
DIETHYLAMINE	73.14	0.7070	20.0	1.3854	-49.8	0.	55.5	3.58	88
ISOBUTYLAMINE	73.14	0.7346	20.0	1.3972	-84.6	0.	67.7	4.43	89
TERT-BUTYLAMINE	73.14	0.6908	25.0	1.3761	-72.7	0.	44.4	0.0	90
DIOXOLANE	74.08	1.0600	20.0	1.3974	-95.0	765.	78.0	0.0	91
ETHYL FORMATE	74.08	0.9289	15.0	1.3625	-79.4	0.	54.1	7.16	92
HYDROXY ACETONE	74.08	1.0824	20.0	1.4295	-17.0	0.	145.5	0.0	93
3-HYDROXYPROPYLENEOXIDE	74.08	1.1110	22.0	1.4350	0.0	0.	166.5	0.0	94
METHYL ACETATE	74.08	0.9342	20.0	1.3614	-98.1	0.	56.3	6.68	95
PROPANOIC ACID	74.08	0.9880	25.0	1.3843	-20.7	0.	140.8	3.44	96
1-BUTANOL	74.12	0.8097	20.0	1.3993	-88.6	0.	117.7	17.51	97
1-BUTANOL	74.12	0.8060	20.0	1.3973	-88.6	0.	117.7	17.51	98
2-BUTANOL	74.12	0.8026	25.0	1.3950	-114.7	0.	99.6	16.56	99
DIETHYL ETHER	74.12	0.7138	20.0	1.3526	-116.2	760.	34.5	4.34	100
ETHYL ETHER	74.12	0.7076	25.0	1.3495	-116.3	0.	34.5	4.34	101
2-METHYL-1-PROPANOL	74.12	0.7978	25.0	1.3939	-108.0	0.	107.7	17.93	102
2-METHYL-2-PROPANOL	74.12	0.7808	25.0	1.3878	25.5	0.	82.2	1.77	103
METHYL PROPYL ETHER	74.12	0.7380	20.0	1.3579	0.0	760.	38.9	0.0	104
TERT BUTYL ALCOHOL	74.12	0.7887	20.0	1.3878	25.5	0.	82.2	1.77	105
1,2-PROPANEDIAMINE	74.13	0.8584	25.0	1.4492	0.0	760.	120.5	0.0	106
NITROETHANE	75.07	1.0446	25.0	1.3897	-89.5	0.	114.1	28.06	107
1-AMINO-2-PROPANOL	75.11	0.9730	18.0	1.4500	-1.0	750.	160.0	0.0	108
3-AMINO-2-PROPANOL	75.11	0.9824	26.0	1.4570	11.0	756.	187.0	0.0	109
2-METHOXYETHANOL	76.10	0.9602	25.0	1.4002	-85.1	0.	124.6	16.93	110
METHYLAL	76.10	0.8665	15.0	1.3563	-105.2	0.	42.3	2.65	111
1,2-PROPANEDIOL	76.10	1.0362	20.0	1.4329	-60.0	0.	187.6	32.00	112
1,3-PROPANEDIOL	76.10	1.0538	20.0	1.4396	-26.7	0.	214.4	35.00	113
CARBON DISULFIDE	76.14	1.2700	15.0	1.6319	-111.6	0.	46.2	2.64	114
1-PROPANE THIOL	76.17	0.8411	20.0	1.4380	-113.3	760.	67.5	0.0	115
2-PROPANE THIOL	76.17	0.8143	20.0	1.4255	-130.5	760.	52.6	0.0	116
3-CHLOROPROPENE	76.52	0.9442	15.0	1.4181	-134.5	0.	45.1	8.20	117
CIS-PROPENYL CHLORIDE	76.53	0.9347	20.0	1.4055	-134.8	760.	32.8	0.0	118
TRANS-PROPENYL CHLORIDE	76.53	0.9350	20.0	1.4054	-99.0	760.	37.4	0.0	119
BENZENE	78.12	0.8790	20.0	1.5011	5.5	0.	80.1	2.28	120
BENZENE	78.12	0.8737	25.0	1.4979	5.5	0.	80.1	2.28	121
DIMETHYLSULFOXIDE	78.13	1.0958	25.0	1.4773	18.5	0.	189.0	46.68	122
ETHANOL-1-THIOL-2	78.13	1.1143	0.0	1.4996	0.0	13.	55.0	0.0	123
ACETYL CHLORIDE	78.50	1.1050	20.0	1.3898	-112.0	0.	51.5	15.00	124
1-CHLOROPROPANE	78.54	0.8909	20.0	1.3879	-122.8	0.	46.6	7.70	125
2-CHLOROPROPANE	78.54	0.8617	20.0	1.3777	-117.2	0.	35.7	9.02	126
2-CHLOROPROPANE	78.54	0.8491	30.0	1.3711	-117.2	0.	35.7	9.82	127
PYRIDINE	79.10	0.9782	25.0	1.5075	-41.6	0.	115.3	12.40	128
PYRIDAZINE	80.09	1.1035	23.0	1.5231	-8.0	0.	208.0	0.0	129
PYRIMIDINE	80.09	0.0	0.0	0.0	22.0	0.	123.7	0.0	130
2-CHLOROETHANOL	80.52	1.2019	20.0	1.4438	-67.5	0.	128.6	25.80	131
1-METHYL PYRROLE	81.11	0.9145	15.0	1.4899	0.0	748.	114.5	0.0	132
1-METHYL IMIDAZOLE	82.10	1.6325	21.0	1.4924	-6.0	0.	198.0	0.0	133
CYCLOHEXENE	82.15	0.8061	25.0	1.4438	-103.5	0.	83.0	2.22	134
1,5-HEXADIENE	82.15	0.6923	0.0	1.4044	-141.0	0.	60.0	0.0	135
N-ME-ALANINE NITRILE	83.11	0.8992	20.0	1.4312	0.0	22.	82.0	0.0	136
VALERONITRILE	83.13	0.7950	25.0	1.3951	-96.2	0.	141.3	19.71	137
CYCLOPENTANONE	84.11	0.9509	0.0	1.9366	-51.3	0.	130.7	0.0	138
THIOPHENE	84.14	1.0649	20.0	1.5289	-38.2	0.	84.2	2.71	139
CYCLOHEXANE	84.16	0.7786	20.0	1.4262	6.6	0.	80.7	2.02	140
CYCLOHEXANE	84.16	0.7739	25.0	1.4235	6.6	0.	80.7	2.02	141
1-HEXENE	84.16	0.6685	25.0	1.3850	-139.8	0.	63.5	2.05	142
METHYL CYCLOPENTANE	84.16	0.7489	20.0	1.4096	-142.4	0.	72.1	1.98	143
DICHLOROMETHANE	84.93	1.3148	25.0	1.4211	-95.1	0.	39.8	8.93	144
ACETONE CYANOHYDRIN	85.11	0.9320	19.0	1.3996	-190.0	0.	82.0	0.0	145
2-PYRROLIDINONE	85.11	1.1070	25.0	1.4860	25.0	0.	245.0	0.0	146
PIPERIDINE	85.15	0.8616	20.0	1.4525	-10.5	0.	106.4	5.80	147
ALLYL FORMATE	86.09	0.9498	18.0	1.3980	0.0	0.	83.0	0.0	148
CIS-2-BUTENOIC ACID	86.09	1.0267	20.0	1.4483	15.5	760.	169.3	0.0	149
BUTYROLACTONE	86.09	1.1254	25.0	1.4348	-43.5	760.	204.0	39.00	150
METHACRYLIC ACID	86.09	1.0153	20.0	1.4314	15.0	760.	160.5	0.0	151
METHYL ACRYLATE	86.09	0.9547	18.0	1.4117	-75.0	0.	80.2	0.0	152
VINYL ACETATE	86.09	0.9312	20.0	1.3959	-92.8	0.	72.5	0.0	153
ALLYL ETHYL ETHER	86.13	0.7597	25.0	1.3861	64.0	0.	0.0	0.0	154
2-PENTANONE	86.13	0.8124	15.0	1.3895	-77.8	0.	102.0	0.0	155
3-PENTANONE	86.13	0.8095	25.0	1.3900	-39.0	0.	102.0	17.00	156
1-PENTENE-3-OL	86.13	0.8395	22.0	1.4183	0.0	0.	115.0	0.0	157
TETRAHYDROPYRAN	86.13	0.8772	25.0	1.4195	-45.0	0.	88.0	5.61	158
TRI-ME ACETALDEHYDE	86.13	0.7927	17.0	1.3791	6.0	0.	75.0	0.0	159
VALERALDEHYDE	86.13	0.8095	20.0	1.3944	-91.5	0.	102.5	10.00	160
2,2-DIMETHYL BUTANE	86.17	0.6492	0.0	1.3687	-99.9	0.	49.7	0.0	161
2,3-DIMETHYL BUTANE	86.17	0.6616	0.0	1.3749	-128.5	0.	57.9	0.0	162
HEXANE	86.17	0.6548	0.0	1.3723	-95.3	0.	68.7	1.89	163
HEXANE	86.17	0.6594	0.0	1.3749	-95.3	0.	68.7	1.89	164
2-METHYL PENTANE	86.17	0.6532	0.0	1.3714	-153.7	0.	60.3	0.0	165
3-METHYL PENTANE	86.17	0.6643	0.0	1.3765	0.0	0.	63.3	0.0	166
N,N-DIMEACETAMIDE	87.12	0.9366	25.0	1.4356	-20.0	0.	166.1	37.78	167
N-ME PROPIONAMIDE	87.12	0.9305	25.0	1.4345	-30.9	0.	148.0	172.20	168
MORPHOLINE	87.12	1.0050	25.0	1.4573	-3.1	0.	128.9	7.42	169
ETHYL ISOTHIOCYANATE	87.14	0.9990	20.0	1.5130	-5.9	760.	131.5	0.0	170

Note: Missing data is indicated by 0, 0., or 0.0.

(continued)

Name	Mol. Wt.	Density	Temp.	Refract. Index	Melting Point	Pressure	Boiling Point	Dielec. Constant	List No.
1-AMINOPENTANE	87.17	0.7547	20.0	1.4118	-55.0	760.	104.4	0.0	171
ETHYLENE CARBONATE	88.06	1.3208	25.0	1.4250	36.4	0.	238.0	89.60	172
PYRROLINE	88.06	1.2272	0.0	1.4138	13.6	0.	115.0	0.0	173
PYRUVIC ACID	88.06	1.2272	20.0	1.4280	13.6	760.	165.0	0.0	174
ALDOL	88.10	1.1030	20.0	1.4497	0.0	12.	79.0	0.0	175
CIS-2-BUTENE-14-DIOL	88.11	1.0740	20.0	1.4793	11.0	0.	235.0	0.0	176
TRN-2-BUTENE-14-DIOL	88.11	1.0685	20.0	1.4779	27.3	0.	132.0	0.0	177
BUTYRIC ACID	88.11	0.9532	25.0	1.3958	-5.2	0.	163.3	2.97	178
1.3-DIOXANE	88.11	1.3042	20.0	1.4165	-42.0	755.	105.0	0.0	179
P-DIOXANE	88.11	1.0280	25.0	1.4203	11.8	0.	191.3	2.21	180
ETHYL ACETATE	88.11	0.9006	20.0	1.3724	-83.9	0.	77.1	6.02	181
ETHYL ACETATE	88.11	0.8946	25.0	1.3698	-83.9	0.	77.1	6.02	182
ISOBUTYRIC ACID	88.11	0.9682	20.0	1.3930	-46.1	0.	154.7	2.73	183
METHYLPROPIONATE	88.11	0.9151	20.0	1.3779	-87.5	0.	78.7	5.50	184
PROPYL FORMATE	88.11	0.9111	15.0	1.3790	-92.9	0.	80.8	7.72	185
VALERONITRILE	88.13	0.8034	15.0	1.3991	-96.2	0.	141.3	19.71	186
ETHYL-N-PROPYL ETHER	88.15	0.7330	0.0	1.3695	-79.0	0.	63.6	0.0	187
2-METHYL-1-BUTANOL	88.15	0.8152	25.0	1.4087	-70.0	760.	128.7	14.70	188
3-METHYL-1-BUTANOL	88.15	0.8071	25.0	1.4052	-117.2	60.	130.5	14.70	189
2-METHYL-2-BUTANOL	88.15	0.8050	25.0	1.4024	-8.8	0.	102.0	5.82	190
3-METHYL-2-BUTANOL	88.15	0.8138	25.0	1.4075	0.0	0.	111.5	0.0	191
METHYL-N-BUTYL ETHER	88.15	0.7443	0.0	1.3736	-115.5	0.	71.0	0.0	192
1-PENTANOL	88.15	0.8115	25.0	1.4079	-78.2	0.	137.8	13.90	193
2-PENTANOL	88.15	0.8054	25.0	1.4044	0.0	0.	119.0	13.82	194
3-PENTANOL	88.15	0.8160	25.0	1.4079	0.0	0.	115.3	13.02	195
TETRAHYDROTHIOPHENE	88.17	0.9938	25.0	1.5257	-96.2	0.	120.9	0.0	196
1-NITROPROPANE	89.10	1.9961	25.0	1.3996	-104.0	0.	131.2	23.24	197
2-NITROPROPANE	89.10	0.9829	25.0	1.3924	-91.3	0.	120.3	25.52	198
2-AMINO-1-BUTANOL	89.14	0.9162	20.0	1.4489	-2.0	0.	178.0	0.0	199
3-AMINO-2-BUTANOL	89.14	0.9299	25.0	1.4502	19.0	745.	159.5	0.0	200
DIMETHYL ETHANOLAMINE	89.14	0.8866	20.0	1.4300	0.0	760.	134.0	0.0	201
2-ETHYLAMINOETHANOL	89.14	0.9140	20.0	1.4440	-9.0	760.	169.5	0.0	202
3-CHLOROPROPIONITRILE	89.53	1.1375	0.0	1.4380	0.0	0.	58.0	0.0	203
DIMETHYL CARBONATE	90.08	1.0694	20.0	1.3687	3.0	0.	90.5	0.0	204
LACTIC ACID DL	90.08	1.2060	25.0	1.4392	18.0	12.	119.0	0.0	205
METHOXYACETIC ACID	90.08	1.1768	20.0	1.4168	0.0	760.	213.0	0.0	206
METHYL GLYCOLATE	90.08	1.1677	18.0	0.0	0.0	760.	151.1	0.0	207
1.2-BUTANEDIOL	90.12	1.0059	20.0	1.4375	0.0	0.	193.0	0.0	208
1.3-BUTANEDIOL	90.12	1.0053	20.0	1.4410	77.0	0.	207.5	0.0	209
1.4-BUTANEDIOL	90.12	1.0171	20.0	1.4460	20.1	0.	235.0	0.0	210
2.3-BUTANEDIOL	90.12	0.9872	20.0	1.4306	34.0	760.	181.0	0.0	211
1.2-DIMETHOXYETHANE	90.12	0.8629	20.0	1.3796	-58.0	760.	83.5	0.0	212
2-ETHOXYETHANOL	90.12	0.9252	25.0	1.4057	-90.0	0.	135.6	29.60	213
1-METHOXYPROPANOL-2	90.12	0.9620	20.0	1.4070	0.0	0.	118.3	0.0	214
1-METHOXYPROPANOL-2	90.12	0.9620	20.0	1.4070	0.0	0.	118.3	0.0	215
1-BUTANETHIOL	90.19	0.8416	20.0	1.4429	-115.7	0.	98.4	5.07	216
ETHYL SULFIDE	90.19	0.8312	25.0	1.4402	-103.9	0.	92.1	5.72	217
CIS-1-CL-1-BUTENE	90.55	0.9153	15.0	1.4194	0.0	760.	63.5	0.0	218
TRANS-1-CL-1-BUTENE	90.55	0.9205	15.0	1.4225	0.0	760.	68.0	0.0	219
2-CHLORO-1-BUTENE	90.55	0.9107	15.0	1.4115	0.0	760.	58.7	0.0	220
3-CHLORO-1-BUTENE	90.55	0.8978	20.0	1.4149	0.0	766.	64.5	0.0	221
4-CHLORO-1-BUTENE	90.55	0.9211	20.0	1.4233	0.0	773.	75.0	0.0	222
CIS-1-CL-2-BUTENE	90.55	0.9426	20.0	1.4390	0.0	758.	84.1	0.0	223
TRANS-1-CL-2-BUTENE	90.55	0.9295	20.0	1.4350	0.0	752.	84.8	0.0	224
CIS-2-CL-2-BUTENE	90.55	0.9239	20.0	1.4240	-117.3	760.	70.6	0.0	225
TRANS-2-CL-2-BUTENE	90.55	0.9138	20.0	1.4190	-105.8	760.	628.0	0.0	226
1-CL-2-ME-PROPENE-1	90.55	0.9250	16.0	1.4221	0.0	775.	68.0	0.0	227
3-CL-2-ME-PROPENE-1	90.55	0.9250	20.0	1.4270	0.0	0.	72.0	0.0	228
2-NITROETHANOL-1	91.07	1.2700	15.0	1.4438	-80.0	765.	194.0	0.0	229
1.2.3-PROPANETRIOL	92.10	1.2613	20.0	1.4746	18.2	0.	290.0	42.50	230
TOLUENE	92.14	0.8669	20.0	1.4969	-94.9	0.	110.6	2.38	231
TOLUENE	92.14	0.8623	25.0	1.4941	-94.9	0.	110.6	2.38	232
CHLOROACETONE	92.53	1.1500	20.0	0.0	-44.5	0.	119.0	0.0	233
EPICHLOROHYDRIN	92.53	1.1807	20.0	1.4380	-57.2	0.	116.1	22.60	234
TERT-BUTYL CHLORIDE	92.57	0.8420	20.0	1.3857	-25.4	760.	52.0	0.0	235
1-CHLOROBUTANE	92.57	0.8862	20.0	1.4021	-123.1	0.	78.4	7.39	236
2-CHLOROBUTANE	92.57	0.8732	20.0	1.3971	-140.5	0.	68.3	7.09	237
1-CL-2-METHYLPROPANE	92.57	0.8773	20.0	1.3980	-130.3	0.	68.8	6.49	238
2-CL-2-METHYLPROPANE	92.57	0.8420	20.0	1.3857	-25.4	0.	50.7	9.96	239
ANILINE	93.13	1.0217	20.0	1.5863	-6.0	0.	184.4	6.89	240
2-METHYLPYRIDINE	93.13	0.9497	15.0	1.5029	0.0	0.	128.8	9.80	241
3-METHYLPYRIDINE	93.13	0.9613	15.0	1.5043	0.0	0.	143.5	9.80	242
GLUTARONITRILE	94.12	0.9911	15.0	1.4295	-29.0	0.	286.0	0.0	243
PHENOL	94.12	1.0576	41.0	1.5428	40.9	0.	181.8	9.78	244
1.2-DIHYDROTOLUENE	94.16	0.8354	0.0	1.4763	0.0	0.	110.0	0.0	245
1-CHLORO-2-PROPANOL	94.54	1.1100	20.0	1.4392	0.0	762.	126.5	0.0	246
3-CHLORO-1-PROPANOL	94.54	1.1309	0.0	1.4450	0.0	0.	161.5	0.0	247
PYRROLE-2-CARBOXALDEHYDE	95.10	0.0	16.0	1.5939	46.5	0.	218.0	0.0	248
2.5-DIME-PYRROLE	95.14	0.9353	20.0	1.5025	0.0	765.	171.0	0.0	249
1-ETHYL-PYRROLE	95.15	0.9009	20.0	1.4841	0.0	0.	164.0	0.0	250
1.4-PYRONE	96.08	1.1900	0.0	1.5238	32.5	742.	216.0	0.0	251
2-FURALDEHYDE	96.09	1.1598	20.0	1.5261	-36.5	0.	161.8	38.00	252
FLUOROBENZENE	96.10	1.0309	15.0	1.4684	-42.2	0.	84.7	0.0	253
2.5-DIMETHYL FURAN	96.14	0.8883	20.0	1.4363	-62.8	760.	93.5	0.0	254
2.4-HEPTADIENE	96.17	0.7384	0.0	1.4578	0.0	0.	108.0	0.0	255

Note: Missing data is indicated by 0, 0., or 0.0.

(continued)

Name	Mol. Wt.	Density	Temp.	Refract. Index	Melting Point	Pressure	Boiling Point	Dielec. Constant	List No.
1-HEPTYNE	96.17	0.7338	20.0	1.4084	-81.0	760.	100.0	0.0	256
1,1-DICHLOROETHYLENE	96.94	1.2132	20.0	1.4247	-122.6	0.	31.6	4.60	257
CIS-1,2-DICLETHYLENE	96.94	1.2837	20.0	1.4490	-80.0	0.	60.6	9.20	258
TRANS1,2DICLETHYLENE	96.94	1.2547	20.0	1.4462	-49.8	0.	47.7	2.14	259
CAPRONITRILE	97.16	0.8052	20.0	1.4069	-80.3	0.	163.6	17.26	260
4-MEVALERONITRILE	97.16	0.7993	25.0	1.4040	-51.1	0.	154.0	15.50	261
FURFURYL ALCOHOL	98.10	1.1238	30.0	1.4801	-29.0	0.	170.0	0.0	262
2-METHOXY FURAN	98.10	1.0646	25.0	1.4468	0.0	0.	110.5	0.0	263
PROPARGYL ACETATE	98.10	0.9982	20.0	1.4187	0.0	0.	121.5	0.0	264
PROPARGYL ACETATE	98.10	0.9982	20.0	1.4187	0.0	0.	121.5	0.0	265
3,4-DIMETHYL FURAZAN	98.11	1.0528	14.0	1.4237	-7.0	744.	156.0	0.0	266
ALLYL ACETONE	98.14	0.8470	0.0	1.4917	0.0	0.	128.0	0.0	267
ALLYL ETHER	98.15	0.8006	25.0	1.4141	0.0	0.	0.0	0.0	268
CYCLOHEXANONE	98.15	0.9510	25.0	1.4520	-32.1	0.	155.6	18.30	269
MESITYL OXIDE	98.15	0.8653	20.0	1.4440	-52.9	760.	129.8	0.0	270
2-METHYLTHIOPHENE	98.17	1.0193	20.0	1.5203	-63.4	760.	112.6	0.0	271
3-METHYLTHIOPHENE	98.17	1.0218	20.0	1.5204	-69.0	760.	115.4	0.0	272
METHYL CYCLOHEXANE	98.18	0.7694	0.0	1.4231	-126.6	0.	98.2	0.0	273
1-HEPTENE	98.19	0.6970	20.0	1.3998	-118.9	760.	93.6	2.07	274
1,1-DICHLOROETHANE	98.96	1.1680	25.0	1.4138	-97.0	0.	57.3	10.10	275
1,2-DICHLOROETHANE	98.96	1.2458	25.0	1.4421	-35.7	0.	83.5	10.36	276
METHYLCYANOACETATE	99.09	1.1225	25.0	1.4166	-13.1	0.	205.1	29.30	277
1-ME-2-PYRROLIDINONE	99.13	1.0279	25.0	1.4680	-24.4	760.	202.0	32.00	278
1-ME-2-PYROLIDONE	99.13	1.0279	25.0	1.4680	-24.4	10.	79.0	32.00	279
ALLYLISOTHIOCYANATE	99.16	1.0126	20.0	1.5306	-80.0	760.	152.0	17.20	280
N-METHYL PIPERIDINE	99.17	0.8159	0.0	1.4355	0.0	0.	107.0	0.0	281
CYCLOHEXYLAMINE	99.18	0.8671	20.0	1.4592	-17.7	0.	134.8	4.73	282
2,4-DIME-PYRROLIDINE	99.18	0.8297	20.0	1.4325	0.0	753.	116.0	0.0	283
ALLYL ACETATE	100.12	0.9280	20.0	1.4040	0.0	0.	104.0	0.0	284
ETHYL ACRYLATE	100.12	0.9234	20.0	1.4068	-71.2	0.	99.5	0.0	285
METHYLMETHACRYLATE	100.12	0.9433	20.0	1.4146	-48.2	0.	100.3	2.90	286
2,3 PENTANEDIONE	100.12	0.9565	19.0	1.4014	0.0	0.	108.0	0.0	287
2,4 PENTANEDIONE	100.12	0.9721	25.0	1.4541	-23.0	734.	139.0	0.0	288
GAMMA-VALEROLACTONE	100.12	1.0520	25.0	1.4320	-37.0	0.	206.0	0.0	289
CYCLOBUT-CARBOXYLIC ACID	100.13	1.0599	20.0	1.4400	-2.0	754.	190.0	0.0	290
BUTYLVINYL ETHER	100.16	0.7727	25.0	1.3997	-92.0	0.	93.8	0.0	291
CYCLOHEXANOL	100.16	0.9684	25.0	1.4648	25.1	0.	161.1	15.00	292
2-HEXANONE	100.16	0.8116	0.0	1.4015	-57.0	0.	126.0	0.0	293
METHYL-T-BUTYL KETONE	100.16	0.8016	0.0	1.3952	-52.5	0.	106.0	0.0	294
3-ME-2-PENTANONE	100.16	0.8181	14.0	1.4002	0.0	0.	118.0	0.0	295
4-METHYL-2-PENTANONE	100.16	0.8008	20.0	1.3957	-84.0	0.	116.5	13.11	296
HEPTANE	100.19	0.6836	20.0	1.3876	-90.6	0.	98.4	0.0	297
HEPTANE	100.19	0.6795	25.0	1.3851	-90.6	0.	98.4	0.0	298
2-METHYL HEXANE	100.19	0.6744	25.0	1.3848	-118.3	0.	90.1	0.0	299
3-METHYL HEXANE	100.19	0.6829	25.0	1.3886	-119.4	0.	91.9	0.0	300
2,3-DIMETHYL PENTANE	100.21	0.6909	25.0	1.3920	0.0	0.	89.8	0.0	301
2,4-DIMETHYL PENTANE	100.21	0.6683	25.0	1.3814	-119.2	0.	80.5	0.0	302
3,3-DIMETHYL PENTANE	100.21	0.6933	0.0	1.3909	-135.0	0.	86.1	0.0	303
2,2,3-TRIMETHYLBUTANE	100.21	0.6901	20.0	1.3894	-25.0	0.	80.9	0.0	304
N-METHYLMORPHOLINE	101.15	0.9051	20.0	1.4332	0.0	750.	115.0	0.0	305
DIISOPROPYLAMINE	101.19	0.7153	20.0	1.3924	-96.3	0.	83.9	0.0	306
DIPROPYLAMINE	101.19	0.7375	20.0	1.4043	-63.0	0.	109.2	3.07	307
TRIETHYLAMINE	101.19	0.7230	25.0	1.3980	-114.7	0.	89.5	2.42	308
ACETIC ANHYDRIDE	102.09	1.0871	15.0	1.3930	-73.1	0.	140.0	20.70	309
4-METHYL DIOXOLANE	102.09	1.2069	20.0	1.4189	-46.8	760.	242.0	0.0	310
BUTYL FORMATE	102.13	0.8917	20.0	1.3890	-90.0	0.	106.6	2.43	311
ETHYL PROPIONATE	102.13	0.8957	15.0	1.3864	-73.9	0.	99.1	5.65	312
ETHYL PROPIONATE	102.13	0.8899	20.0	1.3839	-73.9	0.	99.1	5.65	313
ISOBUTYL FORMATE	102.13	0.8853	20.0	1.3855	-94.5	0.	98.4	6.41	314
ISOPROPYL ACETATE	102.13	0.8717	20.0	1.3773	-73.4	0.	88.2	0.0	315
ISOVALERIC ACID	102.13	0.9308	20.0	1.4064	-29.3	0.	176.5	2.64	316
METHYL-N-BUTYRATE	102.13	0.8984	20.0	1.3870	-95.0	0.	102.6	5.60	317
4-ME-1,3-DIOXANE	102.13	0.9953	20.0	1.4168	0.0	0.	114.0	0.0	318
PROPYL ACETATE	102.13	0.8938	15.0	1.3866	-92.5	0.	101.5	6.00	319
TETRA H FURFURYL ALC	102.13	1.0420	25.0	1.4599	0.0	0.	178.0	13.61	320
VALERIC ACID	102.13	0.9345	35.0	1.4060	33.7	0.	185.5	2.66	321
BUTYL ETHYL ETHER	102.18	0.7448	25.0	1.3793	-103.0	0.	92.2	0.0	322
2-ETHYL-1-BUTANOL	102.18	0.8295	0.0	1.4205	-114.4	0.	146.5	6.09	323
1-HEXANOL	102.18	0.8159	25.0	1.4161	-44.6	0.	157.0	13.30	324
ISOPROPYL ETHER	102.18	0.7182	25.0	1.3655	-85.5	0.	68.3	3.88	325
2-METHYL-2-PENTANOL	102.18	0.8350	20.0	1.4125	108.0	0.	121.5	0.0	326
3-METHYL-2-PENTANOL	102.18	0.8235	25.0	1.4179	0.0	0.	134.3	0.0	327
4-METHYL-2-PENTANOL	102.18	0.8075	20.0	1.4100	-90.0	760.	133.5	0.0	328
3-METHYL-3-PENTANOL	102.18	0.8237	20.0	1.4180	-38.0	0.	121.0	0.0	329
PROPYL ETHER	102.18	0.7419	25.0	1.3780	-123.2	0.	89.6	3.39	330
BENZONITRILE	103.12	1.0006	25.0	1.5259	-12.8	0.	191.1	25.20	331
METHYL URETHANE	103.12	1.0350	15.0	1.4200	0.0	760.	170.0	0.0	332
1-NITROBUTANE	103.12	0.9880	0.0	1.4103	0.0	0.	153.0	0.0	333
DIETHYLENETRIAMINE	103.17	0.9586	20.0	1.4810	-39.0	760.	207.0	0.0	334
METHYL LACTATE	104.12	1.0857	26.0	1.4131	-66.0	760.	144.8	0.0	335
STYRENE	104.14	0.9012	25.0	1.5440	-30.6	0.	145.2	2.43	336
DIETHOXYMETHANE	104.15	0.8319	20.0	1.3748	-665.0	0.	88.0	0.0	337
N-PROPYL NITRATE	105.09	1.0580	20.0	1.3976	0.0	0.	110.5	0.0	338
DIETHANOLAMINE	105.14	1.0899	30.0	1.4747	28.0	0.	268.4	2.81	339
BENZALDEHYDE	106.12	1.0447	20.0	1.5455	-26.0	0.	178.9	17.80	340

Note: Missing data is indicated by 0, 0., or 0.0.

(continued)

Name	Mol. Wt.	Density	Temp.	Refract. Index	Melting Point	Pressure	Boiling Point	Dielec. Constant	List No.
DIETHYLENE GLYCOL	106.12	1.1164	20.0	1.4475	-10.5	0.	244.8	31.69	341
METHOXYMETHOXYETHANOL	106.12	1.0385	25.0	1.4100	-70.0	0.	167.5	0.0	342
ETHYL BENZENE	106.17	0.8626	25.0	1.4932	-94.9	0.	136.2	2.40	343
O-XYLENE	106.17	0.8759	25.0	1.5029	-25.2	0.	144.4	2.57	344
M-XYLENE	106.17	0.8599	25.0	1.4946	-47.8	0.	139.1	2.37	345
P-XYLENE	106.17	0.8611	20.0	1.4958	13.3	0.	138.3	2.27	346
P-XYLENE	106.17	0.8567	25.0	1.4933	13.3	0.	138.3	2.27	347
BENZYLAMINE	107.15	0.9813	20.0	1.5402	10.0	770.	185.0	0.0	348
2,4-DIMETHYL PYRIDINE	107.15	0.9271	25.0	1.4984	0.0	0.	159.2	0.0	349
2,5-DIME PYRIDINE	107.15	0.9261	25.0	1.4982	-15.0	0.	156.8	0.0	350
2,6-DIME PYRIDINE	107.15	0.9200	25.0	1.4953	-5.0	0.	143.0	0.0	351
3,5-DIME PYRIDINE	107.15	0.9385	25.0	1.5032	0.0	0.	171.6	0.0	352
3,4-DIME PYRIDINE	107.15	0.9537	25.0	1.5099	-12.0	759.	178.8	0.0	353
METHYL ANILINE	107.15	0.9891	0.0	1.5702	-57.0	0.	196.3	5.97	354
O-TOLUIDINE	107.16	0.9984	20.0	1.5725	-16.1	0.	200.4	6.34	355
M-TOLUIDINE	107.16	0.9930	15.0	1.5704	-30.4	0.	203.4	5.95	356
P-TOLUIDINE	107.16	0.9659	45.0	1.5540	43.8	0.	200.6	4.98	357
ADIPONITRILE	108.14	0.9510	19.0	1.4597	2.0	0.	180.0	0.0	358
ANISOLE	108.14	0.9893	25.0	1.5143	-37.5	0.	153.8	4.33	359
BENZYL ALCOHOL	108.14	1.0454	20.0	1.5403	-15.3	0.	205.4	13.10	360
BENZYL ALCOHOL	108.14	1.0413	25.0	1.5384	-15.3	0.	205.4	13.10	361
O-CRESOL	108.14	1.1350	25.0	1.5442	30.9	0.	191.0	11.50	362
M-CRESOL	108.14	1.0302	25.0	1.5396	12.2	0.	202.2	11.80	363
P-CRESOL	108.14	1.0178	41.0	1.5311	34.7	0.	201.9	9.91	364
1,3-PROPANEDITHIOL	108.23	1.0783	20.0	1.5403	-79.0	0.	172.9	0.0	365
ETHYL CHLOROFORMATE	108.53	1.3577	20.0	1.3955	-80.6	760.	95.0	0.0	366
METHYL CHLOROACETATE	108.53	1.2337	20.0	1.4218	-32.1	760.	129.8	0.0	367
BROMOETHANE	108.97	1.4708	15.0	1.4276	-118.6	0.	38.4	9.39	368
O-FLUOROTOLUENE	110.13	1.0027	15.0	1.4716	-62.0	0.	114.4	4.22	369
M-FLUOROTOLUENE	110.13	0.9974	20.0	1.4691	-87.7	760.	116.5	0.0	370
P-FLUOROTOLUENE	110.13	0.9975	20.0	1.4688	-56.8	760.	116.6	5.86	371
BENZENETHIOL	110.18	1.0727	25.0	1.5872	-14.9	0.	169.1	4.38	372
THIO-PHENOL	110.18	1.0728	25.0	1.5879	70.5	0.	169.5	0.0	373
2,3-DICHLOROPROPENE	110.98	1.2040	25.0	1.4600	0.0	0.	94.0	0.0	374
ACETAZINE	112.17	0.8422	20.0	1.4535	-125.0	0.	133.0	0.0	375
2-METHYLCYCLOHEXANONE	112.17	0.9240	20.0	1.4493	0.0	757.	165.0	0.0	376
THIOXENE	112.19	0.9956	0.0	1.5130	0.0	0.	137.8	0.0	377
P-DIMETHYLCYCLOHEXANE	112.21	0.7827	0.0	1.4253	-87.0	0.	124.6	0.0	378
ETHYL CYCLOHEXANE	112.21	0.7839	25.0	1.4330	-111.3	0.	131.8	0.0	379
OCTENE-1	112.21	0.7149	20.0	1.4087	-101.7	0.	121.3	0.0	380
DI-ISO-BUTYLENE	112.22	0.7122	25.0	1.4090	0.0	0.	101.0	0.0	381
CHLOROBENZENE	112.56	1.1117	15.0	1.5275	-45.6	0.	131.7	5.62	382
1,2-DICHLOROPROPANE	112.99	1.1560	20.0	1.4394	-100.4	760.	96.4	0.0	383
1,3-DICHLOROPROPANE	112.99	1.1878	20.0	1.4487	-99.5	760.	120.4	0.0	384
2,2-DICHLOROPROPANE	112.99	1.1120	20.0	1.4148	-33.8	760.	69.3	0.0	385
1,1-DICHLOROPROPANE	112.99	1.1321	20.0	1.4289	0.0	760.	88.1	0.0	386
ETHYL CYANOACETATE	113.12	1.0614	20.0	1.4155	-22.5	0.	206.0	26.70	387
CHLOROACETYL CHLORIDE	113.94	1.4202	20.0	1.4541	0.0	760.	107.0	0.0	388
TRIFLUOROACETIC ACID	114.02	1.4890	20.0	1.2850	-15.3	0.	71.8	8.55	389
ALLYL PROPIONATE	114.14	0.9037	25.0	1.4110	0.0	0.	124.0	0.0	390
2,5-HEXANEDIONE	114.14	0.7370	0.0	1.4232	-5.5	754.	194.0	0.0	391
2,4-DIME-3-PENTANONE	114.18	0.8062	20.0	1.4001	0.0	0.	124.0	0.0	392
2-HEPTANONE	114.18	0.8111	20.0	1.4116	-35.0	0.	151.0	0.0	393
3-HEPTANONE	114.18	0.8183	20.0	0.0	39.0	0.	50.0	0.0	394
4-HEPTANONE	114.18	0.8174	20.0	1.4073	-33.0	0.	144.0	0.0	395
CYCLOHEXYLMETHYL ETHER	114.19	0.8790	20.0	1.4355	-74.4	760.	133.0	0.0	396
1-METHYLCYCLOHEXANOL	114.19	0.9251	24.6	1.4587	26.0	0.	157.0	0.0	397
2-METHYLCYCLOHEXANOL	114.19	0.9254	20.0	1.4610	0.0	0.	167.6	13.30	398
CIS-2-ME CYCLOHEXANOL	114.19	0.9360	20.0	1.4640	7.0	0.	165.0	0.0	399
TRN-2-ME CYCLOHEXANOL	114.19	0.9247	20.0	1.4616	-4.0	0.	166.5	0.0	400
3-METHYLCYCLOHEXANOL	114.19	0.9168	20.0	1.4576	0.0	0.	172.0	12.30	401
CIS-3-ME CYCLOHEXANOL	114.19	0.9155	20.0	1.4572	-5.5	0.	168.0	16.47	402
TRN-3-ME CYCLOHEXANOL	114.19	0.9214	20.0	1.4580	-0.5	0.	84.0	8.05	403
4-METHYLCYCLOHEXANOL	114.19	0.9122	20.0	1.4565	0.0	763.	172.0	13.30	404
5-METHYL-3-HEXANONE	114.19	0.8150	17.0	1.3970	0.0	735.	134.0	0.0	405
ISO OCTANE	114.22	0.6918	0.0	1.3915	-107.4	0.	99.2	0.0	406
OCTANE	114.22	0.7025	20.0	1.3974	-56.8	0.	125.6	1.95	407
OCTANE	114.22	0.6985	25.0	1.3951	-56.8	0.	125.6	1.95	408
2,2,4-TRIME PENTANE	114.22	0.7078	0.0	1.3914	-107.4	0.	98.2	1.94	409
2,2,3-TRIME PENTANE	114.22	0.7121	25.0	1.4006	-112.3	0.	109.8	1.96	410
ACETONYLUREA	116.12	0.8018	4.0	0.0	-41.0	0.	82.0	0.0	411
METHYLACETOACETATE	116.12	1.0747	20.0	1.4186	-80.0	0.	171.7	0.0	412
1-UREIDO-2-PROPANONE	116.12	0.8018	4.0	0.0	-41.0	0.	82.0	0.0	413
BETA-ACETOPROPIONIC ACID	116.13	1.1335	20.0	1.4396	37.2	0.	245.8	0.0	414
INDENE	116.15	0.9915	0.0	1.5642	-2.0	0.	182.2	0.0	415
4-ME-2-PENTANONE-4-OL	116.15	0.9385	0.0	1.4235	-44.0	0.	166.0	0.0	416
AMYL FORMATE	116.16	0.8926	15.0	1.3992	-73.5	0.	132.1	0.0	417
BUTYL ACETATE	116.16	0.8713	30.0	1.3827	-73.5	0.	126.1	5.01	418
SEC BUTYL ACETATE	116.16	0.8720	20.0	1.3894	0.0	0.	112.3	0.0	419
CAPROIC ACID	116.16	0.9230	25.0	1.4148	-3.9	0.	205.7	2.63	420
DIACETONE ALCOHOL	116.16	0.9342	25.0	1.4213	-44.0	12.	168.1	18.20	421
ETHYL BUTYRATE	116.16	0.8791	20.0	1.3928	-98.0	0.	121.6	5.10	422
ETHYL ISOBUTYRATE	116.16	0.8693	20.0	1.3903	-88.2	0.	111.0	0.0	423
ISOAMYL FORMATE	116.16	0.8820	20.0	1.3476	0.0	0.	124.2	0.0	424
ISOBUTYL ACETATE	116.16	0.8695	25.0	1.3880	-98.8	0.	118.0	5.29	425

Note: Missing data is indicated by 0, 0., or 0.0.

(continued)

Name	Mol. Wt.	Density	Temp.	Refract. Index	Melting Point	Pressure	Boiling Point	Dielec. Constant	List No.
N-PROPYL PROPIONATE	116.16	0.8330	0.0	1.6015	-76.0	0.	122.5	0.0	426
1122TETRAME UREA	116.16	0.9654	25.0	1.4493	-1.2	0.	175.2	23.06	427
2-HEPTANOL	116.21	0.8171	20.0	1.4210	0.0	0.	159.7	9.21	428
PHENYLACETONITRILE	117.14	1.0155	0.0	1.5233	-23.8	0.	233.5	0.0	429
TOLUIC NITRILE	117.15	1.0125	25.0	1.5209	-23.8	0.	233.5	18.70	430
1-AMINO-2-ME-2-PENTANOL	117.19	0.9081	20.0	1.4510	0.0	15.	83.0	0.0	431
2-BUTYLAMINO ETHANOL	117.19	0.8907	20.0	1.4437	-3.5	760.	199.5	0.0	432
DIMETHYL OXALATE	118.09	1.1716	60.0	1.3790	5.4	760.	164.5	0.0	433
GLYCOL DIFORMATE	118.09	1.1930	1.0	1.3580	-10.0	0.	174.0	0.0	434
ACETAL	118.12	0.8213	25.0	1.3682	0.0	0.	103.6	3.80	435
DIETHYL CARBONATE	118.13	0.9693	25.0	1.3829	-43.0	0.	126.8	2.82	436
ETHYL LACTATE	118.13	1.0328	20.0	1.4124	-26.0	0.	154.5	13.10	437
ETHYL LACTATE	118.13	1.0272	25.0	1.4127	-26.0	36.	69.5	13.10	438
2-MEOXYETHYLACETATE	118.13	1.0049	20.0	1.4022	-65.1	0.	144.5	8.25	439
DIETHOXYETHANE	118.17	0.8341	0.0	1.3819	0.0	0.	102.2	0.0	440
ALPHA-METHYL STYRENE	118.17	0.9140	0.0	0.0	-20.0	765.	165.0	0.0	441
2-BUTOXYETHANOL	118.18	0.9008	20.0	1.4198	0.0	760.	170.2	9.30	442
2-METHYL-2,4-PENTANEDIOL	118.18	0.9254	17.0	1.4250	-40.0	760.	197.0	0.0	443
CHLOROCYCLOHEXANE	118.61	1.0000	20.0	1.4626	-43.9	0.	143.0	0.0	444
PHENYL CARBONIMIDE	119.12	1.0960	0.0	1.5350	0.0	0.	165.8	0.0	445
TRIAZOBENZENE	119.13	1.0880	20.0	1.5589	-27.0	11.	70.C	0.0	446
CHLOROFORM	119.38	1.4799	25.0	1.4429	-63.5	0.	61.1	4.81	447
ACETOPHENONE	120.15	1.0281	20.0	1.5342	19.6	0.	202.0	17.39	448
GLYCEROL DIMETHYL ETHER	120.15	1.0085	0.0	1.4192	0.0	0.	169.0	0.0	449
2-(2-MEOETO)ETHANOL 328	120.15	1.0167	25.0	1.4245	-70.0	0.	194.1	0.0	450
PHENYL ETHYLENE OXIDE	120.15	1.0469	25.0	1.5350	-35.6	0.	191.5	0.0	451
STYRENE OXIDE	120.15	1.0469	25.0	1.5350	-35.6	0.	191.5	0.0	452
SULFOLANE	120.17	1.2614	30.0	1.4820	28.5	0.	287.3	43.30	453
CUMENE	120.19	0.8618	20.0	1.4915	-96.0	0.	152.4	2.38	454
O-ETHYLTOLUENE	120.19	0.8870	0.0	1.5042	-17.0	0.	164.9	0.0	455
ISOPROPYLBENZENE	120.19	0.8575	25.0	1.4889	-96.0	0.	152.4	2.38	456
1-PHENYL PROPANE	120.19	0.8620	0.0	1.4920	-99.5	0.	159.2	0.0	457
1,2,3 TRIME BENZENE	120.19	0.8944	0.0	1.5139	-25.5	0.	176.0	0.0	458
1,2,4-TRIME BENZENE	120.19	0.8890	4.0	1.5030	-60.5	0.	169.3	0.0	459
1,2,5-TRIME BENZENE	120.19	0.8642	0.0	1.4998	-52.7	0.	164.7	0.0	460
PROPYLBENZENE	120.20	0.8620	0.0	1.4900	-99.5	0.	159.2	0.0	461
1,3,5-TRIMETHYLBENZENE	120.20	0.8642	20.0	1.4998	-52.7	0.	164.7	2.27	462
1-CHLOROHEXANE	120.62	0.8784	0.0	1.4240	-83.0	0.	132.0	0.0	463
2-CL-HEXANE	120.62	0.8694	21.0	1.4142	0.0	0.	0.0	0.0	464
3-CL-HEXANE	120.62	0.8700	20.0	1.4163	0.0	0.	0.0	0.0	465
1-BRUMO-1-PROPENE	120.99	1.4133	20.0	1.4519	-116.6	0.	59.5	0.0	466
O-ETHYL ANILINE	121.18	0.9830	22.0	1.5584	-5.0	0.	209.0	0.0	467
O-METHYL TOLUIDINE	121.18	0.9769	20.0	1.5649	0.0	0.	206.0	0.0	468
2,4,6-TRIMETHYL PYRIDINE	121.18	0.9166	22.0	1.4959	0.0	754.	172.0	0.0	469
SALICYLALDEHYDE	122.13	1.1525	25.0	1.5702	-7.0	0.	196.7	13.90	470
O-METHYLANISOLE	122.16	0.9850	15.0	1.5199	0.0	0.	170.3	0.0	471
PHENETOLE	122.17	0.9605	25.0	1.5049	-29.5	0.	170.0	4.22	472
DIETHANOL SULFIDE	122.19	1.1793	25.0	1.5146	-10.0	0.	282.0	0.0	473
2-CHLOROETHYLACETATE	122.55	1.1783	6.0	1.4215	-20.0	0.	145.0	0.0	474
ETHYL CHLOROACETATE	122.55	1.1144	20.0	1.4215	-26.0	740.	144.0	0.0	475
1-CHLORO-3-PENTANOL	122.60	1.0327	25.0	1.4660	0.0	0.	173.0	0.0	476
ACETYL BROMIDE	122.96	1.6630	0.0	1.4538	-96.5	0.	76.7	0.0	477
1-BROMOPROPANE	123.00	1.3452	25.0	1.4317	-109.8	0.	71.0	8.09	478
2-BROMOPROPANE	123.00	1.3060	25.0	1.4221	-89.0	0.	59.4	9.46	479
NITROBENZENE	123.11	1.2082	15.0	1.5546	5.7	0.	210.8	34.82	480
DIETHYL ZINC	123.50	1.1820	18.0	1.4954	-28.0	0.	118.0	0.0	481
M-THIOCRESOL	124.21	1.0526	12.0	1.5752	-20.0	0.	195.4	0.0	482
BETA-CLETHYLCELLOSOLVE	124.57	0.0	19.0	1.4505	0.0	760.	182.0	0.0	483
3-CL-2-CLME-PROPENE	125.00	1.1782	20.0	1.4754	-14.0	0.	138.3	0.0	484
CAPRYLONITRILE	125.21	0.8097	25.0	1.4182	-45.6	0.	205.2	13.90	485
DIMETHYL SULFATE	126.13	1.3283	20.0	1.3874	-31.8	760.	188.5	0.0	486
2,5-DIME-CYCLOHEXANONE	126.19	0.9025	20.0	1.4446	0.0	0.	178.0	0.0	487
2-NONENE (TRANS)	126.23	0.7540	0.0	1.4191	0.0	0.	148.5	0.0	488
1-NONENE	126.24	0.7253	25.0	1.4133	-81.4	0.	146.9	0.0	489
BENZYL CHLORIDE	126.58	1.1000	0.0	1.5391	-39.0	0.	179.4	23.00	490
M-CHLOROTOLUENE	126.58	1.0722	0.0	1.5214	-47.8	0.	162.0	5.55	491
P-CHLOROTOLUENE	126.58	1.0697	0.0	1.5199	7.5	0.	162.0	6.08	492
1,2-DICL ISOBUTANE	127.01	1.0930	20.0	1.4370	-130.0	0.	108.0	0.0	493
1,1-DICHLOROBUTANE	127.02	1.0863	0.0	1.4355	0.0	763.	114.5	0.0	494
1,2-DICHLOROBUTANE	127.02	1.1116	25.0	1.4474	0.0	0.	124.0	0.0	495
1,4-DICHLOROBUTANE	127.02	1.7598	12.0	1.4566	-38.7	0.	162.0	0.0	496
2,3-DICHLOROBUTANE	127.03	1.1134	20.0	1.4420	-80.0	760.	116.0	0.0	497
1,1-DICH-2-ME PROPANE	127.03	1.0111	20.0	1.4330	0.0	760.	105.5	0.0	498
1,2-DICH-2-ME PROPANE	127.03	1.0930	20.0	1.4370	-130.0	760.	108.0	0.0	499
1,3-DICH-2-ME PROPANE	127.03	1.1325	25.0	1.4488	0.0	760.	134.6	0.0	500
1-ACETYLPIPERIDINE	127.18	1.0112	9.0	0.0	108.0	0.	226.5	0.0	501
O-CHLOROANILINE	127.57	1.2125	20.0	1.5881	-1.9	0.	208.8	13.40	502
3-ME-HEPTANONE-2	128.21	0.8318	20.0	1.4172	0.0	0.	167.0	0.0	503
2-ETHYLCYCLOHEXANOL (CIS)	128.22	0.9274	21.0	1.4655	0.0	12.	74.0	0.0	504
OCTANONE-2	128.22	0.8185	0.0	1.4161	-20.9	0.	173.0	0.0	505
OCTANONE-3	128.22	0.8220	20.0	1.4156	0.0	738.	169.0	0.0	506
ISO-NONANE	128.25	0.7134	0.0	1.4032	-80.5	0.	142.8	0.0	507
4-METHYL OCTANE	128.25	0.7199	0.0	1.4061	-119.1	0.	142.4	0.0	508
NONANE	128.25	0.7138	25.0	1.4054	-53.5	0.	150.8	1.97	509
2,2,5-TRIME HEXANE	128.25	0.7032	25.0	1.3997	-105.8	0.	124.1	0.0	510

Note: Missing data is indicated by 0, 0., or 0.0.

(continued)

Name	Mol. Wt.	Density	Temp.	Refract. Index	Melting Point	Pressure	Boiling Point	Dielec. Constant	List No.
DICHLOROACETIC ACID	128.94	1.5634	20.0	1.4658	10.8	0.	192.5	8.20	511
1,3-DICL-2-PROPANOL	128.99	1.3506	17.0	1.4837	0.0	760.	176.0	0.0	512
N-ACETYL MORPHOLINE	129.16	1.1165	20.0	1.4383	14.0	50.	152.0	0.0	513
CINNAMONITRILE	129.16	1.0283	20.0	1.6043	22.0	0.	263.8	0.0	514
QUINOLINE	129.16	1.0977	25.0	1.6293	-14.9	760.	237.1	9.00	515
DIBUTYLAMINE	129.25	0.7619	20.0	1.4177	-62.0	0.	159.6	2.98	516
ETHYLACETOACETATE	130.15	1.0222	20.0	1.4192	-40.0	0.	180.8	15.70	517
PROPIONIC ANHYDRIDE	130.15	1.0110	20.0	1.4045	-43.0	0.	169.0	18.30	518
N-BUTYL PROPIONATE	130.18	0.8818	0.0	1.3982	-89.5	0.	145.5	0.0	519
HEPTANOIC ACID	130.18	0.9185	0.0	1.4216	-10.0	0.	223.0	0.0	520
ISOBUTYLPROPIONATE	130.18	0.8876	20.0	1.3975	-71.4	0.	136.8	0.0	521
4(2-AMINOET)MORPHOLINE	130.19	0.9915	20.0	1.4715	25.6	50.	116.0	0.0	522
AMYL ACETATE	130.19	0.8753	20.0	1.4028	-100.0	0.	149.2	4.75	523
ETHYL ISOVALERATE	130.19	0.8652	20.0	1.3962	-99.3	0.	134.7	4.71	524
2-OCTANOL(DL)	130.22	0.8193	0.0	1.4203	-38.5	0.	180.0	0.0	525
2,2,4-TRI ME PENTANOL-4	130.22	0.8270	0.0	1.4293	-16.7	0.	194.5	0.0	526
BUTYL ETHER	130.23	0.7641	25.0	1.3968	-95.2	0.	142.2	0.0	527
2-ET-1-HEXANOL	130.23	0.8291	25.0	1.4292	-76.0	0.	184.4	4.41	528
2-ME-HEPTANOL-2	130.23	0.8142	20.0	1.4279	0.0	0.	156.0	0.0	529
3-ME-HEPTANOL-3	130.23	0.8282	20.0	1.4238	-83.0	0.	163.0	0.0	530
1-OCTANOL	130.23	0.8221	25.0	1.4275	-15.0	0.	195.2	10.34	531
P-FLUORO CHLORO BENZENE	130.55	1.2260	0.0	1.4990	-28.3	757.	130.0	0.0	532
TRIMETHYL BORATE	130.92	0.9200	23.0	1.3568	-34.0	0.	69.0	0.0	533
ISOAMYL ACETATE	130.98	0.8664	25.0	1.3984	78.5	0.	42.0	4.63	534
4(B-HYDROXYET)MORPHOLINE	131.17	1.0710	20.0	1.4780	0.0	757.	227.0	0.0	535
1-NITROHEXANE	131.17	0.0	0.0	1.4229	0.0	0.	0.0	0.0	536
TRICHLOROETHYLENE	131.39	1.4762	15.0	1.4800	-86.4	0.	87.2	3.42	537
DIMETHYLMALONATE	132.11	1.1544	0.0	1.4140	-80.0	0.	183.0	0.0	538
CELLOSOLVE ACETATE	132.16	0.9730	20.0	1.4050	-61.7	0.	156.3	7.57	539
CINNAMALDEHYDE	132.16	1.0497	20.0	1.6195	-75.0	760.	253.0	16.90	540
1234TET H NAPHTHALENE	132.21	0.9702	20.0	1.5414	-35.8	0.	207.6	2.77	541
1234TET H NAPHTHALENE	132.21	0.9662	25.0	1.5492	-35.8	0.	207.6	2.77	542
1-FLUORO OCTANE	132.22	0.8103	0.0	1.3935	0.0	0.	142.5	0.0	543
ETHYLDIETHANOLAMINE	133.19	1.0135	20.0	1.4663	-50.0	760.	247.0	0.0	544
1,1,1-TRICHLOROETHANE	133.41	1.3376	20.0	1.4379	-30.4	0.	74.0	7.53	545
1,1,2-TRICL ETHANE	133.41	1.4405	20.0	1.4706	-37.4	0.	113.0	0.0	546
3-BR PROPIONITRILE	133.98	1.6152	20.0	1.1470	0.0	25.	92.0	0.0	547
ALLYL PHENYL ETHER	134.18	0.9788	20.0	1.5200	0.0	18.	85.0	0.0	548
CARBITOL	134.18	0.9881	0.0	1.4273	0.0	0.	201.0	0.0	549
CINNAMYL ALCOHOL	134.18	1.0440	20.0	1.5819	33.0	0.	257.5	0.0	550
2-(2-ETOETO)ETHANOL	134.18	0.9814	25.0	1.4254	0.0	0.	202.0	0.0	551
BIS(2-MEOET) ETHER	134.18	0.9440	25.0	1.4043	0.0	0.	159.8	0.0	552
1-PHENYL-2-PROPANONE	134.18	1.0157	20.0	1.5168	27.0	0.	216.5	0.0	553
PROPIOPHENONE	134.18	1.0105	20.0	1.5269	18.6	0.	218.0	0.0	554
O-DIETHYLBENZENE	134.21	0.8800	0.0	1.5035	-20.0	0.	183.0	0.0	555
1,3-DIETHYLBENZENE	134.21	0.8639	20.0	1.4955	-20.0	0.	181.0	0.0	556
1,4-DIETHYLBENZENE	134.21	0.8620	20.0	1.4967	-35.0	0.	182.0	0.0	557
P-ISOPROPYL TOLUENE	134.21	0.8569	20.0	1.4904	0.0	0.	177.0	0.0	558
1234 TETRA-ME-BENZENE	134.21	0.9010	0.0	1.5187	-64.0	0.	203.5	0.0	559
1235 TETRA-ME-BENZENE	134.21	0.8906	0.0	1.5134	-24.0	0.	196.0	0.0	560
BUTYL BENZENE	134.22	0.8561	25.0	1.4874	-87.9	0.	183.3	2.36	561
SEC-BUTYL BENZENE	134.22	0.8580	25.0	1.4878	-75.5	0.	173.3	2.36	562
P-CYMENE	134.22	0.8533	25.0	1.4885	-67.9	0.	177.1	2.25	563
TERT-BUTYL BENZENE	134.22	0.8624	25.0	1.4902	-57.9	0.	169.1	2.37	564
4-CHLOROCYCLOHEXANOL	134.61	1.1435	17.0	1.4930	0.0	14.	106.0	0.0	565
1-CHLOROHEPTANE	134.65	0.8810	0.0	1.4248	-69.0	0.	158.0	0.0	566
2-CL-HEPTANE	134.65	0.8725	15.0	1.4221	0.0	19.	46.0	0.0	567
3-CL-HEPTANE	134.65	0.8960	20.0	1.4237	0.0	751.	144.0	0.0	568
4-CL-HEPTANE	134.65	0.8710	20.0	1.4237	0.0	758.	144.0	0.0	569
4-BROMO-1-BUTENE	135.01	1.3230	20.0	1.4622	0.0	758.	98.5	0.0	570
BENZEDRINE	135.20	0.9400	15.0	1.5463	0.0	0.	203.5	0.0	571
PHENYL ACETATE	136.14	1.0730	25.0	1.5051	0.0	765.	195.8	18.40	572
BENZYL FORMATE	136.15	1.0817	25.0	1.5121	0.0	20.	93.0	0.0	573
METHYL BENZOATE	136.15	1.0790	30.0	1.5123	-12.1	0.	199.5	6.59	574
N-PROPYL PHENYL ETHER	136.19	0.9530	15.0	1.5011	0.0	0.	189.5	0.0	575
BENZYL ETHYL ETHER	136.20	0.9446	25.0	1.4934	0.0	0.	185.0	3.90	576
3-PHENYL-1-PROPANOL	136.20	1.0080	20.0	1.5357	-18.0	750.	236.5	0.0	577
ALPHA PINENE	136.24	0.8539	25.0	1.4632	-64.0	0.	156.9	2.26	578
BETA PINENE	136.24	0.8667	25.0	1.4768	-61.5	0.	166.0	2.50	579
ETHYL CHLOROGLYOXYLATE	136.54	1.2226	20.0	0.0	0.0	760.	135.0	0.0	580
6-CL-HEXANOL-1	136.62	0.0	0.0	1.4531	0.0	12.	107.0	0.0	581
BROMOBUTANE	137.03	1.2764	20.0	1.4389	-112.4	0.	101.3	0.0	582
2-BROMO-2-ME PROPANE	137.03	1.2220	20.0	1.4283	-20.0	0.	73.3	0.0	583
1-BROMO-2-ME PROPANE	137.03	1.3356	25.0	1.4348	0.0	0.	91.0	0.0	584
DIETHYL SELENIDE	137.06	1.2300	20.0	1.4768	0.0	0.	108.0	0.0	585
ANILINOETHANOL	137.18	1.1129	25.0	1.5749	0.0	0.	286.0	0.0	586
METHOXY BENZYL ALCOHOL	138.16	1.0430	25.0	1.5490	0.0	0.	249.0	0.0	587
2-PHENOXYETHANOL	138.17	1.1020	22.0	1.5340	14.0	0.	237.0	0.0	588
VERATROLE	138.17	1.0819	25.0	1.5323	22.5	0.	206.3	4.09	589
DIETHYL SULFITE	138.19	1.0829	20.0	1.4144	0.0	768.	157.0	0.0	590
ISOPHORONE	138.21	0.9229	20.0	1.4759	-8.1	754.	214.0	0.0	591
DECAHYDRONAPHTHALENE	138.24	0.8789	25.0	1.4758	-125.0	0.	191.7	0.0	592
CIS-DECAHYDRONAPHTHALENE	138.24	0.8967	0.0	1.4811	-43.3	0.	195.7	0.0	593
TRAN-DECAHYDRONAPHTHALENE	138.24	0.8700	0.0	1.4696	-32.5	0.	187.3	0.0	594
BROMOACETIC ACID	138.95	1.9335	50.0	1.4804	5.0	760.	208.0	0.0	595

Note: Missing data is indicated by 0, 0., or 0.0.

(continued)

Name	Mol. Wt.	Density	Temp.	Refract. Index	Melting Point	Pressure	Boiling Point	Dielec. Constant	List No.
1-PHENYL-2-PROPANOL	139.20	0.9727	19.0	1.5314	36.0	0.	217.0	0.0	596
PELARGONIC NITRILE	139.23	0.7860	16.0	1.4235	-34.2	0.	224.0	0.0	597
DECAHYDROQUINOLINE	139.24	0.9426	20.0	1.4926	-40.0	20.	90.0	0.0	598
TRIMETHYLPHOSPHATE	140.08	1.2144	20.0	1.3967	-46.0	760.	197.2	0.0	599
2-ACETYL CYCLOHEXANONE	140.18	1.0782	20.0	1.5138	0.0	18.	112.0	0.0	600
2-ISOPROPYLCYCLOHEXANONE	140.23	0.9270	20.0	1.4538	0.0	76.	199.0	0.0	601
BUTYLCYCLOHEXANE	140.27	0.8178	20.0	1.4400	-78.6	0.	179.0	0.0	602
1-DECENE	140.27	0.7369	25.0	1.4191	-66.3	0.	170.6	0.0	603
1,5-DICHLOROPENTANE	141.04	1.1006	20.0	1.4564	-72.8	760.	180.0	0.0	604
IODOMETHANE	141.94	2.2649	25.0	1.5270	-66.5	0.	42.4	7.00	605
1-ACETYL CYCLOHEXANOL	142.20	1.0257	20.0	1.4726	0.0	50.	125.0	0.0	606
1-METHYL NAPHTHALENE	142.20	1.0202	20.0	1.6170	-22.0	760.	244.6	0.0	607
METHYL HEPTYL KETONE	142.23	0.8185	0.0	1.4161	-20.9	0.	173.0	0.0	608
4-NONANONE	142.24	0.8370	20.0	1.4210	0.0	0.	187.5	0.0	609
5-NONANONE	142.24	0.8270	20.0	1.3980	-4.8	0.	188.4	0.0	610
3,3,5-TRIME CYCLOHEXANOL	142.24	0.9006	16.0	1.4550	37.3	750.	202.0	0.0	611
DECANE	142.27	0.7262	25.0	1.4119	-29.7	0.	174.1	1.99	612
METHYL-DICHLORO ACETATE	142.97	1.3774	20.0	1.4429	-51.9	760.	142.8	0.0	613
BIS(2-CL ETHYL)ETHER	143.01	1.2192	20.0	1.4575	-46.8	0.	178.8	21.20	614
1,3-DICL-2-ME-2-PROPANOL	143.02	1.2758	20.0	1.4744	0.0	0.	174.5	0.0	615
2-METHYL QUINOLINE	143.18	1.0585	0.0	1.6126	-1.0	0.	246.5	0.0	616
TRI-N-PROPYL AMINE	143.27	0.7530	25.0	1.4176	-93.5	0.	156.0	0.0	617
1-BR-2-CL-ETHANE	143.43	1.7392	20.0	1.4917	-16.7	0.	107.0	0.0	618
DIMETHYL MALEATE	144.13	1.1513	20.0	1.4422	-17.5	0.	200.4	0.0	619
TETRAHYDROFURFURYL ACET	144.17	1.0672	25.0	1.4352	0.0	14.	85.0	0.0	620
4(3-AMINOPROP)MORPHOLINE	144.21	0.9872	20.0	1.4749	-15.0	50.	134.0	0.0	621
AMYLPROPIONATE	144.21	0.8760	20.0	1.4096	-73.1	760.	168.7	0.0	622
BUTYLBUTYRATE (N)	144.21	0.8717	20.0	1.4045	-91.5	760.	166.6	0.0	623
ISOBUTYL-N-BUTYRATE	144.21	0.8364	18.0	1.4030	0.0	0.	157.0	0.0	624
N-PROPYL VALERATE	144.21	0.8888	0.0	1.4057	0.0	0.	167.5	0.0	625
CAPRYLIC ACID	144.22	0.9106	20.0	1.4280	16.5	0.	239.9	2.45	626
2-ETHYLBUTYL ACETATE	144.22	0.8790	20.0	1.4109	-100.0	760.	162.5	0.0	627
HEXYL ACETATE	144.22	0.8779	15.0	1.4092	-80.9	760.	171.5	0.0	628
ISOBUTYLISOBUTYRATE	144.22	0.8742	0.0	1.3999	-80.7	0.	147.5	0.0	629
ISOBUTYLISOBUTYRATE	144.22	0.8489	25.0	1.3981	-80.7	0.	175.6	0.0	630
NONYL ALCOHOL	144.25	0.8273	0.0	1.4333	-5.0	0.	212.0	0.0	631
2,6-DIMETHYL-4-HEPTANOL	144.26	0.8090	21.0	1.4242	0.0	760.	176.5	0.0	632
SILICONANE	144.34	0.7682	0.0	1.4243	0.0	0.	154.7	0.0	633
4-PHENYLBUTRONITRILE	145.21	0.9762	0.0	1.5150	0.0	0.	139.0	0.0	634
3,3,3-TRICHLOROPROPENE	145.43	1.3690	20.0	1.4827	30.0	0.	104.5	0.0	635
DIETHYL OXALATE	146.14	1.0669	30.0	1.4060	-40.6	0.	185.4	1.80	636
ETHYLENE DIACETATE	146.14	1.1043	20.0	1.4159	-41.5	0.	190.9	10.00	637
BUTYL LACTATE DL	146.19	0.9803	22.0	1.4217	-43.0	13.	83.0	0.0	638
2-ETHYL-1,3-HEXANEDIOL	146.23	0.9325	22.0	1.4497	-40.0	760.	244.0	0.0	639
HEXYL CELLOSOLVE	146.23	0.8894	20.0	1.4291	-45.1	760.	208.0	0.0	640
TRIETHYLENETETRAMINE	146.24	0.9820	20.0	1.4971	12.0	760.	266.5	0.0	641
O-DICHLOROBENZENE	147.01	1.3059	20.0	1.5515	-17.0	0.	180.5	9.93	642
M-DICHLOROBENZENE	147.01	1.2884	20.0	1.5459	-24.8	0.	173.0	5.04	643
P-DICHLOROBENZENE	147.01	1.2416	60.0	1.5285	53.1	0.	174.1	2.41	644
KAIROLINE	147.21	1.0220	0.0	1.5082	0.0	758.	0.0	0.0	645
1-PHENYL-PYRROLIDINE	147.22	1.0260	25.0	1.5803	0.0	9.	113.0	0.0	646
TRICHLOROACETALDEHYDE	147.40	1.5120	0.0	1.4557	-57.5	0.	98.0	0.0	647
1,1,1-TRICHLOROPROPANE	147.43	1.2870	23.0	0.0	0.0	760.	107.5	0.0	648
1,1,2-TRICHLOROPROPANE	147.43	1.3720	25.0	0.0	0.0	760.	140.0	0.0	649
1,1,3-TRICHLOROPROPANE	147.43	1.3557	20.0	1.4718	-59.0	760.	145.6	0.0	650
1,2,2-TRICHLOROPROPANE	147.43	1.3180	25.0	1.4609	0.0	762.	124.0	0.0	651
1,2,3-TRICHLOROPROPANE	147.44	1.3940	0.0	1.4858	-14.7	0.	156.0	0.0	652
BUTYROPHENONE	148.20	0.9880	0.0	1.5320	11.0	727.	231.0	0.0	653
1 PHENYLPENTANE	148.24	0.8662	0.0	1.4943	-78.3	0.	205.3	0.0	654
1-CHLORO OCTANE	148.67	0.8748	0.0	1.4306	0.0	765.	181.0	0.0	655
1-CL-2,5-DIME HEXANE	148.68	0.8476	18.0	1.4232	0.0	14.	44.0	0.0	656
2-BR-3-ME-1-BUTENE	149.04	1.2328	20.0	1.4504	0.0	757.	105.0	0.0	657
1-BR-3-ME-2-BUTENE	149.04	1.2819	20.0	1.4930	0.0	40.	50.5	0.0	658
2-BR-3-ME-2-BUTENE	149.04	1.2773	20.0	1.4738	0.0	766.	119.5	0.0	659
TRIETHANOLAMINE	149.19	1.1196	25.0	1.4835	21.6	0.	335.4	29.36	660
N-BUTYLANILINE	149.24	0.9323	20.0	1.5341	-14.4	760.	241.6	0.0	661
2,2,2-TRICHLOROETHANOL	149.42	1.5521	25.0	1.4890	17.8	737.	151.0	0.0	662
BENZYL ACETATE	150.18	1.0550	20.0	1.5232	-51.5	0.	215.5	5.10	663
ETHYL BENZOATE	150.18	1.0511	15.0	1.5075	-34.7	0.	212.4	6.02	664
TRIETHYLENE GLYCOL	150.18	1.1274	15.0	1.4578	-4.3	0.	288.0	23.69	665
N-BUTYL PHENYL ETHER	150.22	0.9351	0.0	1.4969	-19.0	0.	210.0	0.0	666
2-ME-1-PHENYLPROPANOL-1	150.22	0.9869	14.0	1.5193	0.0	0.	223.0	0.0	667
2-ME-1-PHENYLPROPANOL-2	150.22	0.9774	19.0	1.5201	24.0	0.	215.0	0.0	668
4-CHLOROBUTYL ACETATE	150.61	1.0759	0.0	1.4340	0.0	0.	83.0	0.0	669
1-BROMO-3-ME BUTANE	151.05	1.2609	20.0	1.4420	-112.0	0.	120.0	0.0	670
1-BROMOPENTANE	151.05	1.2177	0.0	1.4413	-95.0	0.	129.7	0.0	671
ET-2-PYRIDINECARBOXYLATE	151.16	1.1194	20.0	1.5104	1.0	0.	243.0	0.0	672
METHYL SALICYLATE	152.15	1.1831	20.0	1.5365	-8.6	0.	233.3	9.41	673
2-BENZYLOXYETHANOL	152.20	1.0640	20.0	1.5233	-19.0	760.	256.0	0.0	674
CINNAMYL CHLORIDE	152.63	0.0	0.0	0.0	0.0	0.	214.0	0.0	675
BROMOMETHYLACETATE	152.99	1.6560	12.0	1.5619	10.5	750.	130.0	0.0	676
O-NITROANISOLE	153.14	1.2527	20.0	1.4631	-23.0	0.	265.0	0.0	677
CARBON TETRACHLORIDE	153.82	1.6037	15.0	1.4574	-23.0	0.	76.7	2.24	678
CARBON TETRACHLORIDE	153.82	1.5844	25.0	1.4004	-24.5	0.	76.7	2.24	679
DIETHYL SULFATE	154.19	1.1774	20.0			0.	208.0	0.0	680

Note: Missing data is indicated by 0, 0., or 0.0.

(continued)

Name	Mol. Wt.	Density	Temp.	Refract. Index	Melting Point	Pressure	Boiling Point	Dielec. Constant	List No.
1,8-CINEOLE	154.25	0.9192	25.0	1.4555	1.3	0.	176.0	4.57	681
2-NBUTYLCYCLOHEXANONE	154.28	0.9350	20.0	1.4545	0.0	2.	7.0	0.0	682
1-UNDECENE	154.29	0.7506	0.0	1.4261	-50.0	0.	193.0	0.0	683
SUCCINYL CHLORIDE	154.98	1.3748	20.0	1.4683	20.0	760.	193.3	0.0	684
1,6-DICL-HEXANE	155.08	1.0677	20.0	1.4572	0.0	0.	202.4	0.0	685
IODOETHANE	155.97	1.9357	20.0	1.5133	-111.1	0.	72.3	7.82	686
3-CYCLOHEXPROPANOIC ACID	156.22	0.9966	20.0	1.4658	16.0	750.	193.0	0.0	687
1-ETHYL NAPHTHALENE	156.23	1.0082	20.0	1.6062	-13.9	760.	258.7	0.0	688
2-ETHYL NAPHTHALENE	156.23	0.9922	20.0	1.5999	-7.4	760.	257.9	0.0	689
DECANONE-2	156.26	0.8230	22.0	1.4621	14.0	767.	210.5	0.0	690
2-BUTYLCYCLOHEXANOL	156.27	0.9020	20.0	1.4641	0.0	3.	75.0	0.0	691
3-ISOPRO-2-HEPTANONE	156.27	0.8195	20.0	1.4250	0.0	0.	78.0	0.0	692
UNDECANE	156.30	0.7401	0.0	1.4172	-25.0	0.	195.5	2.01	693
BETA-CHLOROPHENETOLE	156.61	1.1443	25.0	1.5328	0.0	0.	0.0	0.0	694
P-CHLOROPHENETOLE	156.61	1.1231	20.0	1.5227	21.0	0.	213.0	0.0	695
2-CLETHYLCHLOROACETATE	157.00	1.3600	25.0	1.4619	0.0	760.	202.0	0.0	696
2,3-DICHLORODIOXANE	157.00	1.4680	20.0	1.4928	30.0	10.	81.0	0.0	697
ETHYLDICHLOROACETATE	157.00	1.2821	20.0	1.4386	0.0	0.	157.0	10.00	698
BROMOBENZENE	157.02	1.4882	25.0	1.5571	-30.8	0.	155.9	5.40	699
DIPENTYLAMINE	157.30	0.7771	20.0	1.4272	-70.0	760.	202.5	0.0	700
ALLYLIDENE DIACETATE	158.15	1.0749	20.0	1.4193	-37.6	0.	180.0	0.0	701
TETRAHYDROFURFURYL PROP	158.19	1.0321	25.0	1.4380	0.0	18.	101.0	0.0	702
BUTYRIC ANHYDRIDE	158.20	0.9668	20.0	1.4127	-66.7	0.	199.5	12.90	703
HEXYL PROPIONATE	158.23	0.8698	20.0	0.0	57.5	0.	190.0	0.0	704
ISOAMYL ETHER	158.27	0.7777	20.0	1.4085	0.0	0.	173.4	0.0	705
AMYL ETHER	158.28	0.7790	25.0	1.4098	-69.4	0.	186.9	0.0	706
DIETHYL MALONATE	160.17	1.0549	20.0	1.4136	-48.9	0.	199.3	7.87	707
CYCLOHEXYLBENZENE	160.26	0.9387	25.0	1.5239	7.0	0.	240.1	0.0	708
1-CHLOROPENTANE	160.60	0.8818	20.0	1.4120	-99.0	0.	107.7	6.60	709
BENZAL CHLORIDE	161.03	1.2557	14.0	1.5502	-16.4	0.	205.2	0.0	710
N-BUTYLDIETHANOLAMINE	161.25	0.9692	20.0	1.4625	-70.0	741.	274.0	0.0	711
1,2,3-TRICHLORO BUTANE	161.46	1.3164	20.0	1.4790	0.0	725.	166.5	0.0	712
2-ME-1,2,3-TRICL PROPANE	161.47	1.3012	25.0	1.4765	0.0	0.	162.0	0.0	713
ISOSAFROLE	162.18	1.1140	25.0	1.5740	0.0	0.	0.0	0.0	714
SAFROLE	162.18	1.0950	25.0	1.5383	11.2	0.	233.5	0.0	715
DIETHYLENEGLYCDIET ETHER	162.22	1.9063	0.0	1.4115	0.0	0.	188.0	0.0	716
BUTYLCARBITOL	162.23	0.9553	20.0	1.4321	-68.1	760.	231.0	0.0	717
1,2,4-TRIETHYL BENZENE	162.27	0.8738	0.0	1.5024	0.0	0.	218.0	0.0	718
1,3,5-TRIETHYL BENZENE	162.27	0.8621	0.0	1.4958	-66.5	755.	216.0	0.0	719
1-PHENYLHEXANE	162.28	0.8540	0.0	1.4860	-62.0	760.	226.0	0.0	720
2-CHLORONAPHTHALENE	162.61	1.1377	71.0	1.6079	59.0	0.	256.0	0.0	721
1-CHLORONAPHTHALENE	162.62	1.1938	20.0	1.6332	-2.3	0.	259.3	5.04	722
TRICHLOROACETIC ACID	163.39	1.6218	64.0	1.4603	5.8	760.	197.6	0.0	723
2-CHLOROQUINOLINE	163.60	1.2464	25.0	1.6259	37.5	751.	275.0	0.0	724
DICHLOROBROMOMETHANE	163.85	1.9800	20.0	1.4964	0.0	742.	89.5	0.0	725
4-PHENYL-1,3-DIOXANE	164.19	1.1038	20.0	1.5306	0.0	0.	245.0	0.0	726
EUGENOL	164.20	1.0664	20.0	1.5410	9.2	760.	255.0	0.0	727
PROPYL BENZOATE	164.21	1.0232	20.0	1.5003	-51.6	0.	231.2	0.0	728
BENZYL BUTYL ETHER	164.25	0.9407	20.0	1.4970	15.0	744.	220.5	0.0	729
NITROTRICHLORO METHANE	164.38	1.6566	20.0	1.4622	-64.5	760.	111.8	0.0	730
1-BROMOHEXANE	165.08	1.1763	20.0	1.4478	-85.0	0.	154.0	0.0	731
2-BR-HEXANE	165.08	1.1658	20.0	1.4432	0.0	0.	144.0	0.0	732
3-BR-HEXANE	165.08	1.1799	20.0	1.4486	0.0	744.	144.0	0.0	733
TETRACHLOROETHYLENE	165.83	1.6311	15.0	1.5076	-22.3	0.	121.2	2.30	734
BICYCLOHEXYL	166.31	0.8862	20.0	1.4800	3.6	0.	239.0	0.0	735
2-BROMOETHYLACETATE	167.02	1.5140	20.0	1.4547	-13.8	0.	162.5	0.0	736
1122-TET CL ETHANE	167.35	1.5786	30.0	1.4868	-43.8	0.	146.2	8.20	737
1112 TET CL ETHANE	167.86	1.5532	0.0	1.4821	-68.1	0.	129.5	0.0	738
3-IODOPROPENE	167.99	1.8454	22.0	1.5540	-99.3	0.	102.5	0.0	739
O-PHENYL TOLUENE	168.23	1.0100	0.0	1.6824	45.0	0.	262.5	0.0	740
2-BENZYL PYRIDINE	169.23	1.0670	20.0	1.5785	12.0	742.	276.0	0.0	741
1-IODOPROPANE	170.00	1.7394	25.0	1.5028	-101.3	0.	102.4	7.00	742
2-IODOPROPANE	170.00	1.6946	25.0	1.4961	-90.0	0.	89.5	8.19	743
1-ACETONAPHTHANE	170.21	1.1336	20.0	1.6280	12.0	0.	297.0	0.0	744
DIPHENYL ETHER	170.21	1.0661	20.0	1.5763	26.9	0.	258.3	3.69	745
DODECANE	170.34	0.7487	20.0	1.4216	-9.6	0.	216.3	2.01	746
BENZYL CHLOROFORMATE	170.60	1.1950	25.0	1.5160	0.0	0.	0.0	0.0	747
M-BROMOTOLUENE	171.04	1.4019	20.0	1.5510	-39.8	0.	183.7	5.36	748
P-BROMOTOLUENE	171.04	1.3898	20.0	1.5490	28.5	0.	184.5	5.49	749
B.B'DICL-DIISOPROP ETHER	171.07	1.1030	20.0	1.4505	0.0	0.	187.0	0.0	750
DIETHYL MALEATE	172.18	1.0687	20.0	1.4400	-8.8	0.	225.3	8.58	751
1-ETHOXY NAPHTHALENE	172.23	1.0600	20.0	1.5953	5.5	0.	280.5	0.0	752
HEPTYL PROPIONATE	172.26	0.8679	20.0	0.0	-50.9	0.	209.0	0.0	753
2-ETHYLHEXYL ACETATE	172.27	0.8718	20.0	1.4204	-93.0	0.	198.6	0.0	754
UNDECYL ALCOHOL	172.30	0.8298	20.0	1.4392	19.0	760.	243.0	0.0	755
2-UNDECANOL	172.31	0.8270	20.0	1.4369	12.0	760.	225.4	0.0	756
O-BROMOPHENOL	173.02	1.4924	20.0	1.5892	5.6	0.	194.5	0.0	757
METHYLENE BROMIDE	173.85	2.4970	20.0	1.5420	-52.6	760.	97.0	7.50	758
DIETHYL SUCCINATE	174.19	1.0406	20.0	1.4201	-22.0	0.	217.7	0.0	759
N-DIPROPYL OXALATE	174.19	1.0169	20.0	1.4168	-46.3	0.	214.5	0.0	760
DIMETHYL ADIPATE	174.20	1.0600	20.0	1.4283	10.3	13.	115.0	0.0	761
DIBUTOXYETHANE	174.28	0.8370	0.0	1.4131	-69.0	0.	203.0	0.0	762
P-FLUORO BROMOBENZENE	175.01	1.4946	20.0	1.5604	-17.0	764.	152.0	0.0	763
N-OCTYL NITRATE	175.22	0.9750	0.0	0.0	0.0	20.	111.0	0.0	764
2-(2-ETOETO)ETACETATE	176.21	1.0096	20.0	1.4213	-25.0	0.	217.4	0.0	765

Note: Missing data is indicated by 0, 0., or 0.0.

(continued)

Name	Mol. Wt.	Density	Temp.	Refract. Index	Melting Point	Pressure	Boiling Point	Dielec. Constant	List No.
ETHYL CINNAMATE	176.21	1.0494	20.0	1.5598	6.7	0.	272.7	0.0	766
1-CHLORODECANE	176.73	0.8683	0.0	1.4373	0.0	0.	222.5	0.0	767
ISOAMYLISOVALERATE	177.26	0.8583	18.7	1.4130	0.0	0.	194.0	3.62	768
METHYL TRICL ACETATE	177.43	1.4890	19.2	1.5250	-17.5	765.	152.5	0.0	769
BENZYL BUTYRATE	178.23	1.0140	19.0	1.4920	9.0	0.	109.0	0.0	770
TETRAMETHYL TIN	178.83	1.3140	0.0	1.4386	-54.8	0.	78.0	0.0	771
1-BROMOHEPTANE	179.11	1.1384	20.0	1.4505	-58.0	0.	179.0	0.0	772
HEXAME PHOSPHORAMIDE	179.20	1.0270	20.0	1.4588	7.2	0.	233.0	30.00	773
(ISO)PROPYL SALICYLATE	180.20	1.0729	15.0	1.5065	0.0	0.	241.0	0.0	774
1,4-(BIS CL ME)CYC HEX	181.11	1.1180	25.0	1.4908	15.0	0.	121.0	0.0	775
1,2,4-TRICL BENZENE	181.45	1.4542	20.0	1.5717	17.0	760.	213.5	0.0	776
1112TETRACLPROPANE	181.89	1.4695	22.0	1.4855	-64.0	0.	152.5	0.0	777
1,1-DIPHENYLETHANE	182.27	0.9875	20.0	1.5761	-215.0	0.	286.0	0.0	778
1,2-DIPHENYLETHANE	182.27	0.9950	20.0	1.5338	0.0	0.	285.0	0.0	779
TERT-BUTYL IODIDE	184.02	1.5445	20.0	1.4918	52.2	0.	0.0	0.0	779
1-IODOBUTANE	184.03	1.6123	0.0	1.5000	-103.1	0.	130.0	0.0	780
N-TRIDECANE	184.37	0.7559	0.0	1.4233	-5.5	0.	243.0	0.0	781
BENZYL CHLOROACETATE	184.63	1.2223	4.0	1.5426	10.0	0.	133.3	0.0	782
TRI-N-BUTYLAMINE	185.36	0.7771	20.0	1.4297	-70.0	760.	213.0	0.0	783
TRIBUTYL AMINE (ISO)	185.36	0.7640	20.5	1.4252	-21.8	0.	191.5	0.0	784
CIS-1,2-DIBR ETHYLENE	185.80	2.2464	20.0	1.5428	-53.0	760.	112.5	0.0	785
TRANS-1,2-DIBR ETHYLENE	185.80	2.2308	20.0	1.5505	-6.5	760.	108.0	0.0	786
2-BROMO-4-ME ANILINE	186.06	1.4745	25.0	1.6012	15.0	0.	0.0	0.0	787
HEXAFLUOROBENZENE	186.06	1.6182	20.0	1.3781	5.1	0.	80.3	0.0	788
OCTYL PROPIONATE	186.29	0.8663	0.0	0.0	0.0	0.	0.0	0.0	789
DIHEXYL ETHER	186.34	0.7936	0.0	1.4204	-43.0	768.	223.0	0.0	790
BIS(2-CL ET) CARBONATE	187.02	1.3444	25.0	1.4595	10.0	0.	117.0	0.0	791
112TRIFL-122TRICLETHANE	187.38	1.5635	25.0	1.3557	-36.4	0.	47.7	0.0	792
1,1-DIBROMOETHANE	187.87	2.0554	0.0	1.5122	-63.0	0.	110.0	0.0	793
1,2-DIBROMOETHANE	187.87	2.1687	25.0	1.5360	9.8	0.	131.4	4.78	794
1245 TETRAETHYL BENZENE	190.32	0.8788	0.0	1.5054	10.0	0.	250.0	0.0	795
ETHLYTRICL ACETATE	191.44	1.3826	20.0	1.4507	0.0	0.	167.5	7.80	796
ETHYL BENZOYL ACETATE	192.21	1.1220	20.0	1.5312	0.0	14.	165.0	0.0	797
3-BROMOMETHYL HEPTANE	193.13	1.1227	25.0	1.4548	0.0	0.	57.0	6.00	798
1-BROMO OCTANE	193.13	1.1180	0.0	1.4527	-55.0	0.	202.3	0.0	799
METHYL BENZOPHENONE	193.24	1.1464	0.0	1.6738	-18.0	0.	325.0	0.0	800
DIMETHYL PHTHALATE	194.18	1.1905	20.7	1.5150	0.0	0.	283.0	0.0	801
METHYL PHTHALATE	194.18	1.1890	25.0	1.5150	0.0	734.	283.0	0.0	802
TETRAETHYLENE GLYCOL	194.22	1.1285	0.0	1.4598	-6.2	0.	328.0	0.0	803
PHENYL CHLOROFORM	195.48	1.3800	0.0	1.5011	-4.8	0.	220.7	0.0	804
TETRANITROMETHANE	196.04	1.6372	0.0	1.4398	13.0	0.	126.0	0.0	805
1-TETRADECENE	196.38	0.7852	0.0	1.4932	-12.0	0.	246.0	0.0	806
DIBENZYLAMINE	197.28	1.0256	22.0	1.5143	-26.0	250.	270.0	3.60	807
1-IODOPENTANE	198.06	1.5170	0.0	1.4955	73.1	0.	155.0	0.0	808
BENZYL ETHER	198.27	1.0428	20.0	1.5406	36.0	0.	288.3	0.0	809
TRICHLOROBROMOMETHANE	198.30	2.0120	0.0	1.5061	-21.0	0.	104.0	0.0	810
N-TETRADECANE	198.40	0.7627	0.0	1.4290	6.0	0.	253.5	0.0	811
PHENYL-N-PROPYL BROMIDE	199.09	1.3098	19.0	1.5517	0.0	0.	121.5	0.0	812
(3-BROMOPROPYL)BENZENE	199.10	1.3098	19.0	1.5517	0.0	0.	121.5	0.0	813
BIS(4-CL BUTYL) ETHER	199.12	1.0796	25.0	1.4800	0.0	10.	130.0	0.0	814
2,3-DIBROMOPROPENE	199.88	2.0346	25.0	1.5416	0.0	760.	141.0	0.0	815
TRIBUTYL CARBINOL	200.35	0.8408	0.0	1.4445	20.0	0.	230.0	0.0	816
O-BROMOPHENETOLE	201.07	1.4105	25.0	1.5532	0.0	22.	124.0	0.0	817
P-BROMOPHENETOLE	201.07	1.4031	25.0	1.5498	11.0	0.	233.0	0.0	818
1,1-DIBROMOPROPANE	201.91	0.0	0.0	1.5100	0.0	74.	135.4	0.0	819
1,2-DIBROMOPROPANE	201.91	1.9366	0.0	1.5206	-55.5	0.	141.4	0.0	820
1,3-DIBROMOPROPANE	201.91	1.9893	0.0	1.5230	-34.2	0.	0.0	0.0	821
ACETYLDIMETHYLMALONATE	202.20	1.0830	26.0	1.4374	0.0	0.	120.0	0.0	822
DIBUTYL OXALATE	202.25	0.9873	20.0	1.4234	-30.5	773.	242.0	0.0	823
DIETHYLADIPATE	202.25	1.0076	20.0	1.4272	-19.8	760.	245.0	0.0	824
PENTACHLOROETHANE	202.30	1.6881	15.0	1.5054	-29.0	0.	162.0	3.73	825
1122TETCLDIF ETHANE	203.83	1.6252	35.0	1.4083	26.0	0.	92.8	2.52	826
IODOBENZENE	204.01	1.8230	25.0	1.6197	-31.4	0.	189.0	0.0	827
N,N-DIBUTYLANILINE	205.34	0.9037	20.0	1.5186	-32.2	760.	274.8	0.0	828
1-BROMONAPHTHALENE	207.08	1.4834	20.0	1.6580	6.2	0.	281.1	4.83	829
1-BROMONONANE	207.16	1.0851	25.0	1.4520	0.0	5.	84.0	0.0	830
2-BROMO-NONANE	207.16	1.0810	0.0	1.4519	0.0	767.	208.5	0.0	831
1,10-DICHLORODECANE	211.18	0.9936	0.0	1.4600	0.0	0.	148.0	0.0	832
BENZYL BENZOATE	212.25	1.1121	25.0	1.5681	19.4	0.	323.5	4.90	833
N-PENTADECANE	212.41	0.7689	0.0	1.4315	10.0	0.	270.5	0.0	834
3-BRPROPYL PHENYL ETHER	215.10	1.3650	16.0	0.0	11.0	18.	127.0	0.0	835
2-NITRO-DIPHENYL ETHER	215.21	1.2539	22.0	1.5750	-20.0	8.	184.0	0.0	836
1,2-DIBROMOBUTANE	215.94	1.7950	0.0	1.5500	0.0	0.	166.3	0.0	837
1,4-DIBROMOBUTANE	215.94	1.8080	0.0	1.5175	-26.0	0.	197.5	0.0	838
2,3-DIBROMOBUTANE	215.94	1.7830	0.0	1.5133	-70.3	0.	161.0	0.0	839
1,2-DIBR-2-ME PROPANE	215.94	1.7590	20.0	1.5090	10.5	760.	150.0	0.0	840
METHYLENE IODIDE	217.87	3.3254	0.0	1.7559	6.0	0.	181.0	0.0	841
2,3-DIBR-1-PROPANOL	217.90	2.0739	20.0	1.5466	0.0	17.	118.0	0.0	842
O-IODOTOLUENE	218.05	1.7130	0.0	1.6090	0.0	0.	211.5	0.0	843
GLYCEROL TRIACETATE	218.21	1.1562	20.0	1.5064	3.2	0.	259.0	0.0	844
PENTAETHYL BENZENE	218.37	0.8985	19.0	1.5127	-20.0	0.	277.0	0.0	845
ISOBUTYL TRICL ACETATE	219.50	1.2550	25.0	1.4456	0.0	0.	188.0	0.0	846
TETRAETGLYCOL DIME ETHER	222.29	1.0132	20.0	1.4336	-21.4	760.	275.8	0.0	847
1-HEXADECENE	224.42	0.7825	0.0	1.4441	4.0	0.	155.0	0.0	848
2-CLETHYL TRICL ACETATE	225.89	1.5357	20.0	1.4813	0.0	766.	217.0	0.0	849

Note: Missing data is indicated by 0, 0., or 0.0.

(continued)

Name	Mol. Wt.	Density	Temp.	Refract. Index	Melting Point	Pressure	Boiling Point	Dielec. Constant	List No.
TRICHLOROACETYL BROMIDE	226.29	0.1900	15.0	0.0	0.0	760.	143.0	0.0	851
TRI-N-PENTYLAMINE	227.44	0.7907	20.0	1.4367	-70.0	0.	242.5	0.0	852
DIBUTYL MALEATE	228.29	0.9950	20.0	1.4454	-80.0	0.	280.0	0.0	853
1,5-DIBROMOPENTANE	229.96	1.7060	18.0	1.5091	-39.5	0.	222.2	0.0	854
TRIBUTYL BORATE	230.16	0.8580	20.0	1.4092	-70.0	0.	233.5	0.0	855
DIISOPROPYL ADIPATE	230.31	0.9659	20.0	1.4247	-1.1	6.	120.0	0.0	856
DIPROPYL ADIPATE	230.31	0.9790	20.0	1.4314	-15.7	11.	151.0	0.0	857
DIPHENYL SELENIDE	233.17	1.3510	20.0	1.6500	2.5	760.	301.5	0.0	858
TETRAETHYL TIN	234.94	1.1870	23.0	1.4724	-112.0	0.	181.0	0.0	859
O-DIBROMOBENZENE	235.92	1.9557	0.0	1.6081	6.7	0.	221.0	0.0	860
M-DIBROMOBENZENE	235.92	1.9523	0.0	1.6083	-7.0	0.	220.0	0.0	861
1-IODO OCTANE	240.14	1.3297	0.0	1.4890	-45.7	0.	255.5	0.0	862
1,2-DIBROMOHEXANE	243.99	1.5872	15.0	1.5012	0.0	16.	287.0	0.0	863
DIETHYL AZELATE	244.33	0.9729	20.0	1.4351	-18.5	0.	291.5	0.0	864
TRICHLOROIODOMETHANE	245.30	2.3650	0.0	1.5854	-19.0	0.	142.0	0.0	865
4-BR-DIPHENYL ETHER	249.11	1.4225	19.0	1.6088	18.0	0.	305.0	0.0	866
BENZAL BROMIDE	249.95	1.5100	0.0	1.6147	0.0	0.	140.0	0.0	867
TRIBROMOACETALDEHYDE	250.76	2.6650	25.0	1.5939	0.0	760.	174.0	0.0	868
BROMOFORM	252.76	2.8889	20.0	1.5976	8.1	0.	149.6	4.39	869
1-IODO NAPHTHALENE	254.07	1.7399	20.0	1.7026	4.2	0.	302.0	0.0	870
DIBUTYL ADIPATE	258.35	0.9652	0.0	1.4369	-37.0	14.	183.0	0.0	871
DIBUTYL ADIPATE	258.36	0.9652	20.0	1.4369	-37.0	4.	145.0	0.0	872
DIETHYL SEBACATE	258.36	0.9646	20.0	1.4359	5.0	773.	306.0	0.0	873
DI ISOBUTYL ADIPATE	258.36	0.9530	20.0	1.4315	-17.0	760.	282.0	5.19	874
12DIBRTETF ETHANE	259.83	2.1630	25.0	1.3670	-110.5	0.	47.3	2.34	875
HEXACHLOROACETONE	264.77	1.7440	12.0	0.0	-30.0	0.	203.9	0.0	876
TRIBROMOETHYLENE	264.78	2.7080	20.5	1.6345	0.0	0.	163.5	0.0	877
1,1,2,3,4,4-HEXACLBUTANE	264.82	1.6460	20.0	1.5258	0.0	10.	111.0	0.0	878
TRI-N-BUTYL PHOSPHATE	266.32	0.9760	25.0	1.4226	-79.0	0.	289.0	7.96	879
1,1,2-TRIBROMOETHANE	266.79	2.5789	0.0	1.5933	-35.5	0.	188.5	0.0	880
TETRAMETHYL LEAD	267.33	1.9950	20.0	1.5120	-27.5	0.	110.0	0.0	881
DIIODOMETHANE	267.83	3.3078	25.0	1.7380	6.1	0.	182.0	5.32	882
DIBUTYL PHTHALATE	278.35	1.0465	20.0	1.4926	-35.0	0.	340.0	6.44	883
1,1,2-TRIBROMO PROPANE	280.80	2.3548	20.0	1.5790	0.0	760.	200.5	0.0	884
1,2,2-TRIBROMO PROPANE	280.80	2.2985	20.0	1.5670	0.0	760.	190.5	0.0	885
1,2,3-TRIBROMO PROPANE	280.80	2.4209	20.0	1.5862	16.9	760.	222.2	0.0	886
1,1-DIIODOETHANE	281.86	2.8400	0.0	1.6730	2.8	0.	179.5	0.0	887
OLEIC ACID	282.47	0.8870	25.0	1.4582	13.4	0.	360.0	2.46	888
1-BROMO-2-IODOBENZENE	282.92	2.2571	25.0	1.6618	8.0	0.	257.0	0.0	889
1-BROMO-3-IODOBENZENE	282.92	2.2220	25.0	1.6608	-9.0	754.	252.0	0.0	890
1-CHLOROOCTADECANE	288.95	0.8586	25.0	1.4525	19.0	2.	158.0	0.0	891
1-BROMOPENTADECANE	291.32	0.9999	25.0	1.4592	18.6	80.	172.0	0.0	892
123-TRIBR-2-ME PROPANE	294.65	2.1750	0.0	1.5652	0.0	0.	223.5	0.0	893
1,2,4-TRIBROMO BUTANE	294.83	2.1700	20.0	1.5608	-18.0	760.	215.0	0.0	894
2,2,3-TRIBROMO BUTANE	294.83	2.1724	20.0	1.5602	-1.9	760.	206.0	0.0	895
1,2,3-TRIBROMOBUTANE	294.84	2.1938	0.0	1.5680	-19.0	21.	113.5	0.0	896
METHYL OLEATE	296.50	0.8702	25.0	1.4502	19.9	0.	217.0	3.21	897
TRI-CL-ACETIC ANHYDRIDE	308.76	1.6908	20.0	0.0	0.0	760.	169.0	0.0	898
DI-N-BUTYL SEBACATE	314.47	0.9366	20.0	1.4397	1.0	0.	345.0	4.54	899
DIBUTYL SEBACATE	314.47	0.9324	25.0	1.4415	-11.0	0.	345.0	4.54	900
TETRAETHYL LEAD	323.45	1.6590	11.0	1.5915	-136.8	19.	91.0	0.0	901
N-BUTYL OLEATE	338.56	0.8657	25.0	1.4480	-10.0	0.	227.5	4.00	902
BUTYL STEARATE	340.60	0.8540	25.0	1.4422	26.3	0.	222.5	3.11	903
1122-TET BROMOETHANE	345.67	2.9529	25.0	1.6323	0.0	0.	243.5	7.00	904
1,1,1,2-TETRABROMOETHANE	345.70	2.8748	0.0	1.6277	0.0	0.	103.5	0.0	905
BIS(2-ETHOXY ET)SEBACATE	346.46	0.9953	25.0	1.4440	-10.0	0.	0.0	0.0	906
TRI(2-TOLYL)PHOSPHATE	368.36	1.1830	25.0	1.5575	11.0	20.	264.0	0.0	907
DI(2-ET HEX) ADIPATE	370.58	0.9220	25.0	1.4474	-67.8	5.	214.0	0.0	908
BIS(2-ETHEX) PHTHALATE	390.57	0.9843	20.0	1.4859	-50.0	0.	231.0	5.30	909
DIETHYLHEXYL AZELATE	412.66	0.9150	25.0	1.4460	-78.0	5.	237.0	0.0	910
BIS(2-ETHYLHEX)SEBACATE	426.66	0.9120	25.0	1.4510	-48.0	5.	256.0	4.03	911

Note: Missing data is indicated by 0, 0., or 0.0.

References

(1) Mellan, I., *Industrial Solvents,* Reinhold Publishing Corp., New York, NY (1939)

(2) Mellan, I., *Industrial Solvents,* 2nd Ed., Reinhold Publishing Corp., New York, NY (1950)

(3) Mellan, I., *Handbook of Solvents,* Vol. I, "Pure Hydrocarbons," Reinhold Publishing Corp., New York, NY (1957)

(4) Phillips Chemical Co., Bridgestone One, 2600 N. Loop West, Houston, TX 77092

(5) Arizona Chemical Co., Berdan Ave., Wayne, NJ 07470

(6) Chevron Chemical Co., 575 Market St., San Francisco, CA 94105

(7) Mellan, I., *Source Book of Industrial Solvents,* Vol. II, "Halogenated Hydrocarbons," Reinhold Publishing Corp., New York (1957)

(8) Exxon Chemical Americas, P.O. Box 3272, Houston, TX 77001

(9) Exxon, P.O. Box 3272, Houston, TX 77001

(10) Eastman Chemical Products, Inc., Kingsport, TN 37662

(11) E.I. du Pont de Nemours and Co., Inc., Wilmington, DE 19898

(12) Sun Petroleum Products Co., 1608 Walnut St., Philadelphia, PA 19103

(13) Union Chemicals Division, Petrochemicals Group, Union Oil Co. of California, Schaumburg, IL 60195

(14) Shell Chemical Co., One Shell Plaza, Houston, TX 77002

(15) Ansul Specialty Chemicals, Wormald U.S., Inc., One Stanton St., Marinette, WI 54143

(16) Getty Oil Co., P.O. Box 3000, Tulsa, OK 74102

(17) Allied Chemical, P.O. Box 1139R, Morristown, NJ 07960

(18) Hynes Chemical Research Corp., P.O. Box 5273, Columbia, SC 29205

(19) Union Carbide Corp., Chemicals and Plastics, 270 Park Ave., New York, NY 10017

(20) Amoco Chemicals Corp., 200 East Randolph Drive, P.O. Box 8640-A, Chicago, IL 60680

(21) Charter International Oil Co., 8441 Gulf Freeway, P.O. Box 87535, Houston, TX 77287-7535

(22) PPG Industries, Inc., One PPG Place, Pittsburgh, PA 15272

(23) Dow Chemical U.S.A., Midland, MI 48640

(24) Stauffer Chemical Co., Westport, CT 06880

(25) Pfizer Chemical Division, 235 East 42nd St., New York, NY 10017

(26) Humko Chemical Division, Witco Chemical Corp., P.O. Box 125, Memphis, TN 38101

(27) Occidental Chemical Corp., Hooker Industrial & Specialty Chemicals, MPO Box 728, Niagara Falls, NY 14302

(28) Hercules, Inc., 910 Market St., Wilmington, DE 19899

(29) Holson, R.W. et al, *Journal of the American Chemical Society,* 63:2094 (1941)

(30) Diamond Shamrock Chemicals Co., P.O. Box 2300, Irving TX 75061-1433

(31) Mellan, I., *Source Book of Industrial Solvents,* Vol. III, "Monohydric Alcohols," Reinhold Publishing Corp., New York, NY (1959)

(32) Mellan, I., *Polyhydric Alcohols,* Spartan Books, Washington, DC (1962)

(33) Kirk, R.E., and Othmer, D.F., *Encyclopedia of Chemical Technology,* Interscience Encyclopedia, Inc., New York, NY

(34) Angus Chemical Co., 2211 Sanders Road, Northbook, IL 60062

(35) Haas, H.B., and Riley, E.F., *Chemical Review,* 32, No. 373 (1943)

(36) Crown Zellerbach Corp., P.O. Box 4266, Vancouver (Orchards), WA 98662

(37) Pennwalt Corp., Pennwalt Bldg., Three Parkway, Philadelphia, PA 19102

(38) ICI Americas, Inc., Wilmington, DE 19897

(39) Procter & Gamble Co., P.O. Box 599, Cincinnati, OH 45201

(40) Conoco Chemicals Co., 15990 North Barkers Landing Road, P.O. Box 19029, Houston, TX 77224

(41) Eastman Chemical Products, Inc., P.O. Box 431, Kingsport, TN 37662

(42) Celanese Chemical Co., Inc., 1250 W. Mockingbird Lane, Dallas, TX 75247

(43) ICI Mond Division, P.O. Box 19, Runcorn, Cheshire, England

(44) Rember, R.F., *Acetone: Its Production and Uses,* National Wood Chemicals Association, Washington, DC

(45) Northwest Petrochemical Corp., P.O. Box 99, Anacortes, WA 98221

(46) Quaker Oats Chemicals, Inc., Merchandise Mart Plaza, Chicago, IL 60654

(47) BASF Wyandotte Corp., 100 Cherry Hill Road, Parsippany, NJ 07054

(48) Texaco Chemical Co., Box 430, Bellaire, TX 77401

(49) GAF Corp., 140 West 51st St., New York, NY 10020

(50) Mellan, I., *Ketones,* Chemical Publishing Co., New York, NY (1968)

(51) Mapstone, G.E., *Chemical Processing,* April 1967

(52) Rohm and Haas Co., Independence Mall West, Philadelphia, PA 19105

(53) American Cyanamid Co., Wayne, NJ 07470

(54) Sohio Chemical Co., Industrial Chemicals Division, Midland Bldg., Cleveland, OH 44115

(55) *American Ink Maker,* New York, NY

(56) Reilly Tar & Chemical Corp., 1510 Market Square Court, 151 North Delaware St., Indianapolis, IN 46204

(57) *Chemical Processing,* p. 94b, September, 1954

(58) Henkel Corp., Minerals Industry Division, 7900 West 78th St., Minneapolis, MN 55435

(59) Akzo Chemie America, Box 1805, Chicago, IL 60690

(60) Henkel Corp., 425 Broad Hollow Road, Melville, Long Island, NY 11746

(61) Stauffer Chemical Co., Westport, CT 06881

(62) Glyco Inc., 51 Weaver St., P.O. Box 700, Greenwich, CT 06830

(63) Emery Industries, 1300 Carew Tower, Cincinnati, OH 45202

(64) Mobil Chemical Co., 1024 South Ave., Plainfield, NJ 07067

(65) Velsicol Chemical Corp., 341 East Ohio St., Chicago, IL 60611

(66) Olin Chemicals, 120 Long Ridge Road, Stamford, CT 06904

(67) Hatco Polyester Division, W.R. Grace and Co., 1605 West Elizabeth Ave., Linden, NJ 07035

(68) Drew Chemical Corp., One Drew Chemical Plaza, Boonton, NJ 07005

(69) Ashland Chemical Co., Division of Ashland Oil Co., 5200 Paul G. Blazer Memorial Parkway, Dublin, OH 43017

(70) Howard, F.L., Brooks, D.C., and Streits, R.E., National Bureau of Standards Circular 576, Washington, DC

(71) 3M Co., 3M Center, 2501 Hudson Road, St. Paul, MN

(72) "Solvay's Chlorobenzenes," *Chemical Engineering,* April 1954

(73) Liggett, L.M., *Anal. Chem.* 26:748 (1954)

(74) Mobil Oil Corp., 5151 Belt Line Road, Suite 600, Dallas, TX 75240

(75) Penreco, 106 South Main St., Butler, PA 16001

(76) Kendall Amalie Division, Witco Chemical Corp., 77 W. Kendall Ave., Bradford, PA 16701

(77) Stepan Co., Northfield, IL 60093

(78) Publicker International, Inc., 1429 Walnut St., Philadelphia, PA 19102

(79) Arco Chemical Co., 1500 Market St., Philadelphia, PA 19101

(80) Sylvachem Corp., 2110-A West 23rd St., Panama City, FL 32405

(81) Alpha Metals, Inc., 600 Route 440, Jersey City, NJ 07304

(82) Ethyl Corp., Ethyl Tower, 451 Florida, Baton Rouge, LA 70801